ANNOTATED TEACHER'S EDITION
SCIENCE PLUS®
TECHNOLOGY AND SOCIETY

LEVEL BLUE

Project Directors

International: **Charles McFadden**
Professor of Science Education
The University of New Brunswick
Fredericton, New Brunswick

National: **Robert E. Yager**
Professor of Science Education
The University of Iowa
Iowa City, Iowa

Project Authors

Earl S. Morrison *(Author in Chief)*
Alan Moore *(Associate Author in Chief)*
Nan Armour
Allan Hammond
John Haysom
Elinor Nicoll
Muriel Smyth

This new United States edition has been adapted from
prior work by the Atlantic Science Curriculum Project,
an international project linking teaching, curriculum
development, and research in science education.

HOLT, RINEHART AND WINSTON
Harcourt Brace & Company
Austin • New York • Orlando • Atlanta • San Francisco • Boston • Dallas • Toronto • London

Requests for permission to make copies of any part of the work should be mailed to: Permissions Department, Holt, Rinehart and Winston, 6277 Sea Harbor Drive, Orlando, Florida 32887-6777.

Portions of this work were published in previous editions.

Acknowledgments: See page S196, which is an extension of this copyright page.

SCIENCEPLUS is a registered trademark of Harcourt Brace & Company licensed to Holt, Rinehart and Winston.

Printed in the United States of America

ISBN 0-03-095097-X

4 5 6 7 032 00 99

ACKNOWLEDGMENTS

Project Advisors

Herbert Brunkhorst
Director, Institute for Science Education
California State University,
San Bernardino
San Bernardino, California

David L. Cross
Science Consultant
Lansing School District
Lansing, Michigan

Jerry Hayes
Associate Director, Science Outreach
Teacher's Academy,
Mathematics and Science
Chicago, Illinois

William C. Kyle, Jr.
*Director, School Mathematics
and Science Center*
Purdue University
West Lafayette, Indiana

Mozell Lang
Science Education Specialist
Michigan Department
of Education
Lansing, Michigan

Annotated Teacher's Edition Writers

Patricia Barry
Milwaukee, Wisconsin

Carol Bornhorst
Chula Vista, California

Catherine Carlson
Plano, Texas

Sharon Effinger
Evansville, Indiana

Kenneth Horn
Forest Hills, Maryland

Roberta Jacobowitz
Williamston, Michigan

Shirley Key, Ph.D.
Multicultural Consultant and Writer
Missouri City, Texas

Annette Moran
Carrollton, Texas

Lynn Roudabush-Novak
Midlothian, Virginia

Donna Robinson
Rapid City, South Dakota

Pamela Russ, Ph.D.
Multicultural Consultant and Writer
Turlock, California

Sandy Tauer
Derby, Kansas

Feature Writers

Glyn Brice
Wenatchee, Washington

Margy Kuntz
San Anselmo, California

Bruce R. Mulkey
Austin, Texas

David Stienecker
Forest Hills, New York

Teacher Reviewers

Robert W. Avakian
Alamo Junior High
Midland, Texas

Caroline Beller
C. D. York Junior High
Conroe, Texas

Carol Bornhorst
Bonita Vista Middle School
Chula Vista, California

Paul Boyle
Perry Heights Middle School
Evansville, Indiana

Wayne Browning
Clarkstown South High School
West Nyak, New York

Renae Cartwright
Cedar Park Middle School
Cedar Park, Texas

Kenneth Creese
White Mountain Junior High
Rock Springs, Wyoming

Harry Dierdorf
Educational Consultant
Carnegie Science Center
Pittsburgh, Pennsylvania

Leila R. Dumas
LBJ Science Academy
Austin, Texas

Kenneth Horn
Fallston Middle School
Fallston, Maryland

Roberta Jacobowitz
C.W. Otto Middle School
Lansing, Michigan

Pamela Jones
Birmingham Covington School
Birmingham, Michigan

David Negrelli
Miami Lakes Middle School
Hialeah, Florida

Kevin Reel
Thacher School
Ojai, California

Larry Tackett
Andrew Jackson Middle School
Cross Lanes, West Virginia

Donald Yost
Cordova High School
Rancho Cordova, California

Academic Reviewers

David Armstrong, Ph.D.
Department of EPO Biology
University of Colorado
Boulder, Colorado

Bruce Briegleb
National Center for Atmospheric
Research
Boulder, Colorado

Kenneth Brown, Ph.D.
Professor of Chemistry
Northwestern Oklahoma State
University
Alva, Oklahoma

Linda K. Butler, Ph.D.
Division of Biological Sciences
University of Texas
Austin, Texas

Andrew A. Dewees, Ph.D.
Department of Biological Sciences
Sam Houston State University
Huntsville, Texas

Frederick R. Heck, Ph.D.
Associate Professor of Geology
Ferris State University
Big Rapids, Michigan

Arthur Huffman, Ph.D.
Department of Physics
University of California at
Los Angeles
Los Angeles, California

James Kaler, Ph.D.
Professor of Astronomy
University of Illinois
Urbana, Illinois

Gloria Langer, Ph.D.
University of Colorado
Boulder, Colorado

Doris I. Lewis, Ph.D.
Professor of Chemistry
Suffolk University
Boston, Massachusetts

R. Thomas Myers, Ph.D.
Professor of Chemistry Emeritus
Kent State University
Kent, Ohio

Alvin M. Saperstein, Ph.D.
Associate Professor
Department of Physics
Wayne State University
Detroit, Michigan

Thomas Troland, Ph.D.
Associate Professor
Department of Physics
and Astronomy
University of Kentucky
Lexington, Kentucky

Blue-Ribbon Committee

The following teachers constituted a special committee of current *SciencePlus* users who provided information and insights on how to improve the *SciencePlus* program. Their input was invaluable.

Patricia Barry
Milwaukee, Wisconsin

Carol Bornhorst
Chula Vista, California

Catherine Carlson
Plano, Texas

Harry Dierdorf
New Brighton, Pennsylvania

Jeff Felber
Spring Valley, New York

Barbara Francese
Madison, Connecticut

Kenneth Horn
Forest Hills, Maryland

Doug Leonard
Toledo, Ohio

Betsy Mabry
Enid, Oklahoma

Debbie Melphi
Syracuse, New York

Lynn Roudabush-Novak
Midlothian, Virginia

Donna Robinson
Rapid City, South Dakota

Sandy Moerke-Schaefer
Marysville, Washington

Marvin Selnes
Sioux Falls, South Dakota

Patricia Soto
Coral Gables, Florida

Margaret Steinheimer
St. Louis, Missouri

Sandy Tauer
Derby, Kansas

Joy Ward
Chicago, Illinois

Gary Weaver
Vacaville, California

Nancy Wesorick
Longmont, Colorado

Brenda West
Scott Depot, West Virginia

Project Associates

We wish to thank the thousands of science educators, teachers, and administrators from the scores of universities, high schools, junior high schools, and middle schools who have contributed to the success of *SciencePlus*.

CONTENTS ANNOTATED TEACHER'S EDITION

OWNER'S MANUAL

The *SciencePlus* Philosophy T16

The *SciencePlus* Method T19
  Building a Better Understanding of Science T19
  Conceptual Framework T20

Components of *SciencePlus* T23
  Pupil's Edition T23
  SourceBook T25
  Annotated Teacher's Edition T26
  Teaching Resources T29
  Other Teaching Aids T30
  Science Discovery Videodisc Program T31

Using *SciencePlus* T32
  Themes in Science T32
  Integrating the Sciences T34
  Cross-Disciplinary Connections T34
  Science, Technology, and Society T35
  Communicating Science T36
  Journals and Portfolios T37
  Concept Mapping T39
  Cooperative Learning T41
  Process Skills T47
  Critical Thinking T48
  Environmental Awareness T49
  Multicultural Instruction T50
  Meeting Individual Needs T52
  Materials and Equipment T55
  SciencePlus Teacher's Network T55

Assessing Student Performance T56
  A Comprehensive Approach to Assessment T56
  Assessment Aids in *SciencePlus* T56
  Assessment Strategies T58
  Developing Assessment Items T61
  Assessment Rubrics T62

UNIT INTERLEAF INFORMATION

Unit 1 Life Processes 1A
Unit 2 Particles 75A
Unit 3 Machines, Work, and Energy 143A
Unit 4 Oceans and Climates 209A
Unit 5 Electromagnetic Systems 287A
Unit 6 Sound 355A
Unit 7 Light 421A
Unit 8 Continuity of Life 487A

Answers to SourceBook Unit CheckUps S198

Answers to Tables and Diagrams S206

CONTENTS

To the Student xii
About Safety xiv
Concept Mapping xvii

Unit 1 LIFE PROCESSES 2

CHAPTER 1 **Energy for Life** 4
  Lesson 1 Where Does the Energy Come From? 5
  Lesson 2 Where Does the Starch Come From? 10
  Lesson 3 Food Factory Basics 17

CHAPTER 2 **A World of Water** 28
  Lesson 1 The Indispensable Chemical 29
  Lesson 2 The Water Cycle 31
  Lesson 3 Spreading Water Around 35
  Lesson 4 Osmosis: Controlled Diffusion 43
  Lesson 5 Putting Water Where It Counts 47

CHAPTER 3 **Maintaining Life** 54
  Lesson 1 Turning Food Into Fuel 55
  Lesson 2 Unlocking the Energy in Food 57
  Lesson 3 Maintaining the Balance 66

  Making Connections 70

  Eye on the Environment: Watchfrogs of
      the Environment 72
    Science and Technology: Space Food 73
    Science in Action: Eric Pianka, Zoologist 74
    Health Watch: Meatless Munching 75

iv

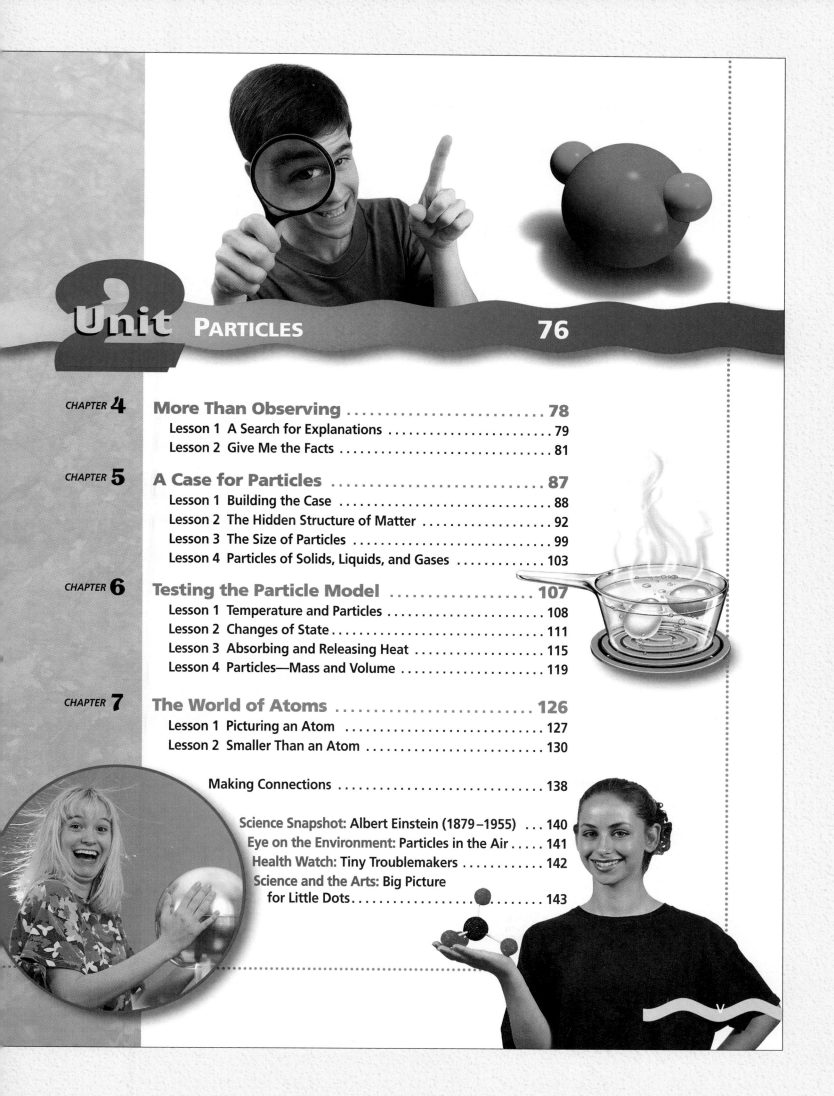

Unit 2 PARTICLES 76

CHAPTER 4 **More Than Observing** **78**
  Lesson 1 A Search for Explanations 79
  Lesson 2 Give Me the Facts 81

CHAPTER 5 **A Case for Particles** **87**
  Lesson 1 Building the Case 88
  Lesson 2 The Hidden Structure of Matter 92
  Lesson 3 The Size of Particles 99
  Lesson 4 Particles of Solids, Liquids, and Gases 103

CHAPTER 6 **Testing the Particle Model** **107**
  Lesson 1 Temperature and Particles 108
  Lesson 2 Changes of State 111
  Lesson 3 Absorbing and Releasing Heat 115
  Lesson 4 Particles—Mass and Volume 119

CHAPTER 7 **The World of Atoms** **126**
  Lesson 1 Picturing an Atom 127
  Lesson 2 Smaller Than an Atom 130

  Making Connections 138

  Science Snapshot: Albert Einstein (1879–1955) . . . 140
  Eye on the Environment: Particles in the Air 141
  Health Watch: Tiny Troublemakers 142
  Science and the Arts: Big Picture
      for Little Dots 143

Unit 3 · MACHINES, WORK, AND ENERGY 144

CHAPTER 8 **Putting Energy to Work** 146

Lesson 1 Machines for Work and Play 147
Lesson 2 The Idea of Work 152
Lesson 3 Work and Energy 158
Lesson 4 Energy Changes 164

CHAPTER 9 **Harnessing Energy** 169

Lesson 1 Lightening the Load 170
Lesson 2 Machines and Energy 180
Lesson 3 Transferring Energy 186
Lesson 4 Mechanical Systems 193
Lesson 5 The Human Machine 198

Making Connections 204

Science and Technology: Micromachines 206
Science in Action: Wheelchair Innovators 207
Weird Science: Unbelievable Bicycles 208
Science and the Arts: Machines as Art 209

Unit 4 OCEANS AND CLIMATES 210

CHAPTER 10 **The Temperate Planet** 212

Lesson 1 An Oasis in Space 213
Lesson 2 The Greenhouse Effect 217
Lesson 3 Land and Sea 226

CHAPTER 11 **Oceans of Water and Air** 237

Lesson 1 Hidden Wonders 238
Lesson 2 The Moving Oceans 243
Lesson 3 The Atmosphere 251

CHAPTER 12 Winds and Currents **259**
    Lesson 1 Pressure Differences 260
    Lesson 2 The Direction of Flow 269
    Lesson 3 Hurricanes 277

    Making Connections 282

    Weird Science: Weather From Fire 284
    Science and Technology: Space-Age Subs 285
    Technology in Society: The Wildest of Waves 286
    Science in Action: From Thermometer to Satellite 287

Unit 5 ELECTROMAGNETIC SYSTEMS 288

CHAPTER 13 Energy in a Wire **290**
    Lesson 1 Indispensable Energy 291
    Lesson 2 What Is Electricity? 300

CHAPTER 14 Sources of Electricity **309**
    Lesson 1 Electricity From Chemicals 310
    Lesson 2 Electricity From Magnetism 315
    Lesson 3 Other Sources of Electricity 322

CHAPTER 15 Currents and Circuits **326**
    Lesson 1 Circuits: Channeling the Flow 327
    Lesson 2 Making Circuits Work for You 331
    Lesson 3 Controlling the Current 335
    Lesson 4 Electromagnets 338
    Lesson 5 How Much Electricity? 343

    Making Connections 350

    Science Snapshot: Paul Chu (1941–) 352
    Eye on the Environment: A Cleaner Ride 353
    Health Watch: Magnets in Medicine 354
    Technology in Society: Riding the Electric Rails 355

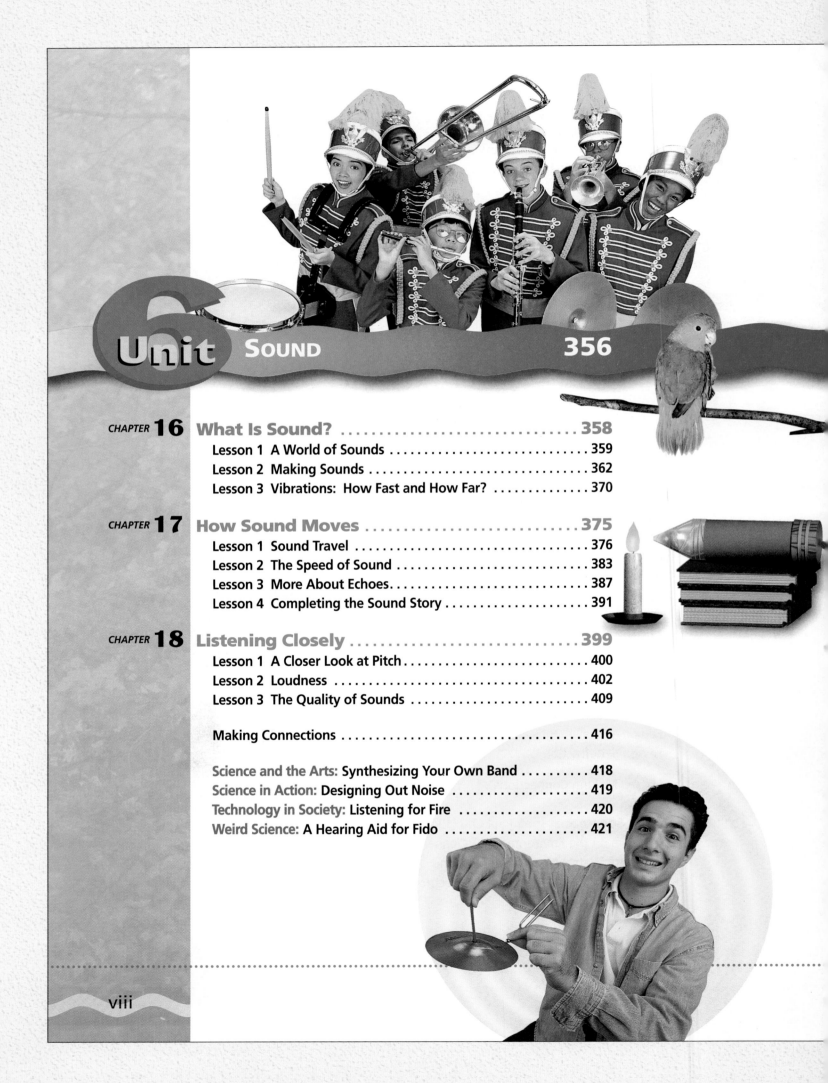

6 Unit SOUND 356

CHAPTER 16   What Is Sound? **358**
　Lesson 1 A World of Sounds 359
　Lesson 2 Making Sounds 362
　Lesson 3 Vibrations: How Fast and How Far? 370

CHAPTER 17   How Sound Moves **375**
　Lesson 1 Sound Travel 376
　Lesson 2 The Speed of Sound 383
　Lesson 3 More About Echoes. 387
　Lesson 4 Completing the Sound Story 391

CHAPTER 18   Listening Closely **399**
　Lesson 1 A Closer Look at Pitch. 400
　Lesson 2 Loudness 402
　Lesson 3 The Quality of Sounds 409

　Making Connections 416

　Science and the Arts: Synthesizing Your Own Band 418
　Science in Action: Designing Out Noise 419
　Technology in Society: Listening for Fire 420
　Weird Science: A Hearing Aid for Fido 421

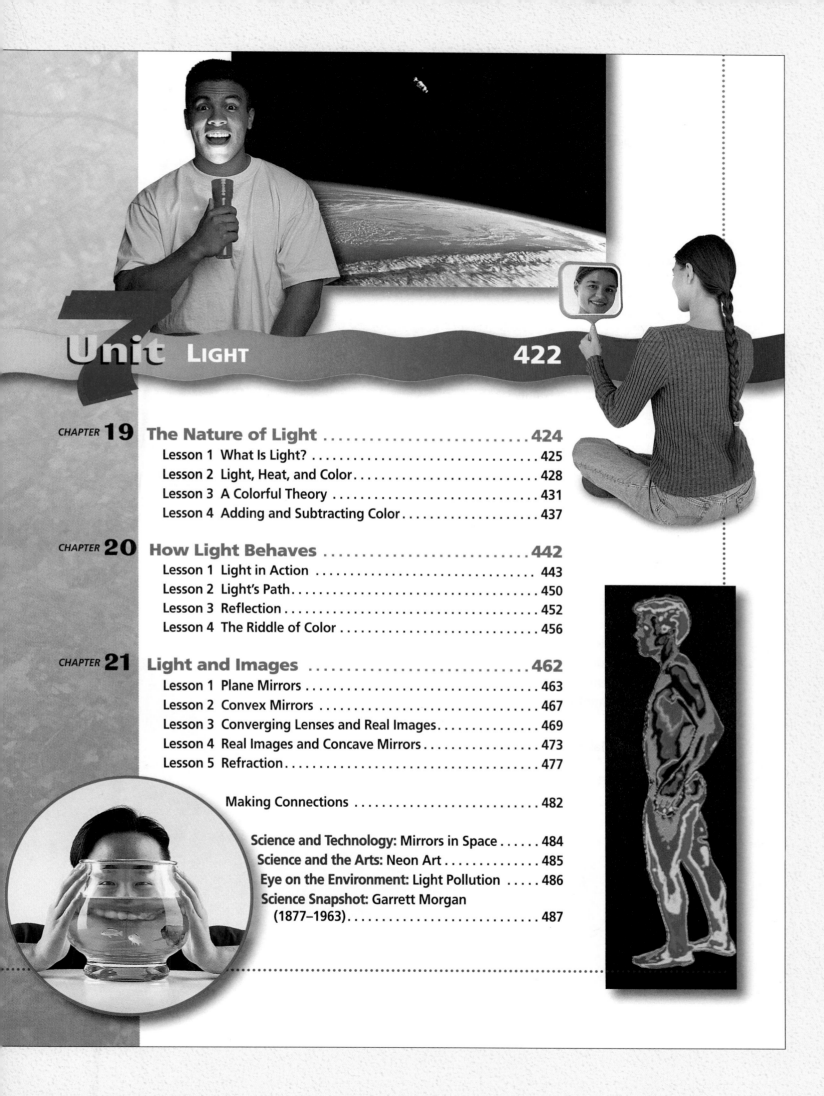

Unit 7 LIGHT 422

CHAPTER 19 **The Nature of Light** **424**
Lesson 1 What Is Light? 425
Lesson 2 Light, Heat, and Color 428
Lesson 3 A Colorful Theory 431
Lesson 4 Adding and Subtracting Color 437

CHAPTER 20 **How Light Behaves** **442**
Lesson 1 Light in Action 443
Lesson 2 Light's Path 450
Lesson 3 Reflection 452
Lesson 4 The Riddle of Color 456

CHAPTER 21 **Light and Images** **462**
Lesson 1 Plane Mirrors 463
Lesson 2 Convex Mirrors 467
Lesson 3 Converging Lenses and Real Images 469
Lesson 4 Real Images and Concave Mirrors 473
Lesson 5 Refraction 477

Making Connections 482

Science and Technology: Mirrors in Space 484
Science and the Arts: Neon Art 485
Eye on the Environment: Light Pollution 486
Science Snapshot: Garrett Morgan
(1877–1963) 487

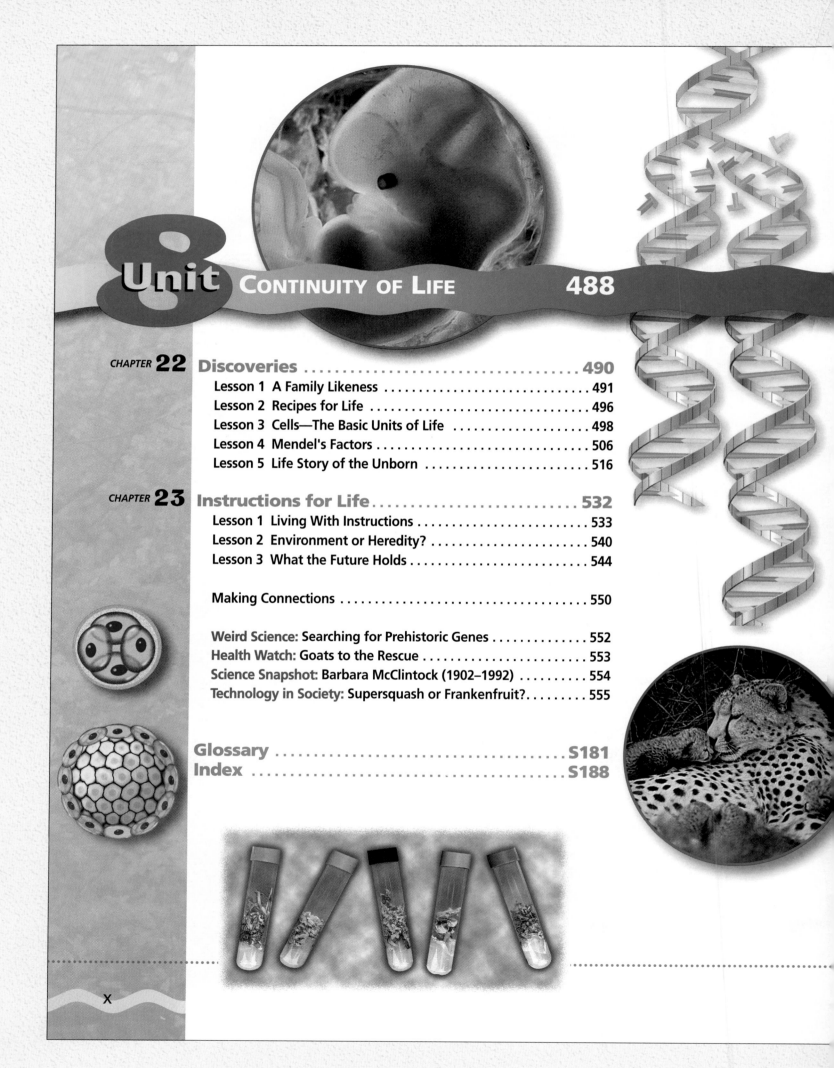

8 Unit CONTINUITY OF LIFE 488

CHAPTER **22** **Discoveries** 490

Lesson 1 A Family Likeness 491
Lesson 2 Recipes for Life 496
Lesson 3 Cells—The Basic Units of Life 498
Lesson 4 Mendel's Factors 506
Lesson 5 Life Story of the Unborn 516

CHAPTER **23** **Instructions for Life** 532

Lesson 1 Living With Instructions 533
Lesson 2 Environment or Heredity? 540
Lesson 3 What the Future Holds 544

Making Connections 550

Weird Science: Searching for Prehistoric Genes 552
Health Watch: Goats to the Rescue 553
Science Snapshot: Barbara McClintock (1902–1992) 554
Technology in Society: Supersquash or Frankenfruit?......... 555

Glossary S181
Index S188

SOURCEBOOK

UNIT 1 Life Processes S1
  The Chemistry of Life . . . S2
  Building Biochemicals
  With Light S9
  Basic Life Processes . . . S13
  Unit CheckUp S19

UNIT 2 Particles S21
  Particles of Matter S22
  Particles in Motion S27
  Particles of Particles . . . S35
  Unit CheckUp S39

UNIT 3 Machines, Work,
  and Energy S41
  Force, Work,
  and Power S42
  Simple Machines S47
  Machines at Work S52
  Unit CheckUp S57

UNIT 4 Oceans and
  Climates S59
  The Atmosphere S60
  Weather and Climate . . S66
  Ocean Resources S75
  Unit CheckUp S79

UNIT 5 Electromagnetic
  Systems S81
  Two Kinds of
  Electricity S82
  Magnets and
  Magnetic Effects S91
  Using Electricity S97
  Unit CheckUp S103

UNIT 6 Sound S105
  Waves S106
  Sound Waves S115
  Sound Effects S121
  Unit CheckUp S125

UNIT 7 Light S127
  Light: A Split
  Personality? S128
  Visible Light S133
  Invisible Light S137
  High-Tech Light S144
  Unit CheckUp S151

UNIT 8 Continuity
  of Life S153
  Foundations of
  Heredity S154
  Chromosomes, Genes,
  and Heredity S161
  DNA—The Material
  of Heredity S167
  Reproduction and
  Development S173
  Unit CheckUp S179

xi

Science Plus Owner's Manual

THE SciencePlus PHILOSOPHY

SciencePlus is lively, engaging, and relevant to the student's world.

Welcome to *SciencePlus*, an innovative approach to science education. *SciencePlus* is unlike any science program you have used before. It is designed from the ground up to teach science in precisely the way that students learn best—by thinking, talking, and writing about what they do and discover. *SciencePlus* is activity- and inquiry-based. In other words, it is both hands-on and minds-on. *SciencePlus* is lively, engaging, and relevant to the students' world. *SciencePlus* is loaded with thought-provoking activities designed to challenge students' thinking skills while introducing them to realistic methods of science.

SciencePlus works for students and teachers alike. Students enjoy and benefit from its varied and active approach, while teachers find it to be teachable under real conditions. Laboratory-type activities require about 30 to 40 percent of class time. The remainder of class time is taken up by a rich variety of learning activities.

SciencePlus is tailored to meet the needs of middle school–aged students. It accommodates every type of student you are likely to encounter: the basic student, who requires substantial guidance; the average student, who is not yet fully equipped with abstract-thinking skills; and the advanced student, who needs only to be pointed in the right direction to succeed.

At every stage, the *SciencePlus* program emphasizes concept and skill development over memorization of facts. By doing science activities and then thinking about the results, students learn the *whys* and *hows,* not just the *whats* and *whens,* of science. Ultimately, students come to see science as a system for making sense of the world.

Origins

The *SciencePlus* program was originally developed by the Atlantic Science Curriculum Project (ASCP) of Canada to replace the traditional recall-based curriculum, which had proven to be ineffective. *SciencePlus* represents a ground-breaking effort, the culmination of many years of labor by dozens of talented, dedicated science educators.

The *SciencePlus* development team was guided every step of the way by the latest insights into how children actually learn. The *SciencePlus* program has been thoroughly tested on real students in realistic settings, refined, and then retested. The result is a program that works! Teachers using *SciencePlus* have reported dramatic gains in student comprehension and retention of scientific concepts. Above all, students enjoy using *SciencePlus* and develop a heightened interest in science.

A Continuing Tradition

This edition of *SciencePlus*, now in its second edition as an American publication, continues the tradition of excellence begun in Canada many years ago. The program has undergone many changes in response to recommendations of curriculum reform movements, such as Project 2061 and the NSTA Scope, Sequence, and Coordination. In addition, the revision of this program was heavily influenced by *SciencePlus* teachers, whose combined and varied experiences helped us make the program even more relevant and easier to implement.

An Interactive, Effective Program

SciencePlus employs proven teaching strategies: guided and open-ended investigations, small-group discussions, exploratory writing and reflective-reading tasks, games, picture and word puzzles, and independent long-range projects. This variety motivates and helps maintain the interest of students and teachers alike.

SciencePlus develops scientific process skills as an essential goal. As the curriculum progresses, the students will master increasingly complex tasks. For example, students will move from directed to open-ended inquiry and from reading and completing tables and graphs to constructing them from experimental data they have collected on their own.

In general, each of the units in *SciencePlus* is self-contained and may be taught as a separate instructional module. *SciencePlus* contains a balance of physical, biological, earth/space, and environmental science topics.

Guiding Principles

The guiding principles of *SciencePlus* are simple and few:

- **Anyone can learn science**
  The image of science as the private domain of the superintelligent is wrong and damaging. Science is for everyone. Children exposed to science for the first time take to it naturally. It is only later, after science explorations have been replaced by fill-in-the-blank worksheets and recall drills, that love for science is replaced by boredom and even dread. *SciencePlus* can rekindle the sense of wonder and fascination that lies dormant within your students.

- **Science is a natural endeavor**
  Whether we realize it or not, each of us applies science nearly every day. Hardly a day passes that we don't ask ourselves, "How does this work?" or "Why does that happen?" or "What happens if . . . ?" Scientists differ from other people only in that it is their profession, rather than their avocation, to figure out "how," "why," and "what if."

Unfortunately, stereotypes about science and scientists abound. Many students feel that only "nerds" or "geeks" enjoy science. This falsehood may do as much to turn people away from science as any curricular shortcoming. *SciencePlus* actively refutes these stereotypes. It portrays science as a rewarding, quintessentially human undertaking. Scientists are portrayed as normal people, not aloof geniuses who talk in equations.

- **Science is its own reward**
  There is no feeling quite like the thrill of discovery or the sense of accomplishment that comes from rising to a difficult challenge. Science can be thought of as a voyage into the unknown. This voyage can be exciting and rewarding for all.

Aims

SciencePlus is designed to help you further develop each of the following in your students:

- Understanding of the interrelationships among science, technology, and society
- Understanding of important science concepts, processes, and ideas
- Use of higher-order thinking skills
- Ability to solve problems and apply scientific principles
- Commitment to environmental protection
- Interest in independent study of scientific topics
- Social skills
- Communication skills

To accomplish these goals, a wide variety of teaching strategies are employed. The common denominator among these is their emphasis on doing. At all times, students are to be active and involved.

> *SciencePlus* can rekindle the sense of wonder and fascination that lies dormant within your students.

Science, Technology, and Society

Science and technology are flip sides of the same coin; each supports the other. Neither should be studied in isolation. *SciencePlus* explores the relationship between science and technology, and the effect of both on our society as a whole. Even people who never again set foot in a laboratory after leaving school can benefit from an understanding of science and technology and how both relate to each other and to society at large.

Our society, complex as it is, will become even more so in the years to come. Science and technology will play roles in every aspect of life. In the future, "high-tech" will be more than a catch phrase—it will permeate every aspect of life. To prepare students for the challenges of the future, they must become science literate. They must be given the tools that will enable them to become responsible and productive individuals in a highly techno-logical world.

Science for All

SciencePlus is designed to put the "process" back into science education and, in so doing, to provide students with the mental skills they need to truly understand and apply science. Today, as never before, a thorough grounding in science is absolutely essential; without it, students—the citizens of tomor-row—cannot expect to be fully conversant in and responsive to the complex issues of the twenty-first century.

No program can teach itself. You, with your energy, enthusiasm, and ability, are the key to a successful outcome. *SciencePlus* will help you help your students develop all of the skills they need to learn independently. *SciencePlus* fosters a spirit of joint explo-ration. Let the journey begin.

To prepare students for the challenges of the future, they must become science literate.

THE SciencePlus METHOD

BUILDING A BETTER UNDERSTANDING OF SCIENCE

The *SciencePlus* program is based on the Constructivist Learning Model (CLM). Constructivism is based as much on common sense as on the results of research. With the CLM, students "construct" an understanding of concepts step by step. Students begin by identifying what they already know about a topic. Any misconceptions they may have about a topic are exposed at this point. Identifying these misconceptions is a critical part of the process. Next, students do hands-on activities to experience the subject matter directly. Their experiences cause them to amend, add to, or scrap altogether the mental model they already have of the subject in question.

Constructivism is based on a few key steps:

1. Invitation
The Invitation stimulates students' curiosity and engages their interest. At this stage, students note the unexpected, pose questions, or define a problem.

2. Exploration
Explorations engage students in the search for solutions or explanations. Students look for alternative sources of information, collect and evaluate data, and clarify their findings through discussion and debate.

3. Proposing explanations and solutions
At the conclusion of the Explorations, students propose their response to the problem or question posed in the Invitation. The class is exposed to a variety of possible responses, and students have the opportunity to consider each.

4. Taking action
Students make decisions about a course of action based on the various proposals offered. If the class reaches consensus, then this stage may bring about closure of the lesson. It may happen, though, that this stage identifies new questions to explore.

Another way to think of the Constructivist Learning Model is in terms of the five *E*s: Engage, Explore, Explain, Elaborate, and Evaluate.

For an in-depth discussion of Constructivism, see "The Constructivist Learning Model," by Bob Yager, The Science Teacher, *September 1991.*

With the Constructivist Learning Model, students "construct" an understanding of concepts step by step.

CONCEPTUAL FRAMEWORK

CONCEPT FOCUS	CONTENT INTEGRATION				PROCESS SKILLS	
	Content Focus	Supporting Content	Thematic Focus	STS	All Major Skills Covered	Process Skills Focus
Unit 6 **Sound** The features and variations of sound, including its production and transmission	Physical science	•Life science •Earth science	•Energy •Systems •Structures	•Using an oscilloscope •Examining how the human ear works •Measuring pitch and loudness in musical instruments	X	•Communicating •Contrasting •Analyzing
Unit 7 **Light** Light as a form of energy, the behavior of light, and how color is determined	Physical science	•Life science •Earth science	•Energy •Changes Over Time •Structures	•Testing a pinhole camera •Developing systems of lenses and mirrors •Invention of the traffic light	X	•Predicting •Measuring •Hypothesizing
Unit 8 **Continuity of Life** Genetics, the role of heredity in life, and the development of human life from conception to birth	Life science	•Earth science •Physical science	•Changes Over Time •Cycles •Structures	•Developing a model of the structure of DNA •Debating the value of genetic engineering •Detecting hereditary traits	X	•Communicating •Classifying •Organizing

COMPONENTS OF SciencePlus

SciencePlus is no ordinary textbook. *SciencePlus* is a student-friendly text: lively; abundantly illustrated with clever, colorful illustrations; and loaded with engaging activities. Every effort has been taken to make this text the sort of book that students will actually want to use.

Units, Chapters, Lessons

SciencePlus contains eight units, which are further divided into chapters and lessons on closely related subject matter. Each lesson includes a wide variety of activities and explorations designed to develop the lesson content.

ScienceLog

Each chapter begins with a ScienceLog page that invites students to express what they already know about the subject of the chapter. Highly graphic and playful questions help expose misconceptions students may have about the topics to be covered. This important process sets the stage for firsthand exploration of the subject area. The answers to the ScienceLog questions are the equivalent of hypotheses or predictions, which are either supported or disproved as the students explore the chapter content. Either way, the students become invested in what they are about to learn.

Explorations

Scattered throughout each unit are a series of Explorations—hands-on, inquiry-based activities. These Explorations allow students to see scientific principles in action. The Explorations are essential for inducing real learning in students. As students do the Explorations, they have the opportunity to compare their mental models of scientific principles with the real things. As weaknesses are exposed, students adjust their thinking to accommodate what they have learned.

Most of the Explorations are designed so they can be done cooperatively in small groups. In this way, the Explorations model real scientific experiences, in which scientists work together to solve problems. By working in cooperative groups, students develop important skills such as communicating ideas and sharing responsibility. And the cooperative groups not only make science more interactive and more fun, but also provide valuable opportunities to develop social skills. They are also more economical to implement.

> ***For more information on cooperative learning, see pages T41 through T46 of this Annotated Teacher's Edition.***

Most of the supplies needed for the Explorations consist of very common equipment and materials. In most cases, they can be gathered from the home. For help in gathering supplies, the *Teaching Resources* package contains a comprehensive *Materials Guide,* which contains both a master materials list and individual unit materials lists. In addition, the second page of the Home Connection parent letter in each *Teaching Resources* booklet provides a checklist of materials that you can use to invite parent donations. Using this portion of the parent letter will help you keep your budget low and at the same time get parents involved in the *SciencePlus* experience.

Assessment

The *SciencePlus* Pupil's Edition contains a variety of methods for checking student learning.

- **Chapter Review—Challenge Your Thinking**
  To make sure your students comprehend the new information, each chapter concludes with Challenge Your Thinking questions. These questions challenge students to apply newly learned material in a variety of ways. Many of the questions are like brain-teasers in that they are unusual and highly creative. Because the questions require more than simple recall, students find them fun to figure out.

  The chapter review pages conclude with an invitation to students to rewrite their answers to the ScienceLog questions located at the beginning of the chapter. This exercise gives students the opportunity to confront and discard any misconceptions they may have had at the outset of the chapter.

- **Unit Review—Making Connections**
  The Making Connections pages consist of two parts: The Big Ideas and Checking Your Understanding. The Big Ideas asks students to formulate their own summary of the unit, using a list of questions as a guide. The Checking Your Understanding poses a selection of comprehensive questions designed to gauge students' understanding of the unit's subject matter.

  Also included on the Making Connections pages is a short description and table of contents of the corresponding SourceBook unit. This reference directs students to the appropriate unit of the SourceBook, where they can extend their knowledge of the concepts they just explored.

> *All answers to the Challenge Your Thinking and Making Connections questions are located in the Wrap-Around Margins of this Annotated Teacher's Edition.*

> *For more information about assessment, see pages T56 through T64 of this Annotated Teacher's Edition.*

ScienceLog

For more information about using the ScienceLog, see pages T37 and T38.

Special Features

A common complaint among students is that the material they learn is not relevant. The end-of-unit features in *SciencePlus* help show students how science is an integral part of their lives.

In keeping with the spirit of *SciencePlus*, all of the features are interactive, with thought-provoking questions and research ideas that encourage students to explore the topic further. Each unit concludes with four features selected from the menu below.

- **Science in Action** features illustrate working scientists today. These features focus on the excitement and variety that characterize science careers. Students will learn about real experiences and what the scientists like most and least about their work. They may even learn about the scientists' personal career aspirations as well as their greatest disappointments.

- **Science Snapshots** are glimpses into the lives of famous scientists. Such scientists are featured to give students a sense of the colorful history of science. Present and future scientists really do stand on the shoulders of giants!

- **Health Watch** features provide insights into the variety of connections between science and health.

- **Eye on the Environment** features focus on the effects that science and technology have on our environment—both good and bad.

- **Science and Technology** features explore the science on which new technologies are based. In many cases, explorations in science open the doors to new technology. Highlighting the impact of this technology on our everyday lives shows students how relevant scientific research can be.

- **Technology in Society** features take the logical step beyond the technology that has been created through the use of science. These features explore the impact of technology on our society and put forth the inevitable question: Do the benefits outweigh the potential problems?

- **Science and the Arts** features show that science and art are not necessarily polar opposites. The connection between science and society is reinforced by showing students how artists have been inspired by nature and how scientific methods or principles have enhanced or empowered their work.

- **Weird Science** features showcase some of the odd, outlandish, and even unbelievable creatures and phenomena that make up our unique world.

SOURCEBOOK

The SourceBook is an in-text science reference located at the back of the Pupil's Edition. This unique reference includes information that is organized to match the units of *SciencePlus*. Each SourceBook unit both reinforces and extends beyond the material presented in the Pupil's Edition, providing an excellent resource that students can use to add depth to their understanding of the topics presented in *SciencePlus*.

Because the SourceBook is not designed for hands-on exploration, it lends itself to reading at home or outside the lab. It's handy for homework assignments or for when you have a substitute teacher. Each unit of the SourceBook ends with a Unit CheckUp for checking students' comprehension. Worksheets are also available in the *Teaching Resources* booklets.

> The SourceBook is referenced in many ways throughout the Pupil's Edition.

The *SciencePlus* Annotated Teacher's Edition will help you achieve the full potential of *SciencePlus*. The Annotated Teacher's Edition consists of two major parts: the Unit Interleaf and the Wrap-Around Margins. Each Unit Interleaf consists of a six-page insert preceding each unit. The Wrap-Around Margins provide on-page annotations and teaching suggestions.

Using the Unit Interleaf

Each Unit Interleaf consists of six pages of information to help you plan your lessons. Each interleaf has the following planning and preparation aids.

- **Unit Overview** is a quick overview of the concepts and content presented in the unit.

- **Using the Themes** provides descriptions of the themes that are relevant to the unit, including short explanations of how they apply and how to integrate them into your teaching.

- **Using the SourceBook** is a concise description of the content in the accompanying SourceBook unit.

- **Bibliography** consists of three subsections categorized for teachers; for students; and for films, videotapes, software, and other media.

- **Unit Organizer** is a comprehensive chart that identifies the objectives, time requirements, and teaching resources that are available for all of the chapters and lessons in the unit.

- **Materials Organizer** is a chart that identifies the materials required to carry out each Exploration and Activity in the unit.

- **Advance Preparation** identifies what you may need to do in advance to prepare for the Explorations and Activities.

- **Unit Compression** provides recommendations on how to compress or reduce the time spent teaching the unit without sacrificing the integrity of the unit.

- **Homework Options** is a chart that lists the various homework opportunities, suggestions, and worksheets available to support each of the chapters in the unit.

- **Assessment Planning Guide** is a chart that identifies the assessment opportunities and worksheets available at the lesson, chapter, and unit levels.

- **Using the *Science Discovery Videodiscs*** provides an overview and barcodes for the Science Sleuth mystery and Image and Activity Bank selections that correspond to the unit.

Using the Wrap-Around Margins

The entire Pupil's Edition of *SciencePlus* has been reduced in size and placed on the pages of this Annotated Teacher's Edition. The margins on the reduced pages have been filled with teaching suggestions and commentary to help you teach each lesson with maximum effectiveness.

UNIT OPENER

Each unit opener has the following teacher's information to help you quickly engage your students in the subject of the unit.

- **Unit Focus** provides interactive suggestions for introducing students to the unit.

- **Connecting to Other Units** is a table that shows at a glance the integration between that unit and other units in the program.

- **Using the Photograph** provides suggestions for using the unit photograph to start students thinking and asking questions about the subject of the unit topic.

- **Answer to In-Text Question** provides the answer to any question that is posed to the students on the unit opener pages.

CHAPTER OPENER

Each chapter has the following information to help you prepare to teach the chapter.

- **Connecting to Other Chapters** is a chart that identifies the basic topics explored in each chapter of the unit. This chart allows you to see the logical progression of concepts through the unit.

- **Prior Knowledge and Misconceptions** describes how to use the ScienceLog questions and suggests other ways to access students' prior knowledge or misconceptions about the chapter topics.

LESSON

Each lesson of the unit has extensive commentary and teaching notes to help you in teaching the material. The lesson information is divided into three parts: Focus, Teaching Strategies, and Follow-Up. In addition, a Lesson Organizer is provided at the beginning of each lesson.

Lesson Organizer

The Lesson Organizer is provided at the beginning of each lesson. The Lesson Organizer provides easy access to the following information about the lesson:
- Time Required
- Process Skills
- Theme Connections
- New Terms
- Materials
- Teaching Resources

FOCUS

The Focus consists of Getting Started and the Main Ideas. Getting Started provides suggestions for an activity and discussion to begin each lesson. The Main Ideas list the main concepts of the lesson.

TEACHING STRATEGIES

Teaching Strategies provide a variety of methods for successfully guiding your students through each lesson. The following categories of strategies are provided to serve your needs.

Meeting Individual Needs

Today's classrooms are places of diversity—populated by students with different ability levels and from a variety of ethnic backgrounds and cultures. The following categories of teaching suggestions are provided to help you tailor your instruction to meet the needs of all of your students.
- Gifted Learners
- Second-Language Learners
- Learners Having Difficulty

CROSS-DISCIPLINARY FOCUS

Suggestions and activities are provided that cross the boundaries between science and other disciplines, such as art, health, mathematics, social studies, music, language arts, foreign language, and industrial arts. By doing these activities, students will experience the interrelatedness of science with other areas of study and will come to understand that science is not just for scientists.

Multicultural Extension

Under this heading you will find activities and suggestions that serve to highlight cultural diversity and show its positive influence on science as well as other disciplines. These activities build on the multicultural elements already included in the Pupil's Edition to ensure ample opportunities to show students how culture and science are integrated to the benefit of us all.

Integrating the Sciences

These strategies help you explore concepts from the standpoint of more than one scientific discipline. In this way, you can help students to see how life, earth, and physical sciences work together to help us understand the world around us.

Theme Connection

Each Theme Connection consists of a Focus question and its answer. By asking the Focus question, you can help students organize their learning within an overarching theme to help them make better sense of the ever-increasing pool of scientific knowledge.

ENVIRONMENTAL FOCUS

Many lessons lend themselves to an environmental focus. In these cases, information is provided both to inform and to invite class discussion.

Homework

Homework suggestions are provided as well as recommendations as to which activities and worksheets make good homework assignments.

Portfolio

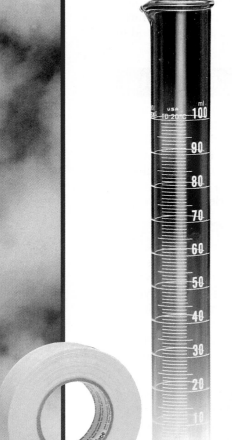

Students are encouraged to choose a portion of their work to include in a personal Portfolio. The suggestions provided in Teaching Strategies help you guide your students in developing this Portfolio. You may then choose to use the Portfolio as part of your final assessment.

Cooperative Learning

Suggestions are provided to help you organize cooperative-learning activities. Grouping strategies, individual and group responsibilities, and other important topics are covered.

Guided Practice

Questions are provided to help you lead group discussions related to the concepts presented in the lesson. The questions help invite and encourage group participation.

Independent Practice

Suggestions are provided for additional activities that students can do independently to help them solidify their understanding of difficult concepts.

Answers to Questions

All questions that have been posed in the course of the lessons, either within the running text or in the review materials, are answered in the Wrap-Around Margins for your convenience. In only a few instances, where the answer requires a graph or chart, will you have to turn pages to refer to the answer. In addition, all questions in the running text have lettered annotations on the reduced Pupil's Edition pages for quick and easy reference.

Did You Know. . .

Science facts are provided to add special insights and interest to your lessons.

Safety Alert/Waste Disposal

Additional information is provided when there are special safety and waste-disposal concerns.

FOLLOW-UP

The Follow-Up consists of four parts:
- Reteaching
- Assessment
- Extension
- Closure

This information provides effective methods to close out the lesson and to evaluate whether students have grasped the main concepts. If you find that your students need additional help, a reteaching strategy is provided. The extension provides an idea or activity for learning more about a concept covered in the lesson.

Review Pages

All questions appearing in each Challenge Your Thinking and Making Connections are fully answered in the Wrap-Around Margins. All answers for the SourceBook Unit CheckUps are provided in the back of this Annotated Teacher's Edition. Annotations in the Wrap-Around Margins reference the appropriate page numbers.

Features

All of the special-interest features at the end of each unit include background information, answers to questions, and a selection of teaching strategies, including ideas for extending, discussing, debating, and analyzing the feature.

TEACHING RESOURCES

All of the worksheets you will need or want are bound into eight convenient unit booklets. Now you do not have to spend precious time compiling worksheets from various sources. Each *Teaching Resources* booklet has all the worksheets you need to teach an entire unit.

Each *Teaching Resources* booklet contains the following types of worksheets:

Home Connection

The Home Connection begins with a two-page parent letter that introduces the unit of study to the parents and includes home activities to encourage parents' participation. The last page lists the supplies needed to do the Explorations and Activities in the unit. This page makes it easy for you to invite donations in order to keep your budget low.

Chapter Worksheets

- **Resource Worksheets** Worksheet versions of charts, graphs, puzzles, and activities from the Pupil's Edition
- **Exploration Worksheets** Worksheet versions of the Explorations in the Pupil's Edition
- **Discrepant Event Worksheets** Demonstrations and activities that spur students' curiosity and motivate further exploration
- **Theme Worksheets** Worksheets that draw thematic connections between concepts
- **Transparency Worksheets** Review, reinforcement, and assessment worksheets that correspond to the overhead transparencies
- **Math and Graphing Practice Worksheets** Worksheets that reinforce important math concepts and graphing skills
- **Review Worksheets** Worksheet versions of the Challenge Your Thinking review pages in the Pupil's Edition
- **Chapter Assessment** Tests consisting of a variety of questioning strategies to check students' comprehension. Challenge questions that require students to use higher-order thinking skills are included.

Unit Worksheets

- **Unit Activity Worksheets** Extension and review activities, usually in the form of a crossword puzzle or word search
- **Unit Review Worksheets** Worksheet versions of the Making Connections review pages from the Pupil's Edition
- **Self-Evaluation of Achievement** Checklist that helps students evaluate their progress throughout the unit
- **End-of-Unit Assessment** Test that consists of a variety of question types, including word usage, short response, correction/completion, short essay, numerical problems, and interpreting illustrations, graphs, and charts. At least two Challenge questions are also provided to check higher-order thinking skills.
- **Activity Assessment** Performance-based test that requires students to use materials and data to solve a problem
- **Spanish Resources** Spanish blackline masters consisting of Home Connection letters, Big Ideas, and unit glossaries

SourceBook Worksheets

- **SourceBook Activity Worksheets** Hands-on laboratory explorations and activities to illustrate and reinforce the content in the SourceBook
- **SourceBook Review Worksheets** Worksheet versions of the Unit CheckUp review pages in the SourceBook
- **SourceBook Assessment** Tests consisting of multiple-choice, true-false, and short-answer questions

The *SciencePlus* program includes the following support materials to make your teaching both effective and efficient.

Teaching Transparencies

Full-color transparencies that highlight key points from each chapter and organize important content with charts, graphs, and puzzles from the Pupil's Edition and *Teaching Resources*

Getting Started Guide
(for Levels Red and Blue)

Blackline master booklet designed to orient both teachers and students who are new to the *SciencePlus* method of teaching and learning science

Assessment Checklists and Rubrics

Blackline master booklet that provides a variety of assessment checklists and rubrics to make your assessment tasks efficient and timely. Electronic versions of these blackline masters are available on the *SnackDisc*.

Materials Guide

Complete listing of all the supplies and equipment needed to teach the entire level, as well as each individual unit, of *SciencePlus*

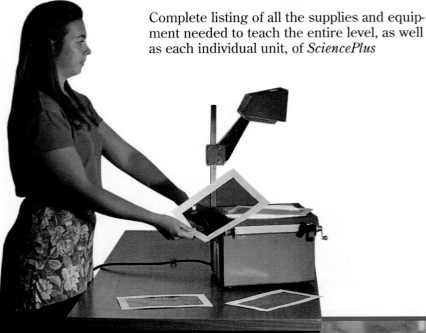

Test Generator

Test item software for Macintosh® and Windows® allows you to create your own tests from an extensive bank of tests and test items. In addition, you can edit the items and even add your own original items. A comprehensive *Test Item Listing* is provided to preview the test items in hard copy.

English/Spanish Audiocassettes

Audiocassette tapes that provide important preview information for each unit in both English and Spanish. Each unit begins with an attention-grabbing skit or story and then takes students on a visual tour of the unit.

Videodisc Resources

Instructions, barcodes, and worksheets for using the *Science Discovery* videodiscs with the *SciencePlus* program. For more information about the *Science Discovery* videodisc program, see page T31 of this Annotated Teacher's Edition.

SnackDisc

CD-ROM disc for Macintosh® and Windows® that contains a vast array of worksheets and information, most of which is customizable to fit your specific needs. The following is a partial listing of the items on the *SnackDisc*.
- Safety First! including safety review, test, and contracts
- Student Review Worksheets, providing review of math, graphing, and the metric system
- Science Sites, showing points of scientific interest on maps of the United States, Canada, and Mexico
- Home Connection letters in English and Spanish
- Assessment Checklists and Rubrics
- Lab Inventory Checklists
- District Requirements Reference Chart
- *SciencePlus* Communicator

Science Discovery is a comprehensive videodisc program that blends imagination and state-of-the-art technology into an exciting program that will add a new dimension to your science teaching. The program includes the following:
- Science Sleuths videodisc
- Science Sleuths Teacher's Guide
- Science Sleuths Resource Directory
- Image and Activity Bank videodisc
- Image and Activity Bank Directory
- Image and Activity Bank Quick Reference Card

Science Sleuths

The Science Sleuths videodisc consists of 24 open-ended science mysteries, one for each unit of the *SciencePlus* textbook series. Each mystery begins with a dramatization in which a problem is presented and the students are asked to serve as consultants (sleuths) to figure it out. Problems include such mysteries as Exploding Lawn Mowers, Dead Fish on Union Lake, and The Misplaced Fossil. The videodisc then becomes a "videophone" that connects students to the laboratory, where they can request information such as the testing of samples, experiments, interviews, tables, newspaper clippings, news broadcasts, and many other bits of information from an extensive menu.

The students work as a class or in small groups to explore the mystery, develop hypotheses, and support their position using the data from the videodisc. They try to solve the mystery using as few of the video segments as possible. In this way they either challenge themselves or other groups to see who can solve the mystery most efficiently.

In working as Science Sleuths, students improve their problem-solving and reasoning skills while applying their science knowledge from the textbook. Students work in a realistic scientific mode in which information comes from a variety of sources. Students must also judge the accuracy of each source and separate raw data from interpretation and inference.

Science Sleuths Teacher's Guide
The Science Sleuths Teacher's Guide contains the teaching plans for using the videodisc.

Science Sleuths Resource Directory
The Science Sleuths Resource Directory contains the barcodes and frame numbers for using each mystery. There are five directory sets so that you can have five working cooperative groups at a time.

Image and Activity Bank

The Image and Activity Bank videodisc consists of a still and motion image database designed to reinforce and extend concepts presented in *SciencePlus*. The still images include hundreds of photographs and computer graphics related to life, earth, and physical sciences. The motion images include demonstrations, experiments, and selected motion footage.

Image and Activity Bank Directory
The Image and Activity Directory provides descriptions, frame numbers, and barcodes for the Image and Activity Bank videodisc. A separate Reference Card provides barcodes and frame numbers for selected topics.

> *For complete directions on how to use* Science Discovery *with* SciencePlus, *see the Videodisc Resources booklet described on page T30.*

USING SCIENCE PLUS

THEMES IN SCIENCE

One subject can be addressed from the viewpoint of many different themes.

The traditional division of science into three branches—life, earth, and physical—often leads students to believe that nature is similarly arranged. Too often, students view science as a system of separate, unrelated abstractions or as a compilation of facts and difficult-sounding terms. But science is simply the study of nature, and there are certain underlying principles, or themes, that unite the study of all areas of science. The themes provided here are not meant to replace the traditional teaching of scientific disciplines, but rather to create a framework for the unification of these disciplines.

The following themes are emphasized in *SciencePlus*. These themes are intended to integrate facts and ideas to provide a context for discussing the textual matter in a meaningful way. You can employ these themes as an organizational tool to reinforce understanding of the subject matter.

Energy

Energy puts matter into motion and causes it to change. Energy is what makes the universe and everything in it dynamic. The study of dynamic systems in any field of science requires an understanding of energy: its origins, how it flows through systems, how it is converted from one form to another, and how it is conserved. Energy provides the basis for all interactions, whether biological, chemical, or physical. Thematically, energy connects all disciplines.

Systems

A system is any collection of objects that influence one another. The parts of a system can be almost anything—planets, organisms, or machines, for example. And a system may be very small, such as a cell nucleus; very large, such as a galaxy; or very complex, such as the human body. All of the science disciplines involve the study of some kind of system or systems. Understanding a system involves knowing what its important parts are and how those parts work together.

Structures

Structure provides a basis for studying all matter, from the most basic forms to the most complex. Structure is closely related to function, so scientists study the structures of things to learn how they work. For example, the structure of an eye reveals much about the process of vision. And the key to a diamond's strength lies in the tight lattice structure of carbon atoms within the diamond.

Changes Over Time

A change over time is not a single alteration, but a progression of alterations that occurs over the continuum of time. Biological evolution is an example of change over time. Evolution has gradually changed the characteristics of organisms ranging from the starfish to the giraffe. Changes over time occur in physical and earth science as well. For example, the chemical change of rust forming on an iron nail can be traced through time, as can the movements of the Earth's plates. An understanding of how changes occur over time allows students to appreciate not only the present state of the world around them but also what it may have looked like in the past and what it may look like in the future.

Cycles

A cycle is a pattern of events that recurs regularly over time or a circular flow of materials in a system. Cycles occur throughout nature and appear in all of the scientific disciplines. The water cycle, the rock cycle, and the life cycle of a particular organism are examples of cycles at work. A defining feature of a cycle is that it has no beginning or end; therefore, the study of any part of a cycle is incomplete without consideration of the other elements in the cycle.

Using the Themes

A major strength of the thematic approach is that seemingly different processes, structures, or systems can be shown to have underlying similarities. Although many thematic organizations are possible, each Unit Interleaf in this Annotated Teacher's Edition suggests at least three major themes that can be discussed in relation to the unit and chapter material. Focus questions are also provided in the Wrap-Around Margins to promote discussion and an understanding of how the themes relate to specific text material.

Although major themes have been identified in each Unit Interleaf and in the Wrap-Around Margins, it is up to you to decide which themes are most appropriate. The direction that your class discussion takes will most likely guide you in your choices.

In your class discussions, use the themes to provide a framework of understanding for your students. For example, whether you are studying photosynthesis or the way in which the forces of nature have shaped the physical appearance of the Earth, the theme of **Energy** can be discussed. Similarly, one subject can be addressed from the viewpoint of many different themes. For example, in discussing an organism such as a zebra, **Energy** can be applied in a discussion of how the zebra takes in food from the environment; **Changes Over Time** can be discussed in relation to how the zebra's structures are adaptations to its environment; and **Cycles** can be introduced in a discussion of how the zebra's migration habits are based on the seasons.

Theme Connection

Changes Over Time
Focus question: When succession occurs at an abandoned farm, in what order would the following plants appear?
a. taller plants—goldenrod and grasses
b. weeds that thrive in sunny locations—wild carrots and dandelions
c. slower growing trees—oak and hickory
d. young saplings—aspens
(*Answer: b, a, d, c*)

In your class discussions, use the themes to provide a framework of understanding for your students.

INTEGRATING THE SCIENCES

Throughout *SciencePlus*, students are asked to look at things from many different viewpoints. Since *SciencePlus* is not organized according to life, earth, and physical sciences, these disciplines are integrated throughout the program. Each discipline comes into play as needed in covering a main concept or theme in *SciencePlus*. In this way, science disciplines are blended together so that the emphasis is on student comprehension of the "big picture" rather than on isolated components of different areas of science.

Additional integration suggestions are provided in the Wrap-Around Margins. Look for the Integrating the Sciences within the Teaching Strategies of the lessons. These suggestions will help you build connections between the science disciplines so that students can get a fuller, more in-depth view of the concept at hand.

Integrating the Sciences

Life, Earth, and Physical Sciences
Every field of science employs its own classification system in order to make sense of diversity. Students will learn later in this unit how biologists apply a classification system to living things. You might want to explain to the class how some other disciplines classify things in their own field.

CROSS-DISCIPLINARY CONNECTIONS

SciencePlus is also integrated in terms of non-science disciplines. For example, students are asked to write a headline for a science story, to write advertising copy highlighting geological features for a travel agency, and even to make dioramas of animal habitats and ways of life. In doing these activities, students see firsthand the connections between science and a variety of other disciplines, including history, geography, social studies, and others. And the students even have fun making the connections.

Although the Pupil's Edition contains a wide variety of cross-disciplinary activities and Explorations, additional suggestions are included in the Wrap-Around Margins of this Annotated Teacher's Edition. Look for the Cross-Disciplinary Focus in the Teaching Strategies of the lessons. Each Cross-Disciplinary Focus provides a suggestion for an activity that highlights the interrelatedness of science and disciplines such as mathematics, social studies, art, music, health, language arts, foreign language, and industrial arts.

CROSS-DISCIPLINARY FOCUS

Foreign Languages
The naming system used by scientists makes use of many Latin words. Have students make a dictionary of some of the commonly used Latin words, examples of each word's use, and the common name for the animal. For example, *Homo* means "man" and is used in such species names as *Homo erectus* and *Homo sapiens*, the name given to modern humans.

SCIENCE, TECHNOLOGY, AND SOCIETY

STS is an approach to teaching science in which the impacts of scientific, technological, and social matters are explored. STS teaches science from the context of the human experience and in so doing leads students to think of science as a social endeavor. STS emphasizes personal involvement in science. Students become active participants in the scientific experience. For this reason, *SciencePlus* incorporates an STS approach.

STS contrasts with the traditional "basic science" approach to science education, in which students follow a carefully sequenced program of the basic models, concepts, laws, and theories of science. The fundamental flaw of the basic science approach is that the student is more or less a passive participant in the learning process.

There are three parts to STS:

 Science concept and skill development, and knowledge of the nature of science

This component of STS introduces science as a system for learning about the natural world and gives students the foundation they need to actually practice science in and out of the classroom. The ultimate goal of STS, and of *SciencePlus*, is to turn students into scientists, at least for the duration of their science education. To accomplish this, students first learn the methods of science and the skills that scientists draw on. Whenever possible, the major ideas of science are introduced from the standpoint of those who developed them. In this way, students come to see the reasoning that went into the development of these ideas.

 Knowledge of the relationship of science and technology, and engagement in science-based problem solving

Students who understand the real-world applications of science are better able to appreciate and enjoy it. This component of STS reinforces the practical value of science. Students see that science is a system for solving practical problems. Students themselves become practical scientists—first identifying problems and then developing solutions to them. Students learn to analyze, to plan, to organize, to design, and to refine models and designs.

 Engagement in science-related social issues and attention to science as a social institution

This component of STS deals with the ways in which science serves human needs. The benefits may be tangible or intangible, but either way, they are real. To emphasize the social responsibility of science, scientists are shown to be concerned about the impact of their work on society as a whole. Advances in science and technology sometimes lead to thorny ethical issues. Such issues are often used as a focus for discussion and investigation. From these investigations, students draw conclusions and form reasoned opinions.

Studies have shown that students begin the study of science full of curiosity and enthusiasm. But after a few years of a traditional curriculum, the curiosity is squelched and the enthusiasm has all but vanished. By contrast, under the STS approach, student interest builds through the years and is maintained long after formal study ends. The key to this success is the active involvement in science that STS imparts. The end products of the STS approach are students who appreciate and understand science and who are equipped to deal sensibly with the complex issues that will become commonplace in the decades ahead.

> *For a fuller discussion of STS, see the NSTA position paper, July 1990.*

STS emphasizes personal involvement in science. Students become active participants in the scientific experience.

One of the most important skills that students can acquire is the ability to communicate what they have learned, both orally and in writing. *SciencePlus* challenges students to develop and communicate their mastery of new ideas in novel ways—for example, by writing for one another or for some audience other than the teacher. Students' comprehension is enhanced when they are called upon to reformulate in their own words what they have learned.

Throughout *SciencePlus*, students are called on to communicate what they have learned in many different ways. Students may be called on to interpret a passage, illustrate a paragraph, write a headline, label a diagram, or write a caption for a photo or drawing. These strategies complement the inquiry approach followed throughout *SciencePlus*.

SciencePlus challenges students to communicate ideas in novel ways.

Reading and Writing in the Classroom

The time spent in helping students prepare to read is critical in fostering comprehension. Two strategies are particularly important in the pre-reading phase of instruction: building on prior knowledge and establishing a purpose for reading.

The Power of Prior Knowledge

The amount of prior knowledge that students have about a topic directly influences their comprehension of that topic. The more students know about something, the easier it is for them to grasp new information about it. Helping students identify the information they already know about a topic assists them in relating the new information to existing knowledge.

Research strongly suggests that students tend to retain misconceptions they may have about a topic. If the text information seems to conflict with their preconceptions, students may ignore or reject new information. It is therefore extremely important to identify these misconceptions so that they may be dispelled.

Reading to Understand

One way to establish a purpose for reading is to make a study guide with questions that students can answer as they read. You can also help students learn to make their own study guides. First teach students to preview a unit by looking at all the unit headings, illustrative material, and terms and phrases in boldface or italics. This technique helps students gain a feel for the unit and helps them build a basic structure for the new information. Then show them how to use the captions, headings, and highlighted words to devise study-guide questions to answer after reading the unit. Following this method, students will read with a purpose in mind, a purpose of their own devising. You may also wish to consider using the English/Spanish Audiocassettes, which are an excellent means for previewing the unit with students.

Writing to Understand

Studies show that writing is an effective tool for improving reading. As students write, they are creating a text for others to read. The most important advice to give students about scientific writing is to strive for clarity and accuracy. These characteristics can often be achieved with simple vocabulary and short sentences. One useful approach might be to have them imagine that they are writing for a younger audience.

Keeping a Journal

One highly successful tool for improving students' performance in science is the journal. The *SciencePlus* version of the journal is called the ScienceLog. The ScienceLog has many functions. First and foremost, it is an ongoing record of students' learning. Students begin the study of a new topic by recording prior knowledge of that topic. Any misconceptions that students may have are thus exposed. As the lesson progresses, students record any and all new findings. In many cases, students find that what they learned in the activities contradicts their preconceptions.

Much of the work that students do should be recorded in their ScienceLog. The ScienceLog is a constant reminder to students that learning is occurring. Students can look back and compare their early work with later work to see and take pride in the progress they have made. To supplement their other work, you may want to ask students to briefly summarize what they have learned each week. This makes a very handy capsule history of their work.

What makes a good ScienceLog? Insofar as is possible, the ScienceLog should be neat and easy to follow. Students may organize their ScienceLogs in any of a number of ways—chronologically, by unit, by chapter, or by lesson, to name a few. Some kind of heading should set off each major entry.

A spiral-bound notebook or hard-bound lab-type notebook makes a good ScienceLog. Or you may make copies of the sample ScienceLog pages in the *Teaching Resources* or *SnackDisc* and distribute them to your students to use if you wish. These ScienceLog pages are located at the beginning of the Unit Worksheets for Unit 1.

Portfolios: What, Why, and How

Definitions and descriptions vary as to what constitutes a portfolio. Very simply, a portfolio is a collection of work that is done by the student during the course of the year. Usually, the students themselves have a say as to what goes into the portfolio, but selections should represent the objectives outlined by the curriculum. The purpose of the portfolio is to represent the student's mastery of skills and knowledge within the subject area. This may sound simple enough, and, in fact, it can be very simple. On the other hand, portfolios can be elaborate collections of work that are limited only by the teacher's and students' imaginations.

Your initial decision to use portfolios in your assessment leads to a host of other decisions that must be made. However, before you launch into making these decisions, it is a good idea to seek input from other teachers as well as from your school administration. You may even want to discuss your ideas with parents and other members of your community. This will make it easier to get feedback later as to the impact portfolios are having on student learning and attitudes.

You will also need to plan a system for organizing and managing the portfolios. You will need to establish guidelines for what types of information will be admitted into the portfolios, how and by whom the selections will be made, and when materials can be added or revised. These decisions will be based on your individual preferences and on the level and attitudes of your students. The following guidelines may help you in developing your plans.

> The most important advice to give students about scientific writing is to strive for clarity and accuracy.

- Allow students to select for themselves the sample materials that best represent their level of understanding and mastery. Although you may require that certain projects or materials be placed in the portfolios, it is advised that the students play an active role in selecting their best work. This gives students ownership and encourages them to take increasing responsibility for the quantity and quality of their work.
- Allow students to revise the selections in their portfolios at any time. The portfolios should evolve as your students' skills improve.
- Identify well in advance when you will be assessing the portfolios. Students should be given plenty of warning before the portfolios are collected for assessment.
- Determine in advance the criteria you will use for grading the portfolios. Share this criteria with the students up front. You may want to make a criteria checklist that each student can place inside his or her portfolio. This checklist would provide both you and your students with a ready reference to the grading criteria.
- You may want to keep the portfolios in the classroom or in some other area, allowing frequent but controlled access to them. This will decrease the chances of portfolios being lost. This may require a significant amount of room, depending on the nature of the portfolios.

- Make available examples of high-quality portfolios so that students can see examples of excellent work. This, of course, may be possible only after you have used portfolio assessment for at least one course.

 as a Portfolio

The ScienceLog provides an excellent opportunity for portfolio assessment. By directing your students to include in their ScienceLogs representative samples of their work, as well as comments and reflections about their samples, you can add depth to your evaluation method. To help you utilize portfolios effectively in your teaching, specific strategies are provided in Portfolio boxes in the Wrap-Around Margins of this Annotated Teacher's Edition.

PORTFOLIO

Students may wish to include their written composition in their Portfolio.

CONCEPT MAPPING

Too often, students are able to master the individual elements of a topic without truly grasping the "big picture." If students fail to understand how the elements fit together or relate to one another, they cannot truly comprehend the topic. Concept mapping is a very effective method of helping students see how individual ideas or elements connect to form a larger whole. Concept maps are a highly effective tool for helping students make those logical connections.

The most effective concept maps are those that students construct on their own. Used in this way, concept maps are both a self-teaching system and a diagnostic tool. To construct a proper concept map, the student must first examine closely his or her mental model of the topic at hand. Any flaws or shortcomings in that model will be reflected in the concept map.

Concept maps are flexible. They can be simple or highly detailed, linear or branched, hierarchical or cross-linked, or they can contain all of these major elements. Students can construct their own maps from scratch or can finish incomplete maps. Concept maps can take almost any form as long as they are logically arranged.

Making Concept Maps

The steps involved in making a concept map are outlined below. To provide guidance to your students in making concept maps, direct them to page xvii in the Pupil's Edition.

1. Make a list of the concepts to be mapped. Concepts are signified by a noun or short phrase equivalent to a noun.
2. Choose the most general, or the main, idea. Write it down and circle it.

3. Select the concept most directly related to the main idea. Place it underneath the main idea and circle it. If two or more concepts bear the same relationship to the main idea, they should be placed at the same level.
4. Draw a line between the related concepts, leaving a space for a short action phrase that shows how the concepts are related. These are linkages.
5. Continue in this way until every concept in the list is accounted for.

The simple concept map below shows the relationship among the following terms: plants, photosynthesis, carbon dioxide, water, and sun's energy. More detailed maps are shown on the next page.

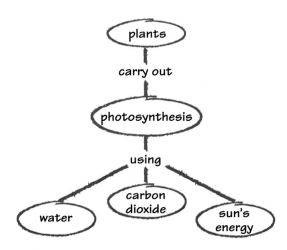

For any given topic, there is no single "correct" concept map. Not all maps are equally valid, however. Good concept maps have most or all of the following characteristics:
- start with a single, general concept—a big idea—and work down to more specific ideas
- represent each concept with a noun or short phrase, each of which appears only once
- link concepts with linkage words or short phrases
- show cross-linkages where appropriate
- consist of more than single path
- include examples where appropriate

> Concept maps are a highly effective tool for helping students make logical connections.

Using Concept Maps

Concept maps can be applied in many ways.

- to gauge prior knowledge of a topic
- as end-of-lesson, chapter, or unit evaluation
- as pre-test review
- to help summarize special presentations, such as films, videos, or guest speakers
- as an aid to note taking
- for reteaching

You may also want to use partially completed concept maps as pop quizzes or as devices for summarizing particularly difficult class sessions. Also, be sure to use the concept map in each Making Connections review.

Evaluating Concept Maps

Again, there is no single correct concept map. However, you should consider the following criteria as you evaluate your students' concept maps.

- how comprehensive the map is (Are all relationships shown?)
- how clearly the concepts are linked (proper relationship between concepts, use of linkage terms between all concepts)
- overall clarity of presentation (Could the map be simpler? Is it redundant? Is it logically arranged? Are linkage terms used properly?)

Used properly, concept maps can increase comprehension, improve retention, and sharpen study skills in your students. They are a valuable addition to any student's arsenal of learning strategies.

> The most effective concept maps are those that students construct on their own.

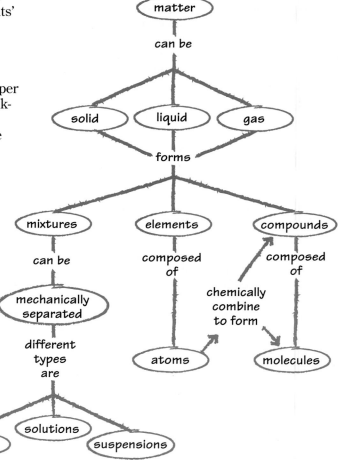

Cooperative learning is a teaching technique that brings students together to learn in small, heterogeneous groups. In these groups, students work interdependently without constant and direct supervision from the teacher. Assignments are structured so that everyone contributes. Challenges as well as rewards are shared. Brainstorming, lively discussion, and collaboration are the hallmarks of the cooperative-learning classroom.

Cooperative learning is an ideal complement to the *SciencePlus* approach. The discussions, explorations, research projects, games, and puzzles of *SciencePlus* can all be used in a cooperative-learning format.

What It's Not!
- Cooperative learning is **not** the same as ability grouping, where a teacher divides up the class in order to instruct students with similar skills.
- Cooperative learning is **not** having students sit side by side at the same table to talk while they complete individual assignments.
- Cooperative learning is **not** assigning a task to a group in which one student does the work and the others get equal credit.

Benefits of Cooperative Learning

Traditionally, teachers structure lessons so that students work individually to achieve learning goals or compete against one another to be "the best." While these formats can be useful, cooperative learning provides an important alternative.

- **Cooperative learning models the scientific experience.**
  Students working in groups learn about the joys as well as the frustrations involved in scientific inquiry. Cooperative learning models real scientific experience, in which scientists work together, not in isolation, to solve difficult problems. With cooperative learning, the classroom becomes a fertile environment for ideas and novel solutions.

- **Cooperative learning empowers and involves students.**
  Cooperative learning raises students' self-esteem because they are learning something on their own through cooperation, rather than being handed prepackaged knowledge. It helps students become self-sufficient, self-directed, lifelong learners. In a cooperative-learning environment, students are less dependent on you for knowledge.

- **Cooperative learning serves the heterogeneous classroom.**
  With group work, everyone has the chance to participate, and everyone has a role to play. As students join forces to achieve a common goal, they come to recognize commonalities that cut across differences related to ethnicity, socioeconomic background, and gender. Likewise, cooperative learning provides an excellent vehicle for students of differing ability levels to work together in a positive way. Basic students can interact successfully with average and advanced students, and in so doing can learn that they, too, have something to offer.

- **Cooperative learning strengthens interpersonal skills.**
  Group tasks are structured so that students must cooperate to succeed. Students quickly understand that they will "sink or swim" together by how constructively they interact. Consequently, students develop important interpersonal and social skills that help them function in a group setting and that will ultimately benefit them socially, at work, and in other situations.

- **Cooperative learning develops appropriate social skills.**
  When doing cooperative group work, students channel their energies into constructive tasks while satisfying their fundamental need for social interaction.

- **Cooperative learning is an effective management tool.**
  Establishing cooperative learning in the classroom requires you to relinquish some control, so the students themselves can become responsible for building their own knowledge. Working in groups to probe and investigate ideas, answer questions, and draw conclusions about observations allows students to discover and discuss concepts in their own language. When students learn through cooperation, the knowledge derived becomes their own, not just a loan of your ideas or those from the textbook.

- **Cooperative learning increases achievement.**
  Since the 1920s, there has been extensive research on cooperative-learning techniques. Results clearly indicate that cooperative learning promotes higher achievement for all grade levels in all subject areas.[1]

[1] Johnson, Johnson, Holubec, and Roy. *Circles of Learning, Cooperation in the Classroom.* Association for Supervision and Curriculum Development. ©1984

Using Cooperative Learning With *SciencePlus*

Cooperative learning is an essential component of the *SciencePlus* philosophy. To help you take full advantage of this important component, several cooperative-learning activities have been highlighted in each unit of this Annotated Teacher's Edition. These activities are broken down to provide suggestions for group size, group goal, positive interdependence, and individual accountability.

This symbol designates cooperative-learning activities in the Wrap-Around Margins of this Annotated Teacher's Edition. Cooperative Learning worksheets are are also provided in the *Teaching Resources* booklets.

Group Size Although group size will vary depending upon the activity, the optimum size for cooperative learning is between three and four students. For students unaccustomed to this learning style, keep the group size to about two or three students.

Group Goal Students need to understand what is expected of them. Identify the group goal, whether it be to master specific objectives or to create a product such as a chart, a report, or an illustration. Identify and explain the specific cooperative skills required for each activity.

Positive Interdependence A learning activity becomes cooperative only when everyone realizes that no group member can be successful unless all group members are successful. The "we're all in this together" part of group work is the positive interdependence. Encourage positive interdependence by assigning each student some meaningful role, or allow students to do this themselves. You can also encourage positive interdependence by dividing materials, resources, or information among group members.

Individual Accountability Each group member should have some specific responsibility that contributes to the learning of all group members. At the same time, each group member should reach a certain minimum level of mastery.

Meeting Individual Needs With Cooperative Learning

Cooperative learning is an effective tool for meeting the individual needs of your students. Cooperative learning builds relationships among students where relationships might not have developed before. Students are forced to interact with each other as individuals with common goals. In so doing, students learn more about each other's personal characteristics, and as a result, many stereotypes are destroyed.

There is no single set of cooperative-learning strategies that will work with all students in all situations. However, the following strategies may provide you with insight and guidance in developing your own set of strategies that will work for your students.

Balance the needs of students of all levels and learning styles
Your ultimate goal will be to ensure that all students are able to work effectively in any group. Initially, however, you may wish to develop special grouping strategies to foster the growth of learners having difficulty and second-language learners, and to assure gifted students that their grade will not be affected by slower learners. More information is provided about grouping strategies on the next page.

Have a clearly defined goal
When you tell students what is expected of them, be sensitive to their special needs. Be sure that each student understands the group goal and his or her own personal responsibility.

It is also important that assignments be specifically appropriate for groups. In other words, simply having students fill in the blanks of a worksheet or answer the end-of-unit questions is not creating an adequate cooperative-learning assignment. Students need tasks that cannot be easily completed alone. Students should see that if they work together, the end product will be better and more complete than if they had worked alone.

Answer questions only when the whole group has the same question. It's a good idea to designate one person per group as the liaison between you and the group.

Praise success
If students seem unmotivated or feel that their individual tasks are unimportant to the success of the group, you may wish to consider offering group rewards. Reward the groups as they successfully complete each activity. Reward successful project results as well as positive interaction and effective group-process skills. However, rewards should not become automatic. They should be used only for the short term. For the long term, students should take pride in their group's achievements and should benefit from the knowledge that these achievements contribute to their success as individuals.

Encourage interpersonal problem solving within groups

Pulling a disruptive student out of a group is sometimes necessary, but be sure that the isolation is only temporary. Difficult students need the support of others. Build into each group the spirit of encouraging each other. Suggest that groups evaluate their own performance after an activity is finished. Encourage students to suggest solutions to problems without criticizing individuals.

Grouping Strategies

With cooperative learning, you can either place students in particular groups or assign students to groups at random. There are advantages to both approaches. At first, however, it is recommended that you assign students to particular groups.

> *The SciencePlus SnackDisc CD-ROM and the SciencePlus Assessment Checklists and Rubrics booklet contain several checklists to help you implement cooperative-learning groups.*

Assigned Grouping

Composing groups yourself lets you create groups that are heterogeneous in terms of academic ability, gender, ethnicity, and cultural background. Heterogeneous groups are preferred because cooperation among diverse students not only teaches the widest range of interpersonal skills but also promotes frequent exchange of explanations and greater perspective in discussions. This increases depth of understanding and retention of concepts.

To create *effective* heterogeneous groups, balance each group with students who have different strengths. First decide who your resource students are. These are students you think will facilitate group work—either because of their academic ability or because of their interpersonal skills. Assign at least one resource student to each group. Distribute students who may be disruptive and students who lack academic skills evenly throughout the groups. Avoid putting close friends together to prevent cliques from disrupting teamwork. Put students who have limited English proficiency in groups with bilingual students who can act as translators.

Random Grouping

Random grouping can be especially effective with experienced cooperative learners or if you plan to change group membership often. To create random groups, you can simply have students count off from 1 to 4. All of the 1s form a group, all of the 2s form another group, and so on.

There are many other fun ways to assign groups randomly. For example, you can hold a lottery in which students pick numbers out of a hat. Numbers 1 to 4 form one group, numbers 5 to 8 form another, and so forth. You can also use cards naming sets of a particular type of item. All students who draw items belonging to the same set form a group. For example, all students whose cards name types of flowers belong in one group, students whose cards name farm animals belong in another group, those with cards naming heavy-metal bands form a third group, etc. Students have a fun and lively time discovering who belongs in the same group.

You can also combine lesson content with assigning groups. First decide how many students you want in each group. For each group, write a different scientific term or principle on a flashcard. Then for each group's term, list on separate cards the definition of the term, a synonym, or an example of what the term means. Mix up the cards and hand them out as students come into the room or once students are seated. Students use the cards as clues to find the others in their group.

Assigning Roles

Assign roles at first; students can choose their own later.

Assigning roles is very important, especially at first. Consider the behavior patterns of students, and assign roles that will complement those patterns. Group work needs to be structured so that everyone has a part to play. In other words, there needs to be positive interdependence. Just as members of a surgical team work together, with each person contributing his or her own special skill, students work effectively in teams when everyone has a unique role that is vital to the group's success.

Some examples of useful roles are listed here. Use as many as you need, modify them, combine them, or invent roles yourself.

- **Facilitator** The facilitator is a leadership role. The facilitator keeps an activity running smoothly by presiding over the work flow. He or she manages the group so that all members have a chance to talk, questions are answered, students listen to one another's ideas, and ideas are substantiated with reasons and explanations.
- **Recorder** The recorder records data and answers questions posed to the group.
- **Reporter** The reporter explains the group's findings to the teacher or the entire class.
- **Safety officer** The safety officer makes sure safety practices are followed and notifies the teacher of any unsafe situations.
- **Checker** The checker makes sure that everyone has finished his or her worksheet or other individual assignment.
- **Materials manager** The materials manager gathers activity materials at the outset, monitors their use during the activity, and organizes the cleanup and return of materials to their proper place after an activity.

Again, assign roles carefully, especially at first, taking into account students' behavior patterns. A shy student might be most comfortable as a recorder, while a student who likes attention might make the best reporter. The facilitator is a role that some students will always want and others will avoid. Be careful not to stereotype. Sometimes the most unlikely students will make the best leaders.

Assessment

Assessment within a cooperative-learning setting is not as difficult as it may seem. Like any other assessment, you must determine in advance what you would like to assess and to what degree. You will also need to develop some slightly different monitoring skills. In addition to the following information, the *SciencePlus* program includes several checklists and rubrics to make this task simple and efficient.

A variety of checklists designed for monitoring and assessing cooperative group work are provided on the SciencePlus *Snackdisc CD-ROM and in the* SciencePlus Assessment and Rubrics *booklet. Both student and teacher checklists are provided.*

Monitor Groups

Resist the temptation to get caught up on paperwork as the groups do their work—this is the time to observe, monitor, and coach. As you monitor the groups, you can reinforce cooperative behavior with a formal observation sheet. Appropriate checklists are available on the *SnackDisc* CD-ROM and in the *Assessment Checklists and Rubrics* booklet. Record how many times you observe each student using a collaborative skill, such as contributing ideas or asking questions.

If a group seems hopelessly confused or "stuck," you can intervene to guide students to a solution. But make sure students have the opportunity to reason through problems themselves first. Consider the following differences between direct supervision and the kind of monitoring that supports cooperative learning.

DIRECT SUPERVISION	SUPPORTIVE MONITORING
Lecturing	Giving feedback
Disciplining	Encouraging problem solving
Telling students what to do	Providing resources
Leading discussions	Observing

What to Assess

What should you assess in a cooperative-learning activity? Individual success? Group success? Cooperative skills? Actually, many teachers find it useful to evaluate all three. And there are many ways to assess each of these areas.

Individual success can be evaluated by asking students to fill out answers to a work-sheet as they progress through an activity; by having them record, analyze, and submit data; or by having them take a quiz. Some activities are structured so that each student turns in a product, such as a report or a poster, that can be individually graded. The individual accountability portions of the cooperative-learning features in this Annotated Teacher's Edition also provide ways to assess individual performance.

Group success is evaluated according to how well the group accomplished its assigned task. Was the task completed? Were the results accurate? If not, were errors explained and accounted for? Criteria such as these provide a framework for group evaluation.

Cooperative skills are evaluated based on your observations of students' behavior in their groups. Evaluating students' use of cooperative skills will motivate students to use them. If you intend to grade cooperative skills, it is helpful to use a formal observation checklist as you monitor students at work. Log the frequency with which group members exhibit cooperative skills or disruptive behavior.

Weigh each of the three grades as you wish to compute a single overall grade. Stress the factors that you consider most important. Use cooperative learning to meet the needs of your students. Use it and enjoy!

PROCESS SKILLS

Process skills are a means for learning and are essential to the conduct of science. For this reason, *SciencePlus* is strongly process oriented.

Perhaps the best way to teach process skills is to let students carry out scientific investigations and then point out to them the process skills they used in the course of the investigations. The Lesson Organizer at the beginning of each lesson identifies the process skills that are emphasized in the corresponding text.

SciencePlus makes regular use of many different process skills, as highlighted below. These and other process skills are called upon in *SciencePlus* in virtually every lesson.

Observing

An observation is simply a record of a sensory experience. Observations are made using all five senses. Scientists use observation skills in collecting data.

Communicating

Communicating is the process of sharing information with others. Communication can take many different forms: oral, written, nonverbal, or symbolic. Communication is essential in science, given its collaborative nature.

Measuring

Measuring is the process of making observations that can be stated in numerical terms. In *SciencePlus*, all measurements are given in SI units.

Comparing

Comparing involves assessing different objects, events, or outcomes for similarities. This skill allows students to recognize any commonality that exists between seemingly different situations. A companion skill to comparing is contrasting, in which objects, events, or outcomes are evaluated according to their differences.

Contrasting

Contrasting involves evaluating the ways in which objects, events, or outcomes are different. Contrasting is a way of finding subtle differences between otherwise similar objects, events, or outcomes.

Organizing

Organizing is the process of arranging data into a logical order so it is easier to analyze and understand. The organizing process includes sequencing, grouping, and classifying data by making tables and charts, plotting graphs, and labeling diagrams.

Classifying

Classifying involves grouping items into like categories. Items can be classified at many different levels, from the very general to the very specific.

Analyzing

The ability to analyze is critical in science. Students use analysis to determine relationships between events, to identify the separate components of a system, to diagnose causes, and to determine the reliability of data.

Inferring

Inferring is the process of drawing conclusions based on reasoning or past experience.

Hypothesizing

Hypothesizing is the process of developing testable explanations for phenomena. Testing either supports a hypothesis or refutes it.

Predicting

Predicting is the process of stating in advance the expected result of a tested hypothesis. A prediction that is accurate tends to support the hypothesis.

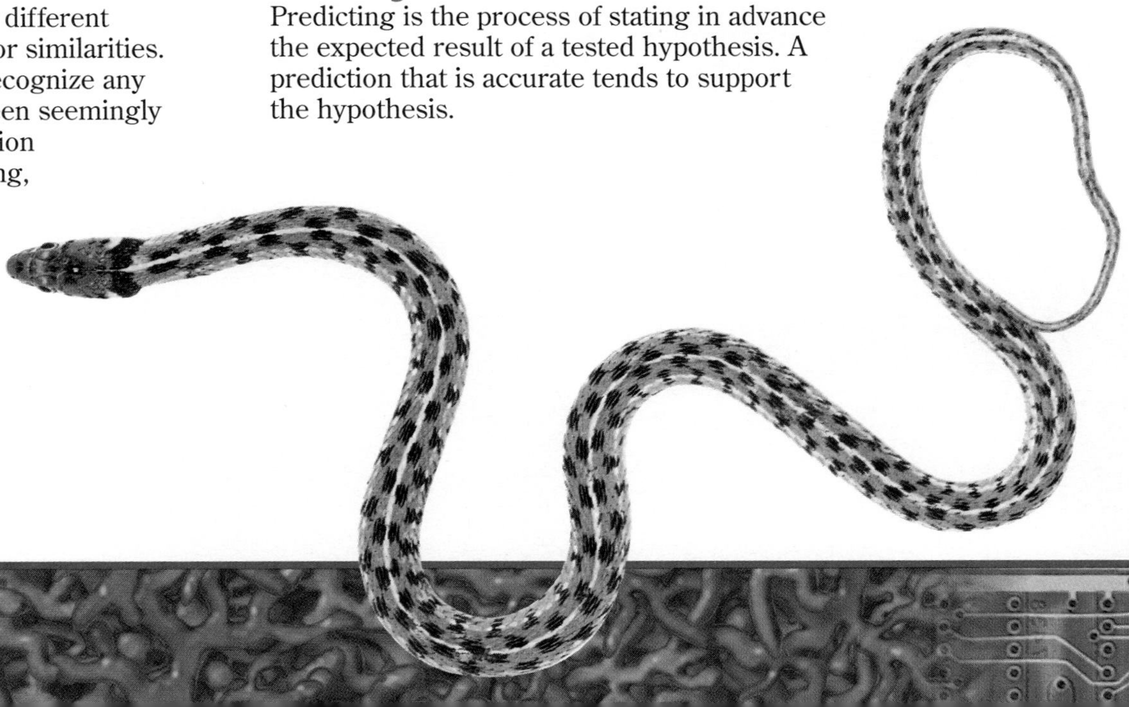

Critical-thinking skills are essential for making sense of large amounts of information. Too often, science lessons leave students with a set of facts and little ability to integrate those facts into a comprehensible whole. Requiring students to think critically as they learn improves their comprehension and increases their motivation.

Loosely defined, critical thinking is the ability to make sense of new information based on a set of criteria. Critical-thinking skills draw on higher-order thinking processes, especially synthesis and evaluation skills. Critical thinking takes a number of different forms.

Validating Facts

This type of critical thinking involves judging the validity of information presented as fact. Too often, people will accept as valid almost any statement, no matter how outrageous, as long as it comes from a supposedly authoritative source. It is important for scientists to treat all untested data with suspicion, no matter how reasonable it may seem.

Students may validate facts in a number of ways: by observing, by testing, or by rigorously examining the logic of the so-called fact. *SciencePlus* presents students with many opportunities to critically evaluate facts and hypotheses.

Making Generalizations

A scientist must often be able to identify similarities among disparate events. Generalizations are drawn based on a limited set of observations that can be applied to an entire class of phenomena. One does not have to test every substance known to make the generalization that solid substances melt when heated. Generalizations allow scientists to make predictions. Once the rule is known, future outcomes can be forecast with a high degree of confidence.

It is important that students base their generalizations on an adequate amount of information. A generalization that is formed too quickly may be wrong or incomplete or may lead the student down a dead-end path.

Making Decisions

Many students would not regard science as a field requiring decision-making skills. But in fact, scientists must make decisions routinely in the course of their work. Any time a scientist works through a problem or develops a model, a whole series of decisions must be made. A single faulty decision can throw the entire process into disarray. Making informed decisions requires knowledge, experience, and good judgment.

Interpreting Information

Having all the information in the world is useless unless one also has the tools to interpret that information. Scientists must know how to separate the meaningful information from the "noise." Information can come in any form detectable by the five senses. It is important that scientists and students alike interpret information to determine its meaning, validity, and usefulness.

> Requiring students to think critically as they learn improves their comprehension and increases their motivation.

ENVIRONMENTAL AWARENESS

No species affects its surroundings as dramatically as does the human species. Thanks to recent highly publicized events—such as Chernobyl, destruction of the rain forests, the depletion of the ozone layer, and the greenhouse effect—people have come to realize the global impact that human actions can have. It is incumbent upon the educational system to promote environmental awareness among students. *SciencePlus* addresses environmental issues in a way that students can easily grasp.

Environmental issues run the gamut from global to local. While large-scale problems get headlines, they can be hard to grasp for many students, who may have never directly observed their impact. In most cases it is best to introduce your students to local issues to start building their awareness. Local issues not only are more relevant to their lives, but also are more likely to lead to direct involvement.

Environmental awareness serves two purposes: it promotes understanding of the living world and the place of humans within it, and it produces a positive change in students' behavior toward the environment. *SciencePlus* pursues both goals. You may involve students directly in environmental issues by using the suggested activities in the Pupil's Edition.

In addition, this Annotated Teacher's Edition contains teaching strategies and special Environmental Focus boxes throughout the Wrap-Around Margins. See the sample below. The Environmental Focus boxes contain relevant environmental information that you can use to add depth to the topic at hand and to encourage discussion or some other action.

> Environmental awareness produces a positive change in students' behavior toward the environment.

ENVIRONMENTAL FOCUS

Many organisms produce poisonous chemicals called *biotoxins* that are used as a defense against predators or in capturing prey. The most-studied biotoxins are snake venoms because they are so easy to obtain. Have students do some research or contact a local government agency to find out about poisonous organisms in their area. Students can then create a booklet of information about biotoxins from poisonous plants and animals found in their area.

> Respect for your students' cultures will go a long way in helping each of your students achieve and maintain a positive self-concept.

The success of our nation would not be possible without the contributions of the many cultures and ethnic groups that make up our country. Multicultural instruction serves to ensure that all students have the chance to learn, to succeed, and to become whoever they would like to become, regardless of race, gender, socioeconomic background, or disability. Multicultural instruction affirms the positive nature of this country's diversity by helping students develop an open mind, a positive self-concept, and a realistic understanding of the world that surrounds them.

Meeting the needs of culturally diverse students is perhaps the most demanding challenge faced by today's teachers. We must constantly strive to arouse adolescent curiosity, minimize risks of failure, and be as relevant as possible to our individual students' needs. The more culturally relevant we make our science programs, the better we will be at serving our changing class populations. These challenges are especially difficult with middle-school students because they are also going through the physical and emotional changes associated with adolescence.

By their very nature, middle-school science programs have the potential for promoting the full development of individual learners—especially when science classrooms are perceived as places of inquiry and discovery. When such environments exist in science classrooms, students become more successful learners.

Relevance, Positive Self-Concept, and Multiculturalism

Let your students help you incorporate multicultural learning into the classroom by allowing them the freedom to express their feelings and attitudes during your classes. With a program tailored specifically to the personal experiences of your students, you will find that your students are more curious about the world around them. With an increased level of relevance, learning is more important to young thinkers.

Likewise, it is very important to be sensitive to the cultural identities of your students. You must view these cultural identities on an equal footing. Having a healthy respect for your students' cultures will go a long way in helping each of your students achieve and maintain a positive self-concept.

Using Multicultural Instruction

A strong program of multicultural instruction can begin by implementing a few of these basic strategies. While none of the strategies are exclusively multicultural, they can provide proper contexts and situations that capitalize on the cultural backgrounds of students.

- Recognize and convey to students that all languages are equally valid. The learning of English, however, increases the range of opportunities available to the students.
- Draw special attention to the diversity of role models in *SciencePlus*. At every opportunity, provide information about past and current scientists from diverse cultures.

  The power of such role models should not be underestimated. Role models may create interest and motivation, and may even influence a student's pursuit of a career.
- Use cooperative learning to diversify student groups. You will find that students develop more of an open-minded awareness as well as more positive, accepting, and supportive relationships with peers. Labels concerning ethnicity, gender, ability, social class, and handicaps cease to exist. For more detailed information about cooperative learning, see pages T41–T46.
- Peer and cross-age tutoring is an excellent strategy for fostering better understanding among individuals. Peer tutoring involves students tutoring students their own age. Cross-age tutoring involves older students tutoring younger students. These strategies are beneficial both to the tutors and to the students being tutored.

  In using either of these strategies, be very careful when pairing students. Although this is an excellent opportunity to integrate students, both the tutor and the student must be willing participants.

- Take every opportunity to relate science to personal experiences. Invite your students to discuss any of their own experiences that may apply. You may discover some very relevant connections and analogies, and the learning process will become more interactive and personalized to the class. This process might also add to the richness of the class by highlighting the cultural differences among your students.
- Allow students to select independent projects relevant to their own world. These projects should permit students to create new, positive avenues of self-expression from their own experiences. Students find opportunities to select, design, and articulate their own interest within science programs while developing their creative-thinking and problem-solving skills. In addition, these activities promote the development of positive attitudes toward general academics, social interactions, and the study of science.

Multicultural Instruction and *SciencePlus*

Science is for everyone, and *SciencePlus* is designed to serve the multiethnic and multicultural classrooms of today. Students, regardless of their ethnic backgrounds, will not have to look hard to find positive role models in the pages of *SciencePlus*. In addition, content that shows events, concepts, and issues from diverse ethnic and cultural perspectives is provided. As students work through *SciencePlus*, they will come to understand that science is a human endeavor that has been advanced by the contributions of many cultures and ethnic groups.

To add depth to your multicultural instruction, the Wrap-Around Margins of this Annotated Teacher's Edition periodically include a feature called Multicultural Extension. This information provides activities to help you focus on cultural diversity, highlighting the individuality and contributions of different ethnic groups.

Meeting the needs of culturally diverse students is perhaps the most demanding challenge faced by teachers today.

Obviously, to teach effectively you must be able to reach every individual in your class. This is seldom easy, given the diverse nature of most of today's classrooms. In addition, certain students present special challenges. Dealing adequately with these students requires special preparation and strategies. In many cases a minimal amount of preparation is sufficient to make the classroom a place where all can learn. Some of the more common situations you are likely to encounter are discussed below.

Learners Having Difficulty

Learners having difficulty are those who, for any of a number of reasons, are liable to perform poorly and who have a high probability of dropping out. *SciencePlus* is engaging and interesting throughout, appealing to all students. Throughout *SciencePlus*, clear easy-to-read prose and straightforward, attractive graphics reduce the potential for students to become bored. The style of *SciencePlus* is intentionally friendly and unintimidating. Field-testing has shown that the performance of students considered at-risk in science increased substantially when using *SciencePlus*.

Additional activities and teaching suggestions for learners having difficulty are provided under the learners Meeting Individual Needs: Learners Having Difficulty heading in the Wrap-Around Margins of this Annotated Teacher's Edition.

Meeting Individual Needs

Learners Having Difficulty

Using a hot plate, heat a few spoonfuls of sugar in a nonstick frying pan, stirring frequently. Once the sugar has caramelized, pour it onto wax paper to cool. Provide students with a cube of sugar and a piece of the caramelized sugar for comparison. Students should describe the metamorphosis that has taken place, compare the appearances of the two substances, and relate the change to the formation of metamorphic rock.

Second-Language Learners

Because *SciencePlus* places so much emphasis on doing science rather than reading about it, the program is ideal for students who are not proficient in English. Science is a universal language—the language of curiosity and logical reasoning. Many *SciencePlus* activities are easy to follow and require a minimum of reading. Lengthy explanations are seldom called for. You need only to get students started in the right direction; thereafter their intuition and common sense take over. The cooperative approach emphasized in *SciencePlus* helps to give second-language learners the extra support they need.

Additional activities and teaching suggestions for second-language learners are found in the Wrap-Around Margins of this Annotated Teacher's Edition. Look for the heading Meeting Individual Needs: Second-Language Learners.

Meeting Individual Needs

Second-Language Learners
Have students tell the story of a volcanic eruption using a comic-strip format. Students can create or gather a series of drawings or photographs to illustrate the sequence of events. Then encourage them to write a simple dialogue or commentary to accompany the pictures.

The Spanish Resources section of the *Teaching Resources* booklets also contains useful information, including blackline masters of parent letters translated into Spanish, as well as unit summaries and unit glossaries in Spanish.

Also available are *English/Spanish Audiocassettes,* which provide important preview information to assist Spanish-speaking students and students who are auditory learners.

Gifted Learners

The difficulty of teaching gifted students lies in keeping them interested, motivated, and challenged. Gifted students who are inadequately challenged may become bored, withdrawn, or even openly disruptive. *SciencePlus* includes many activities suitable for even the most advanced student. Open-ended activities, in particular, are especially suited for gifted students.

The *SciencePlus* approach emphasizes creative problem solving. In many cases there is no single right answer to a problem or question, so students' answers can reflect their individual abilities. This approach is ideal for gifted students, as they may extend the activities to fit their interests and talents.

Additional activities and teaching suggestions for gifted students are found in the Wrap-Around Margins of this Annotated Teacher's Edition. Look for the heading Meeting Individual Needs: Gifted Learners.

Meeting Individual Needs

Gifted Learners
Have students create their own guidebooks of local plants. Students could either sketch or photograph each plant. They can then list the common name of the plant along with its scientific name and some of its characteristics.

Physically Impaired Students

Make your classroom as easy to move about in as possible. Remove or bypass any obvious barriers. Encourage your students to assist physically impaired students. If the student uses a wheelchair, make the aisles wide enough to accommodate the chair. Make sure that the student can reach any equipment he or she needs. You may wish to enlist the aid of other students in the class to assist the disabled student as necessary.

As much as possible, adapt the classroom to make it possible for physically impaired students to engage in the same activities as other students. Use a mobile demonstration table so that it can be moved to different areas of the room for maximum visibility.

Visually Impaired Students

Seat students with marginal vision near the front of the room to maximize their view of both you and the chalkboard, or assign a student to make copies of what you write. You could also assign a student to explain all visual materials in detail as they are presented.

Students who are completely blind should be allowed to become familiar with the classroom layout before the first class begins. Promptly inform these students of any changes to your classroom layout. Whenever possible, provide blind students with Braille or taped versions of all printed materials. Blind students may also use hand-held devices for converting written text into speech.

Hearing-Impaired Students

If you have hearing-impaired students in your class, remember to always face the class while speaking. Minimize classroom noise, and arrange seating in a circle or semicircle so that hearing-impaired students can see others. This arrangement facilitates speech reading. Speak in simple, direct language and avoid digressions or sudden changes in topic. During class discussions, periodically summarize what students are saying and repeat students' questions before answering them. Use visual media such as filmstrips, overhead projectors, and close-captioned films when appropriate. You might arrange a buddy system in which another student provides copies of notes about activities and assignments.

A student who is completely deaf may require a sign-language interpreter. If so, let the student and the interpreter determine the most convenient seating arrangement. When asking the student a question, be sure to look at the student, not at the interpreter. If the student also has a speech impairment, group assignments for oral reports may be advisable.

> In many cases a minimal amount of preparation is sufficient to make the classroom a place where all can learn.

Speech-Impaired Students

Mainstreaming speech-impaired students is generally not very difficult. Patience is essential when dealing with speech-impaired students, however. For example, resist the temptation to finish sentences for a student who stutters. At the same time, do not show impatience. Also pay attention to nonverbal cues, such as facial expression and body language.

Be supportive and encouraging. You need not leave the speech-impaired student out of normal classroom discussions. For example, you may call on a speech-impaired student to answer a question and then allow the student to write out his or her response on the chalkboard or overhead projector. Use multisensory materials whenever possible to create a more comfortable learning environment for the speech-impaired student.

Learning-Disabled Students

Learning disabilities are any disorders that obstruct a person's listening, reasoning, communication, or mathematical abilities, and they range from mild to severe. An estimated 2 percent of all adolescents have some type of learning disability. Learning disabilities are the most common type of disability. Provide a supportive and structured environment in which rules and assignments are clearly stated. Use familiar words and short, simple sentences. Repeat or rephrase your instructions as needed.

Students may require extra time to complete exams or assignments, with the amount of extra time being dependent on the severity of their disability. Some students may need to tape-record lectures and answers to exam questions. For those who have difficulty organizing materials, you might provide chapter or lecture outlines for them to fill in. Having peer tutors work with learning-disabled students on specific assignments and review materials can be effective.

Computer-assisted instruction is an extremely useful tool for some learning-disabled students. This mode of instruction can even help these students develop good learning skills. For learning-disabled students, computers serve as a tireless instructor with unlimited patience. In addition, students receive simplified directions; proceed in small, manageable steps; and receive immediate reinforcement and feedback with computerized instruction.

Students With Behavioral Disorders

Behavioral disorders are emotional or behavioral disturbances that hinder a student's overall functioning. The behaviorally impaired may exhibit any of a variety of behaviors, ranging from extreme aggression to complete passivity.

Obviously, no single teaching strategy can accommodate all behavioral disorders. In addition, behavioral psychologists disagree on the best way to deal with students who have behavioral disorders. As a general rule, try to be fair and consistent, yet flexible, in your dealings with behaviorally disabled students. Make sure to state rules and expectations clearly. Reinforce desirable behavior or even approximations of such behavior, and ignore or mildly admonish undesirable behavior.

Because learning disabilities often accompany behavioral disorders, you might also wish to refer to the guidelines for learning disabilities.

> Computer-assisted instruction is an extremely useful tool for some learning-disabled students.

MATERIALS AND EQUIPMENT

SciencePlus is designed to be teachable even by those with a limited budget for materials. Most activities use common household items that can be brought to class by students or parents or that can otherwise be easily obtained. For a comprehensive listing of the materials and equipment you will need to carry out the activities of *SciencePlus*, see the *Materials Guide*. The *Materials Guide* includes both a master list of materials and a unit-by-unit list.

In addition, the second page of the parent letter contained in the *Teaching Resources* booklets lists the supplies needed for each unit. This page makes it easy for you to invite donations from parents to keep your budget as low as possible.

For teachers who would rather have the convenience of purchasing supplies through the mail, Science Kit is the official *SciencePlus* materials and equipment supplier. For your convenience, Science Kit offers kits that contain the materials and equipment you will need to teach each unit of *SciencePlus*. Or, if you prefer, you can order needed materials and equipment individually as necessary.

Science Kit ordering information is available from your local HRW representative.

SCIENCEPLUS TEACHER'S NETWORK

The *SciencePlus* Teacher's Network (SPTN) is part of the Atlantic Science Curriculum Project (ASCP), linking teaching, curriculum development, and research in science education. The ASCP, which produced the Canadian version of *SciencePlus*, has existed for over 15 years, beginning as a collaborative, grass-roots activity to improve science teaching in middle schools and junior high schools. The goal of the SPTN is to continue this collaborative effort, linking teachers with teachers through newsletters, electronic bulletin boards, conferences, and small-group and regional meetings.

The *SciencePlus Communicator* is the official U.S. publication of the SPTN. The communicator is written by and for *SciencePlus* teachers. It provides a forum for *SciencePlus* teachers to share their classroom experiences with other *SciencePlus* teachers. Each edition includes copy masters, activities, current research, assessment ideas, and much more that can help make your teaching more effective. The *SciencePlus Communicator* links you to a community of like-minded teachers whose breadth of experience can provide encouragement and support.

ASSESSING STUDENT PERFORMANCE

A COMPREHENSIVE APPROACH TO ASSESSMENT

Developing strategies for assessing student progress is an important step in realizing the goals of *SciencePlus*. Students pay the most attention to those aspects of a lesson on which they know they will be graded. Teachers who want their students to be successful should therefore teach with continual assessment in mind.

In *SciencePlus*, there is no distinct boundary between teaching and assessing. You will find that most of the tests and assessment activities in this program are designed to teach as well as to evaluate comprehension and performance. This emphasis can help correct the preoccupation with measuring and sorting students. The suggestions here are intended to aid you in your primary task: teaching.

ASSESSMENT AIDS IN *SCIENCEPLUS*

SciencePlus includes a wide variety of assessment aids to help you measure your students' mastery of the concepts and processes covered in this course. In addition to the assessment materials contained in the Pupil's Edition, including the Challenge Your Thinking chapter reviews, Making Connections unit reviews, and SourceBook Unit CheckUp reviews, the following materials are available for comprehensive and convenient assessment of your students.

In the *Teaching Resources* booklets
Blackline-master tests are available in the *Teaching Resources* booklets in the following categories:
- **Chapter Assessment**
- **Activity Assessment**
- **End-of-Unit Assessment**
- **SourceBook Assessment**

In addition, a checklist is available to help you implement self-assessment into your classroom. The **Self-Evaluation of Achievement** checklist will allow you to add this facet of assessment to your other assessment methods. The checklists are tailored to each unit in the Pupil's Edition.

In the *Assessment Checklists and Rubrics* booklet
This booklet contains over 40 different checklists for student self-evaluation, peer evaluation, and teacher assessment. Checklists are also provided to aid ongoing assessment.

This booklet contains a variety of assessment rubrics that serve as models for grading writing assignments, portfolios, reports, presentations, experiments, and technology projects.

Several progress reports are also provided to help you keep track of your students' progress.

On the *SnackDisc* CD-ROM
All of the checklists and rubrics contained in the *Assessment Checklists and Rubrics* booklet are also available on the *SnackDisc*. This CD-ROM for Macintosh® and Windows® makes these assessment materials extremely easy to access and fully customizable. Now you can quickly and easily create assessment checklists and rubrics that fit your own criteria and class situation. You can add to the recommended criteria or replace the criteria with your own.

In the *Test Generator* and *Test Item Listing* booklet

The *Test Generator* for Macintosh® and Windows® contains five categories of tests for each unit of the Pupil's Edition: Chapter Assessment, End-of-Unit Assessment, SourceBook Assessment, Activity Assessment, and Extra Assessment Items. The Extra Assessment Items consist of questions found only in the *Test Generator* and *Test Item Listing*. All of the other tests have blackline masters in the *Teaching Resources* booklets.

The *Test Generator* is easy to use, includes graphics, and is fully customizable. You can easily change the questions or add questions of your own.

The *Test Item Listing* booklet is a handy way to preview the tests and questions contained in the *Test Generator*.

In the *Annotated Teacher's Edition*

Each Unit Interleaf in this Annotated Teacher's Edition contains a comprehensive **Assessment Planning Guide.** This guide identifies, in chart form, all of the assessment components available in the program. A sample Assessment Planning Guide is shown below.

The Wrap-Around Margins of this Annotated Teacher's Edition also contain valuable assessment options. Each lesson contains a Follow-Up section that includes an optional assessment strategy. See page T28 for an example of these strategies.

Assessment Planning Guide

Lesson, Chapter, and Unit Assessment	SourceBook Assessment	Ongoing and Activity Assessment	Portfolio and Student-Centered Assessment
Lesson Assessment Follow-Up: see Teacher's Edition margin, pp. 303, 313, 323, 338, and 343 **Chapter Assessment** Chapter 14 Review Worksheet, p. 35 Chapter 14 Assessment Worksheet, p. 38* Chapter 15 Review Worksheet, p. 46 Chapter 15 Assessment Worksheet, p. 48* **Unit Assessment** Unit 5 Review Worksheet, p. 52 End-of-Unit Assessment Worksheet, p. 54*	SourceBook Unit Review Worksheet, p. 64 SourceBook Assessment Worksheet, p. 69* 	Activity Assessment Worksheet, p. 59* **SnackDisc** Ongoing Assessment Checklists ♦ Teacher Evaluation Checklists ♦ Progress Reports ♦	Portfolio: see Teacher's Edition margin, pp. 298, 309, 320, 323, and 334 **SnackDisc** Self-Evaluation Checklists ♦ Peer Evaluation Checklists ♦ Group Evaluation Checklists ♦ Portfolio Evaluation Checklists 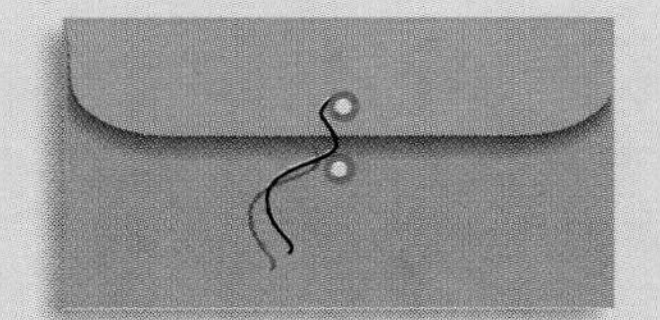 ♦

* Also available on the Test Generator software
♦ Also available in the Assessment Checklists and Rubrics booklet

Assessment should be ongoing and should measure performance on exams and quality of class work. Homework, lab work, and ScienceLog entries should all be factors in assigning grades.

The authors strongly discourage reliance on recall-based assessment strategies. Teachers who currently rely heavily on such assessment strategies may find it difficult at first to adopt new methods of assessment. However, once the transition is made, the reward—in the form of improved student performance and motivation—will more than offset the inconvenience.

ScienceLog Assessment

SciencePlus provides many opportunities for students to demonstrate their understanding of specific concepts. The first page of each chapter is devoted to getting students to think about what they already know about the concepts in the chapter. Students are asked several questions and are encouraged to write down what they already know or think they know. After students complete the chapter, they are

given the opportunity to revise their ScienceLog entries in the Challenge Your Thinking chapter review. Here students will confront any misconceptions they may have had in the beginning. In this way students actually assess their own prior knowledge and make adjustments accordingly.

Although you should not grade these ScienceLog entries beyond checking that students have done them, viewing students' initial entries can give you a good idea of their understanding of the main concepts. In this way, these entries can provide you with an excellent diagnostic tool to determine where to start your teaching and what concepts will need the most emphaisis.

By checking your students' revised ScienceLog entries, you can get a good idea of the progress that has been made as well as what topics may need to be revisited to ensure understanding.

In addition, the ScienceLog serves as a companion notebook for most of the written work that is assigned throughout the program. You may want to have students hand in their entire ScienceLog periodically to check their work and progress.

Portfolio Assessment

SciencePlus is ideally suited to the use of Portfolios as one method of assessing your students' performance and accomplishments. For more information about Portfolios and their use in *SciencePlus*, see page T38 of this Annotated Teacher's Edition.

For help in assessing student Portfolios, several checklists and rubrics have been provided in the *Assessment Checklists and Rubrics* booklet and on the *SnackDisc*.

Assessing Scientific, Psychomotor, and Communication Skills

In *SciencePlus*, knowledge and understanding are closely linked to the development of important process skills such as observing, measuring, graphing, writing, predicting, inferring, analyzing, and hypothesizing. All learning tasks are designed to help develop these skills. The teacher can assess such skill development by inspecting student work and by observing student performance.

The sample tables below are suitable models for evaluating student performance.

Assessing Environmental Awareness

SciencePlus was written with a commitment to environmental awareness. Many activities that promote such awareness are included. The teacher is provided with suggestions on extending this theme through creative projects, cleanup or recycling projects, and so on. Tasks such as these promote environmental consciousness. The care that students take in carrying out these activities is a measure of their awareness of environmental issues.

A successful science program is reflected in a student body with outside interests in science.

ASSESSING SCIENTIFIC BEHAVIOR					
Behavior	Poor	Satisfactory	Good	Very Good	Excellent
• Cooperates with others in small groups					
• Observes and records observations					
• etc.					

ASSESSING TECHNICAL SKILLS			
Task	Yes	No	Uncertain
• Is able to read thermometer correctly			
• Is able to use spring scale to measure force			
• etc.			

Assessing Scientific Attitudes

It can be useful to survey your students about the types of science-related hobbies and interests that they pursue outside of class. In a direct way, this provides feedback on the success of your school's science program. A successful science program is reflected in a student body with outside interests in science. Ask your students to keep a tally of any science-related activities they undertake outside of class. These could include reading or writing about science and technology, science-related projects, visits to museums, attending lectures on scientific and technological topics, and viewing science programs on television.

For developing and assessing individual students' interest in and attitudes toward science, the assignment of elective reading and independent projects is essential. In addition to the numerous project ideas and extension activities included in the units of *SciencePlus*, the Science in Action features provide a range of project ideas from which students may choose. Students may also want to read further about a topic in the SourceBook or in another science-related book or magazine.

Student work on elective projects should count as a significant part of overall assessment. This type of work provides the surest indication of a student's interest and proficiency in science, especially the student's ability to study, plan, and research independently.

A Balanced Assessment

The authors of *SciencePlus* recommend a balance between the different forms of assessment. As a general rule, the following proportions are suggested:

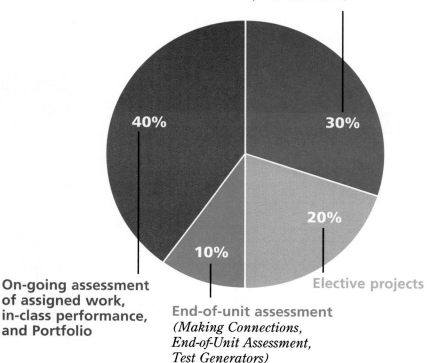

End-of-chapter assessment
(Challenge Your Thinking, Chapter Assesssment, Test Generator)

40%

30%

20%

10%

On-going assessment of assigned work, in-class performance, and Portfolio

End-of-unit assessment
(Making Connections, End-of-Unit Assessment, Test Generators)

Elective projects

Assessing Science Projects

SciencePlus offers students abundant opportunities for independent investigation. Many open-ended, curiosity-stimulating questions are posed to students. Some of these questions are natural starting points for science projects. Students using *SciencePlus* have often developed successful science-fair projects based on questions in the text. The Wrap-Around margins of this Annotated Teacher's Edition also contain additional project ideas where appropriate.

Undertaking a project provides students with a host of positive experiences. Students learn to organize, plan, and piece together many separate ideas and pieces of information into a coherent whole. Undertaking a project also allows students to experience the sense of accomplishment that comes from tackling and completing a difficult task. It has even been argued that no science education is complete without having undertaken and completed a major project.

Many students will resist the idea of undertaking a major project because they feel that it is too much work or that they are simply not up to the task. The following suggestions may help you overcome students' reluctance:

- Allow students select their own project ideas.
- Encourage students to be creative.
- Provide a clear set of guidelines for developing and completing projects.
- Help students locate sources of information, including people in science-related fields who might advise students about their projects.
- Allow students the option of presenting their completed projects to the class.
- Emphasize the satisfaction students will derive from completing their projects.
- Inform students of the general areas on which assessment may be made, such as scientific thought, originality, and presentation.
- Do not emphasize the details of assessment. "Scoring points" should not be a major incentive.

Do not allow preconceived notions about how the project should be done to detract from the students' interest, enjoyment, and satisfaction in doing original work. Rather than forcing all students to fit their project work into a mold suited to scientific research, establish three sets of criteria, as described in the tables on pages T62 to T64.

DEVELOPING ASSESSMENT ITEMS

When developing tests, you should bear in mind the sort of skill required to answer each assessment item. The manner of testing determines what is learned: tests that require tick-mark responses teach tick-marking. Tests that require verbal, graphic, illustrative, and numeric responses develop writing, speaking, graphing, drawing, and mathematical skills. A superior test draws on as many skills as possible.

The assessment-item development model below is designed to help you construct comprehensive tests that will meet the educational goals of *SciencePlus*. The model features four categories (verbal, graphic, illustrative, and numeric) and twelve types of items. The variety of tests and test items that accompany this program have been developed in accordance with this model.

ASSESSMENT-ITEM DEVELOPMENT MODEL

VERBAL

Word Usage: The words given are to be used in a prescribed situation.

Correction/Completion: Incorrect or incomplete sentences and paragraphs are given for correction or completion.

Short Essays: Information is given or a question is posed for short-essay response.

Short Responses: Answers to these questions require a tick mark, a line, or a single word, phrase, or sentence.

ILLUSTRATIVE

Illustrations for Interpretation: Illustrations (drawings or photographs) are presented for interpretation.

Illustrations for Correction or Completion: An incorrect or incomplete illustration is given for correction or completion.

Answering by Illustration: A question requiring a drawing as the expected answer is asked.

GRAPHIC

Graphs for Interpretation: A graph of a relationship between two variables is given for interpretation.

Graphs for Correction or Completion: An incorrect or incomplete graph is given for correction or completion.

Graphing Data: Data are given to be graphed.

NUMERIC

Data for Interpretation: A data table is given for interpretation.

Numerical Problems: A problem requiring a numerical solution is given.

> The manner of testing determines what is learned: tests that require tick-mark responses teach tick-marking.

Rubric for Reports and Presentations

SCIENTIFIC THOUGHT (40 POINTS POSSIBLE)				
40–36	35–31	30–26	25–21	20–10
Complete understanding of topic; topic extensively researched; variety of primary and secondary sources used and cited; proper and effective use of scientific vocabulary and terminology	Good understanding of topic; well-researched; a variety of sources used and cited; good use of scientific vocabulary and terminology	Acceptable understanding of topic; adequate research evident; sources cited; adequate use of scientific terms	Poor understanding of topic; inadequate research; little use of scientific terms	Lacks an understanding of topic; very little research, if any; incorrect use of scientific terms

ORAL PRESENTATION (30 POINTS POSSIBLE)				
30–27	26–23	22–19	18–16	15–5
Clear, concise, engaging presentation; well-supported by use of multisensory aids; scientific content effectively communicated to peer group	Well-organized, interesting, confident presentation supported by multisensory aids; scientific content communicated to peer group	Presentation acceptable; only modestly effective in communicating science content to peer group	Presentation lacks clarity and organization; ineffective in communicating science content to peer group	Poor presentation; does not communicate science content to peer group

EXHIBIT OR DISPLAY (30 POINTS POSSIBLE)				
30–27	26–23	22–19	18–16	15–5
Exhibit layout self-explanatory, and successfully incorporates a multisensory approach; creative use of materials	Layout logical, concise, and can be followed easily; materials used in exhibit appropriate and effective	Acceptable layout of exhibit; materials used appropriately	Organization of layout could be improved; better materials could have been chosen	Exhibit layout lacks organization and is difficult to understand; poor and ineffective use of materials

Rubric for Experiments

SCIENTIFIC THOUGHT (40 POINTS POSSIBLE)

40–36	35–5
An attempt to design and conduct an experiment or project with all important variables controlled	An attempt to design an experiment or project, but with inadequate control of significant variables

ORIGINALITY (16 POINTS POSSIBLE)

16–14	13–11	10–8	7–5	4–2
Original, resourceful, or novel approach; creative design and use of equipment	Imaginative extension of standard approach and use of equipment	Standard approach and good treatment of current topic	Incomplete and unimaginative use of resources	Lacks creativity in both topic and resources

PRESENTATION (24 POINTS POSSIBLE)

24–21	20–17	16–13	12–9	8–5
Clear, concise, confident presentation; proper and effective use of vocabulary and terminology; complete understanding of topic; able to arrive at conclusions	Well-organized, clear presentation; good use of scientific vocabulary and terminology; good understanding of topic	Presentation acceptable; adequate use of scientific terms; acceptable understanding of topic	Presentation lacks clarity and organization; little use of scientific terms and vocabulary; poor understanding of topic	Poor presentation; cannot explain topic; scientific terminology lacking or confused; lacks understanding of topic

EXHIBIT (20 POINTS POSSIBLE)

20-19	18-16	15-13	12-11	10-6
Exhibit layout self-explanatory, and successfully incorporates a multisensory approach; creative and very effective use of materials	Layout logical, concise, and can be followed easily; materials used appropriate and effective	Acceptable layout; materials used appropriately	Organization of layout could be improved; better materials could have been chosen	Layout lacks organization and is difficult to understand; poor and ineffective use of materials

Rubric for Technology Projects

SCIENTIFIC TECHNICAL THOUGHT (40 POINTS POSSIBLE)

40–36	35–31	30–26	25–21	20–10
An attempted design solution to a technical problem; the problem is significant and stated clearly; the solution reveals creative thought and imagination; underlying technical and scientific principles are very well understood	An attempted design solution to a technical problem; the solution may be a standard one for similar problems; underlying technical and scientific principles are recognized and understood	A working model; underlying technical and scientific principles are well understood; model is built from a standard blueprint or design	Model is built from a standard blueprint or design or from a kit; underlying technical and scientific principles are recognized but not necessarily understood	Model is built from a kit; underlying technical and scientific principles are not recognized or understood

PRESENTATION (30 POINTS POSSIBLE)

30–27	26–23	22–19	18–16	15–5
Clear, concise, confident presentation; proper and effective use of vocabulary and terminology; complete understanding of topic; able to extrapolate	Well-organized, clear presentation; good use of scientific vocabulary and terminology; good understanding of topic	Presentation acceptable; adequate use of scientific terms; acceptable understanding of topic	Presentation lacks clarity and organization; little use of scientific terms and vocabulary; poor understanding of topic	Poor presentation; cannot explain topic; scientific terminology lacking or confused; lacks understanding of topic

EXHIBIT (30 POINTS POSSIBLE)

30–27	26–23	22–19	18–16	15–5
Exhibit layout self-explanatory, and successfully incorporates a good sensory approach; creative and very effective use of material	Layout logical, concise, and easy to follow; materials used in exhibit appropriate and effective	Acceptable layout of exhibit; materials used appropriately	Organization of layout could be improved; better materials could have been chosen	Layout lacks organization and is difficult to understand; poor and ineffective use of materials

SCIENCE PLUS

TECHNOLOGY AND SOCIETY

TO THE STUDENT

This book was written with you in mind!

There are many things to try, to create, and to investigate—both in and out of class. There are stories to read, articles to think about, puzzles to solve, and even games to play.

GET INVOLVED!

The best way to learn is by doing. In the words of an old Chinese proverb:

Tell me—*I will forget*
Show me—*I may remember*
Involve me—*I will understand*

The activities in this book will allow you to make some basic and important scientific discoveries on your own. You will be acting much like the early investigators in science who, without expensive or complicated equipment, contributed so much to our knowledge.

What these early investigators had, and had in abundance, was curiosity and imagination. If you have these qualities, you are in good company! And if you develop sharp scientific skills, who knows?— you might make your own contributions to science someday.

Scientists are usually interested in understanding things that happen in nature. However, the discoveries that scientists make are often used by inventors and engineers. Using science in this way has resulted in our most sophisticated technology, including such things as computers, laser discs, nuclear reactors, and instant global communication.

SCIENCE & TECHNOLOGY

There is an interaction between science and technology. Science makes technology possible. On the other hand, the products of technology are used to make further scientific discoveries. In fact, much of the scientific work that is done today has become so technically complicated and expensive that no one person can do it entirely alone. But make no mistake, the creative ideas for even the most highly technical and expensive scientific work still come from individuals.

A built-in reference section is located at the back of this book. It's called the SourceBook. **CHECK IT OUT!**

GO FOR IT!

Science is a process of discovery, a trek into the unknown. The skills you develop as you do the activities in this book—like observing, experimenting, and explaining observations and ideas—are the skills you will need in order to be a part of science in the future. There is a universe of scientific exploration and discovery awaiting those who take up the challenge.

Keep a *ScienceLog*

A journal is an important tool in creative work. In this book, you will be asked to keep a type of journal, called a ScienceLog, to record your thoughts, observations, experiments, and conclusions. As you develop your ScienceLog, you will see your own ideas taking shape over time. This is often the way scientists arrive at new discoveries. You too may log some discoveries as you develop your own journal.

SAFETY FIRST!

The study of science is challenging and fun, but it can also be dangerous. Don't take any chances! Follow the guidelines listed here, as well as safety information provided in the particular Exploration you are doing. Also, follow your teacher's instructions and don't take shortcuts—even when you think there is little or no danger.

Accidents can be avoided. The major causes of laboratory accidents are carelessness, lack of attention, and inappropriate behavior. These things reflect a person's attitude. By adopting a positive attitude and by following all safety guidelines, you can greatly reduce your chances of having an accident. Even a minor accident in a science laboratory can cause major injuries, so be very careful.

SAFETY GUIDELINES

GENERAL

Always get your teacher's permission before attempting any laboratory explorations. Read the procedures carefully, paying particular attention to safety information and caution statements. If you are unsure about what a safety symbol means, look it up here or ask your teacher. You cannot be too careful when it comes to safety! If an accident does occur, inform your teacher immediately, regardless of how minor you think the accident is.

EYE SAFETY

Wear safety goggles when working around chemicals, acids, bases, or any type of flame or heating device, and any other time when there is even the slightest chance that harm could come to your eyes. If any substance gets into your eyes, notify your teacher immediately. Treat any unknown chemical as if it were a dangerous chemical. Never look directly into the sun with an optical device, and never use direct sunlight as a light source for a microscope.

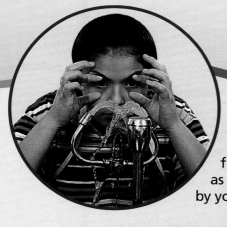

SAFETY EQUIPMENT

Know the location of and how to use the nearest fire alarms and any other safety equipment, such as fire blankets and eyewash fountains, as identified by your teacher.

NEATNESS

Keep your work area free of all unnecessary books and papers. Tie back long hair and secure loose sleeves or other loose articles of clothing such as ties and bows. Remove dangling jewelry. Don't wear open-toed shoes or sandals in laboratory situations. Never eat, drink, or apply cosmetics in a laboratory setting; food, drink, and cosmetics can easily become contaminated with dangerous materials.

SHARP/POINTED OBJECTS

Use knives and other sharp instruments with extreme care. Never cut objects while holding them in your hands. Place objects on a suitable work surface for cutting.

HEAT

Wear safety goggles when using a heating device or a flame. Whenever possible, use an electric hot plate instead of a flame as a heat source. When heating materials in a test tube, always slant the test tube away from yourself and others. Wear oven mitts, when instructed to do so, to avoid burns.

ELECTRICITY

Be careful with electrical wiring. When using a microscope with a lamp, do not place the cord where it could cause someone to trip. Do not let cords hang over a table edge in a way that could cause equipment to fall if the cord is accidentally pulled. Do not use equipment with damaged cords. Be sure your hands are dry and that the electrical equipment is in the "off" position before plugging it in. Turn off equipment when you are done.

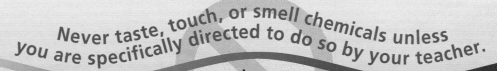

Never taste, touch, or smell chemicals unless you are specifically directed to do so by your teacher.

If you are instructed to note the odor of a substance, wave the fumes toward your nose with your hand. Never put your nose close to the source. Never mix chemicals unless you are told to do so by your teacher.

CHEMICALS

Wear safety goggles when handling any potentially dangerous chemicals, acids, or bases. If a chemical is unknown, handle it as you would a dangerous chemical. Wear an apron and latex gloves when working with acids or bases or when told to do so in the Exploration or Activity. If a spill gets on your skin or clothing, rinse it off immediately with water for at least 5 minutes while calling your teacher.

ANIMAL SAFETY

Always obtain your teacher's permission before bringing any animal into the school building. Handle animals only as your teacher directs. Always treat animals carefully and with respect. Wash your hands thoroughly after handling any animal.

PLANT SAFETY

Do not eat any part of a plant or plant seed used in the laboratory. Wash hands thoroughly after handling any part of a plant.

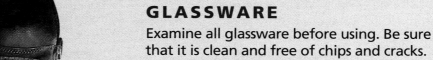

GLASSWARE

Examine all glassware before using. Be sure that it is clean and free of chips and cracks. Report damaged glassware to your teacher. Glass containers used for heating should be made of heat-resistant glass.

CLEANUP

Before leaving, clean up your work area. Put away all equipment and supplies. Dispose of all chemicals and other materials as directed by your teacher. Make sure water, gas, burners, and electric hot plates are turned off. Hot plates and other electrical equipment should also be unplugged. Wash hands with soap and water after working in a laboratory situation.

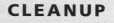

concept mapping

A Way to Bring Ideas Together

What Is a Concept Map?

Have you ever tried to tell someone about a book or a chapter you've just read, and you find that you can remember only a few isolated words and ideas? Or maybe you've memorized facts for a test, and then weeks later you're not even sure what topic those facts are related to.

In both cases, you may have understood the ideas or concepts by themselves, but not in relation to one another. If you could somehow link the ideas together, you would probably understand them better and remember them longer. This is something a concept map can help you do. A concept map is a visual way of choosing how ideas or concepts fit together. It can help you see the "big picture."

How to Make a Concept Map

1. **Make a list of the main ideas or concepts.**

   It might help to write each concept on its own slip of paper. This will make it easier to rearrange the concepts as many times as you need to before you've made sense of how the concepts are connected. After you've made a few concept maps this way, you can go directly from writing your list to actually making the map.

2. **Spread out the slips on a sheet of paper, and arrange the concepts in order from the most general to the most specific.**

   Put the most general concept at the top and circle it. Ask yourself, "How does this concept relate to the remaining concepts?" As you see the relationships, arrange the concepts in order from general to specific.

3. **Connect the related concepts with lines.**

4. **On each line, write an action word or short phrase that shows how the concepts are related.**

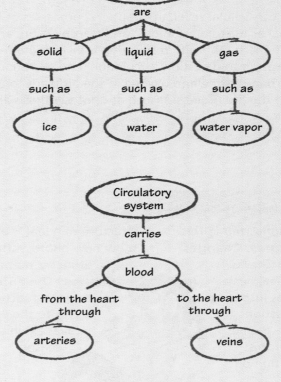

Look at the concept maps on this page and then see if you can make one for the following terms: **plants, water, photosynthesis, carbon dioxide, and sun's energy.**

The answer is provided below, but don't look at it until you try the concept map yourself.

Unit 1  Life Processes

Bibliography for Teachers

Attenborough, David. *The Private Life of Plants.* Princeton, NJ: Princeton University Press, 1995.

Benards, Neal, ed. *The Environment in Crisis: Opposing Viewpoints.* San Diego, CA: Greenhaven Press, Inc., 1991.

Bibliography for Students

Lamb, Marjorie. *Two Minutes a Day for a Greener Planet.* New York City, NY: Harper Paperbacks, 1990.

Newman, Arnold. *Tropical Rainforest.* New York City, NY: Facts on File, Inc., 1990.

The Visual Dictionary of Plants. New York City, NY: Dorling Kindersley, Inc., 1992.

Films, Videotapes, Software, and Other Media

Diffusion and Osmosis
Videotape
Britannica
310 S. Michigan Ave.
Chicago, IL 60604

Green Plants and Sunlight
Videotape
Britannica
310 S. Michigan Ave.
Chicago, IL 60604

Photosynthesis
Software (Macintosh, Windows)
Sunburst
101 Castleton St.
P.O. Box 100
Pleasantville, NY 10570

Plants: Parts and Processes
Film
National Geographic Society
Educational Services
P.O. Box 98019
Washington, D.C. 20090-8019

The Water Cycle
Videotape
Barr Media Group
12801 Schabarum Ave.
Irwindale, CA 91706-7878

Unit Overview

In this unit, students explore some of the processes of living things. In Chapter 1, photosynthesis is examined at the cellular level. In Chapter 2, processes of water movement that affect plant life, such as diffusion, osmosis, and transpiration, are explored. In Chapter 3, the release of energy through respiration is investigated. Students then study the relationship of respiration to digestion, circulation, and excretion. An explanation of the balance between carbon dioxide and oxygen in the air leads to a discussion of the effect of air pollution on the life processes (such as photosynthesis and respiration) of all living things.

Using the Themes

The unifying themes emphasized in this unit are **Energy, Structures, Changes Over Time,** and **Cycles.** The following information will help you incorporate these themes into your teaching plan. Focus questions that correspond to these themes appear in the margins of this Annotated Teacher's Edition on pages 13, 21, 23, 32, and 62.

Energy is a dominant theme in this unit. It is particularly important in Chapter 1, in which students learn about photosynthesis, and in Chapter 3, in which students examine the release of energy from food by respiration.

Students study many examples of **Structures** throughout the unit. In Chapter 1, the basic structure of a leaf is examined in order to show how photosynthesis works.

Students may also consider **Changes Over Time** in Chapter 1 by relating the function and chemistry of leaves to the color changes that leaves exhibit from season to season.

Chapter 2, Lesson 2, illustrates the theme of **Cycles.** Students follow the path of transpired water from plants to water vapor, to clouds and precipitation, and finally to water taken up by plant roots. In Chapter 3, students learn about how deforestation of the world's tropical rain forests disrupts the water cycle.

Using the SourceBook

Unit 1 focuses on the organic chemistry of life, with an introduction to the important groups of biomolecules: carbohydrates, lipids, and proteins. The chemistry and physics of water are then discussed, with an emphasis on water's role in many life processes. The chemistry of other life processes, such as photosynthesis, cellular respiration, diffusion, and osmosis, are also described in detail. The unit ends with an overview of metabolic processes.

Unit Organizer

Unit/Chapter	Lesson	Time*	Objectives	Teaching Resources
Unit Opener, p. 2				Science Sleuths: The Nature House Malady English/Spanish Audiocassettes Home Connection, p. 1
Chapter 1, p. 4	Lesson 1, Where Does the Energy Come From? p. 5	2	1. Explain how all of our food comes, directly or indirectly, from plants. 2. Describe the basic process of photosynthesis.	Exploration Worksheet, p. 3
	Lesson 2, Where Does the Starch Come From? p. 10	3	1. Demonstrate that carbon dioxide, water, and light are necessary for photosynthesis to occur. 2. Explain that plants make sugar and oxygen during photosynthesis and that the sugar is stored as starch.	Exploration Worksheet, p. 6 Exploration Worksheet, p. 8 Exploration Worksheet, p. 9 Transparency Worksheet, p. 11 ▼
	Lesson 3, Food Factory Basics, p. 17	4	1. Describe the different structures that make up the external features of a leaf. 2. Describe the structure and function of the cells that make up the internal structure of a leaf. 3. Explain the function of chlorophyll during photosynthesis.	Image and Activity Bank 1-3 Transparency 2 Exploration Worksheet, p. 13 Transparency 3 Exploration Worksheet, p. 15 Exploration Worksheet, p. 16 Activity Worksheet, p. 18 ▼
End of Chapter, p. 26				Chapter 1 Review Worksheet, p. 19 Chapter 1 Assessment Worksheet, p. 23
Chapter 2, p. 28	Lesson 1, The Indispensable Chemical, p. 29	1	1. Identify several conditions that allow for the formation of liquid water on Earth. 2. Describe several characteristics of water that make it valuable to living things.	Image and Activity Bank 2-1 Transparency Worksheet, p. 25 ▼
	Lesson 2, The Water Cycle, p. 31	2	1. Demonstrate that plants give off water through their leaves. 2. Describe how transpired water fits into the water cycle.	Image and Activity Bank 2-2 Exploration Worksheet, p. 27 ▼
	Lesson 3, Spreading Water Around, p. 35	3	1. Describe how water is absorbed by the roots of plants. 2. Explain how and why diffusion takes place. 3. Explain the difference between permeable, semipermeable, and impermeable membranes.	Image and Activity Bank 2-3 Transparency 7 Exploration Worksheet, p. 30 Exploration Worksheet, p. 32
	Lesson 4, Osmosis: Controlled Diffusion, p. 43	2	1. Describe the relationship between osmosis and diffusion. 2. Explain how osmosis affects the rigidity of plant cells. 3. Identify factors that determine the rate of osmosis in plants.	Image and Activity Bank 2-4 Exploration Worksheet, p. 34 Exploration Worksheet, p. 37
	Lesson 5, Putting Water Where It Counts, p. 47	2	1. Describe how root pressure affects the absorption of water by a plant. 2. Describe different means of water movement, including osmosis, cohesion, transpiration, and capillary action. 3. Explain which types of water movement are significant in plants.	Image and Activity Bank 2-5
End of Chapter, p. 52				Chapter 2 Review Worksheet, p. 40 Chapter 2 Assessment Worksheet, p. 44
Chapter 3, p. 54	Lesson 1, Turning Food Into Fuel, p. 55	1	1. Explain the function of digestion. 2. Describe how the digestion of starch takes place. 3. Identify the main parts of the human digestive system.	Image and Activity Bank 3-1
	Lesson 2, Unlocking the Energy in Food, p. 57	4	1. Explain the process of respiration as it applies to both plants and animals. 2. Describe how human cells receive the oxygen they need for respiration. 3. Explain the effect of physical activity on the rate of respiration. 4. Identify the products of respiration as carbon dioxide and water.	Image and Activity Bank 3-2 Discrepant Event Worksheet, p. 46 Transparency 8 Exploration Worksheet, p. 47 Exploration Worksheet, p. 49 Exploration Worksheet, p. 52
	Lesson 3, Maintaining the Balance, p. 66	1 or 2	1. Describe the interdependence of plants, animals, and the atmosphere. 2. Explain the greenhouse effect. 3. Identify some of the possible consequences of increasing the amount of carbon dioxide in the atmosphere.	Image and Activity Bank 3-3 Theme Worksheet, p. 54 Graphing Practice Worksheet, p. 55
End of Chapter, p. 68				Chapter 3 Review Worksheet, p. 57 ▼ Chapter 3 Assessment Worksheet, p. 59
End of Unit, p. 70				Unit 1 Activity Worksheet, p. 61 Unit 1 Review Worksheet, p. 62 Unit 1 End-of-Unit Assessment, p. 66 Unit 1 Activity Assessment, p. 72 Unit 1 Self-Evaluation of Achievement, p. 75

* Estimated time is given in number of 50-minute class periods. Actual time may vary depending on period length and individual class characteristics.

▼ Transparencies are available to accompany these worksheets. Please refer to the Teaching Transparencies Cross-Reference chart in the Unit 1 Teaching Resources booklet.

Materials Organizer

Chapter	Page	Activity and Materials per Student Group
1	7	**Exploration 1, Part 1:** 2 containers (such as 250 mL beakers or recycled margarine tubs); about 5 mL of cornstarch; about 5 mL of sugar; about 500 mL of water; a few drops of iodine starch-test reagent; eyedropper; stainless steel scoop or metric measuring spoon; glass stirring rod; safety goggles; lab aprons; latex gloves (additional teacher materials: 25 mL of 0.1 M sodium thiosulfate solution; see Advance Preparation on page 1D.); **Part 2:** 2 geranium leaves; materials to test for starch: about 500 mL of water, large beaker, small beaker, about 25 mL of methanol, hot plate, a few drops of tincture of iodine, eyedropper, sheet of white paper, watch or clock, safety goggles, lab aprons, oven mitts, latex gloves (additional teacher materials: 25 mL of 0.1 M sodium thiosulfate solution; a few sheets of old newspaper; 250 mL of water; 600 mL beaker; see Advance Preparation on page 1D.)
	10	**Exploration 2:** geranium plant (grown outdoors); 10 cm × 10 cm piece of thin cardboard; scissors; 4 or 5 straight pins; grow lamp; materials to test for starch from Part 2 of Exploration 1 on page 7 (additional teacher materials: 25 mL of 0.1 M sodium thiosulfate solution; a few sheets of old newspaper; 250 mL of water; 600 mL beaker; see Advance Preparation on page 1D.)
	13	**Testing for Carbon Dioxide:** 2 small jars with lids; 2 small candles; a few matches; about 200 mL of limewater; safety goggles (additional teacher materials for preparing limewater: 4 mL of calcium hydroxide; 200 mL of water per class; see Advance Preparation on page 1D.)
	14	**Exploration 3:** 2 leaves; 2 test tubes; 2 large jars with lids; about 20 mL of 4% sodium hydroxide solution; 1 g of baking soda (sodium bicarbonate); 19 mL of water; wax pencil; lamp with 100 W bulb; materials to test for starch from Part 2 of Exploration 1 on page 7 (additional teacher materials: 1 g of sodium hydroxide and 24 mL of water for preparing sodium hydroxide solution; 50 mL of vinegar; a few strips of pH paper; 25 mL of 0.1 M sodium thiosulfate solution; a few sheets of old newspaper; 250 mL of water; 600 mL beaker; see Advance Preparation on page 1D.)
	15	**Exploration 4:** lamp with 100 W bulb; 2 test tubes; wooden splint; a few matches; 50 g of baking soda (sodium bicarbonate); 950 mL of water; 2 clear plastic containers or 600 mL beakers; 2 clear funnels; elodea sprig; safety goggles
	19	**Exploration 5, Activity 1:** magnifying glass or low-power binocular microscope; a variety of fresh leaves; **Activity 2:** microscope; 2 lettuce leaves; 2 microscope slides and coverslips; transparent metric ruler; eyedropper; a few drops of water
	22	**Exploration 6:** prepared slide of leaf cross section; microscope; sharp, hard pencil; unlined paper
	23	**Exploration 7:** plant with leaves that have white areas, such as coleus or pothos ivy; materials to test for starch from Part 2 of Exploration 1 on page 7 (additional teacher materials: 25 mL of 0.1 M sodium thiosulfate solution; a few sheets of old newspaper; 250 mL of water; 600 mL beaker; see Advance Preparation on page 1D.)
2	31	**Transpiration: A Demonstration:** geranium or other hardy plant; large, clear plastic bag; string or twist tie; about 150 mL of water
	33	**Exploration 1:** about 200 mL of water; large, glass juice bottle; a few extra-long matches
	35	**In-Text Activity:** 2 paper towels; about 100 mL of water; a few radish seeds; optional materials: microscope; microscope slide and coverslip; eyedropper; a few drops of water
	39	**Exploration 2, Activity 1:** uncooked egg; about 50 mL of vinegar; petri dish; materials to support egg, such as tape; wide-mouthed bottle filled with water; needle or probe; 10 cm of thin glass tubing, such as capillary tubing; candle and a few matches or silicone sealant; support stand with clamp; watch or clock; safety goggles; **Activity 2:** support stand with clamp; 20 cm of thin glass tubing; animal membrane; about 200 mL of 5% sugar solution; rubber band; about 600 mL of water; large beaker or other large container (additional teacher materials for preparing sugar solution: 10 g of sugar; 190 mL of water; see Advance Preparation on page 1D.)
	42	**Exploration 3:** 2 lengths of dialysis tubing, 14 to 16 cm each; two 250 mL beakers; 20 mL of 5% cornstarch and water mixture; 20 mL of 5% sugar solution (corn syrup and water); a few drops of iodine starch-test reagent; 2 paper clips; about 300 mL of water; eyedropper; about 30 cm of thread; materials to test for sugar: one 250 mL beaker, 2 mL of Benedict's solution, hot plate, 10 mL graduated cylinder, about 150 mL of water, test tube, test-tube clamp, watch or clock, safety goggles, lab aprons, latex gloves (additional teacher materials: 1 g of corn syrup; 1 g of cornstarch; 38 mL of water; 50 mL of 1 M sulfuric acid solution; six 6d iron nails; steel wool; portable burner; tongs; watch glass; 50 mL of 4% sodium hydroxide solution; pH paper; strainer; 25 mL of 0.1 M sodium thiosulfate solution; a few sheets of old newspaper; see Advance Preparation on page 1D.)
	43	**Exploration 4:** uncooked potato; four 600 mL beakers; 85 g of salt; 1.5 L of tap water; knife; potato peeler; watch or clock
	44	**Exploration 5:** carrot; coring knife; 20 cm glass tube fitted with a one-hole rubber stopper; about 30 mL of molasses; candle and a few matches or silicone sealant; support stand with clamp; 250 mL beaker; about 200 mL of water; graph paper; grease pencil; metric ruler; materials to test for sugar from Exploration 3 on page 42 (additional teacher materials: thick cloth or leather gloves; about 0.5 mL of glycerin; 50 mL of 1 M sulfuric acid solution; six 6d iron nails; steel wool; portable burner; tongs; watch glass; 50 mL of 4% sodium hydroxide solution; pH paper; strainer; see Advance Preparation on page 1D.)
	55	**The Mouth—Where It All Starts:** 1 unsalted soda cracker per student
3	60	**Exploration 1:** stopwatch, clock, or watch with a second hand
	61	**Exploration 2: Test 1:** plastic drinking straw; about 20 mL of limewater; test tube; safety goggles (additional teacher materials for preparing limewater: 4 mL of calcium hydroxide and 200 mL of water per class; see Advance Preparation on page 1D.); **Test 2:** a mirror or shiny piece of metal; piece of cobalt chloride paper
	63	**Exploration 3:** 20 seeds, such as radishes or beans; 4 test tubes; 4 rubber stoppers; wooden splint; 5 mL of limewater; about 10 mL of water; test-tube rack; a few matches; 10 mL graduated cylinder; safety goggles (additional teacher materials for preparing limewater: 4 mL of calcium hydroxide and 200 mL of water per class; see Advance Preparation on page 1D.)
	64	**Hot Plants?** materials for monitoring heat production during plant respiration (See Advance Preparation on page 1D.)

Exploration 3, page 14: To provide each student group with a 4% solution of sodium hydroxide, add 1 g of sodium hydroxide to every 24 mL of water and stir thoroughly. To dispose of the sodium hydroxide solution, you will need 50 mL of vinegar and a few strips of pH paper per student group. See the Waste Disposal Alert on page 14. Make an extra liter of the sodium hydroxide solution for disposal purposes in Explorations 3 and 5 on pages 42 and 44, respectively.

Exploration 2, Activity 2, page 39: Make the 5% sugar solution beforehand by dissolving 10 g of sugar in 190 mL of water per student group.

Exploration 3, page 42: Make the 5% sugar solution and the 5% cornstarch and water mixture beforehand by adding 1 g of corn syrup to 19 mL of water and 1 g of cornstarch to 19 mL of water per student group.

Exploration 5, page 44: To insert the glass tubes into the one-hole stoppers, first make sure that the tubes are fire-polished to remove any sharp edges. Wearing thick cloth or leather gloves and using a small amount of glycerin to lubricate the tubes, insert the tubes into the stoppers with a gentle twisting motion.

Hot Plants? page 64: Students will need to design their own experiments. Useful materials may include a thermometer, test tubes or jars, moist radish or bean seeds, masking tape, and a watch or clock.

Other Explorations: To dispose of iodine-stained materials, you will need about 25 mL of 0.1 M sodium thiosulfate solution per student group for each of the following Explorations: Exploration 1, Part 1, on page 7; Exploration 3 on page 42; Exploration 1, Part 2, on page 8; Exploration 2 on page 10; Exploration 3 on page 14; and Exploration 7 on page 23. For the last four of these Explorations, you will also need some old newspaper, a 600 mL beaker, and about 250 mL of water per student group for disposal purposes. See the Waste Disposal Alerts on pages 7 and 8.

You will need 4 mL of calcium hydroxide and 200 mL of water per class to prepare limewater for the following activities: Testing for Carbon Dioxide on page 13; Exploration 2, Test 1, on page 61; and Exploration 3 on page 63. To prepare the limewater, add 4 mL of calcium hydroxide to 200 mL of water, shake, and let settle.

To dispose of the Benedict's solution in Explorations 3 and 5 on pages 42 and 44, you will need 50 mL of 1 M sulfuric acid solution, six 6d iron nails, steel wool, a portable burner, tongs, a watch glass, 50 mL of the 4% sodium hydroxide solution, pH paper, and a strainer. See the Waste Disposal Alert on page 42.

Chapter 1
See Teacher's Edition margin, pp. 6, 12, 15, 19, 21, 23, and 24
Activity Worksheet, p. 18
SourceBook, p. S2 and S9

Chapter 2
See Teacher's Edition margin, pp. 30, 31, 32, 33, 37, 40, 45, 46, 50, and 51
SourceBook, pp. S6 and S13

Chapter 3
See Teacher's Edition margin, pp. 56, 58, 60, 62, 66, and 67
Exploration Worksheet, p. 47
Theme Worksheet, p. 54
Graphing Practice Worksheet, p. 55
SourceBook, pp. S3 and S16

Unit 1
Unit 1 Activity Worksheet, p. 61
SourceBook Activity Worksheet, p. 76

Unit Compression

If time becomes short, try to preserve the conceptual progression of the unit. Most of the Explorations in Chapter 1 may be performed as teacher demonstrations or partially prepared ahead of time. For instance, for Exploration 5 on page 19, prepare some slides and set up stations so that students can observe different leaf structures. For Chapters 2 and 3, you may wish to omit certain sections as necessary based on your students' prior knowledge of the material.

Assessment Planning Guide

Lesson, Chapter, and Unit Assessment	SourceBook Assessment	Ongoing and Activity Assessment	Portfolio and Student-Centered Assessment
Lesson Assessment Follow-Up: see Teacher's Edition margin, pp. 9, 16, 25, 30, 34, 42, 46, 51, 56, 65, and 67 **Chapter Assessment** Chapter 1 Review Worksheet, p. 19 Chapter 1 Assessment Worksheet, p. 23* Chapter 2 Review Worksheet, p. 40 Chapter 2 Assessment Worksheet, p. 44* Chapter 3 Review Worksheet, p. 57 Chapter 3 Assessment Worksheet, p. 59* **Unit Assessment** Unit 1 Review Worksheet, p. 62 End-of-Unit Assessment Worksheet, p. 66*	SourceBook Review Worksheet, p. 77 SourceBook Assessment Worksheet, p. 81*	Activity Assessment Worksheet, p. 72* **SnackDisc** Ongoing Assessment Checklists ♦ Teacher Evaluation Checklists ♦ Progress Reports ♦	Portfolio: see Teacher's Edition margin, pp. 22, 50, and 56 **SnackDisc** Self-Evaluation Checklists ♦ Peer Evaluation Checklists ♦ Group Evaluation Checklists ♦ Portfolio Evaluation Checklists ♦

* Also available on the Test Generator software
♦ Also available in the Assessment Checklists and Rubrics booklet

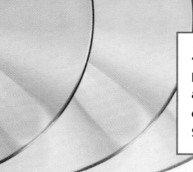

Science Discovery is a versatile videodisc program that provides a vast array of photos, graphics, motion sequences, and activities for you to introduce into your *SciencePlus* classroom. *Science Discovery* consists of two videodiscs: Science Sleuths and the Image and Activity Bank.

Using the *Science Discovery* Videodiscs

Science Sleuths: The Nature House Malady
Side B

Dario, an allergy sufferer, moves into Nature House, a superinsulated condominium, on the advice of his physician. Instead of getting better, his symptoms get worse.

Interviews
1. Setting the scene: Allergy sufferer **17413 (play ×2)**

2. Nature House manager **18677 (play)**

3. Nature House tenant **19313 (play)**

4. Allergist **20160 (play)**

5. Pulmonologist **20990 (play)**

6. Custodial engineer **21446 (play)**

Documents
7. Nature House pamphlet **22046 (step)**

8. Allergist's newsletter **22049 (step ×2)**

Literature Search
9. Search on the words: AIR, FORMALDEHYDE, HOUSE PLANTS, POLLUTION, RADON, ROCKY MOUNTAIN CITY **22053 (step)**

10. Article #1 ("Good Air Into Bad") **22056 (step)**

11. Article #2 ("Do You Have a Hazardous Home?") **22059 (step ×2)**

12. Article #3 ("Formaldehyde") **22063 (step)**

13. Article #4 ("Some Greens to Get the Blues Out") **22066 (step ×2)**

Sleuth Information Service
14. Schematic plan of Nature House **22070**

15. Map of Rocky Mountain City area **22072 (step)**

16. Monthly weather data **22075**

17. Air pollutant levels allowed by the EPA **22077 (step ×3)**

18. EPA drinking water standards **22082 (step ×3)**

Sleuth Lab Tests
19. Medical reports **22087 (step ×2)**

20. Air analysis **22091**

21. Water analysis **22093 (step)**

Image and Activity Bank
Side A or B

A selection of still images, short videos, and activities is available for you to use as you teach this unit. For a larger selection and detailed instructions, see the Videodisc Resources booklet included with the Teaching Resources materials.

1-3 Food Factory Basics, page 17
Leaf, corn; microscopic cross section 4201
A stoma is a small opening in the epidermis of a plant, usually on the leaf. It permits water and gases to pass into and out of the plant.

Plant structure; stomata of the epidermis 3063
Structure of a stoma (breathing pore)

Plant structure; leaf anatomy 3066–3067 (step)
Structure of a leaf in a temperate climate (step) Structure of a leaf in a hot, dry climate

Leaf color; autumn 3045
Leaves lose their chlorophyll in the fall, allowing other pigments to become visible.

◀| Step Reverse   Play ▶   Pause || Step Forward |▶

2-1 The Indispensable Chemical, page 29
States of matter 2355–2357 (step ×2)
Water in its solid state (step) Heat changes water from its solid state to its liquid state. (step) Heat added to liquid water causes it to change to its gaseous state.

2-2 The Water Cycle, page 31
Water and plants 309
Placing plastic bags on a plant demonstrates transpiration.

2-3 Spreading Water Around, page 35
Plant structure; plant roots 3065
Diffusion through plant root hairs

Root, corn; microscopic cross section 4203
Roots hold a plant in place and absorb water and minerals from the soil.

Root and root hairs 4214
A root hair is a microscopic projection of a cell from the surface of a root. A root hair has a thin membrane that, by increasing the surface area of the root, makes it easier for the root to absorb more water and minerals.

Plant structure; diffusion 3064
Water moves from the roots up the plant.

Stem, corn; microscopic cross section 4202
Vascular tissue is a type of tissue found in plant stems. It conducts material through the plant. It contains phloem, which conducts food, and xylem, which carries water and minerals and helps to support the plant.

Movement into and out of cells; lab setup 340–343 (step ×3)
An aerosol can diffuses scent and propellant throughout a room. (step) A celery stalk in red food coloring demonstrates water moving into and out of cells. (step) Use these materials for an experiment involving an apple in salt water and in fresh water. (step) Place slices of apple in fresh water and others in salt water. What happens to each slice?

2-4 Osmosis: Controlled Diffusion, page 43
Osmosis; weighing a dialysis bag filled with red sugar water 40186–41568 (play ×2) (Side B only)
A membrane filled with a red-dyed sugar solution is weighed and then placed in distilled water. After 3 hours the bag weighs more. This is because distilled water passes through the membrane into the sugar solution, which has a lower concentration of water.

Osmometer 39416–39855 (play ×2) (Side B only)
A membrane tube, filled with a thick sugar solution (dyed red) is attached to a long glass tube. The membrane tube is placed in distilled water. The sugar solution climbs up the glass tube, pushed by the osmotic pressure across the membrane.

Osmosis experiment 2523
When celery is placed in salt water, it loses water, causing it to wilt. At the same time, the salt solution is slightly diluted with water from the celery.

2-5 Putting Water Where It Counts, page 47
Plant structure; effects of water on plant life 3070–3072 (step ×2)
Loss of turgor in a dry plant (step) Recovery after watering (step) The movement of water through a plant stem

Water and plants; lab setup 308
This setup demonstrates the capillary action of plants. The four strips (polyester, cotton, linen, and wool) have different rates of water transport.

3-1 Turning Food Into Fuel, page 55
Digestive system, human 4158
Digestion begins in the mouth, where starch is broken into sugars. Proteins are digested by acid in the stomach. The food is neutralized and passed through the intestines, where the nutrients are absorbed into the bloodstream.

Body membranes 4159
Intestinal wall

Capillaries carry food 4161

3-2 Unlocking the Energy in Food, page 57
Circulatory system, human 4160
Oxygen-rich blood flows out of the heart through the red arteries. The oxygen is transferred to the tissues, and the blood flows back through the blue veins. Then the blood is sent through the lungs to pick up more oxygen, and the cycle repeats.

Respiratory system, human 4169
The respiratory system includes the nose, sinuses, mouth, throat, trachea, bronchial tubes, and lungs.

Respiration, external; various organisms 4163–4168 (step ×5)
Earthworms receive oxygen through their skin. (step) Grasshoppers pump air into their bodies through tiny openings on their abdomens called spiracles. (step) Frogs and tadpoles absorb most of their oxygen directly through their skin. Frogs also have lungs, and tadpoles have gills. (step) Fish absorb oxygen from the water through gills. The gills are thin fluffy flaps through

---

which the blood flows, picking up oxygen. (step) Birds breathe with lungs. They also have air sacs and hollow bones that reduce their weight, making flight possible. (step) Humans absorb oxygen and later release carbon dioxide through their lungs. Blood full of carbon dioxide flows through capillaries in the lungs, and the carbon dioxide is replaced by oxygen.

Respiration products; carbon dioxide reacts with limewater to form a precipitate 38960–39415 (play ×2) (Side B only)
Limewater is an indicator for carbon dioxide.

Plant respiration equation 3061
During respiration, sugar is broken down into water and carbon dioxide, releasing energy that converts ADP to ATP. ATP provides energy for most biochemical processes in the cell.

3-3 Maintaining the Balance, page 66
Pollution, air 1457
The buildup of air pollution is common in cities, especially during fair weather when there is little wind or rain to disperse it.

Automobiles on freeway 2609
Collectively, automobiles make a big impact on the Earth's atmosphere because they release pollutants and greenhouse gases into the air.

Smokestacks 2610
Smokestacks release materials into the atmosphere.

Feedlot 2612
Feedlots can release a significant amount of methane, a greenhouse gas, into the atmosphere.

Unit 1 · Life Processes

UNIT FOCUS

Begin the unit by asking students to identify some of the biological processes that keep them alive. Make a list of suggestions on the board. *(Students may suggest processes such as breathing, food digestion, heart beating, or muscle contraction.)* Then ask students to identify some of the processes that keep plants alive. *(Students may suggest making food, absorbing water, or absorbing light.)* Tell them that there are no right or wrong answers at this point. Ask: Do plants need to exist in order for animals to exist? *(Accept all reasonable responses.)* After discussing student answers to this question, point out that this unit explores some of the life processes that all living organisms have in common.

A good motivating activity is to let students listen to the English/Spanish Audiocassettes as an introduction to the unit. Also, begin the unit by giving Spanish-speaking students a copy of the Spanish Glossary from the Unit 1 Teaching Resources booklet.

Unit 1

CHAPTER

1 Energy for Life
1 Where Does the Energy Come From? . . . 5
2 Where Does the Starch Come From? . . . 10
3 Food Factory Basics . . 17

CHAPTER

2 A World of Water
1 The Indispensable Chemical 29
2 The Water Cycle 31
3 Spreading Water Around 35
4 Osmosis: Controlled Diffusion 43
5 Putting Water Where It Counts 47

CHAPTER

3 Maintaining Life
1 Turning Food Into Fuel 55
2 Unlocking the Energy in Food 57
3 Maintaining the Balance 66

2

Connecting to Other Units

This table will help you integrate topics covered in this unit with topics covered in other units.

Unit 2 Particles	Processes such as osmosis and diffusion involve the transport and exchange of particles.
Unit 3 Machines, Work, and Energy	All life processes rely on a constant supply of energy to perform work.
Unit 4 Oceans and Climates	The plant life in a given area is shaped by and helps shape the climate in that area.
Unit 5 Electromagnetic Systems	A leaf is like a chemical cell in that it uses chemical reactions to generate energy.
Unit 7 Light	Most leaves are green because chlorophyll reflects the green component of sunlight.

A mist-shrouded mountainside lies awash in lush tropical greenery. Every square meter, it seems, is occupied by riotous growth. Thanks to nearly optimal conditions, the plants in this picture grow at a furious pace, seeking the sunlight that makes their existence possible. To human eyes, the view is beautiful. But what function underlies such dazzling form?

The reader cannot help but notice how intensely green is this scene. Every conceivable shade of this color, it seems, is represented here. Obviously this is no accident. The "green" in greenery is clearly important, but how? The chemical that is the source of the green color actually enables plants to capture and store the energy of sunlight. This energy is transferred whenever a plant—or the animal that eats it—is consumed by another living thing.

Water also figures prominently in this photograph. Although not itself organic, water is perhaps the most important chemical of life. No living thing can exist without water; its unique chemical properties are critical to life processes. When all of the ingredients of life—water, sunlight, and vital gases—are abundant, as they are here, life flourishes. But remove any one of these ingredients and life quickly ends—or never even begins.

Lush rain forest covers a mountain on the northeastern side of the island of Hawaii.

3

This mountainside is overflowing with sumptuous vegetation because both water and sunlight are abundant, allowing for bountiful growth. In fact, Mount Waialeale on Kauai, another island in the Hawaiian chain, is the wettest place on Earth. It receives almost 12 m of rainfall annually. Much of the humidity and rainfall in a rain forest is due to water released by the trees themselves through *transpiration.*

Have students imagine what life would be like in such a place. Ask: What would you do for food? What do you think plants or animals who might live in this forest do for food? *(Accept all responses.)* When answering these questions, students should consider where energy can be found in the forest. *(Mostly in plants)*

Critical Thinking

The trees in rain forests such as these form such a dense cover of leaves that in some places less than 1 percent of the sunlight reaches the forest floor. The highest layer of trees is known as the *upper canopy.* These trees are able to receive full sunlight. Beneath this layer, smaller trees form one or two *lower canopies,* capturing what light remains. Some plants grow on the branches of trees, where more sunlight can penetrate through the overlying canopies. These plants are called epiphytes, which means "air plants," and they include various kinds of ferns and mosses.

Ask: Where is the upper canopy in this photograph? Are there any trees that are likely to be part of a lower canopy? Can you see the forest floor? *(Much of this photograph is of the upper canopy, although some trees at the bottom right are likely part of a lower canopy. The forest floor is too shaded to be visible at the bottom left.)* Have students point out the various parts of a rain forest as they answer these questions. Then ask: Do the upper layers prevent the underlying layers from receiving rainfall as much as they prevent them from receiving sunlight? *(Since water is absorbed by the roots of plants and not by the leaves, the height of plants is not necessarily a factor in obtaining water.)*

CHAPTER 1
Energy for Life

Connecting to Other Chapters

> **Chapter 1**
> investigates the importance of photosynthesis and the plant structures that make it possible.

> **Chapter 2**
> focuses on water transport, with an emphasis on transpiration, diffusion, osmosis, and cohesion.

> **Chapter 3**
> explores how cells turn food into energy through the process of respiration.

Prior Knowledge and Misconceptions

Your students' responses to the ScienceLog questions on this page will reveal the kind of information—and misinformation—they bring to this chapter. Use what you find out about your students' knowledge to choose which chapter concepts and activities to emphasize in your teaching. After students complete the material in this chapter, they will be asked to revise their answers based on what they have learned. Sample revised answers can be found on page 27.

In addition to having your students answer the questions on this page, you may wish to have them complete the following activity: Ask students to make a labeled drawing that represents the process by which plants make food. Ask students to include water, sun, oxygen, carbon dioxide, and any plant structures that they think might be important, so that their drawings are as detailed as possible. Assure the students that there

CHAPTER 1
Energy for Life

1 Where do plants get their food?

2 Describe a typical plant.

What's inside a leaf?
3

ScienceLog

Think about these questions for a moment, and answer them in your ScienceLog. When you've finished this chapter, you'll have the opportunity to revise your answers based on what you've learned.

4

are no right or wrong answers to this question. Collect the papers, but do not grade them. Instead, use the students' drawings to identify possible problem areas in the chapter. Read the papers to find out what students know about life processes, what misconceptions students may have, and what aspects of this topic are interesting to them.

Where Does the Energy Come From?

The girl in the photograph is eating a popular meal—green leaves, slices of ripened plant ovaries, ground-up seeds, and cow muscle. Actually, most people call it a hamburger! You probably know that hamburgers and other foods provide us with the energy necessary to function. You need energy from food to go about your daily activities, such as walking, talking, thinking—and eating more food!

Now consider the following statement:

Without plants, there would be no food.

Do you think the statement above is true? To help you decide, think about the hamburger in the photograph. The lettuce, obviously, is a plant. The bun is made from wheat, so it is also plant matter. But how about the meat from the cow? Where did the cow get its food? Cows and other plant-eating animals eat plants directly. Humans and other meat-eating creatures eat the animals that ate plants. Somewhere in the food chain there is a connection to plants.

Without plants, there would be no food. How can that be true? If we didn't have plants to eat, we would still have meat, right?

Right, but who wants to eat just meat?

Besides that, where does the meat come from? That animal has to eat something.

Yeah, and most of the animals we use for food eat plants, as far as I know.

5

LESSON 1 ORGANIZER

Time Required
2 class periods

Process Skills
observing, inferring, analyzing

New Term
Starch—the stored form of the food produced by plants in photosynthesis

Materials (per student group)
Exploration 1, Part 1: 2 containers (such as 250 mL beakers or recycled margarine tubs); about 5 mL of corn-starch; about 5 mL of sugar; about 500 mL of water; a few drops of iodine starch-test reagent; eyedropper; stain-

less steel scoop or metric measuring spoon; glass stirring rod; safety goggles; lab aprons; latex gloves (additional teacher materials: 25 mL of 0.1 M sodium thiosulfate solution; see Advance Preparation on page 1D.);
Part 2: 2 geranium leaves; materials to test for starch: about 500 mL of water, large beaker, small beaker, about 25 mL of methanol, hot plate, a few drops of tincture of iodine, eyedropper, sheet of white paper, watch or clock, safety goggles, lab aprons, oven mitts, latex gloves (additional teacher

continued ▶

Where Does the Energy Come From?

FOCUS

Getting Started
Have students construct three food chains: (1) a forest food chain, (2) a marine food chain, and (3) a desert food chain. Ask students what organisms the three food chains have in common. *(Each food chain includes plants. Also, there are probably many nonplant microbes involved.)* Ask: If animals eat plants to obtain energy, do plants also "eat" to obtain energy? If so, what do they eat? How do plants obtain their food? *(Virtually all plants subsist on starch, which they make for themselves using light from the sun and water and nutrients absorbed from the soil.)*

Main Ideas
1. Almost all food for humans and other animals comes, directly or indirectly, from plants.
2. Photosynthesis is the process by which green plants make food.
3. The presence of starch in the leaves of a green plant indicates that food has been produced by the plant.

TEACHING STRATEGIES

Have each student write on a slip of paper the name of a food and an explanation of how it is derived from plants. *(Sample answer: Cheese ultimately derives from plants because it is made from milk, which comes from cows that eat grasses and other plants.)* Collect the papers and discuss a few of the examples with the class. It may be most interesting to discuss examples of foods that have less obvious connections to plants. *(For instance, honey is a substance that bees produce from plant nectar.)*

Okay, but how does the plant get food?

Here is your challenge: Figure out what should go where the question mark is. By thinking and doing the Explorations in this chapter, you will figure out the answer to the question, "How do plants get food?" Get started by forming a group with several other students, and do the following:

1. Read and discuss possible answers to Question 1 on page 4. In talking, you may come up with ideas you hadn't thought of before.

2. Work out an answer that your group considers reasonable. (A)

3. Make a poster that clearly illustrates your group's answer.

4. Decide what you could do to test your answer. On the back of your poster, illustrate your plan with labeled diagrams.

5. Share your ideas with the other groups, explaining both sides of your poster to them.

As you do the Explorations in this chapter, see how closely they resemble the plan you suggested on the back of your poster.

6

Some Food for Thought

Before you get serious about figuring out how plants get food, you have to know what kind of food is in plants. The food in plants is **starch**. The starch in plants provides almost 70 percent of the world's food supply.

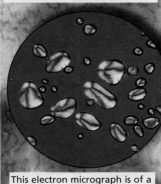

This electron micrograph is of a starch-storage cell in a potato. The sphere-shaped objects are granules of pure starch.

If you look at the photograph above, you will see some starch granules inside a potato plant. The photograph was taken using an electron microscope. (An electron microscope is a powerful instrument that uses a beam of electrons to form an enlarged image.) You can't use an electron microscope in class, but there is another way to tell whether something contains starch, which you will learn in Exploration 1. You will also use the technique later in the unit as you answer the question, "Where do plants get food?"

EXPLORATION 1

The Search for Starch

PART 1

A Simple Test

You Will Need

- 2 containers
- cornstarch
- an eyedropper
- a stainless steel scoop
- a glass stirring rod
- iodine solution
- sugar
- water

What to Do

1. Mix some cornstarch in water, and then add two drops of iodine solution. What happens?

   Be Careful: Iodine is poisonous and stains clothing and skin. Wear a lab apron.

2. Repeat the test with plain water. Then add a little sugar to the water and repeat the test. How do the results of these tests differ from the result of the test with cornstarch?

3. When testing for a substance, a *positive* result signals that the substance is present; a *negative* result means it is not present. What signaled the presence of starch? For each of your tests for starch, was the result positive or negative?

Add iodine to each container to test for starch.

Be sure to save iodine solution for later Explorations.

Exploration 1 continued ▶

Some Food for Thought

You may wish to make students aware that the end product of photosynthesis is actually glucose (a sugar), not starch. Since plants usually produce more glucose than they need, the excess is changed into starch and stored in the plants' cells. The production of starch, instead of sugar, is stressed here because starch is easier to detect, and the presence of starch is an indicator of glucose.

 An Exploration Worksheet is available to accompany Exploration 1 (Teaching Resources, page 3).

Multicultural Extension

The Impact of New World Sugar

Ask students what the sources of common table sugar are. *(Sugar beets and sugar cane)* Explain that during America's colonial period the production of sugar from sugar cane, especially in the West Indies, was a very profitable industry. The sugar industry had an immediate and lasting effect on land use in Latin America and the Caribbean, on the African diaspora, and on the world market. Have students work in groups to report on different ways in which the sugar industry affected all of the cultures involved.

EXPLORATION 1

Cooperative Learning
EXPLORATION 1

Group size: 3 to 4 students
Group goal: to determine whether starch is present in a leaf
Positive interdependence: Assign each student a specific role, such as task leader, safety coordinator, materials organizer, or checker.
Individual accountability: Each student should record his or her own data. The data can be used to complete Explorations 2 through 4.

PART 1

SAFETY ALERT Before students begin the activity, alert them that iodine is poisonous if swallowed. Also point out that it will stain just about anything it comes in contact with, especially clothes and skin. For these reasons, they should be careful when using the iodine starch-test reagent.

WASTE DISPOSAL ALERT Pour the cornstarch-iodine mixture into a labeled container. While stirring, slowly add a 0.1 M sodium thiosulfate solution until the mixture turns white. Pour the resulting mixture down the drain.

Answers to
Part 1

1. The iodine starch-test reagent turns from a golden brown color to a blue-black color when it is added to the mixture of cornstarch and water.

2. When the iodine starch-test reagent is added to plain water and to the sugar solution, it remains a shade of golden brown rather than changing to blue-black.

3. A change to a blue-black color indicates the presence of starch. No change in color indicates the absence of starch. Of the three tests for starch, only the cornstarch mixture showed a positive result.

Exploration 1 continued ▶

EXPLORATION 1, continued

PART 2

You may wish to heat the water yourself and distribute it to students. In this way, only one heat source is needed in the classroom, and you can ensure that the methanol is not used near it. If you are concerned about your students' ability to perform this Exploration in a mature manner, you may wish to remove the chlorophyll from the leaves yourself as a class demonstration. Then distribute the leaves to individuals or groups so that they can finish the Exploration themselves.

SAFETY ALERT Explain to students that methanol is more dangerous than most other chemicals that they will use. Methanol is poisonous and highly flammable; all heat sources in the room should be turned off and completely cooled before the methanol is distributed. Methanol can also be absorbed through the skin; this is why a lab apron, latex gloves, and safety goggles should be worn at all times. Also, be sure the work area is well ventilated, and allow no more than 250 mL of methanol to be in the classroom at one time. Have students use the minimum amount of methanol necessary to cover the leaves. By placing the leaves flat against the bottom of the beaker, as little as 5 mL of methanol may be needed. Distribute the methanol yourself only after all safety precautions have been met. Be sure you are familiar with your school system's policies for handling hazardous materials.

WASTE DISPOSAL ALERT Pour the methanol into a labeled container containing enough water to equal 10 times the volume of the methanol. Then pour the liquid down the drain while flushing the drain with running water.

To dispose of the iodine-stained materials, treat them with enough 0.1 M sodium thiosulfate to decolorize them, wrap them in old newspaper, and put them in the trash. Wrap all other disposable materials from this Exploration in old newspaper, and put them in the trash as well.

PART 2

Starch in Green Leaves

You Will Need

- oven mitts
- water
- a geranium plant
- a large beaker
- a small beaker
- methanol
- a hot plate
- tincture of iodine
- an eyedropper
- white paper
- latex gloves

What to Do

1. Remove two leaves from a healthy plant that has been growing in a sunny spot.
2. Remove the green color from the leaves. To do this, first heat water in a large beaker until it boils.
3. Turn off the heat source. While wearing oven mitts, move the beaker to a place away from any heat sources.
4. Put the leaves into a small beaker and cover them with methanol.

   Be Careful: Methanol is poisonous and highly flammable. Keep methanol away from open flames and other heat sources. Wear a lab apron, latex gloves, and goggles.

5. Put the small beaker into the large beaker containing hot water, as shown in the diagram. The methanol will become warm enough to boil.
6. Soak the leaves in the warm methanol for 5 to 10 minutes. The methanol should become green as it dissolves the coloring of the leaves. Simultaneously, the leaves should become pale.

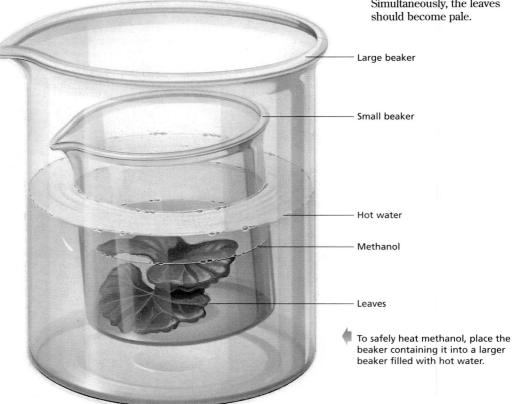

Large beaker

Small beaker

Hot water

Methanol

Leaves

To safely heat methanol, place the beaker containing it into a larger beaker filled with hot water.

8

Answers to Questions, page 9

1. Students may note minor changes in shape, but the most noticeable change should be in the color of the leaves.

2. The iodine starch-test reagent should turn blue-black when placed on the leaf, indicating the presence of starch.

3. Heating the methanol helps to remove more of the chlorophyll from the leaves.

4. The leaf that was not tested with the iodine starch-test reagent was used as a control. (A control is an experimental setup in which nothing is changed; it is used for comparison with another setup in which a variable, or experimental condition, is changed.) The only difference between the two leaves was the application of iodine. Therefore, any difference in the leaves must be due to the iodine.

7. Remove the leaves from the methanol and rinse them with tap water. Spread them face up on a piece of white paper.

8. Treat one leaf with three or four drops of the iodine solution. Then place the leaf in water to *fix* the color (make it permanent). Compare the treated leaf with the untreated leaf. Look for any slight change in color.

Questions

1. Sketch the two leaves in your ScienceLog, noting any changes observed.
2. What evidence shows you that starch is present in a plant leaf?
3. Why was it necessary to heat the methanol that you used to soak the leaves?
4. What was the purpose of using two leaves if only one was treated with iodine?

Starch—Essential for Life

The starch in plants is essential to other living things. Even organisms that do not feed on plants depend on them indirectly for food. Carnivorous (meat-eating) animals, for example, prey upon other animals. But the preyed-upon animals either ate plants or ate animals that ate plants. Consider the statement, "All meat is ultimately grass." What does it mean? Write your explanation in your ScienceLog. **Ⓐ**

Starch is a carbohydrate—a complex organic compound. Learn more about organic compounds on pages S2–S6 of the SourceBook.

Starch—Essential for Life

The lesson concludes with a writing exercise about the meaning of the following statement: All meat is ultimately grass. Students should be encouraged to express their ideas in their own words.

When students have finished writing, provide them with the opportunity to share and discuss their ideas. Monitor their discussions to assess how well students understand the concept that virtually all food for virtually all living things ultimately comes from plants.

Answer to
In-Text Question

Ⓐ The statement refers to the fact that the meat we eat generally comes from animals that eat grasses and other plants.

Reteaching

Have students make poster diagrams that illustrate how all of the items they ate the previous day came, either directly or indirectly, from plants.

Assessment

Suggest that students use an iodine starch-test reagent to test several substances, such as a slice of potato, orange juice, flour, a slice of bread, and a piece of apple, for starch. (*The potato, flour, and bread should test positive for starch, but the apple and orange juice should not.*)

Extension

Point out to students that starch is a nutrient that makes up over 70 percent of the world's food supply. Have students research how the human body uses starch. Have volunteers share what they learn with the class.

Closure

Have students research which foods contain large amounts of starch. They should compile their information in a chart. The chart should include the name of each food item, a specified quantity of each food, the amount of starch in each sample, and what percentage of the USRDA (United States Recommended Daily Allowance) this amount of starch corresponds to.

Meeting Individual Needs

Learners Having Difficulty

Have students make a poster or mobile that depicts how energy moves through a food chain. They may wish to use the food chain that they constructed for the Getting Started activity on page 5. You may wish to have volunteers present their posters to the class.

FOCUS

Getting Started

Have students grow a potato in a glass of water in order to see van Helmont's experiment firsthand. (See Water—How Essential Is It? on page 11.) They should begin the experiment 2 weeks before starting this lesson. Have them weigh the potato, insert tooth-picks into the potato, and then balance it on the rim of the glass with the bot-tom of the potato submerged in the water. After 2 weeks, they should weigh the potato a second time and explain where the extra mass came from. *(The water)*

Main Ideas

1. Sunlight and water are necessary for photosynthesis to occur.
2. Carbon dioxide is consumed and oxygen is released during photosynthesis.
3. Photosynthesis results in the formation of a simple sugar that is later converted into starch.

TEACHING STRATEGIES

EXPLORATION 2

Plants that have been grown under artificial light typically do not show dramatic results for this Exploration.

In order to complete this Exploration in one class period, you may wish to have plants prepared ahead of time. Place a supply of plants in a dark place, such as a closet, for 4 days before they will be needed. Have students cut card-board shapes to suit the sizes of the leaves for step 3. At least one plant should be prepared for step 4. A grow lamp should produce excellent results. Place the lamp as close to the leaves as possible without harming the plant (20 to 25 cm away). Leave the lamp on for 24 hours.

WASTE DISPOSAL ALERT Refer to page 8 for instructions on disposing of the iodine-stained materials.

LESSON

2

Where Does the Starch Come From?

In Explorations 2, 3, and 4 you will be figuring out where the starch in plants comes from. As you do the Explorations, think about what light, water, and carbon dioxide could have to do with the starch in plants.

EXPLORATION 2

Light and Starch

The leaf you tested for starch in Exploration 1 was from a plant that had received plenty of light. If a plant receives no light, will it also have starch in its leaves?

You Will Need

- a geranium plant (grown out-doors)
- thin cardboard
- scissors
- straight pins
- materials and equipment to test leaves for starch

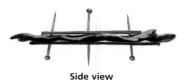

What to Do

1. Put a geranium plant in a dark but warm place for 4 days. The soil should be moist but not too wet.
2. After the 4 days have passed, test some of the plant's leaves for starch, as in Exploration 1. Record your findings in your ScienceLog.
3. Now cover one leaf with card-board, as shown in the illus-trations. (Important—the leaf must still be attached to the plant!) Affix a letter or a

number to the top surface of the leaf, using the method shown in the illustration.

4. Put the plant in bright light, such as under a grow lamp, for 24 hours. Then remove the cardboard pieces and test the covered and uncovered parts of the leaf for starch as you did in Part 2 of Exploration 1. Record your findings.

Questions

1. What effect did darkness have on the amount of starch in the plant?
2. Which part of this Exploration, steps 1 and 2 or steps 3 and 4, suggests that light has an important role in determining the amount of starch in green plants? Why?
3. Could you perform steps 3 and 4 without first doing steps 1 and 2? Explain your answer.

Upper surface of leaf

Cardboard pinned tightly to upper surface of leaf

Cardboard pinned tightly to lower surface of leaf

Side view

LESSON 2 ORGANIZER

Time Required 3 class periods

Theme Connection Energy

Process Skills observing, analyzing, inferring, predicting

New Term
Photosynthesis—the process by which green plants make food

Materials (per student group)
Exploration 2: geranium plant (grown outdoors); 10 cm × 10 cm piece of thin cardboard; scissors; 4 or 5 straight pins; grow lamp; materials to test for starch from Part 2 of Exploration 1 on page 7 (additional teacher materials: 25 mL of

0.1 M sodium thiosulfate solution; a few sheets of old newspaper; 250 mL of water; 600 mL beaker; see Advance Preparation on page 1D.)
Testing for Carbon Dioxide: 2 small jars with lids; 2 small candles; a few matches; about 200 mL of limewater; safety goggles (additional teacher materials for preparing limewater: 4 mL of calcium hydroxide; 200 mL of water per class; see Advance Preparation on page 1D.)
Exploration 3: 2 leaves; 2 test tubes; 2 large jars with lids; about 20 mL of 4% sodium hydroxide solution; 1 g of

continued ▶

Water—How Essential Is It?

When you started this chapter, you may have thought that plants get their food only from the soil. If you did, you are in good company. That's what the great philosopher Aristotle thought too! But a Belgian scientist of the seventeenth century, Jan Baptista van Helmont, questioned this belief and decided to look into the matter further. He decided to find out what role water had in plant growth.

What happens when a plant is deprived of water? You have probably seen what happens when you or somebody else forgets to water a houseplant. Plants that don't get enough water wilt. If they continue to be deprived of water, they eventually die.

Van Helmont performed an experiment that convinced him that water was so important that it was completely responsible for the great change in mass that occurs in growing plants. In other words, he concluded that water was completely responsible for growth in plants.

What do you think? Before you decide, let van Helmont tell you about his experiment, translated from his own words. Don't let the odd wording and spelling throw you. People spelled and spoke quite differently in van Helmont's day. (The masses have been changed to metric units.)

The corn plant on the left is healthy because it received adequate rainfall. The corn on the right is stunted because it did not receive enough water.

Journal

I have learned from this handicraft operation that all vegetables do immediately and materially proceed out of the element of water only. For I took an earthen vessel in which I put ninety kilograms of earth that had been dried in a furnace, which I moystened with rainwater, and I implanted therein the trunk, or stem, of a Willow tree, weighing two kilograms and about two-hundred and fifty grams. At length, five years being finished, the tree sprung from thence did weigh seventy-six kilograms. But I moystened the earth vessel with rain-water, or distilled water (always when there was need) ... and lest the dust should be co-mingled with the earth, I covered the lip or mouth of the vessel with an iron plate covered with tin, and easily passable with many holes. I computed not the leaves that fell off, in the four autumnes. At length, I again dried the earth of the vessel, and there were found the same ninety kilograms, wanting a few grams. Therefore almost seventy-four kilograms of wood, bark, and roots arose out of water only.

Jan Baptista van Helmont

11

 An Exploration Worksheet is available to accompany Exploration 2 on page 10 (Teaching Resources, page 6).

Answers to
Questions, page 10

1. The leaves from a plant left in the dark have little or no starch compared with leaves from a plant that has been in bright light.

2. Both parts suggest that green plants need light to produce starch. In each case the plants (or parts of the plants) that did not receive light had less starch. However, steps 1 and 2 could not be considered without comparing the results with those from Exploration 1.

3. Steps 3 and 4 could have been done without first doing steps 1 and 2. However, steps 1 and 2 reduce the amount of starch in the plant's leaves so that the production of starch in the uncovered parts of the leaves is more apparent.

Water—How Essential Is It?

Before students begin reading, have them discuss what happens when a plant is deprived of water. *(It wilts and eventually dies.)* Help them to conclude that water is essential for healthy growth.

Have students silently read about van Helmont's experiment, or call on a volunteer to read it aloud. Monitor students' understanding of the material by having them discuss and complete the pictorial summary on page 12.

Point out to students that before the 1600s, scientists thought that plants obtained their food directly from the soil. This theory was first proposed by the Greek philosopher Aristotle (384–322 B.C.). This food was believed to have been absorbed directly by the roots and utilized by the plants without much change. Van Helmont was the first to provide experimental evidence that conflicted with this theory.

Primary Source

Description of change: excerpted from early writings by Jan Baptista van Helmont
Rationale: excerpted to focus on the importance of water to plant growth

ORGANIZER, continued

baking soda (sodium bicarbonate); 19 mL of water; wax pencil; lamp with 100 W bulb; materials to test for starch from Part 2 of Exploration 1 on page 7 (additional teacher materials: 1 g of sodium hydroxide and 24 mL of water for preparing sodium hydroxide solution; 50 mL of vinegar; a few strips of pH paper; 25 mL of 0.1 M sodium thiosulfate solution; a few sheets of old newspaper; 250 mL of water; 600 mL beaker; see Advance Preparation on page 1D.) **Exploration 4:** lamp with 100 W bulb; 2 test tubes; wooden splint; a few matches; 50 g of baking soda (sodium bicarbonate); 950 mL of water; 2 clear plastic containers or 600 mL beakers; 2 clear funnels; elodea sprig; safety goggles

Teaching Resources
Exploration Worksheets, pp. 6, 8, and 9
Transparency Worksheet, p. 11
Transparency 1

Answers to
In-Text Questions

Ⓐ The following data complete the summary:
- mass of willow tree at the beginning of the experiment = 2 kg and about 250 g (or about 2.25 kg)
- mass of dried earth = 90 kg
- mass of dried earth at the end of the experiment = 90 kg, minus a few grams
- mass of grown tree = 76 kg

Ⓑ Van Helmont carefully planned his experiment before he began. He took careful quantitative measurements at the beginning and end of the experiment. He was careful to prevent unknown material from entering the earthen vessel he used.

Ⓒ Van Helmont assumed that water was the only substance that entered the tree. He also assumed that the tree absorbed all of its nutrients through its roots. Van Helmont did not realize that the tree gained a nutrient (carbon dioxide) from the air.

CROSS-DISCIPLINARY FOCUS

Mathematics

Have students repeat van Helmont's experiment using some fast-growing plants, such as bean or corn plants. Students should keep careful notes and record their data in the form of graphs. They can then share the results of their work with the class.

Homework

Have students design an imaginary experiment that refutes van Helmont's conclusion that water is solely responsible for plant growth. *(One example is to try to grow a plant in the absence of carbon dioxide, such as underwater or in space.)*

Here is a summary of van Helmont's experiment. Fill in the missing data in your ScienceLog. **Ⓐ**

Mass of dried earth =

Mass of willow tree at the beginning of experiment =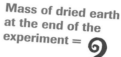

Van Helmont's experiment indicated that water was an important raw material for plants. Somehow, plants used water to build new tissues and structures. It was left to others, however, to find out exactly how the conversion of water into plant matter takes place.

Van Helmont was a careful experimenter. Can you give some examples that show this? But even though he was **Ⓑ** careful, van Helmont made a major mistake when he concluded that water alone was completely responsible for the increase in mass in plants. What do you think was wrong with his conclusion? **Ⓒ**

In the next lesson you will perform Explorations involving gases that van Helmont knew nothing about. You will also be getting more information that will allow you to answer the question, "Where does the starch in plants come from?"

Mass of dried earth at the end of the experiment =

Mass of grown tree =

What About Carbon Dioxide?

Scientists know that starch is composed of three substances: carbon, hydrogen, and oxygen. You cannot see hydrogen or oxygen (they are colorless gases), but you can see carbon. If you cook carrots too long, they eventually turn black. What happens when bread gets stuck in a toaster? It, too, turns black. In fact, any food that is cooked too long will turn black because food is made from living things and all living things contain carbon.

Where do plants get their carbon? The answer is: from the air. As you will learn in a later unit, air is made up of different gases. One of these gases is carbon dioxide, which is produced whenever carbon-containing substances are burned. A burning candle produces carbon dioxide, for example. You can test for the presence of carbon dioxide with the chemical *limewater,* which consists of calcium hydroxide dissolved in water. Limewater turns milky when it comes into contact with carbon dioxide.

Testing for Carbon Dioxide

Try the following activity. Take two small jars with lids and put a small warming candle in each one. Light only one candle. Then put the lids on both containers. After the candle flame goes out, pour limewater into both jars. Remove the candles. Gently swirl the liquid, keeping the lids on the jars. What happens to the limewater? Why is a jar with an unlit candle also used? **ⓓ**

1 **2** **3**

Integrating the Sciences

Earth and Life Sciences
Share the following information with the class: When a plant absorbs carbon dioxide from the air to use in photosynthesis, some of the carbon becomes part of the plant. When the plant dies, it may be buried by sand or silt. Over millions of years, the decayed remains of such plants become fossil fuels, such as oil, coal, or natural gas. When we burn these fossil fuels for energy, we finally release the carbon that was taken out of the air millions of years ago. You may wish to have students construct poster diagrams to illustrate this cycle.

Theme Connection

Energy
Focus question: How is the conversion of energy in photosynthesis similar to the formation of fossil fuels? *(Both photosynthesis and the formation of fossil fuels require energy from the sun, which is converted into chemical energy and stored. As organic matter becomes buried and compressed over time, volatile gases are removed during compression, leaving behind a greater proportion of carbon. The result is fossil fuel, a form of stored chemical energy. During photosynthesis, plants store energy in sugars that are later converted into starch.)*

What About Carbon Dioxide?

Have a student read aloud the first paragraph on this page. Then have students recall the conclusion drawn from van Helmont's experiment. Remind them of the composition of water. *(Hydrogen and oxygen)*

Testing for Carbon Dioxide

Depending on your classroom situation and the availability of supplies, you may wish to have students work in small groups to perform the test for carbon dioxide.

Meeting Individual Needs

Learners Having Difficulty
You may wish to perform the following demonstration to help students visualize the composition of starch: Heat some starch, such as cornstarch or a piece of potato, in a test tube over a flame. Have students look for evidence of carbon in the form of a black substance and for evidence of hydrogen and oxygen in the form of water vapor.

Prior to the Exploration, prepare the 4% sodium hydroxide solution. (See Advance Preparation on page 1D.) The following points should be noted for the successful completion of Exploration 3:

- The leaves need continuous, strong light for at least 3 days.
- Be sure students understand that the sodium hydroxide solution takes carbon dioxide away from the leaves, while the sodium bicarbonate solution (also called sodium hydrogen carbonate solution) supplies carbon dioxide to the leaves.
- In order to conduct the Exploration in one class period, you may wish to have several students set up the materials ahead of time.

SAFETY ALERT Sodium hydroxide is caustic. Sodium hydroxide may not cause pain when it comes in contact with the skin, but it can nonetheless cause irreparable damage to the living tissue. Students should be alert for any amount of solution that comes into contact with their skin and wash the exposed area immediately. Make sure that students also wash their hands thoroughly at the end of the experiment. If you are unsure about any part of the procedure for this Exploration, consult a chemistry teacher. Because sodium hydroxide is a hazardous chemical, you may wish to perform the experiment yourself beforehand to determine the minimum amount of sodium hydroxide solution necessary to complete the experiment properly. As little as 5 to 10 mL of sodium hydroxide solution may be adequate.

WASTE DISPOSAL ALERT Pour the sodium hydroxide solution into a labeled container containing enough vinegar to neutralize all of the sodium hydroxide solution used by the class. Use pH paper or another indicator to verify that the resulting solution is neutral. Then pour the mixture down the drain.

Refer to page 8 for disposal of the iodine-stained materials.

 An Exploration Worksheet (Teaching Resources, page 8) is available to accompany Exploration 3.

Carbon Dioxide and Starch

You Will Need

- sodium hydroxide solution (4%)
- baking soda (sodium bicarbonate) solution (5%)
- a lamp with 100 W bulb
- materials and equipment to test leaves for starch
- 2 large jars with lids
- 2 test tubes
- a wax pencil
- 2 leaves

What to Do

Arrange the two jars as shown in the illustration below. Jar 1 contains a 4% solution of sodium hydroxide, which absorbs carbon dioxide, thus removing it from the air.

Be Careful: Handle sodium hydroxide solution carefully—it can burn your skin!

The second jar holds a 5% baking soda solution. To make a 5% solution, add 1 part baking soda to 19 parts water. Baking soda contains carbon; its chemical name is *sodium bicarbonate*. If all of the carbon dioxide in the air is used up by the plant, the baking soda releases more carbon dioxide into the air.

Label the jars with a wax pencil. Put the jars in strong light for 3 to 5 days. Then test each leaf for starch. Do you observe differing amounts of starch? What do you conclude about the relationship between carbon dioxide and starch in plants? How could you have tested for the presence of carbon dioxide in the two jars? **A**

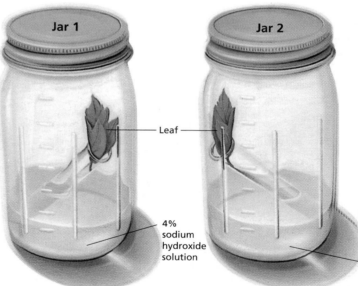

Jar 1

Jar 2

Leaf

4% sodium hydroxide solution

5% sodium bicarbonate (baking soda) solution

Put It All Together!

So now what do you think about how plants get food? If you decided, as a result of doing these Explorations, that plants use light, water, and carbon dioxide to make starch, you are right. Plants actually *make* their own food in their leaves. To do this, they need energy from the sun plus two raw materials: water and carbon dioxide. The process by which plants make food is called **photosynthesis**. Like many scientific terms, this one is a combination of two Greek words: *photo*, which means "light," and *synthesis*, which means "putting together." Given what you learned in doing the Explorations, why do you think the process is called *photosynthesis*? **B**

Ultimately, photosynthesis is responsible for feeding practically all the organisms on Earth! Plants use photosynthesis to make their own food. Animals feed directly on plants, and other animals feed on those animals. At your next meal, think about the fact that none of the food on your plate could exist without photosynthesis.

Now, here is your next challenge:

What do plants give off during photosynthesis?

Answers to In-Text Questions

A The leaves in jar 2 test positively for starch. The leaves in jar 1 should test negatively for starch, although some starch produced by the leaf prior to the experiment may remain. Carbon dioxide is necessary for green plants to make starch. Students can test for carbon dioxide by placing a test tube of limewater in each jar and then replacing the covers. If the limewater turns milky, carbon dioxide is present. If the limewater remains clear, little or no carbon dioxide is present. (You may wish to point out to students that mixing the limewater with the solutions in the jars would test only for the presence of carbon dioxide in the solutions instead of in the air of the jars.)

B Photosynthesis is the process that plants use to "put together," or construct, starch using light, water, and carbon dioxide.

Guess Which Gas!

You Will Need

* a lamp with 100 W bulb
* 2 test tubes
* a wooden splint
* baking soda
* 2 clear plastic containers
* 2 funnels
* an elodea sprig
* water
* matches

What to Do

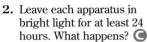

1. Arrange the materials as shown in the illustrations.
2. Leave each apparatus in bright light for at least 24 hours. What happens? **C**
3. If a test tube has collected gas, remove it from the container, making sure to hold your finger over the end of the tube so that the gas does not escape.
4. Insert a glowing (not burning) wooden splint into the test tube. If the gas is carbon dioxide, the splint will go out immediately. If the gas is oxygen, the splint will burst into flames.

Questions

1. What discoveries did you make?
2. Why is a setup without a plant sprig also used?
3. Notice that baking soda dissolved in water was used again in this Exploration. What purpose did this serve?

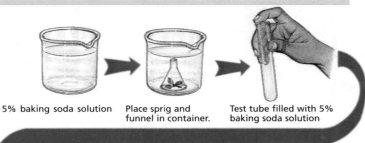

5% baking soda solution   Place sprig and funnel in container.   Test tube filled with 5% baking soda solution

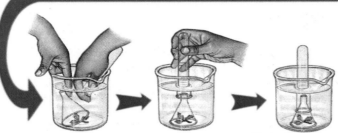

Invert the test tube while holding your finger over its mouth.   Lower the filled test tube over the funnel.   **Wait 24 hours**

Setup With Control

Baking soda in water

Sprig of elodea

Testing for gas   Insert the glowing splint quickly.

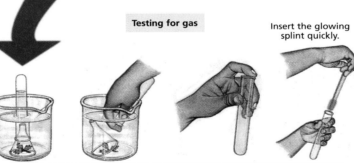

15

The following points should be noted for the successful completion of Exploration 4:

* If elodea is not available, other small aquarium plants, such as hydrilla, may be used.
* Strong sunlight gives good results. However, a lamp with a 100 W bulb or a grow lamp set near the containers produces a satisfactory amount of gas in less than 24 hours. Placing a jar of water between the lamp and the plant helps prevent heat from damaging the plant.
* Be sure that the same concentration and amount of sodium bicarbonate solution are used in both test tubes.
* Point out that the test tubes should be as full of the sodium bicarbonate solution as possible. Any air spaces at the top of the inverted test tubes will make the results less dramatic.
* Be sure that students keep a finger or thumb over the openings of the test tubes as they lift them out of the water at the end of the Exploration.
* The openings should not be uncovered until the test tubes are upright. This will help to prevent any collected gas from escaping, since it is heavier than air.

Answer to
In-Text Question

C The test tube in the container with a plant should collect gas, which can be seen by the lower level of the baking soda solution in the test tube.

Answers to
Questions

1. In the test tube over the funnel containing the plant, students should observe that baking soda solution has been replaced with a gas bubbling up from the plant. When the glowing splint is placed into the test tube, it bursts into flame, indicating the presence of oxygen. The test tube from the container without a plant should not collect any gas. However, if there is an air space at the top of that test tube, a glowing splint placed in the test tube will

grow dim or go out, indicating a lack of oxygen. Students should conclude that the plant gave off oxygen in the presence of light, water, and carbon dioxide from the baking soda solution.

2. The setup without a plant sprig is a control. It ensures that the only variable between the two setups is the plant. Therefore, any differences in the outcomes must be due to the presence of the plant.

3. The baking soda supplies the plant with the carbon dioxide it needs for photosynthesis.

Homework

Hydroponics is the growing of plants in nutrient-rich water without soil. Have students research hydroponic techniques and try to grow a plant without soil. Volunteers may wish to present their findings to the class.

★ **An Exploration Worksheet is available to accompany Exploration 4 (Teaching Resources, page 9).**

Answers to
A Formula for Food

1. Starch (Oxygen is also produced.)

2. Water or carbon dioxide

3. Carbon dioxide or water

4. Light (usually from the sun)

5. Oxygen; most organisms need it for respiration.

 A Transparency Worksheet (Teaching Resources, page 11) and Transparency 1 are available to accompany A Formula for Food.

Answers to
In-Text Questions

Ⓐ The following is a sample description: During the process of photosynthesis, green plants use *light* energy and the *raw materials water* and *carbon dioxide* to *manufacture* the *product, food*. The food is stored in the form of *starch*. *Oxygen* is released as *waste* in the process.

Ⓑ The question marks can be replaced, from left to right, by (2 or 3), (3 or 2), (4), (1), and (5). The arrow means *produces* or *yields*.

A Final Word on Photosynthesis

Be sure students understand that sugar is the first result of photosynthesis and that it is converted into starch later. The process of changing sugar to starch and starch to sugar is going on almost all of the time in the leaves of green plants. You may wish to point out that the name of the simple sugar is *glucose*. *Sucrose* is the compound we refer to as table sugar.

A Formula for Food

Photosynthesis can be likened to the processes that occur in a factory: raw materials come in, energy is used, a finished product results, and wastes are produced. Now, check what you have learned about leaves as "food factories."

1. What important product results from photosynthesis?

2. What is one essential raw material for photosynthesis?

3. What other raw material is required?

4. What is the source of energy for photosynthesis?

5. What is a waste product of photosynthesis? Why is this waste product important to other living things?

Now, write a description of photosynthesis based on what you have learned so far in this unit. Use the following terms in your description: light, oxygen, waste, starch, water, raw material, carbon dioxide, food, manufacture, and product. Your description of photosynthesis could turn out to be quite long. A much shorter description, which still conveys a great deal of information, is the following simple equation: **Ⓐ**

Copy this equation into your ScienceLog. Then replace each question mark with the number of the question above that relates to it. What does the arrow mean? **Ⓑ**

A Final Word on Photosynthesis

You have learned that photosynthesis produces a food called starch. However, when it is first made, the food in the leaf is actually in a form that is chemically less complex than starch—a simple sugar. Many sugar particles join together in a specific way to form the more complex starch particle. The sugar is changed into starch when the plant stores food for future use. When the plant needs food, it changes some of its stored starch back into sugar. If this is so, why didn't you test for sugar? Mostly because it is easier to test for starch than for sugar, and the presence of starch is evidence that food is being manufactured by the leaf.

You can more accurately rewrite the photosynthesis word equation as follows:

FOLLOW-UP

Reteaching

Have students make a diagram that illustrates the process of photosynthesis. Their diagram should show the roles that water, carbon dioxide, sugar, oxygen, and sunlight play in the process. You may wish to have students compare this diagram with the one they made in the Prior Knowledge and Misconceptions activity at the beginning of this chapter.

Assessment

Present students with the following scenario: A natural disaster blocks out half of the light coming from the sun for several months. How does this affect green plants? *(Answers will vary but could include the following effects: fewer or smaller plants, less productive plants, and the death of plants due to climate change and the loss of energy needed to make food. There would also be a drastic effect on the animals that depend on plants for food.)*

Extension

One theory that attempts to explain the disappearance of the dinosaurs involves an asteroid striking the Earth and sending up huge clouds of dust to cover the land and oceans. Have students describe how this event could have affected the Earth's plants, which would have affected the dinosaurs in turn.

Closure

Have students design an experiment to compare the rate of food production in a variety of plants. Do not allow students to conduct any experiment without first checking their proposed procedure with you for safety.

Food Factory Basics

Plant Structure

A well-designed structure is a thing of beauty. Architects and engineers strive to design buildings, bridges, machines, and other structures that are elegant as well as efficient and functional. Structures are also found in nature. Take plants, for example. How do they rate in terms of their efficiency? their functionality? their "ingenuity" of design? **C**

D Quickly draw a plant. What parts did you show—leaves? stem? roots? What other plant parts might you include? How about flowers, fruit, and seeds? When we say the word *plant,* we usually think first of a large, important group that we call the seed plants. This group includes many familiar plants. Look at the drawing of a typical seed plant.

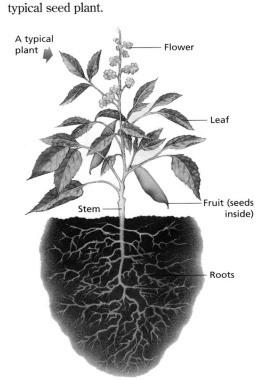

A typical plant
Flower
Leaf
Fruit (seeds inside)
Stem
Roots

As you know, parts of your body that have special functions, such as your heart or your stomach, are called **organs**. Plant parts may also be called organs because they perform special functions for a plant. What are the special functions of each of the plant organs shown in the drawing at lower left?

Plant Organ	Functions
Roots	
Stem	
Leaves	
Flowers	
Fruit	
Seeds	

Listed below are a number of functions (tasks) that plant parts perform. Which plant organ carries out which function? Copy the names of the plant organs into your ScienceLog and then match the letters indicating functions with the appropriate plant organs.

Functions of Plant Organs

More than one organ may perform certain functions, so you may want to use a letter more than once.

a. Make food by photosynthesis
b. Hold plant firmly in the soil
c. Produce the leaves
d. Exchange gases with the air
e. Allow the plant to reproduce
f. Store food for later use
g. Hold leaves up to the light
h. Produce a new plant without using seeds
i. Absorb water and minerals from the soil
j. Hold the developing seeds
k. Let water vapor escape into the air
l. Conduct water and minerals upward to the leaves

LESSON 3 ORGANIZER

Time Required
4 class periods

Theme Connections
Structures, Changes Over Time

Process Skills
observing, analyzing, comparing, classifying

New Terms
Blade—the flattened, green part of a leaf
Chlorophyll—a green pigment in plants; an essential ingredient for photosynthesis

Organ—part of an animal or plant that performs a special function
Petiole—the stalk that fastens a leaf to the stem of a plant
Pigment—a material that gives substances their color by absorbing some colors of light and reflecting others
Stomata—slitlike openings located mainly on the underside of a leaf. They allow water vapor and other gases to pass into and out of the leaf.
Veins—the channels through which materials move in a leaf. They also help strengthen the leaf.

continued ▶

FOCUS

Getting Started
Ask students to draw several types of plants. Then have them identify the name and function of each plant part. *(Accept all reasonable responses.)*

Main Ideas
1. Various plant organs perform distinct functions.
2. A leaf is made up of many different kinds of cells.
3. The upper and lower surfaces of a leaf differ in a number of ways.
4. Gases enter and exit a leaf through structures called stomata.
5. Specific cells inside a leaf are the sites of the production of food.

TEACHING STRATEGIES

Answers to
In-Text Questions

C Students should support their statements with examples. For example, plants are efficient in that they get energy directly from the sun and utilize the animal waste product carbon dioxide.

D Accept all reasonable responses.

Answers to
Functions of Plant Organs

The following table shows how the functions from the list may be incorporated into the in-text table:

Plant organ	Functions
Roots	b, e, f, h, i, l
Stem	a, c, e, g, h, i, l
Leaves	a, d, f, k
Flowers	e
Fruit	e, f, j
Seeds	c, e, f

 Transparency 2 is available to accompany Functions of Plant Organs.

18

Answers to
Out of the Ordinary

Corn
- The brace roots of the corn plant grow above ground.
- They sprout from the lower stalk to provide additional support.

Ivy
- The climbing roots of ivy grow above ground along its winding stem.
- If a root touches fertile ground, it can bud, sprouting new roots.

Raspberry
- Tip layering occurs when a plant's stem bends over, its tip touches the ground, and a new root system sprouts from the tip.
- The tips of the stems provide one way for the raspberry plant to reproduce.

Iris
- The iris's stem, called a rhizome, grows horizontally underground.
- Buds on the rhizome produce leaves above ground, creating a new plant.

Onion
- The stem of the onion grows underground, is short, and provides little structural support.
- The stem is part of the bulb, an unusual leaf system that the plant uses to store food.

Potato
- The stem of the potato grows underground and is very short.
- The tuber, which allows the potato to store food, is an enlarged part of the stem.

Did You Know...

Only black and purple raspberries propagate by tip layering. Red raspberries undergo a process called suckering to generate offspring. A sucker is a shoot from the bud or stem that grows very near the surface. It eventually sprouts its own root system and grows into a new plant.

Out of the Ordinary

You have just identified functions of stems and roots. You also know where to look to find these parts on a plant. Some stems and roots, however, grow in unusual locations on a plant, where they have special functions to perform.

Examine the plants shown at right. Refer to the exercise you have just completed to help you

- identify what is unusual about the location of one part of each plant.
- explain what special function that part might perform because of its location.

Corn | Climbing roots | Purple raspberry
Brace roots | Ivy | Tip layering
Unusual roots ➤

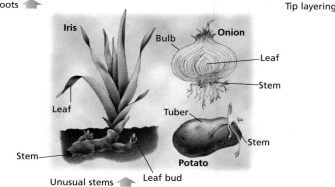

Iris | Bulb | Onion
Leaf | Stem | Leaf | Stem
Stem | Tuber | Potato
Unusual stems ➤ | Leaf bud

Leaves

A leaf is specially structured for its main task—the production of food by photosynthesis. The thin green part is called the **blade**. Each kind of plant has a leaf blade that distinguishes it, by size or shape, from other kinds of plants.

A network of **veins** gives strength to the leaf. The pattern of veins is different for each kind of plant. Materials move into and out of the leaf through the veins.

The leaf is fastened to the stem, usually by means of a stalk called the **petiole**. The leaves of some plants are fastened directly to the stem, without petioles.

Edge
Blade
Vein
Petiole

ORGANIZER, *continued*

Materials (per student group)
Exploration 5, Activity 1: magnifying glass or low-power binocular microscope; a variety of fresh leaves;
Activity 2: microscope; 2 lettuce leaves; 2 microscope slides and coverslips; transparent metric ruler; eyedropper; a few drops of water
Exploration 6: prepared slide of leaf cross section; microscope; sharp, hard pencil; unlined paper
Exploration 7: plant with leaves that have white areas, such as coleus or pothos ivy; materials to test for starch from Part 2 of Exploration 1 on page 7

(additional teacher materials: 25 mL of 0.1 M sodium thiosulfate solution; a few sheets of old newspaper; 250 mL of water; 600 mL beaker; see Advance Preparation on page 1D.)

Teaching Resources
Exploration Worksheets, pp. 13, 15, and 16
Activity Worksheet, p. 18
Transparencies 2–4

Sycamore

Note the different vein patterns in these leaves. How might you describe them? Ⓐ

Elm

Lily

Grass   Beech   Carrot

How might you describe the leaf edges shown here?
Ⓑ

Maple

Mimosa   Oak

Simple or compound? How can you tell? Ⓒ

Although leaves show great variation, they all perform the same basic functions for plants. In the next Exploration you will closely examine the outside surfaces of leaves and their internal structure.

EXPLORATION 5

A Good Look at Leaves

ACTIVITY 1

An Outside View

You Will Need

- a magnifying glass or low-power binocular microscope
- a variety of fresh leaves

What to Do

Choose a leaf. Look carefully at both its upper and lower surfaces. Which surface has more features? Look for patterns of Ⓓ veins, hairlike projections, and other structures. Compare the leaf to leaves from several other plants.

Choose one type of leaf and write descriptions of both of its surfaces. Let a classmate read your description and try to identify the leaf you described.

To get a closer look at a leaf, you will need to use a microscope. You will also need to prepare a slide of leaf tissue that is thin enough for light to pass through. You will do this in the next Activity.

Exploration 5 continued ▶

19

Ⓐ Accept all reasonable responses. You may wish to point out that the vein pattern in a leaf is called *venation*. Leaf venation can be *netted, parallel,* or *palmate.* The elm leaf shows netted venation. It has one large central vein from which smaller veins branch out. The lily has parallel venation because it has many large veins that run parallel to one another, with smaller veins in between. The sycamore has a few large veins, each one running along a different lobe of the leaf. This is palmate venation.

Ⓑ Accept all reasonable responses. Leaf edges, or margins, are generally classified as *entire, dentate,* or *lobed.* An entire leaf has a smooth edge, such as the grass leaf. Dentate leaves have pointed "teeth" and may be called serrated or incised. Examples include the leaves of the beech and the maple. Lobed leaves have rounded intrusions, such as those of the carrot.

Ⓒ The oak leaf is a simple leaf because only one blade attaches to the petiole. The mimosa leaf is compound because it consists of separate leaflets that attach to the petiole independently of one another.

Ⓓ The lower surface of a leaf usually has more features.

EXPLORATION 5

Before students begin, you may wish to have them review the procedures for handling a microscope and for making wet-mount slides.

ACTIVITY 1

For Activity 1, present students with leaves from a variety of plants. You may wish to make wet-mount slides or find prepared slides of lower and upper leaf surfaces for students to compare.

Exploration 5 continued ▶

Homework

To familiarize students with some of the terms used in this chapter, have them make a list of at least 10 words, and their definitions, that contain the roots *photo* or *chlor.* Ask: What do these roots mean? (Photo *means "light," as in photograph or photosynthesis.* Chlor *means "green," as in chlorine or chlorophyll.)*

Meeting Individual Needs

Second-Language Learners

Have students develop a multilingual catalog of the names for various plant parts, along with a definition for each part. Students should point out any similarities in the names between the languages.

 An Exploration Worksheet is available to accompany Exploration 5 (Teaching Resources, page 13).

ACTIVITY 2

You may want to make extra wet-mount slides of the lower and upper surfaces of lettuce leaves for students who have difficulty preparing their own. Suggest that students use a high-power objective lens to estimate the number of stomata because it will reduce the microscope's field of view and thus reduce the number of stomata that the students must count. Also, preview the slides of lettuce leaves to make sure that students will not be required to count an unreasonable number of stomata.

Answer to
In-Text Question

Ⓐ Students should observe slitlike openings of stomata similar to the example shown in the photograph.

Answers to
Interpreting Your Observations

1. The bottom surface contains more stomata.

2. Estimates will vary. Students can determine the diameter of the field of view by using a transparent metric ruler to count the maximum number of millimeters that can be seen. If the diameter of the field of view at high-power (400×) is 1 mm, then the area of this field (determined by the equation $A = \pi r^2$) is about 0.8 mm². Students can count the visible stomata and multiply the total by 125 to estimate the number of stomata in 100 mm², or 1 cm². (You may wish to assist students with their calculations. Typically, 1 cm² of a leaf has about 10,000 stomata. However, this number may vary from 1000 to 200,000 depending on the species and growth conditions.)

3. Accept all reasonable responses. Students may correctly speculate that the stomata allow substances to pass into and out of the leaf.

4. Accept all reasonable responses. Students may correctly speculate that the guard cells change in shape and size to open and close the stomata.

A Close-Up View

ACTIVITY 2

You Will Need

- a microscope
- 2 lettuce leaves
- a microscope slide and coverslip
- a clear plastic ruler
- an eyedropper
- water

What to Do

You are going to make a wet-mount slide of the cells of the lower surface of a lettuce leaf, and then you are going to make a wet-mount slide of the upper surface. For the lower-surface slide, use your fingernail to gently scrape away the upper part of the leaf so that only the thin lower layer remains. Mount the thin leaf section on a slide, as shown in the illustrations at right.

Look for the kinds of cells shown below. Do you see any slit-like openings? Ⓐ

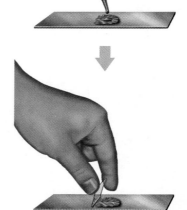

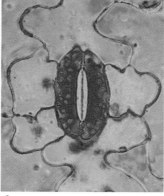

▲ A stoma (with *guard cells* stained purple)

These openings are called *stomata* (singular *stoma*). Study them closely. Then draw and label one stoma and the cells around it. You should see two kinds of cells: guard cells and regular epidermal (surface) cells.

Now follow the same process for the upper surface of a leaf.

Interpreting Your Observations

Working with a classmate, find the answers to the following questions:

1. On which leaf surface (top or bottom) did you observe more stomata?

2. Estimate the number of stomata on one square centimeter of the lower surface of the leaf. (Hint: How much magnification are you getting with the microscope? What do you estimate to be the area of the field of view in the microscope?)

3. What purpose do you think the stomata might serve?

4. How might the guard cells open and close the stomata?

5. Do you think the stomata close at night? Why or why not?

6. What do you predict would happen if grease were smeared on the lower surface of the leaves of a healthy plant and left there for several days? Why would this happen?

In case you still have questions about the purpose that stomata serve, their function will be explained further, beginning on the next page.

---

5. Many stomata close at night because photosynthesis occurs less at lower light levels. Therefore, there is less need for gas exchange through the stomata.

6. The leaves would not be able to make food from carbon dioxide because there would be no gas exchange through the stomata. Eventually, the plant would die.

CROSS-DISCIPLINARY FOCUS

Mathematics

You may wish to reinforce student understanding of the actual sizes of the objects described in this lesson. For instance, have students note the magnification of the onion cell pictured on page 22. *(300×)* At this scale, a typical pencil would appear just over 2 m wide and almost 60 m long! Using this scale, have students measure and calculate the adjusted sizes of several nearby objects, such as a textbook, fingernail, or desk chair.

Stomata Data

As you saw in Exploration 5, the stomata are located mainly on the underside of a leaf. Stomata allow water vapor and other gases to pass into and out of the leaf. Despite their important function, stomata are extremely tiny—the width of a stoma is about $\frac{1}{10}$ the thickness of this page! Each stoma is flanked by a pair of guard cells, which control the size of the opening. When the guard cells absorb water and swell, the stoma is opened. The guard cells relax when water leaves them—this closes the stoma. Did you guess that this is how stomata work? Usually stomata open during the day and close at night. Considering what you know about photosynthesis, why does this make sense?

An Inside View

The drawing below shows a cross section of a typical leaf. Imagine that you could shrink to the size of a water or carbon dioxide particle. You could journey into a leaf and make some interesting discoveries about this "food factory." Study the diagram of the leaf, and then answer the questions below.

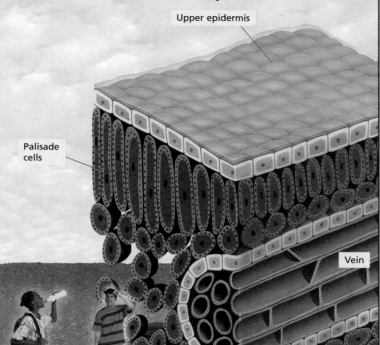

Upper epidermis

Palisade cells

Vein

Guard cells

Stoma

1. In your miniaturized state, what are two ways you could enter the leaf?

2. If you were a carbon dioxide particle, where would you wait, along with large numbers of other gas particles, until you could be used?

3. Where would you find huge, bulging "gates"?

4. Where would you exit the leaf? Would a water particle and a carbon dioxide particle exit at the same place? Why?

5. Where would you find a pipeline system to carry liquids such as water?

6. Can you find the elongated cells near the top of the leaf? Would these cells be well lit? Might these be good places to make food?

7. Food is also made in the closely packed, somewhat rounded cells farther inside the leaf. Would these be as well lit as the elongated cells? Could as much food be made there?

21

Theme Connection

Structures

Focus question: In what ways is the structure of a leaf similar to the structure of human skin? *(The plant leaf and human skin both contain a protective layer, the epidermis. In both, pipelike structures transport water and other chemicals utilized or produced by the organism.)*

Answers to
Stomata Data

Because photosynthesis occurs in the presence of light, it is helpful for the stomata to remain open during the day so that gases can move into and out of the leaves. At night, when less photosynthesis is taking place, the stomata can close to conserve water.

Stomata Data

To promote critical thinking, you might want to point out that the guard cells are the only epidermal cells that conduct photosynthesis. Ask: Why is this significant? *(The opening and closing of the stomata can be stimulated by the presence and absence of light.)*

Answers to
An Inside View

1. You could enter through a vein or through a stoma.

2. You could wait in the air spaces between cells.

3. The guard cells on either side of the stomata are like huge, bulging gates.

4. Water and carbon dioxide particles could both exit the leaf through the stomata because the stomata are not selective in the type of particles that they allow to pass through. Some particles could also exit through the leaf's veins, which carry sugar and starch to other parts of the plant.

5. There are pipelike veins in the leaves that connect to the stems and roots.

6. Because the elongated palisade cells are located under the top surface of the leaf, they receive a great deal of light and would be good places to conduct photosynthesis and make food.

7. These cells would not receive as much light; therefore, they would not be able to produce as much food as the palisade cells. (This layer is called the *spongy layer*.)

 Transparency 3 is available to accompany An Inside View.

Homework

As students complete An Inside View, have them create a travel log that chronicles their imaginary trip through a leaf. Log entries should focus on the leaf's structures and functions, with particular emphasis placed on the materials involved in photosynthesis.

Encourage students to refer to the diagram on page 21 as they study the slides. The criteria listed in Making a Good Microscope Drawing should help students as they complete Exploration 6. They may be able to identify the upper and lower epidermis, palisade cells, veins, stomata, and guard cells.

Answers to
In-Text Questions

a. The epidermal cells prevent the leaf from drying up.
b. Food is made in the palisade cells and in the rounded, loosely packed cells. The layer of palisade cells adds strength to the leaf because these cells are tightly packed.
c. The vein is part of the transportation system of the leaf.
d. The guard cells control the flow of gases into and out of the leaf through the stomata.

An Exploration Worksheet is available to accompany Exploration 6 (Teaching Resources, page 15).

 **PORTFOLIO**
Suggest that students include their diagram from Exploration 6 in their Portfolio. The drawings represent mastery of an important laboratory skill.

Meeting Individual Needs

Gifted Learners
Challenge students to consider whether the elements involved in photosynthesis are "used up" during the process. *(The molecules of water drawn from the soil are split into molecules of hydrogen and oxygen. The hydrogen combines with molecules of carbon dioxide from the air, forming a simple sugar. Oxygen molecules left over from the broken-down water molecules are released into the air. The individual atoms of the elements involved are rearranged but not destroyed.)*

Making a Good Microscope Drawing

You don't have to be an artist to make a good microscope drawing. Below is a drawing of onion-skin cells viewed under a microscope. It is good because

- a sharp, hard pencil was used on plain unlined paper; thick markers, soft pencils, and colored pencils were not used.
- details are shown, and only one cell is shown with all of the details.
- labels are outside the drawing and neatly printed.
- labeling lines end at the feature being identified.
- there is no blurring, shading, or added color.
- the magnification is provided.
- the drawing was done neatly and carefully.

In the next Exploration, you will try your hand at making a microscope drawing.

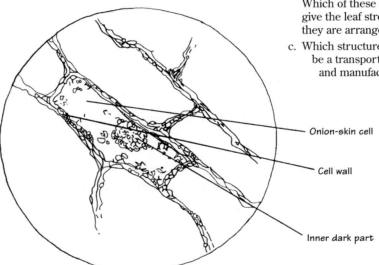

Onion-skin cell
Cell wall
Inner dark part

Onion-skin cell, magnification = 300x

A Microscopic Activity

You Will Need
- a prepared slide of a leaf cross section
- a microscope
- a sharp, hard pencil
- unlined paper

What to Do
In this activity, you will use a microscope to view a cross section of a leaf, and make drawings of what you see.

1. Examine the cross section under a microscope. Use the cross-sectional view of a leaf on page 21 to help you identify the kinds of cells present.
2. Draw a detailed diagram of each type of cell you see. Label the various kinds of cells.
3. Answer each of the following questions, even if you did not see all of the cell types in your cross section.

   a. Which cells protect the leaf, preventing the inside from drying up?
   b. In which cells is food manufactured? Which of these cells can also function to give the leaf strength because of the way they are arranged?
   c. Which structure looks like it might be a transportation system for water and manufactured food?
   d. Which cells control the flow of gases into and out of the leaf?

ENVIRONMENTAL FOCUS
Share the following information with students: About 2.5 billion years ago, the Earth's atmosphere contained very little oxygen. Organisms called *cyanobacteria* did not need abundant oxygen to survive, but they produced it as a result of photosynthesis. Because of the cyanobacteria, algae, and plant life that have appeared since that time, the atmosphere is now about 21 percent oxygen. This has allowed many types of oxygen-dependent organisms to thrive on Earth.

CROSS-DISCIPLINARY FOCUS

Health
Share the following information with students: Plants are not the only organisms that use sunlight for nourishment. When exposed to sunlight, human skin cells produce vitamin D. For this reason, vitamin D is sometimes referred to as the "sunshine vitamin." Too much or too little of this vitamin can cause damage to bones. Have interested students form groups and produce a pamphlet that identifies sources of vitamin D and their benefits.

Why Green?

You have been investigating the process by which plants produce food. You now know about the raw materials plants use, their source of energy, and the products of photosynthesis. You have no doubt noticed that most living plants are green. Obviously this color must be important, but why?

First of all, this green color is due to a substance called *chlorophyll*. When you tested for starch in Exploration 1, you dissolved chlorophyll and removed it from the leaf so that you could see the results of the test for starch. What function do you think chlorophyll performs? The next Exploration will give you a good idea.

EXPLORATION 7

What Does Chlorophyll Do?

To figure out the function of chlorophyll, you need leaves, such as those from a coleus, that have some white areas (indicating a lack of chlorophyll).

You Will Need

• materials and equipment to test leaves for starch

What to Do

Give the plant a good chance to make food (you know the conditions), and then remove a couple of leaves. Test for starch as you did previously. (Remember to heat the methanol safely!)

Make a brief report in your ScienceLog explaining your idea about the function of chlorophyll. Include a diagram showing the areas with and without starch.

Exploration 7 continued ▶

23

Theme Connection

Changes Over Time

Focus question: Why do many leaves change color in the autumn? Have students research what chemical changes are occurring in these leaves and what environmental factors cause the leaves to change color. (*In addition to chlorophyll, many leaves also contain pigments called carotenoids. Carotenoids are usually red, orange, or yellow, but their colors are ususually masked by the plentiful chlorophyll in the leaf. As days get shorter and temperatures become cooler, the metabolism of the leaf slows and its cells gradually die. Since carotenoids tend to withstand these changes longer than chlorophyll, the leaves lose their green coloration and briefly display the vibrant colors often associated with the fall.*)

EXPLORATION 7

Before students begin, remind them that the safety precautions that they took in Exploration 1 must also be taken in this Exploration because many of the same procedures are being repeated. Make sure that students separate the green and white portions of the leaves prior to boiling them in methanol.

By completing this Exploration, students should discover that the parts of leaves without chlorophyll do not produce starch. Conversely, the parts of leaves that do contain chlorophyll also contain starch. Using this information, students should conclude that in order for photosynthesis to take place, chlorophyll must be present.

WASTE DISPOSAL ALERT Refer to page 8 of this Annotated Teacher's Edition for proper disposal of the iodine-stained materials.

Answers to *Crash-Course Quiz*, page 24

1. Carbon dioxide + water + chlorophyll + light energy → sugar (or starch) + oxygen + chlorophyll

2. Chloroplasts (Students are introduced to chloroplasts on page S9 of the SourceBook.)

3. Chlorophyll is found in the palisade cells and in the rounded, loosely packed cells. (Some chlorophyll is also found in guard cells.)

4. The palisade cells are the chief food producers. They receive the most light because they are closest to the leaf's upper surface. Also, the greatest concentration of chlorophyll is found there.

★ An Exploration Worksheet is available to accompany Exploration 7 (Teaching Resources, page 16).

Homework

Have students write a few paragraphs explaining the function of chlorophyll in photosynthesis. Paragraphs should include the following terms: chlorophyll, chloroplasts, palisade cells, loosely packed cells, sunlight, energy, pigment, and photosynthesis.

Answer to
In-Text Question

Ⓐ Chlorophyll reflects green light.

Topics to Discuss, *page 25*

By discussing science questions in small groups, students can share and evaluate their ideas and the ideas of others. After each group is finished with the discussion and research work, suggest that each group choose a spokesperson to report their findings to the class.

Answer to
Topic 1, *page 25*

Trees absorb water from the soil through their roots. The water reaches the leaves by traveling through the stem and veins.

Answers to
Topic 2, *page 25*

a. Leaves at the bottom of a tree may be broader because they are more shaded and require a larger surface area to absorb adequate light.
b. Plants with leaves that are not green still contain chlorophyll. The green is overpowered by other pigments that are yellow, red, or purple. Light absorbed by these pigments is also used by the plant as energy for photosynthesis.
c. Plant leaves contain special chemicals that are very efficient at using the small amount of carbon dioxide that is in the air.
d. The cells in the leaves of evergreen trees have thick cell walls and contain little water. This helps them keep from freezing during the winter.

Answer to
Topic 3, *page 25*

The materials in a closed terrarium are recycled by the organisms that live there, just as water, carbon, oxygen, and nitrogen are recycled in natural ecosystems.

A Crash Course in Colors

Light from the sun seems to have no particular color. But, in fact, sunlight is a mixture of many different colors blended together. We can see the colors that make up sunlight by passing it through a prism, or by looking at a rainbow. If sunlight is white, then why do the things around us appear to have color?

Suppose that a substance absorbed all of the colors of sunlight except red, which it reflected. What color would you perceive the substance to be? Red, of course. Substances appear to have color because they absorb some colors and reflect others.

A *pigment* is a substance that absorbs light. Some pigments absorb all the colors of light, and some absorb certain colors but reflect the rest. In this way pigments give substances their color. Chlorophyll is a pigment that reflects one color of light. What color is this? **Ⓐ**

As you know, sunlight has energy. Chlorophyll absorbs certain colors of light, so chlorophyll absorbs energy. How do you think plants put this energy to work?

The process of photosynthesis puts the energy absorbed by chlorophyll to work, chemically combining carbon dioxide and water to form sugars. Further reactions change these sugars into starches. Now that you know this, you can take the following mini-quiz.

Crash-Course Quiz

1. Write a word equation for photosynthesis, placing *chlorophyll* in the equation.
2. Study the diagram of a leaf cell shown below. Which of the cell bodies contain chlorophyll (and make the plant's food)? Find out what these chlorophyll-containing bodies are called.
3. In which leaf cells is chlorophyll found?
4. Which type of leaf cell would you expect to be the chief food producer? Support your inference with at least two observations you have made.

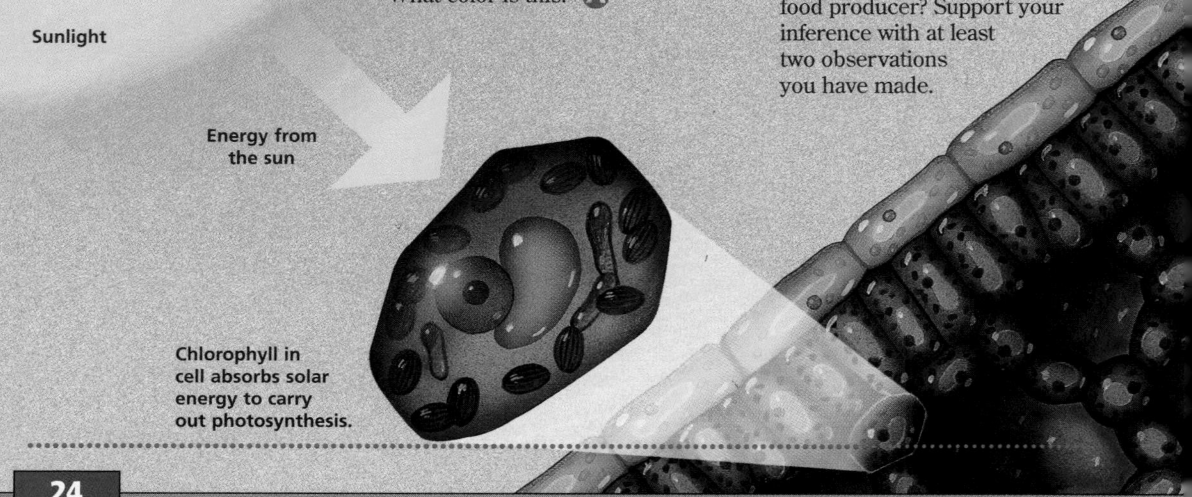

Sunlight

Energy from the sun

Chlorophyll in cell absorbs solar energy to carry out photosynthesis.

24

Homework

Have students define the word *succulent* in the context of leaf structure. (*A succulent leaf is a fleshy, juicy leaf used to store water. It is often protected by spines or thorns.*) What are some examples of succulent plants? (*Common succulent plants include cactuses and aloe.*)

Topics to Discuss

Form small groups and discuss any two of the following topics. You may need to do some research first. Then share your conclusions with the whole class.

Topic 1

A tree releases water through its stomata. How, and from where, did the tree get this water? How do you think the water reaches the leaves?

Topic 2

a. Why are leaves often broader at the bottom of a tree?

b. How do plants without green leaves (such as the purple velvet vine below) make food?

c. Carbon dioxide is fairly scarce, making up only about 0.035 percent of the atmosphere. How do plant leaves get enough carbon dioxide to carry out photosynthesis?

d. How do the leaves of evergreen trees survive cold winters?

Topic 3

How do plants in closed terrariums manage to get raw materials for photosynthesis?

Topic 4

Discuss the design of a specific leaf, taking into consideration the following questions:

a. In what ways is the leaf well suited for its particular environment?

b. Can you find any "design flaws"? If so, identify them and suggest improvements.

Topic 5

Coniferous trees, such as pines, have leaves that are radically different in structure from those of plants such as maple trees, sycamore trees, or orange trees.

a. Identify some of the differences.

b. How might these structural differences reflect differences in each plant's way of life?

Pine   Orange   Maple
Sycamore

25

Answers to
Topic 4

a. Accept all reasonable responses. For example, plants in arid regions have leaves that are thick and fleshy or small and leathery, or these plants have thick, fleshy, green stems with no leaves. These features help these plants reduce water loss and store water while providing adequate surface area to absorb light for photosynthesis.

b. Students should make reasonable suggestions for design flaws and corresponding improvements.

Answers to
Topic 5

a. Leaves of most conifers are small, stiff, needlelike, and evergreen, while most other plants have leaves that are broad, flat, flexible, more succulent, and last only one growing season.

b. Most coniferous trees live in cold, dry climates. The needlelike shape reduces water loss. Having thickened cell walls and cells that contain little water enables the needles to resist freezing. Also, the leaves photosynthesize year-round to provide enough food for growth in cool climates with limited sunlight.

Reteaching

Have students examine some deciduous tree leaves. Ask: Which side is darkest? Relate this to where the palisade cells are found. Then have students predict the number and position of stomata on the leaves of a water lily *(many on the top surface)*, a desert plant *(few in the shade of the lower surface)*, and grass *(uniformly distributed)*. Relate this to the needs for water conservation and gas exchange in the leaf.

Assessment

Ask: What happens to a plant's ability to produce chlorophyll if it does not have access to light? *(Plants cease to produce chlorophyll in the absence of light.)* Suggest that students design an imaginary experiment to answer this question.

Extension

Point out to students that chlorophyll is used in some common everyday products. Have them investigate what some of these products are and why they contain chlorophyll. *(Students may be surprised to know that chlorophyll is used as an air freshener. It is used as a dye in such products as soap, oils, wax, candy, preserves, cosmetics, toothpaste, and chewing gum. Chlorophyll is also sometimes taken as an herbal dietary supplement.)* Suggest that students compile their findings in a chart to display in the classroom.

Closure

Have students work in small groups to make three-dimensional models of the cross section of a leaf. Materials may include clay, papier-mâché, or construction paper. Display the finished models in the classroom.

Homework

You may wish to assign the Activity Worksheet on page 18 of the Unit 1 Teaching Resources booklet as homework. If you use the worksheet in class, Transparency 4 is available to accompany it.

Answers to *Challenge Your Thinking*, pages 26–27

CHALLENGE YOUR THINKING

1. Accept all reasonable responses. Sample answer: Plants, windmills, and hydroelectric dams all convert energy that ultimately comes from the sun. For example, plants convert sunlight into food. Because plants are involved in so many food chains, most living things receive energy from the sun indirectly through the foods they eat. Windmills convert energy from winds, which result from the warming of Earth's atmosphere by the sun. Hydroelectric dams convert energy from Earth's water cycle, which depends on energy from the sun to cause the evaporation of water.

2. **a.** Plants appear green because the chlorophyll in them absorbs energy from all parts of the visible spectrum except green. Green is reflected, causing the plants to appear green.
   b. A plant that received only green light would not grow well because it would reflect most of the light it received.
   c. A plant grown in red or blue light would grow better than a plant grown in green light. However, the plant would grow best in normal sunlight because it would then receive the most energy.

3. Accept all reasonable responses. Answers outlined by students might include the following new ideas: (1) Plants make food for themselves as opposed to being "fed" by fertilizers. (2) Plants make food in their leaves as opposed to taking food in through their roots. (3) Water from the soil and carbon dioxide from the air combine in green leaves to make food as opposed to plants absorbing food that is already made. (4) Sunlight provides energy for food making in plants as opposed to plants needing the sun for warmth only.

   Sample explanation for a fifth-grade student: Plants make food for themselves in a process called photosynthesis. To make food, they use light, carbon dioxide, and water. Water is taken in through the roots and travels to the leaves. Carbon dioxide enters the leaves from the air through openings called stomata. In the leaves, a green substance called chlorophyll traps light energy.

1. How Is a Plant Like a Windmill?

You have been asked to respond to the issue raised in the cartoon below. How would you go about providing evidence to support your response?

2. It's Not Easy Being Green

The diagram at left shows the *absorption spectrum* (pattern of light absorption) for a typical plant. Light that is not absorbed is reflected. Study the diagram and then answer the following questions. Remember that white light (such as that from the sun) is a mixture of all the different colors of light.

a. What causes plants to appear green?

b. If a plant received only green light, how might it grow compared to a plant grown in normal light?

c. Would a plant grown in red light do well? blue light?

Absorption axis: Less — More

Red   Yellow   Green   Blue   Violet

Oxygen is produced during photosynthesis and is released into the air through the plants' stomata. The food produced is sugar or starch.

4. Answers will vary. Students should describe a series of experiments in which they test whether a specific kind of plant will grow without the presence of starch in the soil or water. In the first experiment, one plant could be grown in soil containing starch, while the control is grown in soil without starch. In the second experiment, one plant could be watered with water containing starch, and the other with plain water. Other variables such as the

type of plant, light, temperature, and size should be identical for both plants.

5. Answers will vary according to each student's continuation of the radio report. Students should demonstrate their knowledge of normal plant processes by contradicting what they have learned. For example, students may report that plants are giving off carbon dioxide instead of oxygen or that they are reflecting red light instead of green. Students should also describe some effects, such as malnutrition or a lack of oxygen, that would result from the abnormal plant behavior.

3. Back to the Drawing Board

This chapter began with the question, "How do plants get their food?" Has your thinking changed since then? Here are some suggestions to help you organize your ideas.

a. Begin by writing a brief report about how your thinking has changed.

b. Look at the poster your group made. Discuss and write comments on the poster.

c. Produce a new poster and compare it with the first one.

d. Prepare a way to explain to some fifth-grade students how plants get food.

4. Tony's Hypothesis

Your friend Tony (who isn't taking this class) has a hypothesis about the starch in plants. "The starch in plants comes either from the soil they grow in or the water they use," Tony said. "Plants just won't grow where the soil or water isn't starchy." How could you test Tony's hypothesis?

5. The Day the Earth Stood Still

Imagine that you hear the following report on the radio:

Scientists have noticed that plants all over the world are behaving strangely. The leaves of plants fold up when they are in sunlight and they open up in the dark. The stomata of some plants remain shut all day, and then they open at night . . .

Continue this story and describe other behaviors that would be strange for plants. How would other living things be affected by such behaviors?

Review your responses to the ScienceLog questions on page 4. Then revise your original ideas so that they reflect what you've learned.

 You may wish to provide students with the Chapter 1 Review Worksheet that is available to accompany this Challenge Your Thinking (Teaching Resources, page 19).

Sciencelog

The following are sample revised answers:

1. Most plants make their own food. They use energy from light to make sugar from carbon dioxide and water. The sugar is then converted to starch for storage.

2. A typical plant has leaves for food production, a stem for transporting water and nutrients and for supporting the leaves, and roots for taking up water and minerals from the soil and for anchoring the plant. Many plants also produce flowers, fruit, and seeds for reproduction.

3. A leaf contains a plant's food-making machinery. This includes an epidermis to protect the leaf; palisade cells, where most of the leaf's photosynthesis takes place; rounded, loosely packed cells that provide air space for diffusion of gases and where a lesser amount of photosynthesis takes place; veins that transport water and nutrients to or from the leaf; stomata, which are openings through which carbon dioxide, oxygen, and water vapor enter and exit the leaf; and guard cells that control the opening and closing of the stomata.

Connecting to Other Chapters

Chapter 1
investigates the importance of photosynthesis and the plant structures that make it possible.

Chapter 2
focuses on water transport, with an emphasis on transpiration, diffustion, osmosis, and cohesion.

Chapter 3
explores how cells turn food into energy through the process of respiration.

Prior Knowledge and Misconceptions

Your students' responses to the ScienceLog questions on this page will reveal the kind of information—and misinformation—they bring to this chapter. Use what you find out about your students' knowledge to choose which chapter concepts and activities to emphasize in your teaching. After students complete the material in this chapter, they will be asked to revise their answers based on what they have learned. Sample revised answers can be found on page 53.

In addition to having students answer the questions on this page, you may wish to have them write a paragraph titled "Why Water Is Important to Plants." They should use the knowledge that they have already gained in this unit, as well as giving other reasons why water is important to plants. Assure students that there are no right or wrong answers to this question. Collect the papers, but do not grade them. Instead, read the paragraphs to find out what students know about water, what misconceptions students may have, and what aspects of this topic are interesting to them.

CHAPTER

A World of Water

1 How much of you is water?

2 How does water get from the roots to the top of this tree?

3 How does water move within plants?

4 Plants don't use all the water they get. Where does the extra water go?

ScienceLog

Think about these questions for a moment, and answer them in your ScienceLog. When you've finished this chapter, you'll have the opportunity to revise your answers based on what you've learned.

28

The Indispensable Chemical

If you looked at the Earth from space, what would impress you more than anything else—its smallness in the vast emptiness of space? the lack of national boundaries? How about the colors? Blues and whites totally dominate the view. What is the significance of these colors? They signal the presence of liquid water. Look around you. Water is everywhere. One of the remarkable things about our planet is that conditions are just right for liquid water to exist. If our planet were a little colder or a little warmer, or had a weaker gravitational field, this would not be the case. And without an abundant supply of liquid water, life as we know it could not exist.

Water Facts

What is this stuff we call water, and why is it so important? Here are just a few facts about this remarkable chemical.

- About 70 percent of the Earth's surface is covered by water; that's a total of 1.4 billion km^3. Three percent of this water is fresh (and 75 percent of that is ice).
- All living things are made mostly of water. A chicken is about 75 percent water, an earthworm is about 80 percent water, and a tomato is about 95 percent water. You are about 65 percent water.
- Water can dissolve more substances than any other substance. It is the closest thing there is to a universal solvent.

- Water has a higher *heat capacity* (absorbs more heat without increasing in temperature) than almost any other substance.
- Water is the only substance on Earth that is naturally present in solid, liquid, and gaseous states.
- About 265,000 km^3 of rain or snow fall every year (enough to cover Arizona to a depth of 1 km).
- A person will drink about 60,000 L of water in an average lifetime.

29

LESSON 1 ORGANIZER

Time Required
1 class period

Process Skills
communicating, organizing

New Term
Heat capacity—the amount of heat that a substance can absorb without increasing in temperature

Materials (per student group)
none

Teaching Resources
Transparency Worksheet, p. 25
Transparency 5
SourceBook, p. S6

The Indispensable Chemical

FOCUS

Getting Started
Provide groups of students with at least one fresh produce item. (Remind them never to eat anything in a classroom setting unless specifically instructed to do so.) Have students find the mass of each item and place it in a dehydrator or under a bright lamp. Have students record the mass of each item every day for a week and then calculate what percentage of each item's original mass was lost through evaporation.

Main Ideas
1. More of the Earth's surface is covered by water than by land.
2. The properties and abundance of water make life on Earth possible.

TEACHING STRATEGIES

Display a glass of water. Have students discuss the water's properties, and keep a list of student ideas on the board. *(Sample answer: Liquid water is colorless and odorless, takes the shape of its container, flows, and is wet.)* Ask students what would happen to a plant or animal if it did not receive any water. *(It would become dehydrated and die.)* Help students to conclude that water is necessary for life to exist.

Multicultural Extension

Rain Dance
The American Indians of the Southwest perform a special ceremony known as the rain dance at specific times of the year to ask spirits to send rain for their crops. Each tribe practices these ceremonies in its own way. You may wish to have students of American Indian descent interview family members about rain ceremonies performed by their tribes. If possible, have a guest speaker talk with your class about American Indian customs.

Homework

Have students keep track of how much water they use in a 24-hour period and what they use it for. Afterward, have them compare their results with each other.

FOLLOW-UP

Reteaching

Show students a world map or globe in order to emphasize how much of the Earth's surface is covered by water. Compare the size of the oceans with that of the lakes, rivers, and streams in order to demonstrate how much of the Earth's liquid water is salty.

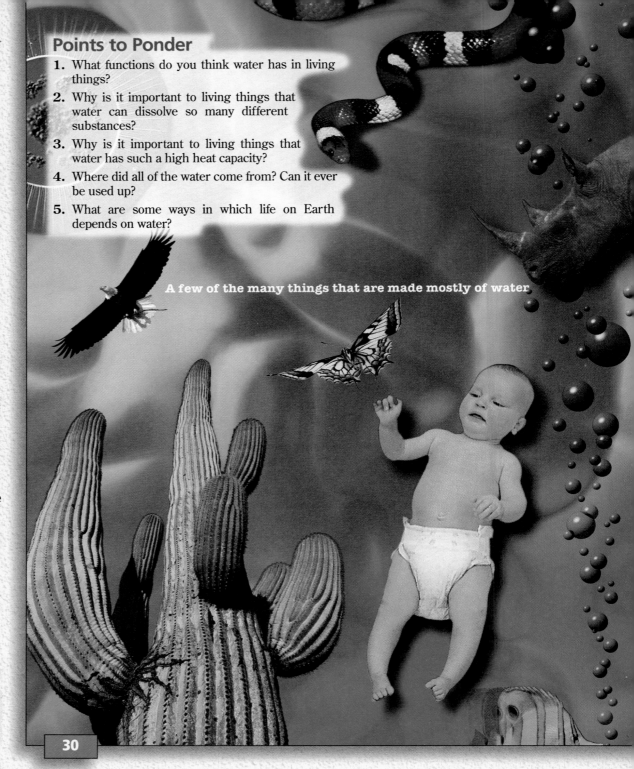

Points to Ponder

1. What functions do you think water has in living things?
2. Why is it important to living things that water can dissolve so many different substances?
3. Why is it important to living things that water has such a high heat capacity?
4. Where did all of the water come from? Can it ever be used up?
5. What are some ways in which life on Earth depends on water?

A few of the many things that are made mostly of water

30

Assessment

Have students design and construct a mobile to illustrate the different roles that water plays in the environment. Suggest that students begin by drawing a concept map and then use it as a model for their mobile.

Extension

Have students research the following question: With so much water on Earth, why are people concerned about a water shortage? *(Answers should*

provide information about sources of water pollution and the difficulty of cleaning up water supplies that have become polluted.)

Closure

Have students make crossword puzzles that incorporate the "water facts" they have learned in this lesson. Be sure that students make an answer sheet to accompany their puzzle. Students can then trade with their classmates and complete one another's puzzles.

The Water Cycle

So far in this unit, you have been looking at various aspects of the food-making process in leaves, which is called photosynthesis. Photosynthesis is not the only process that occurs in leaves, as the following demonstration will show.

Transpiration: A Demonstration

Use a geranium or some other hardy plant. Give the plant some water, and then cover the leaves with a plastic bag and seal the bag carefully around the stem. Observe the setup over a period of 24 hours. Record any changes you see. What do you think causes the change(s)? Ⓐ

31

LESSON
2

The Water Cycle

FOCUS

Getting Started
Have a student exhale on a cool, glassy surface and describe what happens. *(Water vapor condenses on the surface.)* Ask students if they think plants give off water vapor too. *(Accept all responses.)* Point out that in this lesson, they will find out whether their ideas are correct.

Main Ideas
1. Plants give off water through the process of transpiration.
2. Most water vapor evaporates from plants through stomata.
3. As part of the water cycle, transpired water may form clouds that later release precipitation.

TEACHING STRATEGIES

Transpiration: A Demonstration
This activity works well as a teacher demonstration or as a student activity. Discuss the reasons for sealing the bag carefully around the plant stem. *(To ensure that no changes inside the bag are caused by the soil or the air in the room, to ensure that water vapor does not escape, and to provide a surface on which the water vapor can condense)*

Answer to *In-Text Question*

Ⓐ Accept all reasonable responses. Some students may think that the water is condensed water vapor from the air, while others may think that the water originates from the plant.

Homework
With their parents' permission, have students repeat this activity with a plant that has had the underside of its leaves greased. *(A plant with greased leaves should give off little water vapor.)*

LESSON 2 ORGANIZER

Time Required
2 class periods

Theme Connection
Cycles

Process Skills
observing, analyzing, inferring

New Terms
Transpiration—the process by which water vapor is given off through the stomata of leaves
Water cycle—the natural circulation of water on Earth, such as from the ocean, to the atmosphere, to land, and back to the ocean

Materials (per student group)
Transpiration: A Demonstration: geranium or other hardy plant; large, clear plastic bag; string or twist tie; about 150 mL of water
Exploration 1: about 200 mL of water; large, glass juice bottle; a few extra-long matches

Teaching Resources
Exploration Worksheet, p. 27
Transparency 6

Theme Connection

Cycles

Many of the concepts in this unit lend themselves to a discussion of cycles in nature. Examples include the water cycle, the cycle of photosynthesis and respiration, the carbon cycle described in Integrating the Sciences on page 13, or the "life" cycle of a cloud included in the Environmental Focus on this page. Point out one such cycle to students.
Focus question: How do the different stages of this cycle interact? Have students answer this question in the form of a diagram. *(Accept all reasonable responses.)*

ENVIRONMENTAL FOCUS

Students may be interested in further investigating the formation of clouds. Encourage students to find out more about how clouds form, how clouds are classified, and what causes precipitation. Students should present the results of their findings in a diagram or poster. Volunteers may wish to share their findings with the class.

Multicultural Extension

Water Supply and Usage

Rainfall is not distributed uniformly across the globe, and many communities must ration their water during dry periods to keep their supply from becoming depleted. Other communities experience frequent flooding. Have students research the availability of water in a region of the world of their choosing. They should prepare a report analyzing how the culture has adapted to the local water supply.

The process you observed in the previous demonstration, in which water evaporates from the plant's leaves, is called **transpiration.** Transpiration occurs continually. From which part of the leaf do you think the water exits? A

How does water evaporate? Think about how a clothes dryer works. The clothes dryer produces heat. The heat is absorbed by water in the clothes, causing the water to change into a gas—water vapor. This is evaporation.

A similar process causes some of the water in leaves to evaporate as well. You might be surprised to learn that a mature tree gives off as much as 15 L of water daily through its stomata. Imagine how much water an entire forest gives off! Imagine how much heat is absorbed by all this water as it evaporates!

The Brazilian rain forest transpires so much water that rain clouds form almost every day.

Homework

Have students write down three uses for water in each of its three states: liquid, solid, and gas. *(Answers may include sailing, drinking, and bathing for the liquid state; ice skating, sledding, and preserving food for the solid state; and steam cleaning, cooking, and driving a turbine for the gaseous state.)*

CROSS-DISCIPLINARY FOCUS

Language Arts

Have students find out the definition and derivation of the word *xerophyte*, and have them identify some examples of xerophytes. *(A xerophyte is a plant, such as a succulent, that is specially adapted to very dry conditions. Xerophyte comes from words that mean "dry plant.")* You may wish to point out that many botanical words end in *-phyte*. These include *epiphyte, microphyte, sporophyte, chlorophyte,* and *pyrrophyte.*

Where Does the Water Go?

What happens to the water given off by plants? Evidence of transpiration can be seen in the photograph of the rain forest on the facing page. What is the evidence? The clouds you see contain a lot of moisture. How do clouds form? In the transpiration demonstration, you saw water droplets on the inner surface of the plastic covering around the plant. How did the water from the plant get back into the liquid state again? Is there a connection between this phenomenon and cloud formation?

EXPLORATION 1

From a Gas to a Liquid

Look closely at the demonstration that Danny did for his friends to show how fog and clouds develop.

1. Danny poured some water into a large juice bottle, rinsed it around, and poured the water out.

2. He held a match at the mouth of the bottle, lit it, and inserted it inside, holding it there for several seconds before removing it and putting it out.

3. Immediately, Danny put his mouth over the opening and blew into the bottle several times.

4. Then he held the bottle up to show the swirling, whitish (foglike) substance that appeared inside it.

   Try this for yourself. Then try it again, but leave out step 2.

   "This is really mysterious," Hannah said. "I guess the air in the bottle represents the atmosphere, but how did the fog form?"

   Raquel wondered, "There are no plants in the bottle to transpire, but I suppose some water vapor forms from the water poured into the bottle."

   "How does the match help?" Ian asked. "I saw some smoke enter the bottle. Is that important?"

   Perhaps you also have questions about Danny's demonstration. Write down any questions that occur to you. If you have no questions, write your explanation for the formation of the swirling, cloudlike substance in the bottle. **B**

Exploration 1 continued ▶

33

EXPLORATION 1

This Exploration helps students understand how transpired water can later fall as precipitation. The drops of water that collect on the sides of the bottle are analogous to drops of water on the surfaces of leaves after transpiration. To add depth to the class discussion, you may wish to discuss the role of condensation nuclei in the process of cloud formation.

★ **An Exploration Worksheet (Teaching Resources, page 27) and Transparency 6 are available to accompany Exploration 1.**

Answer to
In-Text Question

B Sample explanation: When Danny exhales into the bottle, he adds heat and increases air pressure in the bottle. The temperature in the bottle increases as a result, and water droplets on the sides of the bottle evaporate, forming water vapor. Then, when Danny removes his mouth from the bottle, the trapped gases in the bottle quickly escape, imitating the upward movement of warm air in the atmosphere. This upward movement causes air pressure to decrease. In the bottle, as in the atmosphere, this lower pressure results in a lower temperature. The cooler air cannot hold as much moisture as it could when it was warmer, so the water begins to condense. Condensation is aided in the bottle, as in the atmosphere, by the presence of tiny smoke particles that act as nuclei around which the droplets of water form.

Answers to
Where Does the Water Go?

The water given off by plants enters the atmosphere as water vapor, which forms clouds as it condenses. In the transpiration demonstration, water vapor from the plant condensed on the inside of the plastic covering and formed water droplets, just as water vapor collects in the atmosphere to form clouds.

Homework

Have students find out what common houseplants are related to plants that survive well in a tropical rain forest.

Answer to
In-Text Question

A Student answers will vary. Sample answer: During *evaporation,* the *heat* absorbed by the liquid *water* molecules is converted to kinetic *energy* of the gaseous water molecules. During *condensation,* or *cloud* formation, gaseous water molecules give up kinetic energy to the *atmosphere.* During *transpiration,* water is given off through the *stomata* in the leaves of *plants.*

Answer to
What About the Heat?

1. Precipitation: Water droplets become too large to be supported by air currents and fall as rain, sleet, or snow.
   Runoff: Water flows downhill to a river, lake, or ocean.
   Evaporation: Water goes from a liquid state to a gaseous state and rises into the atmosphere.
   Condensation: Water vapor condenses to form water droplets.

2. Precipitation: Heat energy is released by water droplets.
   Runoff: No appreciable heat exchange occurs.
   Evaporation: Heat energy is absorbed by evaporating water.
   Condensation: Heat energy is released by water vapor.

FOLLOW-UP

Reteaching

Have students illustrate the water cycle by making a small flip chart of images from various sources. Encourage students to include as many different types of environments as they can and to use captions to trace the flow of water within the water cycle. Allow volunteers to present a summary of the water cycle to the class, using their flip chart as a visual aid.

Assessment

Ask students to consider whether desert plants or tropical plants transpire more. *(Desert plants that have small leaves with fewer stomata transpire very little because water is scarce. Tropical plants have large leaves and transpire a lot because water is abundant.)*

Danny produced a copy of a page from a reference book. "I thought you might find that demonstration interesting," he said. "Here's an explanation for cloud formation in the atmosphere. Read it and see if you can pick out the details that correspond to the things I did and the observations you made."

Check to see whether your questions have been answered. If you already wrote an explanation, check it and change it if you find that improvements are needed.

What About the Heat?

What happens to the heat when water changes from the liquid state to the gaseous state, that is, when it evaporates? What happens to the heat when water condenses, as it does during cloud formation?

Summarize your thoughts about these questions. Include the following words: evaporation, condensation, transpiration, water, heat, energy, plants, stomata, atmosphere, and cloud. **A**

The processes of transpiration, evaporation, and condensation are constantly recurring.

When water droplets in the clouds reach a certain mass, they fall, and *precipitation*—rain, sleet, or snow—begins. Precipitation returns water to the Earth, where it is used by plants and animals. Each of the changes you've learned about in this lesson is part of the **water cycle.**

The drawing below illustrates the water cycle.

1. What process takes place at each labeled area?
2. For each process, is heat being absorbed or released?

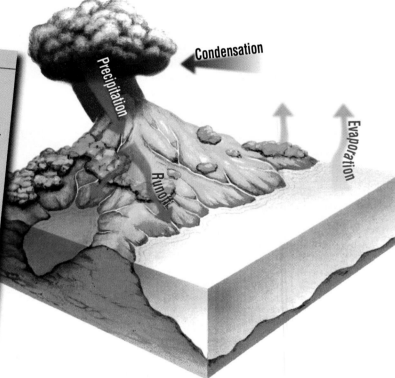

23 How Clouds Form

Heat evaporates millions of tons of water into the air daily. Not only does water evaporate from lakes, rivers, and oceans, but an amazing amount of water transpires from the leaves of plants. As warm moist air rises, there is less pressure on it from above, so it expands. This causes it to fall in temperature. As the air cools, some of the moisture it carries begins to condense. The tiny droplets of water that produce clouds. This process of condensation in the atmosphere is aided by dust or smoke particles, on which the water droplets can form.

Extension

Point out to students that another way that some plants lose water is by a process known as *guttation.* Suggest that they do some research to find out what this process is and how it differs from transpiration. *(In guttation, water leaves the plant in liquid form, not as water vapor. It occurs when water uptake exceeds water loss. Grass leaves may release water through guttation in the spring.)*

Closure

Have students formulate three questions

they have about the topics they have studied in this lesson. Call on students to share their questions and then discuss the answers with the class.

Spreading Water Around

How does a plant obtain the water it uses? As you may already know, this water enters the plant through its roots and travels upward. The water contains dissolved minerals that combine with sugar to make other kinds of materials used by the plant. The plant uses these materials to grow and to maintain itself.

A seedling needs a generous supply of water. Its *root hairs* ensure that this need is met. Just look at the many hairs on a sprouting radish seed. You can sprout radish seeds between two damp pieces of paper towel. Keep the paper towels slightly moist at all times. Once the seeds have sprouted, estimate the number of root hairs on a single seedling. But don't just guess. Use some kind of systematic method. You might want to use a microscope.

Now look at the close-up drawing of a section of a root. Note that each root hair is a projection of one of the cells that make up the outside of the root tip.

Root hairs grow among the tiny particles of soil. Moisture often fills the spaces between soil particles, and this is where plants get most of their water.

Inside the leaf
Stomata
Vein
Underside of leaf
Inside the stem
Inside the root

Closely examine some radish seedlings to see root hairs.

35

LESSON 3 ORGANIZER

Time Required 3 class periods

Process Skills observing, analyzing, predicting, inferring

New Terms
Diffusion—the uniform mixing of the particles of one substance with those of another substance, caused by the motion of both types of particles
Impermeable—describes a substance or surface that prevents other substances from passing through it
Permeable—describes a substance or surface that allows other substances to pass through it
Semipermeable—describes a substance or surface that allows some substances to pass

through while preventing others from doing so

Materials (per student group)
In-Text Activity: 2 paper towels; about 100 mL of water; a few radish seeds; optional materials: microscope; microscope slide and coverslip; eyedropper; a few drops of water
Exploration 2, Activity 1: uncooked egg; about 50 mL of vinegar; petri dish; materials to support egg, such as tape; wide-mouthed bottle filled with water; needle or probe; 10 cm of thin glass tubing, such as capillary tubing; candle and a few matches or silicone sealant; support stand with clamp; watch or clock; safety goggles;

continued ▶

Spreading Water Around

FOCUS

Getting Started
Have students add a small amount of powdered drink mix to a beaker of warm water. Tell them not to stir the mixture. Have them record their observations for a few minutes. Encourage students to discuss how the drink mix moves through the water.

Main Ideas
1. Water carrying dissolved minerals enters plants through root hairs.
2. Root hairs increase the surface area of a root, allowing more diffusion.
3. Particles always diffuse from a region of higher concentration to a region of lower concentration.
4. A semipermeable membrane allows only certain types of particles to pass through.

TEACHING STRATEGIES

You may wish to prepare some seeds a few days ahead of time. Have students examine them first with a magnifying lens and then with a microscope. The root tips can be used to make wet mounts.

Ask students to carefully study the diagrams on this page. Emphasize how the root hairs increase the root's surface area, thereby increasing the amount of water that can be absorbed.

Defining Diffusion, *page 36*
GUIDED PRACTICE   You may wish for volunteers to prepare examples (a) through (c) as demonstrations for the class.

SAFETY ALERT   Hot tap water should be sufficient for demonstrating example (b). To avoid scalding, do not use water that is above 50°C. Use a strong perfume instead of an ammonia solution. Before demonstrating example (c), remind students that they should never directly inhale fumes from laboratory chemicals and that they should treat any unknown substance as toxic. You may wish to take this opportunity to demonstrate the proper procedure for wafting and smelling fumes that are not toxic.

Defining Diffusion, continued

Involve the class in a discussion of what is happening in each case. Have them read the top of page 37 to confirm and revise their ideas.

Encourage students to formulate their own definition of diffusion, such as the uniform mixing of one substance with another. Have them write their ideas in their ScienceLog to review and revise as they continue with the lesson.

Answers to
Defining Diffusion

a. The sugar cube gradually disappears as it dissolves.

b. The food coloring disperses slowly in the cold water, but it disperses quickly in the hot water.

c. The smell of ammonia gas should be detected within a minute.

Answer to
Caption

A Gases and small particles from the trash have become airborne and are spreading uniformly throughout the air.

Integrating the Sciences

Life and Physical Sciences

Have students examine the close-up views of root cells on pages 35 and 38. Ask: How are water and soil nutrients absorbed by a plant's roots? *(Water and nutrients diffuse through the cell walls of the plant's root cells.)* Then ask: Why do some root cells have special cell walls? *(The outer root cells have extensions called root hairs that are able to grow into the small spaces between soil particles where water collects.)* Emphasize to students that the root hairs also increase the surface area of the root cells, thereby increasing the rate at which water can diffuse into the roots. Challenge students to illustrate how root hairs increase a cell's surface area without greatly increasing the cell's volume. *(Students may choose to use different geometric shapes or equal volumes of modeling clay to model the root cells and then calculate the surface-area-to-volume ratios of the different shapes.)* Allow volunteers to present their results to the class.

How do the root hairs provide a plant with its water and mineral requirements? The answer is, through the process of **diffusion.** Under normal conditions, water *diffuses* inward from the soil, through the walls of the root hair, and into the other cells. Now you know *what* happens. However, you still do not know *how* it happens. What causes water and minerals to enter root hairs? To understand this you will need to look more closely at the process of diffusion.

Defining Diffusion

Here are some other situations in which diffusion occurs. Thinking through what happens in each situation will help you form a definition for *diffusion.*

a. A sugar cube is left in a beaker of water for a while. What happens?

b. Several drops of food coloring are placed into a beaker of water at room temperature. What happens? The experiment is then repeated using very hot water. What do you predict will happen?

c. Invisible fumes of ammonia gas rise from a container full of concentrated ammonium hydroxide with the stopper removed. A person stands about 3 m away. How soon will she smell the ammonia gas?

How is diffusion at work here?

ORGANIZER, continued

Activity 2: support stand with clamp; 20 cm of thin glass tubing; animal membrane; about 200 mL of 5% sugar solution; rubber band; about 600 mL of water; large beaker or other large container (additional teacher materials for preparing sugar solution: 10 g of sugar; 190 mL of water; see Advance Preparation on page 1D.)

Exploration 3: 2 lengths of dialysis tubing, 14 to 16 cm each; two 250 mL beakers; 20 mL of 5% cornstarch and water mixture; 20 mL of 5% sugar solution (use corn syrup and water); a few drops of iodine starch-test reagent; 2 paper clips; about 300 mL of water; eyedropper; about 30 cm of thread; materials to test for sugar: one 250 mL beaker, 2 mL of Benedict's solution, hot

plate, 10 mL graduated cylinder, about 150 mL of water, test tube, test-tube clamp, watch or clock, safety goggles, lab aprons, latex gloves (additional teacher materials: 1 g of corn syrup; 1 g of cornstarch; 38 mL of water; 50 mL of 1 M sulfuric acid solution; six 6d iron nails; steel wool; portable burner; tongs; watch glass; 50 mL of 4% sodium hydroxide solution; pH paper; strainer; 25 mL of 0.1 M sodium thiosulfate solution; a few sheets of old newspaper; see Advance Preparation on page 1D.)

Teaching Resources
Exploration Worksheets, pp. 30 and 32
Transparency 7
SourceBook, p. S13

In situation (a), the sugar diffuses throughout the water and seems to disappear, yet you can taste the sugar in every part of the water. In situation (b), the food coloring diffuses throughout the water until the mixture is the same color throughout. The process occurs much more quickly in the hot water. In situation (c), the ammonia gas diffuses through the air, eventually reaching your nose. Soon it can be smelled in every part of the room. Now write a definition of *diffusion*. Can you suggest other examples of diffusion? **B**

You will find additional information to help you define *diffusion* more precisely later in this lesson.

A Theory of Diffusion

As you probably already know, scientists have learned that all matter is composed of incredibly tiny particles. (You will take this up in more detail in Unit 2.) These particles are in constant motion. The hotter the temperature, the faster the particles move. These particles are far too small to be seen, and therefore you cannot directly observe their motion. But you can see indirect evidence of their motion. Diffusion, for example, is evidence of particle motion.

A diagram may help you understand the connection between particles and diffusion. Here is a drawing of situation (a) from the previous page. This is only a model of what happens. You can't really see the sugar and water particles.

● Water particles
◇ Sugar particles
↘ Direction of motion

Refining the Definition of Diffusion

Once you have seen the connection between particles and diffusion in situation (a), you can work in small groups to try to explain how diffusion occurs in each of the other two situations. Make use of the ideas suggested by the theory of particles. After you have finished, flex your creative muscles a little bit with the following exercise: Write a short account of the experiences of a particle undergoing diffusion.

37

A Law of Diffusion

Be sure students understand the law governing the process of diffusion. Review the diagram on page 37. Provide time for students to write or diagram their explanations for the other situations. When they have finished, have them share and discuss their ideas. Drawing diagrams of each situation on the chalkboard may help make the explanations clearer.

A Further Refinement

Review each of the points in the summary with the class. Call on students to cite the evidence they have seen or examined to support each one. Then provide time for students to review and revise their own definitions.

 Transparency 7 is available to accompany The Root of the Matter.

Meeting Individual Needs

Learners Having Difficulty

Have each student blow up a balloon and measure its circumference. Set the balloons aside overnight and have students measure them again the next day. Students should notice that the circumferences have decreased. Explain to them that the air inside the balloon was under pressure, so it was more concentrated than the air outside the balloon. Because the balloon was not airtight, air particles diffused out of the balloon.

A Law of Diffusion

The most important law governing the process of diffusion states:

> **"The particles of a substance always diffuse from a region of high concentration to a region of low concentration."**

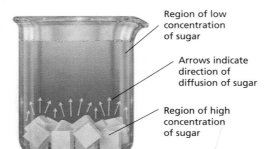

Region of low concentration of sugar

Arrows indicate direction of diffusion of sugar

Region of high concentration of sugar

Do you understand why the diffusion process results in solvent and solute becoming uniformly mixed? This is because diffusion continues until the particles of the substances involved are completely and randomly scattered.

Turn back to the situations discussed earlier. In your ScienceLog, identify for situation (b) the areas of high and low concentration of food coloring and its direction of diffusion. For situation (c), identify the regions of high and low concentration of ammonia gas and its direction of diffusion. Ⓐ

A Further Refinement

Here is a quick summary of what you have learned so far about diffusion.

- One substance intermingles with another.
- The intermingling seems to be uniform.
- The two substances that intermingle do not need to be mixed mechanically; mixing seems to occur naturally.

Using these observations, we can formulate a good working definition of the process of diffusion:

Diffusion is the uniform intermingling of the particles of one substance with those of another substance. This occurs naturally because of the motion of both types of particles.

Compare this definition to your own. Do you need to refine your definition in any way? Ⓑ

The Root of the Matter

Look at the illustration of a plant's root hair. Where is there an area of higher concentration of water particles? an area of lower concentration? In what direction will diffusion take place? If water is to diffuse from the soil into the root-hair cells, the water must pass through both the cell wall and the cell membrane, which enclose the contents of each root-hair cell. Water does, in fact, pass easily through both the cell wall and cell membrane. After this it must go from cell *A* to cell *B*, again through the membrane and wall of each cell. Diffusion through a membrane occurs in living things. In the Exploration that follows, you will study this special kind of diffusion.

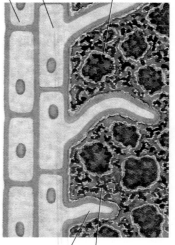

Cell B Cell A

Water around soil particles

Low water concentration

High water concentration

Root hair

Water and air in spaces between soil particles

Answers to In-Text Questions

Ⓐ The colored water is a region of high concentration. The clear water is a region of low concentration. Therefore, the food coloring diffuses throughout the clear water until the entire volume of water is a uniform color.

The ammonia fumes coming from the bottle represent a high concentration of ammonia particles. The surrounding air represents a low concentration of ammonia particles. Therefore, the ammonia fumes diffuse throughout the air.

Ⓑ Answers will vary.

Ⓒ Students should conclude that the higher concentration of water particles is in the spaces between the particles of soil. There is a lower concentration of water particles inside the root hairs. Therefore, water particles will diffuse from the soil into the roots. (In a similar fashion, the water particles in a cell with a high concentration will diffuse into a cell with a low concentration. In this manner, water can be passed from cell to cell throughout the plant.)

Diffusion Through a Membrane

Here are two models of diffusion through root hairs.

ACTIVITY 1

You Will Need

- a support stand and clamp
- an egg
- vinegar
- a wide-mouthed bottle
- a needle
- a thin glass tube (10 cm)
- a candle or silicone sealant
- matches
- a Petri dish
- water

An egg is a single cell—a very large one. This cell, like all cells, is protected from its surroundings by membranes.

What to Do

1. Find a way to support the egg so that only the small end rests in the vinegar. The shell will become very thin in about 24 hours.

2. Fill the bottle with water and place the egg (thinned shell down) on the mouth of the bottle as shown. Use the needle to poke a hole through the top of the egg large enough to accommodate the glass tube.

3. Insert the glass tube. It should just penetrate the inner membrane, which is right under the shell. Seal the glass tube in place with candle wax or silicone.

4. Leave the egg in place for several hours.

5. Record your observations.

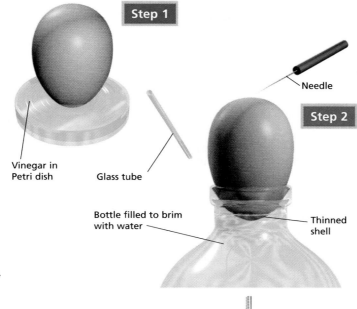

Step 1

Needle

Step 2

Vinegar in Petri dish

Glass tube

Bottle filled to brim with water

Thinned shell

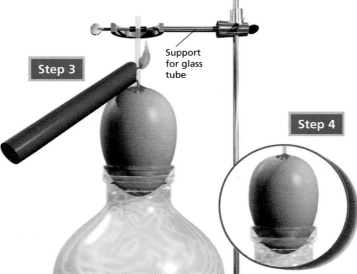

Step 3

Support for glass tube

Step 4

Exploration 2 continued ▶

39

This Exploration illustrates the process of diffusion through a semipermeable membrane. You may wish to set up both Activities in advance so that students can observe the results quickly. For instructions on how to prepare the sugar solution, see Advance Preparation on page 1D.

Have students check for results every 15 minutes or so for both Activities. Eventually, they should note that water is rising in both tubes. In each case, students should recognize that water moves from the container into the membrane's interior because the concentration of water particles in the container is greater than the concentration of water particles inside the membrane.

SAFETY ALERT Make sure that the ends of the glass tubes are fire-polished to remove any sharp edges.

Cooperative Learning
EXPLORATION 2

Group size: 3 to 4 students
Group goal: to observe diffusion through a membrane
Positive interdependence: Assign each student a role such as manager (to read directions), recorder (to collect and report data to other group members), materials organizer (to obtain and return equipment), or technician (to carry out the experiment).
Individual accountability: Have students answer question 4 on page 40 individually and prepare a detailed illustration of what took place at each membrane.

ACTIVITY 1

Caution students not to touch the thinned shell because it is soft and can tear easily. To avoid breaking the prepared egg, you may want to help students pierce the egg with the needle. The glass tube should be inserted so that it just pierces the membrane beneath the shell. It is crucial that there is an airtight seal where the tube is inserted into the shell. Once the glass tube is in place, lightly tapping the top of the tube with a finger may help the liquid begin to rise up the tube.

Some students may wonder why other substances do not pass out of the egg into the bottle. Point out that many membranes allow the passage of only certain substances. In this case, the membrane allows water to pass through, but not the other substances inside the egg. This quality of membranes will be discussed in greater detail later in the lesson.

Students may suggest supporting their egg with modeling clay and toothpicks, or a cardboard tube. A good way to support the egg is to crisscross two strips of tape on top of the egg, and stick the ends of the tape to the underside of the petri dish. Be sure that you approve their plans for safety before they begin.

★ An Exploration Worksheet is available to accompany Exploration 2 (Teaching Resources, page 30).

Exploration 2 continued ▶

ACTIVITY 2

Animal membranes are available from most science supply companies. The membrane should be filled with as much sugar solution as possible. For best results, use tape, candle wax, or silicone sealant to create an airtight seal where the membrane is joined to the glass tube. Be careful when setting up the experiment to avoid spilling any of the sugar solution into the beaker of water.

You may wish to substitute corn-starch for sugar because starch can be detected easily with iodine starch-test reagent. Students can determine empirically whether the solution inside the membrane (cornstarch) diffuses out through the membrane into the beaker.

Answers to *Questions*

1. In both Activities, the level of the liquid rose in the glass tube. (Students may also note a very slight decrease in the water level of the bottle or beaker.)

2. Activity 1: The concentration of water particles is higher in the bottle than in the egg; therefore, water diffuses into the egg. Activity 2: The concentration of water particles is higher in the beaker than in the sugar solution within the membrane sack; therefore, water diffuses into the sack.

3. The concentration of the egg contents was higher inside the egg than in the bottle, and the concentration of sugar was higher inside the membrane sack than in the beaker. The egg contents and the sugar did not diffuse through the membranes. If they had diffused, the level of liquid in the glass tubes may not have risen, or it may have even dropped. Testing the containers of water for the presence of egg or sugar would indicate whether these substances diffused through the membranes. For example, boiling the bottle of water and finding pieces of cooked egg would show that some of the egg's contents had diffused. (Students will learn how to test for sugar in Exploration 3 on page 42.)

ACTIVITY 2

You Will Need
- a support stand with clamp
- a thin glass tube (20 cm)
- an animal membrane
- sugar solution (5%)
- a rubber band
- a large beaker or similar container
- water

What to Do
Arrange the equipment as shown in the illustration, and then leave it for several hours.

Questions
1. What evidence is there in Activities 1 and 2 that water has passed through the membranes?
2. For each of Activities 1 and 2, identify in your ScienceLog the regions of high and low concentrations of water. Identify the direction of diffusion of the water.
3. What can you say about the concentration of the egg contents in Activity 1? the sugar in Activity 2? Did these substances also diffuse through the membranes? How could you find out?
4. Develop an explanation for the results of this Exploration.

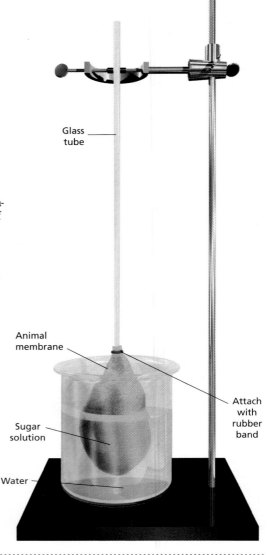

Glass tube

Animal membrane

Attach with rubber band

Sugar solution

Water

4. Accept all reasonable responses. Sample explanation: The membranes of living things allow certain substances to pass through them while acting as a barrier to other substances. Substances that pass through membranes move from areas of higher concentration to areas of lower concentration.

Homework
You may wish to have students answer the questions at the end of Exploration 2 as homework.

Meeting Individual Needs

Gifted Learners
Have interested students repeat Activity 2 with several different concentrations of sugar solution. They will need several membranes of equal volume. *(Students should find that the amount of water entering the tube depends on the concentration of the sugar solution.)*

A Model of Cell Membranes

Exploration 2 demonstrated that not all particles can pass through every type of cell membrane. To better understand the selection process, consider what happens in the following situations, which involve common substances.

- Does water, salt, or sand pass through glass?

- Does water, salt, or sand pass through a fine-mesh screen?

- Does water, a solution of salt and water, or sand pass through filter paper?

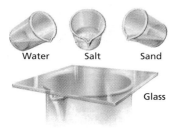
Water   Salt   Sand
Glass

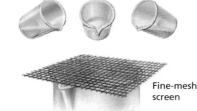
Fine-mesh screen

Filter paper

The word *permeable* means "open to passage," while *impermeable* means "closed to passage." Of the glass, screen, and filter paper, which is permeable to all three substances—water, salt, and sand? Which is impermeable to all three substances? Which is permeable to some substances but impermeable to others? What might you call a material with such a property? **Ⓐ**

The screen is a permeable material, while the glass is an impermeable material. The filter paper is a *semipermeable* material. In general, what do you think determines whether materials are permeable, semipermeable, or impermeable? (Think about what you have learned so far in this unit.) **Ⓑ**

Semipermeable membranes allow some particles to pass through them, depending on the size and shape of the particles. In general, semipermeable membranes prevent larger particles from passing through them.

Suppose, for a moment, that the egg and the animal membrane in Exploration 2 are accurate models for the way diffusion takes place in cells. What kind of membranes do cells have: permeable, semipermeable, or impermeable? What does this model suggest about the way a cell interacts with the surrounding environment? **Ⓒ**

The action of the cell membrane is similar to that of the filter paper, which allowed passage of salt and water particles, but prevented passage of the sand. In other words, the membranes allow some particles to pass through, while acting as a barrier to other, larger particles. Cell membranes are, therefore, semipermeable.

The next Exploration will reveal more about semipermeable membranes.

41

Answers to In-Text Questions

Ⓐ The screen is permeable to all three substances. The glass is impermeable to all three substances. The filter paper is permeable to the water and to the salt solution but impermeable to the sand. Accept all reasonable names for a semipermeable material.

Ⓑ The presence and size of the openings in a material, along with the size of the particles attempting to pass through, determine the permeability of a material to those particles.

Ⓒ Cells have semipermeable membranes. This suggests that cells regulate their internal concentration of certain substances by letting some substances pass through while blocking others.

EXPLORATION 3, page 42

Students should wash their hands before handling the dialysis tubing. Rinsing off the tubing before placing it in the beakers will yield more accurate results. The liquid levels in the dialysis tubing should be even with the levels in the beakers. To prevent the tubing from slipping completely into the water, attach the tubing to the beaker with a clothespin or binder clip.

Only 5 drops of iodine starch-test reagent are needed in beaker A. The starch solution in the dialysis tubing should change color. This indicates that the iodine can pass through the tubing. By contrast, the liquid in the beaker should not change color unless some starch leaked out of the dialysis tubing.

Students will not be able to observe any changes in beaker B until it has been tested with Benedict's solution. The test should turn out positive, indicating that sugar passes through the tubing into the water in the beaker.

Once all of the data from the Exploration has been collected, help students recognize that the tubing is semipermeable because it allows only certain substances to pass through. Thus, the starch particles must be larger than the sugar particles. If necessary, draw a diagram on the chalkboard to demonstrate the concept.

Cooperative Learning
EXPLORATION 3

Group size: 3 to 4 students

Group goal: to determine the effects of semipermeable membranes on the movement of substances

Positive interdependence: Assign each student a role such as task leader, recorder, materials handler, or technician.

Individual accountability: Have each student record the data. Using the setup illustrated on page 42, have students predict the outcome of pouring water, 5% starch solution, and 5% sugar solution through each "membrane." Students should determine what substances each of the beakers would contain and describe how he or she would verify the contents of each one.

Exploration 3 continued ▶

EXPLORATION 3, continued

WASTE DISPOSAL ALERT Pour all solutions containing Benedict's solution into a beaker. While stirring, slowly add 1 M sulfuric acid solution until the pH is between 5 and 6. Scour six 6d iron nails with steel wool until they are shiny. Immerse the nails in the acidified solution overnight or until all the copper has precipitated. Remove the nails, and filter the mixture. Using tongs and a burner, heat the nails and the copper precipitate in a watch glass until all the copper is converted to copper oxide, which is black. Let the copper oxide and the nails cool, and put them in the trash. Treat the remaining filtrate with enough 4% sodium hydroxide solution to raise the pH to 8–10. Filter the mixture again. Let this second precipitate dry, and put it in the trash. Pour the filtrate down the drain.

Refer to Exploration 1, Part 1, on page 7 of this Annotated Teacher's Edition for proper disposal of the iodine-stained mixture.

 An Exploration Worksheet is available to accompany Exploration 3 (Teaching Resources, page 32).

Answers to Questions

1. The dialysis tubing is selective. It allows water, iodine, and sugar to pass through, while preventing starch from doing so.

2. Starch particles are larger than sugar particles.

3. Both are semipermeable.

FOLLOW-UP

Reteaching

Draw a model of a cell on the blackboard, representing it simply as a circle bounded by a broken line. Next, represent starch and sucrose particles as circles too big to diffuse through the openings. Draw water, glucose, and iodine particles as smaller circles that are able to fit through the openings in the cell.

Assessment

Have students draw cartoons to illustrate the role that diffusion plays in making a person cry when he or she peels an onion.

Extension

Suggest that students use diffusion to describe what happens to cigarette smoke in a closed room. How is it possible for a nonsmoker to inhale smoke from someone else's cigarette?

Closure

Ask students to make a list of their favorite foods in one column. In a second column, students should list an example of a diffused substance that adds to the item's taste. *(Examples include strawberries and sugar, steak and marinade, and pickles and vinegar.)*

EXPLORATION 3

Semipermeable Membranes

You Will Need

- two 14 cm to 16 cm lengths of dialysis tubing
- three 250 mL beakers
- 20 mL of cornstarch and water mixture (5%)
- 20 mL of 5% sugar solution (use corn syrup and water)
- iodine solution
- Benedict's solution
- 2 paper clips
- a hot plate
- a graduated cylinder
- water
- a test tube
- an eyedropper
- thread
- test-tube tongs
- latex gloves

What to Do

Arrange the materials as shown in the illustrations below.

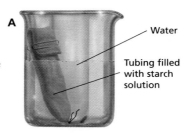

A — Water; Tubing filled with starch solution

B — Water; Tubing filled with sugar solution

Check the liquids in the beakers after about 20 minutes and then again 48 hours later. To determine whether starch particles have passed through the tubing in *A* and whether sugar particles have diffused in *B*, you will have to apply the chemical test for each of these substances to the liquid in the beakers. You already know how to test for starch, shown at left. The test for sugar is shown at right. Report your results in your ScienceLog, and then answer the questions at the end of the Exploration.

Caution: Benedict's solution can irritate the skin. Handle with care.

Iodine

⬆ Testing for starch

A Test for Sugar

Place 2 mL of the solution to be tested for sugar in a test tube. Add an equal amount of Benedict's solution. Put the test tube in boiling water as shown below. Watch carefully for 5 minutes.

When heated, Benedict's solution by itself is blue. Benedict's solution heated with a concentrated sugar solution is brick-red. Benedict's solution heated with a weak sugar solution is greenish yellow.

Questions

1. Is the dialysis tubing selective in what it allows to pass? Which substance(s) does it permit to pass through?

2. What conclusions would you draw about the relative sizes of starch particles and sugar particles?

3. What property do cell membranes and dialysis tubing have in common?

⬆ Testing for sugar

Osmosis: Controlled Diffusion

Diffusion of water (or another liquid) through membranes is called *osmosis*, from the Greek word *osmos*, meaning "push." Osmosis takes place continuously in living things. The cells of all plants and animals are surrounded by membranes that allow smaller particles to pass through but stop larger ones. Even when two different kinds of particles are both small enough to pass through a cell membrane, the smaller ones are likely to get through more quickly. You can see this for yourself in Exploration 4.

EXPLORATION 4

Observing Osmosis

Potatoes have a very high concentration of water, which is why they are used in this Exploration. You are going to study the effect of putting potato slices into different solutions.

You Will Need

- a potato
- 4 beakers
- salt water
- a potato peeler
- a knife

What to Do

Peel a potato and cut four slices that are the same size. Place a slice of potato into each of four beakers that contain either air alone, tap water, a dilute salt solution (20 g/L), or a concentrated salt solution (150 g/L). After about 30 minutes, remove the slices and examine them. Record all observations in your ScienceLog.

Follow-Up

1. Describe the size and rigidity of each of the four slices before and after being placed in a beaker.
2. What seems to be the effect of tap water on a slice?
3. What effect does the concentrated salt solution have on a slice?

Exploration 4 continued ▶

43

LESSON 4 ORGANIZER

Time Required 2 class periods

Process Skills experimenting, observing, inferring, measuring

New Term
Osmosis—the diffusion of a fluid through semipermeable membranes

Materials (per student group)
Exploration 4: uncooked potato; four 600 mL beakers; 85 g of salt; 1.5 L of tap water; knife; potato peeler; watch or clock
Exploration 5: carrot; coring knife; 20 cm glass tube fitted with a one-hole rubber stopper; about 30 mL of molasses; candle and a few matches or silicone sealant; support stand with clamp; 250 mL beaker; about 200 mL of water; graph paper; grease pencil; metric ruler; materials to test for sugar from Exploration 3 on page 42 (additional teacher materials: thick cloth or leather gloves; about 0.5 mL of glycerin; 50 mL of 1 M sulfuric acid solution; six 6d iron nails; steel wool; portable burner; tongs; watch glass; 50 mL of 4% sodium hydroxide solution; pH paper; strainer; see Advance Preparation on page 1D.)

Teaching Resources
Exploration Worksheets, pp. 34 and 37
SourceBook, p. S13

LESSON
4
Osmosis: Controlled Diffusion

FOCUS

Getting Started

Leave two eggs overnight in vinegar. The next day, rub off any shell that has not dissolved. Help students to use water displacement to determine the volume of each egg. Then soak one of the eggs overnight in fresh water and the other in a strong enough salt solution so that the egg floats. The next day, again measure the volume of each egg. Discuss why one egg lost volume while the other gained volume.

Main Ideas

1. Osmosis is the diffusion of a fluid through a semipermeable membrane.
2. The rate of osmosis is equal to the amount of fluid that moves across a semipermeable membrane during a specific amount of time.
3. Osmosis becomes slower and finally ceases as the concentration of a fluid on both sides of a semipermeable membrane equalizes.

TEACHING STRATEGIES

Display a fresh carrot stick and one that is wilted. Ask students to identify how the two are different. Then ask them what they could do to make the wilted stick rigid like the fresh one. *(Place it in water for a few hours.)* Encourage students to use what they know about diffusion to explain why putting the carrot stick in water would make it rigid again.

EXPLORATION 4

In this Exploration, students observe how osmosis affects the rigidity of plant cells. If enough beakers are available, have students work in groups of three or four. After they have had a chance to write their responses to the questions, involve students in a discussion of their observations and conclusions.

Exploration 4 continued ▶

Answers to
Follow-Up, pages 43–44

1. All four potato slices should be about the same size and rigidity before being placed in a beaker. After being placed in the beakers for a while, the size and rigidity of the slices are different depending on what they were immersed in.

2. Tap water makes the potato slice larger and more rigid.

3. The concentrated salt solution causes the potato slice to shrink and become less rigid. If the salt solution is strong enough, the potato slice will become spongy.

4. Beaker of air: The concentration of water is higher in the potato slice than in the air. Water flows out of the potato cells.
   Beaker of tap water: The concentration of water is lower in the potato slice in the beaker. Water flows into the potato cells.
   Beakers of salt solutions: The concentration of water is greater in the potato slices than in the beakers. Water flows out of the potato cells.
   The rule of diffusion explains the results in each case.

5. When a plant is deprived of water, each structure in the plant is affected and becomes less rigid. This causes the leaves and stems to droop, and the whole plant wilts.

6. The slice in tap water became more rigid because it gained water by osmosis. The slices in air and in a dilute salt solution seem to have lost some water because of osmosis. The slice in the concentrated salt solution seems to have lost more water in the same amount of time because of a greater rate of osmosis.

⭐ **An Exploration Worksheet is available to accompany Exploration 4 (Teaching Resources, page 34).**

This experiment takes time and patience to set up correctly but is worth it because the results are so dramatic. Try the Exploration before class to get the conditions just right. When set up correctly, the molasses begins to rise in

---

4. Compare the concentration of water in each potato slice and in each of the beakers. Remembering the rule for diffusion, state the direction of water flow through the cell membranes (i.e., into or out of the cells). Does this explain your results?

5. Plant cells become less rigid (more spongy) when water leaves them. How does this principle explain what happens to plants when they are deprived of water? (They wilt.)

6. Did any slice seem to become more rigid than the others? Which liquid was it in? Explain what happened to this slice. Which slice seems to have lost some water? a lot of water? How do you explain this?

Osmosis Explained

In the beakers containing salt solutions, water particles move out of the cells and into the beaker because initially, the water concentration is greater inside the cells than outside. A kind of pressure forces the water toward the outside until the solution outside has the same concentration as the solution inside. Diffusion always occurs in both directions, from the inside of the cell outward and from the outside of the cell inward. Diffusion occurs fastest, though, where the difference in concentration is greatest. So even though some water diffuses from a salty solution into a cell, it is more than offset by the water diffusing outward from the cell into the salt water. The reverse situation occurs when cells are placed in tap water.

EXPLORATION 5

Osmosis: How Fast?
Purpose: to obtain information about how fast osmosis can occur

You Will Need
- a carrot
- a coring knife
- a metric ruler
- a 20 cm long glass tube fitted with a rubber stopper
- molasses
- candle wax or silicone sealant
- matches
- materials and equipment to test for sugar (see Exploration 3)
- a support stand and clamp
- a beaker
- water
- graph paper
- a grease pencil

What to Do

1. With a coring knife, cut a hole in the carrot about 5 or 6 cm of the way down its length. The hole should be just large enough so that a one-holed stopper will fit tightly.

2. Fill the hole in the carrot with molasses.

3. Push the stopper and glass tube into the hole in the carrot. Seal any openings between the stopper and the carrot with candle wax or silicone sealant.

4. Place the carrot in a beaker filled with water, as shown in the illustration.

5. Mark the initial level of the molasses in the glass tube. This is the level at time zero.

6. Record the level of the molasses in the glass tube several times over the next 3 days, at the same time each day. Before school, during lunch hour, and after school would be good times.

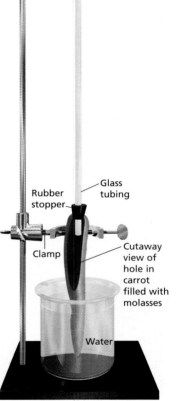

Glass tubing

Rubber stopper

Clamp

Cutaway view of hole in carrot filled with molasses

Water

---

seconds, clearly showing how quickly osmosis can occur. The Exloration may work best as a demonstration.

The carrot should have a large diameter so that the cavity can be carved easily. The stopper must fit the carved opening snugly and should be carefully sealed. A cork stopper may provide a better fit than the rubber stopper. Honey may be substituted if molasses is unavailable. Have student volunteers record the height of the molasses in the tube at the predetermined times.

WASTE DISPOSAL ALERT To dispose of the Benedict's solution, consult the procedure in Exploration 3 on page 42 of this Annotated Teacher's Edition.

 Make sure that the ends of the glass tubes are fire-polished to remove any sharp edges. For instructions on how to insert the tubes into the rubber stoppers, see Advance Preparation on page 1D.

⭐ **An Exploration Worksheet is available to accompany Exploration 5 (Teaching Resources, page 37).**

Graph your findings. Interpret your results using the following questions:

Questions

1. What was the rate of rise of the molasses in millimeters per hour (a) during day 1, (b) during day 2, and (c) during day 3?

2. Use your graph to indicate the level of the molasses at the following points after time zero:
   a. 12 hours
   b. 18 hours
   c. 30 hours
   d. 60 hours

3. Use the sugar test to determine whether any molasses particles went through the carrot into the water.

4. Which do you think is made up of smaller particles, water or molasses? Justify your answer.

5. What happened to the molasses concentration inside the carrot during osmosis?

6. What causes the process of osmosis to slow down? Under what conditions do you think it would stop?

Applied Osmosis

Six examples of applied osmosis are presented here. Can you figure out how osmosis explains each of these common occurrences?

1. When old, spongy potatoes are being prepared for cooking, which would it be better to soak them in, tap water or salt water? Explain.

2. Why is salt sometimes used to kill weeds?

3. To examine blood cells under a microscope, a technologist puts them into an *isotonic* salt solution, which has the same concentration of substances as the blood. Explain why water alone is not used. Predict what would happen if blood cells were dropped into water.

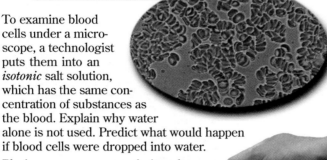

4. Placing sugar on a grapefruit makes the grapefruit become moist very quickly; a sweet syrup forms over its surface. Explain why this happens.

45

Answers to Questions

1. Answers will vary. The fastest rate should occur during the first few hours of the activity. The rate should begin to diminish as the concentration of water inside and outside of the carrot begins to equalize. Because only a few readings will be taken each day, students will need to calculate the average hourly rate. They can do this by dividing the amount that the molasses rises by the number of hours it took to rise. The rate will depend on various factors, including the size of the carrot, the room's temperature, and the humidity.

2. Graphs will differ depending on the setup and the classroom environment. However, graphs should show that the rate at which the molasses rises decreases over time.

3. Molasses particles should not be found in the water unless the water has been contaminated during the Exploration.

4. Water is made up of smaller particles. The cellular membranes of the carrot allow water particles, but not molasses particles, to pass through.

5. The concentration of molasses particles decreased as they mixed with more and more water particles.

6. The process of osmosis slows down as the concentration of water inside the carrot approaches the concentration of water outside the carrot. The process of osmosis would stop if the concentration of substances in the water inside and outside the carrot became equal (reached an equilibrium); however, this is unlikely to happen because there will always be some molasses inside the carrot but not in the water in the beaker.

Answers to Applied Osmosis

1. It would be better to soak them in tap water. Tap water will flow into the potatoes, making them firmer. Salt water would cause water to flow out of the potatoes, causing them to become even spongier.

2. Salt causes water to flow out of the weeds, which then dry up and die.

3. If water alone were used, it would diffuse into the blood cells, where the concentration of water is lower. The blood cells would expand and eventually burst.

4. The sugar dissolves in the liquid on the surface of the grapefruit, making a concentrated sugar solution. The water, which is in higher concentration in the cells of the grapefruit, diffuses out into the sugar-and-water solution, where water is less concentrated.

Answers to Applied Osmosis continued on the next page ▶

Homework

Have students write down two reasons why a semipermeable membrane is necessary to a cell. (*Sample answer: It regulates water pressure and salt content in the cell.*)

5. Too much fertilizer and insufficient water creates a highly concentrated solution around the outside of plant roots. This causes the water in the plants to diffuse into the fertilizer solution. The plants become damaged from dehydration and can eventually die.

6. Place the celery in tap water. Water diffuses into the celery sticks, making them firmer.

Homework

You may wish to assign the questions in Applied Osmosis as homework. Questions 1, 2, 4, and 6 could even be performed as home experiments.

CROSS-DISCIPLINARY FOCUS

Health

A high-salt diet can often lead to high blood pressure in many individuals. This is because salt readily dissolves in the blood. The greater salt concentration causes water to be absorbed into the bloodstream from the neighboring tissues to compensate. The greater volume of fluid in the blood vessels is responsible for the higher pressure.

FOLLOW-UP

Reteaching

Perform the following demonstration: Slice an eggplant in half, sprinkle about 30 mL (2 tablespoons) of salt evenly across the cut surface, and place the eggplant in a colander set inside a large glass bowl. An hour later, ask students to observe the contents of the glass bowl and to write a paragraph explaining what they see. *(Water—with some dissolved material—has drained from the eggplant due to osmosis.)* Cooks sprinkle salt on eggplant before cooking in order to draw out the "eggplant water," which has a bitter flavor.

5. The plants below suffered "fertilizer burn." How does osmosis help explain what happened?

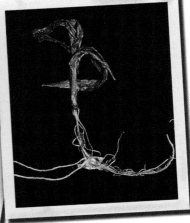

6. Celery becomes limp when it dries out. What might you do to restore its original crispness?

Assessment

Have students write a summary of the lesson. Remind them that their summaries should be a shortened version of what they have read. They should state in their own words the main ideas and the most important facts.

Extension

Explain to students that sea water can be purified into fresh water by a process called *reverse osmosis*. Ask students to research this process and summarize their findings.

Closure

Discuss the practice of putting salt or sugar on a slice of grapefruit. Ask students which substance would cause the most diffusion. *(Since a grapefruit contains much more sugar than salt, the difference in concentration between the grapefruit and the substance on its surface would be greater for a salt solution than a sugar solution.)* Ask students to try putting equal amounts of each on a slice of grapefruit, waiting for 1 hour, and measuring the liquid solution produced in each case. Ask students to explain their observations and to comment on any possible sources of error in their experiment.

5 Putting Water Where It Counts

As you know, water enters the roots of plants by diffusion. But what happens then? How does the water get from the root hairs to the leaves, where it is needed?

The concentration of water around the root hairs is normally higher than the concentration of water inside the root hairs. This difference in concentration results in the diffusion of water into the plant, causing a phenomenon known as *root pressure*. Root pressure forces a steady stream of water into the plant. At the same time, the membranes surrounding the root-hair cells prevent the vital contents of the cells from leaking out. The root hairs pass the water they absorb from cell to cell. Eventually the water reaches special cells that function very much like pipes. These cells carry water to the leaves.

In a living plant there is an unbroken "pipeline" of water from roots to leaves. This becomes pretty impressive when you realize that some trees are over 100 m tall. Can root pressure alone push the water that high, or are there other forces at work? What do you think? Come up with as many ways as you can to explain how plants are able to move water to such great heights.

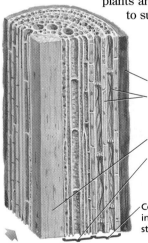

Bark

These long cells, which carry food up and down, are living.

Woody core of the stem

These long cells, which carry water and dissolved minerals up the stem, are hollow, dead cells.

Cells in this region multiply in number, causing the stem to increase in width.

Cutaway view of a woody stem

How does water get all the way from the roots to the tops of these trees?

LESSON 5 ORGANIZER

Time Required
2 class periods

Process Skills
analyzing, classifying, hypothesizing, communicating

New Terms
Capillary action—the climbing of a liquid in a narrow tube due to attractive forces between particles
Cohesion—a strong, attractive force that one particle of a substance has for another particle of the same substance

Root pressure—the pressure, developed in roots and caused by osmosis, that forces water upward in a plant

Materials (per student group)
none

Teaching Resources
SourceBook, p. S6

FOCUS

Getting Started

Ask students to conjecture whether the water-movement mechanism they have studied so far (osmosis) could deliver water to the leaves of plants for photosynthesis. If osmosis would not transport the water, why not? *(Accept all reasonable responses.)* Point out that in this lesson, students will learn about other water-movement mechanisms.

Main Ideas

1. In a living plant, there is an unbroken "pipeline" of water from the roots to the leaves.
2. Several factors contribute to the movement of water in plants.
3. Cohesion is believed to be the most important factor affecting the upward movement of water in plants.

TEACHING STRATEGIES

After students have read page 47, have them examine the diagram of a woody stem and locate the cells forming the "pipelines" that carry food, water, and dissolved minerals up and down the stem of a plant.

On the following two pages, have students review and discuss each of the diagrams. Each example illustrates a mechanism for transporting water up the stems of plants. Allow students to draw their own conclusions about the relative merit of each explanation.

Answer to
In-Text Question

A Accept all resonable responses. Provide students with time to think of ways to explain how plants can move water to such heights. You may wish to have them work in small groups to brainstorm their ideas. Have students state their ideas in the form of hypotheses.

Ⓐ Accept all reasonable responses. The potential of each mechanism for transporting water up the stems of plants is discussed in the corresponding sections on pages 48–49.

Ⓑ Water rises by capillary action because of the attractive forces between the water particles and the inside surfaces of tiny passages. Other examples include the absorbency of sponges or towels and the meniscus that forms at the surface of water in graduated cylinders.

EXAMPLE A

Some students may be skeptical that enough root pressure could be built up to account for the movement of water to great heights. In fact, root pressure can move water considerable distances, but root pressure alone cannot carry water to the tops of tall trees. Students should be able to explain the relationship between osmosis and root pressure. If not, have them review it on page 47.

EXAMPLE B

Students may find the idea of capillary action a plausible explanation for the movement of water up the stems of plants. Most will have observed capillary action in paper towels, bath towels, and other fabrics. They may have also seen it taking place in sand or soil. The water-carrying cells of a plant act as capillary tubes. Point out that for capillary action to be effective, capillary tubes must be very thin. Although small, the diameter of water-carrying cells is not small enough to cause the water in them to rise more than a short distance.

Study the ideas that are presented on these two pages. At one time or another, each has been suggested as the mechanism responsible for delivering water to the tops of the highest trees. Which, if any, seem logical to you? Which, if any, of the following examples do not seem logical? Ⓐ

Example A

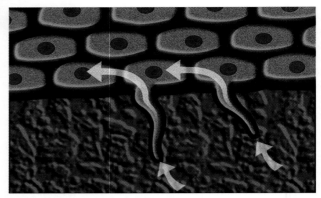

Root pressure caused by osmosis

Example B

Water rises through the process of *capillary action*. Why does it do so? Can you think of other situations in which this occurs? In each case, water rises through tiny passages. Ⓑ

Sugar cube in a little water

Strip of paper towel in water

Glass tube with very small (hair-sized) opening

Did You Know...

An average person who lives for 71 years consumes approximately 60,600 L of fresh water during the course of his or her lifetime.

Example C

Here, atmospheric pressure forces the water up the glass tube of the dropper. The partial vacuum created by water evaporating from the leaves creates a similar situation.

Squeeze the bulb of the eyedropper.

Eyedropper

Place the dropper with the squeezed bulb in water.

Release the bulb, and the water rises.

Example D

Particles of water have a strong attractive force for one another. Notice how droplets of water cling to each other. (This is called *cohesion*.)

Tap turned on slightly

Example E

The strong attraction of water to itself causes water particles to pull each other along. As each water droplet evaporates or drips from the leaf, it "tugs" on the droplet next to it. This tug is transferred back through the unbroken chain of water droplets to the root.

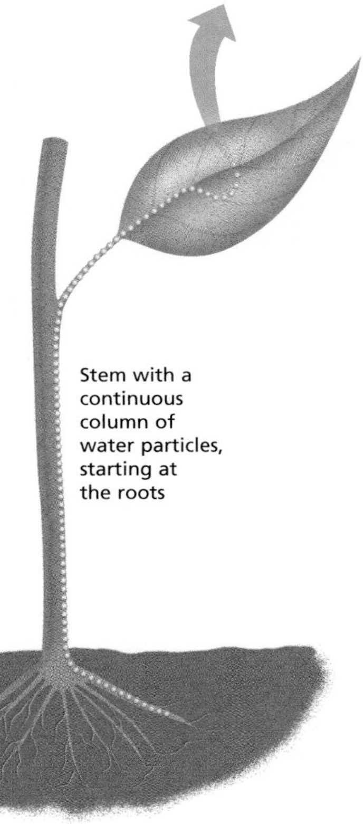

Stem with a continuous column of water particles, starting at the roots

49

EXAMPLE C

Students may recognize this example as similar to what happens during transpiration. As water evaporates from the leaves of a plant, it causes a vacuum in the water-carrying cells. As a result, atmospheric pressure forces more water up these cells. Although students may find this a plausible explanation, transpiration and atmospheric pressure alone do not account for the movement of water up the stems of plants.

EXAMPLE D

The idea of cohesion may be new to students, but they can observe it by watching a rapidly dripping faucet. The drops of water seem to pull each other as they fall from the faucet. Scientists consider cohesion to be a crucial factor in moving water up the stems of plants. Cohesion is discussed on the next two pages.

EXAMPLE E

Students may find this example to be the most logical because it combines transpiration with cohesion. As students will discover by the end of this lesson, this and all of the preceding examples contribute to water transport in plants.

Theories of Water Movement

You may wish to conduct a class survey to determine how students rate each of the explanations for the movement of water up the stems of plants.

Even though the cohesion of water provides the best explanation for the movement of water up the stems of tall plants, it has some shortcomings as well. One of them is the inelastic nature of water. Under laboratory conditions, columns of water under tension are easily broken by a slight movement. In contrast, the water columns within a tree seem to be remarkably elastic and capable of adjusting to very severe conditions, such as the rapid movement of branches in a high wind. Although it seems certain that cohesion is the most important factor in the movement of water in plants, capillary action (adhesion) and root pressure are also involved. Transpiration creates pressure differences and a pulling force that also contribute to the process.

Answer to
In-Text Question

Ⓐ Answers will vary. Accept all reasonable responses. Students should label their diagram with the terms that pertain to the explanation of water movement that they have chosen to illustrate.

PORTFOLIO
Students may wish to include their tree diagrams from Theories of Water Movement in their Portfolio.

Theories of Water Movement

In looking at the examples on pages 48 and 49, you discovered four kinds of forces that could account for the rise of water in the stems of plants. Here are four explanations in more detail:

- **Osmosis** Water particles enter the root cells more quickly than they leave the cell, so pressure in the cells increases, pushing water upward. This is called root pressure.
- **Capillary Action** The narrow passages inside the plant stem cause the water particles to rise by capillary action. Capillary action results from *adhesion* (the attraction between the water and the walls of the passages) and cohesion.
- **Transpiration** Water constantly evaporates through the stomata of the leaves (transpiration). This lowers the pressure in the cells at the top of the plant, allowing atmospheric pressure to push more water upward.
- **Cohesion** The force of cohesion keeps the water column intact. Between the pull of water caused by transpiration and the push caused by root pressure, the water column is able to move upward in a plant.

Which explanation do you think scientists accept today? Draw a diagram of a tree, and label it to show the explanation you find most convincing. Ⓐ

I don't think water can get to the top of the tallest tree by just osmosis. I mean, how far can root pressure push water upward?

I agree. And it doesn't seem as if capillary action or transpiration could do it either—at least not by themselves.

But what about cohesion? Maybe water droplets could pull each other along, all the way up the tree.

Evaporation at the top of the tree—transpiration, I mean—would leave spaces so the water would keep rising . . .

Yeah, as long as there was still a complete column and root pressure. I think cohesion must be the key force involved.

Homework

As a homework assignment, have students revise their tree diagrams from Theories of Water Movement to include all of the mechanisms for transporting water discussed in this lesson.

As it turns out, all four phenomena assist the upward movement of water in plants. However, cohesion is most important in moving water to the tops of the tallest trees. Root pressure can only move water upward a few meters. Capillary action is slow and cannot push water very high at all. Transpiration creates only a slight vacuum, so atmospheric pressure would not be able to push the water very far.

Defying Gravity

Each of the illustrations below shows liquids overcoming gravity. Match each picture with one of the explanations for the upward movement given below.

- attraction between water particles (cohesion)
- atmospheric pressure
- capillary action
- pressure caused by osmosis
- some other force

To find out more about the characteristics of water, see pages S6–S7 in the SourceBook.

A few gravity-defying acts

Starch solution

Water

Wick

Water bulges over top of container without overflowing

51

Answers to
Defying Gravity

The correct explanations for the pictures, from left to right, are atmospheric pressure, pressure caused by osmosis, capillary action, and attraction between water particles (cohesion).

FOLLOW-UP

Reteaching

Point out to students that water tends to flow downward because of the force of gravity. Yet, in plant stems, water flows upward, against gravity. Ask students how they would explain this idea to students who have not studied any science. Ask them to write down what they might say to explain this gravity-defying behavior.

Assessment

Have students select one of the theories of water movement from the lesson and create a poster diagram to illustrate it. Suggest that they use examples other than those presented in the lesson to explain the theories. When the posters are finished, provide an area where they may be displayed.

Extension

Point out to students that trees have annual growth rings from which their age can be determined. If possible, display a cross section of a woody stem and have students identify the annual rings. Then suggest that they research what kind of tissue the annual rings are made of, why they are called annual rings, and what can be learned from them. *(Each annual ring is the seasonal growth of secondary xylem, or water-carrying tissue. The rings provide information about what the growing season was like.)* Have them report their findings to the class.

Closure

Have students make cross-section diagrams of the trunk of a tree. With a little research, their diagrams could include the following labels: bark, phloem, cambium, xylem, and pith. Ask students to add a sentence or two about the function of each of these types of tree tissue. Display the posters for the class.

Homework

The Assessment and Closure activities on this page make ideal homework assignments.

Answers to *Challenge Your Thinking*

1. a. Students should deduce from the drawing that *hypotonic* means the solution is less concentrated than the cell's contents, that *isotonic* means both concentrations are the same, and that *hypertonic* means the solution surrounding the cell is more concentrated. You may need to remind students that a greater concentration of dissolved particles means a lesser concentration of water and vice versa.

b. In the hypotonic solution, water flows into the cell. In the isotonic solution, the flow of water is at an equilibrium. In the hypertonic solution, water flows out of the cell.

c. In the hypotonic solution, the cell might eventually burst if too much water enters. In the isotonic solution, the cell should experience no net change in water content. In the hypertonic solution, the cell might shrink.

2. Because honey is an extremely concentrated solution of sugar and water, it draws the water out of any microorganisms that fall into it; therefore, none survive long enough to reproduce and spoil the honey.

3. a. Particles in a fluid start to diffuse when they are not evenly mixed.

b. Diffusion stops when the particles in a fluid are evenly mixed (or when particles are as evenly distributed as possible across a semipermeable membrane).

c. Particles of matter are constantly moving in all directions. There are more particles available to move from a concentrated area to a less-concentrated area and fewer available to move in the opposite direction. Thus, areas of higher concentration lose more particles than they gain, and areas of lower concentration gain more particles than they lose.

 You may wish to provide students with the Chapter 2 Review Worksheet that is available to accompany this Challenge Your Thinking (Teaching Resources, page 40).

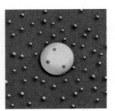

CHALLENGE YOUR THINKING

1. Cell Show and Tell

Casey showed the class the three diagrams below. She said that they show a cell under three different conditions: hypotonic, isotonic, and hypertonic.

Hypotonic solution   Isotonic solution   Hypertonic solution

Then she challenged the other students to answer the following questions:

 Cell

 Dissolved particle

a. What do *hypotonic, isotonic,* and *hypertonic* mean?

b. In which direction are water particles moving in each example?

c. What might eventually happen to each of the cells?

If you were in Casey's class, how would you answer the questions?

2. See You Later, Honey

Honey is basically an extremely concentrated solution of sugar and water. It is also remarkably durable; it has been known to last for centuries without spoiling. How does the principle of osmosis help to explain honey's ability to resist spoilage?

3. Diffusion Confusion

Your 9-year-old brother has just read about diffusion and he has a few questions. Help him answer them.

a. "How does stuff know when to start diffusing?"

b. "How does it know when to stop diffusing?"

c. "When stuff diffuses, why does it always go from a concentrated area to a less-concentrated area? Why doesn't it ever go in the other direction?"

4. How to Kill a Plant Without Really Trying

How does the principle of osmosis help explain why plants wilt when they don't get enough water?

5. Water, Water, Everywhere, But . . .

Why will drinking salt water make you thirstier than not drinking any water at all?

6. How Is a Leaf Like a Wet Towel?

A leaf has been compared to a wet towel on a clothesline.

a. What conditions help a towel to dry?

b. Suggest what you could do to find out whether these conditions would also increase the transpiration rate of a plant.

c. What beneficial effect might transpiration have on plants on a hot day?

7. Keep It Simple

Design a poster to show how a plant gets water and how the water travels through the plant. The design should be large, attractive, and simple enough for a fifth-grade student to understand. Show it to someone. Is the poster clear enough? What changes would help?

ScienceLog

Review your responses to the ScienceLog questions on page 28. Then revise your original ideas so that they reflect what you've learned.

53

ScienceLog

The following are sample revised answers:

1. About 65 percent of a human being is water.

2. Water gets from the roots to the top of a tree because of the cohesion of the water particles, root pressure, adhesion of the water particles to the walls of the cells, and the process of transpiration.

3. Water moves within plants by a process of diffusion called osmosis. Water moves from a cell where it is in greater concentration to a cell where it is in lower concentration. Water flows through cell membranes as long as there is a difference in water concentration between cells.

4. Most of the extra water that a plant takes up from the soil evaporates through the stomata of the plant's leaves (transpiration) and into the atmosphere.

Answers to
Challenge Your Thinking,
continued

4. Plants absorb water through their root cells by osmosis. As plants absorb water, their cells swell against the cell walls, giving the plants structural strength. Without enough water in the soil, the plants may lose more water through transpiration than they gain by osmosis, or the plants may actually lose water to the soil by osmosis. Without enough water, the pressure inside the cells decreases, and the plant sags under its own weight.

5. Body cells that come into contact with the salt water will lose some of their own water through osmosis because the water concentration is greater inside the body cells than in the salt water. This makes the shortage of water in the body's cells even worse.

6. **a.** Hot, dry, windy conditions
   b. Students responses will vary. Possible suggestions include the following:
   (1) Cover a plant with a plastic bag. Compare the water lost through transpiration at different temperatures.
   (2) Put a plant in a room with a dehumidifier. Cover the soil so that no water can evaporate from it. Observe whether the plant requires more water than usual.
   (3) Use a fan to simulate a breeze, and observe how much water a plant needs to keep from wilting. Compare this with a plant not exposed to a breeze. Be sure to cover the soil so that water is lost only through the leaves.
   c. In addition to helping plants transport water up to the leaves, transpiration helps cool the leaves by releasing water that has absorbed heat to become water vapor.

7. In general, the poster should show water entering the roots from the soil and moving upward through the plant to the leaves. Extra water should be shown leaving the plant by transpiration.

Connecting to Other Chapters

> **Chapter 1**
> investigates the importance of photosynthesis and the plant structures that make it possible.

> **Chapter 2**
> focuses on water transport, with an emphasis on transpiration, diffusion, osmosis, and cohesion.

> **Chapter 3**
> explores how cells turn food into energy through the process of respiration.

Prior Knowledge and Misconceptions

Your students' responses to the ScienceLog questions on this page will reveal the kind of information—and misinformation—they bring to this chapter. Use what you find out about your students' knowledge to choose which chapter concepts and activities to emphasize in your teaching. After students complete the material in this chapter, they will be asked to revise their answers based on what they have learned. Sample revised answers can be found on page 69.

In addition to having your students answer the questions on this page, you may wish to have them complete the following activity: Have students create a diagram titled "How the Body Obtains Energy From Food." Tell students that they should include the terms *digestion* and *respiration* in their diagram. Assure the students that there are no right or wrong answers to this question. Collect the papers, but do not grade them. Instead, use the students' diagrams to identify possible problem areas in the chapter. Read the papers to find out what students know about digestion and respiration, what misconceptions students may have, and what aspects of these topics are interesting to them.

CHAPTER 5

Maintaining Life

Animals need plants. Do plants need animals in any way?
1

2 How is the energy in your food released in you?

What keeps heat near the Earth's surface?
3

ScienceLog

Think about these questions for a moment, and answer them in your ScienceLog. When you've finished this chapter, you'll have the opportunity to revise your answers based on what you've learned.

54

Turning Food Into Fuel

You know that the energy needed by living things to carry out life functions comes from food. But exactly how do living things—people, for example—get energy from food?

Food in its raw form is not usable by our bodies. Before our bodies can put food to work it must be converted into a form suitable for absorption by individual cells. It's not just a matter of breaking down the food into tiny pieces, though. It must also be broken down chemically into simpler water-soluble compounds that can pass by diffusion through cell membranes.

The process of breaking down food into substances usable by the body is called **digestion**. Even a simple chemical compound such as starch is too complex to pass through the membranes of individual cells. The starch must be converted into a simpler substance. What do you think that simpler substance is? **A**

The Mouth—Where It All Starts

Put an unsalted soda cracker into your mouth and chew it slowly, letting it soften thoroughly. Hold the food in your mouth and describe any change in taste. Does it taste sweeter or less sweet than when you first put it in your mouth? **B**

Monica did the following demonstration to show how the mouth breaks down starch into a simpler substance.

In one test tube, I put starch, water, and some saliva from my mouth. In the second test tube, I put starch and water. And in the third test tube, I put water and saliva.

After 15 minutes, I checked each test tube for sugar. Here are my results.

Test Tube Number	Substances in Test Tube	Result of Sugar Test
1	Starch, water, and Saliva	Positive
2	Starch and water	Negative
3	Water and Saliva	Negative

Why is it necessary for the starch to be broken down into sugar? What substance breaks down the starch into sugar? **C**

LESSON 1 ORGANIZER

Time Required
1 class period

Process Skills
observing, communicating, hypothesizing, comparing

New Term
Digestion—the process of breaking down food into substances that can be used by the body

Materials
The Mouth—Where It All Starts:
1 unsalted soda cracker per student

Teaching Resources
SourceBook, pp. S3 and S16

Turning Food Into Fuel

FOCUS

Getting Started
Help students recall any knowledge that they may have about the digestive system by asking them what happens to the food they eat. Make a list or flowchart on the chalkboard of the digestive organs they identify. Be sure they understand that digestion involves many organs of the body and that food must somehow enter the cells of the body during digestion. Tell students that in this lesson they will examine the processes that make this possible.

Main Ideas
1. Food particles are converted into water-soluble compounds so that they can move across cell membranes.
2. Digestion usually begins when saliva in the mouth mixes with food to break down starch.
3. Digestion takes place in the mouth, stomach, and small intestine.
4. Ultimately, most food particles are carried through the bloodstream to all parts of the body.

TEACHING STRATEGIES

The Mouth—Where It All Starts
Tell students not to swallow their masticated cracker as they mix their saliva with the starch in the cracker.

Answers to
In-Text Questions

A Accept all reasonable responses. Students may realize that the simpler substance is sugar.

B Students should observe that the cracker becomes sweeter as they chew it.

C The starch must be broken down before it can be absorbed by the body's cells. Saliva is the substance that breaks down starch into sugar.

The Rest of the Story

After students have read the summary of digestion, call on a volunteer to state the law of diffusion from Chapter 2 on page 38.

the law of diffusion from Chapter 2 on page 38.

Answer to
In-Text Question

Ⓐ The food particles move by diffusion from areas of higher concentration to areas of lower concentration. This explains the movement of food particles from the small intestine into the bloodstream and from the bloodstream into the body's cells.

FOLLOW-UP

Reteaching

Ask students to find out what saliva contains and how its contents relate to its functions. Then have students share their results to create as complete a list as possible. *(Saliva is about 99 percent water, which softens food for chewing and dissolves food nutrients. The water also keeps the inside of the mouth moist, which is crucial for speech. Different enzymes in saliva begin the process of digestion by breaking down starches into sugars. Saliva also contains mucus, which is made of proteins called mucins, to lubricate food particles for swallowing.)*

Assessment

Have students write a commentary of Monica's demonstration on page 55. Their commentaries should explain the purpose of the demonstration, the reason for using three different test-tube setups, and the results of the demonstration.

Extension

Point out that certain substances called *enzymes* act as *catalysts* during the process of digestion. Explain that a catalyst is a substance that speeds up a reaction without being changed by the process. Suggest that students do some research to discover what these enzymes are called and why they are considered to be catalysts.

The Rest of the Story

Digestion breaks down starch into a simple sugar called *glucose,* which cells can absorb. Once food is completely digested, it is able to diffuse through cell membranes. In the small intestine, most of the digested food diffuses into the blood vessels. The bloodstream then carries the food to all parts of the body. The small particles of digested food are able to pass from the blood vessels into the body cells by means of diffusion through thin-walled vessels called capillaries. (How does diffusion determine the direction that the

Ⓐ food particles move?) Once the digested food is in the body cells, it is ready for the next process.

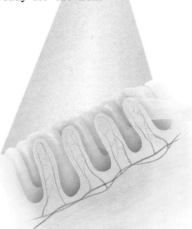

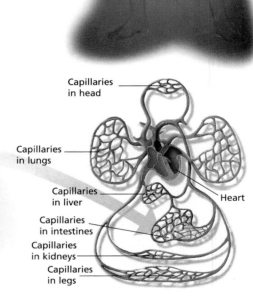

Small intestine

Capillaries in head

Capillaries in lungs

Capillaries in liver

Heart

Capillaries in intestines

Capillaries in kidneys

Capillaries in legs

56

Closure

Have students create a serial cartoon depicting the digestive process from the point of view of a hamburger, an apple, or another food item.

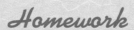

Homework

You may wish to assign the Closure activity on this page as homework. Encourage students to include the instances in which diffusion takes place.

PORTFOLIO

If you have students complete the Closure activity, they may wish to include their cartoon of the digestive system in their Portfolio.

LESSON 2 Unlocking the Energy in Food

A Matter of Energy

Perhaps in earlier studies you burned a peanut (or some other food) to find out how much energy it contained. Where did the energy in the peanut come from? Where does it go when you eat the peanut? How is the energy released? The series of photographs below might give you some ideas. Write a brief caption of about two or three sentences for the series of photographs. **B**

Energy from the sun

Energy stored as food

Energy released and used to activate muscles

How is the energy in gasoline released in an engine? How is the energy in candle wax released as heat and light energy? The energy within digested food in your body is released in a somewhat similar way. Look at the following diagrams. In what ways is the release of energy similar in each case? How does it differ? **C**

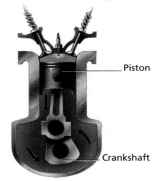

Piston

Crankshaft

A mixture of air and gasoline is ignited. The exploding mixture pushes the piston down, turning the crankshaft.

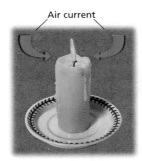

Air current

Oxygen combines with candle wax, releasing heat energy.

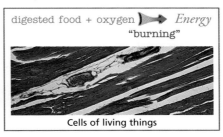

digested food + oxygen ➤ *Energy*
"burning"

Cells of living things

You have probably noticed that living cells release energy differently from the candle or engine. The release of energy that occurs in the cells of living things—both plants and animals —is called **respiration**.

57

LESSON 2 ORGANIZER

Time Required 4 class periods

Theme Connection Energy

Process Skills
observing, analyzing, comparing

New Terms
Circulatory system—the heart and vessels that move blood through the body
Digestive system—the organs associated with the intake, digestion, and absorption of food
Excretory system—the organs associated with collecting and eliminating metabolic waste

Nervous system—the network of structures that control the actions and reactions of the body
Respiration—the release of energy that occurs when digested food combines with oxygen in the cells of living things
Respiratory system—the system of organs that accomplish the exchange of oxygen and carbon dioxide between the body and its surroundings

Materials (per student group)
Exploration 1: stopwatch, clock, or watch with a second hand
Exploration 2: Test 1: plastic drinking
continued ▶

LESSON 2 Unlocking the Energy in Food

FOCUS

Getting Started
Have students exhale on the palm of their hand. Ask them what it feels like. *(It should feel warm and moist.)* Help students conclude that water vapor is exhaled when they breathe. Point out that in this lesson they will learn where the water vapor comes from.

Main Ideas
1. Respiration takes place in the cells of animals and plants.
2. In humans, oxygen is inhaled into the lungs, where a portion of it is absorbed by the blood.
3. Blood carries oxygen through the circulatory system to the cells of the body.
4. Carbon dioxide and water are the waste products of respiration.

TEACHING STRATEGIES

Answers to *In-Text Questions*

B Captions will vary. Sample answer: Plants use and store the sun's energy as they make food. When people eat this food, directly or in the form of meat, the energy is released and used by the body.

C In all cases, energy is released through burning and oxygen is used. In the internal-combustion engine, the burning is explosive. The candle burns at a much slower rate. The "burning" that occurs in cells is the slowest of all.

★ A Discrepant Event Worksheet is available to introduce Lesson 2 (Teaching Resources, page 46).

Ⓐ People are usually unaware of their breathing. How many times someone breathes in a minute depends on factors such as physical exertion and emotional state.

Ⓑ Students are probably more aware of breathing in the winter because the air feels cold as it enters their lungs and because water vapor becomes visible when moist exhaled air cools in the outside air.

Ⓒ Humans get oxygen by inhaling air into their lungs. The oxygen enters the bloodstream through clusters of air sacs.

Fish get oxygen by moving water over their gills. The oxygen dissolved in the water enters their bloodstream through gill filaments.

Grasshoppers take in air through spiracles in their abdomen. The oxygen in the air is distributed within their body by tracheae.

Earthworms absorb oxygen from the air directly through their skin. The oxygen then diffuses into blood vessels.

Integrating the Sciences

Life and Physical Sciences

Inhaled air contains about 21 percent oxygen, and exhaled air contains about 15 percent oxygen. So we use only about a third of the oxygen we breathe for respiration. The remainder is exhaled. At first glance, this seems inefficient. Ask: Why do you think this inefficiency is actually a useful survival mechanism? *(The fact that we do not immediately use all the oxygen in a breath allows us to vary our oxygen consumption when necessary, such as during physical exertion, when we need more oxygen, and while holding our breath, when we receive no additional oxygen at all.)*

Homework

Have students consult a nutritional guide or food labels to determine the energy content of some of their favorite foods. Ask: Which foods provide you with the most energy? *(Answers will vary but should include foods high in fat, such as chocolate, ice cream, or peanut butter.)*

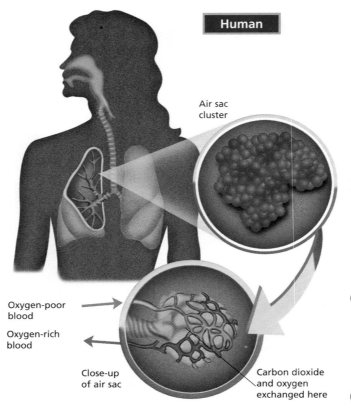

Human

Air sac cluster

Oxygen-poor blood

Oxygen-rich blood

Close-up of air sac

Carbon dioxide and oxygen exchanged here

Fish

External flap covering gill

View of gill filament (under flap)

Water carrying dissolved oxygen

Oxygen-rich blood

Oxygen-poor blood

Expanded view inside gill filament

Grasshopper

Tracheae

Spiracle

How do different animals get the oxygen their cells need? Study the examples on these pages to find out. Ⓒ

Outside

Oxygen

Carbon dioxide

Skin

Blood vessel

Earthworm

The Function of Breathing

Air, which contains oxygen, is necessary for combustion (burning), such as the explosive combustion of gasoline that takes place in a car engine. Likewise, oxygen is essential for respiration in the cells of living things. But how does oxygen reach the cells? In animals, the process begins with breathing.

While you were reading the last few sentences, were you conscious of the fact that you were breathing? How many times a minute do you breathe? Do you have to think about it Ⓐ for it to happen? Unless something goes wrong with your "breathing system," such as a stuffed nose, you are normally not even aware of the flow of air into and out of your lungs. In winter, why are you more likely Ⓑ to be aware of your breathing?

How does a tree get its oxygen? It may surprise you, but plants also "breathe." Their cells must carry on respiration, just as do the cells of animals.

58

ORGANIZER, continued

straw; about 20 mL of limewater; test tube; safety goggles (additional teacher materials for preparing limewater: 4 mL of calcium hydroxide and 200 mL of water per class; see Advance Preparation on page 1D.); **Test 2:** a mirror or shiny piece of metal; piece of cobalt chloride paper
Exploration 3: 20 seeds, such as radishes or beans; 4 test tubes; 4 rubber stoppers; wooden splint; 5 mL of limewater; about 10 mL of water; testtube rack; a few matches; 10 mL graduated cylinder; safety goggles (additional teacher materials for preparing limewater: 4 mL of calcium hydroxide

and 200 mL of water per class; see Advance Preparation on page 1D.)
Hot Plants? materials for monitoring heat production during plant respiration (See Advance Preparation on page 1D.)

Teaching Resources
Discrepant Event Worksheet, p. 46
Exploration Worksheets, pp. 47, 49, and 52
Transparency 8
SourceBook, p. S16

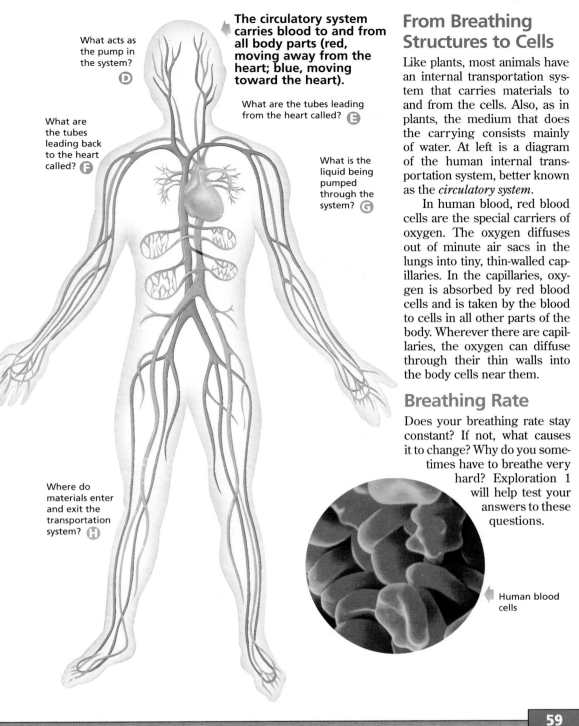

What acts as the pump in the system? **D**

The circulatory system carries blood to and from all body parts (red, moving away from the heart; blue, moving toward the heart).

What are the tubes leading from the heart called? **E**

What are the tubes leading back to the heart called? **F**

What is the liquid being pumped through the system? **G**

Where do materials enter and exit the transportation system? **H**

Human blood cells

From Breathing Structures to Cells

Like plants, most animals have an internal transportation system that carries materials to and from the cells. Also, as in plants, the medium that does the carrying consists mainly of water. At left is a diagram of the human internal transportation system, better known as the *circulatory system*.

In human blood, red blood cells are the special carriers of oxygen. The oxygen diffuses out of minute air sacs in the lungs into tiny, thin-walled capillaries. In the capillaries, oxygen is absorbed by red blood cells and is taken by the blood to cells in all other parts of the body. Wherever there are capillaries, the oxygen can diffuse through their thin walls into the body cells near them.

Breathing Rate

Does your breathing rate stay constant? If not, what causes it to change? Why do you sometimes have to breathe very hard? Exploration 1 will help test your answers to these questions.

59

From Breathing Structures to Cells

Review the diagram of the circulatory system by having students answer and discuss the in-text questions. If students are having difficulty understanding the process by which oxygen gets into cells, suggest that they make a flowchart to show what happens. Their flowchart should convey the following idea:

Oxygen passes
into the cells of the lungs →
into the cells of capillary walls →
into the red blood cells →
to different parts of the body →
out of the red blood cells →
into the cells of capillary walls →
into body cells.

Call on a volunteer to use his or her diagram to describe the process.

Meeting Individual Needs

Second-Language Learners
Have each student create a poster describing the circulatory system and the movement of blood throughout the body. The poster should be labeled in English and in the student's native language. Students may use the illustration on page 59 as a model, but additional research should be encouraged. Display the posters for the class to enjoy.

Multicultural Extension

Charles Richard Drew
Share the following information with students: The African American pathologist Charles Richard Drew was one of the great pioneers of blood preservation. Challenge students to find out more about Drew and his contributions. *(At the request of the British government, Drew set up the first blood bank in England, and during World War II, Drew was appointed director of the American Red Cross blood-donor project. Thousands of soldiers benefited from the blood he helped to collect and preserve.)*

EXPLORATION 1

Answers to
Interpreting Your Data

1. Exercise causes both breathing and pulse rate to increase.

2. The breathing rate is slower than the pulse rate, but when the pulse rate increases, so does the breathing rate. This enables more oxygen to enter the lungs so that enough oxygen is available for diffusion into the increased blood flow. The extra oxygen is needed so that the rate of respiration can keep up with the energy needs of the body.

3. Since breathing and pulse rates may vary slightly from one minute to the next, averaging several readings is more accurate.

4. Breathing supplies oxygen to the circulatory system, which then carries the oxygen to the cells so that respiration can take place. Breathing also rids the body of the wastes produced by respiration. The circulatory system carries the carbon dioxide and water given off by cellular respiration to the lungs, where these waste products are exhaled.

5. Most students will probably believe that the rate of respiration is controlled by the heart. You may wish to point out that respiration is actually regulated by the nervous system.

6. Both respiration and burning give off heat, as indicated by the warming of the body during exercise. Like burning, respiration produces carbon dioxide and water as waste materials.

Homework

You may wish to assign the Exploration Worksheet that accompanies Exploration 1 as homework (Teaching Resoures, page 47).

Testing Your Breathing and Pulse Rates

You will need a partner to help you count and measure time. (Use a stopwatch or a watch with a second hand to measure the time accurately.)

First find out your breathing rate while sitting quietly. Have a partner count the number of breaths you take in 1 minute. Try not to think about your breathing. Repeat the procedure five times and average the results. Then take your pulse using the following methods. Count the number of beats in 30 seconds and multiply by 2 to get your resting pulse rate. Repeat the procedure five times and average the results.

Find out your breathing rate while walking around for 2 minutes. Again, average the results of six readings. Calculate your pulse immediately following this activity using the method outlined above.

Find out how many times you breathe in 1 minute while running. (You may also run in place.) Again, calculate your pulse rate immediately after the activity.

Record all results in your ScienceLog in a data table of your own design.

Interpreting Your Data

1. What effect does exercise have on your breathing rate? on your pulse rate?

2. Is there any relationship between your breathing and pulse rates? If so, what is it? Why do you think this relationship exists?

3. Why did you repeat your measurements to find average breathing and pulse rates?

4. How are breathing and blood circulation related to respiration in the cells?

5. What part of the body do you think might regulate the rate at which respiration occurs?

6. Did you find that you got hot as you exercised? What does this suggest about the relationship between respiration and burning?

CROSS-DISCIPLINARY FOCUS

Health

Students may be interested in knowing that a gram of protein and a gram of carbohydrate both contain about the same amount of energy, 4 dietary calories. In comparison, a gram of fat contains 9 dietary calories, more than twice as much!

Did You Know. . .

In order to keep the body supplied with oxygen, the human heart pumps nearly 7600 L of blood every day. It exerts enough total energy in an hour to lift a small car with a mass of 1400 kg 30 cm off the ground.

The Ins and Outs of Respiration

The process of respiration uses oxygen and releases energy, as does the combustion process in a car engine or the burning of a candle. Each of these processes produces wastes. In Exploration 2 you will learn more about the waste products of respiration.

Be Careful: Do not draw the limewater into your mouth. Blow gently so that the limewater does not splash out of the test tube.

EXPLORATION 2

Products of Respiration

 TEST 1

You Will Need

- a drinking straw
- limewater
- a test tube

What to Do

Blow *gently* through a straw placed into clear limewater in a test tube. What happens? Is there carbon dioxide in the air you breathe out? Might carbon dioxide be a product of the respiration process that takes place in the cells of living things? **Ⓐ**

TEST 2

You Will Need

- a mirror or a shiny piece of metal
- cobalt chloride paper

What to Do

Exhale so that your breath strikes the surface of the mirror or piece of metal. Examine the shiny surface closely for any change in appearance. Repeat this process several times. Wipe the surface with a piece of cobalt chloride paper. This paper is blue when dry but changes to pink when wet. This color change signals the presence of water.

 Exploration 2 continued ▶

61

Answer to *In-Text Question*

Ⓐ Students should observe that the limewater turns cloudy when they blow into it. They should conclude that there is carbon dioxide in the air they exhaled. Help students to recognize, however, that this indicates only that carbon dioxide *might* be a product of respiration. The other possibility is that the limewater is testing positive to carbon dioxide that was already in the air when it was inhaled.

Meeting Individual Needs

Gifted Learners

Students may enjoy preparing a presentation on one of the body systems. The presentation should include a discussion of the purpose and operation of the system and its primary structures. In addition, students should provide a visual aid such as a poster, diagram, or model. You may wish to suggest that they ask the school librarian to help them in their research.

EXPLORATION 2

TEST 1

Make sure the class has an adequate supply of limewater before they begin.

SAFETY ALERT Instruct students not to draw the limewater into their mouth. They should discard the straws as soon as they are finished with them. The test tubes should be cleaned as soon as possible to prevent chemicals from coating the glass. If chemicals do coat the glass, a little dilute acid, such as vinegar, will remove the coating.

TEST 2

You will need to obtain some cobalt chloride paper before you begin this test. Make sure that you store the paper in a capped bottle. Moisture may cause the paper to turn pink, but heating it will turn it blue again.

SAFETY ALERT Remind students never to put laboratory materials in their mouths unless specifically instructed to do so. Also remind students to wash their hands thoroughly with soap and water after each laboratory activity.

⭐ An Exploration Worksheet is available to accompany Exploration 2 (Teaching Resources, page 49).

ENVIRONMENTAL FOCUS

Air pollution is a contributor to breathing problems such as asthma and emphysema. Excessive amounts of carbon monoxide in the air can interfere with the transfer of oxygen from the lungs to body tissues. Have students generate a list of questions about air pollutants and health problems related to them. Then have each student choose one question to try to answer for the class. You may wish to have the class discuss some resources that would be useful.

Exploration 2 continued ▶

1. The mirror appeared foggy, as though moisture had collected on it.

2. The cobalt chloride paper was blue before it was used to wipe the mirror. After the mirror was wiped, the paper turned pink.

3. The substance on the mirror was water.

4. The water was present in the air exhaled from the lungs.

5. The conclusion can be drawn that water is another product of respiration.

Answers to
In-Text Questions

A Students may correctly speculate that the water resulting from the process of respiration is also carried away from cells by the circulatory system. Some of the water is exhaled, and some of it is used by the body in other chemical reactions.

B Student answers will vary. Sample answer: To test for water, expose a piece of cobalt chloride paper to the air for a few minutes. If the paper turns pink, then there is a significant amount of water in the air. To test for carbon dioxide, put some lime-water in a test tube and shake it gently from side to side. This gentle agitation would be enough to turn the limewater milky if there were a lot of carbon dioxide already present in the air in the room.

C Student definitions of respiration will vary but should be similar to the following: Respiration is a process that takes place in the cells of living things. In this process, food (sugar) is broken down in the presence of oxygen to produce carbon dioxide, water, and energy. This definition can be stated as the following equation: digested foods + oxygen → energy + carbon dioxide + water.

   After analyzing the word equation at the bottom of the page, students should conclude that the process of respiration is similar to the burning of any fuel to release energy. Students should recognize that a small amount of energy is needed to begin the burning of a fuel. The same is true for respiration.

Questions

1. What was the appearance of the mirror after you exhaled on it?
2. What color was the cobalt chloride paper before you wiped the mirror? after you wiped the mirror?
3. What was the substance on the mirror?
4. Where did it come from?
5. What conclusion can you draw regarding another product of respiration?

Water and carbon dioxide are produced by the burning of most fuels, for example, gasoline or candle wax. Living things, in the process of respiration, give off these same waste products. Carbon dioxide diffuses into the blood from the cells and is carried back to the lungs to be breathed out. (This is the reverse of the way that oxygen reaches the cells.) What do you think happens to water, the other product of respiration? **A**

It might have occurred to you that the water and carbon dioxide you detected in your breath might have been present *before* you breathed in. How might you determine this experimentally? **B**

Now you are ready to devise a definition of *respiration* that takes into consideration the substances and the energy required, the products, and the location at which the process takes place. When you have written down your definition, complete this word equation describing respiration:

Respiration is part of metabolism. Learn more on pages S16–S17 of the SourceBook.

> **digested foods** + **?** → **energy** + **?** + **?**

Now look at the following simple chemical equation:

> **fuel (gasoline, coal, wood, etc.)** + **oxygen** → **carbon dioxide** + **water** + **energy**

What does this equation tell you about the process of respiration? **C**

Plants Also Respire

Plants must use some of the food they make to carry out their own life processes. Respiration is, therefore, just as essential to their existence as is photosynthesis. However, because photosynthesis is taking place, it is difficult to actually observe a green plant taking in oxygen and giving out carbon dioxide. To observe respiration in plants, it is easier to use germinating (sprouting) seeds, because they have not yet begun to make their own food through photosynthesis. Instead, they are still using food stored inside them. You will observe plant respiration in the next Exploration.

62

Plants Also Respire

Have students suggest some activities of plants that require plants to use their own food. (*Plants use food to grow, reproduce, repair damaged tissues, and respond to the environment.*) Point out that animals eat this food to obtain energy.

Homework

Have students create cartoon strips that could explain the process of respiration to younger students.

Theme Connection

Energy

Focus question: What is the role that oxygen plays in the release of energy during human respiration and during the combustion of wood? (*During respiration, a release of energy occurs when digested food combines with oxygen in the cells. Similarly, oxygen combines with wood, a hydrocarbon, in order to release energy in the combustion of wood.*)

Plant Respiration

You Will Need

- 20 seeds (such as radish or bean seeds)
- 4 test tubes
- 4 stoppers
- a wooden splint
- limewater
- a graduated cylinder

What to Do

1. Put stoppers in two test tubes and set them aside. These test tubes will contain only air.
2. Place 20 moist seeds into the remaining 2 test tubes. (Water helps seeds carry on their life processes.)

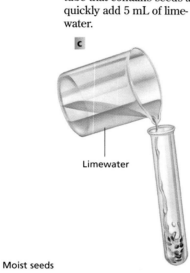

a

Moist seeds

3. Stopper these test tubes tightly and leave them alone for 2 days.
4. After 2 days, remove the stopper from one test tube that contains seeds and quickly insert a lit wooden splint. Did the splint go out? What does this suggest that the seeds might have done? **D**

b

5. Unstopper the second test tube that contains seeds and quickly add 5 mL of lime-water.

c

Limewater

6. Stopper the tube again, and gently swirl it for a minute. Write your observations in your ScienceLog.
7. Perform the same tests on the test tubes that contain only air, and compare the results with those obtained using the test tubes that contain seeds. How can you **E** explain your observations?

What do the results of this Exploration suggest concerning respiration in seeds? Write a summary of your results, along with your conclusions, in your ScienceLog. From your studies so far in this unit, make a list of differences between plant and animal processes. **F**

d

Swirl

Once the seeds have been moistened, do not add additional water. Two additional stoppered test tubes of air should be used as controls. Place the test tubes in a warm, dark place for 2 days. Make sure the students have an adequate supply of limewater before they begin.

Cooperative Learning
EXPLORATION 3

Group size: 3 to 4 students
Group goal: to demonstrate respiration in plants
Positive interdependence: Assign each student a role such as materials handler (to obtain and return equipment), supervisor (to encourage other students to work efficiently), technician (to perform the experiment), or checker (to ask the teacher questions on behalf of the group).
Individual accountability: Have each student design a data chart on which to record the data collected in this investigation. Each student should explain his or her experimental results.

An Exploration Worksheet is available to accompany Exploration 3 (Teaching Resources, page 52).

ENVIRONMENTAL FOCUS

While an increase in carbon dioxide in the atmosphere raises concerns about an increase in the greenhouse effect and global warming, some biologists predict that the greater availability of carbon dioxide to plants and algae would lead to an increase in photosynthetic activity. This would increase the rate at which the carbon dioxide is used, thus reducing the buildup of carbon dioxide in the atmosphere and lowering the risk of global warming. The ultimate outcome is still very much "up in the air." Have students make a diagram that illustrates this concept.

Answers to
In-Text Questions

D When students insert a burning splint into one of the test tubes of seeds, it should go out immediately. This suggests that the seeds used any oxygen that was present in the test tube to conduct respiration.

E Step 7 is a control. When the burning splint is placed in the test tube of air, it continues to burn for a short while, indicating that the amount of oxygen has not decreased. Thus, the oxygen in the test tube containing seeds must have been used by the seeds. The limewater again turns milky, but

after a longer period of shaking than in step 6. This test indicates that less carbon dioxide is present here than in the test tube containing seeds. Therefore, the extra carbon dioxide in the tube containing seeds was produced by the seeds.

F The results of this Exploration suggest that the seeds used oxygen and produced carbon dioxide. The processes of photosynthesis and transpiration are the only processes discussed so far in this unit that plants perform but that animals do not. The processes of diffusion, osmosis, and respiration are carried out by both plants and animals.

Hot Plants?

What do you think? How could you investigate Kimberly's question? Devise an experiment using seeds. **A**

Controlling Respiration

While reading about respiration, did you wonder how your body knows where energy is needed, how much energy is needed, and so on? The answer lies in the glands, the "process controllers" of the body. The glands produce chemicals that regulate or control different functions in the body.

One of the most important glands, the thyroid gland, regulates the rate of respiration in body cells; that is, it controls how quickly food is "burned" in the cells. It does this by means of a chemical substance, *thyroxin,* that diffuses from the thyroid gland into the blood, which carries it to other body cells. The respiration process is monitored and regulated by the nervous system, which consists of the brain and nerves.

Another vital gland is the pancreas. It is tucked between the stomach and the first part of the small intestine. The pancreas makes *insulin,* a chemical that enables sugar to be used as "fuel" for respiration.

These two glands, the thyroid and the pancreas, are directly involved in the process of respiration. There are many other glands in the body as well, not all of which regulate respiration. You will study these in later science courses.

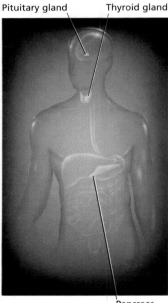

Pituitary gland   Thyroid gland

Pancreas

Respiration: A Summary

Imagine an individual cell in your body carrying on respiration.

1. What does the cell need? How do these materials reach the cell?

2. What does the cell give off? How do these materials leave, first from the cell and then from the body?

3. How do substances get into and out of the cell?

4. Why does the cell carry on respiration? Why is this process so important?

5. Many parts of the body are either directly or indirectly involved with respiration. How do each of the following parts of the body contribute to respiration at the level of individual cells: teeth, heart, skin, kidneys, lungs, salivary glands, thyroid gland, brain, capillaries, pancreas, and stomach?

Body Systems

Question 5 in the summary above named parts of many systems in the body. For example, the teeth and stomach are part of the **digestive system**. The kidneys and skin are two organs in the **excretory system**. The heart and capillaries are part of the **circulatory system**. The lungs are part of the **respiratory system**. The brain is part of the **nervous system**. All of these systems in your body work together to perform respiration and digestion, as well as other life processes.

Look on pages S16–S18 of the SourceBook for more information about body systems.

65

FOLLOW-UP

Reteaching

Have students make concept maps using cellular respiration as the main concept. Words that should be included are *food, oxygen, energy, carbon dioxide,* and *water.* Other suggested words are *blood, diffusion, lungs,* and *breathing.* All words in boxes should be accompanied by verbs or phrases along the lines that join them.

Assessment

Present students with the following scenario: A child decides to hold her breath as she enters a tunnel. She plans to exhale and continue regular breathing when she exits the tunnel. Why is this a bad idea, and what changes will occur in her blood as she holds her breath? *(As respiration continues, the amount of carbon dioxide in her blood will increase, and the amount of oxygen will decrease. Without oxygen, the girl could faint.)*

Extension

Point out to students that gills are the breathing structures in fish. Suggest that students learn how gills function to supply fish with the oxygen necessary for respiration. Call on volunteers to share what they learn by presenting oral reports to the class. Encourage students to use diagrams to make the presentation clear and easier to understand.

Closure

Have students try several physical activities for 2 minutes and measure their breathing rate for 1 minute following each activity. Possible activities include sitting, marching in place, and swinging their arms in large circles. Ask students to list their activities in order of increasing rate of respiration and to explain why the activities produced the varying rates.

- The skin excretes water and heat, two of the products of respiration, through the sweat glands.
- The kidneys remove water and impurities from the blood.
- Oxygen needed for respiration enters the body through the lungs. Carbon dioxide moves from the blood into the lungs and is expelled.
- The salivary glands produce the saliva that helps to break down starch for use in respiration.
- The thyroid gland produces thyroxin, which controls the rate of respiration in cells.
- The brain monitors the needs of different parts of the body and sends out signals that regulate breathing rate, heart rate, production of chemicals, and other activities.
- Capillaries permit easy diffusion of digested food, oxygen, carbon dioxide, and water into and out of the bloodstream.
- The pancreas secretes insulin, which aids in the uptake of sugar by cells so that it can be used as fuel for respiration.
- The stomach helps break down food into simpler substances that can be used by the cells for respiration.

LESSON 3 · Maintaining the Balance

FOCUS

Getting Started

Ask students to write for 5 minutes about the greenhouse effect. Tell students that the point is to brainstorm on paper what they know about this effect as well as to generate questions.

Main Ideas

1. A delicate balance exists between plants and animals on Earth.
2. The burning of fossil fuel releases carbon dioxide into the atmosphere.
3. The use of fossil fuels contributes to the greenhouse effect and could alter life on Earth.

TEACHING STRATEGIES

Answers to
In-Text Questions

Ⓐ The terrarium would require a balanced food web of plants and animals. Outside materials include light, water, soil, and mineral nutrients. Extremes in light and temperature or an introduced disease or pest could destroy the terrarium.

Theme Connection

Cycles

Focus question: How does the destruction of the rain forests threaten our water cycle, which in turn influences our climatic pattern? *(Plants of the rain forests release large quantities of water vapor through transpiration—so much that rains occur daily in most rain forests. The amount of water vapor that can cycle through our atmosphere is reduced as sections of rain forest are destroyed.)*

Homework

You may wish to assign the Theme Worksheet that accompanies the Theme Connection as homework (Teaching Resources, page 54).

LESSON 3 Maintaining the Balance

So far in this unit, we have talked about life processes carried out on a small scale, within a single plant or animal. Let us now turn our attention to life processes carried out on a much larger scale, that of the entire planet.

Our Earth is a very special place. It is the only world we know of where life exists. No other planet in our solar system is even close to being habitable. Life is able to exist on Earth for many reasons, not the least of which is the delicate balance that exists between plants and animals. Animals take in oxygen and give off carbon dioxide, and plants take in carbon dioxide and give off oxygen. Plants and animals need each other.

Consider a sealed terrarium, in which plants and animals live in balance with each other. Once the terrarium is sealed, no additional materials enter and no materials leave. What kinds of plants and animals would you put into a terrarium to make it self-sustaining? What additional materials from outside the terrarium are required? How could the terrarium environment be damaged so that everything in it died? Ⓐ

The Earth is much like a sealed terrarium, only on a much larger scale, of course. Balance must be maintained in order for life to survive. A balance between plants and animals, and between oxygen and carbon dioxide, has existed on Earth for many millions of years. However, it seems now that the actions of human beings can upset this balance.

Work with several others to explore the situation and to write a brief report about it. Your report will consist of six paragraphs. The first sentence of each paragraph is given on the facing page. Complete each paragraph by developing the idea in the opening sentence. Some research will be needed to gather information. Provide a title for the report. (Note: The numbers used to indicate the separate paragraphs should not be used in your report.)

LESSON 3 ORGANIZER

Time Required
1 or 2 class periods

Process Skills
observing, analyzing, hypothesizing

New Terms
none

Materials (per student group)
none

Teaching Resources
Theme Worksheet, p. 54
Graphing Practice Worksheet, p. 55

1. Carbon dioxide is absolutely necessary for plants and animals.

2. At present, it appears that green plants cannot remove the carbon dioxide from the air as fast as humans produce it.

3. An excess of carbon dioxide is a threat.

4. The greenhouse effect is beneficial up to a certain level.

5. It is difficult to predict all the possible effects of a higher concentration of carbon dioxide in the atmosphere.

6. There are ways to lessen the potential problem of an excess of carbon dioxide in the air.

Points for Discussion

Choose one or more of the following topics and discuss them in small groups. Record your findings for presentation to the class.

1. Some people have suggested that in the future, human energy needs could be met by *biomass* alone—that is, energy would come solely from organic materials, such as plants or animal wastes. What are the possible advantages and disadvantages of this approach? Would this be a viable way to meet energy demands? Why or why not?

2. As cities expand, trees often have to be cut down to permit the widening of streets and the building of houses and businesses. Do we need to conserve trees in our cities and industrial areas? Consider these facts: One large tree can absorb 2300 g of carbon dioxide in 1 hour. This is the amount given off by 10 single-family houses. During the same hour, that tree can give out about 1700 g of oxygen. How might large trees help control the level of carbon dioxide in the air around your town or city?

3. So far, you have learned that carbon dioxide gets into the air through cellular respiration and the burning of fuels. Are there other ways? If so, what are some of them? Do a little research. What are some sources of carbon dioxide emission? How many of these carbon dioxide sources might you find in a typical home?

How is the Earth's atmosphere affected by pollutants from automobiles? **B**

by burning forests? **C**

You may wish to use the Graphing Practice Worksheet that accompanies Lesson 3 as a homework assignment (Teaching Resources, page 55).

Answers to
In-Text Questions

B Most automobiles produce carbon dioxide and water vapor, which increase the greenhouse effect and could lead to global warming.

C The burning of forests produces greenhouse gases just as the burning of fossil fuels in automobiles does. More importantly, it reduces the number of plants available to capture the sun's energy through photosynthesis. This reduces the amount of carbon dioxide that is converted into food and oxygen and the amount of water vapor that is released through transpiration.

FOLLOW-UP

Reteaching

Have students do a magazine search and select two contrasting pictures, one showing an environment in balance and the other showing an environment out of balance. Discuss the pictures in a large group, and find differences in how life processes are affected in the balanced and unbalanced environments.

Assessment

Suggest that students make posters to express their concerns about air pollution. Their posters might illustrate some of the causes of, consequences of, and solutions to air pollution.

Extension

Suggest that students do some research to discover why the ozone layer is important, how it is being damaged, and what can be done to stop the damage.

Closure

Have students prepare an oral report on one alternative energy source that might prevent problems related to the greenhouse effect.

Answers to
Points for Discussion

The following information is provided to help you assess student discussions:

1. The main advantage of using biomass is that it is a renewable resource. Also, its production can be carefully controlled. The disadvantages include the vast amounts of land it would take to produce enough biomass to meet energy needs. Burning biomass, like burning most other fuels, would still release excess amounts of carbon dioxide into the atmosphere.

2. Most students will probably agree that trees should be conserved in cities and industrial areas. If enough trees are present, they can help clean carbon dioxide out of the air. By producing oxygen and using carbon dioxide, trees can help maintain the delicate balance between carbon dioxide and oxygen in the urban atmosphere.

3. Carbon dioxide is given off by baking powder used in cooking and by soft drinks. Some fire extinguishers produce carbon dioxide. Dry ice is the solid form of carbon dioxide. When it melts, it releases carbon dioxide gas into the air. Decaying plant and vegetable matter also give off carbon dioxide.

Answers to Challenge Your Thinking, pages 68–69

1.

a. 16	b. 10	c. 4	d. 9
e. 8	f. 5	g. 11	h. 15
i. 13	j. 17	k. 6	l. 3
m. 2	n. 7	o. 18	p. 12

2. Students' answers will vary. The report should include the use of plants as food, as well as some or all of the following uses: furniture, a source of chemicals (oils, turpentine, etc.), shade from the sun, wind-breaks, fuel, ingredients for cosmetics, shelter for animals, building materials, and paper products.

3. Each part of the meal contains energy that was obtained directly or indirectly from the sun: peanut butter is made from the seeds of peanut plants, which use energy from the sun to grow; bread contains wheat and sugar, both of which come from plants that use solar energy to grow; the apple comes from a tree that requires solar energy; and milk is obtained from an animal that gets its energy from plants that use solar energy to make food. The food first undergoes digestion, which breaks it down into glucose that can be distributed to all parts of the body. In this simple form, food can be used by body cells, where it combines with oxygen in the process of respiration. Through this process, food is converted to carbon dioxide, water, and energy.

4. Accept all reasonable responses. The liquid water could be recaptured by condensing the water vapor onto a cool surface. (You may wish to point out that a biosphere is any environment that supports life.)

1. The Magic Square

Match the number of each structure in the illustrations with the letter of the best description, and place the number in the appropriate box. If your choices are correct, the sum of the numbers in each row, column, and diagonal will be the same. The mathematical term for such a box is *magic square*. (Some questions refer to earlier chapters.)

- **a** where diffusion of digested food into the blood primarily occurs
- **B** gland that controls the rate of respiration in body cells
- **C** water pipes in the food factory
- **d** protective waterproof covering for the factory
- **e** the food factory
- **f** air space in the factory
- **g** air pipe with strong supporting rings
- **H** where insulin is made
- **i** center of blood circulation
- **J** where digestion of starch begins
- **K** the regulators for opening and closing the food-factory doors
- **L** cell layer where most of the factory doors are located
- **m** site of the actual manufacture of food
- **n** the service doors of the factory
- **o** capillaries, where diffusion of digested food from the blood into the body cells occurs
- **P** where the blood exchanges gases with the air

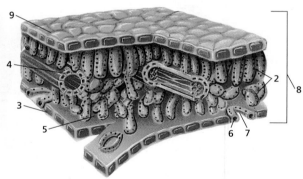

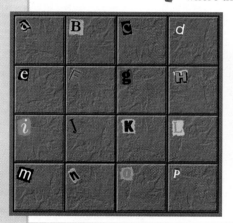

Did You Know...

About 90 percent of the oxygen in the Earth's atmosphere is a product of photosynthetic activity in algae.

★ **You may wish to provide students with the Chapter 3 Review Worksheet (Teaching Resources, page 57) and Transparency 9, which are available to accompany this Challenge Your Thinking.**

2. How Useful?

Write a report on the usefulness of plants. Use all available resources. Be sure to write the report in your own words. Prepare a list of all the resources you used.

3. Lunch to Go

Suppose your lunch consisted of a peanut butter sandwich, an apple, and a tall glass of milk. You would get energy from the meal. The energy would help you to move and to grow. Explain clearly how this energy is linked to the energy from the sun. Then trace the path of this food as it is used as energy by your body.

4. Lost in Space

In a space biosphere, a plant's water supply would have to be maintained. How would you retrieve the water lost by plants through transpiration?

ScienceLog

Review your responses to the ScienceLog questions on page 54. Then revise your original ideas so that they reflect what you've learned.

69

ScienceLog

The following are sample revised answers:

1. Plants use carbon dioxide, which animals release during respiration. Plants and animals rely on one another to maintain a balance between carbon dioxide and oxygen in Earth's atmosphere.

2. Food energy is released to the body through the processes of digestion and respiration. During digestion, complex food molecules are broken down chemically into a simple, water-soluble sugar, glucose. Glucose is then transported to the bloodstream, which carries it to the body's cells. The glucose passes from the bloodstream to the body's cells by diffusion. During respiration, glucose reacts with oxygen to provide the energy needed for life processes.

3. The greenhouse effect is responsible for keeping heat near the Earth's surface. The Earth reradiates much of the energy received from the sun. Greenhouse gases, such as carbon dioxide, in the atmosphere trap part of this radiation and keep it from escaping into space, thus keeping heat near the Earth's surface.

Meeting Individual Needs

Gifted Learners

Have students research the respiratory system of an organism of their choice. They should use their findings to create a poster or model that demonstrates how the organism breathes, and the poster or model should be labeled with or accompanied by an oral description of the respiratory process.

The Big Ideas

The following is a sample unit summary:

Food is manufactured in the leaves of plants. (1) For this reason, leaves are like food factories. They use the sun's energy, chlorophyll, water, and carbon dioxide to make food for themselves in the form of simple sugars. They then store this food in the form of starch. Other organisms eat the plants, and these organisms in turn become food for other organisms. Food provides organisms with the energy and materials they need to live and grow. (2)

Water makes up a large part of most animals and plants. It can carry large amounts of dissolved nutrients and other materials through the environment and within living organisms. (3) Osmosis allows cells to reach equilibrium or near equilibrium with surrounding fluids and provides a mechanism for the movement of particles through cell membranes. (4)

In living things, chemical reactions convert substances that make up food into other substances. The energy released by these chemical reactions is then used by living things. (5) Most living things breathe to obtain oxygen, which is needed to "burn" food for energy. However, of the organisms that require oxygen, not all breathe as most animals do. Some, such as plants, obtain oxygen in other ways. (6)

Respiration is the process of breaking down food to release energy. The process uses oxygen and produces carbon dioxide. (7) Burning is similar to respiration in that oxygen combines with another substance and energy is given off. (8)

Plants and animals each produce gases during respiration. Animals need oxygen and produce carbon dioxide, and plants need carbon dioxide and produce oxygen. (9)

Human activities produce a great deal of carbon dioxide and other pollutants, which have a harmful effect on the global climate. (10)

 You may wish to provide students with the Unit 1 Review Worksheet that is available to accompany this Making Connections (Teaching Resources, page 61).

Making Connections

Unit 1

SourceBook

Look in your SourceBook to read more about the chemistry of living things. The SourceBook also contains additional information about that remarkable substance water, as well as information about the systems of the human body.

Here's what you'll find in the SourceBook:

UNIT 1
The Chemistry of Life S2
Building Biochemicals
   With Light S9
Basic Life Processes S13

The Big Ideas

In your ScienceLog, write a summary of this unit, using the following questions as a guide:

1. Where does food come from?
2. Why are leaves like food factories?
3. Why is water important to living things?
4. What role does osmosis play in living things?
5. How do living things get energy from food?
6. Why do we breathe?
7. What is respiration?
8. How is respiration similar to burning?
9. How do plants and animals depend on each other?
10. How are humans changing the natural environment?

Checking Your Understanding

1. Below are a number of statements about topics you studied in this unit. Suggest ways in which each statement could be scientifically verified.
   a. Plants give off oxygen only during daylight.
   b. Carbon dioxide is a raw material used in the manufacture of food.
   c. Plants exchange gases through their stomata.
   d. Water has an attraction to itself.
   e. Osmosis affects living cells.
   f. Carbon dioxide is a waste product of respiration.
   g. All living things require water.
   h. Temperature affects the rate of respiration in plants.
   i. Plants need sunlight to produce food.
   j. Osmosis creates pressure.
   k. The light of the sun, not its warmth, is what makes life on Earth possible.

2. Read the following statement:

Diffusion begins when you add a substance to water, and it ends when the substance is completely mixed in the water.

Does the statement really tell the whole story, or is there more to it? Explain.

3. Plants are sometimes afflicted with an inherited abnormality (called *albinism*) that results in the plants having no color whatsoever. What kind of problems do you think this might cause for a plant? What would the chances be for this plant to survive and reproduce?

4. Look at the illustrations below. Each tells a story. It is your job to figure out what that story is.

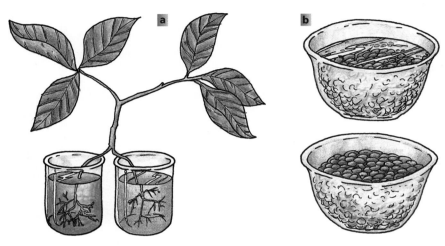

a. After a few days the leaves were mottled with blue and red patches.

b. After a few hours, the water seemed to disappear. What happened?

5. **concept map** Make a concept map using the following terms: carbon dioxide, breathing, water, respiration, oxygen, and energy.

6. How is it that the same water is used over and over again in the natural world?

71

The Unit Activity Worksheet makes an excellent homework assignment (Teaching Resources, page 61).

Homework

The Unit Activity Worksheet makes an excellent homework assignment (Teaching Resources, page 61).

Answers to
Checking Your Understanding,
pages 70 and 71

1. Student answers will vary but should demonstrate a systematic approach to analyzing a situation in a scientific context. Students should propose ways to test these hypotheses. *(For example, (a) could be tested by enclosing a plant in a terrarium and monitoring the amount of oxygen present in the terrarium at several times during the night and during the day.)*

2. Random particle motion is always occurring in matter. Particles reach a state in which their numbers are evenly distributed, but particle motion never begins or ends.

3. In plants, albinism signals a lack of chlorophyll, which is essential for photosynthesis. Without chlorophyll, photosynthesis could not occur and the plant could not survive.

4. a. Transpiration occurs, causing water to be absorbed by the roots. Because the water is colored, it becomes visible in the leaves.
   b. The concentration of water inside the beans is less than that outside the beans. Therefore, water enters the beans by osmosis, causing them to swell.

5. For a completed sample concept map, see page S206.

6. Answers should include an understanding and an explanation of the water cycle. One possible response is as follows: Excess water leaves plants through the stomata by the process of transpiration. The water evaporates into the air and, at some point, condenses to form clouds. This happens as warm, moist air rises and cools. Eventually, rain or another form of precipitation falls to the surface of the Earth. The precipitated water seeps into the ground, where a portion of it is absorbed by plants. Also, water taken in by animals may be released as a waste product that seeps into the ground and is later absorbed by plants. In both cases, plants once again release most of this water into the air through transpiration. Together, all of these processes can be called a water cycle because the water is constantly reused.

Background

In addition to their sensitivity to solar radiation and soil and water pollution, amphibians are good indicators of environmental conditions for several other reasons. First, amphibians are exposed to many different biotic and abiotic aspects of the environment. Amphibian larvae are herbivores until adulthood, when they become carnivores. Likewise, young amphibians live in water, while adults live primarily on land. Because amphibians come in contact with so many different parts of their local environment, the health of an amphibian population is a good indicator of the general health of the local ecosystem. Second, most amphibian populations remain in one area over many generations, making amphibians reliable "witnesses" to changes in the local environment. Finally, because there is so much variation among amphibian species, a problem that is unrelated to environmental conditions, such as disease or predation, would affect only a limited number of amphibian species. Thus, scientists can be fairly certain that a global decrease in amphibian populations is the result of global environmental changes.

ENVIRONMENTAL FOCUS

Have students think about the way an individual amphibian interacts with its environment during the course of its life. Have a volunteer write the interactions on the board. Then ask: How are amphibians important to their ecosystems? *(Many adult amphibians help limit the populations of insects such as mosquitoes and flies by eating them. Amphibian larvae are also a primary source of food for many other organisms. The extinction of amphibians could lead to a serious disruption of many food webs.)*

EYE ON THE ENVIRONMENT

Watchfrogs of the Environment

In 1990 scientists made a troubling discovery. Populations of frogs, toads, salamanders, and other amphibians were decreasing all over the world. Equally troubling were the questions this discovery raised. Why were they dying? Was there a problem with their environment? If there was a problem, could it harm people too?

A Featherless Bird?

Amphibians can be compared to the canaries that miners once carried down mine shafts. If a canary died, the miners knew that poisonous gases were in the air and that the miners were in danger. Similarly, amphibians may be warning us of dangers in our own environment.

Amphibians are particularly sensitive to their surroundings. They live most of their lives underwater or in damp soil. As they breathe, they absorb oxygen through their thin skin. Any pollutants, such as acid rain, industrial chemicals, or pesticides, can easily pass through their skin. Thus, pollutants can drastically affect an amphibian's health. Similarly, only a jellylike coating protects their eggs from the outside world, making the eggs vulnerable to pollutants as well.

No Trees Means No Tree Frogs

What is causing the shrinking of the amphibian population? The likeliest reason is the destruction of their natural habitats as forests are cut down and wetlands are filled in. But there may be another reason: the decrease of *ozone* in the atmosphere. Ozone is an oxygen molecule that absorbs much of the sun's harmful ultraviolet rays before they reach the Earth's surface. Exposure to these rays can damage amphibian eggs, resulting in mutations or even death.

Cleaning Up Our Planet

Perhaps the best way we can help our amphibian friends is to reduce our impact on the environment. For instance, we can recycle paper so that fewer forests need to be cut down. We can buy natural, organic detergents so that when we wash clothes, we do not add pollutants to our water. By carpooling or using public transportation, we can reduce air pollution, too. Little things can add up!

▲ Will pollution, deforestation, or other environmental damage drive amphibians to extinction? Practices such as the one shown above may contribute to their demise.

Take a Breather

Both plants and animals carry out respiration in order to live, but different organisms respire in different ways. Find out how a tree, a fish, an amphibian, and a human respire. How are the processes different for each organism? How are the processes alike?

72

Answers to
Take a Breather

Respiration is the process by which an organism obtains and uses oxygen. Humans accomplish this by inhaling air into their lungs, where gases are exchanged through the thin walls of capillaries. Oxygen is absorbed and carbon dioxide is exhaled. In fish, this gas exchange occurs when water is taken in through the mouth and passed across the gills. Young amphibians also have gills, but adult amphibians have lungs like humans. Amphibians also absorb oxygen through their skin. Plants absorb oxygen primarily through small holes called *stomata* in their leaves. Plants absorb carbon dioxide and release oxygen in the process of photosynthesis.

Space Food

*T*hey say that dining in space leaves a lot to be desired. The menu is limited, the service is nonexistent, and fresh food is out of the question. But don't blame the chef; preparing food for space travel is a difficult task.

Eating Out—Way Out!

Before the first spaceflight, scientists did not know whether eating food in space would be possible without gravity to aid in swallowing. So one of the first things John Glenn did on an American spaceflight in 1962 was try to eat. As Glenn soon discovered, swallowing food is not a problem. However, making space food both nutritious and good to eat is another story.

Food used for space travel must meet several requirements. There are no refrigerators on space shuttles, so the food must be nonperishable and must remain free of harmful bacteria. The stress of space travel is hard on the body, so the food must be rich in minerals and high in calories. Also, limited storage requires dense food and compact packaging. Finally, food must be packaged in special containers that will keep it from floating away.

What's for Dinner?

Early attempts to make space food resulted in small, dried food cubes and tubes of food paste. As you can imagine, these

▲ Astronaut Sally Ride preparing to have a meal on a space shuttle. The food tray is specially packaged to keep the food from floating away.

foods made unappetizing meals. Over time, however, nutritionists developed better methods of packaging tasty, conventional food so that it would meet the astronauts' needs.

Thanks to these advances, a typical meal on today's space shuttle might consist of smoked turkey, mixed Italian vegetables, mushroom soup, strawberries, butterscotch pudding, and tropical fruit punch. The food is canned or specially packaged in foil to solve the problems of storage and weightlessness. Most spices would float away in space, but astronauts can still flavor their food with liquid pepper and other flavorings that will stick to the food. Some of the food is freeze-dried for

easier storage. Before eating this food, astronauts mix it with water that is produced as a byproduct of the shuttle's fuel cells. It's a complicated process, but despite all of the hassle, most astronauts would repeat their dining experiences. After all, the food may not be fancy, but you can't beat the view.

Plan It Yourself

How is planning a backpacking trip like planning for spaceflight? Plan a menu for a three-day backpacking trip. How much water would you need and how would you get it? What other concerns will you need to think about?

73

Extension

Before the twentieth century, ships took months to transport people across the oceans. Ask the class what special nutritional problems people confronted on these voyages. You might have students do some research on this topic. Ask: How does this compare with the difficulties of space travel? *(One common disease was scurvy, caused by a deficiency of vitamin C. The availability of fresh*

water was also a problem during long periods when there was no rainfall or when the ship's water supply became contaminated. Plans for space travel must account for similar limitations on space and fresh supplies. For example, fresh water is collected from the shuttle's fuel cells, which make pure water as a byproduct of their operation.)

Background

In space, the body undergoes many changes due to weightlessness. The heart and leg muscles begin to atrophy, and the mass and strength of bone tissue is gradually reduced. There is also a progressive loss of red blood cells, as well as a loss of calcium and other minerals in the bones. To compensate for some of these changes, the balance of nutrients in an astronaut's diet is different from that found in the diet of someone on Earth. A special mineral-rich diet helps to keep these harmful changes to a minimum. In addition, vigorous exercise on exercise bikes or treadmills helps prevent demineralization and muscular deconditioning.

Discussion Questions

1. How do you think food is preserved without refrigeration? *(Some processes include canning, irradiation, and dehydration.)*
2. Why do you think it is important to have a variety of foods to eat during a mission? *(Most people prefer to eat a variety of foods over an extended period of time. A varied diet also provides the full range of nutrients needed to remain healthy.)*
3. Why do you think astronauts eat tortillas instead of sliced bread in space? *(Sliced bread is too crumbly for the weightless environment; the crumbs could float around the cabin.)*

Answers to
Plan It Yourself

Food supplies for a backpacking trip, just like those used in space travel, must be lightweight, compact, easy to prepare, and resistant to spoiling. If water sources are available on the backpacking trip, carrying extra water would be a waste of energy. However, water taken directly from natural sources must be purified. If water is not available, each person should pack at least 4 L of water for every day that they will be on the trail. Because water is so heavy to carry, this limits the number of days one can backpack safely in the absence of natural water sources.

Background

Zoology dates back to ancient times when the Greek philosopher Aristotle observed and theorized about animal behavior more than 2300 years ago. About 200 years later, Galen, a Greek physician, began dissecting and experimenting with animals. However, there were few advances in the field of zoology until the 1700s and 1800s. During this period, the Swedish naturalist Carolus Linnaeus developed a classification system for animals, and Charles Darwin published his theories on natural selection and evolution.

Today, zoology is divided into many specialized areas. Some areas include comparative anatomy (the study of anatomical structures among many different animals), physiology (the study of how bodies function), genetics (the study of heredity), embryology (the study of embryo development), entomology (the study of insects), ichthyology (the study of fish), and herpetology (the study of reptiles and amphibians).

Meeting Individual Needs

Second-Language Learners

Like zoology, many branches of science include subfields. Learning about the origins of the names can help students understand the differences among these subfields. Provide each student with one related group of subfields. Ask them to identify the larger field of study and to look up each subfield's component root words to find out what they mean. Students should also find examples of simple, common words that have the same root(s). *(For example, the subfields botany, zoology, taxonomy, and ecology would all fall under the field of biology. The word zoo is a simple word that has the same root as the word zoology.)*

Eric Pianka, Zoologist

Eric Pianka became involved in the study of lizards when he was four or five years old. "On a trip across the country with my family, I saw a big, green lizard at a roadside park," Eric explains. "I tried to catch it, but all I got was the tail. At that moment, I knew I had to find out everything I could about the kind of life it led." After years of study, Eric is now a world-famous professor of zoology at the University of Texas.

The Ecology of Desert Lizards

In his research as a zoologist, Eric has focused on the ecology of desert lizards. He goes to a desert, collects lizards, and examines and classifies them. Then he compiles data and interprets it in books or papers. As Eric puts it, "I try to answer questions like, Why are there more lizards in one place than in another? How do they interact with each other and with other species? How have they adapted to their environment?"

Learning From Wildlife

Eric believes that research on lizards and other animals may help protect our environment. "Everyone always asks, 'Why lizards?' I turn that around and say, 'Why you?' The general attitude is that everything on Earth has to somehow serve humans. By looking at how other species have lived and died and changed over millions of years, we can gain a better understanding of the world we live in."

Exploring the Deserts of the World

One of the things Eric likes best about his job is being in the wilderness and seeing things

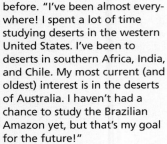

▲ The collared lizard lives in rocky regions of the southwestern United States.

that few people have ever seen before. "I've been almost everywhere! I spent a lot of time studying deserts in the western United States. I've been to deserts in southern Africa, India, and Chile. My most current (and oldest) interest is in the deserts of Australia. I haven't had a chance to study the Brazilian Amazon yet, but that's my goal for the future!"

A Project Idea

Select a common animal that lives in your area and that can be easily observed. Spend a couple of hours watching what it eats, what it does, and where it goes. What other animals does it interact with? What specific features of this animal make it well suited to its environment? Carefully document everything you observe. Did you discover anything you didn't already know?

▼ Eric Pianka with a perentie in Australia's Great Victoria Desert. Perenties are some of the largest lizards in the world, ranging up to 2.4 m in size.

74

A Project Idea

You may wish to have students work in pairs. Remind students not to disturb the animal or its habitat when making their observations. If they have difficulty finding an animal to observe, remind them that insects are interesting subjects for observations. They could also observe their pets.

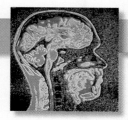

Meatless Munching

What'll it be today, the hamburger special, chicken surprise, or the garden-fresh veggie platter? More and more people are opting for the veggie platter. In fact, research indicates that over 12 million Americans are now eating vegetarian, and this number appears to be growing.

It's Not Just Salads

When you think of a vegetarian diet, you might think only of vegetables. Of course, vegetables are the mainstay of a vegetarian diet, but not all vegetarian diets are alike. Many vegetarian diets include dairy products, eggs, fruits, and nuts. Some vegetarian diets are based solely on plant products, while semivegetarian diets may also include some fish and poultry.

Why the Trend?

Different people have different reasons for going vegetarian. Some people believe that a vegetarian diet is healthy because a decrease in consumption of animal products can lower their intake of saturated fat and cholesterol. Many people also have ecological reasons for eating a vegetarian diet. For example, the grains that are fed to livestock can supply more nutritional value per volume than the meat we get from the livestock. And producing a serving of meat requires more land, water, and chemicals than producing a serving of grain. Finally, many vegetarians believe that it is unethical to kill animals for meat when plant substitutes are available.

▼ All of the foods shown here are the products of plants and are commonly used in vegetarian diets.

Benefits and Risks

A vegetarian diet can be a healthy choice. Recent statistics suggest that a vegetarian diet may reduce the risk of heart disease, adult-onset diabetes, and some forms of cancer. However, this type of diet takes some careful thought. It is not simply a matter of eliminating meat. People may replace the missing meat with too many dairy products and eggs, which are higher in fat than most meats. Others may substitute high-carbohydrate foods (such as pasta) and junk food (such as french fries) for their meat choices instead of increasing their fruit and vegetable intake. This could lead to nutritional deficiencies. The key to consuming the recommended amount of calories and nutrients for a healthy diet is to eat a wide variety of nutritious, low-fat foods. Actually, this is good advice whether you want to decrease your meat intake or not.

Prepare a Healthy Menu

Choose one of the nutrients that meats provide, and do some research to find a vegetable substitute. What difficulties did you encounter in your search? Would you eat the substitute you found?

75

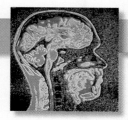
HEALTH WATCH

Background

To be nutritionally complete, our bodies need twenty different amino acids, eleven of which we manufacture ourselves. The other nine are called *essential amino acids.* No food derived solely from plants contains all nine amino acids in large amounts.

By eating a variety of foods, a vegetarian can be assured of getting all of the protein he or she needs. Indeed, many experts consider dietary variety to be the primary factor in achieving a nutritionally complete diet and maintaining good health, regardless of whether a person is a vegetarian.

CROSS-DISCIPLINARY FOCUS

Health

Explain to the class that a protein consists of a long chain of amino acids. Proteins are used in the body to regulate hormones, repair damaged tissue, and help fight infections. Assign each student one of the 20 common amino acids to investigate. Tell them to find out what foods the amino acid can be found in and what function it serves in the body. Then have students write the information in outline form on a small index card along with an illustration of the food or foods. Use the index cards to make a large class poster.

Multicultural Extension

Vegetarian Cultures

Some cultural and religious groups practice vegetarianism. Ask students to research one of these groups. If possible, have students prepare a sample menu of a vegetarian meal from that culture. (*Example: Many Indian dishes and Lenten meals in certain Christian religions are vegetarian.*)

Answers to
Prepare a Healthy Menu

Some nutrients found in meats include vitamin B_{12}, iron, zinc, calcium, and protein. Vitamin B_{12} is not found in plants in appreciable amounts, so many vegetarians must eat foods fortified with B_{12} or take vitamin supplements. To maintain an adequate intake of iron and zinc, a vegetarian could eat dried fruits, beans, spinach, or collard greens. Calcium can be found in milk, tofu, and leafy greens such as broccoli. Finally, many types of plant foods (such as legumes) contain protein. Eaten in the right amounts and combinations, these foods can supply all of the protein necessary for human health.

Unit 2

Unit Overview

In this unit students examine the particle model of matter. In Chapter 4, the concept of size as it relates to the structure of matter is explored, and the use of models to study objects that cannot be seen is introduced. Also, students examine how observations and inferences relate to each other and how they differ. In Chapter 5, students begin to build a particle model for matter based on observation and inference. Additionally, the behavior of particles in solids, liquids, and gases is demonstrated. In Chapter 6, students analyze examples from everyday life to find out what happens as matter absorbs and releases energy. In Chapter 7, students explore the history of the particle model of matter, from Dalton's initial ideas to the current model of the atom.

Using the Themes

The unifying themes emphasized in this unit are **Changes Over Time, Structures,** and **Energy.** The following information will help you incorporate these themes into your teaching plan. Focus questions that correspond to these themes appear in the margins of this Annotated Teacher's Edition on pages 85, 90, 113, 116, 122, 132, 134, and 136.

Changes Over Time can be discussed in terms of how our understanding of the nature of matter has changed through the ages. As students study particles, they should keep in mind that new discoveries and modifications of existing scientific theories are constantly being made.

Structures can be emphasized in this unit in a variety of ways because the physical properties of a substance are determined by the structure of its constituent atoms. Additionally, an accurate understanding of the structure of an atom can be facilitated by means of a number of atomic models.

The random motion of particles in a substance determines the amount of heat **Energy** the substance contains. As the motion of particles within a material decreases, the material cools and solidifies.

Using the SourceBook

In Unit 2, students are introduced to the distinction between molecules and other particles. They then explore the kinetic theory of matter as illustrated by various properties of solids, liquids, and gases. Finally, theories of the forces that hold matter together are discussed. You may notice that many of the SourceBook articles are not direct extensions of the lessons in Unit 2. In this Unit, the SourceBook is best treated as a logical extension of Unit 2 and as a place where students can build on the foundations of their work in Unit 2.

Bibliography for Teachers

Stwertka, Albert. *The World of Atoms and Quarks.* New York City, NY: Twenty-First Century Books, 1995.

Von Baeyer, Hans Christian. *Taming the Atom.* New York City, NY: Random House, 1992.

Bibliography for Students

Darling, David. *Micromachines and Nanotechnology.* Parsippany, NJ: Dillon Press, 1995.

Structure of Matter. Alexandria, VA: Time-Life Books, 1992.

Films, Videotapes, Software, and Other Media

The Atom
   Software (Macintosh, Windows)
   Sunburst Communications
   101 Castleton St.
   P.O. Box 100
   Pleasantville, NY 10570

Matter and Its Properties
   Videotape
   Agency for Instructional Technology
   P.O. Box A
   Bloomington, IN 47402-0120

Matter Is Everything
   Videotape
   Barr Media Group
   12801 Schabarum Ave.
   Irwindale, CA 91706-7878

The Molecular Theory of Matter
   Videotape
   Britannica
   310 S. Michigan Ave.
   Chicago, IL 60604

Particles in Motion: States of Matter
   Film
   National Geographic Society
   Educational Services
   P.O. Box 98019
   Washington, D.C. 20090-8019

Unit Organizer

Unit/Chapter	Lesson	Time*	Objectives	Teaching Resources
Unit Opener, p. 76				Science Sleuths: Fortune or Fraud? English/Spanish Audiocassettes Home Connection, p. 1
Chapter 4, p. 78	Lesson 1, A Search for Explanations, p. 79	1	1. Observe that the sizes of objects can be compared using an exponential scale. 2. Explain how exponential form is used to express very large and very small numbers.	Image and Activity Bank 4-1 Transparency 10
	Lesson 2, Give Me the Facts, p. 81	1 to 2	1. Explain that observations are supported by facts. 2. Explain that inferences are supported by circumstantial evidence. 3. Examine models as representations of events or objects in the real world that can be used to test hypotheses.	Image and Activity Bank 4-2 Resource Worksheet, p. 3 Transparency 11 Exploration Worksheet, p. 4
End of Chapter, p. 85				Chapter 4 Review Worksheet, p. 5 Chapter 4 Assessment Worksheet, p. 8
Chapter 5, p. 87	Lesson 1, Building the Case, p. 88	1 to 2	1. Examine the evidence for the particle nature of matter. 2. Develop a model to explain observations of dissolving and the pouring and mixing of substances.	Image and Activity Bank 5-1 Exploration Worksheet, p. 11 Discrepant Event Worsheet, p. 16
	Lesson 2, The Hidden Structure of Matter, p. 92	2	1. Explain that all matter is composed of atoms. 2. Recognize that there are a finite number of naturally occurring elements. 3. Demonstrate that all atoms of the same element have the same properties. 4. Explain that atoms combine to form molecules. 5. Identify molecules of different kinds of atoms as compounds.	Image and Activity Bank 5-2 Transparency Worksheet, p. 17 ▼ Transparency Worksheet, p. 19 ▼ Transparency 14
	Lesson 3, The Size of Particles, p. 99	2	1. Demonstrate that the molecules of different substances have different sizes. 2. Determine that atoms and molecules are extremely small. 3. Examine how some materials allow certain substances to pass through while blocking others.	Image and Activity Bank 5-3 Exploration Worksheet, p. 21 Exploration Worksheet, p. 22
	Lesson 4, Particles of Solids, Liquids, and Gases, p. 103	2 to 3	1. Explain that the particles of matter are in constant motion. 2. Demonstrate that gas particles are farther apart than are the particles of liquids and solids. 3. Explain that heating a substance causes the particles of the substance to move faster and farther apart.	Image and Activity Bank 5-4 Exploration Worksheet, p. 23
End of Chapter, p. 105				Chapter 5 Review Worksheet, p. 26 ▼ Chapter 5 Assessment Worksheet, p. 29
Chapter 6, p. 107	Lesson 1, Temperature and Particles, p. 108	1	1. Demonstrate that cooling matter causes the particles of matter to slow down and move closer together. 2. Explain that air expands when it is heated. 3. Recognize that not all substances expand and contract at the same rate.	Image and Activity Bank 6-1
	Lesson 2, Changes of State, p. 111	2	1. Demonstrate that melting and freezing are the result of heat energy being added to or taken away from a substance. 2. Explain that the temperature remains constant at a substance's melting point or freezing point until all of the material has either melted or frozen.	Image and Activity Bank 6-2 Exploration Worksheet, p. 31 ▼ Resource Worksheet, p. 32 Graphing Practice Worksheet, p. 35
	Lesson 3, Absorbing and Releasing Heat, p. 115	1 to 2	1. Define endothermic and exothermic changes. 2. Demonstrate that a change of state requires substances to absorb or release heat energy. 3. Construct wet-bulb thermometers to study the process of evaporative cooling.	Image and Activity Bank 6-3 Exploration Worksheet, p. 37 Theme Worksheet, p. 38 Exploration Worksheet, p. 40
	Lesson 4, Particles— Mass and Volume, p. 119	3	1. Investigate the three methods for determining the volume of small solid objects. 2. Demonstrate that different materials with the same volume may have different masses. 3. Explain that the mass of a given volume of a substance is an identifiable property of that substance.	Exploration Worksheet, p. 42 Exploration Worksheet, p. 44 ▼ Transparency 18
End of Chapter, p. 124				Chapter 6 Review Worksheet, p. 46 Chapter 6 Assessment Worksheet, p. 50
Chapter 7, p. 126	Lesson 1, Picturing an Atom, p. 127	1 to 2	1. Investigate Dalton's theory of matter, which states that all matter is made up of atoms. 2. Investigate Thomson's refinement of the atomic theory, which states that the atom contains both a positive and a negative charge.	none
	Lesson 2, Smaller Than an Atom, p. 130	3 to 4	1. Identify the parts of the atom. 2. Investigate Rutherford's discovery of a positively charged nucleus. 3. Investigate Bohr's planetary model of the atom. 4. Discuss the relative sizes of protons, neutrons, and electrons.	Exploration Worksheet, p. 52 Exploration Worksheet, p. 54 Resource Worksheet, p. 56
End of Chapter, p. 136				Chapter 7 Review Worksheet, p. 58 Chapter 7 Assessment Worksheet, p. 61
End of Unit, p. 138				Unit 2 Activity Worksheet, p. 63 ▼ Unit 2 Review Worksheet, p. 64 Unit 2 End-of-Unit Assessment, p. 67 Unit 2 Activity Assessment, p. 73 Unit 2 Self-Evaluation of Achievement, p. 76

* Estimated time is given in number of 50-minute class periods. Actual time may vary depending on period length and individual class characteristics.

▼ Transparencies are available to accompany these worksheets. Please refer to the Teaching Transparencies Cross-Reference chart in the Unit 2 Teaching Resources booklet.

Materials Organizer

Chapter	Page	Activity and Materials per Student Group
4	84	**Exploration 1:** balloon; 5 small plastic beads
5	88	**Exploration 1, Part 2:** drop of red food coloring; stirring rod; 500 mL of water; 100 mL graduated cylinder; 100 mL beaker; lab aprons; **Part 3:** 4 large containers; about 500 mL of sand; about 500 mL of water; about 500 mL of dried peas or beans; **Part 4:** 50 mL of sand; 25 mL of salt; two 100 mL graduated cylinders; 200 mL of water; watch or clock with second hand; 40 mL of rubbing alcohol; stirring rod; funnel; safety goggles
	93	**Time Out for Facts:** 16 g of one color of modeling clay; 2 g of another color of modeling clay; metric balance
	93	**Time Out for Discovery:** model molecule from Time Out for Facts; metric balance
	95	**Time Out for Analysis:** 12 g of one color of modeling clay; 4 g of another color of modeling clay; metric balance
	97	**Even More Models:** five different colors of modeling clay in the following amounts: 160 g, 39 g, 14 g, 196 g, and 32 g; metric balance
	99	**Exploration 2:** jar with lid; 100 mL graduated cylinder; 5 mL of cornstarch; 5 mL of dextrose; hot plate; stirring rod; egg; straight pin; large beaker; 110 mL of water; clock or watch with second hand; 2 test tubes; a few drops of iodine starch-test reagent; 8 drops of Benedict's solution; eyedropper; hot-water bath; safety goggles; lab aprons; test-tube tongs, test-tube clamp, or oven mitts; latex gloves (additional teacher materials: 50 mL of 1 M sulfuric acid solution; six 6d iron nails; steel wool; portable burner; tongs; watch glass; 50 mL of 4% sodium hydroxide solution; pH paper; strainer; 25 mL of 0.1 M sodium thiosulfate solution; a few sheets of old newspaper; see Advance Preparation below.)
	100	**Exploration 3:** clove of garlic; 30 cm of plastic wrap
	102	**Exploration 4:** aluminum foil; scissors; metric ruler; metric balance
	103	**Exploration 5:** plastic disposable syringe without a needle; 20 mL of water; 100 mL graduated cylinder
	103	***Exploration 6, Station 1:** 2 drops of food coloring; 2 small beakers; about 250 mL of cold water; about 250 mL of very hot tap water; lab aprons (additional teacher materials: ice chest with ice); **Station 2:** balloon; 2 L or 3 L plastic soft-drink bottle; 2 large containers or buckets; 5 L of ice water; 5 L of very hot tap water; watch or clock; **Station 3:** 2 drops of rubbing alcohol; eyedropper; 2 microscope slides; a few matches; safety goggles; **Station 4:** 250 mL of ice water; 250 mL beaker; **Station 5:** a cotton ball; metal jar lid; eyedropper; a few drops of perfume; **Station 6:** candle; a few matches; materials to prop up candle, such as a metal jar lid and a small ball of modeling clay; safety goggles
6	111	**Exploration 2:** about 20 mL of stearic acid; scoop; small beaker; 250 mL of hot water (75–80°C); hot plate; alcohol thermometer; wire-loop stirring device; test tube; test-tube tongs or test-tube clamp; test-tube rack; watch or clock with second hand; safety goggles; oven mitts
	115	**Exploration 3:** microscope slide; a small amount of salol; stainless steel scoop; a few matches; magnifying glass or low-power microscope; test-tube tongs or test-tube clamp; tweezers; latex gloves; safety goggles; lab aprons
	117	**Exploration 4:** 3 alcohol thermometers; three 5 cm pieces of hollow shoelace; cup or other small container; 50 mL of water; 30 mL of rubbing alcohol; electric fan; safety goggles
	119	**Exploration 5, Method 1:** overflow can or container with spout or side arm; 100 mL graduated cylinder; small block of wood; straight pin; **Method 2:** block of wood from Method 1; metric ruler; **Method 3:** iron bolt; 250 mL graduated cylinder; 200 mL of water
	121	**Exploration 6:** 250 mL graduated cylinder; metric balance; small iron object; small aluminum object; 200 mL of water
7	133	**Exploration 2, Part 1:** 100 mL graduated cylinder; metric balance; 125 mL of long-grain rice; **Part 2:** two 25 cm pieces of string; a small ball bearing; 2 grains of sand

* You may wish to set up this activity in stations at different locations around the classroom.

Advance Preparation

Exploration 2, page 99: To dispose of the Benedict's solution, you will need 50 mL of 1 M sulfuric acid solution, six 6d iron nails, steel wool, a portable burner, tongs, a watch glass, 50 mL of the 4% sodium hydroxide solution, pH paper, and a strainer. To dispose of the iodine-treated mixture, you will need 25 mL of 0.1 M sodium thiosulfate solution and some old newspaper. See the Waste Disposal Alert on page 100.

Unit Compression

In this unit, students progress from investigating the need for a particle model, to understanding the model itself, and finally to using the model to explain observations and make predictions. When time is short, the unit can be compressed in several ways without sacrificing the integrity of this conceptual progression.

Using what you find out about your students' prior knowledge, you may wish to assign some sections as homework to be read ahead of time. This will help to speed in-class discussion.

Several of the Explorations in this unit, such as Exploration 1 on page 88, can be effectively conducted as teacher demonstrations. Others may be considered optional, including Explorations 2, 3, and 4 on pages 99–102. In addition, the sections that develop the concept of the atom and its internal structure may be omitted if necessary. These sections include Lesson 2 of Chapter 5 and all of Chapter 7.

Homework Options

Chapter 4
See Teacher's Edition margin, pp. 83, 84, and 87
Resource Worksheet, p. 3
Exploration Worksheet, p. 4

Chapter 5
See Teacher's Edition margin, pp. 90, 94, 96, 98, 101, and 102
Exploration Worksheet, p. 22
SourceBook, p. S22

Chapter 6
See Teacher's Edition margin, pp. 110, 112, 113, 117, 121, and 123
Resource Worksheet, p. 32
SourceBook, p. S29
Chapter 6 Review Worksheet, p. 46

Chapter 7
See Teacher's Edition margin, pp. 128, 129, 131, 132, 134, 137, and 138
Exploration Worksheet, p. 52
Resource Worksheet, p. 56
SourceBook, pp. S22 and S25

Unit 2
Unit 2 Activity Worksheet, p. 63
Activity Assessment Worksheet, p. 73
SourceBook Activity Worksheet, p. 77

Assessment Planning Guide

Lesson, Chapter, and Unit Assessment	SourceBook Assessment	Ongoing and Activity Assessment	Portfolio and Student-Centered Assessment
Lesson Assessment Follow-Up: see Teacher's Edition margin, pp. 80, 84, 91, 98, 102, 105, 110, 114, 118, 123, 129, and 135 **Chapter Assessment** Chapter 4 Review Worksheet, p. 5 Chapter 4 Assessment Worksheet, p. 8* Chapter 5 Review Worksheet, p. 26 Chapter 5 Assessment Worksheet, p. 29* Chapter 6 Review Worksheet, p. 46 Chapter 6 Assessment Worksheet, p. 50* Chapter 7 Review Worksheet, p. 58 Chapter 7 Assessment Worksheet, p. 61* **Unit Assessment** Unit 4 Review Worksheet, p. 64 End-of-Unit Assessment Worksheet, p. 67*	SourceBook Review Worksheet, p. 78 SourceBook Assessment Worksheet, p. 82*	Activity Assessment Worksheet, p. 73* **SnackDisc** Ongoing Assessment Checklists ♦ Teacher Evaluation Checklists ♦ Progress Reports ♦	Portfolio: see Teacher's Edition margin, pp. 79, 91, 114, 116, and 128 **SnackDisc** Self-Evaluation Checklists ♦ Peer Evaluation Checklists ♦ Group Evaluation Checklists ♦ Portfolio Evaluation Checklists ♦

* Also available on the Test Generator software
♦ Also available in the Assessment Checklists and Rubrics booklet

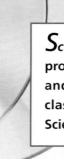

Science Discovery is a versatile videodisc program that provides a vast array of photos, graphics, motion sequences, and activities for you to introduce into your *SciencePlus* classroom. *Science Discovery* consists of two videodiscs: Science Sleuths and the Image and Activity Bank.

Using the *Science Discovery* Videodiscs

Science Sleuths: Fortune or Fraud?
Side B

In the year A.D. 872, King Euripide holds a magic contest in which Mookie the Magnificent appears to turn lead into gold. The greedy King begins buying all the lead in the kingdom. The Fool suspects that it was just a trick.

Interviews

1. Setting the scene: Court jester to the king **44654** **(play ×2)**

2. Magician **45297 (play)**

3. Daughter of the King **45977 (play)**

4. Metallurgist **46689 (play)**

5. Film of magic trick **47614** **(play)**

Documents

6. Correspondence from the King **48813 (step)**

7. Contest notice **48816**

8. Town crier scroll **48818**

9. Contents of magician's bag **48820**

Sleuth Information Service

10. Density of elements table **48822**

11. Conductivity table **48826**

12. Gold **48828 (step)**

13. Lead **48831 (step)**

14. Cold fusion **48834 (step)**

Sleuth Lab Tests

15. Mass of ball before and after trick **48837**

16. Measurement of ball before and after trick **48839 (step)**

17. Powder in vial analysis **48842**

18. Analysis of wand **48844**

19. Rabbit analysis **48846**

20. Cloth analysis **48848**

21. Analysis of foils **48850**

22. pH of liquid **48852**

23. Chemical analysis of precipitate **48854**

24. Conductivity at 20°C **48856**

Still Photographs

25. Ball before and after trick **48858 (step)**

26. Magic wand **48861**

27. Both sides of the cloth **48863 (step)**

28. Vial of powder **48866**

29. Rabbit **48868**

30. Foils **48870**

31. Glass beaker **48872**

32. Precipitate in beaker **48874**

Image and Activity Bank
Side A or B

A selection of still images, short videos, and activities is available for you to use as you teach this unit. For a larger selection and detailed instructions, see the Videodisc Resources booklet included with the Teaching Resources materials.

4-1 A Search for Explanations, page 79
Metric measurements 450
Metric prefixes and symbols and their values

4-2 Give Me the Facts, page 81
Observation and inference 472
Statements of observation and inference

◀| Step Reverse   Play ▶   Pause || Step Forward |▶

5-1 Building the Case, page 88

Salt dissolving; animation 47390–47730 (play ×2) (Side A only)
A cube of NaCl dissolves in water. Notice how the Na ion dissociates from the Cl ion.

5-2 The Hidden Structure of Matter, page 92

Periodic table 2353
A simple periodic table of the elements

Water molecule 2364
A water molecule is made up of one oxygen atom and two hydrogen atoms.

Summary of elements 2366
Properties of elements

Summary of compounds 2367
Properties of compounds

Common compounds 2368 (step)
Common compounds include water, sugar, table salt, chalk, nylon, and baking soda.

All matter 2370
Organizational chart categorizing matter

Elements and compounds 2371 (step)
Representation of separate elements (step) Elements combine to form compounds.

5-4 Particles of Solids, Liquids, and Gases, page 103

Expansion with heat; balloon 45957–46298 (play ×2) (Side A only)
Air expands when heated. Place a balloon over a bottle, and dip the bottle in hot water. The air in the bottle expands, as shown by the swelling balloon.

Changes of state 2354
Terms that describe changes of state

6-1 Temperature and Particles, page 108

State of matter 2355 (step ×2)
Water in its solid state (step) Heat changes water from its solid state to its liquid state. (step) Heat added to liquid water causes it to change to its gaseous state.

Rubber ball in liquid nitrogen 44966–45956 (play ×2) (Side A only)
Though a rubber ball is elastic at room temperature, cooling it with liquid nitrogen causes it to become rigid. Tapping the cooled ball with a hammer shatters it.

6-2 Changes of State, page 111

Melting and boiling points 2362
Table of melting and boiling points of some common substances

Which solid melted? 2359 (step ×2)
Three different solids are heated, and their temperatures are plotted against time. The first solid has a low melting point. (step) The second solid does not melt at all. (step) The third solid has a high melting point.

Mothballs 2320
Mothballs sublimate from a solid state directly into a vapor state.

6-3 Absorbing and Releasing Heat, page 115

Refrigerator 2341
A refrigerator uses electrical energy to pump heat from the inside to the outside. Coils behind or underneath the refrigerator release heat into the air.

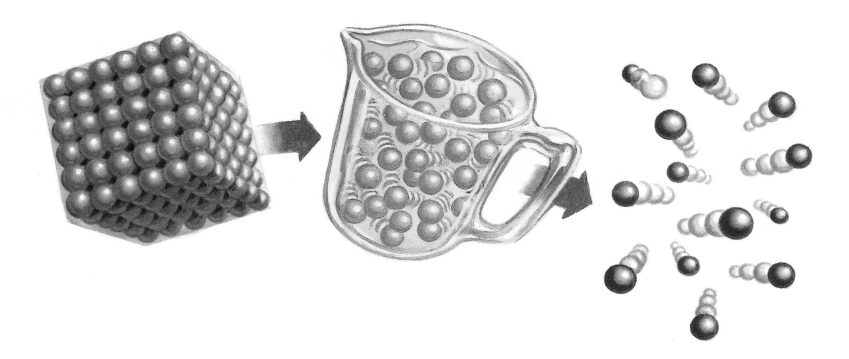

Unit 2 Particles

UNIT FOCUS

Use a discussion of a classroom object such as a window to introduce the concept of particles to students. If students do not have substantial prior knowledge on this subject, you may wish to provide them with the answers to the questions in the following sample discussion. Point to a window and ask: From what kind of matter is the window made? *(The window is made from glass.)* From what is the glass made? *(Silica)* From what is silica made? *(Sand that is weathered quartz, which is made of molecules of SiO_2)* Continue asking similar questions until students either identify *particles, atoms,* or *molecules* or can no longer respond. Point out that in this unit, students will examine evidence that matter is made of particles.

A good motivating activity is to let students listen to the English/Spanish Audiocassettes as an introduction to the unit. Also, begin the unit by giving Spanish-speaking students a copy of the Spanish Glossary from the Unit 2 Teaching Resources booklet.

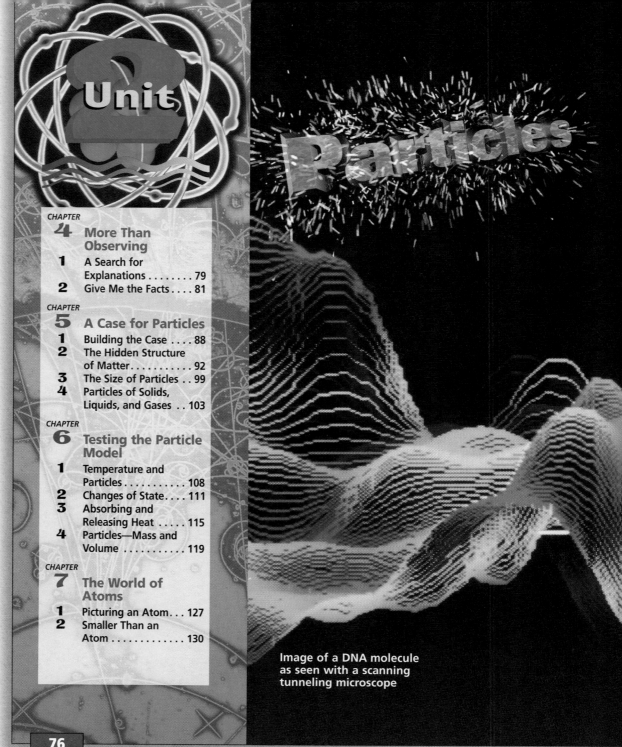

Unit 2

CHAPTER 4 — More Than Observing
1 A Search for Explanations 79
2 Give Me the Facts 81

CHAPTER 5 — A Case for Particles
1 Building the Case 88
2 The Hidden Structure of Matter 92
3 The Size of Particles . . 99
4 Particles of Solids, Liquids, and Gases . . 103

CHAPTER 6 — Testing the Particle Model
1 Temperature and Particles 108
2 Changes of State 111
3 Absorbing and Releasing Heat 115
4 Particles—Mass and Volume 119

CHAPTER 7 — The World of Atoms
1 Picturing an Atom . . . 127
2 Smaller Than an Atom 130

Image of a DNA molecule as seen with a scanning tunneling microscope

Connecting to Other Units

This table will help you integrate topics covered in this unit with topics covered in other units.

Unit 1 Life Processes	All living things are made up of particles and depend on other particles to carry out life processes.
Unit 4 Oceans and Climates	The water cycle involves particles of water in the solid, liquid, and gas states.
Unit 5 Electro-magnetic Systems	An electric current is the flow of negatively charged particles called electrons.
Unit 6 Sound	Sound is propagated by particles in a medium colliding with one another.
Unit 8 Continuity of Life	Hereditary information depends on the sequential arrangement of molecules in our genes.

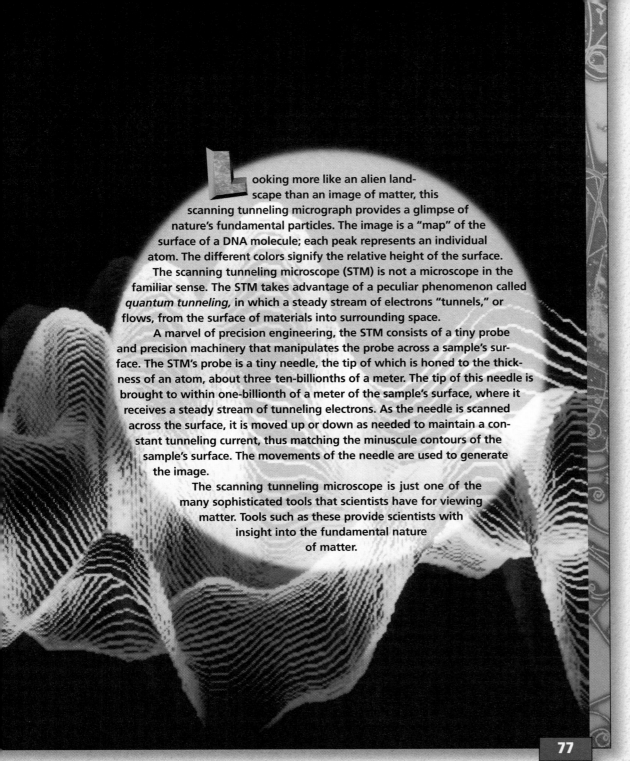

Looking more like an alien landscape than an image of matter, this scanning tunneling micrograph provides a glimpse of nature's fundamental particles. The image is a "map" of the surface of a DNA molecule; each peak represents an individual atom. The different colors signify the relative height of the surface. The scanning tunneling microscope (STM) is not a microscope in the familiar sense. The STM takes advantage of a peculiar phenomenon called *quantum tunneling,* in which a steady stream of electrons "tunnels," or flows, from the surface of materials into surrounding space.

A marvel of precision engineering, the STM consists of a tiny probe and precision machinery that manipulates the probe across a sample's surface. The STM's probe is a tiny needle, the tip of which is honed to the thickness of an atom, about three ten-billionths of a meter. The tip of this needle is brought to within one-billionth of a meter of the sample's surface, where it receives a steady stream of tunneling electrons. As the needle is scanned across the surface, it is moved up or down as needed to maintain a constant tunneling current, thus matching the minuscule contours of the sample's surface. The movements of the needle are used to generate the image.

The scanning tunneling microscope is just one of the many sophisticated tools that scientists have for viewing matter. Tools such as these provide scientists with insight into the fundamental nature of matter.

77

Call on a student to read the text on this page aloud. Explain to students that a computer is used to measure the movements of the needle across the sample. The computer then uses these measurements to generate the image on this page. Ask: Why is studying the structure of DNA particles useful to scientists? *(Accept all reasonable responses. Answers may include the advantages that can be gained in medical research, genetic engineering, anthropology, archeology, or military defense. Help students to realize that the benefits of understanding the structure of matter are endless.)*

Critical Thinking

Ask students to suggest some reasons why scientists develop models to study the physical world. *(Accept all reasonable responses.)* Explain to students that scientists often lack the technology to see the objects they wish to study. Even with modern advancements, objects may be either too small or too far away to see. A model provides scientists with an alternative way of "seeing" objects. A model must demonstrate all of the characteristics of the object it represents. If scientists find that their model does not meet this requirement, they build new models. When technology is finally able to provide scientists with a visual picture, scientists often find that their models are very accurate.

The scanning tunneling microscope was not developed until the 1980s, and scientists have been able to see DNA molecules only in recent years. However, James Watson and Francis Crick constructed a DNA model in 1953 that closely resembles the shape of the molecule in this photograph. Both men were awarded the 1962 Nobel Prize in medicine for their work.

As students proceed through this unit, they will discover other scientists who constructed models of particles centuries before anyone could see the actual particles. You may wish to look for other opportunities to reinforce the theme of models in other units.

Connecting to Other Chapters

> **Chapter 4**
> introduces scientific models and how exponents are used to describe large and small numbers.

> **Chapter 5**
> explores the evidence for an atomic theory of matter and discusses the states and structure of matter.

> **Chapter 6**
> investigates how temperature changes affect matter and explores the properties of matter.

> **Chapter 7**
> surveys the historical development of our understanding of the atom and the particles contained in its nucleus.

Prior Knowledge and Misconceptions

Your students' responses to the ScienceLog questions on this page will reveal the kind of information—and misinformation—they bring to this chapter. Use what you find out about your students' knowledge to choose which chapter concepts and activities to emphasize in your teaching. After students complete the material in this chapter, they will be asked to revise their answers based on what they have learned. Sample revised answers can be found on page 86.

In addition to having students answer the questions on the page, you may wish to have them complete the following activity: Ask students to write these measurements in exponential form: 100 cm, 1000 m, 10,000 m², 0.1 m. Then ask students to name items that could have these measurements. Collect their answers, but do not grade them. Instead, use the answers to identify possible problem areas and to find out what students know about exponents and the

1 **What observations might this scientist be making? To what inferences might these observations lead?**

2 **If you could see individual air particles, what might they look like?**

3 **Some people collect models of cars or trains. How do such models differ from scientific models?**

ScienceLog

Think about these questions for a moment, and answer them in your ScienceLog. When you've finished this chapter, you'll have the opportunity to revise your answers based on what you've learned.

78

size of units, what misconceptions students may have, and what aspects of this topic are interesting to them.

Integrating the Sciences

Earth and Physical Sciences

Point out to students that a diamond and the graphite in a pencil are made of the same material. Ask: What is this material? (Carbon) Why do you think the hardnesses of diamond and graphite are so different? (The carbon atoms are arranged differently in the two materials.)

CROSS-DISCIPLINARY FOCUS

Industrial Arts

In this chapter, students will explore the usefulness of creating physical models. Under the supervision of an industrial-arts teacher, you may wish to have students build a working example of a model that they will learn about or have learned about, such as a model of how the ear works (on page 83) or a model of a cell or cell membrane (in Unit 1). Have an industrial-arts teacher supervise their work, and make sure they have their plans approved by you for safety before they begin construction.

What Is the Smallest Object You Can See?

Angelina's class went to the planetarium. As they sat back in their seats and looked up at the domed ceiling, the lights went out. Stars appeared on the dark ceiling. Angelina felt as if she were in outer space. Then a voice began to speak:

"Imagine that you have journeyed far, far away to another galaxy. You are now on your way home to Earth. You are approaching our galaxy, the Milky Way. The entire Milky Way appears as a mere point of light in space because of the vast distance separating the two galaxies. As you come closer, the Milky Way begins to resolve into billions of stars, one of which is our sun. Move even closer and the planets can be seen shining in the reflected light of the sun. Come closer still and Earth is seen as a sphere of land, ice, and water. When you land on Earth, you find yourself surrounded by a myriad of objects of every shape, size, and color imaginable. With a sharp eye, you can detect tiny particles of dust in the air of a sunlit room. And with the aid of optical instruments, you can view even smaller particles."

What is the limit for detecting smaller and smaller bits of matter? Is there no limit to the size of objects that can be detected? The answers to these questions are important steps in the search for explanations about why matter behaves the way it does. This unit will lead you on a search for these explanations.

79

LESSON 1 ORGANIZER

Time Required
1 class period

Process Skills
inferring, analyzing, comparing, contrasting, measuring

New Term
Exponent—a kind of mathematical shorthand; a number written above and to the right of a base number, denoting the number of times that the base number is used as a factor

Materials (per student group)
none

Teaching Resources
Transparency 10

FOCUS

Getting Started
Ask students to identify as many "size" words as they can. Keep a list of their suggestions on the chalkboard. (*Examples include tiny, little, small, minuscule, minute, petite, large, big, gigantic, huge, immense, and enormous.*) Then ask students to classify the words under the headings Large and Small. Explain that choosing which "size" words to use when comparing or describing objects is important for accuracy. Point out that in this lesson they will explore the concept of size and discover a new way of describing it.

Main Idea
Very large and very small numbers can be expressed in exponential form.

TEACHING STRATEGIES

What Is the Smallest Object You Can See?
Direct students to the images on the right side of the page. Point out that some objects appear small because they are small, and some appear small only because they are far away. Ask: Which appears smaller, the planet Saturn or the Earth? (*Saturn*) Which object is really smaller? (*Earth*) Which of the objects in the imaginary journey seemed small because they were far away and which seemed small because they really are small? (*The Milky Way, stars, and planets are large but often seem small because they are far away. Dust and other smaller particles appear very small to the naked eye because they are very minute.*)

PORTFOLIO
Have students write their own account of an imaginary trip for their Portfolio. Ask students to incorporate concepts about life processes from Unit 1 in their account.

The Sizes of Things

Have students read the paragraph at the top of this page. Then review how to use exponents to express numbers by demonstrating the procedure on the chalkboard with examples from this page. You may wish to have volunteers work a few examples on the board as well. Then help students to understand the following:

- 10 raised to a positive exponent indicates how many times the number 1 is multiplied by 10.
- 10 raised to a negative exponent indicates how many times the number 1 is divided by 10.

Answers to
In-Text Questions

(A) When converting a number to its exponential form, students should count the number of places that the decimal point must be moved to get to the right of the number 1. For very large numbers, the decimal point must be moved to the left, and the exponent should be positive. For very small numbers, the decimal point must be moved to the right, and the exponent should be negative.

(B) • The width of a pencil is less than 10^{-2} m.
- The diameter of a hair is about 10^{-4} m.
- Distances from home to school will vary, but they will generally be between 10^2 m and 10^5 m.
- The diameter of Venus's orbit is about 10^{11} m.

(C) Some suggestions include the distance between two states for 10^5 m and the size of a microorganism for 10^{-5} m.

 Transparency 10 is available to accompany The Sizes of Things.

FOLLOW-UP

Reteaching

Ask each student to use the chart on this page to develop 10 trivia questions about the sizes of various objects. For example, how many powers of 10 larger is the diameter of the solar system than the diameter of Jupiter's orbit? *(One)* Have students quiz one another with their questions.

The Sizes of Things

Some objects and distances are very large. The diameter of the solar system, for example, is about 10,000,000,000,000 m. Other objects and distances are very small. The size of one cell in your body is about 0.000001 m. But it can be very awkward to work with such large or small numbers. One useful way to refer to the sizes of very large and very small things is to use **exponents**. Exponents are a kind of shorthand. Count the number of zeros in the figure that shows the diameter of the solar system. When it is expressed using an exponent, the figure becomes 10^{13} m. The size of a body cell expressed with an exponent is 10^{-6} m.

Now examine the chart below. Explain the rule for using exponents to express both very large and very small numbers. **(A)**

The list at right gives more examples of the sizes of different objects, using exponents. The measurements are approximate, not exact.

Where would you place the following in the list? **(B)**

- width of a pencil
- diameter of a hair
- distance from your home to school
- diameter of Venus's orbit

Suggest other objects that could be added to the list. **(C)**

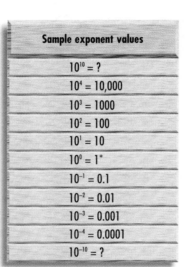

Sample exponent values

$10^{10} = ?$
$10^4 = 10,000$
$10^3 = 1000$
$10^2 = 100$
$10^1 = 10$
$10^0 = 1^*$
$10^{-1} = 0.1$
$10^{-2} = 0.01$
$10^{-3} = 0.001$
$10^{-4} = 0.0001$
$10^{-10} = ?$

* Any number to the power of 0 is always 1.

Meters	
10^{26}	Diameter of the known universe
10^{25}	
10^{24}	
10^{23}	
10^{22}	
10^{21}	Diameter of the Milky Way
10^{13}	Diameter of the solar system
10^{12}	Diameter of Jupiter's orbit
10^{11}	Diameter of Earth's orbit
10^{10}	
10^9	Diameter of the moon's orbit
10^8	
10^7	Diameter of the Earth
10^6	Distance across the continental United States
10^5	
10^4	Deepest part of the Pacific Ocean
10^3	
10^2	Height of a 25-story building
10^1	Length of your classroom
10^{0*}	Height of a young child
10^{-1}	Width of a hand
10^{-2}	Width of a finger
10^{-3}	Diameter of a thread
10^{-4}	Diameter of a fine sand particle
10^{-5}	
10^{-6}	Diameter of a body cell
10^{-7}	
10^{-8}	
10^{-9}	
10^{-10}	Diameter of an atom

Assessment

Have students measure the dimensions of a few objects in the classroom and express the numbers in exponential form. Examples of possible measurements include the size of the room or a desk, the height of the ceiling or a door, and the diameter of a clock. Have students make a chart that lists the objects and their sizes in order.

Extension

Point out that any number can have an exponent. For example, 4^3 is 4 multiplied by itself 3 times, or 64. Challenge students to use a number other than 10 to express a measurement in exponential notation. For example, the number of seconds in an hour can be expressed as 60^2. Have volunteers share their results with the class.

Closure

Have students write a story about how the world would look to them if they were to shrink by a power of 10. *(Accept all reasonable responses.)*

Give Me the Facts

A Question of Proof

You have probably heard people say, "I'll believe it when I see it!" With average vision you can see objects as small as 10^{-4} m. With a microscope, objects as small as 10^{-7} m can be detected. What are the facts about even smaller objects? You cannot see them, so how do you know they exist?

You must rely on *circumstantial evidence*. In scientific terms, that means you must make observations and use those observations (facts) to make inferences.

Observation vs. Inference

How does an inference differ from an observation? The article below offers some clues. While reading the story, try to find at least three inferences made by the defense lawyer and the prosecuting attorney. What facts (observational evidence) support each inference? Are all the inferences true? In your ScienceLog, record your answers in a table similar to the one shown at right.

The AIDS virus (orange dots) is approximately 1.2×10^{-7} m across.

Inference	Supporting observation (evidence)
1.	
2.	
3.	

Jury Still Out

Phoenix, AZ

After a full day of deliberation, the jury has yet to reach a verdict in the trial of Ike Swipe, who is accused of robbing the corner store. Yesterday, District Attorney Ivana Burnham summed up the prosecution's case. She reminded the jury of the facts brought out by the prosecution:

• The accused party was seen in the area of the robbery.
• His blood type was found on the doorknob. He had a cut on his right hand.
• His fingerprint was found on the countertop.
• He was observed spending more than the usual amount of money at the horse races the next day.

• The defendant has a past record of robbery.

Burnham stated that since Swipe was in the area of the crime and has a past criminal record, he must have committed the crime. The blood type found on the doorknob also matches Swipe's, so Swipe cut himself while breaking the store's window. Finally, the prosecutor declared that the money the defendant spent at the races was the money taken during the robbery.

On the other hand, defense attorney Sibyl Quibbler claimed that many people were seen in the area of the robbery and that Swipe is no more

a suspect than any of them. The cut was the result of a dish-washing accident and had nothing to do with the broken window at the store. As for the blood stain on the doorknob, Swipe has a very common blood type. The fingerprint was probably left on the countertop when the defendant bought a paper at the store a few hours earlier. "The evidence is entirely circumstantial," Quibbler noted. Judge Hugo Furst reminded the jury that in order to reach a verdict of guilty, they must find that the evidence shows—beyond a reasonable doubt—that the accused committed the crime.

LESSON 2 ORGANIZER

Time Required
1 to 2 class periods

Process Skills
comparing, contrasting, inferring

New Terms
none

Materials (per student group)
Exploration 1: balloon; 5 small plastic beads

Teaching Resources
Resource Worksheet, p. 3
Exploration Worksheet, p. 4
Transparency 11

Give Me the Facts

FOCUS

Getting Started

Present students with the following scenario: You are walking along a street. Tree branches are scattered on the ground. The roofs of several houses have been damaged. Ask:

• What can you infer about what has happened? (*There was a wind storm.*)
• What are the facts? (*Tree branches are scattered on the ground. Roofs have been damaged.*)
• What are these facts based on? (*Observations*)

Point out that in this lesson students will learn how to analyze a situation in order to discover the difference between fact and inference.

Main Ideas

1. Facts are based on observations.
2. Inferences are based on circumstantial evidence.
3. Models are representations of events or objects in the real world and can be used to test inferences.

TEACHING STRATEGIES

A Question of Proof

The lesson does not begin with the assumption that matter consists of particles—atoms and molecules. Instead, it poses the question that if there are particles that are too small to be seen, how do we know they exist? Although students will want to accept the particle nature of matter as fact, they are challenged to suggest proof for it. In doing so, they learn that often the proof is based on circumstantial evidence.

Throughout this lesson, the terms *facts* and *observations* are used interchangeably. Explanations of the observations are called *inferences*.

Answers to Observation vs. Inference are on the next page. ▶

Answers to
Observation vs. Inference,
page 81

- An inference is a deduction that is drawn from an observation.
- For an example of a completed table, see page S206.
- Help students conclude that facts are based on observations. An inference is a conclusion, drawn from observations, that may or may not be true.

Cooperative Learning
MAX READS FROM HIS SCIENCELOG

Group size: 2 to 3 students
Group goal: to discuss differences between observations and inferences
Positive interdependence: Assign the following roles: reader (reads text material); recorder (writes answers to questions after the group has discussed them); presenter (presents chart to the rest of the class). Each group reads and discusses the questions listed in Max Reads From His ScienceLog and then comes to a consensus concerning the answers to the questions.
Individual accountability: Each student should be prepared to take part in the class. Each student should also write the answers to the questions in Using Models on page 83.

Answers to
Max Reads From His ScienceLog

Some students may agree with Max's opinion for the same reasons that he states. Others may feel that the circumstantial evidence is too strong and that the man is guilty. Max makes a good point in that no one saw the defendant commit the crime; therefore, all of the evidence is circumstantial. An individual must then weigh the amount of circumstantial evidence gathered in order to make a decision about what happened. All inferences are based on circumstantial evidence.

Scientists also weigh direct and circumstantial evidence to determine whether a statement is true. However, scientists are also able to duplicate experimental results to further support their inferences.

★ **Transparency 11 is available to accompany Observation or Inference?**

Max Reads From His ScienceLog

After reading the article titled "Jury Still Out," I think the jury should give a verdict of "not guilty." No one saw the accused at the scene of the crime, and there is no real proof that he did it. I just don't think the circumstantial evidence is good enough to let anyone make an inference that will send that man to prison.

Do you agree or disagree with Max's opinion? Explain why. Are all inferences based on circumstantial evidence? Describe the similarities and differences between the inferences that scientists make and those that the lawyers and Max made.

Observation or Inference?

You live in an ocean of air, but
(A) you cannot see it. Why not? The answer is based on circumstantial evidence. Such an explanation depends on knowledge about the unseen structure of matter. For instance, it is easy for a scientist to observe how liquids or solids behave, but it is more difficult to explain *why* they behave as they do. A scientist must draw inferences based on observations. In this sense, a scientist does exactly what a jury does!

Copy the chart below into your ScienceLog. Working in small groups, examine the statements about air. Do you agree with each one? Place a check mark next to the statements you agree with. Are your decisions based on observations, or are you making inferences? What evidence supports each decision? Discuss your decisions with your classmates. As a group, do you agree or disagree with each statement?

Statements about air	Observation	Inference	Evidence supporting your decision
1. Air can be squeezed into a smaller space.	✔		I can pump air into my bicycle tire.
2. Air is invisible.			
3. Air has volume.			
4. Air has mass.			
5. Air moves.			
6. If we could see air, we would see many particles.			
7. Air behaves like a sponge.			
8. There is water in air.			
9. Create your own statement about air.			

Answer to
In-Text Question

(A) Particles of air are too small to be seen.

Answers to
Observation or Inference?

This exercise may cause considerable debate among students. The evidence that students supply to support their decisions will vary. The following are sample answers:
1. Observation; I can pump air into my bicycle tire.
2. Observation; we cannot see air.
3. Observation; when a glass is held upside-down and pushed into water, water does not go into the glass.
4. Observation; the mass of a balloon full of air is greater than the mass of an empty balloon.
5. Observation; wind is moving air.
6. Inference; air can be compressed, air has volume, and air has mass. Therefore, air must be made up of particles.
7. Inference; particles of air can be compressed like a sponge.
8. Inference; clouds form in air.
9. Answers will vary.

Using Models

Based on your observations and inferences, you can come to some conclusions about air. Do you think air is made of particles, as suggested in statement 6 on page 82? Or is air more like a sponge that can be compressed and expanded, as suggested in statement 7?

These kinds of questions help you develop a mental picture or idea about the structure of air. They help you form a *model* for the concept you are thinking about. So far, you have considered two types of models. In statement 6, the model is an idea: air is made up of particles. You cannot actually see air—you can only observe how it behaves. Therefore, you form an idea about what air is really like based on your observations of its behavior. In statement 7, you compare air to an actual object—a sponge. In what ways is air like a sponge? In what ways is it different? Do you think the particle model is better or worse than the sponge model? Why? How can the models help you understand air and its behavior? **B**

More About Models

Francine created the model below to represent the workings of the inner ear and how sounds are transmitted across the eardrum. Find a diagram of the inner ear in a reference book, and then figure out what each feature of the model represents. Do you find that Francine's model helps explain how the ear works? Do you think it is a good model? Why or why not?

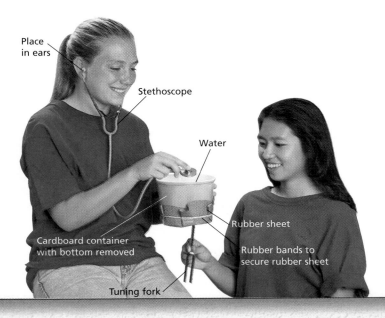

Place in ears

Stethoscope

Water

Cardboard container with bottom removed

Rubber sheet

Rubber bands to secure rubber sheet

Tuning fork

83

Answers to
In-Text Questions

B Most students will recognize that air is like a sponge because it can be compressed and expanded and because it can absorb and hold moisture. Air is unlike a sponge in that a sponge can be seen but air cannot be seen; air does not bounce like a sponge does; and air is evenly distributed, while a sponge is full of irregular holes held together by matter. Encourage students to debate whether air is described better by the sponge or the particle model. Challenge them to support their opinions with the facts they know about air. Do not be surprised at this point if some students prefer the sponge model. Students should recognize that some models can be tested to see if their behavior is the same as the phenomena being modeled.

Answers to
More About Models

The tuning fork represents the bones of the middle ear; the rubber sheet represents the oval window; water in the ice cream container represents fluid in the cochlea; the tubes of the stethoscope represent the auditory nerve.

Most students will agree that the model helps them to understand how the ear works. Encourage students to suggest ways in which the model could be modified to more closely resemble the complex structure of the human ear.

Homework

The Resource Worksheet that accompanies Observation or Inference? on page 82 makes an excellent homework activity (Teaching Resources, page 3).

Cooperative Learning
More About Models

Group size: 3 to 4 students
Group goal: to explain how models can be used to test an inference
Positive interdependence: Assign the following roles: advertiser (to design the advertisement), materials handler (to gather the supplies to be used), discussion leader (to keep the discussion moving and all members on task), and designer (to work with the group to come up with a design for the models). Have group members work together to construct the model described in

Exploration 1 and to answer the in-text questions. The group should then also design a model to describe the particle theory of matter. After the model is complete, students should prepare an advertisement for the model to be entered in a class competition, with the class voting on the best model and advertisement. You may wish to give awards or prizes.
Individual accountability: Each student is responsible for answering the Challenge Your Thinking questions on page 85.

Caution students not to blow up the balloon too much or to rattle the beads excessively. Challenge students to make inferences about the behavior of real air based on their observations of the model.

Answers to
In-Text Questions

Ⓐ Students may not be able to see the model air particles in motion, but they should feel the model air particles just as they can feel real air but cannot see it. The model demonstrates that particles of air move about freely, occasionally bumping into one another as well as into other objects.

Ⓑ The words *representation* and *simulation* may be used to describe a model.
    Examples of other models include models of the solar system, the Biosphere II Project, and a paper airplane.

Answers to
Visualizing, Explaining, and Predicting

Student answers will vary. Encourage students to analyze the usefulness of Francine's and Ramón's models in light of the three bulleted sentences in this section. For instance, students may respond that a successful model should simplify a complex problem by including only the most important features of the problem. Point out that later in this unit, students will be introduced to a number of other models. Encourage them to apply the three tests in this section to these models as well.

Homework

The Exploration Worksheet that accompanies Exploration 1 makes an excellent homework activity (Teaching Resources, page 4).

Making Ramón's Model

Ramón agreed with statement 6 on page 82, so he devised the following model. Now it's your turn to build Ramón's model.

You Will Need

• a balloon
• 5 small plastic beads

What to Do

Place the plastic beads inside the balloon. Then blow up the balloon—but not too much! Tie the balloon closed. The balloon now contains not only real air but also the plastic beads, which represent air particles. Gently shake the balloon until the beads rattle around inside. Can you see the model air particles in motion? Can you feel them in motion? What observation about air does this model explain? Ⓐ

Visualizing, Explaining, and Predicting

Were the models made by Francine and Ramón useful? In your ScienceLog, describe how models help us to do the following things:

• Visualize a complex idea or structure.
• Explain observations and make inferences.
• Make predictions that can be tested through further observations and experiments.

    What word or phrase best describes a model? Suggest another example of a model. Ⓑ

84

FOLLOW-UP

Reteaching

Have students state some of the things they have learned in past science courses about the way the world works. *(For example, students could report that dinosaurs existed on Earth millions of years ago.)* Then ask students to name as many inferences and observations they can think of that led scientists to develop these findings. *(For example, scientists observed fossils and inferred the existence of dinosaurs.)*

Assessment

Have students work together to make one of the models they suggested in Visualizing, Explaining, and Predicting.

Extension

Challenge students to read articles from a newspaper's editorial page and to identify some facts and inferences in the articles.

Closure

Have students take notes during a program they watch on television. A detective or medical program would be best. Students should list inferences made by characters as the plot unfolds.

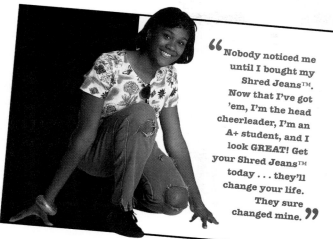

" Nobody noticed me until I bought my Shred Jeans™. Now that I've got 'em, I'm the head cheerleader, I'm an A+ student, and I look GREAT! Get your Shred Jeans™ today . . . they'll change your life. They sure changed mine. "

1. These Jeans Changed My Life!

Obviously, the advertisement above is making some pretty ridiculous claims. It gives several examples of how people draw incorrect inferences from available observations. With another student, make a list of five ridiculous claims that you can support without actually lying. Identify the circumstantial evidence you used to make your claim. Then explain how you can make better inferences based on your observations.

2. This Model Is All Wet

Design a model for water that illustrates the observations you have made about it. What happens when liquid water turns into ice? into steam? Can your model represent these changes? What observations does your model have trouble explaining? After you create your design, summarize its strengths and weaknesses as a model.

85

Theme Connection

Changes Over Time

Since ancient times, researchers have been investigating the structure of matter. Each new particle model improves on the work that came before. **Focus question:** Is our understanding of particles likely to change in the future? *(Even though scientists currently know a great deal about the structure of particles, discoveries in the future may cause researchers to modify their theories. Science is a field of constant discovery and rethinking, and any theory can be modified.)*

Did You Know...

A *googolplex* is the number 1 followed by 100 zeros, written 10^{100}.

Answers to *Challenge Your Thinking*

1. Answers will vary. The following is a sample answer:

Claim: Having plaid sheets on my bed makes me unable to sleep.

Circumstantial evidence: Last week I put plaid sheets on my bed, and I couldn't sleep.

To make a better inference about the cause of my inability to sleep last week, I should make a list of other things that were happening at the same time. For example, it was hot, we had house guests who played the stereo all night, and I had a big test in science class the next day. Perhaps I can try sleeping on the plaid sheets again when there are not so many things on my mind. I could also compare how well I sleep on plaid sheets with how well I sleep on white or pastel sheets.

2. Answers will vary. The following is a sample answer:

Marbles could be used as a model for water. When liquid water turns into ice, the ice is cold and remains solid until it melts. The turning of water into ice could be represented with the marble model by adding a cementing agent to hold the marbles together into a definite shape. When liquid water turns into steam, the steam expands into the surrounding volume of air and is very hot. It would be difficult to represent how liquid water turns into steam or how ice turns into liquid water with this marble model. The model also would have trouble explaining why the steam would expand. One additional weakness of the model is that it would not show the marbles moving around faster in a gas than they would in a liquid, which is what water molecules do. One strength of the model is that it can accurately represent how liquid water moves. The marbles can take the shape of their container and can freely move past one another when shaken or poured, much like the way water flows.

★ You may wish to provide students with the Chapter 4 Review Worksheet that is available to accompany this Challenge Your Thinking (Teaching Resources, page 5).

Answers to
Challenge Your Thinking,
continued

3. Accept all reasonable responses. Students may wish to choose a process that they studied in the last unit, such as the water cycle, photosynthesis, osmosis, or diffusion. They might then choose a model based on one of the previous experiments, such as Exploration 2 on page 39 in which they modeled the process of diffusion through root hairs.

4. Answers will vary. Possible responses are as follows:

Top left:
Observation: There are a number of American Indians gathered together in the photograph.
Inference: The people in the photograph are participating in a cultural celebration.

Top right:
Observation: A band of people on horseback are traveling near a desert canyon.
Inference: The weather is hot and dry.

Bottom:
Observation: The structure (the Leaning Tower of Pisa) is leaning to the right.
Inference: The foundation of the structure has shifted over time.

The following are sample revised answers:

1. The scientist seems to be observing the actions of the chimpanzee. She may be examining the type and speed of the animal's body movements, eye movements, sounds, breathing, and its temperament. From these observations, the scientist may be able to make inferences about the reactions of chimpanzees to different situations.

2. Air is made up of gases consisting of individual molecules, which in turn are made up of atoms that are chemically bonded together. Our conception of how these molecules would look is based on the model used to represent them. Some models, such as the ones shown in the photo, represent molecules and atoms in a greatly simplified manner, while others involve much more complex representations.

3. It's Like This, Kid . . .
Choose a scientific concept or process that you have studied either in this course or in a previous science course. Describe a model that could help a fifth-grade student better understand the concept or process.

4. Road Trip
Choose one of the pictures shown here. Make a list of as many observations as you can about the photograph. Then make as many inferences as you can about the photo, and write them on a separate list.

ScienceLog
Review your responses to the ScienceLog questions on page 78. Then revise your original ideas so that they reflect what you've learned.

3. One big difference between model cars or trains and scientific models is that scientific models are often used to represent ideas instead of concrete objects. Scientific models are also used to make inferences about the ideas they represent.

Meeting Individual Needs

Second-Language Learners
Have students look up the words *observation* and *inference* in the dictionary and write a sentence that explains the relationship between the two words. Students should then record three observations that they made during the course of a school day and three inferences that can be drawn from those observations.

A Case for Particles

1 **If matter is made up of particles, what's in between the particles?**

2 **Why do scientists visualize matter as being made of particles?**

3 **Where does liquid water go when it boils away? How do you know?**

ScienceLog

Think about these questions for a moment, and answer them in your ScienceLog. When you've finished this chapter, you'll have the opportunity to revise your answers based on what you've learned.

87

CHAPTER

5

A Case for Particles

Connecting to Other Chapters

Chapter 4
introduces scientific models and how exponents are used to describe large and small numbers.

Chapter 5
explores the evidence for an atomic theory of matter and discusses the states and structure of matter.

Chapter 6
investigates how temperature changes affect matter and explores the properties of matter.

Chapter 7
surveys the historical development of our understanding of the atom and the particles contained in its nucleus.

Homework

Ask each student to make three observations at home and state what he or she infers from these observations. For example, a student could observe an empty food dish and infer that the dog has eaten its dinner. Or a student might observe the good mood of a sibling and infer that he or she got an A on a test in school that day.

Prior Knowledge and Misconceptions

Your students' responses to the ScienceLog questions on this page will reveal the kind of information—and misinformation—they bring to this chapter. Use what you find out about your students' knowledge to choose which chapter concepts and activities to emphasize in your teaching. After students complete the material in this chapter, they will be asked to revise their answers based on what they have learned. Sample revised answers can be found on page 106.

In addition to having students answer the questions on this page, you may wish to have them complete the following ac-

tivity: Have students describe what would happen if they could break a rock up into gravel, grind the gravel into a powder, and then grind the powder into even smaller pieces. Ask: What would you eventually end up with if you could break the rock up into smaller and smaller pieces?

Assure students that there are no right or wrong answers to this exercise. Collect their answers, but do not grade them. Instead, use the answers to identify possible problem areas and to find out what students know about the particle nature of matter, what misconceptions students may have, and what aspects of this topic are interesting to them.

Building the Case

FOCUS

Getting Started

Have students discuss what the diagram on page 88 shows. *(It shows the relationship between observations and inferences. Observations based on measuring and information from the senses raise questions and lead to inferences, explanations, and models that can be tested.)* Point out that in this lesson students will examine some of the evidence that matter is made of particles.

Main Ideas

1. A particle model of matter can be used to explain observations of dissolving, pouring, and mixing substances.
2. A case favoring the particle nature of matter can be made based on observation and inference.

TEACHING STRATEGIES

EXPLORATION 1

PART 1

Ask students if they have ever noticed that ice cubes seem to shrink if they are left in the freezer for too long. Have them discuss what might have happened to the ice. *(Accept all reasonable responses. Some students may know that the ice has undergone sublimation. Sublimation is the process by which a substance changes directly from a solid to a gas without first becoming a liquid. Point out that they will study such changes of state in Chapter 6.)* Ask students the following questions:

• What is the gaseous state of water called? *(Water vapor)*
• Can you see water vapor? *(No)*
• Is water vapor a part of the air? *(Yes)*
• How do you know? *(Water often condenses out of the air, such as when moisture forms on the outside of a soft-drink can.)*

Building the Case

You are the judge, jury, and attorney in a landmark case—a case that will determine whether all matter is composed of particles. This case may raise as many questions about matter and its behavior as it answers. The following experiments will provide the observations and information you will need to make some important inferences as you prepare your case in favor of the particle theory of matter—or against it.

Observations
Measuring
Hearing
Smelling
Seeing
Feeling

raise

Questions
Why?
How?
What if?

that can lead to

Inferences
Explanations
Models

88

EXPLORATION 1

Making the Case

You Will Need

• red food coloring
• an eyedropper
• a stirring rod
• a 100 mL beaker
• 500 mL of sand
• water
• 500 mL of dried peas or beans
• 40 mL of rubbing alcohol
• 4 large containers
• 25 mL of salt
• 2 graduated cylinders
• a funnel
• a stopwatch or clock

PART 1

A Thought Experiment

What to Do

Read the observation and inferences about liquid and frozen water. Then answer the questions that follow.

Observation

In the freezer, ice cubes become smaller over time.

LESSON 1 ORGANIZER

Time Required
1 to 2 class periods

Process Skills
observing, analyzing, inferring

Theme Connection
Structures

New Terms
none

Materials (per student group)
Exploration 1, Part 2: drop of red food coloring; stirring rod; 500 mL of water; 100 mL graduated cylinder; 100 mL beaker; lab aprons; **Part 3:**

4 large containers; about 500 mL of sand; about 500 mL of water; about 500 mL of dried peas or beans; **Part 4:** 50 mL of sand; 25 mL of salt; two 100 mL graduated cylinders; 200 mL of water; watch or clock with second hand; 40 mL of rubbing alcohol; stirring rod; funnel; safety goggles

Teaching Resources
Exploration Worksheet, p. 11
Discrepant Event Worksheet, p. 16

CHAPTER

5

A Case for Particles

1 **If matter is made up of particles, what's in between the particles?**

2 **Why do scientists visualize matter as being made of particles?**

3 **Where does liquid water go when it boils away? How do you know?**

ScienceLog

Think about these questions for a moment, and answer them in your ScienceLog. When you've finished this chapter, you'll have the opportunity to revise your answers based on what you've learned.

Connecting to Other Chapters

Chapter 4
introduces scientific models and how exponents are used to describe large and small numbers.

Chapter 5
explores the evidence for an atomic theory of matter and discusses the states and structure of matter.

Chapter 6
investigates how temperature changes affect matter and explores the properties of matter.

Chapter 7
surveys the historical development of our understanding of the atom and the particles contained in its nucleus.

Homework

Ask each student to make three observations at home and state what he or she infers from these observations. For example, a student could observe an empty food dish and infer that the dog has eaten its dinner. Or a student might observe the good mood of a sibling and infer that he or she got an A on a test in school that day.

87

Prior Knowledge and Misconceptions

Your students' responses to the ScienceLog questions on this page will reveal the kind of information—and misinformation—they bring to this chapter. Use what you find out about your students' knowledge to choose which chapter concepts and activities to emphasize in your teaching. After students complete the material in this chapter, they will be asked to revise their answers based on what they have learned. Sample revised answers can be found on page 106.

In addition to having students answer the questions on this page, you may wish to have them complete the following activity: Have students describe what would happen if they could break a rock up into gravel, grind the gravel into a powder, and then grind the powder into even smaller pieces. Ask: What would you eventually end up with if you could break the rock up into smaller and smaller pieces?

Assure students that there are no right or wrong answers to this exercise. Collect their answers, but do not grade them. Instead, use the answers to identify possible problem areas and to find out what students know about the particle nature of matter, what misconceptions students may have, and what aspects of this topic are interesting to them.

FOCUS

Getting Started

Have students discuss what the diagram on page 88 shows. *(It shows the relationship between observations and inferences. Observations based on measuring and information from the senses raise questions and lead to inferences, explanations, and models that can be tested.)* Point out that in this lesson students will examine some of the evidence that matter is made of particles.

Main Ideas

1. A particle model of matter can be used to explain observations of dissolving, pouring, and mixing substances.
2. A case favoring the particle nature of matter can be made based on observation and inference.

TEACHING STRATEGIES

EXPLORATION 1

PART 1

Ask students if they have ever noticed that ice cubes seem to shrink if they are left in the freezer for too long. Have them discuss what might have happened to the ice. *(Accept all reasonable responses. Some students may know that the ice has undergone sublimation. Sublimation is the process by which a substance changes directly from a solid to a gas without first becoming a liquid. Point out that they will study such changes of state in Chapter 6.)* Ask students the following questions:
- What is the gaseous state of water called? *(Water vapor)*
- Can you see water vapor? *(No)*
- Is water vapor a part of the air? *(Yes)*
- How do you know? *(Water often condenses out of the air, such as when moisture forms on the outside of a soft-drink can.)*

Building the Case

You are the judge, jury, and attorney in a landmark case—a case that will determine whether all matter is composed of particles. This case may raise as many questions about matter and its behavior as it answers. The following experiments will provide the observations and information you will need to make some important inferences as you prepare your case in favor of the particle theory of matter—or against it.

Observations
Measuring
Hearing
Smelling
Seeing
Feeling

raise

Questions
Why?
How?
What if?

that can lead to

Inferences
Explanations
Models

EXPLORATION 1

Making the Case

You Will Need
- red food coloring
- an eyedropper
- a stirring rod
- a 100 mL beaker
- 500 mL of sand
- water
- 500 mL of dried peas or beans
- 40 mL of rubbing alcohol
- 4 large containers
- 25 mL of salt
- 2 graduated cylinders
- a funnel
- a stopwatch or clock

PART 1

A Thought Experiment

What to Do

Read the observation and inferences about liquid and frozen water. Then answer the questions that follow.

Observation

In the freezer, ice cubes become smaller over time.

LESSON 1 ORGANIZER

Time Required
1 to 2 class periods

Process Skills
observing, analyzing, inferring

Theme Connection
Structures

New Terms
none

Materials (per student group)
Exploration 1, Part 2: drop of red food coloring; stirring rod; 500 mL of water; 100 mL graduated cylinder; 100 mL beaker; lab aprons; **Part 3:**

4 large containers; about 500 mL of sand; about 500 mL of water; about 500 mL of dried peas or beans; **Part 4:** 50 mL of sand; 25 mL of salt; two 100 mL graduated cylinders; 200 mL of water; watch or clock with second hand; 40 mL of rubbing alcohol; stirring rod; funnel; safety goggles

Teaching Resources
Exploration Worksheet, p. 11
Discrepant Event Worksheet, p. 16

Questions

- Where does the ice go?
- How does it disappear?
- Can ice be prevented from disappearing?

Inference and Possible Explanation

- Perhaps ice (water) is made up of particles.
- Maybe some of these particles escaped from the solid state to form a gas, which floated away.

Follow-Up

1. Name another substance that changes directly from a solid into a gas.

2. Could a gas change directly into a solid? If so, think of some examples.

3. Do these observations and explanations support the idea that water is made up of particles? Why or why not?

PART 2

Seeing Red

What is the largest amount of water in which you could dissolve a drop of red food coloring and still detect its color? Here is a way to find out.

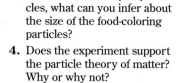

What to Do

Thoroughly dissolve a drop of food coloring in 50 mL of water. Now divide this solution into two equal parts. Wash 25 mL down the sink, and add 25 mL of water to what remains. Once again the total volume of the solution is 50 mL. Is the solution still colored red?

The concentration of the food coloring has been diluted to one-half of the original amount. Repeat the dilution process once more. Can you still see the red coloring in the water? Your beaker now contains one-quarter of the original drop of food coloring. Repeat the procedure—keeping accurate records—until you no longer see the red color.

Before going on to Part 3, discuss the following questions with a partner:

1. Is the color spread evenly throughout the solution, or are bits of food coloring clumped together?

2. Do you think there may be some food coloring left in the solution at the end, even though you cannot see any? How much of the food coloring do you have in the beaker of water at the end of the experiment? How do you know?

3. If matter is made up of particles, what can you infer about the size of the food-coloring particles?

4. Does the experiment support the particle theory of matter? Why or why not?

PART 3

Pour Judgment

What to Do

Fill three large containers with the substances listed below. Do not mix the substances.

- dried peas or beans
- sand
- water

Now pour each substance into an empty container. Did either of the first two substances resemble water in the way they poured? What might you infer about matter from this experiment?

Dried peas

Water

Sand

Exploration 1 continued ▶

 An Exploration Worksheet is available to accompany Exploration 1 (Teaching Resources, page 11). In addition, a teacher demonstration that can be used once students have finished Exploration 1 is described in a Discrepant Event Worksheet (Teaching Resources, page 16).

Answers to Questions

Students will probably not be able to provide answers at this point, although you may wish to discuss the answers with the class.

- The ice evaporates into the air in the freezer.
- The ice changes into water vapor through the process of *sublimation*. Sublimation occurs when a material changes directly from a solid to a gas.
- Ice can be prevented from disappearing by storing it under high pressure.

Answers to Follow-Up

1. Substances may include iodine crystals, solid room deodorizers, mothballs, dry ice, or camphor.

2. Yes, but rarely. Iodine vapor can change directly into iodine crystals, and water vapor changes directly into frost when the temperature is at or below the freezing point.

3. Yes; if part of a solid such as ice can disappear, then the solid may be made of smaller particles that can separate from each other and spread out.

Answers to Part 3

Students will probably agree that the sand and peas or beans all pour like water, but that the sand resembles water the most. They may conclude that the reason the sand pours more like water is because it is made up of very small particles. From this observation, students may infer that water is composed of particles just as sand is, but much, much smaller.

Answers to Part 2

The solution is still colored red after the first two dilutions.

1. The color is spread evenly throughout the water.

2. Students should conclude that there will still be some food coloring left in the solution. The amount of food coloring in the beaker at the end of the experiment will depend on how many times the water has been diluted. After the first dilution, one-half of the food coloring is left. After the second dilution, one-quarter of the food coloring is left. A third dilution will leave one-eighth of the food coloring.

3. Since the food coloring spreads evenly throughout the water, students should infer that the molecules of food coloring are smaller than the spaces between the water particles.

4. Because the food coloring spreads throughout the water and can be divided into smaller and smaller quantities, the experiment supports the particle theory of matter.

In preparation for Part 4, remind
students that alcohol is highly
flammable and should not be handled
around hot plates or flames.

Answers to
Part 4

1. The combined volume is not 100 mL
because water has filled in the
spaces between the particles of
sand.

2. The volume for the salt and water
mixture will be less than 100 mL
because the salt dissolved in the
water. That is, salt filled the spaces
between the water particles.

3. The combination of alcohol and
water is less than 90 mL because the
particles of the liquids are smaller
than the spaces between the particles. Therefore, some particles of
alcohol fill the spaces between the
water particles and vice versa.

Answers to
Drawing Conclusions

Students should recognize that their
observations support the idea that
matter consists of particles. Part 1
shows that parts of a solid (without
obvious particles) can disappear. Part 2
shows that a liquid can spread out
evenly in another liquid and can be
divided into smaller parts. Part 3 shows
that substances that are obviously
made of particles behave in a manner
similar to a substance that is not obviously made of particles. Part 4 shows
that when two substances are mixed
together, their combined volume may
be less than the sum of their separate
volumes, showing that the particles
of one substance can fill the spaces
between the particles of the other
substance.

Theme Connection

Structures
Focus question: In what way is a container of sand similar to a sponge?
*(Sand, like a sponge, contains gaps or
pores where fluids can accumulate. The
pores will hold water until they reach
their capacity. Excess fluid becomes
runoff.)*

PART 4

When $1 + 1 \neq 2$

What to Do

Carry out the following three activities. After making careful observations, use them to develop inferences about the unseen structure of
matter.

1. Pour 50 mL of sand into a 100 mL graduated cylinder. Then pour
50 mL of water into another 100 mL graduated cylinder. Carefully
pour the water into the sand. Record the volume of the mixture.
Suggest an explanation for why the combined volume is not 100 mL.

50 mL of water   50 mL of sand   25 mL of salt   50 mL of water   40 mL of alcohol

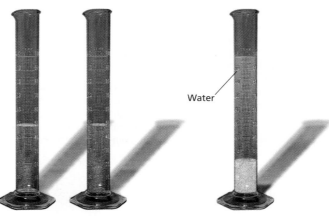

Water

2. Put 25 mL of salt into a graduated cylinder. Add enough water to
bring the combined volume of salt and water to 100 mL. Without
spilling the contents, gently shake the cylinder for a minute or two.
Record the volume after shaking. How do you explain the final volume of salt and water?

To read volume, locate the
curve at the top of the liquid.
Read at eye level the lowest
point of the curve.

3. Pour 50 mL of water into a graduated cylinder. Then pour 40 mL of
alcohol into a second cylinder. Pour the alcohol into the water and
stir. Is the volume of the two combined liquids 90 mL? Explain.

Drawing Conclusions

Do your observations support the idea that matter consists of particles?
Why or why not? In your ScienceLog, summarize your case for or
against the particle theory of matter.

Homework

The writing activity described in Your
Case on page 91 makes an excellent
homework assignment.

Particles on Trial

A ninth-grade class was asked to build a case for or against the particle model. Here are a few of their replies. Imagine you are the teacher—what comments would you write on each report? Has each student built a good case for the existence of particles? How could they improve their case?

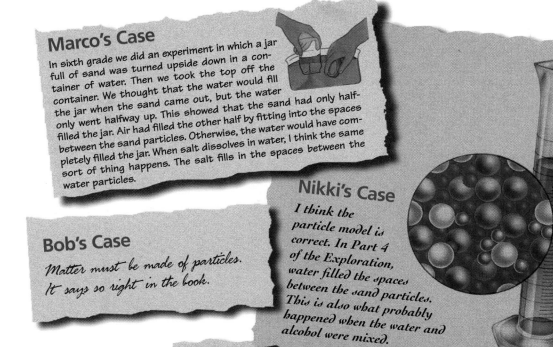

Marco's Case

In sixth grade we did an experiment in which a jar full of sand was turned upside down in a container of water. Then we took the top off the container. We thought that the water would fill the jar when the sand came out, but the water only went halfway up. This showed that the sand had only half-filled the jar. Air had filled the other half by fitting into the spaces between the sand particles. Otherwise, the water would have completely filled the jar. When salt dissolves in water, I think the same sort of thing happens. The salt fills in the spaces between the water particles.

Bob's Case

Matter must be made of particles. It says so right in the book.

Nikki's Case

I think the particle model is correct. In Part 4 of the Exploration, water filled the spaces between the sand particles. This is also what probably happened when the water and alcohol were mixed.

Your Case

In a few paragraphs, revise your own case for or against the particle model. Share it with a few classmates, and ask for their opinions. You should use your summary and evidence from the Exploration, as well as any other evidence you have observed. Now test the strength of your case. Does it explain why you cannot see air? Does it explain why you see sugar when it is in the sugar bowl, but not when it is dissolved in water? How does your case for or against the existence of particles answer these questions?

Answers to
Particles on Trial

Marco's Case
Marco's case is a good model for what happens when salt mixes with water, but it does not demonstrate that water is made up of particles. It demonstrates only that air fills the spaces around particles of sand and that the air has volume. Marco could have made a better case by using the activities in Exploration 1 as examples.

Nikki's Case
Nikki's explanation is a good one. She could have cited the results of the other three parts of Exploration 1 as further evidence supporting her inference.

Bob's Case
Bob has no case at all. Even if it says so in a book, that does not make it true. Bob could have used any or all of the activities in Exploration 1 to make the point that matter is made up of particles.

Your Case
Student answers should demonstrate an understanding of the behavior of matter that indicates it is made of particles. You may wish to have students work in small groups to brainstorm their ideas, take notes, and then write an opinion. Encourage students to share their cases for or against the particle model with the rest of the class.

FOLLOW-UP

Reteaching
Ask students to write a short story about how a magician might use the information found in this lesson to amaze an audience.

Assessment
Have students compare Exploration 1, Part 2 with the diagram on page 88. They should point out how each part of the activity correlates to each step in the diagram. Students could do this in writing, by making a diagram, or by presenting their ideas orally to the class.

Extension
Have students demonstrate other models that might be used to show the particle theory of matter. Such models might include a balloon gradually losing its air, smoke, a chemical change, or odors traveling through the air. Have students present their models to the class.

Closure
Have students brainstorm familiar examples of phenomena that support the particle model. (For example, when sugar is added to water or chocolate syrup to milk, the combined volume does not equal the sum of the separate volumes.)

PORTFOLIO
Have students include their short story from the Reteaching activity in their Portfolio.

Democritus and John Dalton were born more than 2000 years apart in two different countries, Greece and England. But while thinking about the nature of matter, they both separately arrived at the same conclusion: Matter is made up of particles—what Democritus called **atoms**. During the two millennia separating the lives of these men, this idea was largely ignored. How did Dalton arrive at a conclusion that had been neglected for 2000 years? Read Constructing a Particle Theory to find out.

> According to convention, there is a sweet and a bitter, a hot and a cold, and according to convention there is color. In truth there are atoms and a void.

> I should apprehend there are a considerable number of what may properly be called elementary particles, which can never be metamorphosed [changed] one into another.

Constructing a Particle Theory

Recall your study of chemicals from earlier science courses. You found out that all pure substances on Earth can be classified as either *elements* or *compounds*. All matter is made of elements. During chemical changes, elements combine to form compounds, or compounds decompose into elements. Compounds can also change into other compounds.

By Dalton's time, much had been discovered about chemical changes and the elements involved in these changes. Antoine Lavoisier had discovered the role that oxygen plays in combustion. Water had been decomposed into the elements hydrogen and oxygen by passing an electric current through it. Metals could be obtained from ores by chemical changes, and at least 40 different elements had been identified.

Time Out for Facts

- About 2000 years ago, people had already discovered the elements gold, silver, iron, lead, tin, mercury, sulfur, and carbon without understanding that they were elements.

- By 1735, alchemists had added zinc, arsenic, and phosphorus to the list of known elements.

- Hydrogen, oxygen, and nitrogen were discovered between 1765 and 1775. Why do you think elements like gold were discovered long before elements such as oxygen and hydrogen? **A**

FOCUS

Getting Started

Direct students' attention to the lesson title and ask: What do you think it means? Why is the structure of matter referred to as a "hidden" structure? *(You cannot see it.)* Then involve students in a brief review of the previous lessons. Ask: How can you describe the structure of matter if you cannot see it? Help students to recall that a model for matter can be arrived at by drawing inferences from observations. Point out that in this lesson they will learn more about the hidden structure of matter.

Main Ideas

1. All matter is composed of atoms.
2. There is a finite number of elements that occur naturally on the Earth.
3. Atoms of the same element have the same properties.
4. Atoms combine to form molecules.
5. Molecules of different kinds of atoms form compounds.

TEACHING STRATEGIES

Answer to
In-Text Question

A Explain that people knew about elements like gold before discovering elements such as hydrogen and oxygen because elements like gold exist on Earth as pure elements. In their elemental state, oxygen and hydrogen are both invisible gases. They also often combine with other elements to form different substances, such as water.

 Two Transparency Worksheets (Teaching Resources, pages 17 and 19) and Transparencies 12 and 13 are available to accompany Lesson 2.

LESSON 2 ORGANIZER

Time Required 2 class periods

Process Skills
analyzing, inferring, comparing, contrasting

New Terms
Atom—the smallest particle into which an element may be divided and still be the same substance
Compound—a substance consisting of two or more elements that are chemically combined
Element—a substance that consists of only one kind of atom and that cannot be chemically separated into other substances

Molecule—particle of matter that is made of two or more atoms

Materials (per student group)
Time Out for Facts: 16 g of one color of modeling clay; 2 g of another color of modeling clay; metric balance; **Time Out for Discovery:** model molecule from Time Out for Facts; metric balance; **Time Out for Analysis:** 12 g of one color of modeling clay; 4 g of another color of modeling clay; metric balance; **Even More Models:** five different colors of modeling clay in the following amounts: 160 g, 39 g, 14 g,

continued ▶

John Dalton began his scientific investigations at a very early age. When he was 12 years old, he organized a school in an old barn and became a schoolmaster.

As an observant and curious scholar, Dalton explored many scientific questions. One of these explorations led to some rewarding conclusions. Trace Dalton's path of investigation by completing the flowchart below. You'll find the information you need in the next paragraph.

Dalton knew that by passing an electric current through water, the water could be separated into the elements hydrogen and oxygen. He also observed that by identifying the amount of one element in a quantity of water, he could predict the amount of the other element. For example, if 16 g of oxygen were formed by decomposing water, then 2 g of hydrogen would always be formed as well. In other words, the mass of the oxygen was always eight times greater than the mass of the hydrogen. Dalton asked himself, "From this information, what can I infer about the structure of matter?"

Perhaps this activity will help you understand the answer Dalton found. Measure 16 g of modeling clay, and form it into a ball. This ball represents an oxygen atom. Next, form two 1 g balls of clay. These represent two hydrogen atoms.

Attach the model atoms so that they resemble the illustration below. You have just simulated a chemical change between oxygen and hydrogen atoms that forms a new substance—water.

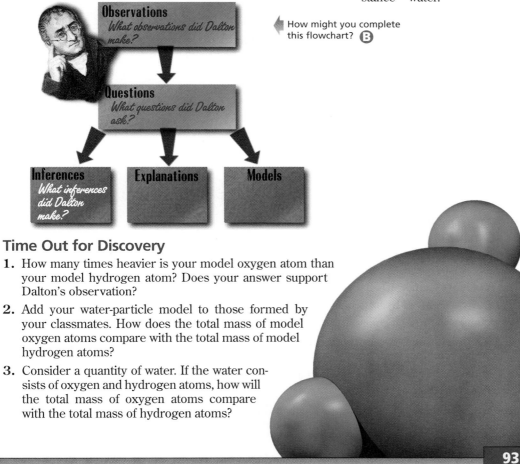

How might you complete this flowchart? **B**

Observations
What observations did Dalton make?

Questions
What questions did Dalton ask?

Inferences
What inferences did Dalton make?

Explanations

Models

Time Out for Discovery

1. How many times heavier is your model oxygen atom than your model hydrogen atom? Does your answer support Dalton's observation?

2. Add your water-particle model to those formed by your classmates. How does the total mass of model oxygen atoms compare with the total mass of model hydrogen atoms?

3. Consider a quantity of water. If the water consists of oxygen and hydrogen atoms, how will the total mass of oxygen atoms compare with the total mass of hydrogen atoms?

93

Answer to
Caption

A The representations show models of how molecules are made up of elements based on a particle theory of matter.

Answers to
In-Text Questions

1. The model oxygen atom was 16 times heavier than the model hydrogen atom.

2. The two model hydrogen atoms were identical.

3. The model oxygen and hydrogen atoms were not altered or destroyed by making the model water molecule.

Answers to
Time Out for Analysis,
pages 94–95

1. Only inference (b) is similar to one of Dalton's. Inferences (b), (d), and (e) can be supported. (The reading above does not indicate that Dalton inferred the formula of a water molecule to be H₂O.) The inferences for which there is no supporting evidence are (a) and (c).

2. Each carbon atom would be 12 times heavier than each hydrogen atom. (3 × 4 = 12) The total mass of the model carbon atom is three times heavier than the total mass of the four hydrogen atoms. (12 g ÷ 4 g = 3) Sample answer: Democritus said that matter is made up of atoms. Dalton said matter is made up of many different kinds of elementary particles, which cannot be changed into other types of particles.

Homework

Have students research the element of their choice to find out when and how that element was discovered. You may wish to have volunteers share their findings with the class.

In 1803 Dalton wrote the following entry in his journal: "An enquiry into the relative weights of the ultimate particles is, as far as I know, entirely new. I have lately been prosecuting this enquiry with remarkable success."

 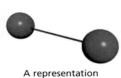

A representation of iodine atoms

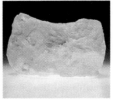 

A representation of sulfur atoms

Iodine (left) and sulfur (right) are examples of elements. What do the representations show? **A**

Of course, the "ultimate particles" Dalton spoke of were much like Democritus's atoms. Although he could not see atoms, Dalton made the following inferences:

1. An atom of one element has a mass that is different from the mass of an atom of any other element. (How much heavier was your model oxygen atom than your model hydrogen atom?)

2. An atom of one element is identical to any other atom of the same element. (Did you make your two model hydrogen atoms the same or different?)

3. An atom of one element cannot be changed into an atom of another element. (Did you alter or destroy your model oxygen and hydrogen atoms by making your model water particle?)

Because Dalton made careful observations about the masses of different elements and asked the right questions, evidence emerged to support a new model for the structure of matter:

Matter is composed of atoms.

Time Out for Analysis

1. What inferences were made in designing the model? Consider the examples below. Not all of these inferences are necessarily true.

   a. Atoms are round.

   b. The two atoms of hydrogen are identical.

   c. Atoms are different colors.

   d. In forming water, two hydrogen atoms combine with one oxygen atom.

   e. An atom of oxygen is 16 times more massive than an atom of hydrogen.

   Which of these inferences are similar to Dalton's? Which of these inferences can you support? Which can't you support?

Meeting Individual Needs

Gifted Learners

Many of the ideas of Democritus contradicted the conventional thinking of the ancient Greeks. Have interested students research the life and work of Democritus. The school librarian should be able to aid students in their search. Have volunteers present their findings to the class.

2. Shown at right are the symbols Dalton used to represent some atoms.

Dalton inferred that in the methane molecule there were four hydrogen atoms for each carbon atom.

Dalton knew that the total mass of the carbon atoms in methane was three times greater than the total mass of the hydrogen atoms. How much heavier is each carbon atom than each hydrogen atom?

Try this: Form a 12 g ball of clay and four 1 g balls of clay. Construct a model of a methane molecule. How does the mass of the model carbon atom compare with the total mass of the four model hydrogen atoms?

This lesson started with quotations from Democritus and Dalton. Rewrite each quotation in your own words.

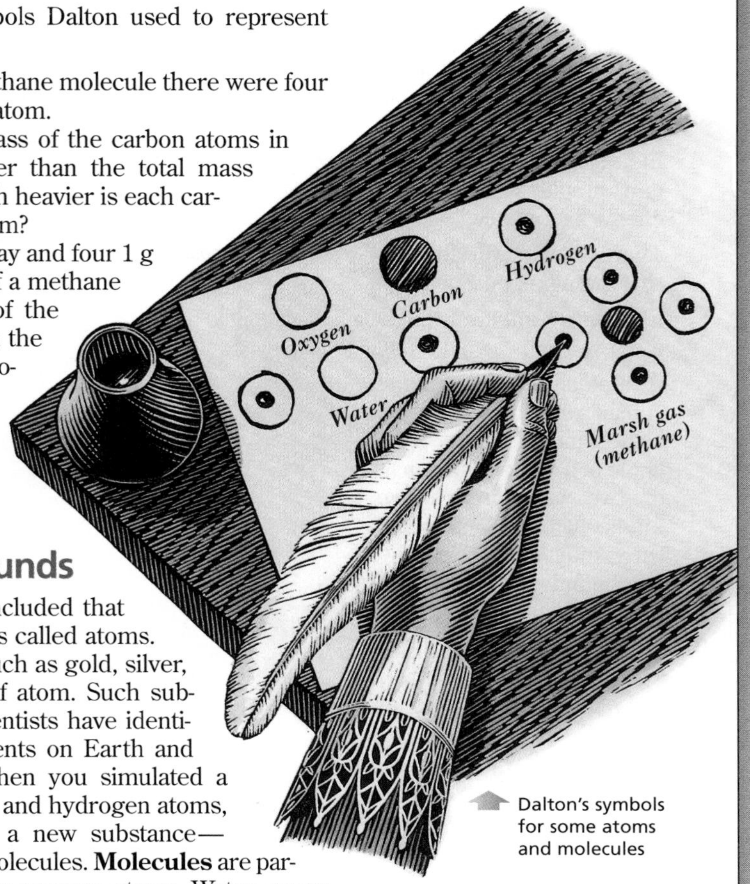

Elements and Compounds

Both Dalton and Democritus concluded that all matter is composed of particles called atoms. It follows that pure substances, such as gold, silver, and sulfur, consist of one kind of atom. Such substances are called **elements**. Scientists have identified 91 naturally occurring elements on Earth and therefore 91 kinds of atoms. When you simulated a chemical change between oxygen and hydrogen atoms, you simulated the formation of a new substance—water. Water particles are called molecules. **Molecules** are particles that are a combination of two or more atoms. Water, sugar, and carbon dioxide consist of molecules made of two or more *different* kinds of atoms. These substances are called **compounds**. The number of existing compounds is practically limitless.

▲ Dalton's symbols for some atoms and molecules

◄ Water and sugar are compounds. Each is made up of more than one type of element.

Elements and Compounds

This material should serve as a review because most students will probably be somewhat familiar with the terms *element, molecule,* and *compound.* To avoid any misconceptions, be sure students understand that the molecules of a compound are made up of atoms from different elements. For example, a water molecule is made up of hydrogen and oxygen atoms. A molecule of sugar is made up of carbon, hydrogen, and oxygen atoms. A molecule of carbon dioxide is made up of carbon and oxygen atoms.

Integrating the Sciences

Earth, Life, and Physical Sciences
All substances—living, nonliving, and dead—are made up of atoms. The smallest molecules, such as carbon monoxide, are composed of two atoms and are called diatomic molecules. Have students suggest examples of simple molecules that are important to consider when studying life, Earth, and physical sciences. Then have students suggest examples of complex molecules. *(Some students may have more background knowledge than others on this topic. An example of an important simple molecule is table salt, or NaCl. Each particle of salt consists of one sodium ion and one chloride ion. Salt is important to the life processes of many organisms, it is a naturally occurring mineral, and it is a reactant or product in many common chemical reactions. One of the most complex molecules can be found in every human cell—deoxyribonucleic acid. DNA is composed of millions of atoms and contains all of the information about an individual's genetic makeup.)*

95

Summing It Up

Call on a volunteer to read aloud the material at the top of page 96. Then involve the class in a discussion of any questions that may arise. Be prepared to help resolve any misconceptions.

Answers to
Testing Your Understanding

1. Encourage students to explain their concept maps either verbally or in writing. The following is one possibility:

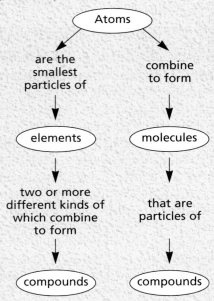

2. **a.** Each molecule of carbon monoxide contains one atom of carbon and one atom of oxygen. Dalton would have represented carbon monoxide by drawing one empty circle and one filled-in circle.
   b. Carbon dioxide has a greater mass than carbon monoxide because it contains one more atom of oxygen.

3. The masses for the clay balls were chosen to be consistent with Dalton's observation that the mass of oxygen in water is eight times the mass of hydrogen. Since a water molecule has twice as many hydrogen atoms as oxygen atoms, then the mass of each oxygen atom must be 16 times greater than that of each hydrogen atom.

Summing It Up

- Elements consist of atoms; each element is made of one kind of atom.
- Elements cannot be divided into simpler substances by chemical changes.
- There are 91 naturally occurring elements on Earth.
- Elements combine with each other to form compounds.
- A molecule of an element consists of two or more of the same kind of atoms. A molecule of a compound consists of two or more different kinds of atoms. A molecule is the smallest particle of a compound that still has the characteristics of the compound.
- Compounds can be broken down into elements by chemical changes.
- The number of compounds that exist or that can be made through chemical changes is essentially unlimited.

Testing Your Understanding

1.  Use the ideas in the summary to draw a concept map showing the relationship between atoms, elements, molecules, and compounds.

2. Both carbon dioxide and carbon monoxide consist of carbon atoms and oxygen atoms. Yet these substances have different properties. Carbon monoxide will burn and is poisonous, while carbon dioxide will not burn and is not poisonous. Here is how Dalton represented carbon dioxide.

   a. How would he have represented carbon monoxide?
   b. How are the masses of these two compounds different?

3. Earlier you made models of water molecules using clay balls to represent oxygen and hydrogen atoms. The model oxygen atom had a mass of 16 g. Each model hydrogen atom had a mass of 1 g. Why do you think these masses were chosen?

I know that atoms make up elements, but don't they make up molecules too?

Molecules can be made up of different kinds of atoms—like carbon, oxygen, and hydrogen. An element has only one kind of atom in it.

If you'd like to know more about molecules and atoms, see pages S22–S26 of the SourceBook.

CROSS-DISCIPLINARY FOCUS

Language Arts

Both Democritus and John Dalton decided to call the particles that made up all matter *atoms.* The word *atom* comes from the Greek word *atomus,* which means "uncut" or "indivisible." Ask students why Democritus and Dalton found this name appropriate. *(Both men believed that atoms were the smallest materials in existence and that they could not be broken into smaller particles.)*

Homework

Although there are only 91 naturally occurring elements on Earth, many elements that do not occur naturally on Earth have been discovered in the twentieth century. Have students research one of these elements to find out how it was discovered.

Even More Models

Here are more models of molecules that you can make. Use clay balls of different colors to represent each element. In order to make your representation fit Dalton's findings, your model atoms should have the masses listed below. (You could also use a fraction of the suggested masses. For example, you can divide all of the masses by two, and your models will still show how much more massive one type of atom is than another.)

For a model carbon (C) atom, use 12 g of clay.
For a model hydrogen (H) atom, use 1 g of clay.
For a model nitrogen (N) atom, use 14 g of clay.
For a model oxygen (O) atom, use 16 g of clay.
For a model sulfur (S) atom, use 32 g of clay.

Compound	Elements Forming Each Molecule	Atoms in the Molecule
water	hydrogen (H), oxygen (O)	H—O—H
hydrogen sulfide	hydrogen (H), sulfur (S)	H—S—H
carbon dioxide	carbon (C), oxygen (O)	O=C=O
methane	carbon (C), hydrogen (H)	H | H—C—H | H
butane	carbon (C), hydrogen (H)	H H H H | | | | H—C—C—C—C—H | | | | H H H H
ammonia	nitrogen (N), hydrogen (H)	H—N—H | H
glucose (sugar in honey)	carbon (C), hydrogen (H), oxygen (O)	H H H | | | O H O H O O | | | | | | H—C—C—C—C—C—C—H | | | | | | H O H O H | | H H
alcohol (from fermentation)	carbon (C), hydrogen (H), oxygen (O)	H H | | H—C—C—O—H | | H H

97

⭐ **Transparency 14 is available to accompany Even More Models.**

Even More Models

You may wish to point out to students that the masses suggested for the model atoms are proportional representations of the masses of the real atoms. For example, the real mass of sulfur is 32 times greater than the mass of hydrogen and twice as great as the mass of oxygen. In fact, each of the masses suggested for the model atoms is based on each atom's atomic mass. The atomic mass for carbon is 12, for hydrogen, 1, and so on.

If students seem somewhat confused by the chart, you may wish to review it with them. Explain that the third column, labeled Atoms in the Molecule, shows one way that scientists represent the atomic structure of molecules. By looking at the diagrams, they can count the number of different atoms in each molecule. They can also tell which atoms are bonded to one another. Be sure students understand that their model molecules must have the same configuration as the molecules in the chart. However, their models do not have to be flat. Provide space in the classroom for students to display their finished models.

 Cooperative Learning
EVEN MORE MODELS

Group size: 2 to 3 students
Group goal: to construct model molecules
Positive interdependence: Assign the following roles: materials coordinator (to gather supplies and return extra materials at the end of class), mass expert (to use the balance to measure the mass of the clay), and model builder (to direct the building of the models). The group should construct the models listed on this page. Other molecules may be added. To save materials, each group can construct one large model and one small one. Once the models are completed, students should present their models to the class for discussion.
Individual accountability: Each student should write the answers to the Questions on page 98.

In Three Dimensions

Have students carefully examine each of the models shown on this page. Call on individual students to describe each one in their own words. You may want to emphasize to students that while each of these models represents some of the properties of the molecule to some degree of accuracy, the molecules do not actually look like the models.

Answers to
Questions

1. The molecules are similar in that they are made of atoms but different because they are each made of different kinds and different numbers of atoms.

2. Carbon dioxide has three atoms; methane has five atoms; and water has three atoms. Each molecule contains two different elements.

3. Individual atoms are not destroyed when they are combined into a molecule.

Homework

The Questions on this page make an excellent homework activity.

FOLLOW-UP

Reteaching

Both water (H_2O) and hydrogen peroxide (H_2O_2) are composed of hydrogen and oxygen. Ask students to draw molecules of each of these substances and to compare the shapes and masses of the two molecules.

Assessment

Have students construct models of molecules and then hang them on mobiles. The model molecules could be made from papier-mâché, pipe cleaners, plastic-foam balls, or clay, or they could simply be poster-board cutouts. The mobiles can be constructed from clothes hangers and thread. Hang several of the finished mobiles in the classroom. Suggest that other students identify the substances that the molecules represent.

Extension

1. Point out to students that all of the elements are arranged in a chart

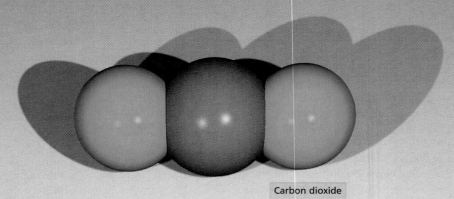

Carbon dioxide

In Three Dimensions

Shown here are three-dimensional models of some fairly common molecules so tiny that even the most powerful microscopes cannot clearly capture their form. Carefully study each model, and answer the questions that follow.

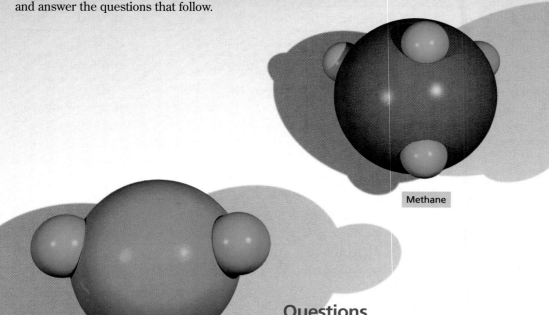

Methane

Water

Questions

1. How are these molecules alike? How are they different?

2. How many atoms are there in each molecule? How many kinds of elements are shown?

3. In forming a molecule of a compound, were the individual atoms of the elements destroyed?

98

called the periodic table of the elements. Have students do some research to find out what information is included for each element on the periodic table and what each piece of information tells about the element. (*Information about each element includes its atomic number, chemical symbol, element name, and atomic mass.*)

2. Point out to students that atoms are not the smallest particles of matter. Atoms themselves can be divided into even smaller particles. Have students research the particles that

make up atoms (*protons, electrons, neutrons*) and share what they learn by making poster diagrams. To extend the exercise, you may wish to have students research other sub-atomic particles, such as *gluons, hadrons, leptons, mesons, neutrinos,* and *quarks.*

Closure

Have students arrange the following items from smallest to largest: ecosystem, atom, molecule, cell, organ, animal.

EXPLORATION 2

Sugar and Starch Molecules

Starch and sugar are two compounds that consist of the same elements, just arranged differently. The molecules of starch and sugar are made up of carbon, hydrogen, and oxygen atoms.

After performing the following experiment, name three differences in the properties of sugar and starch.

You Will Need

- a graduated cylinder
- 5 mL of cornstarch
- 5 mL of dextrose
- a jar with a lid
- a stirring rod
- a large beaker
- 100 mL of hot water
- an egg
- a straight pin
- iodine solution
- Benedict's solution
- a watch or clock
- a hot plate
- a hot-water bath
- 2 test tubes
- an oven mitt or test-tube tongs
- an eyedropper
- latex gloves

What to Do

1. Mix 5 mL of cornstarch with 5 mL of dextrose. (Dextrose is a sugar.) Add this mixture to 100 mL of hot water in a beaker. Stir.

2. Crack an egg in half, and save the larger end of the shell, which contains the air sac.

3. Using a straight pin, carefully remove part of the large end of the shell to expose the air sac. Be careful not to puncture the air-sac membrane. (You will, however, need to break the membrane that lies flush with the eggshell.)

4. Pour 5–10 mL of water into the shell, and float the shell in the sugar-cornstarch-water mixture.

5. After 15 minutes, pour half of the liquid in the eggshell into a test tube. Pour the remaining half into another test tube.

6. Test the liquid in one test tube with a few drops of iodine solution. A blue color indicates the presence of starch. Did starch molecules move through the air-sac membrane into the liquid in the shell?

7. Test the remaining liquid by adding eight drops of Benedict's solution to the second test tube.

Caution: Benedict's solution can irritate the skin. Handle with care.

Heat the liquid *gently* in a hot-water bath to avoid splattering. Use an oven mitt or test-tube tongs to handle the hot test tube. A red or yellow color indicates the presence of sugar. Did sugar molecules pass through the air-sac membrane?

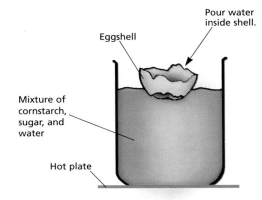

Pour water inside shell.

Eggshell

Mixture of cornstarch, sugar, and water

Hot plate

LESSON 3 ORGANIZER

Time Required 2 class periods

Process Skills analyzing, inferring

New Terms none

Materials (per student group)
Exploration 2: jar with lid; 100 mL graduated cylinder; 5 mL of cornstarch; 5 mL of dextrose; hot plate; stirring rod; egg; straight pin; large beaker; 110 mL of water; clock or watch with second hand; 2 test tubes; a few drops of iodine starch-test reagent; 8 drops of Benedict's solution; eyedropper; hot-water bath; safety goggles; lab aprons; test-tube tongs, test-tube clamp, or

oven mitts; latex gloves (additional teacher materials: 50 mL of 1 M sulfuric acid solution; six 6d iron nails; steel wool; portable burner; tongs; watch glass; 50 mL of 4% sodium hydroxide solution; pH paper; strainer; 25 mL of 0.1 M sodium thiosulfate solution; a few sheets of old newspaper; see Advance Preparation on page 75C.)
Exploration 3: clove of garlic; 30 cm of plastic wrap
Exploration 4: aluminum foil; scissors; metric ruler; metric balance

Teaching Resources
Exploration Worksheets, pp. 21 and 22

FOCUS

Getting Started

Ask students to identify some words that describe extremely large numbers. (*Examples include million, billion, trillion, and quadrillion.*) List the words on the chalkboard, and add to them if necessary. Then call on volunteers to write out each number. (*1,000,000; 1,000,000,000; 1,000,000,000,000; 1,000,000,000,000,000*) Remind students that in Lesson 1 they learned about exponents. Call on volunteers to express each number on the chalkboard using exponents of 10. (10^6; 10^9; 10^{12}; 10^{15}) Point out that in this lesson students will use exponents to represent very small and very large numbers as they explore the size of atoms and molecules.

Main Ideas

1. The molecules of different substances are different sizes.
2. Atoms and molecules are extremely small.
3. Some materials allow certain molecules to pass through while blocking others.

TEACHING STRATEGIES

EXPLORATION 2

Divide the class into small groups and distribute the materials. It may be helpful if you display an eggshell floating in a beaker of water to let students see how their eggs should be prepared so that the eggshells will float in the solution.

 SAFETY ALERT Before students begin the activity, warn them that iodine is poisonous if swallowed. Also point out that it will stain just about anything it comes in contact with, especially clothes and skin. For these reasons, they should be careful when using the iodine starch-test reagent.

★ An Exploration Worksheet is available to accompany Exploration 2 (Teaching Resources, page 21).

Exploration 2 continued ▶

At the end of the activity, students should recognize that the differences in the properties of sugar and starch include the following:

- Sugar molecules are smaller than starch molecules.
- Starch turns blue-black in the presence of iodine, while sugar does not.
- Sugar turns yellow or red in the presence of Benedict's solution, while starch does not.

WASTE DISPOSAL ALERT Pour all solutions containing Benedict's solution into a beaker. While stirring, slowly add 1 M sulfuric acid solution until the pH is between 5 and 6. Scour six 6d iron nails with steel wool until they are shiny. Immerse the nails in the acidified solution overnight or until all the copper has precipitated. Remove the nails, and filter the mixture. Using tongs and a burner, heat the nails and the copper precipitate in a watch glass until all the copper is converted to copper oxide, which is black. Let the copper oxide and the nails cool, and put them in the trash. Treat the remaining filtrate with enough 4% sodium hydroxide solution to raise the pH to 8–10. Filter the mixture again. Let this second precipitate dry, and put it in the trash. Pour the filtrate down the drain.

Pour the iodine-treated mixture into a labeled container. While stirring, slowly add a 0.1 M sodium thiosulfate solution until the mixture turns white. Pour the resulting mixture down the drain.

Answers to
Exploration 2, page 99

6. Starch molecules did not pass through the membrane.

7. Sugar molecules did pass through the membrane.

Something to Think About

1. As a conclusion to the experiment, which of the following statements do you think is most correct? Which statement is an inference?

   a. Sugar molecules have properties that are different from those of starch molecules.

   b. The experiment showed that some molecules pass through an egg membrane.

   c. Since sugar molecules passed through the membranes and starch molecules did not, sugar molecules may be smaller in size.

   d. Iodine solution is a test for starch, while Benedict's solution is a test for sugar.

2. What is the function of the air-sac membrane? What molecules do you think pass through the membrane as the egg develops into a chicken, and why?

Flashback!

You observed the passing of water through a membrane in Chapter 2. As you may recall, this process is called osmosis. Osmosis is an important process in every living organism. Water is transferred by osmosis from cell to cell. Review and explain the observations you made in Exploration 2 of Chapter 2, using the particle model of matter in your explanation. Can you define *osmosis* using the particle model? Ⓐ

A Mini-Experiment— Locking in the Smell

Wrap a clove of garlic in a piece of plastic wrap. Now crush the garlic between your thumb and forefinger. Can you still smell the garlic? How many layers of plastic wrap are needed to seal in the smell? What can you conclude from this mini-experiment?

Answers to
Something to Think About

1. Statement (c) is an inference, and it is the best conclusion to the experiment. All of the others are observations.

2. The air-sac membrane allows molecules of air to enter and leave the egg while preventing the fluids inside the egg from escaping. Obtaining oxygen from the air is essential for the growing embryo, as is eliminating excess carbon dioxide. If the liquids inside the egg were able to escape, the egg would dry out, and the developing embryo would die.

Answer to
In-Text Question

Ⓐ Students should recall from Exploration 4 on page 43 that openings in the cell membranes of the potato slices were just the right size to allow water molecules to pass through, while blocking larger particles such as those of starch and salt. Students' definitions of osmosis will vary but should reflect an understanding that in osmosis, molecules of a certain size are permitted to pass through the openings in a membrane, while other molecules that are too large to pass through are blocked.

Huge Numbers of Tiny Particles

If some particles are so tiny that they can pass through openings invisible to the human eye, then atoms and molecules must be extremely small. Because these particles are so small, a great number of them are needed to make up even a tiny bit of matter. The images on this page illustrate just how small atoms and molecules are.

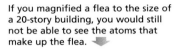

If you magnified a flea to the size of a 20-story building, you would still not be able to see the atoms that make up the flea.

A single drop of water contains approximately 3×10^{21} (3,000,000,000,000,000,000,000) water molecules. If you started to count these water molecules at the rate of one per second, it would take you . . . Well, you do the calculation. How many minutes? How many hours? **B**

Answers to
Exploration 3, page 100

This activity may be done by you as a demonstration, by pairs or small groups of students in class, or by students at home. Students should be able to detect the odor of the garlic even when it is wrapped in plastic wrap. Students will discover that several layers of plastic wrap are needed to block the smell of the garlic. Students should conclude that molecules from the garlic are passing through spaces between molecules of the plastic wrap.

Huge Numbers of Tiny Particles

This short reading reinforces the idea that atoms and molecules are extremely small. Provide students with time to perform the necessary calculations. Then call on volunteers to share and explain their answers.

Answer to
Caption

B 3×10^{21} particles $\div$ 60 sec./min. $\times$ 1 sec./particle = 5×10^{19} min.

5×10^{19} min. $\div$ 60 min./hr. = 833,333,333,333,333,333 hr. (or 8.3×10^{17} hr.) This is equal to 9.46×10^{13} years. (For comparison, the age of the Earth is thought to be about 4.5×10^9 years.)

Homework

The calculation described in the caption on this page makes an excellent homework activity.

Meeting Individual Needs

Learners Having Difficulty
Pose the following scenario to students: You pour muddy water through a filter and clear water comes out. You notice that solid material has collected on the filter. What do these observations tell you about the relative sizes of the particles that make up the mud and the water? *(The pores in the filter must have been large enough to allow water particles to pass through but not mud particles. Students should conclude that a mud particle must be larger than a water particle.)* You may wish to have students recall their experiments with semipermeable membranes in Unit 1.

Answers to
Exploration 4

Answers will vary depending on the type of aluminum foil used.

2. The mass can be found by using a metric balance. A sample mass is 1.5 g.

3. The volume of 1.5 g of aluminum foil is 1.5 g ÷ 2.7 g/cm³ = 0.56 cm³.

4. 10 cm × 10 cm = 100 cm²

5. 0.56 cm³ ÷ 100 cm² = 0.0056 cm, or about 0.006 cm

6. 0.006 cm ÷ 0.00000003 cm/atom = 200,000 atoms

Modern Miracle Fabrics

You might ask students to bring their rain gear to class so that they can compare different types of materials and discuss the benefits of waterproofing that permits "breathing."

Answer to
In-Text Question

Ⓐ Students should recognize that the type of fabric described would be most useful in situations when they want to be protected from rain and moisture, but do not want the buildup of perspiration.

Homework

The Exploration Worksheet that accompanies Exploration 4 makes an excellent homework activity (Teaching Resources page 22).

More Large Numbers!

You know that atoms are small. In fact, each of the aluminum atoms in a piece of aluminum foil has an approximate diameter of 0.00000003 cm (3×10^{-8} cm). How many atoms thick is the aluminum foil?

Here is the equation that you will need to get the answer:

$$\text{Number of atoms} = \frac{\text{thickness of foil in centimeters}}{\text{thickness of one atom in centimeters}}$$

You still have a problem, though. You need to find the thickness of the aluminum foil in centimeters, so try the following exercise:

1. Cut a piece of foil that measures exactly 10 cm × 10 cm.

2. Find its mass in grams.

3. Find its volume in cubic centimeters (cm³) by dividing the mass by 2.7 g/cm³—the *density* of aluminum. (Density is the mass of a substance divided by the volume of the substance. You can learn more about density in Unit 4.)

4. Find its area in square centimeters (cm²):

area = length × width.

5. Divide volume (length × width × thickness) by area (length × width) to find the foil's thickness:

V/A = thickness (cm).

6. Now use the first equation to calculate how many atoms thick a piece of aluminum foil is. Are you surprised?

Modern Miracle Fabrics

Is your jacket water-repellent? If it is made of plastic, it is. Some fabrics, such as nylon, are actually formed of plastic threads and can withstand some rain. Other fabrics are bonded synthetic materials. Gore-Tex® is one such fabric. It contains many tiny pores that are large enough to let individual water molecules through but that prevent the passage of water droplets. As you perspire, the water in your perspiration evaporates and its particles pass through the openings, from inside to outside. This keeps you dry. Can you think of some situations in which you would need such a fabric? Ⓐ

FOLLOW-UP

Reteaching

Remind students that in Unit 1 they learned about diffusion, transpiration, and respiration. Suggest that they review these three life processes and provide an explanation for each based on the particle model of matter.

Assessment

Suggest that students repeat Exploration 3 using different kinds of materials, such as wax paper, aluminum foil, and plastic trash bags. Have them compare how well each material prevents the odor of garlic from escaping. Then suggest that they explain their results in terms of the particle model of matter.

Extension

Explain to students that all atoms are about the same size but vary widely in mass. Suggest that they find out what unit is typically used to express the mass of an atom. *(The atomic mass unit, or amu)* Ask: How many of these units are in 1 g? *(There are 602 billion trillion [6.02×10^{23}] amu in 1 g)*. You may wish to point out that this number has a special name: Avogadro's number. Also, have students make a chart listing the atomic mass of several common elements.

Closure

Have students make a chart with the following columns: Compound name, Mass, Total number of atoms. Students should fill out the chart for each of the compounds found in the table on page 97.

Particles of Solids, Liquids, and Gases

From circumstantial evidence, you have developed a model of the structure of matter. So far, your model includes the following ideas:

1. All matter consists of particles.

2. Particles are different sizes, although all are small.

3. Elements are made up of particles called *atoms*, and compounds are made up of particles called *molecules*.

In the Explorations that follow, you will discover more ideas about the structure of matter to add to these three.

Compressed air in this rocket forces water through the opening at the bottom. As the water is forced downward, the rocket is propelled upward. Why do you think both air and water are needed to make this rocket work? **B**

EXPLORATION 5

Compressing Gases, Liquids, and Solids

Fill a plastic syringe (without the needle) with air. Then, with your finger over the end, push down on the plunger as hard as you can. What does this experiment tell you about the particles that make up a gas?

Now fill the syringe with water, and again try to push down on the piston. What is the difference between the particles that make up a gas and those that make up a liquid?

Finally, consider this: If you could get a solid (such as a piece of chalk) into the syringe, could you compress it? What differences are there between the particles of a solid and those of a liquid or a gas?

To conclude this Exploration, add a fourth statement to the particle model of matter. (See the list of ideas in the first column on this page.)

....

EXPLORATION 6

Particles on the Move

Your model of matter is becoming more and more useful because it can explain more observations. Now you will make a few more observations of the behavior of matter. In each instance, explain your observations in terms of what the particles in the solid, liquid, or gas are doing.

You Will Need

- food coloring
- an eyedropper
- ice water
- hot water
- a balloon
- a plastic soft-drink bottle
- an ice chest with ice
- rubbing alcohol
- 2 microscope slides
- matches
- test-tube tongs
- a beaker
- cotton balls
- a metal lid from a jar
- perfume
- a candle

STATION 1

Place a drop of food coloring into very cold water and another drop into very hot water. Explain the difference in behavior.

Exploration 6 continued ▶

LESSON 4 ORGANIZER

Time Required 2 to 3 class periods

Process Skills analyzing, inferring, comparing, contrasting

New Terms none

Materials (per student group)
Exploration 5: plastic disposable syringe without a needle; 20 mL of water; 100 mL graduated cylinder
Exploration 6, Station 1: 2 drops of food coloring; 2 small beakers; about 250 mL of cold water; about 250 mL of very hot tap water; lab aprons (additional teacher materials: ice chest with ice); **Station 2:** balloon; 2 L or 3 L plas-

tic soft-drink bottle; 2 large containers or buckets; 5 L of ice water; 5 L of very hot tap water; watch or clock; **Station 3:** 2 drops of rubbing alcohol; eyedropper; 2 microscope slides; a few matches; safety goggles; **Station 4:** 250 mL of ice water; 250 mL beaker; **Station 5:** a cotton ball; metal jar lid; eyedropper; a few drops of perfume; **Station 6:** candle; a few matches; materials to prop up candle, such as a metal jar lid and a small ball of modeling clay; safety goggles

Teaching Resources
Exploration Worksheet, p. 23

Particles of Solids, Liquids, and Gases

FOCUS

Getting Started
Ask students what the solid, liquid, and gaseous forms of water are. *(Ice, water, water vapor)* Explain to students that the particles that make up these three states of water are the same. Then invite them to speculate about what causes these three forms of water to be different.

Main Ideas
1. Particles of matter are constantly in motion.
2. Particles are farther apart in gases than in liquids and solids.
3. Heating causes particles in a substance to move faster and farther apart.

TEACHING STRATEGIES

Answers to *Exploration 5*

The particles of a gas can be compressed. The particles that make up a liquid cannot be noticeably compressed. A piece of chalk could not be compressed in the syringe. The particles of a solid are locked in place, while the particles of liquids and gases can move about.

Sample answer: Gas particles are farther apart than are the particles that make up liquids or solids.

Answer to *Caption*

B Air can be compressed, but water cannot be. Once the compressed air is free to expand, this forces the water to shoot out of the rocket and the rocket to move skyward.

★ An Exploration Worksheet is available to accompany Exploration 6 (Teaching Resources, page 23).

Exploration 6 begins on the next page. ▶

EXPLORATION 6, pp. 103–104

Set up five stations in the classroom.

SAFETY ALERT Water above 55°C is scalding. Make sure no open containers of alcohol are in the room while students are lighting their matches. You may also wish to apply the drops of alcohol yourself or have a beaker of water available for disposing of the hot matches.

Answer to
Station 1, page 103

The food coloring diffuses more quickly in the hot water because the heated particles are moving faster.

Answer to
Station 2

When the bottle is placed in hot water, the air particles inside speed up and move farther apart, causing the balloon to expand. When the bottle is placed in the ice water, the air particles slow down and move closer together, causing the balloon to deflate.

Answer to
Station 3

The alcohol evaporates faster on the heated glass because the heated particles of alcohol move faster.

Answers to
Station 4

Water condenses on the beaker. The beaker's cool surface causes particles of water vapor in the breath to slow down and gather in droplets.

Answers to
Station 5

Eventually, the perfume will be smelled from across the room. The particles of perfume are moving farther apart as they change into a gas and diffuse throughout the air.

Answers to
Station 6

The heat causes the wax particles to move freely as a liquid, so the wax melts. When the candle is extinguished and the wax cools, the particles slow down, and the liquid solidifies.

EXPLORATION 6, continued

STATION 2

Place a balloon over the mouth of a 2 L or 3 L plastic soft-drink bottle. Place the bottle into a container of hot water for a few minutes. Now quickly place it into a container of ice water. Use the particle model to explain what happens.

STATION 3

Heat a microscope slide with a match. Then, after extinguishing the flame, place one drop of alcohol on the heated slide and one drop on an unheated slide. Using the particle model, explain the differences you observe.

STATION 4

Pour ice water into a beaker. Now breathe on the side of the beaker. What do you observe? Explain this observation in terms of what you think the water molecules in your breath are doing.

STATION 5

Place a cotton ball on a metal lid. Add a few drops of perfume to the cotton. From how far away can you smell the perfume? What do you think the liquid particles that make up the perfume are doing?

STATION 6

Observe a burning candle. What forms at the top of the candle (not the top of the flame)? What happens after the candle is blown out? Explain these observations in terms of what the particles of wax are doing.

Analysis, Please!

1. Now add at least one more statement to the three given on page 103.
2. Here are six words that help describe the processes you observed in Stations 1–6: condensation, expansion, diffusion, evaporation, melting, and solidification. Which word(s) would you associate with each station?

Expanding the Model

Mr. Chin's class expanded the particle model of matter, as described on page 103, by adding more ideas. If you agree with the statements they added, suggest at least one observation to support each statement.

More Ideas

1. Particles in gases are far apart.
2. Particles that make up liquids and solids must be as close together as possible.
3. Particles move.
4. Particles in a hot substance move faster than particles in a cold substance.
5. The faster gas particles move, the more pressure they exert on the sides of a balloon.
6. Liquid particles can become gas particles, and gas particles can become liquid particles.

Answers to
Analysis, Please!

1. Sample statement: Heating causes the particles of the substance to move faster and farther apart.
2. Condensation—Station 4; expansion—2; diffusion—1 and 5; evaporation—3 and 5; melting—6; solidification—6

Answers to
Expanding the Model

1. Gases can be compressed.
2. We cannot easily compress liquids or solids.

3. Gases diffuse.
4. Raising the temperature of a liquid increases the rate of evaporation.
5. Warming the air inside a balloon causes it to expand, suggesting that the particles exert more pressure on the sides of the balloon.
6. Several changes of state were observed, such as water vapor condensing and liquid perfume evaporating.

The Follow-Up for Lesson 4 is on the next page. ►

CHALLENGE YOUR THINKING

1. Invisible Aerobics

Make a table like the one shown here. The first column lists some words that describe the ways particles may move. Which state of matter—solid, liquid, or gas—is most likely to exhibit each kind of movement? Suggest an everyday event that is similar to the way particles move. One has been done for you.

Word	State of matter	Your analogy
Wriggling	Solid	Like students wriggling while sitting in their seats
Vibrating	?	?
Tumbling	?	?
Bouncing	?	?
Flying	?	?
Shaking	?	?
Whirling	?	?
Sliding	?	?

2. Changes in Behavior

The following pictures illustrate the behavior of particles in solids, liquids, and gases. Write a sentence or two that would explain to a fifth-grader what is happening in each picture.

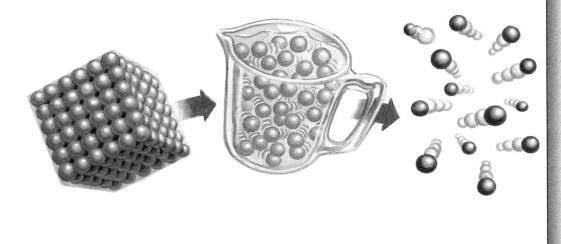

105

FOLLOW-UP

Reteaching

Ask students to write a story from the point of view of a particle in the solid phase as it melts and then evaporates.

Assessment

Have students draw illustrations or cartoons to show the behavior of particles in a solid, liquid, and gas. Then have them draw what happens when the particles change state.

Extension

Inform students that *plasmas* make up a fourth state of matter. Suggest that they do some research on this subject.

Closure

Have students develop three quiz questions (with answers) based on the lesson. You may wish to collect and redistribute the questions or have students trade them.

Answers to *Challenge Your Thinking*

1. Students may have different interpretations of the motion associated with each movement. Accept all reasonable responses. Encourage students to think about the range of motion of the individual particles in each state of matter. For instance, particles in a gas are far apart, move very quickly, and move throughout the atmosphere. *Flying* particles, therefore, are likely to be gas particles. Students may disagree about which state of matter is most appropriate for each movement. For instance, some students could defensibly argue that tumbling refers to molecules in a liquid state rather than a gaseous one. In any case, student analogies should emphasize the motion of the particles in the state of matter that they have chosen.

2. Sample answer: In a solid, the particles can jiggle around, but they are locked in place like soldiers standing in a row.

In a liquid, the particles are able to move about within the boundaries of the liquid in the container and take the shape of the container.

In a gas, the particles are able to go anywhere, like a swarm of bees in the air.

Did You Know. . .

Radial tires on automobiles may be considered full of air even when they look slightly underinflated. This is because tires heat up as they roll along the ground. Heat is transferred to the air in the tires. The air molecules in each tire thus move faster, and the pressure they exert on the inside walls of the tire increases.

★ **You may wish to provide students with the Chapter 5 Review Worksheet (Teaching Resources, page 26). Transparency 15 is also available to accompany this Challenge Your Thinking.**

3. Sample answers:
- The particles in a solid do not move much.
- The particles in a liquid slide over one another, allowing the liquid to flow and take the shape of its container.
- The particles in a gas spread out, or diffuse, until they fill the space they occupy and exert pressure on the walls of the container.

Students' own definitions will vary but should demonstrate that they understand the types of movement that occur in solids, liquids, and gases.

4. a. Real air molecules bounce off each other and hit the sides of the balloon, creating pressure.
b. Shaking the balloon is like heating the air because it causes the beads to move faster and farther apart and to strike the sides of the balloon.
c. Doubling the number of beads in the balloon is like doubling the number of air molecules because it increases the number of collisions between the air particles and the sides of the balloon.
d. This is like cooling the air in that an orderly pattern indicates that the molecules are no longer a gas. Instead they have condensed to form a liquid or have solidified.

3. What's the Matter?

Here are some students' descriptions of solids, liquids, and gases. To which particular characteristic of the particles is each student referring?

Solid—made up of the staying-at-home type of particles

Liquid—consists of particles slip-sliding away

Gas—particles with claustrophobia

Now it's your turn to create an unusual definition of each state of matter. Share it with a friend, and see whether he or she can discover which term you are describing.

4. Air Apparent

In Chapter 4, Ramón used plastic beads in a balloon to simulate particles of air. Explain how each of Ramón's observations (listed below) describes the behavior of air and therefore supports the particle model of matter.

a. "When the balloon was shaken gently, the plastic beads rattled around. I could feel and see them hitting the sides of the balloon."

b. "Shaking the balloon harder caused more frequent and harder collisions."

c. "When I doubled the number of plastic beads in the balloon, the number of collisions with the sides of the balloon increased."

d. "When I stopped shaking the balloon, the plastic beads formed an orderly pattern in the bottom of the balloon."

ScienceLog

Review your responses to the ScienceLog questions on page 87. Then revise your original ideas so that they reflect what you've learned.

ScienceLog

The following are sample revised answers:

1. There is nothing in between the particles of matter except vacuum. A vacuum has nothing in it at all. (Some students may have the mistaken notion that there must be something in between the particles, probably more particles. Point out that this is incorrect.)

2. The scientific particle theory of matter is based on observation and inference. Over the course of hundreds of years, scientists have performed countless experiments investigating matter, yielding a large number of observations about its structure. Drawing inferences from these observations, scientists have come to the conclusion that all matter is composed of particles.

3. The liquid water is converted into water vapor, which diffuses into the air. Evidence to suggest that the evaporated water particles have not simply vanished but are actually in the air includes the condensation of water in clouds and on cool objects.

Why does the thermometer 1
read no higher than 100°C
even though the stove
continues to supply heat to
the water?

100°C

2 How might you
explain this situation?

ScienceLog

Think about these
questions for a moment,
and answer them in your
ScienceLog. When you've
finished this chapter, you'll
have the opportunity to
revise your answers based
on what you've learned.

3
Why might a
person shiver
on a hot day?

107

Connecting to Other Chapters

Chapter 4
*introduces scientific models and how
exponents are used to describe
large and small numbers.*

Chapter 5
*explores the evidence for an atomic
theory of matter and discusses the
states and structure of matter.*

Chapter 6
*investigates how temperature
changes affect matter and explores
the properties of matter.*

Chapter 7
*surveys the historical development of
our understanding of the atom and
the particles contained in its nucleus.*

Prior Knowledge and Misconceptions

Your students' responses to the
ScienceLog questions on this page will
reveal the kind of information—and
misinformation—they bring to this
chapter. Use what you find out about
your students' knowledge to choose
which chapter concepts and activities
to emphasize in your teaching. After
students complete the material in this
chapter, they will be asked to revise
their answers based on what they have
learned. Sample revised answers can be
found on page 125.

In addition to having students
answer the questions on this page, you
may wish to have them complete the following exercise: Students should write a
paragraph to explain evaporation and
condensation, giving several examples
of each and offering an explanation of
what is happening on a particle level
during these two processes. Collect the
papers, but do not grade them. Instead,
use them to find out what students
already know about evaporation and
condensation, what misconceptions they
might have, and what about the topic is
interesting to them.

Temperature and Particles

Temperature and Particles

FOCUS

Getting Started

Remind students of the activity they did in the previous lesson in which they placed a bottle capped with a balloon in hot water and then in ice water. Call on a volunteer to describe what happened. *(The balloon shrank in the ice water and expanded as it warmed up in the room.)* Ask students to explain their observations. *(When heated, the particles absorbed heat energy, sped up, and moved farther apart. In the cold, the air particles lost heat energy, slowed down, and moved closer together.)* Point out that in this lesson they will learn more about how temperature affects the movement of particles.

Main Ideas

1. As matter is heated, the motion of its particles increases, and the particles move farther apart.
2. As matter cools, the motion of its particles slows down, and the particles move closer together.
3. Air expands when it is heated and contracts when it is cooled.
4. Not all substances expand and contract at the same rate.

TEACHING STRATEGIES

EXPLORATION 1

You may wish to encourage students to perform each experiment so that they can verify the results. If so, have students work in pairs or small groups.

For safety, you should insert the glass tubes into the stoppers in advance for students. Be sure to wear thick cloth or leather gloves while doing it. First make sure that the glass tubing has been fire polished to prevent any jagged edges. Then, using a small amount of glycerin as a lubricant, slowly insert the glass tubing into one-hole stoppers with a gentle twisting motion.

You know that matter is made up of molecules that are in constant motion. The particle model of matter suggests that when heat is added to matter (solid, liquid, or gas), the molecules move faster and faster and farther and farther apart. Similarly, as matter cools, the molecules slow down and move closer together. The particle model suggests that this is true for all states of matter— solids, liquids, and gases. The following Exploration shows the way in which two students chose to test these inferences.

EXPLORATION 1

Slowing Down and Speeding Up Particles

Examine the notes jotted down by Sedrick and Leilani on the following two experiments. Then do the following:

1. For each experiment, devise a good title in the form of a question.
2. Draw an appropriate conclusion for each experiment.

Sedrick's Experiment

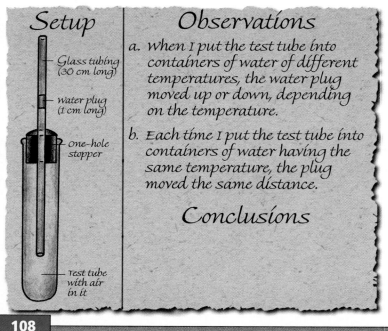

Setup
— Glass tubing (30 cm long)
— Water plug (1 cm long)
— One-hole stopper
— Test tube with air in it

Observations

a. When I put the test tube into containers of water of different temperatures, the water plug moved up or down, depending on the temperature.

b. Each time I put the test tube into containers of water having the same temperature, the plug moved the same distance.

Conclusions

LESSON 1 ORGANIZER

Time Required
1 class period

Process Skills
analyzing, inferring, comparing, contrasting

New Terms
none

Materials (per student group)
none

Teaching Resources

Leilani's Experiment

Setup

In this experiment, I filled two test tubes with two different liquids, glycerin and water.

Here are my two test tubes.

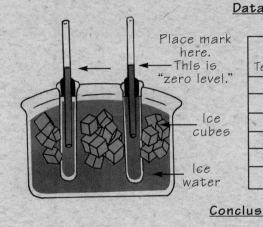

Glass tubing (30 cm long)

Water

One-hole stopper

Glycerin

Procedure

When I put the stoppers in, I had to make sure that the liquids rose about one-third of the way up the glass tubing from the stopper. I also had to make sure there was no air left in the test tubes. Then I put both test tubes in a beaker of ice and water. The water level had to almost cover the test tubes. After the liquids in the glass tubing stopped moving, I marked the level on each piece of glass tubing as "O cm"—my baseline. The baseline for each tube did not have to be at the same level. I then put my test tubes into four more water baths of different temperatures. I measured how far from the baseline the liquid moved up the glass tube each time.

Data

Place mark here. This is "zero level."

Ice cubes

Ice water

Temperature	Height of liquid at different temperatures	
	Glycerin	Water
5°C	O cm	O cm
24°C	0.9 cm	0.4 cm
45°C	3.8 cm	2.6 cm
50°C	4.2 cm	3.1 cm
63°C	5.5 cm	4.1 cm

Conclusions

Sedrick's Experiment, page 108

For Sedrick's Experiment, students can place a finger over one end of the tubing in order to draw in the correct amount of water. The bulb from an eyedropper can then be used to position the water plug. Students should measure the starting and final positions of the water plug.

Have students discuss and compare their data.

Answers to
Sedrick's Experiment, page 108

1. Students' titles will vary but should reflect the main point of the experiment. Examples include How Does Temperature Affect the Volume of a Gas? and What Happens When Air Is Heated and Cooled?

2. Students' conclusions should be similar to the following: When air is heated, the molecules in the air move faster and farther apart, causing the air to expand. When air is cooled, the molecules slow down and move closer together, causing the air to contract.

Answers to
Leilani's Experiment

1. Students' titles will vary but should reflect the main point of the experiment. Examples include Do Different Liquids Expand at the Same Rate? How Does Temperature Affect the Volume of Different Liquids? and How Do the Expansion Rates of Water and Glycerin Compare?

2. Students' conclusions should be similar to the following: When a liquid is heated, the molecules in the liquid move faster and farther apart, causing the liquid to expand. When a liquid is cooled, the molecules slow down and move closer together, causing the liquid to contract. Not all liquids expand and contract to the same extent.

Aisha knows that solids expand when they are heated but that not all solids expand at the same rate. When heated, the jar lid absorbs heat energy and expands faster than the glass jar. This loosens the lid and makes it easy to remove. Encourage students to share other experiences they may have had when a similar technique was successful or unsuccessful.

Answers to
An Eggsperiment

Students should observe that as the water warms, air bubbles escape from the large end of the egg. Students may conclude that as the gases inside the egg expand, some escape through the porous shell. If these gases do not escape fast enough, the shell may crack as pressure builds up inside the egg. When the large end of the egg is punctured, the gases escape more rapidly, preventing pressure from building up inside the egg. Some eggs do not crack because the gases are able to escape fast enough. It is not a good idea to place a raw egg in boiling water because the shell may expand unevenly or the gases inside the egg may expand too quickly, causing the egg to crack.

Homework

An Eggsperiment makes an excellent homework activity.

FOLLOW-UP

Reteaching

Discuss with students how a thermometer works. Encourage them to relate this concept to the particle model of matter. *(The liquid in a thermometer exchanges heat with its surroundings. Thus, when the temperature goes up, the particles in the liquid move faster and spread farther apart. This causes the liquid to move up the thermometer a certain amount, depending on how much temperature rises.)*

Assessment

Suggest that students use the information in Leilani's data table to make a line graph. *(See page S207 for graph.)* Then have them use the graph to

Aisha's Secret

Use the particle model to explain why Aisha's technique works.

An Eggsperiment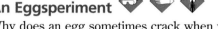

Why does an egg sometimes crack when you boil it? To find an answer, try this activity at home.

Start with a pot of cold water, and heat an egg in it. Watch the large end. What do you observe? How does your observation explain why eggs sometimes crack?

Repeat the experiment with a new egg, only this time pierce the large end with a thumbtack. Start over again with cold water and heat the egg. Again, what do you observe?

Why do some eggs crack when boiled, while others don't? Can you suggest why it is not a good idea to put a raw egg in boiling water, even if you have pierced the large end?

Ask someone in your family how he or she boils eggs without cracking them.

110

answer the following questions:
- Do liquids expand at a constant rate as they are heated? *(No)*
- Which expands faster as the temperature increases: water or glycerin? *(Glycerin)*

Extension

Have students do some research to discover what makes popcorn pop. *(Popcorn kernels pop because they contain a small amount of water that expands when the kernel is heated.)* Invite students to share what they have learned with the class, using diagrams of what is happening inside the popcorn kernel.

Closure

As a demonstration, put a shelled, hard-boiled egg into a boiling flask. The egg will pop into the flask and amaze the students. Wearing oven mitts, first heat the empty flask. Then put the egg firmly on top of the flask. As the flask cools, air molecules come closer together, and pressure inside the flask decreases. Pressure from air outside the flask pushes the egg into the flask. Have students explain what happened with diagrams and written descriptions.

LESSON 2 Changes of State

Does heating a material always cause a temperature change within the material?

"Of course," said Derrick. "When you heat water, it gets hotter, and the pot it's in gets hotter too."

"And," added Wendy, "because the particles making up the water and the pot are vibrating at a faster rate than they were before heating, the particles take up more space, causing the materials to expand."

That explanation seems to make good sense. Now examine what happens when heat is *removed* from a substance. Believe it or not, this will help you answer the question above.

EXPLORATION 2

The Temperature Connection

In this Exploration, you will start with stearic acid at 75–80°C and observe how its temperature changes as it cools for 30 minutes.

You Will Need

- stearic acid
- a scoop
- a beaker of hot water (75–80°C)
- a hot plate
- a thermometer
- a test tube
- test-tube tongs
- a test-tube rack
- a watch with a second hand
- wire-loop stirring device

Be Careful! Do NOT try to pull the thermometer out of the solid; this could break the thermometer.

What to Do

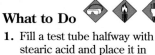

1. Fill a test tube halfway with stearic acid and place it in the hot water.

2. Once the stearic acid has melted, place the test tube in the test-tube rack and measure the temperature of the stearic acid.

3. Slip the wire-loop stirring device over the thermometer. Record the temperature of the molten liquid every 30 seconds. Mix the liquid before each reading by moving the stirring device up and down. Record your data in a table in your ScienceLog.

4. Record the temperature until *all* of the liquid has become solid.

5. Remove the thermometer and stirring device from the frozen stearic acid by melting the solid again, as in step 1. Clean the thermometer if necessary.

How to melt stearic acid

Hot water (75-80°C)

Stearic acid

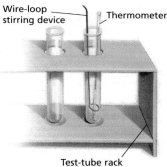
How to cool stearic acid

Wire-loop stirring device

Thermometer

Test-tube rack

111

LESSON 2 ORGANIZER

Time Required
2 class periods

Process Skills
measuring, comparing, contrasting, analyzing

Theme Connection
Energy

New Terms
none

Materials (per student group)
Exploration 2: about 20 mL of stearic acid; scoop; small beaker; 250 mL of hot water (75–80°C); hot plate; alcohol thermometer; wire-loop stirring device; test tube; test-tube tongs or test-tube clamp; test-tube rack; watch or clock with second hand; safety goggles; oven mitts

Teaching Resources
Exploration Worksheet, p. 31
Resource Worksheet, p. 32
Graphing Practice Worksheet, p. 35
Transparency 16
SourceBook, p. S29

LESSON 2 Changes of State

FOCUS

Getting Started

Ask students to define the word *latent.* (Latent *means "present but inactive or hidden."*) Then point out that in the last lesson, students learned that adding heat to a substance increases its temperature. Ask: What do you think the term *latent heat* might mean? What do you think would happen if you added latent heat to a substance? *(Latent heat is heat energy that is absorbed by a substance but does not cause a change in temperature. The heat energy is used to change the state of the substance.)*

Main Ideas

1. Melting occurs as a result of heat energy being added to a substance.
2. Freezing occurs when a substance loses enough heat energy to solidify.
3. The temperature of a substance remains constant at its melting point or freezing point until all of the material has either melted or frozen.

TEACHING STRATEGIES

EXPLORATION 2

To construct the wire-loop stirring device, make a loop slightly smaller than the inside of the test tube at one end of a straightened 25 cm piece of copper wire. Angle the loop so that it is perpendicular to the rest of the wire. At the other end of the wire, make a handle that extends in the opposite direction of the loop. Students will use this device to stir the contents of the test tube by placing the loop around the thermometer and using the handle to move the device up and down.

Exploration 2 continued ▶

⭐ **An Exploration Worksheet (Teaching Resources, page 31) and Transparency 16 are available to accompany Exploration 2.**

EXPLORATION 2, *continued*

SAFETY ALERT Water above 55°C is scalding. Students should not attempt to remove their thermometers from the solid because the thermometer could break.

WASTE DISPOSAL ALERT The scoop is provided so that students can remove the stearic acid from the test tubes without coming into contact with it. Dispose of the stearic acid in an approved incinerator. Discharge, treatment, or disposal methods may be subject to federal, state, or local laws.

Answers to
Analysis, Please!

1. Students' graphs should look similar to the second graph on page 113.

2.

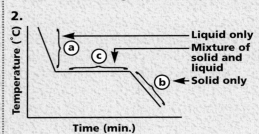

3. All of the graphs should appear similar, with the plateau at 70°C. They should differ only in the length of the plateau.

4. All students should determine that the melting and freezing point is 70°C.

5. During the time interval when the temperature does not drop, heat energy is being released from the stearic acid as it changes from a liquid to a solid. The temperature remains at the freezing point until all of the stearic acid has solidified.

6. After completing the experiment, most students should agree that adding heat does not always cause a temperature change to occur within a material. At the melting point, the temperature does not change. Instead, the added heat energy causes the substance to change state. Discuss with students how their answers to this question differ from their answers from Getting Started.

Analysis, Please!

1. In your ScienceLog, prepare a graph of your results. Place *time* on the horizontal axis and *temperature* on the vertical axis.

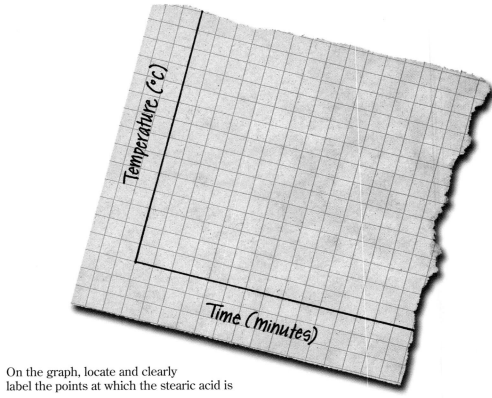

2. On the graph, locate and clearly label the points at which the stearic acid is
   a. liquid only.
   b. solid only.
   c. a mixture of solid and liquid.

3. How does your graph resemble those of other students in the class? How does it differ?

4. What is (a) the melting point and (b) the freezing point of stearic acid? Compare your answer with those of your classmates.

5. The graph should reveal that, at one point, the temperature does not drop as you might expect. Explain this observation.

6. Return to the question that began this lesson: Does adding heat to a material always cause a temperature change within the material? Have you changed your answer?

Cooperative Learning
EXPLORATION 2 AND ANALYSIS, PLEASE!

Group size: 3 to 4 students
Group goal: to conduct an experiment that demonstrates the effects of heating and cooling on a substance
Positive interdependence: Assign students the following roles: materials and safety coordinator (to obtain materials and supervise safety practices within the group), timer (to accurately keep time for the experiment), facilitator (to read the instructions and keep the group on task), and recorder (to keep accurate data). Students will carry out the experiment, discuss the results, and construct graphs.
Individual accountability: Students should construct a graph of their results for use in the rest of this lesson.

Homework

The Resource Worksheet that accompanies the material on this page makes an excellent homework activity (Teaching Resources, page 32).

Derrick's Logbook

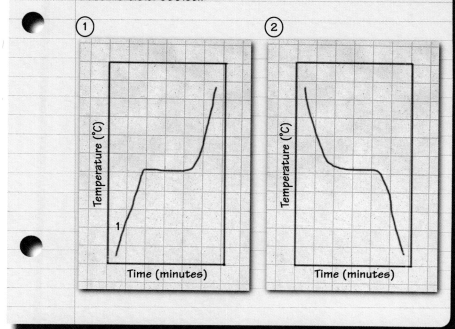

I was surprised by the results of the experiment, although I shouldn't have been. After all, I already knew that you can't cool liquid water below 0°C using only ice, and I knew that you can boil water all day without its temperature going over 100°C.

Here is what I learned by doing this experiment. I drew two graphs to illustrate. The first graph shows what I now predict will happen as stearic acid is heated. The second is what actually happened as the stearic acid cooled.

① Temperature (°C) / Time (minutes)

② Temperature (°C) / Time (minutes)

Derrick's Logbook

Spend a few minutes reviewing Derrick's graphs on this page. Have students compare Derrick's graphs with their own and note the similarities. Ask them to explain what is happening at the point where the temperature remains constant as the stearic acid is heated. (*Heat energy is being used to melt all of the solid. The temperature remains at the melting point until all of the solid has melted.*) Have students copy Derrick's graphs into their ScienceLog, and then have students label their graphs as directed. You may wish for students to tape graph paper into their ScienceLog in order to plot their results.

Answers to
Derrick's Logbook, page 114

You may wish to remind students that each label may be used on only one graph. Sample labeled graphs:

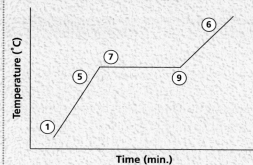

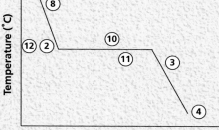

Multicultural Extension

Glass Blowing

One industry that depends on a change in state brought about by heat is glass blowing. Glass blowers heat glass until it becomes a solid-liquid mixture that can be shaped by skillfully blowing air into the glass. Glass blowing was invented by the Phoenicians in the first century B.C., and their exquisite glassware became some of the most admired and sought after products of the ancient world. Have interested students do research on glass blowing and report their findings to the class.

Theme Connection

Energy

As the liquid stearic acid cools and gives up heat, its temperature decreases. At its freezing point, the stearic acid solidifies. Then if the solid is continuously cooled, its temperature will also decrease. **Focus question:** Will the temperature of the cooling, solid stearic acid continue to fall indefinitely? (*At 0 K, or –273°C, all motion ceases. This temperature is called absolute zero. The atoms stop moving so the temperature must remain constant.*)

Homework

For additional practice, have students place the labels from page 114 on the graphs they made for Exploration 2. Point out that not all of the labels will be used.

Answer to *Follow-Up*

Accept all reasonable responses. Students will probably conclude that the force of attraction between molecules is greater in stearic acid than in water. Even though this is technically not true, encourage students to realize that, in general, more heat energy is required to weaken the attractive forces between molecules in substances with a higher melting point. (This is an excellent opportunity to discuss the chemical uniqueness of water. The hydrogen bonds between a negatively charged hydrogen atom in one water molecule and a positively charged oxygen atom in another water molecule make the forces of attraction between water molecules exceptionally strong.)

PORTFOLIO
Have students include their cartoon strip from the Reteaching activity in their Portfolio.

FOLLOW-UP

Reteaching

Have students summarize the lesson by drawing a cartoon strip showing what happens as a substance reaches its melting point or freezing point.

Assessment

Have students write a brief explanation of the following statement: The solid and liquid forms of a substance can exist at the same time. *(The liquid form of a substance freezes at the same temperature at which its solid form melts. Therefore, the freezing and melting points of a substance are the same.)*

Extension

Challenge students to discover how atmospheric pressure affects the melting and freezing points of a substance. *(An increase in atmospheric pressure increases the melting and freezing points of most substances.)* Then have them find out if the freezing and melting points of water are affected by atmospheric pressure in the way they might expect. *(An increase in atmospheric pressure actually lowers the melting and freezing points of water.)*

Closure

Point out to students that the melting of a material may not occur abruptly at a definite temperature; melting depends partly on whether the substance is a single element or compound or a mixture of elements or compounds. Have students suggest a reason for this difference. *(Each element or compound is likely to melt at a different temperature. Thus, a mixture of different elements or compounds will melt slowly over a range of temperatures.)* Have students suggest how this observation could be used to determine how pure a substance is. *(The purer a sample is, the closer its melting point will be to the melting point of the pure substance.)*

Derrick intends to label the parts of his graphs. The labels are listed below in no particular order. First, copy his graphs (on a larger scale) into your ScienceLog. Then, with a classmate, decide where each label belongs. Write the number of each label in the appropriate place on the graphs. Derrick has already placed one label.

1. Heat energy is added to solid stearic acid.
2. Stearic acid starts to form a solid.
3. The particles of solid stearic acid are moving more slowly.
4. The solid stearic acid returns to room temperature.
5. The particles of solid stearic acid are vibrating at a greater rate.
6. The temperature of liquid stearic acid goes up.
7. Melting—no change in temperature
8. The temperature of liquid stearic acid goes down.
9. The stearic acid melts.
10. The particles of stearic acid release heat energy.
11. Freezing—the temperature remains constant.
12. The freezing point of stearic acid

Follow-Up

Ice melts at 0°C. Stearic acid melts at 70°C. In which substance does the force of attraction between molecules appear to be greater? Explain the reason for your choice.

A Graphing Practice Worksheet is available to follow Lesson 2 (Teaching Resources, page 35).

Absorbing and Releasing Heat

When energy is added to or released by a substance, one of two things may happen: the temperature of the substance may change, or the material may undergo a change of state. In your investigations with stearic acid, at what point was there a change in temperature? a change of state? **A**

When enough heat energy is added to a solid such as stearic acid or ice, the solid increases in temperature and then starts to melt. But during the melting process, the temperature of the solid-liquid mixture no longer increases. This change in state causes the plateau you saw on your graph. For stearic acid, the plateau is at 70°C. For water, the plateau is at 0°C. These are the melting points of stearic acid and ice. During melting, heat is *absorbed*, or taken *in*, by the substance being heated. *Endo* means "in." Melting is therefore an example of an **endothermic** change.

When a liquid solidifies, heat is released. During the solidification process, the temperature of the liquid-solid mixture remains constant. This time period of constant temperature corresponds to the plateau on your graph; it is the freezing point for the substance. For stearic acid the plateau occurs at 70°C. For ice the plateau occurs at 0°C. *Exo* means "outside." Because heat energy is released, or given off, freezing is called an **exothermic** change.

Freezing—heat energy being released

Melting—heat energy being absorbed

In the remainder of this lesson, you will be exploring these ideas in more detail.

EXPLORATION 3

Salol and Changes of State

This Exploration involves changes of state, this time for a substance called *salol* (phenyl salicylate). As you follow the procedure, identify the points when

• melting is occurring.
• freezing is occurring.
• energy is being absorbed.
• energy is being released.
• an exothermic change occurs.
• an endothermic change occurs.

You Will Need

• a glass microscope slide
• a test-tube clamp
• a stainless steel scoop
• a pair of tweezers
• a match
• salol
• a magnifying glass or low-power microscope

What to Do

1. Using the scoop, add a small amount of salol to a glass slide.

Be Careful! Salol may cause irritation to the skin. Don't handle it with your bare hands.

Exploration 3 continued ▶

LESSON 3 ORGANIZER

Time Required 1 to 2 class periods

Process Skills
observing, analyzing, inferring

Theme Connection Energy

New Terms
Endothermic—a change, such as melting, that occurs as a result of heat being absorbed by a substance
Exothermic—a change, such as freezing, that occurs as a result of heat being released by a substance

Materials (per student group)
Exploration 3: microscope slide; a

small amount of salol; stainless steel scoop; a few matches; magnifying glass or low-power microscope; test-tube tongs or test-tube clamp; tweezers; latex gloves; safety goggles; lab aprons
Exploration 4: 3 alcohol thermometers; three 5 cm pieces of hollow shoelace; cup or other small container; 50 mL of water; 30 mL of rubbing alcohol; electric fan; safety goggles

Teaching Resources
Exploration Worksheets, pp. 37 and 40
Theme Worksheet, p. 38

Absorbing and Releasing Heat

FOCUS

Getting Started

Display a glass of ice and ask: Is this ice absorbing or releasing heat energy as it melts? *(It is absorbing heat energy.)* When the water froze to form the ice, did it absorb or release heat energy? *(It released heat energy.)*

Main Ideas

1. Heat energy is absorbed during an endothermic change and released during an exothermic change.
2. Substances either absorb or release heat energy as they change from one state to another.
3. Heat energy is continually absorbed and released during the water cycle.

TEACHING STRATEGIES

Answers to
In-Text Questions

A A change in temperature is indicated on the graph by a slope. A change of state is indicated by a horizontal line.

EXPLORATION 3

If an overhead projector is available, place a slide of salol on the screen for the entire class to observe. Student observations should include the following:

• In step 2, melting occurs as heat energy is absorbed. An endothermic change occurs.
• In step 4, students should notice that the salol solidifies, starting near the seed crystal and spreading outward. This is an example of crystallization.

WASTE DISPOSAL ALERT Collect excess salol for disposal by a waste-disposal contractor.

★ An Exploration Worksheet is available to accompany Exploration 3 (Teaching Resources, page 37).

Hot Dogs, Cold Sweats, and Other Phenomena

Ask students to dip a finger in some water and wave it in the air. Have them note the cooling effect. Encourage students to speculate about the cause of the cooling effect. *(As a liquid evaporates, energy is absorbed from the surroundings, thus lowering the temperature of the surroundings.)*

Answers to
Hot Dogs, Cold Sweats, and Other Phenomena

Panting helps a dog to stay cool through the evaporative cooling of saliva from the tongue and mouth. While people perspire through their sweat glands in order to cool themselves, a dog has sweat glands only in its feet, and these glands contribute only a small amount to the cooling of the body.

A refrigerator makes use of evaporative cooling by passing a cool, compressed liquid through the refrigeration compartment, where the liquid absorbs the heat from the compartment's contents. In doing so, the liquid heats up and evaporates. Then the warm vapor leaves the refrigerator and is compressed into a liquid once again and cooled. The cycle then repeats.

 A Theme Worksheet is available to accompany Hot Dogs, Cold Sweats, and Other Phenomena (Teaching Resources, page 38). This worksheet corresponds to the Theme Connection at the bottom of this page.

 PORTFOLIO
Students may wish to draw pictures of the crystal patterns that they observed in Exploration 3 and put the pictures in their Portfolio.

EXPLORATION 3, continued

2. Heat the slide with a match until the salol melts, and then immediately remove the match.
3. Extinguish the match, let the slide cool, and add a small crystal of salol to the slide. This is called a "seed crystal."
4. Using a magnifying glass or low-power microscope, observe what happens.

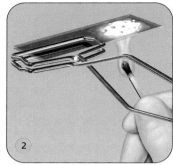

116

Hot Dogs, Cold Sweats, and Other Phenomena

Panting dogs, a sweating person, a refrigerator, and even the Earth itself have something in common. They all use a similar process to exchange heat. Read on to find out how this works.

You've probably experienced the coolness that results when you dip your finger in alcohol, when you sweat, or when you wet your finger to test the wind direction. Why does this occur? What's happening involves a process called *evaporative cooling*. In each example above, a change of state occurs: a liquid evaporates to form a gas. When a liquid evaporates or a solid melts, heat energy is absorbed from the surroundings. As sweat on your skin absorbs nearby available heat (including heat from your body), the sweat changes into water vapor. Since heat has been removed from the skin during this process, you feel cooler. How does this process work for a panting dog or explain how a refrigerator cools our food? Perhaps the following Exploration will help you with these questions.

ENVIRONMENTAL FOCUS

Smog often occurs when moisture in the air condenses onto smoke particles, much in the same way that the salol crystal in Exploration 3 served as a seed crystal. Discuss the effects of smog with students. Smog in heavy concentrations is poisonous. It kills plant life and deteriorates building materials. One incident of smog in London actually killed 4000 people.

Theme Connection

Energy
Focus question: How could the loss of heat energy due to evaporative cooling lead to hypothermia on a cold or windy day? *(The rate of evaporation of sweat depends on the difference in temperature between the surrounding air and the skin. Also, wind can strip away the heated air nearest to our skin, resulting in a faster rate of evaporation. Since evaporation cools the body, this could result in a dangerously low body temperature, called hypothermia.)*

---

116 UNIT 2 • PARTICLES

Cooling Through Changes in State

You Will Need

- 3 thermometers
- 3 pieces of hollow shoelace (at least 5 cm long)
- a fan
- rubbing alcohol
- water

What to Do

1. Record the room temperature.
2. Prepare three *wet-bulb thermometers* by inserting each thermometer bulb into one end of a piece of shoelace. Secure each bulb with masking tape.
3. Wet one bulb in alcohol and the other two in water.
4. Place one of the water-dipped thermometers in a quiet spot on a table top.
5. Place the other water-dipped thermometer and the alcohol-dipped thermometer in front of a fan.
6. Every 30 seconds, record the temperature of each thermometer.

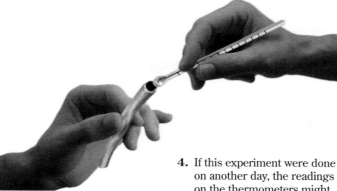

Making Sense of It

1. How did the temperatures change in the three thermometers? Can you explain why?
2. Why did the thermometer that was dipped in alcohol read a lower temperature than the one that was dipped in water?
3. How does this experiment help explain why you feel cool when getting out of the shower? How does it help explain why dogs pant?
4. If this experiment were done on another day, the readings on the thermometers might be higher or lower. Why would a wet-bulb thermometer read lower temperatures on some days than on others?
5. Many of the main points of the particle model are listed on pages 103 and 104. This model is also called the kinetic theory of matter. *Kinetic* means "motion." Why is this a good name for the particle model we have developed?

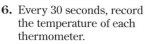 ## Multicultural Extension

Heat From Underground

The ancient Romans used the water vapor produced by hot springs to heat baths and homes. Today, cultures in geothermal regions of the world, including Iceland, Turkey, Italy, and Japan continue to make use of this resource to provide their houses with heat and hot water. Have interested students design houses that make use of hot springs, and have them share their results with the class.

ENVIRONMENTAL FOCUS

As the hot springs of geothermal regions of the Earth undergo evaporative cooling, the steam that they give off can be used to produce electricity for nearby communities. This geothermal energy is seen as an attractive alternative to the burning of fossil fuels, which are increasingly expensive, limited in supply, and harmful to the environment. Have students conduct research on geothermic energy's potential as an alternative to fossil fuels.

 # EXPLORATION 4

SAFETY ALERT Make sure that no flames or hot plates are used near the alcohol.

Answers to *Exploration 4*

1. Students should notice that the temperatures in the three thermometers decreased. This was because of evaporative cooling. However, the thermometer that was not in front of the fan read a higher temperature than the other two thermometers. This was because the breeze increased evaporation and thus further cooled the object.

2. When a wet-bulb thermometer is dipped in alcohol, the temperature is lower than when it is dipped in water because alcohol requires less energy to evaporate—and thus evaporates faster—than water, producing a greater cooling effect.

3. Water absorbs energy as it evaporates, making a person cooler. As saliva evaporates from a dog's tongue, it also absorbs energy and makes the dog cooler.

4. A wet-bulb thermometer reads much lower on dry days than on humid days because the dry air can hold more water vapor than humid air. Therefore, the water evaporates faster, removing more heat from the thermometer.

5. Students should realize that many properties of a material can be explained by modeling the material as a collection of particles in random motion. For example, the temperature of an object is related to how fast the particles are moving, and the state of an object is related to how tightly bound these particles are to one another.

Homework

Have students write a brief statement that describes the relationship between temperature and motion.

 An Exploration Worksheet is available to accompany Exploration 4 (Teaching Resources, page 40).

Answers to
Explanations, Please!

1. a. As the coolant circulates through the walls of the refrigerator, it changes from a liquid to a gas as it absorbs heat (endothermic change). Outside the refrigerator, the gas changes into a liquid as it releases heat into the air (exothermic change). These processes alternate continuously to cool the inside of the refrigerator.

b. It is not possible to cool a room by opening a refrigerator door because the heat absorbed by the coolant inside the refrigerator is released into the room at the back of the refrigerator.

2. Sample answer: As the sun heats the waters of the Earth, evaporation (an *endothermic* change) occurs and heat energy is absorbed. This energy carries the resulting water vapor high into the air. As the water vapor is carried over land, it condenses (an *exothermic* change), releasing rain and heat energy in the process.

Integrating the Sciences

Life and Physical Sciences

Discuss with students why a person can catch a cold by sitting in wet clothes on a cold day. Explain that evaporative cooling occurs as the body's heat causes water to evaporate from the wet clothing. The body must use more energy to produce more heat to compensate for that which is lost in heating up the wet clothing. It is the extra demand for energy that strains the body, leading to illness.

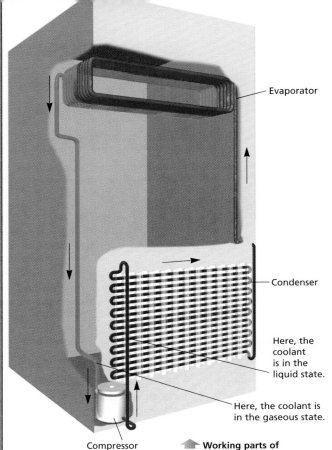

Explanations, Please!

1. a. Did you know that changes of state are involved in maintaining a colder temperature inside your refrigerator than outside? Examine the diagram of a refrigerator. Discuss with a friend why it is colder inside the refrigerator than it is outside.

b. Also discuss the following with a friend: You could warm a room by turning on an oven and opening the oven door. But can you cool a room by opening the door of a refrigerator? (Notice that the back of the refrigerator has pipes containing a coolant. The coolant circulates in these pipes.) If not, why not?

2. The water cycle is illustrated in the diagram at left. Use the diagram to explain how heat is transferred over large distances by the processes of evaporation and condensation. Use the words *endothermic* and *exothermic* in your explanation.

Labels for the refrigerator diagram:
- Evaporator
- Condenser
- Here, the coolant is in the liquid state.
- Here, the coolant is in the gaseous state.
- Compressor
- **Working parts of a refrigerator**

Condensation

Evaporation

FOLLOW-UP

Reteaching

Suggest that students write a story about "a day in the life" of a particle of salol, describing the endothermic and exothermic changes it undergoes. The story should be told from the particle's point of view.

Assessment

Have students make poster diagrams or bulletin-board displays to illustrate endothermic changes (*Melting and evaporation*) and exothermic changes (*Freezing, solidification, crystallization,* and condensation). Suggest that they use pictures from various sources to illustrate these two types of changes.

Extension

Suggest that students do some research to discover what a heat pump is. (*It is a machine that transfers heat from a low-temperature area to a high-temperature area.*) Have students name a common household appliance that uses a heat pump. (*A refrigerator*) Challenge students to discover how the particle nature of matter allows a heat pump to work and to list other uses for heat

pumps and their advantages. (*Heat pumps can cool and heat homes more economically than conventional systems.*)

Closure

Pose the following question: If a refrigerator cannot be used to cool a room, how can an air conditioner cool a room? Challenge students to do some research to discover the answer. Suggest that their answer include a discussion of the absorption and release of heat energy and any useful diagrams.

Identifying Substances

Many everyday observations and phenomena can be explained by the particle model. Perhaps you can help Jon with a question that he brought to class.

Jon found two cubes of the same volume, but their masses were quite different.

He wondered how the particle model explains how two objects with the same volume could have different masses.

If they have the same volume, why not the same mass?

A Comparison of Two Objects Having the Same Volume

Object	Mass (g)	Volume (cm³)
Cube 1	15.2	2.1
Cube 2	3.4	2.1

Write your own explanation based on the particle model. Then **A** try the following Explorations, which will explore this question in more detail.

Measuring Volumes

Here are three methods for determining volume.

METHOD 1

Using an Overflow Can

You Will Need

- an overflow can
- a graduated cylinder
- a small block of wood
- a straight pin

What to Do

1. Fill the can with water so that the water flows out of the spout. Catch and discard the overflow.

2. Stick the straight pin firmly into the block of wood.

3. Holding the block by the pin, carefully submerge it. Avoid putting your fingers into the water. (Why?)

4. Collect the water that overflows in a graduated cylinder. What is the volume of the water? the block of wood?

Exploration 5 continued ➤

LESSON 4 ORGANIZER

Time Required
3 class periods

Process Skills
measuring, observing, analyzing, comparing, contrasting

Theme Connection
Structures

New Term
Density—the mass per unit volume of a material

Materials (per student group)
Exploration 5, Method 1: overflow can or container with spout or side arm; 100 mL graduated cylinder; small block of wood; straight pin; **Method 2:** block of wood from Method 1; metric ruler; **Method 3:** iron bolt; 250 mL graduated cylinder; 200 mL of water **Exploration 6:** 250 mL graduated cylinder; metric balance; small iron object; small aluminum object; 200 mL of water

Teaching Resources
Exploration Worksheets, pp. 42 and 44
Transparencies 17 and 18

FOCUS

Getting Started

Show students an array of objects that have various masses and volumes, such as an eraser, a piece of chalk, a book, a pencil, or a chair. Ask students to list the items in order of increasing mass and then list the items in order of increasing volume. Ask them to give a definition of *mass* and then a definition of *volume*. Clear up any confusion before proceeding with the lesson.

Main Ideas

1. Different materials with the same volume may have different masses.

2. The mass of a given volume of a substance is an identifiable property of that substance.

TEACHING STRATEGIES

Answer to *In-Text Question*

A Student answers will vary. Point out that they will have an opportunity to revise their answers during this lesson.

METHOD 1

Water may be spilled, so students will need a good supply of paper towels or sponges for cleanup. If water is spilled in steps 3 and 4, the results of the experiment will be affected. Emphasize the importance of catching all of the overflow during these steps.

Answers to Method 1 are on the next page. ➤

★ An Exploration Worksheet is available to accompany Exploration 5 (Teaching Resources, page 42).

3. If students put their fingers into the water, they will overestimate the volume of the block because their fingers will cause more water to overflow.

4. The volume of the water collected in the graduated cylinder is equal to the volume of the block of wood.

METHOD 2

As students perform this activity, have them identify sources of error in Method 2. *(Error can be introduced in faulty measurements of height, length, or width; in miscalibration of the ruler; in faulty arithmetic; or by using the wrong formula for volume.)* Then ask students to identify shortcomings of Method 2. *(A major shortcoming is that it can only be used to determine the volume of rectangular objects.)*

Ask students to explain how Method 2 can be applied to determine the volume of other objects. *(The volume of other objects with regular geometric shapes can be determined by taking measurements and using appropriate formulas for volume. For example, the volume of a spherical object can be determined by measuring the diameter, d, and using the formula $\frac{4}{3}\pi\left(\frac{d}{2}\right)^3$.)*

Answer to Method 2

3. Depending on how students round off their measurements, the measured volume may be greater or less than the calculated volume. Students should infer that calculating the volume of the block of wood in cubic centimeters is less exact than measuring the volume in milliliters.

Some students may point out that the calculation of volume would be more precise if the linear measurements were made in millimeters rather than centimeters. (An interesting extension would be to have students follow up with a calculation based on measurements of height, length, and width in millimeters. Students will still probably find that measuring overflow is more precise than using this method.)

EXPLORATION **5**, continued

METHOD 2

Using a Ruler

You Will Need
- the block of wood from Method 1
- a metric ruler

What to Do
The volume of objects with a box-like shape (such as your block) can be determined using the following procedure:

1. Measure the length, width, and height of your block of wood in centimeters.

2. Calculate its volume in cubic centimeters (cm³) using this formula:

Volume = length × width × height

⬆ What is the volume of this block in cubic centimeters? Ⓐ

3. How does the measured volume in milliliters compare with the calculated volume in cubic centimeters? Which do you think provides the better answer?

METHOD 3

Using a Graduated Cylinder

You Will Need
- a graduated cylinder
- an iron bolt

What to Do
A more accurate method for finding the volume of small objects is to use a graduated cylinder. Follow the sequence in these diagrams to determine the volume of an iron bolt.

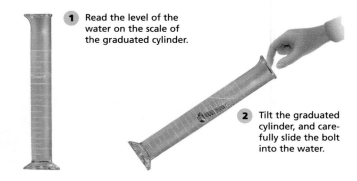

1 Read the level of the water on the scale of the graduated cylinder.

2 Tilt the graduated cylinder, and carefully slide the bolt into the water.

3 Read the new water level. This is the combined volume of the bolt and the water.

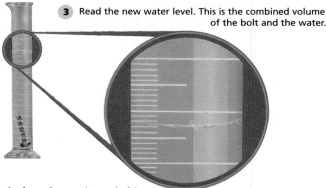

What is the volume of your bolt?
Remember to read the volume as you learned to on page 90.

METHOD 3

Method 3 is often called the displacement method of measuring volume. It is a standard technique in laboratory work and is therefore an important skill for students to master. Conceptually, Method 3 may be more difficult for students to understand than Methods 1 and 2, so extra practice and class discussion may be necessary.

If possible, use plastic graduated cylinders rather than glass ones. Caution students not to drop the bolt into the cylinder because this could break the cylinder. Instead, the bolt should be placed gently in the cylinder.

Answer to Caption

Ⓐ The volume of the block of wood is 5 cm × 5 cm × 5 cm, or 125 cm³.

Mass and Volume of Different Materials

You Will Need

- a graduated cylinder
- a balance
- a small iron object
- a small aluminum object

Note: You may use materials other than iron and aluminum as long as everyone else uses the same materials.

Iron		Aluminum	
Mass (g)	Volume (mL)	Mass (g)	Volume (mL)

What to Do

1. Find the mass and volume of your iron object. Your classmates will likely be using iron objects that have different volumes. In your ScienceLog, record your results in a class data table like the one shown above. Since you will be sharing your results with others in your class, do the measurements carefully.

2. Now find the mass and volume of your aluminum object. Add these values, as well as those of your classmates, to your data table.

3. Prepare a graph by placing *mass* on the vertical axis of the graph and *volume* on the horizontal axis. Choose scales to include the largest of your mass and volume values.

4. Draw a straight line (or line of best fit) through the points that represent the iron objects and another line through the points that represent the aluminum objects. The lines should go through the origin (0, 0). Why?

5. Is it possible to identify a material by knowing only its mass and volume? Could the materials that Jon brought to class be iron or aluminum? How do you know?

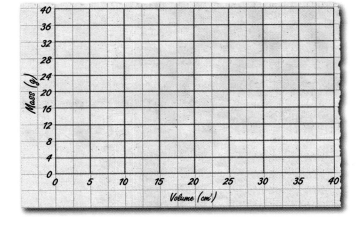

121

 ## Cooperative Learning
EXPLORATION 6

Group size: 2 to 3 students
Group goal: to compare the mass-versus-volume graphs for an iron object and an aluminum object
Positive interdependence: Have one student measure mass and volume for either the aluminum or the iron object. Have another student measure the mass and volume of the remaining object. Have a third student record and graph the data.
Individual accountability: Have students copy their graphs onto the blackboard or an overhead transparency so that the class can compare data and ask questions.

Homework

Have students write a brief statement that describes the difference between mass and weight.

 An Exploration Worksheet (Teaching Resources, page 44) and Transparency 17 are available to accompany Exploration 6.

This Exploration suggests that iron and aluminum objects be used. Other materials such as glass and modeling clay can also be used. The text suggests that each group determine the mass and volume of an object. This data can be shared with the class. Alternatively, each group could determine the mass and volume of several iron and aluminum objects. They could then graph their own data.

Students will observe that the data for all of the iron objects will fall on a straight line that goes through the origin. The data for all of the aluminum objects will fall on a different straight line that goes through the origin. Because different substances produce different mass-versus-volume lines (different *slopes*), substances can be identified by their mass-versus-volume data.

Answers to
Exploration 6

4. The lines should all go through the origin because if the volume of an object were zero, it would no longer contain any material; therefore, the mass would be zero as well. Both the volume and the mass are zero at the origin.

5. It is often possible to identify a material given only its mass and volume because different materials tend to have different mass-to-volume ratios, or densities. The densities of the two cubes that Jon brought to class (see page 119) are 7.24 g/cm³ and 1.62 g/cm³. The density of aluminum is 2.70 g/cm³ and the density of iron is 7.86 g/cm³. So although the density of Cube 1 is close to that of iron and the density of Cube 2 is close to that of aluminum, both cubes are probably made of materials other than iron or aluminum.

   Some students may suggest that the cubes could be a mixture of different materials. Point out that even if the densities were the same, the materials would still not be known because different mixtures of different materials could have the same density.

Answers to
Jon's Class Project

1. Copper is made up of atoms. All copper atoms have the same mass. There are twice as many atoms packed into 2 cm³ as there are in 1 cm³.

2. The masses of equal volumes of copper and water are different because either the particles that make up the objects have different masses or there are more copper atoms than hydrogen and oxygen atoms in a given amount of volume, or both.

CROSS-DISCIPLINARY FOCUS

Mathematics

Volume is expressed in cubic centimeters because objects exist in three dimensions—all have width, height, and depth. Have students complete the following volume problems:

a. Find the volume of a box with a height of 10 cm, a width of 5 cm, and a depth of 3 cm. *(150 cm³)*

b. Find the volume of a cube with a height of 8 cm. *(512 cm³)*

c. Find the height of a box with a volume of 240 cm³, a width of 5 cm, and a depth of 6 cm. *(8 cm)*

Meeting Individual Needs

Gifted Learners

The Greek mathematician Archimedes was once asked to find out whether an object was pure gold or merely gold plated. Because the object was potentially very valuable, Archimedes had to devise an experiment that would not damage the object. Have interested students do research to find out how Archimedes solved this problem. Students should then prepare a visual presentation that demonstrates Archimedes' solution.

 Transparency 18 is available to accompany Jon's Class Project.

Jon's Class Project

Each student in Jon's ninth-grade class collected mass and volume values for a variety of materials. Here is the graph they prepared from the data.

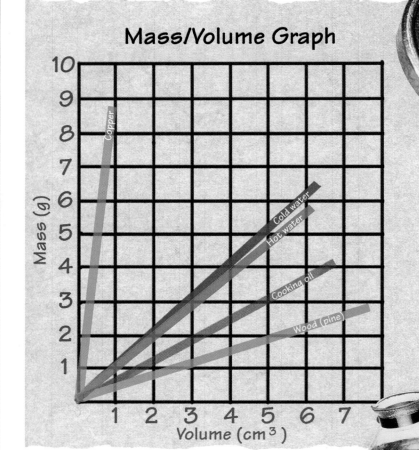

Try to interpret their data in terms of your understanding of the particle model of matter.

1. Why does 1 cm³ of copper have a mass of 8.9 g, and why would 2 cm³ of copper have a mass of 17.8 g?

2. Why would the mass of 5 cm³ of water be less than the mass of the same volume of copper?

Theme Connection

Structures

Osmium is a bluish white metallic element and is the densest substance on Earth. However, individual osmium particles are less dense than individual particles of gold, lead, mercury, or uranium. **Focus question:** How can this be so? *(The close spacing of osmium particles in an osmium crystal gives osmium its great density. More osmium particles fit into a cubic centimeter than do other, more massive and more widely spaced particles of other elements.)*

3. Why would $5 \, cm^3$ of cold water have a greater mass than the same volume of hot water? Be sure to explain this observation based on what you discovered earlier about the behavior of particles.

4. From your experiment and the class's, determine the mass of $1 \, cm^3$ (1 mL) of each of the following materials: iron, aluminum, copper, cold water, oil, and wood. By determining the mass of each of these materials per unit volume, you have found the *density* of each substance. (You will learn more about density in Unit 4.)

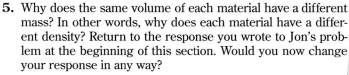

5. Why does the same volume of each material have a different mass? In other words, why does each material have a different density? Return to the response you wrote to Jon's problem at the beginning of this section. Would you now change your response in any way?

6. From your experience, which of the materials in the graph will float in cold water? Which ones will sink? Suggest how you could use the class's graph to predict which materials will float and which will sink.

123

FOLLOW-UP

Reteaching

Have students compare Method 3 in this lesson with Methods 1 and 2. Ask which method is better and why. *(Method 3 is better than Method 2 because the measurements in Method 3 are made in milliliters, thus reducing the amount of rounding error in the answer. Also, the volume of irregularly shaped objects can be easily determined with Method 3 but not with Method 2. Method 3 is better than Method 1 because there is no chance that water will be spilled in Method 3, which would reduce the accuracy of the answer.)*

Assessment

Ask students to investigate the densities of water, ethyl alcohol, and glycerin. Supply them with a metric balance and a graduated cylinder. Then have students provide a description and graph of their results and an answer to the question, Can a liquid substance be identified by its density? *(See the answer to question 5 on page 121.)*

Extension

Point out to students that, unlike most other substances, water expands rather than contracts when it freezes. Have students write a narrative describing what life might be like if water did not exhibit this behavior. *(Sample answer: Ice would not float on water, so the surfaces of lakes would not freeze over in the winter. Many bodies of water would probably freeze solid.)*

Closure

Encourage students to create a visual presentation that demonstrates the relative densities of objects compared with a common substance such as water. Presentations may include posters, models, graphs, or diagrams. Students may wish to include the densities of such materials as air, lead, and the nucleus of an atom.

Answers to
Jon's Class Project, continued

3. As water is heated, its molecules speed up and occupy a greater volume. Therefore, $5 \, cm^3$ of hot water contain fewer molecules than the same volume of cold water.

4. The masses of $1 \, cm^3$ of the various substances are as follows:
   - iron: 7.9 g
   - aluminum: 2.7 g
   - copper: 8.9 g
   - cold water: 1.0 g
   - oil: 0.63 g
   - wood: 0.38 g

5. The same volumes of different materials have different masses because either the particles making up the substances have different masses, or each substance has different numbers of particles packed into the same volume, or both.

6. Wood and oil float in water. On the graph, all materials whose data lines lie below that of water will float.

Homework

The questions in Jon's Class Project on pages 122–123 make an excellent homework activity.

Answers to
Challenge Your Thinking

1. Both examples (a) and (b) can be explained by the particle model of matter. Particles of dust and smoke are extremely small and have very small masses. They tend to move around in air because they are carried by slight currents in the air. In addition, these particles tend to move randomly within the air currents. This haphazard, dancing motion happens because smaller particles of air collide with the dust or smoke particles and cause them to move in different directions. Thus, the smoke particles in example (b) move around even in a box where no air currents are present.

2. For a completed diagram, including a sample title, see page S207.

3. a. Solids 1 and 3 melted when heated. The horizontal section of the line on those graphs indicates that the molecules of those substances were gaining the energy needed to change state rather than becoming hotter.
   b. Solid 2 has the highest melting point because it has not yet melted.
   c. Solid 2, for the same reason as above

CHALLENGE YOUR THINKING

1. A Matter of Particles
Can the particle model of matter explain the phenomena below? If so, how?

a. When a sunbeam shines into a room, you can see dust particles dancing about in the path of light.

b. If you could trap a little smoke inside a transparent box and put it under a microscope, you would see the particles of smoke moving in a haphazard pattern.

2. Name This Diagram!
Copy the diagram at right into your ScienceLog. Then complete it by following the instructions below.

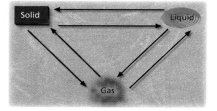

a. Come up with a title for the diagram that summarizes the diagram and is accurate, descriptive, and catchy!

b. Place one of the following words on each arrow:
   - melting
   - freezing
   - condensing
   - vaporizing (or evaporating)
   - subliming (changing directly from solid to gas or vice versa)

c. Label each arrow as signifying either an endothermic or exothermic change.

3. Solid Evidence
Three solids were heated, and their temperatures were plotted against the heating time (shown at left).

a. Which substance(s) melted when heated? How do you know?

b. Which substance has the highest melting point?

c. Which substance would seem to have the strongest forces of attraction between its particles?

Solid 1 — Temperature / Time (min.)

Solid 2 — Temperature / Time (min.)

Solid 3 — Temperature / Time (min.)

 You may wish to provide students with the Chapter 6 Review Worksheet that is available to accompany this Challenge Your Thinking (Teaching Resources, page 46).

Meeting Individual Needs

Gifted Learners
Have students research the freezing and boiling points of a common element or compound. What does this information suggest about how strongly the particles are held together? *(In general, the lower the freezing and boiling points of a material, the more loosely its particles are held together.)*

4. Heavy Subject

The density of an object is the ratio of its mass to its volume ($D = m/v$). Study the diagram and chart below. What kind of wood is the block probably made of?

Wood	Density
balsa	0.13 g/mL
birch	0.64 g/mL
pine	0.42 g/mL

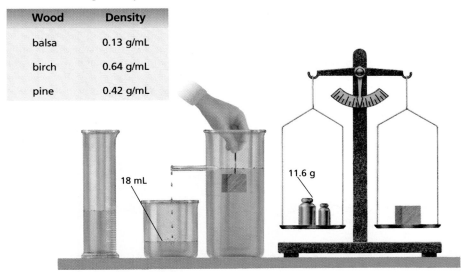

5. Hammering It Home

Janet concluded that according to the particle model of matter, a hot iron nail should have less mass than a cold iron nail. Is she correct in her thinking? Why?

6. Food for Thought

Michael said his data proved that a can of soup would sink in water. Was he right?

Review your responses to the ScienceLog questions on page 107. Then revise your original ideas so that they reflect what you've learned.

Meeting Individual Needs

Learners Having Difficulty
Have students develop a skit that demonstrates to younger students how the motion of molecules determines the state of matter. The skit should include at least one example of each type of change. If possible, have the students perform their skit for elementary students.

---

4. The mass of the piece of wood is 11.6 g. Its volume is 18 mL. Since density is mass divided by volume, the density of the wood is 11.6 g/18 mL, or 0.64 g/mL. Therefore, the wood is probably birch.

5. Janet is incorrect in her thinking. The mass of a substance remains the same no matter how hot or cold the substance is. Janet is probably referring to density rather than mass. Heating a nail causes the nail to expand, increasing its volume. Since density is mass divided by volume and the mass is constant, the density of the nail would decrease when heated.

6. Michael was right. The density of the can of soup is 323 g/284 mL, or 1.14 g/mL. Because this is greater than the density of water, the can of soup would sink in water.

The following are sample revised answers:

1. The heat energy being supplied to the water is used to change the water from a liquid to a gas and not to raise its temperature.

2. Density equals mass divided by volume ($D = m/v$). Another way of stating this is mass equals volume times density ($m = vD$). As indicated by the balance, the masses of the two liquids are the same. Because mercury is much more dense than water, it takes a much smaller volume of it to equal the mass of the greater volume of water.

3. The person might shiver on a hot day because of the effects of evaporative cooling. As the sweat or water on his or her skin evaporates, it draws energy from his or her body, thereby cooling him or her.

Connecting to Other Chapters

> **Chapter 4**
> introduces scientific models and how exponents are used to describe large and small numbers.

> **Chapter 5**
> explores the evidence for an atomic theory of matter and discusses the states and structure of matter.

> **Chapter 6**
> investigates how temperature changes affect matter and explores the properties of matter.

> **Chapter 7**
> surveys the historical development of our understanding of the atom and the particles contained in its nucleus.

Prior Knowledge and Misconceptions

Your students' responses to the ScienceLog questions on this page will reveal the kind of information—and misinformation—they bring to this chapter. Use what you find out about your students' knowledge to choose which chapter concepts and activities to emphasize in your teaching. After students complete the material in this chapter, they will be asked to revise their answers based on what they have learned. Sample revised answers can be found on page 137.

In addition to having students answer the questions on this page, you may wish to have them complete the following activity: Have students draw a picture representing their view of the atom with as much detail as possible. Allow students to share their responses, but do not correct or add to the ideas presented. Assure students that there are no right or wrong answers to this exercise. Collect their pictures, but do not grade them. Instead, use the pictures to identify possible problem areas and to find out what students know about the structure of the atom, what misconceptions students may have, and what aspects of this topic are interesting to them. As students proceed through the chapter, they will compare their view of atoms with those of Dalton and other scientists.

CHAPTER

7

The World of Atoms

1
Do you think there is anything smaller than an atom?

2 How do we know what atoms look like?

3
How small do you think an atom is?

ScienceLog

Think about these questions for a moment, and answer them in your ScienceLog. When you've finished this chapter, you'll have the opportunity to revise your answers based on what you've learned.

126

LESSON 1 · Picturing an Atom

Viewing the Invisible

The particle model explains many observations about the behavior of matter. According to the model, all matter is made up of particles. Dalton called these particles atoms. As you found out in Chapter 6, everything on Earth is made up of 91 different elements. Since there are 91 different elements that exist naturally on Earth, there must be 91 different kinds of atoms.

Imagine one of these atoms magnified large enough for you to see. What would it look like? How does one kind of atom differ from another? Why is an atom of gold different from an atom of oxygen? In this section A you will follow in the footsteps of others who, through their insight and creativity, provided us with a mental picture of the unseen structure of matter and in the process answered these and many other questions.

Before seeing how others pictured an atom, draw one yourself. In your diagram, include your thoughts about the nature of the atom. Does it have a definite shape? Is it made up of identifiable parts, or is it solid like a ball bearing? B Adding labels to your diagram will help you to communicate your ideas and thoughts to others. Then read on as scientists from the past give their views on the nature of the atom.

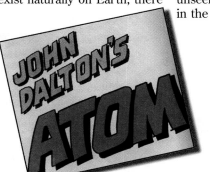

Mr. Dalton, you hypothesize that all matter is made of atoms, do you not?

Quite right. My research has convinced me of this.

Continued

127

FOCUS

Getting Started

Have students imagine the following situation: Imagine that you fall off your chair in slow motion. As you fall, you also shrink in size. You are falling to the floor, but as you approach it you can see that the floor is not solid, as it had appeared when you were bigger. Instead, the floor is composed of hazy blobs of matter. You are small enough now to strike one of the blobs and be crushed by the impact. But before you hit the surface of the particle, you find that it has no solid surface at all. You have fallen into an individual particle of matter—an atom—and you discover that it is mostly empty space.

Explain to students that this lesson will prepare them to investigate the inner structure of the atom.

Main Ideas

1. John Dalton proposed that matter is made up of tiny, indivisible particles called *atoms.*
2. J. J. Thomson conducted experiments that led to his "plum pudding" model, a refinement of Dalton's original particle theory.

TEACHING STRATEGIES

Answers to *In-Text Questions*

A As students will discover by the end of the chapter, an atom is a hazy region consisting of a central nucleus surrounded by moving electrons. Students will also discover that atoms differ from one another in the number of protons contained in the nucleus. Gold atoms contain 79 protons, while oxygen atoms contain 16 protons.

B Atoms do not have definite shapes. Rather, they are hazy regions of space with boundaries that are not sharply defined. Atoms do contain identifiable parts—subatomic particles.

LESSON 1 ORGANIZER

Time Required
1 or 2 class periods

Process Skills
inferring, analyzing, predicting

New Terms
none

Materials (per student group)
none

Teaching Resources

Homework

The questions in Reflecting make an excellent homework activity.

Did You Know...

In 400 B.C., Democritus proposed that all matter is composed of atoms. However, Aristotle's view that matter is continuous prevailed until the seventeenth century. In 1808, Dalton organized the experimental evidence for atoms into his atomic theory. Dalton's statement of the atomic theory finally laid to rest Aristotle's view of continuous matter.

Reflecting

1. What are the main characteristics of atoms as described by John Dalton?

2. How is his picture of an atom similar to yours?

CROSS-DISCIPLINARY FOCUS

Language Arts

Have students compose a short story in which they shrink to the size of a sub-atomic particle and travel through an atom. Remind students of the vast distance between an atom's nucleus and the orbits of its electrons. What would be the advantages and disadvantages of being this small? What would they discover? What might they be able to accomplish? How would this change the way they think about the world?

PORTFOLIO

Have students include their short story about their travel through an atom in their Portfolio. You may wish to have students complete the checklist called About My Portfolio, which is available both in the Assessment Checklists and Rubrics booklet and on the *SciencePlus SnackDisc*.

J. J. Thomson's View of the Atom

The year is 1897. You are listening to a lecture given by Joseph John Thomson, professor of physics at the famous Cavendish Laboratories of Cambridge University in England. He has just shown the audience what he thinks an atom looks like. How is his model different from the one suggested by John Dalton? Ⓐ How does it compare with your diagram of an atom? Ⓑ

I see the atom as being made up of electricity. I call my picture the "plum pudding" model of an atom. Note that in this model negatively charged particles called electrons (the plums) are embedded in a sphere of positive charge (the pudding). A more familiar analogy may be a muffin with raisins. The raisins represent the electrons, while the rest of the muffin represents the positive charge. Read on to find out how my plum pudding model led to important new discoveries about the atom.

Homework

Have students write down two more questions that they would ask John Dalton, and then have them do some research to answer the questions.

FOCUS

Getting Started

Give each student (or group of students) some modeling clay and a different number of washers, bolts, and nuts. Ask students to embed their hardware in the clay and then form the clay into a ball. Students will then trade the clay balls and try to find out what is inside them. They can only use toothpicks to probe the interiors. Discuss the results with the class, and explain that Rutherford's experiment with the atom was similar to the students' experiment with the clay balls.

Main Ideas

1. The atom is mainly empty space with a dense nucleus.
2. The nucleus contains protons and neutrons, while electrons occupy space beyond the nucleus.
3. The nucleus makes up most of the mass of the atom.
4. The Bohr model represented electrons as traveling in discrete orbits around the nucleus.
5. The Bohr model has been modified in light of more recent experiments.

TEACHING STRATEGIES

Throughout this lesson, you may wish to have students refer to the three bulleted items on page 84. They should use these criteria to evaluate the strength of the models they will encounter in the following pages.

Answers to
In-Text Questions

Ⓐ The chief difference is that Rutherford's model includes a nucleus around which electrons are scattered. Students' models will vary but may closely resemble Rutherford's model. However, many students' models may more closely resemble the Bohr model, which features electron orbits.

Rethinking the Plum Pudding Model—Ernest Rutherford's Atom

As a scientist, Ernest Rutherford was very familiar with the "plum pudding" model of the atom, having studied under J. J. Thomson at Cambridge University. However, Rutherford's research led him to another view of the structure of the atom.

Listen in as he describes his atomic model.

Nucleus containing protons

Empty space

Electrons

I suggest that the atom has the following characteristics:

* **It consists of a small core, or nucleus, that contains most of the mass of the atom.**
* **This nucleus is made up of particles called protons, which have a positive charge.**
* **The protons are surrounded by negatively charged electrons, but most of the atom is actually empty space.**

How is Rutherford's diagram different from Thomson's? How does it compare with yours? Ⓐ

Since the atom is far too small to see, how did Rutherford arrive at these surprising conclusions? The answer is that he shot tiny "bullets" at atoms to probe their internal structure. Read on to see how this was done.

130

LESSON 2 ORGANIZER

Time Required 3 or 4 class periods

Process Skills
inferring, analyzing, predicting, comparing, contrasting

Theme Connection
Structures

New Terms
Electron—a particle of an atom that has a negative charge and is located outside the nucleus
Neutron—an electrically neutral part of the atom with a mass similar to that of the proton, located inside the nucleus of the atom

Proton—a particle of the atom that has a positive charge and is located inside the nucleus of the atom

Materials (per student group)
Exploration 2, Part 1: 100 mL graduated cylinder; metric balance; 125 mL of long-grain rice; **Part 2:** two 25 cm pieces of string; a small ball bearing; 2 grains of sand

Teaching Resources
Exploration Worksheets, pp. 52 and 54
Resource Worksheet, p. 56
SourceBook, pp. S22 and S25

Probing the Atom

The "Plum Pudding" Atom

A simulation uses a model to test predictions about the behavior of real objects, processes, and systems. In this simulation, "bullets" will be shot at a model of an atom to find out more about its internal structure. An empty box, open at both ends, will represent the atom as conceptualized by J. J. Thomson. Pieces of popcorn will represent the electrons embedded in a sea of positive charge. The overall atom is neutral.

Predict

What would happen if table-tennis balls were shot at this "atom"? Would they pass straight through, or would they be deflected in different directions?

Observe

Here is what happens.

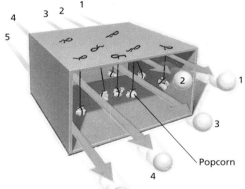

Popcorn

Explain

Why were the balls not deflected by this model of the atom? **B**

Probing Another Structure for the Atom

Predict

This time you will not be able to see inside the box. Can anything be inferred about what is inside the box from the following simulation? Again, balls are shot at the "atom." The diagram below shows what happened.

Observe

Ball 1 passes straight through.
Ball 2 deflects to the right.
Ball 3 bounces back.
Ball 4 passes straight through.
Ball 5 passes straight through.

Explain

How must the arrangement inside this model atom differ from that in Simulation 1? Draw a **C** diagram to show what you infer the inside of this "atom" might look like.

EXPLORATION 1

You may wish to make models of the atom described in the Exploration. If so, you will need two cardboard boxes, open at opposite ends, to represent the atomic models. For Simulation 1, popcorn or cereal can be used to represent the electrons. For Simulation 2, use a massive object such as a small bottle filled with water to represent the nucleus. Secure the bottle in the box so that it does not move on impact.

Have students analyze how Simulation 1 represents Thomson's model. The cereal is distributed more or less evenly throughout the atom, which is how Thomson viewed the atom. The balls pass straight through without being deflected.

In Simulation 2, the balls are sometimes deflected at an angle or even straight back. Only those balls going through the center of the box are deflected. If this happened in a real experiment, it would suggest a model of the atom similar to Rutherford's.

Answers to
In-Text Questions

B The balls were not deflected by the pieces of popcorn (representing the electrons) because the less-massive pieces of popcorn were pushed out of the way by the larger "bullets."

C In this simulation, the inference is that there must be something more massive within the box to deflect the balls than there was in Simulation 1.

Integrating the Sciences

Life and Physical Sciences

The rate at which radioactive carbon-14 atoms decompose into nonradioactive atoms can tell us how long ago an organism lived. The time required for one-half of a sample to decay is known as the *half-life* of that substance. Carbon-14 has a half-life of 5730 years. Plants absorb the radioisotope carbon-14 in the form of carbon dioxide from the atmosphere. Animals get carbon-14 by eating plants or by eating animals that have eaten plants. When an organism dies, the carbon-14 continues to disintegrate, but it cannot be replaced. The age of the remains can therefore be estimated by determining how much carbon-14 is left.

Homework

The Exploration Worksheet that accompanies Exploration 1 makes an excellent homework activity (Teaching Resources, page 52).

ENVIRONMENTAL FOCUS

Nuclear energy is produced by a process called fission, in which a nucleus is split, forming new elements and releasing energy. The energy is used to produce steam that drives an electric generator. Unfortunately, this process produces radioactive waste, and the United States has no facility for permanently disposing of this waste. Have interested students conduct research on radioactive waste disposal and present their findings to the class in the form of an oral report.

The Actual Experiment: Rutherford's Gold Foil Experiment

When Rutherford's team performed the gold foil experiment, they noticed that a number of the "bullets" were deflected at large angles by the atoms making up the foil. This was contrary to the predictions based on Thomson's model.

Answers to *What Rutherford Observed*

1. Simulation 2 produced results similar to those of the gold foil experiment.

2. The fluorescent screen was used to detect the presence of the alpha particles. By observing the flashes, Rutherford's assistants could determine whether the particles were deflected.

3. Most of the alpha particles passed straight through the gold atoms. These are the particles that did not come close to the massive, positively charged gold nuclei.

4. The alpha particles that bounced back probably made direct hits on something more massive than themselves.

5. The alpha particles that were deflected slightly but still hit the screen behind the gold foil probably came close to a positively charged object but did not hit it.

Homework

The questions provided under What Rutherford Observed make an excellent homework activity.

The Actual Experiment: Rutherford's Gold Foil Experiment

Rutherford and his assistants performed similar experiments by shooting atomic "bullets" at real atoms. Instead of table-tennis balls, Rutherford used helium nuclei, each with a positive charge. These charged atoms are called alpha particles. The alpha particles were shot at the gold foil. The gold foil had been made extremely thin so that the alpha particles did not have to pass through great numbers of atoms. What do you think Rutherford predicted before the experiment was performed?

What Rutherford Predicted

Since alpha particles are much more massive than electrons, Rutherford predicted results similar to those in Simulation 1.

What Rutherford Observed

Rutherford was extremely surprised by the results; he compared the experiment to firing a cannon at a piece of tissue paper and having the shells bounce back. Examine the diagram below for his observations.

1. Which simulation produced results most similar to the gold foil experiment?

2. What was the purpose of the fluorescent screens in the gold foil experiment?

3. Which alpha particles passed straight through the gold atoms?

4. Which particles probably made a direct hit on something more massive than themselves?

5. Which particles may have approached, but not directly hit, something with a positive charge?

The Setup for Rutherford's Famous Gold Foil Experiment

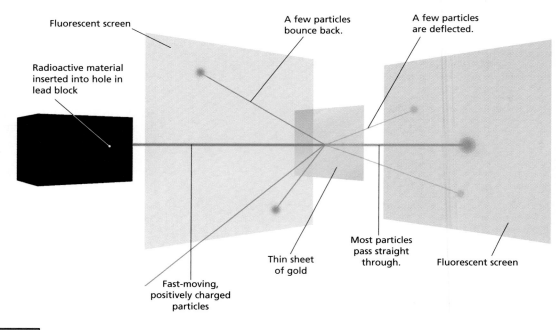

Fluorescent screen

Radioactive material inserted into hole in lead block

A few particles bounce back.

A few particles are deflected.

Thin sheet of gold

Most particles pass straight through.

Fluorescent screen

Fast-moving, positively charged particles

Theme Connection

Structures

Focus question: What features of Bohr's model of the atom are no longer accepted today? *(Bohr's model featured electrons moving in well-defined paths around the nucleus. Today, scientists believe that electrons are found in "clouds"—hazy regions with vague boundaries. Electron clouds have various shapes, but most are not shaped like the elliptical orbits found in Bohr's model.)*

Did You Know...

If a hydrogen atom were 4 mi. (6.4 km) in diameter, its nucleus would be the size of a tennis ball.

Rutherford's Nuclear Atom

Based on his observations, Rutherford arrived at a new model of the atom. The main features were described earlier but are repeated here. Respond to the question that follows each feature.

Atoms consist of a core, or nucleus, that contains most of the mass of the atom. (What evidence from the experiment suggested this?) **A**

The nucleus has a positive charge. (Why did Rutherford infer that the nucleus has a positive charge rather than a negative charge or no charge at all? Hint: What would happen to a positive alpha particle that approached but did not hit a positive nucleus? What would happen if the nucleus had a negative charge?) **B**

Most of the atom is empty space. (On what observation did he base this inference?) **C**

The Bohr Model

In 1913, Niels Bohr, a Danish scientist who worked with Rutherford, developed a theory that electrons travel around the nucleus in orbits like those of the planets around the sun. This theory is called the *Bohr model*. Just as Dalton's theory had to be changed in light of Rutherford's discoveries, Bohr's model has been modified in response to more recent discoveries.

I propose that the electrons of an atom travel around the nucleus in specific orbits. Take a look at my model of a hydrogen atom, shown below. A circular path represents the fast-moving electron, which orbits a nucleus consisting of a single proton.

▲ Even though the Bohr model has been replaced by more accurate models, it is still widely used to represent atoms today.

EXPLORATION 2

Exploring the Size of the Atom

PART 1

Modeling the Mass of Electrons and Protons

You Will Need
- a graduated cylinder
- long-grain rice

What to Do

1. Set one grain of rice aside. This represents the mass of one electron.

2. Measure 125 mL of rice. This will contain approximately 2000 grains of rice and represents the mass of one proton.

Questions

1. Compare the mass of 125 mL of rice with the mass of one grain of rice. According to this activity, how much more massive are protons than electrons?

2. How does this model demonstrate Rutherford's inferences about the structure of the atom?

Exploration 2 continued ▶

133

EXPLORATION 2

PART 1

Rice is used to compare the mass of an electron and a proton. If the mass of an electron is represented by one grain of rice, then the mass of the proton would be the equivalent of about 2000 grains of rice, or about 125 mL. You could use other objects in the same proportion, such as 1 ball bearing to 2000 ball bearings. (The actual ratio of masses is a bit less than 1:2000; it is 1:1856.)

Answers to
Part 1

1. The mass of 125 mL of rice is 2000 times greater than that of one grain of rice. A proton is about 2000 times as massive as an electron.

2. This model demonstrates that an atom's mass is concentrated in its nucleus. Only such a massive object in the atom could deflect alpha particles so effectively.

PART 2 *page 134*

In this simulation, students pace off distances that represent the positions of the electrons closest to the nucleus. This kind of activity is important in helping students obtain a visual picture of the atom and the distances involved.

As an extension after students have paced off the distances involved, have other students represent alpha particles and walk through the "atom." Note the opportunity for collisions. Ask: Which alpha particles will be deflected? Which will strike electrons but keep going because of the greater mass of the alpha particles?

★ An Exploration Worksheet is available to accompany Exploration 2 (Teaching Resources, page 54).

Answers to
In-Text Questions

A Students should recall that because a few particles were deflected by the gold foil, Rutherford concluded that most of the mass of the atom is concentrated in a small, compact core. Only such a massive object could cause the observed deflections.

B The fact that some alpha particles bounced back implies that they were repelled by the nucleus. Because alpha particles are positively charged, the nucleus must be positively charged also. If the nucleus were negatively charged, the alpha particles would have been attracted to the nucleus and far fewer would have bounced back.

C The fact that most of the alpha particles passed straight through the gold foil implies that the atom is mostly empty space.

Answers to
Part 2

1. This simulation illustrates Bohr's idea that the electrons are at fixed distances from the nucleus.

2. Most alpha particles passed through the gold foil because the atom is mostly empty space. If an alpha particle hits an electron, the alpha particle would not be deflected because its mass is greater than that of an electron. Only a direct or near hit on the nucleus would cause the alpha particles to be strongly deflected. If the positive alpha particle merely approached the positive nucleus, there would be some deflection because of the repulsion between like charges.

Theme Connection

Structures
Focus question: Why is Bohr's model of the atom often called the planetary model? *(The Bohr model features electrons that orbit the nucleus like planets orbit the sun.)*

Homework

The Resource Worksheet that accompanies Particle Summary on page 135 makes an excellent homework assignment (Teaching Resources, page 56).

PART 2

Modeling the Space Within Atoms

You Will Need
(for each group of three students)

• two 25 m lengths of string
• a small ball bearing or BB
• 2 grains of sand

What to Do
(This activity is best done outdoors.)

1. Have one person hold the ball bearing. (This ball bearing represents the nucleus of an atom.)

2. The person holding the "nucleus" will also hold one end of each string while the other two students, each holding a grain of sand and the other end of a string, pace out 25 m in different directions. (The grains of sand represent the electrons in an atom.) The position of each student represents the distance of the closest electrons to the nucleus in an atom.

Questions

1. How does this simulation demonstrate Bohr's theory of the structure of the atom?

2. Why did most of the alpha particles in Rutherford's gold foil experiment pass directly through the gold foil, while only a few were deflected back?

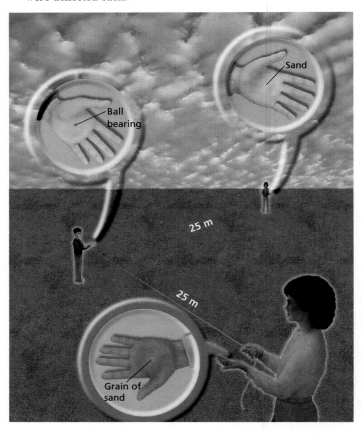

Meeting Individual Needs

Second-Language Learners
Have students design and build or illustrate an atomic model that is labeled in both English and their native language. Tell students that their model does not have to be to scale, and they may build an atom representing the element of their choice. Grade the models on scientific accuracy, with a minimal emphasis on language proficiency. Be sure to check students' designs for safety before allowing them to proceed.

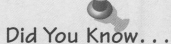

Did You Know. . .

J. J. Thompson won a Nobel Prize in physics in 1906, and James Chadwick won one in 1935. Ernest Rutherford won a Nobel Prize in chemistry in 1908. All of these scientists were awarded this honor for their work on the atomic model.

More Particles in the Nucleus?

James Chadwick, while working with Ernest Rutherford in 1932, discovered that another particle, in addition to the proton, is found in the nucleus of the atom. This particle has no charge and is about the same mass as the proton. It received the name *neutron*.

James Chadwick

To read more about the neutron and other particles smaller than the atom, see pages S35—S38 of the SourceBook.

Particle Summary

Particle	Charge	Mass (compared to electron)	Location
proton	positive (+)	almost 2000 times as massive	in the nucleus
neutron	neutral	almost 2000 times as massive	in the nucleus
electron	negative (–)		outside the nucleus

1. Where is most of the mass of the atom concentrated? What contributes the most to its mass?

2. If the mass of the proton is said to be one atomic mass unit (1 amu), what would be the approximate mass of an atom of lithium, which has a nucleus made up of three protons and four neutrons? (Ignore the comparatively tiny mass of the electrons.)

3. Below are Bohr models for atoms of helium, lithium, and beryllium. Each red orbit represents the path of a single electron.

   a. What do all atoms have in common?

   b. How do atoms of different elements differ?

   c. Draw diagrams for the following atoms:

   • nitrogen (7 protons and 7 neutrons in its nucleus)

   • oxygen (8 protons and 8 neutrons)

   • argon (18 protons and 21 neutrons)

4. By insightful experiments and reasoning, John Dalton discovered that each kind of atom has its own distinctive mass. (For example, oxygen atoms are 16 times more massive than hydrogen atoms.) Of course, Dalton didn't know anything about electrons, protons, and neutrons. How would you explain to him the reasons for his conclusion?

• Protons
• Neutrons
╱ Electron path

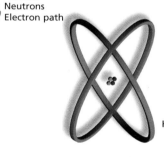

He

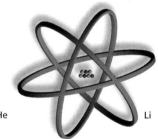

Li

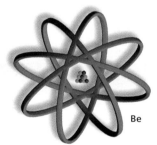

Be

Reteaching

Use the following analogy to further reinforce Rutherford's gold foil experiment: Imagine that there are 50 billiard balls hung from strings at various heights throughout the classroom. A student is blindfolded and given five tennis balls to throw toward the center of the room. Ask students what would happen to the tennis balls? Though some would go straight through, some would be deflected to the side and some would be deflected straight back because they hit a billiard ball.

Assessment

Have students choose either Rutherford's or Bohr's model of the atom. Students should write mock newspaper articles describing the model as if they are reporters covering the story for the first time. All of the important features of the model should be included.

Enrichment

Have students research the currently accepted atomic model, the "electron cloud" model, which is based on quantum mechanics. Have students present their findings to the class in the form of an oral report.

Closure

Ask students to develop analogies for the atomic model. For example, one might say that the atom is like a city because most of the population (mass) tends to be clustered together in the center (nucleus) and diminishes with distance, into the surrounding suburbs (electrons.)

Answers to *Particle Summary*

1. Most of an atom's mass is concentrated in the nucleus. Protons and neutrons make up most of the mass.

2. Lithium atoms have an atomic mass of 7 atomic mass units.

3. a. Sample answer: All atoms have a nucleus, with electrons occupying space outside of the nucleus. The number of protons equals the number of electrons. All atoms except hydrogen have neutrons in the nucleus.

   b. Sample answer: Atoms of different elements differ in the number of protons, neutrons, and electrons. The number of protons in the nucleus is what identifies an atom of a particular element.

   c. Following the examples at the bottom of the page, student drawings of these atoms should each have as many ovals as electrons. At the intersection of the ovals (the center of the atom) students should draw as many dots as the number of protons and neutrons combined. They should distinguish between them by using different colors for the two types of particles. Also, make sure that the number of protons equals the number of electrons (ovals) in the drawings.

4. Sample answer: You might explain to Dalton that each atom is made up of electrons, protons, and neutrons that give it a unique mass. Since oxygen has 8 protons and 8 neutrons in its nucleus, and hydrogen has only 1 proton, oxygen atoms must have 16 times the mass of hydrogen atoms.

Answers to *Challenge Your Thinking*

1. a. This carbon atom has 6 protons and 6 neutrons in the nucleus. It also has 6 electrons occupying the space around the nucleus.

b. There is no charge on this atom because the number of protons equals the number of electrons.

2. a. Both diagrams of hydrogen have 1 proton and 1 electron. However, the first diagram has no neutrons, while the second diagram has 1 neutron in its nucleus. Because of the added neutron, the second atom—known as *heavy hydrogen* or *deuterium*—has approximately twice the mass of the first atom.

b. An atom such as hydrogen is identified by the number of protons in its nucleus. (If an atom has the same number of protons as a certain element but a different number of neutrons, then the atom is a different *isotope* of the element. If instead an atom has the same number of protons as a certain neutral element but a different number of electrons, then the atom is an *ion* of the element.)

3. Students may need to research the contributions of Henri Becquerel and Marie Curie before drawing their time line. They should include the following information on their time line:

- 1808—John Dalton suggests that all matter is made up of small particles called atoms.
- 1896—Henri Becquerel discovers that certain atoms naturally break down, giving off energy in the process. This becomes known as *radioactivity.*
- 1897—J. J. Thomson concludes that every atom contains electrons.
- 1898—Marie Curie discovers two new elements: radium and polonium.
- 1911—Ernest Rutherford suggests that the atom has a dense nucleus.
- 1932—James Chadwick announces the existence of a neutral particle in the nucleus: the neutron.

CHALLENGE YOUR THINKING

1. Neutral Neighbors

Examine the Bohr model of an atom of the element carbon.

a. How many of each kind of particle make up this atom?

b. Is there a charge on this atom, or is it neutral? Why?

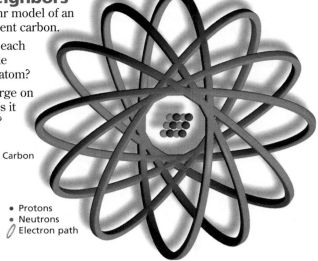

Carbon

- Protons
- Neutrons
- Electron path

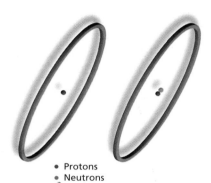

- Protons
- Neutrons
- Electron path

2. Same but Different

Here are Bohr models of atoms representing different forms of hydrogen.

a. How are they similar? How are they different?

b. What identifies an atom as hydrogen and not as some other atom, such as helium?

3. Timely Discoveries

Draw a time line that describes the discoveries made about the nature of the atom. Include the people encountered in this section and their discoveries. Then add two more names to the time line: Henri Becquerel and Marie Curie. What were their contributions? How might you find out?

Theme Connection

Changes Over Time

Focus question: How has our understanding of atomic structure evolved from the theories of Democritus and John Dalton? *(Democritus and Dalton theorized that atoms were indivisible particles. The current atomic model describes the atom as a system in which negatively charged electrons orbit a positively charged nucleus that consists of protons and neutrons. Atoms can, in fact, be split, generating tremendous amounts of energy in the process.)*

 You may wish to provide students with the Chapter 7 Review Worksheet that is available to accompany this Challenge Your Thinking (Teaching Resources, page 58).

4. Ballpark Figure

Here is another analogy for the atom: If the nucleus of an atom were as large as a grape, then the atom itself would be as large as Yankee Stadium. Use this analogy to explain the results of Rutherford's gold foil experiment.

5. Hair-Raising Experience

Here's an activity you can try:

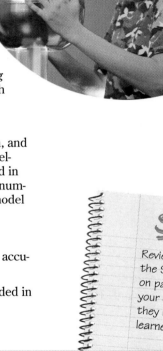

When an object is rubbed against something else, electrons (negatively charged particles) are transferred from one object to another.

a. Describe two or three occasions when this has happened to you.

b. If electrons are added to a plastic strip when it is rubbed with silk, what is the charge on the plastic strip?

c. What happens when this strip is brought close to small pieces of paper? (Try this yourself, using a comb that you have run through your hair.)

6. Piecing It Together

Choose whatever materials you wish, and construct a model of an atom for an element that you have not encountered in this unit. The model must show the number of electrons and protons. Your model could be rated in the following way:

- creative use of materials
- numbers of protons and electrons accurately portrayed
- bonus values if neutrons are included in your model

ScienceLog

Review your responses to the ScienceLog questions on page 126. Then revise your original ideas so that they reflect what you've learned.

137

4. Answers will vary. One possible response is as follows: Suppose that while standing far outside the stadium, one tries to hit the grape with much smaller objects, such as grape seeds. One would notice that most of the seeds fly through the stadium unaffected, while a few seeds bounce off of the grape and are knocked out of the ballpark in different directions. This is similar to Rutherford's experiment with gold foil because the probability of hitting the nucleus of a gold atom with alpha particles is very small. Most of the alpha particles pass undeflected through the gold foil. Additionally, the atom and the stadium consist mostly of empty space.

5. a. Some situations in which objects become electrically charged are the following:
   - rubbing one's feet on the carpet
   - drying clothes in a dryer (static cling)
   - combing one's hair
   - lightning during a thunderstorm
   b. Because the plastic strip will have more electrons than protons, it will have a negative charge.
   c. The small pieces of paper will be attracted to the charged strip.

6. Make sure you approve all materials selected by the students before they proceed. Students' models will vary. The materials that students can choose are numerous, such as clay, clothes hangers, wood, and paper.

The following are sample revised answers:

1. Atoms are composed of smaller particles called protons, neutrons, and electrons. These subatomic particles are in turn composed of even smaller particles.

2. There really is no way of actually "seeing" atoms. Common models of atoms are the result of theories and inferences based on the experiments that scientists perform.

3. An atom is an extremely small object. The typical diameter of an atom is about 10^{-10} m. (Since such a small number is difficult to comprehend, teachers may wish to ask students to consider the following comparisons: The diameter of an atom is less than one billionth of a meter, which is more than one million times smaller than the thickness of a human hair. The smallest speck visible under an ordinary microscope contains more than 10 billion atoms, and the period at the end of this sentence is made up of about 1×10^{18} atoms.)

Homework

Have students find out what *tritium* is. Ask: How is tritium related to hydrogen? (*Tritium is an isotope of hydrogen. Both contain 1 proton, but tritium has 2 neutrons, and hydrogen usually has no neutrons.*)

The Big Ideas

The following is a sample unit summary:

Observations are things that are perceived by the senses, while inferences are conclusions about why or how something occurs. Observations are recorded events or facts, while inferences are attempts to explain observations. (1)

Models are representations or simulations of structures or processes. They are useful in helping to determine whether inferences about the structures or processes are correct. (2)

There is much evidence that matter is made up of particles. When mixed together, two liquids have a combined volume that may be less than the sum of their separate volumes. Different types of matter combine in predictable ratios. Matter can be compressed, and it can change its state. (3) Everyday observations such as condensation forming on a surface, steam rising from boiling water, dust collecting on furniture, smelling scents from across the room, and salt or sugar dissolving in water can all be explained by the particle model of matter. (4)

John Dalton was a teacher who theorized that all matter is made up of "ultimate particles" that are of many different types. He also stated that particles of the same type of matter are identical and cannot be changed into other types. (5)

Solids are made up of particles that are tightly bound together and cannot move relative to one another. Liquids are made of particles that are loosely held together and are able to flow and take the shape of their containers. Gases are made of particles that are not held together at all and so are able to move independently. (6) Increasing the temperature of matter causes its particles to move faster, while decreasing the temperature causes them to move slower. (7) Endothermic changes are changes that require the addition of heat, such as melting and evaporation. Exothermic changes are changes that release heat, such as condensation and solidification. (8)

Objects with the same volume have different masses because either the masses of the particles that make up the objects differ or the spacing among the particles differs, or both. (9)

An atom consists of a massive, positively charged nucleus surrounded by less massive, negatively charged electrons. The nucleus is very dense, contains almost all of the mass of the atom, and is made up of protons and neutrons. In one model of the atom, electrons orbit the nucleus, but more recent models suggest that the nucleus is surrounded by a fuzzy cloud of electrons. Many scientists contributed to the discovery of the structure of the atom, including Dalton, Thomson, Rutherford, and Bohr. (10)

Making Connections

Unit 2

Particles

The Big Ideas

In your ScienceLog, write a summary of this unit, using the following questions as a guide:

1. How do observations differ from inferences?
2. What are models, and why are they useful?
3. What evidence is there that matter is made up of particles?
4. What everyday observations can be explained by the particle model of matter?
5. Who was John Dalton, and what did he conclude about the makeup of matter?
6. How does the particle model explain the properties of solids, liquids, and gases?
7. What is the effect of temperature changes on the particles making up matter?
8. What are some examples of endothermic changes? examples of exothermic changes?
9. Why do many objects with the same volume have different masses?
10. What is the general structure of the atom, and who helped discover this structure?

Checking Your Understanding

1. Why do we feel hotter on hot, humid days than on hot, dry days? Use the particle model of matter to explain your reasoning.

SOURCEBOOK

To find out more about the behavior of particles, look in the SourceBook. You'll discover how scientists measure particles as well as how particles interact. You'll also read about particles even smaller than protons, neutrons, and electrons.

Here's what you'll find in the SourceBook:

UNIT 2
Particles of Matter S22
Particles in Motion S27
Particles of Particles S35

138

Homework

You may wish to assign the Unit 2 Activity Worksheet as homework (Teaching Resources, page 63). If you choose to use this worksheet in class, Transparency 19 is available to accompany it.

 You may wish to provide students with the Unit 2 Review Worksheet that is available to accompany this Making Connections (Teaching Resources, page 64).

2. The following story contains at least five observations that can be explained with the particle model. List them and give explanations for each.

"One more dive and then we gotta go. We can't be late for dinner again. Mom'll get mad." Ben and Josh each dove off the cliff, neatly splitting the water. "It shouldn't take too long to dry in this sun," said Ben. "Oh great," groaned Josh. "My front tire's flat. Guess I should've filled it before we left. We'll have to walk it to the gas station." With the tire pumped up, the brothers raced home to make up for lost time. The breeze felt cool on their damp skin and hair. When they got home, their mother said, "Put your wet things in the dryer and come eat. I want one of you to mow the grass before it gets dark, while it's still dry. There'll be too much dew to mow in the morning." "No prob, Mom," said Josh, "Ben'll do it. Say, dinner smells great."

3. A balloon initially had a mass of 6.2 g and a volume of 2.3 L. What might have been done to the balloon to bring about the following changes?

	Mass	Volume
a.	6.2 g	3.2 L
b.	6.5 g	2.9 L
c.	6.0 g	2.5 L

4. Air fresheners are often placed in different areas of a home, such as kitchens, bathrooms, and basements. Over a period of weeks, the fragrant part of the air freshener gradually disappears. What happens to it? Use the particle model to explain.

5. concept map — Copy the concept map at right into your ScienceLog. Then complete the concept map using the following words: electrons, elements, atoms, nuclei, molecules, matter, neutrons, negative charge, protons, no charge, positive charge, and compounds.

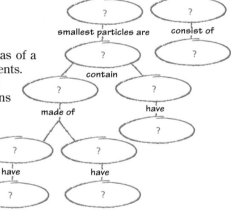

139

Answers to
Checking Your Understanding,
pages 138–139

1. We feel hotter on hot, humid days than on hot, dry days because water does not evaporate from our skin as quickly and does not carry away as much heat.

2. Observations and sample responses include the following:
- "Ben and Josh . . . splitting the water." The particles of water move aside when Ben and Josh dive in.
- "It shouldn't take too long to dry in this sun." Water is heated by the sun, which makes the water particles evaporate.
- "My front tire's flat." Air particles escaped through a small hole or a leaky valve stem.
- "With the tire pumped up . . ." Particles of gas can be compressed, causing the tire to inflate.
- "The breeze felt cool . . ." Wind causes water particles to evaporate more quickly, increasing the cooling effect.

Answers to
Checking Your Understanding,
continued

- "Put your wet things in the dryer . . ." Heat generated by a dryer causes water particles to evaporate, drying the clothes.
- "Dinner smells great." Particles leave the food during the cooking process and diffuse throughout the room, allowing us to smell the food.

3. Possible responses include the following:
a. The balloon has been filled with 0.9 L of air or a similar substance having negligible mass. Or the balloon may have been heated to expand its volume.
b. The balloon has been filled with a substance that has a density of 0.3 g/0.6 L, or 0.5 g/L.
c. The balloon has lost mass but gained volume, so the balloon could have been deflated and then reinflated with a less dense substance. Or the balloon may have lost some of its gas, but the remaining gas could have been heated, thus expanding the balloon's volume.

4. The particles of the fragrant material slowly leave the surface and spread out into the air. Eventually, we can smell the air freshener all over the room. When all the particles of the fragrant material have diffused into the air, the fragrant part of the air freshener will have disappeared.

5. Sample concept map:

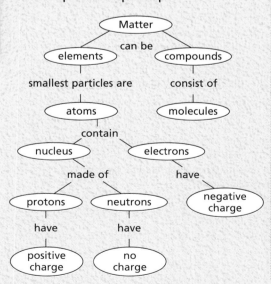

Background

Einstein's unique personality is almost as famous as his scientific accomplishments. It was rumored that he could become so lost in thought that he would forget where he was. He also cared little for money; when asked to name his salary at the Institute for Advanced Study in Princeton, New Jersey, he responded with a very small sum. (He was instead granted a salary five times his request.)

Einstein was born in 1879 in Ulm, Germany, and received his degree from the Swiss Polytechnic Institute in Zurich in 1900. He was awarded the Nobel Prize for physics in 1921 for a paper he had written in 1905 on the photoelectric effect.

The *photoelectric effect* occurs when a beam of light hits a metal object and the metal releases electrons. These electrons can be harnessed in an electric current. Einstein's paper offered the possibility that light, while it exhibits wavelike properties, can also be considered to consist of particles called *photons.* He explained that when these particles strike metal, they force the atoms in the metal to release electrons. In colors of light with too low a frequency, however, the photons do not carry enough energy to produce this effect.

Did You Know...

The Greek philosopher Democritus (ca. 460–370 B.C.) proposed that the building blocks of matter are particles too small to be seen. It was not until the 1600s—about 2000 years later—that the first living cells were observed. Individual atoms (from the Greek word *atomos,* meaning "indivisible") were not detected until the twentieth century.

Science Snapshot
Albert Einstein (1879–1955)

The year 1905 was an extraordinary time in the development of modern physics. In that year, a 26-year-old German patent clerk living in Switzerland published four papers that would forever change our understanding of the physical nature of matter and energy. This young patent clerk's name was Albert Einstein, considered today to be one of the greatest scientists to have ever lived.

The Existence of the Atom

Even though we have only recently developed the technology to see individual atoms, scientists have been collecting evidence for their existence for centuries. In 1827, for example, an English botanist by the name of Robert Brown placed tiny pollen grains on the surface of completely still water. He noticed that even though the water was perfectly still, the pollen grains moved around erratically. Scientists at the time thought that this effect, which is now called Brownian motion, might be caused in some way by living organisms. However, when it was demonstrated that even nonliving particles suspended in a fluid would undergo this motion, scientists

▶ **Albert Einstein overturned our view of reality when he was only 26. He won the Nobel Prize for physics when he was 42.**

began to consider other explanations. Some accepted the view that a fluid must be composed of particles that are in constant motion and that these particles periodically bang into the suspended particles, causing them to move.

Enter Einstein

Many scientists tried to develop theoretical explanations to support Brownian motion, but it was not until Einstein's paper on the subject that Brownian motion was placed upon a firm foundation. Einstein derived the mathematical equations that govern Brownian motion and used these equations to determine the size of the particles.

Einstein's creative mind and incredible mathematical skills

▶ **The zigzag path of a particle executing Brownian motion**

put him in the spotlight as a world-renowned physicist, but Einstein was more than just a scientist. He was also very active in human affairs, and his actions showed a deep concern for the politically and economically oppressed. He also loved classical music and even played the violin.

Einstein moved to the United States in 1933 and became a U.S. citizen in 1940. He died at his home in Princeton, New Jersey, in 1955. On his deathbed, he spoke his last words in German to a nurse who did not understand the language. We will never know what he said.

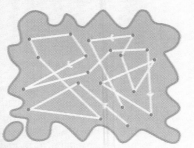

$E = mc^2$

Albert Einstein pushed the boundaries of physics with both philosophical and mathematical reasoning. He is perhaps best known for his theory of relativity and his famous equation, $E = mc^2$. Do some research to learn more about Albert Einstein. And in the process, discover the meaning of his famous equation.

Critical Thinking

Einstein once said that the ideal job for a scientist would be a lighthouse keeper. Ask: What do you think he meant by this? (*Accept all reasonable responses. Students may suggest that scientific investigation requires a lot of solitude and quiet time. They might also explore the idea that a lighthouse shining into the darkness prevents boats from running into unseen rocks. Another interpretation may focus on the lighthouse keeper whose sole purpose it is to keep the lighthouse clean, well oiled, and running.*)

Answers to
$E = mc^2$

Einstein derived this equation when he developed *the theory of special relativity.* The equation may be read "energy equals mass times the speed of light squared." This equation demonstrates that because the speed of light is a constant, any object's total energy is proportional to its mass. For this reason, scientists sometimes refer to mass and energy as being interchangeable.

Particles in the Air

Take a deep breath. You have probably just inhaled thousands of tiny specks of dust, pollen, and other particles. These particles, called particulates, are harmless under normal conditions. But if concentrations get too high or if they consist of harmful materials, they are considered to be a type of air pollution.

Where There's Smoke . . .

Unfortunately, dust and pollen are not the only forms of particulates. Many of the particulates in the air come from the burning of various materials. For example, when wood is burned, it releases particles of smoke, soot, and ash into the air. Some of these are so small that they can float in the air for days. The burning of fuels such as coal, oil, and gasoline also creates particulates. The particulates from these sources can be very dangerous in high concentrations. That's why particulate concen-

trations are one measure of air pollution. Because they are visible in large concentrations, they can also be responsible for the dirty look of polluted air. But don't be fooled—even clean-looking air can be polluted.

Eruptions of Particulates

Volcanoes can be the source of incredible amounts of particulates. For example, when Mount St. Helens blew its top in 1980, it launched thousands of tons of ash into the surrounding air. The air was so thick with ash that the area became as dark as night. For several hours, the

ash completely blocked the light from the sun. When the ash finally settled from the air, it covered the surrounding landscape like a thick blanket of snow. This layer of ash killed both plants and livestock for several kilometers around the volcano.

◄ The burning of most fuels adds harmful particulates to the air.

One theory about the extinction of dinosaurs is that a gargantuan meteorite hit the Earth with such velocity that the resulting impact created enough dust to block out the sun for years. During this dark period, plants were unable to grow to support the normal food chains. Consequently, the dinosaurs died out.

It's a Matter of Health

Because many particulates are so small, our bodies' natural filters, such as nasal hairs and mucous membranes, cannot filter all of them out. When inhaled, particulates in the lungs can cause irritation. Over time, this irritation can lead to diseases such as bronchitis, asthma, and emphysema. The danger increases as the level of particulates in the air increases.

◄ When the ash from Mount St. Helens settled from the air, it created scenes like this one.

Cigarette Smoke

Since the burning of most substances creates particulates, there must be particulates in cigarette smoke. Do some research to find out if the filters on cigarettes are effective at preventing particulates from entering the smoker's body. Your findings may surprise you!

141

Background

Particulate matter is one of the major forms of air pollution, along with carbon monoxide (a toxic gas), sulfur oxides (which contribute to acid rain and human respiratory problems), nitrogen oxides (which contribute to acid rain, smog, and human respiratory problems), and volatile organic compounds, or VOCs (organic chemicals that vaporize and produce toxic fumes).

Particulates are often formed during mechanical processes that break down materials. These include blasting, drilling, and grinding. Some organic matter, such as certain bacteria, are also considered particulates.

The growing problem of particulate pollution has been addressed by the federal government as well as by many local governments and communities. The Clean Air Act, passed by Congress in 1970, sets maximum emission levels for automobiles and industrial sources of pollution. New filtering technology has also helped industries to reduce the amount of particulates being released into the atmosphere.

Did You Know. . .

- In most cases, the majority of the particulates found indoors in dust are particles of discarded human skin.
- Certain types of asbestos are particularly dangerous as particulates. When inhaled, these asbestos fibers scar the lungs, inhibiting breathing and eventually causing cancer.

Meeting Individual Needs

Learners Having Difficulty

Have students collect samples of airborne particulates. Tell them to apply double-sided tape to some microscope slides. They should leave the slides overnight in several locations, both outside and inside. The following day, have them view the slides and compare the types and amounts of particulates seen on the slides.

Answers to
Cigarette Smoke

Cigarette filters absorb some of the particulates found in cigarette smoke, but not all of them. A filter that prevents all particulates from entering the lungs would be much too difficult to inhale through.

Background

The treatment of viral infections is often extremely difficult because the viruses themselves can change from one generation to the next. Any antibodies that the body produces to fight off the infection can become ineffective because they can no longer recognize new strains of the virus. Influenza is an example of such a virus. Every year the vaccines for influenza must be changed in order to account for changes in the virus. Sometimes these changes can occur faster than vaccines can be developed.

Unlike some forms of bacteria, no known virus is beneficial to humans. Also, many viral infections lead to other diseases. For example, the virus hepatitis B has been linked to hepatoma, a type of liver cancer.

Human cells are not the only victims of viruses. A special class of viruses known as *bacteriophages* invade, infect, and eventually destroy bacteria. In fact, virtually every known type of bacteria is preyed on by its own specific virus. Many of these infected bacteria are also harmful to humans, and a single type of these infected bacteria can cause a variety of diseases. For example, one virus-infected bacterium is responsible for both strep throat and scarlet fever, either of which can develop into rheumatic fever.

CROSS-DISCIPLINARY FOCUS

Mathematics

Most viruses have symmetrical, geometric bodies. They are usually either rod-shaped or sphere-shaped. Show some pictures of viruses to the class and have each student draw a picture of a virus. Tell them to indicate the important parts of the virus on the picture, as well as what disease the virus causes and some characteristics of the disease. Ask: What geometric shape is your virus?

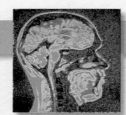

Tiny Troublemakers

You're minding your own business when suddenly your body is infiltrated by invaders so tiny that millions of them could fit on the point of a sharpened pencil. They immediately start using the chemicals and energy inside your cells to make more invaders. Soon thousands of your cells are infected, your nose begins to run, and you start to sneeze. You've just been invaded by a common cold virus!

Thank goodness it was only a cold virus—it could have been worse. Scientists have identified more than 1400 different viruses. Some cause problems such as cold sores and warts. Others cause serious diseases such as polio, measles, chicken-pox, influenza, and acquired immune deficiency syndrome (AIDS). The Ebola virus, for example, can cause severe internal bleeding that can lead to death within three days of being infected.

Living or Nonliving?

Because of their makeup, viruses are hard to classify. Unlike living organisms, viruses are not made up of cells. Instead, a typical virus has only two

▶ **Some viruses look more like machines than living organisms. But once inside a cell, viruses become part of the organism's living system.**

basic components—a core of nucleic acid and an outer coat of protein. On its own, a virus is a lifeless particle that is totally inert. It cannot perform any of the functions of a living organism. It can exist in the environment for weeks, months, or even years without any negative effects or changes.

Once inside a living cell, however, a virus takes on very different characteristics. The virus is no longer inert. It takes over the cell nucleus and redirects the cell to produce proteins that

◀ **Viruses range in size from about 0.01 to 0.03 micron (μm) in diameter. (A micron is 0.001 millimeter!) In contrast, the smallest bacteria are about 0.4 μm in diameter. Shown here are Ebola viruses.**

are necessary for the production of more viruses. The new viruses then burst out of the cell and go on to infect other cells. One virus inside a single cell can produce thousands of new viruses!

Once inside a cell, a virus has many of the characteristics of a living thing—most obviously, it can reproduce. A virus is much like a parasite, using a cell as its host. In fact, some scientists speculate that those viruses which kill their hosts have not fully adapted to their environment. By killing the host, the virus is, in effect, sabotaging own ability to function and reproduce. This is why Ebola epidemics have been short lived.

A Deadly Virus

HIV—the virus that causes AIDS—doesn't actually kill its host directly. Instead, the virus weakens the body's immune system to the point that the body cannot protect itself from other infections. Do some research to find out about how HIV causes AIDS. Also, find out about the role of the drug AZT in the fight against AIDS.

A Deadly Virus

Students may choose to approach this activity in a number of ways. Some may want to illustrate what occurs when HIV attacks a cell and how AZT works in treating AIDS. Others may want to write a report. A report could focus on how infections occur and are transmitted or on where AIDS epidemics are occurring in the world. All students should be aware of the symptoms of the disease.

Big Picture for Little Dots

Take a look at the painting on this page. What do you see? Probably a holiday crowd on an island, right? Take a closer look and you'll see something else—the whole image is created with tiny, uniform dots of single colors! This method of painting is called *pointillism*, and it is as scientific as it is artistic.

Seurat's Dots

Pointillism was developed by the French artist Georges Seurat in the 1880s. Seurat noticed that when two small dots of different colors are placed next to each other, the two colors seem to form a new color when seen from a distance. Seurat then began to develop ways in which different combinations of colored dots could be used to create subtle changes in form and color.

It's Not Connect-the-Dots

Seurat immersed himself in the study of how white light and color interact. As you might know, white light is actually made up of seven colors—red, orange, yellow, green, blue, indigo, and violet—known as spectrum colors. Paints contain tiny grains of colored substances called pigments. When light hits the paint, the pigments absorb some of the colors of the light and reflect the rest. The color you see is actually the color of light that is reflected by the pig-

ments. Mixing different-colored pigments changes which colors of light are absorbed. This type of mixing is called subtractive color mixing because the resulting mixture absorbs, or subtracts, colors from white light. Most of the colors you see work this way. For example, a red shirt appears red because it absorbs all colors of light except red—only red light is reflected. Mixing many pigment colors together will make a dark brown or black paint because together the pigments will absorb almost all light.

Adding It Up

Mixing light is quite different from mixing pigments. Mixing colors of light is called additive color mixing. For example, overlapping a red and a green spotlight will produce a yellow light. Mixing all of the colors of light produces white light because all of the different parts of white light are added together.

Seurat used this information as he painted. He used only pure spectrum colors in his paintings. He placed dots next to each other in such a way that the light reflected by one dot would combine with the light reflected by a second dot to create the color he wanted. It's almost as if

Photograph © 1995, The Art Institute of Chicago, All Rights Reserved

▲ *Sunday Afternoon on the Island of La Grande Jatte* by George Seurat, at the Art Institute of Chicago. Look closely at the inset to see the dots of color.

the light reflected from each dot becomes its own little spotlight of color. In other words, Seurat produced the *effect* of additive light using subtractive pigments. Amazing!

A color television set also uses the principles of pointillism. Look at the screen with a magnifying glass to see for yourself.

Test It Out!

Create a small picture using pointillism. Using only the seven spectrum colors, experiment with different dot combinations to see what other colors you can make.

143

Test It Out!

You may want to provide students with other examples of pointillism or have students find their own examples. You may also wish to display the students' pictures in prominent places around the classroom.

Background

Georges Seurat was born in Paris on December 2, 1859. As a youth, he studied art at the prestigious Ecole des Beaux-Arts and was strongly influenced by the impressionist artists, as well as by the scientific discussions about perception, color, and light that were occurring at the time.

Seurat's work is an example of postimpressionism, a French art movement that developed in the wake of impressionism. Seurat believed that colors obey a scientific system of relations just as musical tones do, and he wanted to apply a scientific method to painting. The painting *A Sunday Afternoon on the Island of La Grande Jatte* is a synthesis of his ideas about color. Pointillism is often referred to as *neoimpressionism* to distinguish it from other postimpressionist techniques.

CROSS-DISCIPLINARY FOCUS

Art

Other postimpressionist artists, such as Vincent van Gogh and Paul Gauguin, experimented further with color. They were concerned with the emotional response that each color elicited in the viewer. Show the class some prints of postimpressionist paintings and ask students what kinds of emotions they think the artists were trying to convey. Have students compare these paintings with some impressionist works by artists such as Renoir or Monet. Point out to the class that the use of color is more realistic in impressionist works than in postimpressionist works.

PORTFOLIO

Have students include their pictures from Test It Out! in their Portfolio.

Unit 3 MACHINES, WORK & ENERGY

Unit Overview

In this unit, students learn that many machines that we use every day are mechanical systems composed of combinations of simple machines. They learn about the scientific concepts of force, energy, work, and power that underlie the operation of machines. They learn about different kinds of energy and also study efficiency, energy transformation, mechanical advantage, and simple machines, such as wheels and axles, pulleys, and gears. Finally, students learn that the human body is also a machine. In several Explorations, students have the chance to develop problem-solving skills by designing and building machines and by performing calculations.

Using the Themes

The unifying themes that appear in this unit are **Systems, Energy,** and **Structures.** The following information will help you incorporate these themes into your teaching plan. Focus questions that correspond to these themes appear in the margins of this Annotated Teacher's Edition on pages 145, 150, 159, 166, 176, 191, and 201.

Systems is an important theme in this unit because students apply their knowledge of energy to the operation of mechanical systems. Students further advance their understanding of systems when they learn about the musculo-skeletal system of the human body in Chapter 9.

Energy is also a theme in this unit. Students investigate the concept of energy explicitly in Chapter 8 when they study the various forms of energy and the transfer and transformation of energy using roller coasters, light bulbs, and other familiar examples. Energy is also a unifying theme because all of the topics in the lesson relate to energy in some way. Work and power, mechanical advantage and efficiency, and the balance of work input and output in simple machines all rest on the basic concept of energy.

Structures is the third theme in this unit. Students analyze the structures of compound machines to find out what simple machines they consist of. Also, students analyze the structures of simple machines. For example, the basic structure of the wheel and axle is studied in depth so that students can understand how this simple machine can make tasks easier.

Bibliography for Teachers

Aaseng, Nathan. *Twentieth Century Inventors.* New York, NY: Facts On File Publications, 1991.

Gonick, Larry, and Art Huffman. *The Cartoon Guide to Physics.* New York, NY: Harper Perennial, 1990.

Ouseph, P. J. *Technical Physics.* New York, NY: John Wiley & Sons, 1991.

William, Trevor I. *The History of Invention: From Stone Axes to Silicon Chips.* New York, NY: Facts On File Publications, 1991.

Bibliography for Students

Ardley, Neil. *Hands-on Science: Muscles to Machines.* New York, NY: Gloucester Press, 1990.

Burnie, David. *Machines: How They Work.* New York, NY: Sterling Publishing Company, 1994.

Macaulay, David. *The Way Things Work.* Boston, MA: Houghton Mifflin Company, 1988.

Films, Videotapes, Software, and Other Media

Energy Does Work
Videotape
Barr Media Group
12801 Schabarum Ave.
Irwindale, CA 91706-7878

Muscular and Skeletal Systems
Videotape
National Geographic Society
Educational Services
P.O. Box 98019
Washington, DC 20090-8019

Simple Machine Series
Film, videotape, or videodisc
Coronet/MTI
108 Wilmot Rd.
Deerfield, IL 60015

Simple Machines
Videotape
Agency for Instructional Technology
P.O. Box A
Bloomington, IN 47402-0120

Work and Machines
Software (Apple, Macintosh, MS-DOS)
Queue, Inc.
338 Commerce Dr.
Fairfield, CT 06430

Using the SourceBook

Unit 3 in the SourceBook further explores force, work, power, and machines. More in-depth discussion and practice with numerical examples are provided. The concepts are presented in the context of everyday examples. The consumption of electrical energy is also presented as students apply their knowledge to household appliances.

Unit Organizer

Unit/Chapter	Lesson	Time*	Objectives	Teaching Resources
Unit Opener, p. 142				Science Sleuths: The Moving Monument English/Spanish Audiocassettes Home Connection, p. 1
Chapter 8, p. 146	**Lesson 1, Machines for Work and Play,** p. 147	1	1. Identify several familiar machines and describe their functions and when they were invented. 2. Distinguish between simple machines and complex mechanical systems.	Image and Activity Bank 8-1
	Lesson 2, The Idea of Work, p. 152	3	1. Define work, in the scientific sense. 2. Calculate work using the formula: work = force × distance. 3. Define joule and demonstrate how large a joule is.	Image and Activity Bank 8-2 Exploration Worksheet, p. 3 Exploration Worksheet, p. 7
	Lesson 3, Work and Energy, p. 158	3	1. Define potential and kinetic energy. 2. Describe how energy is stored and how it is released. 3. Explain the relationship between work and energy. 4. Relate the amount of kinetic energy in a moving body to its mass and speed.	Image and Activity Bank 8-3 Exploration Worksheet, p. 8 Activity Worksheet, p. 12 Activity Worksheet, p. 14
	Lesson 4, Energy Changes, p. 164	2	1. Identify instances in which kinetic energy is converted into potential energy and vice versa. 2. Analyze how energy changes from one form into another. 3. Recognize several energy-conversion devices, and identify the energy changes that occur.	Image and Activity Bank 8-4 Exploration Worksheet, p. 15 Transparency 20 Math Practice Worksheet, p. 20
End of Chapter, p. 167				Activity Worksheet, p. 21 Chapter 8 Review Worksheet, p. 22 Chapter 8 Assessment Worksheet, p. 27
Chapter 9, p. 169	**Lesson 1, Lightening the Load,** p. 170	3 to 4	1. Design and construct a simple machine to make work easier. 2. Identify some simple machines, including levers, pulleys, and inclined planes. 3. Calculate the work input, work output, efficiency, and mechanical advantage of a simple machine. 4. Identify machines that multiply force, multiply distance, or change the direction of a force.	Image and Activity Bank 9-1 Discrepant Event Worksheet, p. 29 Transparency 21 Transparency 22 Theme Worksheet, p. 30 Transparency 23
	Lesson 2, Machines and Energy, p. 180	3	1. Describe how a wheel and axle are used to perform work. 2. Explain the relationship between friction and the total energy output of a machine. 3. Identify the steps used to solve technological problems, and use these steps to design and build a wheel and axle. 4. Give an example of a wheel-and-axle machine that multiplies force and one that multiplies distance.	Image and Activity Bank 9-2 Exploration Worksheet, p. 32 Exploration Worksheet, p. 33 ▼ Exploration Worksheet, p. 35
	Lesson 3, Transferring Energy, p. 186	3	1. Explain how gears, wheels, and belts in machines can be used to transfer energy. 2. Give examples of mechanisms that use gears, wheels, and belts. 3. Explain how gears, wheels, and belts can be used to alter speed. 4. Identify several different kinds of gears, and explain how they transfer kinetic energy.	Image and Activity Bank 9-3 Exploration Worksheet, p. 37 Transparency 25 Transparency 26 Math Practice Worksheet, p. 39 ▼
	Lesson 4, Mechanical Systems, p. 193	3	1. Distinguish between a mechanical system and a simple machine. 2. Identify the subsystems in a given mechanical system. 3. Identify the variables that must be considered when designing a specific mechanical system. 4. Evaluate a mechanical system to determine the effectiveness of its design.	Image and Activity Bank 9-4 Exploration Worksheet, p. 41 Exploration Worksheet, p. 46
	Lesson 5, The Human Machine, p. 198	2	1. Identify structures of the human body that function like simple machines. 2. Describe the motions made possible by the actions of bones, joints, and muscles. 3. Describe how muscles contract, relax, and work in pairs.	Exploration Worksheet, p. 49
End of Chapter, p. 202				Chapter 9 Review Worksheet, p. 54 ▼ Chapter 9 Assessment Worksheet, p. 58
End of Unit, p. 204				Unit 3 Activity Worksheet, p. 61 ▼ Unit 3 Review Worksheet, p. 62 Unit 3 End-of-Unit Assessment, p. 66 Unit 3 Activity Assessment, p. 71 Unit 3 Self-Evaluation of Achievement, p. 74

* Estimated time is given in number of 50-minute class periods. Actual time may vary depending on period length and individual class characteristics.

▼ Transparencies are available to accompany these worksheets. Please refer to the Teaching Transparencies Cross-Reference chart in the Unit 3 Teaching Resources booklet.

Materials Organizer

Chapter	Page	Activity and Materials per Student Group
8	156	***Exploration 1, Task 1:** sheet of paper; scissors; metric measuring tape or meter stick; 100 g, 1 kg, and 5 kg masses; 8 short pieces of masking or transparent tape (See Advance Preparation below.); **Task 2:** chair; metric bathroom scale; meter stick; **Task 3:** metric force meter (or metric spring scale); various objects (door, drawer, book, 1 kg mass, screen or blind); meter stick or metric measuring tape; **Task 4:** meter stick; pull-up bar; metric bathroom scale; **Task 5:** 1 kg bag of gravel (or other 1 kg mass); 2 pulleys; 2–3 m of cord; metric force meter (or metric spring scale); meter stick; support stand with ring clamp; C clamp
	157	**Exploration 2:** meter stick or metric measuring tape; watch or clock with second hand (See Advance Preparation below.)
	163	**Exploration 3, Experience 3:** metal or cardboard trough about 60 cm long; 3 identical books; 3 balls, one twice the mass of the lightest and one three times the mass of the lightest; 2 cm × 15 cm piece of cardboard; meter stick
	164	**Exploration 4:** marble; large bowl; nickel; sheet of paper
9	170	**Helping Tony:** a variety of materials to construct a device for lifting a load, such as the following: support stand with ring clamp; C clamp; single-, double-, and triple-pulleys; metric force meter (or metric spring scale); 1 kg mass; metric ruler; small wedge to be used as a fulcrum; wood block, 10 cm thick; wood board, about 5 cm × 20 cm; 2–3 m of string
	182	**Exploration 2:** a variety of materials to make a simple wheel-and-axle machine, such as the following: spools of various sizes; mailing tubes; film canisters with lids; wire coat hangers; knitting needles; craft sticks; milk cartons; shoe boxes; plastic-foam or paper cups; cardboard; plywood; 1 kg mass; handsaw; white glue; cans; wire; plastic drinking straws; string; pliers; corks; scissors; metric force meter (or metric spring scale)
	184	**Exploration 3, Part 2:** a variety of materials to create an energy source, such as the following: rubber band; water; 500 mL Erlenmeyer flask; one-hole stopper fitted with glass tubing; hot plate; small fan; miniature DC motor with battery; a few meters of copper wire; support stand with ring clamp; C clamp; pulley; 1 kg mass; 2–3 m of string; safety goggles (See Advance Preparation below.)
	187	**Exploration 4:** various unused mechanical devices, such as toys, clocks, or watches that are powered by some form of energy
	189	**A Project—Model Building:** technology construction kit that contains a DC motor with battery, wheels, propellers, and other items needed to transfer energy from the motor to the moving parts
	194	***Exploration 5, Station A:** ballpoint pen; pens other than ballpoint; **Station B:** stapler; staples; 2 pieces of paper; **Station C:** rotary pencil sharpener; **Station D:** bicycles (different makes and models); **Station E:** hand drill without a bit
	196	**One Purpose—Many Designs:** a variety of can openers
	196	**Exploration 6:** materials to build a working model car, such as the following: rubber bands; masking tape; plastic-foam or paper cups with lids; plastic drinking straws; index cards; cardboard tubes; wooden dowels; drawing compass; straight pins; pencils; corks; pliers; scissors; paper; paper clips; white glue; thumbtacks
	200	**Exploration 7, Part 2:** fresh chicken wing; paper towels; small scissors; plastic bag; scalpel

* You may wish to set up these activities in stations at different locations around the classroom.

Advance Preparation

Exploration 1, Task 1, page 156: Each group will need access to an empty wall to perform this task.

Exploration 2, page 157: Each group will need access to a staircase, a steep hill, or a gymnasium rope climb to perform this Exploration.

Exploration 3, Part 2, page 184: Insert fire-polished glass tubing into one-hole stoppers before students enter the classroom. Make sure that the glass tubing has been fire-polished to remove any sharp edges. Wearing leather gloves and using a small amount of glycerin to lubricate the tubing, slowly insert the tubing into the one-hole stoppers with a gentle twisting motion.

Unit Compression

This unit is tightly structured with activities organized in a sequence that develops and consolidates concepts. The conceptual flow should not be interrupted until about two-thirds of the unit is completed. Then some open-ended sections that are concerned with technological problem solving begin. These could be omitted to save time, if necessary. Material that may be omitted includes the following:

- Wheels and Axles, page 182
- Exploration 3, page 184
- Other Wheels and Axles, page 185
- Exploration 4, page 187
- Examining One Model Car Design, page 197

The final lesson in the unit, The Human Machine, could also be omitted without interrupting the flow of ideas.

Homework Options

Chapter 8
See Teacher's Edition margin, pp. 150, 154, 157, 163, 166, and 167
Exploration Worksheet, p. 7
Activity Worksheet, p. 12
Activity Worksheet, p. 14
Math Practice Worksheet, p. 20
Activity Worksheet, p. 21
SourceBook, p. S42

Chapter 9
See Teacher's Edition margin, pp. 173, 176, 177, 181, 182, 184, 187, 191, 192, 195, and 199
Theme Worksheet, p. 30
Exploration Worksheet, p. 32
Math Practice Worksheet, p. 39
SourceBook, pp. S47, S50, S52, and S55

Unit 3
Unit 3 Activity Worksheet, p. 61
Unit 3 SourceBook Activity Worksheet, p. 75

Assessment Planning Guide

Lesson, Chapter, and Unit Assessment	SourceBook Assessment	Ongoing and Activity Assessment	Portfolio and Student-Centered Assessment
Lesson Assessment Follow-Up: see Teacher's Edition margin, pp. 151, 157, 163, 166, 179, 185, 192, 197, and 201 **Chapter Assessment** Chapter 8 Review Worksheet, p. 22 Chapter 8 Assessment Worksheet, p. 27* Chapter 9 Review Worksheet, p. 54 Chapter 9 Assessment Worksheet, p. 58* **Unit Assessment** Unit 3 Review Worksheet, p. 62 End-of-Unit Assessment Worksheet, p. 66*	SourceBook Review Worksheet, p. 77 SourceBook Assessment Worksheet, p. 81* 	Activity Assessment Worksheet, p. 71* **SnackDisc** Ongoing Assessment Checklists ♦ Teacher Evaluation Checklists ♦ Progress Reports ♦	Portfolio: see Teacher's Edition margin, pp. 149 and 193 **SnackDisc** Self-Evaluation Checklists ♦ Peer Evaluation Checklists ♦ Group Evaluation Checklists ♦ Portfolio Evaluation Checklists 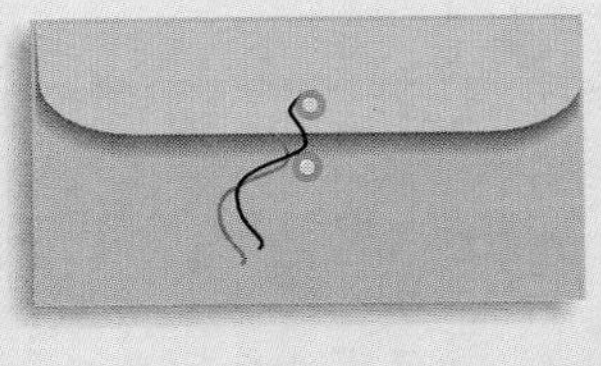 ♦

* Also available on the Test Generator software
♦ Also available in the Assessment Checklists and Rubrics booklet

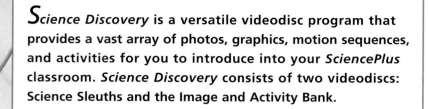

Science Discovery is a versatile videodisc program that provides a vast array of photos, graphics, motion sequences, and activities for you to introduce into your *SciencePlus* classroom. *Science Discovery* consists of two videodiscs: Science Sleuths and the Image and Activity Bank.

Using the *Science Discovery* Videodiscs

Science Sleuths: The Moving Monument
Side B

The city has announced its plans to construct a building on the site of the Arnold Madigli Memorial Monument in Volunteer Park. In order to do so, large machinery was brought in to move the granite monument to a new site in the park. Now, each morning, the monument has been found in a new site, apparently moving back toward its original site. The park is surrounded by concrete traffic barriers, and there is no sign that heavy machinery has been brought in at night. Rumor has it that the heavy rock is moving by itself.

Interviews
1. Setting the scene: City official 22102 (play ×2)

2. Madigli spokesperson 23205 (play)

3. Psychic 24116 (play)

4. Park supervisor 24777 (play)

5. Park maintenance 25783 (play)

6. Reporter 26464 (play)

Documents
7. Memo to Parks Department staff 27073 (step)

8. Letter from Madigli family to mayor 27076 (step ×2)

9. Minutes of Madigli Monument Preservation Society 27080 (step)

Literature Search
10. Search on the words: GHOSTS, GRANITE, MADIGLI, MONUMENT, VOLUNTEER PARK 27083

11. Article #1 ("Old Quarry To Be Flooded") 27085

Sleuth Information Service
12. Map of park 27087

13. Park elevation 27089

14. Scale drawing of monument 27091

15. Objects found in park 27093

16. Top 10 common building materials 27095

17. Lever arm tests 27097 (step)

18. Force required to bend fence poles 27100

19. Sliding granite 27102 (step)

20. Archeological mysteries 27105 (step ×2)

21. Rope strength 27109 (step ×2)

Still Photographs
22. The monument 27113 (step ×3)

23. The monument in different locations 27118 (step ×3)

24. Pile of construction materials 27126

25. Items found in park 27128 (step ×10)

Image and Activity Bank
Side A or B

A selection of still images, short videos, and activities is available for you to use as you teach this unit. For a larger selection and detailed instructions, see the Videodisc Resources booklet included with the Teaching Resources materials.

8-1 Machines for Work and Play, page 147
Machine time line 1837
Machines that have been invented since 3500 B.C.

Piano 1832
A piano converts the pressure on a key to the movement of a lever. The lever then strikes a string, which results in a specific sound.

◀| Step Reverse   Play ▶   Pause || Step Forward |▶

8-2 The Idea of Work, page 152

Work and energy 1868
Work and energy are depicted. The person gains energy from food and uses it to do the work of lifting an iron block. The iron block gains potential energy as it is lifted off the ground.

Newton; unit of force 1739–1741 (step ×2)
The newton is the SI unit for measuring force. (step) Measuring the newton (step)

8–3 Work and Energy, page 158

Kinetic energy; hammering a nail 24720–24877 (play ×2) (Side A only)
The weight of the hammer alone will not drive a nail into the wood. When the hammer is swung, kinetic energy is conveyed from the hammer head to the nail.

Food 1814
Food supplies the energy necessary for humans to perform work and even to live. The nutritional value within food is converted to energy by the body.

8-4 Energy Changes, page 164

Jack-in-the-box 1813
Potential energy is converted to kinetic energy when Jack comes out of the box. Potential energy is built up by winding the spring within the box and is released when the spring pushes Jack out of the box.

Swing; child being pushed 1815
Potential energy and kinetic energy are transformed back and forth in the swinging motion of the child. As the potential energy decreases near the bottom, the child approaches a maximum velocity, or highest kinetic energy.

Energy conversion; ball rolling in pan 28067–28185 (play ×2) (Side A only)
The kinetic energy of the ball is transferred to the pan. The vibrating pan (kinetic energy) creates sound.

Bowling-ball pendulum 29833–29974 (play ×2) (Side A only)
As long as energy is not added to the system, a pendulum does not gain energy in its oscillation. Therefore, it must return to its starting point.

Dam, hydroelectric 1789 (step ×4)
Reservoirs, or large amounts of water, are created behind dams to make the production of electricity easy. The energy gained by moving the water from a higher to a lower elevation is captured in the dam.

Wind farm 1779
Windmills capture the energy of wind. The wind turns the blades. This produces mechanical energy, which is converted into electrical energy to be stored or transported to your home.

Sailboat 2625
The kinetic energy of the wind propels this sailboat.

Steam engine 1780
Steam engines can power car movement. Water heated to a high temperature produces steam. The high-pressure steam moves pistons that, through gears, produce wheel rotation.

Motor, simple DC electric 1852
A motor is an energy converter. Electrical energy is converted to mechanical energy.

9-1 Lightening the Load, page 170

Simple machines; pulleys 1838
Pulleys are simple machines that change the direction of force so that objects can be lifted more easily.

Pulley demonstration 24878–25508 (play ×8) (Side A only)
(play ×2) Single pulley loaded with 5 N weights. A spring scale for measuring newtons is used on the left to measure the force as the weight is lifted. A single pulley reverses the direction once. (play ×2) Two single pulleys reverse the direction twice. (play ×2) This pulley arrangement reverses the direction of the force four times. (play ×2) This arrangement reverses the direction of the force six times.

Work input/output equation 1854
Work input is always greater than work output because of energy losses such as those due to friction.

Ax; splitting wood 1821
An ax is an example of a simple machine. The kinetic energy of the falling ax is used to break apart the wood fibers when splitting wood.

Scissors 1823
Scissors are a simple machine. The high shearing stress at the point of contact cuts the paper fibers.

9-2 Machines and Energy, page 180

Problem solving, technical 478–479 (step)
Model for technical problem solving (step)

Steering wheel; boat 1824
The steering wheel of a boat is a device that converts forward movement to lateral, or side-to-side, movement.

Crank 1825
This crank makes it easier to pull down on a rope to raise a sail.

9-3 Transferring Energy, page 186

Bicycle 1816–1819 (step ×3)
The potential energy of the biker is transformed into kinetic energy through the movement of the biker's legs and the action of a chain and gears. (step) Different gears are used on the front sprocket to change pedaling difficulty for different speeds. A larger gear is used for higher speeds. (step) The rear sprocket changes the pedaling difficulty for different speeds. The higher gear, or higher speed gear, is the smallest one, located at the bottom. (step) Turning the handlebar controls the direction of movement of a bike.

Hoist 1826
A hoist uses gears to lift heavy objects.

9-4 Mechanical Systems, page 193

Clock; inside 1842
A mainspring turns a driving wheel, which turns a pinion. The pinion moves the other gears and wheels.

Pendulum clock 1711
Seconds on a pendulum clock are counted as the time necessary for the pendulum to complete a movement from the left or the right to the center position.

Clock, pendulum 1843–1844 (step)
A pendulum is hung so that it can move back and forth in a regular arc under the influence of gravity.

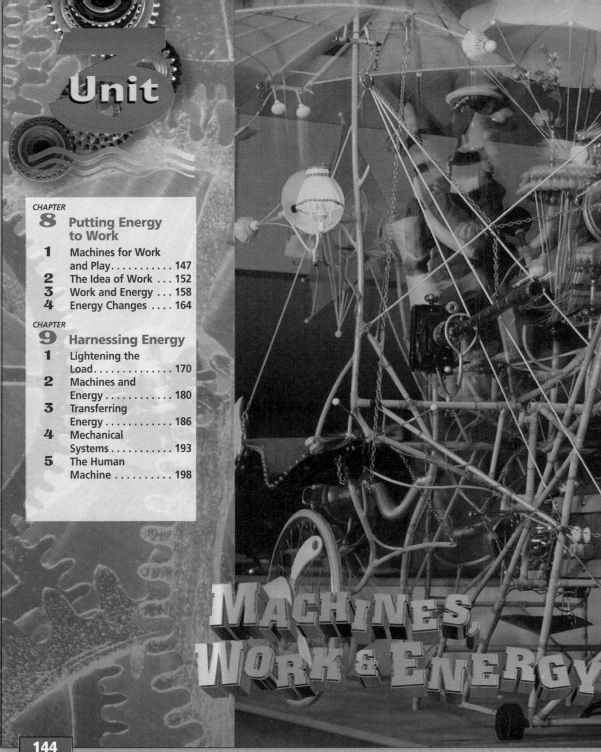

Unit 3

CHAPTER

8 Putting Energy to Work

1 Machines for Work and Play. 147
2 The Idea of Work . . . 152
3 Work and Energy . . . 158
4 Energy Changes 164

CHAPTER

9 Harnessing Energy

1 Lightening the Load 170
2 Machines and Energy 180
3 Transferring Energy 186
4 Mechanical Systems 193
5 The Human Machine 198

144

MACHINES, WORK & ENERGY

UNIT FOCUS

Write the question, What is work? on the board. Invite students to offer their ideas in response to the question. *(Accept all reasonable answers.)* List their answers on the chalkboard under the question. Then have students decide which of their suggestions require energy and which do not. *(Most students will probably agree that energy is needed to complete all of the tasks they have suggested.)* Explain to students that they have correctly pointed out that all work requires energy. Let students know that in this unit they will learn about the relationship between work and energy and about what work is to a scientist.

A good motivating activity is to let students listen to the English/Spanish Audiocassettes as an introduction to the unit. Also, begin the unit by giving Spanish-speaking students a copy of the Spanish Glossary from the Unit 3 Teaching Resources booklet.

Connecting to Other Units

This table will help you integrate topics covered in this unit with topics covered in other units.

Unit 1 Life Processes	The work done by the human body relies on energy supplied by cellular respiration.
Unit 2 Particles	The concepts of physics that explain mechanical systems also explain systems of particles of matter.
Unit 4 Oceans and Climates	Burning fossil fuels to provide energy for machines increases the concentration of carbon dioxide in the atmosphere and contributes to the greenhouse effect.

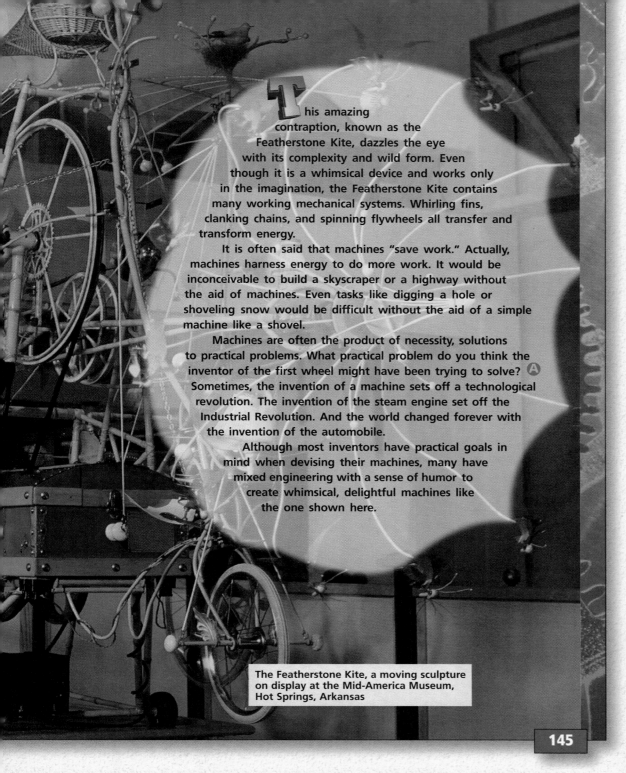

This amazing contraption, known as the Featherstone Kite, dazzles the eye with its complexity and wild form. Even though it is a whimsical device and works only in the imagination, the Featherstone Kite contains many working mechanical systems. Whirling fins, clanking chains, and spinning flywheels all transfer and transform energy.

It is often said that machines "save work." Actually, machines harness energy to do more work. It would be inconceivable to build a skyscraper or a highway without the aid of machines. Even tasks like digging a hole or shoveling snow would be difficult without the aid of a simple machine like a shovel.

Machines are often the product of necessity, solutions to practical problems. What practical problem do you think the inventor of the first wheel might have been trying to solve? Ⓐ Sometimes, the invention of a machine sets off a technological revolution. The invention of the steam engine set off the Industrial Revolution. And the world changed forever with the invention of the automobile.

Although most inventors have practical goals in mind when devising their machines, many have mixed engineering with a sense of humor to create whimsical, delightful machines like the one shown here.

The Featherstone Kite, a moving sculpture on display at the Mid-America Museum, Hot Springs, Arkansas

145

Connecting to Other Units, continued

Unit 5 Electro-magnetic Systems	An understanding of energy, power, and efficiency can be applied to electrical circuits and sources of electrical energy.
Unit 6 Sound	The mechanical basis of sound can be analyzed from the perspective of the work and energy that are involved in sound production.

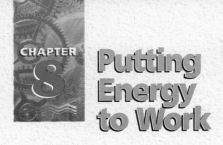

CHAPTER

8 Putting Energy to Work

Connecting to Other Chapters

Chapter 8
presents the scientific meanings of work, energy, and power, along with ways to measure or calculate each.

Chapter 9
examines mechanical systems in terms of efficiency, mechanical advantage, and energy transfer.

Prior Knowledge and Misconceptions

Your students' responses to the ScienceLog questions on this page will reveal the kind of information—and misinformation—they bring to this chapter. Use what you find out about your students' knowledge to choose which chapter concepts and activities to emphasize in your teaching. After students complete the material in this chapter, they will be asked to revise their answers based on what they have learned. Sample revised answers can be found on page 168.

In addition to having students answer the questions on this page, you may wish to assign a "free-write" to assess prior knowledge. To do this, instruct students to write for 3 to 5 minutes on the subject of machines and energy. Tell them to keep their pens moving at all times, writing in a stream-of-consciousness fashion. Emphasize that there are no right or wrong answers in this exercise. It may be best to ask students not to put their names on their papers. Collect the papers, but do not grade them. Instead, read them to find out what students know about machines and energy, what misconceptions they may have, and what about this topic is interesting to them.

CHAPTER

8 Putting Energy to Work

1 What is work? How might it be measured?

2 How is energy converted from one form to another by a roller coaster?

3 What is energy? How is work related to energy?

ScienceLog

Think about these questions for a moment, and answer them in your ScienceLog. When you've finished this chapter, you'll have the opportunity to revise your answers based on what you've learned.

146

LESSON 1

Machines for Work and Play

Can you think of a time when people did not use machines of any sort? If you went back in time 5500 years, you would discover the first plow. How was this invention beneficial? How did it change people's lives? Ⓐ

There were also great construction projects long ago. During the construction of Egypt's Great Pyramids more than 4000 years ago, huge earthen ramps were constructed to help raise enormously heavy stone blocks into position. A ramp is a kind of machine. A pulley is another kind of machine. Pulleys have been in use for over 3000 years. Archimedes is said to have been the first to use a combination of pulleys. These pulleys were used to haul ships ashore.

A sampling of machines

147

LESSON 1 ORGANIZER

Time Required
1 class period

Process Skills
observing, analyzing, inferring

Theme Connection
Systems

New Term
Mechanical system—a machine composed of more than one simple machine

Materials (per student group)
none

Teaching Resources
none

LESSON 1

Machines for Work and Play

FOCUS

Getting Started
Ask students to identify any machines in the classroom, and have them explain what these machines do. Keep track of their suggestions on the chalkboard. (Students may identify such things as clocks, light switches, window and door openers, doorknobs, hinges, window shades or blinds, pencil sharpeners, and computers.) Involve students in a brief discussion of why machines are used. (Machines are used to make jobs easier and life more enjoyable. Some jobs would be impossible without machines.) Point out that in this lesson students will learn more about simple and complex machines.

Main Ideas
1. People have depended on machines for thousands of years to make their lives easier and more enjoyable.
2. Most machines are complex mechanical systems that are designed to perform some overall function.
3. A complex mechanical system is made up of many subsystems composed of simple machines.

TEACHING STRATEGIES
Suggest that students cover page 149 with a piece of paper before they read page 148. Then have them list the machines in order, from the earliest invention to the latest. They can then read page 149 to discover if their answers were correct.

Answer to
In-Text Question

Ⓐ The plow made soil preparation easier and allowed farmers to cultivate larger areas of land in the same amount of time.

A The machines are listed on page 149 of the Pupil's Edition according to the date of their invention. Some students may be interested in the date of invention for the electric circular saw, which is pictured but not listed. Portable electric saws became available to the general public around 1925.

B • A blender mixes and chops food. It uses electrical energy to turn its blades.
• Weighing scales compare the weights of objects. They use energy by moving in response to the weights placed on them.
• A power saw cuts wood. It uses electrical energy to turn its blade.
• A lawn mower cuts grass. It uses the chemical energy in gasoline to turn its blades.
• Gears change the direction or speed of other moving parts. They can also change the amount of force necessary to turn a moving part. Gears use energy to rotate.
• A jet aircraft transports people and cargo through the air. It uses the chemical energy in jet fuel to turn the turbines of its engines.
• An internal-combustion engine can perform a variety of tasks, from generating electricity to turning a vehicle's wheels. It uses the chemical energy in fossil fuels to turn a drive shaft.
• A clothes washer cleans clothes. It uses electricity to spin the clothes.

C • Answers will vary. Accept all reasonable responses. (Point out to students that they will be able to revise their answers as they progress through the unit.) Sample answer: A machine is a device that uses energy to complete a specific task.
• Answers will vary, but most lists should be quite long. Possible answers include a bicycle, car, can opener, pencil sharpener, and even a doorknob (which is a wheel and axle).
• Machines require energy because they do work. Machines get energy from various sources, including electric power plants and the muscular effort supplied by people.
• Hand-operated machines include the shovel, wheelbarrow, ramp, screw driver, eggbeater, nut-cracker, pliers, car jack, and winch.

Try to place the following machines in chronological order. In other words, begin the list with the machine that was invented first, followed by the machine that was invented second, and so on. Also try to determine the approximate century in which each machine was invented. Write your answers in your ScienceLog. **A**

• motorcycle
• internal-combustion engine
• jet aircraft
• steam locomotive
• automatic clothes-washing machine
• CD player
• electric motor
• weighing scales
• gears
• mechanical clocks
• blender
• gasoline-powered lawn mower

148

• Yes, machines do work for people.
• Machines do not reduce the amount of work required to complete a task, but they do make the task easier.
• Accept all reasonable responses. (Students will learn in Lesson 2 that work has a precise scientific definition.)
• Accept all reasonable responses. Sample answer: Energy is the ability to do work. (Students will learn more about the relationship between work and energy in Lesson 3.)

Machines in Your Life

Now compare your answers with the actual dates of invention shown below.

- weighing scales—3500 B.C.
- gears—100 B.C.
- mechanical clocks—1300
- steam locomotive—1804
- electric motor—1830
- internal-combustion engine—1860
- motorcycle—1885
- gasoline-powered lawn mower—1902
- blender—1923
- automatic clothes-washing machine—1937
- jet aircraft—1939
- CD player—1979

Were you surprised by some of the early dates? Notice that several of the machines were invented in the 1800s. At that time, in the wake of the Industrial Revolution, machines were developed quickly and in large numbers.

Take a look at the machines shown on this and the facing page. What does each machine do? How does each use energy? **B**

Now take a moment to consider the following questions: **C**

- How would you define *machine*?
- How many machines have you used so far today?
- Do machines require energy? If so, where do they get it?
- What are some hand-operated machines?
- Do machines do work for you?
- Do machines save work or make it easier?
- What is work?
- What is energy?

You can answer some of these questions now, but you probably cannot answer all of them. This unit will help you answer these remaining questions and, possibly, help you to become an inventor, too!

149

PORTFOLIO

Students may wish to include in their Portfolio a time line that incorporates the inventions listed on this page. Students may also wish to add other inventions of their choice.

CROSS-DISCIPLINARY FOCUS

Language Arts

You may wish to help students familiarize themselves with some of the terms that are associated with mechanical systems in this unit. Sample terms that students could define include *derailleur, sprocket, rack and pinion,* and *flywheel.*

Multicultural Extension

Steel-Driving Man

Many students may be familiar with the folk tale of John Henry, the African American steel driver who raced a mechanical steam drill to prove that a man could dig a tunnel faster than a machine. The story of how John Henry defeated the steam drill and then died of exhaustion is one of the best known folk tales in American culture. Students may not know that the folk tale is based on the life of a real man who did actually race a steam drill. Have interested students consult an encyclopedia for more information on this popular folk tale to find out why the story of how a man could defeat a machine was so popular among people of diverse cultural backgrounds. Encourage students to share their findings with the class.

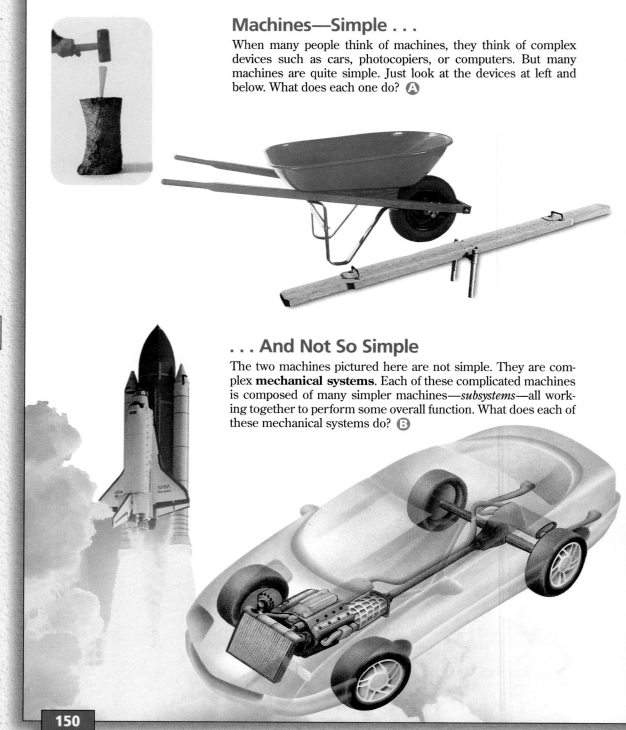

Machines—Simple . . .

When many people think of machines, they think of complex devices such as cars, photocopiers, or computers. But many machines are quite simple. Just look at the devices at left and below. What does each one do? **Ⓐ**

. . . And Not So Simple

The two machines pictured here are not simple. They are complex **mechanical systems**. Each of these complicated machines is composed of many simpler machines—*subsystems*—all working together to perform some overall function. What does each of these mechanical systems do? **Ⓑ**

150

Meeting Individual Needs

Gifted Learners

To emphasize that complex mechanical systems are made up of simpler subsystems, involve students in a discussion of some of the subsystems in space shuttles and cars. The following are some possible responses:

• Space shuttle—launch system, communication system, fuel-storage system, environmental system, waste-disposal system, landing system, food-preparation system

• Car—engine, braking system, exhaust system, climate-control system, suspension system, sound system, seating system, fuel system, ignition system, transmission system, cooling system

Homework

Have students make a list of 5 simple machines and 10 complex machines they have at home. Challenge students to find interesting examples. The next day in class, vote to decide which example was the most surprising or interesting example of a simple machine.

Theme Connection

Systems

Have students examine a mechanical corkscrew and consider it as a system. Have them take it apart, look at its components, and then see how some of the parts work together in subsystems.
Focus question: What subsystems does the corkscrew contain, and what does each subsystem do? *(Sample answer: A corkscrew consists of two major subsystems. The clasp system consists of a screw at the end of an axle. A handle at the other end of the axle allows a person to sink the screw into the cork of a bottle. The extraction system consists of two levers attached to sprockets. When the levers are pushed down, the sprockets turn, raising the axle and extracting the cork from the bottle.)*

Familiar Mechanical Systems

You use mechanical systems all the time. Some, like a bicycle, are complex. Others, like a ballpoint pen, are simpler. But even the pen has a surprising number of parts and subsystems, each of which performs functions that contribute to the pen's overall function. The subsystems of a bicycle are more obvious. The braking subsystem is just one example. How many more subsystems of a bicycle can you identify? What does each do? What parts make up each subsystem? **C**

You are able to move faster on a bicycle than you can move by walking. How does the bicycle make possible **D** this increase in speed? As you progress through this unit, you will be investigating many mechanical systems, their subsystems, and their functions. You will also discover how they can increase either force or speed.

↑ Believe it or not, this is a simple ballpoint pen. And like other pens, it is made up of a number of parts. You can check it out for yourself on page 194.

Work and Energy

Some machines are hand-powered—you have to do some work to use them. A shovel is a very simple kind of machine that requires you to do all the work. If you have to dig a large hole, a shovel helps to make your work easier. Imagine clearing away the dirt without it—using only your hands! Of course, it's even easier to dig a large hole using a backhoe. In this case, though, you don't do the real work—the machine does. The energy to dig up the dirt and to move the backhoe forward comes from the gasoline burned by the machine's engine. You do just a little work in operating the controls.

Notice the words *work* and *energy* in the preceding **E** paragraph. What do they mean? You probably have a fairly clear idea of what work and energy are, whether you realize it or not. The lessons that follow will help you define both terms in a scientific sense.

151

FOLLOW-UP

Reteaching

Have students make cartoon strips to show how machines make life easier and more enjoyable. The cartoons should be specific. That is, they should focus on one task done by hand versus the same task done by machine.

Assessment

Have students make a list of all the machines they use during one 24-hour period. Then, in chart form, have them classify the machines according to their function (e.g., machines for travel, machines for food, writing machines, and entertainment machines). Also suggest that they explain how energy is used in each machine.

Extension

Have students make diagrams of mechanical systems that are familiar to them. Their diagrams should identify several of the subsystems that work together to make up the more complex mechanical system. Display the diagrams for other students to study.

Closure

Have students make a drawing of a machine in use. Each drawing should contain a mechanical impossibility—either the machine featuring parts that it could not realistically contain or the machine being used in a way that is not possible. Students can trade drawings with one another to guess what is wrong with the picture. (*Example: a small person sitting close to the point of support, or fulcrum, of a seesaw and lifting a large person seated on the other end of the seesaw far from the fulcrum*)

FOCUS

Getting Started

Point out to students that work, in the everyday sense, can have several different meanings. Explain, however, that the scientific definition of work is very specific. You might want to emphasize to students that for work to be done, a force must move an object through a distance.

Main Ideas

1. In a scientific sense, work is done only when a force moves an object.
2. Work is calculated by the following formula: work = force × distance.
3. When work is done, energy is transferred from one object to another.
4. Power refers to the rate at which work is done.

TEACHING STRATEGIES

On the chalkboard, list the seven suggested meanings of work found on page 153. Then have students provide examples of each meaning.

Answer to
In-Text Question

Ⓐ Accept all reasonable responses. Various meanings of the word *work* are listed on page 153 of the Pupil's Edition.

LESSON
2

The Idea of Work

What Is Work?

Ken and Monica are having a conversation in which the word *work* is used in several different ways. How many different meanings can you find? Ⓐ

"Hey Ken, guess what—I'm going to *work* part time after school."

"That's great, Monica! What kind of *work* will you do?"

"I'm going to *work* in the stockroom at the Minit Mart."

"That doesn't sound like too much *work*."

"I don't know. I hope it *works* out."

"It's probably not as much *work* as mowing grass."

"No kidding—pushing a mower is really hard *work*."

"It's not as hard as home*work*."

"Yeah, thinking is pretty hard *work*!"

"By the way, what are you *working* on in science class?"

"We've started this really interesting

unit called Machines, *Work*, and Energy."

"Sounds interesting. Speaking of science, I've been *working* on an idea for a science fair project."

"Speaking of *work*, we'd better get back to it or we'll never finish."

LESSON 2 ORGANIZER

Time Required
3 class periods

Process Skills
observing, measuring, comparing

New Terms
Joule—the amount of work done when a force of 1 N is exerted through a distance of 1 m
Power—the rate at which work is done
Work—the result of using force to move an object over a distance

Materials (per student group)
Exploration 1, Task 1: sheet of paper;

scissors; metric measuring tape or meter stick; 100 g, 1 kg, and 5 kg masses; 8 short pieces of masking or transparent tape (See Advance Preparation on page 143C.); **Task 2:** chair; metric bathroom scale; meter stick; **Task 3:** metric force meter (or metric spring scale); various objects (door, drawer, book, 1 kg mass, screen or blind); meter stick or metric measuring tape; **Task 4:** meter stick; pull-up bar; metric bathroom scale; **Task 5:** 1 kg bag of gravel (or other 1 kg mass); 2 pulleys; 2–3 m of cord; metric force meter (or metric spring

continued ▶

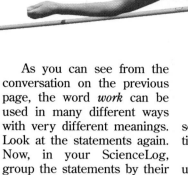

As you can see from the conversation on the previous page, the word *work* can be used in many different ways with very different meanings. Look at the statements again. Now, in your ScienceLog, group the statements by their meanings. Which statements **B** suggest that work is

- a vocation or a job?
- anything that occupies our time?
- the opposite of leisure or recreation?
- a way to earn money?
- connected with forces?
- connected with motion?
- an assignment?

Some statements fit into more than one category.

In science, *work* has a very precise meaning. For example, thinking or preparing for an exam would not be considered work. **Work, in the scientific sense, requires that a force be applied to an object and that the object move.** Try to group the sentences on the previous page and the pictures on these two pages into examples of work in a scientific sense and work in a nonscientific sense. **C**

Another word that is often used with work is *energy*. No doubt you often use this word, but do you really know what energy is? *Energy*, too, can have different meanings. Construct some sentences using the different meanings of *energy*. One way to start would be to look up *energy* in a dictionary. **D**

Work and energy—how are they related? In this unit, you will explore both concepts and learn how one relates to the other.

153

ORGANIZER, continued

scale); meter stick; support stand with ring clamp; C clamp
Exploration 2: meter stick or metric measuring tape; watch or clock with second hand (See Advance Preparation on page 143C.)

Teaching Resources
Exploration Worksheets, pp. 3 and 7
SourceBook, p. S42

Answers to
In-Text Questions

B Answers will vary. Have students justify their responses. Sample answers are as follows:
- A vocation or a job: statements 1, 2, and 3
- Anything that occupies our time: all statements except statement 5
- The opposite of leisure or recreation: all statements (Some students might exclude statements 5 and 12.)
- A way to earn money: statements 1, 2, 3, and 6
- Connected with forces: statements 3, 4, 6, and 7
- Connected with motion: statements 3, 4, 6, and 7
- An assignment: statements 8–13

C Statements 3, 4, 6, and 7 on page 152 involve both force and motion, the criteria for the scientific definition of work. The rest of the statements use the term *work* in a nonscientific sense. Except for the photographs of the man stretching and of the girl reading, all of the photographs show work in the scientific sense because force is being applied to move an object.

D Answers will vary. Accept all reasonable responses. Students' energy statements should reflect some of the common definitions of the term. The following are some examples of how the term *energy* can be used:
- The new power plant produces twice as much *energy* as the old power plant. (electric power)
- By the end of the conference, she had no *energy* left. (pep, vitality)
- Carbohydrates are a good source of *energy*. (food energy, calories)
- His performance conveyed a lot of *energy*. (intensity, vigor)
- He had the *energy* to overpower his opponent. (strength)

Meeting Individual Needs

Second-Language Learners
Briefly discuss the pictures with students, pointing out that in some cases a person is exerting a force and in other cases a machine is exerting a force. In both cases, work is being done.

The Scientific Idea of Work

Encourage students to try to figure out a way to calculate the relative amount of work that each volunteer on this page is doing before reading the discussion of the answers in the second column on this page. If necessary, remind students that they should assume that each book has the same weight.

Answers to
In-Text Questions

Ⓐ • Nicole is doing the least amount of work. She is lifting one book 2 m.
  • Bill is doing the most work because he is lifting three books 3 m.
  • Mark and Melanie are doing the same amount of work. Mark is lifting two books 3 m, and Melanie is lifting three books 2 m.
  • The order of work done from least to most is Nicole, Lynn, Mark and Melanie, and Bill.

Ⓑ Students may suggest multiplying the number of books by the distance the books were moved. Because it is assumed that each book has the same weight, the number of books is proportional to the force exerted to move the books.

Meeting Individual Needs

Learners Having Difficulty
After completing the exercise, most students should recognize that work depends on the distance through which a force is exerted and on the size of the force used. For those who are still having difficulty with the concept, have them perform a book-lifting activity similar to the one illustrated.

The Scientific Idea of Work

Work in Progress

It's moving day in the library. The new shelves have arrived and the librarian has recruited 10 volunteers to help reshelve the books.

"Okay, here's your job," the librarian explains. "The books are stacked on the floor by the shelves. You need to place the books back on the shelves according to the call number found on each book."

Each volunteer has a different task. Look at the illustrations on this page showing the work of five of the volunteers at one specific moment. Then answer these questions:

• Who is doing the least amount of work?
• Who is doing the greatest amount of work?
• Are some volunteers doing the same amount of work?

Ⓐ • Can you place the volunteers in order according to the amount of work each does?

Ⓑ Assume that each book has the same weight.

Lifting a book 2 m obviously requires twice the work needed to lift a book 1 m. Therefore, Kyle must do twice as much work as Marie. What about Sandra? Did you conclude that the amount of work depends on the distance through which the force is exerted?

To lift two books 1 m requires twice the work needed to lift one book 1 m. So Roberta does twice as much work as Marie. When you lift two books, you exert twice the force needed to lift one book. Therefore, the amount of work also depends on the size of the force exerted.

Do you see why Gene does more work than Roberta and Kyle? The order of the volunteers, from least to most work done, is Marie, Kyle/Roberta, Sandra, and Gene.

On the next page is a second set of illustrations to study. Answer the same questions as before. Try to figure out a way to calculate the relative amount of work each student is doing.

154

Homework

Have students list 10 examples of work that they performed during the day. Have students estimate the amount of work involved in each example and rank the examples in order.

Calculating Work

Did you find a good method for calculating how much work each volunteer does? If you multiply the force exerted by the distance through which the force is exerted, you obtain a quantity that indicates the amount of work done:

work done = force × distance

Suppose that each book weighs 10 newtons (N); a newton is a unit of force. How much work does Marie do when she lifts one book up to the second shelf? She exerts an upward force of 10 N to lift the book, and she exerts this force through a distance of 1 m.

work = 10 N × 1 m = 10 Nm
= 10 J

The unit for work is the *joule* (J). It is named after the English scientist James Prescott Joule. A joule is actually a fairly small amount of work, so most work is measured by the *kilojoule* (kJ). How many joules are in a kilojoule? 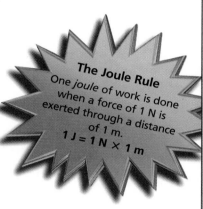 **C**

How much work must Bill do to lift the books to the top shelf? **D**

Now that you know the scientific formula for work, calculate the work being done by each set of volunteers. Do your results verify your earlier rankings of the volunteers in terms of how much work they are doing? **E**

> **The Joule Rule**
> One joule of work is done when a force of 1 N is exerted through a distance of 1 m.
> 1 J = 1 N × 1 m

Did You Know...

One day, a 45 kg (100 lb.) chimpanzee at the Bronx Zoo in New York City lifted 270 kg (600 lb.)—six times its own weight. The champion weight-lifter in the animal kingdom, however, is the ant. An ant can pick up and carry a food crumb 50 times its own weight. That would be like a middle-school student picking up an elephant!

Calculating Work

Review the formula for work with students, and go through some examples. Some students may not understand how to determine the amount of force needed to raise an object. Explain that the force needed to lift an object is equal to the object's weight. Therefore, if a book weighs 10 N, it takes a 10 N force to lift it. If there are two books, each weighing 10 N, the force needed to lift them is double, or 20 N, and so on.

Answers to *In-Text Questions*

D There are 1000 J in 1 kJ.

E Bill must do 90 J of work (10 N × 3 m × 3 books).

F The following calculations should be consistent with students' earlier relative rankings of the volunteers:

Marie: 10 N × 1 m = 10 N•m = 10 J
Kyle: 10 N × 2 m = 20 N•m = 20 J
Roberta: 20 N × 1 m = 20 N•m = 20 J
Nicole: 10 N × 2 m = 20 N•m = 20 J
Sandra: 10 N × 3 m = 30 N•m = 30 J
Lynn: 30 N × 1 m = 30 N•m = 30 J
Gene: 20 N × 2 m = 40 N•m = 40 J
Mark: 20 N × 3 m = 60 N•m = 60 J
Melanie: 30 N × 2 m = 60 N•m = 60 J
Bill: 30 N × 3 m = 90 N•m = 90 J

CROSS-DISCIPLINARY FOCUS

Mathematics

If students need more practice, provide additional examples, such as the following:

- How much work does Cassie do if she raises a 20 N box 4 m? *(20 N × 4 m = 80 N•m = 80 J)*
- How much work does Jeff do if he lifts a rock weighing 10 N a distance of 5 m? *(10 N × 5 m = 50 N•m = 50 J)*
- How much work does Karen do if she lifts a stack of magazines weighing 25 N onto a table 1 m high? *(25 N × 1 m = 25 N•m = 25 J)*

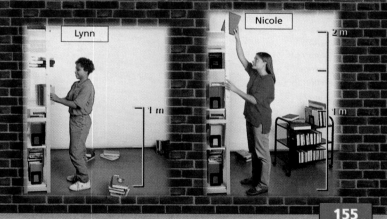

EXPLORATION 1

You may wish to set up this Exploration in stations around the classroom. Have students work in pairs or small groups.

TASK 1

To calculate the height to which the masses must be lifted, students should first express the weight of the masses in newtons. Explain to students that one newton is about equal to the weight of, or gravitational force on, a 100 g mass. *(100 g weighs about 1 N.)*

Answers to
Task 1

2. A complete list of the height required for each label is given in (4) below.

4. 3 m: 3 J (100 g); 30 J (1 kg)
   2 m: 2 J (100 g); 20 J (1 kg); 100 J (5 kg)
   1 m: 1 J (100 g); 10 J (1 kg) 50 J (5 kg)
      To calculate the extra joules needed to lift your body, first determine how much of your body is lifted as you lift the masses. The number of extra joules needed = the mass of that part of your body (in kilograms) × 10 N × the distance (in meters) through which you lift that part of your body.

TASK 2

 SAFETY ALERT Caution students not to tip over the person seated in the chair.

Answers to
Task 2

1. The placement of the scale should only slightly affect the amount of force needed to move the chair. Pushing from a lower position reduces the frictional force between the floor and the chair.

2. Answers will vary. Students should use the following equation to calculate their answer: work (J) = force (N) × distance (0.3 m).

3. No. Lifting the chair would require more work because the full force of gravity must be overcome. To move the chair, only the frictional forces must be overcome.

EXPLORATION 1

What Work Feels Like

Here are a few tasks that will give you an idea of what a joule of work feels like.

TASK 1

A Work-Experience Display

You Will Need
- 1 sheet of paper
- scissors
- measuring tape or a meter stick
- 100 g, 1 kg, 5 kg masses
- masking or transparent tape

What to Do

1. Make a set of labels like the ones below.

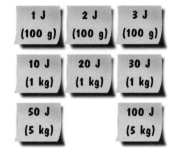

1 J (100 g)	2 J (100 g)	3 J (100 g)
10 J (1 kg)	20 J (1 kg)	30 J (1 kg)
50 J (5 kg)		100 J (5 kg)

2. Choose one label. To what height above the floor must you raise the indicated mass to experience the amount of work shown on the label?

3. Lift the mass to this height and tape the label onto the wall.

4. Repeat for each of the labels.

   You are also lifting part of your body when you do this work. How might you calculate how many extra joules of work are needed?

TASK 2

You're Pushing It!

You Will Need
- a chair
- a bathroom scale

What to Do

1. Hold the bathroom scale up against the back of the chair as shown in the illustration. As smoothly as you can, push the chair with a person seated in it for a distance of 30 cm. Does it matter where you place the bathroom scale as you push?

2. Determine the force needed to move the chair the measured distance. Calculate the work done.

3. Would you do the same amount of work if you lifted the chair and the person to a height of 30 cm? Explain.

4. Will the amount of work done change if a different person sits in the same chair? Explain.

Figuring Force
If the scale measures in kilograms, multiply by 10 to get the force in newtons; if it measures in pounds, multiply by 4.5.

TASK 3

Comparing Different Examples of Work

You Will Need
- a force meter
- various objects

What to Do

Try these activities:

a. Open a door.
b. Pull out a drawer.
c. Lift this book from the floor to a position directly over your head.
d. Pull a 1 kg mass along the floor for 2 m.
e. Pull down a screen or blind.

   Figure out how many joules of work are required to do each task. First design a way to use a spring scale or force meter to determine the size of the force needed for each task. Then measure the distance through which the force is applied.

---

4. Yes; if a person with a different mass sits in the chair, the amount of work done will change because the frictional force between the chair and the floor increases as the mass increases.

TASK 3

If a spring scale is used as a force meter, remind students that 100 g on the scale is equal to 1 N of force.

TASK 4 *page 157*

Because some students may not be able to do pull-ups, you may wish to suggest that one volunteer from each group perform the exercises. Make sure that students do not turn this Exploration into an athletic competition. If you feel your students are incapable of handling themselves in a mature manner, you may wish to have a volunteer perform the exercises as a classroom demonstration.

 An Exploration Worksheet is available to accompany Exploration 1 (Teaching Resources, page 3).

TASK 4

Work While You Exercise

You Will Need

- a meter stick
- a pull-up bar

What to Do

1. Calculate your weight in newtons.
2. Do a pull-up.
3. Have someone measure how high you lift your weight.
4. Calculate the work you do in one pull-up and in five pull-ups. What muscles are doing this work?
5. Repeat this task, this time doing a vertical jump. At the highest point of your jump, have someone measure the distance from your heels to the floor with your legs straight.

TASK 5

Machines Also Work

Answer the following questions about the diagram below:

a. How much upward force must the pulley machine put on 1 kg of gravel to lift it?

b. If the gravel is raised 0.5 m, how much work does the machine do?

You Will Need

- 1 kg bag of gravel (or other 1 kg mass)
- 2 pulleys, with cord
- a force meter or spring scale

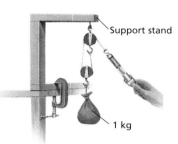

Support stand

1 kg

What to Do

1. Arrange the setup as shown.
2. Measure with a force meter the force needed to operate the machine.
3. Measure how far you need to pull the force meter to lift the gravel 0.5 m.
4. Calculate the work you put on the machine. Compare your work with the work the machine puts on the gravel.

   Why is it helpful to use the machine?

EXPLORATION 2

A Powerful Idea

By now you should have a good feel for the scientific meaning of *work*. But what is the scientific meaning of *power*, an energy-related word we often use? Take a moment to write down your own definition of *power*. Using your definition, which is more powerful, a horse or a car? a diesel locomotive or a jet aircraft? How powerful are you? How would you find out?

What to Do

Find a staircase or a steep hill that isn't too tall. Perhaps your school's gym has a rope climb. Measure the vertical distance to the top. If necessary, ask your teacher how to do this. Find out how much work you would do in climbing to the top. As fast as you can, climb the rope or run to the top of a hill or staircase (don't stumble!). Time how long it takes. To find your power in joules per second, or *watts*, divide the amount of work you did by the time it took to do it. How does the definition of *power* you wrote earlier compare with the definition of *power* that you just learned?

To find out more about power, see pages S44–S45 of the SourceBook.

157

TASK 5

After students have had a chance to experiment with the setup, ask: Why is the force recorded on the force meter one-half of the weight of the 1 kg mass? *(The tension, or force, on one side of a pulley always equals the tension on the other side. Since the mass is suspended by two segments of cord separated by a pulley, each segment of cord supports half of the weight.)*

Answers to *Task 5*

a. The force has to exceed the weight of the bag in order to move it. Since 1 kg weighs about 10 N, the machine must put about 10 N of force on the bag of gravel to lift it.

b. If the bag is raised 0.5 m, the work done will be 10 N × 0.5 m, or 5 J.

2. The force recorded on the force meter should be about 5 N.

3. To lift the bag 0.5 m, the force meter must be moved 1 m.

4. • Student's work = 5 N × 1 m = 5 J
   Pulley's work = 10 N × 0.5 m = 5 J
   (Students may note that they do slightly more than 5 J of work on the pulley system. This extra work is a function of the machine's efficiency.)
   • Using the machine is helpful because it reduces the amount of force required to lift the mass. The machine also allows the user to pull downward to lift the mass, thus changing the direction of the applied force.

EXPLORATION 2

Have students answer the questions in the introductory paragraph prior to doing the activity. *(Accept all reasonable responses.)* Help students to conclude that power is the rate of doing work. Encourage students to define power by writing a formula. *(Power = work ÷ time)*

Homework

The Exploration Worksheet that accompanies Exploration 2 makes an excellent homework activity (Teaching Resources, page 7).

FOLLOW-UP

Reteaching

Have students develop their own activity for demonstrating what work is and how it can be measured.

Assessment

Have students choose one of the tasks in Exploration 1 and submit their results for assessment.

Extension

Have students examine exercise equipment that is available on campus, such as the stations of an exercise course or equipment used in the gymnasium. Have students explain how work is done when each of the devices is used.

Closure

Students often have chores, or work, to do at home. Have them choose some example of work that they do in the scientific sense, determine or estimate the force and distance for each activity, and then calculate the amount of work involved. Have students compile their results in a chart, listing their chores in order of the amount of work done.

FOCUS

Getting Started

Before students write their captions, you may wish to have them briefly review what they learned about food and energy in Unit 1.

Main Ideas

1. When potential energy is released, it is converted into kinetic energy.
2. An object in motion possesses kinetic energy.
3. Work may result in changing potential energy to kinetic energy.

TEACHING STRATEGIES

Answers to
In-Text Questions

A Students' captions may vary, but they should indicate a connection between the energy in food and the energy required for a person to do work. The following is an example of the type of caption students might write: The body uses the energy it gets from food to do work.

CROSS-DISCIPLINARY FOCUS

Language Arts

This Chapter introduces students to the scientific definitions of work and energy. Have students review the differences between the everyday definitions of the terms *work* and *energy* and the scientific meanings of these terms. Then ask students to list other examples of words that have both a scientific and an everyday definition. Possible examples include conductor, pressure, fusion, power, efficiency, weight, and tension.

LESSON 3 Work and Energy

What Do You Need to Do Work?

Look at the illustration shown here. A lot is going on. Write a caption to explain what's happening in terms of work and energy. Where do you think all this energy came from? **A**

The Work/Energy Story that follows will give you many new ideas that relate work to energy. As you read it, see how many blanks you can complete without looking at the answers.

The Work/Energy Story

Energy enables us to do (1). If you lift a 10 N block to a height of 2 m, you will have done (2) J of work on it. If you now let the block fall on a nail, it hits the nail with a (3) and the nail is pushed a short distance into the wood. In other words, the block does (4) on the nail. Since energy enables work to be done, the raised block must have (5). How did the block get it? The block got the energy by the (6) you put on the block to raise it.

The energy of raised objects is called potential (stored) energy. Since you did 20 J of work on the block, the potential energy of the raised block must be (7) J. You, of course, got your energy from the (8) you ate. If you had felt the top of the nail, it would have been warm. As the block did work on the nail, some of the kinetic energy was changed into (9) energy. Note that in every case when work was done, (10) was transferred.

Answers: 1. work, 2. 20, 3. force, 4. work, 5. energy 6. work, 7. 20, 8. food, 9. heat, 10. energy

---

LESSON 3 ORGANIZER

Time Required
3 class periods

Process Skills
observing, analyzing, classifying

Theme Connection
Energy

New Terms
Chemical energy—a form of potential energy that is stored in food, fuels, and other chemicals and that is released by chemical reactions
Kinetic energy—the energy of motion

Potential energy—the energy stored in an object because of its position or chemical composition

Materials (per student group)
Exploration 3, Experience 3: metal or cardboard trough about 60 cm long; 3 identical books; 3 balls, one twice the mass of the lightest and one three times the mass of the lightest; 2 cm × 15 cm piece of cardboard; meter stick

Teaching Resources
Exploration Worksheet, p. 8
Activity Worksheets, pp. 12 and 14

Important Forms of Energy

Potential Energy

Alexander has a lot of *potential* as an archer. What does *potential* mean? It doesn't mean that he's a good archer yet. But he does have the ability to become one.

A raised object has **potential energy**. This does not mean that the raised object is doing work in that position. Rather, it means that the object can perform work on something else if it is released.

When you lift the object, you do work to overcome the gravitational force on the object. The object now has potential energy. When the object is released, gravitational force acts on it, causing it to fall. Because of its motion, the falling object is able to do work on other objects.

A raised object is only one example of potential energy. Potential energy also takes other forms. Work can be done on objects by overcoming other forces, and in so doing, the objects gain potential energy. For example, if you pull back on a bowstring, you are overcoming opposing elastic forces—the tensions of the bow and the bowstring. The work you do against the opposing forces gives potential energy to the bow and bowstring. When the bowstring is released, the elastic forces of the bow and bowstring do work on the arrow, sending it flying.

159

Important Forms of Energy

Provide students with some rubber bands. Ask them if the rubber bands have any energy to do work. *(The unstretched bands do not have energy to do work.)* Have students stretch the rubber bands, and ask them if the rubber bands have energy now. *(Most students will agree that the stretched rubber bands have energy.)* Ask: Where is the energy when the rubber bands are stretched? *(It is in the rubber bands.)* What happens to this energy when the rubber bands are released? *(It is turned into motion.)* Explain to students that this is an example of potential energy being converted to kinetic energy.

Help students understand the following concepts:

- Work must be applied to an object in order for the object to gain energy.
- An object with potential energy has the potential for doing work.
- A raised object's potential energy depends on both its weight and its height.

Integrating the Sciences

Life and Physical Sciences

Hold an arm-wrestling contest between student volunteers. Have all students observe the contests and write a paragraph describing how work and energy are involved in arm wrestling. Note: Before beginning the contest, make sure that students will conduct themselves in a safe and mature manner during this activity.

Meeting Individual Needs

Gifted Learners

Obtain several rubber-band powered, balsa wood airplanes. Have students determine the relationship between flight distance and the number of turns in the rubber band. *(Students should find that there is direct relationship between the amount of potential energy [turns of the rubber band] and the distance the plane travels.)* The best results are accomplished if this activity is performed in an open area such as a long hallway or outdoors.

Theme Connection

Energy

Focus question: How do seat belts protect passengers in a moving vehicle? *(Moving objects, such as a person in a moving car, have kinetic energy. The greater the speed of the car and the mass of the person, the greater the kinetic energy. If the moving car stops suddenly, the passenger continues forward with the same amount of kinetic energy unless restrained by a seat belt.)*

ENVIRONMENTAL FOCUS

Burning fossil fuels releases energy in the form of heat, which provides the energy to produce electricity. Petroleum is the most commonly used fossil fuel in the United States and worldwide. If the current rate of consumption continues, the world's supply of petroleum will be exhausted within the lifetime of an average middle-school student. Discuss with students some alternatives to petroleum. *(Suggestions may include natural gas, solar energy, and wind-generated energy.)*

Group size: 3 to 5 students

Group goal: to describe how potential energy is stored and released

Positive interdependence: Assign each group two or three illustrations from pages 160 and 161 to study. Students should write responses to questions 1–4 on page 160 for each of the illustrations. If you wish, have students make a chart showing their responses. Students should share their answers during a class discussion of the lesson.

Individual accountability: Make each student responsible for listening carefully to the discussion and for analyzing another group's presentation.

Answers to
Potential-Energy Study

There is potential energy in all of the situations shown. Make sure students understand that potential energy does not perform the work directly. Potential energy is put to work only after first being converted to kinetic energy.

a. The potential energy of the chest expander is stored by stretching its springs. The opposing force overcome is the elasticity of the spring. The potential energy is put to work when the force holding the spring in place is removed, and the potential energy is converted to kinetic energy.

b. Potential energy is stored when the racket collides with the ball, compressing the ball and stretching the strings of the racket. The opposing force overcome is the elasticity of the ball and the strings. The potential energy converts to kinetic energy, which returns the ball and the strings to their original shapes.

c. The potential energy of the basketball is stored by pushing the ball under the water. The opposing force overcome is the buoyancy of the water pushing up on the ball. When the hand is removed, the potential energy becomes kinetic energy, and the buoyant force pushes the ball upward.

d. The potential energy of the frozen yogurt cone is stored in the chemical bonds of the chemicals in the yogurt and cone. Molecular forces are overcome to form the bonds. The potential energy becomes kinetic energy when the chemicals in the frozen yogurt and cone are broken down by the cells of the body.

Potential-Energy Study

The illustrations on these pages show objects that have potential energy. Working with a partner, answer the following questions for each situation:

1. Which objects or substances have potential energy?

2. How is the potential energy stored?

3. What opposing force is overcome to store the potential energy?

4. How will the potential energy be put to work?

Now consider the answers to these questions for the bow-and-arrow situation on page 159. The bent bow and stretched bowstring gain potential energy (1) from the archer's work in stretching the bow (2). This work overcomes the elastic forces exerted by the bow and string (3). The stored energy enables the string to do work on the arrow (4).

Note that the energy represented by (d) and (e) is stored in fuel and food, respectively. This energy is released by chemical changes. For this reason, the potential energy in food and fuel is often called **chemical energy**.

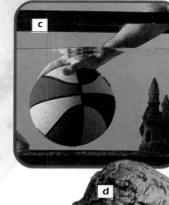

160

e. The potential energy of the gasoline is stored in chemical bonds. Molecular forces are overcome to form the bonds. The potential energy is converted to kinetic energy when the gasoline is burned, breaking the bonds holding the gasoline molecules together.

Answers to Potential-Energy Study continued on the next page. ▶

Integrating the Sciences

Life and Physical Sciences
Point out that energy is transferred in ecosystems. Have students construct a food chain and label the energy transfers and transformations that occur within it.

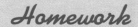

Homework

Questions 1–4 of Potential-Energy Study make excellent homework.

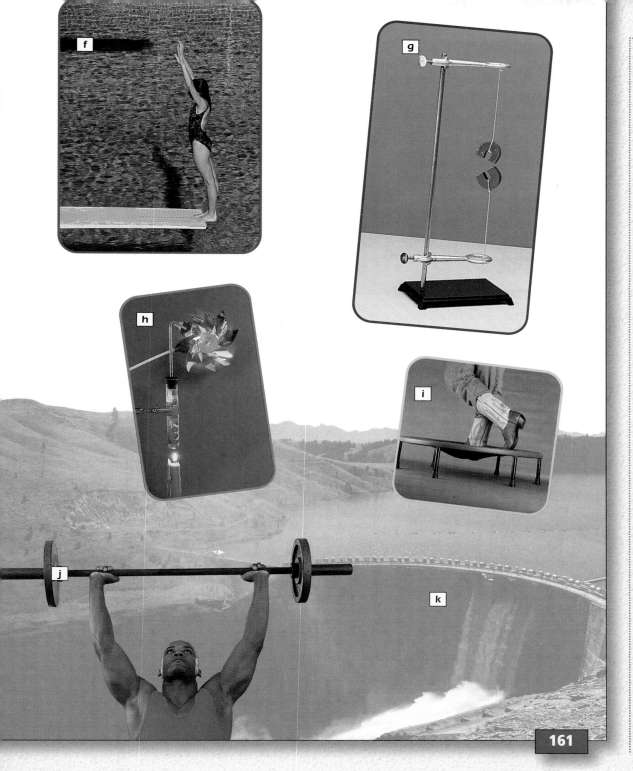

Answers to
Potential-Energy Study, continued

f. Potential energy is stored when the diver climbs up to the board. Some of the diver's energy is transferred to the diving board when she exerts a force on the board, bending the board. The opposing forces overcome are the diving board's elasticity and Earth's gravitational force. The board's potential energy is converted to kinetic energy when the diving board flexes upward.

g. Potential energy is stored in the height of the lower magnet. The opposing force is the force of gravity. The potential energy is converted to kinetic energy when the lower magnet drops.

h. There is potential energy stored in the candle. The opposing force overcome is the force of attraction among the particles of the candle. The potential energy is converted to kinetic energy when the candle burns and steam pushes the pinwheel.

i. Potential energy is stored in the stretched elastic material of the trampoline. The opposing force is the weight of the person. The potential energy will be converted to kinetic energy when the trampoline returns to its unstretched position.

j. Potential energy is stored in the barbell when it is lifted. The opposing force is the downward pull of gravity. The potential energy is converted to kinetic energy when the barbell is allowed to move against the muscles of the weight lifter.

k. Potential energy is stored as the water is held back by the dam. The opposing force overcome is gravity. The potential energy is converted to kinetic energy when the water is allowed to flow and fall.

EXPLORATION 3, *page 162*

You may wish to point out to students that physicists often use simplified models to explain more complex, real-world phenomena. Disregarding forces such as friction and air resistance allows scientists to more easily test hypotheses and to trace the flow of energy in mechanical systems. Students are introduced to the concept of efficiency in Lesson 1 of Chapter 9.

EXPERIENCE 1   *page 162*

After students have read this section, ask: What opposing forces must the bowling ball overcome to do work on the pins? *(Air resistance, friction between the ball and the floor, and the inertia of the pins)* In what other ways does the ball lose energy? *(Some of the ball's kinetic energy is converted to sound energy, especially as it strikes the pins.)*

Answers to
Experience 1, *page 162*

Rodney does 100 J of work (work = force × distance, or 100 J = 100 N × 1 m). Because Rodney does 100 J of work on the bowling ball, it begins moving with 100 J of energy. The ball will be able to do slightly less than 100 J of work because of friction and air resistance and because some of its kinetic energy is converted to sound energy.

EXPERIENCE 2

Help students realize that the potential energy of the brick is calculated in the same way that work and kinetic energy are calculated. If all forces other than gravity are ignored, the brick's initial potential energy would equal its maximum kinetic energy as it hits the nail. In turn, all of this kinetic energy would be converted to work in an ideal system.

Answers to
Experience 2

The brick has 60 J of potential energy (2 kg × 10 N/kg = 20 N; 20 N × 3 m = 60 J). As the brick falls, its potential energy is converted to kinetic energy. The kinetic energy of the brick when it reaches the nail will be slightly less than its initial potential energy (60 J) because of air resistance. The brick's energy results from the force of gravity doing work on the brick, converting potential energy into kinetic energy. The brick will do slightly less than 60 J of work on the nail because of friction and because some of its kinetic energy is converted to sound energy.

Answers to
In-Text Questions

Ⓐ Rodney's energy came from the food he ate.

Ⓑ The force of gravity did work on the brick by changing its potential energy to kinetic energy.

EXPERIENCE 3   *page 163*

You may wish to point out to students that although they are doubling and tripling the potential energy of the ball in steps 2 and 3, its kinetic energy does not necessarily double and triple. This is because the energy used to overcome friction decreases as the angle of the trough increases. Also, as students double and triple the mass of the ball in steps 4 and 5, it is likely that they are also increasing its diameter. If so, the ball's kinetic energy will not double and triple because more energy is contained in the ball's rotational motion instead of in its motion down the incline.

EXPLORATION 3

Another Form of Energy: An Investigation

In this Exploration, you will investigate the following questions through a series of real and imagined experiences.

- Do moving objects do work?
- Do moving objects have energy?
- If moving objects have energy, where do they get it?
- What factors affect the amount of energy that moving objects might have?

The first two experiences are thought experiments; only the last one is a laboratory experience. Put on your thinking cap!

EXPERIENCE 1

Does the moving bowling ball in the illustration below do work when it hits the pins? If the bowling ball does work, what must the moving ball have? Where does this energy come from?

Rodney puts the bowling ball in motion by applying force to it. He does this by swinging the ball forward. If Rodney puts 100 N of force on the ball through a swing of 1 m, how much work does he do? How much energy does the bowling ball have? How much work will the bowling ball be able to do when it strikes the pins?

A simple calculation shows that about 100 J of work was done to the ball. Therefore, the ball was given 100 J of energy, which enabled it to do 100 J worth of work to the bowling pins. This form of energy— the energy of moving bodies—has a name. It is called **kinetic energy**. *Kinetic* comes from an ancient Greek word meaning "motion."

EXPERIENCE 2

A 2 kg brick is raised to a height of 3 m. How much potential energy does it have? When the brick is released, it accelerates downward. What kind of energy does it have now? About how much kinetic energy will it have as it just begins to strike the nail? Where does this energy come from? Does the moving brick do work on the nail? How much work does it do?

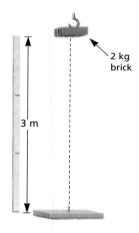

2 kg brick

3 m

- In Experience 1, the source of the kinetic energy was Rodney. Where did Rodney get the energy to do this work? Ⓐ

- In Experience 2, the source of the brick's kinetic energy was the potential energy it had before it was dropped. Whenever energy is transformed from one form into another, work is done. Therefore, work must have been done as the potential energy of the brick was changed into kinetic energy. What did the work on the brick? Ⓑ

Answers to
Experience 3, page 163

1. The rolling ball moves the piece of cardboard. The ball does work on the cardboard. The ball has potential energy before it starts moving; its potential energy is gradually converted to kinetic energy as it speeds up.

2. The speed of the ball is greater than in step 1. The cardboard moves farther. Therefore, the ball in step 2 has more kinetic energy than the ball in step 1. Speed is one characteristic that determines the amount of kinetic energy of the ball.

3. The cardboard moves farther than it did in step 2.

4. The cardboard moves farther than it did in step 1. The increased mass causes the difference. Mass is another characteristic that determines the amount of kinetic energy of the ball.

5. The cardboard moves farther than in step 4, confirming that the mass of a moving object affects the amount of its kinetic energy.

What factors affect the amount of kinetic energy a moving body has?

You Will Need

- a trough (made of metal or cardboard approximately 60 cm long)
- 3 identical books
- 3 balls, one twice the mass of the lightest, and the other 3 times the mass of the lightest
- a piece of cardboard (approximately 2 cm × 15 cm)

What to Do

1. Raise the trough by placing a book under it. Let the lightest ball roll down the trough to strike a piece of cardboard folded as shown. What happens? Is work done by the ball? What kind of energy does the ball have before it starts moving? after it starts rolling? The distance the cardboard moves is an indication of the amount of kinetic energy the rolling ball has. Allow the ball to roll again, and measure the distance the cardboard moves from the starting point.

2. Raise the trough by adding another book. Allow the ball to roll from the same spot on the trough. How does the speed of this ball compare with that of the ball in step 1? Measure and record the distance the cardboard moves. How does the kinetic energy compare in each case? What is one characteristic of the ball that affects how much kinetic energy it has?

3. Try using three books. Record the distance that the piece of cardboard moves.

4. Place the trough on one book again. Allow the ball that has twice the mass as the lightest ball to roll down the trough. Compare the distance the cardboard moves with the distance it moved in step 1. What causes the difference? What is another characteristic that affects the amount of kinetic energy?

5. Check your conclusion from step 4 by using the ball that is 3 times the mass of the lightest ball.

Questions

1. Where does the kinetic energy of the balls come from? How does the potential energy of the ball (before rolling) compare in steps 1, 2, and 3? How does its kinetic energy (just before striking the cardboard) compare in steps 1, 2, and 3?

2. Which factors (variables) do you think affect the amount of kinetic energy a moving body has—color, speed, shape, mass, composition, location, others not listed? Make a list in your ScienceLog.

Answers to *Questions*

1. Gravity does work on the ball, which changes the potential energy to kinetic energy. The potential energy of the ball increases from step 1 to step 3 as the height of the trough increases. The kinetic energy of the ball increases from step 1 to step 3.

2. Students should understand that speed and mass are the main factors affecting the kinetic energy of a moving object. However, these two main factors may be derived from other factors, such as shape, composition, and location.

 An Exploration Worksheet is available to accompany Exploration 3 (Teaching Resources, page 8).

Homework

The Activity Worksheets on pages 12 and 14 of the Unit 3 Teaching Resources booklet make excellent homework activities as students complete Lesson 3.

FOLLOW-UP

Reteaching

Have students make "flip books" to illustrate potential energy changing into kinetic energy. Each page in a small notebook represents a progression of motion. For example, a rock resting at the top of a hill has potential energy. When the rock rolls down the hill, its potential energy changes into kinetic energy. Have students display their books in the classroom.

Assessment

Suggest that students analyze their favorite sporting event to determine when energy is stored as potential energy and when energy is changed into kinetic energy. Have volunteers present their information to the class by giving oral reports. Suggest that they use diagrams or charts to illustrate their ideas.

Extension

Have students collect a set of 10 pictures or photographs that illustrate kinetic or potential energy. (You may wish to provide magazines for students to do this in class.) A brief caption beneath each picture should identify the type of energy and how it is being used. Display the pictures around the classroom.

Closure

Have students write a summary of some ways in which potential energy can be stored. *(Examples include lifting things, stretching an elastic material, bending or twisting an elastic material, chemical bonding, compressing something, or pushing against buoyant forces.)* Also ask students to provide examples of how potential energy can be converted to kinetic energy in each case. *(Dropping a lifted object, letting go of a stretched elastic material, allowing chemicals to react, relaxing compression of an item, removing an object experiencing buoyant force from the fluid in which it is suspended)*

Getting Started

Ask students to rub their hands together quickly. Then have them analyze what happens by responding to the following questions:

• What did you feel when you rubbed your hands together? *(Heat)*
• Is heat a form of energy? *(Yes)*
• What kind of energy produced the heat? *(Kinetic energy)*
• Where did the kinetic energy come from? *(It came from rubbing the hands together.)*
• Was rubbing your hands together work? *(Yes)*
• Why was it work? *(Force and motion were involved.)*

Point out to students that in this lesson they will analyze energy changing from one form into another.

Main Ideas

1. There are many different forms of energy.
2. One form of energy may change into or produce another form.
3. Many types of devices act as energy converters.

Answer to
In-Text Question

 Answers will vary. Examples include a change from chemical energy to kinetic energy in the car and the bowler and from potential energy to kinetic energy in the trampoline and the dam.

Answers to Exploration 4 are on the next page. ►

★ An Exploration Worksheet is available to accompany Exploration 4 (Teaching Resources, page 15).

Energy Changes

How many different types of energy have you seen so far? Can one type of energy change into another type of energy? Look over the last few pages and identify some examples of where this happens. Ⓐ

The following Exploration features a number of investigations. Some are thought investigations that can be performed best by talking about them; others you may actually perform.

164

EXPLORATION 4

An Energetic Discussion

Working in groups, discuss question 1, below, and any two others that interest you. Your group should be prepared to discuss the results with the class.

1. Refer to the Potential-Energy Study on pages 160 and 161 to answer the following questions:
   a. How is potential energy changed to kinetic energy in each situation in the Potential-Energy Study?
   b. When energy is transferred, work is done. What agent is doing the work as potential energy changes into kinetic energy?
   c. How does kinetic energy do work in image (h)? Any moving part of a machine has kinetic energy, which is sometimes called **mechanical energy**. Can you think of other examples?

2. Hold a marble at the rim of a large bowl. Release the marble. What happens? You know that potential energy can change into kinetic energy. Can kinetic energy change into potential energy? When does the marble have the most potential energy? the most kinetic energy?

3. Marta lets go of a ball that is attached to the end of a rope suspended from a support. What happens to the ball? Will the ball swing back and hit Marta if she stands in the same place? What would happen if she moved forward one pace? Is there a "reversible" change of energy going on here?

LESSON 4 ORGANIZER

Time Required
2 class periods

Process Skills
analyzing, predicting, inferring, classifying

Theme Connection
Energy

New Terms
Energy converter—a physical system in which energy changes from one form into another

Mechanical energy—the kinetic energy of the parts of a machine

Materials (per student group)
Exploration 4: marble; large bowl; nickel; sheet of paper

Teaching Resources
Exploration Worksheet, p. 15
Transparency 20
Math Practice Worksheet, p. 20

4. Consider the acrobatic act pictured at right. One acrobat is standing on a raised platform. A second acrobat is standing on one end of a seesaw. Their initial positions are both labeled *1*. The illustration shows two other positions for each acrobat after the one on the platform jumps off (positions *2* to *5*). What happens? Describe the potential and kinetic energy of each acrobat in each position. How do you explain the fact that the second acrobat ends up on a platform higher than the platform used by the first acrobat?

5. Diagram A represents the lifting of a brick through a vertical distance of 6 m, in two steps. The brick has a mass of 2 kg. Diagram B represents the release of the same brick.

   In your ScienceLog, calculate the amount of potential and kinetic energy that you think the brick has in each position in the two diagrams.

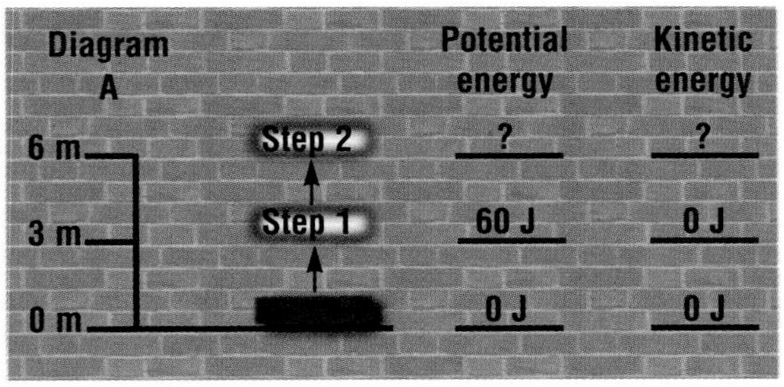

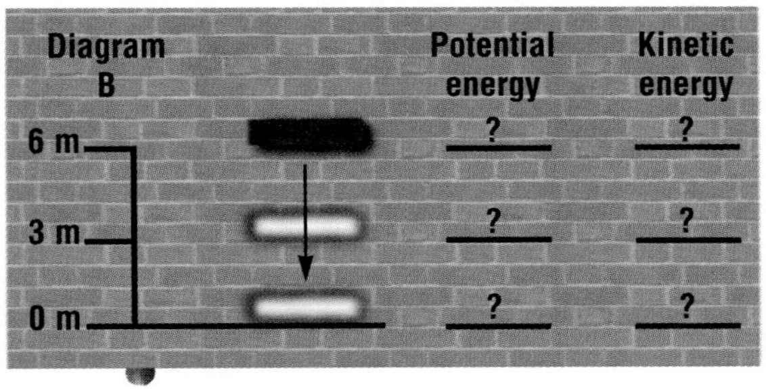

6. Think about dragging a table across the floor. What kind of energy does the table have as it moves? When you stop applying a force to the table, it stops moving. Where does the energy go?

7. Touch a nickel to your chin. Place this nickel flat on a piece of paper. While pressing down hard on it with your fingers, move it back and forth briskly 10 to 20 times. Again touch the nickel to your chin. What do you observe? Explain any energy changes.

8. Look at the stop-action photo below. It shows a bouncing golf ball. When is the potential energy the smallest? Why does the ball stop bouncing? What other type of energy might be present?

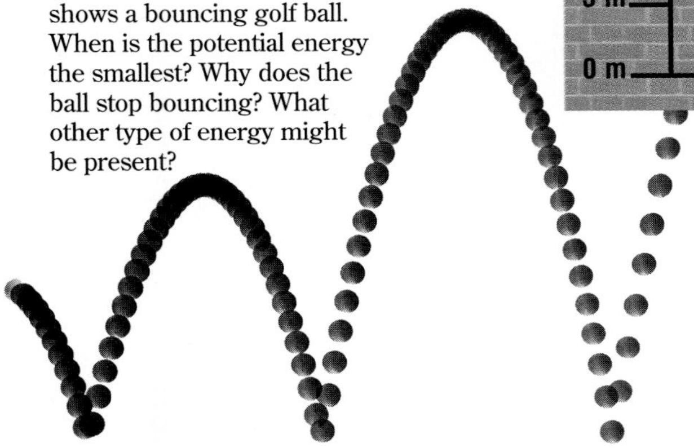

Diagram A		Potential energy	Kinetic energy
6 m	Step 2	?	?
3 m	Step 1	60 J	0 J
0 m		0 J	0 J

Diagram B		Potential energy	Kinetic energy
6 m		?	?
3 m		?	?
0 m		?	?

165

Answers to
Discovering New Forms of Energy

Sample answers:

a. Roller coaster: potential energy → kinetic (mechanical) energy, sound energy, and heat energy → potential energy → kinetic energy, sound energy, and heat energy

b. Ceiling fan: electrical energy → mechanical energy and sound energy

c. Clock: potential energy → kinetic (mechanical) energy and sound energy → potential energy → kinetic energy and sound energy

d. Model plane: potential energy → kinetic (mechanical) energy

e. Light bulb: electrical energy → heat and light energy

 Transparency 20 is available to accompany Discovering New Forms of Energy.

Theme Connection

Energy
Focus question: How is the water flowing through the sluiceway of a dam similar to wind turning blades in a windmill? *(As water flows through the sluiceway of a dam, it turns a turbine, producing mechanical energy. The windmill is similar because the wind causes the windmill's blades to turn, again producing mechanical energy. In both cases, the mechanical energy of the rotating parts is then converted to electrical energy.)*

Homework

You may wish to assign the Math Practice Worksheet on page 20 of the Unit 3 Teaching Resources booklet as homework.

FOLLOW-UP

Reteaching

Have students design a machine that not only performs a particular task (such as heat water or move an object up a ramp) but also features at least five changes of energy from one form to another.

Discovering New Forms of Energy

You have already seen how different types of energy can change from one form to another. Now identify the energy changes in each of the **energy converters** pictured on this page. Look not only for the familiar energy forms you have studied, but also for new ones such as electrical, light, heat, and sound energy. In your ScienceLog, make a table similar to the one shown below, and complete it for energy converters (a) through (d).

Example	Energy converter	Energy change
(e)	Battery	Chemical to electrical

Assessment
Divide the class into two teams. One team should think of an appliance. The other team should identify the energy changes that take place when that appliance is used. Then have teams switch roles.

Extension
Suggest that students research some scientific terms associated with motion, such as velocity, acceleration, and momentum. How are these terms related to each other? Challenge students to draw cartoons that illustrate these concepts.

Closure
Have students bring to class some of their old toys that are either battery- or spring-powered. Allow students to demonstrate and explain how the energy in their toy is transferred and transformed within the toy to make it work.

CHALLENGE YOUR THINKING

1. Pushing the Issue

Suppose your mass is 50 kg. Assume that your hands and arms support half your weight and that in doing one push-up you elevate this weight through a total distance of 40 cm. How many push-ups would you need to do to expend 1 kJ of work?

2. Power Play

a. Assume it took you 0.5 seconds to do a push-up in question 1. Use this to calculate the power you exerted in doing a single push-up, in both watts (W) and kilowatts (kW).

b. A 40 W light bulb uses 40 J of electrical energy each second to produce light and heat. How many such light bulbs would have to burn to equal the rate of work of one of your push-ups?

3. Climbing Potential

a. Mount Everest is 8848 m high. How much work would you have to do to get to the top? How much potential energy would you have at the top?

b. A glass of milk provides 500 kJ of energy, one-quarter of which is available for work. How many glasses of milk would provide you with enough energy to scale Mount Everest? Where does the rest of the energy go?

4. Stop! Look! Listen!

"The one thing constant in this world is change."

For the next 5 minutes, think about this statement and then about your surroundings. Make a list of all the energy converters and energy changes that you recognize. Try to identify where work is being done during each energy change.

167

Homework

The Activity Worksheet on page 21 of the Unit 3 Teaching Resources booklet makes an excellent homework assignment.

★ You may wish to provide students with the Chapter 8 Review Worksheet that accompanies this Challenge Your Thinking (Teaching Resources, page 22).

Answers to *Challenge Your Thinking*

1. One kilogram weighs about 10 N, so the weight of a 50 kg student is 10 N/kg × 50 kg = 500 N. To do a single push-up, the student must apply a lifting force of 250 N through a distance of 0.4 m. Therefore, the work expended to do one push-up is $W = F \times d = 250$ N × 0.4 m = 100 N•m = 100 J. To do 1 kJ of work the student must do 1000 J/100 J per push-up = 10 push-ups.

2. a. Power = work/time, and one push-up requires 100 J and 0.5 s. Power required to do one push-up: $P = W/t = 100$ J/0.5 s = 200 W = 0.2 kW.

b. One light bulb uses 40 W; one push-up requires 200 W. Thus, the number of light bulbs with power equal to that of one push-up is 200 W per push-up/40 W per light bulb = 5 light bulbs.

3. a. Answers will vary with students' weights. A student with a mass of 50 kg weighs 50 kg × 10 N/kg = 500 N.
work = force (student's weight) × distance (height of Mt. Everest)
= 500 N × 8848 m
= 4,424,000 J
= 4424 kJ
The work used to get to the top of Mt. Everest equals the potential energy of the student at the top.

b. One glass of milk provides 500 kJ/4 = 125 kJ; the number of glasses of milk required = 4424 kJ/125 kJ = 35. The rest of the milk's energy would be used to maintain body temperature and to fuel internal body activities.

4. Students' examples will vary. Sample answer:
A jogger converts chemical energy to kinetic energy. The chemical energy supplied by food enables the jogger to move her or his body. Work is done in the jogger's muscles as the muscle fibers are flexed.

Answers to Challenge Your Thinking continued ►

Answers to
Challenge Your Thinking,
continued

5. a. (1) Vic's training activities: riding his bicycle, doing push-ups, pull-ups, and sit-ups, jogging, and lifting weights; (2) Vic's bike ride to the top of the steep hill; (3) the dam and power station; (4) Vic's downhill movement; (5) the evergreen and maple trees; (6) Vic's brake; (7) Vic's watch

b. Energy change (1) involves a change of chemical energy to mechanical (kinetic) and potential energy. Change (2) is a conversion from kinetic to potential energy. Change (3) represents a change from potential energy to kinetic energy to electrical energy. Change (4) is a transformation from potential to kinetic energy. Change (5) represents a change from solar energy to chemical energy. Change (6) is a conversion of kinetic energy to heat energy. Change (7) is a change from chemical energy to mechanical energy.

c. (1) The cells in Vic's body convert chemical energy in his food to the kinetic and potential energy he uses in his workout. The energy converters in (2) are Vic's muscles and his bicycle, which allow him to do work against gravitational force. The converter in (3) is the power station, which converts the kinetic energy of falling water to electrical energy. (4) Vic's bike allows him to convert potential to kinetic energy. (5) The leaves in the trees change solar energy to chemical energy through photosynthesis. (6) Vic's brakes change kinetic energy to heat through friction. (7) The battery-operated components of Vic's watch convert the chemical energy of the battery to mechanical energy.

d. Answers will vary.

The following are sample revised answers:

1. The word *work* can be used in many different ways with very different meanings. In the scientific sense, work is done when force is applied to an object and this force moves the

5. The Race Is On

Excitement was in the air! It was the day of the big Overnight Bike Race. Vic felt confident. He had trained well by riding his bicycle; doing daily push-ups, sit-ups, and pull-ups; jogging; and lifting weights at the gym. Now, 20 minutes into the journey, he reached the top of a steep hill. In the valley below he could see the dam and power station. Water gushed down a sluiceway, making generators hum and electricity surge through the power lines. As Vic sped down the hill, he caught occasional glimpses of evergreens amid colorful maples. "Nature's power stations," he mused.

Vic braked as he approached the first rest station at the bottom of a steep hill. He was the fifth to arrive. He glanced at his watch—he had been racing for 56 minutes.

a. How many energy changes can you identify in the story so far?

b. Name the forms of energy involved in each change.

c. Where possible, name the energy converter.

d. Continue the story. Hide some more energy changes in the narrative, and then have a classmate try to identify them.

ScienceLog

Review your responses to the ScienceLog questions on page 146. Then revise your original ideas so that they reflect what you've learned.

object. Work is measured in joules and can be calculated if the applied force and the distance that the object moves is known: work (J) = force (N) × distance (m).

2. As the cars go downhill on a roller coaster, potential energy (energy of position, or stored energy) is converted to kinetic energy (energy of motion, or released energy). As the cars go uphill, kinetic energy is converted to potential energy. When the cars are moving, some kinetic energy is converted to heat and sound energy as a result of the friction

between the cars' wheels and the roller coaster's track.

3. Energy comes in many different forms, such as light, heat, and electrical energy, and can be converted from one form to another. Stored energy is called potential energy. Chemical energy is one type of potential energy. Energy of motion is called kinetic energy or mechanical energy. Energy enables work to be done. Another way of expressing this relationship is that creating a force to put matter into motion requires energy.

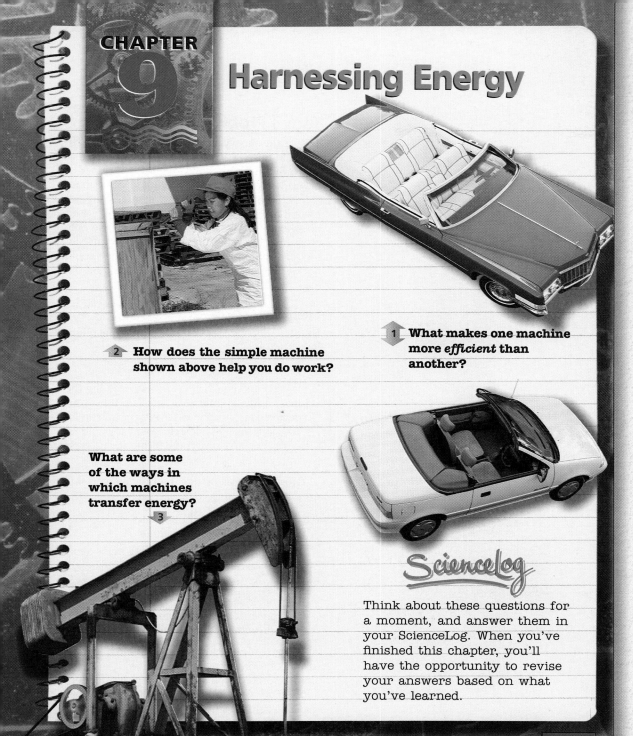

CHAPTER 9

Harnessing Energy

2 How does the simple machine shown above help you do work?

1 What makes one machine more *efficient* than another?

What are some of the ways in which machines transfer energy? **3**

ScienceLog

Think about these questions for a moment, and answer them in your ScienceLog. When you've finished this chapter, you'll have the opportunity to revise your answers based on what you've learned.

169

Connecting to Other Chapters

> **Chapter 8**
> presents the scientific meanings of work, energy, and power, along with ways to measure or calculate each.

> **Chapter 9**
> examines mechanical systems in terms of efficiency, mechanical advantage, and energy transfer.

Prior Knowledge and Misconceptions

Your students' responses to the ScienceLog questions on this page will reveal the kind of information—and misinformation—they bring to this chapter. Use what you find out about your students' knowledge to choose which chapter concepts and activities to emphasize in your teaching. After students complete the material in this chapter, they will be asked to revise their answers based on what they have learned. Sample revised answers can be found on page 203.

In addition to having students answer the questions on this page, you may wish to have them complete the following activity: Students should experiment with pushing open a door, placing their hands at different places on the door each time. Ask: Is it harder or easier to open the door when your hands are far from the hinges? Why do you think this is so? Could this knowledge be applied in any other situations? Where does the energy that you exert go? Remind students that there are no right or wrong answers to the questions at this point. Student responses to these questions should help you evaluate what students understand about machines and energy transfer, what misconceptions they may have, and what is interesting to them about this topic.

FOCUS

Getting Started

The teacher demonsration described in the Discrepant Event Worksheet makes an excellent introduction to this lesson (Teaching Resources, page 29).

Main Ideas

1. Simple machines are used to make work easier.
2. Simple machines reduce the input force required to do work by applying this force over a greater distance.
3. The work input of a machine is always greater than the work output.
4. The mechanical advantage of a machine is the ratio of the force exerted by the machine to the force applied to the machine.

TEACHING STRATEGIES

Tony's Problem

After students have read this section, ask them to brainstorm solutions to Tony's problem. They should realize that it would require four members of Tony's crew to lift the load without the aid of a machine. Thus, their solution should make lifting the load at least four times easier in order for Tony to be able to lift it by himself. It might be helpful to display the following items in the classroom so that students can examine them as they evaluate their ideas: a lever, a single, fixed pulley, a movable pulley, a combination of pulley blocks, and an inclined plane.

Do not instruct students on how to use the equipment at this time. Instead, refer them to the illustrations on page 171, and suggest that they make some of the simple devices shown there.

SAFETY ALERT You should point out to students that they should never lift an object the way Tony does on this page. One should always bend one's legs at the knees, keeping one's back as upright as possible.

LESSON

1

Lightening the Load

Tony's Problem

Tony is the supervisor on a construction site. He needs to move a 100 kg load to a platform that is 1 m high. Tony knows that each member of his crew can safely exert a lifting force of about 250 N. How many of his crew should Tony get to lift the load? (Remember, the weight of a 1 kg mass is about 10 N.) However, Tony is the only crew member who isn't busy. But he can't lift the load by himself. What type of device could he put together so that one person could lift the load onto the platform?

Helping Tony

Here's your chance to design a device that might help Tony.

1. Examine the diagrams on the next page.
2. Design and construct a device for lifting the mass with as little effort as possible. Your device doesn't actually have to be able to lift 100 kg! Instead, design your machine to scale. For instance, it might lift a 1 kg mass with an effort of 2 N.

   Here is some of the information you will need in order to test your machine:

   - the size of the force (in newtons) required to lift the load using your machine
   - the distance through which you must apply that force to raise the load to the desired height
   - the amount of work you do in moving the load
   - the comparison of the amount of work you do with your machine to the amount of work you would do without the machine

   What might you do to improve your machine? Can you lift the load with an even smaller force? If you are able to do this, is there a trade-off with something else? Does your machine actually *save* work? **A**

LESSON 1 ORGANIZER

Time Required
3 to 4 class periods

Process Skills
comparing, contrasting, classifying

Theme Connection
Energy

New Terms
Efficiency—the ratio of the work done by a machine to the work put into it
Mechanical advantage—the factor by which a machine multiplies force

Work input—the work put into or done on a machine
Work output—the work done by a machine

Materials (per student group)
Helping Tony: a variety of materials to construct a device for lifting a load, such as the following: support stand with ring clamp; C clamp; single-, double-, and triple-pulleys; metric force meter (or metric spring scale);

continued ▶

Some Possible Devices to Help Tony

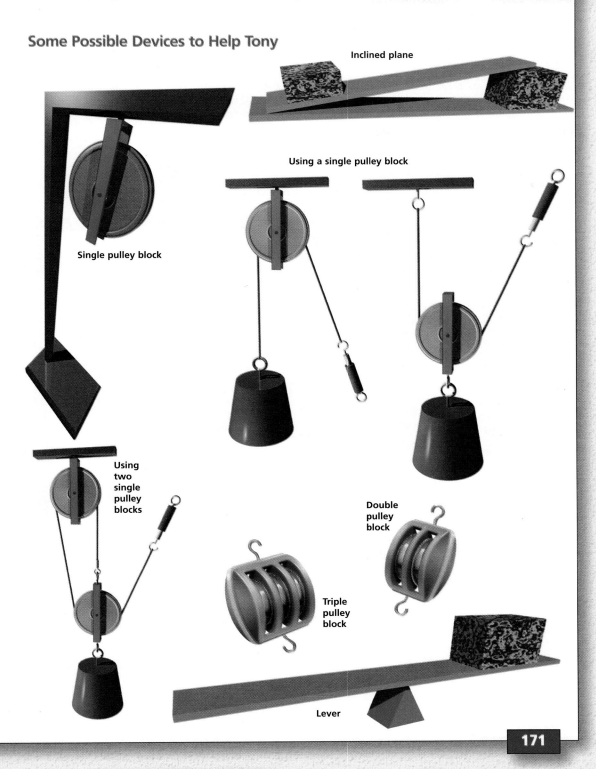

Inclined plane

Single pulley block

Using a single pulley block

Using two single pulley blocks

Double pulley block

Triple pulley block

Lever

171

Cooperative Learning
HELPING TONY, PAGE 170

Group size: 3 or 4 students

Group goal: to design a simple machine to solve a lifting problem

Positive interdependence: Students should choose one of the following roles: construction manager (to direct the building of the machine), builder (to build the machine), materials manager (to gather materials), and illustrator/presenter (to sketch the designs and present the project to the class).

Have students draw a design of their machine and then, using the equipment supplied, construct a machine that can lift the 1 kg mass with no more than 2 N of force. Have students keep their designs secret from the other groups so that the groups can have a contest to create the machine that will do the most work with the least effort. The machines should be shared with the class during class discussion.

Individual accountability: Each student should be able to describe the machine that their group constructed and to defend its benefits and trade-offs.

⭐ **Transparency 21 is available to accompany Helping Tony.**

Answers to
In-Text Questions, page 170

Ⓐ Student suggestions for improving their machine will vary depending on what feature they wish to improve. Accept all reasonable responses. To lift the load with an even smaller force, students may suggest ways to reduce friction, or students may suggest increasing the distance through which the force is applied. The trade-off in reducing the applied force is that the distance through which the load moves will be greater.

A machine cannot save work, but it can reduce the force or distance required to do the work. Students should understand that machines reduce the input force required to do work by increasing the distance over which this force is applied.

ORGANIZER, *continued*

1 kg mass; metric ruler; small wedge to be used as a fulcrum; wood block, 10 cm thick; wood board, about 5 cm × 20 cm; 2–3 m of string

Teaching Resources
Discrepant Event Worksheet, p. 29
Theme Worksheet, p. 30
Transparencies 21–23
SourceBook, pp. S47 and S52

Leona's Machine

If feasible, set up the apparatus as it appears in Leona's report. Encourage students to experiment with the device by lifting loads of various weights. It is likely that they will be surprised at how little force it takes to lift a load using a machine of this type.

Call on a volunteer to read aloud Leona's report. Then have students examine how her machine is assembled. Have them trace the path of the string to discover how it is attached to the machine and to the load. To determine their understanding of the machine's construction, you may wish to ask questions similar to the following:

- How many pulleys are at the top of the machine? at the bottom? *(There are two pulleys at the top and two at the bottom.)*
- Where is the string attached to the machine? *(The string is attached to the bottom of the block containing the top two pulleys.)*
- Where is the load attached to the machine? *(The load is attached to the block containing the bottom two pulleys.)*
- Which of the pulleys are movable? *(The bottom pulleys are movable.)*
- How many sections of string support the movable load? *(Four sections)*

You may wish to explain to students that a combination of pulleys and ropes similar to what Leona put together is called a *block and tackle.* A block and tackle always has movable pulleys in frames called blocks. A single block has one pulley. Double blocks, like the ones Leona used, have two pulleys. A triple block has three pulleys.

A Look at Two Solutions

Leona's Machine

Leona used an arrangement of pulleys to solve Tony's problem. Below is Leona's report about the machine she made. From the information in her report, solve the problems on the facing page.

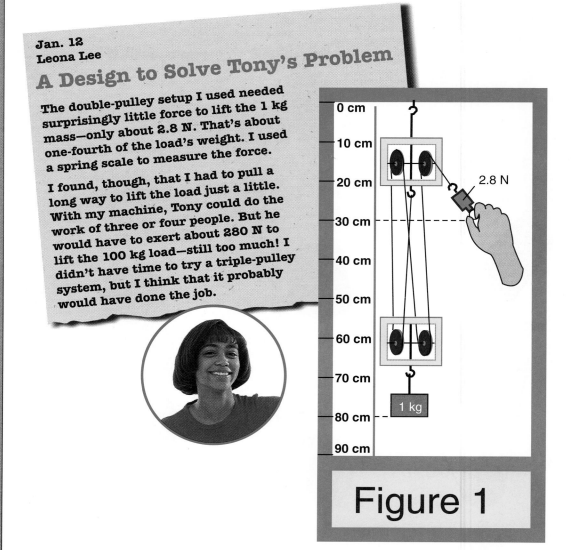

Jan. 12
Leona Lee

A Design to Solve Tony's Problem

The double-pulley setup I used needed surprisingly little force to lift the 1 kg mass—only about 2.8 N. That's about one-fourth of the load's weight. I used a spring scale to measure the force.

I found, though, that I had to pull a long way to lift the load just a little. With my machine, Tony could do the work of three or four people. But he would have to exert about 280 N to lift the 100 kg load—still too much! I didn't have time to try a triple-pulley system, but I think that it probably would have done the job.

Figure 1

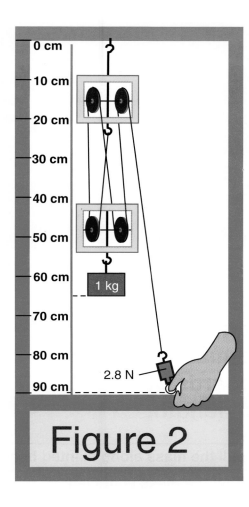

Figure 2

0 cm
10 cm
20 cm
30 cm
40 cm
50 cm
60 cm — 1 kg
70 cm
80 cm — 2.8 N
90 cm

Problem 1

a. How much force did Leona put on the string?

b. How far did she move the string?

c. How many joules of work did Leona do in using the machine?

Machines are commonly used to make work easier. The work put into or done on a machine is called **work input**.

Problem 2

a. What upward force does the machine (pulley and strings) place on the load? (This is the same as the force you would need to lift the load without a machine.)

b. How far is the load raised by the machine?

c. How much work was done on the load by the machine?

The work done by a machine on a load is commonly called **work output**, which is the work you get out of the machine, regardless of the work put into it.

Problem 3

How does the work output in Problem 2 (c) compare with the work input in Problem 1(c)? What do you think accounts for these results? From your own experience, how would you modify the following principle?

The work you put into a machine is (greater than, less than, the same as) the work you get out of it.

Answers to
Problem 1

a. Leona applied an average force of 2.8 N to the string.

b. She moved the string 60 cm (0.60 m).

c. The amount of work done by Leona can be calculated in the following way: 2.8 N × 0.60 m = 1.68 J.

Answers to
Problem 2

a. The upward force placed on the load by the machine is 10 N, which is the amount of force needed to lift a 1 kg load.

b. The load is raised 15 cm (0.15 m) by the machine.

c. The work done by the machine on the load can be calculated in the following way: 10 N × 0.15 m = 1.5 J.

Answers to
Problem 3

The work output in problem 2 is less than the work input in problem 1. In other words, more work was put into the machine than was gotten out of the machine. This is always the case because some of the input work is used to overcome friction. In this machine, friction is created by the pulleys turning on their axles. Also, some work is used to lift the mass of the lower pulley.

The principle at the bottom of the page should read: "The work you put into a machine is greater than the work you get out of it."

173

Homework

You may wish for students to answer the questions found in Problem 1, Problem 2, or Problem 3 as homework. Discuss students' answers in class the following day.

Pam's Machine

Set up Pam's apparatus as it appears in her report. Make sure that students measure the force in the direction parallel to the object's movement. Encourage students to experiment by pulling loads of various weights up the inclined planes. They should immediately recognize that it takes more effort to move a load up an inclined plane than it does to raise the same load with a combination of pulleys and string. Students should also discover that the steeper the inclined plane, the more force it takes to pull a load.

Before students complete Pam's table, you may wish to check their understanding of what Pam did by asking questions similar to the following:

• How many trials did Pam make? *(She made three trials.)*
• How many inclined planes did she use? *(She used three different inclined planes.)*
• How are the inclined planes alike? *(They all rise to the same height.)*
• How are the inclined planes different? *(They are different lengths.)*

Answers to
Pam's Machine

Provide students with time to complete the table in their ScienceLog. Then have them share and discuss their results.

• Inclined plane 2
  FORCE: 3.2 N
  DISTANCE: 0.70 m
  WORK INPUT: 3.2 N × 0.70 m = 2.24 J
  WEIGHT: 10 N
  VERTICAL (HEIGHT): 0.15 m
  WORK OUTPUT: 10 N × 0.15 m = 1.5 J
• Inclined plane 3
  FORCE: 2.1 N
  DISTANCE: 1.3 m
  WORK INPUT: 2.1 N × 1.3 m = 2.73 J
  WEIGHT: 10 N
  VERTICAL (HEIGHT): 0.15 m
  WORK OUTPUT: 10 N × 0.15 m = 1.5 J
• Pam should conclude that for a longer inclined plane, less input force is required. However, for a longer inclined plane, a greater work input is required. For all three inclined planes, the work output remains the same.

Pam's Machine

Pam had a different solution to Tony's problem. She tried three different versions of her idea, which involved using a slanted board. A slanted board is a type of simple machine called an *inclined plane*. Pam tried three boards of different length.

Complete her table in your ScienceLog, and then write some conclusions that she might draw from her solution. The questions below will help you.

Questions

1. Does the length of the board affect the force needed to pull the load? If so, how?

2. Compare the *work output* (the amount of work actually done by the machine) with the *work input* (the amount of work put into the machine) for each of the three boards. What do you notice? Would a long inclined plane and a short inclined plane be equally good at converting work input to work output? How and why might they differ?

3. Compare the work input and work output of Leona's machine with those of Pam's machine. Which of the two machines is better at converting work input to work output?

4. Which machine do you think Tony should use and why?

A SLANTED BOARD
is a type of machine
called an "inclined plane."
I used a mass of 1 kg.
I tried boards of
different length.

FORCE to pull the mass along slanted board
DISTANCE that the mass is moved
WORK INPUT
WEIGHT
VERTICAL (HEIGHT) distance the mass is raised
WORK OUTPUT

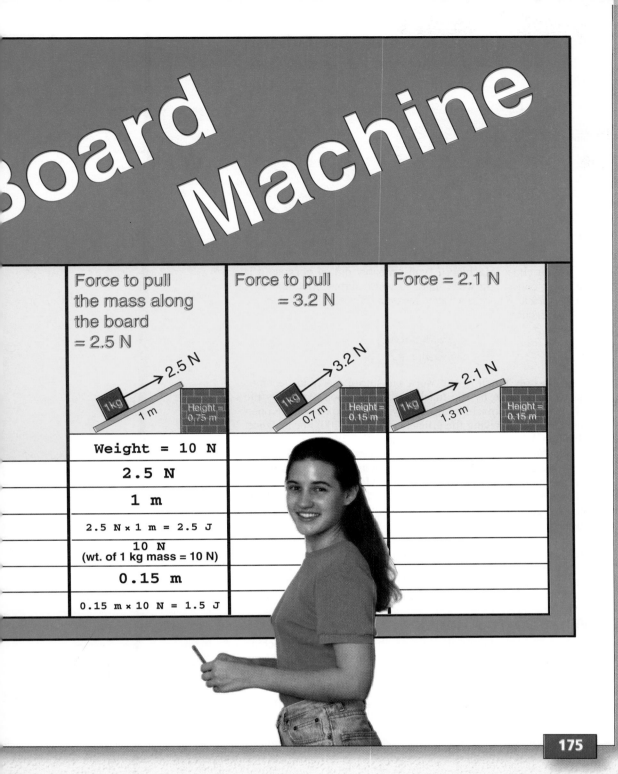

Board Machine

Force to pull the mass along the board = 2.5 N	Force to pull = 3.2 N	Force = 2.1 N
2.5 N	3.2 N	2.1 N
1 m	0.7 m	1.3 m
Height = 0.75 m	Height = 0.15 m	Height = 0.15 m
Weight = 10 N		
2.5 N		
1 m		
2.5 N × 1 m = 2.5 J		
10 N (wt. of 1 kg mass = 10 N)		
0.15 m		
0.15 m × 10 N = 1.5 J		

Answers to
Questions, page 174

1. If the height remains the same, as it does in Pam's machine, then the length of the board does affect the force needed to pull the load. The shorter the length, the greater the force. (The force really depends on the slope of the board. The slope is the vertical distance divided by the horizontal distance.)

2. The ratios of work output to work input are 0.6, 0.67, and 0.55 for the three cases. These ratios vary according to the length of the board. The shorter the board (and the greater the slope), the larger the ratio is and the better the machine is at converting work input to work output. The more vertical the board is, the less friction there is between the load and the board because more of the load's weight is directed *along* the board instead of *against* the board. (Make sure students understand that even though the required force increases with slope, the work input decreases.)

3. The ratio of work input to work output for Leona's machine is 0.89. Thus, Leona's machine is better at converting work input to work output than Pam's machine is.

4. Student answers will vary depending on what feature of the machine is most important for their purposes. For instance, if students desire a small input force, then Pam's third machine would be best. If it is most important to convert the greatest amount of work input to work output, then Leona's machine would be best.

Multicultural Extension

Mechanical Mysteries
Present students with the following scenario: You have been given the assignment of explaining the origin of one of the following artifacts: Stonehenge, the stone statues on Easter Island, or the Pyramids in Egypt. Prepare a presentation for the class explaining what kinds of machines are thought to have been used in the construction of these artifacts.

Some Machines Are Better Than Others

First have students respond to the question at the top of the page. Call on several volunteers to share their ideas with the class, and ask them to give reasons for their choices. *(Accept all responses without comment.)* Then have students continue reading to learn how to calculate the efficiency of a machine. If necessary, review how to calculate the efficiency of machine *A* or *B* on the chalkboard.

Answers to
In-Text Questions

A The efficiency of Machine C is 9 J ÷ 10 J × 100 = 90 percent efficient

The efficiency of Leona's machine is 1.5 J ÷ 1.68 J × 100 = 89 percent efficient.

The efficiency values for Pam's machine are as follows:
- inclined plane 1:
  1.5 J ÷ 2.5 J × 100 = 60 percent efficient
- inclined plane 2:
  1.5 J ÷ 2.24 J × 100 = 67 percent efficient
- Inclined plane 3:
  1.5 J ÷ 2.73 J × 100 = 55 percent efficient

Pam's machines differ in efficiency because the distances over which the mass must be moved are different. The greater the distance that the mass moves along the inclined plane, the greater the amount of force needed to overcome friction.

B Students' answers will vary. A machine is more efficient when the input work is reduced in comparison with the output work. This is best accomplished by reducing the friction in the mechanical system.

Answers to
What Do You Think?

Accept all reasonable responses without comment at this point. As students are about to find out, such a machine is not possible. Pam's and Leona's machines demonstrate that work output is less than work input.

Students should conclude that no mechanical system can have the same work output as work input, or be 100 percent efficient, because some energy is always lost to friction.

Some Machines Are Better Than Others

Which of the following machines would you prefer to use for a job?

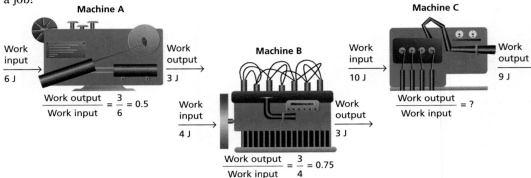

Machine A

Work input 6 J

Work output 3 J

$$\frac{\text{Work output}}{\text{Work input}} = \frac{3}{6} = 0.5$$

Machine B

Work input 4 J

Work output 3 J

$$\frac{\text{Work output}}{\text{Work input}} = \frac{3}{4} = 0.75$$

Machine C

Work input 10 J

Work output 9 J

$$\frac{\text{Work output}}{\text{Work input}} = ?$$

If you chose Machine C, you probably did so because it is better at converting work input to work output. Machine C is more *efficient*. There is a way to measure how efficient a machine is. The fraction

$$\frac{\text{work output}}{\text{work input}}$$

tells you how much of the work put into a machine becomes useful work done by the machine. This fraction expresses the efficiency of the machine. Efficiency can be expressed as a percentage by multiplying the efficiency value by 100. For instance, Machine A is 0.5 × 100 = 50 percent efficient.

Machine B is 75 percent efficient. This means that 75 percent of the work put into Machine B emerges as useful work done by the machine. What is the efficiency of Machine C? What is the efficiency of Leona's machine? How efficient are each of the inclined planes of Pam's machine? Do you observe any differences? Can you explain them? **A**

Now consider the machine you designed to solve Tony's problem. Compute its efficiency. Compare your machine's efficiency with that of each of your classmates' designs. Which machine has the highest efficiency? How do you think your machine might be made more efficient? **B**

What Do You Think?

Look at Machine D at right. Do you think that such a machine is possible? Why or why not? How can Pam's and Leona's results help you reach a conclusion?

In fact, no machine can be 100 percent efficient. Some energy is always "wasted" and does not go into work output. What do you think happens to this energy?

176

Machine D

Work input = 100 J

Work output = 100 J

Theme Connection

Energy

Focus question: How does the use of lubricants affect the overall efficiency of a machine? *(Lubricants reduce the amount of friction between the moving parts of a machine. This, in turn, decreases the amount of energy loss due to the production of heat energy from friction. By reducing the amount of energy loss due to friction, a machine functions more efficiently.)*

Homework

The Theme Worksheet on page 30 of the Unit 3 Teaching Resources booklet corresponds to the Theme Connection at left. This worksheet makes an excellent homework assignment to accompany Some Machines Are Better Than Others.

The Advantage of Machines

If machines do not *save* work, is there any advantage in using them? Indeed there is! With a machine, one person can do the work of many. This is because a small force can be turned into a larger force. Leona, using her machine, can lift the load of 10 N with a force of only 2.8 N. She increases her force almost fourfold! The number of times a force exerted on a machine is increased by the machine is called the **mechanical advantage** of the machine. For Leona's machine, the mechanical advantage is calculated as follows:

$$\text{mechanical advantage} = \frac{\text{force exerted by the machine}}{\text{force exerted on the machine}} = \frac{10 \text{ N}}{2.8 \text{ N}} = 3.6$$

Of course, you never get something for nothing. The smaller force put on the machine must be moved through a longer distance, while the larger force that the machine exerts moves a shorter distance.

A Trade-Off: Force for Distance

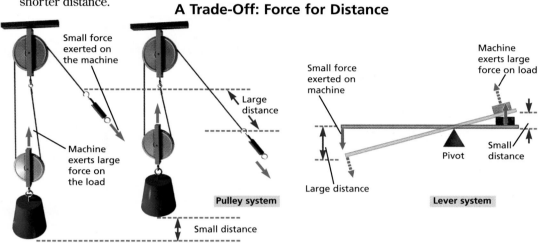

Pulley system

Lever system

Some machines work differently. They multiply distance at the expense of the force. You will see some of these on pages 178 and 179.

1. What is the mechanical advantage of each of Pam's inclined planes? of Leona's machine? of the machine you constructed?

2. Suppose that Leona used a triple-pulley setup and found that a 2 N force could lift the 1 kg mass. What would be the mechanical advantage?

3. Now suppose that Tony used Leona's triple-pulley system to raise his 100 kg load. How much force, in newtons, would be needed to raise the load? How many people would be needed for the job?

Homework

For at-home practice, you may wish to assign the following problems:
- What is the mechanical advantage of a machine that exerts a 33 N force when a force of 3.3 N is applied to the machine?
  (33 N ÷ 3.3 N = 10)
- What is the mechanical advantage of a machine that requires a 25 N force to lift a 65 N load?
  (65 N ÷ 25 N = 2.6)

The Advantage of Machines

Have students silently read the page. Then have them discuss the concept of *mechanical advantage*. If students seem confused about the idea that a machine does not save work but is still worth using, remind them that a machine allows a person to use less force. Help students to recognize the difference between *force* and *work*.

Answers to *The Advantage of Machines*

1. • Pam's machine:
     inclined plane 1: 10 N ÷ 2.5 N = 4
     inclined plane 2: 10 N ÷ 3.2 N = 3.1
     inclined plane 3: 10 N ÷ 2.1 N = 4.8
   • Leona's machine:
     10 N ÷ 2.8 N = 3.6
   • Answers will vary depending on the design of each student's machine.

2. Mechanical advantage = 10 N ÷ 2 N = 5

3. Raising Tony's 100 kg load through 1 m using Leona's triple pulley would require a force equal to one-fifth (mechanical advantage of 5) of the force exerted by the machine: $\frac{1}{5} \times 1000$ N = 200 N. Therefore, one person could raise the load.

⭐ **Transparency 22 is available to accompany The Advantage of Machines.**

CROSS-DISCIPLINARY FOCUS

Mathematics

The *Carnot engine* is a model engine for converting heat energy to mechanical energy. Even this ideal engine is not 100 percent efficient. The formula for the efficiency of a Carnot engine is:

$$\frac{T_{hot} - T_{cold}}{T_{cold}},$$

where T_{hot} is the temperature at which heat is added and T_{cold} is the temperature of the exhaust in kelvins. Ask: What is the efficiency of a Carnot engine that uses steam and whose exhaust is room temperature? *(Assume that the steam's temperature is 100°C [373 K] and the exhaust's temperature is 22°C [295 K]. This gives an efficiency of 0.26, or 26 percent!)*

Which Machines?

You may wish to have students brainstorm their responses for each of the pictures. Then involve class members in a discussion of their ideas. Encourage students to draw diagrams on the chalkboard to illustrate the forces exerted on the machines and by the machines. If you'd prefer to have students work in small groups, see the Cooperative Learning strategy below.

Answers to
Which Machines? pages 178–179

a. The nutcracker is a *lever.* It *multiplies force* at the expense of distance. A force is exerted at the ends of the nutcracker. The nutcracker exerts a force on the nut.

b. The exercise machine is a *pulley.* It *changes the direction* of the force and *multiplies distance* at the expense of force. A sideways force is applied to the padded bars. An upward force is exerted on the masses at the end of the cable.

c. The bar is a *lever.* It *multiplies force* at the expense of distance. An upward force is exerted at one end of the bar. The other end of the bar exerts an upward force on the rock.

d. The arm is a *lever.* It *multiplies distance* at the expense of force. An upward force is applied at the point where the muscle is attached to the bone in the forearm. An upward force is exerted by the hand.

e. The broom is a *lever.* It *multiplies distance* at the expense of force. A sideways force is exerted at the middle of the broom handle. The bottom of the broom exerts a sideways force.

f. The downhill road is an *inclined plane.* It *changes the direction* of the force and *multiplies force* at the expense of distance. The cars exert a downward force on the road. The road exerts an upward force on the bottom of the cars.

Answers to Which Machines? continued on the next page. ▶

 Transparency 23 is available to accompany Which Machines?

Which Machines?

Some familiar machines are shown on these pages. They are all simple machines, although in one or two cases you might not believe it! *Simple* machines make up *complex* machines such as bicycles and backhoes. Your task is to analyze each of the simple machines shown.

1. Classify each machine as one of the following:
   - lever
   - inclined plane
   - pulley
   - some other type of machine

2. Study the machines. Which ones do you think
   - multiply (increase) the force put into them?
   - multiply (increase) the distance put into them?
   - change the direction of the force put into them?

3. Where do you exert a force on each machine?

4. Where does the machine exert a force on something else?

   A sketch showing the forces might be the best way of answering questions 3 and 4. Make sure your sketch shows the forces involved.

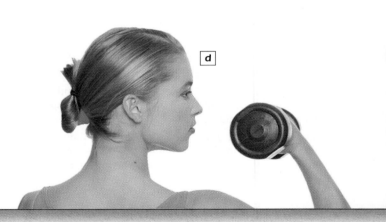

Cooperative Learning
WHICH MACHINES? PAGES 178–179

Group size: The class should be divided into two groups.

Group goal: to analyze and classify simple machines on pages 178 and 179

Positive interdependence: Divide the class into two groups. Prepare letters *a–i.* Place half of the letters in one bag and half in another. Have each student from one group draw from one bag and each student from the other group draw from the other bag. Each student is responsible for sketching and analyzing the picture on page 178 or 179 that corresponds to the letter drawn. They should use the questions on page 178 as a guide. The following day, use the "inside-outside circle" method to share. Place the groups in two concentric circles. The inside circle faces out; the outside circle faces in. Students share their work and then rotate in the circles so that each student gets an explanation for each picture.

Individual accountability: Each student is responsible for analyzing one picture to share with the class. Choose any one of the pictures or provide a new one to analyze for a quiz.

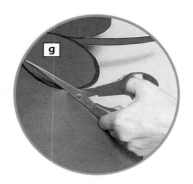

Answers to
Which Machines? *continued*

g. The scissors are an example of a pair of *levers* hinged together. (Each blade is a wedge, which is a form of an *inclined plane.*) They *multiply force* at the expense of distance. The force is applied by pushing the scissor handles together. The force of the scissor blades is exerted on the paper as the blades come together.

h. The steering wheel is a *wheel and axle,* which is a type of *lever.* It *multiplies force* at the expense of distance. The force is applied at right angles to the edge of the steering wheel. The steering wheel exerts a force on the car's front axle and, though a *rack and pinion,* causes the car to change direction.

i. The doorknob is a *wheel and axle,* which is a type of *lever.* It *multiplies force* at the expense of distance. The force is applied at right angles to the outside of the doorknob. The doorknob exerts a force on the rectangular axle attached to it.

179

FOLLOW-UP

Reteaching

Have students use a spring scale to compare the amount of force that is needed to lift an object such as a roller skate from the floor to the seat of a chair with the force that is required to pull the same object up an inclined plane to the seat of a chair. Students should find that the inclined plane greatly reduces the force. Also have students compare the work input to the work output for this simple machine.

Assessment

Have students identify three simple machines found in their homes. Ask them to sketch each machine, identify where the force is applied, and indicate where the machine exerts a force.

Extension

Have students make poster diagrams to illustrate why the *wedge* and the *screw* are considered inclined planes. Their diagrams should indicate what makes a wedge effective *(The small angle between the two surfaces of the wedge)*

and how the mechanical advantage of a screw is determined *(By the ratio of the circumference of the screw to the pitch of the screw [the distance that the screw advances for one revolution]).*

Closure

As a final activity to the lesson on machines, have students design, construct, and demonstrate a catapult used to launch a lightweight foam ball. Students should work in teams and develop creative designs. Approve all designs for safety before students begin.

LESSON 2 — Machines and Energy

FOCUS

Getting Started

You may wish to begin this lesson by showing students a few of the videodisc images, such as the pulley demonstration, from the Image and Activity Bank on pages 143E and 143F. Have the class discuss how energy is transformed and transferred in these machines. *(Accept all reasonable suggestions without criticism.)* Tell students that in this lesson they will study the interaction of machines and energy further.

Main Ideas

1. A wheel and an axle can be combined to create a simple machine.
2. Energy can be transferred and transformed, but the total amount of energy remains the same.
3. Some of the energy put into a machine is converted into heat by friction.
4. A series of steps can be followed to solve technological problems.
5. Either distance or force can be multiplied by a wheel and axle, depending on whether the force is applied to the axle or to the wheel.

TEACHING STRATEGIES

Teaching Strategies begin on the next page. ▶

LESSON 2 — Machines and Energy

You have seen that machines can convert energy from one form to another. You have also seen that machines can multiply force or distance, but never both at the same time. An increase in force always comes at the expense of distance, and an increase in distance comes at the expense of force. What accounts for this?

Efficiency is another matter altogether. As you know, no machine is 100 percent efficient, but where does the "lost" energy go? Does it simply disappear, or is it converted into some unseen form? In this lesson you will tackle these questions and others. You will also learn a bit more about the ways in which machines convert and convey energy.

EXPLORATION 1

Miguel's Machine

Miguel devised a machine that would lift loads easily. With his machine, he found that he could lift a load of 10 N with a force of only 2 N.

Miguel realized, however, that he had to turn the handle a long way to lift the load just a little. In fact, when he turned the handle all the way around once, the load came up by just the distance around the spool.

Evaluate Miguel's machine by completing his evaluation report in your ScienceLog. Could Miguel's machine be used to solve Tony's problem on page 170? Ⓐ

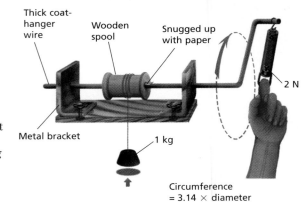

Thick coat-hanger wire — Wooden spool — Snugged up with paper — 2 N — Metal bracket — 1 kg — Circumference = 3.14 × diameter — 5.5 cm — 1 cm — Handle — End of spool

My Machine: An Evaluation	
Distance handle moves in one turn = ____?____	Work input = ____?____
Distance load rises when handle makes one turn = ____?____	Machine's work output = ____?____
Force put on handle = ____?____	Efficiency = ____?____
Upward force put on the load by the machine = ____?____	
Mechanical advantage = ____?____	
Conclusion and Recommendation: This machine would have an even greater mechanical advantage if . . .	

LESSON 2 ORGANIZER

Time Required 3 class periods

Process Skills
analyzing, observing, comparing

New Term
Wheel and axle—a simple machine consisting of a wheel attached to a smaller wheel or shaft

Materials (per student group)
Exploration 2: a variety of materials to make a simple wheel-and-axle machine, such as the following: spools of various sizes; mailing tubes; film canisters with lids; wire coat hangers; knitting needles; craft sticks; milk cartons; shoe boxes; plastic-foam or paper cups; cardboard; plywood; 1 kg mass; hand saw; white glue; cans; wire; plastic drinking straws; string; pliers; corks; scissors; metric force meter (or metric spring scale)
Exploration 3, Part 2: a variety of materials to create an energy source, such as the following: rubber band; water; 500 mL Erlenmeyer flask; one-hole stopper fitted with glass tubing;

hot plate; small fan; miniature DC motor with battery; a few meters of copper wire; support stand with ring clamp; C clamp; pulley; 1 kg mass; 2–3 m of string; safety goggles (See Advance Preparation on page 143C.)

Teaching Resources
Exploration Worksheets, pp. 32, 33, and 35
Transparency 24
SourceBook, pp. S50 and S55

A Closer Look at Efficiency

Miguel put 0.34 J of work into his machine. How much energy did he transfer to the machine? His machine did 0.31 J of work on the load. How much energy did the machine transfer to the load? Was some energy lost? It seems so. Where did the lost 0.03 J of energy go? **B**

Situations like the one above puzzled scientists such as Count Rumford (1753–1814) and James Joule (1818–1889). Through many observations and experiments, the two scientists independently came to the conclusion that the energy apparently lost in machines was actually only diverted; the energy was turned into heat by friction. Their research established that heat was a form of energy and not a form of matter, as was commonly believed at that time. Rumford and Joule reasoned that the amount of work actually done by a machine is equal to the amount of energy put into the machine minus the energy lost to friction. Each scientist concluded that energy is *conserved* in nature. In other words: **Energy can be transferred and transformed, but the total amount of energy in a system always stays the same.**

How would you apply Rumford's and Joule's findings to Miguel's machine? **C**

Count Rumford

James Joule

| **Energy transferred to the machine (? J)** | **=** | **Potential energy gained by the mass raised by the machine (? J)** | **+** | **Heat energy caused by friction (? J)** |

Compute the energy values for one turn of the handle. What would the energy values be for 10 turns of the handle? If the machine had less friction, how would these values change? How would the value for the efficiency change? In your ScienceLog, complete the following principles. **D**

- A machine is a device for transferring or converting __?__ .
- Some of the work (energy) input to a machine is used to overcome friction. This produces __?__ .
- Energy input is equal to energy __?__ plus other forms of energy (such as heat) that are lost to use.

Do you now understand why the energy you put into a machine (work input) is always greater than the energy you get out of the machine (work output)? Why is this so? **E**

How much heat energy was produced as the load was lifted 15 cm by Leona's machine (pages 172–173)? by each of Pam's inclined planes (pages 174–175)? How could both of these machines be made more efficient? What is the relationship between a machine's efficiency and its loss of useful energy? **F**

To find out more about how friction reduces the efficiency of machines, turn to pages S55–S56 in the SourceBook.

181

E̲XPLORATION 1

Students will need to use the formula for finding the circumference of a circle: $C = \pi d$, where C is the circumference and d is the diameter of the circle.

Homework

The Exploration Worksheet on page 32 of the Unit 3 Teaching Resources booklet makes an excellent homework assignment to accompany Exploration 1.

Answers to
In-Text Questions, pages 180–181

A • Distance handle moves in one turn: $C = 3.14 \times 5.5$ cm = 17.3 cm
- Distance load rises when handle makes one turn: $C = 3.14 \times 1.0$ cm = 3.1 cm
- Force put on handle = 2 N
- Upward force put on the load by the machine = 10 N
- Mechanical advantage = 10 N ÷ 2 N = 5
- Work input = 2 N × 0.17 m = 0.34 J

Answers to
In-Text Questions, continued

- Machine's work output = 10 N × 0.031 m = 0.31 J
- Efficiency = 0.31 J ÷ 0.34 J × 100 = 91%

Conclusion and Recommendation: This machine would have an even greater mechanical advantage if Miguel used a spool with a smaller diameter.

Miguel's machine could solve Tony's problem if the machine were built of sturdier materials.

B Miguel transferred 0.34 J of energy to the machine. The machine transferred 0.31 J of energy to the load. Energy was not lost; the missing 0.03 J of energy was transformed to unusable energy.

C Accept all reasonable responses at this point.

D For one turn of the handle, 0.34 J was transferred to the machine, 0.31 J was transferred to the load, and 0.03 J was converted to heat energy. For 10 turns of the handle, 3.4 J was transferred to the machine, 3.1 J was transferred to the load, and 0.3 J was converted to heat energy.

If the friction were less, the work input would be less. The potential energy of the raised load, or work output, would remain the same. As a result, the machine would be more efficient.

E Work input is always greater than work output because some mechanical energy is lost to friction.

F Heat energy produced by Leona's machine = 1.68 J – 1.5 J = 0.18 J. Heat energy produced by Pam's machines:
- Inclined plane 1: 2.5 J – 1.5 J = 1.0 J
- Inclined plane 2: 2.24 J – 1.5 J = 0.74 J
- Inclined plane 3: 2.73 J – 1.5 J = 1.23 J

Leona's and Pam's machines could be made more efficient by reducing friction. The more useful energy a machine loses, the lower the machine's efficiency.

Answers to
A Closer Look at Efficiency

- A machine is a device for transferring or converting energy.
- Some of the work (energy) put in to a machine is used to overcome friction. This produces heat energy.
- Energy input is equal to energy output plus other forms of energy (such as heat) that are given off during use.

A Students should conclude that Miguel's machine is called a wheel and axle because the handle is like a wheel, and the spool is like an axle. Miguel applied his force to the wheel. When Miguel applies a force to the wheel, the machine exerts a force at the axle to lift the load.

B Of the pictures on pages 178 and 179, the doorknob and the steering wheel are examples of wheel-and-axle machines. In the case of the doorknob, a force is applied to the outside of the doorknob (wheel). The doorknob exerts a force on the rectangular axle attached to it. In the case of the steering wheel, a force is applied to the outside of the steering wheel. The steering wheel exerts a force on the steering column (axle).

An Exploration Worksheet (Teaching Resources, page 33) and Transparency 24 are available to accompany Lesson 2.

Homework

The in-text questions in Wheels and Axles can be assigned as homework if class time is short.

Wheels and Axles

Miguel's machine is commonly called a **wheel and axle.** Why? Did Miguel apply his force to the *wheel* or to the *axle* of the machine? Where does the machine apply its force on the load? **A**

Wheels and axles are normally found together—neither is very useful without the other. Now turn back to the pictures on pages 178 and 179. Which ones show a wheel and axle? For each example, determine where the force is applied and where the machine exerts a force on something else. **B**

EXPLORATION **2**

Solving a Technological Problem

You have already encountered a wheel-and-axle machine. Miguel designed such a machine. What steps might he have followed in designing and building it? Try to approach the problem as he might have.

As you follow the steps on the next page, you will be tracing out the steps commonly used in solving technological problems.

You Will Need

Choose from the following readily available materials and tools:

- spools of various sizes
- mailing tubes
- film canisters with lids
- wire coat hangers
- knitting needles
- craft sticks
- milk cartons
- shoe boxes
- plastic-foam or paper cups
- cardboard
- plywood
- a 1 kg mass
- a hand saw
- white glue
- cans
- wire
- straws
- string
- pliers
- corks
- scissors
- a force meter

EXPLORATION **2**

This Exploration uses Miguel's design of a wheel-and-axle machine to introduce students to technological problem solving. Encourage students to think of a variety of original designs, some of which may be preferable to Miguel's.

Allow students as much freedom as possible as they brainstorm new ideas. You may wish to suggest that they spend some time outside of class to collect materials, perfect their designs, and assemble their machines. Make sure that students submit their designs to you for

a safety review and get your approval before building their machines. Also encourage students to be creative in the way they report their results to the class.

You may wish to emphasize to students that the quantitative evaluation of their machines is important because it will help them to think of further improvements, suggestions, and recommendations. As a follow-up, help students recognize that the steps employed in solving a technical problem represent what they might do naturally, logically, and intuitively.

What to Do

1. First review Miguel's problem. He wanted to use simple materials to make a hand-powered wheel and axle strong enough to lift a load of 10 N with a much smaller force.

2. Observe Miguel's machine again. Identify each of its working parts.

3. Make decisions regarding the following:
   a. What will I choose for the spool part of the axle?
   b. What will I use for the wire part of the axle?
   c. How will I construct the wheel?
   d. How can I attach the spool to the wire so that both turn together?
   e. What size would be best for each of these parts?
   f. What will I use for a supporting frame?
   g. How strong should each part be?

4. Make a simple diagram showing your design solution. Name all the materials you plan to use.

5. Construct a model of the wheel and axle according to your design.

6. Test its operation. Are some improvements necessary to make it operate better? If so, make them.

7. Determine your machine's mechanical advantage. Also determine its efficiency and the amount of energy it changes into heat due to friction with every turn.

8. Think of other ways you might critically judge your machine, for example, appearance, strength, and maximum load it can carry.

9. Compare your wheel and axle with those of your classmates. Identify strengths and weaknesses in their wheels and axles as well as in your own.

10. List suggestions to make your machine better.

Congratulations! You have a workable hand-powered wheel and axle of sufficient mechanical advantage to do the job you wanted. You did this by "technological problem solving." The diagram at right shows the steps. Match what you did in this Exploration (steps 1–10) with the steps in the diagram (A, B, C, and D). More than one Exploration step may match each diagram step.

Exploration 3 extends the problem. Your task will be to adapt your machine so that it can be powered by some source of energy other than your hand.

Technological Problem Solving

A — Understanding the Problem

B — Developing a Plan
- Identifying alternative design solutions

C — Carrying Out the Plan
- Constructing a model and troubleshooting

D — Evaluating
- Evaluating your design
- Proposing improvements

Cooperative Learning
EXPLORATION 2, PAGE 182

Group size: 3 to 4 students
Group goal: to use technological problem-solving skills to build a machine with a wheel and axle
Positive interdependence: Assign students the following roles: facilitator (to read aloud all instructions and questions and to keep the group on task), planner (to lead the brainstorming of ideas and to identify alternative design solutions), construction manager (to coordinate the building of the model and to lead the group in troubleshooting), and evaluator (to evaluate the design and to lead the discussion of proposed improvements). Using the materials provided, have students work through the steps of the technological problem that Miguel followed in order to construct a wheel and axle. Students should prepare a detailed sketch, name the materials used, construct and test the model, analyze the mechanical advantage, and compare their machine with others in the class.
Individual accountability: Have students answer the matching section on this page individually.

Answers to *Exploration 2*

1. A
2. A
3. B
4. B
5. C
6. C
7. D
8. D
9. D
10. D

EXPLORATION **3**

Point out to students that the illustrations are meant to show some possible energy sources for their machines. They are not intended to suggest how an energy source might be connected to a machine. As in the previous Exploration, encourage students to be creative. You may wish to have several students work together, allowing them to brainstorm ideas and to help one another.

When the final designs have been completed, call on volunteers to explain their ideas to the class. Then display the designs around the classroom so that students can have an opportunity to compare their ideas with those of their classmates.

If time permits, encourage students to turn their designs into actual models. Make sure that students submit their designs to you for a safety review and receive your approval before building their models. Also specify that if students choose to use electricity in their models, they should use only small batteries. They should not use household current from a wall outlet. Point out that they do not have to make working models. Facsimiles of energy sources can be used instead of the real thing. This is particularly important for students whose designs require the use of combustible fuels or heat sources. Monitor student designs closely to make sure that all necessary safety precautions are taken.

 An Exploration Worksheet is available to accompany Exploration 3 (Teaching Resources, page 35).

Homework

Step 4 of Exploration 3, Part 2, makes an excellent homework activity.

Powering Your Machine

Can you suggest another way to power your wheel-and-axle machine, other than by hand? You might consider using any of the following sources of energy:

- water
- steam
- wind
- electricity
- a rubber band
- a raised weight

How would you connect your energy source to your machine?

PART 1

What to Do

1. Choose an energy source.
2. Decide on a possible way to connect the source to your machine. If you need to modify the machine, determine how you will do it.
3. Check your design. Ask yourself whether you are using simple and readily available materials and equipment. Modify your design if necessary.
4. Draw your design.
5. Show your design to others, and ask for their opinions and advice.

PART 2

What to Do

1. Obtain the necessary materials and equipment, and construct a model of the powered wheel and axle according to your design.
2. Connect the machine to the source of energy you have chosen.
3. Evaluate your model. Does it work? If not, why does it not work? If it works, what size load will it lift? How does it operate compared with the models of others? Propose some improvements to your model.
4. How good a problem solver were you? Exploration 2 showed you a model for solving technological problems. Draw a chart in your ScienceLog like the one on page 183, and complete it by adding details of what you did at each of steps *A–D*.

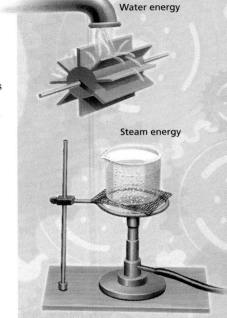

Rubber-band energy

Water energy

Steam energy

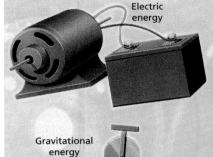

Electric energy

Gravitational energy

Other Wheels and Axles

Let's look at some more examples of wheel-and-axle machines. As you will see, the following machines are different in a key way from those you have seen so far.

Look carefully at the images below. How are these wheels and axles different from Miguel's machine? In each diagram, visualize the circle that is traced out by the handle or pedal as it rotates the machine's axle. How does the circumference of this circle compare with the circumference of the machine's wheel? **B**

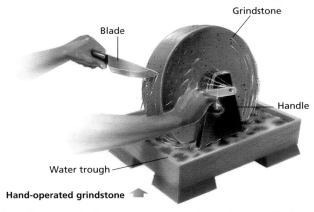

Hand-operated grindstone

In using the grindstone, a person applies a force to the handle, which turns through a smaller circumference than does the rim of the grindstone. How does this wheel and axle affect the speed, distance, and size of the output force? **C**

Analyze the operation of the unicycle shown at right in the same way. Which of these statements is true for these two examples of a wheel and axle? **D**

- Force is multiplied at the expense of distance.
- Distance is multiplied at the expense of force.

Suppose you push the pedals of the unicycle through one complete turn.

1. Through what distance does each pedal move?
2. Through what distance would the wheel move along the road?
3. How much has the pedal distance been multiplied?
4. Suppose you move the pedals through one full turn in 1 second. What is the speed of the pedals in centimeters per second? What is the speed of the unicycle?
5. The speed of the pedals has been multiplied. How did this happen? What's the trade-off?

Diameter of wheel = 60 cm

Diameter of pedal swing = 35 cm

185

Answers to In-Text Questions

A Miguel's machine differs from the grindstone and the unicycle in that the circumference of the driven wheel is much smaller than the circumference of the driving axle.

B The circumferences of the paths made by the wheels of the unicycle and grindstone are much larger than the circumference made by the handle or pedal of the machines.

C In Miguel's machine, force is increased at the expense of the distance moved. The grindstone increases the speed and distance but reduces the force. The output force is exerted by the wheel of each machine. For the grindstone, the operator's force moves through a small distance, while the machine's force, exerted at the wheel's circumference, moves through a much greater distance. As a result, distance and speed are multiplied for the grindstone.

D Distance is multiplied at the expense of force.

Answers to In-Text Questions

1. Distance of pedal in one turn: 2 × 3.14 × 17.5 cm = 109.9 cm

2. Circumference of wheel: 2 × 3.14 × 30 cm = 188.4 cm

3. 188.4 cm ÷ 109.9 cm = 1.7 times

4. Pedal speed is 109.9 cm/s. Unicycle speed is 188.4 cm/s.

5. The speed of the pedals has been multiplied by attaching the pedals to a wheel. Force has been traded for distance and speed. More force must be applied to the pedal than the wheel exerts on the ground.

FOLLOW-UP

Reteaching

Have students make a model to illustrate how the forces exerted by a wheel and axle differ depending on whether force is applied to the wheel or to the axle. Have students use their model to demonstrate the difference to the class. (Hint: A pulley with an extended axle through it is a good starting point.)

Assessment

Have students make a list of all of the wheel-and-axle machines encountered during the last two days. (Point out that sprocket gears are a kind of wheel.) Have students explain how each wheel and axle affects speed, distance, and force.

Extension

Have students do some research to discover the formula used to calculate the mechanical advantage of a wheel and axle. Have them create a poster diagram to illustrate how this formula is used. *(The formula is the following: mechanical advantage = diameter of the wheel ÷ diameter of the axle.)*

Closure

Have students develop three quiz questions (with answers) based on the lesson. Students can trade questions to check one another's understanding.

FOCUS

Getting Started

Display the mechanism of a clock, watch, or other device that uses gears. If such a device is not available, display a picture of one. Ask students to describe what they see. *(An arrangement of gears and springs)* What makes a clock or watch work? *(The way the gears are put together and move)* How is energy transferred in a clock or watch? *(Through the movement of the gears)* Point out that in this lesson they will examine how gears, wheels, and belts transfer energy in machines.

Main Ideas

1. Energy can be transferred by means of gears, wheels, and belts.
2. Gears, wheels, and belts can be used to regulate speed and to change the direction of motion.
3. Meshed gears move in opposite directions; belted wheels move in the same direction.

TEACHING STRATEGIES

Answers to
Getting Into Gear

A. Electricity from the battery causes the motor to turn. The motor causes the belt and pulleys to turn, which cause the axle to turn. As the axle turns, it winds up the string attached to the load.

B. Electricity powers the motor. Energy is transferred through the gear pairs to the drill bit.

C. Heat energy produces the steam. The steam causes the metal strips on the cork to turn, which causes the shaft to turn. As the shaft turns, it winds up the string attached to the load.

D. The energy comes from the person pedaling the bicycle. The circular motion of the pedals turns the front sprocket, which is attached to a chain. The chain turns the rear sprocket, which causes the rear wheel to turn.

Getting Into Gear

Examine the following pictures of machines. Identify how the energy is transferred from its source to the operating parts of each machine.

Model machines made from a construction kit

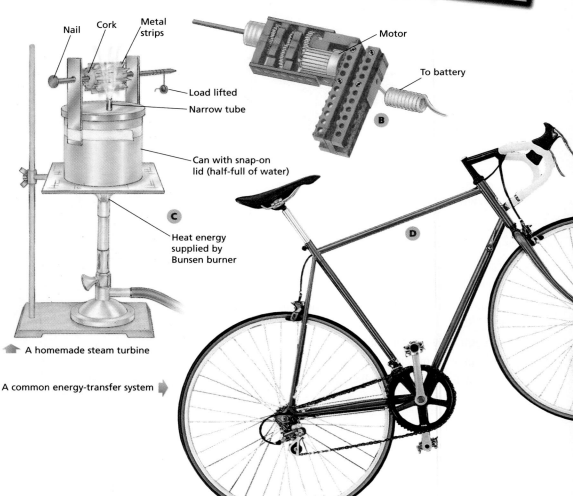

Nail   Cork   Metal strips

Motor

To battery

Load lifted

Narrow tube

Can with snap-on lid (half-full of water)

Heat energy supplied by Bunsen burner

A homemade steam turbine

A common energy-transfer system

LESSON 3 ORGANIZER

Time Required
3 class periods

Process Skills
analyzing, comparing, observing, communicating

Theme Connection
Systems

New Term
Gear—a toothed wheel that meshes with another toothed wheel in order to transmit force

Materials (per student group)
Exploration 4: various unused

mechanical devices, such as toys, clocks, or watches that are powered by some form of energy
A Project—Model Building: technology construction kit that contains a DC motor with battery, wheels, propellers, and other items needed to transfer energy from the motor to the moving parts

Teaching Resources
Exploration Worksheet, p. 37
Math Practice Worksheet, p. 39
Transparencies 25–27
SourceBook, pp. S47 and S52

EXPLORATION 4

What Makes It Tick?

What to Do

To do this activity you will need to collect some unused mechanical devices—such as toys, clocks, or watches—that are powered by some form of energy.

1. Working in small groups, first select a device to observe in detail.

2. Disassemble the device to examine the parts necessary for its operation. For example, if you are examining a toy car, does it have a large wheel that continues to spin for some time once it is set in motion? Such a wheel is a *flywheel*. Of what use is the flywheel? You might also find toothed wheels or cylinders that fit into one another. These are *gears*. What do they do? **F**

3. Identify the sequence of movements as energy is transferred through your machine, from the energy source to the final movement of the object.

4. Sketch your machine, and describe how it works.

187

Multicultural Extension

Blowguns

The blowgun is a machine that efficiently transfers the kinetic energy of air to a small dart. Blowguns are used by hunters in many cultures, including the Jivaro of Ecuador, who use them to hunt birds and other small game. Ask: What are some of the advantages of using blowguns over using firearms? *(Sample answer: Blowguns are silent, have no mechanical parts that can break, and require no fuel.)*

EXPLORATION 4

Divide the class into small groups. Have several mechanisms available, such as clocks, watches, and mechanical toys that have moving parts controlled by gears, wheels, axles, springs, and so on. As the groups disassemble the mechanisms, they should observe how energy and motion are transferred from the power source to each of the moving parts. Suggest that each group decide on a spokesperson to share the group's observations and sketches with the class.

SAFETY ALERT Make sure that students receive safety approval from you before disassembling their devices. Limit the available devices to those that are powered by a spring or a rubber band. Allow disassembly of a spring-powered device only if the spring is completely unwound first. Supervise students closely as they perform this activity.

★ **An Exploration Worksheet is available to accompany Exploration 4 (Teaching Resources, p. 37).**

Answers to
In-Text Questions

E The flywheel is an energy-storage device. It is used to regulate the speed of the machine to which it is attached.

F Gears transmit motion or force in machines. They can also be used to multiply force or distance or to change the direction of a force.

Homework

Have students make a list of five energy transfers that they observe on their way home from school in the afternoon. Many examples are possible, from the kinetic energy they transfer to open a door to the heat energy they transfer to their own clothing.

Joy's Construction Kit

As students consider Joy's questions and examine and experiment with the assemblies she made, they will gain a better understanding of how gears may be used to increase or decrease the speed of rotation. They will also observe how gears can be used to change the direction of motion. You could duplicate Joy's assemblies to help students answer the questions.

Answers to *Assembly 1*

1. The lower gear will turn in the opposite direction.

2. Answers will vary depending on the student's estimate of ratio of the number of teeth in the top gear to the number of teeth in the bottom gear. This ratio determines the speed at which the small gear rotates. For example, if the small gear has 20 teeth and the large gear has 80 teeth, the ratio is 1 to 4. The large gear will make one complete turn in the time that the small gear makes four complete turns.

3. There are no differences when the lower gear is turned. However, when the lower gear drives the larger gear, it reduces the speed of rotation by the same amount as the larger gear increases the speed of rotation when it is used to drive the smaller gear. For example, if the smaller gear had 20 teeth and the larger gear had 60 teeth, the speed of rotation would be reduced by a ratio of 1 to 3. The large gear would make one complete turn for three complete turns of the small gear.

Joy's Construction Kit

Most hobby and toy shops have technology construction kits; LEGO and Meccano are two popular brands. Joy had such a kit. It contained a motor, wheels, propellers, and all the other items needed to get energy from the battery to these moving parts. Joy put together four assemblies to see how everything worked. She also asked herself questions about each assembly. Joy's assemblies are shown on this and the following page.

Assembly 1

1. If I turn the top gear, in what direction does the lower gear turn?

2. How many turns does the lower gear make during one complete turn of the top gear? Is this related to the number of teeth in each gear?

3. What differences would I note if I turned the lower gear instead?

Assembly 2

1. As I turn the small gear, what happens to the large gear?

2. For one turn of the small gear, how far does the large gear move?

3. What differences would I observe if I turned the large gear instead?

Assembly 3

Joy included a worm gear, another gear, and a wheel in this assembly.

1. As I turn wheel *A* through one complete turn, how far does the thread of the worm gear move to the left or right?

2. How far does gear *B* move at the same time?

3. If this system were attached to a motor, how many turns of the motor would be needed to make one turn of gear *B*?

Assembly 4

1. How does the energy get from the battery to the moving parts?

2. For each turn of the wheel on the motor, how much does each successive part listed below turn?

   • the wheel attached to the motor by a belt
   • the small wheel on the other side of the short axle
   • the wheel on the long axle
   • the tractor wheel

ASSEMBLY 1

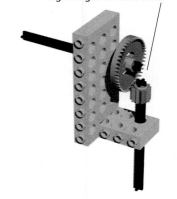

ASSEMBLY 2

Gear wheels of two sizes at right angles to each other

ASSEMBLY 3

Worm gear   Wheel A

Gear B

Elastic band to wheel on motor

Answers to *Assembly 2*

1. As the small gear is turned, the large gear turns more slowly than the small gear. The motion has also changed from horizontal rotation to vertical rotation.

2. Answers will vary depending on the ratio of the number of teeth in the top gear to the number of teeth in the bottom gear.

3. The only difference would be that the large gear is driving the small gear.

Answers to *Assembly 3*

1. The thread of the worm gear appears to make one complete revolution to the left or right.

2. Gear *B* will move a distance equal to that between two adjacent notches on the thread of the worm gear.

3. Answers will vary. The number of times that the motor will have to turn is equal to the number of teeth on gear *B*.

Answers to *Assembly 4*

1. Energy from the battery passes through wires to the motor. The motor wheel is attached by a belt to a larger wheel, which turns a short axle. Another belt connects the short axle to another large wheel, which is attached to the long axle of the tractor wheel.

Answers to Assembly 4 continued ▶

Joy first turned the connecting wheels by hand. She marked the top of each wheel to keep track of its motion.

3. How many turns of the motor's axle are needed to produce one turn of the tractor wheel?

4. How does the speed of the motor compare with the speed of the tractor wheel?

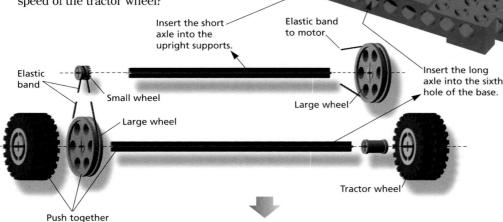

Wires to battery pack

Motor

(ASSEMBLY 4)

Insert the short axle into the upright supports.

Elastic band to motor

Insert the long axle into the sixth hole of the base.

Elastic band

Small wheel

Large wheel

Large wheel

Tractor wheel

Push together

(FULLY ASSEMBLED MODEL)

A Project— Model-Building

Now it's your turn to build a complete, working model from a construction set. It could be a device such as a car or tractor. It should have movable parts and be powered by a motor.

After you have completed your model and have done all the troubleshooting to get it working, observe how the energy is transferred from the battery to the working parts. Make a list of all the links in the energy train. Describe how they function and where changes in speed and direction occur.

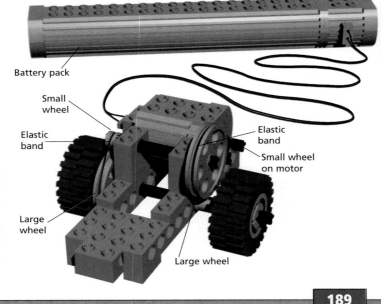

Battery pack

Small wheel

Elastic band

Elastic band

Small wheel on motor

Large wheel

Large wheel

189

Answers to
Situation 1

1. The smaller wheel makes three turns. Students will probably determine the answer by calculating the circumference of each wheel. Then they can divide the circumference of the large wheel by the circumference of the small wheel. At some point, however, you may wish to explain that the rotation of the two wheels can be expressed as a ratio of their diameters, in this case 3 to 1.

2. The speed of the driven wheel's axle is three times as fast as the speed of the driving wheel's axle.

3. Speed has been increased. Multiplying speed is an advantage desired in many mechanical systems where a quickly rotating axle or wheel is needed, such as in a drill.

4. If the driven wheel becomes the driving wheel, the speed of the larger driven wheel is decreased.

Answers to
Situation 2

1. The driving gear has 12 teeth.

2. The driven gear turns clockwise.

3. For every revolution of the driving gear, the driven gear makes only one-third of a turn. The driving gear's axle moves three times as fast as the driven gear.

4. The speed of the driven gear is 10 turns per second. The driven gear exerts less force at its axle because force has been sacrificed for distance.

5. More teeth in the driving gear and fewer teeth in the driven gear will increase the speed of the driven gear. Fewer teeth in the driving gear and more teeth in the driven gear will decrease the speed of the driven gear. You may wish to point out to students that the diameter of the gears would have to be changed to accommodate differing numbers of teeth of the same size.

A Closer Look at Wheels, Belts, and Gears

Here is a chance to see what you have learned about wheels and gears. Analyze the four situations below.

Situation 1
Driving Wheels and Driven Wheels

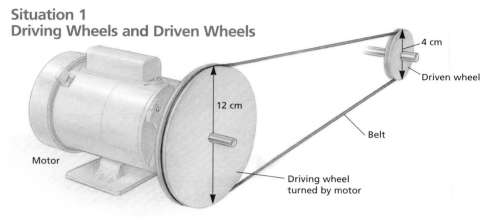

Here, the larger wheel is called the *driving* wheel because it drives the smaller wheel. It is turned by a motor. The belt transfers the energy to the smaller wheel, which is called the *driven* wheel.

1. As the large wheel makes one revolution (turn), how many turns does the smaller wheel make?

2. How does the speed of the axle of the driven wheel compare with the speed of the axle of the driving wheel?

3. Which has been increased in the driven wheel—speed or distance? Of what value could this be?

4. If the driven wheel were to become the driving wheel, what difference would this make?

Situation 2
Different-Sized Gear Pairs

1. The driven gear in the photo at right has 36 teeth. How many teeth are in the driving gear?

2. As the driving gear turns counterclockwise, in what direction does the driven gear turn?

3. For each revolution of the driving gear, how many turns does the driven gear make? Which gear axle moves faster?

4. If the speed of the driving gear is 30 turns per second, what is the speed of the driven gear? Which gear exerts less force at its axle?

5. What changes would you make to the gears to increase the speed of the driven gear? to decrease its speed?

190

ENVIRONMENTAL FOCUS

Wind generators are safe, clean machines that convert the kinetic energy of the wind to electrical energy. In Northern California wind generators produce millions of watts of electricity for nearby residents. However, wind generators are economical only in areas that have consistently high winds. Have interested students find pictures of different types of wind generators to share with the class.

Situation 3
Different Kinds of Gears

1. The diagrams at right show three different kinds of gear pairs that perform in a similar way. How do they work?

2. In diagram (a), when the worm gear moves one turn, how far do you estimate the driven gear will turn? Explain how you know. Is the speed of the driven gear greater than, the same as, or less than the speed of the driving gear? Which gear will exert the greater force? The worm gear is almost always the driving gear. Why do you suppose this is so?

3. Estimate the number of teeth in each of the bevel gears shown in diagram (b). If the driving bevel gear makes 7 turns in 1 second, approximately how many turns will the driven gear make?

4. In diagram (c), when the driving gear makes one turn, how many turns does the smaller driven gear make? Has speed or force been multiplied?

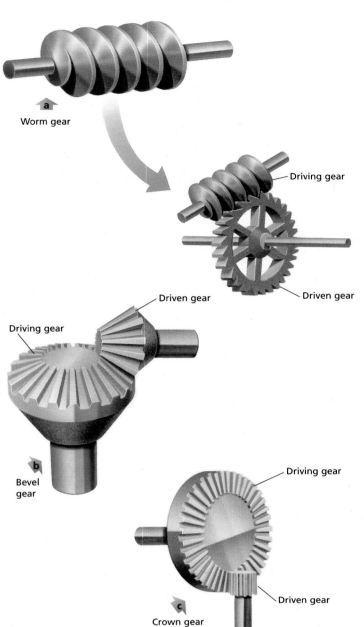

a
Worm gear

Driving gear

Driven gear

Driven gear

Driving gear

b
Bevel gear

Driving gear

Driven gear

c
Crown gear

 Transparency 25 is available to accompany Situation 3.

Homework

Have students conduct a survey at home to discover how many devices work with gears or wheels and belts. Have them share the results of their survey by making a composite chart of all of the different devices they identified.

---

1. In each case, the driving gear rests against the driven gear. Each of the gears changes the direction of motion by up to 90 degrees.

2. As the worm gear revolves once, it moves the teeth of the driven gear the distance between two adjacent threads of the worm gear. Since there are 28 teeth in the driven gear, it will take 28 revolutions of the worm gear to make one complete revolution of the driven gear. Therefore, the driven gear is moving much slower than the worm gear. The driven gear will exert the greater force. The worm gear is usually the driving gear because if it were the driven gear, it would turn so quickly that it would lose too much energy due to friction. This is because of the great increase in speed in the worm gear at the expense of distance.

3. There are about 24 teeth in the driving gear and about 14 in the driven gear. If the driving gear makes seven turns in one second, then 7×24, or 168, teeth mesh with the teeth on the driven gear, causing it to make $168 \div 14$, or 12, turns in one second.

4. The driving gear has about 36 teeth. The driven gear has about 12 teeth. The ratio is 3 to 1. Therefore, the driven gear will rotate three times for each time the driving gear turns. Speed has been multiplied.

Theme Connection

Systems
Focus question: How is a gear-driven mechanical system affected if the teeth on the gears wear down or break with use and age? *(The number of times the teeth mesh between the driving gear and the driven gear will decrease if teeth are missing or broken on the gears. As the number of revolutions per unit of time decreases, so does the speed. It is possible that the mechanical system may be thrown off balance because the gears may not move or turn at a constant rate due to the interruptions in the meshing of teeth between gears.)*

1. Since gear *A* is turned counter-clockwise, gear *D* will turn clockwise. Gear *D* will move with greater speed because it is smaller and has fewer teeth than gear *A*.

2. The speed of the output axle will be greater than it was in question 1 because there are fewer teeth in gear *E* than in gear *D*. Gear *B* appears slightly larger than gear *A*, and gear *E* is smaller than gear *D*. Therefore, the ratio between gear *B* and gear *E* is greater than the ratio between gear *A* and gear *D*.

3. Gear *G* and the output axle will move in a counterclockwise direction because gear *F* reverses the direction a second time.

4. Low: gears *A* and *D*; high: gears *B* and *E*; reverse: gears *C, F,* and *G*. The low gears give more force to the forward motion of the car by sacrificing speed. The high gears give more speed to the forward motion of the car by sacrificing force.

 Transparency 26 is available to accompany Situation 4.

Homework

The Math Practice Worksheet on page 39 of the Unit 3 Teaching Resources booklet makes an excellent homework activity once students have completed Lesson 3. If you choose to use this worksheet in class, Transparency 27 is also available.

Situation 4
A Simple Gearbox

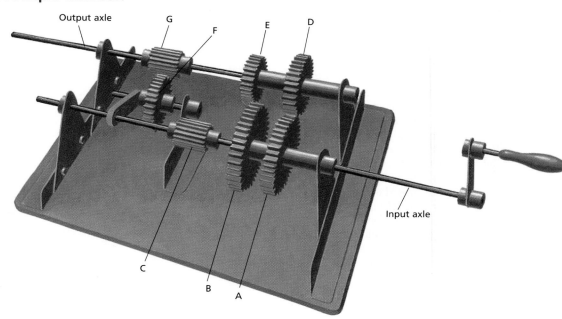

A student made this model of a car transmission (gearbox) from a construction set. The handle can move in and out, causing different gears to engage. There are two "forward speeds" and one "reverse speed." The handle is attached to the input axle. Gears on this axle engage the gears on the output axle.

1. In the illustration, gear *A* on the input axle is engaged with gear *D* on the output axle. Suppose the handle is turned counter-clockwise. In what direction will gear *D* on the output axle turn? Which gear, *A* or *D*, will move with greater speed?

2. Suppose the handle is pushed to the left so that gear *B* now engages gear *E*. Again suppose that the handle is turned counterclockwise at the same speed. How does the speed of the output axle now compare with its speed in question 1? Why?

3. Now suppose that the handle is pushed farther to the left so that gear *C* engages gear *F*. As the handle is turned counterclockwise, in what direction will gear *G* and the output axle move?

4. What pair or group of gears in the model corresponds to low gear? high gear? reverse gear? (In a real car there would be an additional set of gears between the gearbox and the wheels.) Which gear pair gives more force to the forward motion of the car? Which gear pair gives more speed to the car's forward motion?

FOLLOW-UP

Reteaching

Present students with the following scenario: Gear-Go Toys, Inc. wants you, its number one trouble-shooting engineer, to design the new velociraptor toy. The motor that operates the velociraptor must be contained in the velociraptor's belly with the drive shaft parallel to the floor. Also, the velociraptor must be able to lift both legs and to move both arms. Go to it!

Have students draw their designs on butcher paper with markers. Students should present their results to the class.

Assessment

Have students summarize the lesson by making fact sheets of what they learned about gears, wheels, and belts. Suggest that they illustrate each of their fact sheets with a picture or diagram.

Extension

Have students select one of the following devices and do some research to discover how gears are used to make the device work:
- bicycle derailleur
- push mower
- windshield wipers
- hand-operated eggbeater

- speedometer
- lawn sprinkler
- rack-and-pinion steering

Have students make poster diagrams to show what happens in their device. Display the diagrams around the classroom.

Closure

Have several student volunteers bring their bicycles to class. As a class or in small groups, have students calculate the gear ratio(s) on the bicycles. Use what they discover to determine the variation in gear ratios among different types of bicycles.

Mechanical Systems

Most machines that you see and use every day could be properly called *mechanical systems*. This is because they are made up of two or more simpler machines working together. Look back at the mechanical systems depicted throughout this unit. What simple machines do you recognize?

A can opener is one example of a mechanical system that contains several subsystems. Try to identify the following subsystems:

a. a lever that operates like a nut-cracker

b. a kind of gear

c. a wheel and axle

d. a wedge like the blade of an ax

You remember that a mechanical system is designed to do a single task and is made up of groups of parts called *subsystems*. What function does each part of the can-opener system perform? Can you group these four parts into two subsystems? What has been added to one part so that it can serve an additional purpose? **A**

How do you operate the kind of can opener illustrated here? Match the numbered parts in this photo with the lettered subsystems listed above. What simple steps of operation would you describe for someone else to follow? Write them in language that could be understood by an 8-year-old. **B**

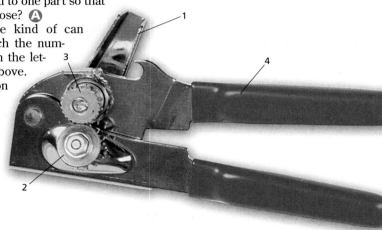

LESSON 4 ORGANIZER

Time Required
3 class periods

Process Skills
analyzing, comparing

New Terms
none

Materials (per student group)
Exploration 5, Station A: ballpoint pen; pens other than ballpoint;
Station B: stapler; staples; 2 pieces of paper; **Station C:** rotary pencil sharpener; **Station D:** bicycles (different makes and models); **Station E:** hand drill without a bit

One Purpose—Many Designs: a variety of can openers
Exploration 6: materials to build a working model car, such as the following: rubber bands; masking tape; plastic-foam or paper cups with lids; plastic drinking straws; index cards; cardboard tubes; wooden dowels; drawing compass; straight pins; pencils; corks; pliers; scissors; paper; paper clips; white glue; thumbtacks

Teaching Resources
Exploration Worksheets, pp. 41 and 46
SourceBook, p. S47

Mechanical Systems

FOCUS

Getting Started

Ask students to identify the simple machines that they have studied so far. Make a list of their responses on the chalkboard. *(Pulley, lever, inclined plane, wheel and axle)* Encourage students to describe the function of each simple machine they identify. Ask students which of these simple machines work together in a pair of scissors *(inclined planes [wedges] and levers)*, in a wheelbarrow *(wheel and axle, lever)*, and in a nail clipper *(lever, inclined plane [wedge])*. If possible, have several can openers, like the one pictured on this page, for students to examine.

Main Ideas

1. Most machines are mechanical systems made up of many subsystems.
2. A mechanical system can be designed in many different ways to perform a specific function.
3. Constructing, testing, and evaluating a working model provides the information needed to improve its design.

TEACHING STRATEGIES

Meeting Individual Needs

Second-Language Learners
For students who have limited English proficiency, have some can openers on hand to demonstrate to the class. Ask students to identify the subsystems of each can opener that are simple machines. Students can draw labeled diagrams of the can openers to show where each of the subsystems is located.

 **PORTFOLIO**
Students may wish to include in their Portfolio sketches of a few of the systems studied in Exploration 5 on the next two pages. Students should identify all of the subsystems in each machine and clearly label all of their parts.

Answers to In-Text Questions are on the next page. ▶

Answers to
In-Text Questions, page 193

A The functions of the four parts of the can opener are as follows:
   a. The handles are two levers that force the blade into the lid of the can and the gears against the top edge of the can.
   b. A kind of gear moves the can, or the can opener, around the edge of the can.
   c. A wheel and axle turn the gear that moves the can opener around the edge of the can.
   d. A wedge acts as the blade that cuts the lid from the can.

   Two subsystems might include (1) the wheel and axle and the gear, which work together to move the can opener around the top edge of the can, and (2) the lever and the wedge, which act together to cut the lid from the can.
   The notch in the handle acts as a bottle opener, allowing the handle to act as a lever to perform a task separate from that of the entire mechanical system.

B The can opener is operated by holding the blade (wedge) against the can with the levers and turning the can with the gears. The numbered parts and the lettered subsystems are related as follows:
   1. C
   2. D
   3. B
   4. A
   The writing exercise following the questions allows students to express themselves in simple language. Suggest that they include the specific function of each subsystem in their instructions.

★ **An Exploration Worksheet is available to accompany Exploration 5 (Teaching Resources, page 41).**

Answers to
Station A

1. Student responses will vary depending on the type of pen they are evaluating.

3. Encourage students to sketch the pen's parts and to write some notes about how the parts fit together and what they do.

4. Answers will vary. You may wish to provide additional pens or have students share pens of their own to examine.

EXPLORATION 5

Analyzing Mechanical Systems

Spend a few minutes at each of the following stations. Record all your answers in your ScienceLog.

STATION A

The Write Stuff

1. Check the operation of the pen. How does it write? Is it retractable? Is it refillable? How?

2. Take apart the pen to see its component parts.

3. How do the parts operate? Sketching the parts in operation may help you explain this.

4. Examine another type of pen. How is it similar? How is it different? Is it a better design? Why or why not?

STATION B

A Fasten-ating Device

1. Staple two pieces of paper together. As you do this, carefully observe each aspect of the stapler's operation.

2. Open the stapler and identify each main part. What is the function of each part?

3. Compare a staple before and after stapling. What causes the staple to bend? Could it be bent in another direction? How could you use the stapler without bending the staples?

4. How do you think a stapler like this compares with one used by carpenters or carpet installers?

To learn more about the simple machines in these mechanical systems, read pages S47–S51 in your SourceBook.

194

Answers to
Station B

2. The stapler has four arms, or levers. The top two arms are connected to a third by a spring. Staples are loaded in the middle arm and are pushed toward the front end of the stapler by a flat piece of metal attached to a spring. A thin, flat metal bar pushes the staple through the paper and against the metal base plate in the lower arm. When stapling has been completed, the thin, flat metal bar retracts, allowing the next staple to be pushed forward.

3. A base plate in the bottom arm bends the ends of the staples. Before, the staple is U-shaped; after, its ends are bent inward. The base plate can be shifted so that the ends of the staple bend outward rather than inward. If the stapler is open when it is used, the staples will not be bent.

4. Sample answer: The carpet installer's device exerts more force than the hand-held stapler. Also, the staples are not bent and are larger and stronger because they are driven into hard objects such as wood.

STATION C

What a Grind!

1. How does this pencil sharpener operate?

2. Identify all of the pencil sharpener's subsystems. Which of these subsystems are simple machines that you studied earlier?

3. How is this system a more effective design than a hand-held sharpener?

4. Suggest a design for a pencil sharpener that is not powered by hand.

STATION D

Wheel You Look at That!

1. Examine a bicycle closely. Observe how each part and subsystem operates.

2. Where do you apply energy to the bicycle? How is the energy transferred?

3. Compare the mechanical systems of different makes and models of bicycles.

STATION E

A Hole-some Machine

1. What is the overall function of a hand drill? How does it work?

2. Identify at least four simple subsystems in this mechanical system.

3. Describe the speed of the drill compared with that of the handle. What accounts for this?

4. How is the drill powered? In what other ways could it be powered?

195

Answers to
Station C

1. As the handle is turned, sharp cutting edges move forward along the tip of a pencil to sharpen it. The hole for the pencil is tapered to produce a sharp point at the end of the pencil.

2. The cutting surfaces of the sharpener are wedges arranged in a manner similar to that of the threads on a screw. The handle and rotating wedges are a wheel and axle.

3. A hand-held sharpener uses a stationary, single blade. The mounted sharpener uses a wheel and axle, which allows a person to apply a small force through a large distance, thus turning the pencil a smaller distance with a greater force.

4. Sample answer: An electric motor could be used to turn the blades.

Answers to
Station D

1. Students should be able to identify simple machines such as levers, cables and pulleys, wheels and axles, and chains and gears (sprockets). The handles and pedal cranks are part of a wheel-and-axle system.

2. Energy is applied by the person riding the bicycle as he or she pedals, turns the handlebars, and uses the brake and gear levers. The energy is transferred from the person's muscles to the pedals, to the forward sprocket, to the chain, to the rear sprocket, and to the wheels.

3. Answers will vary depending on the types of bicycles that are available.

Answers to
Station E

1. A hand drill is used to make holes in wood and metal. The crown gear transmits the motion of the handle to the bevel gears, which rotate the drill.

2. The following illustration identifies the subsystems in a drill:

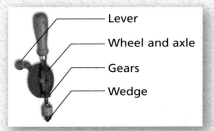

Lever
Wheel and axle
Gears
Wedge

(If a bit is attached, the bit is an inclined plane or screw.)

3. Students should recognize that the speed of the drill is much faster than the speed of the handle because the ratio of the number of teeth on the large driving gear compared with the number on the small driven gear is large.

4. This drill is powered by hand, but many are powered by electric motors.

Homework

You may wish to assign the activity and questions from Station A, B, C, or D as homework.

One Purpose—Many Designs

If possible, display several different kinds of can openers for students to examine, or have students bring can openers from home to share with the class. Then have them discuss their responses to the questions. There are many designs of hand-held and hand-operated can openers available. Some students may have also used electric can openers. Have students compare different kinds of can openers. Suggest that they determine which ones are the safest and the easiest to use. You may wish to have students use different kinds of can openers on the bottom lids of discarded cans. Then have them make an evaluation of which designs work best and are the most practical. Caution students to use care when handling the cut lids.

EXPLORATION 6

This Exploration is intended to be open-ended in that there is no right or wrong design. Allow students time to work together to brainstorm ideas and pique each other's interest. Encourage them to be creative. Be prepared to offer help and suggestions if called on. However, do not solve students' problems for them. Instead, guide them in the right direction by asking questions to point out connections that they may have overlooked. You should have students submit their designs to you for safety approval before beginning to build their machines. Students should not use household current from the wall or an AC or AC/DC converter power source.

In addition to providing an excellent example of technological problem solving, the activity can create a tremendous amount of interest and enthusiasm. You may wish to capture that interest by suggesting that a race be held when the models are finished to determine which car is the fastest and most stable.

Assessing a project of this nature by assigning objective values can be difficult. You may feel that your subjective assessment of each individual is fairer. However, if you desire an assessment checklist, the following points could be considered:

- team work: contribution and cooperation

- clarity and completeness of the design drawings
- quality of the product
- originality of design
- time and effort expended
- the model's performance
- distance traveled, speed, stability, and so on
- completeness of ScienceLog entries
- understanding of the steps involved in solving technological problems

 An Exploration Worksheet is available to accompany Exploration 6 (Teaching Resources, page 46).

One Purpose— Many Designs

The can opener on page 193 is just one device for opening cans. Other devices serve the same purpose.

1. What designs of can openers have you used or seen?

2. Various designs may have different features. What are they?

3. Which designs work best? Which are most practical?

Collect as many can-opening devices as you can. Examine each closely. Identify subsystems and any features that make them useful and effective.

Constructing Your Own Mechanical System

The following Exploration gives you an opportunity to develop a traveling mechanical system. You will design a system and test it.

EXPLORATION 6

Constructing a Model Car

Auto manufacturers are researching different energy sources to power the cars of the future. In this Exploration, you will construct your own car and power it with an unusual alternative energy source. Here is your challenge. Design a model car powered by a rubber band, and construct a prototype of the new car. The car must be able to move in a straight line for a distance of at least 5 m. (Slingshot propulsion is not allowed.)

You have been presented with a problem. How are you going to solve it? First review the steps in Technological Problem Solving on page 183. The following suggestions may also help you.

What You Might Want to Use

- rubber bands
- masking tape
- plastic-foam or paper cups
- cup lids
- drinking straws
- index cards
- cardboard tubes
- wooden dowels
- a drawing compass
- straight pins
- a pencil
- corks
- pliers
- scissors
- paper
- paper clips
- white glue
- thumbtacks
- anything else you can think of

What to Do

1. Record your work in your ScienceLog. Note what you do, why you do it, what problems you encounter, and how you solve the problems.

2. Work with one or two others. Begin by discussing the project. Try to think of as many ideas for solving the problem as you can. Sketch them to help you refine the concepts.

3. Review all of your possible designs. Choose the best one and start to work. Decide what materials to use and how large the car should be. Draw a blueprint of your car that shows exact details, identifies subsystems and individual parts, and indicates the scale. (A blueprint should be detailed enough so that someone could use it to build what the blueprint represents.)

4. Make a list of the materials you will need. Collect the parts needed.

5. Build, test, and evaluate your model. Use the results of your tests to make improvements.

6. Give your product a name. Prepare a promotional brochure designed to convince people to buy the product. Be sure to include technical data in your brochure.

Examining One Model Car Design

Jim Louviere, a science teacher, constructed an "air car" that worked surprisingly well. In order to make it very light, he used nothing but paper, index cards, rubber bands, and paper clips. Examine the diagram of the assembled model below. Discuss the following questions with two other students.

1. How does Jim's model car work? How do you get it to move?

2. What is the source of energy that powers the car?

3. What energy changes take place during its operation?

4. What force moves the car?

5. What causes the car to slow down and eventually stop?

6. How might you measure the energy input into the car?

7. What variables might be important in getting the car to work? You might consider the length of the body, the length of the axles, and other characteristics.

Extending Your Design

You may want to build a car of more durable materials and power it in other ways. What materials might you use? How would you power the car? Draw a blueprint of your design. You might construct this car as a science project, test it, modify it, and evaluate it.

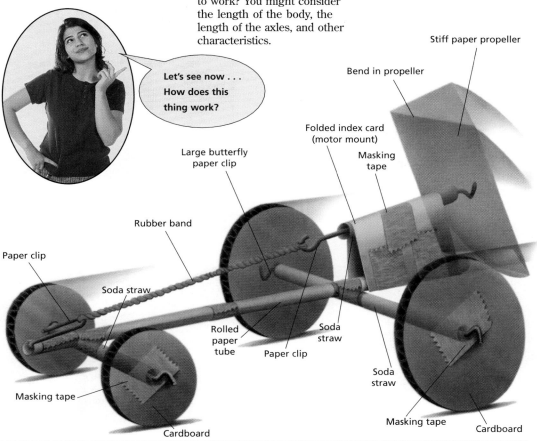

Let's see now . . . How does this thing work?

Stiff paper propeller

Bend in propeller

Folded index card (motor mount)

Masking tape

Large butterfly paper clip

Rubber band

Paper clip

Soda straw

Rolled paper tube

Soda straw

Paper clip

Soda straw

Masking tape

Cardboard

Masking tape

Cardboard

Answers to *Examining One Model Car Design*

1. Jim's model car works by creating an air current that pushes the car forward. The rubber band is wound by means of a handle projecting beyond the propeller. As the rubber band unwinds, the propeller turns, creating the necessary air current.

2. Initially, the source of energy is the person who winds up the rubber band. His or her energy comes from the chemical energy in food. The chemical energy is converted into potential energy in the wound-up rubber band.

3. The potential energy of the rubber band changes into kinetic energy as the propeller is made to spin. The propeller's kinetic energy is transferred to the air.

4. The propeller creates an air current that pushes against the air behind the car, causing the car to move forward.

5. The car slows down as the rubber band unwinds and the propeller stops. The slowing forces include friction between the wheels and the driving surface, friction generated by the wheels turning on the axles of the model, and air resistance.

6. The energy put into the model car could be calculated by determining how much effort it takes to turn the propeller to wind up the rubber band (the distance that the propeller turns times the force applied to the propeller). Each turn would have to be calculated separately because more effort is required as the rubber band tightens.

7. Variables might include the size of the wheels and propeller, the materials used in constructing the model, the size and elasticity of the rubber band, the ease with which the wheels turn, and the strength of the car's frame.

Extending Your Design

Make sure you check all student designs for safety before allowing students to construct their cars.

FOLLOW-UP

Reteaching

Have students list and explain the mechanical systems they use every day.

Assessment

Have students select a mechanical system that they have used. Ask them to analyze the device by identifying each of its subsystems and the simple machines found within each one. They should also describe the function of each subsystem as it relates to the more complex mechanical system.

Extension

Point out to students that automobiles have many mechanical systems. Suggest that they work in small groups to research automobile brake systems.

Closure

Challenge students to design a child's toy that is safe, simple, and attractive and that has one moving part. Ask students to submit their designs and requests for materials to you for approval.

LESSON 5 — The Human Machine

LESSON 5
The Human Machine

FOCUS

Getting Started

Show students an anatomical diagram or model of the human body depicting the muscles. Point out the major muscle groups, including the biceps and triceps, that are discussed in this lesson. Ask students to state whether each of the following statements is true or false:

a. About 40 percent of your body weight is muscle.

b. Muscle tissue is about 75 percent water, 20 percent protein, and 5 percent other materials, such as carbohydrates and minerals.

c. For their size, the muscles that operate the wings of bees, flies, and mosquitoes are stronger than any human muscles.

d. There are more than 600 skeletal muscles in the human body.

Tell students that all of these statements are true!

Main Ideas

1. The musculature of the human body consists of muscles that work together in pairs.
2. Joints, the sites where bones are connected, allow a wide range of motion.
3. Limbs have the mechanical structure of a lever because their motion depends on a fulcrum, load, and applied force.

TEACHING STRATEGIES

Call students' attention to the photographs on this page. Have students state which body parts are acting like a simple machine. Also ask students to identify what type of simple machine each body part resembles. *(All of the photographs show the action of levers.)*

The Human Machine

Leaping, lifting, bending, twisting, stretching, walking, running, swimming—the human body is capable of an amazing variety of motions that no single machine could ever duplicate. Yet the human body is a machine, an extraordinary living machine that not only moves but also transforms energy and does work according to the same principles as the machines you have been studying.

The human machine accomplishes all that it does with the aid of more than 200 bones and 600 muscles. The bones fit together at junctions called joints, where they are held together by ligaments. Muscles run like cables between bones and are attached to the bones by tough tendons.

The human machine's rich variety of motion is possible because of the simple fact that muscles exert force. Even when a person is standing "still," muscles are hard at work, continually readjusting the alignment of bones in response to the brain's commands, which travel to the muscles via nerves.

The following Exploration investigates joints, muscles, and bones—essential components of the human machine.

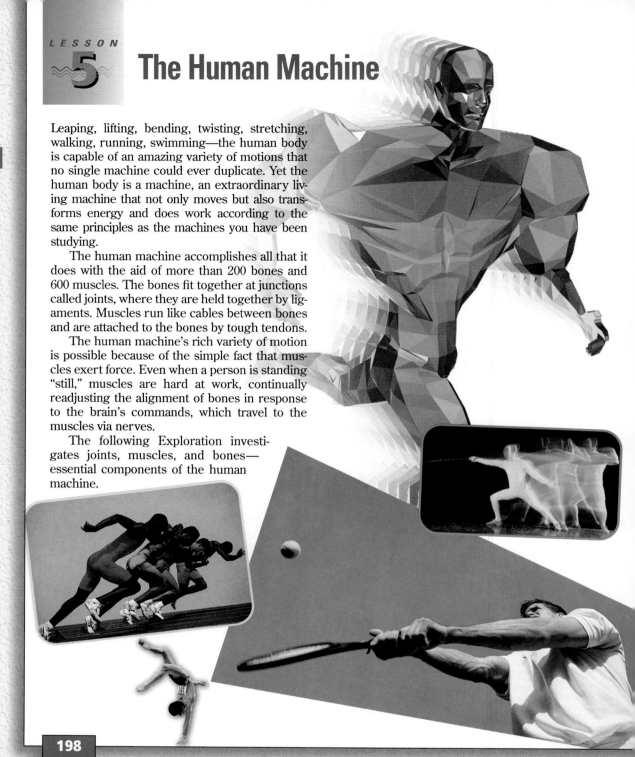

198

LESSON 5 ORGANIZER

Time Required
2 class periods

Process Skills
analyzing, observing, communicating, comparing

Theme Connection
Structures

New Terms
none

Materials (per student group)
Exploration 7, Part 2: fresh chicken wing; paper towels; small scissors; plastic bag; scalpel

Teaching Resources
Exploration Worksheet, p. 49

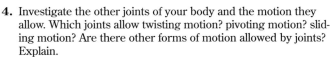

EXPLORATION 7

Motions and Movements

PART 1

Motion Possible

Joints, or sites at which bones are connected to one another, allow a surprising variety of movement. In this part of the Exploration, you will use your own body to investigate how joints work.

What to Do

Get together in small groups with two or three other students. Give yourselves some space in which to move about. The following suggestions will guide you in your investigation of joints and motion.

1. How many joints allow movement of your hand and its parts? Describe the motion that each joint seems to allow.

2. Investigate the possible motions of your forearm. Which joint do you think allows this motion? Is a twisting motion possible? Discuss with the other members of your group how this joint might work. Sketch what you think it might look like.

3. Which joints in your body allow motion through a full circle? Discuss with the other members of your group how this type of joint might work. Sketch what you think it might look like.

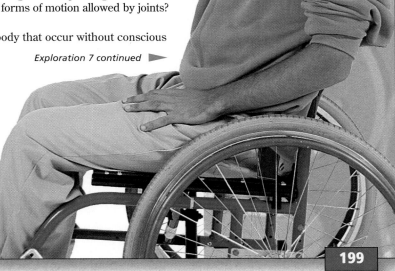

4. Investigate the other joints of your body and the motion they allow. Which joints allow twisting motion? pivoting motion? sliding motion? Are there other forms of motion allowed by joints? Explain.

5. Are there motions in your body that occur without conscious control? If so, explain.

Exploration 7 continued

199

Integrating the Sciences

Life and Physical Sciences

On a chart of the human skeletal system, point out the locations of the joints are that are discussed in this Exploration. You may also wish to point out some examples of unusual joints in the body. For instance, the hip is the only true ball-and-socket joint, and there are joints in the cranium that do not move at all.

Answers to
Part 1

1. The hand contains 14 joints, all of which allow back and forth motion. (You may wish to point out that these joints are the knuckles in the fingers of the hand. Also alert students that the wrist contains numerous small bones that should not be included in the answer.)

2. The elbow allows motion of the forearm. The elbow acts primarily as the fulcrum of a lever. It allows only a small amount of twisting motion.

3. The joints of the ankle, wrist, hip, neck, and shoulder allow motion through a full circle.

4. Students' answers will vary but could include the following: joints at the wrist and ankle allow twisting; the hip and shoulder joints allow pivoting; the shoulder blade allows a sliding motion. Students who are "double jointed" may describe interesting and unusual motions.

5. There are many motions that occur in the body without conscious control. Examples include the heart beating or the pupils dilating.

Homework

If class time is short, students can perform Part 1 of Exploration 7 individually at home. Students should pool results the next day in a class discussion.

★ **An Exploration Worksheet is available to accompany Exploration 7 (Teaching Resources, page 49).**

Answers to
Part 2

4. Tough bands of connective tissue hold the bones together. These bands of tissue are called *ligaments.*

   Students should discover that the motion is smooth. The bones move smoothly because the ends of the bones are covered with smooth, hard tissue that allows the bones to glide past one another.

Meeting Individual Needs

Second-Language Learners

Have students compare the strength of the biceps and triceps muscles. To do this, ask students to use a bathroom scale to measure the force they exert.

- Set the scale on a desktop. While seated, push down on the scale as hard as possible. This measures force exerted by the triceps.
- While seated, push the scale against the underside of the desk. The biceps are the muscles exerting the force in this case.

  Ask students: Why is the biceps muscle stronger than the triceps muscle? *(The biceps is stronger because it is the muscle that contracts when objects are lifted.)*

Integrating the Sciences

Life and Physical Sciences

Have students work in small groups. Each group should select a sport such as football, baseball, or cycling. Group members should prepare a poster representing the motions of the human body when engaged in their sport. They should apply what they have learned in this unit to show how the sport involves energy conversions, simple machines, work, and power.

PART 2

Just Wing It

In this activity, you will look at actual bones and muscles from a chicken to help you understand how they work together.

You Will Need

- a fresh chicken wing
- paper towels
- small scissors
- a plastic bag
- a scalpel

What to Do

1. With your scissors, peel back the skin from a chicken wing, as shown in illustration (a). Identify the transparent skin-like *connective tissue* surrounding the muscle bundles and bone. (You may have to cut some of it away to see the muscles better.) Notice how the muscles are arranged in pairs on opposite sides of the bones.
2. Try squeezing each muscle of the pair, in turn, to see how it moves the end of the chicken wing. Muscles always work in *opposing pairs* to cause motion.
3. Carefully separate the tissue. Identify the tiny white *nerves* that activate the muscles, and identify the blood vessels that bring oxygen and nutrients to the muscles. Identify the durable, white tendons that connect the muscle to the bone.
4. Cut away the remaining tissue to expose a joint. Look at the joint. What holds the bones together? Work the joint back and forth, as shown in illustration (b). Is the motion smooth or do you feel a lot of friction? What accounts for this?

Summarize your observations and write any questions you may have in your ScienceLog. Enclose the remains of the chicken wing in a plastic bag for disposal, and wash your hands thoroughly with soap and water.

Questions

1. What purpose do you think the transparent tissue surrounding the muscle bundles might serve?
2. What do you think is the significance of the arrangement of muscles in opposing pairs?

3. Which of the tissues you examined might carry signals from the chicken's brain?
4. In what ways do tendon tissue and muscle tissue differ? How do these differences reflect the functions of each type of tissue?

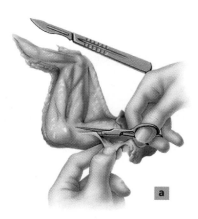

a

b

200

Answers to
Questions

Accept all reasonable responses. The following are sample answers:

1. The transparent tissue protects the muscle fibers. This tissue may also help to reduce friction between moving muscles and other tissues such as skin and bone.

2. When a muscle causes a joint to bend, another muscle is needed to bend the joint back to its original position. Thus, it is useful for muscles to be arranged in opposing pairs.

3. The nerves carry signals from the brain to activate the muscles.

4. Muscles have the ability to cause motion by flexing. Tendons cannot cause motion; their purpose is to connect different types of tissues to one another, such as a muscle to a bone.

Feel the Force

You may think that a chicken and a human aren't very similar. But in fact, they contain similar structures. And their muscles, bones, and joints work in almost exactly the same ways. In this activity you will find out more about opposing muscle pairs and how they work, using the human body as an example.

What to Do

1. While seated, grasp a fixed object such as a desk. Have a partner lightly grasp your upper arm so that he or she can feel the action of your muscles. Pull upward as though to lift the desk. (Make sure the desk is secured to the floor.) Report your observations and those of your partner.

2. Without changing your grip on the desk, pull downward on it. Your partner should continue to observe your muscle action. Report your observations and those of your partner.

3. Switch roles with your partner and repeat the experiment.

4. Read the passage that follows and answer the questions it poses.

5. Summarize your findings in your ScienceLog.

A Lever With Dual Controls

You may not think of your limbs as levers, but in fact, they are. Look at the diagram at right. Like all levers, the arm has a fulcrum (pivot point), a load, and an applied force. Identify each of these. Locate the *biceps* muscle on the upper side of the arm. What happens when this muscle contracts? Now locate the *triceps* muscle on the underside of the arm. What happens when this muscle contracts? What has to happen to the biceps muscle before the triceps muscle can do its job? What do you think the triceps does when the biceps contracts?

You might find it interesting to note that muscles never completely relax. They always exert a little tension. This is called *muscle tone.*

Think again about the lever action of the forearm. Is force or distance multiplied? Explain.

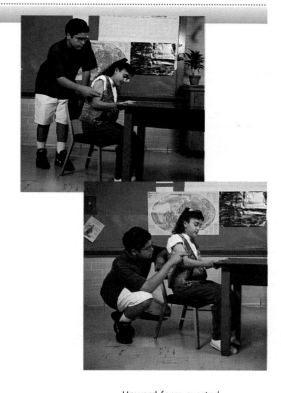

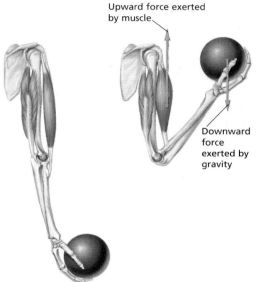

Upward force exerted by muscle

Downward force exerted by gravity

When the biceps muscle contracts, the forearm lifts. When the triceps muscle contracts, the forearm straightens from a bent position. Before the triceps muscle can do its job, the biceps muscle must relax. When the biceps muscle contracts, the triceps muscle relaxes.

Distance is multipled at the expense of force because the applied force in a forearm is relatively large and the distance through which the applied force moves is small.

Theme Connection

Structures

Not only humans, but also other organisms, have bodies that function according to the basic principles that govern mechanical systems. **Focus question:** Describe an anatomical structure of another organism that is essentially a simple machine. *(Sample answer: A woodpecker's beak acts like a wedge.)*

FOLLOW-UP

Reteaching

Bring a poster of the human muscular system to class. Point to a specific muscle and have the class discuss what kind of motion the muscle allows, such as twisting, pivoting, or sliding. (Some muscles may allow more than one type of motion, and others may function without conscious control.) You may wish to have students identify the opposing muscle in the example, if one exists.

Assessment

Have students write a paragraph titled "The Incredible Machine" that summarizes the important points of the lesson. The paragraph should explain how joints, muscles, and bones work together to accomplish a wide range of motion.

Extension

Have students attempt to design a mechanical system that would allow the range of motion achieved by the human arm. Designs should be accompanied by a written or oral explanation of how the system is intended to work, how much weight the mechanical arm should be able to lift, and how life-like its range of motion would be. After checking the designs for safety, encourage interested students to construct their mechanical arm.

Closure

Have students work in small groups to develop skits to present to the class. Each skit should convey how joints and muscles work and why the arm is considered a lever. Encourage creativity and check for accuracy.

Answers to *Challenge Your Thinking*

1. a. Work input = force input × distance the handle was turned through = 12 N × 1.50 m = 18 J

b. The weight of the load is 10 kg × 10 N/kg = 100 N.
Work output = weight (force) of load × distance load was raised = 100 N × 0.02 m = 2 J
Work lost to friction = work input – work output = 18 J – 2 J = 16 J

Zach lost 16 J of work input to friction.

2. Students will have different ideas about which machine is best to use, depending on how each student defines "best." Some students will choose the most efficient machine, some will choose the fastest machine, some will choose the machine requiring the smallest input force, etc. Accept all responses that are correctly based on the data and evidence given.

a. Zach's jack requires the least amount of force; thus, it is the easiest to use.

b. Because Zach's jack requires the least amount of force, it requires this force to be applied through the greatest distance to generate the same amount of work.

c. Efficiency = work output/work input × 100 percent
- stabilizer jack: efficiency = 2 J/18 J × 100 = 11 percent
- block and tackle: efficiency = (100 N × 0.2 m)/(30 N × 0.8 m) × 100 = 20 J/24 J × 100 = 83 percent
- inclined plane and rollers: efficiency = (100 N × 0.2 m)/(25 N × 1 m) × 100 = 80 percent

The block and tackle is the most efficient.

d. The least efficient machine, Zach's jack, loses the most energy to friction.

e. Answers will vary. Assuming an equal rate of input work, the block and tackle would be fastest because it is the most efficient.

Answers to Challenge Your Thinking continued ▶

1. Zach's Jack

Zach suggested another solution to Tony's Problem on page 170. He tried one of the stabilizer jacks used for his family's travel trailer. The measurements he made are shown in his diagram. Zach discovered that when he turned the lever handle through one complete turn, the load was raised a distance equal to the distance between two ridges on the screw bolt.

a. What was Zach's work input for one turn of the lever?

b. How much of Zach's work was lost to friction?

2. You Be the Judge

Which of the three machines is the best—Zach's jack above or one of the two machines below? The diagrams and questions below will help you decide.

a. Which machine would be the easiest to use?

b. Which causes you to work through the greatest distance?

c. Which is the most efficient?

d. Which loses the most energy to friction?

e. Which would be the fastest to use?

Based on your answers to the questions above, which do you think is the best machine?

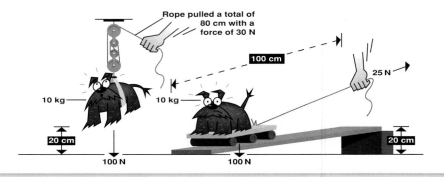

⭐ **The Chapter 9 Review Worksheet** (Teaching Resources, page 54) and **Transparency 28** are available to accompany this Challenge Your Thinking.

Meeting Individual Needs

Learners Having Difficulty

Point out to students that there are six simple machines: the lever, wedge, inclined plane, wheel and axle, screw, and pulley. Have them find examples of each type of machine in this unit. *(Sample answer: On page 150, the wheelbarrow is a lever and a wheel and axle. There is also a hammer striking a wedge. Pulleys are used in Leona's machine on page 172. An inclined plane is used in Pam's machine on page 175. The rotary blades of the pencil sharpener on page 195 consist of screws.)*

3. Time for a Real Challenge

On the right is a simplified diagram of a common mechanical clock. Use the labels and diagram to answer the following questions:

a. How does this device work? (Hint: Start at the mainspring.)

b. Where does the energy to power the device come from? How is it transferred?

c. Why do the hour and minute hands move at different speeds?

d. The hairspring and balance rock back and forth to mark off the seconds. How does this work?

4. Precision Decision

The distributor in a car engine sends a pulse of electric current to each of the spark plugs. It must turn at *exactly* half the speed of the engine. Would gear pairs or a wheel-and-belt system work better? Why?

5. Ship-Shape Machine

The diagram below shows a device for powering a model boat.

a. How does it work?

b. What is the source of energy?

c. How is the energy transferred?

d. How is friction reduced?

e. Is there a trade-off taking place here? If so, what is it?

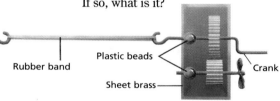

Rubber band — Plastic beads — Sheet brass — Crank

6. Your Invention

Design a mechanical system to perform some task for you. The system must contain three subsystems that are simple machines. Be creative!

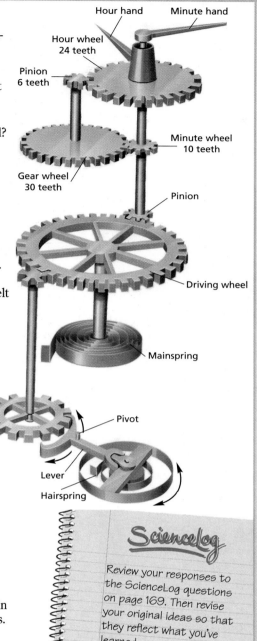

Hour hand Minute hand
Hour wheel 24 teeth
Pinion 6 teeth
Minute wheel 10 teeth
Gear wheel 30 teeth
Pinion
Driving wheel
Mainspring
Pivot
Lever
Hairspring

ScienceLog

Review your responses to the ScienceLog questions on page 169. Then revise your original ideas so that they reflect what you've learned.

203

ScienceLog

The following are sample revised answers:

1. Given equal amounts of work input, the machine that has the greater amount of work output is the most efficient. Efficiency is the ratio of work output to work input and is usually expressed as a percentage: efficiency = work output/work input × 100. Using less energy to overcome frictional forces between its moving parts is one thing that makes a machine more efficient.

2. The person in the photograph is using a lever to pry open a lid. Levers allow a given amount of work to be done with less applied force. A lever consists of a bar that pivots around a fulcrum.

3. Answers will vary. Students should understand that machines can be complex mechanical systems that use combinations of simple machines such as wheels and axles, gears, levers, and pulleys to transfer energy.

---

3. a. The driving wheel engages a small gear, the pinion, on a shaft connected to the minute hand. Another gear on this shaft is connected to a series of gears, which eventually connects to the hour hand.

b. As the mainspring unwinds, its stored potential energy is transferred to the driving wheel, causing the wheel to rotate.

c. The pinion drives the minute hand directly. The two sets of gears slow down the motion of the hour hand to 1/60 the motion of the minute hand. (The product of the gear ratios is 10/30 × 6/30 × 6/24, or 1/60.)

d. The hairspring oscillates, causing the lever to rock back and forth. As the lever rocks, it alternately releases and engages the cogwheel, causing it to advance one cog at a time. Each advance of the cogwheel measures out one second. (Note that without the stopping action of the lever, nothing would prevent the mainspring from unwinding freely.)

4. Gear pairs would work better because the ratio can be determined exactly by the number of teeth in each gear.

5. a. The crank is turned, which twists the rubber band. As the rubber band unwinds, it turns the gears that drive the propeller.

b. The initial source of energy is whatever turns the crank. Then potential energy is stored in the rubber band. When the band is released, its potential energy is converted into kinetic energy.

c. Energy is transferred from the crank to the rubber band, to the driving gear, to the driven gear, and to the propeller.

d. By the use of plastic beads

e. Force is sacrificed to increase the distance that the boat travels in a given time (that is, the speed of the boat).

6. Student designs will vary, but all should contain three subsystems that are simple machines. Make sure you check all designs for safety before allowing students to construct any machines.

The Big Ideas

The following is a sample unit summary:

Work is the force applied to an object times the distance through which the object moves. The work done by a system is equal to the amount of energy that is put into the system. (1)

Work is measured by the force applied to an object (in newtons) times the distance that the object moves (in meters). The unit for measuring force is the joule. The formula for work is $W = f \times d$. (2)

Potential energy is stored energy, such as in a compressed spring. Kinetic energy is the energy due to the motion of an object, such as a falling ball. (3) A mechanical system is a set of simple machines or subsystems which are connected so that energy can be transferred from one part to another to perform a task. A simple machine is a device that does work with one movement. Simple machines include the pulley, lever, wheel and axle, and inclined plane. (4)

Energy is transferred in mechanical systems in many ways, such as through wheels and belts, meshed gears, springs, coils, or rubber bands. (5) Machines do not save work. They can only increase, decrease, or redirect the force put into them. If the force increases, the distance through which it is applied decreases (and vice versa) so that the work done is constant. (6) But because friction is present in all mechanical systems, the work that is put into a machine (work input) always exceeds the work done by the machine (work output). Efficiency equals the work output divided by the work input. Efficiency can never be greater than 100 percent because some energy is always lost due to friction. The more friction there is in a machine, the less efficient the machine is. (7) Machines conserve energy by converting any energy not used to do work into heat energy. (8)

 You may wish to provide students with the Unit 3 Review Worksheet that is available to accompany this Making Connections (Teaching Resources, page 62).

Unit 3

SOURCEBOOK

To find out more about machines and how they make our lives easier, look in the SourceBook. There you will find more information about force, work, and energy and about how even the most complex machines actually consist of many simple machines.

Here's what you'll find in the SourceBook:

UNIT 3
Force, Work, and Power S42
Simple Machines S47
Machines at Work S52

204

The Big Ideas

In your ScienceLog, write a summary of this unit, using the following questions as a guide:

1. What is *work* in a scientific sense? How is it related to energy?
2. How is work measured?
3. How do you distinguish between potential and kinetic energy?
4. How are simple machines and mechanical systems related?
5. How is energy transferred in mechanical systems?
6. Do machines actually save work? Explain.
7. What is efficiency? How is efficiency related to friction?
8. How is energy *conserved* in machines?

Checking Your Understanding

1. Do you agree or disagree with the following statements? Explain your answers.
   a. A machine is a device for converting or transferring energy.
   b. If work is being done, then energy is being converted.
   c. Work is a form of energy.

Answers to *Checking Your Understanding*

1. a. Agree—A machine cannot create or destroy energy, but it can transform energy from one type to another or transfer energy among parts of the machine.
   b. Disagree—Work can be done if energy is simply transferred from one object to another. For example, a car does work on a garage door if the brakes don't work—the kinetic energy of the car is transferred to kinetic energy of the garage door.
   c. Disagree—Work is a mode of energy transfer; work is one way that energy moves from place to place. Strictly speaking, energy is the ability to do work; work is not energy itself.

Homework

You may wish to assign the Unit 3 Activity Worksheet: An Important Message as homework (Teaching Resources, page 61). If you choose to use this worksheet in class, Transparency 29 is available to accompany it.

2. Study the cartoon below. Identify the error or errors in each panel of the cartoon.

"The large and the small bricks are falling at the same rate! Hmmm . . . They must have the same kinetic energy."

"So that's what 200 J of work feels like."

These two students have the same mass.

EARTHQUAKE!

"I've just invented a machine that saves work."

"Sure hope so! Looks like it will get the load all the way up the slope."

The two jars (with identical mass) have the same kinetic energy just before hitting the floor.

"All this energy without putting any work into the bike."

"I've been holding these boxes for 15 minutes! That's the most work I've done in a long time!"

3. Here is an interesting collection of gears.
   a. Trace the motion of the components.
   b. Where does the device change the direction of motion?
   c. Where does a trade-off take place?
   d. Which way does energy probably flow through this system? Explain.
   e. What might you use this system for?

4. (concept map) Make a concept map using the following terms or phrases: mechanical systems, work, energy, machines, and simple machines.

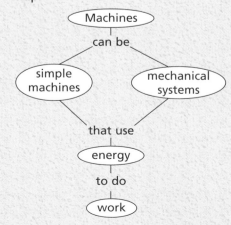

---

2. Starting with the upper left panel and proceeding from left to right, answers are as follows:
   • Because the boy on the left is closer to the fulcrum, he moves a shorter distance and must exert a greater force to lift the girl on the right. Because the two students have the same mass, they should be exerting the same force. Therefore, the boy should be in the air, and the girl should be on the ground.
   • Kinetic energy is dependent on mass and speed. The objects have different masses, so they have different kinetic energies.
   • The problem here is that the weight lifter has calculated the work incorrectly.
   Work = force × distance
   100 kg × 10 N/kg × 2 m = 2000 J
   Also, the weight lifter is not doing any work while he is holding the weight steady over his head. However, he did 2000 J of work in lifting the weights.
   • Although machines can reduce the force needed or change the direction of the force, they cannot save work.

   • The jars are different distances from the floor and therefore have different potential energies. They will not have the same kinetic energy just before hitting the floor.
   • In the top part of the panel, the girl may not be putting any work into the bike, but work is still being done on the bike. Gravity is performing work on the bike to convert the bike's potential energy to kinetic energy as the bike rolls down the hill. In the bottom part of the panel, no work is being done because the force required to hold up the boxes is not causing any motion.

3. a. Answers will vary. You may wish to have students begin at one end of the system and use an arrow to indicate the direction of motion. They should trace the motion through the system, reversing its direction from gear to gear. The direction of motion at the other end of the system will depend on the initial direction chosen. (You may wish to point out that the object at the top of the picture is a type of gear called a *rack.*)
   b. The direction of motion is changed at every point where two gears mesh.
   c. A trade-off takes place anywhere that distance is increased with the transfer of force (for example, from the worm gear to the crown gear).
   d. Energy usually flows from a worm gear to other gears. Therefore, the initial power source probably flows through the bevel gears.
   e. Because the rack is able to convert the rotary motion of the crown gear into a back-and-forth motion, this device might make a good steering mechanism.

4. The following is a sample concept map:

Machines
can be
simple machines
mechanical systems
that use
energy
to do
work

Background

New advances in microtechnology have allowed scientists to achieve impressive results in many fields. For example, medical researchers are working on special pills equipped with sensors, tiny pumps, and drug reservoirs.

Other possible technological advances include microscopic filters and air turbines for controlling the temperature of microchip arrays. One team of scientists has created a molecular "on-off switch" that could be used to store information in computers.

A scanning tunneling microcscope (STM) can be used to study the surfaces of microscopic materials that can carry electric current. As the probe of the STM approaches a material, a current called a *tunneling current* is created between the material and the probe. The strength of the current at different locations allows the STM to create an image of the material's surface. The STM won its inventors the Nobel Prize in physics in 1986.

CROSS-DISCIPLINARY FOCUS

Art

Have students design their own micromachine. Each student should make a poster that shows what his or her micromachine might look like and how it performs its function. Have the class guess what function the micromachine performs, and then have the artist explain his or her idea to the class. You might want students to work in groups for this activity.

Micromachines

The technology of making things smaller and smaller keeps growing and growing. Powerful computers can now be held in the palm of your hand. But what about motors smaller than grains of pepper? Or gnat-sized robots than can swim through the bloodstream? These are just a couple of the possibilities for micromachines.

Minuscule Motors

Researchers have already built gears, motors, and other devices so small that you could accidentally inhale one! For example, one engineer devised a motor so small that five of the motors would fit on the period at the end of this sentence. This micromotor is powered by static electricity instead of electric current, and it spins at 15,000 revolutions per minute. This is about twice as fast as most automobile engines running at top speed.

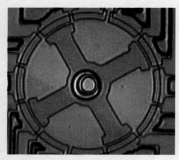

▲ The earliest working micromachine had a turning central rotor.

Small Sensors

So far micromachines have been most useful as sensing devices. Micromechanical sensors can go in places too small for ordinary instruments. For example, blood-pressure sensors can fit inside blood vessels and can detect minute changes in blood pressure within a person's body. Each sensor has a patch so thin that it bends when the pressure changes.

Cell-Sized Robots

Some scientists are investigating the possibility of creating cell-sized machines called nanobots. These tiny robots may have many uses in medicine. For instance, if nanobots could be injected into a person's bloodstream, they might be used to destroy disease-causing organisms such as viruses and bacteria. They might also be used to count blood cells or to deliver medicine.

The ultimate in micromachines would be machines created from individual atoms and molecules. Although these machines do not currently exist, scientists are already able to

manipulate single atoms and molecules. For example, the "molecular man" shown here is made of individual molecules. These molecules are moved by using a *scanning tunneling microscope*.

▼ "Molecular man," drawn with 28 carbon monoxide molecules

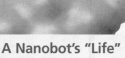

A Nanobot's "Life"

Imagine that you are a nanobot traveling through a person's body. What types of things do you think you would see? What type of work could you do? Write a story that describes what your experiences as a nanobot might be like.

A Nanobot's "Life"

You might want to encourage creativity and scientific accuracy by providing students with a body atlas or similar reference work. You may also wish to tell them that because they are all specialized nanobots, they can only travel through certain systems of the body (such as the circulatory, endocrine, or nervous systems) or certain organs or types of tissue. Ask them what their

environment would look like, what common problems might occur in their environment, and what they could do to help. *(For example, a nanobot inside a lung would see bronchial tubes, alveoli, and capillaries. It could break down contaminants in the air sacs, help fight off infections, or remove fluids in patients who have pneumonia.)*

Wheelchair Innovators

SCIENCE in ACTION

Two recent inventions have dramatically improved the technology of wheelchairs. With these new inventions, some wheelchair riders can control their chairs with voice commands, and others can take a cruise over a sandy beach.

Voice-Command Wheelchair

At age 27, Martine Kemph invented a voice-recognition system that allows people without arms or legs to use spoken commands to operate motorized wheelchairs. Here's how it works: The voice-recognition computer translates spoken words into digital commands, which are directed to electric motors. These commands completely control the operating speed and direction of the motors, giving the operator total control over the chair's movement.

◀ The voice-controlled wheelchair, pictured with its inventor, Martine Kemph, provides more freedom to people who are unable to use their arms or legs.

Kemph's system can execute spoken commands almost instantly, much faster than previous voice-recognition systems. In addition, the system is easily programmed, so each user can tailor the computer's list of commands to his or her individual needs.

Kemph named the computer Katalvox, using the root words *katal*, which is Greek for "to understand," and *vox*, which is Latin for "voice." Katalvox also has great potential for use in surgery, on assembly lines, and in automobiles.

The Surf Chair

Mike Hensler was a lifeguard at Daytona Beach, Florida, when he realized that it was next to impossible for someone in a wheelchair to come onto the beach. Although he had never invented a machine before, Hensler decided to build a wheelchair that could be driven across sand without getting stuck. He began spending many evenings in his driveway with a pile of lawn-chair parts, designing the chair by trial and error.

The result of Hensler's efforts looks very different from a conventional wheelchair. With huge rubber wheels and a thick frame of white PVC pipe, the Surf Chair not only moves easily over sandy terrain, but also is weather resistant and easy to clean. The newest models of the Surf Chair come with optional attachments, such as a variety of umbrellas, detachable armrests and footrests, and even places to attach fishing rods.

▲ Mike Hensler tries out his Surf Chair. People have found the Surf Chair to be practical, fun, and comfortable.

Design One Yourself

Can you think of any other ways to improve wheelchairs? Think about it and put your ideas down on paper. To inspire creative thinking, consider how a wheelchair could be made lighter, faster, safer, or easier to maneuver.

207

Answers to
Design One Yourself

Accept all reasonable designs. (You might want to incorporate this activity into one of the other activities suggested at right. As students develop their designs, be sure to point out that the safety of the user should always be the greatest concern in any design.)

Background

Martine Kemph's Katalvox-driven wheelchair is already being used in medical institutions in Moscow and Paris, at Stanford University Hospital, and at the Mayo Clinic. Also, NASA is testing Katalvox for its ability to control cameras mounted on robotic arms. As an inventor intent on helping people, Kemph has had a very good role model: her father is a polio victim who invented a car that could be driven without the use of legs.

In designing the Surf Chair, Hensler purposefully avoided materials that would make the chair look cumbersome or clinical. Since the beach is a place to relax and have fun, Hensler designed his chair to blend easily into such an environment. This fun and practical wheelchair is now available at many public beaches. Daytona Beach, for example, provides free use of the Surf Chair for those people who need wheelchairs.

Extension

Have students design a device to help mobility-impaired people in the home. You might want to give them a specific goal, such as designing a device to retrieve something from the refrigerator.

Meeting Individual Needs

Learners Having Difficulty

Most devices can be divided into three categories: those that provide humans with an entirely new ability, those that enhance one's ability to do something, and those that partially correct for the loss of an ability. Provide students with the following list: airplanes, Braille, hearing aids, submarines, pacemakers, bicycles, boats, prosthetic arms, telescopes, spacecraft, dialysis machines, and microscopes. Have them place each of these devices into one of the three categories. Students should be able to support their choices with reasonable arguments. Encourage students to come up with their own examples as well.

Background

Bicycles allow riders to move faster than the fastest human can run while requiring less energy to do it. Designers continue to create lighter and more aerodynamic bikes so that riders can go faster and farther than ever before. Sometimes their creations are rather strange, such as a bike that can hold eight riders, a bike that can be pedaled by the hands instead of the feet, and even one with motor-assisted pedals.

One advance in bicycle design is the development of metal matrix composites. MMCs were first employed in the aerospace industry for use in the space shuttle. One type of MMC is a composite of aluminum and silicon carbide fibers. When the fibers are evenly distributed throughout the aluminum, the resulting material is nearly four times stronger, is three times stiffer, and weighs less than the metal by itself.

Extension

Have students make a poster of their own bicycle design. However, explain to them that their bicycle will not be ridden by people. Instead, they are to design it for an animal of their choosing. Some possibilities may be a dog, cat, bird, frog, inchworm, or even a fish!

Multicultural Extension

Cycles Around the World

In many countries, more people ride bicycles than drive automobiles. Ask students: Why are bicycles used more in these places? *(Among other reasons, bicycles are less expensive and require less room.)* What are some advantages of using a bike instead of a car? *(Bikes cause no air or noise pollution. Also, bikes are easier to maintain.)* What are some disadvantages? *(Bikes are slower, have little storage capacity, and do not shelter riders from the weather. Also, many people are not physically able to ride a bike. Finally, some people might need to travel so far every day that riding a bike would be impractical or impossible.)*

Unbelievable Bicycles

Imagine riding a bicycle that can go more than 104 km/h! Or perhaps you'd like a bike frame that weighs only 1.4 kg. How about a bike that you could ride on snow? These aren't bikes from the future—they're being built today. Read on to find out about ways in which bicycles are being made faster, lighter, and easier to handle than ever before.

Light as a Feather

When you ride a bicycle, you are moving not only your own mass, but also the mass of the bike. The less a bike weighs, the less work you have to do.

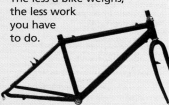

▲ **This bicycle frame was made with MMCs. It weighs only 1.4 kg.**

New, lightweight materials, called metal matrix composites, or MMCs, are now being used to make lighter bicycle frames. MMCs consist of metals that are strengthened and stiffened with nonmetal fibers. MMCs weigh the same as the metals they are made of, but because MMCs are stronger, much less material is needed. Some bicycle frames built with MMCs weigh less than 1 kg.

Through Rain and Sleet and Snow

Riding on slippery surfaces can be both difficult and dangerous. One of the problems is that power is sent only to the back wheel. If the back wheel slips or slides, you lose forward motion.

Inventor Bill Becoat has solved this problem by creating a bike in which both wheels are connected to the drive train. If one wheel slips or slides, the

▲ **This bicycle incorporates Bill Becoat's two-wheel-drive mechanism. A drive cable transmits power from the rear wheel to the front wheel.**

other wheel still receives force from the rider to keep the bike moving.

As Fast as the Wind

Want a faster bike? Maybe you should try lying down! As strange as it may seem, some of the world's fastest bikes are designed to be ridden by someone leaning back or lying down. These positions help reduce air resistance. They also allow the rider to generate more power when pedaling.

◄ **An innovative tandem bicycle**

The Ideal Bike

What's your dream bike? Create a list of at least 10 features that your ideal bike would have. Try to draw a diagram of a bike that has these features.

Did You Know...

- The wheel has been a part of technology since prehistory, but the bicycle was not invented until about 1790.
- The word *bicycle* comes from the Latin prefix *bi*, meaning "two," and the Greek word *kyklos*, meaning "circle."

Answers to
The Ideal Bike

Answers will vary greatly. Innovative features might include a sunshade for the rider or a cooling system. You can encourage creativity by having students think about the comforts offered by other vehicles, such as cars, trains, or boats.

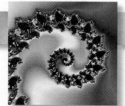

Machines as Art

*W*hat do you consider a piece of art—a painting, a sculpture, or a photograph? How about a blender, a fork, or a pencil sharpener? In fact, blenders, forks, and pencil sharpeners are on display in art museums around the world, as are countless other everyday items and machines that you might not normally think of as art. The people who create these pieces are known as industrial designers. Industrial designers work to make machine-made products, such as small appliances, tools, furniture, and vehicles, both useful and beautiful.

◄ The Waring Blender®, introduced over 50 years ago, combined form and function with an elegant design.

The Introduction of Industrial Design

Until the late 1800s, many machine-made items were simply functional. What a particular item looked like was much less important than what it did. Then, in 1919, German architect Walter Gropius founded a school of architecture and design called the Bauhaus. The Bauhaus became a center for artists and architects who combined sharp, clean, geometric styles with new industrial materials and techniques. Although the school closed in the early 1930s, it changed the way that people viewed machine-made items.

Form and Function

After the introduction of industrial design, manufacturers soon realized that more people might buy an item if it looked appealing. Companies began hiring industrial designers to create products that were functional and that had a graceful shape, color, proportion, and texture. Sometimes the results were less than appealing. For example, following World War I, many products had a number of unnecessary decorations and were odd mixes of styles. At other times, however, the design served to underscore and enhance the function of an item. Many of these latter items were pleasing enough to be

▼ The Tizio table lamp was created by Richard Sapper in 1972. It is now part of the permanent collection of The Museum of Modern Art.

considered works of art. In 1934, the Museum of Modern Art decided to exhibit its first "machine art" collection. This exhibit included objects such as glassware, gears, propellers, springs, pans, valves, irons, axes, racks, and ball bearings.

Today, industrial designers not only design useful and beautiful machines and products, but also work with specialists to ensure that their designs are appropriate for human use.

Create Your Own Machine Art

Find a simple, common machine in your house or at school. Think about ways in which you could change the machine to make it more visually pleasing without changing how it works. Draw a poster that shows the original machine as well as the new machine you designed.

209

Background

The Bauhaus designers emphasized straight lines, simple geometric shapes, and the machine-made nature of industrial objects. They also were known for their use of materials such as glass and steel in their constructions. This Bauhaus philosophy of *industrial design* (a term coined in 1919 by an American named Joseph Sinel) did not become widespread in the design industry until the 1950s.

Some more recent designers employ a principle called *product semantics*. They seek to construct objects in such a way as to trigger a free association of memories and emotions in the viewer. For example, the curves of a stereo may hint at the curves of a musical instrument.

Create Your Own Machine Art

Encourage students to keep their object designs simple. After they have finished their drawings, you may want to show them some pictures of objects made in different styles, such as Art Deco, Art Nouveau, or Rococo. Ask students to observe any similarities between these designs and their own.

Homework

Have students bring to class an object that exhibits an interesting design, such as an old toaster or coffeepot. You might encourage them to ask their parents or grandparents for help in finding unusual machines in household storage areas or even at garage sales. Compare and contrast the different features found in the objects as a group. Be sure to tell students not to bring any object of value to class.

Unit 4 · OCEANS AND CLIMATES

Bibliography for Teachers
DeMillo, Rob. *How Weather Works.* Emeryville, CA: Ziff-Davis Press, 1994.

Wagner, Ronald L., and Bill Adler, Jr. *The Weather Source Book.* Old Saybrook, CT: The Globe Pequot Press, 1994.

Bibliography for Students
Mandell, Muriel. *Simple Weather Experiments With Everyday Materials.* New York City, NY: Sterling Publishing Co., Inc., 1991.
Planet Earth. Alexandria, VA: Time-Life Books, 1992.
Simon, Seymour. *Oceans.* New York, NY: Morrow Publishing Company, Inc., 1990.

Films, Videotapes, Software, and Other Media
The Active Atmosphere Video Series
    Videotape
    Focus Media, Inc.
    P.O. Box 865
    Garden City, NY 11530
Earth's Atmosphere
    Software (Apple, Macintosh, MS-DOS)
    Queue, Inc.
    338 Commerce Dr.
    Fairfield, CT 06430
Oceans and Climate
    Videotape
    Coronet/MTI
    108 Wilmot Rd.
    Deerfield, IL 60015
Planet Earth: The Blue Planet
    Videodisc
    Coronet/MTI
    108 Wilmot Rd.
    Deerfield, IL 60015
Wind and Air Currents
    Film, videotape, or videodisc
    Coronet/MTI
    108 Wilmot Rd.
    Deerfield, IL 60015

Unit Overview

This unit introduces students to oceans and climates and examines the relationship between them. In Chapter 10, students investigate the climates of the inner planets of our solar system, including Earth. Then students explore the evidence for and against global warming and the greenhouse effect. Students also investigate humidity, dew point, and how the differential heating of land and water and the water cycle affect climate. In Chapter 11, students tour the ocean bottom and visit a number of its geological features. They explore deep-ocean currents and the concept of density. The concept of density is then applied to land and sea breezes, rising air over the tropics, and air masses. The chapter concludes with an introduction to weather maps and the distinction between climate and weather. In Chapter 12, students examine the concept of pressure as it relates to both oceans and the atmosphere. Then they discover how the Coriolis effect causes the direction of global winds and ocean currents. The unit closes with an exploration of hurricanes.

Using the Themes

The unifying themes emphasized in this unit are **Energy, Structures, Cycles,** and **Changes Over Time.** The following information will help you incorporate these themes into your teaching plan. Focus questions appear in the margin of this Annotated Teacher's Edition on pages 215, 219, 224, 229, 241, 253, and 255.

When learning about the role of the atmosphere in maintaining a relatively narrow range of temperatures on the Earth's surface, students are asked to consider the theme of **Energy.** For example, in Chapter 10, students explore how atmospheric gases prevent heat energy from escaping into space.

Students encounter the theme of **Structures** in Chapter 11 as they examine the processes that are responsible for the geographical features of the ocean floor.

The theme of **Cycles** plays an important role in this unit as well. In Chapter 10, students investigate the role of the water cycle in determining surface temperatures. Chapter 11 challenges students to explain how the Earth's tilt on its axis causes seasonal changes in weather conditions.

Finally, the theme of **Changes Over Time** is emphasized in Chapter 10 in terms of how the increased burning of fossil fuels may be contributing to global warming. It is also addressed in Chapter 11 when students explore the way meteorologists use records of previous weather conditions to predict future conditions.

Using the SourceBook

In Unit 4, students investigate the evolution of the atmosphere. They also learn about the various layers that make up the atmosphere. The SourceBook also discusses additional factors that cause weather changes. The oceans are identified as a supplier of food, minerals, and energy. The responsibility that humanity shares for the protection of the oceans is also highlighted.

Unit Organizer

Unit/Chapter	Lesson	Time*	Objectives	Teaching Resources
Unit Opener, p. 210				Science Sleuths: The Mystery Fog English/Spanish Audiocassettes Home Connection, p. 1
Chapter 10, p. 212	**Lesson 1, An Oasis in Space,** p. 213	1	1. Compare and contrast Earth with nearby planets. 2. Describe how specific factors affect the Earth's temperature.	Image and Activity Bank 10-1 Exploration Worksheet, p. 3
	Lesson 2, The Greenhouse Effect, p. 217	2	1. Describe the greenhouse effect, its causes, and its possible consequences. 2. Identify the greenhouse gases and their sources.	Transparency 30 Exploration Worksheet, p. 6 ▼ Theme Worksheet, p. 12
	Lesson 3, Land and Sea, p. 226	3 to 4	1. Explain why some regions of the world are warmer or colder than others. 2. Explain how heat is absorbed and released in the water cycle. 3. Calculate dew point given thermometers and dew-point tables. 4. Describe the effect of altitude on temperature.	Image and Activity Bank 10-3 Transparency 33 Transparency Worksheet, p. 14 ▼ Exploration Worksheet, p. 16 Exploration Worksheet, p. 19 Exploration Worksheet, p. 21
End of Chapter, p. 235				Activity Worksheet, p. 23 Chapter 10 Review Worksheet, p. 24 Chapter 10 Assessment Worksheet, p. 28
Chapter 11, p. 237	**Lesson 1, Hidden Wonders,** p. 238	1	1. Describe several geologic features of the ocean bottom. 2. Describe the adaptation of living organisms to ocean ecosystems. 3. Relate global warming to processes occurring in the ocean.	Image and Activity Bank 11-1 Transparency Worksheet, p. 31 ▼
	Lesson 2, The Moving Oceans, p. 243	4 to 5	1. Explain Count Marsili's solution to the Mediterranean puzzle. 2. Calculate the density of a substance using a metric balance and a graduated cylinder. 3. Explain why ocean water varies in density. 4. Describe the causes and effects of ocean currents.	Image and Activity Bank 11-2 Exploration Worksheet, p. 33 Exploration Worksheet, p. 35 Exploration Worksheet, p. 38 Transparency 36 Exploration Worksheet, p. 39
	Lesson 3, The Atmosphere p. 251	2 to 3	1. Describe the relationship between the density currents that occur in air and in water. 2. Describe land and sea breezes and the trade winds in terms of the unequal heating of air that causes density differences. 3. Explain how different air masses form. 4. Describe the characteristics of the air in a given air mass. 5. Interpret weather maps in terms of moving fronts.	Image and Activity Bank 11-3 Transparency Worksheet, p. 40 ▼ Activity Worksheet, p. 42 Transparency 38
End of Chapter, p. 257				Activity Worksheet, p. 43 ▼ Chapter 11 Review Worksheet, p. 44 Chapter 11 Assessment Worksheet, p. 47
Chapter 12, p. 259	**Lesson 1, Pressure Differences,** p. 260	3 to 4	1. Explain the relationship between pressure and altitude in the atmosphere. 2. Explain the relationship between pressure and depth in the ocean. 3. Calculate pressure using the weight of an object and the area on which the force is exerted.	Image and Activity Bank 12-1 Discrepant Event Worksheet, p. 50 Exploration Worksheet, p. 51 Activity Worksheet, p. 55
	Lesson 2, The Direction of Flow, p. 269	3	1. Describe the direction in which moving objects will be deflected in the Northern and Southern Hemispheres. 2. Describe the locations and directions of the major ocean currents and winds. 3. Discuss the causes and effects of El Niño and the Gulf Stream.	Image and Activity Bank 12-2 Transparency Worksheet, p. 56 ▼ Activity Worksheet, p. 58 Transparency 41 Exploration Worksheet, p. 60 Transparency 42
	Lesson 3, Hurricanes, p. 277	1	1. Describe the causes and characteristics of hurricanes. 2. Interpret isobars on a weather map. 3. Predict wind direction and air pressure from weather maps. 4. Explain why winds blow in certain patterns in high- and low-pressure systems.	Image and Activity Bank 12-3 Transparency 43 Graphing Practice Worksheet, p. 62
End of Chapter, p. 280				Chapter 12 Review Worksheet, p. 64 ▼ Chapter 12 Assessment Worksheet, p. 68
End of Unit, p. 282				Unit 4 Activity Worksheet, p. 71 Unit 4 Review Worksheet, p. 72 Unit 4 End-of-Unit Assessment, p. 76 Unit 4 Activity Assessment, p. 82 Unit 4 Self-Evaluation of Achievement, p. 85

* Estimated time is given in number of 50-minute class periods. Actual time may vary depending on period length and individual class characteristics.
▼ Transparencies are available to accompany these worksheets. Please refer to the Teaching Transparencies Cross-Reference chart in the Unit 4 Teaching Resources booklet.

Materials Organizer

Chapter	Page	Activity and Materials per Student Group
10	216	**Exploration 1:** 2 alcohol thermometers, small enough to fit inside a large jar; 2 large glass jars with lids; 2 pieces of transparent tape; lamp with at least a 100 W light bulb (if the weather is not sunny); thin strip of cardboard, about 3 cm × 25 cm
	227	***Exploration 3, Activity 1:** 2 test tubes; test-tube tongs or test-tube clamp; 2 alcohol thermometers; 250 mL beaker; hot plate; test-tube rack; watch or clock with second hand; two small balls of modeling clay; about 10 mL of water; about 10 mL of sand; 100 mL graduated cylinder; safety goggles; **Activity 2:** 2 aluminum pie pans; lamp with at least a 100 W light bulb; about 250 mL of water; about 250 mL of sand; 2 alcohol thermometers; 100 mL graduated cylinder; metric ruler; watch or clock with second hand; optional items: materials to prop up thermometers, such as 2 small balls of modeling clay; **Activity 3:** 2 aluminum pie pans; about 500 mL of sand; 100 mL graduated cylinder; lamp with at least a 100 W light bulb; about 25 mL of charcoal powder; 2 alcohol thermometers; watch or clock with second hand; **Activity 4:** 100 mL graduated cylinder; about 500 mL of sand; 2 aluminum pie pans; about 100 mL of water; 2 alcohol thermometers; lamp with at least a 100 W light bulb; watch or clock with second hand
	229	**Exploration 4, Part 1:** alcohol thermometer; 5 cm piece of shoelace; piece of masking tape; small container of room-temperature water; optional item: small electric fan; **Part 2:** small container of warm water; about 20 mL of rubbing alcohol; safety goggles
	230	**A Mini-Activity in a Bottle:** large, plastic soft-drink bottle; 1 L of water; a few matches; safety goggles
	232	**Exploration 5, Part 1:** 250 mL metal can; several ice cubes; alcohol thermometer; stirring rod; about 50 mL of room-temperature water; **Part 2:** alcohol thermometer; wet-bulb thermometer from Exploration 4, Part 1; small container of water at or above room temperature
	236	**Challenge Your Thinking, question 5:** materials to record dew point from Exploration 5, Parts 1 or 2, on page 232
11	242	**Pictures of the Deep:** poster board; colored markers; construction paper
	244	**Exploration 1:** 150 mL of salt; 5 cm deep aluminum roasting pan (23 cm × 36 cm); 100 mL graduated cylinder; stirring rod; watch or clock; scissors; piece of aluminum foil slightly larger than the roasting pan; about 25 cm of masking tape; a few drops of food coloring; a small amount of pepper; 2 containers, each 1 L or larger; 2 L of water; lab aprons (See Advance Preparation below.)
	245	***Exploration 2, Part 1:** pill bottle or similar small container; about 100 mL of water; 100 mL of each salt solution from Exploration 1; metric balance; 50 or 100 mL graduated cylinder; **Part 2:** See Advance Preparation below; **Part 3:** solutions from Exploration 1; 3 test tubes; 3 pieces of masking tape; permanent marker; a few drops each of 2 different colors of food coloring (both different from the color used in Exploration 1); glass tubing (See Advance Preparation below.) or clear plastic drinking straws; test-tube rack; small beaker or other container; lab aprons
	246	**Exploration 3:** about 1 L of 10% salt solution; 100 mL of cold tap water; 100 mL of hot tap water; few drops each of 2 different colors of food coloring; 2 small, clear jars; large pan or bucket; 4 index cards or stiff pieces of paper; watch or clock with second hand; lab aprons (See Advance Preparation below.)
	250	**Exploration 4:** a variety of materials such as a pencil, modeling clay, plastic drinking straw, cork, thumbtacks, and heavy gauge wire; 100 mL graduated cylinder; 600 mL of water; 12.5 g of salt; permanent marker (See Advance Preparation below.)
12	261	**Exploration 1, Part 2:** brick; 1 m of string; metric spring scale or force meter; metric ruler
	263	***Exploration 2, Activity 1:** large can with a hole punched in it close to the bottom; metric ruler; meter stick; tripod stand; stream table or large pan (about 120 cm × 35 cm) with drain; about 1 L of water (additional teacher materials: hammer; nail; see Advance Preparation below.); **Activity 2:** 2 cans of different diameters, each with a hole punched close to the bottom; meter stick; tripod stand; stream table or large pan (about 120 cm × 35 cm) with drain; about 1 L of water; 100 mL graduated cylinder (additional teacher materials: hammer and nail; see Advance Preparation below.); **Activity 3:** plastic bottle with cap; enough water to fill bottle; small piece of aluminum foil (about 4 cm × 4 cm); watch or clock with second hand
	265	**Try It Yourself:** aluminum can; about 1 L of cold water; hot plate; large container; safety goggles; oven mitts
	266	**Try It Yourself:** large coffee can, open at one end; about 50 cm of masking tape; metric ruler; straight pin; 2 plastic drinking straws; balloon
	267	**Try It Yourself:** index card; drinking glass or plastic cup; enough water to fill the glass or cup; large container
	272	**Exploration 3, Part 1:** ¼ sheet of poster board; **Part 2:** ¼ sheet of poster board from Part 1; scissors; enough baking soda to fill a salt shaker; empty salt shaker; large ball bearing or marble; pushpin or thumbtack; piece of wall paneling or slab of plastic foam larger than the piece of poster board; a few milliliters of water

* You may wish to set up these activities in stations at different locations around the classroom.

Advance Preparation

Exploration 1, page 244: Save 100 mL of each solution for Exploration 2.

Exploration 2, Part 2, page 245: Students will need to design their own experiments.

Exploration 2, Part 3, page 245: If you use glass tubing, make sure that it has been fire-polished to remove any jagged edges.

Exploration 3, page 246: You will need a 10 mL graduated cylinder, salt, water, large beaker or other container, and a stirring rod to make the 10% salt solution prior to the Exploration. Do so by adding 1 mL of salt into the beaker for every 10 mL of water. Stir thoroughly until all the salt is dissolved.

Exploration 4, page 250: You will need a 100 mL graduated cylinder, salt, water, large beaker or other container, balance, and a stirring rod to make the "standard" salt solution prior to the Exploration. Do so by adding 12.5 g of salt into the beaker for every 250 mL of water. Stir thoroughly until all the salt is dissolved. Use either recipe on page 249 to make the artificial sea water.

Exploration 2, page 263: For Activities 1 and 2, use the hammer and nail to punch a hole in each can prior to the Exploration. Cover any jagged edges with masking tape.

Unit Compression

Because the concepts of this unit are interrelated, compression must be done carefully and only if necessary. However, some sections can be omitted without sacrificing development of the most important concepts. The sections that could be omitted include the following:

- Chapter 10, Lesson 2, The Greenhouse Effect
- The Composition and Density of Sea Water in Chapter 11, Lesson 2
- A Climate Sampler in Chapter 11, Lesson 3
- Chapter 12, Lesson 3, Hurricanes

The material that could be omitted is not integral to concept development, but it is highly interesting, and it provides a Science, Technology, and Society focus. Therefore, students should be exposed to as much of this material as possible.

Homework Options

Chapter 10
See Teacher's Edition margin, pp. 215, 218, 221, 224, and 225
Activity Worksheet, p. 23
SourceBook, pp. S60, S66, and S69

Chapter 11
See Teacher's Edition margin, pp. 229, 231, 232, 241, 249, and 253
Activity Worksheet, p. 42
Activity Worksheet, p. 43
SourceBook, pp. S69, S72, and S73

Chapter 12
See Teacher's Edition margin, pp. 260, 262, 264, 271, and 279
Activity Worksheet, p. 55
Activity Worksheet, p. 58
Graphing Practice Worksheet, p. 62
SourceBook, pp. S63, S68, and S74

Unit 4
Unit 4 Activity Worksheet, p. 71
SourceBook Activity Worksheet, p. 86

Assessment Planning Guide

Lesson, Chapter, and Unit Assessment	SourceBook Assessment	Ongoing and Activity Assessment	Portfolio and Student-Centered Assessment
Lesson Assessment Follow-Up: see Teacher's Edition margin, pp. 216, 225, 234, 242, 250, 256, 268, 276, and 279 **Chapter Assessment** Chapter 10 Review Worksheet, p. 24 Chapter 10 Assessment Worksheet, p. 28* Chapter 11 Review Worksheet, p. 44 Chapter 11 Assessment Worksheet, p. 47* Chapter 12 Review Worksheet, p. 64 Chapter 12 Assessment Worksheet, p. 68* **Unit Assessment** Unit 4 Review Worksheet, p. 72 End-of-Unit Assessment Worksheet, p. 76*	SourceBook Review Worksheet, p. 87 SourceBook Assessment Worksheet, p. 91*	Activity Assessment Worksheet, p. 82* **SnackDisc** Ongoing Assessment Checklists ♦ Teacher Evaluation Checklists ♦ Progress Reports ♦	Portfolio: see Teacher's Edition margin, pp. 219, 241, and 268 **SnackDisc** Self-Evaluation Checklists ♦ Peer Evaluation Checklists ♦ Group Evaluation Checklists ♦ Portfolio Evaluation Checklists ♦

* Also available on the Test Generator software
♦ Also available in the Assessment Checklists and Rubrics booklet

Science Discovery is a versatile videodisc program that provides a vast array of photos, graphics, motion sequences, and activities for you to introduce into your *SciencePlus* classroom. *Science Discovery* consists of two videodiscs: Science Sleuths and the Image and Activity Bank.

Using the *Science Discovery* Videodiscs

Science Sleuths: The Mystery Fog
Side B

For about 60 years, the coast near Point Pleasant has been under the influence of a current similar to El Niño. The warm current has recently shifted south, leaving the water at Point Pleasant at a temperature far cooler than it has been in more than 60 years. The wet, oceanic air creates heavy fog when it passes over this cool water.

Interviews
1. Setting the scene: Coordinator of tourist bureau **27146 (play ×2)**

2. Coordinator of Earth! Earth! **28014 (play)**

3. Fisherman **28779 (play)**

4. Mussel farmer **29099 (play)**

5. Atmospheric scientist **29820 (play)**

6. Reporter **30670 (play)**

Documents
7. Environmental group's leaflet **31320 (step)**

8. Letter from power company **31323 (step ×2)**

Literature Search
9. Search on the words: FOG, POINT PLEASANT, POWER PLANT, RADIATION, STEAM VENTS, TOURISM **31327 (step)**

10. Article #1 ("Power Plant Causes Fish Kill in Stuben Lake") **31330 (step)**

11. Article #2 ("London Fog") **31333**

12. Article #3 ("Point Pleasant: Fogged in for Good?") **31335 (step)**

13. Article #4 ("Ocean Currents Run Deep") **31338 (step)**

Sleuth Information Service
14. Map of area **31341**

15. Oceanographic map of bay **31343**

16. Weather map **31345**

17. Temperature maps **31347 (step ×4)**

18. Dew-point chart **31353**

19. Coal-fired power plant **31355**

20. El Niño **31357 (step)**

21. Fog **31360 (step ×2)**

22. Radiation fog **31364 (step)**

Sleuth Lab Tests
23. Simulation with warm air and cool water **31458**

24. Simulation of condensation nuclei **31460 (play)**

Still Photographs
25. Fog in the bay **31641**

26. Fog near power plant **31643**

27. The power plant **31645**

28. Power plant smokestacks **31647**

29. Infrared satellite image of bay **31649 (step ×3)**

Image and Activity Bank
Side A and B

A selection of still images, short videos, and activities is available for you to use as you teach this unit. For a larger selection and detailed instructions, see the Videodisc Resources booklet included with the Teaching Resources materials.

10-1 An Oasis in Space, page 213
Mercury 1580
Mercury is the planet closest to the sun in our solar system.

Venus 1582
Full view

◀| Step Reverse   Play ▶   Pause || Step Forward |▶

Earth 1587
Full view of Earth

Mars 1592
View of the south pole of Mars

Deforestation 24472–24950 (play ×2) (Side B only)
It is estimated that most tropical forests will be gone by the year 2000. This will result in the extinction of millions of species and will contribute to global warming.

Automobiles on freeway 2609
Collectively, automobiles make a big impact on the Earth's atmosphere as they release pollutants and greenhouse gases into the air.

10-3 Land and Sea, page 226
Heat of condensation 1529
As a parcel of air rises, it expands and cools. The moisture in the parcel will condense when it cools to the dew point at that level. As moisture condenses, energy is released as heat into the atmosphere.

Fog, upslope 1477
Fog can form when moisture-laden air moves up a slope and condenses as it cools.

11-1 Hidden Wonders, page 238
Volcanic eruption; Hawaii 952
The streaking arcs are caused by small incandescent fragments (lava bombs).

Volcano, shield; Mauna Loa, Hawaii 963
Mauna Loa fills most of this picture. The crater of the volcano Kilauea is on Mauna Loa's flank in the lower left corner of the image.

Ocean floor; map 1514–1518 (step ×4)
Fracture zones and areas of crustal spreading on the ocean floor (step) Tentative boundaries (step) Ocean trenches (step) Active volcanoes (step) Suggested plate motion

Pillow lava forming underwater 19447–19680 (play ×2) (Side B only)
Pillow lava forms when lava pours into water. The surface of the lava forms a thin shell when cooled by water. The shell fractures, and more lava pours out.

Vents, sea 24171–24381 (play ×2) (Side B only)
Life is scarce at this depth, except near sea vents. Radioactive materials inside the Earth break down to create new oceanic crust and release heat. Sea water reacts with crustal rock and forms hydrogen sulfide.

11-2 The Moving Oceans, page 243
Sea water; composition 1494
The ion concentration of sea water

Buoyancy; egg in solutions 48191–48429 (play ×2) (Side A only)
The egg on the right sinks in water because it is denser than the water. The egg on the left is less dense than the salt solution, so the egg floats.

11-3 The Atmosphere, page 251
Sea breeze 1460
Cool sea breezes blow onshore when sunlight heats the land. As the warm air over land rises, cool sea air takes its place.

Air convection cell 1544
A convection cell is set up to maintain an air balance. This example illustrates how a sea breeze is created.

Front, cold 1538
Cold fronts often bring rain and thunderstorms, followed by cooler and drier air.

Front, warm 1537
Warm fronts often bring continuous, gentle rains with warmer air behind them.

12-1 Pressure Differences, page 260
Cartesian diver; effect of pressure on buoyancy 46299–46757 (play ×2) (Side A only)
Apply pressure to the top of the tube, and the air bubble trapped in the glass diver is also compressed. How does this change the density of the diver?

Air-pressure demonstration; glass filled with water is inverted with a card 12813–13126 (play ×2) (Side A only)
When the cup is covered with a card, air pressure keeps the card pressed against the water. Surface tension prevents the water from seeping out.

Air-pressure demonstration; breaking a ruler 13127–13359 (play ×2) (Side A only)
Air pressure on the paper provides enough weight to resist the applied force and to break the ruler.

12-2 The Direction of Flow, page 269
Wind belts; map 1531
Global map of the wind belts and general location of high- and low-pressure zones

Ocean surface currents 10325–10505 (play ×2) (Side A only)
Currents at the ocean's surface are generated by wind constantly blowing in the same direction. These are called wind-driven currents.

Currents, Atlantic Ocean 1507
Map of the major currents in the Atlantic Ocean

El Niño 1512 (step)
This image shows a normal year. (step) El Niño is caused by a shift in the trade winds that redistributes the warm surface water of the Pacific Ocean.

12-3 Hurricanes, page 277
Typhoon Odessa 1481
Typhoons are intense tropical storms in the western Pacific Ocean. These storms are called hurricanes in the Northern Atlantic Ocean and cyclones in the Indian Ocean.

Hurricane 1533
At the eye of a hurricane, there is downward movement of air, and the winds are calm. The fastest winds occur just outside the eye. The intensity of the storm decreases with height.

Hurricane 15297–15597 (play ×2) (Side A only)
This satellite image shows Hurricane Hugo as it hits Charleston, South Carolina. The red parts indicate where the storm is most intense.

High- and low-pressure systems; maps 1539 (step)
Average locations of high- and low-pressure systems by latitude (step) Isobars

 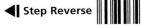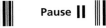

Unit 4 · OCEANS AND CLIMATES

Ask students to name the components that make up the weather. *(Temperature, humidity, precipitation, wind)* Then ask students if they know what causes the weather. Involve them in a discussion of their ideas. Do not attempt to correct them at this point. Tell students that in this unit they will learn about what causes the weather, what causes it to change, and the connection between the atmosphere and ocean currents.

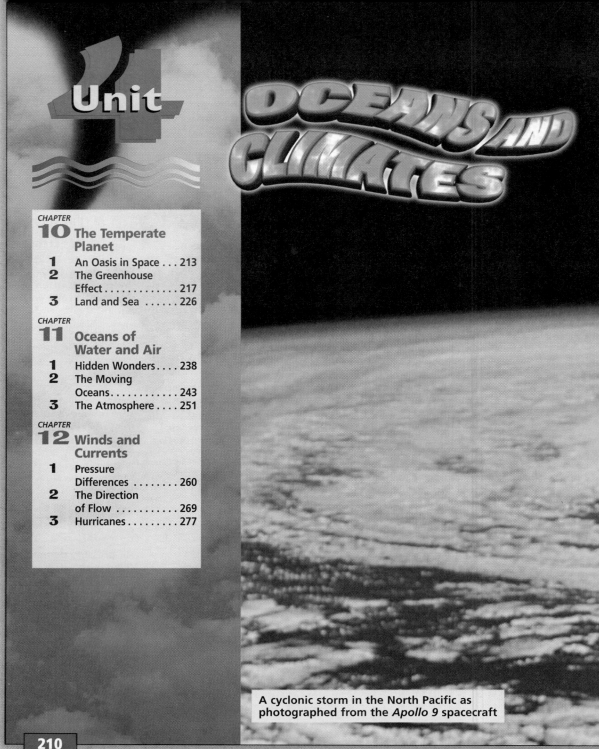

Unit 4 OCEANS AND CLIMATES

CHAPTER
10 The Temperate Planet
1   An Oasis in Space . . . 213
2   The Greenhouse Effect 217
3   Land and Sea 226

CHAPTER
11 Oceans of Water and Air
1   Hidden Wonders 238
2   The Moving Oceans 243
3   The Atmosphere 251

CHAPTER
12 Winds and Currents
1   Pressure Differences 260
2   The Direction of Flow 269
3   Hurricanes 277

A cyclonic storm in the North Pacific as photographed from the *Apollo 9* spacecraft

210

Connecting to Other Units

This table will help you integrate topics covered in this unit with topics covered in other units.

Unit 1 Life Processes	The water cycle is essential for all life processes.
Unit 2 Particles	The relative densities of air and water particles determine the flow of atmospheric and oceanic currents.
Unit 3 Machines, Work, and Energy	Atmospheric and water pressure are discussed in terms of force per unit area.
Unit 8 Continuity of Life	The oceans and the atmosphere make Earth the only planet that can sustain life in this solar system.

giant cyclonic storm rumbles westward across the North Pacific. Driven by the tremendous heat stored in the ocean, cyclonic storms transfer thermal energy from the ocean to the atmosphere on an enormous scale. As the storm sweeps across the North Pacific, it will continue to pick up energy from the warm ocean water. At its peak, the storm may develop winds in excess of 160 km/h and generate huge waves that can travel thousands of kilometers. In the wintertime, when there is a sharp temperature contrast between ocean and air, storms may form one right after another over the North Pacific.

Like hurricanes, which originate in the tropics, cyclonic storms have a distinctive spiral shape. In the Northern Hemisphere, storms spiral counterclockwise. In the Southern Hemisphere, they spiral clockwise. What could cause such a phenomenon? Ⓐ

Beyond the curve of the horizon looms the black emptiness of space. A more startling contrast would be hard to imagine: the living, ever-changing planet below, and the hostile, empty void above. Out of view 150 million kilometers away is the sun, the source of energy that drives the Earth's weather "engine" and makes life on Earth possible. In this unit you will learn about the ways in which the oceans and atmosphere interact to make Earth a unique oasis of life in the void of space.

211

Using the Photograph

Call on a student to read aloud the text on this page. Discuss the answer to the question posed in the second paragraph (see answer below). Choose a volunteer to write student responses to the following question on the board: How does the atmosphere affect life on Earth? *(The atmosphere affects life in many ways: living organisms need oxygen and carbon dioxide for survival; the process of decay requires atmospheric gases; the layers of the atmosphere protect living organisms from harmful radiation and burn up potentially dangerous meteorites as they approach the Earth; the atmosphere helps stabilize global temperatures; atmospheric currents create ocean waves and produce weather that affects most life-forms; air masses interact to produce storms that can benefit or harm life on Earth.)* You may wish to ask students if they have ever seen a tornado. What was the weather like before the tornado formed? How did the weather change as the storm struck? *(Accept all reasonable responses. Students familiar with tornadoes may recall that such storms generally strike during the summer when cool dry winds from the north meet warm, humid air. Dark thunderclouds form and the temperature drops noticeably.)*

Answer to
In-Text Question

Ⓐ Accept all reasonable answers at this point. Tell students that in this unit they will learn about the Coriolis effect, which is caused by the Earth's constant turning on its axis in the same direction, or rotation. When viewed from above the North Pole, the Earth rotates counterclockwise. When viewed from above the South Pole, the Earth rotates clockwise. The Coriolis effect deflects motion to the right in the Northern Hemisphere and to the left in the Southern Hemisphere.

Connecting to Other Chapters

> **Chapter 10**
> offers students the chance to explore the factors that affect the Earth's climate.

> **Chapter 11**
> examines the way density affects the flow of currents in the ocean and in the atmosphere.

> **Chapter 12**
> explores the concept of pressure as it relates to the atmosphere and to the ocean.

Prior Knowledge and Misconceptions

Your students' responses to the ScienceLog questions on this page will reveal the kind of information—and misinformation—they bring to this chapter. Use what you find out about your students' knowledge to choose which chapter concepts and activities to emphasize in your teaching. After students complete the material in this chapter, they will be asked to revise their answers based on what they have learned. Sample revised answers appear on page 236.

In addition to having students answer the questions on this page, you may wish to have them complete the following activity: Have students make two sketches—one of the solar system and one of Earth and its atmosphere. Students should have a basic understanding of the relationship of Earth to the sun, moon, and other planets before beginning the chapter. They should also know that the atmosphere is a layer of gases surrounding Earth. Collect the sketches, but do not grade them. Instead, use them to find out what students know about Earth's atmosphere and what aspects of this topic are interesting to them.

CHAPTER
10
The Temperate Planet

Mars

Earth

Mercury

1 Why is our planet probably the only place in the solar system where life exists?

How is the Earth's **2** atmosphere like a greenhouse?

3 Why does dew form on grass and other objects during the night?

ScienceLog

Think about these questions for a moment, and answer them in your ScienceLog. When you've finished this chapter, you'll have the opportunity to revise your answers based on what you've learned.

212

ENVIRONMENTAL FOCUS

Point out that Earth is the only planet in the solar system with an atmosphere that contains free oxygen. Most of this oxygen has been generated by plants through photosynthesis. During photosynthesis, plants take carbon dioxide from the atmosphere and convert it into food and oxygen. The clearing of tropical rain forests has contributed to the increase of carbon dioxide in the atmosphere, which some scientists suggest is contributing to global warming.

An Oasis in Space

You know that the Earth has many different climates—some cold, some warm, some in between. But even when you consider the extremes of its hottest and coldest places, Earth is still a very pleasant place compared with the other planets and satellites in our solar system.

Take, for example, the moon—our nearest neighbor. The moon's surface temperature soars to above 130°C during the day and plummets to below –170°C at night. Does the temperature in Eureka, California, change that much from day to night? over the whole year? How about Lincoln, Nebraska? How about where you live? Do you think any place on Earth has such a wide range of temperatures? What would explain the very large range of temperatures that the moon experiences? Why does Earth experience a more favorable climate? **A**

Earth is unique. It has an average temperature that allows living things to flourish. Why is this so? Is Earth fortunate enough to be just the right distance from the sun, or are there other factors that contribute to its temperate climate? **B**

Examine the data regarding Earth and its closest neighbors on the pages that follow. Compare these planets. How do their average temperatures vary? What factors seem to determine the temperature ranges of the planets?

One group of students made the suggestions that follow. Can you find data to support each claim? **C**

"Distance from the sun is definitely the most important factor in determining how hot or cold a planet is."

"Another factor is whether or not the planet has an atmosphere."

"I think the kinds of gases that make up the atmosphere affect a planet's temperature."

"Another important thing is how dense or thick a planet's atmosphere is."

What other factors might affect the climate of a planet? Check out the astrocorder readings on pages 214 and 215 for some clues. **D**

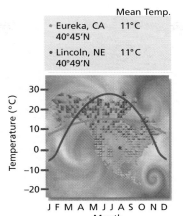

Mean Temp.

- Eureka, CA 11°C
  40°45′N
- Lincoln, NE 11°C
  40°49′N

J F M A M J J A S O N D
Month

Despite the extreme conditions, an afternoon in the Sahara followed by an evening in Antarctica would be pleasant compared with a day and night on the moon.

213

LESSON 1 ORGANIZER

Time Required
1 class period

Process Skills
comparing, contrasting, inferring

Theme Connection
Energy

New Terms
none

Materials (per student group)
Exploration 1: 2 alcohol thermometers, small enough to fit inside a large jar; 2 large glass jars with lids; 2 pieces of transparent tape; lamp with at least a 100 W light bulb (if the weather is not sunny); thin strip of cardboard, about 3 cm × 25 cm

Teaching Resources
Exploration Worksheet, p. 3
SourceBook, p. S60

An Oasis in Space

FOCUS

Getting Started
Write these figures on the board: 58°C and –88°C. Tell students that these are the highest and lowest temperatures ever recorded on Earth. Have them speculate where each temperature was taken. *(The highest temperature was recorded in Al Aziziyah, Libya, in Northern Africa. The lowest temperature was recorded at the Soviet Antarctic station called Vostok.)* Point out that in this lesson students will learn more about the unique features of Earth's climate.

Main Ideas
1. Earth's unique climate allows life as we know it to flourish.
2. Many factors contribute to the temperature of the Earth.

TEACHING STRATEGIES

Answers to
In-Text Questions

A Accept all reasonable responses at this point. Students may understand that one would not find such a range of temperatures on Earth. By looking at the graph, students should realize that although the temperatures in Lincoln, Nebraska, change more over the year than temperatures in Eureka, California, neither location experiences a range of temperatures as large as the moon does in 24 hours. Students may also realize that no place on Earth has this wide a range of temperatures. Some students may understand that the moon has such a dramatic temperature change because there is no atmosphere. The Earth's atmosphere stabilizes the Earth's temperatures by reflecting some heat and trapping some heat.

Answers to In-Text Questions continued ▶

B Many factors determine the Earth's climate.

C Answers will vary.

Distance from the sun is not the most important factor in determining how hot or cold a planet is. The presence of an atmosphere does affect the temperature of a planet. Mercury, which is the only planet in our solar system without an atmosphere, has a wide fluctuation between daytime and nighttime temperatures.

Venus, which has the hottest average temperature, has much more carbon dioxide and a denser atmosphere than does Earth. Mars, despite having an atmosphere that is 95 percent carbon dioxide, is much cooler than Earth because its atmosphere is very thin. (Mars's atmospheric pressure is about 70 times less than that of Earth.)

D The presence of water and living organisms are other factors that may affect a planet's temperature.

Mercury

Be sure students note that Mercury has no significant atmosphere. Students should also note that even though Venus is about twice as far from the sun as Mercury is, the average surface temperature of Venus is higher than that of Mercury.

Mercury has a reflective surface covered with craters, somewhat like that of the moon. Ask: Why does Mercury have so many craters? *(Because meteorites are not burned up by an atmosphere before reaching Mercury's surface)*

Venus

Interested students could find out more about the history of investigating Venus and its atmosphere. *(It is the closest planet to Earth, and it has been explored by the greatest number of spacecraft. Yet it remains one of the most mysterious planets in the solar system. It is a difficult planet to explore because of its harsh climate. The early spacecraft that were sent to Venus were crushed by the planet's atmosphere before they had even landed.*

Research that was initiated in the early 1960s is beginning to find answers to the puzzle of Venus's past and present. The spacecraft Magellan has pro-

Although such devices don't actually exist, these astrocorders provide a quick guide to the four inner planets of the solar system.

MERCURY

Average distance from sun: 58 million km

Average surface temperature: 450°C, day; –173°C, night

Atmosphere: none

Atmospheric pressure: N/A

Water: none

Life: none

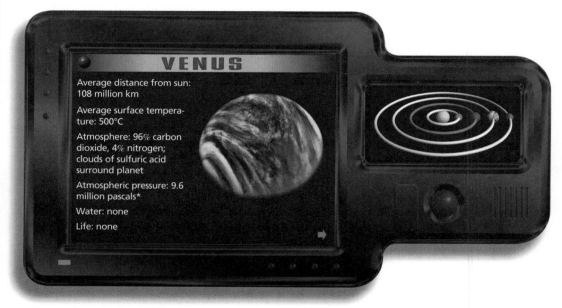

VENUS

Average distance from sun: 108 million km

Average surface temperature: 500°C

Atmosphere: 96% carbon dioxide, 4% nitrogen; clouds of sulfuric acid surround planet

Atmospheric pressure: 9.6 million pascals*

Water: none

Life: none

*Newtons per square meter, a unit of pressure

vided detailed radar images of the planet. These images reveal a new and different world, one which includes craters that look like giant pancakes and crisscrossing ridges and grooves that look like spiders.)

Meeting Individual Needs

Learners Having Difficulty
To be sure that students understand the effect of atmospheric pressure on the temperature of a planet, ask: What accounts for the difference in the aver-

age surface temperature of Mars and Venus? *(Both atmospheres are mostly carbon dioxide—95 percent for Mars and 96 percent for Venus. However, the atmosphere of Venus is much more dense. It has an atmospheric pressure of 9.6 million pascals, while the atmospheric pressure of Mars is only 1500 pascals. It is this difference in density that accounts for the difference in the average surface temperature of the two planets.)*

EARTH

Average distance from sun:
150 million km

Average surface temperature: 15°C

Atmosphere: 77% nitrogen,
21% oxygen, 1% argon,
0%–4% water vapor (variable),
0.03% carbon dioxide; traces
of other gases

Atmospheric pressure:
101,000 pascals

Water: abundant; large oceans,
polar icecaps

Life: extremely abundant and varied

MARS

Average distance from sun:
228 million km

Average surface temperature:
–55°C

Atmosphere: 95% carbon
dioxide, 3% nitrogen

Atmospheric pressure: 1500
pascals

Water: some contained in
polar icecaps and subsurface
permafrost

Life: none detected

Did You Know...

Pluto is nearly 40 times farther from the sun than is Earth. Astronomers, therefore, know very little about Pluto's surface conditions. However, scientists estimate that the temperature may be about –230°C. The planet appears to have a thin atmosphere composed mostly of methane.

Homework
Have students list 10 distinct features of Earth that make it suitable for life.

Meeting Individual Needs

Second-Language Learners
Have students create bilingual astrocorder readings of the planets in our solar system. Display the finished diagrams around the classroom.

Earth

Point out the composition of Earth's atmosphere. Ask: Why is Earth's atmosphere so different from those of Venus and Mars? *(Accept all reasonable responses. Students may note that life processes over the last few billion years are largely responsible for the present composition of Earth's atmosphere.)* You may wish to inform students that by studying this difference in atmospheres, scientists were able to predict that no life would be found on Venus and Mars long before any probes were launched into space.

Mars

Share the following information with students: The first spacecraft to reach Mars was the spacecraft *Mariner 4,* which flew within 10,000 km of Mars in 1965. In 1976, *Viking 1* and *Viking 2* landed on the surface of Mars to analyze the atmosphere and soil. Much of Mars's surface is covered with reddish dust, rocks, and sand. The surface is also marked by craters, canyons, and dry river beds. One goal of the *Viking* probes was to look for signs of life on Mars. Although the probes did not find any evidence, recent studies of a Martian meteorite found in Antarctica may support the theory that primitive life once existed on Mars.

 An Exploration Worksheet is available to accompany Exploration 1 on page 216 (Teaching Resources, page 3).

Theme Connection

Energy
Earth and the moon are nearly the same distance from the sun. Have students answer the following **focus question** in terms of energy: Why does the moon have a much greater range of temperatures than the Earth does? *(The Earth's atmosphere prevents the Earth from overheating by reflecting some solar energy. It keeps the dark side of the planet warm by holding in absorbed solar energy. The moon experiences more extreme temperatures than does Earth because the moon has no atmosphere.)*

Answers to
In-Text Questions

A The size and makeup of a planet's atmosphere affects its surface temperature. As evidence, students may note that the surface temperature of Venus is hotter than that of Mercury even though Mercury is closer to the sun.

EXPLORATION 1

If this Exploration is done indoors, a 100 W or greater light bulb works best. Students should change one factor that simulates some aspect of a planet's atmosphere or surface. For instance, a leaf inside the jar might represent vegetation on a planet's surface.

You may wish to have students work in groups to complete this Exploration.

Answers to
Exploration 1

Changing only one variable at a time ensures a controlled experiment.

1. Answers will vary. Factors that may increase the temperature include changing the jar's inner surface color, adding soil, and filling the jar with carbon dioxide. Factors that may decrease the temperature include increasing the distance from the heat source and adding plants.

2. Answers will vary, depending on the different types and degrees of alteration to the jars.

3. The temperatures in the jars stabilized when equal amounts of heat energy were leaving and entering the jars. This is similar to what happens on Earth because while light energy from the sun is continuously heating the Earth, heat energy in the Earth's atmosphere is continually escaping into space. The result is a relatively stable surface temperature.

4. Sample answer: The model does not show how factors such as oceans, forests, and clouds may interact to heat and cool the Earth.

5. Dark soil absorbs light energy and reemits it as heat.

6. Accept all reasonable responses. Increased carbon dioxide in the atmosphere and deforestation may be contributing to this change.

Understanding Our Planet

Did you conclude that a planet's temperature is determined by more than just its distance from the sun? What other factors play a role? What is the evidence? **A**

In the following Exploration, you will learn about the heating of the Earth by setting up a simulation. As you learned in Unit 2, a simulation is a type of experiment intended to model actual conditions. Scientists often use simulations to study complex problems. In Exploration 1, a glass jar will represent the Earth and a lamp will represent the sun. If the weather is clear, you can do the simulation using the sun instead of a lamp.

EXPLORATION 1

A Simulation

You Will Need

- 2 thermometers
- 2 large glass jars with lids
- a lamp with a 100 W bulb (if not sunny)
- a thin strip of cardboard
- tape

What to Do

Set up two jars, each as shown in the diagram below. One jar is your *control*. The second jar represents Earth and its atmosphere. Your challenge is to modify the second jar in such a way as to cause a change in the temperature. Change anything you want, but change only one variable (such as the distance of the jar from the lamp) at a time. (Why is this necessary?) Now follow the steps below.

1. Get together with two or three other students. Discuss the variables that could influence the temperature in the jar.

2. Choose a variable that your group would like to test, and modify the jar to determine the effect of this variable.

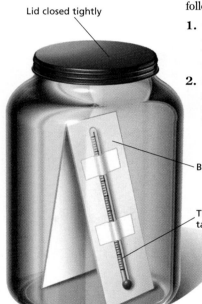

Lid closed tightly

Bent cardboard

Thermometer taped to card

3. Place both jars near the light bulb (or in the sun). Record the temperature changes in both jars.

What did you do to make this a *controlled experiment*?

Conclusions

1. Did your modified jar reach a different temperature than the control did? Was the temperature higher or lower than that of the control? What do you think caused the difference?

2. Make a list of the different modifications that the class made to the simulation, as well as the results obtained with each. Which modifications had the greatest effect?

3. Did the temperatures in your jars stabilize after a time? Why did this happen? How is this similar to what happens on Earth?

4. What are the shortcomings of this model? Compile a list in your ScienceLog.

5. Suppose you add a layer of dark soil to the bottom of the jar. After 20 minutes the temperature in this jar is 1°C higher than the temperature in the control. What might you conclude?

6. Evidence suggests that the average temperature of the Earth has gone up by about 0.5°C during this century. What might be causing this change? Do the results of any of the simulations suggest possible answers? Explain.

FOLLOW-UP

Reteaching

Have students make bar graphs based on the data profiled in the text on the four planets. The graphs could describe data on temperature, pressure, and distance from the sun.

Assessment

Have students summarize the factors that might affect the temperatures of the planets.

Extension

Using what they have learned, ask students to write a weather forecast and a description of the climate for each planet.

Closure

Have students write a one-page advertisement for Earth featuring those characteristics that make it the most desirable planet in the solar system for living things.

LESSON 2 — The Greenhouse Effect

From Today's Headlines

DAILY HERALD
EARTH TO BE A WARMER PLACE IN THE FUTURE

CITY TRIBUNE
GLOBAL WARMING WILL NOT BE AS GREAT AS PREDICTED EARLIER

THE DAILY CHRONICLE
SCIENTISTS DEBATE GREENHOUSE EFFECT-GLOBAL WARMING CONNECTION

THE WORLD REPORTER
OCEAN RISE PREDICTED FROM GLOBAL WARMING

THE JOURNEYMAN'S QUARTERLY
SOME SCIENTISTS SEE RECENT HOT SPELLS AS PROOF OF GLOBAL WARMING

NEWSDAY
GREENHOUSE EFFECT TO COST BILLIONS

Have you seen headlines like those above? What do they imply? It seems that Earth is becoming a warmer place. What could possibly be causing this? Could the sun's energy output be increasing? Could Earth be moving closer to the sun? Could the warming be part of a natural cycle? (After all, throughout most of its history Earth had a much warmer climate than it has now.)

217

LESSON 2 ORGANIZER

Time Required
2 class periods

Process Skills
analyzing, inferring, hypothesizing, organizing

Theme Connections
Energy, Changes Over Time

New Terms
Global warming—the name given to the apparent warming trend affecting Earth's atmosphere in recent years

Greenhouse effect—the warming of Earth or any planet due to the presence of certain gases that trap heat within the atmosphere

Materials
none

Teaching Resources
Exploration Worksheet, p. 6
Theme Worksheet, p. 12
Transparencies 30, 31, and 32
SourceBook, p. S66

LESSON 2 — The Greenhouse Effect

FOCUS

Getting Started

Display newspaper and magazine articles about global warming and the greenhouse effect. Ask students to work in small groups to read and summarize the articles. Have each group present a summary of their article to the class along with a list of the facts and opinions given in the article. Ask students if they agree or disagree with the points of view presented in the articles. Point out that in this lesson they will learn more about global warming and the greenhouse effect.

Main Ideas

1. The greenhouse effect is the warming of a planet due to the presence of certain gases in the atmosphere, including carbon dioxide, water vapor, and methane.
2. The greenhouse effect is essential for maintaining the temperature on Earth's surface.
3. Data suggests that Earth is experiencing a warming trend. However, Earth has gone through a number of warming and cooling periods that have not been caused by human activities.

TEACHING STRATEGIES

From Today's Headlines

Have students examine the graph on page 218. Have students work in small groups to discuss the questions presented on page 218. Emphasize that there are no right or wrong answers to the questions. You may wish to point out that scientists themselves do not agree on the causes and consequences of the greenhouse effect and global warming. Students will be investigating these questions in more depth in Exploration 2.

Ⓐ Students may be aware of the following information: The greenhouse effect is a natural result of atmospheric gases trapping solar energy. Without the greenhouse effect, Earth would experience large daily changes in temperature and would have a much colder average temperature. Gases added to the atmosphere by humans may be trapping more of the sun's energy and raising global temperatures enough to interfere with present ecosystems on Earth.

Ⓑ Some students may note that the prediction does not seem realistic because projected temperatures are drastically higher than current temperatures. Other students may think the prediction is realistic because it shows the effect of a global human population explosion.

Ⓒ Accept all reasonable responses. Students may suggest such possibilities as melting polar icecaps, worldwide flooding, desertification, and drought.

Ⓓ Some students may state that a significant increase in the greenhouse effect could cause global warming.

Homework

Ask students to conduct a survey of friends or family to find out what they know about the greenhouse effect. Students should find out whether the people they interview believe that an increase in the greenhouse effect is causing global warming. Students should also ask those interviewed what their reasoning is for their position. Discuss the results of the surveys in class in order to evaluate the degree to which people use scientific thinking to evaluate the issue of global warming.

You have probably heard a great deal of talk about the *greenhouse effect*. Many people are concerned about the greenhouse effect and blame it for the apparent *global warming* trend. What is causing this greenhouse effect? Are people somehow responsible for it? Ⓐ

With all this talk about global warming, we need to find out more about it so that can we understand it and deal effectively with it.

Look at the graph below. This graph shows the estimated average global temperature over the last century. It also shows the average global temperature through the year 2040 as predicted by some scientists. After examining the graph, think about the questions below.

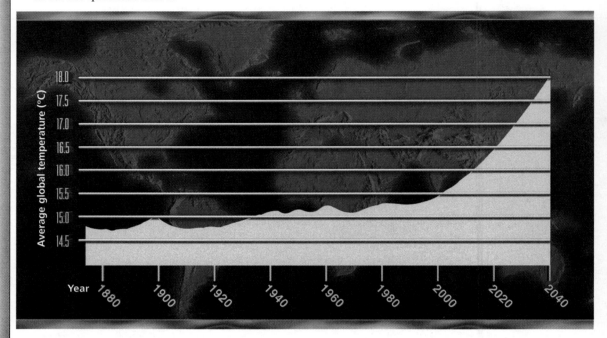

Ⓑ • Is the prediction realistic?

Ⓒ • If the prediction turns out to be right, what might the consequences be?

Ⓓ • How could the greenhouse effect be connected to global warming (if at all)?

Dateline: Earth 2050

Imagine that you could be transported to the year 2050. How might the headlines read then? Make up your own headline and article about the consequences of global warming. Will Earth be a very different and warmer place? Will the icecaps melt, causing the sea levels to rise? Will our temperate Earth become tropical? Perhaps you are skeptical and think that people are overreacting. If that's the case, your headline and article could reflect how all of the predictions about global warming turned out to be wrong.

219

Dateline: Earth 2050

This writing assignment allows students to reconsider their own views and ideas about the consequences of global warming. Ask them to support their opinions. Tell them that they can review and revise their articles as they progress through the unit.

Theme Connection

Energy

Focus question: Why is conserving energy one way to reduce global warming due to an increase in the greenhouse effect? *(Most of the energy consumed worldwide comes from the burning of fossil fuels. A chemical product of the combustion of fossil fuels is carbon dioxide, a greenhouse gas. Therefore, conserving energy reduces the emission of carbon dioxide.)*

Homework

You may wish to assign the creative-writing exercise in Dateline: Earth 2050 as homework.

PORTFOLIO

Suggest to students that they include the creative-writing assignment from Dateline: Earth 2050 in their Portfolio. You may wish to have students complete an About My Portfolio checklist, which is available both in the Assessment Checklists and Rubrics booklet and on the *SciencePlus SnackDisc.*

★ **Transparency 30 is available to accompany the graph on page 218.**

The Story Behind the Greenhouse Effect

You may wish to read these two paragraphs aloud while a volunteer draws a diagram on the board.

Earth's Special Case

If, after answering the in-text questions, students do not understand how the density and composition of the atmosphere affect the temperature of a planet, ask them to diagram the atmospheres of Earth, Venus, and Mars. Have students use one color to represent molecules of carbon dioxide and other colors to represent the other gases. Their diagrams should show the differences in density and in the percentage of carbon dioxide.

Ask: What could happen if the concentration of carbon dioxide increases significantly in Earth's atmosphere? (*Average temperatures on Earth could increase.*)

Explain that water vapor is responsible for most of the greenhouse effect on Earth. Because the amount of water vapor in the atmosphere increases when temperatures rise, any warming caused by increased carbon dioxide will be amplified by an increase in water vapor. This is an example of *positive feedback* in the climate system.

Answers to
In-Text Questions

A In most cases, students should have discovered that the jars trapped heat to some degree. See the Answers to Exploration 1 on page 216 for a more detailed explanation.

B A planet with high atmospheric pressure has a denser atmosphere and therefore a stronger greenhouse effect than a planet with low atmospheric pressure. Venus has a stronger greenhouse effect than the Earth does because Venus's atmosphere is denser. Mars's atmosphere is the least dense and its greenhouse effect is the weakest. Atmospheric composition also influences a planet's greenhouse effect. The larger proportion of carbon dioxide in Venus's atmosphere contributes to a stronger greenhouse effect on Venus than on Earth. However, because Venus and Mars have similar atmospheric compositions, the difference in their greenhouse effects must be caused by a difference in atmospheric pressure, not composition.

The Story Behind the Greenhouse Effect

Without the greenhouse effect, Earth would be a cold, lifeless place.

Here's one headline you probably haven't seen. It suggests an aspect of the greenhouse effect that is rarely noted in news stories about the dangers facing our climate. As you will see, the greenhouse effect is something we all depend on more than we realize.

The term **greenhouse effect** was coined in 1822 by Jean Fourier, a French mathematician. Fourier noted that in many ways the Earth's atmosphere behaves like a greenhouse. The glass of a greenhouse lets solar radiation pass through, but it traps the heat given off by the ground and plants inside as they absorb the sun's energy. The glass walls and roof keep the warm air from blowing away. As a result, the greenhouse warms up. If you've ever been inside a greenhouse on a sunny but cold day, you are probably familiar with this effect. In Exploration 1, did the jar in your simulation behave like a greenhouse? **A**

Earth's Special Case

Certain gases in our atmosphere—principally water vapor, carbon dioxide, and methane—are especially good at blocking the flow of heat from the Earth back into space. These gases together make up a small percentage of the atmosphere, but they have an enormous impact on Earth's climate.

The greenhouse effect is essential for maintaining Earth's moderate temperature. If not for the greenhouse effect, Earth's average temperature would be much colder, at least 33°C colder than it is now—too cold for life as we know it to exist.

Compared with Earth, Venus experiences a very strong greenhouse effect, but Mars experiences a weak greenhouse effect. Look back at the astrocorder readings on pages 214–215. How do you think differences in atmospheric composition and atmospheric pressure might account for the differences in the greenhouse effects on Earth, Venus, and Mars? **B**

The Earth's atmosphere is responsible for the greenhouse effect, which I discovered. To find out more about the atmosphere's origin and composition, turn to pages S60–S65 in the SourceBook.

Jean Fourier
1768–1830

220

Meeting Individual Needs

Learners Having Difficulty

Help students understand the greenhouse effect with a simple model. Students will need two thermometers and two transparent, plastic bags (one large and one small) with twist ties. Have students lay one thermometer inside the small bag. Then inflate the bag by blowing into it and close the bag with the twist tie. Put the inflated bag inside the larger plastic bag. Then inflate the large bag and close it with a twist tie. Place the bags in direct sunlight and lay the second thermometer next to the large bag. Record the temperature on both thermometers after 10 minutes. Move the bag setup and free thermometer to a dark area such as a closet. Record the temperature reading on both thermometers after 30 minutes. Discuss results with the students. (*The double layer of air in the bag acts like a greenhouse or the Earth's atmosphere. The double layer of air allows radiant energy from the sun to enter. The energy is absorbed, converted to heat, and trapped within the bags.*)

Are We Rewriting the Tale?

As you have seen, there is evidence to suggest that Earth is warming up. If this is the case, could human activities be responsible? Are we enhancing the greenhouse effect? (Consider that for at least the last 200 years, humans have had a significant impact on the environment.) Or could the warming trend be attributed to natural causes? Scientists are presently debating these questions.

Form some conclusions of your own. Examine the evidence in the following Exploration. If necessary, do additional research. Work with two or three other students to prepare a position paper presenting your views on the following questions:

- How does the greenhouse effect work?
- Is Earth, in fact, heating up? What evidence supports this interpretation?
- If there is a warming trend, could it be caused by increased amounts of carbon dioxide and other gases in the atmosphere? What evidence is there to support this?
- What else could account for the apparent warming? What evidence suggests this?

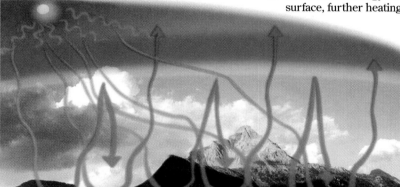
◀ The greenhouse effect illustrated

EXPLORATION 2

Drawing Conclusions

PART 1

How Does the Greenhouse Effect Work?

Refer to the diagram below, which illustrates how greenhouse gases such as water vapor, carbon dioxide, and methane help heat the Earth. Make a rough sketch of this diagram in your ScienceLog and place the letter of each statement below in the appropriate location on the diagram.

a. Sunlight is made up of radiation of different wavelengths.

b. Most of the sun's ultraviolet light and other short-wavelength radiation is absorbed by ozone high in the atmosphere.

c. Radiation with longer wavelengths, such as visible light and infrared light, passes through the atmosphere without being absorbed.

d. This radiation is mostly absorbed by the Earth.

e. The absorbed energy is reradiated at even longer wavelengths (infrared).

f. Some of this radiation escapes back into space.

g. Some of the longer-wavelength radiation is absorbed by greenhouse gases, such as carbon dioxide, water vapor, and methane, thus trapping the energy.

h. Some of this energy is reradiated toward the surface, further heating the Earth.

Exploration 2 continued ▶

221

Are We Rewriting the Tale?

Have a volunteer read this section, stopping at each question to allow for discussion. Explain to students that they will be returning to these questions, and encourage them to keep these questions in mind as they complete the Exploration.

EXPLORATION 2

Divide the class into groups and have students perform Exploration 2 and write their position papers. Explain that the position papers should address all of the questions listed on page 221. Students should be able to find information about the greenhouse effect in books, newspapers, and periodicals.

Remind students that there is still a lot of debate among scientists about these questions. They will need to find out as much as possible and then evaluate the evidence in order to draw their own conclusions. Point out that their conclusions should be supported by convincing arguments.

After students have completed their position papers, have one member of each group present the group's paper to the class.

★ **An Exploration Worksheet (Teaching Resources, page 6) and Transparencies 31 and 32 are available to accompany Exploration 2.**

Answers to
Part 1

For a sample answer, see the labeled illustration on page S208.

Answers to Part 1 continued ▶

Homework

Ask students to explain the greenhouse effect to a family member or friend. Have the students write a summary of the conversation, including a synopsis of the explanation, a list of questions they were asked, and their responses. Have students discuss in class any questions that they were unable to answer.

The glass wall of a greenhouse reflects most of the incoming ultraviolet radiation, whereas the ozone layer of the Earth's atmosphere absorbs it. In addition, the atmosphere varies somewhat in thickness and composition from place to place, whereas the glass of a greenhouse is more uniform.

Answers to
In-Text Questions

Ⓐ A concentration of 350 ppm is a relative measurement indicating that there are 350 carbon dioxide particles in every million particles of air.

Ⓑ The concentration of pepper would be 350 ppm, or 350 grains of pepper in 1 million total grains.

Answers to
Part 2

1. Around 130,000 years ago, the carbon dioxide concentration was at its greatest level. (However, students may note from the graph on the next page that present levels of carbon dioxide are much higher.) The average temperature was highest around 135,000 years ago.

2. There appears to be a connection between high concentration of carbon dioxide in the atmosphere and high temperature. This is demonstrated by the similar patterns of peaks and valleys on the two graphed lines.

3. Looking at the graph, students should observe the low temperatures at about 140,000 years, 60,000 years, 40,000 years, and 15,000 years ago. These low points suggest possible ice ages.

4. Human activities have put large amounts of carbon dioxide into the air. However, students may remember from their history classes that the beginning of the Industrial Revolution occurred less than 300 years ago. Therefore, any large-scale changes in the concentration of carbon dioxide in the atmosphere before that time are unlikely to be due to human activity.

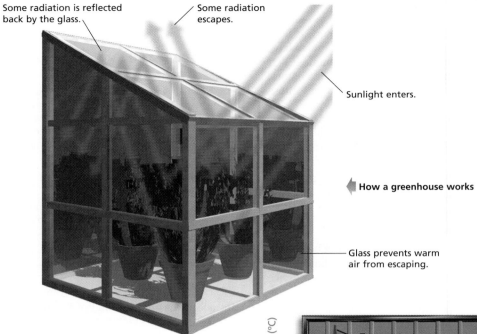

Some radiation is reflected back by the glass.

Some radiation escapes.

Sunlight enters.

◄ How a greenhouse works

Glass prevents warm air from escaping.

An actual greenhouse works a little differently from Earth's "greenhouse." Examine the diagram above and think of some ways in which the two differ. Write down your ideas in your ScienceLog.

PART 2

Temperature Versus Carbon Dioxide Concentration

The concentration of carbon dioxide in the atmosphere is about 350 ppm (parts per million). What does this mean? Ⓐ

Imagine a container filled with a million grains of salt. If you were to replace one of these grains with a grain of pepper, then the concentration of the pepper in the container would be 1 ppm. Imagine replacing 350 salt grains with pepper grains. What would be the concentration of pepper in parts per million? Ⓑ

Using various methods, scientists have been able to estimate the Earth's average temperature, as well as the concentration of carbon dioxide in the atmosphere, over about the last 200,000 years. These are shown in the graph at right.

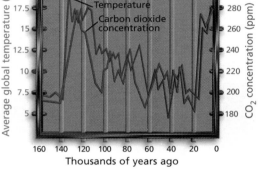

1. When was the carbon dioxide level greatest? When was the temperature highest?

2. Does there seem to be a connection between Earth's temperature and the carbon dioxide concentration in the atmosphere? Explain.

3. At what point during the past 160,000 years do you think there might have been an ice age?

4. Do you think all of the variations in the level of carbon dioxide were caused by human activities? Why or why not?

Carbon Dioxide and Temperature

The graph below contains two sets of data.
Use the data to answer the following questions:

1. What was Earth's average temperature in 1990? in 1890?

2. Does the graph appear to show a relationship between the concentration of carbon dioxide in the atmosphere and the Earth's average temperature?

3. Does the graph *prove* that the change in the atmospheric concentration of carbon dioxide has caused the average temperature to increase? Why or why not?

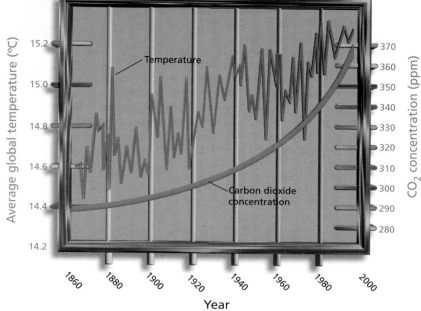

PART 4

Methane and Human Activity

Methane is another greenhouse gas. Methane is present in the atmosphere in smaller quantities than is carbon dioxide. However, a given mass of methane is much more effective as a greenhouse gas than is carbon dioxide. How is the level of methane changing in the atmosphere? Why is it changing? Is there a correlation between methane levels and human population? Using the data at right, prepare a graph in your ScienceLog to help illustrate any possible connection. Then answer the questions on the following page.

Year	Methane concentration (ppm)	Human population (billions)
1990	1.70	5.3
1980	1.50	4.2
1970	1.40	3.5
1960	1.30	3.0
1950	1.25	2.5
1940	1.15	2.0
1900	1.00	1.5
1850	0.85	1.2
1800	0.75	1.0
1750	0.74	0.8
1700	0.72	0.7
1650	0.70	0.6
1600	0.70	0.5

Exploration 2 continued ▶

Answers to
Part 3

1. The average temperature in 1990 was about 15.2°C. In 1890, it was about 14.6°C.

2. Yes; both temperature and carbon dioxide concentration have increased since 1860.

3. No; the graph shows similar trends but does not prove a cause-and-effect relationship. The similarity may be the result of some other factor, or it may be a coincidence.

Answers to
Part 4

The level of methane is increasing. Answers to why it is changing will vary, but students should recognize a correlation between methane levels and human population. For a sample completed graph, see page S208.

Answers to Part 4 continued ▶

ENVIRONMENTAL FOCUS

Point out that it is extremely difficult for scientists to predict the long-term effects of global warming. Such predictions are complicated by a number of interconnected variables. For example, as the Earth's surface temperature increases, more water will evaporate from the oceans, and more clouds will form. Such clouds might decrease the amount of energy that can reach Earth's surface and thus slow the warming trend. However, if partial melting of the polar icecaps occurs, there will be less ice to reflect sunlight away from the Earth's surface, and warming may speed up.

Answers to
Part 4, continued

1. Since 1600, not only has the world population grown, but the rate of that growth has increased. According to the graph, the population in 2000 should be between 6.0 and 6.5 billion.

2. According to the graph, the methane concentration should be between 1.9 and 2.0 ppm.

3. Sources of methane:
   - Cattle—Humans are raising more cattle for food as human populations continue to increase.
   - Rice—Rice is considered a staple in many nations on Earth. As human populations in these nations increase, so does rice production.
   - Decomposition—Growing human populations and the tendency of people in developed nations to consume more goods has led to an increase in the amount of waste deposited in landfills. This material decomposes in the absence of oxygen.
   - Natural gas—As human populations grow and develop new technology, the demand for energy increases. As a result, more natural gas is extracted from the Earth, and leakage from production, processing, and transport is inevitable.
   - Termites—Because human populations are increasing, more and more forests in Earth's warm, moist regions are being cut down to accommodate this population growth. Termites feed on the rotting wood that may be left behind and on structures built from the wood.

EXPLORATION **2,** *continued*

1. From your graph, predict the world's human population in the year 2000.

2. What will be the expected concentration of methane in the year 2000?

3. Listed below are some sources of methane. How is each related to human activities?
   - Cattle—They produce methane as they digest food.
   - Rice paddies—Bacteria in the flooded soil of rice paddies produce methane, which escapes into the air through the plants' hollow stems.
   - Decomposition—The decomposition of organic material in the absence of oxygen produces methane.
   - Natural gas (its main component is methane)—It leaks constantly from the Earth, from pipelines, and from production and processing facilities.
   - Termites—Methane is produced as termites digest wood. Termites are extremely abundant in the Earth's warm, moist regions.

   Return to the questions (on page 221) that preceded this Exploration. Use them to help you prepare your position paper on the greenhouse effect and global warming. What kind of additional data would be helpful? What other questions arose in the course of your study?

Sources of Carbon Dioxide

Carbon dioxide is perhaps the most important greenhouse gas that we add to the atmosphere. It is produced by a wide variety of natural processes and human activities, including the following:

- Burning—Any time organic material (material that contains carbon, such as gasoline) is burned, carbon dioxide is produced.
- Respiration—Carbon dioxide is a byproduct of respiration in both plants and animals.
- Decomposition of organic matter—Decomposition (in the presence of oxygen) is like slow-motion burning and, like all burning of organic matter, produces carbon dioxide.
- Volcanic eruptions—These spectacular events can release huge amounts, sometimes millions of tons, of carbon dioxide into the atmosphere.

   Look at the photographs on this page. Which of these sources of carbon dioxide can people control? Which sources of carbon dioxide can people *not* control?

Theme Connection

Changes Over Time

Focus question: What twentieth-century events could have affected the abundance of greenhouse gases? *(Answers will vary. Burning fossil fuels to power cars and to create electricity releases carbon dioxide into the air. With improved medical care, more people are living longer. The increased population requires more food [such as cattle and rice, which release methane], requires that more land be cleared for agriculture [which removes forests that absorb greenhouse gases], and generates more waste [which releases methane during decomposition]).*

 A Theme Connection Worksheet accompanies this Theme Connection (Teaching Resources, page 12).

Answers to
Sources of Carbon Dioxide

- People can control emissions of carbon dioxide from cars and factories by reducing driving and manufacturing or by using pollution-control devices on cars and factories.
- It is not possible to reduce the carbon dioxide emissions from volcanic eruptions.

- People can control emission of carbon dioxide from rotting wood by implementing reasonable forest-management techniques that leave less waste and reduce the amount of rotting wood (and therefore the amount of carbon dioxide produced).
- People have some control over the number of forest fires. Preventing forest fires can reduce emissions of carbon dioxide. However, not all fires are caused by people. Some fires are caused by lightning or drought. These fires will continue to be a source of carbon dioxide.

Global Warming—How Much?

Perhaps you are not yet convinced that global warming is actually occurring. If so, you are not alone. Many scientists disagree about global warming. Using sophisticated computer models, they try to predict the future behavior of the Earth's climate. Different models have produced very different results. Some scientific models predict a significant rise in global temperatures in the near future. Other models predict little or no warming. None of the models is exactly right. The question is, which model is the most accurate?

The accuracy of a scientific model depends on the reliability of the information used to formulate it. An accurate model of the causes and effects of global warming must answer the following questions. Can you think of other questions? **A**

- Would a higher average temperature change the amount of cloud cover across the Earth? How would a change in cloud cover affect the average temperature of the Earth?

- What is the effect of light-reflecting dust and gases that have been pumped into the atmosphere by human activities and natural processes?

- What is the role of the oceans in absorbing and releasing heat energy and carbon dioxide?

- How would plants respond to a higher level of carbon dioxide in the air? How would other organisms, including people, respond?

If the Earth does warm up by a significant amount, what would be some of the consequences? Consider that a range of possible temperature increases (anywhere from 1.5°C to 5°C) has been predicted. With two other students, make a web of changes and effects that might occur. The web below might give you some ideas. **B**

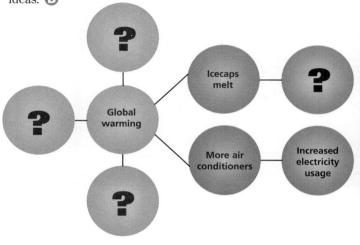

Increased carbon dioxide is absorbed by vegetation.

Warmer temperatures cause faster decay, resulting in faster release of carbon dioxide.

225

Answers to
In-Text Questions

A Students should suggest other questions that a model of Earth should answer. Possible questions include the following: If the amount of pollution was greatly reduced, how would this affect global warming? How would plankton in the oceans respond to higher levels of carbon dioxide in the air?

B Some possible consequences of global warming are given in the sample completed web on page S209.

FOLLOW-UP

Reteaching

Ask students to explain how the greenhouse effect might take part in the following situation: A soda bottle that has been sitting in the sun releases hot air when opened. *(Radiant energy from the sun passes through the glass and heats the soda and air in the bottle. Also, the added carbon dioxide from the soda absorbs more radiation than ordinary air, helping to trap heat inside the bottle.)*

Assessment

Have students discuss ways in which the greenhouse effect is helpful and potentially harmful to living organisms on Earth.

Extension

Have students do some research to discover how scientists determine past temperatures and carbon dioxide concentrations in the atmosphere. *(One way is that scientists bore into a polar ice sheet to recover an ice sample. Then they analyze the air bubbles that are locked in the ice to determine past concentrations of carbon dioxide. Analyses of such samples have revealed that Earth has warmed and cooled many times in the past.)*

Closure

Have students collect and read newspaper and magazine articles related to the greenhouse effect and global warming. Suggest that they summarize the articles and share interesting facts with the class.

Homework

The Extension activity above makes an excellent homework assignment.

LESSON

3 Land and Sea

How much of the sun's energy does the Earth absorb? To find out, turn to page S66 in the SourceBook.

FOCUS

Getting Started

To ascertain students' prior knowledge, ask: Which do you think is warmer in winter, an area that is near the ocean or one that is far away? Why? *(An area that is near an ocean is warmer in winter because water holds heat longer than land does.)*

Main Ideas

1. Land heats up and cools down faster than water does; thus, oceans and lakes moderate air temperatures.
2. Heat is absorbed from the surroundings during evaporation; heat is released during condensation.
3. The dew point is the temperature at which water vapor begins to condense.
4. Temperature generally decreases with increasing altitude.

TEACHING STRATEGIES

Answers to
The Earth's Thermostat

The sun's rays are most direct near the equator. The data suggest that oceans affect how heat is distributed around the globe.

- The hottest temperatures are found in the interior of continents that are near the equator. The coldest temperatures are also found near the poles.
- The image was probably taken during winter in the Northern Hemisphere because the temperatures over both land and water are cooler there than in the Southern Hemisphere.
- Land areas at the same latitude have different temperatures.
- Ocean areas at the same latitude have different temperatures, but they are more similar than land temperatures.
- Students should infer that these areas warm up and cool down differently because continents and oceans absorb and release heat at different rates.

The Earth's Thermostat

Look at the map on this page. This image, taken from space, shows surface temperatures across the Earth at the time the image was made. What does the map tell us about the heating processes of Earth? Find North America on the map. What is the temperature at your location? Where do you think the sun's rays are most direct? What does this data suggest about the role that the oceans play in distributing heat energy across the globe?

Examine the map to help you answer the following questions:

- Where are the hottest temperatures? the coldest temperatures?
- At what time of year was the image taken? How can you tell?
- Do all locations on land at the same latitude (degrees north or south of the equator) have more or less the same temperature?
- Do all locations on the oceans at the same latitude have more or less the same temperature?
- Does the map suggest that continents and oceans warm up and cool down differently? Explain.

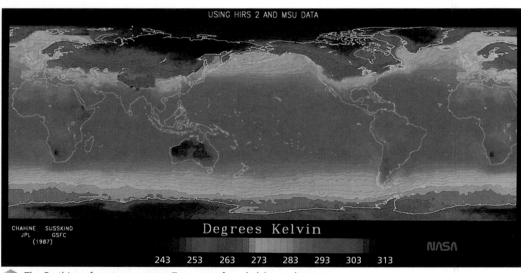

The Earth's surface temperature. To convert from kelvins to degrees Celsius, subtract 273.

LESSON 3 ORGANIZER

Time Required 3 to 4 class periods

Process Skills observing, inferring, predicting, analyzing

Theme Connection Cycles

New Term
Dew point—the temperature at which water vapor begins to condense

Materials
Exploration 3, Activity 1: 2 test tubes; test-tube tongs or test-tube clamp; 2 alcohol thermometers; 250 mL beaker; hot plate; test-tube rack; watch or clock with second hand; two small balls of modeling clay; about 10 mL of water; about 10 mL of sand; 100 mL graduated cylinder; safety goggles; **Activity 2:** 2 aluminum pie pans; lamp with at least a 100 W light bulb; about 250 mL of water; about 250 mL of sand; 2 alcohol thermometers; 100 mL graduated cylinder; metric ruler; watch or clock with second hand; optional items: materials to prop up thermometers, such as 2 small balls of modeling clay; **Activity 3:** 2 aluminum pie pans; about 500 mL of sand; 100 mL graduated cylinder; lamp with at least a 100 W light bulb; about 25 mL of charcoal powder; 2 alcohol thermometers; watch or clock with sec-

continued ▶

Heating of Land and Water

This Exploration is designed to help you answer the questions raised on the previous page. It will be a team effort. Each team will choose an activity, collect the materials, and do the experiment. Later you will give a report on your findings and interpretations. In doing so, make some reference to the photograph of global surface temperatures shown on the facing page. Graph your data to help with your presentation.

You Will Need

- 2 test tubes
- 2 thermometers
- a beaker
- a hot plate or other heat source
- modeling clay
- water
- sand
- aluminum pie pans
- a lamp with a 100 W (or greater) bulb
- charcoal powder
- a graduated cylinder
- test-tube tongs
- a test-tube rack

ACTIVITY 1

Sand Versus Water: Part 1

What to Do

1. Fill one test tube halfway with water. Fill another test tube halfway with sand.

2. Place a thermometer in the sand, and suspend another thermometer at the same depth in the water, using modeling clay to hold each thermometer in place. Place both test tubes in a beaker of hot water. Heat them to a temperature of about 70°C.

3. Remove the test tubes, and record the drop in temperature every 2 minutes for 20 minutes. Did they cool at the same rate?

ACTIVITY 2

Sand Versus Water: Part 2

What to Do

1. Fill one pie pan halfway with water, and fill another pie pan halfway with sand.

2. Place both pie pans an equal distance from the lamp.

3. Every 2 minutes for 20 minutes, record the surface temperature in each pie pan. The thermometer should be laid flat on the sand with the bulb covered by 0.5 cm of sand. The water temperature can be measured by holding a thermometer 0.5 cm below the surface. How did the temperatures change?

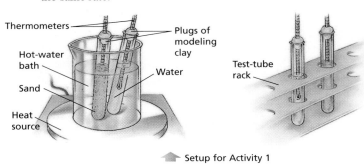

Thermometers — Plugs of modeling clay

Hot-water bath

Sand

Heat source

Water

Test-tube rack

Setup for Activity 1

Water

Sand

Setup for Activity 2

Exploration 3 continued ▶

227

A Transparency Worksheet (Teaching Resources, page 14) and Transparencies 33 and 34 are available to accompany The Earth's Thermostat on page 226. In addition, an Exploration Worksheet is available to accompany Exploration 3 (Teaching Resources, page 16).

EXPLORATION 3

🌐 Cooperative Learning
EXPLORATION 3

Group size: 3 to 5 students
Group goal: to answer the questions raised on page 226 about temperature trends on Earth
Positive interdependence: Divide the class into teams and assign one Activity to each team. Explain that each team will have a chance to make a presentation about their Activity. The presentations should describe the problem investigated, the method used, the data collected, and the conclusions drawn. Assign each of these parts of the presentation to a different team member.
Individual accountability: Have students individually graph their data and display it as part of their presentation.

Answers to Activity 1

Students should find that the sand cools faster than the water does. (When heating the test tubes, students may also observe that the water heats up faster than the sand. This is partly due to the more efficient mixing of the water by convection currents.)

Answers to Activity 2

Students should observe that the temperature of the sand increases faster than the temperature of the water.

ORGANIZER, continued

ond hand; **Activity 4:** 100 mL graduated cylinder; about 500 mL of sand; 2 aluminum pie pans; about 100 mL of water; 2 alcohol thermometers; lamp with at least a 100 W light bulb; watch or clock with second hand
Exploration 4, Part 1: alcohol thermometer; 5 cm piece of shoelace; piece of masking tape; small container of room-temperature water; optional item: small electric fan; **Part 2:** small container of warm water; about 20 mL of rubbing alcohol; safety goggles
A Mini-Activity in a Bottle: large, plastic soft-drink bottle; 1 L of water; a few matches; safety goggles

Exploration 5, Part 1: 250 mL metal can; several ice cubes; alcohol thermometer; stirring rod; about 50 mL of room-temperature water; **Part 2:** alcohol thermometer; wet-bulb thermometer from Exploration 4, Part 1; small container of water at or above room temperature

Teaching Resources
Transparency Worksheet, p. 14
Exploration Worksheets, pp. 16, 19, and 21
Transparencies 33 and 34
SourceBook, pp. S66 and S69

Answers to
Activity 3

The black sand should heat up faster than the light-colored sand. Students may infer that this is because dark colors absorb more heat than light colors do.

Answers to
Activity 4

The dry sand should heat up faster. The wet sand should heat up slowly and very little.

Answer to
Analyzing Activity 4

There are two factors involved in Activity 4. First, water heats up more slowly than sand when exposed to the same heat intensity. Second, evaporation of water from the sand absorbs heat and cools the sand. The heat energy goes into the room.

Answers to
Analyzing and Reporting Your Findings

Students should summarize their findings by concluding that sand heats and cools more quickly than water does. Students will probably suggest that the unequal rate of the heating and cooling of water and land is partly responsible for the unequal heating of the Earth's surface.

The fact that land cools faster than water may explain why the interior of the northern United States is cooler than the ocean at the same latitude. The fact that land heats up faster than water may explain why land temperatures in the Australian interior are warmer than water temperatures at the same latitude.

The map shows land areas that are both hotter and cooler than ocean areas at the same latitude. Those land areas that are cooler than other land areas at the same latitude are farther from bodies of water.

Other questions that could be investigated include the following: How do the temperatures on land and water compare at different times of the year? How do other factors such as kinds of plants or the presence or absence of things such as pavement or buildings affect the heating of the Earth?

(You may wish to point out that altitude also has an effect on average

temperatures. Students will learn on page 233 that the greater the altitude, the lower the temperature. If you choose to provide this information now, you should ask students to point out on the map the locations of as many mountain ranges as they can and comment on how the temperature in these areas differs from that in the surrounding land. Good examples on the map include the Himalayas, the Andes, and the Rocky Mountains. The map on page 256 is a useful aid in locating the major deserts and mountain ranges on Earth.)

ACTIVITY 3
Heating of Different-Colored Sands

What to Do

1. Measure equal quantities of sand into two separate pie pans. Mix one quantity of sand with enough charcoal powder to make it black.
2. Lay a thermometer on top of the sand in each pan. Cover each thermometer bulb with 0.5 cm of sand.
3. Place the pans so that they are equal distances from the lamp.
4. Take readings every 2 minutes for 20 minutes. What differences did you observe?

ACTIVITY 4
Heating of Wet and Dry Sand

What to Do

1. Measure equal quantities of sand into two identical aluminum pie pans.
2. Add water to one so that the sand is quite damp. The water and sand should be the same temperature before they are combined.
3. Bury the bulb of each thermometer about 0.5 cm below the surface, one in each pan.
4. Place both pans an equal distance from the lamp.

5. Record the temperature of each pan every 2 minutes for 20 minutes. Which one warmed up faster?

Analyzing Activity 4

There is more to Activity 4 than first meets the eye. Certainly, water warms up more slowly than sand. However, there is another factor involved. One group carried Activity 4 a step further. At the end of the 20-minute observation period, they switched off the light and observed what happened. After 15 minutes they were surprised to find that the temperature of the wet sand had actually dropped below that of the room. What happened? Where did the heat energy go? The next Exploration will help you solve this mystery.

Analyzing and Reporting Your Findings

What were your findings? What do these findings suggest about the relative effects of water and land in terms of the heating processes of Earth? How does the image on page 226 relate to your findings? What further questions for investigation does this Activity suggest? Summarize your findings in a brief report.

⬆ Setup for Activity 3

⬆ Setup for Activity 4

The Water Cycle and Heat Exchange

In Chapter 6 you learned how to make a wet-bulb thermometer. This type of thermometer is used to measure the amount of moisture in the air. In this Exploration you will review how a wet-bulb thermometer works and will discover how this relates to the water present in our atmosphere.

You Will Need

- a thermometer
- a piece of shoelace 5 cm long
- water (room temperature)
- water (warm)
- masking tape
- rubbing alcohol

PART 1

Making a Wet-Bulb Thermometer

Make a wet-bulb thermometer by putting the bulb of the thermometer into the shoelace as you would put your foot into a sock. Secure the shoelace to the thermometer with masking tape, and then read the temperature registered by the thermometer.

Now wet the shoelace and bulb with water that is at or above room temperature. Wave the thermometer in the air (or hold it in front of a fan) for a minute or so and then read the temperature again. What happened? What might have caused this to happen? (Recall what you learned in Exploration 4 of Chapter 6.)

PART 2

Feeling Cool

Dip your finger into warm water and then wave it in the air. What do you feel? What do you think happened to the water on your finger? Could this help to explain what happens with the wet-bulb thermometer? If you wet your finger with alcohol, will it feel cooler than when you wet it with water? Why or why not? Try it to find out!

What explanation do these activities suggest for why wet sand warmed more slowly than dry sand in the previous Exploration?

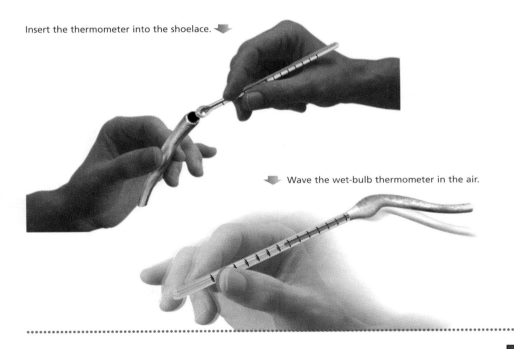

Insert the thermometer into the shoelace. ➡️

➡️ Wave the wet-bulb thermometer in the air.

229

Theme Connection

Cycles
Focus question: How does the water cycle affect temperatures on Earth? *(The water cycle moderates temperatures by cooling organisms, land, and bodies of water through evaporation and transpiration. On cloudy days, the condensation phase of the water cycle warms the air.)*

⭐ **An Exploration Worksheet is available to accompany Exploration 4 (Teaching Resources, page 19).**

Students may find it difficult to grasp the idea that heat is absorbed by water when it evaporates, thus cooling the surroundings. This Exploration provides two experiences that reinforce this concept.

SAFETY ALERT If students wish to perform the activity in Part 2 with alcohol on their skin, make sure that there are no flames, operating hot plates, or other possible sources of ignition in the room.

Answers to
Part 1

Students should read a lower temperature on the wet-bulb thermometer because water absorbs heat from the thermometer as the water evaporates.

Answers to
Part 2

Students should observe that a wet finger waved in the air feels cool regardless of the temperature of the water used.

Students should conclude that the water on their finger is evaporating.

As the water evaporates, heat is absorbed from the finger. Students should recognize that this is what happens with the wet-bulb thermometer.

Students should observe that their finger feels cooler faster when it is wet with alcohol than when it is wet with water. They should conclude that because alcohol evaporates faster than water, the cooling effect is quicker and more noticeable.

These activities suggest that wet sand warms more slowly than dry sand because when the water evaporates from the wet sand, it absorbs heat from the sand.

Life and Physical Sciences

Ask students to explain how evaporation is involved in cooling the body. *(People perspire in hot weather; the perspiration evaporates from the skin and cools the body.)* Have students draw a diagram to illustrate this process.

Answers to
In-Text Questions

Ⓐ Heat energy is absorbed wherever water evaporates. This makes the surroundings cooler.

   Heat is released wherever condensation occurs. This makes the surroundings warmer.

Ⓑ You would expect the greatest amount of water to evaporate from the oceans in the warmer regions—the tropics.

   When the moisture condenses to form clouds, the absorbed heat would be released back into the atmosphere.

Answers to
A Mini-Activity in a Bottle

A cloud forms when the match is dropped in the bottle. When students squeeze the bottle, they should observe that the cloud disappears. The cloud should reappear when the bottle is released.

   When the bottle is squeezed, the air pressure inside the bottle increases. The air warms slightly, allowing it to hold more moisture. When the bottle is released, the pressure decreases, causing the air to cool and the water vapor to condense.

Explaining Your Findings

Your wet finger feels cool. Heat in your finger must have been absorbed by the evaporating water. What happened to this energy? Did it disappear? No, that would be impossible. As you know, energy can never be created or destroyed, only converted from one form to another. What happened was that some of the heat in your finger and its surroundings was absorbed by the individual particles of water as they evaporated. If this evaporated water were to condense, that heat would be released to the surroundings once again.

   Follow the exchange of heat energy in the diagram of the water cycle at right. Where is heat being absorbed from the surroundings? Will this make the surroundings warmer or cooler? Where is heat being released to the surroundings? Does this make the surroundings warmer or cooler? Ⓐ

   Examine again the image showing global temperatures on page 226. Where would you expect to see the greatest amount of evaporation taking place? (Hint: Does more evaporation happen on hot days or cold days?) When would the energy stored in the water vapor be released again? Ⓑ

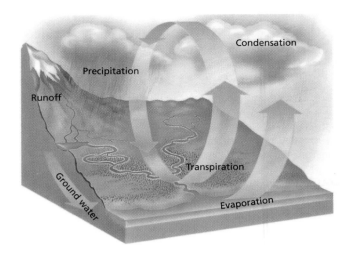

A Mini-Activity in a Bottle

Try this activity. Add some water to a large, plastic soft-drink bottle. Now drop a burning match into the bottle, and screw the cap on tightly. Squeeze the sides of the bottle as hard as you can, and then quickly release it. Do this several times. What do you observe? Explain your observations.

The Dew Point

You may recall from earlier studies that the **dew point** is the temperature at which moisture starts to condense out of the air. The dew point indicates how much moisture there is in the air. The lower the dew point is, the less moisture there is in the air. What happens to the dew point if the amount of moisture in the air increases? The dew point is calculated according to how much a wet-bulb thermometer cools. Can you explain why? Think of a situation in which no amount of shaking of the wet-bulb thermometer would result in a lower reading.

You can calculate the dew point using the table provided below. Find the dry-bulb temperature. Then find the *wet-bulb depression*—the difference between the dry- and wet-bulb temperatures. Use these measurements to find the dew point in the table.

For meteorologists (weather scientists), dew point is actually a very important measurement. In the next Exploration you will use two methods to determine the dew point.

Table for Computing Dew Point

This table allows you to compute the dew point. The dew point is plotted for a dry bulb reading of 22°C and a wet-bulb reading of 19°C.

Something to Think About

Trees cool their surroundings by absorbing solar energy and providing shade. But they also provide a cooling effect in another way. (Recall your study in Unit 1 of processes that occur in plant leaves.) How do you think this works? Why would cutting vast areas of forest have a warming effect on the environment? How might the cutting of trees affect the world climate?

EXPLORATION 5

PART 1

Have students work in groups to complete this activity. Make sure that students' cans have been cleaned with detergent because condensation will be hard to see on a greasy or oily surface. There will probably be variation in students' results because some groups will notice the moisture sooner than others. Also, some groups may not stir the ice and water mixture as much as other groups, which will result in temperature differences between the top and bottom of the can. When all the groups have finished Part 1, have each group record its temperature reading on the chalkboard. Then have students determine an average reading for the dew point.

Answers to
Part 2

Measuring room temperature with a wet-bulb thermometer would give a false reading because evaporation from a wet-bulb thermometer lowers the temperature reading.

Dew-point readings from a wet-bulb thermometer are often higher than those taken with the cold-can method. Possible reasons for the discrepancies include the possibility that students may not have detected the moisture on the can when it initially formed. One weakness of the cold-can method is that students must observe the formation of condensation immediately. A weakness of both methods is that students cannot take readings over an extended period of time without performing the experiment repeatedly.

Answer to
Caption

 Students could use the homemade setup to calculate the dew point the same way they did in Part 2 of this Exploration—by subtracting the wet-bulb temperature from the dry-bulb temperature and using the table on page 231 to determine the dew point. The homemade setup allows students to measure the dew point over an extended period of time.

Measuring the Dew Point

PART 1

The Cold-Can Method

You Will Need

- a metal can (250 mL)
- ice cubes
- a thermometer
- a stirring rod
- water (room temperature)

What to Do

1. Add room-temperature water and a few ice cubes to your can until it is half-full. Do this at your workstation so that you can begin to make observations immediately.

2. Place the thermometer in the water. Using a stirring rod, stir the water continuously while observing the sides of the can for any condensation. Record the temperature at which you first notice moisture (dew) forming on the side of the can. This temperature is the dew point. Compare your results with those of your classmates.

PART 2

The Wet-Bulb Thermometer Method

You Will Need

- a thermometer
- a wet-bulb thermometer

What to Do

1. Measure the temperature of the room. Don't use your wet-bulb thermometer to do this because it will give you a false reading. (Do you know why?)

2. Determine the wet-bulb temperature as you did in the previous Exploration.

3. Find the wet-bulb depression by subtracting the wet-bulb temperature from the dry-bulb temperature.

4. Use the table on page 231 to determine the dew point.

How did the results for the two methods compare? Try to explain any differences. What weaknesses, if any, does each method have?

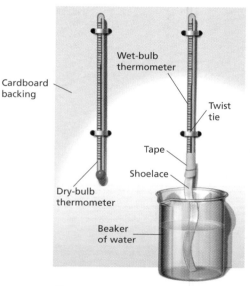

Cardboard backing

Wet-bulb thermometer

Twist tie

Tape

Shoelace

Dry-bulb thermometer

Beaker of water

⬆ How would you use this homemade setup to calculate the dew point?

Homework

Have students write a short paragraph describing how results obtained in Exploration 5, Part 1, depend on the humidity of the environment. *(The dew point is the temperature at which water vapor condenses. Water vapor condenses at a higher temperature when humidity is high. Therefore, dew point indicates humidity.)*

★ An Exploration Worksheet is available to accompany Exploration 5 (Teaching Resources, page 21).

Upward Bound

Imagine taking a trip on the aerial tram in Palm Springs, California. In 15 minutes the tram will travel 4 km and climb from 800 m to 2442 m above sea level. A breeze blows against the mountain and forces air to rise along its slope. As you ascend the mountain, you notice that the air grows cooler and that the vegetation changes dramatically. At the base of the mountain, desert plants prevail. As you climb higher, the desert vegetation is gradually replaced by shrubs and small trees. By the time you reach the summit, you are in an evergreen forest. The pine-scented air is crisp and cool.

Why did conditions change so much from the base of the mountain to the summit? Examine the data below for two trips on the tram, one on a clear day and one on a cloudy day. In both cases the temperature at the base was 30°C.

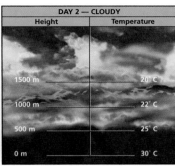
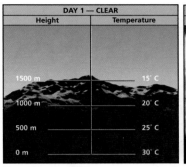

| DAY 1 — CLEAR | | DAY 2 — CLOUDY | |
Height	Temperature	Height	Temperature
1500 m	15° C	1500 m	20° C
1000 m	20° C	1000 m	22° C
500 m	25° C	500 m	25° C
0 m	30° C	0 m	30° C

Base elevation: 800 m

1. By how much did the temperature change on the tram trip on the clear day? on the cloudy day?
2. What was the dew point on the cloudy day?
3. What accounts for the change in vegetation?

As the tram ride clearly shows, the higher you go, the cooler the air becomes. Why does this happen? **Ⓑ**

Raise the Elevation—Lower the Temperature

Let's summarize what you have just learned.

• Air cools as it rises.
• Rising air cools at a slower rate when condensation occurs.

As you may already have realized, these phenomena are very important in shaping the weather. Pause for a moment to answer the following question: When a mass of air sinks, what happens to its temperature? **Ⓒ**

Mountain scenes, from base (bottom) to summit (top)

Integrating the Sciences

Earth and Life Sciences

Share the following information with students: The explosive action of geysers is a result of water expanding when it turns into steam. Underground water is heated by magma, but it cannot boil at first because it is under pressure. Eventually, however, it heats up enough to push up and break through the surface. Once some of the pressure is released, the remaining water changes quickly into steam, resulting in an eruption of steam and hot water. You may wish to have students create a diagram of this process.

Upward Bound

Call on a volunteer to read aloud the top of page 233. Then direct students' attention to the diagram showing the two trips—one on a clear day and one on a cloudy day. Ask them to answer the questions individually. Then involve the class in a discussion of their responses.

Answers to
Upward Bound

1. On the clear day, the temperature changed about 15°C. On the cloudy day, the temperature changed about 10°C.

2. To answer this question, students need to infer that because clouds start forming at a height of 500 m (called "cloud level") at a temperature of 25°C, the dew point must be about 25°C at that altitude.

3. The change in vegetation is due to the change in temperature and available moisture. As the altitude increases, the temperature cools past the dew point, allowing water vapor in the air to condense into a liquid and making it available to plants. Where the temperature is cooler, the rate of evaporation is lower as well.

Answer to
In-Text Question

Ⓑ The air expands as it rises, and it cools as it expands.

Ⓒ Air becomes more compressed as it sinks, causing its temperature to rise.

Raise the Elevation—Lower the Temperature

Call on a volunteer to read this material, pausing at the situations described at the top of page 234. Involve the class in a discussion of these situations, using the information learned so far to explain each.

A • Clouds form around mountain peaks on humid, windy days because the moisture in the air condenses as the air is blown up over the mountains, where temperatures are cooler.
  • The rainiest (or snowiest) places are usually in the mountains because moisture in the air condenses as the air rises over the mountains. This results in heavy precipitation (either rain or snow, depending on the temperature) on the upwind side of the mountain.
  • Areas downwind of mountain ranges often receive little rainfall because the moisture from the air has condensed and fallen as precipitation on the upwind side of the mountains, leaving the air very dry. This air then moves down on the downwind side of the mountains and is warmed.

B The chinook winds are warm, dry winds that descend from mountain slopes. Moist air from the ocean rises up the mountain slope and cools about 5°C per 1000 m, or about 20°C. The moisture condenses, causing clouds to form, and the air loses most of its moisture through precipitation. On the slope facing away from the wind, the air warms by about 10°C per 1000 m of descent, or about 40°C. When the air reaches the base of the mountain, it is drier and about 20°C warmer than the air on the ocean side.

C Vegetation changes reflect a change in climate. The higher the elevation, the cooler and wetter the climate. Vegetation at each level of elevation will be appropriate for that climate.

Keeping Track

Involve the class in a discussion of their ideas to evaluate their insights and identify any misconceptions.

FOLLOW-UP

Reteaching

Have students make a line graph of the average summer and winter temperatures of areas located along a particular latitude in the United States. Have students explain the temperature variations between maritime and inland areas.

As you may have already reasoned, when air sinks, it heats up. This phenomenon is an important factor in shaping weather.

Read the following situations and think about how the phenomena just described explain each situation: **A**

• On humid days when the wind blows, clouds often form around the peaks of mountains.

• The rainiest (or snowiest) places are usually in the mountains.

• Areas downwind of mountain ranges often receive very little rainfall.

Here is one additional mystery for you to ponder:

It is a bitter cold January morning in Billings, Montana. An icy stillness blankets the land. Suddenly, a breath of wind stirs—a warm wind! Within moments a strong westerly breeze is blowing. The temperature soars 5 . . . 10 . . . 20 degrees. Snow and ice begin to melt rapidly. "Chinook!" you hear someone shout. The wind has a name! What on Earth has happened? What is this chinook wind?

You already have part of the answer to this puzzle. Use the following information, along with the diagram provided, to help you solve the puzzle:

• Rising air cools at a rate of about 10°C for every 1000 m of elevation. Sinking air warms at about the same rate.

• Rising air from which moisture is condensing cools at a rate of about 5°C per 1000 m of elevation.

Can you solve the puzzle? Share your explanation of the chinook wind with a classmate. **B**

Climbing a mountain is like taking a journey northward. Climbing 1000 m is roughly equivalent to traveling 1000 km north, in terms of the change in climate. Higher elevations not only are cooler, but also tend to be wetter. How does this help explain the change in vegetation that you saw in the journey on the tram? **C**

Keeping Track

In this section you have been answering many questions and making many discoveries. In your ScienceLog, make a list of your major findings.

Assessment

Ask students to explain the following observation: There are water pipes in a basement. Some of the pipes are dripping with moisture but others are not. (*Students should infer that some pipes are cold and others are hot. Moisture from the air condenses on the cold-water pipes, but not on the hot-water pipes.*)

Extension

Explain that relative humidity is the percentage of moisture in the air compared with how much moisture the air can hold. The amount of moisture the air can hold depends on the temperature. If the temperature increases, the relative humidity decreases, and vice versa. Have students do research to find out how to determine relative humidity using wet- and dry-bulb readings. (*The difference between the temperature readings indicates the relative humidity.*)

Closure

Have students measure the dew point in different locations around the school. Ask them to explain their results.

CHALLENGE YOUR THINKING

1. Country Cool

Towns and cities are generally warmer than the surrounding countryside.

a. What are some reasons for this?

b. What are some things that can be done to help lower the temperature within a town or city?

c. The chart at right shows the average yearly temperature recorded at the Green Acres Weather Station over a 50-year span. Describe the pattern shown by the graph. Explain what might have happened to cause such a pattern. (Hint: When Green Acres was first built, it was in a rural area.)

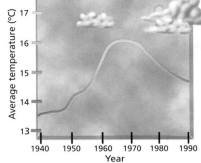

2. Burning Questions

Use your newly acquired knowledge to explain the following phenomena:

a. A car parked in the sun with its windows closed becomes hot very quickly.

b. On a stifling hot, nearly windless day, the slightest breeze feels cool.

c. You are walking barefoot along the waterline at the beach on a summer day. When you walk away from the water, the sand becomes unbearably hot.

3. Care for a Refreshing Swamp Cooler?

Before air conditioners came into widespread use, cooling devices popularly known as "swamp coolers" were widely used. The illustration shows how a swamp cooler works.

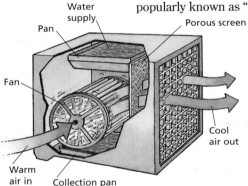

a. Explain the principle of the swamp cooler's operation.

b. Would it work equally well in all climates? Explain.

c. Swamp coolers are still common in certain areas because they are about as effective as air conditioners but much cheaper to operate. Where might you expect to find them in common use?

235

Homework

The Activity Worksheet on page 23 of the Unit 4 Teaching Resources booklet makes an excellent homework assignment once students have completed Chapter 10.

★ You may wish to provide students with the Chapter 10 Review Worksheet that is available to accompany this Challenge Your Thinking (Teaching Resources, page 24).

Answers to Challenge Your Thinking

1. The following are sample answers:
   a. Buildings and pavement in a city absorb the sun's energy and release it as heat. In most rural areas, some of the sun's energy is absorbed by plants and used to drive chemical processes. Transpiration and shade cool most rural areas.
   b. Planting more greenery (especially trees) and building fountains and ponds will help lower temperatures.
   c. Over the 50-year span, the average temperature rose from about 13.5°C to 16°C and then declined to about 14.5°C. The area around the Green Acres Weather Station was probably developed during this time. The increase in temperature was probably due to urbanization, and the later decline in temperature was probably due to the growth of lawns, shrubs, and trees.

2. a. A car warming in the sun is like a greenhouse. Energy from the sun passes through the car's window glass, where it is absorbed and reemitted as heat, which does not easily pass back through glass. Thus, the interior of the car heats up.
   b. People perspire a lot on hot, windless days. The slightest breeze increases the evaporation rate, which cools the body by removing heat from the skin.
   c. The sand at the waterline is cooler because it is saturated with water, which heats up more slowly than dry sand. The wet sand is also cooled as some water evaporates.

3. a. The fan blows air across the porous screen, through which water is flowing. As the water evaporates, it absorbs some heat energy from the air, thus lowering the temperature.
   b. A swamp cooler works most effectively in dry climates where the air has little moisture and works least effectively in humid climates where the air is almost saturated with water.
   c. You would expect to find swamp coolers in hot, dry areas such as deserts.

4. Air expands as it rises, so air at higher elevations is usually cooler. The temperatures on top of Mount Washington are consistently 10°C below those of Montpelier due to the 1500 m higher elevation of Mount Washington.

5. If students note a change in the dew point, they should attribute it to a change in the amount of water vapor in the air. If the dew-point temperature does not vary, it is because the amount of water vapor in the air (the absolute humidity) did not change.

6. Because water changes temperature slowly compared with land, coastal areas tend to have milder climates. Therefore, the interior areas of Pangaea would have colder winters and warmer summers.

   As air moves over areas of high elevation, it cools and its moisture condenses and falls to the ground. Therefore, deserts in Pangaea would most likely be found down-wind of large mountain ranges.

ScienceLog

The following are sample revised answers:

1. Some students may attribute the existence of life on Earth to an optimum distance from the sun. However, the atmosphere and oceans are also important factors in creating conditions that make the Earth suitable for life.

2. On its way from the sun to the Earth's surface, some solar radiation passes through the particles making up the atmosphere just as it passes through the glass walls of a greenhouse. Incoming radiation is absorbed by the Earth's surface or by the floor of a greenhouse, which re-radiates the absorbed energy as heat. This heat is kept from escaping by the gases in the atmosphere just as heat is kept in by greenhouse glass.

3. The air in the atmosphere that surrounds us contains water vapor. The warmer the air is, the more water vapor it can hold. Air is warmest during the daytime and holds a certain amount of vapor at this time. As it

4. New England News

The graph at right shows temperatures for Montpelier, Vermont, and for Mount Washington, New Hampshire, located a short distance away. Explain what is happening and why.

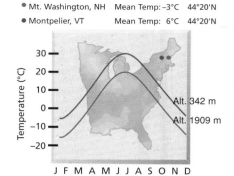

● Mt. Washington, NH   Mean Temp: –3°C   44°20'N
● Montpelier, VT   Mean Temp: 6°C   44°20'N

Alt. 342 m
Alt. 1909 m

5. Something to Dew

Over a few days, keep track of the dew-point temperatures using either of the methods described earlier. Does the dew-point temperature vary from day to day? Why?

6. One Colossal Continent

At the time of the dinosaurs, 200 million years ago, most of the Earth's landmasses were joined together, forming a supercontinent called Pangaea. How would the climate in the interior of this super-continent compare with the climate along its coastlines? If Pangaea had deserts, where would you expect to find them?

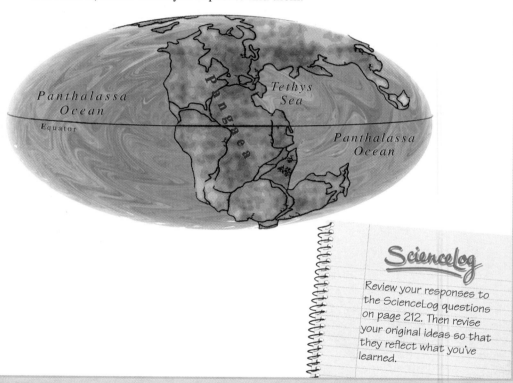

ScienceLog

Review your responses to the ScienceLog questions on page 212. Then revise your original ideas so that they reflect what you've learned.

cools during the night, the air reaches a temperature at which it cannot retain all of this vapor. Some of the vapor leaves the air by condensing on objects in the form of dew.

CHAPTER

11

Oceans of Water and Air

1 If you could tour the ocean floor, what features might you see?

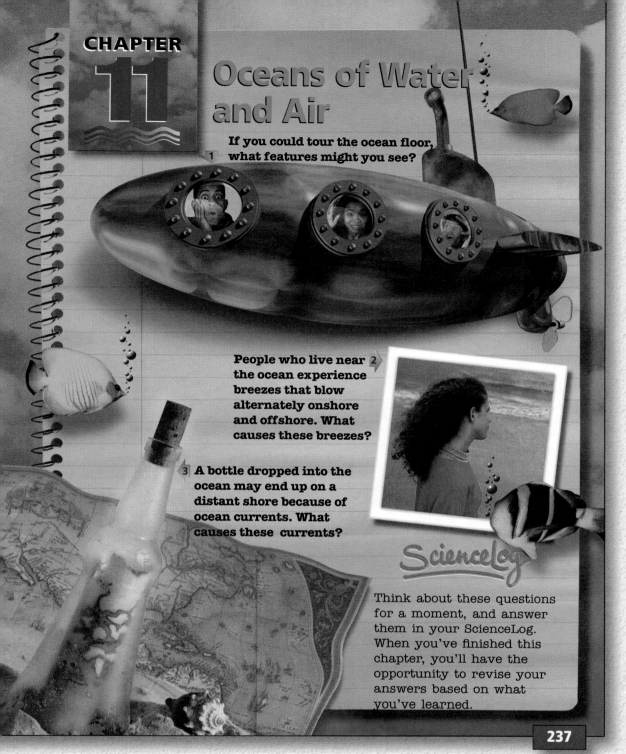

2 People who live near the ocean experience breezes that blow alternately onshore and offshore. What causes these breezes?

3 A bottle dropped into the ocean may end up on a distant shore because of ocean currents. What causes these currents?

ScienceLog

Think about these questions for a moment, and answer them in your ScienceLog. When you've finished this chapter, you'll have the opportunity to revise your answers based on what you've learned.

237

Connecting to Other Chapters

> **Chapter 10**
> offers students the chance to explore the factors that affect the Earth's climate.

> **Chapter 11**
> examines the way density affects the flow of currents in the ocean and in the atmosphere.

> **Chapter 12**
> explores the concept of pressure as it relates to the atmosphere and to the ocean.

Prior Knowledge and Misconceptions

Your students' responses to the ScienceLog questions on this page will reveal the kind of information—and misinformation—they bring to this chapter. Use what you find out about your students' knowledge to choose which chapter concepts and activities to emphasize in your teaching. After students complete the material in this chapter, they will be asked to revise their answers based on what they have learned. Sample revised answers appear on page 258.

In addition to having students answer the questions on this page, you may wish to have them complete the following activity: Have students list 10 or more characteristics of the ocean. Give students a few examples to start their lists, such as the following: the ocean contains currents; the ocean has tides; the ocean contains only salt water and no fresh water. Collect the lists, but do not grade them. Instead, read the lists to find out what students already know about ocean characteristics, what misconceptions they may have, and what about the subject is interesting to them.

Hidden Wonders

LESSON
1

Hidden Wonders

FOCUS

Getting Started

Ask: What do you think the ocean bottom looks like? Is it flat? Is it covered with sand? Are there mountain ranges and canyons under the ocean? *(Accept all reasonable responses.)* Point out that in this lesson students will learn about six of the geographic features found on the ocean bottom.

Main Ideas

1. Many of the geologic features found on land have counterparts under the oceans.
2. The ocean has its own ecosystems, to which living things have become adapted.
3. Processes occurring in the ocean contribute to global warming.

TEACHING STRATEGIES

A Whirlwind Tour

This material introduces students to some of the features of the ocean floor. Ask students to point out as many features as they can on the map. Then call on volunteers to read aloud each of the six stops on the tour. The teaching strategies for each stop provide additional information about each location.

Answer to
In-Text Question

Ⓐ Students may identify abyssal plains, trenches, continental shelves, mountains, and mid-ocean ridges, all of which are visible on this map.

★ A Transparency Worksheet (Teaching Resources, page 31) and Transparency 35 are available to accompany Hidden Wonders.

A Whirlwind Tour

If you were to plan a tour of Earth's great natural wonders, would you include the Grand Canyon? Mount Everest? the Amazon River? All of these sights are truly spectacular. But did you know that some of the world's great natural wonders are hidden from view? These natural wonders are beneath the oceans, which cover 70 percent of the Earth's surface. The oceans conceal huge canyons, enormous peaks, and mountain ranges that are grander than any on land.

The quick six-stop tour on the next few pages will introduce you to some of the world's great hidden wonders. With each stop there is a question to ponder. After reading about each stop on the tour, suggest at least one new question. These questions may lead to research you can do on your own.

First, look at the map shown here. It shows the ocean floor as it would look if all the water were drained away. What "hidden wonders" can you find on your own? Ⓐ

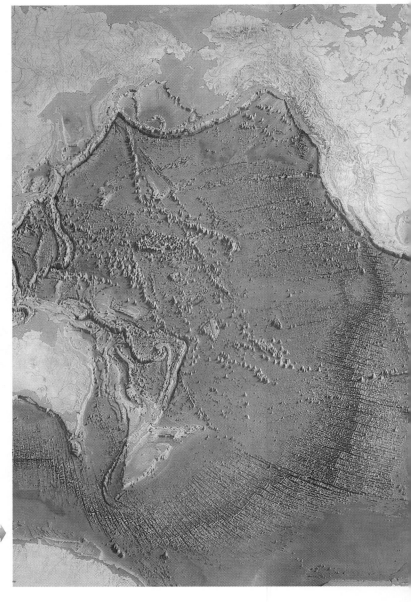

This map shows what the ocean floor would look like if all the water were drained away.

LESSON 1 ORGANIZER

Time Required
1 class period

Process Skills
inferring, comparing, contrasting, analyzing

Theme Connection
Structures

New Terms
Abyssal plains—large, flat areas of the ocean floor
Hydrothermal vents—springs along the ridges that gush hot, mineral-rich waters

Mid-ocean ridges—continuous mountain ranges on the floor of the oceans, varying in height from 500 to 5000 km
Trenches—the deepest parts of the ocean, where the ocean bottom is being pushed downward into the mantle

Materials (per student group)
Pictures of the Deep: poster board; colored markers; construction paper

Teaching Resources
Transparency Worksheet, p. 31
Transparency 35

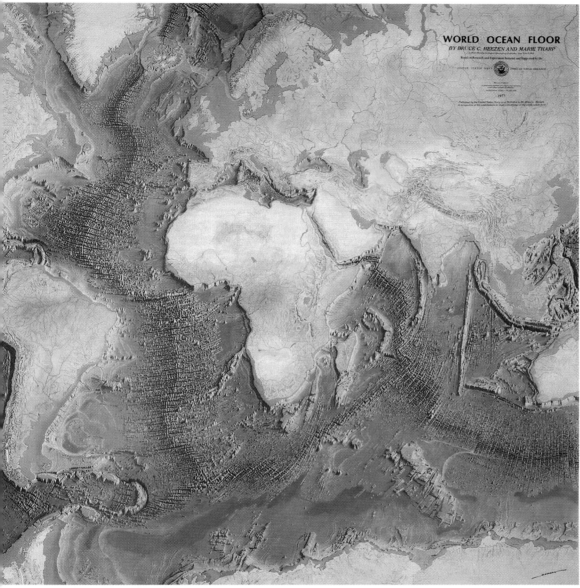

WORLD OCEAN FLOOR
BY BRUCE C. HEEZEN AND MARIE THARP

World Ocean Floor by Bruce C. Heezen and Marie Tharp, 1977

239

Cooperative Learning
A WHIRLWIND TOUR, PP. 238–242

Group size: 2 to 3 students

Group goal: to perform library research in order to identify geological features of the ocean floor and to create a brochure of the ocean bottom

Positive interdependence: Assign roles as follows: primary researcher (to assign research tasks to group members and facilitate library research), designer (to direct the design of the brochure), and presenter (to present the group's brochure to the class). Each group should prepare a brochure for an assigned part of the ocean floor. Each brochure should include important facts about the assigned area and a map showing important features and their location.

Individual accountability: Each student should answer the questions at each stop in the Whirlwind Tour.

Meeting Individual Needs

Second-Language Learners

Have students make a map of the ocean floor that includes labels of some of its hidden wonders. These labels should be given both in the students' native language and in English. You may wish to display the maps around the classroom. Students may be interested to see how the names of the Earth's geological features vary in different languages.

Meeting Individual Needs

Gifted Learners

Have students research a geological feature of the ocean floor. To stimulate interest, you may wish to have students choose which features they would like to investigate. Choices may include deeps, ridges, reefs, or volcanoes. Have students present their findings to the class in the form of an oral report. Encourage students to construct visual aids to accompany their presentations.

Did You Know...

The Southern Hemisphere consists of about 80 percent ocean, and the Northern Hemisphere consists of about 60 percent ocean.

Answer to
Stop 1: Undersea Canyons

Accept all reasonable questions. The following information answers the question posed in the text: For years, the origin of underwater canyons was the subject of debate among geologists. At first, many believed that rivers had carved the canyons during glacial times, when the sea level was much lower. However, surveys revealed that these canyons were much deeper than anyone had predicted. For a river to have carved such a deep canyon in the past, the level of the sea would have to have been incredibly low. Another theory suggested that these canyons had been cut by turbidity (underwater) currents carrying abrasive particles. Turbidity currents form when landslides of various materials run down a slope. These underwater landslides are probably triggered by earthquakes or by the force of gravity.

Stop 2: The Abyssal Plains

The abyssal plains are striking in their flatness. They are flatter than any areas on land—one such plain off the East Coast of the United States varies little more than 30 cm in elevation over a distance of 1.6 km. The plains are covered with hundreds and sometimes thousands of meters of sediments in which evidence of the past are layered. These layers contain the shells, skeletons, and fossils of untold numbers of sea creatures, as well as meteorites and volcanic ash. Also found are rocks, sand, and pebbles carried many kilometers south of their source by great icebergs, silt and other materials dumped into the ocean by rivers, and deposits of sand and gravel from underwater landslides.

Answer to
Stop 2: The Abyssal Plains

Accept all reasonable questions. The following information answers the question posed in the text: The abyssal plains are formed by the constant settling of material from above. The material that drifts down to the depths of the abyssal plains blankets the sea bed, filling up holes and evening out the surface so that it becomes flat.

Stop 1: Undersea Canyons

Everyone has heard of Arizona's awesome Grand Canyon, but not many people realize that only 350 km from New York City is Hudson Canyon, which is every bit as impressive. Some undersea canyons, with walls 5000 m high, would even dwarf the Grand Canyon.

How do you think undersea canyons were formed?

Stop 2: The Abyssal Plains

Much of the deep ocean bottom is made up of *abyssal plains,* vast stretches of featureless wasteland that are flatter and more barren than any place on land. The abyssal plains are forbidding places of eternal darkness, crushing pressures, freezing temperatures, and absolute stillness. Even so, surprising numbers of living things are found there. The remains of tiny dead plants and animals cover the bottom, accumulating with unimaginable slowness—1 cm or less every thousand years. The slightest disturbance raises blinding clouds of mud.

Why do you think the abyssal plains are so flat?

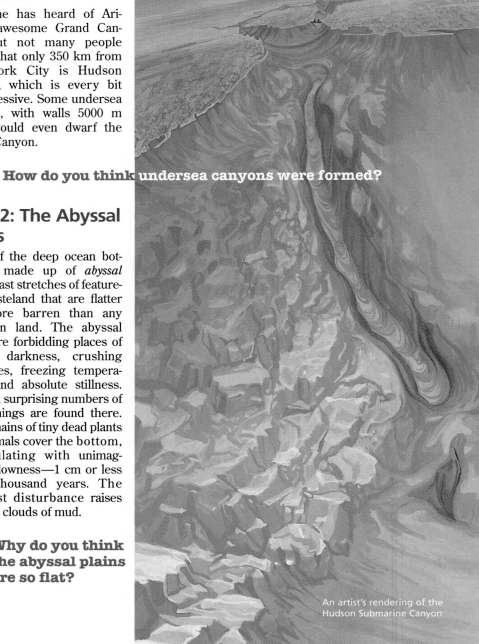

An artist's rendering of the Hudson Submarine Canyon

240

Stop 3: The Mid-Ocean Ridges

Lacing the Earth's surface like the seams on a baseball, the mid-ocean ridges form the world's largest and longest mountain ranges. Hidden beneath the ocean's surface, these ridges consist of harshly rugged landscapes and deep canyons. The mid-ocean ridges form the boundaries between crustal plates. Here, in the middle of the ridges, huge plates are separating, allowing molten rock to flow from the Earth's interior and create new oceanic crust.

Where is the Mid-Atlantic Ridge on this map of the ocean floor?

Stop 4: Hydrothermal Vents

Along the length of the Mid-Atlantic Ridge, volcanoes and hydrothermal vents (springs that gush hot, mineral-rich waters) are common. Clustered about the hydrothermal vents are strange, previously unknown ecosystems. The existence of these ecosystems disproves the long-held assumption that all life-forms depend ultimately on the energy of the sun. Here, where not even the faintest glimmer of light from the surface can penetrate, dwell giant tube worms and clams, snow-white crabs, weird fish and crustaceans, and other bizarre creatures. This ecosystem is based on a species of bacteria that derives energy and nutrients from the mineral-rich waters of the hydrothermal vents.

What would happen to the organisms around hydrothermal vents if the vents were to stop flowing?

241

Stop 5: The Mariana Trench

Explain to students that some of the deepest areas of oceans are located near shorelines. Trenches are located along the island chains of the western Pacific, near the Aleutian Islands, and off the west coast of Central and South America. The Mariana Trench, near Guam, is 11,033 m deep.

Ask students to review the theory of plate tectonics so that they will remember why trenches are found in certain locations. *(Trenches exist where there are breaks in the Earth's crust [separating boundaries] and where the crust is being pushed down into the mantle [converging boundaries].)*

Answer to
Stop 5: The Mariana Trench

Accept all reasonable questions. The following information answers the question posed in the text: Organisms living at the bottom of the Mariana Trench would require adaptations to handle very cold temperatures, sparse food, crushing pressure, and a lack of oxygen.

Stop 6: Hawaiian Volcanoes

Invite interested students to find out more about the volcanoes on the big island of Hawaii. *(They are located in the Volcanoes National Park.)* Mauna Kea, rising 4205 m above sea level, is the world's highest island peak. It was named Mauna Kea, which means "white mountain," because it is high enough to be covered with snow in the winter. Mauna Loa rises 4169 m above sea level. Several eruptions of Mauna Loa have produced tremendous amounts of lava. Kilauea lies on the eastern slope of the larger volcano, Mauna Loa. Kilauea rises to an elevation of 1247 m. Kilauea has erupted many times since the 1950s.

Answer to
Stop 6: Hawaiian Volcanoes

Accept all reasonable questions. The following information answers the question posed in the text: The volcanic islands are forming over a hot spot in the Earth's crust. As plate movement carries the older islands to the northwest, new islands are formed. The islands appear to "trail off" because the older islands have been worn down due to erosion.

Stop 5: The Mariana Trench

Trenches are the sites at which the ocean crust is slowly being pushed, centimeter by centimeter, back into the interior of the Earth. Here the ocean floor dips sharply downward, forming steep-walled valleys that are deeper, in some cases, than Mount Everest is high. An anvil dropped from a boat floating over the deepest part of the Mariana Trench would take 90 minutes to hit bottom. Incredible pressures and an almost total lack of oxygen make life in any ocean trench a difficult proposition.

What special adaptations would an organism living at the bottom of the Mariana Trench have to have in order to survive?

The Mariana Trench, which would swallow Mount Everest, is about 11,000 m deep at its deepest point.

Stop 6: Hawaiian Volcanoes

The Hawaiian Islands are a familiar sight, but did you realize that these islands are only the topmost parts of much larger landmasses? The rest lies beneath the ocean. Measured from its base at the ocean bottom, the island of Hawaii (at 9850 m) would be taller than Mount Everest! Volcanoes such as those that formed Hawaii play a major role in the global cycle of greenhouse gases. Many other mountains of the ocean's floor, known as seamounts, do not reach the ocean's surface.

Why do the Hawaiian Islands "trail off" to the northwest?

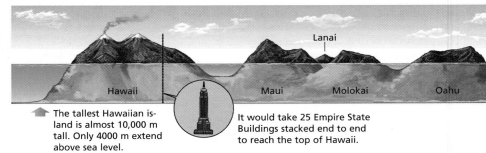

Lanai

Hawaii   Maui   Molokai   Oahu

The tallest Hawaiian island is almost 10,000 m tall. Only 4000 m extend above sea level.

It would take 25 Empire State Buildings stacked end to end to reach the top of Hawaii.

Pictures of the Deep

Have you thought of any questions that you could explore further? Think of yourself as a travel agent, and develop a poster, brochure, or article to inform others about your tour. Try to include an answer to one of your questions in your travel guide. An encyclopedia in your school library is a good place to start your research.

242

FOLLOW-UP

Reteaching

Display a topographic map of the ocean floor. Ask students to make a model using clay or papier-mâché to illustrate as many features mentioned in the lesson as possible. Suggest that they add labels where appropriate.

Assessment

Have students create a picture book that illustrates a cartoon character's underwater expedition to explore the different features of the ocean floor.

Extension

Have students do research to find out more about the organisms that live in one of the zones described in this lesson.

Closure

Have students write an adventure story in which a submarine and its crew encounter the dangers of the deep. Encourage students to include terms they have encountered in this lesson.

The Moving Oceans

The Mediterranean Puzzle

For thousands of years sailors have used winds and ocean currents to propel their ships. Ancient peoples spent much time speculating about causes of winds and ocean currents, but it was left to modern science to solve the puzzle of their origin. What causes these currents in the ocean and atmosphere? Are they somehow related? What role do these currents play in the global climate? These are questions we will investigate in this lesson.

Let's start with an ancient puzzle that was finally solved by Count Luigi Marsili in 1679.

Examine the map of the Mediterranean Sea. The puzzle is this: Sailors had long known that swift currents flowed into the Mediterranean from both the Black Sea and the Atlantic Ocean. Many rivers and streams also empty into it. The Mediterranean has no apparent outlet; therefore, the water level should rise, but it does not. Many explanations were offered—for example, the existence of hidden underground channels to drain the excess water. Can you solve this puzzle?

 The Mediterranean puzzle: Where does the water exit?

Count Marsili thought he could, and so he set up a model of the Mediterranean to test his idea. In the following Explorations you will trace the steps of Count Marsili as you discover the answer to the Mediterranean puzzle. At the end of Exploration 3, be prepared to describe the Mediterranean puzzle and explain its solution.

LESSON 2 ORGANIZER

Time Required
4 to 5 class periods

Process Skills
inferring, predicting, comparing, contrasting

New Terms
Density—the mass-to-volume ratio of a substance
Hydrometer—an instrument that measures the density of a liquid

Materials (per student group)
Exploration 1: 150 mL of salt; 5 cm deep aluminum roasting pan (23 cm × 36 cm); 100 mL graduated cylinder;

stirring rod; watch or clock; scissors; piece of aluminum foil slightly larger than the roasting pan; about 25 cm of masking tape; a few drops of food coloring; a small amount of pepper; 2 containers, each 1 L or larger; 2 L of water; lab aprons (See Advance Preparation on page 209C.)
Exploration 2, Part 1: pill bottle or similar small container; about 100 mL of water; 100 mL of each salt solution from Exploration 1; metric balance; 50 or 100 mL graduated cylinder; **Part 2:** See Advance Preparation on page 209C;

continued

LESSON

2

The Moving Oceans

FOCUS

Getting Started

After students have read this page, direct their attention to the map of the Mediterranean Sea. Ask: Where is it located? *(It is located between Europe and Africa.)* How many openings does it have? *(Two, the Strait of Gibraltar and the opening to the Black Sea)* Then have them identify the many rivers and streams that empty into the Mediterranean. Have students read the material on this page. Ask students to speculate about how water might leave the Mediterranean. Write their responses on the board. Explain to students that they will find answers to these questions when they perform the Explorations on pages 244–246.

Main Ideas

1. To calculate the density of a substance, divide its mass by its volume.
2. Ocean water differs in density because of differences in salinity, excessive evaporation, dilution by runoff, warming or cooling, and formation of ice floes.
3. Hydrometers are used to measure the density of liquids.

TEACHING STRATEGIES

Answer to
In-Text Question

Ⓐ Students will develop answers to this puzzle as they work through Explorations 1–3.

Homework

You may wish to assign Exploration 1 on page 244 as homework.

⭐ An Exploration Worksheet is available to accompany Exploration 1 on page 244 (Teaching Resources, page 33).

EXPLORATION 1

Divide the class into small groups and distribute the materials. Emphasize the importance of making the barrier water-proof. Suggest that they press firmly on the masking tape. Be sure that students gently add liquid to both sides of the container at the same time. Suggest that students write their observations in their ScienceLog. Encourage students to use colored pencils so that their diagram is as clear as possible.

Answers to
Interpreting Your Findings

1. Diagrams of the demonstration should show a surface current flowing from the less salty liquid to the more salty liquid and a deeper current flowing in the opposite direction. Their diagrams of the Mediterranean Sea should show a similar flow—less salty water flowing across the surface from the Atlantic Ocean and the Black Sea into the saltier Mediterranean. Deeper in the water, the saltier water should flow from the Mediterranean to the Atlantic Ocean and Black Sea.

   The demonstration works as a model because it shows how less dense water flows on top of denser water. However, the model is limited in that the currents flow within a closed system, unlike the currents that flow into and out of the Mediterranean Sea.

2. Students should infer that the Atlantic Ocean and the Black Sea are less salty than the Mediterranean Sea. The less salty solution represents the Atlantic Ocean and the Black Sea; the saltier solution represents the Mediterranean.

3. The greater saltiness of the Mediterranean Sea can be attributed to the fact that more water is evaporating from it than is being added to it by rain or runoff. As the water evaporates, the salt is left behind. In contrast, most of the water that evaporates from the Atlantic Ocean returns to the ocean as rain.

4. Sample answer: Marsili discovered that the deeper water is heavier and thus has the greater density (greater mass for the same volume). The next Exploration will confirm this prediction.

EXPLORATION 1

Unraveling the Puzzle

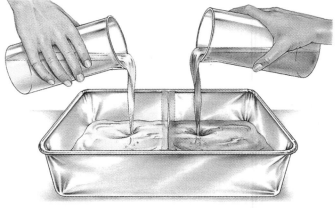

You Will Need

- salt
- an aluminum roasting pan
- aluminum foil
- masking tape
- food coloring
- pepper
- 2 containers that hold at least 1 L of water each
- water
- a sharpened pencil

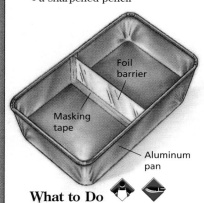

Foil barrier

Masking tape

Aluminum pan

What to Do

1. Cut a piece of aluminum foil slightly wider and taller than the pan. Using masking tape, make a waterproof barrier that divides the pan into two parts.

2. To one container, add 1000 mL of water, 100 mL of salt, and a few drops of food coloring. Stir until the salt is completely dissolved. In the second container, dissolve 50 mL of salt in 1000 mL of water. Do not add food coloring. Save 100 mL of each solution for Exploration 2.

3. With the help of a partner, pour each solution into opposite sides of the pan at the same time (so that the barrier doesn't collapse). Be careful to make the water level on both sides equal. Now sprinkle some pepper into the side containing food coloring.

4. With a pencil point, punch a hole in the aluminum foil just below the surface of the water. Make another hole in the foil near the bottom of the container. Observe the setup for about 10 minutes.

Interpreting Your Findings

1. Make a diagram in your ScienceLog to record what you observed. Use arrows to show current flow. Could this demonstration serve as a model of the Mediterranean Sea? Explain. Using what you have learned, draw a diagram of the Mediterranean showing the inflow and the outflow at both entrances.

2. From your observations, which is saltier—the Atlantic Ocean or the Mediterranean Sea? the Black Sea or the Mediterranean? Which salt solution represents the Atlantic? Which represents the Mediterranean?

3. Why should the saltiness of the Mediterranean differ from that of the Atlantic? (Hint: Most of the Mediterranean basin has a warm and dry climate.)

4. Count Marsili thought that he could explain how the saltiness of ocean water could cause currents. To test his idea he drew water samples from different depths in the Strait of Gibraltar, which connects the Mediterranean and the Atlantic. Using an equal volume of each sample, he found their masses and compared them with the mass of an equal volume of fresh water. What do you think he found?

   Test your prediction by doing the next Exploration.

ORGANIZER, continued

Part 3: solutions from Exploration 1; 3 test tubes; 3 pieces of masking tape; permanent marker; a few drops each of 2 different colors of food coloring (both different from the color used in Exploration 1); fire-polished glass tubing or clear plastic drinking straws; test-tube rack; small beaker or other container; lab aprons

Exploration 3: about 1 L of 10% salt solution; 100 mL of cold tap water; 100 mL of hot tap water; few drops each of 2 different colors of food coloring; 2 small, clear jars; large pan or bucket; 4 index cards or stiff pieces of paper; watch or clock with second hand; lab aprons (See Advance Preparation on page 209C.)

Exploration 4: a variety of materials such as a pencil, modeling clay, plastic drinking straw, cork, thumbtacks, and heavy gauge wire; 100 mL graduated cylinder; 600 mL of water; 12.5 g of salt; permanent marker (See Advance Preparation on page 209C.)

Teaching Resources
Exploration Worksheet, pp. 33, 35, 38, and 39
Transparency 36
SourceBook, pp. S69 and S73

Marsili's Explanation

You Will Need

- a pill bottle or similar small container
- the colored and uncolored salt solutions from Exploration 1
- tap water
- 3 test tubes
- glass tubing or clear drinking straws
- a balance
- a graduated cylinder (50 mL)
- food coloring of 2 different colors
- a beaker or other container
- a pencil or permanent marker

PART 1

Comparing Masses

1. Determine the volume (in milliliters) and mass (in grams) of the pill container.

2. Fill the container with one of your salt solutions or with tap water. Determine the mass of the container and liquid combined. What is the mass of the liquid?

3. Repeat step 2 with the remaining two liquids. Be sure that you are measuring equal volumes of each liquid. How does the mass of the "Mediterranean" solution compare with that of the "Atlantic" solution? How do

both of these masses compare with the mass of tap water?

4. Divide the mass of each solution by its volume. Which solution has the greatest mass per volume? the least mass per volume?

5. Save your solutions for use in Part 3 of this Exploration.

PART 2

Thinking About Density

In step 4 of Part 1 you actually found the density of each solution. As you may recall from Chapter 6, density is mass per unit of volume. Read the following questions about density, and then answer them in your ScienceLog.

Which liquid had the highest density? the lowest density?

The density of any material can be found using this formula:

Density = mass/volume

What are the densities of the following materials?

Material	Mass of 30 mL	Density
water	30.0 g	?
alcohol	21.0 g	?
ice	27.6 g	?
salt water	33.0 g	?
egg	31.5 g	?

Pill bottle — Balance scale

Which materials would float on fresh water? on alcohol? on salt water? Devise an experiment to test your predictions.

PART 3

Layering Liquids

Place three empty test tubes marked *A*, *B*, and *C* in a test-tube rack. Fill test tube *A* with the colored liquid used in Part 1. Color one of the other liquids from Part 1 with a color of food coloring different from that used in liquid *A*, and fill test tube *B* with it. Color the remaining liquid used in Part 1 with the third color of food coloring, and fill test tube *C*. Now use the technique shown below to layer the three liquids in a thin glass tube or drinking straw. Can the liquids be layered in any order?

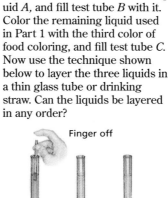
Finger off
A   B   C

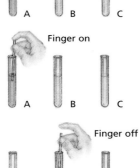
Finger on
A   B   C

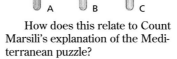
Finger off
A   B   C

A   B   C

How does this relate to Count Marsili's explanation of the Mediterranean puzzle?

PART 2

Be sure that students understand the concept of density before beginning this activity. Also, be sure to check all designs for safety before students begin their investigations.

Answers to
Part 2

Density is mass per unit of volume. The "Mediterranean" solution had the highest density. The tap water solution had the lowest density.

The following are the densities of the materials in the table:

- water: 30.0 g/30.0 mL = 1.0 g/mL
- alcohol: 21.0 g/30.0 mL = 0.70 g/mL
- ice: 27.6 g/30.0 mL = 0.92 g/mL
- salt water: 33.0 g/30.0 mL = 1.1 g/mL
- egg: 31.5 g/30.0 mL = 1.05 g/mL

From the data in the table students can determine that the alcohol and ice should float in fresh water. Nothing shown in the chart would float in alcohol. Water, alcohol, ice, and the egg would float in salt water.

Accept all reasonable experimental designs. To test their predictions, students could devise an experiment in which one material at a time is placed on top of each liquid. Alternatively, students could create a density column in which each liquid is a different color and the most dense liquid is at the bottom.

PART 3

If any solutions from Exploration 1 are used up, it will be necessary to make the additional solutions.

EXPLORATION **2**

PART 1

Make sure that students understand how to determine the mass and volume of their solutions. If necessary, explain that they first need to determine the mass of the empty container. Next, to calculate the mass of the solution, they need to find the container's mass after filling it with salt solution and then subtract the mass of the empty container from the mass of the filled container. The container's volume can be determined by filling the container with

water and then measuring the volume of that water in a graduated cylinder.

Answers to
Part 1

3. A volume of the "Mediterranean" solution has a greater mass than an equal volume of the "Atlantic" solution does. Each of these salt solutions has a greater mass than does an equal volume of tap (fresh) water.

4. The "Mediterranean" solution has the greatest mass for its volume. The tap water has the least mass for its volume.

Answers to
Part 3

The liquids must be layered in order from most to least dense. Students should find that the densest liquid will always be on the bottom.

Count Marsili must have concluded that the denser, saltier water of the Mediterranean Sea flows beneath the less dense, less salty water of the Atlantic Ocean.

★ **An Exploration Worksheet is available to accompany Exploration 2** (Teaching Resources, page 35).

246

Cooperative Learning
EXPLORATION 2, PAGE 245

Group size: 2 to 3 students

Group goal: to determine density experimentally and to observe the effect of layering liquids of different densities

Positive interdependence: Assign the following roles within each group: supervisor (to keep other students on task), materials manager (to arrange equipment for the experiment and to direct cleanup), and recorder (to record and interpret data).

Individual accountability: Each student should write a summary of the Exploration that includes a definition of density, a description of an experimental procedure for determining density, and a calculation of the density of salt water using the density formula.

EXPLORATION 3

SAFETY ALERT

Water over 55°C will scald.

This Exploration serves as a summary for the ideas about density that have been presented so far. After students have completed the Exploration, refer them to the data on global temperatures on page 226, and ask them where mixing of layers of water might occur. To make the 10% salt solution, add 10 g of salt to 100 g of water.

★ **An Exploration Worksheet is available to accompany Exploration 3 (Teaching Resources, page 38).**

Answers to
Exploration 3

The following are sample answers:

Activity 1
- Observation: In (a) the liquids do not mix (the hot water remains on top). When the liquids are placed on their sides in (b), the cold water settles beneath the hot water because cold water is denser than hot water.
- Conclusion: Throughout the oceans, a warmer, less dense top layer of water floats on a colder, more dense bottom layer.

Activity 2
- Observation: When the card is removed, the cold water sinks and the hot water rises. This is because cold

---

EXPLORATION 3

Predict, Observe, Explain

Could density differences in the oceans cause currents similar to those in the Mediterranean? Perhaps the following Exploration will answer this question.

You Will Need
- a 10% salt solution
- tap water (cold and hot)
- food coloring
- 2 small jars
- a large pan or bucket
- index cards or stiff paper

What to Do

In Activities 1–4 you will invert one water-filled jar over another to study how solutions of different densities interact. Be sure each jar is filled to the rim with the correct type of water.

Set up each Activity as shown in the illustration at top right; also follow the additional instructions for each Activity. Note that after setting up the apparatus, you will turn the jars on their side only in Activities 1 and 3.

To catch any spills, carry out this Exploration over a pan or bucket. For each Activity in this Exploration, do the following:

- Predict what you expect to happen, observe what does happen, and finally, give an explanation for your observations.
- Re-examine the map of global temperatures on page 226. Where in the oceans might the phenomenon you observed in this Exploration be occurring?

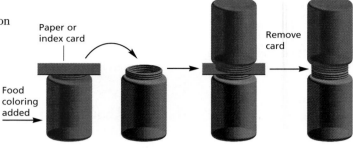
Paper or index card · Food coloring added · Remove card

ACTIVITY 1
a. Allow jars to stand upright for 40 seconds after removing the card or paper.
b. Turn jars on their side while holding them together.

ACTIVITY 2
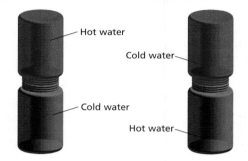
Hot water · Cold water · Cold water · Hot water

What conclusions would you draw from Activities 1 and 2?

ACTIVITY 3
Tap water · Salt water

ACTIVITY 4
Follow the same procedure as in Activity 1.
Salt water · Tap water

---

water is more dense than hot water.
- Conclusion: This occurs in polar regions where cold, dense water from melting ice sinks beneath warmer surface water. This may also occur in places where cold-water currents flow into warm areas.

Activity 3
- Observation: In (a) the liquids do not mix (the tap water remains on top). When the liquids are placed on their sides in (b), the salt water sinks under the fresh water. This is because salt water is more dense than fresh water.
- Conclusion: In regions where rivers and streams empty into the ocean, the fresh water remains above the

---

saltier water. Also, less salty water may float over denser, saltier water when precipitation occurs over salty bodies of water.

Activity 4
- Observation: When the card is removed, the salt water sinks and the fresh water rises. This is because salt water is more dense than fresh water.
- Conclusion: In regions where evaporation occurs, the surface water may become more dense and sink. This effect would be very strong in areas like the Mediterranean that receive little precipitation.

Count Marsili's Letter

After reading the letter at right, write Count Marsili's follow-up letter. Include experiences and diagrams that make your explanation clear. Keep in mind that Marco is less knowledgeable than you about density and currents.

Follow the Flow

How many factors have you discovered that affect the density of ocean water? The previous Exploration suggests two factors. The water of the Mediterranean is denser than that of the Atlantic Ocean because of evaporation. When water evaporates, the salt in it is left behind. Look at the map below. Where else in the world might this be happening? Sea water in the tropics generally has a different density from that of the sea water in cold regions. Would tropical water be more or less dense than polar water? Why?

The Norwegian Sea bordering Greenland and the Weddell Sea bordering Antarctica are both chilled by the cold winds that blow off glacier-covered landmasses. Does this increase or decrease the density of the surface water?

You may be surprised to find that ice formed from sea water is not very salty. Why? The dissolved salt does not fit well into the crystal structure of ice, so it is "squeezed" out as the water freezes. The water left behind becomes a little more salty and therefore more dense.

> Dear Marco,
> I have recently returned from the Mediterranean, where I observed the strong surface currents at both entrances.
> I believe I have found an explanation for the Mediterranean mystery. I shall elaborate in my next letter.
>
> Sincerely,
> Luigi

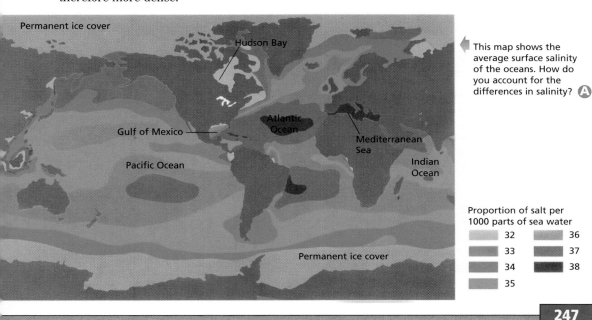

Permanent ice cover

Hudson Bay

Gulf of Mexico

Atlantic Ocean

Mediterranean Sea

Pacific Ocean

Indian Ocean

Permanent ice cover

This map shows the average surface salinity of the oceans. How do you account for the differences in salinity? **A**

Proportion of salt per 1000 parts of sea water

32		36	
33		37	
34		38	
35			

Answer to
Caption

A In their accounts of the differences in salinity, students should include all of the factors discussed on this page, including evaporation, dilution by sources of fresh water, changes in temperature, and the presence of ice. The areas of highest salinity include the Mediterranean Sea and the central Atlantic Ocean; the most important factor affecting the salinity of these areas is probably evaporation. The areas of lowest salinity are cold areas such as Hudson Bay and areas along the edge of continents; the salinity of these areas is probably determined by the presence of ice and incoming currents of fresh water, respectively.

Homework

You may wish to assign Count Marsili's Letter as homework.

 Transparency 36 is available to accompany Follow the Flow.

Count Marsili's Letter

This writing task allows students to consolidate their new ideas and discoveries. Encourage them to include diagrams and experimental results in their papers. Call on volunteers to present their papers to the class.

Answer to
Count Marsili's Letter

Students should explain the mystery by stating that the less salty water from the Atlantic Ocean and the Black Sea flows into the Mediterranean Sea on the surface. The saltier, more dense water from the Mediterranean sinks below the waters of the Atlantic Ocean and the Black Sea and flows into the Atlantic Ocean. Therefore, inward-flowing surface currents are visible at the openings of the Mediterranean Sea, but the average level of the Mediterranean Sea does not increase.

Follow the Flow

Call on a volunteer to read aloud the material on this page. Suggest that students look for at least four processes that affect the density of water.

Answers to
Follow the Flow

Students may identify the evaporation of surface water, the dilution of surface water, the heating or cooling of surface water, and the formation of ice as factors that affect the density of ocean water.

Evaporation would leave salt behind everywhere. This effect would be strongest both in warm, dry areas that receive little precipitation and in areas such as the Arabian Sea that are sheltered from strong currents. Students may also notice the high salinity of the central Atlantic Ocean, a region known as the Sargasso Sea. At this point, accept all student explanations for why this might occur. Students will learn more about the Sargasso Sea on page 273.

The waters of tropical seas are less dense than the waters of polar seas because polar seas are much colder.

Cold winds blowing off glacier-covered land would decrease the surface temperature of the water, making it more dense. The winds would also increase evaporation, making the water have a higher salt concentration. This would also increase the water's density.

Answer to *In-Text Question*

Ⓐ Most deep-ocean currents originate near the poles and flow toward the equator.

Some currents cross each other, so they probably flow at different depths.

Meeting Individual Needs

Learners Having Difficulty

Direct students' attention to the diagram of deep ocean currents on this page. Have them locate a number of deep currents and describe their paths and branches in one or two short paragraphs.

ENVIRONMENTAL FOCUS

The ocean's currents can spread pollution far away from its source. This is one reason why nations must work together to solve the problems of ocean pollution. At least 85 percent of ocean pollution originates on land. Pollutants enter rivers as runoff, and the rivers carry the polluted water to the oceans. Pollutants that are dumped directly into the oceans include garbage and by-products of waste-water treatment. One of the most damaging kinds of pollution is oil that is accidentally spilled from tankers.

Examine the diagram of deep-ocean currents shown below. These currents are caused by density differences in ocean water. Where do they originate? Where do they go? Is there evidence in the diagram that density currents occur at different depths and go in different directions? Ⓐ

These deep-ocean currents play an important role in controlling Earth's temperature. Carbon dioxide dissolves well in water (think of carbonated soft drinks). The surface water of the ocean absorbs huge amounts of carbon dioxide from the atmosphere. When currents push the surface water to the bottom, the carbon dioxide becomes concentrated in the deep-ocean waters. The deep-ocean currents circulate very slowly. The water and carbon dioxide caught in a deep-ocean current may not resurface for centuries. When deep-ocean currents finally resurface, they bring with them nutrients collected over the years. Sites of upwelling deep currents (such as along the west coast of South America) are rich in life and make excellent fisheries.

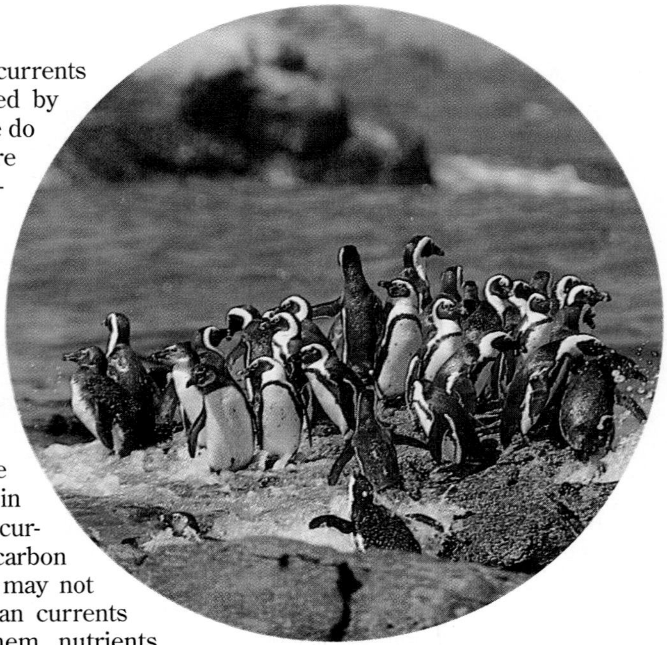

These penguins are able to live very close to the equator thanks to a cold, upwelling current off the western coast of South America.

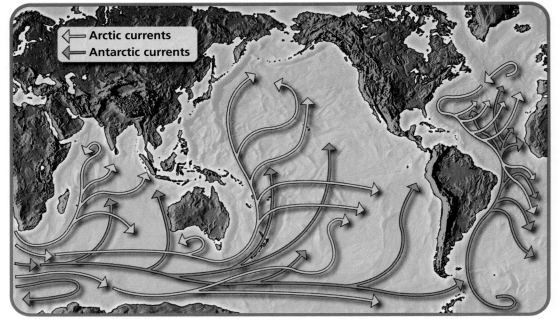

Arctic currents
Antarctic currents

Deep-ocean currents

248

The Composition and Density of Sea Water

The substances that make up sea water remain quite constant no matter where the ocean is sampled. Most sea water contains the substances shown at right.

Here is a simpler recipe for sea water:

Add 9.0 g of salt to 250 mL of water.

This solution will have the same density as the solution made with the more complicated recipe. What is the density of sea water with this salt content? How can a Ⓑ device such as the one in the photos at right be used to determine the density of a salt solution? Such a device Ⓒ is called a *hydrometer*. The hydrometer is calibrated in such a way that it reads the density of the liquid in which it is placed.

Believe it or not, it is easier for you to float in salt water than in fresh water. In fact, some bodies of water such as the Dead Sea and the Great Salt Lake are so salty that a person would find it almost impossible to sink. In the following Exploration you will use this idea to develop your own device for measuring the density of salt water.

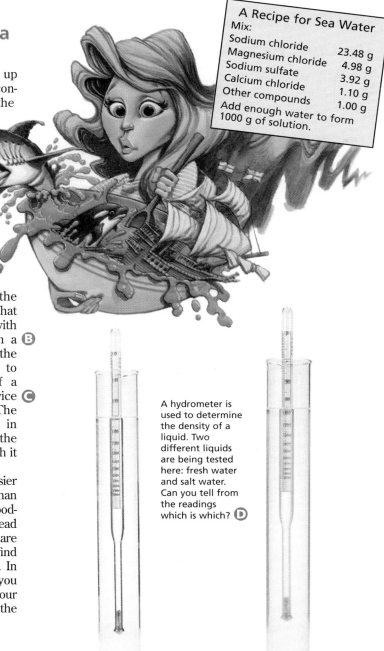

A Recipe for Sea Water
Mix:

Sodium chloride	23.48 g
Magnesium chloride	4.98 g
Sodium sulfate	3.92 g
Calcium chloride	1.10 g
Other compounds	1.00 g

Add enough water to form 1000 g of solution.

A hydrometer is used to determine the density of a liquid. Two different liquids are being tested here: fresh water and salt water. Can you tell from the readings which is which? Ⓓ

249

EXPLORATION **4**

If a real hydrometer is available, demonstrate how it works. If not, direct students' attention to the diagrams of the hydrometer on this page. Answer any questions students may have about how a hydrometer works or how to calibrate one. Then divide the class into small groups and distribute the materials.

Have graduated cylinders containing water and the standard salt solution of known density prepared ahead of time, or allow students to prepare their own. Students can use these solutions to calibrate their hydrometers.

Also have a graduated cylinder of artificial sea water of unknown density prepared. Each group can use their hydrometer to determine the density of the artificial sea water.

Answers to *Exploration 4*

2. Students may assume that the scale can be extended above and below the two marks in increments that are equal to the distance between the two marks. Each mark would indicate a change in density of 0.05 g/mL.

Sensitivity of the hydrometers will vary depending on design.

3. Using the formula on page 245, the density could be calculated once the mass and volume of each sample had been determined.

 An Exploration Worksheet is available to accompany Exploration 4 (Teaching Resources, page 39).

EXPLORATION **4**

The Great Hydrometer Challenge

You Will Need
- an assortment of common materials such as a pencil, modeling clay, a drinking straw, cork, wire, and thumbtacks
- a graduated cylinder
- water
- a "standard" salt solution with a density of 1.05 g/mL (made by adding 12.5 g of salt to 250 mL of water)
- artificial sea water (using either recipe on page 249)
- a permanent marker

What to Do

1. Using the materials listed here or others of your choice, make three hydrometers. Test your hydrometers for their ability to float in a graduated cylinder filled with water.

2. You can *calibrate* your hydrometers by placing them in solutions of different known densities and marking the level to which they sink. First place the hydrometer in tap water and mark the level to which it sinks. Follow this same procedure using the standard salt solution. Using the results of your calibration, can you extend the scale above and below your marks? Which of your hydrometers appears to be the most sensitive?

3. Use your best hydrometer to measure the density of the artificial sea water. How could you confirm this reading?

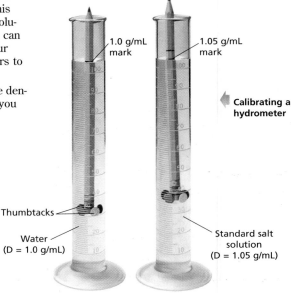

Pencil

Drinking straw

Modeling clay

Thumbtacks

Two possible hydrometer designs

1.0 g/mL mark

1.05 g/mL mark

Calibrating a hydrometer

Thumbtacks

Water (D = 1.0 g/mL)

Standard salt solution (D = 1.05 g/mL)

FOLLOW-UP

Reteaching

As a demonstration, present students with several samples of powdered drinks or non-carbonated juice drinks that have different salt concentrations. Give students a list of salt concentrations, and have them carefully layer the samples in a plastic drinking straw in order to determine the relative concentration of each drink sample. (Have them refer to the layering technique described on page 245.)

Assessment

Have students write a short story in which a powerful alien being has the power to alter the salt content and temperature of the oceans. Encourage students to focus their stories on how these changes in the oceans would affect human and marine life.

Extension

Have students research ways in which marine animals take advantage of ocean currents for migration or other purposes.

Closure

Have students use all of the following phrases in a paragraph describing how ocean waters move: (a) hot water rises, (b) cold water sinks, (c) salty water sinks, (d) less salty water rises, (e) hot, salty water evaporates, leaving salt behind, (f) density increases, and (g) density decreases.

LESSON 3 The Atmosphere

Density Currents in the Atmosphere

Differences in density cause water to flow. Can differences in density also cause air to flow? What would cause density differences to develop in air? Examine the diagram below, which shows a breeze blowing from the ocean to the land during the day. Before reading further, give your explanation of what causes such breezes to blow. Then review the activities you have done in this unit for evidence that each of the following statements is true.

1. In the daytime, land heats up more quickly than does water.

2. The air over land warms and becomes less dense *(A)*. This air rises above (floats on) the denser, surrounding air.

3. Cooler and denser air moves in from the ocean *(B)*, taking the place of the rising air.

4. As air rises, it spreads out and cools *(C)*, thus becoming more dense. The air then sinks *(D)*, completing the cycle.

Draw a similar labeled diagram to show what happens at night, when there is no sun to warm either the land or sea and both begin to cool off.

Storm clouds build over the tropics. Why do these clouds form? Ⓐ

251

LESSON 3 ORGANIZER

Time Required
2 to 3 class periods

Process Skills
inferring, predicting, communicating

Theme Connections
Changes Over Time, Cycles

New Terms
Air masses—bodies of air that cover large areas and have nearly uniform temperature and humidity
Cold front—the leading edge of an advancing cold air mass

Trade winds—almost constant winds that blow toward the equator as a result of unequal heating of the atmosphere
Warm front—the leading edge of an advancing warm air mass

Materials (per student group)
none

Teaching Resources
Transparency Worksheet, p. 40
Activity Worksheet, p. 42
Transparencies 37 and 38
SourceBook, p. S72

LESSON 3 The Atmosphere

LESSON 3 The Atmosphere

FOCUS

Getting Started

Have a volunteer read the opening paragraph aloud, stopping after each question to allow for discussion. To help students grasp the idea of atmospheric density currents, ask them to suggest why most air-conditioning vents are best located near the ceiling. *(Just as cold water sinks in warm water, cold air sinks in warm air.)*

Main Ideas

1. Land and sea breezes and the trade winds can be explained in terms of differences in atmospheric density.

2. An air mass is a volume of air with a specific density that exists over a wide area.

3. Warm fronts occur when a warm air mass invades a cold air mass; cold fronts occur when a cold air mass invades a warm air mass.

TEACHING STRATEGIES

Answers to
Density Currents in the Atmosphere

During the day, the land-surface temperature becomes higher than the water-surface temperature. The warmer air over the land then rises. The cooler air over the ocean moves in to take its place.

Observations that support each statement include the following:

1. Chapter 10, Exploration 3, Activity 2
2. Chapter 11, Exploration 3, Activities 1 and 2
3. Chapter 11, Exploration 3, Activities 1 and 2
4. Chapter 11, Exploration 3, Activity 2

Student diagrams should show that at night, the breeze blows from the land to the ocean.

Answer to
Caption

Ⓐ Storm clouds form over the tropics because there is more water and more heat there and because more evaporation takes place there.

Air Masses

Remind students that in Lesson 2 they learned that bodies of water with different densities, due to either salinity or temperature, do not mix readily. Point out that air behaves in a similar way. Two masses of air with different densities do not mix readily. Just as water that is more dense sinks and pushes up water that is less dense, air that is more dense sinks and pushes up air that is less dense.

Before students respond to the questions, call on a volunteer to summarize the material on this page. (The density of a mass of air depends on what lies under it.)

In order to answer the in-text questions, students will need to know that water vapor is less dense than the other gases that make up air. (This is a result of the differing masses of the gas molecules. For example, one molecule of oxygen gas, O_2, has more mass than one molecule of water vapor, H_2O.)

Answers to
In-Text Questions

Ⓐ Many students will guess correctly that air and water behave similarly in that air or water masses of different densities do not mix readily.

Ⓑ Air over the Sahara tends to be warm and dry, and air over Siberia in winter tends to be cold and dry. Because of the difference in temperature, the air mass over Siberia in winter is far more dense than the air mass over the Sahara.

Ⓒ 1. • *Continental polar*—cold, dry, and very dense because it forms above cold, dry regions
  • *Maritime polar*—cold and moist because it forms over cold waters. It is less dense than cold, dry air.
  • *Maritime tropical*—warm and moist because it forms above tropical seas. It is the least dense of all the types of air masses.
  • *Continental tropical*—warm and dry; less dense than cold air but more dense than warm, moist air

2. Assuming that air masses act like bodies of water, the results from Explorations 1 and 3 in this chapter suggest that different types of air masses will not mix readily.

Where else in the world would you find similar kinds of density currents in the atmosphere, only on a much larger scale? Think about the tropics. What happens to the air over the tropics as it warms? It rises, of course, and as you would expect, cooler air from farther north and south moves in to take its place. This pattern is seen throughout the tropics. The winds that result are called *trade winds*. Trade winds blow steadily from season to season and year to year.

Air Masses

You have seen that bodies of water with different densities do not mix readily.
Ⓐ Would you expect air to behave similarly?

There is constant interaction between a body of air and the land or water it lies over. Air over warm water becomes warm and moisture-laden, while air over a cold landmass becomes cold and dry. What characteristics would air over the Sahara have? How about air over Siberia in the winter? Would these two *air masses* have about the same density? Ⓑ

An **air mass** is a body of air that covers a large area and has nearly uniform temperature and humidity throughout. The diagram at right shows the air masses that affect weather in North America in summer and winter. Use it to answer the following questions: Ⓒ

1. Air masses are either polar or tropical and either maritime or continental. The name of an air mass has two parts, for example, *continental polar*. Here is what each part indicates:

   continental—dry   polar—cold

   maritime—moist   tropical—warm

   Suggest how each type of air mass compares in terms of its temperature, moisture content, and density, based on its location and name. Look back over the last Exploration for observations to support your answer to this question.

2. Would the different types of air masses readily mix together? Which activity that you performed suggests that they do not?

3. What type of air mass is causing your weather today?

4. What type of air mass(es) will probably be causing your weather 6 months from now?

5. What type of air mass do you think you would find over the Atlantic Ocean near the equator?

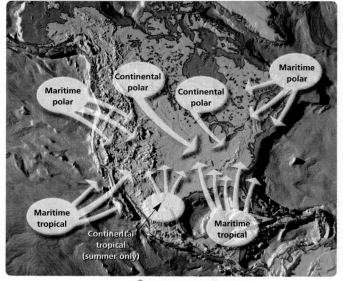

Air masses that affect our weather

To find out more about the behavior of the air and water in our atmosphere, see pages S66–S74 in the SourceBook.

---

3. Answers will vary depending on the location and time of year.

4. Answers will vary depending on the location and time of year.

5. For most of the year, a maritime tropical air mass covers the region.

 A Transparency Worksheet (Teaching Resources, page 40) and Transparency 37 are available to accompany Air Masses.

Air Masses in Motion

If a weather forecaster said, "Look for a Canadian air mass to sweep into our area tomorrow," what kind of weather would you expect? Since much of our weather is the result of collisions between air masses, what would you see if you could see the boundary between air masses of different densities? These boundaries are called *fronts*. The following diagrams show two types of fronts. Answer the questions that go with each diagram.

A Cold Front

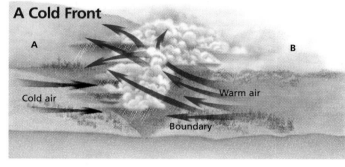

◀ **A cold front** The blue arrows indicate the overall movement of the cold air mass. The red arrows indicate the movement of the air within the warm air mass.

1. Which air mass is more dense?
2. What happens to the warm air as the cold air approaches?
3. What causes the clouds to form?
4. Compare the temperature at *A* with that at *B*. How will the temperature at *B* change shortly?
5. The boundary between the cold air and the warm air is called a cold front, not a warm front. Why?

A Warm Front

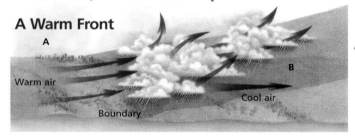

◀ **A warm front** The red arrows indicate the movement of the air within the warm air mass. The blue arrow indicates the overall movement of the cold air mass.

1. In the illustration above, which air mass is moving forward faster—the cold air mass or the warm air mass?
2. What is happening to the warm air as it overtakes the cold air mass?
3. Why are clouds forming?
4. How does the temperature at *A* compare with that at *B*? How will it change?
5. Why is the boundary between the warm air and the cold air called a warm front, not a cold front?

253

The Activity Worksheet on page 42 of the Unit 4 Teaching Resources booklet makes an excellent homework assignment to accompany Air Masses in Motion.

Theme Connection

Changes Over Time

Focus question: How do patterns of change help meteorologists predict the weather? *(Meteorologists analyze weather patterns and predict the weather by comparing past weather changes with current weather patterns. From past experiences, meteorologists know that most weather changes occur when a warm front or cold front moves into an area. By tracking the progress of these fronts, they predict when the weather changes are likely to occur.)*

CROSS-DISCIPLINARY FOCUS

Mathematics

Discuss the "probabilities" that weather forecasters use to describe the chance of oncoming rain. A 10-percent probability of rain means there is a 1 in 10 chance of rain falling during daylight hours at a given location. Ask: If there were a 20-percent chance of rain five days in a row and no rain fell for the first four days, how likely would rain be on the fifth day? *(There would still be a 20-percent, or 1 in 5, chance. Probabilities do not accumulate.)*

Answers to

A Cold Front

1. The cold air mass is more dense than the warm air mass.
2. As cold air approaches, warm air rises above it because warm air is less dense than cold air. Clouds form in the warm air.
3. Clouds form because the warm air cools as it rises, causing water vapor to condense.
4. The temperature at *A* is cooler than the temperature at *B*. As the cold front passes over, the temperature at *B* will drop.
5. The boundary is called a cold front because the cold air mass is advancing.

Homework

Have students answer the questions in A Cold Front and A Warm Front as a homework assignment.

Answers to

A Warm Front

1. The warm air mass is advancing faster.
2. As the warm air overtakes the cold air mass, the warm air rises and flows over the colder, denser air.
3. Clouds form because the warm air cools as it rises, causing water vapor to condense.
4. The temperature at *A* is warmer than the temperature at *B*. The temperature at *B* will rise as the warm front passes over it.
5. The warm air mass is advancing faster than the cold air mass.

Homework

The Activity Worksheet on page 42 of the Unit 4 Teaching Resources booklet makes an excellent homework assignment to accompany Air Masses in Motion.

Answers to
Putting It All Together: Cold Fronts and Warm Fronts

1. The warm front is moving northeast across western Canada and the central United States. The cold front is moving east across the northeastern United States and eastern Canada. By tomorrow, the warm front will have moved across *A*, and the cold front will have moved across *D*.

2. *A* will become warmer while *D* becomes cooler over the next day or two. Area *B* will remain cooler than *C* because the front is stationary in that area.

3. Cold fronts force warmer, less dense air to rise and cool, while warm fronts move in to replace exiting cold air masses.

4. Cold fronts are active in that they move into an area and force warm air to rise and cool. Warm fronts are passive in that they only move in when cooler air leaves.

Answers to
Weather Maps

1. The triangles indicate that the boundary is a cold front. The front is moving in the direction that the triangles are pointing. The half-moons indicate that the air behind the boundary line is warmer than the air in front of it. The front is moving in the direction that the half-moons are facing.

2. *WX* represents a warm front, *XY* represents a stationary front, and *YZ* represents a cold front.

3. On tomorrow's weather map, the front will be farther toward the northeast.

4. A stationary front is indicated by alternating half-moons and triangles on opposite sides of the line.

5. The air-mass diagram demonstrates that the front is stationary at its boundary by showing that both fronts are moving against one another.

Homework

An At-Home Task is designed as a homework activity.

Putting It All Together: Cold Fronts and Warm Fronts

Weather forecasters often make predictions based on their understanding of the movement of cold fronts and warm fronts. Study the diagram at lower left, which shows the interaction of two large air masses, and then answer the following questions:

1. Where is the warm front? the cold front? Approximately where will each front be tomorrow?

2. How will the weather at *A* compare with the weather at *D* for the next day or so? How will the weather at *B* compare with the weather at *C* during the same period of time?

3. How does the interaction between air masses at a warm front differ from the interaction of air masses at a cold front?

4. In many respects, cold fronts are "active" phenomena, while warm fronts are "passive." Explain why these terms apply.

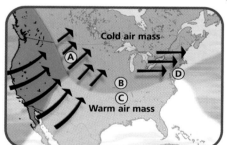

This diagram shows how air masses interact. The black arrows indicate the motion of the air masses.

Weather Maps

Weather maps are simplified representations of the interactions between air masses. Through the use of standard symbols, much information about the weather can be depicted on a single map. Study the weather map below, and then answer the questions that follow.

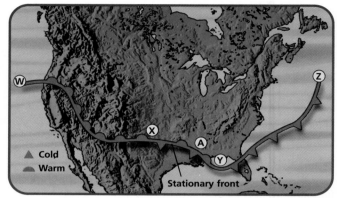

This weather map shows two air masses interacting in three different ways. How does this occur? **A**

1. What do you think the triangles on the line indicate? What do the half-moons indicate?

2. What does the line *WX* on the weather map represent? the line *XY*? the line *YZ*?

3. On tomorrow's weather map, approximately where will the warm front be?

4. As the map shows, one section of the front is not moving. This is called a *stationary* front. How is it represented?

5. How does the air-mass diagram at lower left help explain why the frontal boundary at *A* is stationary?

An At-Home Task

Follow your local weather forecasts for a week. What elements of weather are included in the forecasts? How are forecasters (meteorologists) able to make predictions about future weather?

Answer to
Caption

A The interaction of the air masses depends on the direction of the cold air mass. Cold fronts occur in areas where cold air is moving in; warm fronts occur in areas that are being vacated by cooler air; and stationary fronts occur in areas where there is no cold air movement.

Answers to
An At-Home Task

- Weather forecasts include information about cloud cover and visibility, temperature highs and lows, wind speeds and directions, humidity, and precipitation.

- This information is based on the locations of cold and warm fronts and their directions and speeds. Sometimes satellite pictures that show how clouds and fronts are moving are included in the forecast to explain how and why the weather will change.

A Map Sequence

The weather maps below represent a span of 6 days. Examine the maps. Then answer the following questions:

1. Describe, in general terms, what happens during this sequence. How do the various weather systems move?

2. Identify any fronts. What types of fronts are present?

3. What time of year do you think the maps probably represent?

4. What types of air masses are present in each map?

5. What happens at the boundaries between air masses?

6. What happens to the weather conditions of an area after one air mass overtakes another?

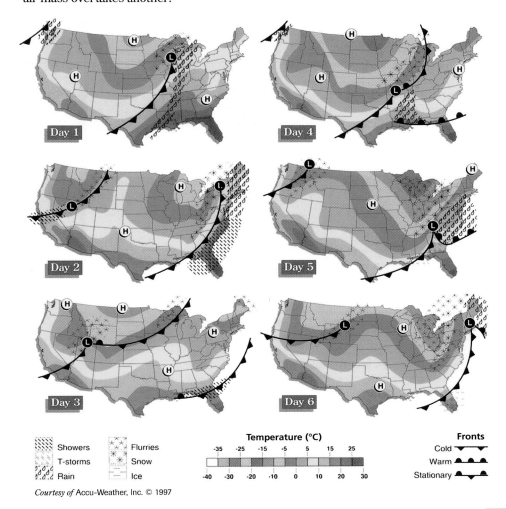

Day 1 | Day 4
Day 2 | Day 5
Day 3 | Day 6

Legend:
- Showers
- T-storms
- Rain
- Flurries
- Snow
- Ice

Temperature (°C)
-35 -25 -15 -5 5 15 25
-40 -30 -20 -10 0 10 20 30

Fronts
- Cold
- Warm
- Stationary

Courtesy of Accu-Weather, Inc. © 1997

255

Meeting Individual Needs

Learners Having Difficulty

To remind students of the effect that temperature has on air movement, have them perform the following activity. Each student will need a ruler, desk lamp, tissue paper, thread, scissors, and cellophane tape. Have each student cut a spiral that is 6 cm in diameter from the tissue paper and a piece of thread that is 15 cm long. Tape one end of the piece of thread to the center of the paper spiral. Turn the desk lamp so that the bulb points upward. Holding the end of the thread, position the paper spiral 10 cm above the bulb. Have students explain their observations. *(The paper spiral should spin. The light energy heats up the surrounding air. As the air heats, it becomes less dense and rises as cooler air moves in. These air movements, called convection currents, are responsible for wind.)*

Theme Connection

Cycles

Focus question: How do the locations of air masses change as Earth revolves around the sun? *(Due to the tilt of the Earth, the direct sunlight that warms the Earth shifts between the tropic of Cancer and the tropic of Capricorn. When direct sunlight is focused on the Southern Hemisphere during its summer, tropical air masses develop in the south. They develop in the north during summer in the Northern Hemisphere.)*

Integrating the Sciences

Earth and Life Sciences

Many animals, especially insects, behave in characteristic ways when rain is coming. Share stories of animal weather predictors with students. After giving each example, have students suggest what might trigger these behaviors. *(Accept all reasonable responses. Bees stay close to their hives when rain is coming. Scientists think the bees' sensitive antennae can detect an increase in humidity, which often signals rain. Ants travel in lines in fair weather and scatter when stormy weather is approaching.)*

Answers to
A Climate Sampler

1. weather
2. weather
3. climate
4. climate, weather

Answers to
In-Text Questions

Ⓐ Students should ascertain from the map that the Canadian city, which is surrounded by land, would be cooler in the winter and warmer in the summer than the Icelandic city, which is on an island.

Ⓑ Climate refers to the average weather conditions of a geographical region, as determined by weather changes over the course of many years. Weather refers to the condition of the atmosphere at a specific place and time.

Ⓒ Students should conclude that latitude, altitude, and proximity to oceans or large bodies of water have the greatest effects on a region's climate. Students should recognize that these factors determine how much heating of the atmosphere will occur, how quickly temperatures will change, how much moisture will enter the air, and how much moisture will leave the air. Thus, these factors determine the characteristics of the air masses that form in those areas.

Ⓓ The two cities have different climates because they are affected by different types of air masses.

Ⓔ Maritime climates are moist because they are located over oceans, and the air absorbs moisture from the water. Continental climates are dry because they are over land and do not absorb as much moisture. Maritime climates show little change in temperature compared with continental climates because water has a moderating effect on climate. That is, water cools down and warms up more slowly than land does .
Exploration 3 of Chapter 10 provided evidence of this phenomenon.

⭐ **Transparency 38 is available to accompany A Climate Sampler.**

Keeping Track
Provide time for students to update their ScienceLog.

A Climate Sampler

Reykjavik, Iceland, and Yellowknife, Canada, are both at about the same latitude, but, as the graph shows, during most months these cities have very different weather. In fact, both cities have very different climates. Why? What clues does the map provide? Ⓐ

Climate is a word we use quite often, but what exactly is a climate? Could you define it? How does it differ from weather? Ⓑ

Write each of the following sentences in your ScienceLog. In each blank, write whichever term best applies, *weather* or *climate*.

1. "What? Rain again! This kind of ____?____ depresses me."

2. "In January we normally have clear, cold ____?____."

3. During the ice age, the Earth had a much colder ____?____.

4. The ____?____ of the Amazon rain forest hasn't changed in millions of years. The ____?____ is almost the same every day.

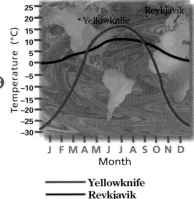

Average Daily Temperature

Yellowknife
Reykjavik

As you probably realize, weather is the condition of the atmosphere at a particular time and place. On the other hand, climate is the average weather conditions over a long period of time—years, decades, or even centuries. The diagram at right shows some of the world's major climatic regions. What seem to be the most important factors in determining climate? Why do you think this is so? Ⓒ

Let's go back to our earlier example. Why do Yellowknife and Reykjavik have such different climates? Ⓓ

Reykjavik has a *maritime climate*, that is, a climate typical of the ocean. Maritime climates are moist and exhibit relatively little change in temperature from day to night and from season to season. Reykjavik's climate is also influenced by the warm waters of the Gulf Stream, which you will learn about later.

Yellowknife has a *continental climate*. Continental climates are drier than maritime climates and typically are marked by large daily and seasonal temperature ranges. Why are continental and maritime climates so different? Where else in this unit have you seen evidence of this? Ⓔ

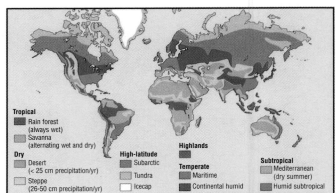

Major climatic regions of the world

Keeping Track

What have been your major findings in this chapter? Continue the record that you started in the last chapter.

256

FOLLOW-UP

Reteaching

Have students design their own experiment to determine the air temperature, dew point, and wind direction for their local weather. They can compare their results with readings from a local news source. Approve student designs for safety before they begin.

Assessment

Provide students with weather maps that show warm and cold fronts. Have them write forecasts for different areas.

Extension

Set aside time for students to learn about careers related to oceans and weather. Possibilities include a meteorologist, oceanographer, sailor, and member of the Coast Guard.

Closure

Have students write a short description of a particular day's weather conditions. Have students switch papers. Based on the information in the paragraph they receive, students should describe the movement of the air masses and winds involved in the description.

CHALLENGE YOUR THINKING

1. Earth's Tragic Thaw?

The Arctic Ocean is covered year-round by a layer of ice. If this ice melted, would you expect sea levels to rise worldwide? Why or why not?

2. The Case of the Baffling Bath Water

Robert filled a tub with water. The water ran cold for about a minute before it finally got hot. Robert expected the water in the tub to be uniformly warm throughout, but instead the hot water floated on top and the cold water stayed underneath. He had to slosh the water around before it felt comfortable.

a. Robert thought, "This is like the Mediterranean effect." What did he mean by this?

b. How do you explain this occurrence?

3. Saline Sailing

Ships loaded to the limit in cold, northern waters would be sailing into danger if they headed for the tropics without first unloading some cargo. To prevent this problem, international marine regulations require most ships to bear *Plimsoll marks*. Plimsoll marks are a kind of scale that shows the maximum safe load for a given water condition. The diagram below shows a typical set of Plimsoll marks. Freshwater marks are on the left, and saltwater marks are on the right.

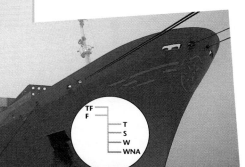

a. Why are the saltwater marks lower than the freshwater marks?

b. Why is tropical fresh (TF) the highest mark?

c. Why is winter north Atlantic (WNA) the lowest mark?

d. In January, a captain plans to sail his fully loaded ship from Oslo, Norway, across the Atlantic to Manaus, Brazil, 1500 km up the Amazon River. How should the captain determine how much cargo the ship should carry?

257

Homework

The Activity Worksheet on page 43 of the Unit 4 Teaching Resources booklet makes an excellent homework assignment once students have completed Chapter 11. If you choose to use this worksheet in class, Transparency 39 is available to accompany it.

Multicultural Extension

Names for the Wind

Wind is a climatic condition found virtually everywhere in the world. Different cultures have different names for wind. The following are examples: *elephanta*, *bhoot* (India), *haboob* (Sudan), *zonda* (Argentina), *williwaw* (Alaska), *xlokk* (Malta), *simoon* (North Africa), and *kwat* (China). Have students pick one of these cultures and find out how that culture distinguishes between different types of air currents.

Answers to *Challenge Your Thinking*

1. Sea levels would not rise. The ice in the Arctic Ocean floats because it is less dense than the sea water. As the ice floats, it displaces a volume of sea water that has the same mass as itself. If the ice melted, its mass would stay the same, but its volume would decrease (that is, its density would increase). The melted ice would take up only as much space as the ice that had been underwater previously, and sea levels would stay the same.

2. **a.** The Mediterranean effect occurs when water of lower density floats on water of higher density, or when water of higher density flows under water of lower density. In both cases, the waters with different densities do not mix.

   b. The hot water is less dense than the cold water. Thus, it remains in a separate layer on top of the cold water.

3. **a.** Because salt water is more dense than fresh water, objects float higher in salt water than in fresh water.

   b. Tropical (warm) fresh water has the lowest density. Thus, a ship would float lowest in tropical fresh water.

   c. Because the northern Atlantic Ocean is cold and salty during the winter, it would have the highest density. Thus, the ship would float highest in this water.

   d. The captain should load the ship to the WNA mark, the highest safe level for the conditions that the ship will encounter after it departs. Loading the ship so that it floats at any higher mark would mean that the ship would sink beyond the safe limit as it headed south into warmer, fresh water that is less dense.

 You may wish to provide students with the Chapter 11 Review Worksheet that is available to accompany this Challenge Your Thinking (Teaching Resources page 44).

4. For this project, students can consult the newspaper, the local TV news, a weather station on cable TV, or the Internet. Or they can determine the characteristics of the air themselves. A thermometer will give them the temperature, and a wet-bulb thermometer will provide the dew-point temperature.

5. Density equals mass divided by volume. The density of the object is 46 g/42 mL or 1.09 g/mL. Because this is greater than the density of water (1 g/mL), the object will sink.

6. Both air and water are materials that flow. In many instances, flow is created by density differences. Also, for both water and air, masses of different densities tend not to mix. In addition, both the water of oceans and lakes and the air surrounding us are solutions. (In the following chapter, students will investigate another similarity between air and water: they both exert pressure.)

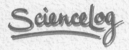
ScienceLog

The following are sample revised answers:

1. Some students may believe that the ocean floor is essentially flat, but this is a misconception. Although portions of the ocean floor are flatter than any area on land, the ocean floor has many physical features similar to those on land, such as mountains, canyons, and volcanoes.

2. Breezes are caused by uneven warming of air over land and water, which creates density differences in the atmosphere. During the day, the land-surface temperature is higher than the water-surface temperature of oceans or other large bodies of water. As the warmer air over the land rises, the cooler, denser air over the ocean moves in to take its place. This creates an onshore breeze. During the night, the land surface cools more quickly than the water surface. As the warmer air over the water rises, the cooler air over the land moves in to take its place, creating an offshore breeze.

4. Massive Project
Find out as much as you can about the air mass presently over your area. What is the temperature of the air mass? Is it a tropical air mass or a polar air mass? Is it a dry air mass, or is the humidity high? Will the same air mass be over your area tomorrow?

5. A Mysterious Matter
An object of unknown material has a mass of 46 g and a volume of 42 mL. Will this object float or sink in water? How do you know?

6. Surprising Similarities
Air and water seem to be completely different substances, but in some ways they are very much alike. Suggest ways in which the two substances are similar.

ScienceLog

Review your responses to the ScienceLog questions on page 237. Then revise your original ideas so that they reflect what you've learned.

258

3. Deep-ocean currents are caused by density differences in ocean water. Some students may incorrectly equate currents with tides because both involve movement of water. However, tides are the result of gravitational effects.

CHAPTER 12

Winds and Currents

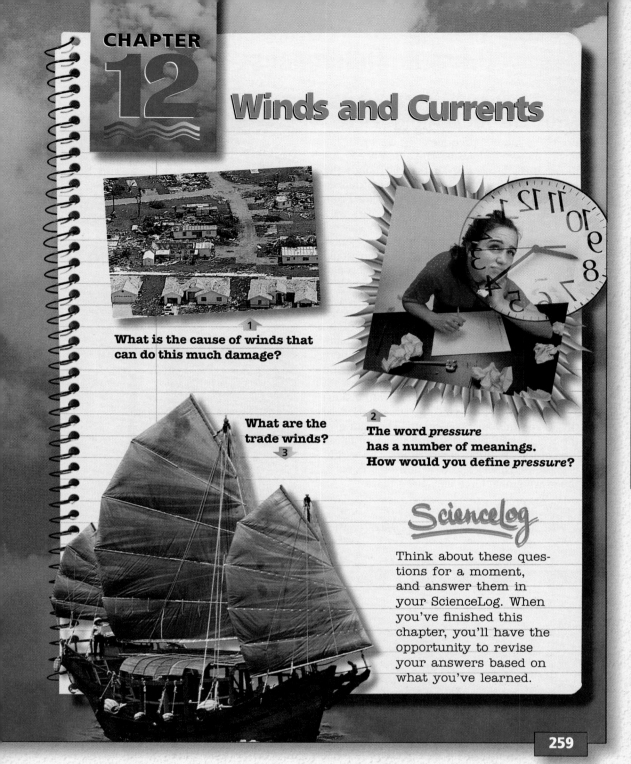

1

What is the cause of winds that can do this much damage?

2

The word *pressure* has a number of meanings. How would you define *pressure*?

What are the trade winds?

3

ScienceLog

Think about these questions for a moment, and answer them in your ScienceLog. When you've finished this chapter, you'll have the opportunity to revise your answers based on what you've learned.

259

Connecting to Other Chapters

Chapter 10
offers students the chance to explore the factors that affect the Earth's climate.

Chapter 11
examines the way density affects the flow of currents in the ocean and in the atmosphere.

Chapter 12
explores the concept of pressure as it relates to the atmosphere and to the ocean.

Prior Knowledge and Misconceptions

Your students' responses to the ScienceLog questions on this page will reveal the kind of information—and misinformation—they bring to this chapter. Use what you find out about your students' knowledge to choose which chapter concepts and activities to emphasize in your teaching. After students complete the material in this chapter, they will be asked to revise their answers based on what they have learned. Sample revised answers appear on page 281.

In addition to having the students answer the questions on this page, you may wish to have them complete the following written assignment: Ask students to imagine that it is 1519 and Ferdinand Magellan has asked them to accompany him on his historic trip around the world. What advice would you offer him about which route to take? Which areas are likely to produce dangerous storms? Where might he be able to benefit from strong winds or ocean currents? Encourage students to

use their imagination if they are unable to answer these questions with certainty. Have the students prepare their recommendations in the form of a letter to Magellan. Collect the letters, but do not grade them. Instead, read them to find out what students know about winds and currents, what misconceptions they may have, and what about the topic is interesting to them.

LESSON 1 Pressure Differences

FOCUS

Getting Started

The Discrepant Event Worksheet on page 50 of the Unit 4 Teaching Resources booklet describes a teacher demonstration that makes an excellent introduction to this lesson.

Main Ideas

1. Pressure is the amount of force exerted on a given area.
2. Atmospheric pressure is the weight of the atmosphere (in newtons) exerted on a given area.
3. Atmospheric pressure decreases with altitude; water pressure increases with depth.
4. The metric unit for pressure is the pascal; one pascal is equal to a force of one newton exerted on an area of one square meter.

TEACHING STRATEGIES

Answers to *In-Text Questions*

Ⓐ A sample graph appears on page S209. Students should discover that half of the atmosphere (exerting a pressure of 50,000 N/m²) is below 5.6 km. By graphing the data, students can estimate the altitude at which three-quarters of the atmosphere is below the shuttle. This altitude is approximately 12 km.

Homework

Students can make graphs to answer in-text question A as homework.

LESSON 1 Pressure Differences

Structure of the Atmosphere

From the moon, space looks black. From the Earth, the sky looks blue. The difference is due to the thick blanket of air that surrounds the Earth. It extends upward with decreasing density until it merges with space at about 500 km. From a space shuttle, the atmosphere appears as a hazy blue band against the blackness of space.

Imagine taking a trip through the atmosphere aboard a space shuttle. The trip would take about 10 minutes. You would pass through the various layers shown in the diagram at right.

The boundaries between layers are not as distinct as those shown here, nor are the elevations of the layers the same in all parts of the world. As you travel in the shuttle, at what altitude do you think one-half of the atmosphere would be beneath you? At what altitude would three-quarters of the atmosphere be beneath you? Use the data in the table below to make a graph that will give you the answer. Record Ⓐ this graph in your ScienceLog. The data give the weight of the atmosphere pressing on an area of 1 m² at different elevations. The weight is expressed in *newtons* (one newton is equal to the gravitational force on a 100 g mass).

Altitude (km)	Weight per square meter (N/m²)
0 (sea level)	100,000
5.6	50,000
16.2	10,000
31.2	1000
48.1	100
65.1	10
79.2	1
100	0.1

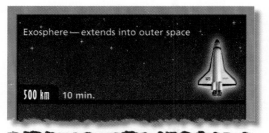

Exosphere—extends into outer space

500 km   10 min.

Thermosphere—northern lights (aurora borealis) occur here

85 km   2 min.

Mesosphere—meteorites start to burn up here

50 km   75 sec.

Stratosphere—contains ozone that filters out most ultraviolet radiation

15 km   30 sec.

Launch

Troposphere—weather occurs here

LESSON 1 ORGANIZER

Time Required
3 to 4 class periods

Process Skills
organizing, classifying, measuring, communicating

New Terms
Pascal (Pa)—a unit of pressure equal to one newton per square meter; normal atmospheric pressure is about 101,000 N/m², or 101,000 Pa
Pressure—the amount of force exerted on a given area

Materials (per student group)
Exploration 1, Part 2: brick; 1 m of string; metric spring scale or force meter; metric ruler
Exploration 2; Activity 1: large can with a hole punched in it close to the bottom; metric ruler; meter stick; tripod stand; stream table or large pan (about 120 cm × 35 cm) with drain; about 1 L of water (additional teacher materials: hammer; nail; see Advance Preparation on page 209C.); **Activity 2:** 2 cans of different diameters, each with a hole punched close to the bot-

continued ▶

Understanding Pressure

You are beginning to work with the concept of **pressure**. The pressure of the atmosphere at sea level is about 101,000 N/m². This means that a force of 101,000 N presses down on every square meter of the Earth at this elevation. Is this pressure higher or lower at 10 km above sea level? Review the graph you made using the data in the table on the previous page. What would be the atmospheric pressure at 10 km? at 50 km? How does pressure vary with altitude?

How would you define *pressure* now? How is pressure related to force? After completing the next Exploration, return to these questions. Then provide a scientific definition of pressure.

EXPLORATION 1

Pressure Situations

PART 1

Talk About It

Pressure is an important factor in the following situation. With a classmate, discuss the explanation for this situation.

You can press on a balloon with your finger, and the balloon bends but does not break. But when you apply the same amount of force to the balloon with a needle, the balloon pops. Why?

Now share a pressure situation of your own with other students.

Exploration 1 continued ▶

261

PART 2

Divide the class into small groups and distribute the materials. Have students predict which faces of the brick will exert the least pressure and the greatest pressure. Then have them calculate the force exerted by the brick in newtons. They can do this by lifting the brick with the spring scale or force meter. Then they can calculate the surface area of each face of the brick in square meters. The pressure exerted by each face of the brick is the force it exerts in newtons divided by the surface area of that face. (For the purpose of this Exploration, students may ignore the fact that the holes in the brick alter its surface area.)

Answers to
Questions

1. Pressure is measured in newtons per square meter (N/m²).

2. The weight of the brick does not change.

3. The pressure of the brick changes depending on the face resting on the table because the area over which the force is exerted changes. The least pressure is exerted when the brick is resting flat, as shown in the bottom photograph. The most pressure is exerted when it is standing on end.

4. It would sink to the greatest depth if it were placed on its smallest face.

Homework

Have students calculate how much pressure they exert on the floor. To complete this assignment, students will have to think of a way to measure the surface area of the soles of their feet. One method is to trace an outline of each sole on a piece of graph paper, count the number of squares within the outlines, and then multiply this number by the area of a single square.

PART 2

A Brick Trick

You Will Need

- a brick with a string (1 m long) tied securely around it
- a spring scale or force meter
- a metric ruler

What to Do

Predict which face the brick must rest on in order to exert (a) the least pressure against a table and (b) the most pressure. Test your predictions. Determine the pressure of the brick against a table when it is lying on each of its faces. Here is the information you will need.

What is the force in newtons?

> Area of a face of a brick = length × width
> Weight = reading on the spring scale in newtons
> Pressure = force (weight of brick) ÷ area

Questions

1. In what units is pressure measured?
2. Does the weight of the brick change when the face it rests on changes?
3. Does the pressure that the brick exerts change depending on which face is resting on the table? Why or why not?
4. If the brick were placed on a piece of foam rubber, on which of its faces, large or small, would it sink deepest?

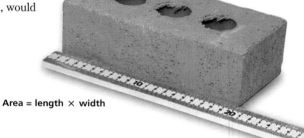

Area = length × width

Pressure in the Ocean

How does water pressure change as you travel deep into the ocean? Is it similar to the changes in atmospheric pressure at different altitudes? The following Exploration focuses on the way in which pressure is exerted by liquids.

ACTIVITY 1

Water Pressure and Depth

You Will Need

• a large can with a hole punched in it close to the bottom
• a meter stick
• a tripod stand
• water
• a stream table
• a ruler

What to Do

Place the can on a stand in a setup like the one shown below. Fill the can with water and observe what happens. Record the height of the water in the can and the distance that the water squirts.

Thinking About It

1. The distance that a volume of water squirts from the hole at the bottom of a can is a measure of the pressure exerted by water. How can you use this information to compare the pressures of different depths of water?

2. Place your data in a table similar to the one shown. Prepare a graph of your data.

Height of water in can (cm)	Distance water squirts (cm)

3. Compare the graph you just made with the graph of atmospheric pressure that you made at the beginning of this lesson. Do you think there is an altitude where the atmospheric pressure is equal to zero? Is there a depth of water for which the pressure in the can is equal to zero? Can you think of an important way in which the air in the atmosphere is different from the water in the can? What is it?

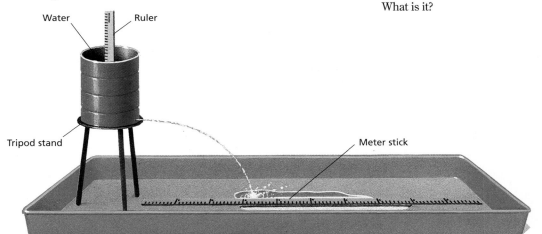

Water · Ruler

Tripod stand

Meter stick

Exploration 2 continued ▶

263

In this Exploration, students discover that there is a direct relationship between depth and water pressure—doubling the depth doubles the pressure exerted by the water.

The three Activities can be done in any order. Prior to the Exploration, you should prepare the cans for Activity 1 by punching holes in them with a hammer and nail. If sufficient materials are available, set up two stations for each activity, and have groups of three or four students work at each station.

★ **An Exploration Worksheet is available to accompany Exploration 2 (Teaching Resources, page 51).**

Integrating the Sciences

Life and Physical Sciences

Have students bend over, dangle their hands next to their toes for 30 seconds, and notice how the veins in the back of their hands stand out. Then, for contrast, have students hold their hands over their head and watch as their veins "disappear." Ask students to explain how this activity demonstrates that liquid pressure depends on depth. *(When hands are held low, pressure exerted by blood in the veins is at a maximum, causing the veins to bulge. When hands are held high, the liquid pressure in them reaches a minimum.)*

Answer to *Thinking About It*

1. By measuring several depths of water and the distance water squirts out of the hole each time, students should discover a relationship between depth and pressure.

2. Answers to the table and graph will vary depending on a number of variables, such as the type of can used and the size of the nail hole.

3. Unlike the graph students just completed, the graph of atmospheric pressure at the beginning of the

lesson shows that atmospheric pressure approaches, but never equals, zero at high altitudes. The graph from Exploration 2 should show that the pressure will equal zero at the point where the depth of water in the can is zero (that is, at the surface of the water).

Students should realize that water is a liquid and air is a gas. There is no well-defined surface to the air in the atmosphere, but there is a definite surface to the water in the can. (You may wish to point out, however, that in this Exploration, students are measuring

gauge pressure and not total pressure. Gauge pressure is defined as the total pressure minus the atmospheric pressure. Thus, at the surface of the water, the gauge pressure is zero, but the total pressure is one atmosphere. See page 268 for a discussion of units of pressure.)

Answers to
Thinking About It (Activity 2)

1. The water pressure is greater in the can with the smaller diameter. Students should conclude that it is not the volume of water in the container that determines pressure, but the height of the column of water.

2. Water at the same depth exerts the same pressure; therefore, water pressure in the cans would be the same.

3. There would be no difference in water pressure. The pressure depends on the depth of the water, not the volume, because the pressure depends on the weight of the water directly above the swimmer, not on how much total water is present.

ACTIVITY 3

When the student drops the foil ball into the bottle, about half of the ball should rise above the water's surface. The student may need to wad the foil more tightly or loosely to achieve this result. Once this is done, the student may proceed with the Exploration. The Exploration can only be performed for a limited time with the foil ball, because eventually the ball will not respond to changes in pressure. A new foil ball may need to be used if results become erratic over time.

Answers to
Thinking About It (Activity 3)

1. Regardless of where the bottle is squeezed or how the bottle is oriented, the foil ball can be controlled equally well. Students should infer that pressure is exerted equally in all directions in a fluid such as water.

2. Sample answer: When the bottle is squeezed, the increased water pressure compresses the ball slightly. This compression reduces the ball's volume, increasing its density and making it denser than water. Thus, the ball sinks. When the pressure on the bottle is released, the ball returns to its previous volume and therefore its previous density. Thus, the ball rises. By varying the amount of pressure on the bottle—and therefore the ball's density—the movement of the ball can be controlled.

ACTIVITY 2
The Effect of Shape and Volume

You Will Need
- 2 cans of different diameters, each with a hole punched close to the bottom
- a meter stick
- a tripod stand
- a stream table
- water
- a graduated cylinder

What to Do
Arrange the materials as shown below. Measure and record a volume of water. Place one of the cans on a tripod stand and pour in the water. Measure the distance the water squirts. Replace the first can with the second can and repeat the experiment, using the same volume of water. Record and then compare your results.

Thinking About It
1. What conclusions can you make from this Activity?
2. What would you observe if you filled both cans with water so that the water was at the same depth in each can?
3. Would there be any difference in water pressure if you swam 2 m below the surface of the water in a swimming pool as opposed to 2 m below the surface in a large lake? Does pressure depend on the volume or the depth of water? Explain your reasoning.

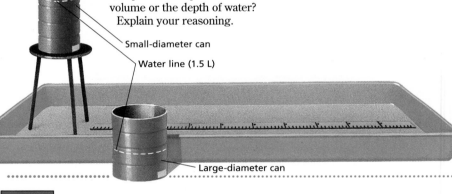

Small-diameter can

Water line (1.5 L)

Large-diameter can

ACTIVITY 3
Is Water Pressure Exerted Equally in All Directions?

You Will Need
- a plastic bottle
- a small piece of foil (4 cm × 4 cm)
- water

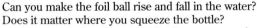

Foil ball

Cap screwed on tightly

Water

What to Do
Fill the plastic bottle all the way to the top with water. Wad the piece of foil into a small ball, and drop it into the container. Screw the cap on tightly. Wait 20 seconds and then squeeze the bottle. Can you make the foil ball rise and fall in the water? Does it matter where you squeeze the bottle?

Now lay the bottle on its side. Can you control the movement of the foil ball? Does it matter where you squeeze the bottle?

Thinking About It
1. How would you answer the question asked in the title of this Activity?
2. Explain the role of the foil ball's density in this Activity.

Integrating the Sciences

Earth and Physical Sciences
Pose the following question to students: One lake is 5 m deep and has a surface area of 5 km². Another lake is 10 m deep and has a surface area of 1 km². A dam will be built on each lake to hold back the water. Which lake will require the stronger dam? Why? *(The smaller but deeper lake will require the stronger dam. The water pressure is greater at the bottom of the deeper lake, not necessarily the lake with the most water.)*

Homework
Exploration 2, Activity 3, can be assigned as homework, especially if class time is short.

Historical Flashbacks

Here are three episodes from history. Each illustrates something about the pressure exerted by the atmosphere. Test your understanding by responding to the questions and completing the tasks that follow each episode.

Holy Hemispheres! The Magdeburg Experiment

During the 1650s, Otto von Guericke was both an inventive scientist and the mayor of the German town of Magdeburg. In 1652 Emperor Ferdinand III heard of his experiments and asked to see them. So Otto von Guericke gave him a dramatic example. He placed two copper hemispheres together so that they formed a hollow sphere with a diameter of about 45 cm. He removed as much air from inside the sphere as he could by using a crude vacuum pump. After creating a partial vacuum inside the sphere, how did the air pressure inside the sphere compare with the air pressure outside?

A depiction of Otto von Guericke's most famous experiment

Now comes the part that astonished the emperor. Two teams of horses—one team attached to each hemisphere—could not pull the two hemispheres apart! What does this tell you about the magnitude of atmospheric pressure? **Ⓑ**

Consider this: Air moves from regions of high pressure to regions of low pressure. Where have you heard of high- and low-pressure areas before? What do you think would happen in von Guericke's experiment if a small hole were punched through the copper sphere? **Ⓓ**

Try It Yourself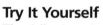

The following simulation will help you understand the effect of air-pressure differences in the atmosphere. Pour a small amount of water into an aluminum can. Heat the can until you see steam escaping from it. Using oven mitts, turn the can upside down in a large container of cold water. What happens? How do you explain this in terms of pressure?

Historical Flashbacks

On pages 265–267, students read three accounts of historical experiments in order to learn more about pressure. Each account is followed by an activity. These could be performed as teacher demonstrations or as student activities after each account has been read and discussed.

Holy Hemispheres! The Magdeburg Experiment

This flashback illustrates the following facts about atmospheric pressure:

- The pressure exerted by the atmosphere is extremely large.
- Air tends to move from regions of higher atmospheric pressure to regions of lower atmospheric pressure.

Answers to
In-Text Questions

Ⓐ When the air was pumped out of the sphere, air pressure inside the sphere was less than the air pressure outside.

Ⓑ Atmospheric pressure outside the sphere pushed the two hemispheres together so tightly that two teams of horses could not pull them apart.

Ⓒ Students may have heard about high- and low-pressure areas in relation to the weather. Wind results when air flows from areas of high pressure to areas of low pressure.

Ⓓ Since air moves from areas of high pressure to areas of low pressure, air would flow into the sphere if it were punctured. The vacuum would be broken, and the hemispheres would fall apart.

Try It Yourself

You may want to do this activity as a demonstration and then involve students in a discussion of their observations.

Answers to
Try It Yourself

Steam expels air from the can. When the can is plunged into cold water, the steam condenses, greatly reducing the pressure inside the can. The greater atmospheric pressure pressing against the outside of the can causes it to collapse.

Blazing a Scientific Trail: Pascal's Experiment

Mercury rather than water was used in early barometers because mercury is 13.6 times more dense than water. Therefore, the atmosphere at sea level can support a column of mercury that is only about 76 cm high. In contrast, the atmosphere can support a column of water 13.6 times higher than the column of mercury. However, it would be difficult to make and measure a column this high. Note: In Pascal's experiment, the mercury column at the foot of the mountain was only 71.1 cm high. This was probably because the foot of the mountain was above sea level.

Try It Yourself

SAFETY ALERT Do not allow students to make a mercury barometer; mercury fumes are poisonous.

Consider building the barometer as a class project. A barometer is a good instrument for measuring relative changes in atmospheric pressure. The barometer can be calibrated by comparing its readings with those of a weather forecaster. The rubber from the balloon is made more sensitive by stretching it. Each reading must occur at the same temperature because an increase in temperature will result in the expansion of the air inside the jar. Cooling the jar results in contraction of the air in the jar. Both of these occurrences will alter the reading.

Encourage students to observe and record atmospheric pressure, cloud cover, precipitation, temperature, and humidity for 2 weeks. You may wish to have them make a classroom chart to record their observations. At the end of 2 weeks, involve the class in a discussion to identify any patterns on the chart.

Blazing a Scientific Trail: Pascal's Experiment

Just before von Guericke performed his demonstration, Blaise Pascal, a French scientist and mathematician, expanded the study of atmospheric pressure. He used a new invention created by the Italian scientist Evangelista Torricelli—the mercury **barometer**. Torricelli had filled a tube with mercury and turned it upside down in a pool of mercury. Not all of the mercury ran out. Instead, a column 76 cm high remained, supported by atmospheric pressure. Whenever the pressure of the atmosphere changed, so did the height of the mercury column.

In 1648, Pascal had his brother-in-law carry a barometer to the top of a 1500 m mountain. At the bottom of the mountain, the column of mercury in the barometer was 71.1 cm high. At the top of the mountain, the mercury column was 62.6 cm high. Why was the measurement at the bottom of the mountain different from the one at the top? What had Pascal proved? Ⓐ

Because of Pascal's contribution to science, he had a unit of measurement named after him. A pressure of 1 N/m² is equal to 1 **pascal** (Pa). Normal atmospheric pressure at sea-level is about 101,000 N/m², or 101,000 Pa.

Grazie, Evangelista! (*thanks)

Prego, Blaise! (*no problem)

Try It Yourself

Below is one design for a homemade barometer that will detect daily changes in atmospheric pressure. Build this barometer or one of your own design, and keep a record of the increase or decrease in atmospheric pressure each day. Also, keep a record of the kind of weather you experience each day. Is it sunny or cloudy? hot or cold? humid or dry? Can you find a relationship between pressure changes and changes in the weather?

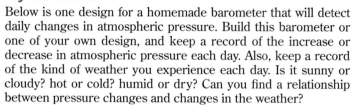

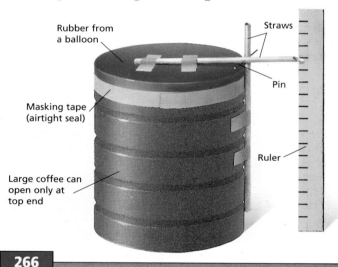

Rubber from a balloon

Straws

Masking tape (airtight seal)

Pin

Large coffee can open only at top end

Ruler

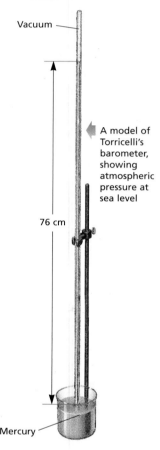

Vacuum

A model of Torricelli's barometer, showing atmospheric pressure at sea level

76 cm

Mercury

Pumped Up: Boyle's Experiment

In 1667 Robert Boyle made an improved pump that created a nearly perfect vacuum in a container. When he pumped air out of a pipe that was placed in a container of water, he was able to raise the water in the pipe a little over 10 m—but not any higher.

- What caused the water to move up the pipe?
- Why was he able to lift about 10 m of water this way? (Hint: Water is 13.6 times less dense than the mercury used in the barometer on the previous page.)

Try It Yourself

Place an index card over a full glass of water. Turn the glass upside down over a bucket or sink, and remove the hand holding the card in place. Why does the water stay in the glass? According to Boyle's experiment, what is the maximum height that a water column can reach in an experiment of this kind?

Maximum water level—10 m

Robert Boyle was able to pump water to a height of only 10 m.

267

Pumped Up: Boyle's Experiment

Another way to show that air supports water in a tube is to lower a straw into some water. Then place your finger on the top of the straw, and lift the straw from the jar. Students should observe that air pressure supports the water inside the straw.

Answers to
Pumped Up: Boyle's Experiment

- Water moved up the pipe to replace the vacuum created above it. However, the water was actually raised by the weight of the atmosphere pushing down on the water in the container.
- Boyle discovered that he could pump water to a height of only 10 m because the weight of the atmosphere could not support a higher column of water. (If 101 kPa supports a column of mercury 76 cm high, then it will support a column of water with a height of 76 cm × 13.6, or over 10 m.)

Try It Yourself

You may wish to recommend that students use a plastic cup instead of a glass to help prevent accidental breakage. When doing this activity, students could try placing the inverted glass on a smooth tabletop and carefully sliding it off the cardboard onto the tabletop. Then they could move the glass slowly over the tabletop. They should observe that the water stays inside the glass even when it is moved along the tabletop.

Answers to
Try It Yourself

The water stays in the glass because air pressure outside the glass pushes on the index card and holds it against the glass.

According to Boyle's experiment, it is possible to support a column of water up to 10 m high, but no higher.

Answers to
In-Text Questions

Ⓐ Students should realize from Robert Boyle's experiment that 10.4 m of water exerts about the same pressure as one atmosphere.

Ⓑ If a diver dove 300 m beneath the surface of the ocean, he or she would experience about 300 m × 1 atm/10 m, or 30 atm, of pressure.

Answers to
Under Pressure

1. Giant squids live at about 1000 m.
1000 m × 1 atm/10 m = 100 atm
Viper fish live at about 3000 m.
3000 m × 1 atm/10 m = 300 atm

2. At the surface, sperm whales experience 1 atm.
1500 m × 1 atm/10 m = 150 atm
150 atm – 1 atm = 149 atm change in pressure

3. The *Titanic* rests at 3660 m.
3660 m × 1 atm/10 m = 366 atm
366 atm × 101 kPa = 36,966 kPa
36,966 kPa × 1000 Pa/kPa = 36,966,000 Pa

4. 11,000 m × 1/10 atm = 1100 atm
1100 atm × 101 kPa = 111,100 kPa
111,100 kPa × 1000 Pa/kPa = 111,100,000 Pa
Area of the sphere: $4 \times \pi \times (1\ m)^2$ = 12.5 m²
111,100,000 Pa = 111,100,000 N/m²
111,100,000 N/m² × 12.5 m² = 1,388,750,000 N

Homework

The Activity Worksheet on page 55 of the Unit 4 Teaching Resources booklet makes an excellent homework assignment to accompany Under Pressure.

FOLLOW-UP

Reteaching

Ask students to summarize the main ideas in this lesson. Have students share their summaries with each other.

Assessment

Ask students to explain how pressure makes it possible to use the following tools: thumbtack, can opener, eyedropper, and aerosol pump. *(The thumbtack and*

Under Pressure

In addition to pascals (Pa), another unit commonly used to measure pressure is the *atmosphere* (atm). One atmosphere is the amount of pressure exerted by the Earth's atmosphere at sea level. You can use the following formula to convert from pascals to atmospheres:

$$101\ kPa = 1\ atm$$

As you discovered in Exploration 2, pressure in the ocean increases with depth. In fact, it increases by 1 atmosphere for every 10.4 m of depth. How does this relate to Boyle's experiment on page 267? Ⓐ

Imagine that you dove 200 m beneath the surface of the ocean. How much pressure would you experience? How much pressure do deep-
Ⓑ dwelling organisms experience? Study the diagram at right, and then answer the following questions:

1. How much pressure do giant squids normally experience? viper fish?

2. The sperm whale spends most of its time near the surface of the ocean, but on occasion it dives to depths of over 1500 m. What change in pressure does the whale experience?

3. While crossing the Atlantic Ocean on its maiden voyage, the passenger ship *Titanic* struck an iceberg and sank. What is the water pressure in pascals and atmospheres at the *Titanic*'s final resting place, nearly 3700 m beneath the surface of the ocean?

4. The deepest part of the Mariana Trench, about 11,000 m deep, was visited by the bathyscaph (deep-water vessel) *Trieste* in 1960. At this depth, how great was the pressure experienced by the *Trieste*? The crew compartment of the *Trieste* was a sphere about 2 m in diameter. What was the *total* force on the crew compartment? (The formula for the area of a sphere is $4\pi r^2$.)

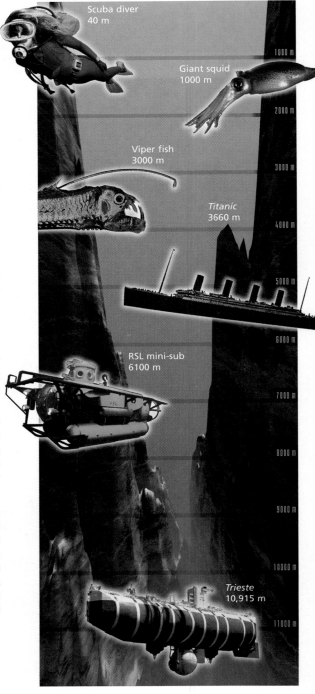

Scuba diver 40 m

1000 m

Giant squid 1000 m

2000 m

Viper fish 3000 m

3000 m

Titanic 3660 m

4000 m

5000 m

6000 m

RSL mini-sub 6100 m

7000 m

8000 m

9000 m

10000 m

Trieste 10,915 m

11000 m

can opener concentrate the force applied on them, thus increasing pressure. Liquid in an eyedropper and an aerosol pump is pushed up by air pressure.)

Extension

Have students create a diagram that explains how pressure is used in a hydraulic press, a whistling teapot, or a bellows. Students may need to do some research to complete this assignment.

Closure

Ask students to design a fictitious machine that uses both air pressure and water pressure to complete a common household task. Encourage creativity.

PORTFOLIO
Suggest that students include their summary from the Reteaching activity on this page in their Portfolio.

The Direction of Flow

Mystery 1

On August 3, 1492, Christopher Columbus lifted anchor and set sail across the Atlantic. He stopped first at the Canary Islands, located just off the coast of Africa. There he knew he could catch steady, northeasterly winds—these later earned the name *trade winds.* It was his idea to use these winds to blow him all the way to China. About 1 month later, he reached the Caribbean and the New World instead.

Portrait of Christopher Columbus

After 3 months, Columbus sailed north hoping to find suitable winds to carry him home. Luckily, he found the *westerlies,* which carried him back to Spain. No previous explorer had documented the great winds that blow in different directions across the oceans.

What was the mystery? No one knew what caused the winds or why they blew in such a consistent and predictable manner. What do you think causes the trade winds?

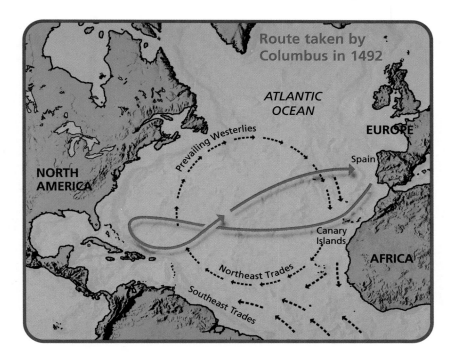

Route taken by Columbus in 1492

ATLANTIC OCEAN

EUROPE

Spain

Prevailing Westerlies

NORTH AMERICA

Canary Islands

AFRICA

Northeast Trades

Southeast Trades

LESSON 2 ORGANIZER

Time Required
3 class periods

Process Skills
hypothesizing, inferring, predicting, contrasting

New Term
Coriolis effect—the deflection of a moving object to the left in the Southern Hemisphere and to the right in the Northern Hemisphere, caused by Earth's rotation
Doldrums—regions of the ocean near the equator that are characterized by calm weather and very light winds

Horse latitudes—regions of the ocean at about 35° N and 35° S that are known for the lack of winds

Materials (per student group)
Exploration 3, Part 1: ¼ sheet of poster board; **Part 2:** ¼ sheet of poster board from Part 1; scissors; enough baking soda to fill a salt shaker; empty salt shaker; large ball bearing or marble; pushpin or thumbtack; piece of wall paneling or slab of plastic foam larger than the piece of poster board; a few milliliters of water

continued ▶

The Direction of Flow

FOCUS

Getting Started

Present students with the following situation: Rhonda was riding on a merry-go-round, watching things pass by her from the left to the right. As she was doing this, her lucky marble slipped out of her pocket and fell to the floor of the merry-go-round. When she reached down to get it, the marble was rolling away. Ask: In which direction was it rolling? *(Out and to the right)* Tell students that in this lesson they will learn about how the Earth's rotation affects the motion of the wind and water.

Main Ideas

1. The major surface currents in the ocean are wind-driven.
2. Both winds and water currents are influenced by the Coriolis effect, which is the deflection of a moving object due to Earth's rotation on its axis.
3. The direction of winds is determined by the existence of high- and low-pressure areas and the Coriolis effect.

★ **A Transparency Worksheet (Teaching Resources, page 56) and Transparency 40 are available to accompany The Direction of Flow.**

TEACHING STRATEGIES

Answer to
In-Text Question

 Accept all responses at this point without supplying the answer. Students will develop an answer to this question as they progress through this lesson.

Group size: 4

Group goal: to prepare a presentation (with illustrations) that explains the cause of the trade winds and Gulf Stream

Positive interdependence: Group members will work together to solve the two mysteries presented in the text. Assign the following roles to each group: chief investigator (to read the text material aloud and keep all group members focused on the group goal), secretary (to take notes, record data, and monitor other group members so that no one looks on page 271 until the discussion is completed), art supervisor (to make sketches and gather materials for the demonstration), and star witness (to present the clues needed to solve the mystery).

Individual accountability: Each student should individually answer the questions in A Windy Solution on page 271.

Mystery 2

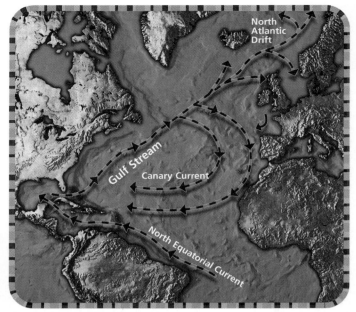

The path of the Gulf Stream

In the mid-1700s another maritime mystery unfolded. Ships sailing from England to New York took much longer to make the journey than ships sailing from England to Rhode Island. And the two destinations were only a single day's sailing apart! Benjamin Franklin solved the mystery by talking to Atlantic whalers. They told him about a strong surface current that flowed across the Atlantic Ocean. This current was named the Gulf Stream because it was believed to originate in the Gulf of Mexico.

While traveling between Europe and America, Franklin discovered that this flow of water was warmer than the water surrounding it. He used this information to draw the first map that included the Gulf Stream. With Franklin's map, sailors could make use of the stream— or avoid it.

This satellite image shows the temperature of the Gulf Stream and surrounding waters.

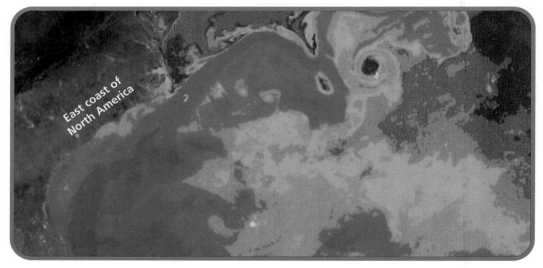

ORGANIZER, continued

Teaching Resources
Transparency Worksheet, p. 56
Activity Worksheet, p. 58
Exploration Worksheet, p. 60
Transparencies 40–42
SourceBook, p. S68

A Windy Solution

How are these two mysteries connected? It turns out that the same winds that blew Columbus back to Europe are also responsible for pushing the ocean water toward Europe. Thus, the ocean's surface currents are wind-driven.

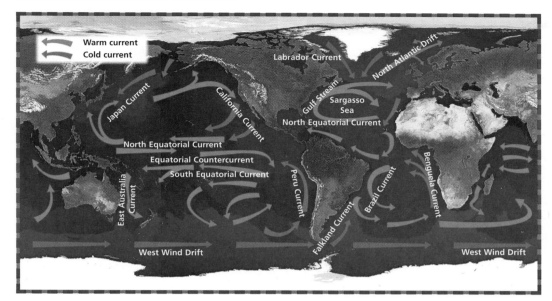

Examine the map above, which shows the surface currents of the world's oceans. What do you notice about the direction of flow of the major currents north of the equator? What about the currents south of the equator? If surface currents such as the Gulf Stream are created by winds, what can you infer about the direction of these winds? Can you think of any reason for the circular motions of the ocean currents?

In 1835, a French scientist, Gustave-Gaspard de Coriolis, published a paper that solved this mystery. He determined that the circular motions of the currents are related to the rotation of the Earth. In the next Exploration you will see exactly how this works.

> I, Gustave-Gaspard de Coriolis, solved the mystery that had sailors and scientists alike scratching their heads! Read on to find out more.

271

Answers to
A Windy Solution

In the Northern Hemisphere, the currents flow clockwise. In the Southern Hemisphere, the currents flow counterclockwise. The currents must flow in approximately the same direction that the winds blow. Accept all reasonable explanations for the circular currents. Have students read the last paragraph on page 271 to discover if their responses were correct.

⭐ **Transparency 41 is available to accompany A Windy Solution.**

CROSS-DISCIPLINARY FOCUS

Social Studies

The ancient Egyptians were the first culture to use air currents to propel their boats. By 3000 B.C., the Egyptians had developed sails that they attached to masts with ropes. Like those of modern pleasure boats, these sails could be manipulated according to which direction the sailors wished to go. Egyptian engineering was so advanced that they constructed huge barges to bring stone for their pyramids from quarries along the Nile River. Such boats could carry as much as 680 metric tons of cargo. Challenge students to find evidence of Egyptian boats in photos of ancient Egyptian artwork.

Homework

The Activity Worksheet on page 58 of the Unit 4 Teaching Resources booklet makes an excellent homework activity to accompany A Windy Solution.

Multicultural Extension

Cultural Exchange

Ask students to speculate about the following questions. Share the answers with them and encourage discussion. How did Benjamin Franklin solve the mystery of the Gulf Stream? *(He asked Atlantic whalers, who were much more familiar with the ocean than he was.)* How can learning about other cultures lead us to new discoveries in science? *(Groups of people who have experiences different from our own often have knowledge and expertise that we do* not.) Challenge students to provide examples of such discoveries. *(The field of medicine has benefited greatly from the practices of different cultures. For example, the Samoan healers use herbal remedies to treat and prevent a variety of diseases. Researchers have been studying these practices in order to find more natural and effective medicines.)*

EXPLORATION 3

To be sure that students understand how the poster-board model represents the Earth, have them observe a rotating globe and the poster-board model from above the North Pole and then from above the South Pole.

In this Exploration, challenge students to discover Coriolis's solution to the mystery of global winds and ocean currents. Remind students to place their poster board on a table before performing Part 2 of the Exploration.

Answers to
Part 1

3. From above the South Pole, the rotation is clockwise.

Answers to
Part 2

3. The path is straight.

4. The path should curve to the right of its initial direction of movement.

5. The path always swings to the right.

6. The South Pole

7. The path should curve to the left when the board is rotating in a clockwise direction.

 An Exploration Worksheet is available to accompany Exploration 3 (Teaching Resources, page 60).

Meeting Individual Needs

Learners Having Difficulty
To reinforce Exploration 3, have students make another model that demonstrates the Coriolis effect. Have each student cut a circle 3 cm in diameter from construction paper and push the point of a pencil through the circle so that the pencil spears the center of the paper circle. Place a drop of water on top of the paper near the pencil. Then have students hold the pencil in their palms and twirl the pencil in a counterclockwise direction. Students should notice that the drop swirls around the paper in a clockwise direction. The rotating circle represents the rotating Earth, and the drop represents Earth's oceans and atmosphere.

EXPLORATION 3

The Coriolis Effect

You Will Need
- $\frac{1}{4}$ sheet of poster board
- baking soda (in a saltshaker)
- a large ball bearing or marble
- a pushpin or thumbtack
- a piece of wall paneling or a slab of plastic foam (larger than the piece of poster board)

PART 1

Direction of Rotation

What to Do

1. In the center of the poster board, write the letter *N* as a symbol for the North Pole of Earth. On the backside of the poster board, write an *S* to symbolize the South Pole.

2. The Earth rotates counterclockwise when viewed from above the North Pole. Slowly rotate your poster-board model of Earth to simulate this.

3. Continue rotating the model in this direction, but raise it so that you can see the backside of the poster board. You are now viewing the Earth's rotation from above the South Pole. In what direction is it rotating now?

PART 2

Coriolis's Explanation
What to Do

1. Near the edge of your poster board, cut a hole large enough for your finger to fit.

2. For backing, place the poster board on the piece of paneling or plastic foam. Locate the center of the poster board. Push a pin or thumbtack through this spot into the backing. Lightly dust the poster board with baking soda.

3. Without rotating the model, roll a wet marble across it. Describe its path.

4. Now place your finger in the hole you cut earlier and spin the poster board (not too fast!) to simulate the Earth's rotation as observed from above the North Pole. As the poster board is rotated, have someone roll the wet marble across its surface, as shown below. (The throw must be timed so that your hand doesn't get in the way as you turn the poster board.) Describe the path the marble takes.

5. Try rolling the marble in different directions across the rotating model. As you look in the direction of the marble's motion, does its path always swing in one particular direction?

6. Rotate the model Earth clockwise. Now what pole does the center of the poster board represent?

7. Roll the marble across the model Earth while it is rotating clockwise. As you look in the direction of the marble's motion, does its path swing one way or another?

Your Analysis

From your observations in this Exploration, find proof for the following conclusions. Then write a statement in your ScienceLog that explains how you came to each conclusion.

1. When viewed from above the North Pole, the Earth rotates in a counterclockwise direction.

2. When viewed from above the South Pole, the Earth rotates in a clockwise direction.

3. A moving object in the Northern Hemisphere swings to the right as an observer looks in the direction of the object's motion.

4. A moving object in the Southern Hemisphere swings to the left as an observer looks in the direction of the object's motion.

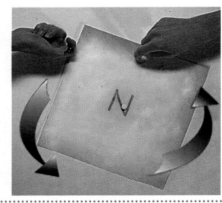

Dust the poster board with baking soda, and rotate the poster board to simulate the Earth's rotation. Roll a wet marble across the poster board's surface and observe its path.

Answers to
Your Analysis

1. When the model Earth is rotated in the proper direction, this direction is seen as counterclockwise when viewed from above the North Pole.

2. When the rotation is maintained as before but the observer changes position to above the South Pole, the direction of rotation is seen as clockwise.

3. The path of the rolling ball bearing or marble deviated to the right when the model Earth was rotated counterclockwise.

4. The path of the rolling ball bearing or marble deviated to the left when rolled across a model Earth rotating clockwise.

Explaining Your Discoveries

If you found evidence to support the statements on the previous page, congratulations! You have rediscovered what Coriolis first explained in 1835—the **Coriolis effect**.

Return to the map of the ocean's surface currents on page 271. Apply your knowledge of the Coriolis effect to an analysis of the Gulf Stream. Can you now explain why the current in the Atlantic Ocean moves in a clockwise pattern? Why do currents in the Southern Hemisphere follow a counterclockwise pattern? Remember that the major factor that drives surface currents is the wind. What can we infer about the direction of the winds from the direction of the surface currents?

⬆ Whereas currents in the North Atlantic flows clockwise, currents in the South Atlantic flow counterclockwise.

The Sargasso Sea

As the currents of the North Atlantic Ocean slowly flow in their clockwise pattern, a calm area forms at the center of the rotation. This area is called the Sargasso Sea. Under constantly clear skies, the sea is warmed by the sun, which increases the evaporation of water in the area. Thus, the Sargasso Sea is saltier than the surrounding ocean waters.

The area has its own ecosystem. A seaweed called sargassum weed has adapted to this unique environment. It floats in the warm water, often forming patches that cover large areas of open sea. The Sargasso Sea is also home to a number of unique animals. One of these, the silver eel, lives in fresh water but

Sargasso Sea life ⬆

returns to the Sargasso Sea to breed. Find out more about the eel and other unique organisms that live in the Sargasso Sea. Then write a paragraph or two in your ScienceLog summarizing what you've learned. Ⓐ

Have students refer to the map on page 271

Explaining Your Discoveries

Have students refer to the map on page 271 to examine the direction of the motion of the Gulf Stream. Ask: In what direction is the Gulf Stream moving? *(They should observe that the Gulf Stream moves in a clockwise rotation.)*

Explain that these motions are caused not only by the Coriolis effect acting on the moving waters, but also by the Coriolis effect acting on the winds. Remind students that wind is the principal cause of the currents.

Answers to
Explaining Your Discoveries

Surface currents in the Atlantic Ocean move in a clockwise direction because the Coriolis effect deflects their motion to the right in the Northern hemisphere.

Currents rotate counterclockwise in the Southern Hemisphere because the Coriolis effect deflects motion to the left in this hemisphere.

At this point, students should be able to infer that the wind moves in the same direction as the currents.

Answer to
In-Text Question

Ⓐ Aristotle was the first person to observe the European eel (*Anguilla anguilla*) migrating to the sea and the young eels returning to the rivers. Their destination in the sea was a mystery until 1856, when a German naturalist discovered a flat and nearly transparent sea creature that resembled a leaf. He named it *leptocephalus*, which means "slender head" in Greek. Later, at the turn of the century, scientists who were studying some specimens of leptocephalus in an aquarium were astonished to see them metamorphose into young eels. It took a Danish oceanographer 18 years and many trips across the ocean to locate the breeding ground of the European eel. His method was to look for younger and younger specimens. The youngest specimens were found in the Sargasso Sea. Other unique organisms include species of shrimp, crab, and barnacle. Students may also research the sargassum fish, which has colorings that make it hard to distinguish from the surrounding sargassum weed.

Integrating the Sciences

Earth and Life Sciences

Tell students about the European eel's adaptations to ocean surface currents. By the millions, 7- to 14-year-old European eels leave American and European rivers. As they travel toward the Sargasso Sea, their reproductive organs mature, their skin toughens, their color changes, and their eyes enlarge. (The color in their retinas also changes so that they are better able to absorb the blue light of the deep sea water.) Then they cease eating and their teeth fall out. In the Sargasso Sea, some

5000 km from the rivers they left, the females lay their eggs. Then the mature males and females die, and the eggs hatch. (Perhaps the eggs are forced out of the females by the water pressure as they descend deeper into the ocean.) The leaflike, 5 cm leptocephali float toward the surface of the Sargasso Sea. There, they float on the ocean currents to the coasts of Europe and North America. However, before entering the rivers, the leptocephali undergo metamorphosis. At this point, it is time for the young eels to swim up the rivers.

So far, students have learned that air moves from more-dense to less-dense areas. In this Exploration, they will learn that air also moves from high-pressure to low-pressure areas. This is not a contradiction because less-dense air weighs less and therefore exerts less pressure.

Answers to
Exploration 4

1. Air masses at warmer temperatures have lower pressures because warmer air is less dense than cooler air. Therefore, a low-pressure area exists at the equator, where the sun's energy causes high temperatures. A high-pressure area exists at each of the poles because the air is cold and, therefore, more dense.

2. The United States is located between 30 degrees and 50 degrees north latitude.

3. The following is a sample completed diagram:

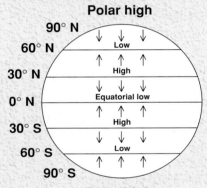

4. The following is a sample completed diagram:

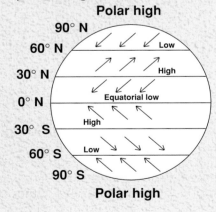

World Winds Explained

What causes the winds that helped Columbus reach America and return home to Spain? These same winds also drive the currents in the Atlantic Ocean. In what direction do they blow? Why? The answers lie in ideas that you have already investigated. First, you know that air moves from areas of high pressure to areas of low pressure. Second, the movement of air is influenced by the Coriolis effect. Keep these ideas in mind as you complete the Exploration.

What to Do

In small groups, discuss the following information. Copy each diagram into your ScienceLog and respond to the questions.

1. Permanent high- and low-pressure areas exist at certain latitudes because of temperature differences. Why is a low-pressure area located at the equator? Why are high-pressure areas located at the poles?

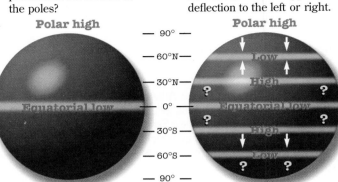

2. Other factors, especially the rising and sinking of air at certain latitudes, create other permanent high- and low-pressure areas.

   What is the approximate line of latitude for the United States?

3. Winds blow from areas of high pressure to areas of low pressure. If Earth did not rotate and there were no Coriolis effect, winds would blow from high- to low-pressure areas with no apparent deflection to the left or right.

A few wind patterns are shown on the diagram above. Sketch the diagram and draw in the missing arrows.

4. Because the Earth rotates, the winds are deflected to the right or left, depending on the hemisphere. A few wind directions are shown below. Sketch the diagram in your ScienceLog and draw in the missing information.

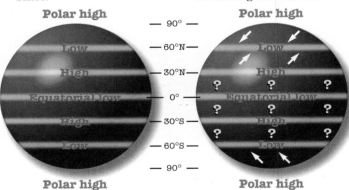

"Into That Silent Sea . . ."

Down dropt the breeze, the sails
    dropt down,
'Twas sad as sad could be;
And we did speak only to break
The silence of the sea!

All in a hot and copper sky,
The bloody Sun, at noon,
Right up above the mast did stand,
No bigger than the Moon.

Day after day, day after day,
We stuck, nor breath nor motion;
As idle as a painted ship
Upon a painted ocean.

Samuel Taylor Coleridge
The Rime of the Ancient Mariner,
Part 2

Coleridge published this poem in 1798. At that time sailing ships were the only means of travel across the vast oceans. How did he refer to the characteristics of winds and currents in his poetry? Where do you think the sailors are stranded? (Hint: See the second stanza, which describes the position and appearance of the sun.) Did sailors really have cause to worry that the wind would stop blowing? **A**

During the time when ships were powered by the wind, certain regions of the ocean presented serious dangers. Near the equator is an area where the air rises due to intense heating, so there is very little wind. Ships could sit for days or even weeks with no wind to fill their sails. This region became known as the **doldrums.**

And how would you like to be stuck in the **horse latitudes**? This is a region of calm located at the northern edge of the northeast trade winds. In the 1700s, ships sailing through this area often carried horses. If a ship was slowed or stranded and supplies ran low, the horses were thrown overboard in order to conserve drinking water—hence the name horse latitudes.

Multicultural Extension

Indian Rain

India has a rainy season that lasts from the middle of June through September. During this period, monsoons (seasonal winds) blow across the Indian Ocean, picking up moisture and dropping it on the Indian countryside. Have students find out more about how monsoons affect Indian life. *(One-third of the Indian economy is based on agriculture, which depends on the rains that the monsoons bring. Sometimes the monsoons arrive later in the season, and crops fail as a result. Some monsoons bring too much rain, destroying crops and causing destructive floods.)*

El Niño

What is El Niño? Consider the following news items, and then answer the questions below.

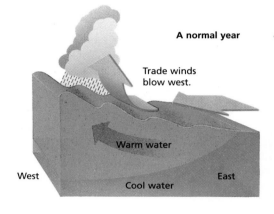

Experts Ponder Impact of El Nino Phenomenon

North America could be facing dramatic weather changes due to El Nino, a variation of temperature and pressure patterns over the eastern Pacific Ocean.

This phenomenon occurs every three to five years — its severity changing from year to year. Weather patterns worldwide are affected. The strong El Nino of 1982-83 was blamed for a devastating drought in Africa and Australia. Also, severe winter storms lashed California, and parts of South America were deluged with torrential rains. The El Nino in 1986-87, however, was barely noticed.

What Causes El Nino?

El Nino (Spanish for "the Christ Child") occurs in the equatorial waters off the west coast of South America around Christmas time. Normally, trade winds push surface waters away from the coastline of South America, causing cold water to well up from below. The cold surface waters suppress the formation of clouds and rain. Every few years these winds die down and then reverse, causing warm water to be pushed toward the South American coast. The warmer waters produce abundant rainfall. At the same time, normal weather patterns are disrupted over a wide area.

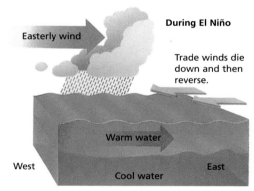

A normal year

Trade winds blow west.

Warm water

West

East

Cool water

During El Niño

Easterly wind

Trade winds die down and then reverse.

Warm water

West

East

Cool water

 In a normal year (top), westward-blowing winds push warm waters toward the western Pacific, where they "pile up" (although only to a height of a meter or two). When El Niño occurs (bottom), these winds reverse, causing warm waters to pool along the western coast of South America.

Questions

1. The El Niño effect illustrates the link between oceans and the atmosphere. Use information from the articles above to describe this link.

2. Examine the map on page 271 that shows the ocean's surface currents. What is the name of the surface current that flows up the coast of South America?

3. Examine the picture of global temperatures on page 226. What evidence suggests that this colder current flows up the west coast of South America?

4. What kind of damage can El Niño cause? What regions suffer the most? How are plant and animal life affected? Research the El Niño phenomenon to learn more about its effects.

276

Have you been following daily pressure changes on your homemade barometer? What kind of pressure change is associated with calm, sunny weather? with unsettled, stormy, or cloudy weather? Did you find that calm weather is usually associated with increasing atmospheric pressure and that stormy weather is associated with decreasing pressure? What is the reason for this? Through the diagrams and pictures that follow, you'll examine the relationship between low atmospheric pressure and stormy weather.

⬆ A hurricane, seen from space

Picture Study 1

The satellite picture above shows an unusual weather system—a *hurricane*. Hurricanes vary in size, but most are 150 to 450 km across. Some have winds up to 300 km/h.

1. Assuming that there is a low-pressure area at the center of the hurricane, which way are the winds blowing—inward or outward?

2. Are the winds moving in a straight line, or are they spiraling around the hurricane's center? How does the photograph suggest an answer to this question?

3. Is the wind moving clockwise or counterclockwise? Why does it blow in this direction? (In the Southern Hemisphere, the wind direction around low-pressure areas is the opposite of that in the Northern Hemisphere.)

277

LESSON 3 ORGANIZER

Time Required
1 class period

Process Skills
predicting, inferring, comparing

New Terms
Hurricane—a cyclonic storm of extreme low pressure and winds of 118 km/h or greater
Isobars—lines on a weather map that connect points of equal pressure

Materials (per student group)
none

Teaching Resources
Graphing Practice Worksheet, p. 62
Transparency 43
SourceBook, p. S74

FOCUS

Getting Started

From newspapers, cut out weather maps that show low-pressure systems. Display these and ask: Can you see a pattern in the direction that the wind blows around the low-pressure systems? *(Students may be able to identify that the wind always blows toward the center of a low-pressure system.)* Point out that in this lesson they will find out why this happens and why the wind blows so fiercely during hurricanes.

Main Ideas

1. Winds blow from regions of higher pressure to regions of lower pressure.
2. Air flowing into a low-pressure system rises, and, as a result, cooling takes place and clouds form.
3. Hurricanes are weather systems with extremely low pressure and winds reaching 118 km/h or greater.

TEACHING STRATEGIES

Picture Study 1

Direct students' attention to the photograph on this page. Ask if any students have personally witnessed a hurricane or seen the destruction that one can cause. Allow students time to share any relevant accounts.

Answers to
Picture Study 1

1. The winds in a hurricane blow inward because air always moves from areas of higher pressure to areas of lower pressure.

2. The swirling configuration of the clouds suggests that the winds are spiraling around the hurricane's center.

3. The wind is flowing counterclockwise around the low-pressure system. This is due to the Coriolis effect. In the Southern Hemisphere, the wind flows clockwise around a low-pressure system.

Answers to
Picture Study 2

1. The trade winds that carry hurricanes toward the coast of the United States are also called the steering winds because they were used to "steer" sailing ships westward. Westward-flowing winds are labeled *1*.

2. The energy is stored in the water molecules as they change from liquid to gas. The surface winds are labeled *2*.

3. As the air rises, it cools to the dew point, and condensation occurs. Warm, moisture-laden air is labeled *3*.

4. High-altitude winds are labeled at *4* spiraling counterclockwise. (Students may notice that the photo shows the hurricane spiraling counterclockwise and inward. High-altitude winds also spiral counterclockwise, but they spiral outward, pushed up and out by the storm. If students become confused, suggest that they focus on the low-altitude air masses, which created the storm and which spiral inward toward a center of low pressure.)

5. Descending air does not form clouds because air warms as it descends, so the air temperature is above the dew point.

 Transparency 43 is available to accompany Picture Study 2.

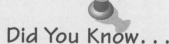

Did You Know . . .

The windiest day on record was April 12, 1934. On this day, winds at the peak of Mount Washington in New Hampshire reached 371 km/h (231 mph).

Picture Study 2

Compare the satellite picture on page 277 with the cross-sectional view of a hurricane shown below. As you read the following, match the italicized terms with the numbers in the diagram and answer the accompanying questions.

1. Most hurricanes are carried toward the eastern coast of the United States by *westward-flowing winds*.

What are these winds called?

2. The *surface winds* pick up heat energy from the warm waters they pass over. In addition, a large amount of water evaporates—another mechanism of gathering energy that will later be released.

How is heat energy stored by the evaporation process?

3. As the *warm, moisture-laden air* spirals toward the center of the hurricane, it also rises. Condensation creates clouds. This process releases heat energy.

Why does moisture condense out of the air as the air rises?

4. Some of the rising air is carried away by *high-altitude winds*.

Which way do these high-altitude winds spiral?

5. Some of the rising air is drawn back down into the *eye of the hurricane*. Air in the eye sinks, warming as it does so and causing clear skies and calm conditions. Why doesn't the descending air form clouds like moist, rising air does?

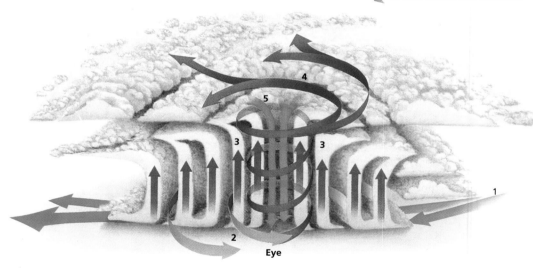

A cross section of a hurricane

Eye

Meeting Individual Needs

Learners Having Difficulty

Students may have difficulty understanding how the Coriolis effect determines the direction of winds around hurricanes or other low-pressure disturbances. Use the following diagram to explain that as winds move inward toward an area of low pressure (in the Northern Hemisphere), they are deflected to the right because of the Coriolis effect. The convergence of these winds as they flow inward and to the right creates a counterclockwise rotation of winds around the area of low pressure.

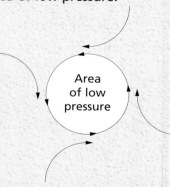

Area of low pressure

Picture Study 3

A hurricane is an extreme example of a low-pressure system. Normal atmospheric pressures are close to 100 kPa. The eye of a hurricane deviates sharply from this norm.

The diagram at right shows a hurricane as it might be symbolized on a weather map. A series of solid lines called *isobars* connect points of equal pressure. The same pressure difference separates each isobar. Wind directions have been included on this map as well.

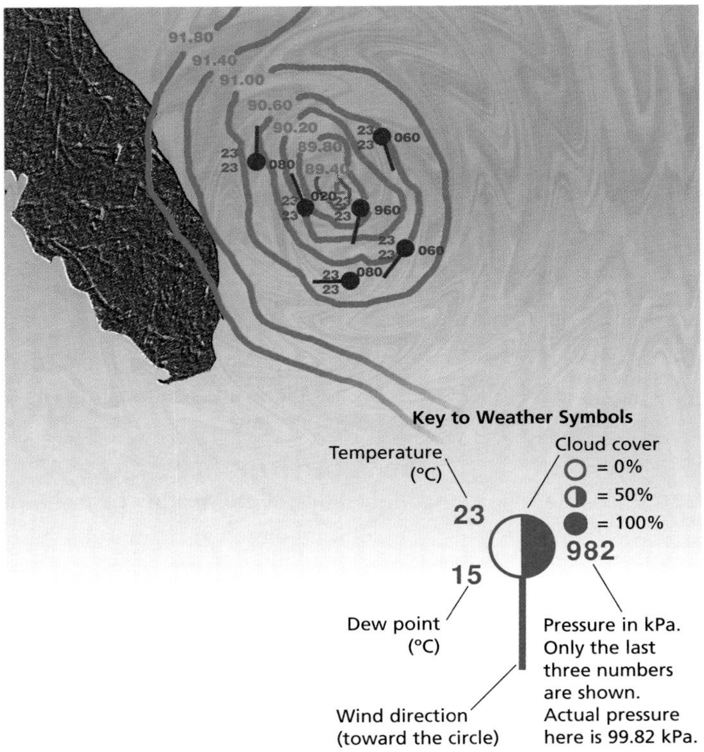

1. What is the wind-direction pattern around a low-pressure system such as a hurricane—clockwise or counterclockwise?

2. How does the pressure change as you progress farther from the eye of a low-pressure system?

3. At each station, what number was dropped from the complete pressure reading? (Why?)

4. Why are the dew point and the air temperature readings the same?

Keeping Track

Make a list of the major ideas and findings you have encountered in this chapter. What have you discovered about pressure in the atmosphere and ocean? What causes global ocean currents and wind? How does a hurricane form? Your list should answer these questions and more. Compare your list with one from a classmate.

Key to Weather Symbols

Temperature (°C)

Cloud cover
○ = 0%
◑ = 50%
● = 100%

23

15

982

Dew point (°C)

Pressure in kPa. Only the last three numbers are shown. Actual pressure here is 99.82 kPa.

Wind direction (toward the circle)

Answers to *Picture Study 3*

1. Wind direction around a low-pressure system is counterclockwise in the Northern Hemisphere and clockwise in the Southern Hemisphere.

2. As you progress farther away from the eye of a low-pressure system, the pressure rises.

3. At each station, the number 8 or 9 was dropped from the pressure reading to allow space for all of the readings. (Most low-pressure systems have pressures around 96 to 99 kPa. Therefore, on most weather maps the tens digit will be dropped. If the pressure reading is above 100 kPa, the hundreds and tens digits are dropped.)

4. Dew point and air temperature are the same because the air is saturated with moisture.

Reteaching

Provide students with a diagram of the cross section of a hurricane. Ask them to draw and identify steering winds, surface winds, warm, moisture-laden air, high-altitude winds, the eye, and clear skies.

Assessment

Ask students to draw a weather map containing the following information:
a. cold front moving in from the north
b. a low-pressure system on the West Coast (the pressure at the center of the low is 98.60 kPa, the pressure difference between isobars is 0.4 kPa)
c. a weather station that has reported a pressure of 99.44 kPa, winds from the south, 100 percent of cloud cover, air temperature of 21°C, and dew point of 15°C

Extension

Have interested students do some research on the causes of tornadoes, or suggest that students do some research on a specific hurricane. Have them find out about the storm's path, the damage it caused, and its wind speeds.

Closure

Have students write a story in which they experience a hurricane. Then have them rewrite the story as it would be told by a meteorologist.

Homework

The Graphing Practice Worksheet on page 62 of the Unit 4 Teaching Resources booklet makes an excellent homework assignment once students have completed Lesson 3.

Meeting Individual Needs

Gifted Learners

Have students create an anemometer (a device for measuring the speed of the wind) using household materials. Be sure you approve all student designs for safety before they proceed. (*Accept all reasonable designs. Some students may design a device that resembles a pinwheel, using revolutions per minute to quantitatively measure wind speed.*)

Answers to *Challenge Your Thinking*

1. Pressure equals force per unit area. Jack's weight, or the force he exerts on the snow, is 600 N; this force is applied over the area of his snowshoes, 1.5 m². Therefore, Jack exerts a pressure of 600 N/1.5 m² = 400 N/m² (or 400 Pa) on the snow.

2. a. A hurricane is a low-pressure system, so it is a cyclone.
b. In the Northern Hemisphere, cyclones move counterclockwise around a center of low pressure, and anticyclones move clockwise around a center of high pressure. In the Southern Hemisphere, cyclones move clockwise around a center of low pressure, and anticyclones move counterclockwise around a center of high pressure.

3. a. From the northeast (This is due to the clockwise rotation around a center of high pressure.)
b. Point *B* is cooler than point *C*.
c. From the northwest (This is due to the counterclockwise rotation around a center of low pressure.)
d. Along the fronts and around the low-pressure system
e. Point *A* and within isobars marked *H*
f. Point *D* and within the isobars marked by *L*
Pressure readings in Boston will drop and so will the temperature. Wind direction will move from southerly, to southwesterly, to westerly, to northwesterly.

Multicultural Extension

Great Winds

Share the following information with students: Hurricanes are the same thing as typhoons or cyclones. When cyclones occur in the China Sea, they are called typhoons. When they occur in the Indian Ocean, they are called cyclones. The word *hurricane* comes from *Hunraken,* the storm god of the Mayas of Central America and Mexico. The word *typhoon* comes from *ty fung,* which is Chinese for "great wind."

CHALLENGE YOUR THINKING

1. Treading Lightly
Jack's mass is 60 kg (600 N). His snowshoes cover an area of 1.5 m². How much pressure does he exert against the snow?

2. Pressure Points
Another name for a low-pressure system is a cyclone. High-pressure systems are called anticyclones.

a. Is a hurricane a cyclone or an anticyclone?

b. Describe the airflow around a cyclone and around an anticyclone.

3. Weather Forecast
Looking at the weather map shown below, identify the following:

a. wind direction at point *A*

b. temperature differences between points *B* and *C*

c. wind direction at point *D*

d. regions of cloudy skies and rain

e. areas of high pressure

f. areas of low pressure

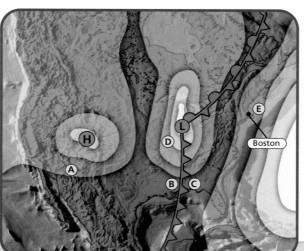

Over the next day the low-pressure system will move over point *E* (Boston). How do you think the pressure readings in Boston will change? What about the temperature? How will wind direction change?

★ You may wish to provide students with the Chapter 12 Review Worksheet that is available to accompany this Challenge Your Thinking (Teaching Resources, page 64). Transparency 44 is also available for your use.

4. The Motion's Over the Ocean

Once a hurricane reaches land, it rapidly loses energy. Can you think of a reason for this?

5. Pressing Issues

Here are some everyday experiences and observations. They all have something to do with atmospheric pressure. What is the connection?

a. On Nora's last plane trip, she noticed that the tops of the small sealed creamer containers bulged outward. But in restaurants on the ground they do not.

b. While drinking a milkshake through a straw at the ice cream shop, Janis wondered exactly how the straw worked.

c. Rick always had trouble getting frozen orange-juice concentrate out of the can, until someone suggested punching a small hole in the bottom of the can.

6. Global Confusion

What do you think would happen to the pattern of global winds and currents if the Earth were suddenly to start spinning in the opposite direction—from east to west instead of from west to east? How would low- and high-pressure systems behave differently?

7. Touchy Question

When you place your finger over the end of a plastic syringe, it becomes very difficult to pull out the plunger. Why is this?

Review your responses to the ScienceLog questions on page 259. Then revise your original ideas so that they reflect what you've learned.

281

ScienceLog

The following are sample revised answers:

1. Students may simply state that hurricanes or tornadoes caused the damage. Others may be able to expand on this answer by explaining how the winds are caused by severe pressure differences.

2. Students should be encouraged to explore the everyday meanings of the word *pressure* as well as its scientific meaning. One example of an everyday meaning is stress, as in "being under pressure." Students may relate pressure to adding air to bicycle or car tires. Point out that pressure in this case is often measured in pounds per square inch (psi). This will help communicate the scientific definition of pressure: force per area.

3. The trade winds are the winds that blow in a westerly direction between 0 degrees and 30 degrees on both sides of the equator.

4. Hurricanes gain heat energy from the warm waters of the ocean. This energy is stored when large amounts of water evaporate. When the moisture condenses, the energy is released again. After moving onto land, there is no more water for evaporation; therefore, the stored energy of the hurricane diminishes rapidly.

5. a. At high altitudes, such as those reached by an airplane, atmospheric pressure drops. The result is that there is less outside pressure against the creamer container when it is in the air than when it is on the ground. Thus, the container on the plane bulges out.

   b. Atmospheric pressure supports the liquid in a straw. By sucking on the straw, Janis creates a vacuum inside the straw that allows the milkshake to move upward through the straw as air pressure pushes down on the milkshake in the glass.

   c. The small hole in the juice can causes the air pressure inside and outside the can to equalize. This way the juice in the can flows out freely instead of being kept inside by atmospheric pressure.

6. Wind directions are determined by the Coriolis effect, which is determined by the Earth's rotation. Therefore, changing the direction of Earth's spin would change the direction of any phenomena dependent on the Coriolis effect. For example, the Westerlies would be the Easterlies, and the winds around low-pressure systems in the Northern Hemisphere would flow in a clockwise direction instead of a counterclockwise direction.

7. There is a partial vacuum in the syringe. The resistance felt as the plunger is pulled out is caused by the higher atmospheric pressure outside the syringe pushing against the plunger. In order to pull out the plunger, a force greater than that exerted by the atmosphere must be supplied in the opposite direction.

The Big Ideas

The following is a sample unit summary:

The greenhouse effect is the trapping of heat by certain gases in the atmosphere. An increase in the greenhouse effect—from human-made or naturally occurring causes—could result in increased global warming. Carbon dioxide, methane, and water vapor are the primary gases that contribute to the greenhouse effect. (1)

Oceans moderate the climate of Earth by storing and slowly releasing great amounts of heat. (2)

Dew point is the temperature at which condensation would occur in a given air mass. (3)

Most ocean surface currents are caused by winds blowing along the surface of the ocean. Other ocean currents, such as those found in the deep ocean and in the Mediterranean Sea, are caused by density differences.

Winds are caused by density differences in the atmosphere, which result from the uneven heating and cooling of the underlying land or ocean. (4)

Density differences drive winds and many ocean currents. Denser air or water tends to sink, and lighter air or water tends to rise. This motion causes convection currents to form. (5)

Air masses are large bodies of air with essentially the same temperature and moisture content throughout. (6)

Winds and surface currents move in accordance with density differences and are guided by the Coriolis effect, which is caused by the Earth's rotation on its axis. (7)

Pressure, by itself, does not directly influence weather. Differences in pressure are caused by differences in density, which cause winds to flow. High pressure indicates an area of dense, sinking air, resulting in clear skies and relatively mild weather conditions. Low pressure indicates an area of rising air, resulting in clouds and precipitation. (8)

 You may wish to provide students with the Unit 4 Review Worksheet that is available to accompany this Making Connections (Teaching Resources, page 72).

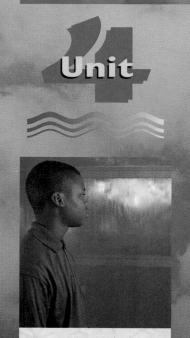

Making Connections

Unit 4

SOURCEBOOK

Look in the SourceBook to read more about the atmosphere and oceans that make Earth a temperate planet. The SourceBook also contains information about a subject that's always topical—the weather.

Here's what you'll find in the SourceBook:

UNIT 4
The Atmosphere S60
Weather and Climate S66
Ocean Resources S75

The Big Ideas

In your ScienceLog, write a summary of this unit, using the following questions as a guide:

1. What is the greenhouse effect, and what gases are responsible for it? How might it relate to global warming?
2. What influence do the oceans have on the climate of the Earth?
3. What is the dew point?
4. What causes ocean currents? What causes winds?
5. What role does density play in driving winds and ocean currents?
6. What are air masses, and how do they form?
7. Why do winds and the ocean's surface currents move in predictable patterns?
8. How does pressure affect the weather?

Checking Your Understanding

1. Every summer, the average concentration of atmospheric carbon dioxide drops significantly in the Northern Hemisphere. What reasons can you suggest for this?

Answers to *Checking Your Understanding*

1. Carbon dioxide levels drop because of all the seasonal vegetation that uses carbon dioxide during photosynthesis. Densely forested areas would be more likely to have lower levels of carbon dioxide than areas without a lot of vegetation, such as deserts.

2. The diagram at right shows the path of a sailplane (a type of glider) on a sunny day.

   a. Use your understanding of air currents to explain the glider's movements.

   b. How would the glider's path differ on a cloudy day? at night? over a city?

   c. How do the effects shown here relate to weather and climate?

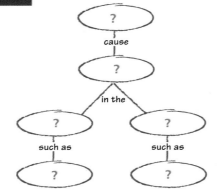

Wind direction

Flight path

Plowed field

Mountain

Forest

Lake

3. Examine the simplified diagram of deep-ocean currents shown below.

   a. What causes these currents?

   b. How do these currents help to absorb the world's "excess" carbon dioxide?

   c. If glaciers melted, what effect might this have on deep-ocean currents?

| North Pole | 45°N | Equator | 45°S | South Pole |

Arctic   Warm-water currents   Antarctic

Upwelling

Antarctic Bottom Current

4. **concept map** Copy the concept map at right into your ScienceLog. Then complete the map using the following words and phrases: currents, oceans, atmosphere, trade winds, Antarctic currents, and density differences.

?

cause

?

in the

?   ?

such as   such as

?   ?

---

   b. On a cloudy day, the glider would probably not experience much up-and-down motion. At night, the glider would probably fall gradually until it came to the lake, where warm air rising would cause it to rise slightly. If the glider flew over a city on a sunny day, it would rise because of the rapid heating of roofs, parking lots, roads, and buildings.

   c. Uneven heating of the Earth's surface causes air to rise or sink and winds to form.

3. a. Ocean surface water cools at the poles, becomes more dense, and sinks to the bottom. From there it flows mainly toward the equator. At places where winds remove surface waters, upwelling occurs, bringing the cold bottom waters to the surface.

   b. Surface waters absorb large amounts of carbon dioxide. When this surface water cools, it sinks to the ocean bottom and carries the carbon dioxide with it. Thus, much carbon dioxide is taken out of circulation.

   c. If all the glaciers melted, the deep-ocean currents would be disrupted. Ocean surface water would be flooded with fresh water, lowering its density. Thus, the surface water might not be dense enough to sink. If the polar waters stopped sinking, deep-ocean currents would change.

4.

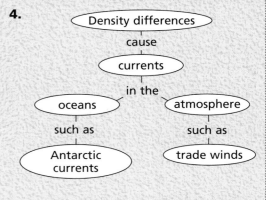

Density differences

cause

currents

in the

oceans   atmosphere

such as   such as

Antarctic currents   trade winds

Homework

The Unit 4 Activity Worksheet on page 71 of the Unit 4 Teaching Resources booklet makes an excellent homework activity after students have completed Unit 4.

Answers to
Checking Your Understanding,
continued

2. a. The glider gains or loses altitude depending on whether the air it passes through is rising or falling. The glider falls over the forest and the lake because they are cooler than their surroundings; therefore, the air above them sinks. The glider rises over the field because it is hotter than its surroundings; therefore, the air above it rises. The glider rises rapidly over the mountain because the mountain deflects the wind

Background

For massive plumes of flame and hot air to develop, a fire must have a wealth of fuel, such as very dry twigs and leaves. There must be warm, rising currents of air, such as those found in unstable atmospheres. But the wind is by far the most important factor in the creation of plumes. A strong crosswind can inhibit the formation of a plume.

Forest fires break out thousands of times a year around the world. For example, in 1985, a total of 81,622 fires destroyed about 1.2 million hectares of forest in the United States alone. In 1971, fires in Wisconsin and Michigan killed 1500 people and burned 1.7 million hectares of land—an area more than five times the size of Rhode Island!

There is another type of interaction between fires and the weather that occurs in confined regions such as valleys. As the upper atmosphere cools after sunset, a *temperature inversion* is set up in which warm air is trapped beneath the denser, cooler air above it. A fire in the valley gradually uses up all of the oxygen in the trapped warm air and eventually starts to smolder, heating up the unburned material surrounding it. When the sun rises and heats the overlying cool air, the oxygen-depleted air escapes and is replaced with oxygen-rich air. The result can be an explosively renewed forest fire.

Extension

Have students find out what types of forest-fire guidelines are offered to the public by local recreation and parks departments. Students can present their findings in an oral or written report.

Answers to
Think About It

Answers will vary, but make sure that each student supports his or her answer with clear and logical reasoning. Students will probably agree that an approach between the two extremes, in which the largest fires are contained but smaller ones are allowed to burn themselves out, is best. (Students may be interested to know that the policy of the National Park Service is to allow most naturally caused fires to burn.)

Weather From Fire

In 1993 a devastating wildfire took place near Santa Barbara, California. As the fire burned, huge storm clouds formed in an otherwise cloudless sky. Fast-moving fiery whirlwinds danced over the ground. The fire wasn't only destroying everything in its path—it was creating its own weather!

Fire-Made Clouds

Hot air rising from a forest fire can create tremendous updrafts. Surrounding air rushes in to take the place of the rising air, stirring up huge plumes of ash, smoke, hot air, and noxious gases. Cool, dry air normally sinks down from above the fire and stops these plumes from developing further. But if the conditions are just right, a surprising thing happens.

If the upper atmosphere contains warm, moist air, the moisture begins to condense on the ash and smoke particles. This creates droplets that can develop into clouds. As the clouds grow, the droplets begin to collide and combine until they are heavy enough to fall as rain. The result is an isolated rainstorm complete with thunder and lightning.

Whirlwinds of Fire

Forest fires can also create whirlwinds. These small, tornado-like funnels can be extremely dangerous. Whirlwinds are similar to dust devils that dance across desert sands. They are created by upward-moving currents of air. Their circular motion is started by an updraft that is forced to curve after striking some obstacle such as a cliff or slope of land. Whirlwinds move across the ground at 8 to 11 km/h, sometimes growing up to 120 m high and 15 m wide.

Most whirlwinds last no longer than a minute, but within this time they can cause some big problems. Firefighters caught in the path of whirlwinds have been severely injured and even killed. Also, if a whirlwind is hot enough, it can suck up tremendous amounts of air into its vortex. The resulting updraft can pull burning debris, called firebrands, up through the whirlwind. In some cases, the updraft has been so strong that burning trees have been uprooted and shot into the air above. When such firebrands land, they often start new fires hundreds of meters away.

▲ A towering whirlwind sucks burning debris from a forest fire in Idaho

Think About It

Many scientists agree that fires are a natural part of the growth of a forest. For example, some tree seeds are released only under the extreme temperatures that occur in a fire. Some people argue that forest fires should be allowed to run their natural course. Others argue that the effects of a forest fire are just too damaging and that forest fires should be put out as soon as possible. Which side do you agree with? Why?

284

Space-Age Subs

*T*he Mariana Trench in the Pacific Ocean near Guam plunges to a bone-crushing depth of 11 km. This is the deepest point of the ocean, and the water pressure at this depth is over 110,000 kilopascals (kPa). This is enough pressure to collapse a conventional submarine. In fact, due to these great pressures, this dark ocean abyss remains shrouded in mystery. But a new generation of space-age submersibles may change all that!

Deep-Diving Robots

One approach to exploring the deepest parts of the ocean is to use unmanned submersibles. Japan's Marine Science and Technology Center has perfected such a vessel, called *Kaiko*.

Kaiko is a 5.4 ton, 3.5 m aquatic robot that operates at more than 10,000 m below the ocean's surface. It is equipped with high-tech TV cameras, sonar, and a pair of 1.8 m robotic arms. However, like most robotic submersibles, *Kaiko* must be tethered to a research ship by 11 km of power and communications cables.

As *Kaiko* sinks to within a few meters of the ocean floor, a small robotic vehicle separates from the bottom of the main craft. Although attached to the main craft by a control cable, the robotic vehicle descends still deeper to explore the ocean floor, transmitting pictures to

▼ The *Kaiko* submarine. *Kaiko* means "trench" in Japanese.

▲ The *Deep Flight* submarine. The pilot lies in this sub in a prone position.

the surface and gathering samples. When its work is done, the robotic vehicle returns to the main craft and is taken back to the surface.

Nothing Beats Being There

The Ocean Exploration Group in Richmond, California, has designed a submersible that looks more like a fighter plane than a submarine. Unlike conventional submersibles, which slowly sink to the bottom, this one-person sub, called *Deep Flight*, is designed to "fly" to the sea floor under it own power. The pilot determines the sub's flight path by adjusting its wing flaps and tail fins. The craft could dive vertically, but this would be uncomfortable for the pilot. Instead, it descends at an angle of about 30 degrees.

Deep Flight's 3.5 m hull is made of a hard ceramic material that is extremely strong under pressure. In addition to the pilot, the vessel houses the controls, an instrument panel, a life-support system, and a 24-volt, battery-operated power supply. The batteries feed a pair of electric motors at the back of the craft that propel it up to 25 km/h (14 knots).

Currently we know more about the surfaces of other planets than we know about our own ocean floor. But with submersibles like *Kaiko* and *Deep Flight*, we may some day unlock the mysteries of the deep.

Find Out for Yourself

Scientists divide the ocean into several layers, or zones—the sublittoral zone, bathyal zone, abyssal zone, and hadal zone. Find out how each of these zones is defined and what its characteristics are.

285

Answers to
Find Out for Yourself

The sublittoral zone runs from just below the surface at low tide to the edge of the continental shelf. It is characterized by abundant oxygen and nutrients. From the edge of the continental shelf (usually 100–200 m below the surface) to a depth characterized by a temperature of 4°C (1 to 3 km below the surface) is the bathyal zone. In this zone there is an absence of light and no photosynthesis. The abyssal zone runs from a depth of about 2 km to 6 km below the surface. And beneath the abyssal zone is the deepest region, known as the hadal zone. These deeper zones have very distinct and unique life-forms.

Background

The first submarine to successfully move underwater was introduced in 1620. Built by the Dutch scientist Cornelius van Drebbel, the submarine was little more than a small wooden boat covered with animal hides. It wasn't until the 1960s that the first deep-sea submarines were constructed. These early subs were both costly and unwieldy. They were constructed of solid steel about 10 cm thick and had huge buoyancy tanks. Designers soon turned to making subs that were more maneuverable and less expensive. These new subs were able to descend to about 6 km, making about 98 percent of the ocean floor accessible to humans.

Scientists have also been interested in studying the 2 percent of the ocean floor below 6 km—which includes parts of the Indian Ocean and Caribbean Sea—for a number of reasons. In these regions, the ocean floor is covered with a variety of life-forms that may rival the diversity of tropical rain forests. Many scientists also study the deep-ocean floor in the hopes of finding unusual geologic features and marine ecosystems not present anywhere else.

Integrating the Sciences

Earth and Life Sciences

Have students discuss what life must be like on the ocean floor. Point out that in this environment there is no sunlight, there are no green plants, and there is tremendous water pressure. Have a volunteer write down the class's ideas. Then direct students to do some research to find out about creatures that inhabit the deep-ocean floor. What adaptations do they have that help them to cope with this unique environment? You may wish to emphasize some unique features of deep-sea organisms, such as bioluminescence.

Background

Tsunamis travel at an average speed of about 800–900 km/h until they reach the shore. As a tsunami approaches the shore, the slowing of the wave forces it to become taller. Although extremely tall tsunamis are very dangerous, shorter tsunamis can be just as destructive. By monitoring earthquakes on the ocean floor, seismologists can usually predict the arrival of a tsunami.

Extension

The Britannica videotape entitled *Waves on Water* is an excellent teaching resource. It uses a wave tank to demonstrate surface water waves, the motion of water molecules in the wave, the generation of tsunamis, and the behavior of water waves as they near the shore. The 16-minute videotape can be ordered from Encyclopaedia Britannica Educational Corporation, ATTN: Customer Service, 310 South Michigan Avenue, Chicago, IL 60604-9839. Ask for catalog number 69603.

CROSS-DISCIPLINARY FOCUS

Social Studies

Divide the class into groups and have each group complete a mini-project on areas that have been hit by severe tsunamis. The projects should include a short report on the event, the scale of the destruction to property, lives, and health, and the effect of that destruction on the local communities. As group members present their report, have them mark the location of their tsunami on a world map. After every group has presented, lead students in a discussion to examine any patterns that they see among the geographic locations. *(Examples of hard-hitting tsunamis include the 1946 tsunami in Hilo, Hawaii, and the 1992 tsunami in Nicaragua. The class may notice that most tsunamis occur in the Pacific Ocean.)*

The Wildest of Waves

*I*magine a foamy wave sloshing onto a peaceful beach. Now imagine the same wave three stories high and moving as fast as a truck on a highway. Sound impossible? It's not. It's called a *tsunami*, and it's very real and very dangerous. *Tsunami* (soo NAH me) means "harbor wave" in Japanese.

The Birth and Death of Giants

Where do these immense waves come from? Tsunamis are created far out in the ocean. A tremor on the ocean floor caused by an underwater landslide or earthquake can make a huge volume of water rise and fall quickly, creating a wave. Out in the middle of the ocean, the wave may not look very large or powerful. However, as the wave nears the shore and the water gets shallower, the wave changes. It slows down and gets taller and taller, sometimes reaching as high as 10 m by the time it crashes onto the shoreline. A large tsunami can demolish trees, houses, cars, and anything else in its path. One tsunami hit Sanriku, Japan, in 1896 and killed between 22,000 and 27,000 people.

▲ Monster waves are well known in many communities along the Pacific coast. Hiroo Kanamori (inset) thinks that mud can make a difference.

Predicting Tsunamis

Since most tsunamis are caused by underwater earthquakes, scientists can monitor earthquakes to predict when and where a tsunami will hit land. But the predictions are not always accurate. Some earthquakes are very weak and should not create powerful tsunamis, yet they do. The reason for this has scientists puzzled.

A scientist named Hiroo Kanamori has an idea. He thinks that when the ocean floor shifts, sediment can sometimes act as a lubricant, allowing a smooth movement of the floor instead of an abrupt jolt. Thus, even if the earthquake does not create a big jolt, it may still create a powerful tsunami. Kanamori calls these special events *tsunami earthquakes*. By modifying traditional measuring devices to account for this effect, he has had more success in predicting the birth of a tsunami. Accurate early warnings can give people enough time to get safely out of the tsunami's way.

Wavy Water

The speed of water waves, including tsunamis, depends on the depth of the water. Check it out for yourself by filling a tub that is about 0.5 m long and at least 5 cm deep with about 1 cm of water. Then nudge the tub. How long does it take for the wave to go back and forth? Add more water, and nudge the tub again. How fast does the wave move the second time? What do you think would happen to the speed of the wave if you made the water deeper still? Try it!

Answers to
Wavy Water

As the water gets deeper, the speed of the wave will increase. For a quantitative approach, you might want to have students verify that the speed of the wave varies with the square root of the water depth.

SCIENCE in ACTION

From Thermometer to Satellite

Predicting floods, observing the path of a tornado, watching the growth of a hurricane, and issuing flood warnings are all in a day's work for Cristy Mitchell. As a meteorologist for the National Weather Service, Cristy spends each working day observing the powerful forces of nature.

Forces in the Atmosphere

When asked what made her job interesting, Cristy Mitchell replied, "There's nothing like the adrenaline rush you get when you see a tornado coming! I would say that witnessing the powerful forces of nature is what really makes my job interesting."

Meteorology is the study of natural forces in the Earth's atmosphere. Perhaps the most familiar field of meteorology is weather forecasting. However, meteorology is also an important aspect of air-pollution control, agricultural planning, and air and sea transportation. Meteorologists also study trends in the Earth's climate, such as global warming and ozone depletion. In fact, anything that relies on or relates to weather is a meteorologist's concern.

Collecting the Data

Meteorologists collect data on air pressure, temperature, humidity, and wind velocity.

▲ Photograph of Hurricane Elena taken from the space shuttle *Discovery* in September 1985. Cristy Mitchell (inset) uses computer technology to analyze the weather.

Then, by applying what they know about the physical properties of the atmosphere and analyzing the mathematical relationships in the data, they are able to forecast the weather.

The data that a meteorologist needs in order to make accurate weather forecasts comes from a variety of sources. When asked what kinds of instruments she uses, Cristy said, "The computer is an invaluable tool for me. Through it, I receive maps and detailed information, including temperature, wind speed, air pressure, and general sky conditions for a specific region."

In addition to computers, Cristy also relies on radar and satellite imagery to show the "whole picture" of a region and the nation's weather. Meteorologists also use sophisticated computer models of the world's atmosphere to help forecast the weather and weather trends.

Find Out for Yourself

Using the barometer you made in Chapter 12, Lesson 1, graph the barometric pressure for 7 days. At the same time, graph the barometric pressure reported by a local meteorologist. How do the graphs compare? What kind of forecast would you make based on the results of your barometer? How does your forecast compare with that of the meteorologist?

Background

Weather prediction is a very difficult task, and a meteorologist's predictions are not always accurate. This is because the Earth's weather system is *chaotic*. A chaotic system tends to be unpredictable because it is very sensitive to its initial conditions. For example, if you started with two identical environments and changed just one small element of one of them, after a while you might find that the small difference had grown so large that the two environments were completely different.

Meteorologists cope with this uncertainty by giving probabilities of expected weather. They use sophisticated computer models to help them assign accurate probabilities.

Discussion Questions

1. Weather stations operate around the clock, seven days a week. Do you think you would enjoy the night work and rotating shifts that are part of a meteorologist's job? *(Some students may think it is exciting, while others may prefer a 9–5 job.)*

2. There is an old saying: "You can talk about the weather, but you can't do anything about it." How might this relate to a meteorologist's job? *(Although scientists like Cristy Mitchell observe the powerful forces of nature, they cannot do anything to stop them. However, issuing accurate weather warnings can save thousands of lives.)*

Critical Thinking

Tell the students to imagine that they are meteorologists studying another planet in the solar system. Ask: As a meteorologist, what features would you look for to get information about the climate and the weather? *(Students should recognize that climates are strongly influenced by major geographic features. A planet's rotation affects prevailing wind patterns. Other features to look for include mountains, deserts, and large bodies of water.)*

Answers to *Find Out for Yourself*

The graphs are likely to agree, at least roughly. Students should find that changes in weather, especially rainfall, occur when the pressure drops and that stable weather occurs when the pressure is high. (This is because air masses always move from high-pressure zones to low-pressure zones. Low local pressure is a sign that different weather is on its way.)

Unit 5 — Electromagnetic Systems

Unit Overview

In this unit, students explore the relationship between electricity and magnetism. In Chapter 13, students are introduced to some basics about the way electricity works in everyday life. Electricity is then explored on the subatomic level. In Chapter 14, students look at some of the sources of electricity, including chemical cells, solar cells, magnetism, thermocouples, and piezoelectricity. Students also build their own dry cells. In Chapter 15, students explore conduction and resistance as they experiment with electric circuits. They learn to represent circuits with circuit diagrams, and they study how current is controlled with switches, bimetallic strips, and magnets. The unit concludes with a discussion of current, voltage, and charge.

Using the Themes

The unifying themes emphasized in this unit are **Energy, Systems,** and **Structures.** The following information will help you incorporate these themes into your teaching plan. Focus questions that correspond to these themes appear in the margins of this Annotated Teacher's Edition on pages 292, 305, 319, 339, 344, 345, and 348.

Energy is a fundamental part of any discussion of the properties of electricity and magnetism because both electricity and magnetism are forms of energy. In electromagnetic devices, both electricity and magnetism

are converted into other forms of energy. This concept is discussed throughout the unit.

The unit relates to the theme of **Systems** in several ways. Systems that produce electricity, such as electrical generators, wet and dry cells, and electromagnets, are discussed. Also, electric circuits are presented as systems that consist of several basic parts arranged in a systematic way.

Structures are important when electricity is explained in terms of the atomic model of matter. Students learn that electricity is the movement of electrons and that an electron is one of the particles that make up the structure of the atom.

Using the SourceBook

In Unit 5, students learn about the differences between static electricity and electric current. Then they explore how voltage, current, and resistance are related. Magnetism is described in terms of poles, forces, and fields. Students also learn how devices that utilize or generate electricity operate. In their study of electrical devices, students take an in-depth look at alternators, direct-current generators, motors, and transformers.

Bibliography for Teachers

Billings, Charlene W. *Superconductivity: From Discovery to Breakthrough.* New York City, NY: Cobblehill Books, 1991.

Gardener, Robert. *Electricity and Magnetism.* New York City, NY: Twenty-First Century Books, 1994.

Vecchione, Glen. *Magnet Science.* New York City, NY: Sterling Publishing Co., Inc., 1991.

Bibliography for Students

Ardley, Neil. *Electricity: The Way It Works.* New York City, NY: New Discovery Books, 1993.

Tomecek, Steve. *Simple Attractions.* New York City, NY: W.H. Freeman and Co., 1995.

VanCleave, Janice. *Electricity.* New York City, NY: John Wiley and Sons, Inc., 1994.

Zubrowski, Bernie. *Blinks and Buzzers: Building and Experimenting With Electricity and Magnetism.* New York City, NY: Morrow Junior Books, 1991.

Films, Videotapes, Software, and Other Media

Build a Circuit
Software (Apple II family)
Sunburst
101 Castleton St.
P.O. Box 100
Pleasantville, NY 10570

Electricity and Magnetism
Filmstrip or videotape
Coronet/MTI
108 Wilmot Rd.
Deerfield, IL 60015

How to Produce Electric Currents With Magnets
Videotape
Britannica
310 S. Michigan Ave.
Chicago, IL 60604-9839

Static and Current Electricity
Film or videotape
Coronet/MTI
108 Wilmot Rd.
Deerfield, IL 60015

Unit Organizer

Unit/Chapter	Lesson	Time*	Objectives	Teaching Resources
Unit Opener, p. 288				Science Sleuths: The Crashing Computers English/Spanish Audiocassettes Home Connection, p. 1
Chapter 13, p. 290	Lesson 1, Indispensable Energy, p. 291	2	1. Describe how electricity functions in a variety of electrical devices on which students depend. 2. Identify examples in which electrical energy is converted into light or heat energy. 3. Construct a simple electric circuit. 4. Explain how electricity can produce magnetic energy.	Image and Activity Bank 13-1 Exploration Worksheet, p. 3
	Lesson 2, What Is Electricity? p. 300	4 to 5	1. Explain electricity in terms of the theory of charged particles. 2. Identify electrons as having negative electrical charge and protons as having positive electrical charge. 3. Demonstrate how an object responds to electrical charges. 4. Explain why electricity flows easily through a conductor but does not flow easily through an insulator. 5. Describe an electric circuit. 6. Describe how a cell produces a continuous electric current.	Image and Activity Bank 13-2 Discrepant Event Worksheet, p. 8 Resource Worksheet, p. 9 Resource Worksheet, p. 10 Exploration Worksheet, p. 12
End of Chapter, p. 307				Chapter 13 Review Worksheet, p. 13 Chapter 13 Assessment Worksheet, p. 15
Chapter 14, p. 309	Lesson 1, Electricity From Chemicals, p. 310	3	1. Explain how a chemical cell produces electricity. 2. Compare a wet cell with a dry cell. 3. Explain how the current in a dry cell or a wet cell can be increased.	Image and Activity Bank 14-1 Exploration Worksheet, p. 17 Exploration Worksheet, p. 21 ▼
	Lesson 2, Electricity From Magnetism, p. 315	3	1. Demonstrate what happens when a magnet moves through a wire coil and when a wire coil moves through a magnetic field. 2. Compare alternating current with direct current. 3. Describe how generators produce alternating current.	Image and Activity Bank 14-2 Exploration Worksheet, p. 26 Transparency Worksheet, p. 30 ▼ Transparency 47 Transparency 48
	Lesson 3, Other Sources of Electricity, p. 322	2	1. Identify several sources of electrical energy. 2. Describe how a solar cell converts light energy into electrical energy. 3. Construct a thermocouple and explain how it produces electricity from heat energy. 4. Explain the piezoelectric effect and how it is used to produce electricity.	Image and Activity Bank 14-3
End of Chapter, p. 324				Chapter 14 Review Worksheet, p. 32 Chapter 14 Assessment Worksheet, p. 35
Chapter 15, p. 326	Lesson 1, Circuits: Channeling the Flow, p. 327	3	1. Draw simple circuit diagrams using conventional symbols. 2. Describe several factors that affect the electrical conductivity of a wire. 3. Explain what happens when a wire resists the flow of current. 4. Identify several practical applications of resistance.	Image and Activity Bank 15-1 Exploration Worksheet, p. 37 Graphing Practice Worksheet, p. 42
	Lesson 2, Making Circuits Work for You, p. 331	2	1. Identify two different types of circuits: series and parallel. 2. Compare the amount of current in series and parallel circuits. 3. Construct and diagram simple parallel and series circuits. 4. Explain the effect of increasing the number of cells connected in series.	Image and Activity Bank 15-2 Transparency Worksheet, p. 44 ▼ Exploration Worksheet, p. 46 Transparency 50
	Lesson 3, Controlling the Current, p. 335	2	1. Identify a variety of mechanisms that are used to control the flow of electric current. 2. Explain how various kinds of switches and dimmers operate. 3. Describe several different kinds of switches.	Image and Activity Bank 15-3 Transparency 51 Transparency 52 Activity Worksheet, p. 50 ▼
	Lesson 4, Electromagnets, p. 338	3 to 4	1. Construct an electromagnet, identify its circuit components, and trace its current. 2. Explain how the strength of an electromagnet can be increased. 3. Explain how electromagnets function in a variety of devices.	Image and Activity Bank 15-4 Exploration Worksheet, p. 51 Transparency 54 Transparency 55
	Lesson 5, How Much Electricity? p. 343	2	1. Describe the flow of electricity in a circuit in terms of current, voltage, and charge. 2. Explain how wattage expresses the relationship between current and voltage. 3. Demonstrate a mathematical understanding of amperes, volts, and coulombs. 4. Determine the effects of various quantities of applied current on the human body.	Image and Activity Bank 15-5 Activity Worksheet, p. 52 ▼
End of Chapter, p. 348				Theme Worksheet, p. 53 Activity Worksheet, p. 54 ▼ Chapter 15 Review Worksheet, p. 55 Chapter 15 Assessment Worksheet, p. 59
End of Unit, p. 350				Unit 5 Activity Worksheet, p. 61 Unit 5 Review Worksheet, p. 63 Unit 5 End-of-Unit Assessment, p. 66 Unit 5 Activity Assessment, p. 71 Unit 5 Self-Evaluation of Achievement, p. 74

* Estimated time is given in number of 50-minute class periods. Actual time may vary depending on period length and individual class characteristics.
▼ Transparencies are available to accompany these worksheets. Please refer to the Teaching Transparencies Cross-Reference chart in the Unit 5 Teaching Resources booklet.

Materials Organizer

Chapter	Page	Activity and Materials per Student Group
13	296	***Exploration 1, Activity 1:** 40 cm of magnet wire; piece of sandpaper, about 10 cm × 10 cm; D-cell; rubber band; flashlight bulb; **Activity 2:** two lengths of magnet wire, 30 cm each; piece of sandpaper, about 10 cm × 10 cm; stick of modeling clay; clothespin; 2 D-cells; wide rubber band; 10 cm strand of steel wool; 10 cm strip of aluminum foil; 10 cm length of thin nichrome wire; **Activity 3:** D-cell; compass; 2 thumbtacks or screws; small wooden block; paper clip; 15 cm and 25 cm lengths of magnet wire; piece of sandpaper, about 10 cm × 10 cm; rubber band; **Activity 4:** 2 D-cells; paper clip; 2 lengths of magnet wire, 10 cm and 250 cm long; piece of sandpaper, about 10 cm × 10 cm; bar magnet; support stand with ring clamp; cork; wide rubber band; small wooden block; strip of cardboard, 2 cm × 30 cm; 2 thumbtacks or screws
	300	**A Classical Current Demonstration:** plastic strip, about 10 cm long, or plastic ruler; wire clothes hanger; paper cup; large nail, about 8 cm long; wheat puff; 10 cm of thread; piece of plastic wrap, about 10 cm × 10 cm; vinyl strip, about 10 cm long; piece of flannel cloth, about 10 cm × 10 cm
	301	**A Theory of Charged Particles:** plastic strip, about 10 cm long, or plastic ruler; wire clothes hanger; paper cup; large nail, about 8 cm long; wheat puff; 10 cm of thread; piece of plastic wrap, about 10 cm × 10 cm; a variety of materials for testing conductivity, such as a glass rod, a wooden stick, and a piece of copper wire or magnet wire, each about 8 cm long
	306	**Exploration 2:** small magnetic compass; 1 m of insulated wire or magnet wire; other materials to hold parts in place, such as a plastic-foam cup and a rubber band or masking tape; straightened paper clip; 10 cm of magnet wire with sanded ends; small cup; 50 mL of concentrated lemon juice
14	310	**Exploration 1, Experiment 1:** commercial galvanometer or homemade galvanometer from page 306; 40 cm of sanded magnet wire from Exploration 1 on page 296; a few pieces of masking tape; 2 to 3 rubber bands; 2 strips of zinc, 3 cm × 8 cm each; 2 strips of copper, 3 cm × 8 cm each; several strips of blotting paper or filter paper, 3 cm × 8 cm each; 50 mL of 0.5 M calcium chloride; 250 mL beaker or other container to hold calcium chloride; pair of forceps; safety goggles; lab aprons; latex gloves (additional teacher materials: bottle; large beaker; 100 mL of 0.5 M sodium hydroxide; a few sheets of newspaper; see Advance Preparation below.); **Experiment 2:** zinc strip, 2 cm × 15 cm; copper strip, 2 cm × 15 cm; 250 mL beaker; 250 mL of water; about 0.5 mL of salt; two 20 cm lengths of magnet wire with sanded ends; 2 alligator clips or clothespins; commercial galvanometer (additional teacher materials: bottle; large beaker; 100 mL of 0.5 M sodium hydroxide; a few sheets of newspaper; see Advance Preparation below.)
	315	**Exploration 3:** commercial galvanometer; 150 cm length of magnet wire; cardboard tube; strong bar magnet; piece of sandpaper, about 10 cm × 10 cm
	321	**Alternating Current:** meter stick or metric ruler; fluorescent or neon light
	322	**Exploration 4, Project 1:** solar cell; light source; meter stick; **Project 2:** 50 cm of uninsulated copper wire; 50 cm of uninsulated iron wire; a few matches; commercial galvanometer; safety goggles
15	329	***Exploration 1, Part 1:** 3 equal lengths of magnet wire, thin (gauge 18 or 20) nichrome wire, and thick (gauge 28 or 30) nichrome wire, about 20 cm each; piece of sandpaper, about 10 cm × 10 cm; wooden dowel; D-cell; flashlight bulb; rubber band; **Part 2:** thin nichrome wire from Part 1; wooden dowel; 3 D-cells; rubber band; coin; **Part 3:** length of nichrome wire from Part 1; wooden dowel; 3 D-cells; rubber band; few sheets of newspaper; **Applications of Resistance:** graphite pencil with some graphite exposed; flashlight bulb; D-cell; 3 lengths of magnet wire with sanded ends; small knife; rubber band (See Advance Preparation below.)
	331	**Exploration 2, Part 1:** six 10 cm lengths of copper wire or magnet wire with sanded ends; several small pieces of masking tape; 3 D-cells; 3 flashlight bulbs in holders; 2 contact switches; **Part 3:** 6 light bulbs; 3 contact switches; 3 D-cells; four 20 cm lengths of uninsulated copper wire; four 10 cm lengths of copper wire or magnet wire with sanded ends; wire cutters
	337	**Exploration 3, Part 2:** materials to construct one of the circuits listed on page 337 (See Advance Preparation below.)
	338	**Exploration 4:** paper clip; several washers; about 1 m of light, insulated wire or magnet wire with sanded ends; 2 D-cells; iron spike or bolt; switch; a variety of materials for increasing the electromagnet's strength (See Advance Preparation below.)
	342	**A Model Motor:** about 1 m of thin insulated copper wire or magnet wire with sanded ends; cork; small knife; short knitting needle; 4 straight pins; 2 thumbtacks; 3 D-cells; stick of modeling clay; 2 block magnets

* You may wish to set up these activities in stations at different locations around the classroom.

Advance Preparation

Exploration 1, page 310: The additional teacher materials will be used in the disposal of the calcium chloride and the salt solution; for detailed instructions, see the Waste Disposal Alert on page 310.

Exploration 1, Applications of Resistance, page 329: Students will need to design their own experiments. You may wish to cut out the pencils before-hand. If you decide to have students do it in class, make sure they handle their knives with extreme care.

Exploration 3, Part 2, page 337: Students will need to design their own circuits.

Exploration 4, page 338: Students will need to come up with their own materials.

Unit Compression

This unit has two areas of emphasis: the conceptual study of electromagnetic systems and the technological applications of such systems. The conceptual sections should be considered core material and should not be skipped. In some sections, the technological applications are integrated within the conceptual framework and so should also be considered core material. Other sections, however, are almost completely technological. These technology-oriented sections can be considered optional and can be skipped if time is short. Such sections include the following:

- Chapter 13, Lesson 1: Exploration 1 and the section titled Electricity and You: Case Studies
- Chapter 14, Lesson 2: the sections titled Alternating Current and Direct Current
- Chapter 14, Lesson 3
- Chapter 15, Lesson 3
- Chapter 15, Lesson 5

Homework Options

Chapter 13
See Teacher's Edition margin, pp. 293, 295, 296, 297, 301, 303, and 307
Resource Worksheet, p. 10
SourceBook, pp. S82 and S86

Chapter 14
See Teacher's Edition margin, pp. 312, 316, 318, 319, and 323
SourceBook, pp. S91 and S97

Chapter 15
See Teacher's Edition margin, pp. 327, 329, 332, 340, 344, 346, 349, and 351
Graphing Practice Worksheet, p. 42
Activity Worksheet, p. 50
Activity Worksheet, p. 52
Activity Worksheet, p. 54
SourceBook, pp. S89 and S101

Unit 5
Unit 5 Activity Worksheet, p. 61
SourceBook Activity Worksheet, p. 75

Assessment Planning Guide

Lesson, Chapter, and Unit Assessment	SourceBook Assessment	Ongoing and Activity Assessment	Portfolio and Student-Centered Assessment
Lesson Assessment Follow-Up: see Teacher's Edition margin, pp. 299, 306, 314, 321, 323, 330, 334, 337, 342, and 347 **Chapter Assessment** Chapter 13 Review Worksheet, p. 13 Chapter 13 Assessment Worksheet, p. 15* Chapter 14 Review Worksheet, p. 32 Chapter 14 Assessment Worksheet, p. 35* Chapter 15 Review Worksheet, p. 55 Chapter 15 Assessment Worksheet, p. 59* **Unit Assessment** Unit 5 Review Worksheet, p. 63 End-of-Unit Assessment Worksheet, p. 66*	SourceBook Review Worksheet, p. 76 SourceBook Assessment Worksheet, p. 80*	Activity Assessment Worksheet, p. 71* **SnackDisc** Ongoing Assessment Checklists ♦ Teacher Evaluation Checklists ♦ Progress Reports ♦	Portfolio: see Teacher's Edition margin, pp. 299, 314, 330 **SnackDisc** Self-Evaluation Checklists ♦ Peer Evaluation Checklists ♦ Group Evaluation Checklists ♦ Portfolio Evaluation Checklists ♦

* Also available on the Test Generator software
♦ Also available in the Assessment Checklists and Rubrics booklet

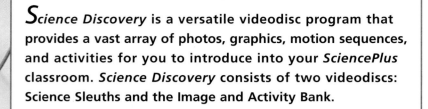

Science Discovery is a versatile videodisc program that provides a vast array of photos, graphics, motion sequences, and activities for you to introduce into your *SciencePlus* classroom. *Science Discovery* consists of two videodiscs: Science Sleuths and the Image and Activity Bank.

Using the *Science Discovery* Videodiscs

Science Sleuths: The Crashing Computers
Side B

Computers in the new offices of Royal Electromag are crashing; the screens go blank and the files being worked on are lost. The vice-president of marketing suspects a rival company of inserting a computer virus into the Royal Electromag computer network.

Interviews

1. Setting the scene: Vice-president of marketing 31658 (play ×2)

2. Executive director of sales 32452 (play)

3. Word processor 33692 (play)

4. Chairman of the board 34052 (play)

5. Computer maintenance 34605 (play)

6. Computer-chip engineer 35235 (play)

7. Guy struck by lightning 35976 (play)

Documents

8. List of computer crashes 37098 (step ×3)

Literature Search

9. Search on the words: COMPUTER, VIRUS, CRASH, ELECTRICITY, LIGHTNING 37103 (step)

10. Article #1 ("The Deadly Virus Is Not So Deadly") 37106 (step)

11. Article #2 ("Lightning") 37109 (step)

12. Article #3 ("Electrical Repairs") 37112

13. Article #4 ("More Things to Worry About") 37114 (step)

Sleuth Information Service

14. Office wiring floor plan 37117

15. Weather graph 37119

Sleuth Lab Tests

16. Computer diagnostics 37121

17. Building power check 37123

18. Electroscopic tests on employees 37125 (step ×7)

Still Photographs

19. Office computers 37134 (step)

20. Surge protector 37137

Image and Activity Bank
Side A or B

A selection of still images, short videos, and activities is available for you to use as you teach this unit. For a larger selection and detailed instructions, see the Videodisc Resources booklet included with the Teaching Resources materials.

13-1 Indispensable Energy, page 291
Switch construction; lab setup 371–375 (step ×4)
Use these materials to construct a switch.

Magnetic field 34144–34290 (play ×2) (Side A only)
The compass needle responds to the magnetic field surrounding a bar magnet. Opposite ends of the needle are attracted to opposite ends of the magnet.

13-2 What Is Electricity? page 300
Electrostatic energy; electroscope 33022–33206 (play) (Side A only)
Charges from the cloth are rubbed onto the glass rod. The separation of the leaves of the electroscope indicates the presence of electrical charges.

Electrostatic energy; Van de Graaff generator 33207–33812 (play ×2) (Side A only)
The blue discharge is ionized air.

14-1 Electricity From Chemicals, page 310
Battery; mercury cells 2122
A battery generates an electric current through a chemical reaction.

◀️ Step Reverse   Play ▶️   Pause ⏸️  Step Forward ▶️

Battery, rechargeable 2123

Rechargeable batteries convert electrical energy back into chemical energy when they are recharging. Disposable batteries are toxic, so the use of rechargeable batteries cuts the amounts of toxic materials in dumps and landfills.

Battery, car 2124

The lead-acid automotive battery stores chemical energy. When charging, the chemical reaction is pushed one way. While discharging, the opposite reaction occurs, and electricity is released.

14-2 Electricity From Magnetism, page 315

Magnetically induced current 32745–33021 (play ×2) (Side A only)

The voltmeter shows a current generated by a constantly changing magnetic field.

Dam, hydroelectric 1789–1793 (step ×4)

Reservoirs, or large amounts of water, are created behind dams to make production of electricity easy. The energy released by moving the water from a higher to a lower elevation is captured in the dam. (step) Large turbines are installed within generators to capture the energy of the moving water. The rushing water pushes against the turbine blades, which turn the turbine, thus creating mechanical energy that can be converted to electricity. (step) The rotating turbine produces large amounts of electricity. Generators convert the mechanical energy of the rotating turbine into electricity. (step) Water is brought down from the reservoir at high pressure to turn the turbines. (step) The electricity captured by the generators is stored for future use or for immediate use in the surrounding areas.

14-3 Other Sources of Electricity, page 322

Solar cell 2126

This solar cell can be used in most lighting conditions, including under fluorescent or tungsten light bulbs.

Solar panel 2127

Entire fields of solar panels such as this one generate great amounts of energy for municipal use.

15-1 Circuits: Channeling the Flow, page 327

Electrical switches 2117

Several scissor switches and a push-button switch are shown. Power flows from the central contact to either of the side contacts. The handle of the switch is insulated.

15-2 Making Circuits Work for You, page 331

Circuits 2128–2135 (step ×7)

Cells and current (step) The circuit with the most cells has the most current. (step) Resistance and current (step) The bulb is the brightest in the circuit with the least resistance. (step) Series vs. parallel circuits (step) The bulbs are the brightest when they are in a parallel circuit. (step) Series and parallel circuits (step) The same amount of current runs through the resistors in the series circuit on top and through the top resistor in the bottom circuit. The two resistors in parallel have slightly less current.

15-3 Controlling the Current, page 335

Dimmer switch 2114

A dimmer switch is a variable resistor. For the same current, different values of resistance will produce different brightnesses of light.

Thermostat 2115

A thermostat is a variable resistor. Keeping the thermostat turned down in the winter and up in the summer is a good way to conserve energy.

15-4 Electromagnets, page 338

Electromagnet construction; lab setup 375–377 (step ×2)

Construct an electromagnet. Coil the wire around the nail, and connect it to the battery and the switch.

Electromagnetic strength; lab setup 378–379 (step)

Test the strength of an electromagnet by using different batteries.

Electromagnet 33813–34143 (play ×2) (Side A only)

Hundreds of turns of wire are inside the lower, blue core. A strong magnetic current is generated when the battery is connected.

Electromagnet; wrecking yard 2120

A current passed through a coil generates a magnetic field. This powerful magnet moves large masses of metal from one place to another.

Motor, electric 1852

A motor is an energy converter. Electrical energy is converted to mechanical energy.

15-5 How Much Electricity? page 343

Appliance wattage; table 1865

List of the wattages for several common appliances

Energy use of appliances; table 1863–1864 (step)

Average energy use of several common appliances

5 Unit Electromagnetic Systems

Ask students to look around the class-room and identify the things they see that are operated electrically. Keep track of their responses on the chalkboard. *(Possible responses include lights, clocks, calculators, watches, and computers.)* Ask: Is there any relationship between electricity and magnetism? *(Accept all reasonable responses.)* Tell students that in this unit, they will learn about electricity, how it is produced, and how it is related to magnetism.

A good motivating activity is to let students listen to the English/Spanish Audiocassettes as an introduction to the unit. Also, begin the unit by giving Spanish-speaking students a copy of the Spanish Glossary from the Unit 5 Teaching Resources booklet.

Unit 5

CHAPTER
13 Energy in a Wire
1 Indispensable Energy 291
2 What Is Electricity? . . 300

CHAPTER
14 Sources of Electricity
1 Electricity From Chemicals 310
2 Electricity From Magnetism 315
3 Other Sources of Electricity 322

CHAPTER
15 Currents and Circuits
1 Circuits: Channeling the Flow 327
2 Making Circuits Work for You 331
3 Controlling the Current 335
4 Electromagnets 338
5 How Much Electricity? 343

Electromagnetic Systems

288

Connecting to Other Units

This table will help you integrate topics covered in this unit with topics covered in other units.

Unit 2 Particles	The sources of electricity and magnetism are the presence and motions of charged particles.
Unit 3 Machines, Work, and Energy	Electromagnetic energy is often used or produced by machines to perform work.
Unit 4 Oceans and Climates	Just as pressure differences drive winds, voltage differences drive electric currents.
Unit 6 Sound	The propagation of electric current, like sound, depends on the material through which it flows.
Unit 7 Light	Electricity flowing through a filament causes an incandescent bulb to give off light.

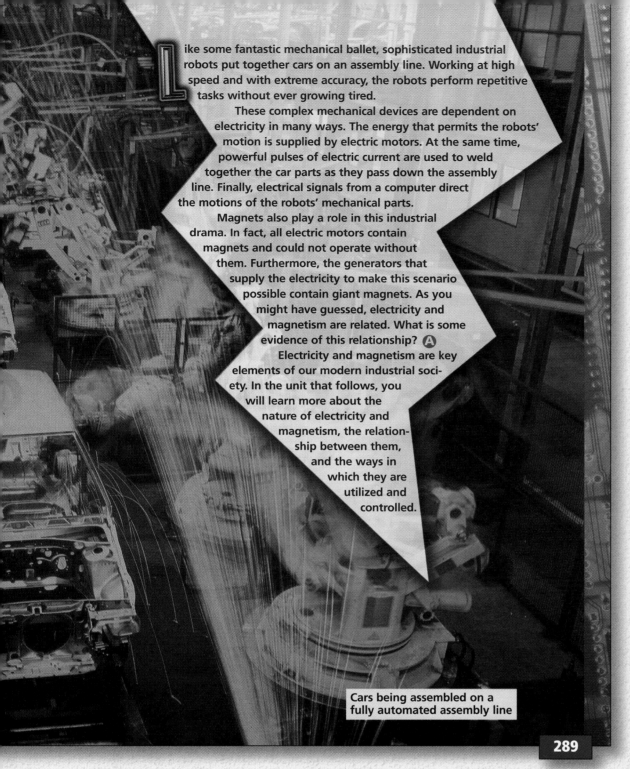

like some fantastic mechanical ballet, sophisticated industrial robots put together cars on an assembly line. Working at high speed and with extreme accuracy, the robots perform repetitive tasks without ever growing tired.

These complex mechanical devices are dependent on electricity in many ways. The energy that permits the robots' motion is supplied by electric motors. At the same time, powerful pulses of electric current are used to weld together the car parts as they pass down the assembly line. Finally, electrical signals from a computer direct the motions of the robots' mechanical parts.

Magnets also play a role in this industrial drama. In fact, all electric motors contain magnets and could not operate without them. Furthermore, the generators that supply the electricity to make this scenario possible contain giant magnets. As you might have guessed, electricity and magnetism are related. What is some evidence of this relationship? Ⓐ

Electricity and magnetism are key elements of our modern industrial society. In the unit that follows, you will learn more about the nature of electricity and magnetism, the relationship between them, and the ways in which they are utilized and controlled.

Cars being assembled on a fully automated assembly line

289

As the photograph illustrates, automobile production is dependent on electricity in many ways. For example, electricity is often used to help extract materials from metal-rich ores through a process called *electrolysis.* In this process, an ore containing the desired metal is dissolved in a solution. Then the solution is placed in an electrolytic cell and a voltage is applied across the cell so that a current flows, which causes the metal to separate out of the solution. Aluminum is recovered in this way and is widely used in the manufacturing of automobiles. Copper, a metal used widely in electric circuits, is refined through the process of electrolysis.

The automobile industry uses electricity in other ways as well. The assembly line in the photograph is devoted to welding automobile parts together. This computer-controlled process is just one example of computer-aided manufacturing, or CAM. Cameras and lasers are sometimes used to perform inspections of the finished product. Also, the automobile itself is equipped with an elaborate electrical system. All of these functions require a constant supply of electricity to the manufacturing plant.

Critical Thinking

Ask students to name as many different forms of energy as they can. Have a volunteer keep a list on the board. Then ask: How many examples of these forms of energy can be found in the photograph? *(Student answers are likely to include heat energy in the sparks, light energy in the sparks and overhead lights, kinetic energy in the automated machinery and the moving cars, and electrical energy in the wires.)* Point out that many of these forms of energy are produced from electrical energy. *(The sparks are produced by welding machines that are powered by electricity. The overhead light comes from bulbs powered by electricity. The mechanical arms in the photograph are moving parts of electrically powered machines.)*

Answer to
In-Text Question

Ⓐ Accept all reasonable responses. Students are not expected to know the answer to this question at this time. The question is meant to stimulate critical thinking. Point out to students that a major goal of this unit is to discover how electricity and magnetism are related.

CHAPTER 13 Energy in a Wire

Connecting to Other Chapters

> **Chapter 13**
> *introduces students to electricity, circuit theory, and the connection between electricity and magnetism.*

> **Chapter 14**
> *introduces alternating current and ways of producing electricity, such as with chemical cells and generators.*

> **Chapter 15**
> *introduces circuit symbols, resistance, parallel and series circuits, electromagnets, and electrical energy.*

Prior Knowledge and Misconceptions

Your students' responses to the ScienceLog questions on this page will reveal the kind of information—and misinformation—they bring to this chapter. Use what you find out about your students' knowledge to choose which chapter concepts and activities to emphasize in your teaching. After students complete the material in this chapter, they will be asked to revise their answers based on what they have learned. Sample revised answers can be found on page 308.

In addition to having students answer the questions on this page, you may wish to have them complete the following activity: Have each student write down three questions and three facts about electricity. Then have students trade papers with a partner to exchange information and discuss each other's questions. Afterward, you may wish to have volunteers discuss one question and one fact of their choice.

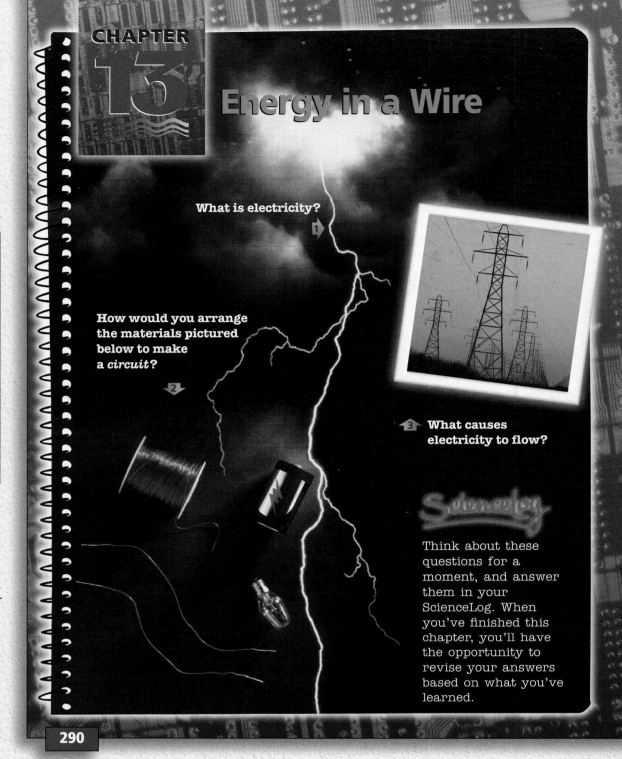

CHAPTER 13 Energy in a Wire

What is electricity?

1

How would you arrange the materials pictured below to make a *circuit?*

2

3 What causes electricity to flow?

Think about these questions for a moment, and answer them in your ScienceLog. When you've finished this chapter, you'll have the opportunity to revise your answers based on what you've learned.

290

Assure students that there are no right or wrong answers to this exercise. Collect the papers, but do not grade them. Instead, read them to find out what students know about electricity, what misconceptions they may have, and what aspects of electricity are interesting to them.

LESSON 1 — Indispensable Energy

Your school is planning an exhibition called the Museum of Today. The idea behind the exhibition is a simple one: to introduce the technology of today to imaginary visitors from the past. It is your job to plan the exhibition. Include as many examples as you can of the technology and gadgetry that make life what it is today.

Where do you start?

You could start with simple things like electric lights. We take them completely for granted, yet we could hardly imagine life without them. Electric lights were a revolutionary development. With their invention, the 24-hour-a-day city suddenly became a possibility. Think of the tremendous variety of different types of lights and their uses. What are some of them? **Ⓐ**

Then, of course, there's the telephone. Who could imagine life without the telephone? A hundred and fifty years ago, to be able to punch a few buttons and talk to someone far away was inconceivable.

And don't forget radio. How wonderful it is to be able to flip a switch and hear music, talk shows, or news—all of it originating many kilometers away.

Television! Now there's an invention! Life would be far different without television. Who would have imagined that you could send pictures through the air and have them appear on a screen? That would seem like magic to someone from the past.

Here are a few examples of technology to get you started. What other examples would you include in your exhibit? **Ⓑ**

291

LESSON 1 — Indispensable Energy

FOCUS

Getting Started

Have students silently read the text on this page. Ask: What do all of these technological devices have in common? *(They all run on electricity.)* Ask students whether they think electricity is a form of energy. Also, encourage them to speculate about what an electric current is and how it is related to an electric circuit. Suggest that they write their ideas in their ScienceLog. Then point out that they will refine their understanding of the technological uses of electrical energy in this lesson.

Main Ideas

1. Much of modern technology depends on electricity to perform a variety of functions.
2. Electrical energy can be changed into light and heat energy.
3. Electricity flows through a circuit.
4. An electric current produces a magnetic field.

TEACHING STRATEGIES

Answers to
In-Text Question and Caption

Ⓐ Accept all reasonable responses. Sample answers include the following: fluorescent lights are used for brightly illuminating large areas; incandescent lamps are used for lighting smaller areas in homes; sodium vapor lights are used as fog lamps on automobiles and street lights; neon signs are used for advertising.

Ⓑ Answers will vary. You may wish to encourage student creativity by having them consider what technologies they use in their own daily lives.

LESSON 1 ORGANIZER

Time Required
2 class periods

Process Skills
observing, analyzing, inferring

Theme Connection
Energy

New Term
Circuit—a continuous path for the flow of electrons

Materials (per student group)
Exploration 1, Activity 1: 40 cm of magnet wire; piece of sandpaper, about 10 cm × 10 cm; D-cell; rubber band; flashlight bulb; **Activity 2:** two lengths of magnet wire, 30 cm each; piece of sandpaper, about 10 cm × 10 cm; stick of modeling clay; clothespin; 2 D-cells; wide rubber band; 10 cm strand of steel wool; 10 cm strip of aluminum foil; 10 cm length of thin nichrome wire; **Activity 3:** D-cell; compass; 2 thumbtacks or screws; small wooden block; paper clip; 15 cm and 25 cm lengths of magnet wire; piece of sandpaper, about 10 cm × 10 cm; rubber band; **Activity 4:** 2 D-cells; paper clip; 2 lengths of magnet wire, 10 cm
continued ➤

The Museum Is Open!

Have students silently read the text on this page. When they have finished, direct their attention to the illustration and photographs on pages 291–293. Call on individuals to identify the function of each electrical device and to explain where the electricity to operate it comes from. Ask them to name any additional examples of technology that they would include in the Museum of Today exhibition. Keep track of their suggestions on the chalkboard. If time permits, you may wish to have groups of students brainstorm some ideas for exhibits. Suggest that they make diagrams of their exhibits to share with the class.

Have students write their ideas about the concepts of electricity and magnetism in their ScienceLog. Students should write definitions of the terms *electricity* and *magnetism* and speculate about how the two terms might be related. Explain that they should review and revise their ideas as they progress through the unit.

Answer to
In-Text Question

Ⓐ As students are about to learn in this chapter, when a button is pushed, a switch is flipped, or a knob is turned in order to turn on an electrical device, electrical contact is made between parts of a circuit, and a current flows through the circuit.

Theme Connection

Energy

Energy can be defined as the ability to make matter move or change. **Focus question:** Does electricity fit this definition? Have students provide at least one example to support their answer. *(Sample answer: Yes, electricity is a form of energy because all of the devices in the Museum of Today move or change in response to the electricity flowing through them.)*

The Museum Is Open!

As you might imagine, most of the exhibits in the Museum of Today would perplex visitors from the past. In fact, the lights, sounds, and moving parts of the items on display might even scare them—a situation that could lead to chaos, since even the idea of turning something "off" would be unfamiliar to most of your guests.

Of course today everybody knows how to flip a switch on and off. But how many people stop to consider what actually happens when something is turned on or off? What happens when that button is pushed or that switch is flipped or that knob is turned? Ⓐ Fortunately, we do not need to know the answer to these questions in order to use our microwave ovens, home computers, lamps, dishwashers, power tools, and so on.

ORGANIZER, continued

and 250 cm long; piece of sandpaper, about 10 cm × 10 cm; bar magnet; support stand with ring clamp; cork; wide rubber band; small wooden block; strip of cardboard, 2 cm × 30 cm; 2 thumbtacks or screws

Teaching Resources
Exploration Worksheet, p. 3

An Electrical Connection

What do all the devices mentioned so far in this chapter have in common? They all use electricity. Most of the devices use magnetism as well. Obviously, electricity and magnetism serve us in **(B)** many ways. Can you identify some of these ways?

Clearly, life without electricity would be very different. **(C)** Imagine going an entire day without it. How would you get by? Write a scenario about a day in your life without electricity.

What exactly is electricity? How would you explain it to your **(D)** visitors from the past? Although you can see what electricity does, you can't see electricity itself. In this unit you will be introduced to this mysterious, indispensable form of energy and its close relative, magnetism.

EXIT

We hope you enjoyed the display!

293

Homework

You may wish to have students read pages 292–293 and answer the in-text questions as a homework assignment.

Did You Know...

Benjamin Franklin was one of the most important early scientists who investigated electricity. In his famous experiment, he hung a key from a kite in a thunderstorm and showed that the key conducted electricity. Franklin was lucky that this experiment did not kill him; it did kill others who tried it. Franklin turned his discovery into an invention—the lightning rod.

Answers to
In-Text Questions

(B) Students' answers will vary, and there is a nearly endless list of possible responses. Students should name devices that use electricity supplied by power stations through electrical outlets and devices that are powered by solar cells, batteries, simple magnetic generators, or even piezoelectric crystals.

(C) Students' responses will vary. Encourage student creativity by having students review any lists they have made of electrical devices around them. Have them consider useful nonelectrical substitutes, such as a bicycle for a train.

(D) Accept all reasonable responses at this point. Students will probably not have a clear idea of what exactly electricity is, but they will clarify their ideas as they progress through the unit.

ENVIRONMENTAL FOCUS

Generating electricity takes tremendous energy; most of the energy required to produce electricity comes from fossil fuels. About 25–30 percent of all energy consumed in the United States is used to generate electricity, depending on the year and season. Discuss with students some alternatives to fossil fuels for producing electricity. *(Alternatives include nuclear power, hydro-electric power, solar power, etc.)*

Meeting Individual Needs

Second-Language Learners

Have students find out the equivalent of each of the following words in their native language: *electron, electricity, switch,* and *circuit.* Students should then prepare a poster labeled in both English and their native language that demonstrates what each term means. You may wish to display the finished posters around the room for interested students to examine at their leisure.

Electricity and You: Case Studies

It takes only a little knowledge and skill to make use of many common electrical devices. It takes more knowledge to understand how they work—and still more to design them. With one or two classmates, discuss how electricity works in one of the following situations. Then make up a case study of your own.

Case Study A

"Let's see," thought Bill. "Colors go in cold water and whites go in hot—I think." Bill has many options for setting the automatic washer. What are some of them? How does he control them? How many different kinds of functions does the machine perform or control? For example, if Bill had only a small load of laundry to do, could the electrical system control the amount of water needed? Does he need to stay beside the washer throughout the cycle? Why or why not? What takes place during the wash cycle? How do you think these events occur? Are there any safety features? Sometimes a system provides you with information as it operates. This is called feedback. Can you identify any controls that serve as feedback mechanisms? What is the source of electricity for the washer?

Case Study B

Jenny arrives at the hospital to visit her friend Kim, who is on the fourth floor. Jenny uses the elevator. What information can she get from the elevator lights? What control does Jenny have over the elevator's operation? Are there other controls in addition to those that she uses? For example, are there safety controls to prevent people from walking into an empty shaft? What features of the electrical design are specifically for the passengers' benefit? for their safety? What feedback systems are there? The elevator is a complex electrical system. Although you cannot see the subsystems, you know that they exist because of the functions they perform.

sensor touches someone or something, a mechanism that controls the speed of the elevator and causes the elevator to slow down as it approaches a floor, and a device that opens and closes the elevator doors at the correct times.

For the passengers' benefit, there are interior lights and air conditioning. Safety features include a mechanism that keeps the elevator doors closed until the elevator stops at a floor, a phone or bell to summon help in case of

an emergency, and an automatic braking system in case the elevator cage breaks free of its cables and begins to fall.

Feedback mechanisms include a device that allows the elevator to align itself with the outside floor before the doors open, a device that sounds an alarm if the elevator is overloaded, and the lights that indicate which floor the elevator is at or near.

Case Study C

With much anticipation, Dylan turned on his computer. After it started up, he inserted the CD-ROM he'd just bought. The computer beeped and whirred. Then the words *American History* appeared on the screen while music played. After a few clicks of the mouse, Dylan was listening to a speech given by John F. Kennedy almost four decades ago. After a few more clicks of the mouse, he was watching Neil Armstrong walk on the moon. Then Dylan used the computer to send a message to his schoolmate across town: "Come on over—I got some great reference material for our history project."

Electricity is certainly at work in a computer. What are some things that a computer can do for you? What are some ways in which a computer provides feedback? The computer is quite complex. However, you do not need to know how it works in order to operate it. How do you control the computer's various operations?

Case Study D

"Drivers start your engines!" Alex turned the key, expecting to feel her race car's powerful engine roar to life. Instead she heard only the "clunk" that signals a dead battery. Later, Alex lifted the hood and stared at the engine. "What could possibly be wrong?" she wondered. "It started fine in practice an hour ago . . ."

What could be wrong with Alex's car? What components make up a car's electrical system? Where does the power needed to start a car come from? What kind of electrical feedback or control systems do cars have? How are different forms of energy converted to electricity, and vice versa, in a car?

Your Own Case Study

In your ScienceLog, describe a situation in which electricity performs some sort of function for you. Describe any devices or processes involved, what takes place, how the device or process is controlled, any energy conversions that take place, and so on.

295

Your Own Case Study

Allow students to choose their own electrical system to analyze. Since this is only an introduction to the unit, the details of each electrical system should not be expected at this time. Instead, emphasis should be on the variety of functions accomplished by electricity, the sequence of those functions in a particular device or mechanism, and the other forms of energy into which electricity may be converted.

Homework

Students can do Your Own Case Study on this page as a homework assignment.

Answers to *Case Study C*

Different computers have different capabilities, but a typical home computer can function as a calculator, word processor, calendar, and stereo. A modern home computer can also provide games, store photographs or video clips, and provide access to the Internet.

A computer gives feedback by providing error messages, supplying prompts, and displaying processed data in graphs, in tables, or in other forms. The computer may provide feedback visually, with sounds, or with an automated voice.

The user controls a computer's operations by inputting commands on a keyboard or with a mouse, by writing programs in a computer language and running the programs, and by installing programs on floppy disks and CD-ROM discs.

Answers to *Case Study D*

Many things could be wrong with Alex's car, but it is probably a faulty component in the electrical system of the car. Such components include the battery, the alternator, the ignition system, the starter, the voltage regulator, the distributor, and the spark plugs. Components such as the horn, heater, air conditioner, and windshield wipers all use electricity but are not considered parts of the electrical system.

The power needed to start a car comes from a battery. Once the car is started, the battery is recharged by the alternator. Electrically operated feedback or control systems include lights that indicate when the oil pressure is low, the emergency brake is on, or the engine is too hot; a sound that goes off if safety belts are not fastened or doors are not closed; and fuses that protect the electrical system by shutting it down if something goes wrong.

Chemical energy stored in the car's battery is used to start the car. Electricity from the spark plugs and chemical energy in the fuel is converted into heat and the mechanical energy of the moving pistons. Mechanical energy from the engine is used to turn the alternator, which charges the battery. Electricity from the battery may be changed into light, sound, heat, or mechanical energy.

Explain to students that magnet wire has a thin varnish coating that must be removed from the ends of the wire.

Answers to
Activity 1

2. Allow students, through trial and error, to find the arrangements that light the bulb. Their sketches should be similar to the following:

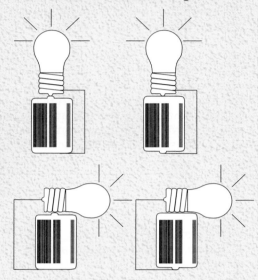

3. Electricity from the battery produces light energy in the bulb and heat energy in the bulb and wire.

4. Other examples of electricity used in this way include electric signs, glowing heat elements, street lights, automobile lights, flashlights, and stadium lights.

5. The electric circuit is made up of a D-cell, wire, and a flashlight bulb.

   The term *circuit* comes from a Latin word that means "to go around." It is an appropriate term because electricity follows, or goes around, a circuit.

   Sample definition of an electric circuit: a loop through which electricity can move in a certain direction.

 An Exploration Worksheet is available to accompany Exploration 1 (Teaching Resources, page 3).

Electricity Working for You

You have probably seen warnings like the one below on appliances or power tools. As useful as it is, electricity in large amounts is deadly and must be carefully controlled.

   The familiar situations that you have been analyzing in the Case Studies involve complex electrical parts and arrangements. Large quantities of electricity are used in the devices in Case Studies A and B. This is true for the operation of most appliances in homes, stores, or industry. A moderate amount is used in the devices in Case Studies C and D. Care must be exercised in using these amounts of electricity.

   Most of the Explorations in this unit require only a small amount of electricity. They are quite safe to do. Here are four experiments in which a small amount of electricity works for you by producing other forms of energy. Watch for these energy forms.

   In these experiments the electricity is supplied by dry cells such as those used in flashlights.

296

ACTIVITY 1
Shedding a Little Light

You Will Need
- a length of magnet wire
- sandpaper
- a flashlight bulb
- a D-cell
- a rubber band

What to Do

1. Begin by sanding the enamel off the last few centimeters of each end of the wire to expose the copper underneath. Then, using only the other items listed, find all of the different arrangements that will light the bulb.

2. Sketch each arrangement.

3. What form(s) of energy does the electricity produce?

4. What are other examples of electricity being used in this way?

5. You have constructed an electric *circuit*. What parts make up this circuit? Check the dictionary to find out the origin of the word *circuit*. How is it significant? What would you say an electric circuit is?

Integrating the Sciences

Life and Physical Sciences
Explain that electricity can run through the human body, just as it can run through the light bulb, once the body becomes part of a circuit. Have students discuss experiences when they have been shocked. Analyze with students what composed the circuit that they were part of when they were shocked.

Homework
Have students write their own definitions of an electric circuit. (See Question 5 above.) Refrain from giving them an exact definition at this time. *(You might expect students to write such phrases as "the path along which electricity flows" or "a cell and a wire to carry electricity." Each is acceptable at this point in the unit.)*

The Heat Is On

You Will Need

- two 30 cm lengths of magnet wire
- sandpaper
- modeling clay
- a clothespin
- 2 D-cells
- a wide rubber band
- a strand of steel wool
- aluminum foil
- a thin nichrome wire (10 cm long)

What to Do

1. Sand the enamel off of the last 5 cm of the ends of each length of magnet wire. Then make a small loop at one end of each wire.

2. Bend the wires and support them with modeling clay, as shown.

3. Attach the ends of the wires to the D-cell, securing them with a wide rubber band.

4. Place a strand of steel wool through the loops and let it rest in contact with each loop. What do you observe? Repeat with a rolled-up length of aluminum foil and then with a length of nichrome wire. Bring your hand close to each piece being tested, but do not touch any of them. What do you feel?

5. What form of energy does the electricity produce?

6. What are some examples of devices in which this type of electrical energy is used?

7. Do Activities 1 and 2 demonstrate the same principle? Explain.

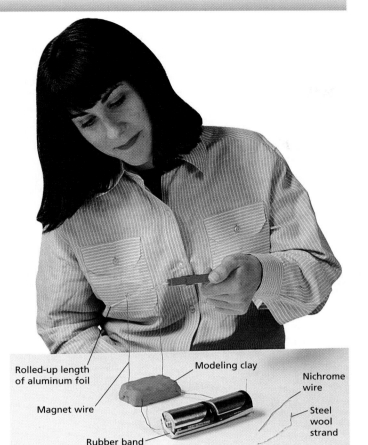

Rolled-up length of aluminum foil

Magnet wire

Modeling clay

Nichrome wire

Steel wool strand

Rubber band

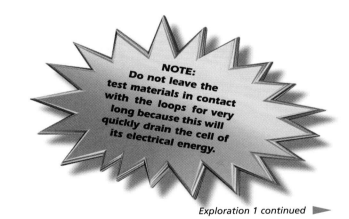

NOTE: Do not leave the test materials in contact with the loops for very long because this will quickly drain the cell of its electrical energy.

Exploration 1 continued ▶

297

Multicultural Extension

Electrical Devices in Other Cultures

Have students choose several electrical devices that they use and depend on daily. Then ask students to research other cultures that do not use electricity. Students should find out how people of the other culture accomplish the task or fulfill the need for which the student uses the electrical device.

Answers to *Activity 2*

4. Students should observe that the strand of steel wool completely or partially melts and that the aluminum foil and nichrome wire become very hot. Number 32 nichrome (an alloy of nickel and chromium) will glow with the electrical energy supplied by a single D-cell. You may wish to dim the lights in order to see the glowing strand better.

   Students should feel some heat radiating from the steel wool, the foil, and the nichrome wire.

5. In this activity, heat energy is produced from electrical energy.

6. Some examples of electrical energy being used to produce heat energy include heating elements in stoves, heaters, toasters, and toaster ovens; light bulb filaments; and car windshield defrosters.

7. Activities 1 and 2 both demonstrate an electric circuit because electricity is flowing in a circuit through wires connected to a D-cell. The steel wool, aluminum foil, and nichrome wire from Activity 2 are similar to the filament in the flashlight bulb from Activity 1.

Meeting Individual Needs

Gifted Learners

Suggest that students complete Activity 2 of Exploration 1 using one D-cell instead of two. Ask them to take careful notes of their observations and to report back to the class. Have them explain each observation. *(By using fewer cells, students will generate a weaker current, and the wool, wire, and foil will become less hot and heat up more slowly. Number 32 nichrome wire will glow less brightly as well.)*

Homework

Have students describe three electrical devices that take advantage of the principle students observed in Exploration 1, Activity 2. *(Possible answers include an electric blanket, toaster, and hair dryer.)*

ACTIVITY 3

Call on a volunteer to describe the flow of electrical charges in the circuit shown in the illustration. *(Electrical charges, or electrons, flow from the negative electrode [or bottom of the battery], through the wire, through the closed switch, over the compass, and back to the battery [at the positive electrode].)* Ask: How can you control the flow of electrical charges through the circuit? *(Opening the switch by releasing the paper clip prevents the flow of charges; closing the switch by pressing on the paper clip allows the charges to flow through the circuit.)* Ask: Why does closing the switch allow the charges to flow and opening the switch prevent them from flowing? *(Closing the switch completes the circuit. Opening the switch breaks the circuit.)*

SAFETY ALERT The amount of time that the switch is closed should be limited because the wire will become hot. Prolonged contact will also run down the battery.

Answers to
Activity 3

2. When the circuit is closed, the compass needle will be deflected to the right or left, depending on which poles of the D-cell the wires are attached to. (You may wish to suggest to students that they reverse the wires on the D-cell and observe what happens to the compass needle. *[It will reverse direction.]* Some students might conclude from their observations that an electric current has a directional flow.)

3. Students should conclude that in this Activity, electricity produces magnetic energy and heat energy (from the wire).

4. Answers will vary. Sample answer: Speedometers, odometers, and fuel gauges make use of electricity in this way.

As you may have guessed from the title of this unit, there is a connection between electricity and magnetism. The following Activities will help illustrate that connection.

ACTIVITY 3

The Electricity-Magnetism Connection

You Will Need

- a D-cell
- a compass
- 2 thumbtacks or screws
- a wood block
- a paper clip
- 2 lengths of magnet wire (15 cm and 25 cm long)
- a rubber band
- sandpaper

What to Do

1. Sand the enamel off of the last 2 or 3 cm of the ends of each wire. Then set up the apparatus as shown. Align the compass needle, and place the wire over the compass in a north-south direction so that the wire lines up with the compass needle.

2. Close the electrical circuit by pressing the contact switch. What happens?

Caution: Don't keep the switch closed for very long.

3. What kind of energy does the electricity produce in this experiment?

4. Can you think of any everyday applications that make use of electricity in this way?

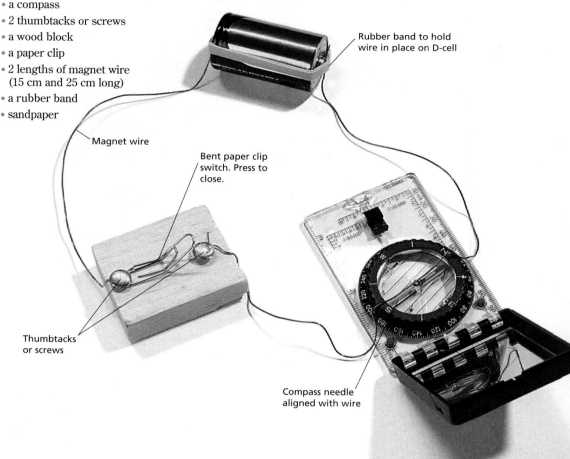

Rubber band to hold wire in place on D-cell

Magnet wire

Bent paper clip switch. Press to close.

Thumbtacks or screws

Compass needle aligned with wire

Multicultural Extension

The First Compass

Even though many ancient cultures were aware of the magnetic properties of the lodestone (the mineral magnetite), the Chinese general Haung Ti was the first to use the lodestone as a compass, in 376 B.C. The Chinese military continued to use the compass on land for centuries, but it wasn't until the thirteenth century that Chinese sailors began to use the compass at sea. Arab and western European nations acquired this device by trade. Have interested students research other inventions that originated in China and present their findings to the class in the form of an oral report.

Let's Get Moving!

You Will Need

- 2 D-cells
- a paper clip
- 2 lengths of magnet wire (10 cm and 250 cm long)
- bar magnet
- a support stand with ring clamp
- a cork
- a wide rubber band
- a narrow strip of cardboard (2 cm × 30 cm)
- 2 thumbtacks or screws
- a wood block
- sandpaper

What to Do

1. Sand the enamel off of the last 2 or 3 cm of the ends of each wire. Then assemble the circuit as shown in the illustration.

2. Have one person hold a strong magnet near the cork while the other person presses the contact switch to complete the circuit. Observe what happens. Open the switch. What happens?

Caution: Don't keep the switch closed for very long.

3. Using the other end of the magnet, repeat step 2. What happens?

4. What kind of energy is produced by the electricity?

5. What are some practical examples of how energy works for you in this way?

A Home Project

Make a battery tester. Use one of the arrangements you discovered in Activity 1 of Exploration 1. Devise a tester that consists of a light bulb with two wires connected to it. Touch the ends of the wire to the battery. The brightness of the light bulb will indicate the strength of the battery.

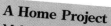

Have students complete the Activity and compare their results. Then ask: How are the functions of the circuits in Activities 3 and 4 similar? *(In each circuit, electrical energy is used to create magnetic energy.)*

Answers to *Activity 4*

2. When the switch is closed, the coil of wire will be repelled or attracted by the magnet, depending on which end of the bar magnet is held near the coil. When the switch is open, the coil of wire will return to its original position. By opening and closing the switch, the coil of wire can be made to swing back and forth.

3. When the magnet is turned around and the switch is closed, the coil is attracted to or repelled by the magnet, depending on which end of the bar magnet was initially held near the coil. When the switch is open, the coil moves back to its original position. Students should recognize that this is the reverse of what happened in step 2.

4. The electricity is producing magnetic energy, which in turn produces energy of motion, or kinetic energy.

5. Sample answer: Doorbells and electric motors use the magnetic field associated with electricity to create movement.

A Home Project

Make sure you check student designs for safety before allowing students to make their battery testers.

FOLLOW-UP

Reteaching

Have students make poster diagrams to illustrate how electricity flows from a wall outlet through a lamp. Call on volunteers to explain their diagrams to the class.

Assessment

Ask students to make a list of all the electrical devices in their homes. Students should list any energy changes that take place when these devices are used. Students may want to compile their findings into a composite wall chart.

Extension

Have students research what makes a light bulb work, including why a light bulb glows, what kind of wire is used in the filament, and why the filament is enclosed in a glass bulb. Have volunteers share what they learned by giving an oral report to the class.

Closure

Have each student make a chart listing electrical devices that they use each day. Next to each item, students should indicate whether there is a way to perform the same task without using electricity. Students should then weigh the advantages and disadvantages of the electrical device and the non-electrical means of performing the task.

 PORTFOLIO
Students may wish to include their diagram from the Reteaching activity or their chart from the Closure activity in their Portfolio.

FOCUS

Getting Started

The Discrepant Event Worksheet on page 8 of the Unit 5 Teaching Resources booklet describes a teacher demonstration that makes an excellent introduction to this lesson.

Main Ideas

1. All matter contains negatively charged electrons and positively charged protons.
2. Like charges repel each other; unlike charges attract each other.
3. Charged particles move readily through conductors and move much less easily through insulators.
4. An electric current is created when electrons flow through material.

TEACHING STRATEGIES

A Classical Current Demonstration

Students can work in groups to complete this demonstration. Afterward, allow students to speculate about what causes the wheat puff to respond as it does. Accept all reasonable suggestions. An in-depth explanation is provided on the following page. Note: If students wish to repeat the demonstration, have them first remove the charge on the wheat puff by touching it.

Answers to
A Classical Current Demonstration

2. The wheat puff moves toward the pointed end of the nail and then "jumps" away from it.

3. The wheat puff moves away from the nail.

4. The wheat puff moves toward the nail.

 A Resource Worksheet is available to accompany A Classical Current Demonstration (Teaching Resources, page 9).

A In Exploration 1 you used electricity to produce other forms of energy. What were they? Later in the unit you will discover that each of these energy forms can, in turn, produce electricity. Remembering that energy can be transformed from one form into another, does it seem that electricity itself must be a form of energy, just as heat and light are? **B**

Electricity accomplishes these changes in energy when an **electric current** is flowing in a circuit. But what is a current? What flows or moves in the circuit? **C**

Have you ever gotten a shock when you touched someone or something after walking across a carpet? If so, you have become *electrically charged*. Have you ever rubbed an inflated balloon on your hair and noticed how your hair sticks to it? In this case, the balloon has become electrically charged. This charge can be so strong that you can even stick the balloon to a wall. What might these charges be? In the demonstration that follows you will see evidence of these charges in action.

A Classical Current Demonstration

Here is a demonstration similar to one first done hundreds of years ago. Make a setup like the one shown below, and try it yourself. Follow the steps closely.

1. Vigorously rub a plastic strip, such as a plastic ruler, with plastic wrap.

2. Quickly touch the strip or ruler to the end of the nail. What happens to the wheat puff?

3. Rub the plastic strip again and bring it close to the end of the nail without quite touching it. Observe the wheat puff.

4. Repeat the experiment using a vinyl strip rubbed vigorously with flannel cloth. What happens to the wheat puff now?

How do you explain the events in this demonstration? **D**

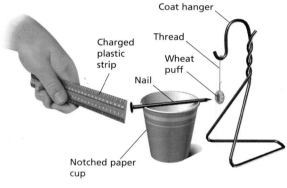

Coat hanger
Charged plastic strip
Thread
Wheat puff
Nail
Notched paper cup

LESSON 2 ORGANIZER

Time Required 4 to 5 class periods

Process Skills
inferring, hypothesizing, observing

Theme Connection Systems

New Terms
Chemical cell—a device in which chemical energy is converted into electrical energy
Conductor—material through which electric current can easily pass
Electric current—the flow of electrons from one place to another

Electrolyte—a solution that conducts electricity
Electron—a negatively charged particle found in atoms
Galvanometer—a sensitive current detector
Insulator—material through which electric current cannot easily pass
Proton—a positively charged particle found in atoms

Materials (per student group)
A Classical Demonstration: plastic strip, about 10 cm long, or plastic ruler; wire clothes hanger; paper cup; large

continued ▶

A Theory of Charged Particles

Scientists explain events such as those seen in the previous demonstration in this way:

1. When one material is rubbed against another, friction causes charged particles to move from one material to the other. The accumulated charge is indicated by the charged material's ability to attract lightweight or finely powdered substances. When did this happen in the previous demonstration?

2. There are two kinds of charged particles: positively charged particles and negatively charged particles. Why might you conclude that the charges on the plastic strip are different from those on the vinyl strip?

3. Objects that have the same kind of charge tend to repel one another. Objects that have different charges tend to attract one another. When did you observe these effects in the demonstration?

4. Charged particles pass easily through certain materials, called **conductors**, but pass with difficulty or not at all through other materials, called **insulators** (nonconductors). What evidence is there that iron is a conductor of electricity? You might experiment by replacing the nail in the demonstration with a glass rod, a wooden stick, a copper wire, or objects made of other materials. Which are conductors? Which are insulators?

The drawings below apply the theory just stated to the previous demonstration. Express in your own words what is taking place in each drawing. **E**

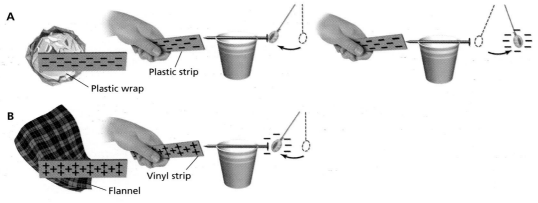

A

Plastic strip

Plastic wrap

B

Vinyl strip

Flannel

The theory we have been considering is an interesting one. It helps to explain many observations about electricity. But how do scientists know about these charged particles? Read the following account from the pages of history.

301

Answers to
In-Text Questions, pages 300–301

A Students produced light energy, heat energy, magnetic energy, and kinetic energy.

B Electricity is a form of energy in the same sense that heat and light are forms of energy.

C A current is the flow of charged particles. Electrons flow in the circuit.

D When the electrically charged plastic strip touches the nail, it transfers its charge to the nail. The nail attracts the uncharged wheat puff and transfers its charge to the wheat puff. After the wheat puff becomes charged, it is repelled by the nail because they are similarly charged. The wheat puff is attracted to the nail when it is charged with the vinyl strip because the two are oppositely charged.

(You should point out to students that even uncharged objects are sometimes attracted or repelled by charged objects. This occurs through *polarization,* a process in which individual charges within the object migrate to opposite sides of the object. Thus, the initially uncharged wheat puff is attracted to the negatively charged nail because the wheat puff has a net positive charge on the side nearest to the nail and a net negative charge on the side farthest from the nail, even though the overall charge on the wheat puff is zero.)

E Students should reword their answers to A Classical Current Demonstration to demonstrate an understanding of the transfer of charged particles through a conductor.

Homework

The Resource Worksheet that accompanies A Theory of Charged Particles makes an excellent homework activity (Teaching Resources, page 10).

A Theory of Charged Particles

SAFETY ALERT Do not allow students to experiment with other objects before you have checked the items for safety. Good choices include a plastic ruler, paper, cardboard, aluminum foil, a steel bolt, a stone, and a pen.

ORGANIZER, *continued*

nail, about 8 cm long; wheat puff; 10 cm of thread; piece of plastic wrap, about 10 cm × 10 cm; vinyl strip, about 10 cm long; piece of flannel cloth, about 10 cm × 10 cm

A Theory of Charged Particles: plastic strip, about 10 cm long, or plastic ruler; wire clothes hanger; paper cup; large nail, about 8 cm long; wheat puff; 10 cm of thread; piece of plastic wrap, about 10 cm × 10 cm; a variety of materials for testing conductivity, such as a glass rod, a wooden stick, and a piece of copper wire or magnet wire, each about 8 cm long

Exploration 2: small magnetic compass; 1 m of insulated wire or magnet wire; other materials to hold parts in place, such as a rubber band or masking tape and a plastic-foam cup; straightened paper clip; 10 cm of magnet wire with sanded ends; small cup; 50 mL of concentrated lemon juice

Teaching Resources
Discrepant Event Worksheet, p. 8
Resource Worksheets, pp. 9 and 10
Exploration Worksheet, p. 12
SourceBook, pp. S82 and S86

1. This occurred when the plastic strip rubbed with plastic wrap and the vinyl strip rubbed with flannel both caused the wheat puff to move toward the nail.

2. When the plastic strip gave its charge to the nail, and the nail passed the charge on to the wheat puff, the wheat puff moved away from the nail. When the vinyl strip gave its charge to the nail, the wheat puff was attracted to the nail. This demonstrates that the charge on the vinyl strip must be opposite that on the plastic strip.

3. After the wheat puff made contact with the nail, they were both negatively charged, so the wheat puff was repelled. Then the nail was positively charged by the vinyl strip, so the wheat puff was attracted.

4. Iron must be a conductor of electricity because the charge seemed to travel through the iron nail. Wood and glass do not conduct electricity and are therefore insulators. Copper wire is a very good conductor.

The Discovery of Charged Particles

Have students read about J. J. Thomson and the discovery of charged particles. When they have finished, check their understanding by involving them in a discussion of the main points made in the article. Be sure students understand that electrons are negatively charged particles (which move through conductors) and protons are positively charged particles. Some students may recognize that in an electric current, such as the one they observed in Exploration 1, electrons flow through a circuit to produce electricity.

Answer to
In-Text Question

Ⓐ Sample answer: An electric current is the flow of negative particles called electrons.

The Discovery of Charged Particles

It is the year 1900 at Cambridge University in England. J. J. Thomson is talking to a small group of students at the famous Cavendish Laboratory. Listen to what he might be saying:

"We believe that we have discovered the smallest particle of negative electricity. We have obtained a stream of identical negative particles from many metals. In fact, we believe that every bit of matter contains these same particles. We call them **electrons.**

"Obviously, most substances are uncharged. Therefore, most substances must have some kind of particle that neutralizes the charge of the electron. We have discovered just that kind of particle in other experiments, and we call it a **proton.** Each proton has a positive charge that exactly counteracts an electron's negative charge.

"It appears that the fundamental particles of matter are composed of electrons and protons in equal amounts so that no overall charge occurs. Electrons are relatively light and move freely in conductors. Protons, being much heavier, remain fixed in their positions."

J. J. Thomson and his colleagues expanded the theory of charged particles. Did you follow what he was saying to the students? Can you relate it to what an electric current is—such as the current you got when you correctly connected the dry cell, wires, and flashlight bulb in Exploration 1? Ⓐ

Don't forget who coined the terms *positive* and *negative* for nature's basic charges—me, Ben Franklin. To find out more about my work with electricity, see page S82 of the SourceBook.

302

Meeting Individual Needs

Gifted Learners

Some people define the direction of electric current to be the direction that negative charge flows, while others define it to be the direction that positive charge flows. Ask: What problems might this difference have for someone working with electricity? *(As long as one is aware of what convention is being used, this should not cause any problems. Some circuit components, such as resistors, will work no matter what direction the current flows through them. Other components, such as batteries and diodes, are polarized, which means that the current must flow through them in only one direction in order for them to work properly.)*

Charged or Uncharged?

In terms of the theory of charged particles, what are positively charged, negatively charged, and uncharged objects? What makes them that way? Note that in the illustrations on page 301, there are no charges shown on the plastic wrap or the flannel. **B** Should there be? The illustrations at right give a more complete picture of what happens when a plastic ruler is rubbed with plastic wrap. Count the number of positive and negative charges in the "Before" and "After" situations.

a. Explain the movement of the electrically charged particles.

b. Now make before and after drawings that illustrate what happens when you rub a vinyl strip with flannel.

c. Each of the balloons shown below is charged differently. Which balloon is slightly negative? strongly negative? uncharged? slightly positive? strongly positive? What remains the same in each balloon? Why?

Before

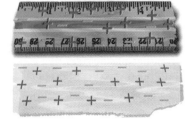

After

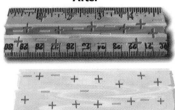

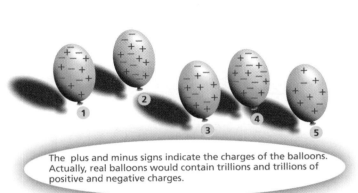

The plus and minus signs indicate the charges of the balloons. Actually, real balloons would contain trillions and trillions of positive and negative charges.

Charges on the Move

The buildup of electric charges on surfaces is called *static electricity*. You have seen several examples of this phenomenon on the previous pages. For example, when a person walks across a carpet, charges accumulate on the surface of his or her body and create an *electrostatic* charge. But what happens to this static electricity when the person nears a conductor like a metal doorknob? Something shocking! The built-up charges move from the person, through the air, and into the metal knob, causing a momentary flow of charges, or **electric current**.

Answers to *In-Text Questions*

B Positively charged objects are objects that have fewer electrons than protons. Negatively charged objects are objects that have more electrons than protons. Uncharged objects have equal numbers of protons and electrons. There should be charges on the plastic wrap and flannel shown on page 301.

Answers to *Charged or Uncharged?*

a. Before rubbing, both the plastic ruler and the plastic wrap have an equal number of electrons and protons. Therefore, neither carries a net charge. They are said to be uncharged, or electrically neutral. After rubbing, electrons have moved from the plastic wrap to the plastic ruler. The plastic wrap now has more protons than electrons and is positively charged. The plastic ruler has more electrons than protons and is negatively charged.

b. Student drawings should show that the electrons have moved from the vinyl strip to the flannel. The number of protons should remain the same in both objects before and after rubbing.

c. 1—Uncharged or neutral
2—Slightly negative
3—Slightly positive
4—Strongly negative
5—Strongly positive

The number of protons remains the same in each diagram because protons are heavier and tend to remain fixed, while electrons are much lighter and tend to move freely in conductors.

Integrating the Sciences

Earth and Physical Sciences

Rub a plastic comb through your hair to charge it, and then hold the comb over a few small pieces of paper scattered on a desktop. The students should see the paper "jump" toward the comb. Help students understand how this demonstration shows that the electrical force is stronger than the gravitational force. *(Earth's gravity pulls the paper toward the desktop, but the comb pulls the paper away from the desktop. The electrical force on the small comb is much stronger than the gravitational force of the Earth, so the paper jumps toward the comb.)*

Homework

Students can answer the questions in Charged or Uncharged? on this page as homework.

Multicultural Extension

Granville Woods

Have students do research on the telephone transmitter and other inventions devised by Granville Woods, an African American, self-taught scientist and inventor who lived in the late nineteenth and early twentieth centuries. Students can write a report about their findings, indicating what led to each invention, how each invention works, and the relationship of the invention to electricity and magnetism.

Answers to
In-Text Questions

Ⓐ The role of electrons in currents will be explored further in Lesson 5 of Chapter 15, where students will draw an analogy between the flow of electrons in a circuit and the flow of water in a pipe. They may also discuss this analogy in reference to the Theme Connection on page 305. For now, simply point out that differences in charge on a conductor (or on conductors in contact with one another) will cause electrons to move, producing a current.

Ⓑ The charged particles of the puff were repelled by like charges on the nail.

Current Thinking

Have students read the top of this page. Check their understanding by asking questions similar to the following:

• What is an electric current? *(The passing, or flow, of electrons through an object)*

• When does a current stop flowing? *(When there is nowhere for the electrons to go)*

• What properties do electrons have that cause them to flow in an electric current? *(They are relatively light and move freely in conductors. Because they are like charges, they repel one another.)*

Other possible models you may wish to use to demonstrate how an electric current flows include Newton spheres (pictured below), a spring toy, and the movement of a transverse pulse in a length of stretched rope or rubber tubing.

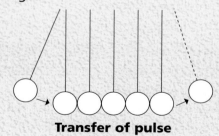

Transfer of pulse

Current Thinking

What causes electrons to move in a conductor to produce a current? Ⓐ

Do you recall the wheat puff in the classical demonstration? You brought a plastic ruler that you had rubbed with plastic wrap near the end of a nail. The wheat puff at the other end of the nail moved away. The puff moved because charged particles passed through the nail to the puff, causing the puff to jump away. The passing of the particles through the nail created a momentary electric current. But how, exactly, did it happen? Ⓑ

Consider the labeled diagram below. First, electrons in the plastic ruler repel (push away) electrons in the point of the nail, *A*.

Moving to the right, these electrons repel electrons at *B*. The electrons at *B* move to the right, repelling electrons at *C*, and so on all the way through the nail to the other end, *Z*.

The electrons at *Z* are forced onto the wheat puff. When the puff is sufficiently charged, the repelling force becomes strong enough to push it away from the nail. The current then stops flowing because there is nowhere for the electrons to go.

Think of electrons as behaving somewhat like falling dominoes. Each electron pushes the one next to it. If there is a gap in the chain of electrons, no electric current can flow.

What causes a continuous flow of current?
Let's summarize. We need two things for a continuous current: a continuous supply of charges and an uninterrupted conducting pathway to carry the charges. Compare current flow to the flow of water to your home. For water to get to your home, you need a source (a reservoir) and a conducting path (pipes).

304

CROSS-DISCIPLINARY FOCUS

Art

Have students develop a comic strip to illustrate how current flows through a circuit or electrical device. You may wish to encourage student creativity by having them form pairs to discuss their ideas with one another and to confirm that their comic strips are scientifically accurate. You may wish to have volunteers present their illustrations to the class.

An electrical conducting path usually includes wire made of a metal such as copper and at least one device that makes use of the electricity. Trace the conducting path in the diagram, starting at the cell and coming back to the cell.

A complete circuit consists of the following:

a. a source of electric charges

b. a conducting path

c. a device that uses the electrical energy

Match (a), (b), and (c) with the numbered labels in the diagram. What would you add to the circuit so that you could control the continuous flow of charges, that is, start and stop the flow at will? What does it mean to make, or close, a circuit? What does it mean to break, or open, a circuit? **C**

A Continuous Supply

One way to obtain a continuous supply of charges is to use a **chemical cell**. People often call a cell a battery. However, a battery is actually a group of connected cells. A car battery, for example, is made up of six separate but connected cells.

As you have observed, a cell always has positive and negative components. These are the *electrodes* of the cell. They are also conductors. The two electrodes of a cell are made of different materials and are in contact with a solution of chemicals called an **electrolyte**.

The electrodes and the electrolyte interact chemically with each other. The result is that electrons accumulate on one electrode and are removed from the other electrode. The electrode with the accumulation of electrons becomes negatively charged. (Why?) The electrode from which electrons are removed **D** becomes positively charged. (Why?) In a typical dry cell, a zinc casing constitutes the negative electrode. The positive electrode is a rod of carbon (graphite) in the center of the cell.

When the negative electrode is connected to the positive electrode by a conducting path, electrons repel each other along that path toward the positive electrode. The positive electrode helps by adding an attractive force on the moving electrons. Thus, a current flows. The chemicals in the cell keep taking electrons from the positive electrode and giving electrons to the negative electrode. The cell, therefore, causes a continuous current to flow.

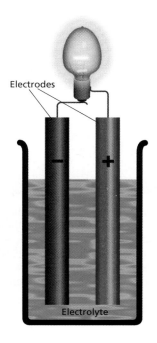

Electrodes

− +

Electrolyte

A Continuous Supply

Before students begin reading, display several different kinds of cells (AA, AAA, C, and D). Pass them around for students to examine. Call on volunteers to describe where the negative and positive electrodes are on each of the cells. Point out that the negative electrodes give up the electrons that flow through a circuit to the positive electrode of the cell, which accepts electrons. Help students to conclude that all cells have similar negative and positive components. Then have them read this section to find out how cells are made and how they work.

Answers to
In-Text Questions

C **a.** The battery, number 3, is the source of electrical charges.
b. The wire, number 1, is the conducting path.
c. The bulb, number 2, is the device that uses the electrical energy.

Most students will probably conclude that they can add a switch to the circuit to start or stop the flow of electrical charges. To make, or close, a circuit means to create a flow of electrical charges or to create an electric current. To break, or open, a circuit means to stop the flow of electrical charges, or to turn the current off.

D The electrode with more electrons becomes negatively charged because it has more negative charges than positive charges. The electrode with fewer electrons becomes positively charged because it has more positive charges than negative charges.

Meeting Individual Needs

Learners Having Difficulty

If students are having trouble understanding how an electric current is set up in a cell, have them trace the path of the electrons in the illustration at the bottom of the page. Help students to recognize that electrons flow from the negative electrode, through the circuit, to the positive electrode, through the electrolyte, and back to the negative electrode.

Theme Connection

Systems
Focus question: How is the flow of current in a wire like the flow of water in a plant? *(In a circuit, as long as there is a sufficient supply of electrons and a pathway with low enough resistance, current will continue to flow. In a plant, water will flow as long as there is a sufficient supply of water and open vascular tubes.)*

ENVIRONMENTAL FOCUS

Used batteries are a disposal hazard because the chemicals in them are toxic and must be treated as hazardous waste. Rechargeable batteries are better for the environment because they last longer and thus produce less waste. You may wish to have students develop an advertising campaign for the school that encourages the use of rechargeable batteries.

Answers to
In-Text Questions

Ⓐ The current deflected a compass needle in Activity 3 of Exploration 1.

EXPLORATION 2

If time permits, have students repeat Activity 4 in Exploration 1 using a coil with more turns of wire. They should discover that the more turns there are in the coil, the stronger the coil's magnetic force becomes. As students experiment with their galvanometers, they should also realize that the more turns of wire there are, the more sensitive their galvanometers become. Note: About 50 turns of wire work well for the purposes of this activity. If a commercial galvanometer is used, break the circuit if the needle goes off the scale. This will prevent damage to the galvanometer.

Students will use a galvanometer frequently in this unit. You may wish to have students learn about other devices used to measure current or resistance, such as the ammeter or ohmmeter. Point out that an ammeter measures electric currents that are typically much larger than those for which galvanometers are designed. An ohmmeter measures electrical resistance.

⭐ **An Exploration Worksheet is available to accompany Exploration 2 (Teaching Resources, page 12).**

Answers to
Exploration 2

5. The compass needle is deflected, indicating that an electric current is flowing.

6. Students should discover that the stronger the electric current is, the farther the compass needle is deflected.

FOLLOW-UP

Reteaching
Have students make a poster diagram of a circuit that includes two flashlight bulbs and two switches. Their diagrams should show the circuit closed and open.

Detecting Electric Current

What these students seem to be saying is, "You can't see electricity, but you know that it's there by its effects." You can feel the heat produced in an electric stove and you know that an electric current is flowing. Likewise, when you see the light given off by a flashlight bulb, you know that an electric current is flowing. In this case, a small amount of current is enough to produce a visible effect.

Later, you will generate even smaller amounts of electricity. How will you detect them? Look back at Activity 3 of Exploration 1 on page 298. What effect did a current of electricity **Ⓐ** produce in that Activity? This effect can be increased many times if you wrap the wire (which must be insulated) around the compass. This involves the same principles of operation as sensitive current detectors called **galvanometers**.

306

EXPLORATION 2

Constructing a Current Detector

Your task is to design and construct a homemade galvanometer. The device consists of a small magnetic compass, insulated wire, and anything else you need to hold the parts in place.

Hints

1. Remember the results of Exploration 1, Activity 3.

2. Try coiling the wire around the compass. Use different numbers of turns and observe the effect.

3. Leave the two ends of the wire free so that you can attach them to the source of the small current.

4. Position the galvanometer so that the compass needle and the coil of wire are parallel to one another.

5. Test your galvanometer using a small current, which can be obtained from a lemon-juice cell. To make a lemon-juice cell, place a straightened paper clip and a piece of sanded magnet wire into a small cup of concentrated lemon juice. The paper clip and the wire are the electrodes, and the lemon juice is the electrolyte. Hook the free ends of the galvanometer wire to the electrodes. What happens?

6. Will your galvanometer be able to give you any information about the size of the current? Explain.

Assessment
Challenge students to design a demonstration in which two objects with the same charge repel each other and two objects with opposite charges attract each other. One example is to use two inflated balloons on strings. By rubbing both balloons with a piece of flannel or wool, the balloons will repel each other. By rubbing one balloon with wool and the other with plastic wrap, the balloons will attract each other.

Extension
Have students do some research to discover as much information as they can about electrons and protons. Have volunteers share what they learn with the class.

Closure
Challenge students to design a game that uses an electric circuit. After you have checked the designs for safety, have students make their games and display them for their classmates to play.

1. All Charged Up

a. One by one, a negatively charged plastic ruler is brought near three light, foil-covered spheres suspended by nylon threads. The ruler repels sphere *A* and attracts spheres *B* and *C*. Sphere *A* attracts sphere *B*, and sphere *C* attracts sphere *B*. Do you have enough information to determine the charges on spheres *B* and *C*? Why or why not?

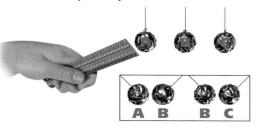

b. Jeff rubbed two pieces of plastic wrap with a sock and then suspended them (below left). Note what he observed.

c. Jeff then brought one of the pieces near a table (below right). Again observe what he saw.

d. Explain what is happening. Then try Jeff's experiment yourself.

2. Sure Shot

When spray painting a screen, the screen is given an electric charge. Why doesn't the spray paint go through or around the screen?

Look, no paint!

307

Answers to *Challenge Your Thinking*

1. a. No. Although sphere *A* must be negatively charged because it is repelled by the negatively charged ruler, sphere *B* and *C* could be either un-charged or positively charged. If sphere *B* is uncharged, then sphere *C* must be positively charged, and vice versa. (You may wish to remind students about the role that polarization plays in attracting or repelling uncharged and charged objects [see A Classical Current Demonstration on page 300]. You may also wish to point out that these conclusions would be different if the spheres were allowed to touch because charge could then be transferred from one sphere to the other.)

b. The two pieces of plastic wrap moved away from each other.

c. The piece of plastic wrap moved toward the table.

d. When the two pieces of plastic wrap are rubbed with the sock, they acquire the same kind of electric charge, so they repel each other. Because of polarization, there is an attraction between the charged plastic wrap and the uncharged table. The plastic wrap has very little mass, so it is pulled toward the more massive table.

2. A charged object attracts oppositely charged objects. A charged object can also attract an uncharged object by polarizing the charges in the uncharged object. The paint spray does not go through or around the screen because the charged screen attracts the less massive, uncharged paint particles in the spray.

★ **You may wish to provide students with the Chapter 13 Review Worksheet that is available to accompany this Challenge Your Thinking (Teaching Resources, page 13).**

 Homework

Give each student two balloons and two lengths of thread (about 15 cm long) to take home. Instruct students to inflate the balloons and then hang them beside each other. Students should rub the balloons together and record and explain their observations. *(The balloons become similarly charged when rubbed together, so they repel one another.)*

 Did You Know...

In metals, currents are carried by electrons. In some other materials, especially liquids and gases, currents are carried by both positive and negative ions.

Answers to
Challenge Your Thinking,
continued

3. a. conductor
b. insulator
c. electron
d. negative
e. proton, positive
f. uncharged
Mystery statement: current =
charges in motion

4. The stretching of the wrap causes
the wrap to become charged. The
attractive forces between the
charged wrap and the uncharged
container cause the wrap (which
has a very small mass) to adhere to
the container.

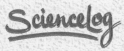

The following are sample revised
answers:

1. Students' responses will vary. Some
may describe electricity as a form of
energy transported by a flow of negative particles (electrons) in a conductor. Others may think of electricity in
terms of charged particles called protons and electrons; electricity makes
an object positive if the object has a
shortage of electrons or negative if it
has an excess of electrons. Electricity
may also be viewed as a form of
energy that can cause things to happen through its conversion. Examples
of this include making a light bulb
glow, causing a material to produce
heat, or causing a material to produce a magnetic field.

2. A sample sketch is provided below.
Make sure student sketches include a
wire running from one electrode of
the power source to the light bulb
and another wire running from the
light bulb to the other side of the
power source.

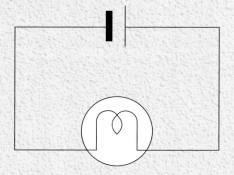

3. Current Puzzle

Copy the puzzle at right into your ScienceLog. Then complete the
statements below, and locate the answers in the puzzle by searching
horizontally, vertically, or in a combination of both directions. Cross
out the letters of each answer in the puzzle. The letters that remain
will tell you something about an electric current.

a. A material that allows charges to go through it is a(n)
_____?_____.

b. A material that does not allow charges to go through
it is a(n) _____?_____.

c. J. J. Thomson discovered a charged particle that
moves readily; it is called a(n) _____?_____.

d. This type of particle has a(n) _____?_____ charge.

e. The other charged particle in materials, which is more
massive and does not move, is called a(n) _____?_____
and has a(n) _____?_____ charge.

f. If a material has equal numbers of these two kinds
of particles, the material is _____?_____.

E	N	N	O	R	T	C	U
G	R	E	L	E	C	R	E
A	T	I	V	E	N	T	R
T	C	U	D	N	O	C	O
O	I	N	S	U	L	A	T
R	=	C	H	I	T	I	V
P	A	R	G	S	U	E	E
R	O	T	O	O	N	D	E
S	I	N	N	P	C	M	G
O	T	I	O	N	H	A	R

4. That's a Wrap

Some kinds of plastic wrap can be
stretched tightly over a container
and down its sides. The plastic
wrap sticks to the sides of the
container. Why?

*Review your responses to
the ScienceLog questions
on page 290. Then revise
your original ideas so that
they reflect what you've
learned.*

308

3. When a circuit is made, or closed,
electricity flows, creating an electric
current. In a temporary current, this
flow is caused by a buildup of excess
electrons at one end of a conductor.
The mutual repulsion of the similarly
charged electrons forces them to
move throughout the conductor. The
flow eventually ceases, however,
because there is no place for the
excess electrons to go. In a continuous
current, excess electrons flow from a
continuous source, through an uninterrupted conducting pathway, and to a
region of perpetual electron shortage.
In this case, the mutual repulsion among
the electrons and the attraction of the
electrons to the excess protons cause
electricity to flow.

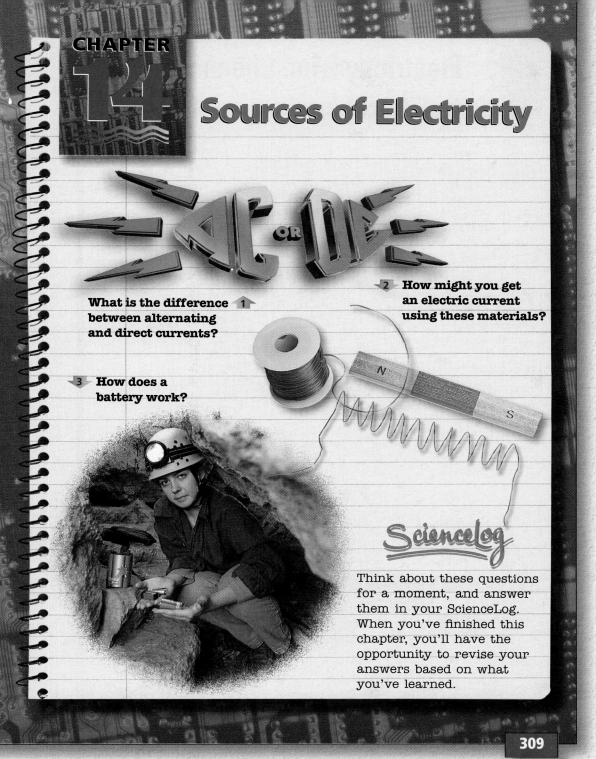

CHAPTER 14

Sources of Electricity

What is the difference 1
**between alternating
and direct currents?**

2 **How might you get
an electric current
using these materials?**

3 **How does a
battery work?**

ScienceLog

Think about these questions
for a moment, and answer
them in your ScienceLog.
When you've finished this
chapter, you'll have the
opportunity to revise your
answers based on what
you've learned.

309

Connecting to Other Chapters

Chapter 13
*introduces students to electricity,
ciruit theory, and the connection
between electricity and magnetism.*

Chapter 14
*introduces alternating current and
ways of producing electricity, such as
with chemical cells and generators.*

Chapter 15
*introduces circuit symbols, resistance,
parallel and series circuits, electromagnets, and electrical energy.*

Prior Knowledge and Misconceptions

Your students' responses to the
ScienceLog questions on this page will
reveal the kind of information—and
misinformation—they bring to this chapter. Use what you find out about your
students' knowledge to choose which
chapter concepts and activities to
emphasize in your teaching. After students complete the material in this
chapter, they will be asked to revise
their answers based on what they have
learned. Sample revised answers can be
found on page 325.

In addition to having students
answer the questions on this page, you
may wish to have them write an explanation of how electric current is generated and share their explanation with a
younger student. Assure students that
there are no right or wrong answers to
this exercise. Collect the papers, but do
not grade them. Instead, read them to
find out what students know about
sources of electricity, what misconceptions they may have, and what aspects
of this subject are interesting to them.

Electricity From Chemicals

FOCUS

Getting Started

Before students begin this lesson, have them briefly survey its contents. Use the illustrations and headings to help students develop an outline for the lesson. Encourage them to include only the information that they feel is important to their understanding of the sources of electricity.

Main Ideas

1. Chemical cells convert the energy stored in chemicals into electricity.
2. Increasing the number of cells in a battery increases the amount of current that can be produced.
3. The amount of current that a cell will provide can be increased by increasing the amount of electrolyte in the cell.

TEACHING STRATEGIES

EXPLORATION 1

WASTE DISPOSAL ALERT   Collect the used calcium chloride and salt solutions and pour them both into a single bottle. Collect the used blotting or filter papers in a large beaker. Pour the contents of the bottle into the beaker on top of the blotting or filter papers. Slowly add 0.5 M sodium hydroxide solution to the beaker while stirring until the mixture reaches a pH of 9. Pour the liquid down the drain. Wrap the blotting or filter papers in old newspaper, and put them in the trash.

 An Exploration Worksheet is available to accompany Exploration 1 (Teaching Resources, page 17).

LESSON 1

Electricity From Chemicals

A surprising number of things that we use every day are powered by cells or batteries. These devices convert the energy stored in chemicals into a form of energy we can use—electricity. What are the advantages of chemical cells? the disadvantages? Are all chemical cells alike, or are some better than others? **A**

You have already made one kind of chemical cell—from lemon juice (page 306). In the Explorations that follow, you will construct several chemical cells, which you can test using a galvanometer.

EXPLORATION 1

Chemical Cells

EXPERIMENT 1

Dry Cells

You Will Need

- a homemade or commercial galvanometer
- a 40 cm length of magnet wire
- masking tape
- rubber bands
- 2 zinc strips (3 cm × 8 cm)
- 2 copper strips (3 cm × 8 cm)
- blotting paper or filter paper
- calcium chloride solution
- a container to hold the calcium chloride solution
- forceps
- latex gloves

What to Do

Caution: Wear goggles and latex gloves when working with calcium chloride.

1. Make a single-cell sandwich like the one shown below.
2. Measure the amount of deflection this sandwich produces in your galvanometer.
3. Now make a double-cell sandwich.
4. Measure the deflection it produces in the galvanometer. How does it compare with the deflection caused by the single-cell sandwich?

Because the double-cell sandwich consists of more than one cell, it is known as a **battery**.

Questions

1. What accounts for the difference in the galvanometer readings for the single-cell and double-cell sandwiches? What would be the effect of adding more layers to the sandwich?

Single-Cell Sandwich

Tape
Rubber bands
Homemade galvanometer
Magnet wire
Plastic-foam cup

Copper (becomes positively charged)
Blotting paper soaked in calcium chloride solution
Zinc (becomes negatively charged)

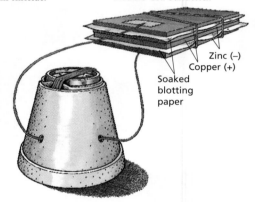

Double-Cell Sandwich

Zinc (−)
Copper (+)
Soaked blotting paper

310

LESSON 1 ORGANIZER

Time Required   3 class periods

Process Skills
observing, analyzing, inferring

New Term
Battery—a group of connected chemical cells that convert chemical energy into electrical energy to produce an electric current

Materials (per student group)
Exploration 1, Experiment 1: commercial galvanometer or homemade galvanometer from page 306; 40 cm of sanded magnet wire from Exploration 1 on page 296; a few pieces of masking

tape; 2 to 3 rubber bands; 2 strips of zinc, 3 cm × 8 cm each; 2 strips of copper, 3 cm × 8 cm each; several strips of blotting paper or filter paper, 3 cm × 8 cm each; 50 mL of 0.5 M calcium chloride; 250 mL beaker or other container to hold calcium chloride; pair of forceps; safety goggles; lab aprons; latex gloves (additional teacher materials: bottle; large beaker; 100 mL of 0.5 M sodium hydroxide; a few sheets of newspaper; see Advance Preparation on page 287C.); **Experiment 2:** zinc strip, 2 cm × 15 cm; copper strip,
continued ►

2. Which electrodes were linked together in converting a single-cell sandwich to a double-cell sandwich?

EXPERIMENT 2

Wet Cells

You Will Need

- a commercial galvanometer
- a zinc strip (2 cm × 15 cm)
- a copper strip (2 cm × 15 cm)
- salt solution
- a 250 mL beaker
- 2 pieces of magnet wire (each 20 cm long)
- 2 alligator clips or clothespins

What to Do

1. Make a setup like that shown below.
2. Add enough salt solution to cover about half of the metal strips. Connect the strips to the galvanometer as shown.

3. Observe the galvanometer. Record the highest reading reached. What happens to the reading? Observe each electrode carefully. What happens to each electrode?
4. Add enough salt solution to fill the beaker.
5. Record the galvanometer reading once again.

Questions

1. How would you connect two wet cells to get more current? Draw a sketch showing your answer.
2. What is one way of increasing the current in a wet cell? How would you explain this?
3. What other factors might be altered in a wet cell to increase its current output?
4. What energy changes take place in the operation of dry and wet chemical cells?

Chemical Cell Technology

You have made some simple chemical cells. The current produced was very small—enough to be detected by a sensitive galvanometer, but not enough to light a bulb. Many different types of chemical cells have been invented—some tiny, and some large and powerful. Chemical cells provide the small amounts of current needed to run calculators, radios, flashlights, pacemakers, hearing aids, and portable telephones. Chemical cells are also used to provide larger amounts of current to operate systems in cars and spaceships. You will find out about how cells are constructed and used in the next Exploration.

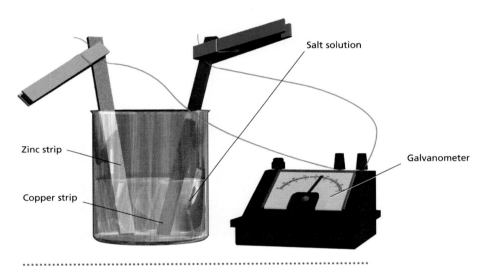

Salt solution

Zinc strip

Copper strip

Galvanometer

ORGANIZER, continued

2 cm × 15 cm; 250 mL beaker; 250 mL of water; about 0.5 mL of salt; two 20 cm lengths of magnet wire with sanded ends; 2 alligator clips or clothespins; commercial galvanometer (additional teacher materials: bottle; large beaker; 100 mL of 0.5 M sodium hydroxide; a few sheets of newspaper; see Advance Preparation on page 287C.)

Teaching Resources
Exploration Worksheets, pp. 17 and 21
Transparency 45

Answer to
In-Text Question, page 310

Ⓐ Answers will vary. Accept all reasonable responses.

Answer to
What to Do, page 310

4. The deflection for the double-cell sandwich should be about twice as much as that for the single-cell sandwich.

Answers to
Experiment 1, pages 310–311

1. The amount of chemicals available for conversion from chemical energy to electrical energy is double. Adding additional layers would increase the current even more.

2. Copper (+) and zinc (−) are placed together when the double sandwich is formed.

EXPERIMENT 2

If a commercial galvanometer is used, leave one of the wires free so that it can be touched to the electrode for only an instant. (Make sure that the ends of the wire are uninsulated.)

Answers to
What to Do

3. The highest reading occurs when the salt solution is added to the beaker. However, the reading on the galvanometer decreases almost immediately, indicating that the current is weakening. As more salt solution is added, covering more of the electrodes, the current increases. Students should observe that a gas is produced and collects in small bubbles on the copper strip (electrode).

Answers to Questions are on the next page. ▶

1. Student sketches should show one wire running from the zinc electrode in one cell to the copper electrode in the other cell. The output wires connect the free electrodes in each cell to the galvanometer.

2. One way to increase the current is to increase the amount of salt solution. This covers more of the electrodes and increases the number of electrons available in the solution.

3. Answers will vary. Possibilities include increasing the number of cells, increasing the size of the zinc and copper strips, and increasing the concentration of the salt solution.

4. The conversion of chemical energy into electrical energy takes place in both types of cells.

EXPLORATION 2

Cooperative Learning
EXPLORATION 2

Group size: 2 to 3 students

Group goal: to compare wet and dry cells and to analyze a variety of dry cells

Positive interdependence: Assign students roles such as artist (to sketch and label the poster), recorder (to write the group's responses), and director (to read material aloud, communicate with the teacher, and present the group's results to the class). Have each group create a poster that compares wet cells with dry cells. Then have them answer the questions in the Exploration as a group. Finally, have each group present their poster and answer some of the questions for the entire class.

Individual accountability: Have each student write a summary of the activity to explain what he or she learned and what questions he or she still has about cells.

 An Exploration Worksheet (Teaching Resources, page 21) and Transparency 45 are available to accompany Exploration 2.

EXPLORATION 2

Commercial Electric Cells

In this Exploration you will examine some common chemical cells. You have probably seen most of them. You may have even wondered how they work. Here's your chance to find out.

PART 1

More Dry Cells

Dry cells were invented to overcome the disadvantages of wet cells. However, dry cells are not really dry. Rather, the solution in them (the electrolyte) is blended with other substances to make it thick and pasty.

1. Look at the cells pictured at right. Which would you use in a standard-sized flashlight? in a penlight? in a watch?

2. The voltage of each cell is marked. Notice how several cells of different size have the same voltage. How can that be? What does *voltage* mean to you?

Check out my lemon-powered flashlight! The lemons are like dry cells, so I don't have to worry about spilling any juice.

Several types of chemical cells and batteries

312

Answers to
Part 1

1. Two D-cells would be used in a standard-sized flashlight; a AAA-cell would be used in a penlight; and a mercury cell would be used in a watch.

2. Different-sized cells with the same voltage can be made by varying the amount and kinds of chemicals used in the cells. Students will probably say that the voltage indicates the strength of a battery.

Homework

Have students write a "consumer fact sheet" in which they organize the information they learn about batteries in Exploration 2. Fact sheets should explain what the practical differences are among different types of cells.

Answers to Part 1, continued ▶

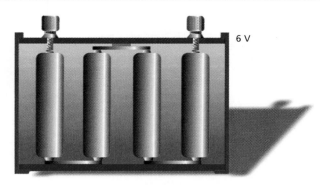

An inside view of cells and batteries. Compare the AAA, AA, and D-cells with the 9 V and 6 V batteries. What differences do you see? **Ⓐ**

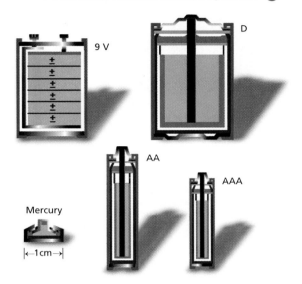

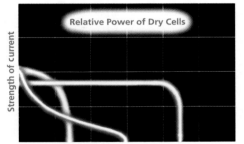

3. Look at the graph below. Which type of cell gradually "winds down"? Which type loses power quickly? Which type of cell would you probably use to power devices that require a steady current?

4. In your ScienceLog, sketch one of the cells and one of the batteries shown on this page. Then use the descriptions below to help you label their parts.

   Like all chemical cells, the ordinary dry cell has two electrodes, or conductors, and an electrolyte solution. The *positive electrode* has two parts—a *graphite rod* in the center of the cell and a mixture of *manganese oxide* and *powdered carbon* surrounding the graphite rod. The *negative electrode* is zinc; it makes up the sides and bottom of the cell. The *electrolyte* fills the space between the electrodes. It consists of *ammonium chloride* paste. At the top of the cell is an *insulator*. *Batteries* consist of at least two *individual cells* joined together by *conducting strips*.

   In the mercury cell, the *positive electrode* consists of a small block of zinc. The *negative electrode* is a layer of *mercury oxide*. The *electrolyte* is *potassium hydroxide*.

5. The *alkaline cell* differs from an ordinary dry cell in two major ways. First, the negative electrode is made of spongy zinc. Second, the electrolyte is potassium hydroxide, a strong base. What effect do these differences have on the power output of the alkaline cell?

Exploration 2 continued ▶

3. The output of the alkaline cell gradually decreases. The ordinary cell loses power rather quickly. The mercury cell would probably be used when a steady current is required.

4. A sample sketch is given below.

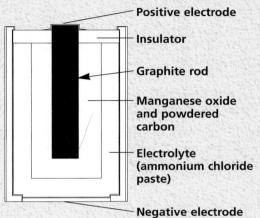

- Positive electrode
- Insulator
- Graphite rod
- Manganese oxide and powdered carbon
- Electrolyte (ammonium chloride paste)
- Negative electrode

5. From the graph, students may infer that these differences make the alkaline cell last longer and provide a steadier current.

Answer to
Caption

Ⓐ The AAA-, AA-, and D-cell batteries are single cells; the 9 V and 6 V batteries are made of multiple cells.

Integrating the Sciences

Life and Physical Sciences

Have students research ways that animals and plants use electricity for purposes such as defense, communication, and predation. *(Sample answer: Sharks, rays, and dolphins communicate with electrical signals, the electric eel can deliver an electric shock large enough to kill a horse, and plant root cells have a small voltage with which they attract positive mineral ions to their root systems.)*

Did You Know...

The lead-acid cell now used in car batteries was invented by the French physicist Gaston Plante in 1860. The dry cell was invented in 1866 by the French chemist George Leclanche.

FOLLOW-UP

Reteaching

Have students repeat the experimental procedure from Exploration 1, Experiment 2, substituting various citrus fruits for the salt solution. Remind students not to ingest any materials. Have them suggest why different fruits produce different amounts of electricity. *(Different fruits contain juices of different acidities and so react with the metal strips at different rates.)*

Assessment

Present students with the following task: Design an advertisement for chemical cells. Display the finished advertisements for all students to enjoy.

Extension

Have students design an activity to compare the cost-effectiveness of a regular cell, an alkaline cell, and a mercury cell. The results should be shown as the average cost of operation per minute for each type of cell.

Closure

Have students make a cell from a lemon or banana by inserting a paper clip and a copper wire into the fruit about 2 cm apart. Remind students not to ingest any materials. Ask students to hold the ends of the wires close together, but not touching, and to place the ends gently on their tongue. Ask students to explain what they feel. *(A slight tingle will be felt and a metallic taste will be experienced. Students close the circuit with their tongues, and a small, harmless current runs through them.)*

 **PORTFOLIO**

Students may wish to include their advertisements from the Assessment activity in their Portfolio.

Answers to
Part 2

1. **a.** The negative electrode is made of spongy lead; the positive electrode is made of lead oxide; the electrolyte is sulfuric acid.
   b. A large battery is needed for a car because a large surge of electricity is needed to start the car's engine.

PART 2

Other Cells

1. A powerful surge of electric current is needed to crank an automobile engine. This surge of current is provided by a group of cells joined together in a battery. Study the drawings at right to discover or infer the answers to these questions:
   a. What substances make up (1) the two electrodes and (2) the electrolyte in a car battery?
   b. Why is such a large battery needed for a car?

2. Automobile batteries have a limited life span, and not all batteries last the same amount of time. Why do batteries wear out? Why do some wear out sooner than others?

3. Research "maintenance-free" batteries. How do they work? How are they different from standard batteries?

4. For many applications, the *nickel-cadmium* cell is replacing both lead-acid batteries and dry cells. Find out how this type of cell works.

5. Unlike dry cells, nickle-cadmium cells and lead-acid batteries can be *recharged*. What does this mean? How is this property useful?

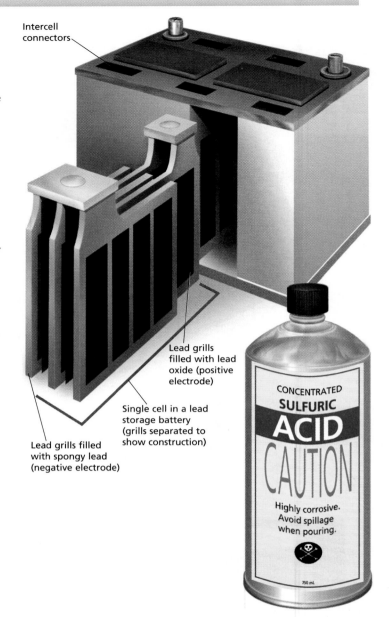

Intercell connectors

Lead grills filled with lead oxide (positive electrode)

Single cell in a lead storage battery (grills separated to show construction)

Lead grills filled with spongy lead (negative electrode)

CONCENTRATED SULFURIC ACID CAUTION

Highly corrosive. Avoid spillage when pouring.

750 mL

2. Batteries wear out because the electrodes are worn away or damaged by the chemical reactions continually taking place in the battery. Some batteries wear out quickly because they are made from less durable materials or because they are used more often.

3. Maintenance-free batteries use a sulfuric acid gel as an electrolyte rather than a sulfuric-acid solution made with distilled water. Also, maintenance-free batteries do not require the addition of distilled water because they are sealed to prevent evaporation of the electrolyte.

4. The nickel-cadmium cell works in a manner similar to a lead-acid cell, but it uses nickel oxide for the positive electrode and cadmium for the negative electrode. A potassium hydroxide solution is the electrolyte.

5. The ability to supply electricity can be restored after the cell or battery has run down. This property is useful because the same battery can be used over and over again.

LESSON 2 Electricity From Magnetism

You have discovered that magnetic effects can be caused by a current flowing in a wire or coil. Could the reverse be true? Could a magnet produce electrical effects in a wire? Try the following Exploration to find out. The Exploration has three parts—an activity that you can do and two completed experiments for you to analyze.

EXPLORATION 3

Moving Magnets and Wire Coils

PART 1

Building Your Own

You Will Need

- a commercial galvanometer
- a 150 cm length of magnet wire
- a cardboard tube
- a strong bar magnet
- sandpaper

What to Do

1. Sand the enamel off of the last 2 or 3 cm of the ends of the magnet wire. Wrap the magnet wire around the tube to make a coil as illustrated below. Attach the bare ends of the wire to a commercial galvanometer.

2. While watching the galvanometer, move a bar magnet into the coil, hold it there for a moment, and then remove it. Is the galvanometer needle affected?

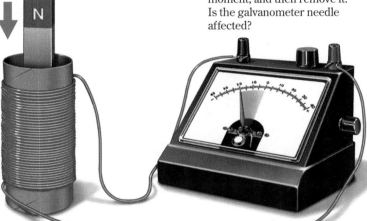

3. Repeat step 2 several times, moving the magnet at different speeds. What do you observe? Does moving the magnet into the coil have a different effect on the galvanometer than moving the magnet out of the coil? What might this suggest about the direction of current flow through the coil?

4. Disconnect the galvanometer and move the magnet to see whether the magnet itself is affecting the galvanometer.

5. Connect the galvanometer to the coil again. This time, hold the magnet still, but pass the coil over the magnet. What do you observe?

6. Use your observations to answer the following questions:

   a. How can a magnet help produce electricity?

   b. How is the direction of the current affected by the motion of the magnet?

   c. How does the speed of the magnet's motion affect the amount of electricity generated?

   d. What changes in forms of energy occur in this investigation?

   e. Could a stationary magnet ever produce electricity? Explain.

   f. Would current still be generated in the wire if the wire were broken at some point? Why or why not?

Exploration 3 continued ▶

315

LESSON 2 ORGANIZER

Time Required
3 class periods

Process Skills
observing, analyzing, inferring, hypothesizing

Theme Connection
Structures

New Terms
Alternating current—a type of electric current in which the direction of the current continually changes
Direct current—a type of electric current that flows in only one direction

Materials (per student group)
Exploration 3: commercial galvanometer; 150 cm length of magnet wire; cardboard tube; strong bar magnet; piece of sandpaper, about 10 cm × 10 cm
Alternating Current: meter stick or metric ruler; fluorescent or neon light

Teaching Resources
Exploration Worksheet, p. 26
Transparency Worksheet, p. 30
Transparencies 46–48
SourceBook, pp. S91 and S97

FOCUS

Getting Started
Remind students that in Chapter 13, Lesson 2, they discovered that an electric current generates a magnetic field. Ask them if they think the reverse could be true, and encourage them to explain their responses. *(Accept all reasonable responses without comment.)* Then direct students' attention to the lesson title, Electricity From Magnetism. Point out that in this lesson they will learn more about the connection between electricity and magnetism.

Main Ideas
1. A magnet moving through a wire coil or a wire coil moving through a magnetic field will generate an electric current.
2. In an alternating current, the direction of the current is continually switching back and forth.

TEACHING STRATEGIES

Cooperative Learning
EXPLORATION 3

Group size: 2 to 3 students
Group goal: to demonstrate how a current can be created by the relative motion of magnets and wire coils
Positive interdependence: Assign students the following roles: coil monitor (to prepare the wire coil and read all instructions), magnet mover (to carry out instructions for moving the magnet), and meter reader (to read and record data from the meter).

Students should complete the experiment and write answers to the in-text questions as a group. Have students design an experiment to test step 6(c).
Individual accountability: Randomly choose a presenter from each group. The presenter should share the group's experimental design for step 6(c) with the class.

Answers to Part 1 are on the next page. ▶

2. While the magnet is moving, students should notice that the needle is deflected.

3. The faster the magnet is moved, the more the needle is deflected. Changing the direction that the magnet moves changes the direction that the needle is deflected. This suggests that the direction of the current has also reversed.

4. The magnet alone will not produce an effect.

5. Moving the wire coil over the magnet will also cause the galvanometer's needle to be deflected.

6. a. A magnet can produce electricity by moving through a coil of wire.
  b. When the motion of the magnet is reversed, the direction of the electric current is also reversed.
  c. When the speed of the magnet's motion increases, the amount of electricity that is produced also increases.
  d. Chemical energy in the person moving the magnet is changed into kinetic energy as the magnet is moved back and forth. Moving the magnet changes the kinetic energy into electrical energy.
  e. A stationary magnet can produce electricity if a coil of wire surrounding the magnet moves.
  f. Initially, current would be generated, but it would cease to flow at the point where the circuit is broken.

 An Exploration Worksheet is available to accompany Exploration 3 on page 315 (Teaching Resources, page 26).

Homework

Students can answer questions 1–4 in Exploration 3, Part 2, on this page as homework.

PART 2

Francesca's Experiment

Francesca devised an experiment to answer questions raised by Part 1 of this Exploration. She started with a wire coil of 15 turns. Illustrations (a) through (c) show the galvanometer readings she recorded. Then she used a coil with twice as many turns. Illustrations (d) and (e) show these readings.

Francesca tried each part of this experiment three times and obtained similar results each time. What conclusions do you think she drew for each part? The illustrations provide some hints.

Analysis

1. When a magnet is moved inside a coil of wire, ___?___ is detected in the wire, which ___?___ its direction when the magnet is moved in the opposite direction inside the coil.

2. A larger current is produced if ___?___ or if ___?___.

3. Suppose Francesca moved the magnet into and out of the coil 15 times in a minute. What would happen to the current? How many times per minute would the current go first in one direction and then in the opposite direction?

4. What do you think would happen if Francesca held the magnet stationary and moved the wire coil instead? Why?

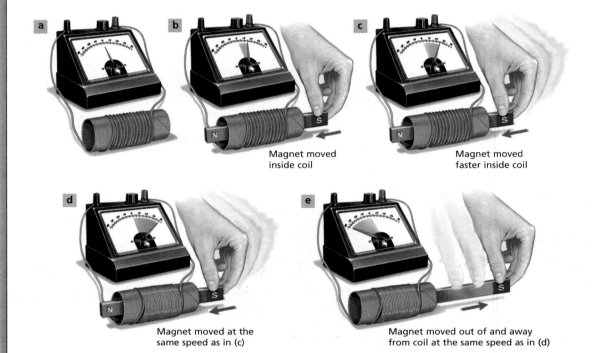

a

b
Magnet moved
inside coil

c
Magnet moved
faster inside coil

d
Magnet moved at the
same speed as in (c)

e
Magnet moved out of and away
from coil at the same speed as in (d)

Answers to
Part 2

1. When a magnet is moved inside a coil of wire, *an electric current* is detected in the wire, which *changes* its direction when the magnet is moved in the opposite direction inside the coil.

2. A larger current is produced if *the magnet moves more quickly* or if *the number of coils of wire is increased.*

3. The current would alternate direction. If Francesca moved the magnet in and out of the coil 15 times per minute, the current would move in one direction 15 times and in the opposite direction 15 times per minute.

4. If Francesca held the magnet stationary and moved the wire instead, an electric current would be produced because the magnet is still moving in and out of the coil.

PART 3

A Related Experiment

Francesca made an important discovery: When a magnet is moved through a coil of conducting wire, electricity is generated. Both the number of coils and the speed of movement of the magnet affect the amount of current produced. Francesca also discovered that the direction in which the magnet was moved made a difference. If the magnet was moved in one direction, current flowed one way. If the magnet was moved in the other direction, current flowed the other way. Let's examine the findings of a related experiment.

But before you begin, think a little bit about how a magnet exerts its influence. Does the magnet have to touch something to have an effect, or does its force act through space? Look at the photo at the upper right. It shows a magnet on which iron filings have been sprinkled. Do you see evidence that (a) the iron filings have been attracted and that (b) the *magnetic force* is exerted through space along curved paths? We call these paths 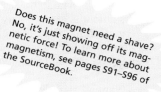 *magnetic lines of force.*

Look at the series of illustrations at right, which represent the results of the experiment. The wire is being moved while the magnet is held stationary. The arrows between the north and south poles of the magnet represent the lines of magnetic force.

Analysis

1. How do the results compare with those of Francesca's experiment?
2. What happens when the wire is momentarily motionless as it changes direction, as in (a)?
3. What happens when the wire is moved parallel to the magnetic lines of force, as in (d)?
4. What role do the magnetic lines of force appear to play in the generation of electricity?

Does this magnet need a shave? No, it's just showing off its magnetic force! To learn more about magnetism, see pages S91–S96 of the SourceBook.

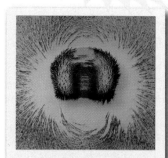

a

b

c

d

317

Did You Know...

Joseph Henry and Michael Faraday discovered in 1831 that the relative motion between a magnetic field and a wire produces a current in the wire. Faraday published his results first and investigated the subject in more detail, so he is usually credited with the discovery. The scientific term for the phenomenon is electromagnetic induction.

Integrating the Sciences

Earth and Physical Sciences

One of the most important instruments of geophysics, the *seismograph,* is based on the principle of electromagnetic induction. A seismograph contains a magnet and a coil of wire. Relative motion between the coil and magnet during an earthquake starts a current flowing in the coil and its conducting wires, indicating a seismic event. Have students research the seismograph and explain its operation in a written or oral report.

EXPLORATION 3

PART 3

In this part of the Exploration, students observe that an electric current is produced when a magnet is stationary and a conducting wire moves through its magnetic field. They should also realize that an electric current is produced only when the wire crosses the lines of magnetic force. Have students respond to the questions after each step of the activity has been done.

Answers to
Part 3

1. The results are similar. A current is produced whether the wire is moving or the magnet is moving.

2. No current is produced when the wire is momentarily motionless just before it changes direction.

3. No current is produced when the wire moves parallel to the magnetic lines of force.

4. The magnetic lines of force appear to cause an electric current to flow in a wire when the wire is crossing the lines and not moving parallel to the lines of force.

Generators—Small and Large, *page 318*

After students have read the introduction to generators, encourage them to discuss whether the electricity generating systems that they have seen so far in the unit could meet the electrical demands of a city. Encourage students to speculate about how any of the systems could be adapted to large-scale energy production.

Answers to
In-Text Questions

Ⓐ Students should have observed that the magnetic force can act through space. The diagram illustrates this. The iron filings are more concentrated near the magnet, suggesting that they are attracted to it. Also, the arrangement of the iron filings traces out the magnetic lines of force. (You may wish to challenge students by asking them to elaborate on this observation.)

Tiny Dynamo

Explain to students that *dynamo* is another name for *generator.* If possible, have a disassembled bicycle dynamo available in the classroom for students to examine. Encourage them to discuss among themselves how the dynamo works and to trace the flow of current through the dynamo, to the light bulb, and back again. Then check students' understanding by involving them in a discussion of the questions that follow on the next page.

Homework

Have students do research to find out where the electricity they use at home is produced. Students can start by making phone calls to the power company that sends electric bills to their home each month.

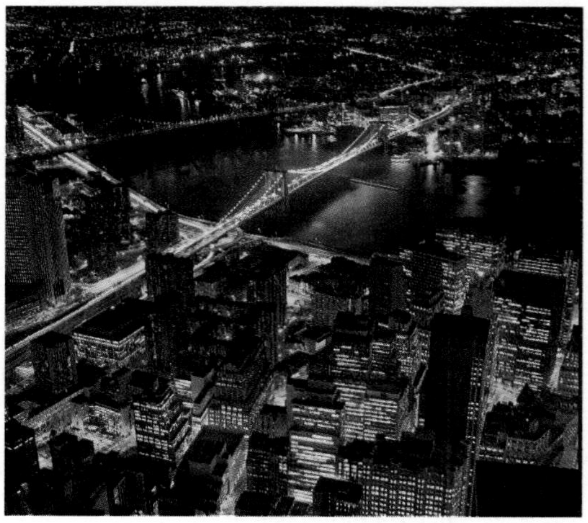

Generators—Small and Large

Our way of life requires large amounts of energy. For example, a medium-sized city requires enormous amounts of electricity to operate normally. Do you think that the electricity-generating systems you have seen so far in this unit could meet such demands? Could they be adapted to do so? Ⓐ

Examine the electricity-generating systems on this and the next page. Although they vary greatly in size, the operating principles of each are essentially the same.

Tiny Dynamo

A bicycle generator is a practical application of the *electromagnetic principle* that you discovered in Exploration 3. Whenever a magnet's influence sweeps across a wire that is part of a closed circuit, an electric current is generated. It does not matter whether the magnet or the wire moves to cause this to happen; the effect is the same. In Exploration 3 you saw electricity generated by back-and-forth motion. However, it is easier to generate electricity by rotating either the magnet or the wire coil. In the generator shown here, look for a magnet that rotates near a coil of wire. The generator is shown with its parts separated to help you see how it works.

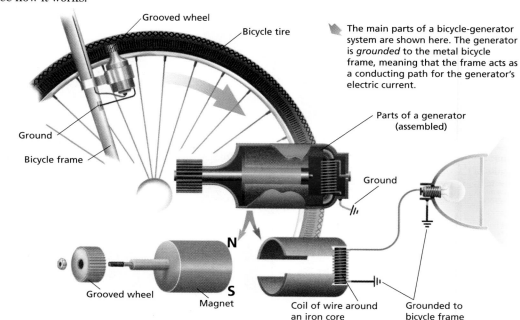

Grooved wheel

Bicycle tire

The main parts of a bicycle-generator system are shown here. The generator is *grounded* to the metal bicycle frame, meaning that the frame acts as a conducting path for the generator's electric current.

Ground

Bicycle frame

Parts of a generator (assembled)

Ground

N

S

Grooved wheel

Magnet

Coil of wire around an iron core

Grounded to bicycle frame

Answers to
Tiny Dynamo, page 319

1. The magnet is the central, rotating drum. The bicycle tire causes the grooved wheel of the dynamo to turn, which spins the magnet.

2. The conducting wire can be seen on the bottom right in the diagrams.

   The circuit has the following connections: top of wire coil → conducting wire → base of flashlight bulb → filament of bulb → side casing of light bulb → conducting wire → bottom of wire coil.

3. Factors that account for the larger current include the speed of the magnet's rotation, the strength of the magnet, and the number of turns in the coil of wire. Parts (b), (c), and (d) of Francesca's Experiment support the conclusion that increasing the speed of the motion of the magnet and increasing the number of coils of wire increase the current.

4. Answers will vary. A sample answer is as follows: The bicycle dynamo is a small electrical generator that is attached to a bicycle tire and sup- plies power to light the headlamp of a bicycle. The dynamo consists of a magnet that is connected to the bicycle tire by means of a grooved wheel; when the tire turns the grooved wheel, the magnet rotates. The magnet is surrounded by a metal tube; the tube transfers the influence of the moving magnet to a coil wrapped around an iron core. The motion of the magnet with respect to the coil-wrapped iron core causes a current to flow through the coil. The coil is attached to conducting wires that deliver the current to the headlamp. The current lights the lamp.

Analyze the generator's construction and operation.

1. Locate the magnet in the generator. How does it move? What causes it to move?

2. Locate the conducting wire that is wound around an iron core. The core is attached to curved metal plates, which help transmit the effect of the magnet coil just as if the magnet were actually moving into and out of the coil. Trace the complete electric circuit.

3. The current Francesca got was very small—it could be detected only by a sensitive galvanometer. The current developed in the generator is hundreds of times greater. What factors in the generator design could account for the larger current? Which of Francesca's experiments support your answers?

4. Write an entry for a student encyclopedia explaining how the bicycle generator works to power the headlight. Write it in a style understandable to a sixth-grader.

Large Generators

Generators usually use moving magnets to generate electricity in coils of wire. Study the illustration, which shows a large hydroelectric generator (*hydro* means "water"). What features of the generator account for the great amount of electric energy it can produce? How do Francesca's results support your answer?

Tracing the Flow of Energy

Use the diagrams on this page to complete the following energy-flow story.

Water in the reservoir has ___1___ energy. As it flows down, this energy changes into ___2___ energy of the moving water. The moving water forces the ___3___ to turn, providing it with ___4___. The attached ___5___ turn inside a stationary ___6___, in which ___7___ is produced. The resulting energy of the generator operation is ___8___ energy.

What is the energy story suggested by this diagram?

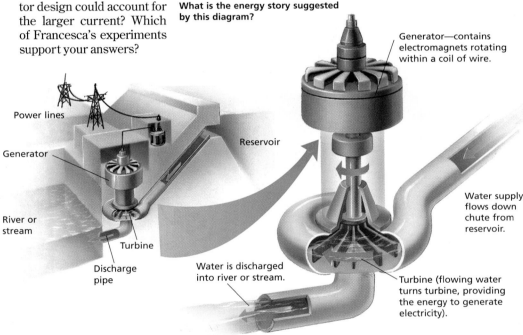

Power lines

Generator

River or stream

Turbine

Discharge pipe

Reservoir

Generator—contains electromagnets rotating within a coil of wire.

Water supply flows down chute from reservoir.

Water is discharged into river or stream.

Turbine (flowing water turns turbine, providing the energy to generate electricity).

319

Answers to
Alternating Current, pages 320–321

1. The current moves from one end of the wire loop to one of the contacts and then travels through the circuit to the other contact through conducting wires.

2. We know the current changes direction because a wire coil moving back and forth through a magnetic field creates an alternating current. The change in direction can also be detected by the galvanometer as the needle moves back and forth. (Remind students that the current reverses direction each time the loop completes one half of a revolution perpendicular to the magnetic field lines. Some students may wonder why the current in the loop in illustration (b) appears to be moving in the same direction as the current in the loop in illustration (d). You may wish to perform a teacher demonstration to show how an object rotating clockwise when viewed from one side rotates counterclockwise when viewed from the other side. After one half-turn, the loop is "upside down," and a current that did not change direction would appear to have reversed its flow. Because the current has reversed direction, it may at first give the illusion of not having changed direction at all. Have students compare the direction of the current at Contacts A and B in illustration (b) with the direction of the current at Contacts A and B in illustration (d). Make sure that students use their finger to carefully trace the current through the system. They will find that the current in the coil has indeed reversed direction.)

3. In illustration (b), the current in the coil enters from Contact A. In illustration (d), the current enters from contact B. It differs because the wire loop has rotated a half-turn.

4. It changes direction twice for every complete turn of the loop.

5. The direction of the current does not matter. The periods of no current occur very quickly, so the bulb appears to be continuously lit.

6. No current is generated because the loop is moving parallel to the magnetic lines of force.

7. The current flows in each direction five times when the handle is rotated five times. A cycle is one com-

Alternating Current

The principles shown by the systems on the previous page can be used to generate a special kind of electric current called **alternating current** (AC for short). The device below is an example of a simple AC generator. The handle sets the device in motion. As the wire loop moves through the magnetic field, electric current is generated.

Carefully study the diagram to figure out how the device works. In your ScienceLog, write a description of what is happening in the diagram. Then answer the following questions:

1. How is current conducted from the moving wire loop to the rest of the circuit, which includes the lamp and galvanometer?

2. As the device rotates, the current changes direction. Why? How do we know this?

3. In what direction is the current flowing in illustration (b)? in illustration (d)? Why does it differ?

4. How often does the current reverse direction with each complete turn?

5. Why is the light bulb unaffected by the change in current direction?

6. What's happening in illustration (c)? Why does this happen? (Remember the third experiment in Exploration 3.)

7. Suppose that you rotated the handle 5 times per second for 1 second. How many times does the current go first in one direction and then in the other? This is a current of 5 cycles per second. What do you think a cycle is?

8. What happens to the current output if you turn the handle faster and faster?

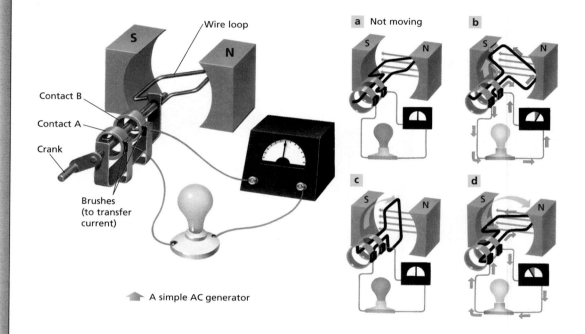

Wire loop

S   N

Contact B

Contact A

Crank

Brushes
(to transfer
current)

a Not moving

b

c

d

A simple AC generator

plete change in current direction—flowing first one way and then reversing to flow the opposite way.

8. As the handle is turned faster and faster, the current increases.

9. The device differs from the system in Parts 1 and 2 of Exploration 3 because there the magnets move within the wires. The device on page 320 is similar to the system in Part 3 of Exploration 3 because the wire moves through the magnetic field in both cases. All the systems are alike in that they produce an alternating current and generate electrical energy from magnetic energy.

10. This type of current is called *alternating current* because its flow alternates between one direction and another.

⭐ **Transparency 48 is available to accompany Alternating Current.**

Answers to
In-Text Question

Ⓐ Accept all reasonable descriptions of the sequence of events shown in the illustrations.

9. How is this device different from the two electricity-generating systems shown in Exploration 3? How is it similar to each system?

10. Why is the type of current produced by this device called *alternating current*?

Normal house current makes 60 complete cycles every second. Have you ever noticed "60 Hz" marked on tools or appliances? Look at the label from the electric drill shown below. The abbreviation *Hz* stands for *hertz*, a unit meaning "one cycle per second." This mark on a device means that the device is designed to run on 60-cycles-per-second alternating current.

Every time alternating current switches direction, for an instant no electric current flows. If this is so, why don't we notice it? We don't notice it

because it happens so quickly that our senses can't detect it. There are ways to detect this change indirectly, though. Here is one way. Wave a meter stick back and forth in the light from a single fluorescent (tube-type) or neon light. An ordinary incandescent light will not work. The room should be dark except for the single light. It also helps to face away from the light. As you sweep the meter stick quickly back and forth, you should see several repeated images of the meter stick. Each image represents the time during which the light is on as the current flows in one direction or another. Every time the current drops to zero as it changes direction, the light actually goes off for an instant. When this happens, you see no image.

Many electrical devices work with either alternating current or *direct current* (current that flows in only one direction). Why? What kind of systems do you think produce direct current? **B**

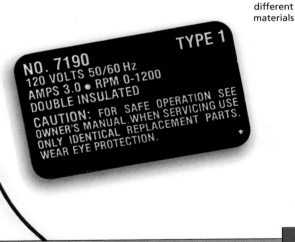

TYPE 1

NO. 7190
120 VOLTS 50/60 Hz
AMPS 3.0 • RPM 0-1200
DOUBLE INSULATED
CAUTION: FOR SAFE OPERATION SEE
OWNER'S MANUAL. WHEN SERVICING USE
ONLY IDENTICAL REPLACEMENT PARTS.
WEAR EYE PROTECTION.

Direct Current

Previously, you were introduced to chemical cells. Why does the current produced by a chemical cell go in just one direction? Study the diagram **C** below to help you answer this question.

Chemical cells produce **direct current** (DC), or current that flows in one direction. Direct current is needed instead of AC for many circuits, such as those in a car. Why can't you charge a battery with alternating current? **D**

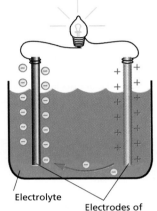

Electrolyte

Electrodes of different materials

Meeting Individual Needs

Second-Language Learners
Read the material on page 320 on alternating current aloud. Display a portable radio or cassette player that operates on either house current or dry cells, and invite students to examine it. Point out that such devices are built with mechanisms that change one current into the other. Help students recognize that chemical cells produce direct current.

FOLLOW-UP

Reteaching

Have students work in small groups to make models of a hydroelectric generator. They may wish to use materials such as modeling clay, plastic foam, poster board, pipe cleaners, and papier-mâché. The finished models should be displayed with explanations of how they work.

Assessment

Suggest that students write a summary of the main ideas in the lesson.

Extension

1. Point out to students that just as there are alternating-current generators, there are also direct-current generators. Suggest that they do some research to discover how the two types of generators are alike and how they are different.

2. Remind students that they examined how a hydroelectric generator operates. Explain that there are also thermoelectric and nuclear-powered generators. Have students do some research to discover how these other systems work.

Closure

Have students connect the ends of a coil of wire to a flashlight bulb (rather than to a galvanometer). Then have students pass a magnet into and out of the coil to see the flashlight bulb light up. Challenge students to make the bulb stay lit. For more fun, have students test magnets of different shapes (round, horseshoe, donut-shaped, etc.) with either the galvanometer or the flashlight bulb in the circuit to see which shape produces the most current.

LESSON
3
Other Sources of Electricity

FOCUS

Getting Started

Ask the class what they think light, heat, and pressure have to do with electricity. *(They can all be used to generate electricity.)* Then have the class suggest methods of generating electricity that have not yet been discussed in class. *(Students may suggest nuclear reactors or geothermal energy.)* Tell students that in this lesson they will learn about three new ways to generate electricity.

Main Ideas

1. Electrical energy can be produced from light energy by solar cells.
2. A thermocouple is used to produce electrical energy from heat energy.
3. Mechanical pressure on certain crystals causes the piezoelectric effect, which produces electrical energy.

TEACHING STRATEGIES

Answers to
In-Text Questions

Ⓐ Students should discover that the closer a solar cell is to a light source, the more electricity it produces.

Ⓑ Power generation by solar cells on the roofs of individual homes may be feasible as an energy source, but solar panels are best used in conjunction with other energy sources.

Solar panels are made up of an array of solar cells, each of which consists of two different types of silicon arranged in layers. As light strikes one layer, electrons are released and flow to the next layer, generating an electric current.

Answers to In-Text Questions continued ▶

So far you have produced electricity on a small scale from (a) kinetic energy alone, (b) chemical energy, and (c) a combination of kinetic and magnetic energies. You have also investigated applications of (b) and (c). In this lesson you will find that electricity can also be produced from light energy, heat energy, and mechanical energy (pressure).

You may not have heard of these last two methods of generating electricity. In fact, these methods can generate only tiny amounts of electric current. Nevertheless, they have specialized applications, as you will see.

EXPLORATION 4

Research Projects

PROJECT 1

Solar Cells

Solar panels generate electricity from light energy. You may be familiar with solar-generated electricity. Many calculators are powered by light energy alone, for example.

Try some experiments with a solar cell using different levels of light at various distances from the solar cell. Check on the amount of electricity being produced in each case. Ⓐ

Generating large amounts of electricity using only solar energy is difficult for many reasons. First of all, solar panels take up a great deal of space. Advances in technology will not shrink them beyond a certain size because there is only so much energy available in a given amount of sunlight.

Investigate sites where solar generation of electricity occurs. What are some advantages and disadvantages of each site? Would power generation by solar panels on the roofs of individual homes be feasible? Find out where this is being done and how solar panels work. Do you think that solar energy is the answer to some of our energy needs? Why is solar power a good source of energy for a satellite or space station? Ⓒ

Solar cells (above) are commonly used to provide electricity for satellites and space probes. Rows of solar cells (at left) produce electricity for a small desert community.

322

LESSON 3 ORGANIZER

Time Required
2 class periods

Process Skills
comparing, contrasting, analyzing

New Terms
Piezoelectric effect—the production of electric currents by certain crystals when they are squeezed or stretched
Thermocouple—a device that converts heat energy into electrical energy

Materials (per student group)
Exploration 4, Project 1: solar cell; light source; meter stick; **Project 2:** 50 cm of uninsulated copper wire;

50 cm of uninsulated iron wire; a few matches; commercial galvanometer; safety goggles

Teaching Resources

Thermocouples

A *thermocouple* converts heat energy into electricity. It consists of wires of two different metals joined together. When heat is applied to the point where the metals are joined, an electric current flows. Make a simple thermocouple using the diagram below as a guide. Apply heat to one end of it (the other end is not heated), and check the current output with a galvanometer.

Thermocouples have only limited use as sources of usable electric energy. They are most commonly used as high-temperature thermometers or thermostats (devices for maintaining a set temperature). How do you think they work? How do you think industries might make use of these devices? **D**

Piezoelectricity

Certain types of crystals produce electric currents when squeezed or stretched. This is called the *piezoelectric effect*. Piezoelectric crystals are useful for turning vibrations into electrical signals.

A classic example of the usefulness of such crystals is the record player, which was the standard home-audio device before compact-disc technology was developed. The needle of a record player uses a tiny quartz crystal. This crystal converts the vibrations created by the grooves in a record into an electrical signal suitable for amplification. How does it do this? As the needle rides along in the groove of a record, the tiny bumps in the groove, which represent the recorded sound, cause the crystal to be squeezed and stretched. Thus, the crystal produces an electric current—a current that mirrors the original recorded sound. Piezoelectric crystals also vibrate when an electric current is passed through them.

Computers, radios, watches, microphones, and many other devices could not operate without piezoelectric crystals. Research how piezoelectric crystals are used in some of these devices. **E**

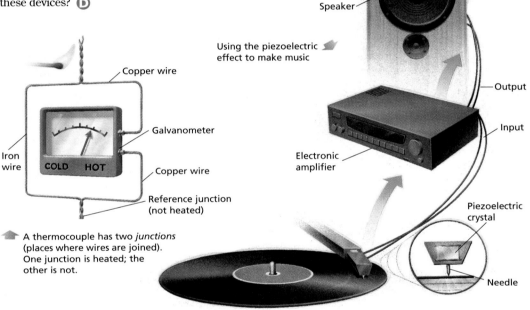

Using the piezoelectric effect to make music

Speaker

Output

Input

Electronic amplifier

Piezoelectric crystal

Needle

Copper wire

Galvanometer

Iron wire

COLD HOT

Copper wire

Reference junction (not heated)

A thermocouple has two *junctions* (places where wires are joined). One junction is heated; the other is not.

Answers to
In-Text Questions, pages 322–323

C Answers will vary. Students may point out the shortcomings of solar energy mentioned on page 322. However, solar-powered technology has become more efficient since its initial development.

Answers will vary. Possible answers include the following: Because the station or satellite is above the atmosphere, it receives a large amount of sunlight. Solar energy is also an inexhaustible energy source.

D Heat causes electrons to flow in the metals of the two wires. The interaction between the electrons in the two metals creates an electric current.

E Although student answers will vary depending on what devices they decide to research, the piezoelectric crystals in most devices generally function as *transducers*. A transducer converts energy from one form into another. For instance, a microphone uses a piezoelectric crystal to convert sound—a mechanical vibration in the air—into an electrical signal.

Homework

You may wish to assign one or more of the research topics described in Exploration 4 as homework.

Reteaching

Have students make a chart to evaluate each of the electricity sources studied. Students should include chemical and magnetic sources as well as the sources explored in this lesson. For each source, students should summarize its availability, reliability, cost, and impact on the environment.

Assessment

Suggest that students organize an "Electrical Energy Fair." The fair should consist of hands-on activities that demonstrate the various ways of producing electricity. Provide time for students to set up and enjoy their fair.

Extension

Point out to students that solar energy may become an important source of electricity in the future. Have them do some research to discover other alternative energy sources for the production of electricity. Suggest that they investigate wind power, wave power, and tidal power.

Closure

Invite a spokesperson from a local utility company to class to discuss the various methods that are used to generate electricity. Most utility companies have speakers available through their public relations departments who often have interesting slide presentations or demonstrations.

Answers to *Challenge Your Thinking*

1. Students may wonder why the current flows through the circuit at all, given that both contacts are located on the same copper ring. Point out to students that the black lines that divide the copper ring represent gaps through which the current cannot pass. The current is therefore forced through the circuit.

a. As the wire loop turns through the magnetic field, a current is produced in the wire. The ends of the wire are connected to two different segments of the armature. As the current flows in one direction, this armature segment touches one contact. As the current flows in the other direction, the other armature segment touches the other contact.

b. This generator produces direct current.

c. Be prepared for various responses. This question requires high-level reasoning skills. The following is a sample answer: The contacts for this generator are on opposite sides of the same copper ring, while the contacts for the generator on page 320 are two separate copper rings. Each leg of the loop in the generator on page 320 is connected to its own specific brush. As the loop rotates through the magnetic field, the current in the loop reverses direction, so the current in the circuit also reverses direction. This type of current is called an alternating current. On this page, even though the flow of the current in the loop also reverses direction, the current's direction in the circuit does not reverse. During the course of one complete rotation, each leg of the loop makes contact first with one brush and then with the other brush on the opposite side. This has the effect of reversing the current in the circuit. This reversal combines with the reversal of direction caused by the loop's half-turn. The net effect is that the two reversals cancel each other out, and the direction of the current remains constant. A current that always flows in one direction throughout a circuit is called a direct current.

d. In setup (a), the coil is not mov-

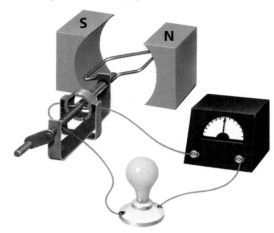

1. Generator X

The sequence of pictures below shows a type of generator in action. Use the pictures to help you answer the questions that follow.

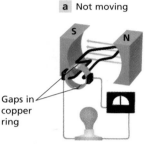 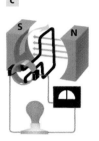 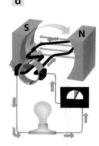

a Not moving   **b**   **c**   **d**

Gaps in copper ring

a. Examine the construction of this generator. How does this generator work?

b. Study the galvanometer readings. What kind of current does this generator produce?

c. How does this generator differ from the generator shown on page 320?

d. Explain what is happening in each illustration in the sequence.

ing, so no current is produced. In setups (b) and (d), the coil is moving through the magnetic lines of force, causing a current to flow. In setup (c), the parts of the coil in the magnetic lines of force are moving parallel to the lines of force, so no current flows.

(Answers will vary. Possible combinations include two 9 V batteries, three 6 V batteries, or twelve 1.5 V batteries.)

★ **You may wish to provide students with the Chapter 14 Review Worksheet that accompanies this Challenge Your Thinking (Teaching Resources, page 32).**

CROSS-DISCIPLINARY FOCUS

Mathematics

Tell students that you have a flashlight that requires 18 V to operate. Ask: What combination of batteries could I use?

2. Current Events

Electricity is related in some way to each of the following energy forms: light, heat, magnetic, chemical, kinetic, and vibrational. Identify the relationship among the energy forms in each of the following converters: dry cell, solar cell, wet cell, light bulb, generator, piezoelectric crystal, and thermocouple.

3. Play It Either Way

A light bulb can use either AC or DC. Hypothesize why this is so.

4. Irregular Exercise

Stan made a jump rope out of a loop of wire and then performed the activity pictured. The galvanometer showed that a current was being generated.

a. Explain what happened. (Hint: What makes a compass work?)

b. Did Stan generate direct current or alternating current? Explain.

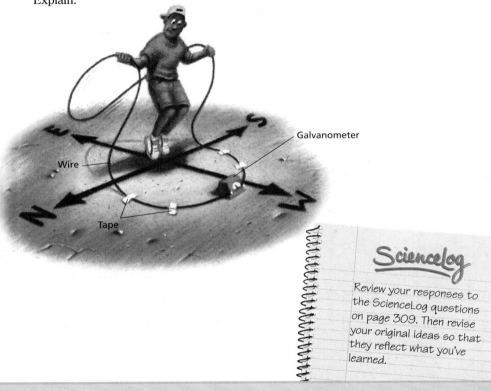

Wire

Tape

Galvanometer

ScienceLog

Review your responses to the ScienceLog questions on page 309. Then revise your original ideas so that they reflect what you've learned.

Answers to Challenge Your Thinking, *continued*

2. • Dry cell—chemical energy to electrical energy
   • Solar cell—light energy to electrical and heat energy
   • Wet cell—chemical energy to electrical energy
   • Light bulb—electrical energy to heat and light energy
   • Generator—mechanical energy and magnetic energy to electrical energy
   • Piezoelectric crystal—mechanical, or vibrational, energy to electrical energy
   • Thermocouple—heat energy to electrical energy

3. The filament within a light bulb acts as a resistor. As current passes through the filament, it heats up until it glows (a process called *ohmic heating* because an ohm is a measure of resistance). A light bulb can use either AC or DC current to heat the filament because a resistor is an unpolarized device. It will heat up and glow regardless of the direction of the current flowing through it.

4. a. Moving the wire through the magnetic field of the Earth caused a current to flow in the wire.

   b. Stan generated alternating current because as the wire moved upward, it generated a current in one direction, and as the wire moved downward, it generated a current in the opposite direction.

ScienceLog

The following are sample revised answers:

1. A direct current (DC) is a current in which the flow of charged particles occurs in one direction only, such as in the currents produced by cells and batteries. An alternating current (AC) is one in which the flow of charged particles reverses its direction, usually many times per second. Household currents are generally alternating currents.

2. Moving a magnet back and forth over or within a coil of wire that is part of a closed circuit will produce an electric current.

3. Remind students that a battery consists of a number of cells connected by conductors. Therefore, it is the cell that is the basic electrical source of a battery. A typical cell contains two different metals immersed in a chemical solution. The chemical solution interacts with the immersed metals, creating a surplus of electrons on one metal and a shortage of electrons on the other. If the metals are connected by a conducting material, electrons will flow from the metal with a surplus of electrons (the negative electrode) to the metal with a shortage of electrons (the positive electrode), creating an electric current that can be used to light a light bulb, power a radio, etc.

CHAPTER 15
Currents and Circuits

CHAPTER 15
Currents and Circuits

Connecting to Other Chapters

Chapter 13
introduces students to electricity, circuit theory, and the connection between electricity and magnetism.

Chapter 14
introduces alternating current and ways of producing electricity, such as with chemical cells and generators.

Chapter 15
introduces circuit symbols, resistance, parallel and series circuits, electromagnets, and electrical energy.

Prior Knowledge and Misconceptions

Your students' responses to the ScienceLog questions on this page will reveal the kind of information—and misinformation—they bring to this chapter. Use what you find out about your students' knowledge to choose which chapter concepts and activities to emphasize in your teaching. After students complete the material in this chapter, they will be asked to revise their answers based on what they have learned. Sample revised answers can be found on page 349.

In addition to having students answer the questions on this page, you may wish to have them complete the following activity: Have students write a paragraph explaining why it is important to learn about electric circuits. Look for areas of interest to highlight as you progress through the chapter. Assure students that there are no right or wrong answers to this exercise. Collect the papers, but do not grade them. Instead, read them to find out what students know about circuits, what misconceptions students may have, and what aspects of this subject are interesting to them.

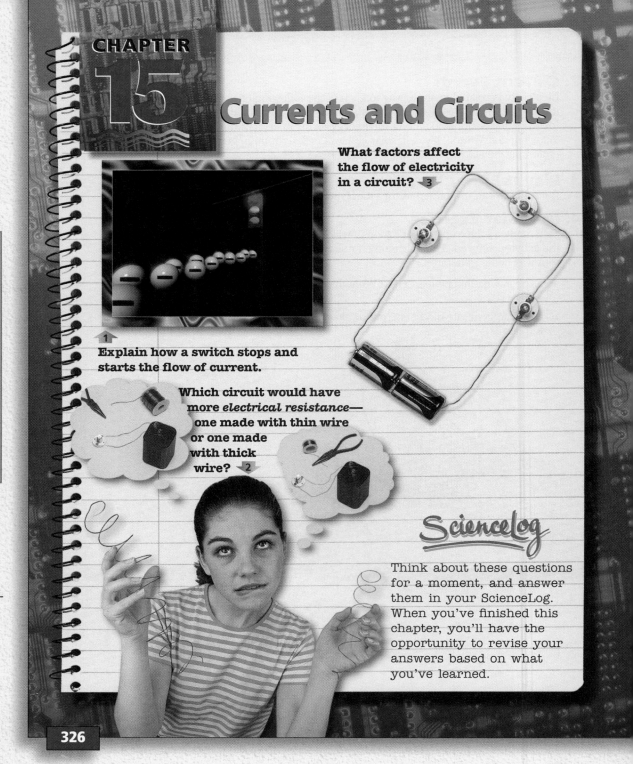

What factors affect the flow of electricity in a circuit? 3

1 Explain how a switch stops and starts the flow of current.

Which circuit would have more *electrical resistance*— one made with thin wire or one made with thick wire? 2

ScienceLog

Think about these questions for a moment, and answer them in your ScienceLog. When you've finished this chapter, you'll have the opportunity to revise your answers based on what you've learned.

326

Making Circuits Work for You

How have you made use of electric circuits in the last 24 hours? How have they worked for you? **E**

In the following Exploration, you will build and test a number of simple yet functional circuits.

Dry cells are quickly drained if left in a closed circuit without something to provide resistance (for example, a bulb). Placing a switch in the circuit and keeping it open until you check the circuit's operation will help to conserve the cells' energy.

EXPLORATION 2

Constructing Circuits

You Will Need

- 6 pieces of copper wire (each 10 cm long)
- masking tape
- 3 D-cells
- 3 flashlight bulbs in holders
- 2 contact switches

PART 1

Exploring

You can devise any circuits you wish. Use some or all of the equipment listed above to construct your circuits. Make any arrangements desired. If you want to use more than one dry cell and do not have a holder, you can use masking tape to hold them together. If the bulbs light up, you have complete circuits. After constructing your circuits, draw them in your ScienceLog using circuit symbols. Suppose that a certain number of electrons flow out of the cell(s) in a given time. Describe the path(s) taken by these electrons.

PART 2

Solving Circuit Problems

The following are a series of circuit problems. What arrangement of circuit components would you make to accomplish the functions described in each problem?

First, make a circuit diagram of your proposed solution. Then construct each circuit.

1. A circuit that lights two bulbs, A and B, when the switch is closed. If either bulb burns out or is unscrewed, the other

bulb goes out too. This type of circuit is called a **series circuit**.

2. A circuit that lights two bulbs, A and B, when the switch is closed. If either bulb burns out or is unscrewed, the other bulb stays lit. This type of circuit is called a **parallel circuit**.

3. A circuit that contains three bulbs, A, B, and C. If A is unscrewed, then B and C go out. If B is unscrewed, A and C stay lit. If C is unscrewed, A and B stay lit.

4. A circuit with two bulbs and two switches, P and Q. When P and Q are closed, both bulbs light up. If either switch is opened, neither bulb lights up.

5. A circuit that contains two switches and two bulbs. If both switches are open, neither bulb lights up. If either one of the switches is closed, both bulbs light up.

6. Analyze your findings.

   a. Identify the series and parallel circuits in each of your designs.

   b. Is there any parallel circuitry in the room where you are now? How could you find out without having to expose any wiring?

Exploration 2 continued ▶

331

Making Circuits Work for You

FOCUS

Getting Started

Draw the circuit symbols from Chapter 15, Lesson 1 (page 328), on the chalkboard and call on volunteers to identify and label each one. Then ask students to draw a simple circuit diagram using some or all of the symbols. When students have finished, ask: How could you determine if your circuit works? *(By constructing and testing it)* Point out that they will construct and test many circuits in this lesson.

Main Ideas

1. There are two main types of circuits: series and parallel.
2. The arrangement of resistors in a circuit affects the amount of current flowing through the circuit.

TEACHING STRATEGIES

Answers to
In-Text Question

E Students answers will vary. Possible answers include appliances such as washers and driers; electronic devices such as televisions, radios, CD players, or computers; automobiles; and battery-operated items such as watches, cameras, or flashlights.

EXPLORATION 2

PART 1

Give students time to experiment with constructing different kinds of circuits. During the process, some students may discover *series* and *parallel* circuits. If this occurs, do not identify the two types of circuits, but ask students to explain what they think is happening.

PART 2

You may wish to have students work in groups of three or four to brainstorm their ideas about how the circuits should be constructed.

LESSON 2 ORGANIZER

Time Required 2 class periods

Process Skills comparing, contrasting, inferring, observing

New Terms
Parallel circuit—a circuit in which the current divides into two or more branches
Series circuit—a circuit in which the components are arranged one after the other with only one path through which the current can flow

Materials (per student group)
Exploration 2, Part 1: six 10 cm lengths of copper wire or magnet wire with sanded ends; several small pieces of masking tape; 3 D-cells; 3 flashlight

bulbs in holders; 2 contact switches;
Part 3: 6 light bulbs; 3 contact switches; 3 D-cells; four 20 cm lengths of copper wire or magnet wire with sanded ends; four 10 cm lengths of copper wire or magnet wire with sanded ends; wire cutters
Exploration 3, Part 2: materials to construct one of the circuits listed on page 337 (See Advance Preparation on page 287C.)

Teaching Resources
Transparency Worksheet, p. 44
Exploration Worksheet, p. 46
Transparencies 49–50
SourceBook, p. S101

Answers to
Part 2, page 331

1. See sample diagram on page S210.
2. See sample diagram on page S210.
3. See sample diagram on page S210.
4. See sample diagram on page S210.
5. See sample diagram on page S210.
6. **a.** Circuit 1 is a series circuit. Circuit 2 is a parallel circuit. (In circuit 3, bulb *A* is in series with bulbs *B* and *C*, which are in parallel. In circuit 4, the switches are in series with other bulbs, but the bulbs could be either in series or parallel with each other. In circuit 5, the switches are parallel with each other, but the bulbs could be either in series or parallel.)
   b. Answers will vary. Sample answer: Yes, the lights in the classroom are wired in parallel. You could find out by unscrewing one bulb to see whether the others stay lit.

Answers to
Part 3, pages 332–333

1. Result: The bulb with two cells was brighter than the bulb with one cell. The bulb with three cells was brighter than the bulb with two cells. Conclusion: (b) If the number of cells is increased in the circuit, the amount of current is increased.

2. Result: When two bulbs were connected in series, they were dimmer than when there was only one bulb. When three bulbs were connected in series, they were dimmer than when there were only two bulbs. Conclusions: (a) Connecting bulbs one after another in a circuit decreases the amount of current flowing in the circuit. (c) If more bulbs are connected in series in a circuit, the resistance of a circuit is increased.

3. Result: The two bulbs in the parallel circuit were brighter than the two bulbs in the series circuit. Conclusions: (e) The current flowing through each bulb in a parallel circuit is greater than the current flowing through each bulb in a series circuit. (f) The resistance of a circuit is decreased when the bulbs are placed in parallel.

4. Result: All of the bulbs were dimmer when they were connected in series than when they were connected in parallel.

 PART 3

Current Questions to Investigate

Remember: The brightness of the light bulb is a measure of the amount of current flowing.

You Will Need
- 6 light bulbs
- 3 contact switches
- 3 D-cells
- copper wire
- wire cutters

What to Do
Construct each of the circuits shown in the table, and record the results in a similar table in your ScienceLog. Make certain that switches are included in the circuits you construct.

Question	Experimental design	Results/conclusions of experiment (Select from the list on the next page.)
1. How is the amount of current affected by the number of cells in a circuit?	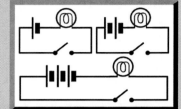	
2. How is the amount of current affected by the number of bulbs connected *in series*, that is, one right after the other?	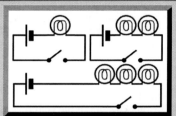	
3. How does the current flowing through each bulb in a *parallel*, or branched, circuit compare with the current flowing through each bulb in a series circuit?	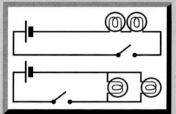	
4. What difference is there between the current flowing through the battery when two bulbs are connected in series and the current flowing through the battery when two bulbs are connected in parallel?	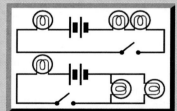	

Conclusions: (d) If two bulbs are placed in a branched circuit rather than in an unbranched circuit, the current through the battery is increased. (g) The resistance of a circuit is more with three bulbs in series than with two bulbs in parallel connected to a third in series.

 A Transparency Worksheet (Teaching Resources, page 44) and Transparency 49 are available to accompany Making Circuits Work for You on page 331. Additionally, an Exploration Worksheet is available to accompany Exploration 2 (Teaching Resources, page 46).

Homework

Ask students to think of careers in which a knowledge of circuits and circuit diagrams is crucial. Then have them interview an appropriate professional to discover what types of devices are used to analyze, build, or repair electronic circuits.

Putting It Together

Below are a number of statements with options. Each statement, with the correct option, is a valid conclusion for one of the four experiments you just did. Choose appropriate conclusions for each experiment.

a. Connecting bulbs one after another in a circuit (decreases, increases) the amount of current flowing in the circuit.

b. If the number of cells is increased in the circuit, the amount of current is (decreased, increased).

c. If more bulbs are connected in series in a circuit, the resistance of a circuit is (decreased, increased).

d. If two bulbs are placed in a branched circuit rather than in an unbranched circuit, the current through the battery is (decreased, increased).

e. The current flowing through each bulb in a parallel circuit is (greater than, less than) the current flowing through each bulb in a series circuit.

f. The resistance of a circuit is decreased when the bulbs are placed in (series, parallel).

g. The resistance of a circuit is (more, less) with three bulbs in series than with two bulbs in parallel connected to a third in series.

How Bright Are You?

How well can you apply what you discovered in Exploration 2? In the circuits shown on the next page, choose the correct brightness (*S* for standard brightness, *L* for less than standard, *M* for more than standard) for each of the 20 numbered bulbs. Assume that all bulbs are identical.

First, note the three illustrated degrees of brightness: *L, S,* and *M.* The standard brightness, *S,* to which *L* and *M* are compared, is the brightness of a single bulb connected to a single dry cell.

In your ScienceLog, make a table similar to the one below. Record your choices in it. Be prepared to defend your choices. The first situation is done for you.

S—standard brightness

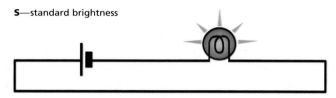

L—less than standard brightness

M—more than standard brightness

Bulb	L, S, or M
1	S
2	
3	
4	

SCORECARD

Over 15	You dazzle me!
10 – 15	You are bright!
Under 10	You need enlightenment

333

How Bright Are You?

A simple rule of thumb that you may wish to use in analyzing most of the circuits on the next page is as follows: To determine how bright a bulb would be in a circuit, you should compare the number of cells in the circuit with the number of bulbs in the circuit, but count only the bulbs along one path through a parallel circuit, not all of the bulbs. Thus, three bulbs in series would count as three, but three bulbs in parallel would count as only one. Numbers 15 and 16 on the next page are the exception to this rule.

Circuits can be better analyzed with an understanding of voltage and power and how they are related to current and resistance. These issues will be explored in Lesson 5 of this chapter.

Invite students to consider what might happen if a bulb were removed from one of the branches of a parallel circuit, such as bulb 7 on page 334. Point out that electric current tends to follow the path of least resistance. Because the circuit wire offers only minimal resistance, nearly all of the current will flow through the branch containing no bulbs. In such cases, the bulbs in the other branches, such as bulb 8, will draw no current and thus will no longer light up. We say that these bulbs have been *short-circuited.*

Cooperative Learning
HOW BRIGHT ARE YOU?

Group size: 3 to 5 students
Group goal: to compare parallel and series circuits and to predict the brightness of bulbs in various circuits
Positive interdependence: Prepare a grab bag with the numbers 1–20 on slips of paper. Have each group draw four numbers that correspond to the circuits on page 334. Each group will fill out the chart on page 333 for each of the circuits they picked. Once each group has filled out the chart, they should build their circuits to test their predictions. Have each member of the group analyze a different circuit and contribute their data to the group's chart.
Individual accountability: Have each student draw a circuit number from another group to analyze and explain to the entire class.

Answers to
How Bright Are You? *page 334*

1. S—One cell lights up one bulb, so it emits a standard brightness.

2. L—In this series circuit, one cell lights up two bulbs, so they are dimmer.

3. See (2) above.

4. L—In this series circuit, one cell must light up two bulbs, so they are dimmer, even though the locations of the bulbs in the circuit differ from the example above.

5. See (4) above.

6. L—In this series circuit, one cell lights up three bulbs, so they are dimmer.

7. S—In this parallel circuit, one cell lights up each bulb individually, so each bulb emits the standard brightness.

8. See (7) above.

9. S—In this parallel circuit, one cell lights up each bulb individually, so each bulb emits the standard brightness.

Answers to How Bright Are You? continued on the next page ▶

10. M—Two cells light up one bulb, so it is brighter.
11. S—In this series circuit, two cells light up two bulbs, so they each emit a standard brightness.
12. M—In this parallel circuit, two cells light up each bulb individually, so they are brighter.
13. M—In this series circuit, three cells light up two bulbs, so they are brighter.
14. S—In this series circuit, three cells light up three bulbs, so they emit a standard brightness.
15. M—Students should treat this bulb as the first in a series of two bulbs (the bulbs in 16 are in a parallel circuit and effectively act as one bulb).
16. S—This is a very difficult question, so you may wish to tell students that the bulb in 15 is operating at exactly double the standard brightness. Because this requires the energy of two of the three cells, students should realize that the electrical energy of one cell is devoted to each of the two bulbs in parallel. The two bulbs therefore demonstrate standard brightness.
17. M—In this parallel circuit, three cells light up each bulb individually, so they are brighter.
18. L—In this series circuit, three cells light up four bulbs, so they are dimmer.
19. See (18) above.
20. M—Again, this circuit is neither a purely series circuit nor a purely parallel circuit. It consists of two parallel sets of two bulbs that are in series. Thus, the three cells light up each set of two bulbs individually, so the bulbs are brighter.

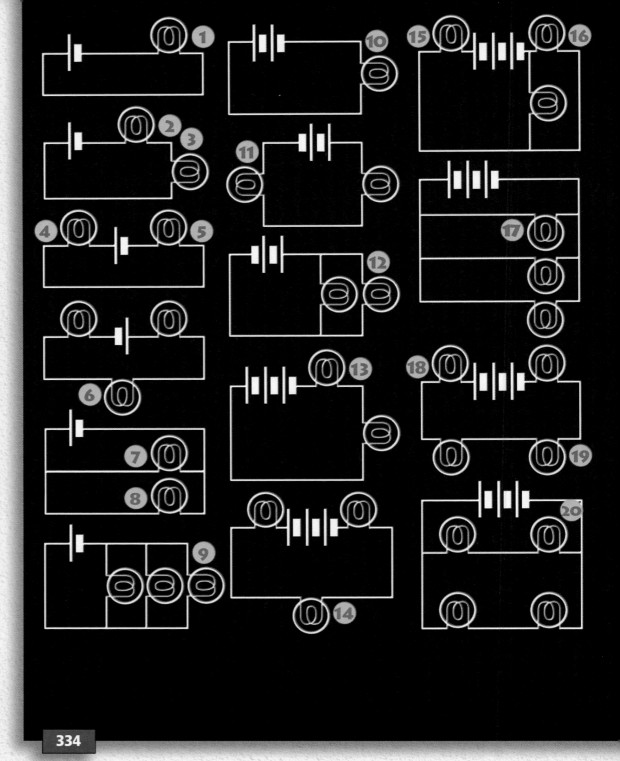

334

FOLLOW-UP

Reteaching

Suggest that students make a simple diagram of how their home might be wired. Their diagrams should include at least one lamp, one outlet, and one appliance.

Assessment

Have students make charts to summarize what they have learned about series and parallel circuits. For each circuit, students should make a circuit diagram and provide practical examples of the circuit in use.

Extension

Point out to students that many electronic devices, such as televisions and stereo components, come with circuit diagrams. Have students look for circuit diagrams they might have at home, such as in owner's manuals or on the backs of appliances. Ask them to analyze the diagrams to discover what the symbols mean. Display some of the circuit diagrams around the classroom.

Closure

Display two types of holiday lights—a string of lights with bulbs connected in series (usually these are older or less expensive lights) and a string of lights with the bulbs connected in parallel. Take one bulb out of each string and show the class the results. (None of the bulbs in the series string will light; the remaining bulbs in the parallel string will light.)

LESSON

3

Controlling the Current

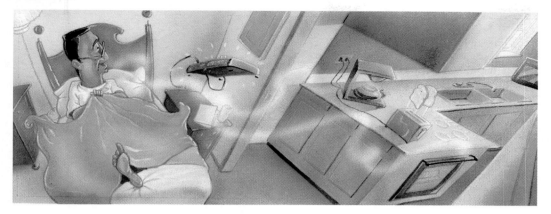

The radio blared suddenly to life, jolting Nick from a sound sleep. He sat up, looked at the time on the clock radio that had just come on, and jumped out of bed. Nick could smell the aroma of coffee coming from the kitchen, indicating that the timer on his new automatic coffee maker had worked properly. "Right on time," he thought to himself, looking at his watch as he entered the kitchen. Nick put some bread in the toaster. After a few moments, a couple of pieces of golden brown toast popped up. He then set the oven to come on at 4 P.M. to cook his casserole for 1 hour, knowing he would come home about 5:30. "I'm hungry already," he said to himself as he pulled out of the driveway. When Nick returned home that night, his apartment was already lit up. A light-sensitive switch had triggered the lights to come on at sunset. Best of all, there was a delectable smell in the air. He dimmed the kitchen lights and sat down to a scrumptious meal.

Automation! People depend on various types of circuits, switches, and controls to make life easier and more convenient. Although you may not know how they work, you are still able to make use of them.

1. How many automatic devices are mentioned in this story?
2. Which of these devices have switches that merely turn electricity on or off?
3. What are some of the kinds of controls that affect what is being done by the appliances and electrical devices?
4. Which of these switches or controls operate by mechanical means? by some other means?

335

LESSON 3 ORGANIZER

Time Required
2 class periods

Process Skills
analyzing, inferring, observing, predicting, measuring

New Terms
none

Materials (per student group)
Exploration 3, Part 2: materials to construct one of the circuits listed on page 337 (See Advance Preparation on page 287C.)

Teaching Resources
Activity Worksheet, p. 50
Transparencies 51–53

Getting Started
Remind students that they have made circuits in which they controlled the flow of electric current. Ask them if they can recall what they did. *(In Chapter 13, Lesson 1, pages 298 and 299, students built circuits with simple on/off switches. In Chapter 14, Lesson 2, they also learned how circuit components can affect the flow of current.)* Then encourage students to discuss why it is important to be able to control the flow of electricity. *(Accept all responses.)*

Main Ideas
1. The flow of electric current can be controlled in a variety of ways.
2. There are many different kinds of switches, including on/off switches, dimmer switches, and time-delayed switches.

TEACHING STRATEGIES

Answers to
In-Text Questions

1. At least six automatic devices are mentioned in this story, including a radio-alarm clock, automatic coffee maker, toaster, oven timer, light-sensitive switch, and dimmer switch. Some students may include Nick's wristwatch or his automobile in this list as well, but these devices do not make explicit use of timers, switches, or other controls in this story.

2. All of the devices except the toaster and the dimmer switch simply turn themselves on and off.

3. Timers are used in the radio-alarm clock, automatic coffee maker, and oven timer to control when they turn on. A thermostat controls how long the toaster will toast the bread. A light meter controls when the lights turn on. A variable resistor in the dimmer switch controls how bright the lights are.

Answers to In-Text Questions continued ▶

Students may also point out that the radio has a volume control and that the oven has a thermostat, but these are not explicitly mentioned in the story.

4. All of the devices, except for the light-sensitive switch, operate by mechanical means. The light-sensitive switch responds to light, so it is an electromagnetic switch, not a mechanical switch. Some students may consider thermostats to be heat switches and not mechanical switches because they respond to heat. You may wish to point out that in these devices, a bimetallic strip is mechanically bent as it absorbs heat.

⭐ **Transparencies 51 and 52 are available to accompany Exploration 3.**

Answer to
In-Text Question
Ⓐ

Answers will vary, but students are likely to use all of these types of controls at home.

Answers to
Part 1, pages 336–337

Circuit 1: This circuit contains two switches, represented by the metal strips on the wooden blocks. In order for the bulb to light, both switches must be closed. One use of this circuit is for safety: by opening one switch, the other switch is disabled. This kind of circuit is used in many automobiles, where it allows the driver to disable other switches that control functions such as unlocking the doors or rolling down the windows. In the story, this type of circuit is probably used in both the coffee machine and the oven, where one switch turns the device on and off and the other switch is controlled by a timer.

Circuit 2: The wire wrapped around the wooden dowel functions as a *variable resistor*. Sliding the contact to the left or right alters the resistance of the coil. The variable resistor controls the amount of current flowing through the circuit and thus the brightness of the bulb. Nick used a dimmer switch to lower the kitchen lights.

Circuit 3: The stiff metal wire in this

EXPLORATION 3

Switched On!

PART 1

Under Control

Team up with two classmates. Together, study and discuss the circuits shown here and on the following page. Identify the switches in each circuit.

Can you determine how each switch works to control the current? In your ScienceLog, write down what you think is happening. Share your responses with other groups, and discuss any questions you still have about the operation of the circuit switches.

What practical use might be made of each circuit? Identify where Nick made use of similar current controls in the preceding story. Which current controls do you use at home? Ⓐ

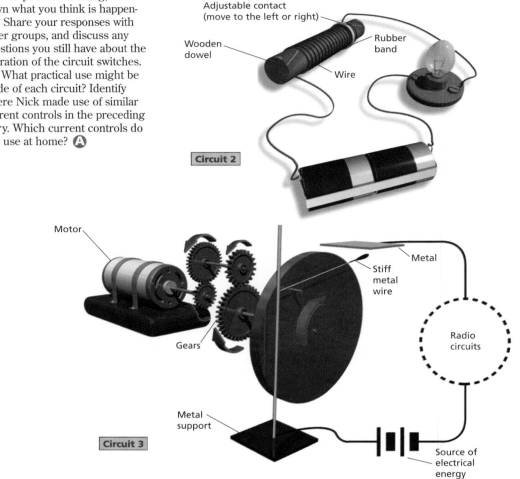

336

---

circuit is part of a timer switch. The motor turns the gears that slowly turn the large wheel. As the wheel turns, a rod attached to the wheel slowly moves upward, raising the metal wire. Eventually, the wire contacts the metal plate, closing the circuit and turning on the radio. Nick used a timer switch such as this to wake up in the morning.

Circuit 4: The bimetallic strip in this circuit operates as a heat switch. As current flows in the circuit, the resistance wire heats up. Some of this heat is transferred to the bimetallic strip, which bends in response. The wire contact represented by the arrow acts as an on/off switch: when it is in contact

with the bimetallic strip, the circuit is closed. As the strip heats and bends, it will eventually lose contact with the arrow, stopping the flow of current and the heating of the resistance wire.

Also, the left end of the bimetallic strip is held in place so that only the right end is free to move. By changing where the arrow touches the bimetallic strip, the circuit controls how much heat is required to shut off the circuit. In the story, such a circuit was used in the toaster to control how long the bread was toasted.

Answers to Part 1 continued ▶

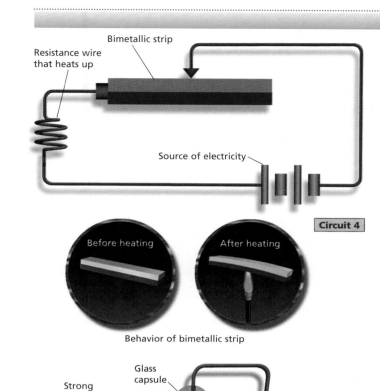

Resistance wire that heats up

Bimetallic strip

Source of electricity

Circuit 4

Before heating

After heating

Behavior of bimetallic strip

Strong magnet

Glass capsule

Circuit

Circuit 5

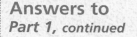

PART 2

To the Drawing Board!

Applying what you have learned so far, design and construct two circuits from the following list. Construct a circuit that

a. has two lamps and two switches. Each switch operates only one lamp.

b. has two lamps and three switches. One switch turns on both lamps; the other two switches can turn off the lamps one at a time.

c. has a lamp, a switch, and a brightness control.

d. is a model circuit of a doorbell operated by two different push buttons.

e. contains a simple switch that could be operated by a magnet.

f. closes when a given temperature is reached.

For each circuit, follow these steps:

1. Brainstorm possible solutions with your two partners.
2. Choose the best solution.
3. Draw your design.
4. Assemble the materials and construct a model of your design.
5. Suggest and try possible improvements.
6. Decide on a good application for the circuit.

Answers to
Part 1, *continued*

Circuit 5: The glass capsule in this circuit functions as a magnetic switch. As the magnet gets closer to the capsule, one of the metal strips in the capsule is magnetized and attracts the other metal strip. When the two strips touch, the circuit is closed and current flows.

Answers to
Part 2

Student answers will vary. Sample circuits are as follows:

a. See page S211 for sample circuit diagram.

b. See page S211 for sample circuit diagram.

c. See page S211 for sample circuit diagram.

d. A diagram of a single-button doorbell circuit can be found on page 339.

e. See circuit 5 on this page.

f. See circuit 4 on this page.

Meeting Individual Needs

Learners Having Difficulty
Have students use the circuit symbols that were introduced in Chapter 15, Lesson 1, to redraw the first two circuits discussed in Exploration 3, Part 1. See S211 for sample sketches.

★ **An Activity Worksheet (Teaching Resources, page 50) and Transparency 53 are available for use after completing Lesson 3.**

337

FOLLOW-UP

Reteaching

Have a contest to see how many different kinds of time-delay switches students can invent. For example, students might make switches from timing devices such as alarm clocks, balloons, or dripping water. Encourage students to have fun with the activity.

Assessment

Have students make a list of the switches they use every day. Then ask them to make a chart to classify the switches as either on/off, dimmer, or time-delay switches and to indicate the purpose of each switch.

Extension

Point out to students that magnetic switches are often more useful than mechanical switches. Have them do some research to discover some of the advantages and uses of magnetic switches. They should share what they learn by making poster diagrams for the classroom. *(Because the metal contacts are sealed in an airtight container, free from water or other contaminants, magnetic*

switches are more reliable in adverse environments where mechanical switches would quickly deteriorate. They are also faster and require less power to operate. Such switches are used in computers and automobiles.)

Closure

Bring to class a variety of devices that use different kinds of current controls. For example, you could provide a buzzer, thermostat, or ignition switch for students to examine.

LESSON 4 Electro-magnets

FOCUS

Getting Started

Display a bar magnet and a coil of wire, and ask students to recall what they have observed about the relationship between electricity and magnetism. *(An electric current flowing though a wire creates a magnetic field. A magnet moving through a coil of wire creates an electric current.)* Ask students to consider how they might be able to make a magnet from a coil of wire with a current flowing through it. *(Accept all reasonable responses.)* Point out that in this lesson they will have a chance to test their ideas.

Main Ideas

1. Electromagnets are used in a variety of everyday devices.
2. The strength of an electromagnet depends on the strength of the current and the number of turns in the coil.

TEACHING STRATEGIES

For Exploration 4 on this page and A Model Motor on page 342, you should use the wire that you sanded down for use in Exploration 1 on page 296. If you did not perform that Exploration, you will need to sand the enamel off the last few centimeters of each end of the wire.

Electromagnets are magnets created by flowing electric currents. You made a kind of electromagnet when you constructed your galvanometer. The current-bearing coil of wire became magnetized and deflected the magnetic compass needle. Look at the circuit in the diagram below. This arrangement of a spike and a coil of wire is a simple electromagnet.

A typical electromagnet consists of a core of iron or soft steel surrounded by a coil of insulated wire. (Why must the wire be insulated?) 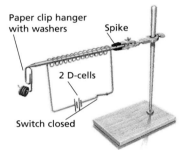 The magnet wire you have been using in the Explorations in this unit is an insulated copper wire that is often used in making electromagnets. When the current flows through the wire, the core quickly becomes a temporary magnet. When the current is interrupted, though, the core loses its magnetism. Note that there is a switch in the circuit that allows the circuit to be easily disconnected.

EXPLORATION 4

Constructing an Electromagnet

You Will Need

- a paper clip
- washers
- some light, insulated wire
- 2 D-cells
- an iron spike
- a switch

What to Do

Get together with one or two classmates. Your task will be to make a functioning electromagnet and then to determine how its strength can be increased.

Making the Electromagnet

Use the materials listed above to make an electromagnet capable of supporting a paper clip from which several washers are hanging. Identify the parts of the circuit, and trace the path of the current. Are the spike, paper clip, and washers part of the circuit? How many washers can your first design hold? **B**

Paper clip hanger with washers
Spike
2 D-cells
Switch closed

Increasing the Electromagnet's Strength

How can you make your electromagnet hold more washers? Take some time to discuss the following:

- factors or variables that might be altered
- ways to measure the magnet's strength

- safety precautions
- the apparatus

Get your design approved by your teacher. Then assemble the necessary apparatus, do the experiment, record the results, and draw conclusions based on your results. Share your results with others.

How does this electromagnet work? **D**

EXPLORATION 4

Provide students with time to read the introduction, study the illustrations, and discuss what they are to do. Point out to students that if the switch is closed for too long, the cell will run down, and the current will be too weak to create a useful electromagnet.

⭐ **An Exploration Worksheet is available to accompany Exploration 4 (Teaching Resources, page 51).**

LESSON 4 ORGANIZER

Time Required
3 to 4 class periods

Process Skills
analyzing, observing, hypothesizing

Theme Connection
Structures

New Terms
Electromagnet—a magnet made from a current-bearing coil of wire wrapped around an iron or steel core

Materials (per student group)
Exploration 4: paper clip; several washers; about 1 m of light, insulated wire or magnet wire with sanded ends; 2 D-cells; iron spike or bolt; switch; a variety of materials for increasing the electromagnet's strength (See Advance Preparation on page 287C.)
A Model Motor: about 1 m of thin insulated copper wire or magnet wire with sanded ends; cork; small knife; short knitting needle; 4 straight pins; 2 thumbtacks; 3 D-cells; stick of modeling clay; 2 block magnets

Teaching Resources
Exploration Worksheet, p. 51
Transparencies 54–55

Figuring Out Electromagnetic Circuits

No doubt you have used diagrams when putting something together or figuring out how something works. A diagram can show how something operates or how parts should be assembled. The following are diagrams of circuits containing electromagnets. Working in small groups, determine how several of these circuits work. A few hints are provided in some of the drawings.

A good plan is to begin with the source of the electricity and follow the path of the complete circuit back to its source. For example, in the doorbell circuit, when the button is pushed down, the circuit is completed. Electrons flow along the parts of the circuit: battery, switch, contact screw, springy metal strip, coil, and back to the battery. Now retrace the path and think about what is happening in each part of the circuit, especially the electromagnet.

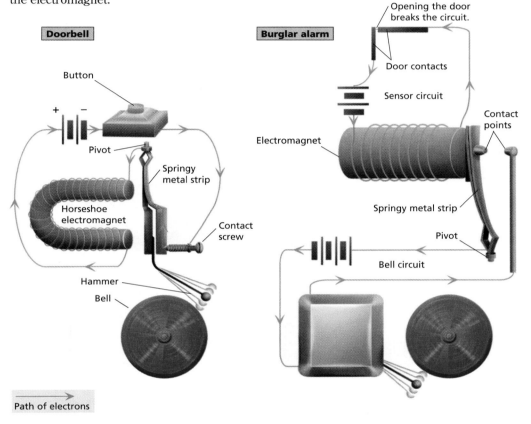

Doorbell

Button
+ –
Pivot
Springy metal strip
Horseshoe electromagnet
Contact screw
Hammer
Bell

Path of electrons

Burglar alarm

Opening the door breaks the circuit.
Door contacts
Sensor circuit
Electromagnet
Contact points
Springy metal strip
Pivot
Bell circuit

339

Answers to
In-Text Questions and Caption, page 338

Ⓐ After examining the illustration of the electromagnet, students should be able to conclude that the wire must be insulated to prevent the current from flowing into the iron spike. If current flows into the iron spike, the spike will become part of the circuit and will not be magnetized.

Ⓑ The parts of the circuit include the cell and the insulated copper wire. The current travels from the cell through the wire and back to the cell again. The iron spike, the paper clip, and the washers are not part of the circuit because current does not flow through them.

The number of washers that the paper clip can hold depends on the strength of the electromagnet.

Ⓒ Allow students to brainstorm their ideas for increasing the strength of the electromagnets. Inference and experimentation should lead them to the conclusion that increasing the strength of the current or increasing the number of turns in the coil will increase the strength of the electromagnet. One simple way to test their ideas is to see how many washers the iron spike will support after each change is made in the circuit. Make sure you check student ideas for safety.

Ⓓ This type of strong electromagnet is extremely massive and highly coiled and has a large flow of electrical energy running through it.

Theme Connection

Structures

Focus question: In what way is a resistor like a tourniquet? *(Both items are used to control flow. A tourniquet is used to control bleeding by slowing the rate of blood flow to a body extremity. A resistor is used to control the flow of current through a wire.)*

Answers to
Figuring Out Electromagnetic Circuits

Students may have used diagrams for such activities as assembling furniture or building toy models.

Doorbell: Pushing the button on the doorbell closes the circuit. Electrons flow from the battery, through the button switch, along the wire, through the contact screw, the springy metal strip, and the wire coil, and back to the battery. When electricity flows through the coil, it activates the electromagnet, which pulls the springy metal strip away from the contact screw, breaking the circuit. With the circuit broken, the electromagnet is deactivated, and the springy metal strip springs back to the contact screw, closing the circuit again. This process of breaking and closing the circuit occurs many times a second and causes the hammer attached to the springy metal strip to vibrate against the bell.

Answers to Figuring Out Electromagnetic Circuits continued on the next page ▶

Answers to
Figuring Out Electromagnetic Circuits, *continued*

Burglar alarm: When the door is closed, the electrons in the sensor circuit flow from the power source, through the wire coil, through the wire and door contacts, and back to the power source. The electromagnet attracts the springy metal strip as long as the current flows in the sensor circuit. When the door is opened, the door contacts and the circuit are broken, deactivating the electromagnet. The springy metal strip springs away from the electromagnet, closing the bell circuit and activating the bell. The electrons in the bottom circuit flow from the power source, through the bell, through the contact points and springy metal strip, through the wire, and back to the power source.

Telephone: When a person speaks into the receiver, the sound waves cause a thin sheet called a *diaphragm* to vibrate. These vibrations compress a block of carbon, changing its resistance and thus changing the amount of current flowing through the circuit. The varying current flows through an electromagnet in the earpiece. The electromagnet attracts a second diaphragm, and the strength of the attraction varies with the strength of the current. The second diaphragm vibrates in response to the changing strength of attraction. This vibration causes the air near the diaphragm to vibrate, reproducing the original sound signal so that it can be heard by a listener whose ear is near the earpiece.

Car starter: When a person turns the key in the ignition, a circuit is closed, causing electrons to flow from the battery through an electromagnet. The small current in the electromagnet causes it to pull on a pivot. The pivot rotates, closing a second circuit. A large current then flows in the second circuit from the battery through the starter motor, which turns a wheel that powers the car's engine.

 Transparencies 54 and 55 are available to accompany Figuring Out Electromagnetic Circuits on pages 339 and 340.

Homework

Assign one or more of the devices in Figuring Out Electromagnetic Circuits as homework.

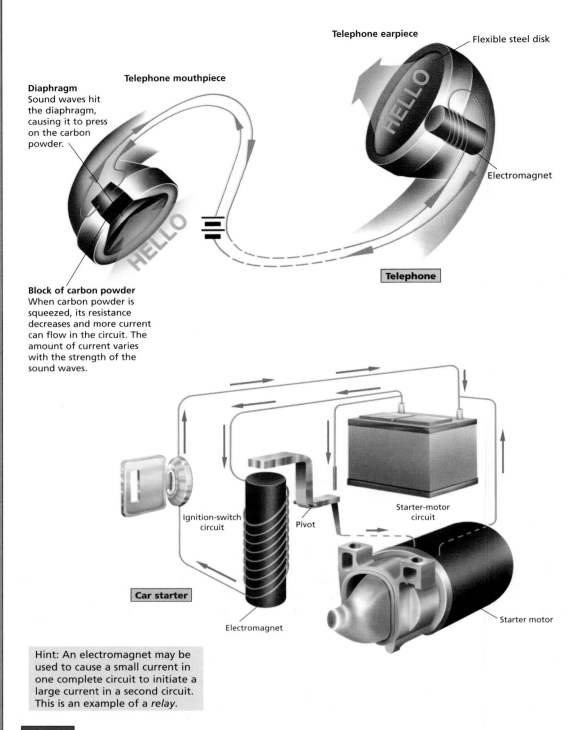

Telephone earpiece — Flexible steel disk

Telephone mouthpiece

Diaphragm
Sound waves hit the diaphragm, causing it to press on the carbon powder.

Electromagnet

Block of carbon powder
When carbon powder is squeezed, its resistance decreases and more current can flow in the circuit. The amount of current varies with the strength of the sound waves.

Telephone

Ignition-switch circuit

Pivot

Starter-motor circuit

Car starter

Electromagnet

Starter motor

Hint: An electromagnet may be used to cause a small current in one complete circuit to initiate a large current in a second circuit. This is an example of a *relay*.

340

Did You Know. . .
Hans Christian Oersted discovered the electromagnet in 1819 in a lucky accident. He had intended to demonstrate the "toaster effect" in which a current warms the wire through which it flows. As he flipped the switch to start the current flowing, he noticed that a nearby compass needle was deflected, showing that a magnetic field had been produced by the current.

Meeting Individual Needs

Gifted Learners
Have students learn about *thermistors* and *transistors,* two other elements of circuits. Ask: What are the functions of these devices in electric circuits? How did they get their names? Do you think their names are appropriate? *(A thermistor is an object whose resistance varies with temperature. The word is a combination of the words* thermal *and* resistor. *A transistor is a device that is used to control the flow of electricity. The word is a combination of the words* transfer *and* resistor.*)*

Moving Coils—Revolutionary!

One of the most common applications of electromagnets is the motor. Can you imagine life without motors? The invention of the motor was revolutionary in more than one way.

Take a look at an electric motor, such as the one in a washing machine or furnace blower, and examine the metal tag attached to it. You may see information similar to that shown here. Some of these words and symbols may be unfamiliar to you. What do you think "RPM" means? "RPM 3450" means that in 1 minute the motor makes 3450 complete turns, or *revolutions,* about a central axis.

What part does an electromagnet play in a motor? What part of a motor revolves? To find out, assemble the device below and close the circuit. What happens? **Ⓐ**

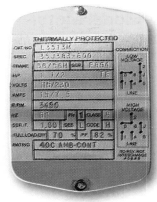

This motor will make 3450 revolutions in 1 minute.

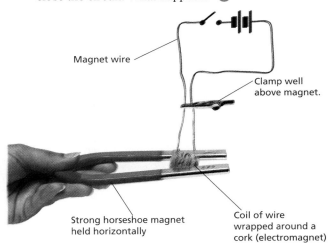

Magnet wire

Clamp well above magnet.

Strong horseshoe magnet held horizontally

Coil of wire wrapped around a cork (electromagnet)

The coil of wire acts like a magnet. When a current flows, the coil has north and south poles. If, as in the photo at right, you curl the fingers of your left hand in the direction of the flow of electrons through the wire coil, the position of your thumb gives you the north pole of the coil.

Locate the N and S poles of the coil above. What will be the interaction between the N pole of the magnet and the N pole of the coil? between the S pole of the magnet and the N pole of the coil? What other interactions are there? You should realize that the coil will rotate in the presence of the horseshoe magnet when a current is flowing. Will it turn clockwise or counterclockwise? Can the coil make a complete turn? If you reverse the coil's connections with the battery after the coil has turned halfway, what **Ⓑ** happens? What causes a continuous rotary motion in a motor? Constructing a motor will help you see the importance of coil connections with the wires of the circuit.

The Polarity Rule

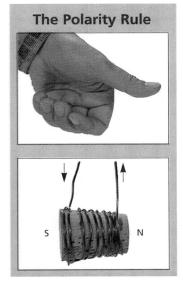

S   N

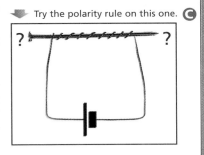

Try the polarity rule on this one. **Ⓒ**

?   ?

Moving Coils—Revolutionary!

This demonstration provides another example of the magnetic effect created by an electric current. Demonstrate the polarity rule for students and be sure they understand it.

Answers to
In-Text Questions and Caption

Ⓐ Students should notice that the electromagnet will turn. This rotational motion can be used to turn another object such as a wheel or pulley.

Ⓑ Using what students know about magnetic properties, they should conclude that the north pole of the coil is repelled by the north pole of the magnet and attracted to the south pole of the magnet. Likewise, the south pole of the coil will be repelled by the south pole of the magnet and attracted to the north pole.

When students close the switch on the circuit, they should observe the coil turning in a clockwise direction.

The coil will make a half-turn so that the north and south poles of the coil (the electromagnet) are closest to the opposite poles of the magnet.

If students reverse the connections with the battery after the coil has turned halfway, they should observe that the coil either reverses its direction or completes the turn. They should conclude that reversing the battery connections changes the polarity in the coil, causing it to turn as it realigns itself with the opposite poles of the magnet.

The electromagnet in a motor exhibits continuous rotary motion because the motor is run by an alternating current that is constantly changing direction. This is similar to changing the polarity of the battery after every half-turn of the coil. Before students make their model motors, challenge them to use what they have learned from the demonstration to predict which part of a motor revolves and what causes this to happen.

Ⓒ Students should determine that the head of the nail is the north pole.

CROSS-DISCIPLINARY FOCUS

Industrial Arts
Have students design and build a lamp. Students should specify what electrical parts will be needed and should brainstorm design ideas. Make sure that they submit a design proposal and have it approved for safety before building their lamp. Display finished lamps around the room.

Meeting Individual Needs

Gifted Learners
Suggest that students design and build an electric motor of their own. When the motors have been built, allow all students to determine which motor is the best—the most powerful, the longest running, the fastest, or a combination of all three factors. For safety reasons, be sure to approve all student designs before the motors are built.

Answers will vary. The best responses will include thoughtful suggestions for improving designs.

Answers to
More on Motors

Possible answers include an electric blender, power saw, lawn mower, drill, garbage disposal, refrigerator, garage-door opener, VCR, electric razor, and cassette deck.

Answers to
Captions

Ⓐ The electrons flow from the negative side of the battery in order through the points A, B, C, and D and back to the positive side of the battery. This determines which side of the wire coil is the south pole and which side is the north pole. The south pole of the coil is attracted to the north pole of the magnet, causing the cork to rotate. After rotating through a half-turn, the pins will switch their contacts with the stationary wires and the flow of electrons through the coil will reverse.

Ⓑ After a half-turn, the poles of the armature are aligned with the opposite poles of the magnet. If the current did not reverse at this point, the coil would stop turning. But now the electrons flow in order through the points D, C, B, and A. This changes the polarity of the coil, causing it to rotate again as it continually realigns itself with the magnet's fixed poles.

A Model Motor ◆

Before building this motor, read the labels and comments in each step.

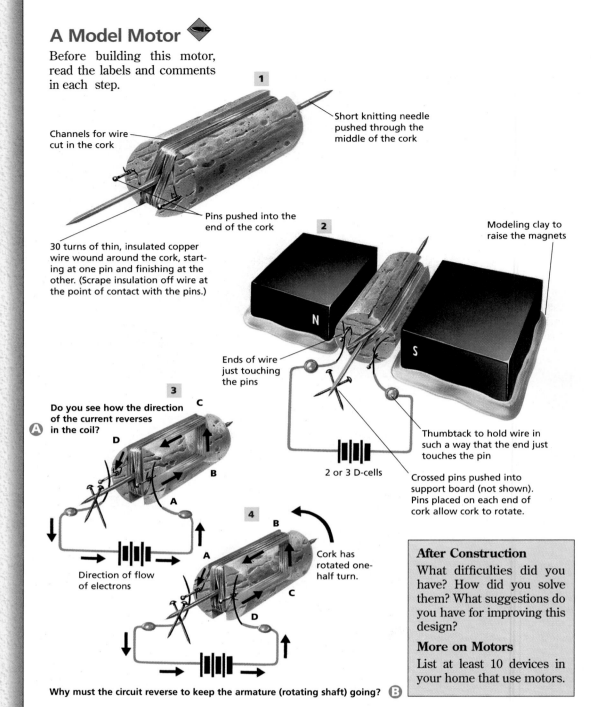

1

Channels for wire cut in the cork

Short knitting needle pushed through the middle of the cork

Pins pushed into the end of the cork

30 turns of thin, insulated copper wire wound around the cork, starting at one pin and finishing at the other. (Scrape insulation off wire at the point of contact with the pins.)

2

Modeling clay to raise the magnets

N

S

Ends of wire just touching the pins

Thumbtack to hold wire in such a way that the end just touches the pin

2 or 3 D-cells

Crossed pins pushed into support board (not shown). Pins placed on each end of cork allow cork to rotate.

3

Do you see how the direction of the current reverses Ⓐ in the coil?

C
D
B
A

Direction of flow of electrons

4

B
A
C
D

Cork has rotated one-half turn.

Why must the circuit reverse to keep the armature (rotating shaft) going? Ⓑ

342

FOLLOW-UP

Reteaching

Bring to class an example of one of the circuits described in this lesson. Disassemble the device for the class, and ask students to identify the important components of the device.

Assessment

Suggest that students have a contest to see who can build the strongest electromagnet. Have a panel of students determine the contest rules, such as the kinds of materials that can be used, how many cells can be in a circuit, and how much time will be allowed for the construc-

tion. The panel should also decide on a way to evaluate the strength of each electromagnet.

SAFETY ALERT Approve all student designs before students construct their electromagnets. A switch should be included in each device so that it can be turned off quickly if it overheats.

Extension

Have students research the motor to discover what kinds of devices use motors and to find out how motor-driven machines affect the everyday lives of people living in different parts of the

world. Students could also research how the invention of the motor allowed new technologies to be developed (such as the automobile) that have had historical significance and have affected human events on a large scale.

Closure

Involve students in a discussion of how a motor works. While doing so, have them trace the flow of electrons through the model motor that they built in this lesson.

After Construction

What difficulties did you have? How did you solve them? What suggestions do you have for improving this design?

More on Motors

List at least 10 devices in your home that use motors.

Look again at the motor tag pictured on page 341. This tag bears information about the motor. We already know what RPM and Hz stand for, but what do the terms *amps* and *volts* mean? These terms relate to certain characteristics of electricity. In this case, these terms and the numbers that accompany them indicate how much electricity the motor needs to run properly.

C How would you go about measuring electric current? Electric currents are often compared to flowing water. We can use this as a model.

How Much Charge?

If you were to talk about amounts of water, you might use words like *liters, cubic centimeters,* or *milliliters.* In describing amounts of electricity, you might say, "How many

electrons?" However, electrons are far too small to be easily used as a unit of measurement. Charge is measured in *coulombs* (pronounced KOO lahms). One coulomb (C) is the charge carried by 6.24 quintillion (billion billion) electrons.

The more charges you have, the more current that can flow. How would you measure the amount of electricity available from a given source? Let's use the water analogy. Which of the containers shown below do you think contains the greater "charge"? How did you arrive at this decision? **D**

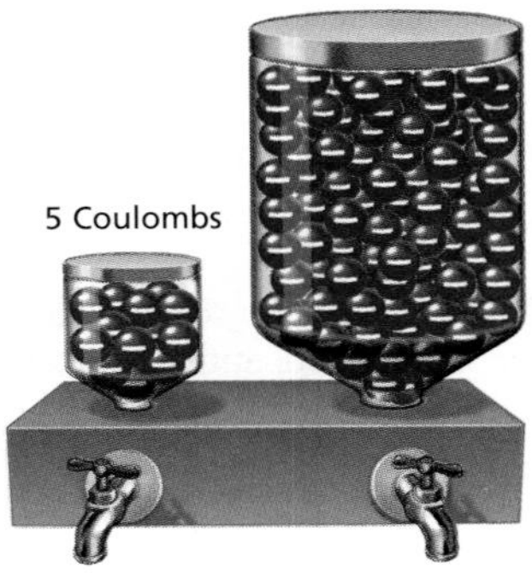

How Much Current?

Look at the diagram at right. Which hose has a greater rate of flow? That is, which hose would carry a greater amount of water in a given amount of time? What is your clue? **E**

The electrical equivalent of flow rate is *current.* Current is measured in *amperes* (A), or *amps* for short. A current of 1 ampere is the rate of flow at which 1 coulomb of charge

passes a given point in 1 second. The more coulombs passing a given point in 1 second, the greater the rate of electrical flow, or amperage. Which hose has the greater "amperage"? What is your clue? **F**

How Much Energy?

Have you ever tried to hold your thumb over the mouth of a garden hose with the water turned on full blast? Most likely, the water pushed your thumb out of the way and you got wet. Why was it so hard to close off the hose with your thumb? The answer, of course, is pressure.

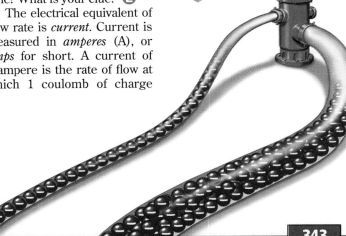

Current is like the rate of flow through a hose.

50 Coulombs

5 Coulombs

⬆ Charge is like a quantity of water.

343

LESSON 5 ORGANIZER

Time Required
2 class periods

Theme Connection
Systems

Process Skills
measuring, classifying

New Terms
Coulomb—unit of electrical charge equivalent to the charge of 6.24 quintillion (billion billion) electrons
Current—the rate of flow of electrical charge, measured in amperes (A)

Voltage—a measure of the energy, or "pressure," given to the charge flowing in a circuit, measured in volts (V)
Watt—unit of power, or rate of energy transport. One watt (W) equals one joule per second.

Materials (per student group)
none

Teaching Resources
Activity Worksheet, p. 52
Transparency 56

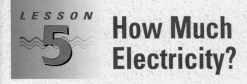
FOCUS

Getting Started

Display a light bulb. Ask students to respond by a show of hands if they have ever read what is printed on the top of a light bulb. Ask them if they can recall the information that was identified there. *(Watts, amps, and volts)* On the chalkboard, write the information that appears on the top of the bulb. Invite students to speculate about what each of the quantities refers to. Point out that in this lesson they will learn more about what these words mean and how they are related.

Main Ideas
1. Current, voltage, and electrical charge are measurable quantities of an electric current and are used to determine electrical energy.
2. The relationship among current, voltage, and electrical charge can be expressed in mathematical terms.

TEACHING STRATEGIES

Answers to
In-Text Questions

C Encourage students' speculations at this point to pique their interest. They will learn how electricity is measured as they continue through the lesson.

D You would measure the amount of electricity available from a given source by "counting" the number of electrons available in the source. One crude way to accomplish this would be to multiply the current provided by a source by the time it takes to run down, assuming that the current is constant. The larger container has the greater "charge" because it contains more electrons.

E The larger hose would carry more water in a given amount of time than the smaller hose, as suggested by the greater number of particles per row.

Answers to In-Text Questions continued ▶

Answers to
In-Text Questions, continued

F The larger hose would have a greater amperage than the smaller hose. The larger number of electrons depicted in the larger hose helps to make this point. To reinforce this analogy, point out that the ampere is used to measure the flow of electricity in much the same way that liters per second is used to measure the flow of water.

Putting It All Together

You may wish to review this analogy step by step with students. Point out that each numbered step in the first part corresponds to each numbered step in the second part. If students have difficulty understanding this analogy, ask them to draw a diagram of the electric circuit and compare it with the diagram of the water circuit.

Answers to
In-Text Questions

A Students should conclude that voltage and current are directly proportional. As one increases, so does the other.

B The hose connected to the high-speed pump has a higher voltage than the hose connected to the low-speed pump. The clue in the illustration is the fact that the electrons in the hose on the right are more energetic than those in the hose on the left.

Homework

The Activity Worksheet that accompanies the material on page 343 makes an excellent homework assignment (Teaching Resources, page 52). If you choose to use this worksheet in class, Transparency 56 is available to accompany it.

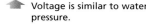

High-speed pump

Low-speed pump

▲ Voltage is similar to water pressure.

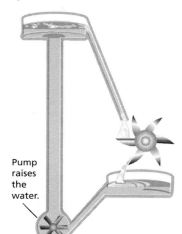

▼ A "water circuit"

Pump raises the water.

Pressure is given to the water by a pump or by a reservoir in an elevated position. The greater the force given to the water, the greater the pressure. Electrical **voltage** is somewhat like water pressure. Voltage is measured in *volts* (V). Voltage is a measure of the energy given to the charge flowing in a circuit. The greater the voltage, the greater the force, or "pressure," that drives the charge through a circuit. How do you think voltage and current are related? **A**

Which hose in the illustration at left has the higher "voltage"? What is your clue? **B**

Putting It All Together

Charge, current, voltage—how do they all work together? So far, our water model has worked well. The flow of electricity behaves remarkably like that of water. However, we have to complete the analogy by considering that electricity flows in circuits. Look at the diagram below, which shows a "water circuit." Let's analyze how it works. (A mathematical formula is provided for key steps.)

1. The pump raises a quantity of water to the top. In doing so, it boosts the water's potential energy level (J/L).

2. As the water flows downward, its potential energy is converted to kinetic energy in the form of moving water—a "water current" (L/s).

3. At the water wheel, work is done as most of the energy in the falling water is converted to mechanical energy.

4. The water returns to the pump. The pump once again puts energy into the water, beginning the cycle again.

Let's compare our water circuit to an electric circuit.

1. A cell, battery, or generator provides a voltage, which boosts the potential energy (J/C or V) of the negative charges, or electrons, to a high level.

2. Closing the circuit allows the potential energy of the electrons to be converted to kinetic energy. The moving electrons create a current (C/s or A).

3. As the current flows through circuits, the electrons do work. The kinetic energy of the electrons is changed into other forms of energy.

4. The electrons, their energy expended, return to the electrical source, where their energy is boosted and the cycle begins again.

Theme Connection

Systems

Focus question: How are the electrical and plumbing systems in a home or apartment building similar? *(Both the flow of water and the flow of electricity can be controlled as they move through a home or apartment building. The resistance to the flow of water can be increased by narrowing the pipes or by the presence of debris or ice. Resistance to the flow of electricity can be introduced by making the connecting wire longer and thinner.)*

Using Electrical Units

Here is a summary of all you need to know to make practical use of electrical units.

- The quantity of electric charge is measured in coulombs.
- The rate of electric flow, or current, is measured in amperes. One ampere of current is the same as one coulomb of charge passing a given point in one second.
- Voltage is the electric energy given to a unit of charge flowing in a circuit. One volt is the same as one joule of electric energy given to one coulomb of charge.
- Electricity, like other forms of energy, is measured in joules.
- Power, measured in watts, is the rate at which electric energy is used. One watt is one joule of electric energy used in one second.

You will draw on these definitions to complete the activities that follow.

Understanding Amperes, Volts, and Coulombs

1. Imagine using a toaster. Suppose that the toaster draws 5 A of current and that it takes 3 minutes to toast a slice of bread.

   a. How many coulombs flow through the toaster in 1 second? during the time it takes to toast the bread?

   b. To calculate the number of coulombs, multiply the number of ___?___ by the number of ___?___. This can be represented by the following word equation:
   coulombs = amperes × ___?___.

2. The toaster is connected to a house's 110 V electrical source.

   a. The house's electrical source supplies ___?___ J to 1 C of charge, and ___?___ J to the 900 C required to toast the bread.

   b. The calculation of the number of joules of energy used may be represented by the following word equation:
   joules = ___?___ × ___?___.

Understanding Watts

The rate at which appliances use energy is measured in watts. This is also known as the power of the appliance.

3. If something uses 1 W, it uses ___?___ J of electric energy every second.

4. An 800 W iron uses ___?___ J of electric energy every second. If the iron is used for 15 minutes, it uses ___?___ J of electric energy.

One coulomb of charge is like a bag containing 6.24 billion billion electrons.

10 C of electrons passing a point in 1 s equals 10 A of current.

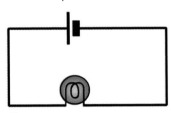
1.5 J of electric potential energy given to 1 C of charge by the cell

Reddy Kilowatt says if you want to read more about amperes, volts, and electric currents, turn to pages S86–S88 of the SourceBook.

Reddy Kilowatt®, The Reddy Corporation, International, 1996

345

Answers to
Understanding Watts, continued

5. seconds; seconds

6. If joules = volts × coulombs, and joules = watts × seconds, then watts × seconds = volts × coulombs. Also, because coulombs = amps × seconds, watts × seconds = volts × amps × seconds, or watts = volts × amps.

7. 12; 1320 W ÷ 110 V = 12 A
   watt—unit used to measure the rate of energy use
   volt—unit used to measure the amount of pressure, or potential energy, in a current
   ampere—unit used to measure the rate of current flow
   In other words, an electric heater that uses 1320 J of energy per second requires an energy source that applies a pressure of 110 V at a rate of 12 A (coulombs per second).

Answers to
Power by the Hour

8. 2

9. 10

10. 36,000,000
    J = W × sec.
    Energy = 2000 W × (60 sec./min. × 60 min./hr. × 5 hr.)
    Energy = 2000 W × 18,000 sec.
    Energy = 36,000,000 J

11. 3,600,000

12. It is much easier to calculate energy consumption in kilowatt-hours because the numbers are smaller.

Homework

Have students develop five circuit problems (with answers) to bring to class. In class, have students trade papers with a partner to solve one another's problems and discuss solutions.

5. To calculate the number of joules of energy used, multiply the number of watts by the number of ___?___. This can be represented by the following word equation:
$$\text{joules} = \text{watts} \times \underline{\quad?\quad}.$$

6. In questions 2b and 5, you found that total electrical energy, measured in joules, is the product of two different measures of electrical energy. By setting these two different measures equal to one another, see whether you can show that watts = volts × amps.

7. A 1320 W electric heater is plugged into a 110 V household circuit. A current of ___?___ A flows through the heater. Explain in your own words what each of the numbered quantities in this question means.

Power by the Hour

As you saw earlier, multiplying power by time gives you the total amount of energy used. For example, watts multiplied by seconds equals joules. Utility companies use another unit to measure the energy used by consumers—the kilowatt-hour (kWh). It is obtained by multiplying the power in kilowatts by the time of energy usage in hours.

Imagine that you ran a 2000 W electrical heater for 5 hours.

8. A kilowatt is equal to 1000 W. The power of the heater is ___?___ kW.

9. In 5 hours an electric heater uses ___?___ kWh.

10. The heater used ___?___ J of energy. (Hint: First figure out how many seconds are in 5 hours.)

11. In questions 9 and 10 you found the energy used by the heater in both joules and kilowatt-hours. Therefore, 1 kWh = ___?___ J.

12. Why do you think kilowatt-hours, instead of joules, are commonly used to measure energy usage?

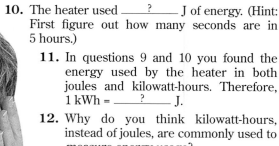

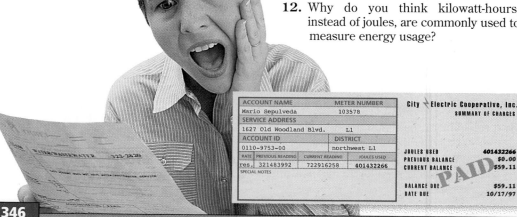

Volts, Amperes, and the Human Body

You probably don't worry about electric shock when you use dry cells. Why not? First of all, dry cells cannot produce enough current to be very harmful. Second, the resistance of the skin is great enough to prevent the dry cell's small current from flowing through it. What happens if the resistance is decreased? Look at the graph to the left. What happens when the skin gets wet? What implications does this have for working with electricity? How high should the bar for the resistance of dry skin be if the bar for wet skin is 5 cm high? Ⓐ

Now look at the graph at right to see what amount of current is safe for the body to receive. Are you surprised at the small amount of current that our bodies are able to receive internally without causing havoc? The resistance of our skin is very important when you consider the amount of current flowing through most household electric devices.

What about voltage? The voltage of a dry cell is 1.5 V, that of a car battery is 12 V, and the voltage of household current is generally 110 V. What is safe? Generally, any voltage above 30 V can overcome the skin's resistance. If the supply of current is large, higher voltages can force enough current through the skin to cause shock or more serious damage. What kinds of procedures and protection would you suggest for people who work with high-voltage circuits? Ⓑ

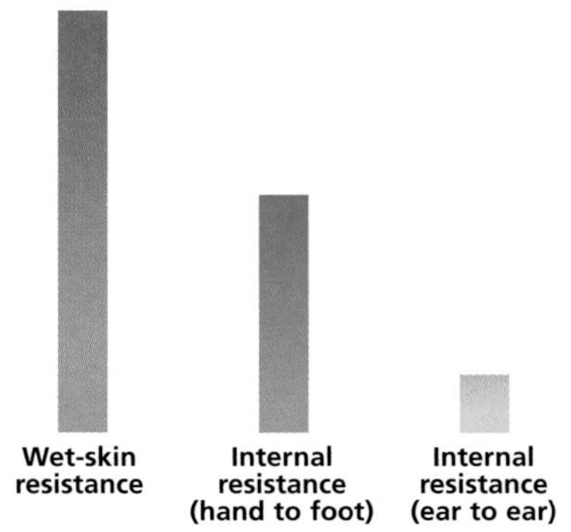

1 A

Heart convulsions

50 mA

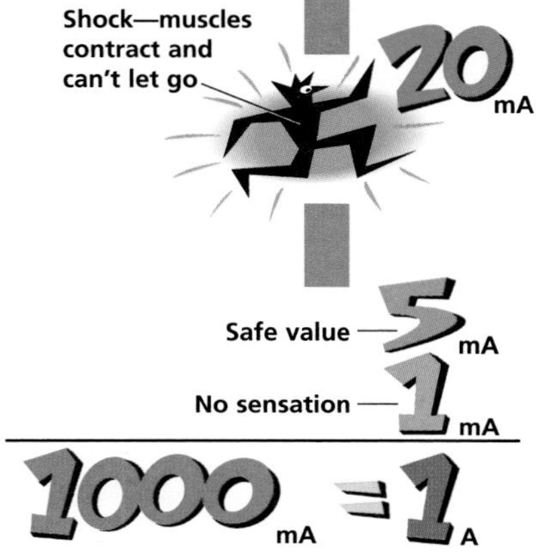

Shock—muscles contract and can't let go

20 mA

Safe value — 5 mA

No sensation — 1 mA

1000 mA = 1 A

Dry-skin resistance | Wet-skin resistance | Internal resistance (hand to foot) | Internal resistance (ear to ear)

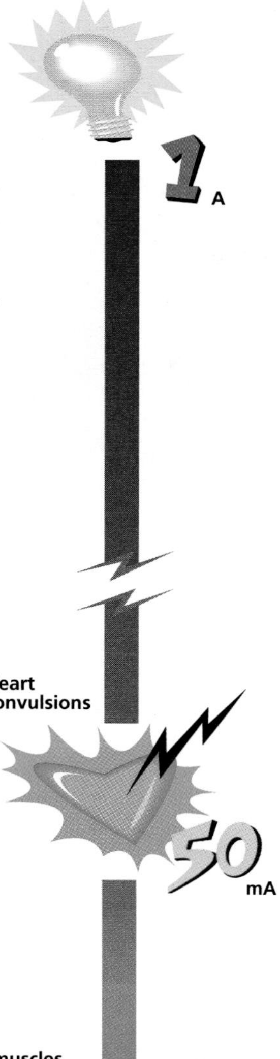

347

Answers to In-Text Questions

Ⓐ If the resistance is decreased for a given amount of voltage, the current will increase. The graph shows that the skin's resistance decreases dramatically when wet. This reinforces that rule that one should not work with electricity in the presence of water. The wet-skin bar should be about 1.5 cm high.

Ⓑ First point out that the best protection when working with high-voltage circuits is to make absolutely sure that the circuit is open and that all voltages are zero. All voltage sources should be turned off and unplugged. Never touch a live electrical device.

When working with high-voltage circuits that cannot be disconnected, one should wear nonconducting gloves and shoes to increase resistance and should stay away from water (a dry object has greater resistance than a wet object). Additionally, when touching any electrical device, never grip it with your hand. In case of shock, the muscles of the hand might contract, preventing you from releasing the object. Instead, if you must touch an electrical device, always touch it with the back of the hand.

FOLLOW-UP

Reteaching

Suggest that students make charts to compare a water circuit and an electric circuit. The charts should compare the following items and quantities: what flows (water and electrons); the measurement of quantity (liters and coulombs); the measurement of rate of flow (liters per second and amperes); the measurement of potential energy (joules per liter and volts); and the measurement of the rate of energy use or power (watts and watts).

Assessment

Have students find the solution to the following problem: Two 1100 W projectors are operated 4 hours a day at the local movie theater.

a. Which costs more to operate, the projectors or the 800 W sign that comes on automatically at dusk and stays on for 10 hours? Both are operated from 110 V electrical outlets. *(The two projectors use 8.8 kWh and the sign uses 8 kWh, so the projectors cost more.)*

b. What is the current in each? *(The current flowing in the projector is 10 A, and the current flowing in the sign is 7.3 A.)*

Extension

Have students do some research on safety procedures to follow when working with electricity and electrical equipment. Then have them make posters to illustrate what they learned.

Closure

Suggest that students ask their parents or other adults if they can use last year's monthly electric bills to make a bar or line graph. They should compare the amount of electricity in kilowatt-hours that was used each month. Ask: During which months was the most energy consumed?

Answers to *Challenge Your Thinking*

1. Sample answer: When the button is pressed, current flows through the circuit. The current flowing through the wire causes the bar inside the wire coil to become magnetized. The electromagnet pulls down the arm attached to a pivot. This removes the lock from the door.

2. Sample answer: The battery supplies energy to the electrons on the negative electrode. As the current of electrons moves through the circuit, it feels resistance first from the wire and then from the resistor, and some of its electrical energy is changed into heat energy. As the electrons flow through the highly resistant filament of the lamp, more of their energy is lost in the form of heat and light energy. Most of the remaining energy in the current is changed into the mechanical energy of the motor. Depleted of their energy, the electrons return to the battery, where they are recharged with more energy to begin the cycle again.

   Starting at the upper right and going clockwise, possible energy labels are heat, heat, heat or light, and mechanical.

3. A possible arrangement is as follows:

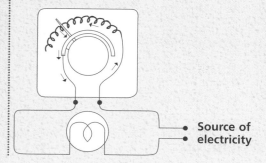
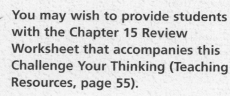

   • Source of electricity

⭐ You may wish to provide students with the Chapter 15 Review Worksheet that accompanies this Challenge Your Thinking (Teaching Resources, page 55).

CHALLENGE YOUR THINKING

1. Safe Circuit

Many apartment buildings have security doors that can be opened from each apartment. The diagram below shows the circuitry of one such security door. Explain how it works.

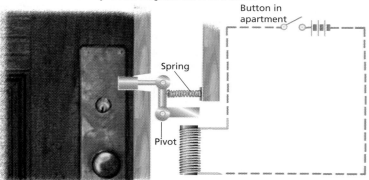

Button in apartment
Spring
Pivot

2. Go With the Flow

This drawing from a sixth-grade science book illustrates a lesson about electricity. Write a paragraph to go with this lesson. What words should be used to fill in the blanks?

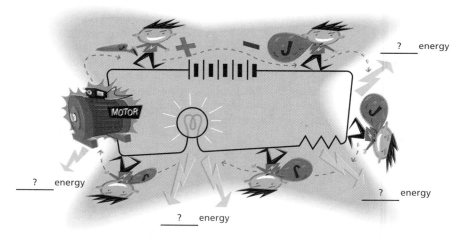

? energy

MOTOR

? energy

? energy

? energy

3. The Inside Story

A common type of switch is the *dimmer switch*. Turn the knob clockwise, and the light becomes brighter. Turn the knob counterclockwise, and the light becomes dimmer. Draw a sketch showing the circuit that may be inside.

348

Theme Connection

Systems

Focus question: How is current through a circuit similar to blood flow through the human body? *(The heart is a pump that pushes blood cells through the arteries. The chemical cell pushes electrons through the circuit. The flow of blood is like a direct current in a parallel circuit—it flows in one direction* along several different paths through the body. Valves in the heart operate like switches, controlling the flow of blood through the veins.) *You may wish to have students complete the Theme Worksheet that accompanies this Theme Connection as a homework activity (Teaching Resources, page 53).*

4. You Be the Teacher

Ms. Alvarado asked her class to design a circuit containing two switches (S_1 and S_2), two light bulbs (L_1 and L_2), and one dry cell. She asked them to arrange the circuit so that it does the following:

a. If only S_1 is closed, L_1 will light.

b. If only S_2 is closed, nothing will happen.

c. If S_1 and S_2 are closed, both lamps will light.

Their work is shown below.

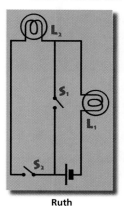

Ruth

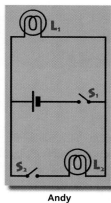

Andy

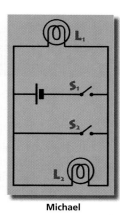

Michael

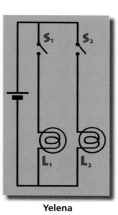

Yelena

Play the role of Ms. Alvarado, and grade the students' work. Indicate how you know whether each design is right or wrong.

5. The Ol' Double Switcheroo

In the circuit shown at right, the upper switch moves from contact A to contact C and the lower switch moves from contact B to contact D. Suggest a function for this circuit.

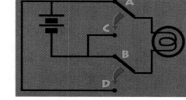
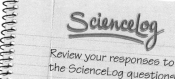

6. Safe at Home

Fuses are designed to protect household circuits against damage due to electrical overload. Fuses are made of metal alloys that have low melting points. How do you think fuses work? (Hint: Think about the effects of resistance.)

ScienceLog

Review your responses to the ScienceLog questions on page 326. Then revise your original ideas so that they reflect what you've learned.

Homework

The Activity Worksheet on page 54 of the Unit 5 Teaching Resources booklet makes an excellent homework assignment once students have completed Chapter 15. If you choose to use this worksheet in class, Transparency 57 is available to accompany it.

349

Answers to Challenge Your Thinking, continued

4. Before answering this question, you may wish to review the concept of a *short circuit* with the class. See page 333 for a discussion of short circuits.

   Only Andy's circuit met all three requirements. In Ruth's circuit, only (a) is met. Her circuit fails requirement (b) because L_1 and L_2 would light, and it fails requirement (c) because only L_1 would light. In Michael's circuit, requirement (b) is met but his circuit fails to meet requirement (a) because both L_1 and L_2 would light, and it fails requirement (c) because neither L_1 nor L_2 would light. Requirements (a) and (c) are met in Yelena's circuit, but her circuit fails requirement (b) because L_2 will light.

5. Answers may vary. One use for this circuit would be to allow two different switches to control one light.

6. Sample answer: Fuses are made of materials that offer high resistance to the current passing through them. Therefore, some of the electrical energy is converted into heat energy. When this buildup of heat energy becomes too great, the fuse melts, breaking the circuit.

ScienceLog

The following are sample revised answers:

1. In order for current to flow in a circuit, electrons must have a complete loop around which to travel. Opening or closing a switch can break or connect the loop, causing the electrons to start or stop flowing.

2. The only difference between the two setups is the thickness of the wire used in the circuits. The setup on the right would offer greater resistance because the wire used is thinner.

3. The flow of electricity in a circuit is affected by a number of variables, including the voltage and the type and shape of the material through which the current is flowing.

The Big Ideas

The following is a sample unit summary:

Electricity is the movement of electrons. Electricity is produced by a number of methods. For example, when two different materials are rubbed together, electrons are transferred from one material to another. Another method is to move a coil of conducting material through a magnetic field. Chemical cells, solar cells, piezoelectric crystals, and thermocouples also produce electricity. (1)

A cell is a source of electrical charge that consists of two electrodes made of two different materials in a solution known as an electrolyte. Due to a chemical reaction, one electrode loses electrons to the electrolyte, while the other electrode gains electrons. One electrode thus becomes positively charged and the other becomes negatively charged. Electrons will flow through a wire connecting the two electrodes, creating a current. A battery consists of two or more cells connected together. (2, 3)

There are two different types of current: direct current and alternating current. Direct current flows in only one direction. Direct current is produced by cells and batteries. In an alternating current, the current flow changes direction. Alternating current is produced when a wire loop rotates within a magnetic field. As the wire rotates, the current changes direction. (4)

Factors that affect the size of currents in circuits include how many cells are placed in the circuit and how much resistance there is in the circuit. (5)

Series circuits are those in which there is only one path for the current to follow. In parallel circuits, there is more than one path that the current can take. (6)

Electromagnets are temporary magnets created by the flow of electricity through a wire. Their strength can be increased by increasing the number of turns of the coil or by increasing the current passing through the coil. (7)

The flow of water is similar to the flow of electricity in that electrons flow through a substance in a way similar to that in which water flows through a

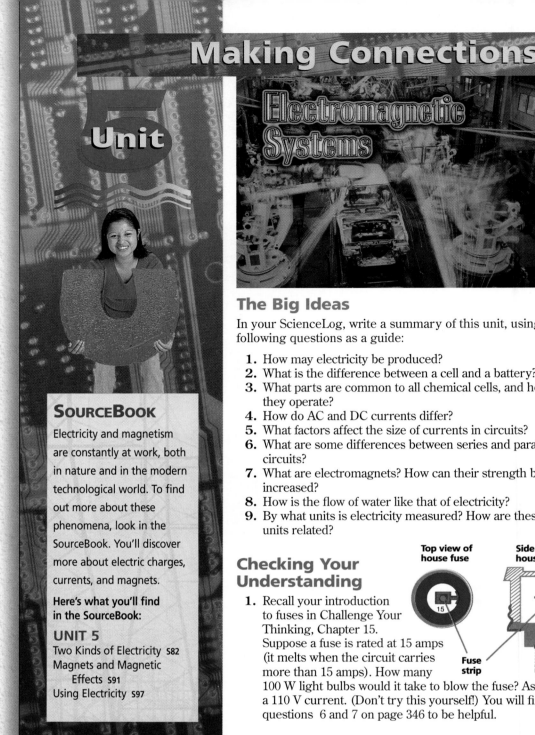

Unit 5

Electromagnetic Systems

The Big Ideas

In your ScienceLog, write a summary of this unit, using the following questions as a guide:

1. How may electricity be produced?
2. What is the difference between a cell and a battery?
3. What parts are common to all chemical cells, and how do they operate?
4. How do AC and DC currents differ?
5. What factors affect the size of currents in circuits?
6. What are some differences between series and parallel circuits?
7. What are electromagnets? How can their strength be increased?
8. How is the flow of water like that of electricity?
9. By what units is electricity measured? How are these units related?

Checking Your Understanding

Top view of house fuse

Side view of house fuse

Fuse strip

1. Recall your introduction to fuses in Challenge Your Thinking, Chapter 15. Suppose a fuse is rated at 15 amps (it melts when the circuit carries more than 15 amps). How many 100 W light bulbs would it take to blow the fuse? Assume a 110 V current. (Don't try this yourself!) You will find questions 6 and 7 on page 346 to be helpful.

SOURCEBOOK

Electricity and magnetism are constantly at work, both in nature and in the modern technological world. To find out more about these phenomena, look in the SourceBook. You'll discover more about electric charges, currents, and magnets.

Here's what you'll find in the SourceBook:

UNIT 5
Two Kinds of Electricity S82
Magnets and Magnetic Effects S91
Using Electricity S97

350

pipe. Water flow, like current flow, is affected by pressure and resistance. In both water and electricity, potential energy can be stored and then converted into kinetic energy. (8)

Electricity is measured by a combination of units. Voltage is a measurement of the amount of energy available to move electrons. It is expressed in joules per coulomb, or volts. Current, measured in amperes, is the amount of charge flowing past a given point on a

circuit per second. One coulomb of charge passing a given point in one second equals one ampere. The power of an electric current is measured in watts. One watt equals one joule per second, or one ampere times one volt. The electrical equivalent of pressure is voltage; the electrical equivalent of flow rate is current; the electrical equivalent of power is watts. (9)

2. At right is a diagram of a circuit breaker, a device that mechanically performs the same task as a fuse, breaking the circuit when too much current flows through it. How does this device work?

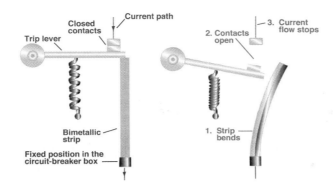

Trip lever
Closed contacts
Current path
Bimetallic strip
Fixed position in the circuit-breaker box

2. Contacts open
3. Current flow stops
1. Strip bends

A. Normal current

B. Overload current

3. Electricity is often compared to flowing water. Read the following examples and decide whether each suggests high or low amperage, high or low voltage, or any combination of these.
  a. the Mississippi River
  b. Niagara Falls
  c. the blast from the spray-gun nozzle at a do-it-yourself carwash
  d. a dripping faucet

4. The diagrams below show two different types of microphones. Use the diagrams and your knowledge of the principles of electricity to explain how each of these microphones works.

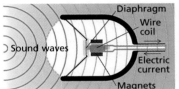
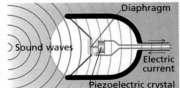

Diaphragm
Wire coil
Sound waves
Electric current
Magnets

Diaphragm
Sound waves
Electric current
Piezoelectric crystal

5. (concept map) Make a concept map using the following terms: volts, current, electricity, amperes, voltage, coulombs, and charge.

 You may wish to provide students with the Unit 5 Review Worksheet that accompanies this Making Connections (Teaching Resources, page 63).

Homework

The Activity Worksheet on page 61 of the Unit 5 Teaching Resources booklet makes an excellent homework assignment.

Answers to
Checking Your Understanding,
pages 350–351

1. Watts are equal to volts times amperes.
110 V × 15 A = 1650 W
1650 W ÷ 100 W/light bulb = 16.5 light bulbs
Therefore, 17 light bulbs would be needed.

2. Sample answer: As the current flows through the circuit, the bimetallic strip offers some resistance to the current flow, causing the strip to heat up due to friction. Because the two heated metals expand at different rates, the strip bends. If the current is great enough, the strip bends enough so that it no longer holds the contacts together, and the circuit is broken.

3. a. High amperage, low voltage
  b. High amperage, high voltage
  c. Low amperage, high voltage
  d. Low amperage, low voltage

4. In the microphone on the left, the vibrating diaphragm causes a wire coil to move across a magnetic field. As the wire cuts through the field, a current is produced in the wire.
  In the microphone on the right, sound waves hit the diaphragm, causing it to vibrate. This vibration squeezes the crystals, which produces an electric current. This current corresponds to the original sound.

5. See sample concept map below.

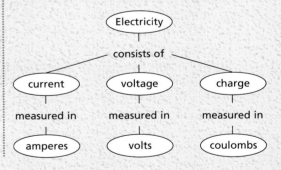

Electricity
consists of
current — voltage — charge
measured in — measured in — measured in
amperes — volts — coulombs

Science Snapshot

Background

All matter can be classified as a conductor, semiconductor, or insulator, according to what it is made up of. Point out to students that a typical conductor has a billion times less resistance than a typical semiconductor, and a typical semiconductor has 100 billion times less resistance than an insulator.

Electrical resistance is a measure of how much a substance opposes the flow of electricity. Resistance reduces the efficiency with which a material conducts electricity by causing some of the electrical energy to change into heat. If electricity is the flow of electrons, then resistance can be thought of as the amount of friction experienced by the moving electrons. Just as friction produces heat when you rub your hands together, resistance produces heat when electrons collide with the fixed particles of the conductor. In superconduction, a material loses all of its electrical resistance.

In addition to developing higher-temperature superconductors, scientists continue to investigate the possible uses of this technology. For example, superconducting switching devices could be used to control computer circuits extremely quickly and without generating heat. Superconductors could make electric motors and generators more compact and efficient. Superconducting power lines could save a lot of energy by carrying current over long distances without losing energy to heat.

Some of the most promising applications of superconductors involve the production of powerful electromagnets. With coils made of superconducting material, electromagnets can produce fields with 200,000 times the strength of Earth's magnetic field. In addition to being used in maglev trains, electromagnets made with superconducting materials could be used in particle accelerators, in devices that measure magnetic fields for medical diagnosis, and in supercooled refrigerators.

Science Snapshot
Paul Chu (1941–)

Imagine a puck whizzing around on an air-hockey table. The puck glides effortlessly on little jets of air. Now imagine a train moving just like the hockey puck, zooming along a frictionless track. Paul Chu and other physicists are investigating ways to take friction out of the transportation picture.

The Future in a Deep Freeze

Chu is a physicist who works with *superconductors*, materials that have no electrical resistance. Electric currents traveling through a superconductor can speed along forever. This sounds perfect, but there's a catch. Superconductors have to be cooled to extremely low temperatures in order to work. That makes them too expensive to be manufactured on a large scale. Chu has made a discovery that may help solve this problem.

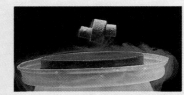

▲ A magnet levitating over a superconducting material. One day scientists might be able to levitate trains over superconducting tracks.

Cheap as a Can of Cola

Chu was born in China and grew up in Taiwan. He came to the United States and earned a doctorate from the University of California at San Diego. He then went to work at the University of Houston, where he and his colleagues began to test ceramics for superconductivity. In the mid-1980s, they finally created a ceramic material, called compound 1-2-3, that superconducts at a then record-high temperature of –175°C.

Having to cool compound 1-2-3 to only –175°C means that scientists can use liquid nitrogen, a resource that is as cheap as a can of cola, as a coolant.

Chu, the director of the Texas Center for Superconductivity at the University of Houston, has discovered superconductors at temperatures as high as –109°C. Yet he continues to work toward the ultimate goal: finding a material that superconducts at room temperature. If Chu were to find such a material, levitated trains and electric cars could soon become common modes of transportation.

The Path of Least Resistance

Model the passage of an electric current through a normal conductor by randomly sticking 10–15 pushpins through the bottom of a shoe box. The pushpins represent the molecules of the conductor. Now lift up one end of the box and roll a few marbles—which represent the electric current—down the slope. Resistance occurs when the marbles hit the pushpins and lose energy. Now model a superconductor by moving the pushpins into two rows running down the slope, and again roll the marbles down the slope. Why is this situation more efficient in terms of energy loss?

▶ Japanese researchers have developed working models of superconducting levitated trains called *maglev* trains. These models have reached speeds over 480 km/h!

Extension

Have interested students extend their understanding of superconductors by doing additional reading. You may wish to have students summarize one application of superconductor technology for the class or have them illustrate concepts related to electrical resistance and conduction. Encourage students to use diagrams or photographs to illustrate their presentations. (For instance, students who study maglev, or magnetic levitation, trains should find that superconducting magnets may be used to suspend these trains about 3–8 cm above the tracks. Because these trains run on electricity, the trains themselves produce little air pollution.)

Answers to
The Path of Least Resistance

Students should conclude that their model superconductor is more efficient than their conductor because the marbles experience fewer collisions and thus lose less energy to resistance.

A Cleaner Ride

At least half the air pollution in this country comes from automobile exhaust. Many state governments are now creating new regulations to reduce the amount of air pollution generated by cars. For example, by 2003, 10 percent of all new cars sold in California must produce no exhaust gases. Government officials in other states are looking at similar measures. But how can automobile companies produce cars that don't pollute? One answer may be electric cars.

Batteries Included

Electric cars are powered by electric motors that get their energy from a large bank of on-board batteries. This means that electric cars are quiet and exhaust-free. In addition, electric cars are more dependable, more durable, and cheaper to operate than gasoline-powered cars.

The batteries in electric cars are similar to the batteries in gasoline-powered cars, but are much bigger and much more expensive. And, like all batteries, electric-car batteries go dead. Currently, the batteries must be recharged every 190 km (120 mi.) or so. This process can take 3 hours or more. The batteries also have a limited life span and must be replaced every 40,000 km (25,000 mi.).

▼ Some electric cars use electric power in combination with solar power. The solar cells on top of the car help recharge the batteries.

Using Energy Wisely

At present, batteries provide very little energy for their weight. Therefore, electric-car manufacturers must use that limited energy wisely. One electric-car model has a number of special features designed to help make the most of the batteries' energy. For example, the car uses a special braking system called a *regenerative braking system*. When the accelerator is released, the car's motor becomes a generator. The turning of the car's wheels drives the generator, which both produces energy and helps slow down the car. The electrical energy produced by the generator helps recharge the batteries.

Dreams and Drawbacks

Widespread use of battery-operated cars could result in many positive changes. For example, city skies would be much cleaner, and traffic noise would be greatly reduced. It sounds exciting, doesn't it? Electric-car manufacturers think so too! But electric cars are far from perfect. Currently, electric cars are expensive to produce. Also, the electricity used to power these cars usually comes from fossil fuels, which also pollute the environment. In addition, because they are large and must be replaced frequently, electric-car batteries will be costly and will pose disposal problems. Finally, many of the materials used in existing batteries are harmful to the environment. However, automobile companies are hopeful that new batteries will be more powerful, smaller, and less toxic.

Think About It

Believe it or not, at one time in the United States, more electric cars were sold than any other type of car. Do a little research about this time in the late 1800s. Write a short paper on what you find out.

Background

The electric car has been around for a while. In fact, the first electric cars were developed about the same time as gasoline cars. Eventually, however, the electric car lost its popularity to the gasoline car.

Interest in electric cars reemerged in the 1970s, mostly due to the demands of gasoline-powered cars on fossil-fuel supplies and to the worsening effects of air pollution. Today car manufacturers and designers all over the world are creating their own versions of the electric car. Some of these cars run solely on electricity, while others use electric power in combination with some other source of energy, such as gasoline, alcohol, or solar power.

ENVIRONMENTAL FOCUS

Battery-driven devices, like all technologies, have some negative effects on the environment. Have students research another energy source that could be used to power vehicles. Then lead a discussion with the class about the advantages and disadvantages associated with each energy source. Energy sources to consider include ethanol, steam, and the sun.

Discussion Questions

- Are electric cars the best solution to the air pollution problem? Why or why not? What are some alternative solutions? *(Answers will vary. Accept any opinion as long as it is supported by evidence. Alternatives to electric cars include walking, solar-driven cars, and bicycles.)*
- Would you want to drive an electric car? Do you think the benefits outweigh the problems associated with them? *(Answers will vary. Accept all thoughtful responses.)*

Think About It

Students should discuss why electric cars were popular and why they were eventually replaced by gasoline-powered automobiles. *(People liked electric cars because they were easy to operate, were quiet, and produced no fumes. However, the cars could travel only about 34 km/h and had to be recharged every 80 km. Gasoline cars were more powerful and cheaper to run.)* You might also have them research the role of transportation in the westward expansion and industrialization of the nation.

Background

MRI has been used as a valuable diagnostic tool since the early 1980s. An MRI scan is usually done as an outpatient procedure. During the scan, the patient must lie very still inside a huge, hollow tube for approximately half an hour while the scan is taking place. The patient is exposed to an extremely powerful magnetic field. Although the magnetic field is extremely powerful, so far no research has shown MRI to be dangerous to human health.

One of the advantages of MRI is that it can provide clear images of parts of the body that are surrounded by dense bone tissue. As a result, MRI is particularly valuable for studying the brain and spinal cord. MRI is also useful for examining joints and soft tissues, as well as the heart, blood vessels, eyes, and ears.

Discussion Questions

1. What does MRI stand for? *(MRI stands for magnetic resonance imaging.)*
2. What does it do? *(It enables doctors to view the internal structure of the human body at work.)*
3. How is using the MRI scanner like producing an animated cartoon? *(An animated cartoon is a series of drawings linked together to create the effect of movement. Similarly, multiple MRI scans can be linked together and viewed in succession to produce the illusion of motion.)*

Answers to
Picture This

Accept all reasonable responses. Students may realize that the different colors represent tissues of different density and chemical composition.

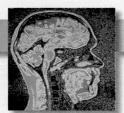

HEALTH WATCH
Magnets in Medicine

Think about what it would be like to peer inside the human body to quickly locate a tumor, find tiny blockages in blood vessels, or even identify damage to the brain. Medical technology known as magnetic resonance imaging (MRI) allows doctors to see these things and more. MRI is a quick, painless, and safe way for doctors to look inside the body to diagnose a variety of ailments.

Magnetic Images

Like X rays, MRI creates pictures of a person's internal organs and skeleton. But MRI produces clearer pictures, and it does not expose the body to potentially harmful radiation as X rays do. Instead, MRI uses powerful electromagnets and radio waves to create the images.

The patient is placed in a large, tunnel-shaped machine that contains four electromagnetic coils and a device that transmits radio signals. When electric current flows through the coils, a powerful magnetic field is created around the patient. Because the human body is made mostly of water, there are a lot of hydrogen atoms in the body. The magnetic field in the machine causes the hydrogen atoms to line up. At the same time, a brief radio signal is transmitted, knocking the atoms out of alignment. After

the signal ends, a computer measures the activity of the hydrogen atoms as they return to their original lineup. The computer analyzes this activity to produce an image.

A Diagnostic Device

MRI is particularly useful for locating small tumors, revealing subtle changes in the brain, pinpointing blockages in blood vessels, and showing damage to the spinal cord. This technology also allows doctors to observe the function of specific body parts, such as the ear, the heart, muscles, tendons, and blood vessels.

Researchers are experimenting with more powerful magnets that work on other types of atoms. This technology is known as magnetic resonance spectroscopy (MRS). One current use of MRS is to monitor the effectiveness of chemotherapy in cancer patients. Doctors analyze MRS images to find chemical changes that might indicate whether the therapy is successful.

◀ This color-enhanced MRI image of a brain shows a tumor (tinted yellow). The tumor was removed, and the patient resumed a healthy life.

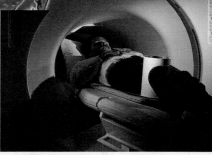

▲ This patient, entering an MRI tunnel, is wearing a cylinder on his knee to provide greater detail of the bone and soft tissue in that area.

Picture This

The images of an MRI scan appear on the screen as cross sections. The doctor can adjust the point of view and depth of the cross section to focus on a specific organ or tissue. Look again at the color-enhanced MRI image on this page. What do you think the different colors represent? What do you think produces the differences in color?

CROSS-DISCIPLINARY FOCUS

Health

Share the following information with students: Both MRI and X-ray imaging form pictures by sending electromagnetic radiation into a patient's body. MRI is safer than X-ray imaging because it uses radio waves, which do not cause any known damage to human tissue. An overdose of X rays, on the other hand, can cause cancer and damage or even destroy body tissue.

One advantage of X-ray imaging is that it is less expensive than MRI. Also, even though X rays can cause cancer, X rays can be used to treat some types of cancer. X rays can also be used to sterilize medical equipment.

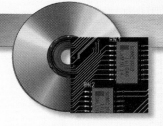
Riding the Electric Rails

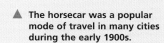

For more than 100 years, the trolley, or streetcar, was a popular way to travel around a city. Then, beginning in the 1950s, most cities ripped up their trolley tracks to make way for automobiles. Today, trolleys are making a big comeback around the world.

From Horse Power to Electric Power

In 1832, the first trolleys, called *horsecars,* were pulled by horses through the streets of New York City. Soon horsecars were used in most large cities in the United States. However, using horses for power presented several problems. Among other things, the horses were slow and required special attention and constant care. So inventors began looking for other sources of power.

In 1888, Frank J. Sprague developed a way to operate trolleys with electricity. These electric trolleys ran on a metal track and were connected by a pole to an overhead power line. Electric current flowed down the

pole to motors in the trolley. A wheel at the top of the pole, called a *shoe,* rolled along the power line, allowing the trolley to move along its track without losing contact with its power source. The current passed through the motor and then returned to a power generator by way of the metal track.

Taking It to the Streets

By World War I, more than 40,000 km of electric-trolley tracks were in use in the United States. The trolley's popularity helped shape American cities because businesses were built along the trolley lines. But competition from cars and buses grew over the next decade, and many trolley lines were abandoned.

By the 1980s, nearly all of the trolley lines had been shut down. But by then people were looking for new ways to cut

▲ The horsecar was a popular mode of travel in many cities during the early 1900s.

down on the pollution, noise, and traffic problems caused by automobiles and buses. Trolleys provided one possible solution. Because they run on electricity, they create little pollution, and because many people can ride on a single trolley, they cut down on traffic.

Today, a new form of trolley is being used in a number of major cities. Called light-rail transit vehicles, these trolleys are quieter, faster, and more economical than the older versions. They usually run on rails alongside the road and contain new systems such as automated brakes and speed controls.

◄ Many cities across the country now use light-rail systems for public transportation.

Think About It!

Just because trolleys operate on electricity, does this mean that they don't create any pollution? Explain your answer.

Background

The first modern light-rail transit (LRT) systems were developed in the 1980s. LRTs have since been established in more than 18 cities in the United States, including Los Angeles, Philadelphia, Cleveland, and Portland, Oregon.

In addition to their more streamlined look, LRT vehicles differ from earlier trolleys in several other ways. The LRT vehicles can operate singly, in short trains of two or three cars, or as a longer series of units that are hinged together to allow for sharp turns. They can operate at higher speeds, are quieter, and can carry up to 16,000 more passengers per track per hour than conventional streetcars. Most of the vehicles also rely on computers to control acceleration and braking. In addition, the new LRT tracks now run along the roadside instead of down the middle of the street. As a result, LRT vehicles are more accessible to passengers and interfere less with automobile traffic.

Discussion

Have students compare LRTs with other forms of mass transportation such as subways, buses, and trains. What are the benefits and drawbacks of each form of mass transportation? You might divide the class into groups and assign each group a form of mass transportation to research. They should consider elements such as the vehicles' construction and maintenance costs, convenience, environmental effects, and aesthetic appeal.

ENVIRONMENTAL FOCUS

Much of the drive to develop LRTs arises from the need to clean up the environment and reduce traffic congestion. But this is not the only alternative form of transportation. Have the class think about other clean forms of transportation. Ideas might include walking, cycling, skating, or skateboarding. Have students suggest specific ways in which they could alter their daily routine to incorporate such changes. Also point out that these alternatives provide an additional benefit—physical exercise.

Answers to *Think About It!*

Running electric-powered vehicles is a clean process, but the production of electricity often is not. Electricity is often generated by burning fossil fuels, which pollutes the environment. (You may wish to point out that this is true for electric cars as well. The batteries used in electric cars also pose disposal problems. They quickly wear out and have to be replaced. They also contain chemicals that are harmful to the environment.)

Unit 6 Sound

Unit Overview

In this unit, students learn what sound is, how it is created, and how it travels. Chapter 16 introduces the distinctive characteristics of sounds, including pitch, loudness, duration, and quality. In this chapter, students see how vibrating objects produce sound. In Chapter 17, students investigate the relationship among wavelength, frequency, and the speed of sound. Students then use echoes to estimate the speed of sound, and they summarize research on echolocation. The structure and function of the human ear and how sound travels through different materials are also discussed. In Chapter 18, students learn how the pitch of a stringed instrument is related to the length, thickness, and tension of its strings; how the loudness of sound is measured; how the loudness of sound can be increased by effects such as forced vibration or resonance; and how a sound can be identified by its shape on an oscilloscope screen. Students also learn about the dangerous effects of loud sounds.

Using the Themes

The unifying themes emphasized in this unit are **Energy, Structures,** and **Systems**. The following information will help you incorporate these themes into your teaching plan. Focus questions that correspond to these themes appear in the margins of this Annotated Teacher's Edition on pages 365, 368, 378, 394, and 405.

Energy is the dominant theme in this unit. In Chapter 16, students discover that sound is produced when energy causes an object to vibrate. In Chapter 17, students investigate how the energy of a vibrating object is transmitted by the particles of a medium to a receiver. In Chapter 18, students examine the relationship between energy and loudness, and they see how converting sound energy into light energy with an oscilloscope helps them study the wave characteristics of sounds.

Structures can be emphasized in Chapter 17 by describing the structures of the human ear in terms of the ear's ability to convert sound waves into electrical impulses that the brain can interpret. Analyzing structures provides a way to compare both the sound-making and hearing abilities of different animals. Throughout the unit, students refine their concept of sound by explaining how the structures of musical instruments contribute to the sounds they make.

The theme of **Systems** is used in Chapter 18 to show how stringed instruments regulate pitch. The use of this theme could be expanded to show how the various parts of an instrument make up a system for regulating all of the different dimensions of sound.

Using the SourceBook

In Unit 6, students review the characteristics of waves in general and of sound waves in particular. Next, students learn about infrasonic and ultrasonic effects of sound. Finally, the intensity of sound is viewed in terms of its potential dangers to human hearing.

Bibliography for Teachers
Gardner, Robert. *Experimenting With Sound.* New York City, NY: Franklin-Watts, 1991.
Lampton, Christopher F. *Sound: More Than What You Hear.* Hillside, NJ: Enslow Publishers, Inc., 1993.
Pierce, John. *The Science of Musical Sound.* New York City, NY: W. H. Freeman, 1992.

Bibliography for Students
Buchanan, Yvonne. *The Science of Music.* New York City, NY: Crowell Collier Press, 1989.
Gibson, Gary. *Science for Fun: Hearing Sounds.* Brookfield, CT: The Millbrook Press, 1995.
Parsons, Alexandra. *Make It Work! Sound.* New York City, NY: Aladdin Books, 1993.
Ward, Alan. *Experimenting With Sound.* New York City, NY: Chelsea House, 1990.

Films, Videotapes, Software, and Other Media
Invention: Mastering Sound
Videodisc
Coronet/MTI
108 Wilmot Rd.
Deerfield, IL 60015
Sound: A Microcomputer-Based Lab
Software (Apple, MS-DOS)
Queue, Inc.
338 Commerce Dr.
Fairfield, CT 06430
Sound, Acoustics, and Recording
Film or videotape
Coronet/MTI
108 Wilmot Rd.
Deerfield, IL 60015
Sound Science: You Can Do It
Videotape
Britannica
310 S. Michigan Ave.
Chicago, IL 60604
The World of Sound Energy
Film or videotape
Focus Media, Inc.
P.O. Box 865
Garden City, NY 11530

Unit Organizer

Unit/Chapter	Lesson	Time*	Objectives	Teaching Resources
Unit Opener, p. 356				Science Sleuths: Noises in School English/Spanish Audiocassettes Home Connection, p. 1
Chapter 16, p. 358	Lesson 1, A World of Sounds, p. 359	2	1. Recognize the number and variety of sounds in our environment. 2. Describe how sounds affect humans. 3. Recognize the differences between various sounds. 4. Classify sounds by their pitch, loudness, duration, and quality.	Image and Activity Bank 16-1 Resource Worksheet, p. 3
	Lesson 2, Making Sounds, p. 362	2	1. Identify or infer what is vibrating to produce various sounds. 2. Understand that sound is a form of energy. 3. Identify what makes a sound loud or soft and high or low. 4. Describe how different animals make sounds.	Image and Activity Bank 16-2 Exploration Worksheet, p. 4 Exploration Worksheet, p. 9 ▼
	Lesson 3, Vibrations: How Fast and How Far? p. 370	1 to 2	1. Distinguish between the frequency and amplitude of a vibrating object. 2. Identify the relationships among the length of a pendulum, the frequency of vibration, and the amplitude of vibration. 3. Define vibration.	Image and Activity Bank 16-3 Exploration Worksheet, p. 11
End of Chapter, p. 373				Chapter 16 Review Worksheet, p. 13 Chapter 16 Assessment Worksheet, p. 16
Chapter 17, p. 375	Lesson 1, Sound Travel, p. 376	2	1. Explain how sound waves travel. 2. Describe how sound travels from a sound source to the ear. 3. Describe the relationship among wavelength, frequency, and the speed of sound waves. 4. Compare how sound waves are transmitted through different media.	Image and Activity Bank 17-1 Transparency Worksheet, p. 18 ▼ Exploration Worksheet, p. 20 Exploration Worksheet, p. 29
	Lesson 2, The Speed of Sound, p. 383	2	1. Calculate the speed of sound, given the distance a sound wave travels and the time it takes to travel that distance. 2. Compare the speed of sound with the speed of light. 3. Describe how echoes are produced.	Image and Activity Bank 17-2 Exploration Worksheet, p. 31
	Lesson 3, More about Echoes, p. 387	1	1. Explore how scientists investigate problems. 2. Describe how blind people use sound to "see." 3. Describe how bats and dolphins use echolocation to detect objects.	Image and Activity Bank 17-3 Math Practice Worksheet, p. 33
	Lesson 4, Completing the Sound Story, p. 391	1	1. Identify and label the structures of the human ear. 2. Describe the path that sound vibrations take when traveling from the outer ear to the brain, where they are interpreted as specific sounds. 3. Examine definitions of sound.	Image and Activity Bank 17-4 Resource Worksheet, p. 34 ▼ Transparency Worksheet, p. 35 ▼
End of Chapter, p. 397				Chapter 17 Review Worksheet, p. 37 Chapter 17 Assessment Worksheet, p. 41
Chapter 18, p. 399	Lesson 1, A Closer Look at Pitch, p. 400	1 to 2	1. Describe how changes in the tension, length, and thickness of vibrating strings affect the pitch of the sounds they produce. 2. Explain how an octave is produced.	Image and Activity Bank 18-1 Exploration Worksheet, p. 44 Theme Worksheet, p. 48
	Lesson 2, Loudness, p. 402	3	1. Identify several ways of making sounds louder. 2. Recognize the relationship between the loudness of a sound and its decibel level. 3. Describe the effects of noise pollution on human health.	Image and Activity Bank 18-2 Discrepant Event Worksheet, p. 50
	Lesson 3, The Quality of Sounds, p. 409	2	1. Describe how an oscilloscope gives a visual shape to a sound and how it represents pitch, loudness, and quality. 2. Describe how the vibration patterns of strings and columns of air can vary and how these patterns affect the sound produced.	Image and Activity Bank 18-3 Exploration Worksheet, p. 51 ▼ Transparency 63
End of Chapter, p. 414				Chapter 18 Review Worksheet, p. 54 Chapter 18 Assessment Worksheet, p. 58
End of Unit, p. 416				Unit 6 Activity Worksheet, p. 62 ▼ Unit 6 Review Worksheet, p. 64 ▼ Unit 6 End-of-Unit Assessment, p. 66 Unit 6 Activity Assessment, p. 72 Unit 6 Self-Evaluation of Achievement, p. 75

* Estimated time is given in number of 50-minute class periods. Actual time may vary depending on period length and individual class characteristics.
▼ Transparencies are available to accompany these worksheets. Please refer to the Teaching Transparencies Cross-Reference chart in the Unit 6 Teaching Resources booklet.

Materials Organizer

Chapter	Page	Activity and Materials per Student Group
16	360	**Describing Sounds:** audiotape of various sounds or a tape recorder and blank audiotape
	362	***Exploration 1, Tuning-Fork Sounds:** a few tuning forks of different sizes; rubber stopper; glass container of any size filled with water; sheet of paper; plastic cup or drinking glass; rubbing alcohol; a few cotton balls; miscellaneous objects to test the tuning forks; **Ruler Sounds:** wooden ruler; metal ruler; **Rubber-Band Sounds:** small cardboard box; rubber band; **Straw Sounds:** plastic drinking straw; scissors; **Simulated Voice Sounds:** balloon; 12 cm cardboard tube; rubber band; scissors
	367	**Exploration 2, Part 1:** meter stick
	370	**Exploration 3:** heavy button, washer, or large paper clip; metric ruler; watch or clock with a second hand; 25 cm of fine thread; scissors
17	376	***Exploration 1, Experiment 1:** 6 pennies; **Experiment 2:** small ball of modeling clay; one-hole rubber stopper; wire clothes hanger; wire cutters; pliers; jingle bell; borosilicate boiling flask; about 20 mL of water; hot plate; safety goggles; oven mitts; **Experiment 3:** balloon; mailing tube; several rubber bands; piece of aluminum foil, about 15 cm²; scissors; materials to support candle, such as a small ball of modeling clay and jar lid; candle; a few matches; a few books; wastebasket; safety goggles; **Experiment 4, Part 1:** Slinky spring toy; 20 cm ribbon; **Experiment 5:** syringe without a needle; small ball of modeling clay; **Experiment 6:** scissors; piece of cardboard, about 10 cm × 20 cm
	381	**Sound Thinking:** metric ruler
	382	***Exploration 2:** 30 cm wooden rod, at least 2 cm in diameter; thumbtack; tuning fork; various plastic, metal, and paper objects such as a plastic ruler, metal ruler, and spiral notebook; 2 metal spoons; 8 to 10 L of water; small aquarium or other very large container; a large and small funnel; balloon or piece of cellophane, about 25 cm × 25 cm; a few rubber bands; 1 m of rubber tubing; scissors
	385	**Exploration 3:** meter stick or metric measuring tape; watch or clock with second hand (See Advance Preparation below.)
	387	**A Research Project:** one index card per student
	395	**Questions to Think About:** stethoscope; about 1 L of water; large paper container with the bottom removed; large balloon or rubber sheet, about 30 cm × 30 cm; a few large rubber bands; tuning fork
18	400	**Exploration 1:** materials to construct apparatus for investigating pitch, such as the following: three 4 L plastic jugs; three 2 m lengths of string or nylon fishing line of different thicknesses; 3 short nails; hammer; 40 cm wooden dowel; 2 bricks; 60 cm × 40 cm × 1 cm sheet of plywood; box of strong cardboard or plastic foam, about 60 cm × 40 cm; 6 L of water; 1000 mL beaker; sheet of paper; about 10 cm of tape; safety goggles
	402	**Some Solutions to the Problem:** tuning fork; variety of materials of different sizes to test, such as a large and small glass container, a large and small hollow wood block, a rubber-headed mallet, a large and small metal cymbal, and a large and small plastic cup; sheet of paper; cardboard tube, at least 40 cm long; scissors; metric ruler; (See Advance Preparation below.)
	412	**How a String Vibrates:** rubber tubing, 5 to 10 m in length, or a Slinky spring toy
	413	**Locating the Higher Sounds:** violin or guitar (See Advance Preparation below.)

* You may wish to set up these activities in stations at different locations around the classroom.

Advance Preparation

Exploration 3, page 385: Each student group will need access to a wall to perform this Exploration.
Some Solutions to the Problem, page 402, and Locating the Higher Sounds, page 413: You may wish to contact a music teacher to borrow musical instruments.

Unit Compression

Sound is a highly interesting, interdisciplinary topic. It incorporates concepts of physics as well as aspects of physiology, zoology, mathematics, and psychology. In addition, this unit ventures into research methodology and environmental issues. If time is short, these extensions can be omitted although the broader context of sound will be sacrificed. Extensions that can be omitted include Chapter 16, Lesson 2, Exploration 2 on page 367; all of Chapter 17, Lesson 4; and Chapter 18, Lesson 3, Noise Pollution—The Effect of Sound on Health on pages 407–408.

Chapter 18 is a reconsideration of topics treated in Chapter 16, such as pitch, loudness, and sound quality, but these topics are covered in more depth, a few new topics are added, and musical applications are introduced. The essential science content is thus mostly complete in the first two chapters. Dropping Chapter 18 would reduce the time spent on the unit by at least a quarter, although depth would be greatly sacrificed.

Homework Options

Chapter 16
See Teacher's Edition margin, pp. 360, 361, 364, 366, 369, and 372
Resource Worksheet, p. 3
SourceBook, pp. S106, S109, S121, and S124

Chapter 17
See Teacher's Edition margin, pp. 375, 380, 385, 390, 392, and 395
Math Practice Worksheet, p. 33
Resource Worksheet, p. 34
SourceBook, pp. S107, S111, S112, S115, S117, and S121

Chapter 18
See Teacher's Edition margin, pp. 407 and 413
SourceBook, pp. S116, S117, and S124

Unit 6
Unit 6 Activity Worksheet, p. 62
Activity Assessment Worksheet, p. 72
SourceBook Activity Worksheet, p. 76

Assessment Planning Guide

Lesson, Chapter, and Unit Assessment	SourceBook Assessment	Ongoing and Activity Assessment	Portfolio and Student-Centered Assessment
Lesson Assessment Follow-Up: see Teacher's Edition margin, pp. 361, 369, 372, 382, 386, 390, 396, 401, 408, and 413 **Chapter Assessment** Chapter 16 Review Worksheet, p. 13 Chapter 16 Assessment Worksheet, p. 16* Chapter 17 Review Worksheet, p. 37 Chapter 17 Assessment Worksheet, p. 41* Chapter 18 Review Worksheet, p. 54 Chapter 18 Assessment Worksheet, p. 58* **Unit Assessment** Unit 6 Review Worksheet, p. 64 End-of-Unit Assessment Worksheet, p. 66*	SourceBook Review Worksheet, p. 78 SourceBook Assessment Worksheet, p. 82*	Activity Assessment Worksheet, p. 72* **SnackDisc** Ongoing Assessment Checklists ♦ Teacher Evaluation Checklists ♦ Progress Reports ♦	Portfolio: see Teacher's Edition margin, pp. 364, 389, 390, and 411 **SnackDisc** Self-Evaluation Checklists ♦ Peer Evaluation Checklists ♦ Group Evaluation Checklists ♦ Portfolio Evaluation Checklists ♦

* Also available on the Test Generator software
♦ Also available in the Assessment Checklists and Rubrics booklet

Science Discovery is a versatile videodisc program that provides a vast array of photos, graphics, motion sequences, and activities for you to introduce into your *SciencePlus* classroom. *Science Discovery* consists of two videodiscs: Science Sleuths and the Image and Activity Bank.

Using the *Science Discovery* Videodiscs

Science Sleuths: Noises in School
Side B

The school has made a great many sacrifices in order to purchase word processors. Now the students complain that noise in the classroom prevents them from concentrating enough in word-processing class to learn anything. They never had these problems with typewriters. It is suspected that extra noise is caused by the gym teacher moving his classes closer to the word-processing classroom.

Interviews

1. Setting the scene: School principal **40107 (play ×2)**

2. Word-processing instructor **41014 (play)**

3. Student in 9 A.M. class **41558 (play)**

4. Student in 2 P.M. class **42134 (play)**

5. Coach **42769 (play)**

6. Custodian **43791 (play)**

Documents

7. Memo to custodial staff **44576 (step)**

8. PTA minutes **44579 (step)**

9. Memo to Department of Highways **44582**

10. Letter from furnace and fuel company **44584 (step)**

Literature Search

11. Search on the words: BUDGET CUTBACKS, NOISE, NOISE REDUCTION, NORWOOD HIGH, VANDALISM, WORD PROCESSING **44587 (step)**

12. Article #1 ("Noise Reduction") **44590 (step)**

13. Article #2 ("The Dolby Wave") **44593 (step)**

14. Article #3 ("CQC Wants Quieter Traffic") **44596**

15. Article #4 ("Several Stations Give All-Day Traffic Reports") **44598**

16. Article #5 ("Football Player Pleads Guilty to Vandalism") **44600**

17. Article #6 ("Editorial: Computers vs. Football?") **44602 (step)**

18. Article #7 ("Word Processing: The Wave of the Present") **44605 (step)**

Sleuth Information Service

19. Map of school **44608**

20. Map of school area **44610**

21. Weather for September 1–18 **44612**

22. Steam furnaces **44614 (step)**

23. Traffic **44617**

Sleuth Lab Tests

24. Sonogram of typing room (morning) **44619**

25. Sonogram of typing room (afternoon) **44621**

26. Sonogram of typing room (night) **44623**

27. Sonogram of school courtyard (morning) **44625**

28. Sonogram of school courtyard (afternoon) **44627**

29. Sonogram of school courtyard (night) **44629**

30. Sonogram of word processor **44631**

31. Sonogram of typewriter **44633**

32. Sonogram of football field during practice **44635**

33. Sonogram of bandroom during rehearsal **44637**

34. Sonogram of boiler room (off) **44639**

◀‖ Step Reverse   Play ▶   Pause ‖   Step Forward ‖▶

Homework Options

Chapter 16
See Teacher's Edition margin, pp. 360, 361, 364, 366, 369, and 372
Resource Worksheet, p. 3
SourceBook, pp. S106, S109, S121, and S124

Chapter 17
See Teacher's Edition margin, pp. 375, 380, 385, 390, 392, and 395
Math Practice Worksheet, p. 33
Resource Worksheet, p. 34
SourceBook, pp. S107, S111, S112, S115, S117, and S121

Chapter 18
See Teacher's Edition margin, pp. 407 and 413
SourceBook, pp. S116, S117, and S124

Unit 6
Unit 6 Activity Worksheet, p. 62
Activity Assessment Worksheet, p. 72
SourceBook Activity Worksheet, p. 76

Assessment Planning Guide

Lesson, Chapter, and Unit Assessment	SourceBook Assessment	Ongoing and Activity Assessment	Portfolio and Student-Centered Assessment
Lesson Assessment Follow-Up: see Teacher's Edition margin, pp. 361, 369, 372, 382, 386, 390, 396, 401, 408, and 413 **Chapter Assessment** Chapter 16 Review Worksheet, p. 13 Chapter 16 Assessment Worksheet, p. 16* Chapter 17 Review Worksheet, p. 37 Chapter 17 Assessment Worksheet, p. 41* Chapter 18 Review Worksheet, p. 54 Chapter 18 Assessment Worksheet, p. 58* **Unit Assessment** Unit 6 Review Worksheet, p. 64 End-of-Unit Assessment Worksheet, p. 66*	SourceBook Review Worksheet, p. 78 SourceBook Assessment Worksheet, p. 82*	Activity Assessment Worksheet, p. 72* **SnackDisc** Ongoing Assessment Checklists ♦ Teacher Evaluation Checklists ♦ Progress Reports ♦	Portfolio: see Teacher's Edition margin, pp. 364, 389, 390, and 411 **SnackDisc** Self-Evaluation Checklists ♦ Peer Evaluation Checklists ♦ Group Evaluation Checklists ♦ Portfolio Evaluation Checklists ♦

* Also available on the Test Generator software
♦ Also available in the Assessment Checklists and Rubrics booklet

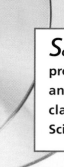

Science Discovery is a versatile videodisc program that provides a vast array of photos, graphics, motion sequences, and activities for you to introduce into your *SciencePlus* classroom. *Science Discovery* consists of two videodiscs: Science Sleuths and the Image and Activity Bank.

Using the *Science Discovery* Videodiscs

Science Sleuths: Noises in School
Side B

The school has made a great many sacrifices in order to purchase word processors. Now the students complain that noise in the classroom prevents them from concentrating enough in word-processing class to learn anything. They never had these problems with typewriters. It is suspected that extra noise is caused by the gym teacher moving his classes closer to the word-processing classroom.

Interviews

1. Setting the scene: School principal **40107 (play ×2)**

2. Word-processing instructor **41014 (play)**

3. Student in 9 A.M. class **41558 (play)**

4. Student in 2 P.M. class **42134 (play)**

5. Coach **42769 (play)**

6. Custodian **43791 (play)**

Documents

7. Memo to custodial staff **44576 (step)**
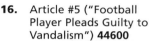

8. PTA minutes **44579 (step)**

9. Memo to Department of Highways **44582**

10. Letter from furnace and fuel company **44584 (step)**

Literature Search

11. Search on the words: BUDGET CUTBACKS, NOISE, NOISE REDUCTION, NORWOOD HIGH, VANDALISM, WORD PROCESSING **44587 (step)**

12. Article #1 ("Noise Reduction") **44590 (step)**

13. Article #2 ("The Dolby Wave") **44593 (step)**

14. Article #3 ("CQC Wants Quieter Traffic") **44596**

15. Article #4 ("Several Stations Give All-Day Traffic Reports") **44598**

16. Article #5 ("Football Player Pleads Guilty to Vandalism") **44600**

17. Article #6 ("Editorial: Computers vs. Football?") **44602 (step)**

18. Article #7 ("Word Processing: The Wave of the Present") **44605 (step)**

Sleuth Information Service

19. Map of school **44608**

20. Map of school area **44610**

21. Weather for September 1–18 **44612**

22. Steam furnaces **44614 (step)**

23. Traffic **44617**

Sleuth Lab Tests

24. Sonogram of typing room (morning) **44619**

25. Sonogram of typing room (afternoon) **44621**

26. Sonogram of typing room (night) **44623**

27. Sonogram of school courtyard (morning) **44625**

28. Sonogram of school courtyard (afternoon) **44627**

29. Sonogram of school courtyard (night) **44629**

30. Sonogram of word processor **44631**

31. Sonogram of typewriter **44633**

32. Sonogram of football field during practice **44635**

33. Sonogram of bandroom during rehearsal **44637**

34. Sonogram of boiler room (off) **44639**

◀◀ Step Reverse ‖‖‖‖   Play ▶ ‖‖‖‖   Pause ‖‖ ‖‖‖   Step Forward ‖▶ ‖‖‖‖

35. Sonogram of boiler room (on) **44641**

Still Photographs
36. Classroom this year **44643 (step)**

37. Classroom last year **44646**

Image and Activity Bank
Side A or B

A selection of still images, short videos, and activities is available for you to use as you teach this unit. For a larger selection and detailed instructions, see the Videodisc Resources booklet included with the Teaching Resources materials.

16-1 A World of Sounds, page 359
Steel drum band 38725–40227 (play ×2) (Side A only)
A steel drum band performs. A steel drum plays a scale.

Musical instrument 2176
Musical instruments produce sound waves that travel through the air to our ears.

Pitch 2236
Is the pitch of a snare drum or of a bass drum higher?

Loudness 2229
Which is louder, a cannon or a rifle?

Duration 2227
Which sound lasts longer, a gong being struck or a nail being hammered?

Quality 2228
Which bird sound is "sweeter" to your ears?

16-2 Making Sounds, page 362
Vibration and pitch 2233
Faster vibrations (higher frequencies) produce higher pitched sounds.

Frog, Pacific tree 3647
Pacific tree frogs are the ones that make a "ribbet" sound. Male frogs inflate their vocal pouches to give their songs resonance when they call mates. Frogs sing in choruses so loud that they can be heard at great distances. Each species has its own song.

Windpipe; bird 4183 (step ×2)
Windpipe of a bird, showing the bronchus and the syrinx (including the air sac and membranes) (step) Air rushes through the syrinx, vibrating thin muscles and membranes so that they create sound. (step) Different notes can be made by altering the size of the air passage and the tension of the membranes.

16-3 Vibrations: How Fast and How Far? page 370
Waves 2220
The parts of a transverse and longitudinal wave are labeled.

17-1 Sound Travel, page 376
Waves, sound 2219
Sound waves are longitudinal and are transmitted in the air. The amount of compression of the wave affects the air pressure. These small changes in air pressure have an effect on the eardrum.

17-2 The Speed of Sound, page 383
The speed of sound 2234
The speed of sound in various materials at 0°C

17-3 More About Echoes, page 387
Bat echolocation 2224 (step)
Bats use sound waves to locate their prey. (step) The bat judges the location of its prey by the characteristics of the returning wave, or echo.

Whale, beluga; sonogram of song 52189–52304 (play ×2) (Side B only)
Beluga whales are called sea canaries by fishermen because they chirp and squeak so much. They make the sounds by squeezing air through tight passages in their heads. They communicate, echolocate, and stun fish with the sounds.

17-4 Completing the Sound Theory, page 391
Ear, human 4180 (step ×2)
Outer ear, middle ear (eardrum and bones), inner ear (cochlea and semicircular canals), auditory nerve, and brain (step) Sound vibrations are transmitted through the eardrum to the inner ear. (step) Sound vibrations are translated by the inner ear into nerve impulses that are then transmitted to the brain.

Ear, human; model 4117 (step ×7)
Surface of external ear with pinna (outer funnel-shaped structure that picks up sound waves) (step) Exterior of ear showing auditory canal (step) Exterior of ear showing auditory canal with tympanic membrane at the end of the canal (step) Opening in skull where external ear attaches (step) Middle ear (left) and inner ear (right) shown by removal of surrounding bones (step) Cross section of inner ear with the semicircular canals on the left and the cochlea on the right (step) Location of tympanic membrane in bone; three bones of the middle ear (step) Details of the structures of the inner and middle ear and semicircular canals (left); cochlea (middle); tympanic membrane, malleus, incus, and stapes (right)

18-1 A Closer Look at Pitch, page 400
Pitch; three glasses of water 38183–38302 (play ×2) (Side A only)
Wave forms of three glasses filled with different amounts of water. The more water, the lower the pitch.

Xylophone; relation between length and pitch 38432–38724 (play ×2) (Side A only)
As you play a shorter length of tube, the pitch rises.

18-2 Loudness, page 402
Decibel meter 2178
A decibel meter measures the intensity of a sound wave.

Decibel scale 2238
Common noises measured in decibels

Sound, high-frequency 38303–38429 (play ×2) (Side A only)
The resonant frequency of the beaker is the same as the pitch of the sound. As the beaker begins to vibrate from the sound, the beaker breaks.

18-3 The Quality of Sounds, page 409
Waveforms; harmonics 40228–40578 (play ×4) (Side A only)
(play ×2) An oscilloscope displays the wave form of a violin being played. (play ×2) An oscilloscope displays the wave form of a trumpet being played.

Waveforms; sine waves 40579–40759 (play ×4) (Side A only)
(play ×2) An oscilloscope shows the wave form of an F tuning fork as it is being played. (play ×2) An oscilloscope shows the wave form of an A tuning fork as it is being played.

◀ Step Reverse   Play ▶    Pause ❚❚   Step Forward ❚▶

Unit 6 Sound

UNIT FOCUS

To emphasize the importance of sound, have students try to imagine what their world would be like if they were hearing impaired. Ask: What would it be like to talk in sign language? What would it be like if you could not hear warning devices such as automobile horns? How could you talk on the telephone? What would it be like to live in a world in which you could feel music but could not hear it? If possible, invite a hearing-impaired person to share with the class what his or her world is like without sound and how he or she has adjusted to it. You may also wish for volunteers to experience what it might be like to live without sound by having them wear headphones or earplugs for a designated amount of time. Tell students that in this unit, they will learn about the relationship between hearing and sound.

A good motivating activity is to let students listen to the English/Spanish Audiocassettes as an introduction to the unit. Also, begin the unit by giving Spanish-speaking students a copy of the Spanish Glossary from the Unit 6 Teaching Resources booklet.

CHAPTER
16 What Is Sound?
1 A World of Sounds . . 359
2 Making Sounds 362
3 Vibrations: How Fast and How Far? 370

CHAPTER
17 How Sound Moves
1 Sound Travel 376
2 The Speed of Sound 383
3 More About Echoes 387
4 Completing the Sound Story. 391

CHAPTER
18 Listening Closely
1 A Closer Look at Pitch. 400
2 Loudness 402
3 The Quality of Sounds. 409

When a firecracker explodes, the rapid release of energy produces a flash of light and an invisible pulse, which we hear as a loud "pop."

356

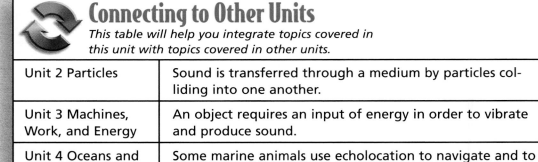

Connecting to Other Units

This table will help you integrate topics covered in this unit with topics covered in other units.

Unit 2 Particles	Sound is transferred through a medium by particles colliding into one another.
Unit 3 Machines, Work, and Energy	An object requires an input of energy in order to vibrate and produce sound.
Unit 4 Oceans and Climates	Some marine animals use echolocation to navigate and to find prey.

KABOOM!
The ignition of each of these colorful energy converters sparks a spectacular transformation. In an instant, the chemical energy stored within the firecracker's powder is transformed into an ear-battering bang accompanied by a tiny tempest of light, heat, and blackened bits of paper. Some of the released energy even causes motion in the remaining firecrackers, which recoil with each blast.

It would be dangerous to be close to this explosion. Among other dangers, the intensity of such a loud sound could cause pain and even damage to your ears. Although you cannot always prevent loud noises, you can play a part in determining their effect on you. What are some factors that you think affect the loudness of a sound? Ⓐ

Even from a safe distance, the noise of an exploding firecracker may cause a feeling of pressure in your ears. This has to do with the way in which sounds are created, transmitted, and detected. In this unit, you'll discover how sounds originate and how they travel, which will help you understand why you'd actually see the flash of one of these explosions before hearing the noise it produced.

357

Using the Photograph

Share the following information with students: In the photograph, ignited black powder in the firecrackers is causing them to explode. Black powder was invented in China over 1000 years ago.

Answer to
In-Text Question

Ⓐ Accept all reasonable responses. Some factors that affect the loudness of a sound include the actual volume of the sound, distance from the source, direction of the sound relative to our ears, the material through which the sound is traveling, and whether the surroundings reflect or absorb sound energy.

♺ Connecting to Other Units, continued

Unit 5 Electro-magnetic Systems	A microphone is a device that converts sounds into electrical signals.
Unit 7 Light	Our ears, like our eyes, can detect waves only within a certain range of frequencies.
Unit 8 Continuity of Life	Genetic variation has allowed species to adapt their hearing to suit their environments.

Connecting to Other Chapters

Chapter 16
shows how vibrations produce sounds and discusses ways of distinguishing different sounds.

Chapter 17
explores how the particles of a medium transmit sound energy and how our ears receive that energy.

Chapter 18
invites students to reexamine pitch, loudness, and sound quality in a quantitative way.

Prior Knowledge and Misconceptions

Your students' responses to the ScienceLog questions on this page will reveal the kind of information—and misinformation—they bring to this chapter. Use what you find out about your students' knowledge to choose which chapter concepts and activities to emphasize in your teaching. After students complete the material in this chapter, they will be asked to revise their answers based on what they have learned. Sample revised answers can be found on page 374.

In addition to having students answer the questions on this page, you may wish to assign a "free-write" to assess prior knowledge. To do this, instruct students to write for 3 to 5 minutes about the characteristics of sound that help us to distinguish one sound from another. Tell students to keep their pens moving at all times, writing in a stream-of-consciousness fashion. Emphasize that there are no right or

wrong answers in this exercise. It may be best to ask students not to put their names on their papers. Collect the papers, but do not grade them. Instead, read them to find out what students know about sound, what misconceptions they may have, and what about the topic interests them.

CHAPTER 16
What Is Sound?

1 How will the sound from the banjo change if the player moves her finger halfway up the neck of the banjo?

2 How would you describe the difference between music and noise?

3 What causes sound?

ScienceLog

Think about these questions for a moment, and answer them in your ScienceLog. When you've finished this chapter, you'll have the opportunity to revise your answers based on what you've learned.

358

LESSON 1

A World of Sounds

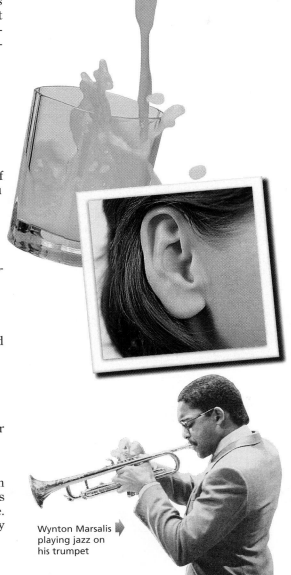

Sounds All Around

You are surrounded by a sea of sound. There is a tremendous variety of sounds in this sea. At times, some of them become part of the background. For example, did you *really* hear the following sounds this morning?

- the hiss of water running from the tap
- the orange juice splashing into the glass
- the rattle of dishes
- the slam of the door
- the din of distant traffic

How many other sounds can you think of that are part of your everyday life but that you don't really notice? Why do you think we filter out certain sounds? **A**

Investigating Sound

Studying sound raises many interesting questions. Think about the following questions:

- What was the loudest sound ever heard?
- Are there sounds you can't hear?
- How can a faint whisper sometimes be heard many meters away?
- What is an echo? What causes it?
- Why is it so silent after a snowfall?
- How do blind people use sounds to "see"?
- How do musical instruments make their sounds?
- How do music and noise differ?

Can you add to this list of questions? With your friends, generate a list of other questions about sound that you would like to investigate. At the end of your study of sound, see how many you have answered.

Wynton Marsalis
playing jazz on
his trumpet

LESSON 1 ORGANIZER

Time Required
2 class periods

Process Skills
observing, analyzing, comparing, contrasting, communicating

New Term
Pitch—the "highness" or "lowness" of a sound

Materials (per student group)
Describing Sounds: audiotape of various sounds or a tape recorder and blank audiotape

Teaching Resources
Resource Worksheet, p. 3
SourceBook, p. S124

LESSON 1

A World of Sounds

FOCUS

Getting Started

Have students shut their eyes and be still and silent for 3 minutes. Tell them to concentrate on what they hear. Have students open their eyes and list the noises they heard. Ask students to think about why they do not normally notice these sounds. *(Accept all reasonable responses.)* Tell the class that they will find out about background noise in this lesson.

Main Ideas

1. We are surrounded by a variety of sounds, but we are often unaware of them.
2. Both too much and too little sound can have harmful effects on humans.
3. The dimensions of sound are loudness, pitch, duration, and quality.

TEACHING STRATEGIES

Investigating Sound

Have students work in small groups to read and discuss the list of questions about sound, most of which will be answered in this unit. Then ask them to generate more questions that they would like to investigate. Have them discuss possible answers to the questions. Emphasize that they are not expected to know the correct answers at this point.

Answer to
In-Text Question

A Accept all reasonable responses. By the end of the lesson, students should understand that if they were not able to filter out certain sounds in the environment, they would be overwhelmed by the many sounds around them and would probably have difficulty functioning normally.

The Sound of Silence

After students have read this paragraph, you may wish for them to make a stethoscope out of two funnels and rubber tubing by attaching a funnel to each end of the tubing. Have them use the stethoscope to listen to the sounds produced in their own body.

 ### Cooperative Learning
THE SOUND OF SILENCE AND NOISE

Group size: 2 to 3 students

Group goal: to read, discuss, and answer the in-text questions on this page in order to appreciate the number and variety of common sounds

Positive interdependence: Assign the following roles: reader (to read aloud all text and questions), investigator (to serve as the liaison between the teacher and the student group and to manage materials), and director (to keep discussion going and to make sure everyone participates). After the material on the page has been read, have each student write one of the in-text questions on an index card and place the card in a grab bag. Each group member should draw one card from the bag and, after a group discussion, write the group's response to the question on the back of the card.

Individual accountability: Collect the index cards and call on students from each group at random to discuss their group's responses to each of the questions.

Answers to
In-Text Questions

Ⓐ Students should be able to hear a number of sounds made by their bodies, such as their heart beating, stomach growling, lungs breathing, or esophagus swallowing.

Ⓑ Some sources of offending sounds in cities include traffic, sirens, machinery, people, and jets. Some possible harmful effects of excess noise include hearing loss and headaches.

Ⓒ Just as white light is a mixture of all colors of the spectrum, white noise is a mixture of sounds of many different frequencies. (Students will learn more about white light in Unit 7.)

Describing Sounds

Have students work in small groups to generate lists of descriptive words. This

The Sound of Silence

Is there such a thing as total silence? For instance, if you were put in a completely sound-proof room, would you hear anything? Actually, you would. People placed in soundproof rooms hear two distinctly different sounds. Sound engineers have discovered that one sound is made by the person's nervous system and the other sound is made by the person's blood circulating. Can you hear *your* body's sounds right now? Ⓐ

This specially designed room eliminates almost all sound. It is being used to check the amount of static transmitted over a telephone system. →

Noise

In many cities today, the continuous background of sound is increasing to alarming levels. Studies show that many people suffer physically and mentally when exposed to excessive noise. What are some sources of offending sounds? What are some possible harmful effects? Ⓑ

At the other extreme, people can become nervous and irritable when there is too little sound. Modern buildings are often so quiet that even ordinary noises can startle people working in them. To overcome this problem, sound is piped back into the rooms! It sounds something like the hiss of escaping steam and is called "white noise." Why do you suppose it is given this name? Ⓒ

Describing Sounds

Make or obtain an audiotape of various sounds. Have others guess what is making each sound. Then brainstorm to think of words or word combinations that describe each sound, such as *harsh*, *loud*, *sweet*, *shrill*, *grating*, *high*, or *low*. Afterward, think of all the ways you might classify these word descriptions.

Yikes! Sometimes sounds can be too loud. For information about sound and safety, see page S124 of the SourceBook.

360

activity gives students their own vocabulary for distinguishing sounds, a vocabulary that they will expand as the unit proceeds. Have students classify their words under categories of their own choice.

Homework

Have students write three questions that interest them about sound. Discuss the questions in class the next day.

CROSS-DISCIPLINARY FOCUS

Social Studies

Have students create a *soundscape*. A soundscape is an illustration of a specific environment and all of the sounds common to that environment. Assign a different soundscape to each student, such as an urban, rural, mountain, or desert environment. Ask them to compare the different soundscapes in order to increase their awareness of the differences among sounds in different types of environments.

Classifying Sounds

As you read the following article, write a descriptive heading for each paragraph in your ScienceLog. Use no more than four words for each heading.

(a)

Sounds have certain characteristics that can be thought of as dimensions. We also speak of dimensions when measuring objects. One such dimension is height. In a way, sounds also have height. A sound may be high like the chirp of a cricket, lower like the croak of a frog, or still lower like the grunt of a pig. The "highness" or "lowness" of a sound is called its pitch.

(b)

Sounds like rain falling may be quite soft. Other sounds, like a door slamming, may be louder. A speeding train is louder still. Loudness is another dimension of sound.

(c)

The rat-a-tat-tat of a jackhammer is a sequence of very short sounds; each lasts a fraction of a second. Other sounds linger, such as that made by a large bell. Are musical sounds generally long or short? The length of time, or duration, for which a sound can be heard is another dimension of sound.

(d)

When we try to further classify sounds, we may use words like sweet, dull, bright, blaring, or harsh. We may also use words such as crackling, buzzing, clanging, and tinkling. We use these descriptive words to describe another dimension of sound—its quality. A bell makes a sound of a different quality from that of a car horn, even if the loudness, pitch, and duration of both sounds are the same.

(e)

In music, tonal quality is called timbre, or tone. Why is there such variation in tonal quality? When you mix paints, the shade you get depends on the colors you start with and how much of each color you use. This is also true of sounds. The quality you hear depends on the mix of component sounds that form the overall sound.

(f)

What is music, and what is noise? Think of the musical tones of the flute, the less musical sound of a ringing telephone, and the totally unmusical noise of the lawn mower. Musical sounds have fewer component sounds; they are "purer."

The dimensions of sound are pitch, loudness, duration, and quality. Changes in these dimensions create the incredible variety of sounds that surround us.

Which sound has a higher pitch? **D**

Which sound is louder? **E**

SQUAWK!

Which sound is "sweeter"? **F**

Why is this music, rather than noise? **G**

361

FOCUS

Getting Started

Ask for a few volunteers and give each volunteer a different classroom object. Tell them to produce a sound with the object. Then ask: How does the object produce the sound? How can you stop the sound? Can you make the sound higher or lower and softer or louder? *(Accept all responses.)* Point out to students that in this lesson they will learn new ways to modify sounds.

Main Ideas

1. Sounds are produced by vibrating objects.
2. Sound is a form of energy.
3. The larger the vibrations of an object, the louder the sound; the faster the vibrations, the higher the pitch of the sound.
4. Animals produce sounds in many different ways and hear different sounds than humans.

TEACHING STRATEGIES

EXPLORATION 1

This Exploration works well in stations. Students should record specific observations from each experiment in their ScienceLog.

 SAFETY ALERT Each student should clean the tuning fork with rubbing alcohol after using it. Make sure there are no operating hot plates, open flames, or other sources of ignition nearby when using alcohol. Also, to avoid chipping a tooth, students should be very careful when touching the base of the tuning fork to the base of their teeth. For Activity 1, make sure that you approve all miscellaneous materials for safety before students test them with the tuning fork.

★ An Exploration Worksheet is available to accompany Exploration 1 (Teaching Resources, page 4).

EXPLORATION 1

A Symphony of Sound

It's not difficult to make sounds, but it is sometimes difficult to see what is happening when sounds are made. Do several of the Activities that follow. For each Activity, record answers to the following questions:

1. How do I make the object produce the sound?
2. What is the object doing as it produces the sound?
3. How long does the sound last?
4. How can I stop the sound?
5. Can I change any characteristics (dimensions) of the sound, such as loudness and pitch? If so, how?

You Will Need

- tuning forks of different sizes
- a rubber stopper
- a glass container of water
- 2 rulers—one wooden, one metal
- a drinking glass
- a cardboard box
- 2 rubber bands
- a pencil
- a plastic drinking straw
- scissors
- a balloon
- a cardboard tube (12 cm long)
- rubbing alcohol
- cotton balls
- paper
- miscellaneous objects

ACTIVITY 1

Tuning-Fork Sounds

1. Strike a tuning fork on the edge of a rubber stopper.
2. Hold the tuning fork close to your ear. What do you observe?
3. Lightly touch the prongs to various parts of your body.

Caution: Do not touch the prongs to your eyes or eyeglasses. Clean the tuning fork with rubbing alcohol and a cotton ball before another student uses it.

After striking the fork again, touch the prongs to the surface of a glass container of water and then to a loosely held piece of paper. What do you observe?

4. Next, while the tuning fork is sounding, touch its base to your teeth, to the table, to a cup held over your ear, and to other objects.
5. Try some of these same experiments with a tuning fork of a different size.

ACTIVITY 2

Ruler Sounds

1. Hold one end of a wooden ruler firmly on the edge of a table. Push down on the other end, and then let it go. Try this several times, with various lengths of the ruler extending over the end of the table. What do you observe?
2. Substitute a metal ruler for the wooden one, and repeat the activity. What do you observe now?

LESSON 2 ORGANIZER

Time Required 2 class periods

Process Skills observing, comparing

Theme Connections Energy, Structures

New Terms
Larynx—the sound-producing organ in many vertebrates; the "voice box"
Syrinx—the sound-producing organ in birds; the "song box"

Materials (per student group)
Exploration 1, Tuning Fork Sounds: a few tuning forks of different sizes; rubber stopper; glass container of any size filled with water; sheet of paper; plastic cup or drinking glass; rubbing alcohol; a few cotton balls; miscellaneous objects to test the tuning forks; **Ruler Sounds:** wooden ruler; metal ruler; **Rubber Band Sounds:** small cardboard box; rubber band; **Straw Sounds:** plastic drinking straw; scissors; **Simulated Voice Sounds:** balloon; 12 cm cardboard tube; rubber band; scissors
Exploration 2, Part 1: meter stick

Teaching Resources
Exploration Worksheets, pp. 4 and 9
Transparency 58
SourceBook, pp. S106 and S121

ACTIVITY 3

Rubber-Band Sounds

1. Pluck a rubber band that has been stretched across a cardboard box. Listen carefully and observe what happens.
2. Tighten the part of the rubber band that is on the top of the box. Pluck it again. Is there any difference in sound? Why or why not?
3. Put a pencil across the top of the box (the short way), under the rubber band, and pluck again. Do you hear or see any differences? Why or why not?

ACTIVITY 4

Straw Sounds

1. Make a "straw saxophone," as shown below.
2. Adjust the position of the "sax" in your mouth until a steady sound is produced. Try producing different sounds by blowing in different ways.

Be careful: Use only your own straw, and dispose of it after completing the Exploration.

3. While blowing a steady sound, use scissors to cut the straw shorter and shorter. What happens?

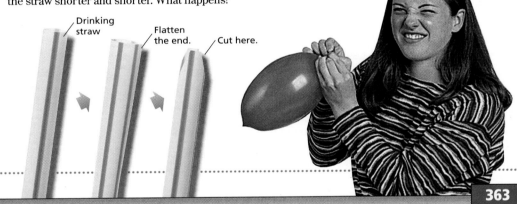

Drinking straw   Flatten the end.   Cut here.

ACTIVITY 5

Simulated Voice Sounds

1. Place your fingers on your neck near your vocal cords. Say "Ah" loudly. What do you feel? Try some high sounds and some low sounds.
2. Blow up a balloon. Make the balloon squeal as you slowly release air from it. Try for variations of sound. What must you do to get a higher sound? a lower sound? This is similar to what happens in your throat when you speak.
3. Make a working model of human vocal cords. Stretch a piece of balloon over the end of a cardboard tube, but not too tightly. Cut a narrow slit in the piece of balloon, and blow into the tube from the opposite end. Now tighten the balloon so that the slit becomes longer, and blow again. How does increasing the tension affect the pitch?

Caution: Blow only into your own tube. Dispose of tube after use.

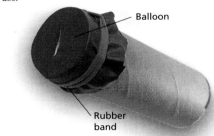

Balloon

Rubber band

363

Answers to
A Symphony of Sound, page 362

The following are sample responses:

1. All of the objects can be made to produce sound by adding energy to them. It takes the energy of an applied force doing work on an object to produce sound.
2. The object is vibrating as it produces the sound.
3. The sound lasts for as long as the object vibrates. This may vary from less than a second to many seconds.
4. The sound can be stopped by "damping" the object (absorbing its energy) so that it stops vibrating.
5. The characteristics of the sound can be changed by changing the characteristics of the vibrating object, the object's surroundings, or the size of the applied force. For example, shortening the length or increasing the tension of the vibrating object raises the pitch of the sound. Applying more force to the object or placing it against certain materials increases the sound's loudness.

Answers to
Activity 1, *page 362*

2. The tuning fork vibrates, producing sound.
3. Students should feel the movement of the tuning fork. When the tuning fork touches the container of water, ripples form on the surface of the water. When the tuning fork touches the paper, the paper vibrates.

Answers to
Activity 2, *page 362*

1. There is a relationship between pitch and how much of the ruler extends over the edge of the table—the shorter the length of ruler hanging over the table, the higher the pitch; the longer the length, the lower the pitch.
2. The pitch of the metal ruler will still get higher as the length hanging over the table gets shorter. However, the sound made by the metal ruler will differ in loudness and quality from that made by the wooden ruler.

Answers to
Activity 3

1. The rubber band vibrates and produces a sound. The box makes the sound louder. (In Chapter 18, students will examine factors that affect loudness, such as forced vibration and resonance.)
2. Yes; the pitch is higher because the tension in the rubber band has increased.
3. Yes; the sound has a higher pitch because the pencil shortens the length and increases the tension of the rubber band.

Answers to
Activity 4

3. Shortening the straw produces a higher pitch.

Exploration 1 continued

364

Answers to
Activity 5, page 363

1. The vocal cords vibrate. Fast vibrations produce high (or high-pitched) sounds; slow vibrations produce low (or low-pitched) sounds.

2. To get a higher sound, decrease the size of the balloon's opening. To get a lower sound, increase the size of the opening. To produce a louder sound, push the air out of the balloon more quickly.

3. Tightening the balloon (lengthening the slit) raises the pitch.

Answers to
Making Sense of Your Observations

1. Vibrations occur in the tines of the tuning fork, the ruler, the stretched rubber band, the end of the straw, the column of air within the straw, the tissue of the vocal cords, the columns of air in the larynx and in the cardboard tube, and the stretched rubber of the balloon.

2. Energy was provided by striking the tuning fork, bending and releasing the ruler, plucking the rubber band, blowing air into the straw, blowing air into the balloon, blowing air through the vocal cords, and blowing air through the cardboard tube. Blowing gives energy to the air particles, causing them to move.

3. A louder sound was produced by providing more energy by hitting harder, by plucking farther, and by blowing more air into the straw, balloon, or vocal cords.

Answers to Making Sense of Your Observations are continued on the next page. ▶

Meeting Individual Needs

Learners Having Difficulty
For students who need more reinforcement to understand the term *pitch*, ask students to make a list of high-pitched and low-pitched sounds. Examples of high-pitched sounds include chalk (or fingernails) scraping a board, a rusty hinge, and a soprano voice. Examples of low-pitched sounds include a bass guitar, distant, rumbling thunder, and motors of large vehicles such as boats and buses.

Making Sense of Your Observations

After doing Exploration 1, you now know the following facts about sound:

1. *Sound is caused by vibrating objects.* What vibrates in each of the experiments in Exploration 1?

2. *Energy must always be added to an object in order for sound to be produced.* A force is exerted on a part of an object to move it a certain distance and to make it start vibrating. This means that work is done on the object. Whenever work is done on an object, energy is given to it. In the case of sound sources, this energy causes a vibration, which results in sound—one of the forms of energy produced by the vibrating object.

How did you provide energy in each of the Activities in Exploration 1?

3. *If more energy is added to an object, the object passes through a greater distance as it vibrates, and the sound is louder.*

Ruler motionless—no sound produced

Ruler pushed down and released—sound produced

Ruler pushed farther down and released—louder sound produced

How did you produce a louder sound in each experiment in Exploration 1?

 PORTFOLIO
You may wish for students to keep their results from Exploration 1 and their responses to Making Sense of Your Observations in their Portfolio as a record of their initial work with sound. Later, to illustrate what they have learned, students could also include in their Portfolio results from the more sophisticated laboratory activities near the end of this unit.

Homework
Have students experiment at home with making sounds by blowing across the tops of soda bottles. Tell students to try bottles with different amounts of water to see how the sound varies with the water depth. The next day, have students report their findings to the class.

4. *The more often an object vibrates within a given period of time, the higher the pitch of the sound produced.*

Vibrates slowly—low sound

Vibrates faster—higher sound

Vibrates even faster—even higher sound

How do our vocal cords produce higher-pitched sounds? How did you get higher-pitched sounds with the rubber band? How did you get higher-pitched sounds with the straw saxophone?

5. *Many vibrating objects have a natural rate of vibration, which depends on their length or size.* Can you use this idea to relate the sound of an object to its length or size? How did the length and size of the tuning fork affect the sound it produced? How did the length of the straw affect the sound? If you dropped a long pencil and a short one onto a hard surface, how would the sounds made by each pencil differ? Try it. Then try dropping both a large book and a small one. What would you do if you wanted to keep the loudness of their impacts approximately the same?

➤ Which pencil would make a higher sound when it hits the floor?

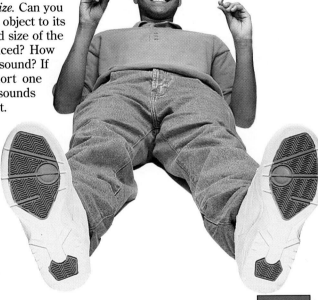

Theme Connection

Energy
Focus question: How can different forms of energy be used to produce sound? *(Accept all reasonable responses. Examples include chemical energy in fireworks and animals, electrical energy in radios and televisions, potential energy in raised objects, and heat energy in whistling kettles.)* Encourage students to think of as many different forms of energy as possible. Then ask: What type of energy must be present when sound is produced? *(Kinetic energy)*

Answers to
Making Sense of Your Observations, *continued*

4. Vocal cords create high pitches when the membranes are tight. More relaxed membranes create lower pitches.
    The rubber band produced a higher pitch when the tension in the rubber band was greater or when the rubber band was shorter.
    Decreasing the length of the air column in the straw produced higher pitches.

5. From the results of these experiments, students should conclude that the larger or longer an object, the slower its natural rate of vibration.
    The larger tuning fork, with its larger and longer tines, produced a lower pitch. In fact, the slower vibration rate in the larger tuning fork can be observed visually.
    The shorter the straw, the higher the pitch and the greater the vibration rate of the column of air in the straw.
    The longer pencil would produce a lower pitch; a shorter pencil would produce a higher pitch.
    To keep the loudness of the books' impacts approximately the same, drop the smaller book from a greater height or throw it to the ground. This would give the smaller book a greater velocity than the larger book. Because the smaller book is likely to have less mass, its greater velocity will result in both books having approximately the same energy and thus the same loudness at impact.

Meeting Individual Needs

Learners Having Difficulty
Perform the following demonstration: Bring a bicycle to class. Turn it upside down. Spin the wheel while holding an index card against the wheel. Ask students what is vibrating to make the sound. *(The card)* Spin the wheel faster and ask students what happens to the sound and why. *(The pitch gets higher because the card vibrates faster.)* Ask students where the card gets the energy it needs to vibrate and make a sound. *(The bicycle wheel transfers its kinetic energy to the card, which converts the energy to sound energy.)*

1. What is vibrating:
   a. Soft drink bottle—The column of air inside the bottle vibrates.
   b. Tearing paper—Strands of material in the paper vibrate when they are broken apart by the tearing action.
   c. Whistle—The column of air between the lips vibrates.
   d. Shutting a door—The door, door jam, and wall vibrate.
   e. Listening on the phone—The diaphragm of the speaker in the earpiece vibrates.
   f. Tap-dancing—The floor and the shoes vibrate as the toes and heels of the shoes strike the floor.

2. What is vibrating:
   a. Piano—Hammers hit the strings, causing them to vibrate. This vibration is transferred to the soundboard, which is connected to the strings.
   b. Harmonica—A series of graduated metal reeds vibrate as air is blown or sucked across them.
   c. Trumpet or bugle—Your lips vibrate, causing a column of air to vibrate when it is forced through the trumpet or bugle.
   d. Drum—The skin of the drum and the column of air inside the drum vibrate when struck with a stick or mallet.
   e. Upright bass—The strings vibrate when they are bowed or plucked.
   f. Flute—Air blown over the mouthpiece opening causes the column of air in the instrument to vibrate.

Homework

The questions in Some Puzzlers make an excellent homework activity.

Some Puzzlers

1. What is vibrating when you perform each of the following actions?
   a. blow over the mouth of a soft-drink bottle
   b. tear a piece of paper
   c. whistle
   d. shut a door
   e. listen on a phone
   f. tap-dance on an uncarpeted floor

2. What is vibrating when you play each of the following musical instruments?
   a. a piano
   b. a harmonica
   c. a trumpet or bugle
   d. a drum
   e. an upright bass
   f. a flute

 Cooperative Learning
EXPLORATION 2, *PAGE 367*

Group size: 3 to 4 students
Group goal: to design an experiment to find out what features of vibrations make them audible
Positive interdependence: Assign the following roles: recorder (to write the group's responses), reader (to read the directions for Part 1 aloud), manager (to make sure all group members participate and understand and to serve as the group's liaison with the teacher), and equipment manager (to obtain materials and direct cleanup). Students should complete Part 1 of Exploration 2 according to the directions. The manager should present the group's experimental design to the teacher before proceeding with the experiment.
Individual accountability: Students should complete Part 2 of Exploration 2 individually.

 An Exploration Worksheet (Teaching Resources, page 9) and Transparency 58 are available to accompany Exploration 2 on page 367.

EXPLORATION 2

Sound Questions to Investigate

PART 1

Design an experiment that uses a meter stick to answer the following questions:

- Must vibrations occur at a certain rate to be audible (capable of being heard)?
- Must vibrations move a certain distance to be audible?
- Is a minimum amount of energy necessary in order for sound to be audible?

Write your procedure in your ScienceLog, and then perform the experiment. The diagrams in Making Sense of Your Observations (pages 364 and 365) may give you some ideas.

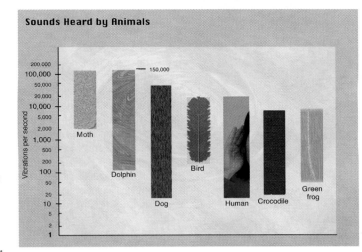

Sounds Heard by Animals

PART 2

The graph above shows the approximate vibration rates required to produce sounds that various kinds of animals can hear. From the graph, find the answers to the questions below. (Hint: First write down the approximate range of vibration rates heard by each animal.)

1. Note that a vibrating object must move back and forth at least 20 times per second to produce a sound that humans can hear. Above what rate of vibration are sounds inaudible to humans?

2. Which animal(s) can hear higher sounds that aren't audible to people? Which animals can hear lower sounds?

3. Which animal can hear the widest range of vibration rates?

4. What sounds can be heard by a green frog but not by a bird?

5. Which animal can't hear most of the sounds that a human can hear? What range of vibration rates can both humans and that animal hear?

6. Wanda blows a whistle. She hears nothing, but her dog begins to howl. Why?

367

EXPLORATION 2

PART 1

Divide the class into small groups and distribute the meter sticks. Students should have little difficulty in designing simple experiments to answer the three questions. Suggest that one person from each group be in charge of holding the meter stick firmly against the table; another student should record the observations; and the remaining students should carry out the experiment.

Answers to
Part 1

- Yes; for humans the rate of vibration must be between 20 and 20,000 vibrations per second.
- Yes, there is a minimum distance that the ruler must move back and forth for the vibrations to be audible.
- Yes, there is a minimum amount of energy needed to make the ruler vibrate so that the sound is audible.

PART 2

Student observations from Part 1 should help them to realize that their ability to hear sounds may be different from that of many other organisms.

Note: Most animal species listed on this page fall within the frequency ranges shown. However, there are some species whose hearing ranges fall outside of those depicted.

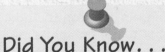

Did You Know...

Mosquitoes, crickets, moths, grasshoppers, and butterflies are among the few kinds of insects that have the ability to hear.

Answers to
Part 2

Answers will be approximations of the actual values. Sample answer:

Moth:	2000–150,000
Dolphin:	150–150,000
Dog:	15–50,000
Bird:	250–21,000
Human:	20–20,000
Crocodile:	30–6000
Green frog:	50–8000

1. Humans cannot hear sounds above about 20,000 vibrations/second.

2. Dogs, moths, birds, and dolphins can hear higher sounds. Dogs can hear slightly lower sounds.

3. The dolphin can hear the widest range of vibrations.

4. Sounds in the range of 50–250 vibrations/second can be heard by a green frog but not by a bird.

5. The crocodile cannot hear most of the sounds that a human can hear. (Among the animals listed, crocodiles share the smallest range of vibration rates with humans.) Both humans and crocodiles can hear sounds between 30 and 6,000 vibrations/second.

6. The whistle's rate of vibration is too high to be audible to Wanda, but it is audible to the dog.

Sounds, Naturally

Have students read the material on animal sounds individually, or call on volunteers to read it aloud. Before students begin reading, ask them to keep the following questions in mind:

- What do some frogs use to amplify the sound from their vocal cords? *(Balloonlike air sacs)*
- What are the loudest sounds made by any animal? *(Songs of humpback whale)*
- How do humpback whales produce sounds if they have no vocal cords? *(By forcing air back and forth through the larynx, or voice box)*
- What is a syrinx, and how is it used to produce sounds? *(It is an organ located at the bottom of a bird's windpipe. It contains membranes that vibrate when air passes over them. Muscles attached to the membranes change the pitch of the sound.)*
- Name some ways that insects produce sounds. *(Grasshoppers rub their legs against their wing covers. Mosquitoes flap their wings in the air hundreds of times a second. Crickets move a toothed file located under one wing against a scratcher located under the other wing.)*

CROSS-DISCIPLINARY FOCUS

Mathematics

Students may be interested to know that the temperature of the air can be estimated by counting the number of times a male North American field cricket chirps in 14 seconds. Add 40 to this number to get the temperature in degrees Fahrenheit. To convert this to degrees Celsius, subtract 32 and multiply by 5/9.

Meeting Individual Needs

Second-Language Learners

Students who have a tape recorder and access to an area frequented by birds can perform a study of bird songs. They should record the sounds of birds in the area and, with the help of a bird identification guide, try to identify the bird species from the songs.

Sounds, Naturally

Animals produce sounds of many kinds and make them in many ways. A dog, for instance, whines, whimpers, barks, growls, and howls. To make these sounds, it uses vocal cords, just as humans do. In fact, this is true of most mammals as well as other animals, including frogs. Many kinds of frogs also have balloon-like air sacs that amplify the sound made by their vocal cords.

Some other mammals (as well as humans) actually make sounds that we might describe as songs. Perhaps the most interesting songster is the humpback whale. Roger Payne, an American naturalist, investigated whale sounds with underwater microphones. Imagine his surprise when he discovered that humpbacks sing, producing what he described as "hauntingly beautiful sounds." Lasting from 6 to 30 minutes, the songs consist of low moaning, wailing, and lowing (cow-like) sounds, as well as high-pitched whistles and screeches.

Humpbacks' songs are the loudest sounds made by any animal. Scientists infer that before the time of steam- and oil-powered ships with noisy propellers, the song of a whale could be heard by other whales as far away as 1600 km. Even today, despite the continuous background noise from ships' propellers, a whale can hear another whale's song up to 160 km away.

Humpback whales have no vocal cords. Their sounds, or songs, are probably produced by forcing air back and forth through the **larynx,** or "voice box."

The most familiar animal songs are those produced by birds. Bird songs are made up of chirps, trills, whistles, and, sometimes, quick mechanical sounds made by clicking the beak. The nonmechanical sounds are produced by an organ called the **syrinx,** or "song box." The syrinx is located at the bottom of the windpipe (the tube connecting the mouth to the lungs). It contains a pair of membranes that vibrate when air passes over them, producing sound. Muscles attached to the membranes change the pitch of the sound.

◀ Humpback whale   ▲ Eastern meadowlark

Theme Connection

Structures

Focus question: How are the structures that animals use to make sound similar to musical instruments? *(Wind instruments depend on columns of air vibrating within them, which is similar to the windpipe and larynx or syrinx of some animals. Stringed instruments are plucked or bowed, which is similar to insects rubbing their legs or wings together. Percussion instruments are struck by a stick or other object to make sounds. Some animals thump the ground or water for a similar effect.)*

Integrating the Sciences

Life and Physical Sciences

Discuss with students why humans cannot produce the high-pitched sounds that birds are capable of producing. *(The membranes of our vocal cords cannot vibrate fast enough.)* Point out that we can imitate the high-pitched sounds of birds by whistling. When we whistle, we cause air to vibrate at a very high speed.

Grasshoppers produce their mechanical sounds by rubbing, or "bowing," their long, hard, rough legs against their hardened wing covers. Mosquitoes produce their high-pitched whine by flapping their wings in the air many hundreds of times per second. All male crickets and females of some species chirp by moving a body part that looks like a toothed file, and is fastened under one wing, against a scratcher located under the other wing. There are approximately 4000 kinds of crickets, each producing its own distinct pattern of chirps. This is how members of the same species recognize each other. Even if 50 kinds of crickets are singing in a meadow, female crickets can recognize the call of the male crickets of their species.

Remember that sound is energy—and that it takes energy to make sound. For each chirp of a cricket, 19 pairs of muscles are doing work! Where does the energy come from when *you* make sounds? Ⓐ

369

Answer to
In-Text Question

Ⓐ The energy to make sounds ultimately comes from the food we eat. (Chemical energy is converted into mechanical energy.)

ENVIRONMENTAL FOCUS

Have students research how animals not mentioned in the section titled Sounds, Naturally produce sounds for communication. (*Examples include the lobster, which clacks parts of its exoskeleton; the male cod fish, which grunts to female cod fish; and the satin fish, which can produce purring sounds.*)

FOLLOW-UP

Reteaching
Show a large stereo speaker without its cover. Play music with bass notes so that students can watch the diaphragm vibrate. Increase the volume and have students explain what happens. (*The amplitude of the vibrations increases. They may even be able to feel the vibrations through the air if they place their hand near the diaphragm.*) Warn students not to put their ears too close to the speaker.

Assessment
Give pairs of students a tuning fork and a pith ball or small ball of aluminum suspended from a thread. Have one member of the pair strike the tuning fork while the other tries to pass the pith ball between the tines. (This will be impossible as long as the fork is vibrating strongly.) Have students explain their observations. (*The pith ball is pushed away by the sound waves produced by the tines.*)

Extension
Point out to students that human communication relies as much on how people manipulate the sounds of a language as it does on the literal meanings of words. For example, a person can change a statement into a question by simply raising the pitch, or frequency, of his or her voice at the end of the statement. Challenge students to think of other ways in which the sound qualities of speech affect meaning. If you have time, allow students to demonstrate how sound alone can convey meaning and emotion or how changing various qualities of speech can drastically affect meaning.

Closure
Ask students to write a short explanation of how sound could indicate the efficiency of a machine, such as a bicycle or a car engine. (*The louder the sound from the machine is, or the more sounds produced, the less efficient the machine is.*)

Homework
You may wish to have students complete the short writing assignment in Closure as homework.

Vibrations: How Fast and How Far?

LESSON 3

Vibrations: How Fast and How Far?

When a bird sings, the membranes of its syrinx may vibrate as many as 20,000 times per second. When you pluck middle G on a guitar, it vibrates about 400 times in 1 second. The larynx of the humpback whale vibrates about 20 times per second when making its lowest sound.

Objects that produce sound generally vibrate too quickly for the vibrations to be seen. How might you observe the individual vibrations? One way is to take pictures of them with a high-speed motion picture camera and then project the pictures at a slow speed. You would then see the vibrations in slow motion.

In the following Exploration, you will analyze a slow, "sound-less" vibrating situation. This will help you understand the much faster vibrations that produce sound.

FOCUS

Getting Started

Show students a swinging pendulum. Have them estimate how many times the pendulum moves back and forth in one second. Explain that if the pendulum produced sound, the pitch would be too low for humans to hear. Ask: How many times per second would the pendulum have to vibrate to produce a sound that we could hear? *(At least 20 times per second)* Point out that in this lesson they will experiment with a pendulum to understand how vibrations produce sound.

Main Ideas

1. A vibration consists of one complete back-and-forth motion.
2. The frequency of vibration is the number of times an object vibrates in one second.
3. The frequency of vibration of a pendulum does not change when the amplitude of the vibration is changed.
4. Decreasing the length of a pendulum increases the frequency of vibration.

EXPLORATION 3

Investigating an Object Vibrating Very Slowly

You Will Need

- a heavy button, washer, or large paper clip
- a metric ruler
- a watch or clock with a second hand
- fine thread
- scissors

What to Do

1. Make a pendulum by hanging a button from a thread that is 25 cm long. Pull the button 10 cm to one side and let it go.

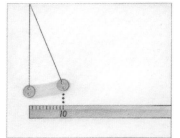

2. A **vibration** is defined as one complete back-and-forth movement of an object, from one side to the other and back again. Approximately how long does one vibration take? Does the time seem to change from one swing to the next?

Devise a way to find the time one vibration takes. About how many vibrations are there in 1 second?

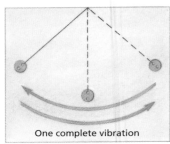

One complete vibration

TEACHING STRATEGIES

EXPLORATION 3

Because experimental errors may occur, tell students to take several measurements and average the results. Also, suggest that students tap the table when each vibration is complete. They should realize that the time between taps remains constant for a given length of the pendulum. The time it takes for a pendulum to complete one vibration is called the *period* and does not vary with small changes in amplitude.

 An Exploration Worksheet is available to accompany Exploration 3 (Teaching Resources, page 11).

LESSON 3 ORGANIZER

Time Required
1 to 2 class periods

Process Skills
observing, hypothesizing, analyzing, inferring, measuring

New Terms
Amplitude—the size of a pendulum swing; the greatest distance that a vibrating object moves from its rest position
Frequency—the number of vibrations in one second
Vibration—one complete back-and-forth movement of an object

Materials (per student group)
Exploration 3: heavy button, washer, or large paper clip; metric ruler; watch or clock with a second hand; 25 cm of fine thread; scissors

Teaching Resources
Exploration Worksheet, p. 11
SourceBook, p. S109

3. Repeat step 2, but pull the button 20 cm to one side. The size of the swing—the horizontal distance from the side position to the central position, as shown below—is called the **amplitude**. Does the time for one vibration appear to change when the button swings twice as far?

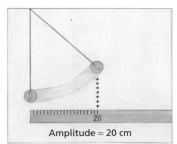

Amplitude = 20 cm

4. Shorten the thread to 6 cm. Pull the button 3 cm to one side and let it go. How much time does one vibration take? How many vibrations are there in 1 second? The number of vibrations in 1 second is called the **frequency** of vibration.

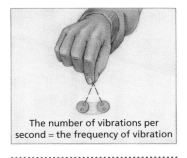

The number of vibrations per second = the frequency of vibration

Vibration
Amplitude
Frequency?

Mason's Problem

Suppose you had to explain the meaning of *vibration, amplitude,* and *frequency* to Mason, a fourth-grader. How would you do this, using a metal ruler as your teaching tool? In your ScienceLog, write down what you would say.

Check It Out

Review what you discovered in doing the previous Exploration. From the statements below, identify those that you agree with, those that you disagree with, and those that you are unsure about.

1. Changing the amplitude of the swing did not change the time needed for one swing very much. If I could do the experiment accurately enough, it probably wouldn't change the time at all.

2. If I double the distance the button swings, I can make a big change in the time it takes for one vibration to occur.

3. If I were to start the swing only 5 cm out from the center, there would be fewer vibrations in 1 second.

4. If I shorten the thread, the time needed for one vibration to occur will be reduced.

5. If I shorten the thread, the frequency will be greater (the button will complete more swings in 1 second).

6. If I use about one-quarter of the length of thread, the time it takes for one vibration to occur will be about half that required when using the whole length of thread.

7. Changing the length of the thread does not change the number of vibrations that occur in 1 second.

8. If I were to make the thread 100 cm long, the time of one vibration might be 2 seconds.

 Cooperative Learning
EXPLORATION 3

Group size: 2 to 3 students
Group goal: to determine the relationship between the length of a pendulum and the amplitude and frequency of its vibration
Positive interdependence: Assign the following roles: architect (to read the directions to the group and build the pendulum), timer (to hold the pendulum and time the swings), and recorder (to count and record the number of swings). The groups should conduct the Exploration using a pendulum system that they construct themselves. Each group should prepare a data chart and graph. To improve accuracy, each group should conduct three trials for each activity and average the results. The group should then work together to answer the in-text questions.
Individual accountability: Each student should individually answer the questions from Check It Out on this page.

Answers to
Exploration 3, pages 370–371

2. One vibration should take about 1 s. The time does not change appreciably from one swing to the next. To find the time for one vibration, measure the time required for several vibrations and then divide it by the number of vibrations. For a 25 cm pendulum, there is about 1 vibration/second.

3. No, the time should not appear to change.

4. One vibration takes about 0.5 seconds (a frequency of 2 vibrations/second).

Answers to
Check It Out

1. Agree; the period does not depend on the amplitude of the swing.

2. Disagree; the period does not depend on the amplitude of the swing.

3. Disagree; the frequency of vibration does not depend on the amplitude of the swing.

4. Agree; the period depends on the length of the pendulum.

5. Agree; if the time for one vibration (the period) is decreased, the frequency is increased.

6. Agree; this follows from the results of Exploration 3.

7. Disagree; the frequency of vibration depends on the length of the pendulum.

8. Agree; based on the results of Exploration 3, lengthening the string by 4 times should double the period.

Answer to
Mason's Problem

Student answers will vary. You may wish to discuss the meaning of the three terms before students write their explanations. Then have volunteers present their answers to the class, and discuss which responses are most accurate and which teaching methods might be most effective. Students will explore these terms further in A Sound Experiment on page 372.

Reteaching

Attach a small weight to a piece of string about 1 m long and swing it from side to side. Remind students that this is a simple pendulum. Identify the *frequency* by counting the number of back-and-forth motions in a given time period and dividing by the time. Identify the *period* by timing the duration of a number of back-and-forth swings and dividing by the number of swings. Point out that the frequency multiplied by the period is always 1—they are reciprocals of one another. Repeat the demonstration, but vary the amplitude to show that the period and frequency do not change from trial to trial.

Assessment

Using thread and a washer, challenge students to make a timer that will accurately measure 4 s. *(Students will need to find the proper length of thread so that the period of the pendulum is 1 s or a multiple of 1 s. For instance, 25 cm of thread will give a period of 1 s, 1 m of thread will give a period of 2 s, and so on.)*

Extension

Have students research the role of the pendulum in a grandfather clock. Or suggest that students visit a music store to find out how frets and bridges in guitars and other stringed instruments affect the sounds produced by those instruments.

Closure

Bring an encyclopedia to class and read the entry on Alexander Graham Bell. After reading the material and discussing how Bell invented the telephone, ask students to write mock transcripts of an interview with him. Tell students that their interviews should include the terms *pitch*, *frequency*, and *amplitude*.

Homework

The root word *phon* comes from the Greek language and means "sound." Challenge students to list as many words containing this root as they can. Then discuss the definitions of the words that they find. *(Likely words include phonetics, cacophony, homophone, and telephone.)*

A Sound Experiment

Jennifer wanted to count the vibrations of some metal rulers. She took pictures of them with a high-speed movie camera and then projected the film at a much slower speed—about 1/200 of the original speed. Suppose one back-and-forth motion (one vibration) took 1 second when her film was shown at this slower speed. Satisfy yourself that the ruler's actual frequency of vibration would be 200 vibrations per second.

Jennifer counted the number of vibrations in a given time from her screen projections, as observed in slow motion.

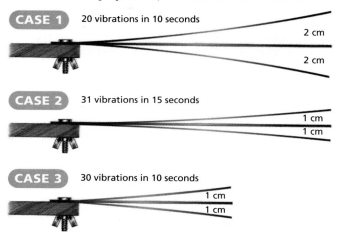

CASE 1   20 vibrations in 10 seconds   2 cm / 2 cm

CASE 2   31 vibrations in 15 seconds   1 cm / 1 cm

CASE 3   30 vibrations in 10 seconds   1 cm / 1 cm

For Case 1

What is the frequency in slow motion? Try calculating the actual frequency.

For Case 2

a. What is the frequency in slow motion? in real motion?

b. How do the amplitudes in Cases 1 and 2 compare?

c. Does the amplitude affect the frequency of vibration?

For Case 3

a. What is the frequency in slow motion? in real motion?

b. How do the lengths of the two rulers in Cases 1 and 3 compare?

c. Do their lengths affect the frequency of vibration?

Conclusions

a. In which Case is the loudest sound produced?

b. In which Case is the highest-pitched sound produced?

372

Answers to *A Sound Experiment*

Case 1
The frequency in slow motion is 20 ÷ 10, or 2, vibrations/second. Therefore, the frequency in real motion is 200 × 2, or 400, vibrations/second.

Case 2
a. The frequency in slow motion is 31 ÷ 15, or about 2, vibrations/second. Therefore, the frequency in real motion is about 200 × 2, or 400, vibrations/second.

b. The amplitude in Case 1 is two times the amplitude in Case 2.

c. The amplitude does not appreciably affect the frequency of vibration.

Case 3
a. The frequency in slow motion is 30 ÷ 10, or 3, vibrations/second. Therefore, the frequency in real motion is 200 × 3, or 600, vibrations/second.

b. The ruler is shorter in Case 3 than in Case 1.

c. Yes; as the length is shortened, the frequency increases.

Conclusions
a. Case 1 has the largest amplitude and therefore the loudest sound.

b. Case 3 has the greatest frequency and therefore the highest pitch.

CHALLENGE YOUR THINKING

1. Silent Communication

Sara is totally deaf and has never heard a sound. You are able to communicate with her only by writing. What would you write to help Sara understand the ideas of high- and low-pitched sounds, soft and loud sounds, and how we distinguish one sound from another?

2. Liner Notes

Ryan attached a piece of nylon fishing line to a nail at one end of a board. He stretched it along the board and hung a pail of sand from the other end. When he plucked the string, he got noise. When he slipped a wooden dowel between the string and the board, though, he got a musical sound. Why? What different ways could he devise to make the musical sound (a) higher pitched? (b) louder? (c) of a different quality?

3. Sound Reasoning

How would you demonstrate or explain convincingly each of the following to someone who has never studied science before?

a. Energy is needed to produce sounds.

b. Sound is produced from vibrating objects.

c. Not all vibrating objects produce sounds that you can hear.

d. Some sounds are too high or low for humans to hear.

e. The vibrations that produce noise and music are different.

373

Did You Know...

The word *stethoscope* means "to look at the chest" even though the device is used to listen to the chest. Some doctors have suggested that people call the device by the more accurate name of *stethophone*.

Answers to *Challenge Your Thinking*

1. Responses will vary. Sample answer: Sounds that we hear come from vibrations, like those you might feel from a motor or an earthquake. How quickly an object vibrates, or an object's frequency of vibration, determines pitch. Pitch refers to how high or low a sound seems. Rapid vibrations produce high-pitched sounds, and slow vibrations produce low-pitched sounds. The loudness of a sound depends on the size, or amplitude, of the vibrations. Just as a vibration has to be large or strong enough to be felt, a sound has to be loud enough to be heard.

Pitch and loudness are two dimensions of sound that help us distinguish one sound from another. We also distinguish sounds by their duration and quality. The duration of a sound is the length of time that the sound can be heard. The quality of a sound relates to the mixture of sounds that contribute to the over-all sound. Two sounds that have the same pitch, loudness, and duration can be distinguished from one another by their sound quality.

2. Without the wooden dowel, the string was not free to vibrate back and forth to produce a simple musical tone. Instead, it slapped against the board in various places and created a noisy mixture of sounds. With the wooden dowel under the string, the string vibrated freely and produced a purer sound (containing fewer component sounds).

a. To make the sound higher pitched, Ryan could move the dowel farther away from the pail to shorten the vibrating portion of the string, or he could increase the weight of the pail to increase the tension of the string. Both of these changes would increase the frequency of vibration of the string. A string with a smaller diameter would also produce a higher-pitched sound.

b. To make the sound louder, Ryan could pluck the string harder, thereby giving more energy to the string and increasing its amplitude of vibration. Changing the material of the board, the dowel, or the string could also make the sound louder.

c. To change the quality of the sound, Ryan could pluck the string with different materials, or he could use a bow to vibrate the string instead of plucking. Ryan could also change the material and shape of the board, the dowel, or the string to produce a sound of a different quality.

3. Answers will vary. The following are sample explanations or demonstrations:

a. A drum produces no sound until it is struck.

b. When a tuning fork is struck, it produces sound while it vibrates.

Challenge Your Thinking continued ▶

Answers to Challenge Your Thinking, continued

c. A dog whistle vibrates when blown, but humans can't hear the sound.

d. A speaker producing a sound too high or low for humans to hear could be placed near a bowl of water so that ripples are produced in the water.

e. Noise from a lawn mower is a mixture of sounds resulting from the vibrations of many different moving parts. Music from a violin is a mixture of fewer component sounds resulting from the vibrations of a few, carefully controlled, parts.

4. Possible answers include the *pitter-patter* of rain on the roof, the *swish* of water against galoshes, the *lapping* of water on the seashore, the *drip* of water from a tap, the *bubbling* of water in a kettle, the *rushing* of water down an incline, the gentle *murmuring* of water in a brook, and the *roar* of a waterfall.

5. **a. s**lowly
b. lower
c. quality
d. energy
e. loud**er**
The mystery word is *sound.*

★ **You may wish to provide students with the Chapter 16 Review Worksheet that is available to accompany this Challenge Your Thinking (Teaching Resources, page 13).**

ScienceLog

The following are sample revised answers:

1. When the player moves her finger up the neck of the banjo, the vibrating part of the banjo string becomes shorter, causing the string's frequency of vibration to increase and the pitch produced to become higher. If the player moves her finger halfway up the neck from the original position of her finger, a note that is exactly twice the frequency (also referred to as one *octave*) will result when she plucks the string.

2. Music is composed of repeating patterns of sound waves. A vibrating object that produces music, such as a musical instrument, vibrates as a whole in a relatively simple manner and with some duration before the vibration dies out. Noise, on the other hand, consists of sound waves that are irregular, with no identifiable pattern.

3. Sound is produced when energy is transferred to an object, causing the object to vibrate. The vibrations are then transferred to the air (or another medium, such as water) in the form of sound waves. When these sound waves strike a receiver (for example, a human ear), they cause the receiver to vibrate in response. These vibrations are then sent to a processor (for example, the human brain) that interprets the vibrations. The result is the detection of sound.

Multicultural Extension

Onomatopoeia in Different Languages

Explain that *onomatopoeia* is the use of a word that sounds like what it describes, such as buzz, hiss, and splash. Have students find out some onomatopoeic words in other languages. (Words describing animal sounds are often good examples.)

---

4. I Call This One "Kerploosh"

Words often sound like the sounds they describe. For example, the words *gurgling* and *splash* convey certain sounds made by water. How many more watery-sounding words can you think of? What are they? Coin some new words that describe the action of water in various situations.

5. Acrostics

Copy the acrostic puzzle shown at right into your ScienceLog. Can you fill in the blanks with the five appropriate words and then determine the mystery word? The questions should be answered in order.

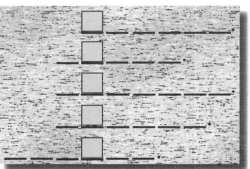

a. An object whose frequency is small vibrates ____?____.

b. If a violin string is slightly loosened, the musical tone will be ____?____.

c. Two sounds of the same pitch and loudness may still differ in ____?____.

d. The loudness of a sound depends on the ____?____.

e. If the amplitude of a vibrating object is increased, its sound will be ____?____.

Now make an acrostic puzzle of your own using the ideas you have learned about sound so far. Give it to someone else to solve.

ScienceLog

Review your responses to the ScienceLog questions on page 358. Then revise your original ideas so that they reflect what you've learned.

CHAPTER

17

How Sound Moves

10:06:17

1 You see a flash of lightning, and several seconds later you hear the roll of thunder. How do you explain this?

10:06:19

BOOM!

2 Is there sound in outer space?

3 How does sound travel from source to listener?

ScienceLog

Think about these questions for a moment, and answer them in your ScienceLog. When you've finished this chapter, you'll have the opportunity to revise your answers based on what you've learned.

375

Homework

Have students try the following experiment at home and report their findings the next day: Ask students to repeatedly dip a finger into a wide pan of water to make a circular wave on the surface. What happens to the distance between waves if the finger is dipped more frequently? How is this like sound? *(The distance between water waves becomes smaller as the frequency of dipping is increased. This corresponds to the distance between sound waves [compressions] becoming smaller as the frequency of the vibrating source increases. The result is sound with higher pitch.)*

Connecting to Other Chapters

> **Chapter 16**
> *shows how vibrations produce sounds and discusses ways of distinguishing different sounds.*

> **Chapter 17**
> *explores how the particles of a medium transmit sound energy and how our ears receive that energy.*

> **Chapter 18**
> *invites students to reexamine pitch, loudness, and sound quality in a quantitative way.*

Prior Knowledge and Misconceptions

Your students' responses to the ScienceLog questions on this page will reveal the kind of information—and misinformation—they bring to this chapter. Use what you find out about your students' knowledge to choose which chapter concepts and activities to emphasize in your teaching. After students complete the material in this chapter, they will be asked to revise their answers based on what they have learned. Sample answers can be found on page 398.

In addition to having students answer the questions on this page, you may wish to have them complete the following activity: Draw a cartoon picture of a face and a ringing bell on the board. Write the title "Hearing the Sound of a Bell" over the drawing. Ask students to copy the illustration and add labels and additional features as needed to explain how sound travels and how we hear sound. Collect the drawings, but do not grade them. Instead, use the drawings to find out what students know about how sound moves and how we hear sounds, what misconceptions they may have, and what aspects of this topic are interesting to them.

FOCUS

Getting Started

Have each student work with a partner to complete this short activity. Partners should stand at opposite ends of a table. First one student drops a pencil on the table while the other student listens. Then the second student puts one ear against the table while the pencil is dropped again. Ask the students to trade roles and repeat the activity. Ask: What differences do you observe between the two trials? *(The sound of the pencil hitting the table is louder the second time.)* Why is this so? *(Sound travels as the vibrations of particles of matter. Sound travels more easily through a solid than a gas because the particles of a solid are much closer together than those of a gas.)*

Main Ideas

1. Energy can be transmitted through the particles of a substance.
2. Sound cannot be transmitted in the absence of particles.
3. Sound waves are alternate compressions and expansions caused by the back-and-forth motion of the particles of a medium.
4. Sound travels not only through air but also through materials such as wood, metal, water, and bone.

TEACHING STRATEGIES

Answer to
Caption

Ⓐ The potential energy stored in the suspended ball is turned into kinetic energy as the ball descends. This energy is transferred through each of the balls until it reaches the last ball, pushing this ball away from the other balls.

★ A Transparency Worksheet (Teaching Resources, page 18) and Transparency 59 are available to accompany the material on this page.

LESSON
1 Sound Travel

How does sound energy travel from the sound source to your ear? Do you have a theory that might explain this process? What evidence do you have for your theory? The six experiments in the next Exploration will provide evidence to help you build a theory of sound travel.

EXPLORATION 1

Transmitting Sounds

EXPERIMENT 1

How are other forms of energy transmitted?

1. Place five pennies flat on the table.
2. Flick a sixth penny so that it hits penny *E*.
3. What happens to penny *A*? What kind of energy was transmitted from *E* to *A*? What was the energy transmitted through?

Scientists tell us that air is made of tiny particles. Is sound energy transmitted through these particles in the same way that kinetic energy (energy of motion) is transmitted through the pennies?

Have you ever experimented with a device like this one? It operates on a principle similar to that demonstrated by the pennies. Can you explain what happens?

LESSON 1 ORGANIZER

Time Required 2 class periods

Process Skills hypothesizing, inferring, observing, analyzing

Theme Connection Energy

New Terms
Medium—a substance, such as air or water, through which a wave travels
Wavelength—the distance between two identical points on neighboring waves

Materials (per student group)
Exploration 1, Experiment 1: 6 pennies; **Experiment 2:** small ball of modeling clay; one-hole rubber stopper; wire clothes hanger; wire cutters; pliers; jingle bell; borosilicate boiling flask; about 20 mL of water; hot plate; safety goggles; oven mitts; **Experiment 3:** balloon; mailing tube; several rubber bands; piece of aluminum foil, about 15 cm²; scissors; materials to support candle, such as a small ball of modeling clay and jar lid; candle; a few matches; a few books; wastebasket; safety goggles; **Experiment 4, Part 1:** Slinky spring toy; 20 cm ribbon; **Experiment 5:** syringe without a needle;
continued ➤

EXPERIMENT 2

If there were no air (or anything else) between a sound source and your ear, would sound still be transmitted to you?

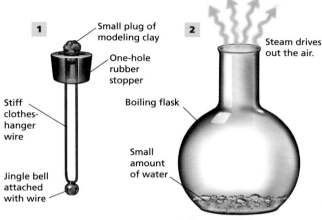

1
- Small plug of modeling clay
- One-hole rubber stopper
- Stiff clothes-hanger wire
- Jingle bell attached with wire

2
- Steam drives out the air.
- Boiling flask
- Small amount of water

Boil water and then remove from heat source. Do NOT boil away all the water.

3 Quickly insert the bell assembly. Let flask cool and then shake.

What do you observe?

4 Remove the clay plug to allow air to enter. Then replace plug and shake.

What do you observe?

EXPERIMENT 3

Does the air between a sound source and your ear move?

1. Touch the piece of stretched balloon lightly with your finger as you hold the mailing tube steady. What happens? Is air moving in the tube?

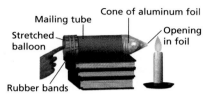

- Mailing tube
- Cone of aluminum foil
- Stretched balloon
- Opening in foil
- Rubber bands

2. Make a loud sound near the piece of stretched balloon by hitting the bottom of a wastebasket with your hand. Observe the flame.

3. Try counting out loud next to the piece of stretched balloon. What happens?

4. As a control, remove the apparatus and repeat steps 2 and 3.

When sound energy moves from a vibrating object through the air to the ear, what do you think happens to the air?

EXPERIMENT 4

PART 1

A Model of Air

How is energy transmitted through a coiled spring, for example, a Slinky?

1. Hold one end of a spring and have a classmate hold the other end. Send pushes and pulls along the spring. Do you see energy being transmitted?

Exploration 1 continued ▶

ORGANIZER, continued

small ball of modeling clay; **Experiment 6:** scissors; piece of cardboard, about 10 cm × 20 cm
Sound Thinking: metric ruler
Exploration 2: 30 cm wooden rod, at least 2 cm in diameter; thumbtack; tuning fork; various plastic, metal, and paper objects such as a plastic ruler, metal ruler, and spiral notebook; 2 metal spoons; 8 to 10 L of water; small aquarium or other very large container; a large and small funnel; balloon or piece of cellophane, about 25 cm × 25 cm; a few rubber bands; 1 m of rubber tubing; scissors

Teaching Resources
Transparency Worksheet, p. 18
Transparency 59
Exploration Worksheets, pp. 20 and 29
SourceBook, pp. S107 and S115

EXPLORATION 1

Except for Experiment 2, which may be done as a teacher demonstration, all of the activities in this Exploration can be set up in stations at different locations around the classroom. Make sure students wear oven mitts when handling the cooling flask.

Answers to
Experiment 1, page 376

3. Penny *A* slides away from the other pennies. Kinetic energy was transmitted from penny *E* to penny *A*. The energy was transmitted through the pennies.
   Sound is transmitted by air particles in the same way that kinetic energy is transmitted through the pennies.

Answers to
Experiment 2

3. If a good vacuum is created in the flask, students should hear very little sound, if any, when the bell is shaken.

4. When air enters the flask, the bell can be heard much more clearly. Students should conclude that air particles are needed for sound to be transmitted.

Answers to
Experiment 3

1. The flame flickers, indicating air movement. Touching the balloon pushes air down the tube. Air is funneled toward the flame (through the hole in the foil), which then flickers.

2. When a loud sound is made near the stretched balloon, the flame should flicker.

3. When students count out loud next to the balloon, the flame should flicker.

4. Students should notice a decreased effect or no effect at all on the candle flame.
   When sound energy moves from a vibrating object through the air, the particles of air vibrate.

Exploration 1 continued ▶

 An Exploration Worksheet is available to accompany Exploration 1 (Teaching Resources, page 20).

 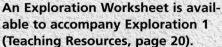

Answers to
Experiment 4, Part 1, pages 377–378

1. The compressions and expansions of the spring show that energy is being transmitted.

2. When a push or pull reaches the students' hands, they should feel a slight push or pull on their hands.

3. Answers will vary. Accept all reasonable responses.

4. A ribbon tied to part of the spring will show a back-and-forth motion as energy travels through the spring. This is evidence that energy is transferred by a back-and-forth motion of the particles of the medium.

Answers to
Experiment 4, Part 2, page 379

1. The coils of wire near Mario's end of the spring are compressed.

2. The compression travels along the spring as subsequent coils compress and preceding coils resume their original position.

3. Kate feels a push on her hand when the compression reaches her. The energy given to the spring by Mario travels along the spring to Kate and is transferred to Kate's hand.

4. When Mario pulls the spring toward himself, the coils of wire expand. After he stops pulling, the expansion travels along the spring from Mario to Kate.

5. Mario gives energy to the spring every time he produces a vibration.

6. The energy is passed along the coils of wire in the spring by a series of alternating compressions and expansions.

7. Each coil of wire goes back and forth as a compression and expansion pass through it.

8. As a tuning fork vibrates, it causes air particles to move back and forth and the air to alternately compress and expand. The compressions and expansions travel through the air, transferring the sound energy to the ear.

2. What do you feel when the "push" or "pull" reaches your hand?

3. How fast is the energy transferred along the spring?

4. Tie a ribbon to the middle of the spring. What happens to the ribbon as energy passes along the coiled spring?

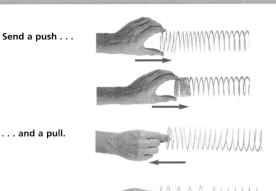

Send a push . . .

. . . and a pull.

PART 2

Spring Fling

The steps below match the diagram at lower right.

a. Mario and Kate hold the stretched spring on the floor.

b. Mario gives a quick push on the coiled spring, toward Kate. He then pulls it back quickly to its original position. Do you see the energy being transferred? **A**

c. When the coiled spring is still again, Mario quickly pulls the spring toward himself. He then returns it quickly to its original position. Do you observe the energy being transferred? **B**

d. Mario pushes his end of the spring quickly. He then pulls the spring back quickly as far past the starting position as he had pushed it forward. He then quickly pushes the spring back to the starting position. Mario has produced one complete vibration of the end of the spring. Kate says, "I think this is like what happens when a sound is made. One compression and one expansion together create a sound wave."

e. Mario produces two complete vibrations of the end of the spring, one right after the other.

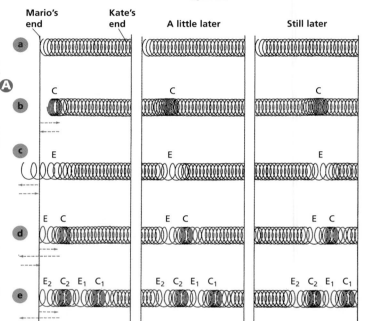

Key
C = Compression of the spring. The coils or turns of the spring are closer together than normal.

E = Expansion of the spring. The coils or turns are spread farther apart than normal.

Answers to
In-Text Questions

A The energy that travels from Mario to Kate is represented by the compression of the coils of wire.

B The energy being transferred from Mario to Kate is represented by the expansion of the coils of wire.

Theme Connection

Energy
Focus question: How is the electrical energy from a lightning bolt transformed into the sound energy of thunder? *(The electrical energy from lightning creates heat energy in the air through which it passes. This heat expands the air rapidly, thereby creating a compression wave in the air particles. The compression wave travels faster than the speed of sound and creates a sonic boom.)*

Use what Kate and Mario observed on page 378 to help you answer the following questions:

Questions

1. When Mario pushes, as in (b), what happens to the coils of wire?

2. What happens to the spring compression after Mario stops pushing?

3. When Mario pushes on the end of the spring, as in (b), he gives energy to the spring. What happens to Kate's hand when she receives the compression?

4. When Mario pulls, as in (c), what happens to the coils of wire? What happens to the spring expansion after he stops pulling?

5. Every time Mario makes his end of the wire go through one vibration, as in (d), what is given to the coiled spring?

6. How is the energy that Mario gives to his end of the spring passed along to Kate?

7. What is happening to each coil of wire as Mario sends a series of "waves" along the spring? (Review your observations of the ribbon.)

8. Suppose that air particles act like the turns of wire in a coiled spring. How would sound energy from a vibrating tuning fork pass through the air to the ear?

EXPERIMENT 5

Elastic Air

Are air particles elastic? That is, do they act like the turns of wire in a coiled spring?

1. Fill a syringe halfway with air, and plug its tip with modeling clay.

Trapped air

2. Push down on the piston, compressing the trapped air. Let the piston go. What happens?

3. Now pull up on the piston, letting the trapped air expand. Let the piston go. What happens?

4. Try compressing and extending a ball of clay. Which behaves more like a spring, clay or air?

EXPERIMENT 6

Viewing Moving Waves

Here's another way to view what may be happening in air. Cut a narrow slit 10 cm long and 1–2 mm wide in a piece of cardboard. Place the slit over the diagram on the next page and move it down at a constant speed.

Questions

1. In which direction does the compression move? (It's difficult to see expansions move.)

2. How can you make compressions move in the opposite direction?

3. How can you make compressions move faster?

4. How many compressions can you see at one time?

5. What do you think the **wavelength** is? Measure it.

6. As compressions move along, what is happening to each line as the card goes down the page?

Exploration 1 continued ▶

2. If air is compressed in a syringe and the piston is released, the piston will be pushed back to its original position.

3. If air is expanded by raising the piston and the piston is released, the piston will be pulled back to its original position.

4. Air behaves more like a spring because air returns to its original volume after being compressed or expanded, just as a spring returns to its original length. The clay, however, retains its new shape and will not return to its original shape.

Answers to *Experiment 6*

1. A series of four compressions move from right to left.

2. Moving the card up the page makes the compressions move in the opposite direction.

3. Moving the card faster makes the compressions move faster.

4. Two compressions, at most, can be seen, but usually only one is apparent.

5. The *wavelength* is the distance from the beginning of one compression to the beginning of the next. In this case, the wavelength is about 6.75 cm.

6. The lines are moving back and forth, but each line does not move at the same time as its adjacent line. There is a slight lag in motion (or position) from one line to the next.

Meeting Individual Needs

Learners Having Difficulty ◀

Provide students with a piece of carbon paper and a tuning fork with pieces of stiff wire attached to the prongs. Students should place the carbon paper on a blank sheet of paper and strike the tuning fork so that it vibrates. Instruct students to point the prongs of the tuning fork away from them and let the tips of the wires just touch the paper as they pull the tuning fork down the page. Ask students how the pattern produced on the blank sheet of paper helps explain sound waves. *(The back-and-forth motion of the prongs of the tuning fork compress and expand the air between the prongs. These alternating compressions and expansions of air particles create sound waves.)*

Drawing Conclusions

Have students work in small groups to construct their theories of how sound energy is transmitted. Encourage group members to discuss their ideas with one another. These ideas should be based on evidence gathered in the experiments. Emphasize that there may be more than one theory supported by the available evidence. The answers below provide a list of experimental evidence gained from Exploration 1.

Answers to *Drawing Conclusions*

1. Experiment 2 showed that sound cannot be transmitted through a vacuum.

2. Experiment 5 showed that air, like a spring, is elastic.

3. Experiments 1 and 4 showed that energy often uses a medium for its transmission.

4. Experiment 3 showed that when sound moves through air, the air moves.

5. Experiments 4 and 6 showed that one way in which energy may be transmitted is in the form of waves of compression and expansion.
   ScienceLog question 3 on page 375 asks students to provide an explanation of how sound travels. The sample response on page 398 also serves as a good model for how sound energy is transmitted.

EXPLORATION 1, *continued*

Drawing Conclusions

From which experiment might you have drawn the following conclusions?

1. Sound energy is not transmitted through a vacuum. In other words, sound cannot be transmitted where there is no air.

2. Air is elastic, like a spring.

3. Energy often uses a **medium** (consisting of particles) for its transmission.

4. When sound moves through air, the air moves.

5. One way in which energy may be transmitted is in the form of waves of compression and expansion.

Using this evidence, construct a theory of how sound energy is transmitted through air from a sound source to your ear.

On the next page is a series of diagrams that will help you visualize what is happening in the air around a vibrating object. Later, you will be able to compare the theory that these diagrams suggest with the theory you constructed here.

Homework

Have students survey two of their friends or family members to find out how others explain the transmission of sound from a source to a listener's ears. Have students give an oral report on the various explanations given. You may wish to have students critique the survey responses, outlining what they believe are the strengths and weaknesses of each explanation.

Sound Thinking

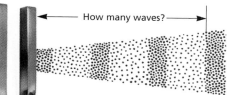

How many waves?

Figure 1

Figure 2

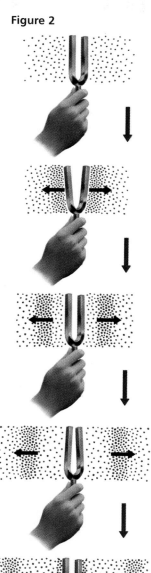

1. Here are illustrations of what you might see if the air around a vibrating tuning fork were visible. Locate the places where the particles are pushed together (compression) and where they are much farther apart (expansion). In Figure 1, how many waves are shown in the indicated region? How many times has the fork vibrated within that region?

2. Measure the wavelength in Figure 1 with a ruler. For waves of the length shown, how many centimeters would a compression move to the right when the tuning fork vibrates once? when it vibrates five times?

3. Figure 1 is a scale drawing of what would happen in air with a tuning fork of a certain frequency. In the figure, 1 cm along the wave represents approximately 40 cm. How far would the sound travel during four complete vibrations of the tuning fork? during one vibration?

4. The tuning fork in the illustration vibrates 440 times in 1 second. How far (in centimeters) would a wave travel through the air in 1 second? How many meters would this be?

5. In question 4 you calculated the distance that the wave traveled through the air in 1 second. In so doing, you estimated the speed of the sound wave. Complete the following equation:

   wavelength × ____?____ = speed of sound waves

6. Suppose a tuning fork vibrates at the rate of 220 times per second. What would be its wavelength, given the value for the speed of sound that you obtained in question 5? How would an illustration of the sound waves produced by this tuning fork differ from Figure 1 above?

7. Figure 2 shows what happens during one complete vibration of the fork. Add four more drawings to show what happens during the second vibration of the fork. Include the motion of the first vibration in your new drawings.

381

Sound Thinking

As students answer each question, they should discover the relationships among the wavelength, frequency, and speed of sound. You may wish to remind students that one complete wavelength includes both a compression and an expansion.

Answers to Sound Thinking

1. Three waves are shown. Therefore, the fork must have produced three complete vibrations.

2. The wavelength is about 2 cm. This means that the compression moves to the right a distance of one wavelength, or 2 cm, for each complete vibration of the fork. Therefore, the compression moves to the right 2 cm × 5, or 10 cm, for five complete vibrations of the fork.

3. In the drawing, 1 cm represents 40 cm. In 4 vibrations of the fork, sound travels a distance of 4 × 2 cm × 40 = 320 cm. In 1 vibration of the fork, sound travels a distance of 2 cm × 40 = 80 cm.

4. For 440 vibrations of the fork, the sound would move 440 × 2 cm × 40 = 35,200 cm = 352 m.

5. Wavelength × frequency = speed of sound waves

6. Wavelength × 220 vibrations/s = 35,200 cm/s. Thus, the wavelength is 35,200 cm/s ÷ 220 vibrations/s = 160 cm. In Figure 1, the distance between compressions would be 160 cm ÷ 40 = 4 cm. Therefore, the figure would show half the number compressions it currently has.

7. The drawings should show the generation of a second wave consisting of a compression and an expansion as well as the outward movement of the first wave.

CROSS-DISCIPLINARY FOCUS

Mathematics

In question 5 of Sound Thinking, students derived a formula for the speed of sound (speed = wavelength × frequency). Give students more practice with the formula by asking them to calculate the wavelength of a 200-wave-per-second tuning fork, a 500-wave-per-second tuning fork, and an 800-wave-per-second tuning fork. Students should assume that sound has the same speed that they determined in question 4 (352 m/s). Ask students which of these tuning forks would have the highest pitch and which would have the lowest pitch. *(The wavelengths are 1.76 m, 70.4 cm, and 44 cm, respectively. The first fork would have the lowest pitch, and the third fork would have the highest pitch.)*

EXPLORATION 2

Make sure students understand that an efficient transmitter of sound is one that transmits sound loudly. The quality of the sound need not be considered in this Exploration.

Answers to
Exploration 2

1. The scratching noise is louder when wood is the only medium the sound travels through.

2. Metal is the most efficient at conducting sound energy. All of the materials transmit some sound.

3. The sound of the spoons is more easily heard in water than in air.

4. Students should find that water transmits sound more efficiently than air does and that they hear sounds best when their ears are in the same medium as the sound source. Students may infer that some of the sound is reflected when it reaches a different medium.

5. Student solutions will vary. Students should discover that sound travels more efficiently through the floor than through the air.

6. The sound is louder because it is traveling through the bones and muscles of the head rather than through the air. Plugging the ears makes the sound louder by blocking out other sounds.

 An Exploration Worksheet is available to accompany Exploration 2 (Teaching Resources, page 29).

FOLLOW-UP

Reteaching

Present students with the following scenario: You have been asked to write an entry for a children's encyclopedia to explain how sound travels from a source to a listener. Use illustrations and examples to clarify your explanations, and remember to use words the students can understand.

Assessment

Have students compare the sound from a vibrating tuning fork that is brought near their ear with one that touches the

bone of their skull just behind the ear. Have students explain the difference in sound. *(Sound travels through bone more efficiently than it does through air.)*

Extension

Have students try this experiment to find out if sound can be channeled and if it can travel around bends. They will need a garden hose with a funnel attached to each end. Using this apparatus, ask them to find out how the loudness of a sound traveling through the hose compares with the loudness of a sound traveling through the open air. Have them use this same apparatus to

EXPLORATION 2

Investigations of Sound Transmission

Is sound transmitted through wood, metal, cardboard, plastic, and other materials? Find out by doing the following activities.

1. Hold one end of a wooden rod (at least 2 cm in diameter) close to your ear. While doing so, scratch the other end with a tack. Can you hear the sound through the wood? Could you hear the same sound in air?

2. Hold a vibrating tuning fork against the end of the rod away from your ear. Then repeat the experiment, substituting plastic, metal, and paper objects for wood.

Be Careful: Use objects whose diameters are larger than that of your ear canal. Carefully place each object so that it is very close to the outside of your ear canal but not touching.

   Which material transmits sounds most loudly? In other words, which material is most efficient in conducting sound energy? Is there a material that will not transmit sound?

3. Which transmits sounds more efficiently—air or water? Try the following activity: Hit two spoons together in air and then in water, listening carefully to the sounds each time. (Use a homemade

stethoscope like the one shown below to listen to the sound in water.)

4. Try the following combinations at home in the bathtub.

   Caution: Do not submerge your head under water, and keep all electrical devices away from the water.

   a. sound source in water—ears in air, ears in water

   b. sound source in air—ears in air, ears in water

5. You've probably seen a movie in which someone put his or her ear to the ground to hear the sound of horse hoofs before the sound could be heard through the air. Design an experiment to test this idea, using the floor. You will need a very quiet room.

6. Do some humming. While you hum, plug your ears. What difference(s) do you notice in the sound? Why does this occur?

find out if sound travels around bends inside the hose.

Closure

Ask students to find out how far sound can be channeled through different materials. They will need a watch that ticks audibly and various materials to test, such as a long board or a metal curtain rod. They should then place the watch at one end of the board or rod. Ask them to safely place their ear at different positions along the object (and above and below it) to hear how well and how far the sound travels.

LESSON 2 • The Speed of Sound

LESSON

2

The Speed of Sound

What Is Speed?

Whew. That was fun. How'd I do, Travis?

Is that a good speed?

Wow, Lena, you lapped the track in only 1 minute. That's 3 kilometers in 60 seconds.

Well, you traveled 3000 meters in 60 seconds, so your speed was 50 m/s. That means you broke the track record of 48 m/s. Congratulations, Lena!

How did Travis calculate the speed? What facts did he need to know? Travis divided the distance in meters by the time in seconds and found the speed in meters per second. Lena's motorcycle speed tells us that she traveled 50 m in 1 second. (To change meters per second to kilometers per hour, which is how speed is customarily measured, multiply by 3.6.) How would you define *speed*? Ⓐ

An Unusual Race!

Was Lena's speed really that good? Suppose Lena raced for 5 seconds with each of the following:

a. a cheetah (the world's fastest animal sprinter)

b. a peregrine falcon (the world's fastest bird)

c. the fastest human runner

d. the *Bluebird* (a record-holding race car)

e. the sound from a rifle shot

f. the Concorde (a supersonic jet airplane)

g. a space shuttle

h. the flash of light from a rifle shot

i. the bullet from a rifle

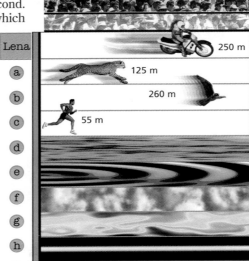

Scale: 1 cm = 50 m
Drawn to scale, the distances shown represent how far each competitor went in 5 seconds.

Lena

a — 250 m

b — 125 m

c — 260 m

d — 55 m

e

f

g

h

i

Starting Line

383

LESSON 2 ORGANIZER

Time Required
2 class periods

Process Skills
observing, measuring, analyzing

New Term
Sonic boom—a loud noise produced by the shock wave of an object traveling faster than the speed of sound

Materials (per student group)
Exploration 3: meter stick or metric measuring tape; watch or clock with second hand (See Advance Preparation on page 355C.)

Teaching Resources
Exploration Worksheet, p. 31
SourceBook, p. S111

FOCUS

Getting Started

Speed is a difficult concept for some students to grasp. To help students understand this concept, draw a simple diagram on the chalkboard. Use a meter stick to mark off intervals of 10 cm, from 0 to 100 cm. Draw a bug at 0 cm. Tell students that a bug starts crawling at this point. Then draw the bug at 10 cm and label this point "2 s." Tell students that the bug reaches this point after 2 seconds. Draw the bug at 10 cm intervals, and label the times 4 s, 6 s, 8 s, and so on. Ask: How fast (at what speed) is the bug moving? *(The bug moves 10 cm every 2 seconds, or 5 cm/s.)*

Main Ideas

1. Sound travels through air at a speed of slightly more than 330 m/s (at 0°C).

2. Light travels much faster than sound does.

3. Sound waves are reflected by obstacles in their path and produce echoes.

TEACHING STRATEGIES

What Is Speed?

First ask students to write a definition of speed in their ScienceLog. Then call on a volunteer to read aloud the introductory material for this lesson, which reviews the concept of speed. Ask: Why does multiplying by 3.6 convert meters per second to kilometers per hour? *(You must multiply 60 sec./min. by 60 min./hr. to find the total number of seconds traveled, which is 3600 sec. Then you must divide meters by 1000 to express the distance in kilometers.)*

Answer to
In-Text Question

Ⓐ Student answers will vary but should arrive at a definition of speed as distance traveled per unit of time.

Answers to
An Unusual Race! pages 383–384

1. Cheetah and the fastest human runner

2. Everything but the cheetah and the fastest human runner

3. Light

4. The flash of light arrives first; the sound of the shot arrives last.

5. 1000 m ÷ 5 s = 200 m/s

6. 55 m ÷ 5 s = 11 m/s

7. 1650 m ÷ 5 s = 330 m/s (The speed of sound is actually 331.5 m/s at 0°C.)

8. The *Concorde* and the space shuttle are the only objects that will cause a sonic boom. A bullet is too small to cause a sonic boom, and light does not compress the air in front of itself, so it generates no shock wave. You may wish to point out that in space, the space shuttle would not generate any sound because there is no medium through which sound can travel.

EXPLORATION 3, *page 385*

Before students perform this activity, locate a wall that reflects sound well. You may wish to review the procedure step by step as well as the sample calculation with Blake's data.

Cooperative Learning
EXPLORATION 3, PAGE 385

Group size: 3 to 4 students
Group goal: to calculate the speed of sound
Positive interdependence: Assign the following roles: recorder (to measure and record the distance), timer (to determine the time between the clap and its echo), mathematician (to calculate the speed using the data obtained), and clapper (to provide the sound to be measured). Each student should record all data and answers to the questions. The group should then complete Some Sound Puzzlers on page 386 using the "jigsaw" method: one student begins reading and continues until coming to a question. The student then gives the common answer sheet to the student on his or her left. This student writes the question and an answer. The group should discuss and agree on the answer.

1. Which of these would Lena outrace?

2. Which would outrace Lena?

3. Which of all these has the fastest speed?

4. A rifle is fired 100 m away from a target. Which reaches the target first—the bullet, the flash of light, or the sound of the rifle shot? Which arrives last?

5. What is the speed record (in meters per second) for a race car?

6. What is the speed of the fastest human runner in meters per second?

7. What is the speed of sound?

8. Objects that approach the speed of sound compress the air ahead of them until the air is almost like a solid wall. At the speed of sound, a shock wave is created that makes a bang that sounds like a clap of thunder. Such a bang is known as a **sonic boom**. Which of the race contestants might cause a sonic boom?

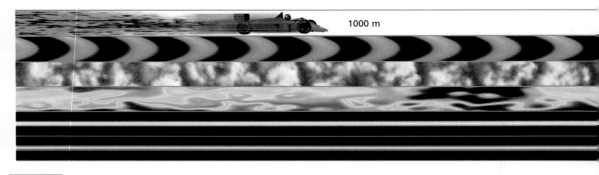

1000 m

The group continues until all information has been read and all questions have been answered.

Individual accountability: Present students with the following scenario, and have them turn in their answers individually: It's 25°C outside. You see a lightning bolt and hear the thunder 10 seconds later. How far away is the lightning? *(10 s × 345 m/s = 3450 m)* One minute later, you see another lightning bolt and hear the thunder 5 seconds later. How far away is the lightning now? *(5 s × 345 m/s = 1725 m)* How fast is the thunderstorm approaching? *(The*

storm moved 3450 m – 1725 m, or 1725 m, in 60 s. So it is approaching at a speed of 1725 m ÷ 60 s = 28.75 m/s.)

 An Exploration Worksheet is available to accompany Exploration 3 on page 385 (Teaching Resources, page 31).

An Echo Experiment to Measure the Speed of Sound

You Will Need

- a meter stick or tape measure
- a watch or clock with a second hand

What to Do

1. Stand about 50 m from the wall. Clap your hands. Listen for the echo. Clap your hands twice. Listen for two echoes.

   Try clapping your hands at such a speed that you begin a new clap just when you hear the echo from the previous clap. You won't hear the echoes any more.

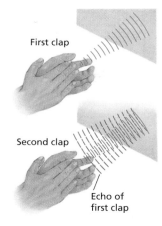

First clap

Second clap

Echo of first clap

What does the time between claps tell you?

2. Have someone time how long it takes to clap 20 times in such a way that you cannot hear the echo. (Don't count the first clap.) Try this several times.

   What is the average time between claps? Where has sound traveled during this time?

3. Measure the distance to the wall, in meters, by pacing the distance (once you've found the length of your pace). Have two or three people check the distance. What is the total distance traveled by the sound of the clapping—the distance from your hand to the wall and back to your ear?

4. From the distance and time of travel, calculate the sound's speed.

   Blake used the above procedure to calculate the speed of sound. Here are his results.

   20 claps in 8 seconds
   115 paces to wall
   10 paces are 6 m long
   Distance sound traveled
   = 138 m
   Time per clap = 0.4 seconds
   Speed of sound = 345 m/s

   Did you notice that the speed of sound calculated by Blake was considerably higher than the speed you calculated from the race on page 383? Blake performed his experiment at 25°C. The speed from the race is what it would have been at 0°C. What effect does temperature appear to have on the speed of sound? What is the average increase in the speed of sound per degree Celsius?

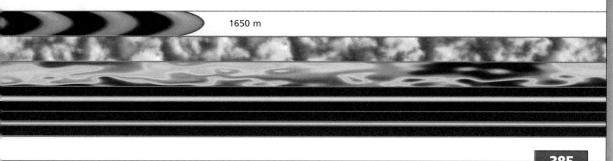

1650 m

Did You Know...

Sound travels about 870,000 times slower in air than light does, but in glass, sound travels only about 34,000 times slower than light—an increase of a factor of 27.

ENVIRONMENTAL FOCUS

You can use the fact that sound is slower than light to estimate how distant a storm is. Just count the number of seconds from when you see the lightning to when you hear the thunder. The storm is about 1 km away for every 3 sec., or about 1 mi. away for every 5 sec. Have students perform some practice calculations.

Answers to Exploration 3

1. The time between claps tells you how long it takes for the sound to reach the wall and return to you because you clap every time an echo reaches you. Students should note that how quickly they must clap is a good indication of how fast sound travels.

2. The average time between claps is the total time for 20 claps (this will vary with each student) divided by the number of claps (20). The sound travels to the wall and back in the time between claps.

3. Student answers will vary depending on how far from the wall they stand. The total distance traveled by the sound of a clap is twice the distance to the wall.

4. Answers will vary but should be approximately 345 m/s. Blake calculated the distance that the sound traveled by doubling the distance to the wall: 115 paces × 6 m ÷ 10 paces × 2 = 138 m. He then calculated the average time per clap to determine the time it took for the sound to travel 138 m: 8 seconds ÷ 20 claps = 0.4 seconds/clap. Blake was able to calculate the speed of sound using the formula, speed = distance ÷ time: 138 m ÷ 0.4 seconds = 345 m/s.

   The speed of sound in air increases as the temperature increases. At 25°C, the speed of sound is 345 m/s. At 0°C, the speed of sound is approximately 330 m/s. The average increase in speed is calculated as follows:

$$\frac{(345 \text{ m/s} - 330 \text{ m/s})}{(25°C - 0°C)} = \frac{15 \text{ m/s}}{25°C} = 0.6 \text{ m/s} \cdot °C$$

For every Celsius degree of increase, speed increases about 0.6 m/s.

Homework

For practice with the concept of speed, have students measure their speed while walking, crawling, and running. Students should make a chart of their results.

1. As sound travels through pockets of warmer and cooler air, the sound can speed up and slow down. The long rolling effect may also be accentuated by the sound echoing from hills or other structures.

2. For an observer very close to where lightning strikes, the light and sound of thunder arrive at almost the same time. As an observer moves farther away, the time for the sound to arrive is noticeably greater than the time for the light to travel the same distance. Because light travels so fast, we can simplify the calculation by assuming that the light reaches the observers instantaneously. Then the distance to the lightning is simply the speed of sound multiplied by the time between the flash and the thunder. Because sound travels 345 m/s at 25°C, the two middle observers must be about 345 m (1 s × 345 m/s) and about 1380 m (4 s × 345 m/s), respectively, from where lightning struck. The last observer did not hear the sound of thunder because it was too weak to be detected by the human ear at that distance.

3. Yes. When you pass a building, sound from the car is reflected by the building, making the sound of the car louder. In a space with no buildings, the lack of reflected sounds makes the sound of the car much less noticeable. An intersection, with its numerous angles and objects, would probably produce and reflect a variety of sounds.

4. The sound waves travel faster than the boat and are trapped and reflected by the walls of the canyon, allowing them to travel farther than they would in a more open area.

5. Snow absorbs (rather than reflects) most sound waves.

6. Sounds are reflected off the shower walls, and the echoes give your voice depth.

7. Assume that it takes 0.5 s to utter "hello." In 0.5 s, the sound will have traveled 172.5 m because the speed of sound is 345 m/s at 25°C. If one stands more than 86.25 m from a building—half of 172.5 m— one will hear an echo right after the original sound.

Some Sound Puzzlers

1. What causes thunder to rumble, sometimes for quite a long while?

2. The speed of sound at 25°C = 345 m/s, and the speed of light = 300,000,000 m/s. Can you explain the different observations below? How far from the lightning bolt are the two people in the middle of the drawing? Why doesn't the last person hear the thunder?

> That sure was close—a lightning flash and a thunderclap at almost the same time!

> I wonder how far away the lightning is. I heard the thunder about a second after the flash.

> The time between the flash and the thunder was about 4 seconds. The distance must be . . .

> No thunder. The storm must be pretty far away.

3. Suppose you are riding in a car down a city street. Close your eyes. Can you tell when you are passing a building, passing a space where there are no buildings, or passing an intersection? How does the sound of the car help you?

4. The Colorado River winds its way through the deep gorge of the Grand Canyon. In this canyon, an outboard motor on a raft can be heard 15 minutes before the raft actually arrives. Why?

5. Why is it so silent after a snowfall?

6. Why does your singing in the shower sound so great?

7. A challenge! How far away from a wall should you be if you want to hear a complete, distinct echo of your voice saying "hello"? (Hint: To solve this, you'll need to measure how long it takes to say "hello.")

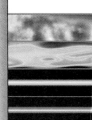

Why do you sound better in the shower?

2900 m

FOLLOW-UP

Reteaching

Have students make recordings of sound effects that illustrate the important concepts of the lesson.

Assessment

Giving the approximate speed of sound at 25°C as 345 m/s, have students determine the distance to a wall using echoes. (After finding the time between echoes, students can find distance by using the equation, speed = distance ÷ time.)

Extension

Have students research the use of sonar (reflected sound) to locate fish, map the ocean floor, and find sunken objects.

Closure

Have students graph the distance traveled by sound, light, and some of the objects listed in the text over a given period of time. Students should plot distance on the *y*-axis and time on the *x*-axis. In order to compare speeds, students should compare the slopes of the lines, which indicate the speed of the objects (distance/time).

More About Echoes

How Do the Blind "See"?

Ask a blind person how he or she is able to sense and avoid obstacles in his or her path. You may get answers like these:

"I feel the presence of the object."
"There are pressures on my skin that tell me an object is there."
"I sense danger."
"I have facial vision."

There is a story about a six-year-old blind boy who learned how to ride his tricycle on the sidewalks near his home. He never had an injury or an accident. He could veer around people, and he knew when to turn corners without going into the street. How did he "see"?

For years, there was a blind cyclist who rode his bike in downtown Toronto, Canada. With less than 10 percent vision, he could not see traffic lights, jaywalkers, or road repair crews. How did he "see"?

How might blind individuals use the tapping sound of their cane to identify objects around them? Ⓐ

A Research Project

A Cornell University professor and two of his students, one of whom was blind, wondered how blind people are able to avoid obstacles even though they can't see them. The research team designed a project to find out.

Think like a researcher: What steps might you follow? The questions below will assist you. Each question is followed by an explanation of what the researchers actually did. Cover these explanations with a card, revealing them only after you have suggested a possible answer.

The researchers were particularly interested in testing the following hypotheses:

- *Skin-Pressure Hypothesis:* Obstacles send out signals that produce pressure on the skin, enabling the blind person to be aware of the obstacles.

- *Sound Hypothesis:* Sounds hit obstacles and bounce back, alerting the blind person to their presence.

a. What kind of people would you use for the experiment—blind, blindfolded, or both?

The Cornell team decided to use both blind and blindfolded subjects. They gave special training to the blindfolded people. After a little practice, they too could avoid obstacles placed in their path.

4000 m

387

LESSON 3 ORGANIZER

Time Required
1 class period

Process Skills
hypothesizing, inferring, analyzing, communicating

New Term
Echolocation—using echoes to detect the presence and location of objects

Materials (per student group)
A Research Project: one index card per student

Teaching Resources
Math Practice Worksheet, p. 33
SourceBook, pp. S112 and S121

LESSON

3

More About Echoes

FOCUS

Getting Started

Have students read the introduction, How Do the Blind "See"? Then ask them to construct their own theories about how blind people could sense and avoid obstacles. Give students the opportunity to debate their theories with one another.

Main Ideas

1. Blind individuals can detect objects in their surroundings by listening to echoes of sounds reflected by these objects, a process known as echolocation.

2. Some animals such as bats and dolphins detect the presence of objects by using echolocation.

TEACHING STRATEGIES

Answer to
Caption

Ⓐ The quality of the sound that echoes from a tapped object will be different for different objects. A blind person identifies objects by tapping them and listening to the sound quality of the echoes.

Meeting Individual Needs

Gifted Learners

Ask students if they have ever noticed how the sound of a car changes as it approaches and then passes by. Explain to students that this change in pitch and frequency is known as the *Doppler effect*. Have students research what causes this effect, and illustrate their findings on a poster board. If possible, have students record the sound of a train or car as it passes by. They should then make a second recording from within the moving vehicle. Ask students to play their recordings for the class in order to compare the sounds.

A Research Project,
pages 387–388

Tell students that people often jump to conclusions without having solid evidence to support their beliefs. Ask students if they can think of examples when they have done this. Explain that scientists must train themselves to be as objective as possible and to try not to jump to conclusions prematurely. The research project that they will read about will give them practice in carefully considering evidence before drawing conclusions.

This research project is adapted from the book, *Echoes of Bats and Men* (Doubleday Science Study Series). The research project was set up to determine how blind people avoid obstacles.

The article allows students to act as researchers and to speculate on how they would carry out each step of the research described in the article. Questions are provided to help them in this role. Students should provide answers to the questions before reading about what the researchers actually did. One way to accomplish this is to use an index card to cover the discussions in the blue print as suggested in the text. You may wish to divide the class into research teams of two or three students. After they discuss each question, have students read from the text to discover what the researchers actually did.

After students have finished reading what the researchers actually did, use the following questions to aid in your discussion of the article: What was the most surprising part of the story? If the students had been in the position of the researcher, when might they have concluded their experimentation? What were the steps researchers followed in conducting their research?

Allow students to try echolocation for themselves. Have them work in pairs. One student in each pair should be blindfolded while the other student acts as a spotter. Tell them to check for a change in the sound of their heels on the floor as they walk down a corridor and pass a classroom with an open door. Then have them compare the sounds of a meter stick hitting the floor far away from a wall and when close to a wall. For safety reasons, you may wish to set this up as a demonstration or allow only a few students to do this activity at one time.

b. What kind of experimental situation would you set up to test the avoidance of obstacles?

The experimenters walked down a long hallway toward a fiberboard screen. The experimenters changed the position of the screen from trial to trial. Before long, each subject could walk down the hallway, detect the presence of the screen, and stop. On average, subjects detected the screen when they were about 2 m away from it.

c. Which hypothesis would you test first?

The researchers tested the hypothesis that blind people mention most often, the skin-pressure hypothesis.

d. What would you do to the subjects to prevent their skin from being affected by the screen?

The subjects had to wear an "armor" of thick felt over their heads and shoulders and heavy leather gloves on their hands.

While clothed this way, they couldn't even feel the air from an electric fan, but they could still hear sound.

e. If they bumped into the obstacle when wearing the armor but didn't bump into the obstacle when they were without it, what would you conclude? If they avoided the obstacle when wearing the armor, what would you conclude?

The subjects wearing the armor avoided the obstacle. However, they tended to approach it more closely, on average, than they did before. They stopped 1.6 m away, as compared with 2 m without the armor. The researchers concluded that the skin-pressure hypothesis was not correct. Blind people do not "feel" the pressure of obstacles with their hands and face.

f. What would you do to test the sound hypothesis?

Ear coverings of wax, cotton, earmuffs, and padding were worn by the subjects. They could not even hear their own footsteps. Their faces and hands were left uncovered.

g. If they bumped into the screen while their ears were plugged, what would you conclude? If they stopped short of the obstacle, what would you conclude?

Spectacular results! Both blind and blindfolded people bumped into the obstacle. The blind people said that all sensation of "feeling" had gone. The sound hypothesis best explained the researchers' observations.

h. Some scientists objected to this conclusion. Perhaps the ear coverings changed the pressure on the ear and ear canal—pressure that would have resulted from the obstacle's presence. What could you do now to determine that avoidance of the obstacle was due to sound rather than to some sort of pressure on the ears?

Cooperative Learning
A RESEARCH PROJECT

Group size: 2 to 3 students
Group goal: to develop hypotheses and interpret data from actual scientific research studies
Positive interdependence: The group will work together to read, discuss, and answer the questions featured in A Research Project. Assign roles as follows: reader #1 (to read each of the questions [a]–[m]), reader #2 (to read the text describing the researcher's actions in [a]–[m]), and recorder (to record the group's responses).

Individual accountability: Have each student role-play the part of a researcher by writing a letter to a friend in order to describe his or her research on echolocation.

The subjects were placed in a soundproof room some distance away from a hallway where an experimenter walked with a microphone. The sounds of the experimenter's footsteps were picked up by the microphone and carried by a telephone line to the subjects. The subjects were asked to locate the screen by listening to the sounds picked up by the microphone.

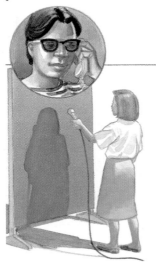

i. What results would you expect if the sound hypothesis were true?

The subjects were able to detect the screen from the sounds picked up by the microphone. The closest approach was 1.9 m. That's more proof for the sound hypothesis! The ability of blind people to avoid obstacles is apparently due to sound—not to pressure on the ear.

j. Some scientists saw weaknesses in this approach as well. Can you think of any?

The experimenter's pace and breathing might have given the subjects hints about the obstacle's presence.

k. What other experiment might you design to check this possible weakness?

The experimenters decided to use a motor-driven cart to carry the microphone toward the screen, and a loudspeaker to make sounds. The movements of the cart were controlled remotely by the subjects in the soundproof room.

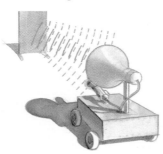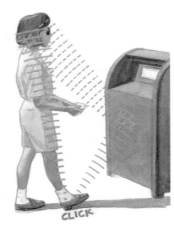

l. What conclusions would you draw if the subjects detected the screen?

Although the average closest approach was a little closer than in previous experiments, the subjects always detected the screen. Therefore, signals from the breathing or pace of the walker could not explain why the subjects were able to detect the screen. The experimenters made three general conclusions.

1. Blind people locate obstacles by sound and by the reflection of this sound—that is, by echoes from the obstacles.

2. With practice, sighted people can also learn to detect obstacles by this means.

3. We are often not aware of how our senses and brain are working. Blind people sense that they "feel" the presence of objects around them. But scientific experiments show that they are, instead, hearing changes in sound as they approach obstacles.

m. What further experiments might you now do?

CLICK

The researchers are still interested in sending different kinds of sounds through the loudspeaker. Through experimentation, they may discover what sounds are best suited for the **echolocation** of objects.

389

e. If the subjects bumped into the obstacle while wearing the armor but avoided the obstacle while not wearing it, students should suggest that the experiment supported the skin-pressure hypothesis. If the subjects avoided the obstacle in both cases, students should suggest that the experiment did not support the hypothesis. (Make sure students understand that errors can occur in every experiment and that one test cannot be considered conclusive.)

g. If the subjects bumped into the obstacle while their ears were plugged, students should suggest that the experiment supported the sound hypothesis. If the subjects avoided the obstacle, students should suggest that the experiment did not support the hypothesis. (Make sure students understand that errors can occur in every experiment and that one test cannot be considered conclusive.)

CROSS-DISCIPLINARY FOCUS

Music
Have students do biographical research on blind musicians (for example, Stevie Wonder or Ray Charles) to find out how their lack of sight has affected their sensitivity to sound.

PORTFOLIO
You may wish for students to keep the results of A Research Project in their Portfolio as evidence of their ability to apply the scientific method in a realistic research setting.

Reporting Your Findings

Encourage students to make their letter as interesting, thorough, and accurate as possible. You may wish to have some students read their letter to the rest of the class.

PORTFOLIO

Students may wish to include their letter from Reporting Your Findings in their Portfolio.

Echolocation at Its Best!

Have students read this section silently. Interested students can pick one of the animals listed and do additional research to find out more about how it uses echolocation to survive.

Homework

After students have read Echolocation at Its Best!, you may wish to assign the Math Practice Worksheet on page 33 of the Unit 6 Teaching Resources booklet as homework.

Reporting Your Findings

Often, the final step of a research project is a written report in a scientific journal. In addition, scientists, like many people, enjoy sharing their interests with friends by writing letters. (Galileo, for instance, wrote many fascinating letters about his work to his friends.) Write a letter to a friend, telling him or her of your exciting discoveries in your role as a researcher into how blind people "see."

Echolocation at Its Best!

Whales and dolphins make clicks and other noises that reflect off objects as echoes. Using echolocation, river-dwelling dolphins can thread their way among logs and fallen trees in muddy rivers. In the open sea, dolphins accurately echolocate the fish they eat.

Bats, too, use echolocation to find food. They hunt flying insects in the dark by sending out pulses of high-frequency sound waves (about 45,000–100,000 vibrations per second—too high to be audible to humans). The waves bounce off of the insects, and the bats find their prey by listening for the echoes. In one experiment, a North American brown bat was able to echolocate and catch 175 mosquitoes in 15 minutes. If a bat were unable to echolocate, it might have to fly all night with its mouth open before catching one mosquito.

A bat usually chirps about 20 times per second. This number increases to 200 shorter chirps per second as it approaches its prey. How far away from its prey could a bat be and still hear a separate echo? Ⓐ

Many ships have a navigation system that uses echoes to find the depth of the water, the presence of schools of fish, and even the structure of the rock on the ocean bottom. This system is called *sonar*, which stands for **so**und **n**avigation **a**nd **r**anging. Sonar works in much the same way that echolocation works for a bat. The sonar device sends short pulses of sound waves through the water. When the sound waves hit the ocean floor, some of the waves are reflected back as an echo. The echo is then detected by a receiver. The gathered information can then be displayed on video monitors.

Shown here is a screen image from a fisherman's depth finder. This device uses the principles of echolocation to map the ocean bottom and to spot schools of fish.

Dolphins are champions at echolocation.

FOLLOW-UP

Reteaching

Ask students to write a letter to a second-grader in order to explain how bats use echolocation to navigate and to catch insects. Encourage students to include pictures with their explanations.

Assessment

Ask students how they would design an experiment to determine whether a certain fish uses echolocation to navigate. Have students submit summaries of their hypothetical experiment.

Extension

Have interested students find out more about how geophysicists use echolocation to study the structure of the crust and upper mantle of the Earth.

Closure

Have the class read a newspaper or magazine article that discusses a recent scientific breakthrough. Students should evaluate the research described in the article by writing their own interpretation of the hypotheses that were tested,

the observations that were made, and the conclusions that were reached. You may wish to give each student an opportunity to read aloud from the article.

Completing the Sound Story

LESSON

4

Completing the Sound Story

You know that sounds start with vibrating objects, which start sound waves in a medium (such as water, wood, or air). But the sound story isn't complete until sound waves activate your ear or some other *receiver*. You will now complete the sound story by looking at what happens between the time that sound waves reach a receiver and the time that the receiver registers, or "hears," the sound.

Designing an Instrument for Hearing

Can you design a device that can

- detect air particles moving back and forth in a sound wave,
- detect particles hitting it with only 0.000001 J of energy (by comparison, a square centimeter of paper dropped from a height of 50 cm hits with 0.001 J of energy),
- function when subjected to very great air pressure (very loud sounds)—up to 5 J of energy,
- detect air particles vibrating as slowly as 20 times per second and as quickly as 20,000 times per second,
- distinguish between air particles vibrating 500 times per second and those vibrating 505 times per second (compare this with the difference in vibrations per second between notes B and C on the piano keyboard at right),
- change the information it has detected into electrical impulses and carry this information to a central recording place,
- and detect its own orientation in space—whether level, on a slant, or upside down?

If your device can do all this, congratulations! You have designed the human ear!

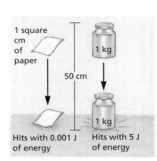

1 square cm of paper

1 kg

50 cm

1 kg

Hits with 0.001 J of energy

Hits with 5 J of energy

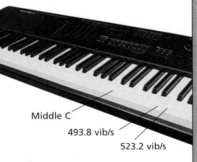

Middle C

493.8 vib/s

523.2 vib/s

LESSON 4 ORGANIZER

Time Required
1 class period

Process Skills
inferring, analyzing, communicating

Theme Connection
Structures

New Term
Hertz—the unit of measurement for frequency; one hertz (Hz) equals one complete vibration, or sound wave, per second

Materials (per student group)
Questions to Think About:
stethoscope; about 1 L of water; large paper container with the bottom removed; large balloon or rubber sheet, about 30 cm × 30 cm; a few large rubber bands; tuning fork

Teaching Resources
Resource Worksheet, p. 34
Transparency Worksheet, p. 35
Transparencies 60 and 61
SourceBook, p. S117

FOCUS

Getting Started

Tell students that their ability to tell the direction a sound comes from depends on having two ears. To demonstrate this, have a group of students sit in a circle around a blindfolded student. Students sitting in the circle should take turns tapping two pencils together. The blindfolded listener should try to locate the direction of the sound and point to it. Then have the listener plug one ear and repeat the activity. Have different students take turns being the blindfolded listener.

Main Ideas

1. Sound waves reaching the ear set up vibrations in the ear that are translated into electrical impulses, which travel to the brain.
2. Sound exists if there is a vibrating object, a medium to carry the sound wave, and a receiver to hear the sound.

TEACHING STRATEGIES

Designing an Instrument for Hearing

Have a volunteer read the bulleted items that describe the instrument for hearing. After the volunteer has read the bulleted statements, ask another volunteer to interpret the two illustrations on the right. One shows a large variation in the amount of energy (air pressure) that the instrument can receive; the other shows a small difference in the number of vibrations per second (notes B and C on the piano) that the instrument can distinguish. Make the point that this hearing instrument (the human ear) can receive even larger variations in energy and can distinguish even smaller differences in frequency. The ear must be really amazing to be able to perform both kinds of tasks.

The Human Ear

Point out to students that this labeling exercise is not a memorization activity. To help students gain an understanding of how the ear works, have them read the story and label the diagrams at the same time. The questions following The Story of Hearing will help students read reflectively.

The following is some additional background information that you may want to share with your students:

Many animals have pinnas that are much larger than those of humans. You may wish to have students try a simple activity to demonstrate the advantage of larger pinnas. Tell students to make large model animal ears out of cardboard and hold them next to their ears. They should be able to hear fainter sounds.

The eardrum is only 10 mm in diameter. Wax helps protect the eardrum by trapping dirt that would otherwise lodge itself against the eardrum. Sometimes deposits of wax build up in the ear and can impede hearing. These should be removed by a doctor. Emphasize to students that they should never try to remove wax themselves because they could easily puncture the eardrum.

Students may also be interested to know that the stirrup is the smallest bone in the human body. It is smaller than a grain of rice.

The Eustachian tube is closed most of the time. It opens when you swallow, yawn, or blow your nose. During those times, the air pressure on both sides of the eardrum is equalized. Ask students if they have ever experienced their ears popping when landing or taking off in an airplane. The popping sensation is the Eustachian tube opening, allowing air to escape from (when ascending) or to enter (when descending) the middle ear.

Explain that sound waves can also be conducted to the inner ear through the bones of the skull. Some of the sound produced by your voice travels to your inner ears in this way.

Ask students if they have ever suffered from motion sickness when traveling by boat, automobile, train, or airplane. Explain that motion sickness is caused by excessive stimulation of the nerves in the semicircular canals. When traveling by car, you can help prevent motion sickness by keeping your eyes on the horizon in front of you, by not

The Human Ear

Here are three diagrams of an amazing instrument—the human ear. Figure 1 shows the whole ear, while the two smaller figures show sections of the ear, enlarged for easier viewing. You'll understand how the ear works after reading The Story of Hearing on the next two pages.

Notice that The Story of Hearing is divided into two parts. The left-hand column describes the structures of the ear, while the right-hand column describes the journey of a sound wave through the ear. In your ScienceLog, match the italicized terms in the story with the numbered items in Figures 1, 2, and 3, using the information provided. For each term, write a brief description summarizing its function.

When you have finished, you'll use your understanding of how we hear to answer some questions.

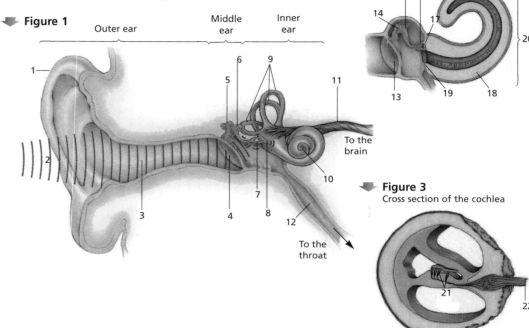

Figure 1

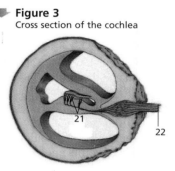

Figure 2
Close-up of middle ear and part of inner ear

Figure 3
Cross section of the cochlea

reading and by not looking through side windows. When spinning, dancers and ice skaters prevent motion sickness by keeping their eyes on the same spot for as long as possible during each turn.

Homework

You may wish to assign the activity described on this page as homework. A Resource Worksheet is available to accompany the activity (Teaching Resources, page 34). If you choose to do this activity in class, Transparency 60 is available for your use.

Answers to *Figure 1*

A discussion of each structure's function appears in The Story of Hearing on pages 393–394.

1. Pinna
2. Sound waves
3. Canal
4. Eardrum
5. Hammer
6. Anvil
7. Stirrup
8. Oval window
9. Semicircular canals
10. Cochlea
11. Auditory nerve
12. Eustachian tube

The Story of Hearing

In Figure 1, the brackets mark off the three regions of the ear.

Vibrating source of sound

The story of hearing tells how sound waves are passed from one region of the ear to the next. The story begins when a vibrating object, for example, a tuning fork, produces vibrations in the air particles surrounding it.

The outer ear (which is what people are usually referring to when they say "ear") includes the large external flap of cartilage and skin called the *pinna*. It leads into a narrow tube called the *canal*.

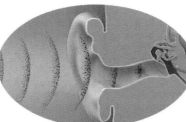

Sound waves enter the ear.

The pinna catches the sound waves and directs them into the canal.

At the end of the canal is the *eardrum*. This is the beginning of the middle ear. The middle ear is connected to the throat by the *Eustachian tube*.

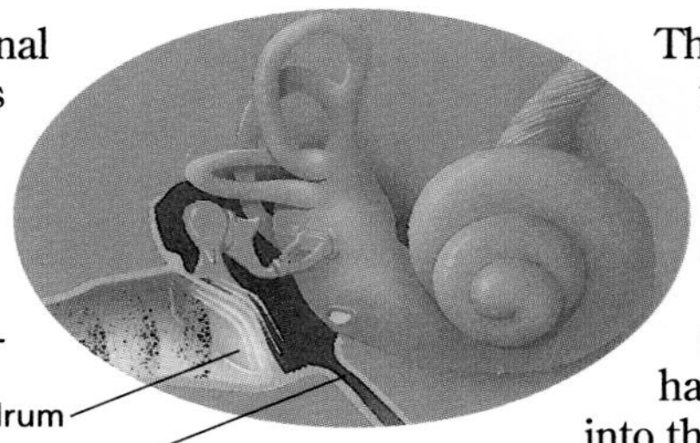

Eardrum

Eustachian tube

Sound waves cause the eardrum to vibrate.

The vibrations of the air particles cause the eardrum to vibrate. The energy of the sound waves determines how far and how fast the eardrum vibrates. Sometimes the eardrum is pushed far into the middle ear by the excessive pressure of a loud sound. When this happens, some of the pressure is released into the throat through the Eustachian tube.

Next to the eardrum are the three bones of the middle ear: the *hammer*, the *anvil*, and the *stirrup*. They fit snugly into each other. Pictured above the stirrup are the three *semicircular canals* of the inner ear.

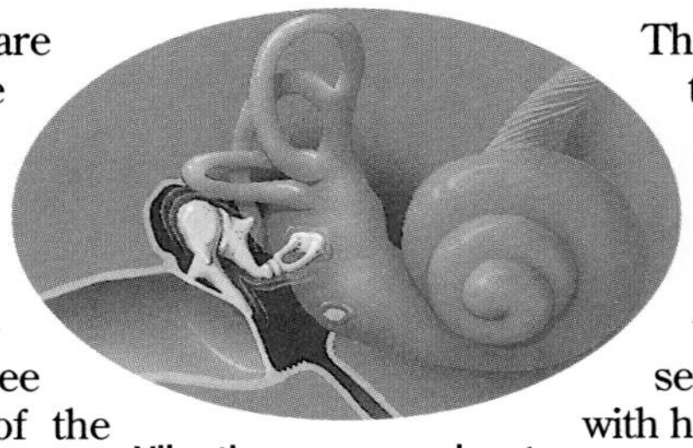

Vibrations are passed on to the bones of the middle ear.

The eardrum passes the vibrations on to the three bones of the middle ear, which in turn vibrate. The bones act like levers, multiplying the force of the vibrations (at the expense of distance). The stirrup then passes the vibrations on to the inner ear. The semicircular canals have nothing to do with hearing. They are the organs of balance.

Answers to
Figures 2 and 3, page 392

13. Eardrum
14. Hammer
15. Anvil
16. Stirrup
17. Oval window
18. Watery liquid
19. Round window
20. Cochlea (stretched out)
21. Hair cells
22. Auditory nerve

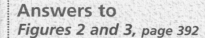

Integrating the Sciences

Life and Physical Sciences
Ask students to compare how the human ear works with how a microphone or the transmitter portion of a telephone works. *(Both the ear and the microphone or telephone convert sound waves into vibrations that are then converted into electrical impulses.)* Encourage students to use diagrams or illustrations to make their comparison clear, and have them label analogous parts where possible.

★ **A Transparency Worksheet (Teaching Resources, page 35) and Transparency 61 are available to accompany Lesson 4.**

Structures

Focus question: How are the structures of the ear similar to the structures that make up the eye? *(Both the ear and the eye have structures that receive waves [the outer ear receives sound waves, and the pupil receives light waves]; both have structures that concentrate the wave signal [the eardrum and bones of the middle ear, and the lens]; both have structures that translate wave signals into nerve signals [the hair cells and the retina]; and both have structures that carry nerve impulses to the brain for interpretation [the auditory nerve and the optic nerve].)*

The stirrup fits into the small *oval window* in the inner ear (see Figures 1 and 2). The inner ear is a cavity in the bone of the skull that is filled with *watery liquid.* The cavity in the bone is made up of the semicircular canals as well as a snail-shaped part called the *cochlea.* (A close-up view of the cochlea is shown in Figure 2.) Located below the oval window is the *round window,* another membrane-covered opening of the inner ear.

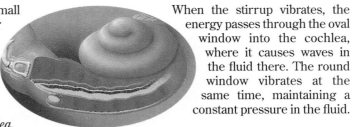

Stirrup transfers the vibrations to the liquid of the inner ear.

When the stirrup vibrates, the energy passes through the oval window into the cochlea, where it causes waves in the fluid there. The round window vibrates at the same time, maintaining a constant pressure in the fluid.

The cochlea contains a membrane that runs the length of the tube, down its middle. Located on this membrane are a large number of little *hair cells.* Figure 3 shows a cross section of the cochlea.

Vibrating hair cells stimulate the nerve endings of the inner ear.

Movement of the liquid in the cochlea causes the little hairs there to vibrate, stimulating nerve endings within the inner ear. The short hairs at the beginning of the cochlea vibrate in response to sound waves of high frequency (which are caused by high-pitched sounds). Farther down the cochlea, longer hairs respond to sound waves of lower frequency.

Located at the base of the hairs in the cochlea are nerve cells that join together to form the *auditory nerve,* which is shown in Figure 1. The auditory nerve connects the ear to the brain.

Nerve impulses are sent to the brain.

The nerve cells change the movements of the hair cells into electrical impulses (which are much like vibrations). The nerve impulses travel through nerve fibers to the auditory nerve, which carries the impulses to the brain. The brain then interprets the impulses as various kinds of sound. It is the brain that ultimately allows us to hear.

Questions to Think About

1. When a compression in a sound wave in the air hits the eardrum, in which direction does the eardrum move? In which direction does the eardrum move when an expansion of a sound wave arrives?

2. If you hear a bird sing a note with a frequency of 2000 vibrations per second, how many times per second does each air particle vibrate? How many compressions (and expansions) reach the eardrum per second? How many times does the eardrum vibrate per second?

3. How does the ear's response to a loud sound differ from its response to a soft sound? to a high sound and a low sound?

4. How does the ear strengthen the sound waves so that they will be strong enough to affect the liquid of the inner ear?

5. You turn around fast a few times and find that you can hardly stand up. Suggest what might be happening in your ear to "upset" you. (Hint: Swirl water in a glass and then set it down. What happens?)

6. Francine's model of the inner ear is shown in the photograph below. Which part of the inner ear is represented by each of these items?

   a. the tubes of the stethoscope

   b. the water in the container

   c. the stretched rubber sheet

   d. the tuning fork

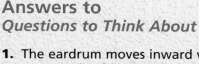

Although they can make you feel like dancing, musical instruments actually produce *standing sound waves*. To find out more about standing waves, see page S117 of the SourceBook.

Does Francine's model work? Try it! Ⓐ

395

Answers to *Questions to Think About*

1. The eardrum moves inward when a compression hits it. It moves outward when an expansion strikes it.

2. 2000; 2000; 2000

3. The eardrum vibrates with greater amplitude (it stretches farther) for a loud sound. The eardrum vibrates with greater frequency (more vibrations per second) for a higher sound.

4. The bones of the middle ear, acting as a system of levers, strengthen (increase) the force of the vibration.

5. The liquid in your inner ear continues to move after you have stopped moving. Your balance is affected because it is controlled by the fluid in the three semicircular canals.

6. **a.** Auditory nerve
   b. Liquid in cochlea
   c. Oval window
   d. Stirrup against oval window

Answer to *Caption*

Ⓐ Students should find that Francine's model does work. You may wish to perform this as a demonstration.

A Centuries-Long Debate, *page 396*
Sound can be considered just the waves produced by a vibrating object. Another definition of sound includes a vibrating source, sound waves, and a receiver. Finally, sound can be considered from the human viewpoint: that is, sound exists only within the range of frequencies and loudness detectable by the human ear.

Homework

You may wish to assign any or all of the Questions to Think About for homework.

Answers to
More Items for Debate

Whether students feel there is sound in each situation will depend on how they define sound: as the waves produced by a vibrating object, or as the detection of the waves by a receiver.

1. Waves from a vibrating object, no—there is no medium through which any sound waves could travel; detection by receiver, no

2. Waves from a vibrating object, yes; detection by receiver, yes

3. Waves from a vibrating object, yes; detection by receiver, no—insufficient sound energy to activate the human ear

4. Waves from a vibrating object, yes; detection by receiver, yes—but often insufficient sound energy to activate the human ear

5. Waves from a vibrating object, yes; detection by human receiver, no

6. Waves from a vibrating object, no; detection by receiver, no

7. Waves from a vibrating object, yes; detection by receiver, yes (bats), but inaudible to humans

8. Waves from a vibrating object, yes; detection by receiver, no—frequencies greater than can be detected by the human ear

9. Waves from a vibrating object, yes; detection by receiver, yes—but insufficient sound energy to activate the human ear

10. Waves from a vibrating object, yes; detection by receiver, yes—but heard after a short delay

A Centuries-Long Debate

An ancient philosophical question asks, If a tree falls in the forest when no one is around to hear it, is there a "sound"? Write your opinion, and back it up with some principles and facts that you have learned in this unit, such as the following:

- Sounds are produced by vibrating objects.
- Sounds are transferred from place to place by vibrating particles.
- Sounds are heard as a result of vibrations transmitted through parts of the ear and the nerves to the brain.
- Dogs hear sounds that people cannot hear.
- Bats make sounds that they can hear but that people cannot hear.

More Items for Debate

Is sound produced in each of these situations? Give reasons for your answers.

1. Huge explosions occur on the sun's surface.
2. In the middle of a quiet night, you are in a house all alone.
3. A tiny square of tissue paper falls to the rug.
4. Your heart beats.
5. There is a running brook in the middle of the woods.
6. An electric bell rings in an airless bell jar.
7. Moths make clicks to confuse the echolocation sounds of bats. These clicks are inaudible to humans.
8. An electronic audio oscillator vibrates at 25,000 vibrations per second (25,000 Hz).
9. Particles of the gases in air constantly hit all of the objects around us.
10. A plane moving faster than the speed of sound has just passed you.

An explosion on the sun's surface

You already know that the *frequency* of vibration is the number of vibrations produced by an object in 1 second. Frequency is measured in units called hertz. One hertz (Hz) is the frequency of one complete vibration, or wave, per second.

The space shuttle would travel 50,000 m in 5 seconds. That's another 53 book pages. But what about the flash of light? This book would need another 20,000,000 pages!

50,000 m

1,500,000 m

FOLLOW-UP

Reteaching

Have students write three possible definitions for the word *sound*. Then ask them to give an example of each definition.

Assessment

Ask students to write a short report comparing how the parts of the ear function when loud, soft, high, and low sounds enter it. (*Answers should reflect differences in frequency and amplitude of vibration of the ear's structures.*)

Extension

Have students research ultrasonic waves and write a report on the scientific or medical applications of such waves. (*Ultrasonic waves have frequencies that are greater than can be detected by the human ear. They are used for measurement and detection because the bones, fluids, and tissues of the human body each reflect ultrasonic waves a little differently. Physicians are able to use these echoes to detect brain tumors and other conditions. Ultrasound is also used to monitor the growth of unborn babies.*)

Closure

Have interested students design their own model to simulate the workings of the human ear. They can use Francine's model on page 395 as a starting point. Ask them to add labels and explanations of how each part of the ear functions. You may wish to have students construct working models and display them around the classroom.

CHALLENGE YOUR THINKING

1. Listen Up

How might you explain each of the following?

Thread

Light paper ball

a. When you strike tuning fork *A*, the paper ball jumps off its resting spot on an identical tuning fork, *B*. Why?

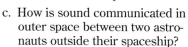

A   B

b. A grasshopper produces sound by rubbing its rough legs against its hard wing covers. This sound can be heard almost 100 m away. There are over 1000 tons of air in 1 cubic hectometer of air (1 hectometer = 100 m). How does a grasshopper have the strength to move that much air as it chirps?

Wooden box

c. How is sound communicated in outer space between two astronauts outside their spaceship?

d. The sound of a dentist's drill is louder to the patient than to the dentist. Why?

e. In the mountains, why does the echo of a shout last so much longer than the original shout?

2. Beat It!

Yale was watching Janelle play her bass drum on the football field. Janelle kept a steady rhythm going—about 20 beats every 10 seconds. When Yale backed away from Janelle to a distance of 85 m, he noticed that each *Boom!* came precisely when the drumstick was farthest away from the drum.

a. How do you explain this phenomenon?

b. From the mathematical information given, find (1) the time between drumbeats, (2) the time it takes for the drumstick to get to the farthest point from the drum, and (3) the speed of sound in air.

85 M

397

⭐ **The Chapter 17 Review Worksheet on page 37 of the Teaching Resources booklet is available to accompany this Challenge Your Thinking.**

Answers to *Challenge Your Thinking*

1. a. Tuning fork *A* vibrates, causing waves of vibrating air particles to pass to tuning fork *B*. Tuning fork *B* then vibrates, and the paper ball jumps.

b. This is essentially a chain reaction. The grasshopper has to move only the air particles closest to itself. This, in turn, moves the adjacent air in ever-widening circles.

c. Accept all reasonable responses. Sound waves are converted into radio waves that are transmitted across space to a receiver. Unlike sound waves, radio waves do not need a medium for transmission.

d. The sound travels through the teeth and the bones of the patient's jaw as well as through the air. Bones, being a solid material, transmit sound better than the air does.

e. The echo is actually a series of overlapping echoes. The sound waves from the shout spread out in many directions, and they are reflected by different surfaces at various distances from the source. The reflected sound waves form a series of overlapping echoes as they return to the source of the shout. The overlapping sound makes the echo last longer than the shout.

2. a. This phenomenon results from the tremendous difference between the speed of sound and the speed of light. In the time it takes for the sound to reach Yale, Janelle is able to complete another half of a drumbeat.

b. (1) The time between drumbeats = 10 s ÷ 20 = 0.5 s
(2) One half of a drumbeat = 0.25 s
(3) Speed of sound = distance ÷ time = 85 m ÷ 0.25 s = 340 m/s

Challenge Your Thinking continued ▶

Answers to
Challenge Your Thinking,
continued

3. Answers will vary in complexity. Sample answer: When Jim blows into the pitch pipe, the column of air inside the pitch pipe vibrates at a frequency corresponding to the note of A. The vibrations from the pitch pipe cause the surrounding particles of air to vibrate at the same frequency and to form waves of compressions and expansions that travel through the air. When the sound waves reach the singers, the waves are channeled through the ear canal to the eardrum and cause the singers' eardrums to vibrate. The vibrations of the eardrum travel through the middle ear bones to the inner ear, where they create waves in the fluid of the inner ear. The movement of fluid causes tiny hairs to vibrate, stimulating the nerve endings in the inner ear. The nerve cells convert the vibrations of the hairs into electrical impulses that travel to the brain. The brain interprets the nerve impulses as the note A. Through training, each singer is able to produce a vibration in his or her vocal cords that corresponds to this note. The vibrations of the singers' voices then travel to the conductor's ears in the same manner, and he is pleased by the note that he hears.

4. Answers will vary. Student definitions of sound may refer to either the transmission of particle motion from a vibrating object through a medium or to the detection and recognition of particle waves by a receiver.

5. The row of dominoes is a good model of wave motion because it shows that energy travels by way of the motion of the particles of a medium. The model shows that the motion of one particle of the medium sets into motion the particle next to it. The model also shows that a wave travels at a constant speed determined by the medium through which it is traveling.

To depict a louder sound, larger dominoes could be used to represent the greater compressions and expansions of the particles of the medium. Dominoes cannot depict a sound of longer duration, however, because the dominoes would need to return to their upright position after every fall to mimic the back-

3. A Simple Tune

Jim blows the note "A" on his pitch pipe. The barbershop singers all hum the note. "Right on key," says the conductor.

Can you describe the whole story of what is going on, from the note Jim blew to the hum that pleased the conductor?

4. One Hand Clapping in a Forest?

After considering the debate questions on page 396, you may have realized that your answers depended on how you defined *sound*. Complete the sentence "Sound is . . . " in as many words as you wish, in order to create a definition that satisfies you. State why this is *your* definition.

5. A Model With Impact

Suppose you had a row of dominoes and you hit the first domino so that it fell backward. What would happen? Is this a good model of wave motion? Why? How would you depict a louder sound? a sound of longer duration?

Sciencelog

Review your responses to the Sciencelog questions on page 375. Then revise your original ideas so that they reflect what you've learned.

and-forth motion of the particles of a medium.

Sciencelog

The following are sample revised answers:

1. Light travels though air nearly 1 million times faster than sound does. Although the thunder and lightning are produced at the same time, the light travels to the observer almost instantaneously, while the sound of the thunder takes much longer to reach the observer.

2. There is no sound in outer space because there is no medium of particles to transmit sound waves.

3. By vibrating, a source causes the particles of a medium near it, such as air, water, or wood, to vibrate. These vibrating particles, in turn, cause the particles next to them to vibrate. This chain reaction continues until the particles next to a receiver, such as an ear, are vibrating, thus transmitting the vibrations, or sound, to the listener.

CHAPTER
18
Listening Closely

1 All other factors being equal, these instruments will play notes of different pitch. What accounts for this?

2 Will everyone in the audience hear the sound at the same volume? Why or why not?

3 Why do a violin and a trumpet playing the same note at the same loudness sound different?

ScienceLog

Think about these questions for a moment, and answer them in your ScienceLog. When you've finished this chapter, you'll have the opportunity to revise your answers based on what you've learned.

399

Connecting to Other Chapters

> **Chapter 16**
> *shows how vibrations produce sounds and discusses ways of distinguishing different sounds.*

> **Chapter 17**
> *explores how the particles of a medium transmit sound energy and how our ears receive that energy.*

> **Chapter 18**
> *invites students to reexamine pitch, loudness, and sound quality in a quantitative way.*

Prior Knowledge and Misconceptions

Your students' responses to the ScienceLog questions on this page will reveal the kind of information—and misinformation—they bring to this chapter. Use what you find out about your students' knowledge to choose which chapter concepts and activities to emphasize in your teaching. After students complete the material in this chapter, they will be asked to revise their answers based on what they have learned. Sample revised answers can be found on page 415.

In addition to having students answer the questions on this page, you may wish to have them complete the following activity: Draw a sine wave on the board to represent one complete vibration of a sound wave. Using a tape recorder, play a selection of sounds for the class. Ask students to draw waves similar to the one on the board to depict what they hear. Collect the papers, but do not grade them. Instead, examine the drawings to find out what students know about measuring sound, what misconceptions they may have, and what about the topic is interesting to them.

Homework

You may wish to have students research how musicians use the effect of beat frequencies to tune their instruments to a proper pitch. *(When a note whose frequency is close to a reference frequency is played, the sound of the two notes together seems to pulse, or beat, over time. A musician can adjust his or her instrument until this beating no longer occurs.)* You may wish to demonstrate this effect in class with a guitar or other string instrument.

Multicultural Extension

Musical Instruments
Divide the class into small groups and assign each group a culture. Have each group investigate the contributions of the culture to our present understanding of sound. Representative scientists from Europe, India, and many Arabic cultures made major contributions. You may wish to have one member of each group present the group's findings to the class.

FOCUS

Getting Started

Review the term *variable*. Later, as students complete Exploration 1, point out how the term applies to the Exploration.

Main Ideas

1. Increasing the thickness of a vibrating string lowers its pitch.
2. Increasing the tension of a vibrating string increases its pitch.
3. Decreasing the length of a vibrating string increases its pitch.

TEACHING STRATEGIES

EXPLORATION 1

Answers to
What to Do, pages 400–401

1. Using the same amount of tension ensures that any difference in pitch among the strings is caused only by their different thicknesses.

2. The greater the thickness of the string, the lower the pitch.

4. Increasing the mass of the jug with water increases the tension, which raises the pitch.
   Try this: Adding 3 kg of water to the jug makes the total mass of the water and jug equal to 4 kg. The new tone produced should be close to high doh *(doh')*.

5. The shorter the length, the higher the pitch.

6. Halving the length doubles the frequency.

Answer to
Caption, page 401

Ⓐ Thickness, length, and tension of the individual strings affect their pitch. The bass produces a lower range of pitches than the violin does.

 An Exploration Worksheet is available to accompany Exploration 1 (Teaching Resources, page 44).

EXPLORATION 1

An Investigation: Strings and Pitch

Before you do this Exploration, think about these questions:

- Many musical instruments use strings to produce their sound. What are some of these instruments?
- How can you make high or low musical sounds (tones) with these instruments?
- What characteristics of the strings can be changed to make musical tones higher or lower?

You Will Need

The diagram below shows one setup that will enable you to do this Exploration. You may, however, find other materials that work just as well. Try to design your own apparatus.

Be Careful: Do not let the plywood extend over the table's edge.

What to Do

Consider the following questions:

> Does the thickness of an instrument's string affect its pitch?

1. Stretch the three different strings or fishing lines using the same amount of tension. (Why?) You can do this by adding the same amount of water to each container.

2. Pluck the strings. What do you observe? By sounding the musical scale (*doh, re, mi*, etc.) you may be able to locate the tones produced by these strings on the scale. What conclusions can you draw?

> Does a string's tension (the amount it is stretched) affect pitch?

3. Test this hypothesis by using the strongest of the three nylon strings (a 9-kg-test fishing line works well). While one person plucks the string, another person can gradually add water to the jug.

4. What do you observe? What conclusions can you draw from your observations?

Try this: Add enough water to a jug so that the jug and water have a total mass of 1 kg. Hang the jug from the strongest string, and pluck the string. Sing this tone as *doh*. Then sing the scale *doh-re-mi-fa-sol-la-ti-doh'*. (The scale from *doh* to *doh'* forms an **octave**.) Try to keep in mind the sound of *doh'* (or record it).

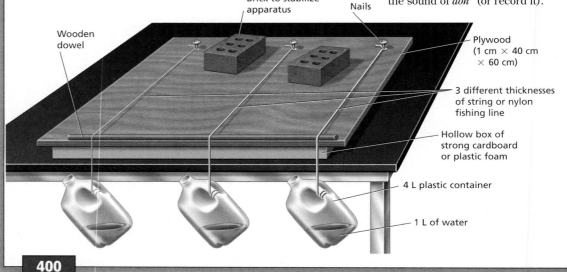

Brick to stabilize apparatus

Nails

Wooden dowel

Plywood (1 cm × 40 cm × 60 cm)

3 different thicknesses of string or nylon fishing line

Hollow box of strong cardboard or plastic foam

4 L plastic container

1 L of water

400

LESSON 1 ORGANIZER

Time Required
1 to 2 class periods

Process Skills
observing, analyzing, classifying

New Term
Octave—the interval between a given musical tone and one with double or half the frequency; the musical scale from doh to doh'

Materials (per student group)
Exploration 1: materials to construct apparatus for interpreting pitch, such as the following: three 4 L plastic jugs; three 2 m lengths of string or nylon fishing line of different thicknesses; 3 short nails; hammer; 40 cm wooden dowel; 2 bricks; 60 cm × 40 cm × 1 cm sheet of plywood; box of strong cardboard or plastic foam, about 60 cm × 40 cm; 6 L of water; 1000 mL beaker; sheet of paper; about 10 cm of tape; safety goggles

Teaching Resources
Exploration Worksheet, p. 44
Theme Worksheet, p. 48
SourceBook, p. S116

1 All other factors being equal, these instruments will play notes of different pitch. What accounts for this?

2 Will everyone in the audience hear the sound at the same volume? Why or why not?

3 Why do a violin and a trumpet playing the same note at the same loudness sound different?

ScienceLog

Think about these questions for a moment, and answer them in your ScienceLog. When you've finished this chapter, you'll have the opportunity to revise your answers based on what you've learned.

399

Homework

You may wish to have students research how musicians use the effect of beat frequencies to tune their instruments to a proper pitch. *(When a note whose frequency is close to a reference frequency is played, the sound of the two notes together seems to pulse, or beat, over time. A musician can adjust his or her instrument until this beating no longer occurs.)* You may wish to demonstrate this effect in class with a guitar or other string instrument.

Multicultural Extension

Musical Instruments
Divide the class into small groups and assign each group a culture. Have each group investigate the contributions of the culture to our present understanding of sound. Representative scientists from Europe, India, and many Arabic cultures made major contributions. You may wish to have one member of each group present the group's findings to the class.

CHAPTER
18
Listening Closely

Connecting to Other Chapters

> **Chapter 16**
> *shows how vibrations produce sounds and discusses ways of distinguishing different sounds.*

> **Chapter 17**
> *explores how the particles of a medium transmit sound energy and how our ears receive that energy.*

> **Chapter 18**
> *invites students to reexamine pitch, loudness, and sound quality in a quantitative way.*

Prior Knowledge and Misconceptions

Your students' responses to the ScienceLog questions on this page will reveal the kind of information—and misinformation—they bring to this chapter. Use what you find out about your students' knowledge to choose which chapter concepts and activities to emphasize in your teaching. After students complete the material in this chapter, they will be asked to revise their answers based on what they have learned. Sample revised answers can be found on page 415.

In addition to having students answer the questions on this page, you may wish to have them complete the following activity: Draw a sine wave on the board to represent one complete vibration of a sound wave. Using a tape recorder, play a selection of sounds for the class. Ask students to draw waves similar to the one on the board to depict what they hear. Collect the papers, but do not grade them. Instead, examine the drawings to find out what students know about measuring sound, what misconceptions they may have, and what about the topic is interesting to them.

LESSON

1

A Closer Look at Pitch

FOCUS

Getting Started
Review the term *variable*. Later, as students complete Exploration 1, point out how the term applies to the Exploration.

Main Ideas
1. Increasing the thickness of a vibrating string lowers its pitch.
2. Increasing the tension of a vibrating string increases its pitch.
3. Decreasing the length of a vibrating string increases its pitch.

TEACHING STRATEGIES

EXPLORATION 1

Answers to
What to Do, pages 400–401

1. Using the same amount of tension ensures that any difference in pitch among the strings is caused only by their different thicknesses.

2. The greater the thickness of the string, the lower the pitch.

4. Increasing the mass of the jug with water increases the tension, which raises the pitch.
   Try this: Adding 3 kg of water to the jug makes the total mass of the water and jug equal to 4 kg. The new tone produced should be close to high doh *(doh')*.

5. The shorter the length, the higher the pitch.

6. Halving the length doubles the frequency.

Answer to
Caption, page 401

Ⓐ Thickness, length, and tension of the individual strings affect their pitch. The bass produces a lower range of pitches than the violin does.

★ An Exploration Worksheet is available to accompany Exploration 1 (Teaching Resources, page 44).

EXPLORATION 1

An Investigation: Strings and Pitch
Before you do this Exploration, think about these questions:

- Many musical instruments use strings to produce their sound. What are some of these instruments?
- How can you make high or low musical sounds (tones) with these instruments?
- What characteristics of the strings can be changed to make musical tones higher or lower?

You Will Need
The diagram below shows one setup that will enable you to do this Exploration. You may, however, find other materials that work just as well. Try to design your own apparatus.

Be Careful: Do not let the plywood extend over the table's edge.

What to Do
Consider the following questions:

> Does the thickness of an instrument's string affect its pitch?

1. Stretch the three different strings or fishing lines using the same amount of tension. (Why?) You can do this by adding the same amount of water to each container.

2. Pluck the strings. What do you observe? By sounding the musical scale (*doh, re, mi,* etc.) you may be able to locate the tones produced by these strings on the scale. What conclusions can you draw?

> Does a string's tension (the amount it is stretched) affect pitch?

3. Test this hypothesis by using the strongest of the three nylon strings (a 9-kg-test fishing line works well). While one person plucks the string, another person can gradually add water to the jug.

4. What do you observe? What conclusions can you draw from your observations?

Try this: Add enough water to a jug so that the jug and water have a total mass of 1 kg. Hang the jug from the strongest string, and pluck the string. Sing this tone as *doh*. Then sing the scale *doh-re-mi-fa-sol-la-ti-doh'*. (The scale from *doh* to *doh'* forms an **octave**.) Try to keep in mind the sound of *doh'* (or record it).

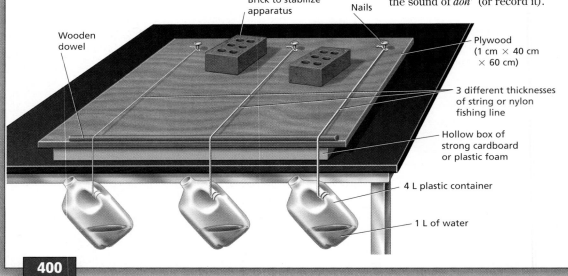

Brick to stabilize apparatus

Nails

Wooden dowel

Plywood (1 cm × 40 cm × 60 cm)

3 different thicknesses of string or nylon fishing line

Hollow box of strong cardboard or plastic foam

4 L plastic container

1 L of water

400

LESSON 1 ORGANIZER

Time Required
1 to 2 class periods

Process Skills
observing, analyzing, classifying

New Term
Octave—the interval between a given musical tone and one with double or half the frequency; the musical scale from doh to doh'

Materials (per student group)
Exploration 1: materials to construct apparatus for interpreting pitch, such as the following: three 4 L plastic jugs;

three 2 m lengths of string or nylon fishing line of different thicknesses; 3 short nails; hammer; 40 cm wooden dowel; 2 bricks; 60 cm × 40 cm × 1 cm sheet of plywood; box of strong cardboard or plastic foam, about 60 cm × 40 cm; 6 L of water; 1000 mL beaker; sheet of paper; about 10 cm of tape; safety goggles

Teaching Resources
Exploration Worksheet, p. 44
Theme Worksheet, p. 48
SourceBook, p. S116

Now add 3 L of water to the jug (that is, 3 kg of water). What is the total mass of the jug and water now? Again, pluck the string. Does the tone sound like the note you sang earlier? If not, can you tell how far away it is on the scale?

You may have discovered an important idea. If the tension of a string is quadrupled, the frequency is doubled; in other words, *doh* becomes high *doh*, or *doh'*.

Does the length of a string affect pitch?

5. You can make a vibrating string shorter by using a pencil as shown below. How does the musical tone change as you shorten the vibrating part of the string?

6. Suppose the tone produced by the whole string vibrating is *doh*. Find the length of string that gives *doh'* (twice the frequency). How does this length compare with the whole length?

7. By moving the pencil along one string, locate the tones of the entire octave. Label them *d, r, m, f, s, l, t, d'* on a piece of paper taped under the string, as shown above.

Analysis

1. What effect does the thickness of a string have on the pitch of a sound?

2. What effect does the tension of a string have on the pitch of a sound?

3. What change in tension causes the frequency to double?

4. What effect does the length of a string have on pitch?

5. What change in the length of a string causes the frequency to double?

6. What change in length might cause the frequency to be halved?

An Orchestra

Find out more about the main groups of instruments in an orchestra—strings, woodwinds, and brass.

- How is sound made in each musical instrument?
- How are sounds made higher? lower? louder?
- How is the size of each instrument related to the pitch of its musical tones?

What are some factors that explain the different pitches of these instruments? Ⓐ

1. The thicker the string, the lower the pitch, assuming that the tension remains the same.

2. Increasing tension increases pitch.

3. Quadrupling the tension

4. The shorter the length, the higher the frequency.

5. Halving the length

6. Doubling the length

Answers to
An Orchestra

- In stringed instruments, sound is made by bowing or plucking the strings. In woodwind instruments, sound is made by blowing across one or two reeds to vibrate the reeds or by blowing across an opening to vibrate the air column inside. In brass instruments, sound is made by blowing through pursed lips to vibrate the lips and the air column inside the instrument.
- In stringed instruments, pressing on the strings shortens the vibrating portion of the strings and makes the sounds higher. Lifting fingers off the strings lengthens the vibrating portion of the strings and makes the sounds lower. Bowing or plucking with greater force makes the sounds louder. In woodwind instruments, uncovering holes in the instrument shortens the vibrating column of air inside and makes the sounds higher. Covering the holes lengthens the column and makes the sounds lower. In brass instruments, tightening the lips makes the sounds higher and loosening the lips makes the sounds lower. Brass instruments also have valves or slides that adjust the pitch further. In both woodwind and brass instruments, blowing with greater force makes the sounds louder.
- The larger the instrument, the lower the range of pitches it can produce.

FOLLOW-UP

Reteaching

Ask student volunteers to demonstrate how a stringed instrument such as a violin or cello is played. Have students discuss how each instrument makes sound, how its sounds can be made louder, and how its pitch can be varied.

Assessment

Ask students to make a poster that represents the most important concepts of the lesson. Have each student explain his or her poster to the class.

Extension

Have students investigate a stringed instrument that has not been discussed in this unit. Students should report on the design of the instrument and on how its features relate to the sounds it makes.

Closure

Have students make stringed instruments using common materials such as rubber bands, pie tins, and dowels. Each student should explain to the class how pitch can be changed on his or her instrument.

LESSON 2 — Loudness

FOCUS

Getting Started

The Discrepant Event Worksheet on page 50 of the Unit 6 Teaching Resources booklet describes a teacher demonstration that makes an excellent introduction to this lesson.

Main Ideas

1. A sound can be made louder by adding energy, by decreasing the distance to the sound source, by directing the sound energy, by forced vibration, and by causing another object to resonate.
2. Loudness can be measured by a sound meter in units called decibels.
3. An increase in loudness of 10 dB is equivalent to a tenfold increase in sound energy.
4. Loud noises constitute a type of pollution that can affect human health.

TEACHING STRATEGIES

Making Sounds Louder: A Problem

Divide the class into small groups, and have them think of as many ways as they can to make the sound of a tuning fork louder. After you have reviewed student ideas for safety, have each group demonstrate their ideas using a tuning fork.

Answers to
Making Sounds Louder: A Problem

Accept all reasonable speculations at this point. Students will discover in this lesson, for example, that the tuning fork can be made louder by creating forced vibrations.

Answers to
Some Solutions to the Problem

Darrel correctly reasoned that more energy produces a greater amplitude of vibration and thus a louder sound.

Making Sounds Louder: A Problem

You are given a tuning fork. When you strike it, you can scarcely hear it. How can you make the sound of the fork louder? Before you read any further, write down some suggestions in your ScienceLog.

Some Solutions to the Problem

Darrel came up with a simple solution. He asked himself a question: How is the loudness of a sound related to the energy put into the object that produces the sound? His answer can be seen in the illustrations at right.

Nicole also found a simple solution. She wondered about the relationship between the loudness of the sound and the distance from the ear to the source of the sound. Her answer can be seen in the illustrations below, which she found in a reference book.

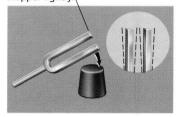

Hit the rubber stopper lightly.

Small amplitude

Hit the rubber stopper strongly.

Large amplitude

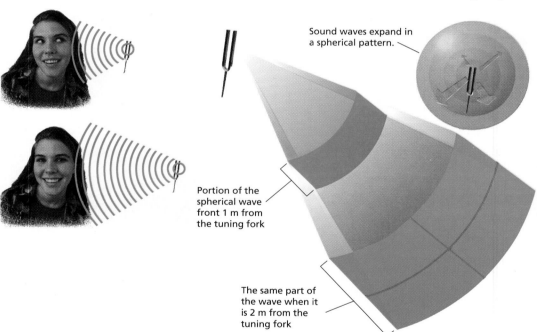

Sound waves expand in a spherical pattern.

Portion of the spherical wave front 1 m from the tuning fork

The same part of the wave when it is 2 m from the tuning fork

402

LESSON 2 ORGANIZER

Time Required 3 class periods

Process Skills communicating, analyzing, inferring, comparing

Theme Connection Systems

New Terms
Decibel—a unit of measure of the loudness of sound
Forced vibration—the vibration produced in an object when it comes in contact with a vibrating object
Megaphone—a cone-shaped device used to direct or amplify the voice
Resonance—a phenomenon in which an object vibrates in response to the

vibrations of another object

Materials (per student group)
Some Solutions to the Problem:
tuning fork; variety of materials of different sizes to test, such as a large and small glass container, a large and small hollow wood block, a rubber-headed mallet, a large and small metal cymbal, and a large and small plastic cup; sheet of paper; cardboard tube, at least 40 cm long; scissors; metric ruler (See Advance Preparation on page 355C.)

Teaching Resources
Discrepant Event Worksheet, p. 50
SourceBook, p. S124

Study the illustration that Nicole found. Notice that when the distance of the sound wave from the tuning fork doubles, the wave spreads out over four times as much area. How loud will the sound be to a listener who is 1 m from the tuning fork, as compared with the loudness at 2 m? at 3 m? Try to derive a formula showing the relationship between the area of the sound energy and its distance from the sound source. Write this formula in your ScienceLog. **Ⓐ**

Diana discovered that placing the base of the tuning fork on a tabletop or a hollow wooden box makes the sound louder. Doug discovered that he could get a similar effect by using a cymbal rather than a box.

Diana and Doug have discovered **forced vibration**—getting an object to vibrate by touching it with an already vibrating object. What is the loudest sound you can get from a tuning fork by using forced vibration? To find out, try touching the fork to objects made of various materials, such as glass, metal, and plastic. How important is the size of the object? the material it is made of? **Ⓑ**

Examples of forced vibration

Yolanda found that the cone-shaped apparatus shown below helped make the sound of the tuning fork louder. How did it work? She concluded that the cone shape prevented some of the sound waves from spreading out in all directions. The cone concentrates the sound energy in a specific direction. Therefore, the sound appears louder to a listener in the path of the sound energy.

Have you ever used a **megaphone**? What are some situations in which a megaphone is used? Design an experiment to show how much farther a sound can travel when a megaphone is used. **Ⓒ**

A simple megaphone

403

Answers to *In-Text Questions*

Ⓐ From the illustration at the bottom of page 402, students should conclude that the sound energy reaching a given surface area twice as far from the fork is spread over four times as much surface area. Hence, the sound will be only one-fourth as loud. The same sound energy will be spread out over nine times the original surface area when it is three times farther away. Hence the loudness of the sound will be only one-ninth as loud. Students may determine that the area of the sound energy depends on the square of the distance to the source.

Ⓑ Students should discover that large objects should make louder sounds than smaller objects of the same material. Metal objects produce the loudest sounds. Plastic-foam objects produce the least sound.

Ⓒ Megaphones are sometimes used by cheerleaders, police officers directing traffic, and other people who must communicate outdoors where there is no electricity for a microphone.

To design and perform their experiment to find out how much farther a sound can travel using a megaphone, students would need to control the amount of energy put into the sound-producing object (e.g., by tapping a tuning fork with exactly the same force each time). The distance from the sound source to the listener, with and without a megaphone, must also be controlled.

CROSS-DISCIPLINARY FOCUS

Industrial Arts

If your school facilities permit, have interested students make a steel drum. Make sure you check all designs for safety before allowing students to construct the drums. This might be a good opportunity to involve the industrial arts and music departments with the science program. Students will need to apply what they have learned about pitch to tune the steel drum. Suggest that they listen to a recording of a steel-drum band or do some research on the construction and use of steel drums.

Multicultural Extension

Musical Scales

Have students compare the musical scales of different musical traditions. Do all cultures use the same set of scales? Do all cultures use the same set of notes? *(Western musical traditions recognize 12 separate tones. Some musical traditions, such as those of Japan, China, India, and some American Indian cultures, actually use more or different tones. These* "extra" tones vibrate at frequencies in between the frequencies of the 12 Western tones. Also, the scales of two different cultures may include the same notes, but because they are used in different patterns, the resulting music can sound very different.)

Answers to
In-Text Questions

A Accept all reasonable answers. Students may be aware that when one object causes another object to resonate, the effect can intensify the magnitude of the vibrations, in this case producing a louder sound.

B The piano string and the natural frequency of the object in the room are the same. Another example of resonance occurs when a note of a certain frequency causes a wine glass to crack.

C In resonance, sound waves from a vibrating object cause another object to begin vibrating. In forced vibration, an object is caused to vibrate when touched with an already vibrating object.

D The length of the tube should be 17,200 cm ÷ 512 = 33.6 cm.

E When you close one end of the tube by placing it on the table, the tube should not resonate with the tuning fork.

F When the tube is cut in half, students should observe that the two halves will resonate with the tuning fork only if one end is closed. From this, students should realize that the tube length necessary to cause resonance depends on whether the column is closed at one end or open at both ends. Students should realize that by closing one end of the tube, they have changed the tube's natural rate of vibration.

Answers to
Who Said It?

a. Nicole
b. Leonard
c. Yolanda
d. Darrel
e. Diana or Doug

Leonard discovered that he could make the sound louder by holding the tuning fork next to a cardboard tube. If the air column within the tube is just the right length—so that it has the same natural rate of vibration as the tuning fork—it will start to vibrate along with the fork. As a result, the sound is louder. The air column is vibrating in **resonance** with the tuning fork. What do you think causes this to happen? **A**

Resonance also occurs when you strike a certain key on the piano and something in the room begins to vibrate. What would you say about the frequency of the piano string and the natural frequency of the object in the room? What other experiences have you had that involved resonance? Try writing the meaning of *resonance* in your own words. How is *resonance* different from *forced vibration*? **C**

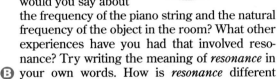 Leonard's design for making the tuning fork louder is quite simple. To try it, cut a cardboard tube to the right length using this approximate formula:

$$\text{length of tube in centimeters} = \frac{17,200}{\text{frequency of fork (Hz)}}$$

For a 512 Hz tuning fork, to what length would you cut the tube? You might want to cut the tube **D** a little long and then shorten it as needed until it is just right.

After you get the tube to work, place it vertically on the table so that one end is closed. Next, hold the vibrating fork above the tube. Does the air column still vibrate in resonance with the fork? **E**

Now cut the tube in half and try again, first with the tube open at both ends and then with it closed at one end. What do you observe? Why do you think this happens? **F**

Who Said It?

Which of the students might have made each of the following comments based on his or her findings?

a. As more and more air particles are affected, their back-and-forth movement decreases.

b. If two objects are vibrating at the same frequency (their natural frequency), the sound will be louder.

c. If the disturbance of the particles is directed, their energy lasts longer.

d. If the air particles vibrate a large amount, the sound is loud.

e. The more air particles that are set into vibration, the more likely they are to hit the eardrum.

404

Meeting Individual Needs

Gifted Learners
Some students might like to make a *telescoping tube* to find resonance lengths for tubes closed at one end, as well as for tubes open at both ends. The formulas for these resonating lengths are as follows:
- For tubes open at both ends, the length of the tube must be (17,200 cm × *n*) ÷ frequency of fork, where *n* = 1, 2, 3, 4, and so on.

- For tubes open at only one end, the length of the tube must be (8600 cm × *n*) ÷ frequency of fork, where *n* = 1, 3, 5, 7, and so on.

Measuring Loudness

The loudness of a sound can be measured with a *sound meter*. The scale on such a device is marked in units called **decibels (dB)**. On the meter shown below, what would happen to the reading if the meter were moved closer to the source of sound? farther from the source?

Estimating Loudness

How sensitive are your ears? How well can you tell whether one sound is louder than another? On the next page, List A contains a number of sounds. List B gives the approximate loudness of each of these sounds but not in an order that corresponds to List A. Match each sound with its loudness. Then examine the answers at the bottom of the next page to check your work.

405

Theme Connection

Systems
Use a guitar to show students how some stringed instruments have a system of metal bars, or frets, to help musicians produce specific pitches. **Focus question:** Why does the distance between the frets on a guitar decrease as you go up the neck? *(Shortening the vibrating portion of a string by a certain percentage increases its pitch by a set amount.*

For example, halving the length doubles the frequency. Because the vibrating portion of the string gets smaller as you go up the neck, the distance between frets must also get smaller if they are to represent the same percentage of the vibrating portion of the string.) A Theme Worksheet is available to accompany this Theme Connection (Teaching Resources, page 48).

Answer to *Measuring Loudness*

If the meter were moved closer to the source of sound, the decibel rating would be higher, and if moved farther away, the decibel rating would be lower.

Estimating Loudness

Be sure that students understand what they are to do in this matching exercise. (Students arrange the 15 sounds on page 406 in order from softest to loudest. Then the values from 0 to 150 dB can be assigned.) Divide the class into small groups and have them discuss the answers. If List A is written on the chalkboard prior to the class, students will not be tempted to look at the key provided at the bottom of the page. After each group has arrived at an answer, reassemble the class, and ask a spokesperson from each group to present the group's list of sounds in increasing order of loudness with the matching decibel level. Resolve any differences that may arise.

Integrating the Sciences

Life and Physical Sciences
Have students research the biological hazards of loud sounds. Allow students to develop their own research topic or make assignments from the following list:

- What kinds of noise are the most hazardous to your ears?
- What have scientists discovered about how hearing loss corresponds to age?
- Is there a significant difference between the amount of hearing loss in industrial areas and in rural areas?
- What parts of the ear can be damaged by loud sound?
- What precautions can people take to minimize the risk of hearing loss due to exposure to loud sound?

The Decibel— An Unusual Unit

Call on a volunteer to read aloud the material. Then direct students' attention to the Sound Energy and Decibels table. Allow time for students to study it. Students should discover that each rating on the decibel scale represents 10 times the energy of the rating before it. (This is a logarithmic relationship.) For example, 20 dB has 10 times as much energy as 10 dB, and 30 dB has 10 times as much energy as 20 dB and 100 times as much energy as 10 dB.

Ask students to fill in the omissions on the table. (*The answers are as follows: 10,000 × sound energy, 40 dB more; 1,000,000 × sound energy, 60 dB more.*)

Answers to
A Puzzler

400 = 4 × 100

From the table:

4 × sound energy = 6 dB more
100 × sound energy = 20 dB more
Total = 26 dB more

One sound has a loudness of 30 dB. The second sound, 400 times louder, has a loudness of 30 dB + 26 dB, or 56 dB.

Answer to
Caption

Ⓐ The sound of a ringing telephone is about 70 dB.

CROSS-DISCIPLINARY FOCUS

Mathematics

For mathematically inclined students, explain that the decibel scale is logarithmic. Because it is a logarithmic scale, an increase of 10 dB means that sound intensity increases by a factor of 10. So a sound of 10 dB is 10 times as intense as 0 dB. But 20 dB is not 20 times as intense as 0 dB, as might be expected. It is 10 times louder than 10 dB, or 100 times as intense as 0 dB. For practice, ask students: How much more intense is a 60 dB sound compared to a 30 dB sound? (*A 60 dB sound is 1000 times as intense as a 30 dB sound.*)

List A

1. circular saw (nearby)
2. leaves rustling in a breeze
3. telephone ringing
4. water at the foot of Niagara Falls
5. jet taking off (nearby)
6. jackhammer breaking concrete (nearby)
7. rock-and-roll band at a concert
8. high-powered rifle shot (nearby)
9. quiet restaurant
10. ordinary conversation
11. quiet neighborhood at night
12. silence in a soundproof room
13. vacuum cleaner
14. whisper
15. ordinary breathing

List B

0 dB, 10 dB, 20 dB, 30 dB, 40 dB, 50 dB, 60 dB, 70 dB, 80 dB, 90 dB, 100 dB, 110 dB, 120 dB, 130 dB, 140 dB

Here is the correct order for the Estimating Loudness activity: 1. 100 dB, 2. 20 dB, 3. 70 dB, 4. 90 dB, 5. 140 dB, 6. 110 dB, 7. 120 dB, 8. 130 dB, 9. 50 dB, 10. 60 dB, 11. 40 dB, 12. 0 dB, 13. 80 dB, 14. 30 dB, 15. 10 dB

The Decibel—An Unusual Unit

The decibel scale of loudness is an unusual scale. For example, the sound of a person breathing registers about 10 dB, while the rustle of a newspaper registers about 30 dB. Yet the sound energy provided by the newspaper is about 100 times that of breathing. The table below shows the relationship between sound energy and decibels.

Sound Energy and Decibels—Some Simple Rules	
Change in sound energy	Increase in loudness
2 × sound energy	3 dB more
4 × sound energy	6 dB more
8 × sound energy	9 dB more
10 × sound energy	10 dB more
100 × sound energy	20 dB more
1000 × sound energy	30 dB more
? × sound energy	40 dB more
1,000,000 × sound energy	? dB more

A Puzzler
One sound has a loudness of 30 dB. Another sound has 400 times as much energy. What is the loudness of the second sound, in decibels? (Use the table to find the answer.)

How loud is a ringing telephone in decibels? Ⓐ

Meeting Individual Needs

Second-Language Learners

To provide students with an opportunity to practice their English skills, ask them to give a simple demonstration of different ways to change the pitch and amplitude of a sound. They could use an instrument such as a guitar, or they could use an open shoebox with rubber bands stretched across the top. Be sure to check students' proposed procedures for safety before allowing them to proceed.

Integrating the Sciences

Earth and Physical Sciences

Have students research the Richter scale, which is used by geophysicists to measure the magnitude of earthquake tremors. Point out that, like the decibel scale, the Richter scale is also a logarithmic scale. Ask: Why do you think a logarithmic scale is used to measure vibrations in the Earth? (*The Richter scale goes from 0 to 10, and an increase of 1 corresponds to a tenfold increase in magnitude. For convenience, scientists often use such scales when dealing with phenomena that involve very large numbers.*)

Noise Pollution—The Effect of Sound on Health

As you read Now Hear This! find clues that will help you answer these questions:

1. What is crammed into the small space of your ear? (You know this from material you studied earlier in this unit.)
2. What does the author think are the most damaging sounds to a person's ear?
3. As you grow older, what kind of change will occur in your hearing range?
4. What kinds of sounds will be harder for you to hear as you get older?
5. What causes ringing in the ears?
6. What causes middle-ear infections?
7. What can you do to protect your hearing now?
8. What are four types of noise pollution that can really hurt your ears?

▲ The human ear extends deep into the skull. Its main parts are the outer ear, the middle ear, and the inner ear.

Now Hear This!

Got your headphones on? Going to the concert and gonna stand right in front of the speakers? Got the TV, the radio, the stereo, or the CD player turned all the way up? Well you'd better hear this—that's right, I'm talking to YOU! And just who am I? I am your high-powered, anatomically amazing right ear.

I think you and I need to have a little talk. I mean, we've hung out together all our lives, but how much do you really know about me? Did you know that there's enough circuitry packed inside me to light a city? Or at least to give it phone service. And that's no joke!

Without me, you'd be missing out on the world of sound—the sounds of leaves rustling, waves crashing, music playing, and friends talking. And I do all of this in a space so small that I rival computer chips. But our lines of communication seem to be breaking down. Those tiny parts and delicate pieces that make me so incredible are beginning to deteriorate. So you gotta take care of me if you want to keep me working for you.

I was at my best the day you were born. From that time on, my abilities have been on the decline. What causes this? Loud sounds, mainly. Sounds that are too loud, occur too often, and last too long. As I get older, my tissues begin to lose their flexibility. We may be pretty young, but just 10 or 15 years down the road, we're going to feel the effects of any abuse now. Besides the loss of elasticity, hair cells begin to fall apart, and deposits of calcium build up. I'm sure you didn't know it, but that's all going on right now, right here, inside me.

When you were born, I had a hearing range of 16 to 30,000 cycles (vibrations) per second. But by now, my upper limit is just about 20,000 cycles per second. By the time we get to be as old as our grandparents, I could be down to as low

Noise Pollution—The Effect of Sound on Health

Have students read this story individually and look for answers to the questions.

This would be an excellent time to invite an audiologist to visit the class. Have students write questions for the speaker on slips of paper beforehand. The speaker could draw questions randomly from a box or sort through the questions to find those that he or she has time to address. Also, students should be encouraged to relate their own experiences with ear problems.

Answers to
Noise Pollution—The Effect of Sound on Health

1. The parts of the ear include the pinna, ear canal, eardrum, hammer, anvil, stirrup, oval window, cochlea, round window, semicircular canals, and auditory nerve.
2. High-pitched noises like electric guitars or loud noises like jet airplanes
3. As a person ages, he or she becomes less sensitive to high frequency sounds.
4. As a person ages, it becomes more difficult for him or her to detect weak sound signals and higher-pitched sounds.
5. Ringing in the ears can be caused by drugs, fever, circulation changes, or tumors on the acoustic nerve.
6. Middle-ear infections are usually caused by microbes in the throat that have easy access to the ear through the Eustachian tube.
7. To protect your hearing, you can wear earplugs and avoid loud, high-pitched sounds and other forms of noise pollution.
8. Answers will vary. The sources of noise pollution mentioned in the story include rock music, headphones, factory noises, and jet airplanes.

Homework

Have students research why sound intensity is often measured using a logarithmic scale such as decibels. (*Our ears sense sounds logarithmically. For example, a 40 dB sound seems twice as loud as a 20 dB sound to our ears, but it is 100 times as intense.*)

Multicultural Extension

Musical Instruments of Other Lands

Have interested students find out about musical instruments from other lands. For example, they could research the *zither* from Arabia or the *acarina* from Latin America. They should discover what these instruments look like, how they are played, what they sound like, what the vibrating parts are, and how each instrument amplifies the sound made by the musician.

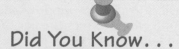
ENVIRONMENTAL FOCUS

Because sounds are waves, they can sometimes cancel each other out, an effect known as *destructive interference.* Have students research this effect, and then share the following information with them: One way of coping with noise pollution is to produce a signal in a set of headphones that uses destructive interference to cancel out the undesirable noise. Such a device is called an *interference filter.*

Did You Know...

If all of the sound energy produced each second by a 75-member orchestra playing at full volume were converted to electrical energy, it would only be enough to power a 70 watt light bulb.

as 4000 cycles per second. Think of all the sounds—good and bad—you'd be missing out on.

I tell you I'm cool, but I'm also fragile. Drum punctures happen very easily. Fortunately, these usually heal themselves. Ringing in here is another problem—the doctors call it tinnitus, and it can come from almost anything: drugs, fever, circulation changes, or tumors on my acoustic nerve. Sometimes the ringing can be stopped—but not always. I can also get infected, usually inside the middle ear. The Eustachian tube exposes me to those nasty infections. The Eustachian tube goes from the middle ear to the throat. In the throat there are a lot of microbes. Those microbes can make their way up to me, causing a painful middle-ear infection.

Another way I get damaged is by an overgrowth of bone from my middle ear. This freezes the motion of the bones, causing conduction deafness. If you have conduction deafness, a hearing aid might help. A surgeon can also go in and replace the stirrup bone with a metal duplicate. This is effective about 80 percent of the time. But bone overgrowth is not the most common cause of hearing

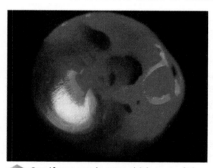

⬆ See if you can locate which part of the middle ear shown here could be surgically replaced by stainless steel. Ⓐ

loss. I think you know what that is. You got it—LOUDNESS!

When you crank up the volume, my tiny muscles tighten up. I'm better at standing up to sounds like thunder or a bass guitar. But high-pitched, piercing sounds from things like jet airplanes, factory machines, and lead guitarists in rock bands destroy my delicate hairs and wreck my tiny muscles. I mean, when I am at my best, I am a finely tuned organ. But when you make me listen to high-pitched, glass-shattering screeches over and over again—well, then I'm done for.

So that's why I'm asking you to listen to me. We've got to do something to protect ourselves. Hearing is too precious for you to lose—take care of your ears. Keep them away from damaging sounds—plug 'em if you have to. Don't use cotton, though; it doesn't work. Use special earplugs instead. Hey, I'm the only right ear you got. And I'll keep hearing for you if you look out for me. Deal?

Effects of Noise

Listed below are the decibel levels at which you'd experience certain effects.
70 dB: difficulty in hearing conversation
80 dB: annoyance at the noise level
90 dB: hearing damage if noise is continuous over time
120 dB: permanent hearing loss if noise is sustained
130 dB: beginning of pain in the ear
140 dB: sharp pain in the ear
170 dB: total deafness

FOLLOW-UP

Reteaching

Have students experiment with different objects that resonate. For example, suggest that they hold up different-sized beakers, bottles, or jugs to their ears. Ask them to compare the sounds that they hear. Suggest that they sing the doh-re-mi scale into a large jar to see if resonance occurs while singing any particular notes.

Assessment

Ask students to write a story about different methods to make sounds louder. Stories should be written from the point of view of the vibrating particles and should describe what happens to the particles when each method is used.

Extension

Ask students to do some research to decide if and how legislation should protect an individual's rights in the case of noise pollution.

Closure

If a decibel meter can be obtained, have students check loudness levels in various places in their community. Then encourage them to suggest actions that city officials might take in regard to persistent or loud sounds in such places as stores, factories, construction sites, malls, and dance clubs.

"Seeing" Sounds

Take a look at the paint smears on the palette. In what way are they similar? In what way do they differ? Although all of them are blue, each represents a different shade of the color blue. Which might be described as navy blue? ocean blue? Just how many shades of blue might there be?

Just as there are shades of color, there are "shades" of sound as well. Your ear (and brain) can distinguish between the different sounds produced by a trumpet, a piano, a violin, and a guitar all playing middle C at about the same loudness. But what makes this difference? If you could see sounds the way you can see different shades of color, you would better understand what gives each of these instruments a unique sound.

There is, in fact, a way to change sound waves into shapes that can be seen using a device called an **oscilloscope**. Study this method by examining the images below.

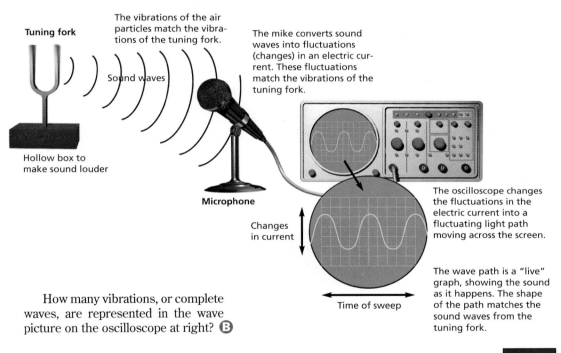

Tuning fork

The vibrations of the air particles match the vibrations of the tuning fork.

Sound waves

Hollow box to make sound louder

Microphone

The mike converts sound waves into fluctuations (changes) in an electric current. These fluctuations match the vibrations of the tuning fork.

Changes in current

The oscilloscope changes the fluctuations in the electric current into a fluctuating light path moving across the screen.

Time of sweep

The wave path is a "live" graph, showing the sound as it happens. The shape of the path matches the sound waves from the tuning fork.

How many vibrations, or complete waves, are represented in the wave picture on the oscilloscope at right? Ⓑ

409

LESSON 3 ORGANIZER

Time Required
2 class periods

Process Skills
observing, comparing, classifying

New Terms
Harmonic—a component of a musical tone whose frequency is a multiple of the fundamental frequency
Oscilloscope—an instrument that visually represents sound waves

Materials (per student group)
How a String Vibrates: rubber tubing, 5 to 10 m in length, or a Slinky spring toy
Locating the Higher Sounds: violin or guitar (See Advance Preparation on page 355C.)

Teaching Resources
Exploration Worksheet, p.51
Transparencies 62 and 63
SourceBook, pp. S116 and S117

LESSON

3

The Quality of Sounds

FOCUS

Getting Started

For a demonstration, set up an oscilloscope with a speaker connected to the input jack. Invite student musicians to play different notes into the microphone, and point out how pitch corresponds to the frequency of waves on the screen. Also ask students to vary the loudness of notes. Point out how loudness corresponds to the amplitudes of the waves on the screen. For more complex waveforms, have students speak into the microphone or play recorded music into it.

Main Ideas

1. An oscilloscope is an instrument that visually represents sound waves.
2. The relative height of the wave on an oscilloscope is a measure of a sound's loudness or amplitude.
3. The relative number of waves on the screen is a measure of the sound's frequency.
4. A string can vibrate in many different ways, each way producing a different sound with a different pitch and loudness.

TEACHING STRATEGIES

"Seeing" Sounds

Have students follow the sequence of drawings and explanations that show how a sound is given a visual shape on an oscilloscope. Stress that the shapes and sizes of the sound pictured on an oscilloscope accurately represent the specific characteristics of a given sound wave. Make sure students understand that each complete wave on the screen represents one vibration.

....

Answer to
In-Text Question

Ⓑ Two vibrations or complete waves are represented in the wave picture on the oscilloscope.

Answers to
In-Text Questions

Ⓐ If the time to sweep across the screen is 0.01 s and two complete waves are shown, then the object vibrates twice every 0.01 s, or 200 times per second. Thus, the frequency is 200 Hz.

Ⓑ If the frequency of the sound waves were 300 Hz, then there would be three complete waves on the screen. If the frequency of the sound waves were 550 Hz, then there would be 5.5 waves on the screen.

Ⓒ If the time of sweep on an oscilloscope is set at 1/50 s and the frequency of a sound source is 200 Hz, then there would be four complete waves on the screen. (0.02 second × 200 vibrations/second = 4 vibrations)

EXPLORATION 2

Divide the class into pairs, and have them work through the Exploration together. Suggest that they discuss all of the questions and analyze the pictures before coming to any conclusions.

Answers to
Exploration 2

1. Frequency = 3.25 vibrations ÷ 0.01 s = 325 Hz.

2. The height of the waves indicates a sound's loudness.

3. The number of waves on the screen indicates a sound's pitch, or frequency.

 An Exploration Worksheet (Teaching Resources, page 51) and Transparency 62 are available to accompany Exploration 2.

As the tuning fork sounds, the light path sweeps from left to right. Suppose that the oscilloscope was set in such a way that one complete cycle of the light path crossed the screen in $\frac{1}{100}$ of a second. You wouldn't see the path as it was being traced out because it would happen too fast. Instead, you would see a steady wave picture; this picture would continue as long as the tuning fork was sounding.

If two complete waves appear on the screen in $\frac{1}{100}$ of a second, how many vibrations does the fork make in 1 second? Ⓐ

As you can see, the frequency of a sound can be determined by knowing how long it takes the light path to sweep across the oscilloscope screen and by knowing the number of waves appearing on the screen. How many complete waves would appear on the screen if the frequency of the sound were 300 vibrations per second (300 Hz)? 550 Hz? Ⓑ

If the time of sweep on an oscilloscope were set at $\frac{1}{50}$ of a second and the frequency of the sound were 200 Hz, how many complete waves would be pictured on the screen? Ⓒ

EXPLORATION 2

Analyzing Wave Pictures on an Oscilloscope Screen

What to Do

Discuss the following questions with a partner. Be sure to write your conclusions in your ScienceLog.

1. This oscilloscope has its time of sweep set at $\frac{1}{100}$ of a second. What is the frequency of the pictured sound?

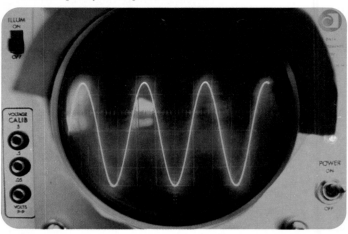

2. How does a wave picture on an oscilloscope show the loudness of sounds? Study these pictures and form your own conclusions.

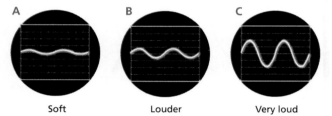

A	B	C
Soft	Louder	Very loud

3. How does a wave picture show the pitch of sounds?

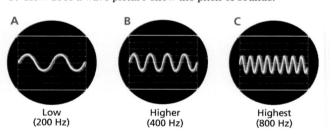

A	B	C
Low (200 Hz)	Higher (400 Hz)	Highest (800 Hz)

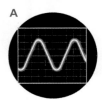 A
 B
 C
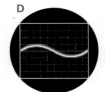 D
 E

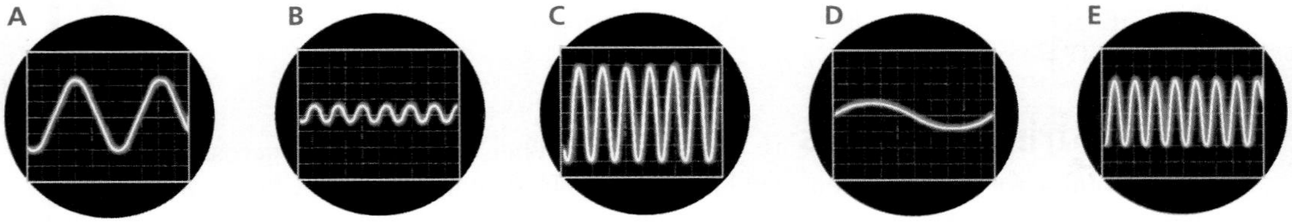

4. Check your conclusions to questions 1 and 2 by studying the five screen displays shown above and answering the following questions:

a. Match the screen displays above with each of the following:
- a sound that is both soft (quiet) and high-pitched
- a sound that is both loud and low-pitched
- a sound that is both loud and high-pitched
- a sound that is both soft and low-pitched

b. Which two screen displays show sounds of about the same loudness? of the same pitch?

c. Which screen display shows the lowest sound? the softest sound?

d. What are the frequencies of the sounds in *D* and *E* above if the time of sweep on the oscilloscope is $\frac{1}{100}$ of a second?

5. Below are the screen displays of five different sounds that you would probably recognize. The sounds differ primarily in quality.

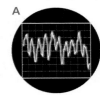 A
Jazz organ

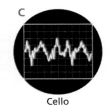 B
Voice

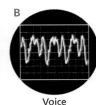 C
Cello

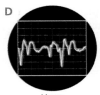 D
Horn

 E
Piano

a. Do you recognize any similarities in the displays? What are they?

b. What differences are there among the wave patterns? Suggest some reasons for these differences.

6. Now consider the screen displays of two common noises. Note the haphazard displays of these nonmusical sounds.

Based on what you've observed, can you identify some of the differences between noise and music as shown by an oscilloscope? Describe these differences in your ScienceLog.

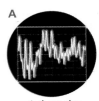 A
A sharp slap

 B
Office noise

Answers to
Exploration 2, continued

4. a. • B
- A
- C
- D

b. *A* and *C* are almost the same loudness; *B* and *C* are the same pitch.

c. *D* is the lowest sound; *B* is the softest sound.

d. *D* has 1 vibration per screen (every 0.01 s), giving a frequency of 100 Hz. *E* has 7.5 vibrations per screen (every 0.01 s), giving a frequency of 750 Hz.

5. a. All have approximately the same pitch and loudness.

b. Answers will vary. The shapes are different because of differences in the quality of the sounds produced by each instrument. Variations in quality arise from differences in how each instrument vibrates.

6. Students should understand that the largest difference between music and noise is the relative simplicity and repetitive nature of the wave form of music.

PORTFOLIO

Students may wish to keep a summary of their observations from Exploration 2 in their Portfolio. Suggest to students that they use illustrations similar to those on pages 410–411 to explain how an oscilloscope displays the different dimensions of sound.

How a String Vibrates

This material begins with the question, "What accounts for the different qualities of sounds that you have seen pictured on the oscilloscope?" By doing this demonstration, students learn that a string can vibrate as a whole, in halves, thirds, quarters, and so on, or in all these ways simultaneously. Students learn that each vibration produces a different sound with its own pitch and loudness. These different sounds are superimposed on the oscilloscope screen, forming the complex patterns shown.

Divide the class into pairs and distribute the materials. Suggest that students begin the demonstration by generating these kinds of waves with rubber tubing or a stretched spring. The kinds of waves being formed are called *standing waves*. Standing waves are really the result of two waves traveling in opposite directions, each wave being reflected from either end of the tubing or spring.

A string vibrating in two parts produces a sound that has twice the frequency of the sound produced by the string vibrating as a whole, and it is musically one octave higher. If the string is vibrating in thirds at the same time, it adds a sound that has three times the frequency of the sound produced by the string vibrating as a whole. The sound is about one-and-a-half octaves above the sound of the string vibrating as a whole.

 Transparency 63 is available to accompany How a String Vibrates.

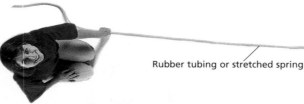

Rubber tubing or stretched spring

How a String Vibrates

What accounts for the different qualities of sounds that you have seen pictured on the oscilloscope? The screen displays on the previous page suggest an answer. Also, recall that most things vibrate in a complex manner, causing a "mix" of many sounds. Do you see how the pitch and loudness of each of these component sounds could affect the quality of the sound you hear? Ⓐ

Examining how a string on a stringed instrument vibrates will make this clearer. First do the following activity with a partner:

Have one person hold steady one end of a long piece of rubber tubing or a stretched spring. The other person should slowly move the other end back and forth rather widely until the tubing or spring vibrates as a whole, as shown in (a) at right.

Now gradually add more energy; that is, make the tubing or spring move faster and faster until it assumes a more stable pattern of vibration. What happens? Were you able to produce the pattern shown in (b)? With even more energy, can you get other patterns of vibration, as Ⓑ shown in (c)? How many patterns did you get?

When you run a bow across a violin string, the string vibrates as a whole, just as the stretched tubing or spring did at first in the demonstration. However, at the same time, the violin string also vibrates slightly in various patterns, which consist of a varying number of parts. The second drawing below shows the string vibrating as a whole and in two parts at the same time. You know that the length of a vibrating string affects the frequency and pitch of the sound it makes. Therefore, when the violin string vibrates as a whole, it makes one sound. When it vibrates in two parts, it makes a higher sound. When it vibrates in three parts, it makes a still higher sound, and so on.

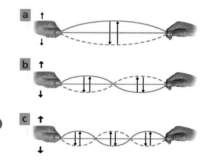

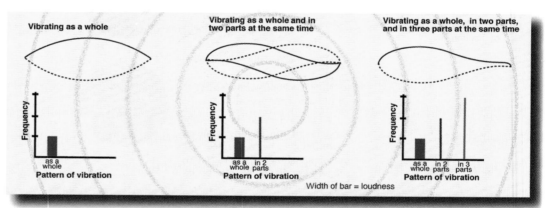

412

Answers to In-Text Questions

Ⓐ Vibrating strings have more than one pattern of vibration at one time. The main vibration of the string (the whole) is generally the loudest and essentially determines the pitch of the sound. The other vibrations occur at the same time, adding softer sounds that are usually higher in pitch. The combination of all the sounds together results in the characteristic tone, or quality, of the sound.

Ⓑ Students should note that as they make the tubing or spring move faster, the number of distinct parts in which the string vibrates also increases. A specific amount of energy is required to produce each vibration pattern. Challenge students to produce a wave with as many parts as they can.

In reality, a violin string vibrates in many more than three parts. Generally, the lowest vibration as a whole is the most vigorous, so this sound is loudest. The higher sounds are softer. These higher sounds add the "wiggles" to the waves you see on the second and third oscilloscope screens shown at right. Put together all of the sounds made by the string's different vibrations, and you have the quality of the sound produced by the violin string.

The pleasant effect of certain vibrating strings has been known for centuries. Turn to page S116 of the SourceBook to find out more.

Wave Pictures of a Vibrating String

String vibrating as a whole

String vibrating as a whole and in two parts

String vibrating as a whole, in two parts, and in three parts

Locating the Higher Sounds

1. Pluck the lowest-pitched string of a violin (or guitar) (a).

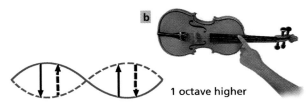

2. Then lightly touch the string exactly at its center to stop it from vibrating as a whole (b), and listen for a higher note. (This may take a little practice.) The note is the sound made by the string when it is vibrating in two parts. Does it sound one octave higher than the first note? Describe the quality of this sound.

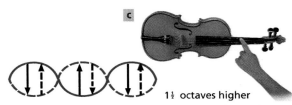

1 octave higher

3. Pluck the string again. Now touch it one-third of the way along its length. With a little skill, you can get the next highest **harmonic** (component tone) (c). This is the sound made by the string as it vibrates in thirds.

1½ octaves higher

4. Try to get even higher harmonics by touching the string closer and closer to one end. Could you get the string to vibrate in four parts? five parts? more parts?

413

FOLLOW-UP

Reteaching
Play different notes on a piano or guitar. Play a note loudly or softly. Ask students to draw a picture of how the sounds would appear on an oscilloscope screen.

Assessment
Present students with pairs of pictures illustrating how different sounds appear on an oscilloscope screen. Ask them to identify the differences in the sounds.

Extension
Have interested students try the following piano experiment: Press down the middle C key on a piano carefully, without making a sound. (This removes the felt dampening the strings for middle C, leaving them free to vibrate.) Now have them strike the C key one octave below middle C and release it. Ask: Do you hear the middle C string vibrating? What does this tell you? *(Yes; some part of the tone from the lower C is vibrating at the same frequency as middle C. This sets up resonance in the strings of middle C. This phenomenon is called sympathetic vibration.)*

Closure
Have students work in groups with an oscilloscope to design an investigation to answer the question, How do the sound waves of music differ from the waves of noise? Have each group turn in a report featuring the group's findings as well as a description of the experimental procedure. Check each group's proposed procedure for safety before allowing students to perform any experiments.

Locating the Higher Sounds
You may want to perform this demonstration yourself, depending on the number of violins or guitars available. Playing harmonics on a violin or guitar may take some practice. For best results, pluck the string forcefully and then place your finger very lightly on the string over the fingerboard. This demonstration should convince students that a string vibrating in two parts produces a note that is one octave higher than when it is vibrating in one part, and that a string vibrating in three parts produces a note that is one-and-a-half octaves higher.

A similar demonstration may be done with a handbell from a handbell choir. Ring the handbell and you will hear a beautiful loud sound. Touch the lower edge of the bell and the main sound stops. However, a higher sound can still be heard.

Homework

Have students research and explain *overtones* in terms of harmonics. Ask students to relate overtones to the characteristics of pitch, loudness, and sound quality (timbre).

Answers to *Challenge Your Thinking*

1. a. Answers will vary. We are likely to miss many common sounds around us every day because we are so accustomed to them that they become part of the background of a given soundscape.

b. Each person produces sounds with his or her voice by using air to vibrate the vocal cords in his or her throat. Not only are each person's vocal cords physically unique, but each person activates his or her vocal cords in a unique way. Thus, the vibration of these cords results in sounds of quality, pitch, and loudness that are unique to each person.

c. Answers will vary. Situations in which sounds are uncomfortably loud may include attending a rock concert, standing near an airport, and operating construction equipment. Ways of reducing the discomfort caused by loud sounds include wearing earplugs, directing your ears away from the sound source, and walking away from the sound source. All three of these methods reduce the amount of sound energy that is able to reach your ear. Turning down the volume is also an option with electronic devices such as radios and televisions; this would help by decreasing the amplitude of the sounds produced.

2. Answers will vary. Accept all reasonable responses. Sample answers:

a. The sounds of different musical instruments can be distinguished by the *quality* of the sounds they produce. Sounds that have the same *frequency* and *amplitude* of vibration may exhibit different *patterns of vibration* because the shape and composition of the instruments affect the *resonance* of the sounds produced.

b. The *pitch,* or *frequency* of vibration, of a string depends on the string's thickness, *length,* and *tension.* The thicker the string is, the lower its pitch. The longer the string is, the lower its pitch. The less tension a string has, the lower its pitch.

c. Sound pollution occurs when the *loudness* of sounds is so great

CHALLENGE YOUR THINKING

1. Hear, There, and Everywhere

Consider each of the following situations:

a. What sounds, although they are loud enough to be heard, do you often miss? Why?

b. You can easily recognize a person just by the sound of his or her voice. Explain this in scientific terms.

c. What are some situations in which sounds are uncomfortably loud? Suggest ways to help yourself in these situations. Explain why each suggestion works.

2. Sound Words

> tension, frequency, amplitude, energy, loudness, pitch, length, quality, resonance, distance, patterns of vibration

Use several appropriate words from the list above to write one or two sentences about each of the following:

a. distinguishing sounds of different musical instruments

b. sounds of vibrating strings

c. sound pollution

3. Steely Band

Pierre selected six different pieces of steel wire. He hung different numbers of identical masses from the wires, as shown at right. Then he plucked the strings and arranged them in order of pitch, from lowest to highest. Predict this order (and learn something about yourself)!

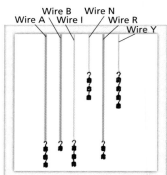

Wire A Wire B Wire I Wire N Wire R Wire Y

that it interferes with a person's activities or health. Some loud sounds even have enough *energy* to break glass that is vibrating in *resonance* with them. Increasing the *distance* to the sound source and decreasing the *amplitude* of vibration of the sound source are two ways of avoiding sound pollution.

3. If the wires are put in the correct order, the letters will spell out *brainy.* The following factors raise the pitch: more tension, shorter wire, thinner wire.

★ **You may wish to provide students with the Chapter 18 Review Worksheet that is available to accompany this Challenge Your Thinking (Teaching Resources, page 54).**

4. A Broad Scope of Sounds

Alejandro connected a microphone to an oscilloscope to see what sounds looked like. Here are some of the pictures that he saw.

a. Which screen shows the noisiest sound?

b. Which shows the purest musical sound?

c. Which shows the highest sound?

d. Which shows the loudest sound?

e. Which three sounds are the same pitch but were made by different instruments?

f. Which two sounds were made by the same instrument but are almost an octave apart?

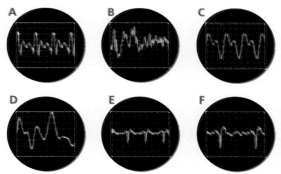

5. Instrumental Instruction

a. Marie blows into the trombone with the slide (1) all the way in and then (2) all the way out, with equal loudness both times. Which tone is lower? Why?

b. Joy plays the cello and Suzanne plays the upright bass. Which instrument is likely to produce higher sounds? Why?

c. Mrs. Cleaver's favorite music CD ends with a loud, sustained chord. She notices that when she plays the CD, her piano sounds the exact same chord for a few seconds after the music stops. Explain what is happening.

ScienceLog

Review your responses to the ScienceLog questions on page 399. Then revise your original ideas so that they reflect what you've learned.

415

ScienceLog

The following are sample revised answers:

1. The range of pitches that an instrument produces corresponds to the size of the instrument. Large instruments like an upright bass vibrate slowly, producing a range of low-pitched notes, whereas small instruments like a violin vibrate more quickly, producing a range of high-pitched notes. The design and material of an instrument will also affect the range of pitches produced.

2. In general, the loudness of a sound varies with the distance from the vibrating source. Therefore, people closer to the band will tend to hear the music more loudly. However, because sound waves are reflected by objects such as the wall behind the stage, certain members of the crowd in line with the reflected sound waves will hear the music more loudly than people closer to the stage.

3. The different tones produced by a violin and a trumpet are due to differences in sound quality. The quality of each sound depends on the mix of component sounds making up the overall sound. This mix is the result of the various vibrations that occur when each overall note is produced. The vibrating section of string, in the case of the violin, and of the air column, in the case of the trumpet, may vibrate as a whole or in many parts at the same time with varying degrees of loudness. The pattern of vibration for a note differs considerably from one instrument to another.

Answers to
Challenge Your Thinking,
continued

4. **a.** *B*
   b. *C*
   c. *A, C,* or *E*
   d. *D*
   e. *A, C,* and *E*
   f. *E* and *F*

5. **a.** The tone produced with the slide all the way out is lower because the vibrating column of air is longer.

b. The cello produces higher sounds because it has thinner, shorter, and tighter strings, which vibrate at a higher frequency.

c. The chord from the CD is amplified sufficiently to cause the corresponding piano strings to vibrate in resonance.

Making Connections

The Big Ideas

The following is a sample unit summary:

Sound is produced when energy is added to an object, causing it to vibrate. (1) Sound can be characterized by four different dimensions: loudness, duration, pitch, and quality. The quality of a sound relates to its purity because the quality depends on the mix of the component sounds produced by an object. For example, musical tones have fewer component sounds than noises have. (2)

When an object vibrates, the particles within the object also vibrate. Any particles of matter that come into contact with this vibrating object also begin to vibrate. As the particles are set into motion, they bump into each other, transferring vibrations among one another. (3)

Some animals have hearing organs that can detect vibration rates above or below the range that human ears can detect. (4)

The speed of sound can be calculated with the following formula: speed of sound = distance ÷ time. (5)

The loudness of a sound can be increased by increasing the energy used to create the vibration. This can be done by forced vibration, by decreasing the distance to the sound source, or by increasing the amplitude through resonance. A megaphone can also be used to direct the energy of a sound to one place. Pitch can be increased by increasing the frequency of the vibration. This can be achieved by increasing the tension, shortening the length, and decreasing the thickness of the vibrating object. (6)

The greater the frequency, the higher the pitch. The greater the amplitude, the louder the sound. (7)

Microphones and oscilloscopes translate sounds into electrical impulses. (8)

An oscilloscope represents pitch by the number of waves, loudness by the amplitude of a wave, and quality by the combination of waves pictured on the screen. (9)

A guitar string vibrates as a whole and, at the same time, in slightly different patterns to create its own distinctive quality. (10)

Unit 6

SOUND

SourceBook

Without the air around us, there would be no sound waves produced by vibrating objects, and our ears would detect no sound at all. To find out more about the role of air in hearing and about how sound waves interact with each other, look in the SourceBook.

Here's what you'll find in the SourceBook:

UNIT 6
Waves S106
Sound Waves S115
Sound Effects S121

416

The Big Ideas

In your ScienceLog, write a summary of this unit, using the following questions as a guide:

1. What causes sound?
2. What are some characteristics of sound?
3. How is sound transported from a sound source to a sound receiver?
4. Why can some animals hear sounds that people can't hear?
5. How can you calculate the speed of sound in air?
6. In what ways can you increase the loudness of a sound? the pitch of a sound?
7. How are the terms *frequency* and *amplitude* related to the pitch and loudness of sound?
8. How do microphones and oscilloscopes "hear" sounds?
9. How does an oscilloscope represent the pitch, loudness, and quality of different sounds?
10. How could you use a vibrating violin string to illustrate what gives rise to the distinctive quality of the sound it makes?

Checking Your Understanding

1. Because you have two ears, you can tell the direction from which sounds are coming. Your brain can detect the slight difference in arrival time between ears. Calculate the difference in the time taken by each ear to hear a sound, if the sound source is on your left.

Answers to *Checking Your Understanding*

1. If the distance between two ears is 15 cm, then the sound will get to the left ear about 0.15 m ÷ 345 m/s, or 0.00043 s, earlier.

2. What's wrong in the illustration shown at right?

3. A haiku is a Japanese nature poem consisting of three lines of five, seven, and five syllables, respectively. Read the examples that follow. They describe sounds, with a little science added too!

> *Stirrings in the night*
> *Quickly, quietly moving*
> *Tiny night noises*
>
> *Leaves lightly rustling*
> *Sound's volumes vary vastly*
> *Thunder deafening*

Write your own haiku about the sounds of nature. Be sure to include some of the scientific concepts you have learned.

4. **concept map** Copy the concept map into your ScienceLog. Then complete the map using the following words: sound, vibrations, frequency, amplitude, pitch, loudness, waves.

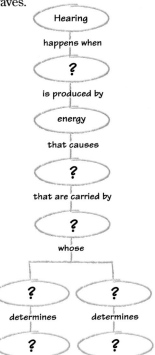

Hearing
↓ *happens when*
?
↓ *is produced by*
energy
↓ *that causes*
?
↓ *that are carried by*
?
↓ *whose*
? *determines* **?**
? — **?**

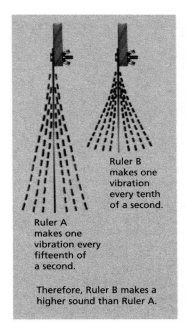

Ruler B makes one vibration every tenth of a second.

Ruler A makes one vibration every fifteenth of a second.

Therefore, Ruler B makes a higher sound than Ruler A.

 Homework

You may wish to assign the Unit 6 Activity Worksheet as homework (Teaching Resources, page 62). If you use the worksheet in class, Transparency 64 is available to accompany it.

Answers to *Checking Your Understanding,* continued

2. Ruler A has a higher frequency of vibration. Therefore, Ruler A should make a higher sound than Ruler B. (Some students may point out that if the rulers are made of the same material, Ruler B should vibrate more quickly and should make a higher sound because it is shorter.)

3. Answers will vary. Students should try to include concepts related to the dimensions of sound, how sound travels, or how sounds are heard.

4. Sample concept map:

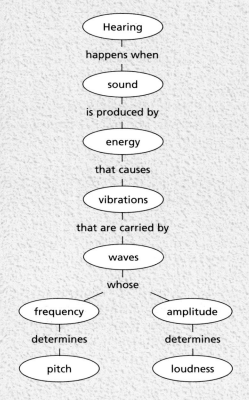

Hearing
↓ *happens when*
sound
↓ *is produced by*
energy
↓ *that causes*
vibrations
↓ *that are carried by*
waves
↓ *whose*
frequency — amplitude
↓ *determines* ↓ *determines*
pitch — loudness

★ **You may wish to provide students with the Unit 6 Review Worksheet that accompanies this Making Connections (Teaching Resources, page 64). Transparency 65 is also available for your use.**

Background

Some modern synthesizers use a computer-connection system called MIDI, which is short for musical instrument digital interface. With MIDI, the musician can feed sounds into a synthesizer from various instruments to create a variety of new effects. MIDI can also be used to write notes directly into a musical score just by playing them. Before synthesizers were widely available, special-effects artists employed common objects to reproduce sounds.

Extension

Live radio actors used to crinkle cellophane to simulate the sound of a crackling fire. Have students experiment with sound by using some everyday objects to reproduce certain sounds. They might attempt sounds such as thunder, an opening door, footsteps, leaves, wind, or animal sounds. You might want to bring a special effects book to class for ideas.

CROSS-DISCIPLINARY FOCUS

Music

In the past 30 years, many composers have written music expressly for the synthesizer. You might want to play some samples of electronic music for students and ask them what they think about the sounds. Are they reminded of sounds they have heard before, or are the sounds completely new? Artists you might consider include Philip Glass, Tangerine Dream, Jean-Michel Jarre, and Tomita. How do the selections compare with recordings that use nonelectronic instruments?

SCIENCE and the *Arts*

Synthesizing Your Own Band

When you hear a musical instrument, your ear is responding to the specific frequencies, or vibrations, it creates. Most musical instruments create only a narrow range of frequencies, giving each instrument its distinctive sound. Not so for a synthesizer! A synthesizer can produce a range of frequencies so wide that it can it be used to imitate the instuments of an entire orchestra.

How Does It Work?

If you could look beneath the rows of keys, buttons, and knobs, you would find a generator, various filters, and an amplifier. The generator converts electricity from a wall plug into a vibrating electric signal. The rate of the vibration depends on which key on the keyboard is pressed. The electric signal then passes through a series of electronic filters that change the quality of the vibration according to the effect that is desired. Finally, the refined signal passes to an amplifier, where its strength is boosted to operate a speaker.

Any Sound at Your Fingertips

By adjusting the controls on the synthesizer, you can adjust the electronic filters to get just the sound you want. In this way, musicians can achieve the sounds of a piano, a violin, a trumpet, and even a snare drum. But synthesizers don't stop there. Some synthesizers can also duplicate natural sounds such as rainfall, wind, surf, and thunder. The synthesizer can also be used to make entirely new sounds. In fact, many musicians play synthesizers to make unique sounds rather than to mimic conventional instruments.

Some synthesizers allow you to play many different parts of a song separately and then blend them together into a finished song. A microprocessor stores all the notes you played and even remembers how hard you pressed the keys. When you have finished recording each instrument's part, you can use the synthesizer to combine the parts and play the final composition.

Science Meets Music

The first commercially available synthesizer was created by an American physicist, Robert A. Moog, in 1964. Since then, synthesizers have become very sophisticated. If you were to go to a music store, you may even find synthesizers bearing Moog's name. Now you know where this name came from!

Find Out for Yourself

Go to a music store and try out a synthesizer. See for yourself the range of possibilities that this instrument provides. Try playing a violin sound. Do you think it is as rich sounding as a real violin? Why do you think this is so?

▲ Musicians can use synthesizers to create special sounds that would be difficult to achieve any other way. The first portable synthesizer, shown above with inventor Robert Moog, went on the market in 1970.

Find Out for Yourself

Some state-of-the-art synthesizers create sounds that are realistic enough to be almost indistinguishable from the real thing. A real sound consists of numerous overtones, many of which change over time. It requires very sophisticated—and expensive—electronics to reproduce these overtones. You might have students compare the violin sounds produced by different synthesizers, from the least expensive to the most expensive. They should notice a difference in quality.

Designing Out the Noise

Who hasn't covered their ears as a huge airliner passed overhead with an earth-shaking rumble? "Why can't they make those things quieter?" you ask. Richard Linn has done just that. As an aeronautical engineer, his job is to do something about that high-flying noise.

▲ Richard Linn reviewing an aerial photo of Dallas–Fort Worth International Airport

Turning Down the Volume

Aeronautical engineers specialize in the design, construction, or testing of aircraft. Also called aerospace engineers, they are responsible for designing anything that flies, from airplanes to spacecraft.

Richard has applied his 35 years of experience in aeronautical engineering to noise control. He analyzes and offers solutions for the noise created by the jet engines that send modern airplanes into the blue.

When asked what it was like when he got started, Richard replied, "Back in the 1960s the noise of airplanes really began bothering people. It became a social problem that needed attention."

The attention it received involved various kinds of research on how to reduce the noise from jet engines. At the same time, homeowners and political groups began putting pressure on Congress to do something about the problem. As a result, laws were passed that forced airplane manufacturers to make quieter jet engines.

As the years passed, however, air traffic increased, and the constant noise of airplanes flying overhead became just as bothersome as the louder jets had been. To attack this problem, aeronautical engineers turned to the high-tech world of computers. Richard explains the process: "We have new computer technology that allows us to use what we call *noise footprints*. These are airplane-simulation programs that help us determine the effect of engine noise on a community."

A Joint Effort

Richard feels that noise control should be a joint effort between the airlines and the community. He says, "It's frustrating to realize that in spite of the well-known noise problems, cities

▲ Airport employees who work on the runway must wear protective headgear to prevent hearing loss.

often zone or rezone the land around airports for homes, churches, and schools. The airlines can do only so much by themselves—without the community's help in zoning, we'll never completely solve the problem."

When asked how it feels to watch the first flight of a brand-new airplane soaring over a neighborhood, Richard responds, "It's great knowing that all your research and experimentation has paid off and you can demonstrate to the community that you have new, quieter technology."

Think About It

Do you think airports and airplane noise affect the animals and insects that live in the surrounding area? Discuss your thoughts with your classmates.

419

Background

Aerospace research ranges from designing quieter and more fuel-efficient aircraft to developing new materials that can withstand the extreme conditions of spaceflight.

Aerospace engineers must have a working knowledge of aerodynamics in order to design strong, safe, and efficient aircraft. They must also have a thorough understanding of the physical properties of the materials they are working with and be able to predict how these materials will behave during flight. Aerospace engineers also work closely with other engineers, such as electrical, mechanical, and civil engineers.

Cooperative Learning
DESIGNING OUT THE NOISE

Group size: 4 to 5 students
Group goal: to develop a new airport
Positive interdependence: Assign each student a role such as land developer, environmentalist, local resident, head of an airline company, or aerospace engineer.
Individual accountability: The problem is that there are three possible sites for the new airport. An endangered species of squirrel lives on the first site. The second site is near a quiet residential community. The third site has limited space, so little future expansion of the airport would be possible. Ask the groups to discuss some ways in which they can compromise so that everyone is satisfied.

Think About It

To facilitate critical thinking, you may want to pose the following questions: Should the effect on wildlife be taken into account when designing a new airport? What are some species that might be affected? Students may discover that the airport's flight paths often pass over sensitive areas, such as residential communities, parks, or environmentally sensitive regions. If so, you may wish to have students interview residents in the flight path. Students could ask questions such as the following: How does the noise affect your life? Which aircraft are the most noisy? Should there be stricter laws to prevent noise pollution?

Did You Know. . .

The principal habitat of an endangered butterfly, the El Segundo blue, is in the dunes that neighbor the runways of Los Angeles International Airport. Because of the noise, humans and other intruders tend to stay away. The butterflies are not affected, however, because they have no ears!

Background

Piezoelectric crystals are used as converters or *transducers* in many common devices. A transducer—which literally means "to lead across"—is any device that converts energy from one form to another. In sonar systems, for example, crystal transducers convert the mechanical energy of an underwater sound wave to an electrical signal. Piezoelectric crystals can also be found in phonograph needles and microphones.

The human ear can sense vibrations with frequencies as low as 20 Hz and as high as 20,000 Hz, or 20 kHz. The expansion of a heated material may cause it to emit sounds at frequencies of up to 500 kHz, well beyond the range of our hearing.

Extension

Energy can exist in a variety of forms, such as mechanical, thermal, sound, chemical, and electrical. Write these terms on the board and explain to the class the role of machines and appliances in converting one form of energy to another. Then draw a line between any two of them and ask the class for an example of a machine or device that can convert energy from one of these forms to another. *(Generators and turbines convert mechanical to electrical, motors convert electrical to mechanical, car brakes convert mechanical to thermal, steam engines convert thermal to mechanical, car engines convert chemical to mechanical, speakers convert electrical to sound, and ears convert sound to mechanical and then mechanical to electrical.)*

Find Out for Yourself

Students should observe the largest cone of a speaker. This cone is known as the *woofer* and generates the lowest pitches. The surface of the woofer should oscillate in step with the music and should move more when the volume is increased. Also, the lower the pitch, the more the woofer should vibrate. Higher pitches can be generated by smaller cones known as *tweeters.* You might ask students where they think these names came from. *(They are reminiscent of the pitches of sounds made by dogs and birds.)*

Listening for Fire

*M*any homes today are fitted with smoke detectors, but smoke detectors have their limitations. If a fire starts behind a wall, for example, by the time the smoke reaches a detector, the fire has already spread. However, it might not be long before there's a device that can "hear" a fire before it has had a chance to burst into flames.

▲ When the materials that make up this house burn, they produce high-frequency sound waves.

The Sounds of Heat

When a material heats up, it expands. If the material is made up of different components, each component may expand at a different rate. When a beam of wood is heated, for example, the cellulose fibers expand at a different rate than does the sap. When the wood fibers finally rupture from the heat, or when the sap boils, high-frequency sounds are emitted.

Piezo What?

William Grosshandler, a mechanical engineer at the National Institute of Standards and Technology, has developed an electronic system that "listens" for these high-frequency sounds. Grosshandler's system uses small sensors, each containing a piezoelectric crystal. These crystals are similar to the crystals used in quartz watches and clocks. Piezoelectric crystals produce an electric charge when they vibrate. In a watch, power from a small battery makes the crystal vibrate to produce a pulse of current at the precise rate of one pulse per second.

Sound Waves to Electric Charges

In Grosshandler's system, sound waves from a heated material cause the piezoelectric crystals to vibrate at a certain frequency, generating a specific electric signal. The electric signal is amplified, filtered, and routed to a microprocessor that is programmed to analyze the signal and determine whether or not it was caused by a heated material. For example, the microprocessor is programmed to differentiate between signals caused by wood fibers that are breaking and those caused by wood fibers that are about to burst into flames. Otherwise, the device would create false alarms.

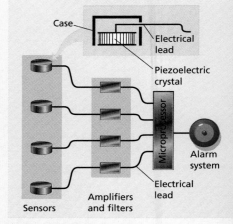

Case — Electrical lead — Piezoelectric crystal — Microprocessor — Alarm system — Electrical lead — Amplifiers and filters — Sensors

▲ In Grosshandler's system, sensors in walls send signals to a microprocessor before a fire gets out of control.

Grosshandler's system still needs some fine-tuning to improve its ability to recognize different signals in order to prevent false alarms. But Grosshandler believes it could be an effective way of detecting fires before they become dangerous.

Find Out for Yourself

You can see the vibrations caused by sound. Look at the surface of a speaker the next time you see one without its cover. Note how the speaker's surface vibrates to different frequencies of sound. Can you recognize a specific sound by the way the speaker vibrates? This is precisely the function of the microprocessor in Grosshandler's system!

A Hearing Aid for Fido

Normally, dogs have a keen sense of hearing. But many dogs, like humans, lose some of their hearing as they grow older. Thanks to dog's best friend, however, help may be on the way.

Testing a Dog's Hearing

Curtis Smith, an audiologist at Auburn University in Alabama, became interested in canine hearing loss when a friend's dog started going deaf. He contacted A. Edward Marshall, a researcher at Auburn's college of veterinary medicine, who had been working on a way to test dogs for deafness.

Marshall's hearing test involved fitting a dog with lightweight headphones. Electrodes were then attached to the dog's head to monitor its brain waves. Short sounds of different intensities were then played through the headphones. On a computer screen, Marshall could observe the response of the dog's brain to the sounds. By changing the frequency of the sounds, Marshall could determine the limit of the dog's auditory range and discover how much hearing the dog had lost.

A Canine Hearing Aid

How do you fit a dog with a hearing aid? Using a conventional hearing aid turned out to be impractical. The dog simply pawed its ear until the device

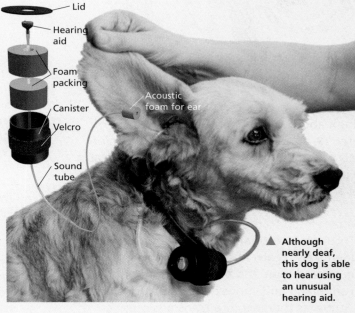

Lid — Hearing aid — Foam packing — Canister — Velcro — Sound tube — Acoustic foam for ear

▲ Although nearly deaf, this dog is able to hear using an unusual hearing aid.

fell out. The hearing aid that Marshall and Smith developed involves attaching the hearing aid hardware to the dog's collar. The amplified sound is channeled through hollow, lightweight tubes to the dog's ears. The tubes are held in place with plastic foam.

Still Room for Improvement

The hearing aid for dogs still has some problems. Although the collar system is tolerated by most dogs, they don't like it. Another problem is related to the dog's outer ear, or pinna. In humans, the pinna remains stationary. In dogs, however,

the outer ear moves in many directions to pick up incoming sounds. This movement can cause the tubes of the hearing aid to work loose. Still another problem is that the device amplifies sounds only in the 500 to 6000 Hz range. These frequencies are much lower than a dog's normal hearing range.

Find Out for Yourself

Different animals hear sound in different frequency ranges. Do some research and make a chart to show what the range of sound is for several different animals. Don't forget to include yourself.

421

Background

Dogs, like humans, would not be able to sense sound without the *cochlea,* the hollow, snail-shaped structure in the inner ear. As an animal gets older, some of the sound-sensing hair cells in the cochlea die, and the auditory signals to the brain fade.

Dogs rely heavily on their ability to hear. The cochlea in their ears have more turns than those in human ears. More turns result in more hair cells, and more hair cells result in a wider range of hearing. For example, the average dog can detect sounds between 30 and 50,000 Hz, while the human range is between 20 and 20,000 Hz.

Extension

You might wish to invite a hearing specialist, or *audiologist,* to come and address the class on hearing and hearing loss. He or she should emphasize the impact of the environment on our hearing. Suggest that students prepare questions ahead of time that they would like to have answered.

Find Out for Yourself

A bar graph will work best for this activity. The needed information will most likely be found in an encyclopedia or biology textbook. Keep in mind that a graph like this one appears in Exploration 2 on page 367. Students should use animals other than those included on page 367. Students should include familiar animals, such as cats, along with less familiar animals that have specially adapted hearing, such as fish and bats. You might have students make note of each animal's range of hearing and environment. To facilitate critical thinking, you may wish to ask the following questions: Can all water creatures hear higher pitches than land creatures? Why can't all animals hear sounds at any pitch? Would this be useful?

Unit 7 LIGHT

Unit Overview

This unit gives students the opportunity to think about the fundamental properties of light and its interaction with matter. In Chapter 19, students explore light as a form of energy, investigating high-temperature objects and the light they produce. Students also recreate some of Newton's experiments with prisms and the separation of white light into its component colors. Finally, the chapter shows students how colors of light can be combined to make new colors of light.

In Chapter 20, students examine phenomena related to the behavior of light, including scattering, transmission, and absorption. Through investigations of shadows, students discover that light travels in straight lines. Students also learn about diffuse and specular reflections and the difference between mixing colored light and mixing colored paints.

In Chapter 21, students study reflection from plane and convex mirrors, from which they develop the concept of a virtual image. Students then explore transmission phenomena, using double-convex mirrors and concave mirrors to study the production of real images. The unit concludes with a discussion of refraction and total internal reflection.

Using the Themes

The unifying themes emphasized in this unit are **Energy, Structures,** and **Changes Over Time.** The following information will help you incorporate these themes into your teaching plan. Focus questions that correspond to these themes appear in the margins of this Annotated Teacher's Edition on pages 427, 448, 454, 465, 476, and 480.

Energy is a fundamental theme in any discussion about properties, characteristics, and uses of light because light itself is a form of energy. For example, light from the sun may produce heat when it strikes an object. Likewise, an object may produce light when it is heated.

Structures is another important theme. Students learn about the structures of optical devices such as cameras. They also learn about the structure of the visible spectrum.

The theme of **Changes Over Time** appears in Chapter 19, in which students observe how objects change color as they become hotter. In Chapter 21, the theme can be observed in the behavior of light as it is reflected or refracted by different materials.

Using the SourceBook

In Unit 7, students consider how light is regarded as being composed of both particles and waves. They also learn how light originates as electrons undergo de-excitation. They explore the electromagnetic spectrum and the properties exhibited by various frequencies of electromagnetic radiation. Students also learn about characteristics of light such as speed and polarization. Finally, they are introduced to technological applications such as lasers, spectroscopy, and fiber optics.

Bibliography for Teachers
Cobb, Vikki, and Josh Cobb. *Light Action! Amazing Experiments With Optics.* New York City, NY: Harper Collins Publishers, 1993.
Fowler, Shannon A. *The Equations of Light.* New York City, NY: Carlton Press, Inc., 1989.
Tomecek, Steve. *Fun With Bouncing and Bending Light.* New York City, NY: W.H. Freeman and Co., 1995.

Bibliography for Students
Ardley, Neil. *The Science Book of Color.* New York City, NY: Gulliver Books, Harcourt Brace Jovanovich, 1991.
Catherall, Ed. *Exploring Light.* Austin, TX: Steck-Vaughn Company, 1990.
Cobb, Vicki. *Light Action!* New York City, NY: HarperCollins Publishers, 1994.
Wood, Robert W. *Physics for Kids: 49 Easy Experiments With Optics.* Blue Ridge Summit, PA: TAB Books, Inc., 1990.
Zubrowski, Bernie. *Mirrors: Finding Out About the Properties of Light.* A Boston Children's Museum Activity Book. New York City, NY: Morrow Junior Books, 1992.

Films, Videotapes, Software, and Other Media
How to Bend Light
Videotape
Britannica
310 S. Michigan Ave.
Chicago, IL 60604
Light, Color, and the Visible Spectrum
Film or videotape
Coronet/MTI
108 Wilmot Rd.
Deerfield, IL 60015
Optics on Computer
Software (Apple)
Focus Media, Inc.
P.O. Box 865
Garden City, NY 11530

Unit Organizer

Unit/Chapter	Lesson	Time*	Objectives	Teaching Resources
Unit Opener, p. 422				Science Sleuths: The Fogged Photos English/Spanish Audiocassettes Home Connection, p. 1
Chapter 19, p. 424	Lesson 1, What Is Light? p. 425	2	1. Identify several ways that light affects matter. 2. Describe light as a form of energy.	Image and Activity Bank 19-1
	Lesson 2, Light, Heat, and Color, p. 428	1	1. Give examples of heat and light occurring together. 2. Describe how the colors emitted from a glowing object relate to the object's temperature.	Image and Activity Bank 19-2 Exploration Worksheet, p. 3 Resource Worksheet, p. 5
	Lesson 3, A Colorful Theory, p. 431	3	1. List the colors of the spectrum in the order in which they occur. 2. Describe how a prism separates light into its component colors. 3. Explain how a filter changes the color of light that passes through it.	Image and Activity Bank 19-3 Exploration Worksheet, p. 6 Transparency Worksheet, p. 12 ▼ Transparency Worksheet, p. 14 ▼ Math Practice Worksheet, p. 16
	Lesson 4, Adding and Subtracting Color, p. 437	1	1. Identify the primary color components of white light. 2. Explain how to obtain secondary colors from primary colors.	
End of Chapter, p. 440				Activity Worksheet, p. 18 ▼ Chapter 19 Review Worksheet, p. 20 Chapter 19 Assessment Worksheet, p. 23
Chapter 20, p. 442	Lesson 1, Light in Action, p. 443	4	1. Describe the behavior of light in terms of reflection, scattering, transmission, and absorption. 2. Explain the difference between translucent and transparent objects.	Image and Activity Bank 20-1 Exploration Worksheet, p. 25
	Lesson 2, Light's Path, p. 450	2	1. Derive evidence that light travels in straight lines by observing shadows. 2. Derive evidence of light's extreme speed. 3. Use a pinhole to study the path of light.	Exploration Worksheet, p. 30
	Lesson 3, Reflection, p. 452	2	1. Explain how a flat mirror reflects a beam of light. 2. Explain why smooth surfaces make better reflectors than rough ones. 3. Describe how light reflects from colored objects. 4. Compare specular and diffuse reflection.	Image and Activity Bank 20-3 Transparency 69
	Lesson 4, The Riddle of Color, p. 456	2	1. Explain how transmitted color is obtained by using colored filters. 2. Compare mixing light and mixing paints. 3. Identify the primary colors of paint.	Exploration Worksheet, p. 31
End of Chapter, p. 460				Activity Worksheet, p. 32 ▼ Chapter 20 Review Worksheet, p. 33 ▼ Chapter 20 Assessment Worksheet, p. 37
Chapter 21, p. 462	Lesson 1, Plane Mirrors, p. 463	2	1. Identify the characteristics of images reflected by plane mirrors. 2. Distinguish between virtual and real images.	Image and Activity Bank 21-1
	Lesson 2, Convex Mirrors, p. 467	1	1. Describe the structure of a convex mirror and the characteristics of the images it forms. 2. Compare convex and plane mirrors. 3. Identify some uses of convex mirrors.	
	Lesson 3, Converging Lenses and Real Images, p. 469	2	1. Describe a converging lens. 2. Explain how virtual and real images are formed by a converging lens. 3. Describe the focal point and focal length of a converging lens. 4. Identify some uses of converging lenses.	Image and Activity Bank 21-3 Discrepant Event Worksheet, p. 40 Exploration Worksheet, p. 41 Transparency 72 Transparency Worksheet, p. 45 ▼ Transparency Worksheet, p. 47 ▼
	Lesson 4, Real Images and Concave Mirrors, p. 473	2	1. Describe a concave mirror. 2. Explain how concave mirrors produce virtual and real images. 3. Compare images produced by plane, convex, and concave mirrors and converging lenses. 4. Describe how concave mirrors can concentrate light energy.	Image and Activity Bank 21-4 Exploration Worksheet, p. 49 Theme Worksheet, p. 52
	Lesson 5, Refraction, p. 477	2	1. Describe and explain the phenomenon of refraction. 2. Diagram the path of light as it passes from one medium to another. 3. Explain total internal reflection.	Image and Activity Bank 21-5 Exploration Worksheet, p. 53 Transparency 75
End of Chapter, p. 480				Chapter 21 Review Worksheet, p. 56 Chapter 21 Assessment Worksheet, p. 59
End of Unit, p. 482				Unit 7 Activity Worksheet, p. 62 ▼ Unit 7 Review Worksheet, p. 63 ▼ Unit 7 End-of-Unit Assessment, p. 66 Unit 7 Activity Assessment, p. 71 Unit 7 Self-Evaluation of Achievement, p. 74

* Estimated time is given in number of 50-minute class periods. Actual time may vary depending on period length and individual class characteristics.
▼ Transparencies are available to accompany these worksheets. Please refer to the Teaching Transparencies Cross-Reference chart in the Unit 7 Teaching Resources booklet.

Materials Organizer

Chapter	Page	Activity and Materials per Student Group
19	426	**Exploration 1: Activity 1:* a few sheets of newspaper; scissors; lamp with a 100 W light bulb; **Activity 2:** sheet of white paper; sheet of black paper; 2 alcohol thermometers; several small pieces of tape; watch or clock with second hand; **Activity 3:** flashlight; metric spring scale or metric balance; watch or clock; **Activity 4:** about 50 cm of copper wire; sheet of blueprint paper; watch or clock; large container of water (optional item: sunlamp); **Activity 5:** lamp with a 100 W light bulb; radiometer; watch or clock with second hand (See Advance Preparation below.)
	428	**Exploration 2, Part 1:** 3 D-cells, 1 weak and 2 fully charged; about 20 cm of uninsulated copper wire; flashlight bulb; about 10 cm of masking tape; **Part 2:** clothespin; piece of copper wire, about 10 cm long; portable burner; striker; safety goggles; **Part 3:** hot plate or electric stove
	431	**Exploration 3, Experiment 1:* flashlight; 15 cm × 15 cm piece of aluminum foil; a few small pieces of tape; prism; piece of white cardboard, about 30 cm × 30 cm; **Experiment 2:** flashlight; 15 cm × 15 cm piece of aluminum foil; a few small pieces of tape; prism; red filter; piece of white cardboard, about 30 cm × 30 cm; sheet of white paper; blue filter (See Advance Preparation below.); **Experiment 3:** flashlight; red filter; piece of white cardboard, about 30 cm × 30 cm; green filter; **Experiment 4:** 3–6 red filters; flashlight; piece of white cardboard, about 30 cm × 30 cm; 3–6 green filters; 3–6 dark blue filters; **Experiment 5:** flashlight; 15 cm × 15 cm piece of aluminum foil; a few pieces of tape; 2 prisms; piece of white cardboard, about 30 cm × 30 cm; red filter; dark blue filter
	439	**Filter Fun:** overhead projector; filters of various colors, such as red, blue, green, cyan, yellow, and magenta
20	444	**Building a Light Box:** large shoe box or other cardboard box, about 35 cm × 15 cm × 10 cm; metric ruler; scissors; piece of aluminum foil, about 8 cm × 8 cm; roll of masking tape
	444	**Exploration 1, Part 1:* light box from Building a Light Box; small flashlight; piece of rough black paper, about 8 cm × 8 cm; sheet of white paper, about 8 cm × 8 cm; a few sheets of other colored paper, about 8 cm × 8 cm each; wooden splint; a few matches; small ball of modeling clay; safety goggles; **Part 2:** light box from Building a Light Box; small flashlight; piece of window glass, about 8 cm × 8 cm; small ball of modeling clay; wooden splint; a few matches; protractor; safety goggles (additional teacher materials: 50 cm of masking tape; see Advance Preparation below.); **Part 3:** light box from Building a Light Box; small flashlight; flat piece of paraffin wax, about 8 cm × 8 cm; small ball of modeling clay; wooden splint; a few matches; protractor; safety goggles (additional teacher materials: knife; see Advance Preparation below.); **Part 4:** light box from Building a Light Box; small flashlight; small beaker of water; a few drops of milk; eyedropper; stirring rod; **Part 5:** large beaker or jar of water; small flashlight; a few drops of milk; eyedropper; stirring rod
	450	**Light's Path:** flashlight; sheet of white cardboard; plastic comb
	451	**Exploration 2:** several index cards with pinholes of different sizes; candle; small ball of modeling clay; small jar lid; a few matches; metric ruler; 2 clothespins; lined white index card; safety goggles
	452	**Exploration 3, Part 1:** light box from Building a Light Box; small flashlight; small ball of modeling clay; flat mirror, about 8 cm × 8 cm; wooden splint; a few matches; protractor; safety goggles; **Part 2:** light box from Building a Light Box; small flashlight; small ball of modeling clay; 15 cm × 10 cm pieces of white, green, and other colored cardboard; wooden splint; a few matches; protractor; safety goggles
	457	**Exploration 4:** cardboard tube; scissors; sheet of white paper; red, yellow, blue, and purple crayons; 2 pieces of dark blue cellophane, about 5 cm × 5 cm each; 2 pieces of dark red cellophane, about 5 cm × 5 cm each; tape
	458	**A Lesson in Art:** samples of red, blue, and yellow watercolor paint; about 20 mL of water; stirring rod (optional materials: several sheets of thick paper; paintbrush)
21	463	**Exploration 1:** 3 sheets of graph paper; plane mirror, about 8 cm × 8 cm; small ball of modeling clay; flat piece of colored glass, about 8 cm × 8 cm; metric ruler; index card
	467	**Exploration 2, Part 1:* large, shiny spoon; plane mirror; **Part 2:** light box from Building a Light Box; small ball of modeling clay; small convex mirror; wooden splint; a few matches; small flashlight; safety goggles; **Part 3:** candle; jar lid; a few matches; convex mirror; 2 small balls of modeling clay; sheet of paper; safety goggles; **Part 4:** convex mirror
	470	**Exploration 3, Part 1:* book or piece of paper with large print; double-convex lens; **Part 2:** light box from Building a Light Box; wooden splint; a few matches; small double-convex lens; small ball of modeling clay; small flashlight; safety goggles; **Part 3:** double-convex lens; candle; jar lid; 2 small balls of modeling clay; a few matches; blank white index card; safety goggles
	473	**Exploration 4, Part 1:* large, shiny spoon; **Part 2:** light box from Building a Light Box; small concave mirror; small ball of modeling clay; wooden splint; a few matches; small flashlight; safety goggles; **Part 3:** sheet of paper; concave mirror; candle; jar lid; 2 small balls of modeling clay; a few matches; safety goggles
	477	**Exploration 5, Part 1:* plastic cup filled with water; 2 pennies; **Part 2:** 3 plastic cups, one filled with water, one with rubbing alcohol, and one with vegetable oil; 2 pennies; felt-tip marker; **Part 3:** 600 mL (or 1 L) beaker; 400 mL (or 650 mL) of water; **Part 4:** 600 mL (or 1 L) beaker; 400 mL (or 650 mL) of water; a few drops of milk; eyedropper; stirring rod; small piece of cardboard, about 10 cm × 10 cm; hole punch; flashlight; about 5 mL (or 1 teaspoon) of chalk dust; **Part 5:** 600 mL (or 1 L) beaker; 400 mL (or 650 mL) of water; a few drops of milk; eyedropper; stirring rod; flashlight; rubber band; 15 cm × 15 cm piece of aluminum foil; about 5 mL (or 1 teaspoon) of chalk dust; protractor

* You may wish to set up these activities in stations at different locations around the classroom.

Advance Preparation

Every Exploration that requires colored filters may also work with colored cellophane. Try these experiments in advance before substituting cellophane.

Exploration 1, page 426: For Activities 1, 2, and 5, you will need a window that is exposed to direct sunlight for at least part of the day. Activity 4 also requires direct sunlight or a sunlamp.

Exploration 1, Part 2, page 445: You may wish to tape the edges of the pieces of glass for safety reasons.

Exploration 1, Part 3, page 445: You will need to use a knife to shave the paraffin wax pieces to approximately the same thickness as the pieces of glass used in Part 2.

Unit Compression

The lessons in this unit are assembled so that they form a logical sequence of concepts and activities. In general, you can save time by having students read certain sections ahead of time or as homework. This will help to speed in-class discussion. Also, there may be questions and other thought exercises that can be assigned as homework.

If time is short, care should be taken to include all of the material that is essential to the coherence of the unit. The lessons that could most easily be omitted are the following:

- Chapter 19, Lesson 4, Adding and Subtracting Color
- Chapter 20, Lesson 2, Light's Path
- Chapter 21, Lesson 5, Refraction

Other modifications may be made depending on your preferences and the characteristics and previous knowledge of your students.

Homework Options

Chapter 19
See Teacher's Edition margin, pp. 425, 429, 433, 435, 436, 439, and 440
Resource Worksheet, p. 5
Math Practice Worksheet, p. 16
Activity Worksheet, p. 18
SourceBook, pp. S128, S133, S134, and S139

Chapter 20
See Teacher's Edition margin, pp. 446, 448, 449, 454, 457, and 461
Activity Worksheet, p. 32
SourceBook, p. S132 and S137

Chapter 21
See Teacher's Edition margin, pp. 465, 466, 470, 475, and 478
SourceBook, pp. S135, S136, and S142

Unit 7
Unit Activity Worksheet, p. 62
SourceBook Activity Worksheet, p. 75

Assessment Planning Guide

Lesson, Chapter, and Unit Assessment	SourceBook Assessment	Ongoing and Activity Assessment	Portfolio and Student-Centered Assessment
Lesson Assessment Follow-Up: see Teacher's Edition margin, pp. 427, 430, 436, 439, 449, 451, 455, 459, 466, 468, 472, 476, and 479 **Chapter Assessment** Chapter 19 Review Worksheet, p. 20 Chapter 19 Assessment Worksheet, p. 23* Chapter 20 Review Worksheet, p. 33 Chapter 20 Assessment Worksheet, p. 37* Chapter 21 Review Worksheet, p. 56 Chapter 21 Assessment Worksheet, p. 59* **Unit Assessment** Unit 7 Review Worksheet, p. 63 End-of-Unit Assessment Worksheet, p. 66*	SourceBook Review Worksheet, p. 77 SourceBook Assessment Worksheet, p. 81*	Activity Assessment Worksheet, p. 71* **SnackDisc** Ongoing Assessment Checklists ♦ Teacher Evaluation Checklists ♦ Progress Reports ♦	Portfolio: see Teacher's Edition margin, pp. 436, 447, and 475 **SnackDisc** Self-Evaluation Checklists ♦ Peer Evaluation Checklists ♦ Group Evaluation Checklists ♦ Portfolio Evaluation Checklists ♦

* Also available on the Test Generator software
♦ Also available in the Assessment Checklists and Rubrics booklet

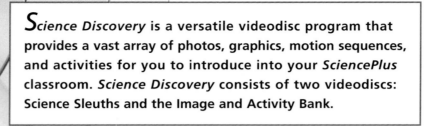

Science Discovery is a versatile videodisc program that provides a vast array of photos, graphics, motion sequences, and activities for you to introduce into your SciencePlus classroom. Science Discovery consists of two videodiscs: Science Sleuths and the Image and Activity Bank.

Using the *Science Discovery* Videodiscs

Science Sleuths: The Fogged Photos
Side B

For the past three months, the photographs taken at an observatory have been fogged. The director of the observatory suspects that the new camera is flawed.

Interviews

1. Setting the scene: Observatory manager **37145 (play ×2)**

2. Astronomer **37609 (play)**

3. Engineer **38500 (play)**

4. Reporter **39047 (play)**

Documents

5. Observatory log **40046 (step)**

Literature Search

6. Search on the words: AIRPORT, CAREY COUNTY, LIGHTNING, HIGHWAY, OBSERVATORY, X RAYS **40049 (step)**

7. Article #1 ("The X-Ray Machine and You") **40052 (step ×2)**

8. Article #2 ("Ammonia Spill on Highway 61") **40056**

9. Article #3 ("Council Member Proposes New Lights") **40058 (step)**

10. Article #4 ("Lights Cast a Shadow") **40061**

Sleuth Information Service

11. Schematic diagram of telescope **40063**

12. Current map of area **40065**

13. 10-year-old map of area **40067**

14. 25-year-old map of area **40069**

15. Precipitation table **40071 (step ×2)**

16. Lunar phases of the past year **40075**

17. Electromagnetic spectrum chart **40077**

18. Spectrum analysis **40079 (step ×4)**

19. Visible light spectrum **40085 (step)**

20. Radar **40088 (step ×2)**

Still Photographs

21. Recent observatory photo **40092**

22. Last year's observatory photo **40094**

23. Recent short-exposure photo **40096**

24. Photo taken with dirty lens **40098**

25. Photo taken with a jerky telescope motor **40100**

26. Photo taken with old film **40102**

27. Light-leak test **40104**

Image and Activity Bank
Side A or B

A selection of still images, short videos, and activities is available for you to use as you teach this unit. For a larger selection and detailed instructions, see the Videodisc Resources booklet included with the Teaching Resources materials.

19-1 What Is Light? page 425
Solar cell 2126
This solar cell can be used in most light conditions, including those provided by fluorescent or tungsten light bulbs.

Light bulb 1845 (step ×3)
Power off (step) Power on—electricity moves to the coil (step) Light energy (step) Heat output

◀| Step Reverse   Play ▶   Pause ‖   Step Forward |▶

19-2 Light, Heat, and Color, page 428

Neon sign 1803
Electricity can be used to excite neon gas molecules. The molecules within the bulb quickly capture and release energy. As they release the energy, visible light is emitted.

Steel, molten 2170
Molten steel gives off infrared radiation (heat).

Red-hot molten iron 41307–41524 (play ×2) (Side A only)
Molten iron emits infrared radiation (heat) and visible light.

19-3 A Colorful Theory, page 431

Electromagnetic spectrum 2239
The electromagnetic spectrum includes gamma rays, X rays, ultraviolet light, visible light, infrared radiation, microwaves, and radio waves.

Prism 2180
This crystal breaks white light into its individual colors. Each color of light has a different wavelength.

Wavelength of light 2240 (step)
A prism is able to bend the different wavelengths of light at different angles, creating the spectrum. (step) The specific wavelengths of visible light for the visible spectrum

20-1 Light in Action, page 443

Sky, blue 1448
Atmospheric gases and particles scatter the shorter wavelengths of light to produce a blue sky.

Sunlight in the atmosphere 2248
The shorter wavelengths of visible light are scattered by gases and particles in the atmosphere that have sizes similar to the wavelengths of blue light. This is why the sky appears blue.

Sunset 2183
At sunset, the longer wavelengths of light (red and orange) are seen more than those of blue light.

Rainbow 2181
Sunlight (white light) is broken into its individual colors by rain falling from the clouds. Each color of light has a different wavelength. Note the dimmer second rainbow. It is due to a second refraction off water droplets in the air.

20-3 Reflection, page 452

Reflection of light 2249 (step ×7)
How large is angle *X*? (step) The angle of incidence of light is equal to the angle of reflection. (step) Which is the correctly reflected beam? (step ×5)

21-1 Plane Mirrors, page 463

Image, inverted; candle 2189
A real image created by a single lens will always be inverted.

21-3 Converging Lenses and Real Images, page 469

Lenses 2257
Shapes of double-convex and double-concave lenses

Convex lens 43517–43693 (play ×2) (Side A only)
The farther away a convex lens is from an object, the greater the magnification will be.

Magnifying glass 2186
A magnifying glass, which is usually a curved, double-convex lens, makes objects appear larger than they really are.

Water drop; magnifier 2188
A water drop magnifies type on paper.

Lenses 2258
A double-convex lens projects an image outside the lens. The point of projection is called the focal point.

Camera 2259
Diagram of a camera

Binoculars 2260
Diagram of binoculars and the path of light through the binoculars

21-4 Real Images and Concave Mirrors, page 473

Concave mirror 2262
A concave mirror concentrates light at the focal point.

Concave lens 43694–43881 (play ×2) (Side A only)
The farther away a concave lens is from the object, the smaller the object appears.

Telescopic images and the eye 2261
How images are seen through a telescope

Solar cooker 2052
The energy of the sun can be captured to produce enough heat to cook objects.

Solar cooker, parabolic 1801
The high-energy radiation from the sun can be used to heat materials. Greater amounts of heat can be obtained if the light is concentrated, or focused. The parabolic surfaces focus light.

21-5 Refraction, page 477

Bent light 2263 (step)
The eye views an object along a light path. This may make an object appear to be in a place other than its actual location. (step)

Reflection, total internal 2265
Total internal reflection occurs at an angle equal to or greater than the critical angle. The critical angle of a substance, such as water, depends on its density.

Refraction 41525–41683 (play ×2) (Side A only)
Water bends light, so the penny is no longer visible.

Refraction; pencil in water 41177–41306 (play ×2) (Side A only)
Water refracts light. The pencil appears to be in a position above the water that is different from its position in the water.

Unit 7 LiGHT

UNIT FOCUS

Write the following statement on the chalkboard: At noon tomorrow there will no longer be any light. Ask students to consider its implications, and call on volunteers to offer their ideas. *(Without light, for example, no one could see, plants could not make food, and the Earth would become extremely cold.)* Next, ask students to brainstorm some ideas about what light is. *(Accept all reasonable responses.)* Then have students suggest how they use and enjoy light and why it is important to them. Keep track of their ideas on the chalkboard. This exercise will give you an opportunity to determine students' ideas about light. Explain that in this unit they will learn about the characteristics and behavior of light.

A good motivating activity is to let students listen to the English/Spanish Audiocassettes as an introduction to the unit. Also, begin the unit by giving Spanish-speaking students a copy of the Spanish Glossary from the Unit 7 Teaching Resources booklet.

CHAPTER
19 The Nature of Light
1 What Is Light?...... 425
2 Light, Heat, and Color 428
3 A Colorful Theory . . . 431
4 Adding and Subtracting Color . . . 437

CHAPTER
20 How Light Behaves
1 Light in Action...... 443
2 Light's Path 450
3 Reflection 452
4 The Riddle of Color.. 456

CHAPTER
21 Light and Images
1 Plane Mirrors....... 463
2 Convex Mirrors 467
3 Converging Lenses and Real Images 469
4 Real Images and Concave Mirrors 473
5 Refraction 477

422

Connecting to Other Units
This table will help you integrate topics covered in this unit with topics covered in other units.

Unit 1 Life Processes	Light energy is used by plants to make food through the process of photosynthesis.
Unit 2 Particles	Light interacts with particles in the atmosphere to make blue skies and red sunsets.
Unit 3 Machines, Work, and Energy	Solar cells convert light energy into electrical energy, which enables machines to do work.

These natural, prism-like quartz crystals took thousands of years to form. From this you might infer that they are rare and valuable, like diamonds. But this is not true—quartz is one of the most common minerals on Earth. Nonetheless, quartz crystals are certainly beautiful and, due to their shape and composition, can create dazzling displays of light.

The large crystal shown here is actually colorless, so it is not the crystal itself that is beautiful, but rather the crystal's effect on the light that strikes it. If you could look straight through a section of this quartz, it would be similar to looking through thick window glass. However, a description of a crystal such as this one would likely include statements about color. How is it possible for a clear material like this quartz crystal to pro-

Ⓐ duce colors? You will learn more about this topic later in the unit.

You will also learn more about lenses and how they affect light. You will learn that lenses are really nothing more than transparent material shaped in specific ways for specific purposes. In fact, certain types of quartz crystals are ground into lenses for high-quality microscopes and telescopes. Although the quartz crystal shown here has not been shaped into a lens, let it focus your attention on the fascinating subject of light.

This transparent quartz crystal reflects light in such a way that it appears to have color.

423

Using the Photograph

Most students are probably familiar with prisms and their effect on light. Explain to students that the quartz crystal is a natural prism. Ask: What is a prism? *(Students may know that a prism is a transparent object whose cross section is a polygon and whose sides meet at edges that are parallel to each other.)* What does a prism do? *(Accept all reasonable responses at this time. Sample response: A prism splits light into different colors.)* What does this tell us about the nature of light? *(Accept all reasonable responses.)* Explain to students that in this unit they will use prisms to explore the nature of light and the relationship between light and color. Ask students to find examples of other natural prisms or of other instances in which light is separated into different colors. *(Examples might include water droplets, rainbows, icicles, and other types of crystals.)* Use students' examples to create a display to introduce the unit.

Answer to
In-Text Question

Ⓐ Answers will vary. Accept all reasonable responses at this time. Light consists of energy of different wavelengths. The clear crystal produces color by physically separating the different wavelengths of the light.

Connecting to Other Units, continued

Unit 4 Oceans and Climates	Solar (light) energy drives many of the climatic processes in our atmosphere.
Unit 5 Electro-magnetic Systems	Electricity flowing through a resistor generates heat and light energy.
Unit 6 Sound	Light, like sound, consists of waves that provide us with information about our environment.

Connecting to Other Chapters

> **Chapter 19**
> *introduces light as a form of energy, the relationship among light, heat, and color, and the theory of color mixing.*

> **Chapter 20**
> *explores diffuse and specular reflection and the scattering, transmission, and absorption of light.*

> **Chapter 21**
> *analyzes lenses and mirrors, their uses, and how they produce real and virtual images through the refraction of light.*

Prior Knowledge and Misconceptions

Your students' responses to the ScienceLog questions on this page will reveal the kind of information—and misinformation—they bring to this chapter. Use what you find out about your students' knowledge to choose which chapter concepts and activities to emphasize in your teaching. After students complete the material in this chapter, they will be asked to revise their answers based on what they have learned. Sample revised answers can be found on page 441.

In addition to having students answer the questions on this page, you may wish to have them complete the following activity: Have each student write down three statements and three questions about light and then trade papers with a partner to exchange information and discuss each other's questions. As a class, discuss one question and one statement from each student pair. Assure students that there are no right or wrong answers to this exercise. Collect the papers, but do not grade them. Instead, read the papers to find out what students know about light, what misconceptions they may have, and what aspects of this topic are interesting to them.

CHAPTER 19 The Nature of Light

1 Why do some stars appear reddish, some white, and some bluish white?

ScienceLog

Think about these questions for a moment, and answer them in your ScienceLog. When you've finished this chapter, you'll have the opportunity to revise your answers based on what you've learned.

2 How are rainbows produced?

3 Is light matter or energy? Explain.

424

Answers to
In-Text Questions, page 425

A Without light to find their way, students would have to rely on the other senses, such as hearing or touch. Items that are likely to be useful in darkness would be ones that produce light, such as a candle, lamp, or matches. One item that is often useful that does not produce light is a walking cane or stick.

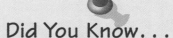

Did You Know...

Scientists once thought that light waves traveled like water waves. Because water waves move through water, astronomers reasoned that light waves must also move through a substance, called ether. Scientists now realize that light is a regularly varying pattern of electric and magnetic fields and that it can travel through empty space.

What Is Light?

Imagine that you've taken a shortcut home and you're walking down a back road on a moonless night. Looking around, you can see nothing but blackness. Without light to aid you, how would you find your way? What kind of items could you have carried to help you make your way through the darkness?

The items you named are probably objects that produce light. Light is something most of us take for granted. Indeed, you have probably never really had to walk home in total darkness because houses, apartments, and streetlights usually light up the night.

It would be hard to imagine life without light. But while you probably realize its importance, how much do you know about what light is? Here are some questions that may help direct your thinking about the nature of light.

1. When a flashlight is turned on, it gives off light. Does the flashlight lose mass as a result?

2. Why is a room with light-colored walls brighter than a room with dark walls?

3. Why is the sky blue when seen from the Earth but black when seen from space?

4. How do we see? Why can we see an object when a light is on but not when a light is off?

Make a list of other puzzling questions about light that you would like answered. By the time you have finished this unit, you'll be able to shed some light on the subject!

425

LESSON 1 ORGANIZER

Time Required
2 class periods

Theme Connection
Energy

Process Skills
observing, analyzing, inferring

New Terms
none

Materials (per student group)
Exploration 1, Activity 1: a few sheets of newspaper; scissors; lamp with a 100 W light bulb; **Activity 2:** sheet of white paper; sheet of black

paper; 2 alcohol thermometers; two small pieces of tape; watch or clock with second hand; **Activity 3:** flashlight; metric spring scale or metric balance; watch or clock; **Activity 4:** about 50 cm of copper wire; blueprint paper; watch or clock; large container of water (optional item: sunlamp); **Activity 5:** lamp with a 100 W light bulb; radiometer; watch or clock with second hand (See Advance Preparation on page 421C.)

Teaching Resources
SourceBook, p. S128

What Is Light?

FOCUS

Getting Started

Ask students to recall the definition of mechanical energy from Unit 3. *(Mechanical energy is the energy of the moving parts in a machine.)* Remind students that mechanical energy depends on the mass and motion of the parts. Explain that in this lesson, students will be introduced to a type of energy that has motion but no mass—light.

Main Ideas

1. Like other kinds of energy, light has no mass and does not occupy space.
2. Light may be produced from and changed into different forms of energy.

TEACHING STRATEGIES

Answers to
In-Text Questions

1. A flashlight does not lose mass when it gives off light.

2. A room with light-colored walls is brighter because the light-colored walls reflect more light around the room.

3. On Earth, the sky appears blue because the sun's light is scattered when it is reflected by water vapor and gas particles in the atmosphere. In space, the sky appears black because there is no atmosphere to reflect and scatter light.

4. Light that is produced or reflected by an object enters the eye and stimulates special cells on the retina at the back of the eye. These cells send electrical signals to the brain describing what the eye sees. When the light is off, the cells on the retina are not stimulated, so we see nothing.

Homework

Have students perform the activity Light and Paper on page 426 at home over a weekend to see more dramatic results.

Be sure to have students read the directions for developing the blueprint paper in Activity 4. Because the paper is sensitive to light, only the portion covered by the copper wire should remain unexposed. Be sure that students do not expose the paper before they are ready to place objects on it.

Cooperative Learning
EXPLORATION 1

Group size: 2 to 3 students
Group goal: to explore the properties of light and determine whether light is matter or energy
Positive interdependence: Assign the following roles to group members: primary investigator (to read the activity to the group and serve as a liaison between the teacher and the group), recorder (to organize data charts for recording information at each station), and manager (to make sure that each student has a job to do and is contributing at each station).
Individual accountability: Call on students at random to explain their responses to the questions in the Exploration.

Answers to
Activity 1

Half a day of direct sunlight will noticeably yellow the color of newspaper. A strong light bulb, such as a flood lamp, will have the same effect but will take longer. This Activity shows that light can change the color of newspaper.

Answers to
Activity 2

The temperature of white paper increases when exposed to light. The temperature of black paper increases more than the temperature of white paper when exposed to light. Light energy will heat up dark paper more than light paper.

Answers to
Activity 3

The mass of the flashlight does not change by any detectable amount. Students should conclude that light has no mass and does not occupy space.

Light Brigade

Each of the Activities in this Exploration will give you a better understanding of what light is.

You Will Need

- a newspaper
- a light bulb or light source
- 2 thermometers
- white paper
- black paper
- tape
- scissors
- a radiometer
- blueprint paper
- a spring scale or balance
- copper wire
- a flashlight

What to Do

Form small groups. Perform at least three of the Activities, dividing them among the groups. While you're waiting for some of the light effects to occur in the longer Activities, you can start on a different Activity. Keep a record in your ScienceLog of what you find out about light. Later, you can report what you discovered to the other groups.

ACTIVITY 1

Light and Paper

Take a piece of newspaper 10 cm square or larger and put it in a window where the sun will shine on it. Leave it there for half a day. Compare its appearance with a piece of newspaper that has not been exposed to sunlight. Will a bright light bulb produce the same effect? Try it. What does this Activity tell you about light?

ACTIVITY 2

Light, Color, and Temperature

Tape a piece of white paper and a piece of black paper to the inside of a window that faces the sun. Then tape a thermometer to each piece, as shown below. Record the

temperature shown on both thermometers every minute for 5 minutes. What do your findings tell you about light and white paper? about light and black paper? What effect do you think light has on the temperature of different colored papers?

A radiometer

ACTIVITY 3

Flashlights and Mass

Find the mass of a flashlight. Then turn it on for 10 minutes. Now check its mass again. Is there any change? What does this tell you about light? Do you think light occupies space?

ACTIVITY 4

Lightly Done

Shape some copper wire into a flat design and place it on the surface of a piece of blueprint paper. Be sure to keep the paper covered and away from light until you are ready to perform the experiment. Expose the paper to direct sunlight or a sunlamp for 5 to 10 minutes. Then immerse the paper in water. Allow it to dry. What does your "photograph" look like? How was it made? What did light have to do with it?

ACTIVITY 5

The Light Windmill

Examine a light windmill, also called a *radiometer*. Put the radiometer in direct sunlight for a minute or so. What happens? Place it in the dark. What happens? Place it at different distances from a bright light bulb. What do you notice? In which direction does the radiometer rotate? What does the radiometer show you about light?

Answers to
Activity 4

The copper-wire design forms a white "negative" image on the blueprint paper. The design was made by shading parts of the paper from the light and exposing other parts. The light caused a chemical change in the special coating on the paper. Washing the paper removes this special coating and prevents changes to the image produced.

Answers to
Activity 5

The radiometer rotates in the presence of light, but it does not rotate in the dark. Students should observe that the speed of rotation of the radiometer increases with the intensity of the light or the closeness of the light source. Students should observe that the black sides of the vanes always move away from the light. The radiometer shows that light energy can cause some objects to move.

Light—A Quick-Change Artist!

Light obviously has many properties. Use what you learned about light in Exploration 1 to answer the following questions. Answer either *yes, no,* or *not certain.*

1. Can light make things move?
2. Can light cause the color of some things to change slowly?
3. Can light cause the color of some things to change rapidly?
4. Can light cause things to heat up?
5. Does light have mass?
6. Does light take up space?

What do you think? If you answered *yes* to the first four questions, you are correct. The Activities in Exploration 1 show that light can do two things: it can make things move, and it can cause changes in color and temperature. Isn't this what you would expect of *energy*? And would you expect light, as a form of energy, to have mass or occupy space as matter does? How did you answer questions 5 and 6 above? If you answered *no* to both questions, you are right again.

As the photographs on this page show, light is actually involved in many energy changes. All of the items shown are converting energy. That is, they are involved in energy changes. For each example, indicate the energy conversion taking place. Ⓐ

a

b

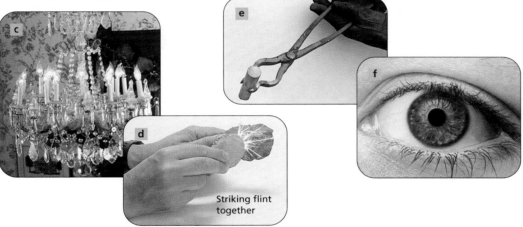

c

e

d

Striking flint together

f

Light—A Quick-Change Artist

These paragraphs provide a rationale to help students arrive at the conclusion that light is a form of energy and not a form of matter. It may be helpful to have students review what they learned about energy in previous units, including heat energy (Unit 3), electrical energy (Unit 5), and sound energy (Unit 6).

Answers to
In-Text Questions

Ⓐ
- **a.** Chemical to light
- **b.** Light to chemical (photosynthesis)
- **c.** Electrical to light
- **d.** Kinetic to light
- **e.** Thermal to light
- **f.** Light to chemical (neural response to light)

Theme Connection

Energy

Focus question: Why do we say that light energy is a kind of electromagnetic energy? *(What people perceive as light is just the visible part of the electromagnetic spectrum, which also includes X rays, infrared radiation, and ultraviolet radiation. Electromagnetic waves carry electromagnetic energy, so light energy is a form of electromagnetic energy.)*

FOLLOW-UP

Reteaching

Have each student illustrate four or more events in which a transformation of light energy occurs.

Assessment

Ask students to make a list of all the light-emitting objects they can find during a 24-hour period.

Extension

Suggest that students do some research to learn about photons. Have them address specific questions such as What are photons? and How do they behave? Have students share what they learn with the class.

Closure

Have students write a response to the following puzzle: Most of the light on the Earth comes from the sun. Is the light that reaches the Earth from the sun the same thing as "solar energy"? Why or why not? *(Students should understand that solar energy includes electromagnetic waves [such as ultraviolet] in addition to visible light.)*

Light, Heat, and Color

Light, Heat, and Color

FOCUS

Getting Started

Direct students' attention to the lesson title and encourage them to speculate about what it means. Ask them if they know what the expression "red-hot" means. *(Very hot)* Then invite students to describe how the color of some familiar objects changes as they become hot. *(Objects that students might mention include the filament of a light bulb, the burner on an electric stove, and burning coals.)* Point out that in this lesson they will learn more about the relationships among light, heat, and color.

Main Ideas

1. The color of light emitted from a glowing object is often an indication of how hot it is.
2. As a solid object becomes hotter, the light it emits changes from red to yellow to white.

TEACHING STRATEGIES

EXPLORATION 2

SAFETY ALERT Students should not touch the wire in Part 2 or the burner in Part 3. Also, before performing step (c) in Part 1, test the light bulb for your students to be sure that it can withstand the current supplied.

Answers to
Part 1

a. The filament should be dimly lit and glowing red. The bulb and wire may feel slightly warm to the touch. The wire will not change color.
b. The filament should appear yellow and brighter than in (a). The bulb and wire will feel warm. The wire will not change color.
c. The filament should appear white and give off more light than in (a) or (b). The bulb and wire will be even warmer. The wire will not change color.

Think of several things that produce light. What general observation could you make about the temperature of these things? Are they hot or cold? Is there a relationship between the light that an object gives off and the object's temperature? In the following Exploration you will take a closer look at this question.

EXPLORATION 2

Observing Hot Solids

You Will Need

- 3 D-cells—1 weak and 2 fully charged
- a copper wire (uninsulated, approximately 20 cm long)
- a flashlight bulb
- masking tape
- a clothespin
- a Bunsen burner
- a hot plate or electric stove

PART 1

Starting with setup (a), arrange the D-cells as shown below. Be sure to complete each circuit by touching the wire to the base

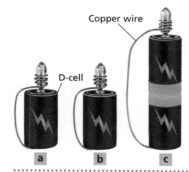

Copper wire

D-cell

a   b   c

of the appropriate cell. Feel the wire and bulb and observe the color in each case. Write down your observations in your ScienceLog.

Be Careful: The flashlight bulb may be hot.

a. A weak cell—just strong enough to make the filament start to glow
b. A strong cell
c. Two strong D-cells held together with masking tape

PART 2

Using a clothespin, hold a piece of copper wire in a flame until it glows (gives off light). What color do you see? If you were to place your hand close to the wire, what would you feel?

Be Careful: Do not touch the hot wire!

PART 3

Turn an electric stove element to low and gradually increase the heat to high. Observe the color of the burner and the heat it gives off in each case.

Be Careful: Keep your hands away from the heated surface!

PART 4

What is the color of the molten steel in the photo below? What do you think its temperature is?

As you probably realize, most things that produce light are hot. There are exceptions, however. Fluorescent lights and neon signs, for example, are relatively cool—even when lit for a long time. And fireflies don't get very hot either.

LESSON 2 ORGANIZER

Time Required
1 class period

Process Skills
observing, analyzing, inferring

New Terms
none

Materials (per student group)
Exploration 2, Part 1: 3 D-cells, 1 weak and 2 fully charged; about 20 cm of uninsulated copper wire; flashlight bulb; about 10 cm of masking tape;

Part 2: clothespin; piece of copper wire, about 10 cm long; portable burner; striker; safety goggles; **Part 3:** hot plate or electric stove

Teaching Resources
Exploration Worksheet, p. 3
Resource Worksheet, p. 5

A Light Quiz

Consider what you learned in Exploration 2. Then complete the following statements by selecting the correct word or phrase from the choices provided. Record your choices in your ScienceLog.

1. Heat and light (rarely, often) occur together.
2. When electrical energy passes through a wire, the wire (gets cold, stays the same temperature, gets hot).
3. When a small amount of electrical energy passes through a light bulb, (white, reddish) light is produced.
4. When a large amount of electrical energy passes through a light bulb, (white, reddish-orange) light is produced.
5. When a small amount of electrical energy passes through an electric stove element, the element gets hot and (gives off, does not give off) light.
6. When a large amount of electrical energy passes through an electric stove element, the element gets hot and (gives off a reddish light, gives off a white light, does not give off light).
7. As an electric stove element becomes hotter, it gives off light that gets (brighter—turning from red to white, darker—turning from white to red).
8. Heating a wire in a hot flame (does not affect its appearance, causes it to glow).
9. When metal is heated to a very high temperature, it gives off (red, white) light.
10. The color of light given off by a substance often indicates (how large, how hot, what shape) the substance is.

How might you summarize your findings from Exploration 2? In general, is there a relationship between the temperature of an object and its color? Write down your ideas in your ScienceLog. Ⓐ

Ⓑ What evidence do you see here to support the fact that the sparks are cooling as they fly through the air?

429

A Light Quiz

The answers to the statements provide a detailed summary of Exploration 2. Suggest that students write the complete sentences in their ScienceLog. Then provide class time for them to discuss their responses.

Answers to
A Light Quiz

1. often
2. gets hot (Students may respond that it stays the same temperature because the heating effect may be so slight that they cannot detect it.)
3. reddish
4. white
5. does not give off
6. gives off a reddish light
7. brighter—turning from red to white
8. causes it to glow
9. white
10. how hot

Homework

You may wish to assign A Light Quiz as homework. A Resource Worksheet is available to accompany the activity (Teaching Resources, page 5).

Answers to
In-Text Questions

Ⓐ Answers may vary. Students should realize that when solids are heated enough, they glow red. With more heat, the light they emit becomes yellowish and then white.

Ⓑ The light given off by the sparks changes color from white to yellowish orange as the sparks fly through the air.

Answers to
Part 2, page 428

The copper wire should turn from its natural color to red and then to orange or yellow as it becomes hotter and hotter. Students should suggest that they would feel heat if they were to place their hands close to the wire.

Answer to
Part 3, page 428

The stove element should change from black, to red, and to orange as it becomes hotter and hotter. Students

should be able to feel the amount of heat increase as the color changes.

Answers to
Part 4, page 428

Most students will probably recognize that the bright yellow color of the molten steel indicates that it is very hot. The steel is probably around 1500°C.

⭐ An Exploration Worksheet is available to accompany Exploration 2 (Teaching Resources, page 3).

Light and Temperature

The emphasis of this exercise is on students' awareness of temperature differences and the relationship of temperature to color. Encourage the class to have fun matching the different substances with the correct temperatures. Guessing may be necessary.

What Color Is Hot?

The questions at the end of this section lead into the next lesson, A Colorful Theory. The questions are not meant to be answered completely at this time. Instead, you may wish to involve students in a discussion to determine what they already understand about color.

Answers to
In-Text Questions

Ⓐ Encourage discussion of any answers that may puzzle students. For instance, students may not have realized that a light-bulb filament is so hot.

Ⓑ You may wish to briefly explain to students that the color of a glowing object comes from light energy emitted by the object. This energy, in turn, comes from other energy sources, such as electrical energy in a resistor or heat energy from a flame. White is a combination of all colors of light, while black is the absence of all colors of light.

FOLLOW-UP

Reteaching

Have students make poster diagrams to illustrate the relationships among light, heat, and color. Suggest that they use one of the examples in the lesson: flashlight bulb, heated wire, stove burner, molten steel, flashbulb, or an example of their own, such as stars, burning logs, or charcoal. Display the finished posters in the classroom for all to enjoy.

Assessment

Have students observe the flame of a burning candle. Students should draw the flame, label the colors, and indicate which part of the flame is the hottest and which is the coolest. Have them give reasons for their answers.

Light and Temperature

In your ScienceLog, place the items listed below in order of increasing temperature.

a. the surface of the sun

b. the human body

c. boiling water

d. the glowing filament in a 100 W bulb

e. red-hot iron

f. solid carbon dioxide (dry ice)

g. melting ice

h. white-hot iron

Now place the corresponding temperature next to each item on the list: –80°C, 0°C, 37°C, 100°C, 500–650°C, 1500°C, 2200–2700°C, 5000–6000°C. Compare your lists with those of other students. Are you surprised at the results? Ⓐ

What Color Is Hot?

You have seen that the light given off by an object can indicate its temperature. Recall the color change in copper wire held over a flame. When the wire is red-hot, it is hotter than when it is dull red. When the wire is white-hot, it is hotter than when it is red-hot.

As an object becomes hot, it usually gives off light. The color of the emitted light changes as the temperature of the object rises. Certain colors generally represent certain temperatures. To understand the relationship between temperature and the colors of light, you need to consider the following:

• Where does color come from?

• Is white a color?

• Is black a color?

• What is the relationship between white light and other colors of light? Ⓑ

You may be surprised by how much you already know about color!

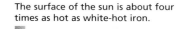

The surface of the sun is about four times as hot as white-hot iron.

Answers to Light and Temperature:
(f) –80°C, (g) 0°C, (b) 37°C, (c) 100°C,
(e) 500–650°C, (h) 1500°C, (d) 2200–
2700°C, (a) 5000–6000°C.

Extension

Have students consult an astronomy textbook to find the names of the brightest stars in the sky. Students should find out the color and temperature of each of these stars. Ask: Is there a correlation between stellar color and temperature? What is it? *(The color of a star depends on its surface temperature. Red stars, such as Betelguese, are the coolest stars, and blue stars, such as Rigel, are the hottest.)* If possible, students may want to try and identify a few of these stars in the sky.

Closure

Have students research the relationship of heat and light to the electromagnetic spectrum. *(Students should explain to the class that heat is infrared radiation and light is visible radiation. Stated simply, heat is radiation of longer wavelengths and lower energy than visible light, which is radiation of shorter wavelengths and higher energy.)*

A Colorful Theory

In 1666 Isaac Newton, age 24, was experimenting with prisms. People had known about and used prisms since the time of Aristotle (about 350 B.C.), but Newton made some remarkable new discoveries. In Exploration 3, you can relive a great moment in science—one of Newton's first discoveries.

EXPLORATION 3

Lights, Prisms, and Filters

This series of experiments works best in a darkened room. The activities should be done in the order given. Use the questions provided to help you think about and develop your own "theory of color."

You Will Need

- 2 prisms
- a flashlight
- white cardboard
- 3–6 red filters (or colored cellophane)
- 3–6 blue filters
- 3–6 green filters
- a piece of white paper
- aluminum foil
- tape

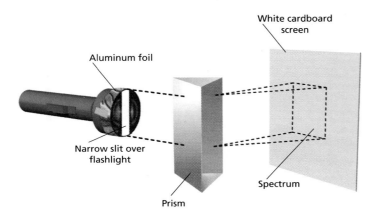

White cardboard screen

Aluminum foil

Narrow slit over flashlight

Spectrum

Prism

EXPERIMENT 1

Light and Prisms

Following the setup shown below, shine a narrow beam of light on a prism in such a way that a rainbow of colors (called a **spectrum**) forms on the white cardboard. Notice how the colors blend into one another. How many different colors do you see on the screen? What are the colors at the ends of the spectrum? Which color seems to come through the thickest part of the prism?

Before Isaac Newton, people believed that the thickness of the prism determined the colors of the spectrum. According to this theory, white light passing through the thinnest part of the prism changes to red; white light passing through the thickest part of the prism changes to blue; and white light passing through a medium thickness of the prism changes to green. Do you agree with this explanation? Why or why not? How else could the rainbow of colors be produced from white light?

Isaac Newton didn't agree with the common explanation of his day. After completing the following experiments, you will understand Newton's reasoning.

Exploration 3 continued

431

LESSON 3 ORGANIZER

Time Required 3 class periods

Process Skills
predicting, analyzing, hypothesizing, observing, communicating

New Term
Spectrum—a continuous range or sequence, such as the pattern of colored bands produced by passing white light through a prism

Materials (per student group)
Exploration 3, Experiment 1: flashlight; 15 cm × 15 cm piece of aluminum foil; a few small pieces of tape; prism; piece of white cardboard, about

30 cm × 30 cm; **Experiment 2:** flashlight; 15 cm × 15 cm piece of aluminum foil; a few small pieces of tape; prism; red filter; piece of white cardboard, about 30 cm × 30 cm; sheet of white paper; blue filter (See Advance Preparation on page 421C.); **Experiment 3:** flashlight; red filter; piece of white cardboard, about 30 cm × 30 cm; green filter; **Experiment 4:** 3–6 red filters; flashlight; piece of white cardboard, about 30 cm × 30 cm; 3–6 green filters; 3–6 dark blue filters; **Experiment 5:** flashlight; 15 cm × 15 cm

continued

A Colorful Theory

FOCUS

Getting Started

Project light through a red filter onto a screen. Invite students to speculate about why the light on the screen is red. *(Accept all reasonable responses.)* Then have them think of some devices that use filters to display a certain color of light. *(Examples include traffic signals and stage lights.)* Point out to students that in this lesson they will use colors to develop a theory about the nature of light.

Main Ideas

1. White light consists of all of the colors of the spectrum, while black is the absence of all light.
2. Using a prism, white light can be separated into its component colors.
3. A filter allows a specific color of light to pass through, while absorbing all other colors of light.
4. When white light passes through a filter, some of its energy is absorbed.

TEACHING STRATEGIES

EXPLORATION 3

🔄 Cooperative Learning
EXPLORATION 3

Group size: 3 to 4 students
Group goal: to analyze the spectral components of white light
Positive interdependence: Have students complete the Exploration in a round-table format. They should take turns reading the directions for each of the experiments and supervising the activity to ensure that it is performed correctly. Each student should continue reading and supervising the activity until he or she comes to an in-text question, at which time another group member becomes the reader and supervisor.
Individual accountability: Each student should write answers to the in-text questions. Collect and evaluate individual answer sheets.

Exploration 3 continued

Answers to
Experiment 1, page 431

Students will probably be able to identify four or five colors. Red appears on
the right end of the spectrum, and
violet or blue appears on the left end.
Violet or blue comes through the thickest part of the prism.

Students may infer correctly that the
prism does not change the color of
white light but rather separates it into
its component colors. The remaining
experiments in this Exploration should
provide students with evidence to support this inference.

Answers to
Experiment 2

When the red filter is placed between
the flashlight and the prism, the color
spectrum on the screen is reduced to
red light. The color of the light
between the red filter and the prism is
also red. Students should conclude that
the prism, regardless of thickness, does
not change the color of the red light.

When students replace the red filter
with the blue filter, they should
observe only blue light on the screen
and between the filter and the prism.
Again, the prism does not cause any
color change in the blue light.

Sample summary: White light is
made up of different colors of light.
Prisms separate the different colors of
light. Filters allow specific colors to
pass through while absorbing all other
colors of light.

Answers to
Experiment 3

The red filter allows only red light to
pass through. The red filter absorbs all
colors of light except red. The green
filter absorbs all colors of light except
green.

Answers to
Consider This

By considering the analogy of the
rubber ball, students should conclude
that just as the ball loses energy when
it hits a puddle of water, light loses
energy when it passes through a
colored filter.

 **An Exploration Worksheet is avail-
able to accompany Exploration 3
(Teaching Resources, page 6).**

EXPERIMENT 2
Light and Filters

Return to the setup from
Experiment 1. Place a red filter
between the flashlight and the
prism. Observe the screen.

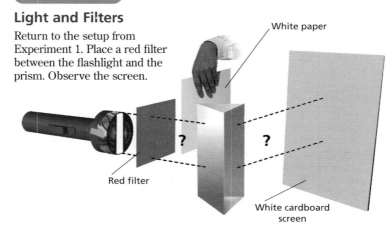

White paper

Red filter

White cardboard
screen

How is the spectrum different
from before?

Place a piece of white paper
between the filter and the prism.
What is the color of the light
between the filter and the prism?
Did the prism change this color
of light? Shine the light through
the thinnest part of the prism, the
thickest part of the prism, and
several areas in between. Did the
various thicknesses of the prism
alter the red light in any way?

Repeat this experiment, substituting a blue filter for the red
one. Again, observe the color of
the light. What is the color of the
light between the blue filter and
the prism? Did the prism cause
any color change?

Now summarize the infor-
mation you have acquired
about light and color in your
ScienceLog. State what you think
white light may be made of, what
you think prisms do to light, and
what you think filters do.

EXPERIMENT 3
More About Filters

Prepare the setup shown in the figure below. What color of light
comes through the filter? Does
the red filter add something to
white light, or does it take something away from it?

Repeat this experiment using
a green filter in place of the red
filter. Does the green filter add
something to white light or take
something away from it?

Consider This

Here is a good analogy for light
passing through a filter: When a
ball moving along the floor hits a
puddle of water, does the ball
gain or lose energy? Remember
that light is energy. When light
passes through the red filter,
does it lose energy, gain energy,
or retain its original energy?

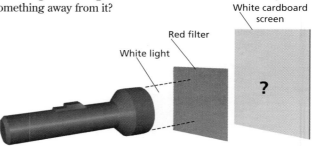

White cardboard
screen

Red filter

White light

432

ORGANIZER, continued

piece of aluminum foil; a few pieces of
tape; 2 prisms; piece of white cardboard, about 30 cm × 30 cm; red filter;
dark blue filter

Teaching Resources
Exploration Worksheet, p. 6
Transparency Worksheets, pp. 12
  and 14
Transparencies 66 and 67
Math Practice Worksheet, p. 16
SourceBook, pp. S133 and S139

EXPERIMENT 4

A Combination of Filters

White cardboard screen

Return to the setup from Experiment 3. One by one, add more red filters, as shown below.

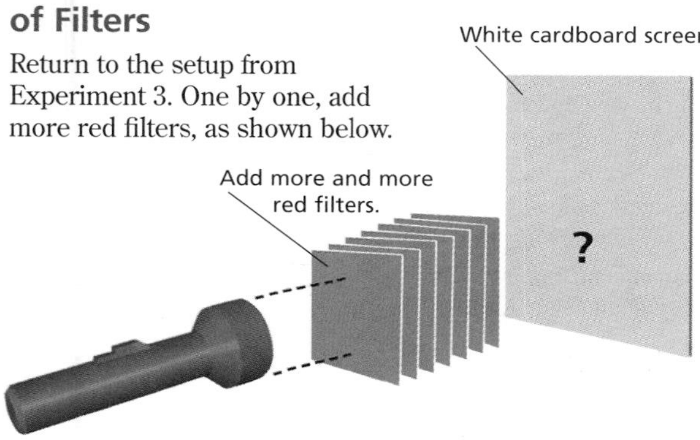

Add more and more red filters.

?

What happens to the white light as you add more red filters? Do the filters remove light energy? Could you add enough filters so that no light energy got through? How many red filters would it take? If no energy got through, what color would you see on the screen? In other words, what color is "no light"?

Repeat this activity using green filters, and consider the same questions.

Now shine the light through a single red filter, and then add either a dark blue or a dark green one, as shown below. Was the energy loss similar? Recall how many red filters you had to use so that no light energy reached the screen. With a single red filter, how many blue or green filters did you have to use to get the same effect? Why do you think this is so?

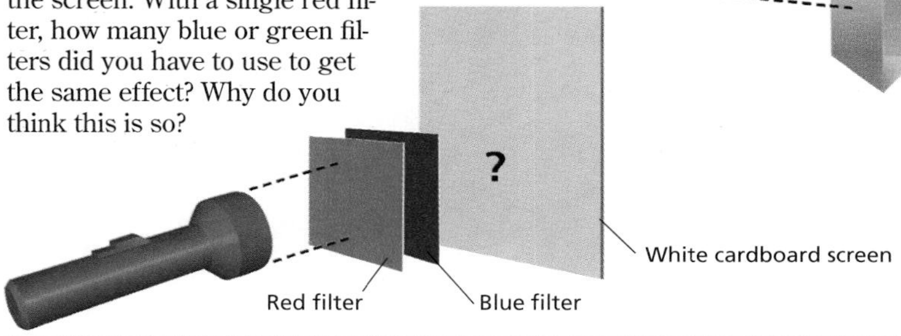

After considering the results of Experiments 3 and 4, would you agree that filters absorb some of the light energy in white light? In the following experiment, you will discover the nature of the energy loss caused by filters.

EXPERIMENT 5

Disappearing Colors

Shine white light through a prism and obtain a spectrum on the white screen.

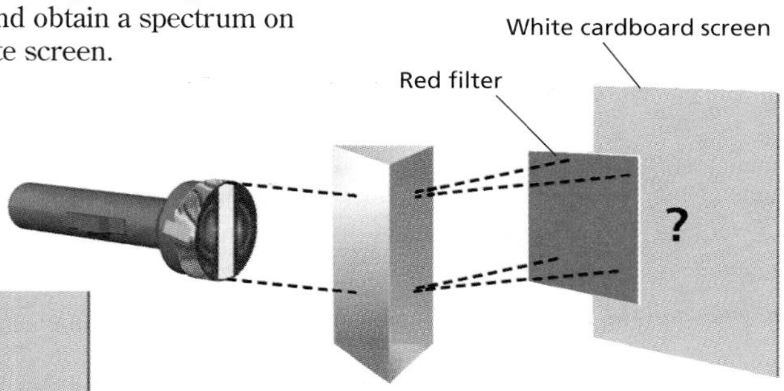

Red filter

White cardboard screen

?

Red filter   Blue filter

White cardboard screen

?

Now place a red filter between the prism and the screen. What colors seem to have disappeared? In other words, what colors are absorbed by the filter? What color(s) pass(es) through the red filter? If only red light were shone on this filter, what would happen to the light? What effect would this filter have if green light were shone on it? If blue light were shone on it?

Repeat the experiment, this time using a dark blue filter between the prism and the screen. Consider the same questions.

Predict what would happen if you shined white light through two prisms (with the second prism facing in the opposite direction from the first one). Try it.

Your Theory

Think through all that you have discovered, and write down your own "theory of color." Then see how your theory compares with Isaac Newton's explanation, which is described next.

433

Answers to Experiment 5

Placing a red filter between the prism and the screen results in only red light appearing on the screen. All of the other colors are absorbed by the filter. The filter allows only red light to pass through. If only red light were shone on the filter, it would pass through to the screen. If green or blue light were shone on the filter, little or no light would pass through to the screen.

The results are the same for the dark blue filter except that only blue light appears on the screen.

White light results from passing white light through the two prisms. The first prism separates the white light into the colors of the spectrum. The second prism recombines the colors of the spectrum back into white light. Note: The second prism must be placed precisely for the white light to emerge. For a demonstration of how to set up the two prisms, see page S212.

⭐ **Two Transparency Worksheets (Teaching Resources, pages 12 and 14) and Transparencies 66 and 67 are available for use after students have completed Exploration 3.**

Your Theory

Allow students time to write and discuss their "theory of color." Their theories should indicate an understanding of the following points:

- White light is composed of many colors, which make up the spectrum.
- Colored filters absorb all colors except the color of the filter itself.
- The colors of the spectrum may be recombined to create white light.
- The thickness of glass does not change white light into different colors.
- Filters do not add color to white light.
- The absence of light energy results in blackness.

Homework

You may wish to assign Your Theory as a homework activity.

Answers to Experiment 4

As more red filters are added, the intensity of the light on the screen decreases. Each red filter removes some light energy. If enough red filters are added, eventually no light energy will pass through, and the screen will become black. (The number of filters needed may vary depending on the quality of the filters and the intensity of the white light.) Black, therefore, is the absence of light energy.

Repeating the experiment with green filters should result in similar observations and conclusions.

Theoretically, when a blue or green filter is placed in front of a red filter, students should observe that all of the light is blocked and the screen becomes black. This occurs because the blue or green filter absorbs all of the red light, leaving no light to pass through. However, depending on its quality, the red filter may not be able to prevent some blue or green light from passing through, or vice versa. If so, it may take more than one blue or green filter to absorb all of the remaining light energy.

Newton's Bright Ideas

Call on a volunteer to read this material aloud. Then involve the class in a discussion of Newton's theory of color. Encourage students to compare Newton's description with their own discoveries in the experiments they performed.

Answer to
In-Text Question

Ⓐ Students should recognize that the colors of the spectrum are red, orange, yellow, green, blue, indigo, and violet (ROY G. BIV).

Answers to
Things to Think About

1. Red, green, and blue light each bend a different amount when passing through a prism.

2. Newton passed the individual spectral colors of light through a prism and discovered that they did not change.

3. Student answers may vary, but possible responses include the following:
- Some scientists may have been envious of Newton's discoveries.
- New ideas or theories in science are often not quickly accepted.

4. The red filter removes all light except red. The blue filter permits only blue light to pass through, but no blue light is present because it was absorbed by the red filter. Therefore, no light reaches the screen, and the screen appears black.

Answers to Things to Think About continued ▶

Newton's Bright Ideas

Consider Newton's theory about light and color. Newton said that light is not changed when it goes through a prism. Instead, he said, it is physically separated. Newton reasoned that, if the old theory were true—that is, if the thickness of a prism did change the color of white light—then another prism should change the resulting colors again. This did not happen when he tried it. What he found was that once the colors of white light were separated, they could not be changed any further. Newton called this the "critical experiment."

In addition, Newton noticed that the spectral colors always occurred in the same order. Thinking back to your investigations, can you recall what the order is? Some people remember the order of the colors of the spectrum by thinking of "Mr. Color," ROY G. BIV. What does each letter stand for? **Ⓐ**

Newton also noticed another interesting phenomenon about the spectral colors. He found that by using a second prism, the colors could be blended to form white light again.

Newton's theory of light and color was a totally new one. Many scientists of his day disputed it and attacked his ideas. Newton eventually wrote a letter saying that he was so harassed by arguments against his theory that he regretted losing his peace of mind in order to "run after a shadow" of an idea.

Isaac Newton
(1642–1727)

Things to Think About

1. How do red, green, and blue lights differ when passing through a prism?

2. How do you think Newton proved that any color produced by passing white light through a prism cannot be changed to any other color?

3. Why do you think other scientists disputed Newton's theory so strongly?

4. Using the illustration below and your own theory of light and color, predict the color of the light that enters the blue filter. Why do you think the screen appears black?

White screen appears black

Performed in a darkened room

Transparent blue filter

Transparent red filter

White light

?

434

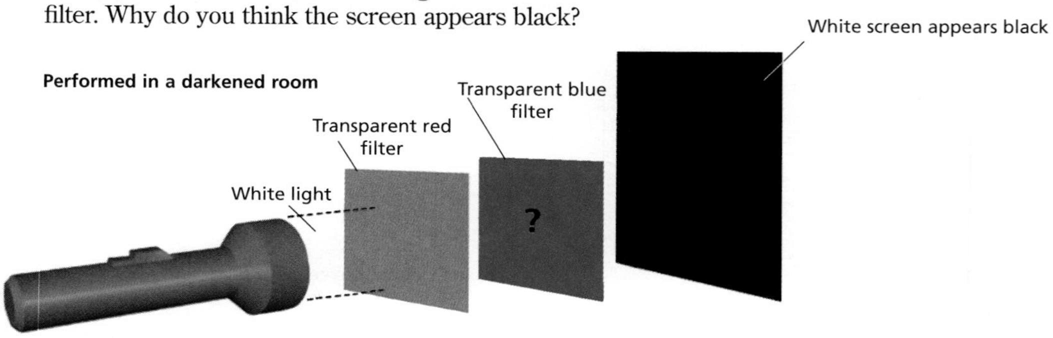

Multicultural Extension

Natural Dyes

Many cultures produce and use natural dyes for fabrics and other objects. Ask students to find out about some of these dyes. (*Indigo, a deep blue dye, is an Indian dye made from the indigo plant. Many Middle Eastern cultures use saffron, a yellow dye extracted from the crocus. Carmine is a red dye from Mexico that is made from the bodies of certain insects.*) Students may wish to bring in samples of materials colored with natural dyes.

ENVIRONMENTAL FOCUS

Point out to students that most of the energy the Earth receives from the sun is in the form of visible light but that some energy also comes in invisible forms such as infrared and ultraviolet rays. Have students write a short report on the beneficial and harmful effects of the sun's ultraviolet rays and on how pigments found naturally in skin can protect people from the harmful effects. As an extension, have interested students also report on the problem of ozone depletion in the Earth's atmosphere.

5. Predict what you would see in the situation shown here, and explain your prediction.

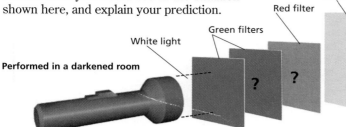

Performed in a darkened room

White light

Green filters

Red filter

White cardboard screen

6. Write a letter to a friend relating your personal discoveries about light. Be factual but expressive.

Activities to Try at Home

Each activity below shows a way of separating white light into its colors.

1. Dip the mouth of a jar into soapy water. Look at the light reflected by the soap film covering the mouth of the jar.
2. Observe the light reflected off of the surface of a compact disc.
3. With the sun behind you, look at a fine spray of water from the nozzle of a hose pointing toward the ground.
4. Look through a nylon stocking at the light coming through a small hole.
5. Observe the film that forms after a drop of oil hits the surface of a beaker of water.

Nylon stocking

Hole

Clear light bulb

Light

435

FOLLOW-UP

Reteaching

Point out to students that a rainbow may include all of the colors of the visible spectrum. Have them draw a colored illustration of a rainbow showing the bands of color in the correct order. Have students include a sentence or two that explains how light from a rainbow is similar to light passing through a prism. *(Both the drops of water in the atmosphere and the prism separate white light into its spectral colors.)*

Assessment

Have students work together to create imaginative posters, bulletin-board displays, or murals to illustrate the theory of light and color that they have investigated in this lesson. Display their work in the classroom or school library for others to see.

Extension

Explain to students that the band of colors produced when white light passes through a prism is called the *visible spectrum.* Suggest that students make poster diagrams to illustrate the visible spectrum. Their diagrams should identify each color and its wavelength. Display the diagrams around the classroom.

Closure

Pose the following scenario to the students: In a restaurant, you order white fish with green beans and tomatoes. The full-spectrum white light bulb at your table burned out, and the manager replaced it with a red light bulb. How will your meal look? Should you send it back to the cook? *(The only color that can be reflected is red, so the green beans will appear black, the fish will look red, and the tomato will look dark red. The meal only looks bad, so there is no need to send it back. A better solution would be to ask for a new light bulb.)*

Light and Color

You now have a theory of light: *White light is composed of, and can be separated into, light of different colors.* The opposite of this is also true. Light of these different colors can be recombined to make white light. How does this theory explain the color of hot objects?

Which end of the spectrum do you think has the most "energetic" light? First recall the observations you made of the flashlight bulb in Exploration 2.

- A flashlight bulb attached to a nearly dead D-cell emits (gives off) reddish light. In this case the energy supplied to the bulb is small. The temperature of the bulb's filament is relatively low. What can you infer about the energy of red light?

- When the cell is stronger, the bulb emits a more yellowish light. The hotter filament gives off not only red light, but also orange and yellow light. What can you infer about the energy of orange light? of yellow light?

- If the bulb is connected to a new cell, even more energy is present. The filament is hotter. It emits not only red light, but the other colors of the spectrum as well, all of which combine to produce white light.

Based on these findings, how would you compare the energy of each color of the spectrum (ROYGBIV)?

My photo was taken with infrared light. To find out more about this form of light, read page S139 in the SourceBook.

R

Nearly dead cell

White light

G

Y   B

O   I

R   V

New cell

Answers to *Light and Color*

- Red light has the lowest energy of visible light.
- Orange light is more energetic than red light, and yellow light is more energetic than orange light.
- Moving from red to violet in the spectral sequence ROYGBIV, each color has progressively more energy.

PORTFOLIO

Suggest to students that they include their statement of the theory of light in their Portfolio along with an explanation of how the activities in Exploration 3 support the theory.

Homework

The Math Practice Worksheet makes an excellent homework assignment after students have completed the lesson (Teaching Resources, page 16).

LESSON 4 — Adding and Subtracting Color

Adding Colored Lights

As you know, Newton was able to recombine the colors of the spectrum to produce white light. Did you observe a similar phenomenon in Exploration 3? (What happened when you shone white light through two prisms, one facing in the opposite direction from the other?) Ⓐ

Like Newton's experiment, the demonstration shown here involves mixing light of different colors. In this case, however, only red, green, and blue light are mixed. What do you predict will happen when these three colors of light are combined? when any two colors are combined? Ⓑ

In the pages ahead, you will have the chance to find out for yourself what happens when different colors of light are combined.

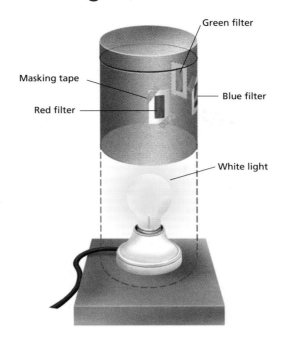

Green filter
Masking tape
Red filter
Blue filter
White light

Performed in a darkened room

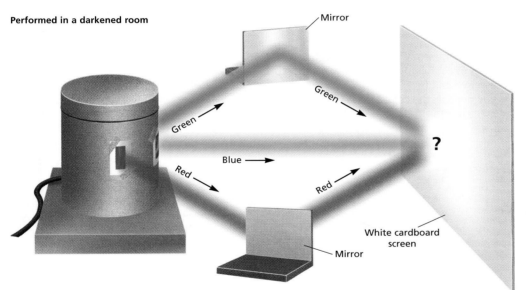

Mirror
Green
Green
Red
Blue
Red
?
White cardboard screen
Mirror

437

LESSON 4 ORGANIZER

Time Required
1 class period

Process Skills
observing, communicating, analyzing, predicting

New Terms
Complementary colors of light— specific pairs of colored lights, consisting of one primary and one secondary color, that produce white light when mixed
Primary colors of light—the three colors of light (red, green, and blue) that can be mixed in various combina-

tions to form different colors or that can be mixed together equally to form white light
Secondary colors of light—the three colors of light (cyan, yellow, and magenta) produced by mixing two primary (light) colors in equal quantities

Materials (per student group)
Filter Fun: overhead projector; filters of various colors, such as red, blue, green, cyan, yellow, and magenta

Teaching Resources
SourceBook, p. S134

FOCUS

Getting Started
Before students begin the lesson, have them mix red, blue, and yellow paint on a piece of butcher paper. Ask the class what color results. *(Brown, gray, or black will result.)* Then write the following question on the blackboard: If you mix red, blue, and yellow light, what color do you get? *(Red)* Have students write down their responses. Discuss the responses to pique student interest but do not answer the question yet. Tell students that they will learn the answer in this lesson.

Main Ideas
1. Colors of light combine according to the rules of color addition.
2. Primary colors of light combine to produce white light.
3. A secondary color of light combined with a primary color of light produces white light.
4. Filters alter colors according to the rules of color subtraction.

TEACHING STRATEGIES

Answers to
In-Text Questions

Ⓐ Students should have observed that the spectra produced by two prisms recombine to produce white light.

Ⓑ Accept all reasonable predictions at this point without giving away the answer.

Meeting Individual Needs

Second-Language Learners

Have students copy the drawing of three intersecting circles below. Have them use colored pencils to represent the primary and secondary colors obtained by mixing colored light. Reinforce the idea that the colors obtained by mixing light are not the same as those obtained by mixing paint. They will learn more about mixing paint in Lesson 4 of Chapter 20.

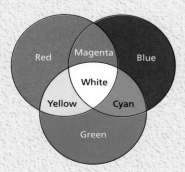

Color-Conscious Calculations

If you were to shine three flashlights covered with red, green, and blue plastic, respectively, on a white surface in a dark room, you would see the same results as if you carried out the demonstration on page 437.

With the help of the diagram at right and the results of the colored-light activity, you can do a little color math! What would you replace the question marks with in the following statements?

1. Red light + Blue light = ? light
2. R + ? = Y
3. B + G = ?
4. R + G + B = ?

Here are a couple of trickier equations:

5. Y + B = ?
6. Cyan + ? = White

Red, green, and blue are three special colors of light because they combine to make *white* light. They are called **primary colors of light**. When you mix any two primary colors of light together, you get **secondary colors of light**. How many secondary colors of light must there be? What are they? **A**

In equations 5 and 6, did you observe that when a particular secondary color of light and a particular primary color of light are mixed, white light is produced? Pairs of colored lights that produce white light are called **complementary colors of light**. What is another pair of complementary colors of light other than those in 5 and 6? **B**

Do you need all the colors (ROYGBIV) of the spectrum to produce white light? Explain. **C**

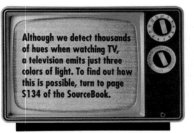

Although we detect thousands of hues when watching TV, a television emits just three colors of light. To find out how this is possible, turn to page S134 of the SourceBook.

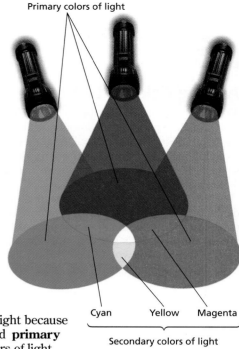

Primary colors of light

Cyan   Yellow   Magenta

Secondary colors of light

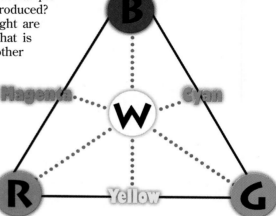

What information can you get from this light triangle? **D**

Did You Know...

Essentially the same process of color mixing by addition that is demonstrated in this lesson is used by television sets. On a color television screen, red, green, and blue light are emitted by dots that are energized by electrons. The red, green, and blue dots mix to produce different colors.

Integrating the Sciences

Earth and Physical Sciences

Have students review the photograph and the introduction to Unit 6 on pages 356 and 357. Remind them that the light given off by the exploding firecrackers is evidence of the energy changes involved in chemical reactions. Ask students to find out what chemicals or elements are added to fireworks to produce certain colors. *(For example, some fireworks contain barium compounds that produce blue or green light.)*

Filter Fun

You have just been examining *additive* light phenomena. What do you think this term means? Review the results of Experiments 2, 3, and 4 of Exploration 3, in which you made use of filters. Are these experiments also additive light activities? Do filters "add" something to white light, or do they "subtract" from it? For example, when white light shines through a red filter, what color of light does the filter let through? Some colors apparently do not go through the filter; that is, they are subtracted from the white light.

Here is some "filter fun" you can do as a class experiment with an overhead projector. Place a colored filter or combination of filters on the projector and observe the light on the projection screen. Then deduce which colors are transmitted through the filter(s) and which are subtracted from the white light by the filter(s). In your ScienceLog, write what you think is happening in each case.

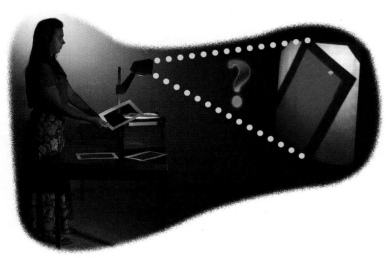

You should have a selection of colored filters: red, blue, green, yellow, cyan, magenta, and others. Below and at right are some suggestions. Add them to your own arrangements.

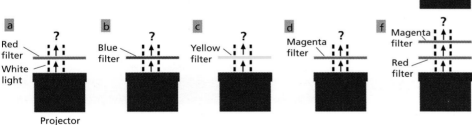

439

Homework

The questions in Color-Conscious Calculations on page 438 make an excellent homework activity.

FOLLOW-UP

Reteaching

Have students write an imaginary letter to a second-grade student explaining the main points of this lesson. Encourage students to illustrate their letter with diagrams.

Assessment

Have students write a paragraph explaining how a stage-lighting director can use three colors of spotlights to produce any color of light on stage. Students should include an illustration with their paragraph.

Extension

Have students research the phenomenon of colorblindness. Students can perform library research or conduct a survey of neighbors and friends. Students should find out what primary colors colorblind people cannot see and, consequently, what secondary colors they also cannot see.

Closure

Have students draw a concept map using the following terms: white light, colors of light, primary colors, secondary colors, red, yellow, green, blue, magenta, cyan. Then have students construct a color-coded mobile of their concept maps.

Filter Fun

You may wish to place the various filters on the projector yourself to help guide class discussion. Before you place the filters on the projector, have students predict what the result will be for each different combination. To demonstrate the difference between additive and subtractive color phenomena, you may wish to set up a second projector focused on the same area of the screen as the first projector. Two filters on separate projectors will produce a different color on the screen than they will when placed on the same projector.

Answers to *Filter Fun*

The use of filters is not an additive light activity because filters subtract colors of light from the white light that shines through them. Students should recognize that in each case, the filters will allow only the light energy of its specific color to pass through.

Students should demonstrate an understanding that good colored filters subtract all colors of light except their own.

CROSS-DISCIPLINARY FOCUS

Language Arts

Challenge students to find synonyms for the terms *additive* and *subtractive* as they are used in the context of color mixing. *(Answers will vary but should mention that additive implies the emission of light while subtractive implies the absorbtion of light. Additive color mixing could also be called color mixing by emission of light, while subtractive color mixing could be called color mixing by absorbtion of light.)*

Answers to *Challenge Your Thinking*

1. Students' answers will vary. Some possible responses are as follows:
- Open the curtains, and let the lightning light up your room.
  Energy change: electrical energy → light energy + heat energy
- Light a candle or kerosene lamp.
  Energy change: chemical energy → light energy + heat energy
- Generate electricity to power the electric lights by turning a hand generator.
  Energy change: mechanical energy → electrical energy → light energy
- Turn on a flashlight.
  Energy change: chemical energy → electrical energy

2. Because the light from an ordinary light bulb has less energy (is less hot) than the light from a flash-bulb, there is more red and orange light in it than blue or violet light. This gives photos a reddish tint. A flash gives off higher-energy light, which has more blue and violet light. A blue filter placed over the camera lens reduces the amount of red light that passes through. When the blue filter removes (absorbs) some of the excess red light, the camera picks up light with a more even amount of red and blue.

3. a. R + B + G = W
   b. B + G = C
   c. R + B = M
   d. C + M = (W + B) = B
   e. Y + M = (W + R) = R
   f. M + G = W
   g. M + C + Y = W

 You may wish to provide students with the Chapter 19 Review Worksheet that is available to accompany this Challenge Your Thinking (Teaching Resources, page 20).

Homework

You may wish to assign the Activity Worksheet that accompanies this Challenge Your Thinking as homework (Teaching Resources, page 18). If you choose to do this activity in class, Transparency 68 is available for your use.

CHALLENGE YOUR THINKING

1. Lights Out!

It is nighttime and a storm has caused the electrical power in your house to fail. How many different ways can you find to light your room so that you can see your way around? For each way, identify the energy change involved.

2. Photo Facts

Photographers observe that when pictures are taken with the light from ordinary incandescent light bulbs, the developed photos have a reddish tint. Why? If a flash is used, the color is better. Why? Another way to improve the color of the photo is to place a blue filter over the camera lens. How does the blue filter help?

Photo taken without flash

Photo taken with flash

3. Color "Math"

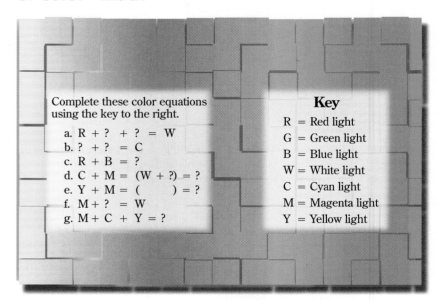

Complete these color equations using the key to the right.

a. R + ? + ? = W
b. ? + ? = C
c. R + B = ?
d. C + M = (W + ?) = ?
e. Y + M = (   ) = ?
f. M + ? = W
g. M + C + Y = ?

Key

R = Red light
G = Green light
B = Blue light
W = White light
C = Cyan light
M = Magenta light
Y = Yellow light

Meeting Individual Needs

Learners Having Difficulty

Students may need extra time to determine the results of mixing secondary colors of light in questions 3(d)–3(f) and question 4 of this Challenge Your Thinking. It may help to remind students that each secondary color contains two primary colors and that mixing all three primary colors produces white light. In question 3, have students rewrite the left side of the equations in terms of primary colors before solving the equations. You may need to point out that adding white light to a primary or secondary color of light does not change that color. In question 4, students are asked to predict all of the possible outcomes of mixing secondary colors. You may wish to discuss the sample answer on page 441 of this Annotated Teacher's Edition with students who are still unclear about the additive properties of light. Students may wish to review the material on page 438 of the Pupil's Edition to help them answer this question.

4. Color Commentary

Michelle asked, "What colors do you get when you mix different secondary colors of light together?" What do you predict? What color might each question mark represent in the illustration at right? Explain your prediction. You might test your prediction in an experiment.

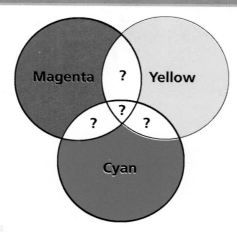

5. A Light Lunch

Millie says, "Plants change *light* into food. Therefore, light must be matter." Do you agree or disagree? Support your answer with a good argument.

6. Color Combo Quiz

Are you a color expert? Consider the illustrations shown below. Each one involves light of various colors. It's up to you to identify the colors you would expect to see at each labeled point, *A–F*. Remember to write in your ScienceLog, not in this book.

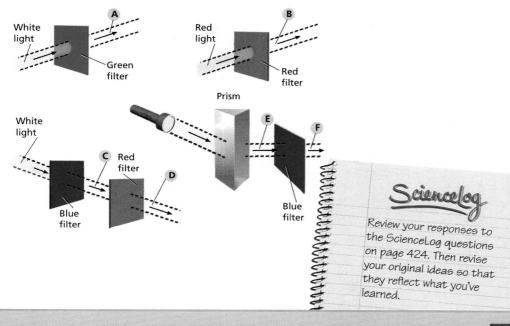

Review your responses to the ScienceLog questions on page 424. Then revise your original ideas so that they reflect what you've learned.

441

The following are sample revised answers:

1. The color of light that a star emits depends on its temperature. The order of colors produced, from lowest temperature to highest temperature, is the following: red, orange, yellow, green, blue, indigo, and violet. Red stars, therefore, are low-temperature stars. White stars are high-temperature stars that combine all of the high-energy colors with all of the low-energy colors. Bluish white stars are even hotter and are dominated by high-energy blue light.

2. Rainbows are formed when water droplets in the atmosphere act like prisms, separating the sun's white light into the colors of the spectrum.

3. Light cannot be classified as matter because it has no mass and takes up no space. However, light does produce energy such as heat and can itself be produced from other forms of energy. Light is therefore described as energy.

Answers to Challenge Your Thinking, continued

4. Magenta light mixed with yellow light will produce red light; yellow light mixed with cyan light will produce green light; cyan light mixed with magenta light will produce blue light; and magenta, yellow, and cyan light mixed together will produce white light. In the diagram, the question marks represent red (top), white (center), green (right), and blue (left). Students' explanations may vary but should include the following points:
   - Each secondary color of light comes from a combination of two of the three primary colors of light: magenta contains red and blue light; cyan contains blue and green light; and yellow contains red and green light.
   - Two secondary colors always have one primary color in common. Also, each secondary color in a pair contains the primary color that complements the other secondary color. (A complementary pair of colors contains all three primary colors and combines to form white light.)
   - Mixing two secondary colors of light will result in the primary color of light that the two have in common because there is twice as much of this color and because the other primary colors of light contained in the mixture produce white light.
   - Mixing all three secondary colors of light will produce white light because it combines equal parts of all three primary colors of light.

5. Disagree. Sample argument: Light can't be matter because it has neither mass nor volume. Plants do not actually change light into food. Instead, light provides plants with the energy that they need to change carbon dioxide and water into food. This process converts light energy into chemical energy and demonstrates that light is just one form of energy.

6. A. Green
   B. Red
   C. Blue
   D. No light
   E. All (spectrum)
   F. Blue

Connecting to Other Chapters

Chapter 19
introduces light as a form of energy, the relationship among light, heat, and color, and the theory of color mixing.

Chapter 20
explores diffuse and specular reflection and the scattering, transmission, and absorption of light.

Chapter 21
analyzes lenses and mirrors, their uses, and how they produce real and virtual images through the refraction of light.

Prior Knowledge and Misconceptions

Your students' responses to the ScienceLog questions on this page will reveal the kind of information—and misinformation—they bring to this chapter. Use what you find out about your students' knowledge to choose which chapter concepts and activities to emphasize in your teaching. After students complete the material in this chapter, they will be asked to revise their answers based on what they have learned. Sample revised answers can be found on page 461.

In addition to having students answer the questions on this page, you may wish to have them complete the following activity: Have students make a labeled drawing showing the process by which a person sees an object. The drawing should feature the following elements: the sun (the light source), a tree (the object perceived), and the eyes of a viewer. Although some students will understand that light from the sun is reflected off the tree and into the viewer's eyes, diagrams may show that many students do not understand this idea. Alternatively, you may wish to have students produce a labeled diagram that shows how color is perceived.

1 **Why is the sky blue?**

2 **How do we see objects that do not produce light?**

3 **What determines an object's color?**

ScienceLog

Think about these questions for a moment, and answer them in your ScienceLog. When you've finished this chapter, you'll have the opportunity to revise your answers based on what you've learned.

442

This drawing should include a light bulb (the light source), a red rose, and the eyes of a viewer. Few students are likely to understand that the rose absorbs all of the colors of white light except red, which is reflected into the eyes of the viewer. Collect the drawings, but do not grade them. Instead, use the drawings to identify possible problem areas in the chapter and to find out what students know about the behavior of light, what misconceptions students may have, and what aspects of this topic are interesting to them.

ENVIRONMENTAL FOCUS

The whitish gray haze in large cities is caused by *particulates* in the air that are mainly the result of combustion of fossil fuels. An unpolluted sky is blue because the air molecules scatter blue light. In polluted skies, however, particulates scatter other colors of light as well, so that the sky is white rather than blue. The gray color is caused by particulates that absorb rather than scatter light. Have students design posters about air pollution to identify the types of particulates associated with haze in large cities.

LESSON 1 · Light in Action

Perhaps you've heard the words *scattering, transmission,* and *absorption* before. These terms are commonly used by scientists when describing light. What do you think these words mean? Read the following statements, which use forms of these words. Rewrite the sentences and replace the words in boldface type with your own words to convey what you think these special light words mean. Ⓐ

- When a light beam shines on a mirror, it bounces off in a specific direction. However, when a light beam hits a white piece of paper or the white wall of a room, the light **scatters**.
- When a white light shines on a blue piece of paper, the blue color in the white light is **scattered** by the paper, while all the other colors are **absorbed**.
- When white light shines on a red filter, only the red color is **transmitted** through the filter. All the other colors are **absorbed** by the filter.
- When a beam of white light shines through smoky air, part of the light is **transmitted** through the smoke. The rest of the light is **scattered** by the smoke particles. Look at the photo at right.

A Light Box

By performing experiments with a *light box,* you can get a better understanding of scattering, transmission, and absorption. A light box shuts out most of the unwanted light in a room, making it easier for you to see the light used in the experiments. Follow the instructions on page 444 to build your own light box.

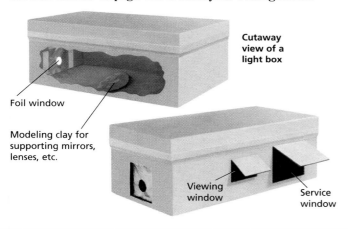

Cutaway view of a light box

Foil window

Modeling clay for supporting mirrors, lenses, etc.

Viewing window

Service window

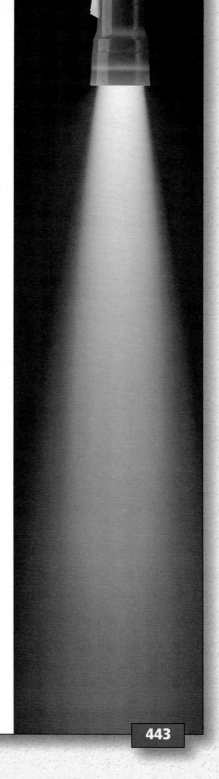

LESSON 1 ORGANIZER

Time Required 4 class periods

Theme Connection Structures

Process Skills
observing, analyzing, hypothesizing, predicting, communicating

New Terms
Absorption—the intake of light without reflection or transmission
Scattering—the reflection of light in random directions
Translucent—term used for a material that allows most light to pass through while scattering the rest so that objects

are not clearly visible through the material
Transmission—the passing of light through an object
Transparent—a term used for a material that allows light to pass through without being scattered

Materials (per student group)
Building a Light Box: large shoe box or other cardboard box, about 35 cm × 15 cm × 10 cm; metric ruler; scissors; piece of aluminum foil, about 8 cm × 8 cm; roll of masking tape

LESSON 1 · Light in Action

FOCUS

Getting Started
Display a hand mirror and use it to reflect some light around the classroom. Ask students to describe what is happening. *(Light is reflected from the mirror.)* Then display some cardboard and a piece of glass. Ask students to describe what happens when light strikes these materials. *(Light that strikes glass passes through it; light that strikes cardboard does not.)*

Main Ideas
1. When light strikes an object, it may be reflected, transmitted, scattered, or absorbed.
2. A dark-colored surface absorbs more light than a light-colored surface, while a light-colored surface reflects more light than a dark-colored surface.
3. A transparent object transmits most of the light that strikes it, while a translucent object scatters most of the light that strikes it.
4. Objects can be seen clearly through a transparent material but cannot be seen clearly through a translucent material.

TEACHING STRATEGIES

Have students silently read the bulleted statements on this page. Then have them substitute their own words and phrases for the words *scattered, absorbed,* and *transmitted.* Make sure that students understand that scattering is the reflection of light in all directions.

Answers to
In-Text Questions

Ⓐ Possible responses are as follows:
- reflects in many directions
- reflected; retained by the paper
- allowed to pass through; retained
- able to pass through; forced to bounce off

continued ▶

443

Building a Light Box

Construction of the light boxes can be done by groups of students either as a classroom activity or as an at-home project. Check the completed boxes to make sure that they meet the required specifications. The holes in the boxes should be small enough so that the boxes will be effective in a well-lit room.

You may wish to consider several different light sources for the boxes. Normally, a flashlight will do, but a filmstrip, slide, or overhead projector can provide a more concentrated beam of light. Bright sunlight is an even better option. Mirrors may be used to focus the sunlight where it is needed.

The piece of paraffin wax for Part 3 should be about the same thickness as the window glass used in Part 2. You may need to shave the piece of wax to match the thickness of the glass, a process which requires a sharp knife and may take up to 20 minutes per piece. After each experiment is completed, encourage students to write a brief statement in their ScienceLog describing what they have observed and concluded.

Cooperative Learning
EXPLORATION 1

Group size: 3 to 4 students
Group goal: to apply knowledge of light to further investigate various properties of light
Positive interdependence: Divide the class into five groups. Assign each group one of the five parts of Exploration 1. Each group should develop a creative way to review the findings of their part of the Exploration. Also, they should develop a short quiz to test the class for understanding of their part. Give each group time to collect and evaluate the quizzes as well as time to reteach their part to the class if necessary.
Individual accountability: Each student should be able to answer the questions found in Interpreting Your Experiments.

⭐ An Exploration Worksheet is available to accompany Exploration 1 (Teaching Resources, page 25).

Building a Light Box

Here is how to make a light box from simple, readily available materials. You will use this light box for many experiments in this unit.

Start with an empty cardboard box about 35 to 45 cm long, 15 to 20 cm wide, and 10 to 15 cm high. A large shoe box will work.

On one end, near the bottom, cut out a square window about 6 cm on each side. Then cut a square of aluminum foil about 8 cm on each side. About 3 cm from the bottom of the foil square, centered between the left and right edges, cut a round hole about 1.5 to 2.0 cm in diameter. Tape the square of foil over the window at the end of the box, as shown in the illustration on page 443. Be sure that you can see through the hole into the box's interior.

Cut two flaps in one side of the box—one near the center, the other one lower and to the right, at the end of the box away from the foil window. The flap near the center is for looking into the box. It should be about 6 cm × 4 cm. The flap to the right is for putting things into or removing things from the box. It should be about 8 cm on each side. You may find it easier to move objects into and out of the box by simply removing the lid. Be sure to replace the lid before you do any experiment.

EXPLORATION 1

Enlightening Experiences

You Will Need

- a light box
- a flashlight
- rough, black paper
- modeling clay
- white paper
- colored paper
- a wooden splint
- matches
- a piece of window glass
- a protractor
- a piece of paraffin wax
- a small beaker of water
- milk
- an eyedropper
- a stirring rod

PART 1

A Black-Surface Experiment

Set up the light box that you made. Position the flashlight so that its light shines through the foil window. Look through the viewing window.

What to Do

1. There is now some light in the box. Where does most of the light come from? Is it from the beam as it passes through the air or from light scattered when the beam hits the end of the box?

2. Put a piece of rough, black paper in the path of the beam near the end of the box. Prop it up with modeling clay, as shown below. What happens to the brightness of light in the box when the black surface is added? What happens to most of the light when it strikes the black surface?

Black paper

ORGANIZER, continued

Exploration 1, Part 1: light box from Building a Light Box; small flashlight; piece of rough black paper, about 8 cm × 8 cm; sheet of white paper, about 8 cm × 8 cm; a few sheets of other colored paper, about 8 cm × 8 cm; wooden splint; a few matches; small ball of modeling clay; safety goggles; **Part 2:** light box from Building a Light Box; small flashlight; piece of window glass, about 8 cm × 8 cm; small ball of modeling clay; wooden splint; a few matches; protractor; safety goggles (additional teacher materials: 50 cm of masking tape; see Advance Preparation on page 421C); **Part 3:** light box from

Building a Light Box; small flashlight; flat piece of paraffin wax, about 8 cm × 8 cm; small ball of modeling clay; wooden splint; a few matches; protractor; safety goggles (additional teacher materials: knife; see Advance Preparation on page 421C.); **Part 4:** light box from Building a Light Box; small flashlight; small beaker of water; a few drops of milk; eyedropper; stirring rod; **Part 5:** large beaker or jar of water; small flashlight; a few drops of milk; eyedropper; stirring rod

Teaching Resources
Exploration Worksheet, p. 25

3. Replace the black paper with a sheet of white paper. What does this do to the amount of light in the box? Which color of paper—black or white—absorbs more light? Which color reflects more light? Try other colors of paper as well.

4. Add some smoke from a smoldering wooden splint.

Caution: Beware of fire.

The particles of smoke scatter some of the light in all directions, allowing you to see the path it takes. Be careful not to use too much smoke. You must be able to see through the smoky air.

5. You should now be able to complete the following statements. Write them in your ScienceLog, and fill in the answers.

a. As a beam of light passes through the air, it (lights up, does not light up) the box.

b. When the beam of light hits the end of the box, it (lights up, does not light up) the box.

c. When the beam of light hits the black screen, the box is lit up (less than, more than, to the same extent as) in step (b).

d. White light falling on a black surface is absorbed (less than, more than, to the same extent as) when it falls on a white surface.

e. White light falling on a black surface is scattered (less than, more than, to the same extent as) it is scattered by a white surface.

f. A colored surface scatters (more light than, less light than, the same amount of light as) a white surface and (more light than, less light than, the same amount of light as) a black surface.

g. A green surface scatters (what color?) light from its surface. Therefore, it must absorb (what colors?) of light.

h. A black surface absorbs (what colors?) of light.

PART 2

Doing Windows

Replace the black paper that is in the path of the light beam with a square of window glass. Place the glass near the center of the box, just behind the viewing window. Add smoke so that you can see the light beam. Place the glass at various angles and observe the results.

What to Do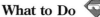

1. Place the glass at a 45° angle to the light beam, and then answer the following questions:

a. Does the beam of light scatter as it hits the glass, or does it bounce off (reflect) in a certain direction?

b. Can you see light pass through the glass? In other words, can you see any transmitted light?

c. Which seems brighter—the reflected beam or the transmitted beam? Why?

2. Repeat step 1 again, first placing the glass at a steep angle such as 80° (measured with respect to the horizontal) and then at a shallow angle such as 10°. For each angle, answer the questions in step 1.

3. Now write several statements like those you completed after Part 1 of this Exploration. They should describe what you have learned by doing this experiment.

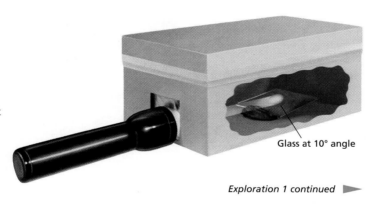

Glass at 10° angle

Exploration 1 continued ▶

Answers to
Part 2

1. a. Most of the light bounces off the glass.
   b. Some of the light is transmitted.
   c. The reflected beam is brighter because more light is reflected than transmitted.

2. With the glass at 80°, some of the light bounces off the glass, but most is transmitted. The transmitted beam is brighter because most of the light is transmitted. With the glass at 10°, most of the light bounces off the glass, but some is transmitted. The reflected beam is brighter because most of the light is reflected.

3. Students' statements will vary but should reflect the following ideas:
   • Glass transmits light.
   • When light strikes glass, some of the light is reflected from the surface of the glass.
   • When a beam of light strikes a piece of glass that is nearly parallel to the path of the beam, most of the light is reflected rather than transmitted. When a beam of light strikes a piece of glass that is nearly perpendicular to the path of the beam, most of the light is transmitted rather than reflected.

Answers to
Part 1, pages 444–445

1. Most of the light in the box comes from the light that is scattered when the beam hits the end of the box.

2. The inside of the box becomes darker because most of the light that strikes the black surface is absorbed by it.

3. The inside of the box becomes brighter. Black paper absorbs more light than white paper. White paper reflects more light than black paper or brown cardboard. As students try different colors of paper, they should observe that the darker colors absorb more light and that the lighter colors reflect more light.

5. a. does not light up
   b. lights up
   c. less than
   d. more than
   e. less than
   f. less light than; more light than
   g. green; all colors except green
   h. all colors

Answers to
Part 3

2. The paraffin appears cloudy in the beam of light while the glass appears clear. The light passes through the glass at a fixed point, while the light seems to be scattered, or diffused, throughout the paraffin. The box should appear brighter when the glass is in place.

3. Students' summaries will vary but should reflect the following ideas:
- When you shine light at glass, the light passes through.
- When you shine light at paraffin, most of the light scatters in all directions within the paraffin. Only a little of the light passes through.
- You can see objects clearly through glass but not through paraffin.

Answers to
Part 4

1. Students should not be able to see the beam of light through the water.

2. When the milk is added to the water, the beam of light becomes visible because the milk particles are reflecting the light. Students should observe that some of the light is transmitted through the slightly cloudy water. The whitish color is evidence that light is scattered by cloudy water.

3. Students' conclusions will vary but should reflect the following ideas:
- Some of the light that strikes slightly cloudy water is *transmitted*.
- Some of the light that strikes slightly cloudy water is *scattered*.

Homework

Part 5 of Exploration 1 makes an excellent homework activity.

PART 3

Wax Facts

What to Do

1. Replace the glass with a piece of paraffin wax. Hold the wax at various angles in the light beam. Be sure to use smoke to help you see the beam.

2. How does the appearance of the paraffin compare with that of the glass when each is in the beam? Is the box brighter when the paraffin is in place or when the glass is present?

3. Record your findings in a series of simple statements.

PART 4

Light and Water

What to Do

1. Put a small beaker nearly filled with water in the light box. Turn on the flashlight and shine a light beam through the water. Can you see the beam in the water?

2. Add a drop or two of milk to the water and stir. Is the beam now visible in the water? What are the milk particles doing to the light? Is there evidence that light is transmitted through the slightly cloudy water? Is there evidence that light is scattered by the cloudy water?

3. Write a conclusion for this experiment using the words *transmitted* and *scattered*.

PART 5

Another Angle

What to Do

1. Take a large beaker or jar of tap water, and hold a small flashlight against one side of it. See the illustration below. Look at the light from the opposite side of the container as well as at right angles to the light beam. You can obtain the best results in a darkened room. What color is the light? What color, if any, is the water when viewed at right angles to the beam?

2. Add a few drops of milk to the water and stir. What color does the light seem to be now? What color, if any, is the water?

3. Repeat step 2, adding more milk until faint color effects are observed.

4. Describe your results in this experiment.

Beaker filled with water

Water with a few droplets of milk

Answers to
Part 5

1. The light will appear to be white. The water has no color when viewed at right angles to the beam.

2. The light beam becomes visible as a whitish color. The water should still appear to be colorless or faintly cloudy.

4. Students' descriptions will vary but should reflect the following ideas:

- The milk particles cause the light in the water to scatter.
- Seen from the front (or opposite the flashlight), the light beam becomes orange-red in color as it passes through the water and strikes the milk particles.
- Seen from the side, the light beam becomes blue-white in color as the light is scattered by the milk particles in the water.

Interpreting Your Experiments

Recall the experiments you did in Exploration 1 of this chapter. Compare your concluding statements for each part of the Exploration with the explanations below.

Part 1

The inside of the light box was darker when the black paper was used than when the white paper was used. The reason for this has to do with how the light entering the light box interacts with each piece of paper. The black surface absorbs most of the light, turning this light energy into heat. The white surface, on the other hand, reflects most of the light and its energy. This is something to think about the next time you dress for a hot summer day. Should you wear light or dark colors? Why?

White surfaces reflect most of the light that strikes them.

Part 2

Glass is said to be **transparent**. How would you define this term? You might say that light can pass through a transparent material and that you can see objects on the other side of this material. However, not all of the light that strikes the glass passes through. Instead, a small amount of light is reflected at the surface of the glass. The amount of reflected light depends on the angle at which the light meets the glass. For example, when light strikes glass at a small angle such as 10°, more light is reflected than if the angle were increased to 70° or 80°. Taking this into consideration, how would you revise your definition of transparent?

How does the surface of a still pond or puddle resemble the window glass in Part 2 of Exploration 1? Ⓐ

Now consider this situation. Melissa sits in front of her bedroom window, and she can see the shrubs and trees outside. At the same time she can also see her own reflection. Why?

Under what conditions would Melissa be able to see her own reflection but not be able to see outside? Under what conditions would Melissa be able to see outside but not see her reflection?

Melissa is seeing reflected and transmitted images at the same time.

447

Answer to Caption

Ⓐ Students should recognize that the surface of a still pond transmits and reflects light in the same way that a window does. Depending on the size of the angle created by the line of sight and the surface of the pond, a person will be able to see through the surface of the water, to see things reflected by it, or both.

Interpreting Your Experiments

This material consolidates the results of the experiments that students completed in Exploration 1 and provides some additional ideas for them to consider.

Answer to Part 1

Students should conclude that light-colored clothing would be cooler and more comfortable than dark-colored clothing because the dark colors absorb more light energy, which would then be converted into heat energy. In contrast, light-colored clothing would reflect much of this light energy.

Answers to Part 2

Students should convey the idea that transparent materials allow light rays that are perpendicular to their surface to pass through. Sample answer: Melissa can see the shrubs and trees outside because the light reflecting off the shrubs and trees is transmitted by the window glass. Sunlight that is transmitted by the glass, in addition to light from inside her bedroom, reflects off Melissa. Some of this light reflecting off Melissa is reflected by the glass and allows Melissa to see herself in the window. Melissa would be able to see her reflection but not see outside if her bedroom was lighted but it was dark outside. Melissa would be able to see outside but not see her reflection if it was light outside but no light was reflecting off her.

Multicultural Extension

Scattering
In 1930, Indian physicist Sir Chandrasekhara Venkata Raman (CHUHN druh SHAY kuhr uh VEHNG kuh tuh RAH muhn) was awarded the Nobel Prize in physics for discovering that a beam of light scatters and changes frequency when it passes through a liquid or gas. Ask students what impact this might have had on the scientific community. *(The way that light scatters and the different frequencies produced are determined by the molecules that the light passes through. By studying the color of scattered light, scientists can learn more about the substance it passes through.)*

Answer to
Part 3

Some possible translucent materials include frosted glass, wax paper, tissue paper, ice cubes, and dense fog.

Answer to
Parts 4 and 5

The light beam became visible.

Parts 4 and 5

Encourage students to review their results in Part 4 of Exploration 1. Then have them compare their observations with the illustration at the bottom of this page. If students indicate that their results differed from those in the illustration, help them to identify the variables that may account for the difference. For example, if tap water was used or if there were tiny air bubbles in the water, students may have been able to see the beam of light before any milk was added to the water. This problem could be corrected by using distilled water or by letting the water sit in its container for a few minutes and then checking it with a beam of light.

Have students review their results for Part 5 of Exploration 1. Help them recognize that the color of the light beam and the color of the cloudy water are the result of different colors of light being scattered by the particles of milk in the water. To check their understanding, ask: What color would the light beam be if a little cornstarch were added to the water instead of milk? What color would the cloudy water be? *(Students may recognize that the results would be the same whether milk or cornstarch is used.)*

Encourage students to brainstorm ideas in response to the questions on page 449. Accept all reasonable suggestions. The answers will become apparent as students continue reading. Have students write their ideas in their ScienceLog to review and revise as they continue working through the lesson.

Part 3

How is the block of paraffin different from either the glass or the white paper? When you shine a light beam at glass, most of the beam passes through. When you use white paper, the light bounces off in all directions. But when you use paraffin, something different happens. Much of the light is scattered in all directions *within* the block itself, while a little of the light passes through.

Paraffin is **translucent**. It allows some light to pass through it, but scatters most of the light within it. You cannot see things clearly through a translucent material. Can you think of other materials that are translucent?

Glass and paraffin wax are positioned between you and a light source. Observe how only some of the light passes through the paraffin wax.

Parts 4 and 5

The experiments with the milk and water illustrate something else about how light behaves. In Part 4, the presence of a few milk particles in water caused light to scatter. What effect did this have on the light beam when viewed from a position at right angles to it?

When you added a little more milk in Part 5, you observed something different. Looking directly at the beam of light through the slightly milky water, you saw faint colors. The beam, viewed head-on, looked faintly yellowish red, but when viewed from the side, the surrounding water looked faintly blue. Why?

The reason is that the milk particles scatter mostly blue, violet, and indigo light. This is why the water around the beam looks bluish. The light coming out of the water looks yellowish red, however, because the beam's red, orange, and yellow light are not scattered.

The light that scatters from a beaker of milky water is of different colors. What color you see depends on your position around the beaker.

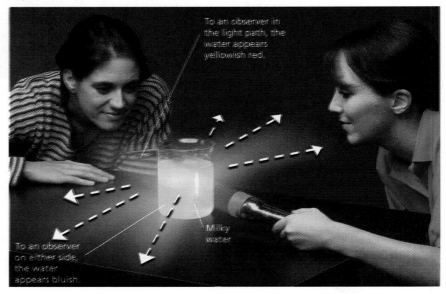

To an observer in the light path, the water appears yellowish red.

To an observer on either side, the water appears bluish.

Milky water

Theme Connection

Structures

Imagine a planet with an atmosphere containing molecules that scatter red light better than blue light. **Focus question:** What would the sky look like on such a planet? What would the sunset look like? *(The sky would appear red, and the sunset would be blue-violet.)*

Homework

Ask students to bring three objects to class: one transparent object, one translucent object, and one opaque object. Set up a display in class.

The results of these experiments will help answer some interesting questions. Examine the images at right. Can you provide answers to the questions accompanying them? Write your answers in your ScienceLog.

Why is space black? Ⓐ

Why is a sunset red? Ⓑ

Why does the ocean look blue? Ⓒ

Reteaching

Have students make posters with captions and illustrations to represent what they discovered in each part of Exploration 1.

Assessment

Point out to students that they encountered many new words in this lesson. Suggest that they compose some definitions for these terms and include them in a Dictionary of Light. Some of the terms from this lesson include the following: scattering, transmission, absorption, translucent, and transparent. Suggest that they include illustrations. Display the completed dictionaries for others to read.

Extension

Ask students to investigate the northern lights, or the aurora borealis. Students should find out what this phenomenon is, where it occurs, and how it is produced. *(The sun is constantly ejecting a stream of charged particles from its surface that are captured in the magnetic field, or Van Allen belts, surrounding the Earth. The magnetic field pulls these charged particles down to the poles of the Earth. While doing so, the particles strike and excite electrons in the upper atmosphere, which emit light energy in the process. We call this light an aurora.)*

Closure

Have students ask friends or family members each of the questions on page 449. Students should write a summary of their findings and provide corrections to any incorrect survey responses.

Answers to
Captions

Ⓐ Beyond the atmosphere, space is black because there are no air molecules to scatter light.

Ⓑ The reddish color of a sunset is also caused by light scattering. When the sun is low in the sky, its light shines through a much thicker layer of atmosphere than when it is high in the sky. As the light travels through this thicker layer, the colors at the blue end of the spectrum are scattered out of the light beam, while the colors at the red end of the spectrum continue on a straight path. So when we look directly at a sunset, we see more red light than blue light.

Ⓒ The ocean looks blue primarily because the ocean reflects the blue color of the sky to our eyes, just as the colors of trees and mountains are reflected by the surface of a lake.

Homework

You may wish to have students find a few photographs of the aurora borealis as a homework activity.

LESSON

2
Light's Path

FOCUS

Getting Started

Discuss the term *light ray* with your students. Explain that a light ray is a symbol that represents the path taken by a group of light waves. The arrow on a light ray shows the direction in which the light waves move. Draw several luminous objects on the blackboard, such as a candle, the sun, or a lamp, and draw the rays emanating from them to show how light rays represent the path of light.

Main Ideas

1. Light travels very quickly.
2. Light travels in straight lines.
3. Shadows are evidence that light travels in straight lines.

TEACHING STRATEGIES

Answers to
In-Text Questions

Ⓐ Students probably noticed that the edges of the beam were fairly well defined and that the beam's direction was in a straight line. When the smoke disappears, however, the path of the light beam is no longer visible, even though the light is still present. Students may wish to include their analysis of Exploration 1 as evidence to support their answer.

Ⓑ Answers will vary, but students should realize that they would see a relatively sharp shadow of the comb on the screen. As the comb moves out of the path of the light beam, the shadow becomes fuzzier.

Ⓒ Since the time lag between the motion of the comb and the motion of its shadow is undetectable, students should conclude that light moves quickly. The fact that the teeth of the comb are visible in the shadow suggests that light travels in straight lines.

Ⓓ The shadow on the top screen is larger but slightly fuzzier than the shadow on the bottom screen.

Through your light-box experiments you were able to determine some characteristics about light. For example, you observed that light traveling through space is invisible, but you can see its path through air if smoke is present or through water if a bit of milk is added.

Now take a look at the diagram on page 444. Think about the path of the light beam you studied in Exploration 1. What did you observe about the edges of the beam? What was the beam's direction? When you observe a light beam passing through smoky air, what happens when the smoke completely disappears? Is the light still there? List some evidence in your ScienceLog to support your answer. Ⓐ

Examine the diagram below. What would you expect to observe on the white cardboard screen? Try setting up this experiment. Observe Ⓑ the screen. Then move the comb into and out of the path of the light beam, watching the screen as you do so.

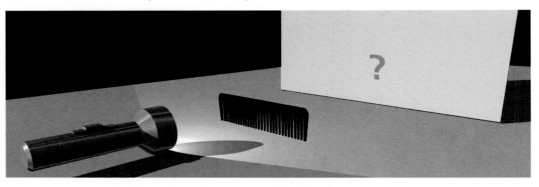

Based on the comb's shadow on the screen and the observations you made during the light-box experiments, what can you say about the way light travels? Does it move slowly or quickly? What evidence suggests that light moves in straight lines? Ⓒ

Light Lines

Study the shadows of a ball illuminated by two different-sized light sources. Examine the screens closely. How are the shadows different? Ⓓ

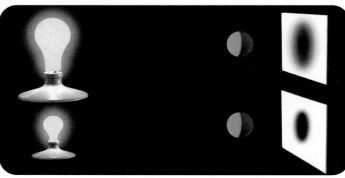

LESSON 2 ORGANIZER

Time Required
2 class periods

Process Skills
observing, analyzing, inferring, contrasting, measuring

New Terms
none

Materials (per student group)
Light's Path: flashlight; sheet of white cardboard; plastic comb

Exploration 2: several index cards with pinholes of different sizes; candle; small ball of modeling clay; small jar lid; a few matches; metric ruler; 2 clothespins; lined white index card; safety goggles

Teaching Resources
Exploration Worksheet, p. 30
SourceBook, p. S132

Sketch the light-bulb diagrams in your ScienceLog. Using a ruler, draw two straight lines from the top of the large light bulb—one to the top of the ball casting the shadow, the other to the bottom of the ball. Now draw two lines from the bottom of the large light bulb—one line to the top of the ball, the other to the bottom. Draw similar lines on the diagram with the small light bulb. Do the lines suggest an explanation for the two types of shadows cast on the screens? Write your explanation in your ScienceLog. Look again at the diagrams in your ScienceLog. Would shadows occur if light did not travel in straight lines? **E**

In the next Exploration you will further investigate "light lines" by examining light as it passes through a pinhole. As you perform the experiment, look for evidence that light travels in straight lines.

EXPLORATION 2

Pinhole Images

You Will Need

- a lined white index card
- a metric ruler
- 2 clothespins
- a candle
- modeling clay
- matches
- several index cards with pinholes of different sizes
- a small jar lid

What to Do

1. Perform the experiment in a darkened room. Arrange the apparatus as shown in the diagram.

2. Start with the pinhole about 3 cm from the candle flame and the screen 3 cm from the pinhole. You should get a good image of the flickering flame. Describe the image.

Can you explain its appearance? (Hint: Make a drawing of the setup as shown below. Then draw thin "lines of light" from the top and bottom of the flame through the pinhole and to the screen.)

3. Observe the size of the image as the screen is moved (a) closer to the pinhole and (b) farther from the pinhole. Now examine the image as the pinhole is moved (c) closer to the flame and (d) farther from the flame. Make drawings of (a) through (d) using "light lines" to show how each image is formed.

4. Write a letter about light lines to a younger sibling or friend. Tell him or her how to obtain pinhole images and explain how these images suggest that light travels in straight lines.

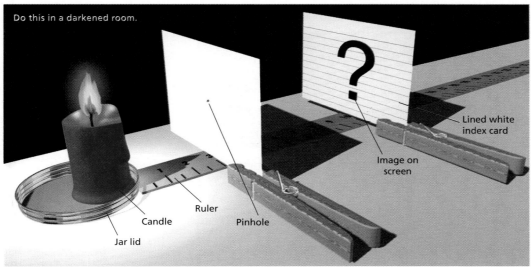

Do this in a darkened room.

Lined white index card

Image on screen

Ruler

Candle

Pinhole

Jar lid

451

FOLLOW-UP

Reteaching

Draw ray-tracing diagrams of the candle, pinhole, and screen on the blackboard, and discusss the diagrams with students. Have students make similar drawings to trace the paths of rays using an arrow as the object.

Assessment

Ask students to develop three quiz questions (with answers) based on Exploration 2. Collect the questions and generate a quiz for the class.

Extension

Have students research the first camera, the *camera obscura*. Students can include diagrams that show how the camera worked and make a model of it.

Closure

Have students use their pinhole-image makers to examine other small, brightly lit objects around the classroom. For example, have students observe how the writing on a light bulb is magnified as they move the image maker closer to the bulb.

EXPLORATION 2

Students will need to try several pinholes of different sizes to produce an adequate image. Warn students that the image produced will be quite dim and indistinct.

Answer to
Exploration 2

2. The image of the candle flame is the same size as the actual flame but is inverted. The angle formed by the lines of light from the top and bottom of the flame is the same on both sides of the pinhole (forming congruent, vertical angles). Because the lined card is the same distance from the pinhole as the pinhole is from the flame, the image is the same size as the actual flame. The image is inverted because light travels in straight lines. Thus, light from the top of the flame can pass through the pinhole only at a downward angle, and light from the bottom of the flame can pass through only at an upward angle.

3. The image gets smaller as the screen is moved closer to the pinhole (a) and gets larger as the screen is moved farther from the pinhole (b). The image gets larger as the pinhole is moved closer to the flame (c) and gets smaller as the pinhole is moved farther from the flame (d).

4. Answers will vary but should include details similar to those discussed in steps 2 and 3.

 An Exploration Worksheet is available to accompany Exploration 2 (Teaching Resources, page 30).

Answers to
In-Text Questions

E The different sizes of the shadows result from the different sizes of the light bulbs, and the shadows would not occur at all if light did not travel in straight lines. (Note: For best results, students should extend the lines in their diagrams from the light bulb all the way to the screen.)

The borders of both shadows are fuzzy in part because the light source does not come from one point. At the edges of the shadow, some light will be blocked and some will not be. At the center, virtually all light is blocked.

FOCUS

Getting Started

Write the word *ambulance* backward on the blackboard. Then have students explain why the lettering on the front of the vehicles that bear this label is often backward. *(It is backward so that its reflection will be readable from a rearview mirror.)* Tell students that they will learn more about reflections in this lesson.

Main Ideas

1. Every visible object reflects light.
2. The angle between a flat mirror and an incident beam is equal to the angle between the mirror and the reflected beam.
3. When a beam of light is reflected from a rough surface, it is scattered in all directions, but when a beam of light is reflected from a smooth surface, it is reflected in one direction.
4. The light reflected from a colored object is the same color as the object.

TEACHING STRATEGIES

Pass out index cards and small mirrors or pieces of Mylar to each student or group. Have volunteers read the introductory text. Try the activities described and discuss students' observations. Allow time for students to figure out what the next symbol in the series should be and what it should look like.

Answers to
In-Text Questions

Ⓐ By the time they have finished reading the introduction, most students will have figured out that the symbols were made from mirror images. It may take more time, however, for them to recognize that the symbols stand for the numbers 1, 2, 3, and 4, with the next two symbols in the series being 5 and 6.

Ⓑ Students must place the mirror to the right of each symbol to complete a letter of the alphabet. The phrase is *WOW WHAT A HIT.*

LESSON 3 Reflection

Look at the symbols below. Are they hieroglyphics? some kind of code? Can you guess what they are? Try drawing the symbols that come next in the Ⓐ upper series. Having difficulty? Maybe a flat mirror will help. Cover up the right half of each of the symbols for a hint as to where to place the mirror, as well as to what the series of symbols is. Now draw the next two symbols in the series.

Can you use the mirror to crack the code in the lower Ⓑ series? You know that mirrors are good reflectors of light. But do you realize that *every* object you see reflects light? If it didn't, you would not be able to see the object. The activities in Exploration 5 will help you better understand the reflection of light.

EXPLORATION 3

Reflection Inspection

You Will Need

- a flashlight
- a light box
- modeling clay
- a flat mirror
- a protractor
- matches
- a wooden splint
- white cardboard (15 cm × 10 cm)
- green cardboard (15 cm × 10 cm)
- cardboard of various colors (15 cm × 10 cm)

PART 1

Light Reflected From a Flat Mirror

What to Do

1. Shine the flashlight into the light box. Place the mirror in the path of the light beam. Prop the mirror up using modeling clay. Use a smoldering splint to add smoke to the box, and then observe the light beam. On its way to the mirror, the light beam is called the **incident beam**. After it bounces off the mirror, it is called the **reflected beam**.

LESSON 3 ORGANIZER

Time Required 2 class periods

Process Skills hypothesizing, predicting, analyzing, comparing

Theme Connection Changes Over Time

New Terms

Diffuse reflection—reflection of light in more or less random directions

Incident beam—a beam of light that strikes an object

Reflected beam—a beam of light after it has bounced off an object

Specular reflection—reflection of light off a smooth surface such as a mirror

Materials (per student group)

Exploration 3, Part 1: light box from Building a Light Box; small flashlight; small ball of modeling clay; flat mirror, about 8 cm × 8 cm; wooden splint; a few matches; protractor; safety goggles;

Part 2: light box from Building a Light Box; small flashlight; small ball of modeling clay; 15 cm × 10 cm pieces of white, green, and other colored cardboard; wooden splint; a few matches; protractor; safety goggles

Teaching Resources Transparency 69

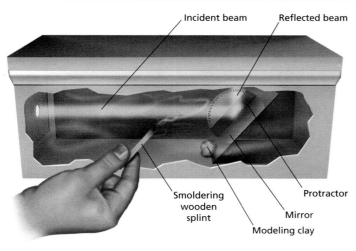

Incident beam   Reflected beam

Smoldering wooden splint

Protractor

Mirror

Modeling clay

2. Place a protractor with its base against the mirror in such a way that it lies in both the incident beam and the reflected beam at the same time. The center of the two beams, where the light strikes the mirror, should be at the midpoint of the base of the protractor. What is the size of the angle between the incident beam and the mirror? between the reflected beam and the mirror?

3. Rotate the mirror to a different position. Again measure the angles between the incident beam and the mirror and between the reflected beam and the mirror.

4. What is the rule about the direction a reflected beam takes after leaving a flat mirror?

5. Suppose that a beam of light hits a mirror in a light box. The mirror is rotated backward by 20°. How many degrees will the reflected beam rotate? If necessary, use the light box and the figure below to help you answer this question.

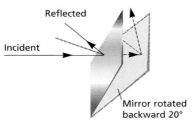

Reflected

Incident

Mirror rotated backward 20°

PART 2

Light Reflected From Cardboard

What to Do

1. In this activity, you will use white cardboard and green cardboard instead of a mirror. Use a smoldering splint to add some smoke to your light box. Place the white cardboard in the incident beam. Describe the appearance of the beam reflected by the white cardboard. How does this beam differ from the beam reflected by the mirror?

2. Rotate the white cardboard in the light beam. In which position does the cardboard reflect the brightest beam? the weakest beam?

3. Describe the appearance of the reflected beam at the point where it hits the top of the light box.

4. Repeat steps 1 through 3 using the green cardboard in place of the white cardboard. Answer the same questions as you go along.

5. How does the appearance of the beam reflected from the green cardboard differ from that of the beam reflected from the white cardboard?

6. Repeat the experiment using different colors of cardboard.

Exploration 3 continued ▶

453

Answers to
Part 2

1. Students should observe that when the beam of light is reflected from the white cardboard, it is scattered in many directions, so the reflected beam appears blurred or fuzzy. By comparison, a beam of light reflected from a mirror is reflected in only one direction with little scattering.

2. The cardboard reflects the brightest beam when it is perpendicular to the incident beam. As the cardboard is rotated so that the incident beam strikes it at smaller and smaller angles, the beam of reflected light is dimmer.

3. The reflected beam appears blurry or fuzzy where it hits the top of the light box. It is more spread out and less focused.

4. Students should observe that when the light beam strikes the green cardboard, it is reflected in the same way as it did when it struck the white cardboard.

5. The beam of light reflected from the green cardboard is green instead of white, and it is less bright.

6. As students try different colors of cardboard, they should discover that the results are the same as before except that the reflected light is the color of the cardboard being used and that it varies in intensity depending on the darkness of the color.

Did You Know. . .

The lenses in polarized sunglasses are a different type of light filter than the colored type used in this unit. To demonstrate this, shine a light through two pairs of polarized sunglasses that are facing each other. As you rotate one pair of sunglasses 90°, the amount of light that passes through both pairs gradually decreases to zero. Polarizing filters are discussed on page S144 of the SourceBook.

Answers to
Part 1, page 452

2. Students' measurements of the angles will vary. Point out to students that the protractor must be positioned so that the beam of light strikes the middle of the base line.

3. Answers will vary.

4. The angle between the incident beam and the mirror is equal to the angle between the reflected beam and the mirror. (The diagram on this page illustrates this concept.)

5. The reflected beam will rotate 20°, satisfying the rule discovered in (4). Encourage students to verify their answer with the light box.

Reflection Reflections

This exercise consolidates the two previous activities by asking students to think about their observations. Two new terms are introduced: *specular reflection* and *diffuse reflection*. Mirrors exhibit specular reflection. This means that the reflection occurs in a single direction. A white sheet of paper exhibits diffuse reflection because the reflected light is scattered in all directions.

 Cooperative Learning
REFLECTION REFLECTIONS

Group size: 2 to 3 students
Group goal: to compare specular and diffuse reflection by completing a set of thought experiments
Positive interdependence: Have students work together to read the explanations and then discuss and write answers to Parts 1 and 2 of Reflection Reflections. After completing the questions, students should list 20 objects that produce specular reflection and 20 objects that produce diffuse reflection.
Individual accountability: Each student should organize his or her list into a chart and be ready to discuss the objects listed during class discussion.

⭐ **Transparency 69 is available to accompany Reflection Reflections.**

Answers to
Part 1

a. 25°
b. *Q*
c. *S*
d. *Q* is the incident beam; *S* is the reflected beam.
e. In specular reflection, the angle between a flat mirror and the incident beam is *equal to* the angle between *the mirror and the reflected beam.*
f. Specular reflection requires an extremely smooth surface. Student suggestions might include chrome fenders, spoons, metal doorknobs, television screens, or the surface of still water.

 Homework

If class time is limited, you may wish to have students complete Reflection Reflections individually as homework.

Reflection Reflections

PART 1

What type of reflection occurred when the mirror was used? The incident light beam was reflected from the mirror without being scattered. Reflection from very smooth surfaces is called **specular reflection.**

Here is a review to check what you discovered about light beams reflected from a mirror. Write your answers in your ScienceLog.

▶ **a** How large is angle *X*?

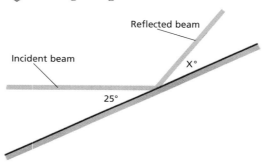
Reflected beam
Incident beam
X°
25°

▶ **b** Which is the correct reflected beam, *P*, *Q*, *R*, or *S*?

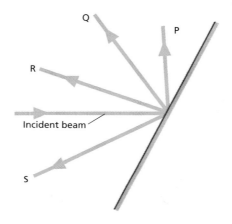
Q
P
R
Incident beam
S

▶ **c** Now that the incident beam has changed its position, which is the correct reflected beam?

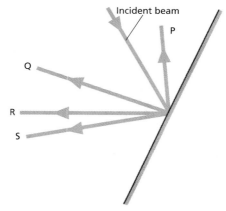
Incident beam
P
Q
R
S

▶ **d** Here the mirror is placed in a different position. In this case, which is the incident beam and which is the reflected beam?

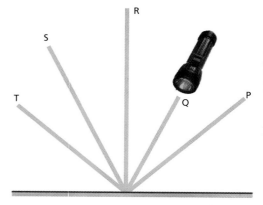
R
S
T
Q
P

▶ **e** Devise a rule for the reflection of light from a flat mirror. Here is a start: In specular reflection, the angle between a flat mirror and the incident beam is (equal to, greater than, less than) the angle between ____?____ .

▶ **f** List some other surfaces on which specular reflection occurs.

Meeting Individual Needs

Gifted Learners
Have students draw a magnified rough surface on the blackboard and a magnified smooth surface. Ask them to apply what they have learned about the angle of reflection to show why specular reflection occurs from smooth surfaces and diffuse reflection occurs from rough surfaces. *(Students should show reflected rays from the smooth surface as parallel lines. Reflected rays from the rough surface should diverge in many directions.)*

Theme Connection

Changes Over Time
Focus question: At night, have you seen your reflection in a living room window? Why don't you see your reflection in the window during the day? *(At night, there is no light coming through the window from the outside. During the day, the light reflected from yourself is much less intense than the light coming in the window from outside, so the reflected image is obscured by the light outside.)*

How did the white cardboard reflect the beam of light? Was it specular, as in the case of the flat mirror, or more scattered—reflected in all directions? **A**

The scientific term for "scattered reflection" is **diffuse reflection**. What causes diffuse reflection? **B** Examine the two diagrams below to help you understand this question.

Specular Reflection

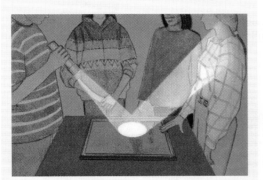

Beam of light

Surface of mirror is smooth (even when magnified)

Diffuse Reflection

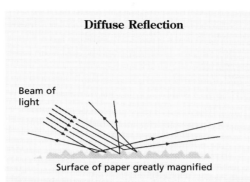

Beam of light

Surface of paper greatly magnified

Diffuse reflection is more common than specular reflection. Why do you think this is so? C

Was the reflection from the green cardboard specular or diffuse? What color was the light scattered from the green cardboard? What color was the smoke? When the light entered the light box, it was white. What colors found in white light were not visible? The following illustration suggests what happened. **D**

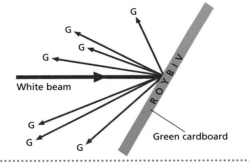

White beam

R O Y B I V

G

G

G

G

G

G

G

Green cardboard

Answers to
In-Text Questions and Caption

A The reflected beam was generally not specular. The beam reflected from the white cardboard generally followed one direction, but it was more scattered than the beam reflected by the mirror.

B Diffuse reflection is caused by an uneven or rough surface. Because the surface is uneven, even parallel light rays that strike the surface very near each other will strike the surface at different angles. The result is diffuse reflected light instead of a focused beam.

C Diffuse reflection is much more common than specular reflection because there are many more rough surfaces than smooth ones.

D The reflection from the green cardboard was diffuse. The color of the scattered light was green and caused the smoke and walls of the box to appear green as the light was reflected from them. The red, orange, yellow, blue, indigo, and violet colors of white light were not visible because they were absorbed by the green cardboard, which reflected only green light.

FOLLOW-UP

Reteaching

Suggest that students try making a mirror. They should start with a piece of clear plastic or Plexiglas and then experiment with different coverings to see which works best. For example, they might try covering one side of the plastic glass with aluminum foil or different colors of paper or try painting the back of the plastic glass black or white. Have students discuss their results and determine which materials make the best mirrors.

Assessment

Ask students to conduct a survey to identify objects that produce specular reflection and objects that produce diffuse reflection. Have them organize their data into a chart to show the difference between the two kinds of reflection.

Extension

Suggest that students do some research to discover how mirrors are made. Their research should answer questions such as the following: What makes the surface of a mirror reflective? How is the reflective material applied to a mirror? Have students share what they learn by presenting oral reports to the class.

Closure

Have students set up two pocket mirrors at right angles to each other and place a coin between them so that they see four coins. Have students change the angle of the mirrors and note how many coins they then can see. You may wish to point out that every time a light beam is reflected off a mirror, the beam loses about 10 percent of its brightness. Ask: Based on this fact, how many images would you predict that you would see in the mirrors? *(Answers will vary depending on the brightness of the initial image.)*

FOCUS

Getting Started

Have students use a magnifying glass to examinine the colored illustrations in this book. Ask students if they can identify what basic colors were used to make the dots that, when combined, produce all of the colors of the illustrations. Guide students to observe closely. You may wish to provide the hint that there are three basic colors plus black that are used in "four-color printing" and that the colors are not the same as the primary light colors studied earlier. Tell students they will find out why as they proceed through the lesson.

Main Ideas

1. We see objects by the color of light they reflect.
2. Mixing light of different colors tends toward white; mixing paint of different colors tends toward black.
3. The primary colors of the paint that artists use are red, blue, and yellow.

TEACHING STRATEGIES

Answers to
In-Text Questions

Ⓐ Students should recall that a white surface reflects white light, a black surface reflects no light, and red cardboard reflects red light. Students should thus label the reflected light in the diagram as red light and the absorbed light as all light but red.

Ⓑ Student answers will vary but should reflect the questions answered in the first paragraph on this page.

Ⓒ In red light, for example, only the red portions of a sweater would be unchanged. Other colors would tend to appear black. White would appear red.

The Riddle of Color

When you shone light on the green cardboard in Exploration 3, what color did the cardboard reflect? This is how we perceive colors in objects—by the colored light that they reflect. What colors does a white surface reflect? a black surface? If you substituted red cardboard for the green cardboard in the light-box experiment on page 453 and shone white light on it, what do you think would happen? How would you label the colors in the diagram at right, which shows this experiment? Ⓐ

Look at the illustration below, and then explain why Lisa sees the color she does. Ⓑ

Do objects have the same color when they are viewed under white light as they do under light of another color? For example, if Lisa viewed her friend's sweater in the dim red light of a photographer's developing lab, would the sweater be the same color? Perhaps you have noticed that sometimes a sweater bought in a store illuminated by artificial light seems a different color when you look at it in sunlight or that the color of a jacket at night under certain street lights looks different from when the jacket is viewed in sunlight. Does the light shining on an object have an effect on the color you see? In the following Exploration you'll have the chance to find out. Ⓒ

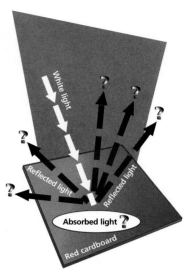

White light
Reflected light
Reflected light
Absorbed light ?
Red cardboard

Wow, Kenny, I really like that blue sweater!

Thanks, Lisa. Blue is my favorite color.

456

LESSON 4 ORGANIZER

Time Required
2 class periods

Process Skills
observing, comparing, inferring

New Term
Primary colors of paint—the three colors of paint—red, blue, and yellow—from which all other colors of paint can be created

Materials (per student group)
Exploration 4: cardboard tube; scissors; sheet of white paper; red, yellow,

blue, and purple crayons; 2 pieces of dark blue cellophane, about 5 cm × 5 cm each; 2 pieces of dark red cellophane, about 5 cm × 5 cm each; tape
A Lesson in Art: samples of red, blue, and yellow watercolor paint; about 20 mL of water; stirring rod (optional items: several sheets of thick paper; paintbrush)

Teaching Resources
Exploration Worksheet, p. 31
SourceBook, p. S137

Changing Colors

You Will Need

- a cardboard tube
- a white piece of paper
- red, yellow, blue, and purple crayons
- tape
- 2 pieces of dark blue cellophane
- 2 pieces of dark red cellophane

What to Do

1. Cut out a small window at the bottom of the tube to construct a cardboard viewer like the one shown here.

2. Cut the paper into four rectangular pieces. Make a heavy mark on each with one of the colored crayons.

3. Tape the blue cellophane over the window of the tube, as shown below.

Now view each mark with this blue filter in place. What is the color of each mark? of the background? Cover the window with a second piece of blue cellophane. Is there any difference?

4. Suppose step 3 were repeated using red filters. Predict the colors of the crayon marks and the background. Check your predictions.

5. Use your viewer to look at the color photos in this unit under different lights. How good have you become at predicting the results of color mixing?

Did you know that visible light is part of a much bigger spectrum? To find out more, read pages S137–S140 in the SourceBook.

457

 An Exploration Worksheet is available to accompany Exploration 4 (Teaching Resources, page 31).

Meeting Individual Needs

Learners Having Difficulty

For extra reinforcement, you may wish to have students repeat Exploration 4 using green cellophane.

Homework

Have students make cellophane viewers in class by taping a piece of red, blue, or green cellophane to a window cut out of an index card. Have students use the viewer while at home and note any reactions they have to objects being colored differently. They should report their findings to the class on the next day. *(For example, students may find that food appears unappetizing under filtered light.)*

Give the students at least 30 minutes to make their observations. Caution the students to make accurate observations, looking not only for the principal effect, but also for secondary effects. A student may view a red mark under a blue filter and report a purple color, when the principal effect is actually black with a slight purplish tinge. The students think they see dark purple because there may be some residual blue reflected to the eye as well as some red due to the extraneous white light.

Answers to Exploration 4

3. The red mark is black, the blue mark is blue, the yellow mark is black, the purple mark is dark blue, and the background is blue. With a second piece of blue cellophane, the blue mark, the purple mark, and the background are darker blue, and the other two marks are black.

4. Predictions will vary but should be clear and logical. Students should find that with a red filter the red mark looks red, the blue mark looks black, the yellow mark looks red, the purple mark looks dark red, and the background looks red.

Integrating the Sciences

Life and Physical Sciences

Place a red and a blue filter side by side on an overhead projector and project the light onto a white screen. Ask students to stare at the screen for a minute or two and then to describe the image they see when they close their eyes. *(Students should be able to see a ghost image that has the same shape as the projected image but different colors. The ghost image, or afterimage, should be green where the original image was red, and yellow where the original was blue.)* Have students research *successive contrast* and design experiments like the one mentioned above to find out more about this phenomenon. Be sure to check students' proposed procedures for safety before allowing them to proceed with any experiment.

Answers to
In-Text Questions, pages 458–459

A Mixing red, blue, and green paints together will produce black. Mixing red, blue, and green light together will produce white. Paints show different colors by absorbing some colors of light and reflecting others. Blue, red, and green paints each absorb different parts of white light. In general, red paint absorbs green and blue light and reflects red. Likewise, blue paint reflects blue and absorbs red and green, and green absorbs red and blue and reflects green. When mixed together, the three colors of paint absorb all colors of white light to produce paint that looks black.

B The diagrams on page 438 show that the primary colors of light (red, blue, and green) combine to form white. In contrast to mixing colors of paint, mixing colors of light is called additive color mixing.
   This may be a good time to explain the term *pigment*. A pigment is a substance that reflects some colors and absorbs others. Paint is a type of pigment.

C Most students will probably observe that green has been replaced by yellow, but that the other two colors remain the same. This may be a good time to explain the difference between pure and compound colors. Some students may be aware that printing presses use magenta, yellow, and cyan as the primary colors of pigment. Explain that printing presses use "pure" pigments. The true primary colors of pigment are cyan, magenta, and yellow (the secondary colors of light). Artists use paints that are not pure pigments, but are "compound" pigments. Using compound pigments for mixing produces different results: although pure blue (absorbing red and green) and pure yellow (absorbing blue) would produce black, compound blue and compound yellow produce green.

D Red and yellow paint make orange paint.

E Mixing red, yellow, and blue paint will produce black or dark brown because together the paints subtract all or most colors of light.

F Other paint mix equations could include the following: green + orange = reddish brown; yellow + green = yellowish green; green +

Mixing Paint

What happens when you mix blue, red, and green paint together? Do you get the same result as when you mix these colors of light? You can check the result for light on page 438. The diagram below shows what happens when you mix blue, red, and green paint together. Why the difference? Before reading further, discuss with a friend a possible reason for this difference. **A**

When you mix various colors of light, you are "building" white light. In other words, you are moving toward "lightness," or "whiteness." But colored paint, like colored objects, absorbs the white light that shines on it. If you mix paint of one color with paint of another color, the resulting blend absorbs more from white light than either paint would absorb separately. Less light means more "dark." Thus, mixing paint tends to increase "darkness," or "blackness."

How do the diagram on page 438 and the color-mixing schemes shown below support these ideas? **B**

A Lesson in Art

If you have a set of watercolor paints, you can test other color-mixing experiences. Mix the dry paint with a little water. The more water you use, the "lighter" and more dilute the color will be. You won't need all the colors in your set—only three! Artists find that they can make all the colors they need if they start with *blue*, *yellow*, and *red*. In fact, artists call these the three **primary colors of paint**. How are these colors different from the three primary colors of light? **C**

purple = bluish brown; and orange + green + purple = black.
   Students may be interested to learn the following basic equations for mixing colors of "pure" pigment: yellow + magenta = red; magenta + cyan = blue; cyan + yellow = green; magenta + cyan + yellow = black; and red + green + blue = black.

Meeting Individual Needs

Gifted Learners
Explain to students that a laser produces a beam of light that contains only a few wavelengths and whose waves are all in step with each other. Lasers can be used to burn a hole through 2 cm of steel and perform delicate surgery. Suggest that students do research to discover how light from a laser differs from other light, how a laser beam is produced, or how laser light is used with mirrors and lenses to produce holograms. Have students share what they learn by making bulletin-board displays for the classroom.

Here is an example. What color would you expect when blue and yellow paint are mixed together? Examine the diagram at right. How do you explain the color of this mix? Actually, the results of mixing paint lead us to more discoveries about the light reflected from objects. When light shines on colored paint, it reflects not only its own color, but small amounts of its neighboring colors in the color spectrum (ROYGBIV) as well.

Take a look at the diagrams at right. Notice that blue paint *absorbs* red, orange, yellow, and violet and *reflects* not only blue, but also a little green and indigo—blue's neighbors in the spectrum. Yellow paint absorbs red, blue, indigo, and violet and reflects yellow as well as its neighbors, green and orange. When blue and yellow paint are combined, together they absorb red, orange, yellow, blue, indigo, and violet. The only color reflected by the paint mixture is green, which is the color that you see.

D Which two primary colors of paint do you think make the color orange? Sketch the situation in your ScienceLog, showing which colors of white light are absorbed and reflected by each primary color and by the mixture.

What happens when you mix all three primary colors of paint together? Try this using equal amounts of paint and not too much water. How do you explain the result? **E**

Examine the color triangle below. What information does it contain? Some hints are in the paint mixes to the right of the triangle. Add a few of your own paint-mix equations with the help of the color triangle. You might even want to test your paint-mix equations using real paint. **F**

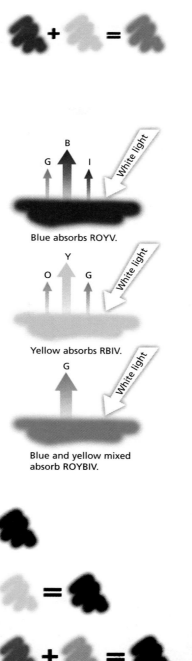

Blue absorbs ROYV.

Yellow absorbs RBIV.

Blue and yellow mixed absorb ROYBIV.

459

Color Choices
Societies throughout history have used colors to convey emotions and have attached cultural or historical significance to certain colors. Ask each student to research a culture of their choice and to explain the culture's use of colors to the class. Afterward, involve students in a discussion of the similarities and differences among the cultures presented.

Meeting Individual Needs

Second-Language Learners
Have students make a drawing of intersecting color circles for pigments similar to the one they may have made for light on page 438. Reinforce the idea that the colors obtained by mixing pigments are not the same as those obtained by mixing light. Have them review what they learned about mixing colored light in Lesson 4 of Chapter 19 if necessary.

Cross-Disciplinary Focus

Art
Invite an art teacher to come to class to discuss the color wheel and the mixing of colored paints. Students will better understand why artists need to be careful about which exact shades of the primary colors they use in order to achieve mixed colors that go well together.

FOLLOW-UP

Reteaching
Have students make a top by cutting out a cardboard disk a few centimeters in diameter and punching a short stick through the center. They should then color half of the disk blue and the other half yellow with markers or paints. Instruct students to spin the top and record and explain their observations. *(The disk will appear whitish because the eye sees both halves of the spinning disk simultaneously and the halves reflect complementary colors of light.)* You may wish for students to experiment further with other colors and to comment on the results.

Assessment
Have students write stories explaining how they might use the information from the lesson if they had one of the following occupations: cosmetics salesperson, interior decorator, grocery-store manager, or stage-lighting director.

Extension
Have students research an artist's color wheel. Students should make posters showing the wheel and the relationships among the colors on it. Students should also include a paragraph explaining how artists use the wheel in color mixing.

Closure
Set up a colorful still life of flowers for students to paint in watercolors. Provide each student with three primary colors of pigment only. Instruct students to mix colors in order to paint the still life in realistic hues. Encourage creativity as well as accurate color mixing.

CHALLENGE YOUR THINKING

1. Brent saw a low-intensity, or grayish, white light. Each segment reflects one of the colors contained in white light and absorbs the other colors (subtractive). Spinning the disk causes all of the reflected colors to be perceived as one. Because all of the colors of the spectrum combine to form white light, Brent saw white light reflected from the spinner (additive). The light is of a low-intensity because the spinner absorbs more energy than it reflects. Brent would see the same results if the segments were yellow and blue because yellow and blue are complementary colors of light. If the segments were red and blue, Brent would see a low-intensity magenta color because the reflected red and blue light would combine to form magenta.

2. Answers will vary but could include the following points:
- When smoke or other particles in the air reveal a beam of light, the path of the beam is a straight line.
- Shadows would not occur if light did not travel in straight lines.
- Light can be reflected by a mirror at predictable and measurable angles.

3. The colored stripes in Joseph's coat would change to blue and black stripes. The red, yellow, green, and orange stripes would look black because they reflect only red, yellow, green, and orange light, respectively, and absorb all other colors of light, including blue. The blue stripes would look blue because they reflect blue light. The purple stripes would look very dark blue because they reflect red and blue light. The white stripes would look blue because they reflect all colors of light.

4. This question is open to the students' ingenuity. The answer to the reflection question presented is *NATURE.* It differs from the puzzles presented earlier in the chapter in that the plane of reflection is horizontal instead of vertical.

5. Transparent: plastic wrap, sunglasses, salt water
Translucent: wax paper, oiled brown paper, a single ply of tissue paper, ice cube, frosted glass
Opaque: brown paper, five sheets of tissue paper, newspaper

CHALLENGE YOUR THINKING

1. Colorful Moves

Brent made a color spinner like the one shown below. He wound the cord tightly and started the disc spinning by pulling back and forth on the cord. What did he see? (Both additive and subtractive color phenomena occur.) What would Brent see if the colored segments were yellow and blue only? red and blue? Explain.

2. Straighten Her Out

Pam made this remark in class: "When you flip a light switch, light suddenly appears everywhere. How can you say that light travels in straight lines?"

How many different reasons could you give Pam to help her understand that light really does travel in straight lines? Describe them.

3. A Staged Question

A local production of Andrew Lloyd Webber's *Joseph and the Amazing Technicolor Dreamcoat* is in progress. At the front of the stage is Joseph in his brilliant coat of red, yellow, green, blue, orange, purple, and white stripes, all the brighter under the high-powered white stage lights. As the scene nears its end, the white lights are suddenly subdued and strong blue stage lighting predominates. Have the colors of Joseph's coat changed? Predict what you would see.

460

6.	Specular reflection	• sunglasses on each of the outside diners • cafe window (some of the light is reflected, some is transmitted) • metal strip framing cafe window • cream container on table • metal parts on wheelchair • tops of salt and pepper shakers • drinking glass on table (some light is reflected, some is transmitted)
	Diffuse reflection	• most of the objects in the photo, including the people and their clothes, the white concrete plant holder, the plants, the table, the coffee cup, the matte parts of the wheelchair, etc.
	Transmission of light	• cafe window • sunglasses • drinking glass and water in glass
	Reflection of blue light	• boy's shirt • blue back of the wheelchair seat • blue blanket (barely visible) in the wheelchair, next to the girl's lap
	Scattering	• All diffuse reflection is scattering; the white objects in the photo are the only ones that scatter *all* light (roygbiv) that strikes them.
	Absorption of all colors	• boy's black hair • black tag on the wheelchair • waiter's black pants and black hair

4. Time for Reflection

Devise more "reflection puzzles" like those on page 452, and then give them to a classmate to solve. In turn, solve some puzzles created by your classmate. Here is one to begin with. What is it? How does it differ from the first one on page 452?

5. Light Going

The amount of light that passes through a material depends on whether that material is transparent, translucent, or opaque (allows no light to pass through). Classify each of the following materials as one of the above types: wax paper, plastic wrap, brown paper, oiled brown paper, newspaper, a single ply of tissue paper, five sheets of tissue paper, an ice cube, sunglasses (not mirrored), salt water, a frosted glass.

6. Describe the Scene

Locate in the photo below where each of the following occur: specular reflection, diffuse reflection, transmission of light, reflection of blue light, scattering, and absorption of all colors.

ScienceLog

Review your responses to the ScienceLog questions on page 442. Then revise your original ideas so that they reflect what you've learned.

461

The following are sample revised answers:

1. When the white light from the sun hits microscopic particles located in the atmosphere, much of the blue portion of the white light is scattered at right angles by the particles. The redirected blue light then enters the students' eyes, making the sky appear blue. The other colors in the sunlight are either not scattered or are not scattered enough to affect the color perceived by the students.

2. We see objects, such as buildings, that do not produce their own light because light from other sources, such as the sun or a lamp, is reflected off the objects and into our eyes. Each point on a particular object reflects light in all directions so that people in different locations can see the same object.

3. An object's color refers to the specific color or combination of colors of light that the object reflects when exposed to white light. Nonwhite light, however, does not contain all of the colors of the spectrum. When viewed in nonwhite light, therefore, an object's color may appear distorted because the object is not reflecting all of its component colors.

★ **You may wish to provide students with the Chapter 20 Review Worksheet that is available to accompany this Challenge Your Thinking (Teaching Resources, page 33). Transparency 71 is also available for your use.**

Homework

The Activity Worksheet that accompanies this Challenge Your Thinking makes an excellent homework assignment (Teaching Resources, page 32). If you choose to do this activity in class, Transparency 70 is also available for your use.

ENVIRONMENTAL FOCUS

Have students research solar eclipses in terms of the shadows that they cast on the Earth's surface. They should relate total, partial, and annular eclipses to the umbra and penumbra cast by the moon onto the Earth. *(A total eclipse only occurs for an observer in the umbra of the moon's shadow. Partial and annular eclipses occur when the observer is in the penumbra.)* You may wish to have students draw and label diagrams representing each of these cases.

Connecting to Other Chapters

Chapter 19
introduces light as a form of energy, the relationship among light, heat, and color, and the theory of color mixing.

Chapter 20
explores diffuse and specular reflection and the scattering, transmission, and absorption of light.

Chapter 21
analyzes lenses and mirrors, their uses, and how they produce real and virtual images through the refraction of light.

Prior Knowledge and Misconceptions

Your students' responses to the ScienceLog questions on this page will reveal the kind of information—and misinformation—they bring to this chapter. Use what you find out about your students' knowledge to choose which chapter concepts and activities to emphasize in your teaching. After students complete the material in this chapter, they will be asked to revise their answers based on what they have learned. Sample revised answers can be found on page 481.

In addition to having students answer the questions on this page, you may wish to have them complete the following activity: Divide the class into groups. Tell each group that they represent a team of designers who must create a hall of mirrors for an amusement park that is opening soon. What type of effects do the mirrors need to create? What kinds of mirrors will be needed? Encourage students to be creative in their designs. Collect the designs, but do not grade them. Instead, use them to determine what students know about mirrors, what misconceptions they may have, and what about the topic is interesting to them.

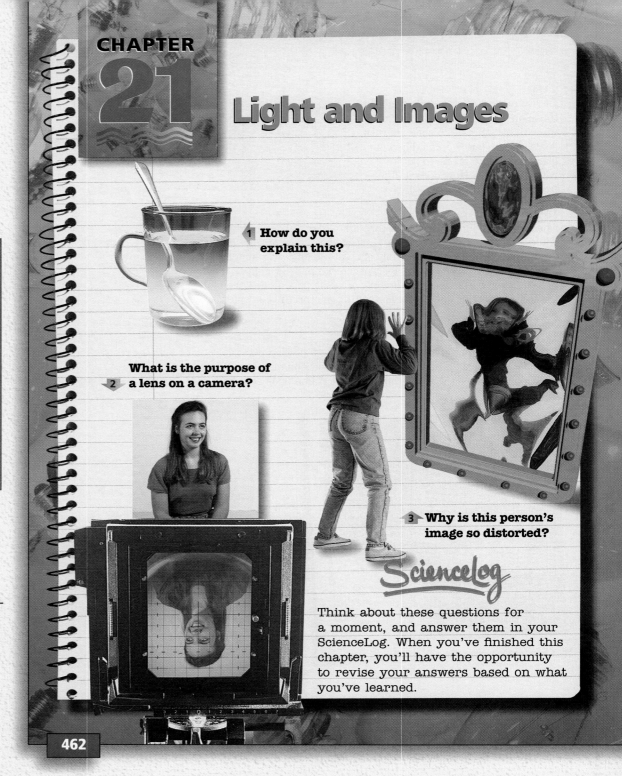

LESSON 1 — Plane Mirrors

Consider the Plain Plane Mirror

Flat mirrors are also known as **plane mirrors**. Plane mirrors are what most people mean when they say "mirror." An image in a plane mirror obviously resembles the object that formed it, but is the image an exact duplicate? If not, how does it differ?

What's Your Image?

You read earlier that if an object did not reflect light, you could not see it. In fact, every object reflects at least some light. The fact that people standing at different places in a room can all see a particular object shows that light is being reflected in all directions from the object. Light bounces off of you in all directions.

If you stand in front of a plane mirror, some of the light reflecting off of you strikes the mirror and is reflected from it back to your eyes. What you see in the mirror is your image. Your body, standing in front of the mirror, is the *object;* the body that seems

to be behind the mirror is your *image.* Are the images in a plane mirror exactly the same as their objects? Hold a watch or a book in front of a mirror to find out. In fact, an object and its plane-mirror image are *not* identical. The following Exploration will help you learn more about images in a plane mirror.

EXPLORATION 1

Plane-Mirror Insights

You Will Need

- graph paper
- a plane mirror
- a pencil
- a flat piece of colored glass
- modeling clay
- a metric ruler
- an index card

What to Do

1. Make a dark line over one of the horizontal lines on a piece of graph paper. Stand the plane mirror straight up along the line you made. Use modeling clay to stand the mirror up along that line.

2. Put your pencil in front of the mirror. Where in the mirror is the image of the pencil?

3. Now place the pencil six squares in front of the mirror. Where is the image now? How many squares is it behind the mirror? Place the pencil in another position, and check the location of the image again.

4. Replace the plane mirror with a piece of colored glass. The glass reflects light just as a plane mirror does. However, you can also see through the glass. Both of these properties will help you make your plane-mirror insights.

5. Draw a straight line across the center of a new sheet of graph paper. Place the glass along the line. Use modeling clay to hold the glass upright. Draw a triangle in front of the glass, near the middle of that half of the page. Look at the image of the triangle. While looking

Exploration 1 continued ▶

463

LESSON 1 ORGANIZER

Time Required
2 class periods

Theme Connection
Energy

Process Skills
comparing, contrasting, predicting, inferring, measuring, communicating

New Terms
Image—the visual impression of an object produced by reflection in a mirror or refraction by a lens
Plane mirror—a flat mirror; reflects light to produce virtual images

Real image—an image that forms where light rays coming from an object converge and that can be projected onto a screen
Virtual image—an image that is not formed by converging light rays and that cannot be projected onto a screen

Materials (per student group)
Exploration 1: 3 sheets of graph paper; plane mirror, about 8 cm × 8 cm; small ball of modeling clay; flat piece of colored glass, about 8 cm × 8 cm; metric ruler; index card

Teaching Resources
none

LESSON 1 — Plane Mirrors

FOCUS

Getting Started

Call on a volunteer to read aloud the introductory material in the section titled Consider the Plain Plane Mirror. Ask students to explain how a mirror works. (Accept all reasonable answers.) Then direct students' attention to the photograph. Ask: Which is the object and which is the image? (The real person is the object, and the face that seems to be behind the mirror is the image.)

Main Ideas

1. Plane mirrors exhibit reflections that are proportional in size to the original object's distance from the mirror.
2. The image reflected by a plane mirror appears to be the same distance behind the mirror as the object is in front of the mirror.

TEACHING STRATEGIES

Answers to *In-Text Questions*

Ⓐ Images in plane mirrors are not exact duplicates of objects. The images are smaller and reversed and are two-dimensional representations of objects that occupy space. (Make sure that students understand that images in plane mirrors appear life-size only when the objects are very close to the mirror. Images decrease in size proportionally as the object's distance from the mirror increases.)

EXPLORATION 1

In step 2, have students place the pencil directly against the mirror. In steps 4 through 8, some of the light will be reflected from the surface of the glass, and some of the light will pass through it, allowing students to see through the glass and to see images reflected in it. In steps 9 and 10, students explore symmetry and reversal in reflected images.

Answers to
Exploration 1, pages 463–464

2. The pencil's image appears to be behind the mirror.

3. The image of the pencil appears to be as many squares behind the mirror as the actual object is in front of the mirror.

6. The image distance and the object distance are the same. A line that joins a point on the object to a corresponding point on its image is perpendicular to the surface of the glass. (It forms a 90-degree angle with the glass.)

   The image of an object placed in front of a plane mirror appears to be the same distance behind the mirror as the object is in front of the mirror. The image is in the same position relative to the mirror as the object is.

7. The image reflected in a plane mirror is smaller than the object in front of the plane mirror. However, the image's smaller size is exactly proportional to the object's distance from the mirror.

8. The image car turns to its right as the object car turns left and vice versa.

9. The words should appear upside down. The words KID and OXO appear the same because the letters are symmetrical about a horizontal axis. The words WOW and POP appear as MOM and bOb in the reflected image because the letters are not symmetrical about a horizontal axis.

10. The order of the letters in KID is reversed in the reflected image. WOW and OXO appear the same as before because the letters are symmetrical about a vertical axis. The letters in KID and POP also face the opposite direction because they are not symmetrical around a vertical axis. When the words are turned upside down, the image is the same as that in step 9.

---

through the glass, trace the lines of the image on the paper behind the glass. Now remove the glass and modeling clay.

6. Draw a line from one point on the triangle to the corresponding point on its image. Measure the distance along this line

   a. from the glass to the image (called the *image distance*) and

   b. from the glass to the object *(object distance)*.

   How do the two distances compare? What is the angle between the glass and the line joining the points on the object and image? What conclusion can you make about where you will see the image of an object that is placed in front of a plane mirror?

7. Measure the sides of the object triangle and those of the image triangle. How do the sizes of the object and the image compare?

8. Draw a center line on a new piece of graph paper. Then set the glass vertically on the line as shown below. Put your pencil on the paper and slowly move it toward the glass, drawing a line as you go. Imagine that your pencil point is a car on a road driving toward the glass.

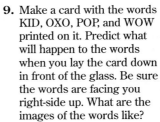

Observe the image of the pencil and line. Now, gradually curve your pencil line to the left. What direction does the image turn? Now trace over the image line that you see on the paper behind the glass. As you follow the line toward the glass and around the curve, in which direction is the imaginary car turning?

9. Make a card with the words KID, OXO, POP, and WOW printed on it. Predict what will happen to the words when you lay the card down in front of the glass. Be sure the words are facing you right-side up. What are the images of the words like?

10. Next, predict what will happen when you hold the card upright in front of the glass. How are the images of the words different? Turn the card so that the words are upside down. What does the image in the glass look like now? Where have you seen that image before?

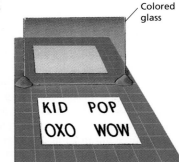

Colored glass

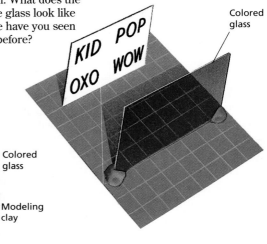

Colored glass

Colored glass

Modeling clay

CROSS-DISCIPLINARY FOCUS

Language Arts

Have students define the word *palindrome. (A palindrome is a word that is the same whether read backward or forward.)* Then have students find as many palindromes as they can. *(Examples include wow or madam.)* Ask: If a palindrome is reflected about a vertical axis, is it still a palindrome? *(Students may answer yes, but they should be aware that although the order of the letters will be correct, the letters themselves will be backward.)*

Looks Like Leonardo's

Try writing your name on a card in such a way that it can be read correctly in a mirror. Interestingly, this is how the famous Renaissance scientist Leonardo da Vinci habitually wrote his notes. He did this so that other people couldn't read them.

Checking the Facts

You have seen lots of plane mirrors, including the ones you used in Exploration 1 of this chapter. Some of the following statements are true; others are false. Decide which statements you agree with, disagree with, or are uncertain about.

1. The image in a plane mirror is the same size as the object.

2. Images of objects in plane mirrors appear to be on the surface of the mirror.

3. If an object moves farther away from a mirror, the image seems to move backward in the mirror.

4. If an object moves farther away from a mirror, its image appears to grow smaller.

5. The perpendicular distance from the object to the mirror is the same as the apparent perpendicular distance from the image to the mirror.

6. If a point on an object and the corresponding point on the image are joined by a straight line, the line intersects the mirror at 90°.

7. You need a plane mirror half your size to see all of yourself at one time.

8. If a moving object in front of a plane mirror turns right, the image seems to turn left.

9. The images of objects held at a right angle (90°) to a mirror are upside down.

10. The images of objects held in front of a plane mirror are reversed left to right.

11. The images of letters *H, O,* and *X* look the same as the objects in a plane mirror, no matter how the letters are held in front of it.

12. The images in a plane mirror aren't where they seem to be.

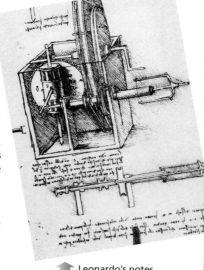

Leonardo's notes for a yarn-spinning machine

465

Real or Virtual?

This material is intended only as an introduction to the concepts of real and virtual images. Both concepts are covered in more depth in the next three lessons, which deal with the topics of convex and concave mirrors and converging lenses.

Have students read the material silently and formulate their own definitions of real and virtual images. Then call on several volunteers to share their ideas with the class. If disagreements occur, involve the class in a discussion of the concepts so that students can resolve their differences.

You may wish to point out to students that virtual images are always in focus, while real images are not.

Homework

Have students provide two symmetrical "half-words" that become full words when reflected by a mirror. (See page 452.) One word should be reflected about a vertical axis, and the other about a horizontal axis.

FOLLOW-UP

Reteaching

Suggest that students think of as many words as they can that read the same way when they are laid down in front of a mirror. (*BOXED, HEEDED, CHIDED, HIKED*) Suggest that students combine their lists and display the words for others to try.

Real or Virtual?

Images are representations of objects. Your bathroom mirror *reflects* light to produce an image of your face. A movie screen *projects* light to produce an image on a screen from a frame of film. Obviously, both the movie projector and the bathroom mirror give you images. But one of these images is **real** while the other image is **virtual**. Why? (Look up *virtual* if you are uncertain about its meaning.) What is the difference between a real image and a virtual one? **A**

The image on a movie theater screen is made by light that is projected onto the screen. If you stood in front of the screen, the image would be projected onto you. Look at the photo to the right. You could place a light meter at the location of the screen image and measure its brightness. For these reasons, the movie-projector image is a real image.

Now think about a mirror image. Where does the image seem to be located? If you answered, "the same distance behind the mirror as the object is in front of the mirror," you are correct. Imagine putting a screen at the location behind the mirror where the image appears to be. Would you get an image? Would a light meter record a reading for the brightness of the image?

The answer to these questions is no. This is why the image in a mirror is virtual, not real. Virtual images cannot be formed on a screen. They are a sort of optical illusion having an apparent location, at which there is no actual light present. On the other hand, real images are formed on a screen by the actual presence of light.

Look at the examples of images provided on this page. Which are real? Which are virtual? How do you know? **B**

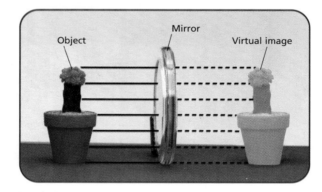

Object   Mirror   Virtual image

Assessment

Have students summarize the lesson by making fact sheets that list the characteristics of images reflected in plane mirrors.

Extension

Have students investigate what a periscope is and how two plane mirrors can be used to make a periscope. You may wish to have students draw illustrations of what they learn. (*A periscope is a device that is used to view objects behind an obstruction. A simple mirror periscope can be made from two parallel mirrors that are placed at an angle to the viewer.*)

Closure

Have students experiment with the funny effect they can make by standing at the edge of a large mirror. (*For example, by placing one's face in the right position, one can make one's nose "disappear."*) Have each student demonstrate one reflection "trick" to the class and explain how it works. You may also wish to bring an assortment of curved mirror surfaces to class for the students to experiment with.

LESSON 2 Convex Mirrors

Not all mirrors are flat. Sometimes the reflective surface is curved in some way. How does this affect the image? You have probably seen each of the items at right behaving as a mirror. What do the images reflected in these mirrors look like? What is similar about each of these mirrors? **C**

Each object's surface is a **convex mirror**. Convex mirrors curve outward toward the viewer. How do you think convex mirrors differ from plane mirrors? What are objects and images like in convex mirrors? Consider the following questions about convex mirrors:

- How do the images compare in size with their objects?
- Are the images *erect* (right-side up) or *inverted* (upside down)?
- Are the images behind or in front of the mirror?
- Are the images reversed from left to right?
- Are the images real or virtual?

First, try answering these same questions with respect to plane mirrors. If you can do so, you will have a head start on understanding convex mirrors. But to really understand convex mirrors, do the following Explorations. **D**

EXPLORATION 2

Exploring Convex Mirrors

You Will Need

- a large, shiny spoon
- a light box
- mirrors (1 plane, 1 convex)
- a wooden splint
- matches
- a flashlight
- a pencil
- a candle
- a jar lid
- modeling clay
- a sheet of white paper

PART 1

Comparing Mirror Images

Examine your image on the back of a spoon where the curve is greatest.

a. How do the size of the image and the size of the object compare?

b. Is the image right-side up or upside down?

c. Is the image reversed?

d. How does the field of view in this convex mirror compare with that in a plane mirror?

Exploration 2 continued ▶

467

LESSON 2 ORGANIZER

Time Required
1 class period

Process Skills
predicting, hypothesizing, measuring, comparing, contrasting

New Term
Convex mirror—a mirror in which the reflective surface curves outward toward the viewer

Materials (per student group)
Exploration 2, Part 1: large, shiny spoon; plane mirror; **Part 2:** light box from Building a Light Box; small ball of modeling clay; small convex mirror;

wooden splint; a few matches; small flashlight; safety goggles; **Part 3:** candle; jar lid; a few matches; convex mirror; 2 small balls of modeling clay; sheet of paper; safety goggles; **Part 4:** convex mirror

Teaching Resources
none

LESSON 2 Convex Mirrors

FOCUS

Getting Started

Direct students' attention to the lesson title, and ask students to explain the meaning of the word *convex*. *(A convex surface curves outward like the outer surface of a sphere.)* Ask students to suggest how the image formed by a convex mirror differs from that of a plane mirror. *(Accept all reasonable responses.)*

Main Ideas

1. A convex mirror has an outwardly curving surface.
2. Images reflected by a convex mirror are upright, smaller than the object, and virtual.
3. Light hitting the surface of a convex mirror is reflected in many directions.
4. The image in a convex mirror appears to be behind the mirror.

TEACHING STRATEGIES

Answers to
In-Text Questions

C The images in such mirrors are often distorted. They are also smaller than the object and are erect. Each of these mirrors exhibits a convex curved surface.

D In plane mirrors, images are erect, appear behind the mirror, are reversed from left to right, and are virtual.

EXPLORATION 2

Answers to
Part 1

a. The image is smaller than the object.
b. The image is right-side up.
c. The image is reversed.
d. The field of view in a convex mirror is greater than in a plane mirror.

Exploration 2 continued ▶

Answer to
Part 2

Diagrams should show that when light strikes a convex mirror, it is reflected in many directions.

Answer to
Part 3

A real image cannot be made to appear on the paper; thus a convex mirror reflects a virtual image.

Answers to
Part 4

When the pencil moves toward the mirror, its image seems to move toward the viewer. When the pencil moves away, its image also seems to move away.

Images in convex mirrors appear to be behind the mirror, are virtual, and seem smaller than the object.

Answers to
Uses of Convex Mirrors

Convex, side-view mirrors have a larger field of view. Because the images are smaller, objects look farther away than they really are.

Large convex mirrors are used to make more of the store visible to security personnel.

Possible responses: They are used to improve vision at intersections and in subway stations.

CROSS-DISCIPLINARY FOCUS

Art

Have students draw a picture of a reflection from a convex mirror. Have them trade drawings with a partner and see if the partner can identify the subject of the drawing.

FOLLOW-UP

Reteaching

Have students make videos of their actions reflected in convex mirrors. Videos should not only be entertaining, but also provide explanations of how convex mirrors create the effects seen in the videos.

EXPLORATION 2, *continued*

PART 2

How Convex Mirrors Affect Light

Use modeling clay to support a convex mirror in your light box. Insert a smoking splint, and then shine a flashlight or projector beam through the hole in the foil window. Draw a diagram to show what happens to the light that strikes the convex mirror.

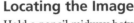

Smoking splint — Convex mirror

PART 3

Classifying Convex-Mirror Images

Recall that plane mirrors produce virtual images. Which type of images do convex mirrors produce—real or virtual? The following activity will help you answer this question.

In a darkened room, place a lit candle near one end of a table.

Place a convex mirror supported with modeling clay at the other end of the table, facing the candle. Hold a piece of paper between the candle and the mirror, but slightly to one side so as not to block the light from the candle. Move the paper slowly from side to side and back and forth to see whether you can focus an image of the candle on it, that is, obtain a real image. Are you able to obtain a real image?

PART 4

Locating the Image

Hold a pencil midway between your face and a convex mirror. Now move the pencil toward the mirror. How does the image seem to move? Move the pencil away from the mirror. How does the image seem to move? Where does the pencil seem to be?

Summary

Based on the observations you made in this Exploration, are images in a convex mirror behind or in front of the mirror? real or virtual? smaller, larger, or the same size as the object?

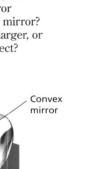

? — Convex mirror

Uses of Convex Mirrors

Consider the following applications of convex mirrors, and think of the characteristics that make them so useful.

- A car's passenger-side rearview mirror is usually convex. Why is this so? Such mirrors usually have a warning attached: "Objects may be closer than they appear." Why?

- Large convex mirrors are found in many stores. Why?

How many more uses for convex mirrors can you find?

Assessment

Have students make drawings showing a reflected object and its image in a plane mirror and convex mirror.

Extension

Have students construct a ray-tracing diagram that illustrates why a convex mirror provides a greater field of view than a plane mirror does. Ask: As the curvature of the convex mirror increases, does the field of view increase, decrease, or stay the same? *(The field of view increases.)*

Closure

Have students make a list of all the convex reflecting surfaces they can find at home. Have students describe how the images they obtain from the surfaces vary. Also ask students to explain why different convex surfaces produce different images. *(Different images are produced due to different amounts of curvature.)*

Converging Lenses and Real Images

What do a camera, binoculars, eyeglasses, and an animal's eyes have in common?

If you said, "They all contain lenses," you are correct. A **lens** is a transparent solid with curved surfaces. Lenses that are thicker in the center than at the edges are called converging lenses. What do you think a **converging lens** does to light? **A**

As the name implies, converging lenses cause light rays to come together. Examine the cross sections of lenses shown below.

Which would you classify as converging lenses? *Farsightedness* is a condition in which things far away are in focus but things nearby are not. One of these converging lenses is used to correct farsightedness. Which do you think it is and why?

The lens shown at the far left of the diagram is called a double-convex lens. What do you think it could be used for? **B**

469

FOCUS

Getting Started

Display a pair of eyeglasses and a magnifying glass. Ask: What do these items have in common? *(Both contain lenses.)* Ask students to think of uses for lenses. *(Cameras, projectors, microscopes, telescopes, binoculars)* Point out that in this lesson they will learn about a kind of lens called a converging lens.

Main Ideas

1. Double-convex lenses are a type of converging lens.
2. Real or virtual images can be formed by converging lenses, depending on how far the object is from the lens.
3. Real images formed by converging lenses are always upside down and on the opposite side of the lens from the object.
4. Virtual images formed by converging lenses are always right-side up and on the same side of the lens as the object.

TEACHING STRATEGIES

Answers to *In-Text Questions*

Ⓐ Converging means coming together at a point. Students may therefore conclude that a converging lens causes light passing through it to come together at a point.

Ⓑ Students should be able to identify the first, fourth, and sixth lenses from the left in the diagram as converging lenses because they are thicker at the center than at the edges.

Students may recognize that the fourth lens is the type used in eyeglasses for farsighted vision. Students may also recognize that a magnifying glass uses a double-convex lens.

★ A Discrepant Event Worksheet is available to introduce Lesson 3 (Teaching Resources, page 40).

LESSON 3 ORGANIZER

Time Required
2 class periods

Process Skills
predicting, hypothesizing, measuring, classifying, inferring, communicating, comparing, contrasting

New Terms
Converging lenses—lenses that are thicker in the center than at the edges
Focal length—the distance from the center of a converging lens to its focal point

Focal point—the point at which light converges after it passes through a converging lens
Lens—a transparent solid with curved sides

Materials (per student group)
Exploration 3, Part 1: book or piece of paper with large print; double-convex lens; **Part 2:** light box from Building a Light Box; wooden splint; a few matches; double-convex lens; small ball of modeling clay; small flashlight; safety

continued ▶

Before students begin the Exploration, you may wish to point out that the type of converging lens they are using is a double-convex lens.

Answers to
Part 1

2. When the lens and eye are farthest from the print, the print appears inverted and smaller than the print's actual size.

3. When the lens was only a few centimeters from the print, the image was larger, clearly in focus, and erect.

4. As the lens is moved slowly toward the print, the print becomes larger but is still inverted. Then, as the lens continues to move toward the print, the print becomes very large, fuzzy, and distorted. When the lens moves even closer, the image becomes slightly smaller (though still larger than the print), very clear, and erect.

5. a. inverted
b. larger
c. distorted
d. erect; larger
e. decreases

 An Exploration Worksheet is available to accompany Exploration 3 (Teaching Resources, page 41).

Have students identify all of the items they can find at home that contain converging lenses. You may wish to present a prize to the student who creates the longest list.

Converging Lens Experiences

You Will Need

- a book or piece of paper with writing on it
- a double-convex lens
- a light box
- modeling clay
- a candle
- a jar lid
- a screen
- a wooden splint
- matches
- a flashlight
- a blank index card

PART 1

Strange Sights

1. Place a book or a piece of paper with large print on it about 1 m from your face. Hold a double-convex lens about halfway between you and the book. Slowly move both the lens and your eyes closer to the book. Continue until the lens touches the book.

2. Describe the appearance of the print when you first looked at it through the lens. Was the print erect or inverted? Determine whether the image of the print was larger than, smaller than, or the same as the print's actual size.

3. Describe the appearance of the print when the lens was only a few centimeters away from it.

4. Repeat the procedure slowly, making notes in your Science-Log of every change that takes place in the "image" of the print as you bring the lens closer and closer to the print.

5. On the basis of your observations, choose the word that best fits in each of the following items. You may have to double-check your observations.

   a. When the converging lens is far away from the object, the image of the object formed by the lens is (erect, inverted).

   b. When the converging lens is moved slowly toward the object, at first the image of the object gets (smaller, larger).

   c. As the converging lens moves closer to the object, at one point the image becomes (clear, distorted).

   d. When the converging lens gets even closer to the object, the image is changed. It is now (inverted, erect). It is also (smaller, larger) than the object.

   e. As the converging lens moves still closer to the object, the image (increases, decreases) in size.

ORGANIZER, continued

goggles; **Part 3:** double-convex lens; candle; jar lid; 2 small balls of modeling clay; a few matches; blank white index card; safety goggles

Teaching Resources
Discrepant Event Worksheet, p. 40
Exploration Worksheet, p. 41
Transparency Worksheets, pp. 45 and 47
Transparencies 72–74
SourceBook, p. S135

How Converging Lenses Affect Light

1. Set up your light box, and insert a smoking wooden splint for a few seconds. Mount a double-convex lens on some modeling clay so that the lens lies in the beam of light from the flashlight.

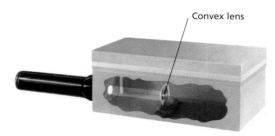

Convex lens

2. Draw a diagram of the beam of light before and after it reaches the lens. Is there any difference in the behavior of a beam of light passing through a double-convex lens and a beam bouncing off a convex mirror? If so, describe the difference.

3. When the beam of light passes through a converging lens, it comes to a point at a position known as the *focal point* of the lens. Label the focal point in your diagram.

   Estimate how far, in centimeters, the focal point is from the lens. This distance is called the *focal length* of the lens.

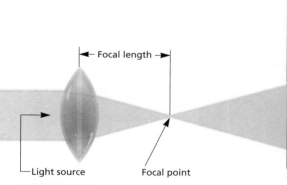

Focal length

Light source   Focal point

PART 3

Converging-Lens Images

1. Using a double-convex lens, find the image of a lighted candle on a blank white index card. The card should be on the side of the lens opposite the candle. Discover in what positions the lens, candle, and screen should be located in order to obtain the following:

   a. an image of the object larger than the object itself

   b. an image the same size as the object

   c. a smaller image

   It is sometimes easier to leave the lens in one position. Start with the object fairly far from the lens. Try to locate its image on the screen. Then systematically change the position of the object.

2. For each situation in step 1, draw a diagram in your ScienceLog.

3. You have been looking at *real images* made by a double-convex lens. Why are these images real and not virtual? Do converging lenses ever form virtual, erect images? (See Part 1.) How can you make this happen? Position the candle and lens in such a way that you get this type of image. Where does the image seem to be? How does it compare in size with the object? Why couldn't you use the screen to locate the virtual image?

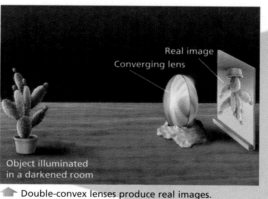

Real image

Converging lens

Object illuminated in a darkened room

▲ Double-convex lenses produce real images.

471

Answers to
Part 2

2. For a sample diagram, see page S212. Students should discover that light seems to be concentrated, or focused, at one point after it passes through a double-convex lens, while light spreads out in many directions when it bounces off a convex mirror.

3. Results may vary due to differences in lenses. (Explain to students that beyond the focal point, the light diverges again, as shown in the drawing at the bottom left of this page.)

Answers to
Part 3

1. **a.** When the lens is closer to the lighted candle than to the screen, a large, inverted image is formed on the screen.

   b. When the lens is halfway between the candle and the screen, the image is the same size as the object. (At this point, the object and the screen are at a distance from the lens equal to twice the focal length.)

   c. When the lens is closer to the screen than to the candle, the image is smaller than the object.

2. Sample diagrams:

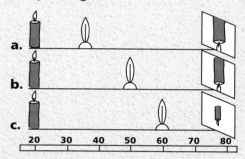

a.

b.

c.

| 20 | 30 | 40 | 50 | 60 | 70 | 80 |

3. These images are real because they are formed by light that is projected onto a screen. Converging lenses can form virtual, erect images when the distance of the object to the lens is very small (less than one focal length). As a person looks through the lens, the image will appear to be behind the lens and larger than the object. A screen cannot be used to locate the virtual image because there is actually no projected light where the image appears to be.

Life and Physical Sciences

When students have finished reading the sections A Special Converging Lens and Eye Exam on page 472, you may wish to check their understanding by asking questions similar to the following:

- What kind of a lens is used in a camera to focus an image onto a piece of film? (*A double-convex lens*)

- Why is a double-convex lens used for this purpose? (*It bends the light coming from an object so that an image of the object can form on the film.*)

Answers to
Picture This

Other uses of converging lenses may include telescopes, microscopes, and projectors.

Encourage students to make their lesson plans. Some students may be interested in doing additional research to find out more about how a camera works. When students finish their plans, call on volunteers to share what they have written with the class.

Answers to
A Special Converging Lens

* Cornea
* Sclera
* Lens
* Retina
* Iris

Answer to
Eye Exam

1. The iris is analogous to the shutter. The retina is analogous to the film.

2. The lens of the eye is analogous to the lens of a camera.

3. Accept all reasonable responses. One difference is that the human eye has a blind spot on the retina, while a frame of film does not.

4. The eye forms real images because the image is projected onto the back of the eye.

 Two Transparency Worksheets (Teaching Resources, pages 45 and 47) and Transparencies 72, 73, and 74 are available for use after completing Lesson 3.

Picture This

As you might imagine, converging lenses have many uses. One common device that uses a converging lens is a camera. Can you name some others?

Study illustrations *A* and *B* below, which show how a converging lens helps a camera produce a photograph. Imagine you have been asked to tell a fifth-grade class how the camera produces an image. Make a lesson plan, as if you were their teacher. Write what you would say or do to help the class understand how the device works.

A Special Converging Lens

You have probably heard the human eye compared to a camera. How accurate is this analogy? Study the diagram of the human eye shown below. In your ScienceLog, match the labeled parts of the eye to the descriptions below. Then answer the questions that follow.

* Transparent front part of the eye; refracts (bends) the light entering the eye
* Layer made of tough tissue that forms the protective outer layer of the eye
* Disk-shaped soft, transparent structure; attached to muscles that can change its shape
* Layer containing millions of light-sensitive cells
* Adjustable diaphragm that alters the size of the eye's light-admitting aperture

Eye Exam

1. Which part of the eye is analogous to the shutter in a camera? Which part is analogous to the film?

2. Which part of the eye is analogous to the camera lens?

3. In what, if any, ways is the eye *unlike* a camera?

4. Does the eye form real or virtual images? Explain.

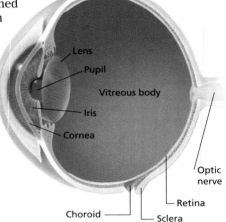

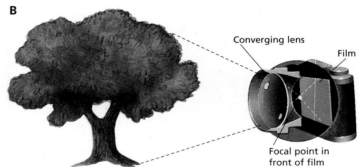

A

Image smaller than object, and inverted

Focal points on each side of lens

Object

B

Converging lens

Film

Focal point in front of film

Photo after development

472

FOLLOW-UP

Reteaching

Provide students with two different sizes of double-convex lenses. Be sure that the curvatures of the lenses are distinctly different. Have students use their light boxes to discover the focal length of each lens. Ask them to share what they learn with the class. (*The lenses should have different focal points and focal lengths.*)

Assessment

Using the diagrams created to answer question 2 from Part 3 of Exploration 3 on page 471, have students explain why the images are real or virtual. (*The images are real because they are projected onto a screen.*)

Extension

Supply students with two convex lenses that have different focal lengths, and ask them to make a telescope with the two lenses. One lens should be held near their eye while the second lens is held farther away. Ask: How far apart do the two lenses need to be to see a sharp image? (*Students should calculate that the distance separating the two lenses needs to be equal to the sum of their focal lengths.*)

Closure

Present students with the following scenario: Steve wanted to start a campfire to cook lunch, but he did not have any matches. However, he did have a magnifying glass and some paper. Have students use their knowledge of convex lenses to explain how a magnifying glass could be used to start a campfire. (*The convex lens focuses sunlight on the flammable material. The concentrated sunlight makes the paper hot enough to ignite.*)

LESSON 4 Real Images and Concave Mirrors

At right are two candles and their images. Which image do you think was produced by a plane mirror and which was produced by a convex mirror? **Ⓐ**

You have studied mirrors that curve outward. You are now going to take a look at mirrors that curve inward. Such mirrors are called **concave mirrors**. A spoon is a simple example of both a concave mirror and a convex mirror. Which side is which? **Ⓑ**

EXPLORATION 4

Discovering Concave Mirrors

You Will Need
- a large, shiny spoon
- a concave mirror
- a metric ruler
- a pencil
- a light box
- a wooden splint
- matches
- modeling clay
- a flashlight
- a candle
- a jar lid
- a sheet of paper

PART 1

A Concave-Spoon Mirror

Look at your image in the concave side of the spoon, holding it 40 cm or so from your eye. What do you see? Now move the spoon very close to you. How has your image changed?

Now, with the other hand, bring a pencil 5–10 cm in front of the spoon. Move it sideways and up and down. What do you observe? Now move the pencil slowly toward the spoon, noting every change in its image. Continue until the pencil touches the spoon.

Exploration 4 continued ▶

473

LESSON 4 ORGANIZER

Time Required
2 class periods

Process Skills
analyzing, measuring, inferring, comparing

Theme Connection
Energy

New Term
Concave mirror—mirror with an inwardly curving surface

Materials (per student group)
Exploration 4, Part 1: large, shiny spoon; **Part 2:** light box from Building

a Light Box; small concave mirror; small ball of modeling clay; wooden splint; a few matches; small flashlight; safety goggles; **Part 3:** sheet of paper; concave mirror; candle; jar lid; 2 small balls of modeling clay; a few matches; safety goggles

Teaching Resources
Exploration Worksheet, p. 49
Theme Worksheet, p. 52
SourceBook, p. S136

FOCUS

Getting Started
Show a large, shiny spoon to the class. Tell students that a spoon possess two types of mirrors: convex and concave. Ask students to speculate about what kind of image a concave mirror produces. *(Concave mirrors produce real, inverted images or virtual, upright images.)* Explain to students that a concave mirror is a mirror that "caves in."

Main Ideas
1. A concave mirror has an inwardly curving surface.
2. Concave mirrors focus a light beam to a point in front of the mirror.
3. Concave mirrors can form either real inverted images or upright virtual images, depending on the distance of the object from the mirror.
4. Concave mirrors are used to collect and reflect light energy.

TEACHING STRATEGIES

Answers to *In-Text Questions*

Ⓐ The image on the left was made by a plane mirror; the image on the right was made by a convex mirror.

Ⓑ The concave side is the side that is used to scoop; the convex side is the back of the spoon.

EXPLORATION 4

PART 1

This simple activity is intended to set the stage for a more in-depth analysis of images produced by concave mirrors.

★ **An Exploration Worksheet is available to accompany Exploration 4 (Teaching Resources, page 49).**

At arm's length, students should see an inverted image of themselves reflected in the concave surface of the spoon. When the spoon is held very close, the image should appear right-side up and larger.

The pencil's image is inverted and moves in the opposite direction from the pencil at a distance of 5–10 centimeters. As the pencil moves closer to the spoon, the image becomes larger than the pencil and is right-side up.

PART 2

The purpose of this activity is to show that light reflected from a concave mirror converges at a focal point. With a small amount of smoke in the light box, students should observe the reflected light forming a focal point in front of the mirror. Students should note that the distance from that point to the center of the mirror's surface is the focal length of the mirror.

PART 3

Before students begin this activity, you may wish to demonstrate how a real image can be obtained from a concave mirror. Darkening the room somewhat will make it easier for students to see the image on the sheet of paper. When the demonstration is over, have students work in small groups to complete each of the steps in Part 3.

Answers to
Part 3

a. A clear, inverted image should appear when the paper is the correct distance between the mirror and the candle. The location of the clearest image in relation to the mirror and the candle will depend on the curvature of the mirror (its focal length) and the distance of the candle from the mirror.

b. The images formed in step (a) are real images. The images are real because they can be projected onto a screen.

c. The real images produced by concave mirrors appear to be in front of the mirror.

EXPLORATION **4,** *continued*

The following activities will help you to understand why concave mirrors form the images they do.

PART 2

Light and Concave Mirrors

Set up the light box as before, but this time include a concave mirror supported by a piece of modeling clay. Use a smoldering wooden splint to put smoke into the light box. Aim the flashlight through the hole and at the concave mirror.

a. Draw a diagram in your ScienceLog to show what happens to the light when it strikes the concave mirror.

b. Label the focal point on your diagram. Label the focal length (distance between the focal point and the mirror) on your diagram.

Concave mirror

474

PART 3

Concave-Mirror Images

Hold a sheet of paper between a concave mirror and a candle. The paper should be slightly off to one side, so that light from the flame can reach the mirror. Move the mirror and its base slowly back and forth toward the paper. Look for an image on the paper.

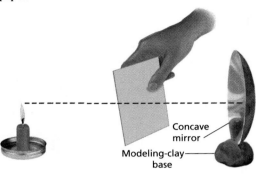

Concave mirror

Modeling-clay base

a. Was a clear image of the candle formed? If so, draw a diagram to show the locations of the candle, the paper, and the concave mirror when the image was clearest.

b. Are the images formed by concave mirrors real or virtual? How can you tell?

c. Do the images appear to be behind the mirror or in front of it?

d. Gradually move the candle toward the mirror. Keep the paper between the candle and the mirror, but adjust the paper to get a sharp image. What change takes place in the location of the image? Is the size of the image always the same?

e. Continue the experiment until you can construct three diagrams. The first should show where the candle, paper, and mirror are placed if a very small image is desired. The second should show the setup that produces an image the same size as the object. The third diagram should show how the candle, paper, and mirror should be arranged to produce an image larger than the object.

f. Now place the candle very close to the mirror. Look into the mirror. What do you see? What kind of image is it—virtual or real? How can you tell?

---

d. As the candle is slowly moved toward the mirror, the image will be farther away from the mirror, and the image will increase in size.

e. • To obtain a small image of the candle, the candle should be more than two focal lengths from the mirror, and the paper screen should be between one and two focal lengths from the mirror.

• To obtain an image the same size as the object, the candle and paper should be the same distance from the mirror—usually twice the focal length.

• To obtain an image larger than the object, the candle should be between one and two focal lengths from the mirror, and the paper screen should be more than two focal lengths from the mirror.

• These three setups are illustrated on page S213.

f. When the candle is placed very close to the mirror (less than one focal length), an upright virtual image is visible in the mirror. Like all virtual images, it appears to be behind the mirror and cannot be projected onto a screen.

Comparing Lenses and Mirrors

1. Using a candle as an object, you were able to get images with a concave mirror that looked like those at right. Which images were real? Which were virtual? What other optical device gave you the same kinds of images?

2. When an image is real, what characteristic do you expect the image to have? When an image is virtual, what other characteristic do you expect it to have?

3. If a real image is smaller than the object, which is located closer to the device forming the image—the object or the image?

4. All of the mirrors shown below form virtual images.

Candle images

A   B   C

Which mirror(s) give(s) only virtual images? Which mirror(s) can give both virtual and real images? Do virtual images appear in front of the mirror, on the mirror, or behind the mirror? Do real images appear in front of the mirror, on the mirror, or behind the mirror?

5. Which mirror or lens would you use to
   a. magnify your finger?
   b. examine the tip of your nose?
   c. see yourself life-size and right-side up?
   d. see yourself smaller than you are, but right-side up?
   e. see yourself smaller than you are and upside down?

Convex lens   Plane mirror   Convex mirror   Concave mirror

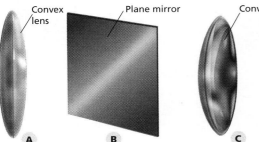  

A   B   C   D

475

Ⓐ Any light placed at the focal point is reflected away from the mirror as parallel beams of light. Flashlights, searchlights, car headlights, and film projectors use this feature of concave mirrors to concentrate and reflect light in a particular direction. In a flashlight, the concave mirror focuses and reflects the light from the bulb.

Ⓑ In a solar cooker, the source of light is at a great distance from the concave mirror. The light collects at the focal point, where the sun's energy is so concentrated that enough heat for cooking is produced.

In a searchlight, a light placed at the focal point of the mirror is reflected away from the mirror as parallel beams of light.

Ⓒ A large main mirror will catch more light, producing a brighter image.

Theme Connection

Energy

Focus question: Can a concave mirror be used to increase the intensity of other forms of energy in addition to visible light? *(Other types of energy, such as sound energy and infrared energy, travel in the same way that light does and can therefore be focused at a particular point by a concave mirror.)* Encourage students to sketch the path along which television signals travel when they are received by a dish-shaped antenna. *(Sketches should resemble the diagram of a solar cooker on this page.)*

 A Theme Worksheet is available to accompany Uses of Concave Mirrors (Teaching Resources, page 52).

FOLLOW-UP

Reteaching

Provide students with two different concave mirrors. Be sure that the curvatures of the mirrors are distinctly different. Have students use their light boxes to find the focal length of each one. Ask them to explain why the focal lengths of the two mirrors are different. *(Because the curvatures of the surfaces of the two lenses are different)*

Uses of Concave Mirrors

If you look around, you'll probably notice that most of the curved reflecting surfaces you see are convex. But there are surprisingly many uses for concave mirrors. For example, what is the purpose of the concave mirror in the device at right? **Ⓐ**

Recall that light passing through a converging lens comes together at a focal point. Similarly, when a beam of light is aimed at a concave mirror, almost all of the light goes to the focal point.

The opposite is also true; if you place the filament of a light bulb at the focal point of a concave mirror, the light reflects away from the mirror as a beam. Suggest other examples of concave mirrors that work on the same principle as a flashlight reflector.

Study the diagrams below of a solar cooker and a searchlight. How do the two devices differ from each other? How are they the same? **Ⓑ**

Some reflecting telescopes have concave mirrors that measure 5 or 6 m in diameter. What purpose could such a large mirror serve? Why would a telescope with a large main mirror work better than a telescope with a small main mirror? **Ⓒ**

As you probably guessed, the bigger the concave mirror in a telescope, the more light it can collect from very faint stars. The best telescopes can intensify the available light a billionfold. Study the photograph and the diagram shown below to determine how a reflecting telescope works.

Most flashlights have a concave mirror. Do you know why?

A solar cooker. Where is the focal point?

A searchlight

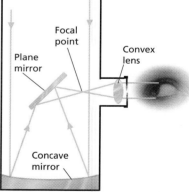
When you look into a reflecting telescope, you see the image at the focal point of the large concave mirror.

Focal point

Plane mirror

Convex lens

Concave mirror

Assessment

Have students make a list of the various kinds of reflective surfaces they notice in a 24-hour period. Have them classify the reflective surfaces as plane, convex, or concave mirrors. Suggest that the class make a composite chart of all the data accumulated from the lists.

Extension

Explain to students that just as there are convex lenses, there are also concave lenses. Suggest that they do some research or perform experiments to discover how concave lenses affect light and what kinds of images they form. Have them share their results with the class.

Closure

Suggest that students draw some diagrams to show how light is reflected from concave mirrors to produce the following images:

- a real, inverted image smaller than the object
- a magnified, inverted, real image
- a virtual image

The Bending of Light Beams

Light travels through empty space at a speed of about 300,000 km/s. Moving at this speed, someone could travel around the world about $7\frac{1}{2}$ times in 1 second! Light travels slightly slower in air than through empty space. In water, light travels even slower—at about 225,000 km/s. Light's speed in glass is only about 200,000 km/s. That's still pretty fast, though!

When light passes through one material into another, its speed changes and some unusual things happen. Complete the following experiments to find out more about this phenomenon.

EXPLORATION 5

Light-Bending Experiences

You Will Need

- 2 pennies
- water
- a plastic cup
- a pencil
- a felt-tip marker
- rubbing alcohol and vegetable oil
- a protractor
- a 600 mL or 1 L beaker
- milk
- a flashlight
- chalk dust
- a small piece of cardboard
- aluminum foil
- a rubber band

PART 1

Which Penny Is Closer?

Fill a plastic cup with water until nearly full. Place a penny inside the container, up against the side. Put another penny on the table outside the cup, close to the penny inside the cup. Look directly down at the pennies.

a. Describe what you see.

b. Which penny seems closer?

c. Move your head slowly from side to side while observing the two pennies from above. Does either penny appear to move across the background of the table? If so, which one?

Water

Remember, reflected light from the outside penny comes from the penny, through the air, and to your eye. Reflected light from the penny in the water first passes through the water and then through the air to reach your eye. What effect occurs when light must pass through two materials to reach your eye?

PART 2

Measuring Apparent Depths

Place a penny in a cup filled with water. Look down at the penny. Indicate where the penny appears to be by making a mark on the side of the cup while still looking down. Compare the apparent depth of the penny with the real depth.

Substitute alcohol and then vegetable oil for the water in the cup. Compare the depths at which you see the pennies.

Water, alcohol, or vegetable oil

Exploration 5 continued ▶

477

LESSON 5 ORGANIZER

Time Required
2 class periods

Process Skills
comparing, contrasting, measuring, analyzing, inferring, observing

New Terms
Refraction—the bending of a beam of light as it passes from one material into another

Total internal reflection—the complete reflection of a beam of light by the inner surface of a medium so that none of the light leaves the medium

Materials (per student group)
Exploration 5, Part 1: plastic cup filled with water; 2 pennies; **Part 2:** 3 plastic cups, one filled with water, one with rubbing alcohol, and one with vegetable oil; 2 pennies; felt-tip marker; **Part 3:** 600 mL (or 1 L) beaker; 400 mL (or 650 mL) of water; **Part 4:** 600 mL (or 1 L) beaker; 400 mL (or 650 mL) of water; a few drops of milk; eyedropper; stirring rod; small piece of cardboard, about 10 cm × 10 cm; hole punch; flashlight; about 5 mL (or

continued ▶

FOCUS

Getting Started

Ask students if they have ever noticed that stars "twinkle" in the night sky. Explain that this happens because light from the star is bent as it passes through the atmosphere. Because the atmosphere is in continuous motion, the amount that the light is bent is constantly changing, which results in a twinkling effect. Ask: If you were in space, would you still see stars twinkling? Why or why not? *(No, because there is no atmosphere to bend the light)*

Main Ideas

1. The speed of light changes when it passes through different kinds of transparent materials.

2. Light travels more slowly in water than in air, and more slowly in glass than in water.

3. The refraction, or bending, of light occurs when light passes from one material into another.

TEACHING STRATEGIES

EXPLORATION 5

Answers to
Part 1

a. The penny in the water may appear slightly larger. Otherwise, the two coins should appear the same.

b. The penny observed through the water appears closer than the one placed outside the container.

c. The penny in the water seems to move across the background of the table.

When light must pass through two materials before it reaches the eye, it bends. In this case, the bending of the light causes the penny in the water to appear closer.

★ An Exploration Worksheet is available to accompany Exploration 5 (Teaching Resources, page 53).

EXPLORATION 5, continued

Answers to
Part 2, page 477

Students should observe that the apparent depth of the penny is less than the real depth. The penny seems slightly closer to the surface when placed in the alcohol and even closer when placed in the vegetable oil.

Answers to
Part 3

a. The pencil appears to be bent or broken at the water's surface for all angles except 90 degrees (when the pencil is perpendicular to the water's surface).

b. When seen from above, the pencil is bent upward (toward the surface) below the surface of the water.

Answers to
Part 4

a. The beam of light appears to bend downward (away from the surface) as it enters the water.

b. The beam of light is going into the water. When you observe the underwater part of the pencil, you see it as a result of light coming to your eyes from the pencil.

c. The part of the pencil that was underwater appeared to bend upward (toward the surface) because the light was moving from the water into the air. The light beam from the flashlight seemed to bend downward (away from the surface) where it entered the water because the light was moving from the air into the water and to the eye.

Answers to
Part 5

a. The beam of light appears to bend downward as it comes out of the top of the water.

b. Total internal reflection will occur at an angle of about 41.5 degrees between the incident beam and the water's surface.

PART 3

Bent Objects

Fill a large container, such as a 600 mL or 1 L beaker, two-thirds full of water. Put a long pencil into the container. Observe the pencil from all sides and from above. Save this setup for Parts 4 and 5.

a. Describe the appearance of the pencil from different angles.

b. When seen from above, is the part of the pencil that is below the surface apparently bent upward or downward?

PART 4

Bent-Light Paths

Use the setup from Part 3 (minus the pencil). Stir in 2 or 3 drops of milk—just enough to give the water a faint milky appearance. Cover the top of the setup with a piece of cardboard that has a hole about 0.5 cm in diameter at the center. Darken the room. Shine a flashlight through the hole, at an angle, into the water. Observe the light from the side. Shake a small amount of chalk dust into the air above the water so that the beam of light is visible above the water, as well as in the water.

a. How does the beam appear to bend when it goes into the water? Look at the beam through the side of the container. Draw a diagram showing how the beam bends.

b. Is the beam of light going into or coming out of the water? When you observed the underwater part of the pencil in

Part 3, did you see it as a result of light going *from* your eyes to the pencil or light coming *to* your eyes from the pencil?

c. Why did the part of the pencil that was underwater appear to bend upward, while the light beam seems to bend downward where it enters the water?

PART 5

Total Internal Reflection

Darken the room. Move the container near the edge of the table and direct the flashlight beam upward from below the water level through the side of the container. Use the flashlight setup shown in the diagram below. Again shake a bit of chalk dust over the water to make the beam visible as it comes out of the water.

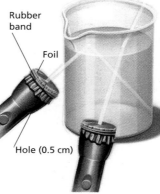

Rubber band

Foil

Hole (0.5 cm)

a. Do you observe bending? Make a sketch showing how the beam bends. Slowly raise the flashlight so that the angle between its beam and the surface becomes smaller. What happens to the beam in the air? At one particular angle the

beam stops coming out of the surface. At all smaller angles the beam is reflected back into the water as though there were a mirror at the surface. This phenomenon is called *total internal reflection*. Total internal reflection occurs only when light reaches a boundary between one material and a second material through which light can travel at a faster speed. Thus, light traveling in water can be totally internally reflected when it comes to a boundary between the water and air—provided the light hits the boundary at a sufficiently small angle.

b. Using a protractor, measure the angle between the light beam and the water surface when total internal reflection is first observed.

> Light has some peculiar characteristics. To learn more about light, turn to pages S142–S143 of the SourceBook.

ORGANIZER, continued

1 teaspoon) of chalk dust; **Part 5:** 600 mL (or 1 L) beaker; 400 mL (or 650 mL) of water; a few drops of milk; eyedropper; stirring rod; flashlight; rubber band; 15 cm × 15 cm piece of aluminum foil; about 5 mL (or 1 teaspoon) of chalk dust; protractor

Teaching Resources
Exploration Worksheet, p. 53
Transparency 75
SourceBook, p. S142

Interpreting Your Findings

Check your understanding of light bending by answering the following questions in your ScienceLog.

1. Does a penny viewed in water appear deeper, shallower, or at the same depth as it really is?

2. In the case of question 1, is light going by the path eye → air → water → penny or by the path penny → water → air → eye?

3. In situation A, which path will the light beam most likely follow—1, 2, or 3?

4. a. In situation B, light is traveling from the water to the air to the person's eye. Which beam of light—1 or 2—will most likely reach the person's eye?

   b. Which beam of light—1 or 2—will probably not exit the water, but will instead be reflected as R?

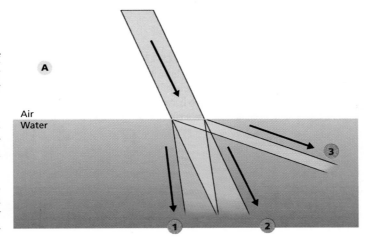

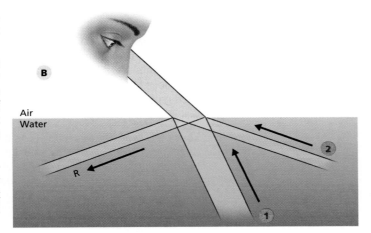

In Conclusion

As you have seen, light bends when it passes from one medium to another. The scientific term for this phenomenon is **refraction.** What other examples of refraction can you think of?

479

Transparency 75 is available to accompany Interpreting Your Findings.

Answers to
Interpreting Your Findings

1. A penny in water appears shallower (closer to the surface) than it really is.

2. Penny → water → air → eye

3. The light beam will most likely follow path number 1.

4. **a.** Light beam number 1 will most likely reach the eye.
   b. Light beam number 2 will most likely be reflected back into the water.

Answer to
In Conclusion

Encourage students to realize that refraction occurs any time that light changes mediums, including when light passes through a lens.

Answers to Challenge Your Thinking

1. • (top) The image should be larger than the object.
   • (bottom) The image is virtual and should not be located on the screen. It should be on the same side of the lens as the object, but farther away.

2. Dr. Hughes uses a concave mirror to examine her patients' teeth because it magnifies objects that are placed near it. Her motorcycle has a convex mirror because this type of mirror produces images that are upright. Dr. Hughes can also see a wide-angle view of the scene behind her so that she can operate her motorcycle more safely.

3. The illustration shows the path that light actually takes—it bounces off the pencil, travels upward through the water, travels through the air, and reaches your eye. However, as the light passes from one material into another, its speed changes. This causes the refraction, or bending, of the light. In the picture, the light's path was bent, giving the illusion that the pencil is bent instead.

Meeting Individual Needs

Learners Having Difficulty
Tell each student to place a dime in an empty test tube and then to sketch a circle that is the apparent size of the dime when they look straight down into the test tube. Then have students add water to the test tube until it is about two-thirds full. Again, have students draw a circle that is the apparent size of the dime as they look straight down into the test tube. *(The surface of the water will bulge downward so that the water acts as a concave lens. This will cause the dime to appear smaller when viewed from above.)* Finally, have students pour water into the test tube until it overflows at the rim. Have students again draw a circle the size of the dime. *(The surface of the water will bulge upward so that the water acts as a convex lens. This will cause the dime to appear larger when viewed from above.)*

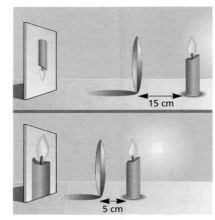

1. Help the Artist
Explain what's wrong in each of these pictures. Then redraw each one so that it's correct. The lens has a focal length of 10 cm.

15 cm

5 cm

2. The Right Mirror for the Job
Dr. Hughes uses a small, slightly concave mirror when examining her patients' teeth. Her motorcycle has a convex rearview mirror. Why does Dr. Hughes use these different types of mirrors? How is each type of mirror appropriate for its situation?

3. Eye Fooled You!
Your eye always views an object as if its light path never changed direction. This explains why a pencil immersed in water appears to bend upward. Write a brief explanation of the accompanying illustration based on what you know about light paths.

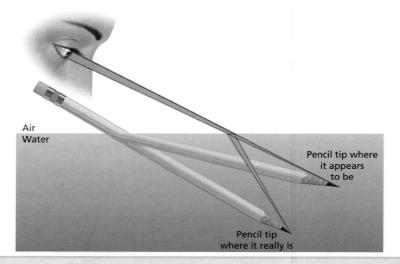

Air
Water

Pencil tip where it appears to be

Pencil tip where it really is

Theme Connection

Changes Over Time
Refraction is the change in the direction of light at the boundary of two materials, such as air and glass. Thus, refraction is a change over space. **Focus question:** In what way is refraction also a change over time? *(Refraction occurs because the speed of light changes and depends on the medium through which it is traveling. Because refraction is caused by changes in the speed of light, it is a change that takes place over time.)*

You may wish to provide students with the Chapter 21 Review Worksheet that is available to accompany this Challenge Your Thinking (Teaching Resources, page 56).

4. Mirror, Mirror

As the photos show, a reflective cylinder produces a different image depending on whether it is held vertically or horizontally. Explain how the cylinder is able to produce such different images. What kind of mirror is the cylinder—convex, plane, concave—or something else? Explain.

5. You Make the Call

You know that optical devices such as plane mirrors, convex mirrors, and converging lenses produce images. You also know that some of the images produced by these devices are real, and some are virtual. The illustration below shows an object and several images of it that were produced by the optical devices just mentioned. Which are real and which are virtual? Which optical device could have produced each image? Where would the object have to be placed in relation to the optical device to get a given result?

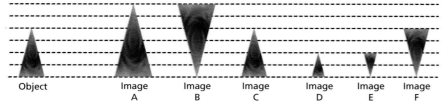

| Object | Image A | Image B | Image C | Image D | Image E | Image F |

6. Ponder This

Use the principles of light refraction to explain why a clear, still pond seems to be shallower than it really is.

Review your responses to the ScienceLog questions on page 462. Then revise your original ideas so that they reflect what you've learned.

481

ScienceLog

The following are sample revised answers:

1. The various effects of the image shown in the photo are due to refraction—the bending of light as it passes from one medium to another (such as from water to air, from water to glass, and from glass to air).

2. A camera lens is a converging lens. Its purpose is to make the light reflecting off an object converge onto the film, creating a smaller, real image of the object that is chemically recorded and developed to produce a photograph.

3. The person's image is distorted because the reflecting surface of a funhouse mirror is not flat, as that of a normal plane mirror would be. Although it does contain some flat areas, a funhouse mirror consists mostly of a variety of convex and concave reflecting surfaces of various curvatures. The result is an image with many different areas of diverging and converging light. Thus, some parts of the image appear larger than the object, some smaller, some stretched out, some squashed, etc.

Answers to Challenge Your Thinking, continued

4. When the cylinder is on its side, it is a convex mirror in the vertical direction and a plane mirror in the horizontal direction. A convex mirror produces an upright image that is smaller than the object, and a plane mirror produces an upright image that is about the same size as the object when the object is close to the mirror. Therefore, the image is about as wide as but shorter than the object. When the cylinder is upright, it is a plane mirror in the vertical direction and a convex mirror in the horizontal direction. Therefore, the image is about as tall as but thinner than the object.

5. Sample answers are as follows. Other solutions are possible. Have students sketch their answers to test their accuracy.

Image	Type	Optical device	Placement
A	virtual	concave mirror	very close
B	real	converging lens	between one and two focal lengths from the lens; the image appears on the opposite side of the lens
C	virtual	plane mirror	in front
D	virtual	convex mirror	in front
E	real	converging lens	more than two focal lengths from lens
F	real	converging lens	exactly twice the focal length from lens

6. A still pond seems to be shallower than it really is because the light reflected from the bottom of the pond bends as it passes from the water to the air. In this case, the bending of the light makes the bottom of the pond appear closer than it actually is.

Integrating the Sciences

Earth and Physical Sciences

Explain to students that mirages are atmospheric effects caused by refraction. Suggest that they do some research to find out how refraction from a hot surface creates a mirage. Then have them make poster diagrams to illustrate how a mirage is formed.

The Big Ideas

The following is a sample unit summary:

Light has no mass and does not take up space. It is readily converted into other forms of energy, and other energy forms can be converted into light. Thus, you can infer that light is energy. (1)

Colored light can be produced by passing white light through filters of various colors. A red filter, for example, will absorb all of the colors of white light except red. The red light will pass through the filter. To produce every color found in white light, white light is passed through a prism. Because the prism separates the colors in white light, a spectrum of all the colors will appear. (2)

Most objects that emit light are hot. The temperature of an object determines the color of light it emits. Relatively cool stars, for instance, emit red light, warmer stars emit yellow light, and very hot stars emit white to bluish light. (3)

A filter is a device that absorbs certain colors of the spectrum while allowing others to pass through. A red filter, for instance, absorbs all the colors of light except red, which is transmitted through the filter. (4)

Objects both absorb and reflect light. The light that they reflect determines their color. For example, a blue car absorbs all the colors of light except blue, which is reflected. Passersby see the reflected blue light and characterize the car as blue. (5)

Light always travels in a straight line through any given medium. When light passes into a different medium (as from air to water), it changes speed and bends. It then continues in a straight line until it changes mediums again. (6)

Mixing all the colors of light will produce white light. The color of paint, however, depends on which color of light it reflects (the remaining colors are absorbed). Mixing colors of paint, therefore, will eventually result in a substance that absorbs all the colors of light and reflects none of them, thus appearing black. (7)

Reflection is the throwing back of light from the surface of a substance. Refraction is the bending of light as it passes from one medium to another. (8)

A plane mirror produces virtual images that are proportional in size to the object and appear to be the same distance away from the mirror. A convex mirror produces virtual images that are smaller than the object. A converging (convex) lens and a concave mirror produce both real and virtual images, depending on the distance from the object to the lens or mirror. If the distance is large, the image is real and smaller than the object. If the distance is small, the image is virtual and larger than the object. All virtual images are upright, and all real images are inverted. For mirrors, real images are on the same side as the object, and virtual images appear to be on the opposite side. For lenses, real images are on the opposite side of the lens from the object, and virtual images are on the same side. (9)

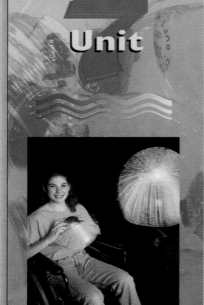

Making Connections

Unit

LIGHT

SOURCEBOOK

You've learned quite a bit about light's properties and behavior. To find out more about light and related phenomena, read the SourceBook. You'll find that there's more to light than meets the eye.

Here's what you'll find in the SourceBook:

UNIT 7
Light: A Split
        Personality? S128
Visible Light S133
Invisible Light S137
High-Tech Light S144

482

The Big Ideas

In your ScienceLog, write a summary of this unit, using the following questions as a guide:

1. How can you tell whether light is matter or energy?
2. How can colored light be produced from white light?
3. What is the relationship between the temperature of an object and the color of the light it emits?
4. What are filters, and how do they work?
5. What gives an object its color?
6. Along what kind of path does light travel?
7. How is mixing colored lights different from mixing colored paints?
8. What is reflection? refraction?
9. How do the various types of lenses and mirrors differ in terms of the images they produce?

Checking Your Understanding

1. Objects lit by a strong, nearby light source cast shadows that expand very quickly, unlike objects that are lit by a faraway light source. Use your understanding of light to explain this.

Homework

The Unit 7 Activity Worksheet makes an excellent homework activity (Teaching Resources, page 62). If you choose to complete the activity in class, Transparency 76 is also available.

2. Why do athletes, such as baseball players, often smear a dull, black substance below their eyes on sunny days?

3. All lenses cause a prismatic effect; that is, they bend each component color of light by a slightly different amount, causing a distorted image. For this and other reasons, the best telescopes always use mirrors rather than lenses.
   a. Why do mirrors not exhibit the problem of the prismatic effect?
   b. In what other ways are mirrors advantageous?
   c. How might you correct the prismatic effect using lenses alone?

4. How could you use two mirrors to see the top of your head?

5. What's inside each mystery box? The only clues you have are the light beams that enter and leave each box.

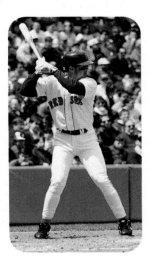

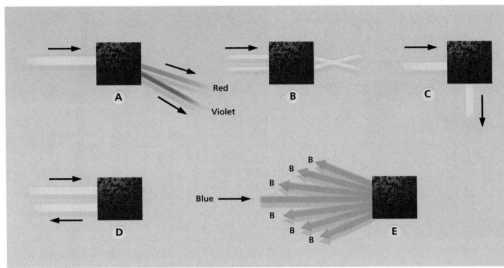

6. concept map Draw a concept map showing the relationship between these words or phrases: converging lens, reflected, real image, refracted, virtual image, light, and plane mirror.

★ **You may wish to provide students with the Unit 7 Review Worksheet that is available to accompany this Making Connections (Teaching Resources, page 63). If you decide to complete this worksheet in class, Transparency 77 is also available.**

Answers to *Checking Your Understanding, pages 482–483*

1. Light radiates from the light source at diverging angles. This light expands very quickly as it moves away from its source. Objects that are close to the light source will therefore block a greater proportion of those rays than will objects that are far away. The shadows of objects close to the light source expand rapidly because they are preventing the light from expanding.

2. Some sports players put dark smears under their eyes to absorb light, thus preventing light from reflecting off their cheeks into their eyes.

3. **a.** Because light is reflected off a mirror and does not pass through the material making up the mirror, the beam is not bent. Thus, no prismatic effect occurs.
   b. Answers may vary. Possible responses include the following: Mirrors are less massive than lenses. Because most incident light reflects off a mirror instead of passing through it, as with a lens, a mirror absorbs much less of the light's energy, and it is possible to use less-intense light beams. Because mirrors provide only one surface of incidence, as opposed to the multiple surfaces provided by lenses, it is easier to focus a light beam with a mirror.
   c. Teachers should note that this is a challenging question. Accept all reasonable responses. Some students may realize that the prismatic effect produced by passing a light beam through one lens can be countered by using another lens to bend the light in the direction opposite to that produced by the first lens.

4. Hold one mirror in front of your face and the other above your head. By orienting the mirrors at slight angles toward one another, you should be able to see the top of your head.

5. *A*—Prism
   B—Double-convex lens
   C—Plane mirror
   D—Two plane mirrors at right angles to one another
   E—Convex mirror

6.

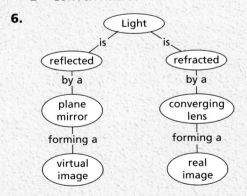

Background

The Russian companies that built *Banner* originally intended it as a solar sail that would use light from the sun to propel spacecraft. Light consists of small bundles of energy called *photons.* These photons would strike the sail and transfer energy to the spacecraft, pushing it forward. When interest in solar sailing waned, the Russians realized that their sail could be used as a mirror instead.

Extension

There are many kinds of orbits, but most human-made satellites travel in one of three types: low altitude, geosynchronous, and sun synchronous. Low-altitude orbits lie in the upper layers of the Earth's atmosphere so that there is almost no air to slow the craft. Satellites in high-altitude geosynchronous orbits move in synchronicity with the Earth's rotation. From the Earth's surface, these satellites appear to be stationary. Finally, satellites in sun-synchronous orbits revolve around the Earth in such a way that they pass over the same place on the Earth at the same time every day.

Ask students to suggest reasons why different types of satellites might travel in each orbit. *(Accept all reasonable responses. Most deep-space observatories, such as the Hubble Space Telescope, operate in low-altitude orbits because of the relatively low launching expenses. Most Earth-observation satellites, such as some weather satellites, follow sun-synchronous orbits in order to record daily changes, such as weather, over as much of the Earth's surface as possible. Most communication satellites are put into high-altitude geosynchronous orbits because they need to relay communication between two fixed locations on the Earth's surface.)*

SCIENCE & TECHNOLOGY

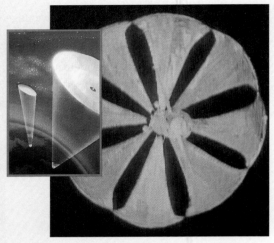

Mirrors in Space

*H*ow would you like to live in continuous daylight? Would you enjoy the extra time to garden or to play outdoors, or would it disrupt your normal cycle of work, play, and sleep? These questions are not as strange as they may seem now that Russian scientists have put a mirror in space.

Shedding New Light

Usually, only the moon reflects a noticeable amount of the sun's light to our planet's dark side.

▲ The moon acts much like a mirror, reflecting light from the sun back to the Earth's surface.

But on February 4, 1995, the Russian space mirror, named *Banner*, illuminated a small area of predawn Europe to about the brightness of a full moon. *Banner* was released from the Russian space station, *Mir*, 362 km above the Earth's surface. The cosmonauts unfolded the 4 kg space mirror like a fan. Its very thin aluminum-coated plastic created a disk 19.5 m in diameter. For 8 minutes the mirror was able to reflect a stream of sunlight 4 km wide! The beam of light moved across the Atlantic and parts of Europe before disappearing into the sunlit side of the globe.

Commercial Possibilities

The success of this experiment convinced some Russian scientists that a series of space mirrors could illuminate large areas of the Earth at night. They speculate that such a system could save billions of dollars in electrical lighting costs, could reduce the use of natural resources, and could allow outdoor projects such as planting and harvesting crops to continue around the clock. However, these predictions may be optimistic considering the dimness of the reflection. Also, to accomplish any of this, scientists must first overcome a technological challenge: the mirrors must be able to focus light on a specific location while orbiting at thousands of kilometers per hour. Presently, the reflected beam of light moves as *Banner* moves. However, the scientists are hopeful

▲ The space mirror *Banner*, or *Znamya* in Russian, unfolds like a huge umbrella. An artist's conception (inset) shows two space mirrors in action.

that they can create space mirrors that will continuously project a beam of light onto one area.

Enlightened Predictions

Many scientists in Russia and around the world are concerned about the idea of perpetual sunlight. How would animals that depend on darkness for hunting and protection survive? How would the growth cycle of plants be affected? How would humans be affected physically and emotionally? Choose one of these questions, and make predictions about how our world could change.

Did You Know. . .

Although some natural satellites follow orbits that move against the rotation of the Earth, virtually all human-made satellites orbit with the Earth's rotation because it takes much less energy to launch a spacecraft using the Earth's rotation.

Enlightened Predictions

Responses will vary depending on the subject chosen. You might want to stimulate critical thinking by having students do some preliminary research on their topic. They should take notes or make an outline of what they find and use this information to support their predictions. Then group students by chosen subject and have them share their ideas with each other.

Neon Art

In 1898 two chemists made an exciting discovery. They passed an electric current through a gas they had collected while experimenting with liquid air. The gas began to glow with a fiery, reddish orange light. The chemists, Sir William Ramsay and Morris Travers, had discovered a new element! They named the element *neon*, from the Greek word meaning "new."

A New Art Form

Not long after the discovery of neon, people began to think of artistic and decorative uses for it. By the 1920s, neon signs were in wide use. Today neon signs are so common that we are likely to take them for granted. But have you ever wondered how they are built or how they work?

▼ **Making neon signs requires skill and patience. The artisan must blow air through the tube as it is heated to prevent the tube from collapsing as it is bent.**

Tubes of Gas

The first step in creating a neon sign or sculpture is to shape the glass. To do this, a glass tube is heated until it becomes flexible, and then it is gently bent into shape. Next, electrodes are installed in each end, and a vacuum pump is used to pump the air out of the tubing. Finally, the tube is filled with invisible neon gas.

Color Magic

When an electric current is applied to neon gas, a reddish orange light is produced. But if a few drops of mercury are added, the light changes to a brilliant blue! The color changes because the atoms of different gases give off different-colored light when they are energized. However, this is not the only

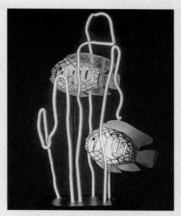

▲ **The discovery of neon led to new art forms.**

way to change the color of a neon light. Green, yellow, and orange light can be created by coating the inside of the glass tubes with powdered phosphors. When ultraviolet rays from the mercury vapor strikes the phosphors, they glow, giving off the desired color. Still other colors can be achieved by replacing the neon gas with argon gas or by using different colors of glass. With all of these techniques at their disposal, neon artists have a complete palette of glowing colors to work with!

Think About It!

When you see a green neon sign or a red stoplight, the color you see is the color of the light itself. How is this different from what you see when you look at a green leaf or a bright red T-shirt?

Answers to *Think About It!*

When you look directly at a light source, you are directly viewing colors of light. However, when the color of an object is viewed, you are seeing only the colors of light that are reflected, or not absorbed, by the object. For a more thorough explanation, see the sections on additive and subtractive color in this unit.

Background

Neon, a relatively rare element, is an odorless, colorless gas found naturally in the Earth's atmosphere. Ordinarily, gases are poor conductors. However, they can be made to conduct electricity in a tube by decreasing the pressure inside the tube and applying a voltage across the ends of the tube. This causes electrons to stream from one end of the tube to the other and to collide with the electrically neutral gas atoms. The collisions cause electrons in these atoms to gain energy, and the electrons are pushed to higher energy levels within the atom. When these electrons return to their original levels, they give off the extra energy in the form of light. Because atoms of different elements have different electron-level structures, different colors of light can be produced by using different gases.

Extension

You can separate the colors of light with a diffraction grating, which is easily available from most science supply houses. Provide students with a diffraction grating, and, if any neon signs are available, have them look at the light through the grating and explain what they see. They can also look at sunlight (but not directly at the sun!), a normal light bulb, a fluorescent light, and a street lamp. What will they see? *(They will either see separate lines of color or continuous bands of color, depending on the light source.)*

Meeting Individual Needs

Learners Having Difficulty

Ask a local merchant whether the class may borrow a small neon sign to display in the classroom for a week. Have students explain in their own words how the neon sign works. If the sign has different colors, have the students do some research and explain to the class why the colors are different. It might be helpful to know that in a neon sign, mercury glows blue, neon glows red, and hydrogen glows purplish red. All three gases are commonly found in neon tubes.

Background

Other ways in which an individual can help reduce light pollution include installing low-wattage bulbs in porch lights and using outdoor light sources only when necessary. One can also join an organization such as the International Dark-Sky Association, which helps educate the public through lectures and newspaper articles.

Visible light pollution is not the only form of pollution that affects astronomy. So far, radio astronomers have not had to contend much with the problem of light pollution because it does not interfere with observations of radio signals. However, as more appliances and telecommunications systems generate more radio noise, radio signals from space become lost in the noise. One solution is to put observatories in orbit around the Earth, but launching satellites is very expensive, and the observatories can be subject to collisions with orbiting debris. Additionally, communications satellites already in orbit could also produce radio noise.

Extension

Many of the observatories in the United States are in danger of becoming inoperative if nearby cities continue to grow. Provide the class with a map of the United States. Point to an observatory site, and ask the class what major city is nearest to the site. You might want to have them find out the city's rate of population growth as well as the kind of astronomical research that is done at the observatory. Important observatories include Lick Observatory near San Francisco, Mount Wilson Observatory in the Los Angeles area, and Kitt Peak National Observatory near Tucson.

EYE ON THE ENVIRONMENT

Light Pollution

Lights from cities can be seen from space, as shown in this photograph taken from the space shuttle *Columbia*. Bright, uncovered lights (inset) create a glowing haze in the night sky above most cities in the United States.

Have you ever seen a large city at night from a distance? Soft light from windows forms the outlines of office buildings. Bright lights from stadiums and parking lots shine like beacons. Scattered house lights twinkle like sparkling jewels. The sight can be stunning!

Unfortunately, these lights are considered by astronomers to be a form of pollution. This pollution, called light pollution, is reducing our ability to see beyond our own cosmic neighborhood. In other words, the light pollution from cities and other light sources is making it difficult to see the night sky. Astronomers around the world are losing their ability to see through our atmosphere into space.

Sky Glow

Twenty years ago, stars were very visible above even large cities. The stars are still there, but now they are obscured by the glow of city lights. This glow, called sky glow, is created when light reflects off of dust and other particulates suspended in the atmosphere. Even remote locations around the globe are affected by the cumulative effects of light pollution. The sky glow affects the entire atmosphere to some degree.

The majority of light pollution comes from outdoor lights such as automobile headlights, street lights, porch lights, and the bright lights illuminating parking lots and stadiums. Other sources include forest fires and gas burn-offs in oil fields. Air pollution makes the situation even worse, adding more particulates to the air so that there is even greater reflection.

A Light of Hope

Unlike other kinds of pollution, light pollution has some simple solutions. In fact, light pollution can be cleaned up in as little time as it takes to turn off a light! While turning off most city lights is impractical, several simple strategies can make a surprising difference. For example, using covered outdoor lights instead of uncovered ones keeps the light angled downward, preventing most of the light from reaching particulates in the sky. Also, using motion-sensitive lights and timed lights helps eliminate unnecessary use of light. Many of these strategies also save money by saving energy.

Astronomers hope that public awareness will help improve the dwindling visibility in and around major cities. Some cities, including Boston and Tucson, have already made some progress in reducing light pollution. It has been projected that if left unchecked, light pollution will affect every observatory on Earth within the next decade.

See for Yourself

With your parents' permission, go outside at night and find a place where you can see the night sky. Can you count the number of stars that you see? Now turn on a flashlight or porch light. How many stars can you see now? Compare your statistics. By what percentage was your visibility reduced?

486

Discussion

Even with low levels of light pollution, the light from astronomical objects is often distorted by the atmosphere. The strength of the distortion is determined by phenomena such as atmospheric dispersion, absorption, reddening, and refraction. Ask students to give some common everyday examples of these effects. (*Examples include mirages, rainbows, the twinkling of stars, and the redness of the sunset.*)

Answers to
See for Yourself

Student responses are likely to vary dramatically depending on how dark the sky is. In a medium to large city, it is not unusual to be able to see only a handful of stars, and a nearby light will make this number even less. In the countryside, students will likely see a sky full of stars. (Ironically, brightly lit skies can aid some observations because then only the very brightest stars are visible, thus aiding their identification.)

Science Snapshot

Garrett Morgan (1877–1963)

One day in the 1920s an automobile collided with a horse and carriage. The riders were thrown from their carriage, the driver of the car was knocked unconscious, and the horse was fatally injured. One witness of this scene was an inventor named Garrett Morgan, and the accident gave him an idea.

A Bright Idea

Morgan came up with an idea for an electrical and mechanical signal to direct traffic at busy intersections. Morgan's idea for the signal included electric lights of different colors. These lights ensured that the signal could be seen from a distance and could be clearly understood.

Morgan patented the first traffic light in 1923. This traffic light looked very different from those used today. Unlike the small, three-bulb signal boxes that now hang over most busy intersections, the early versions were T-shaped and had the words *stop* and *go* printed on them.

Morgan's traffic light was operated by a preset timing system. An electric motor turned a system of gears, which operated a timing dial. As the timing dial turned, it caused switches to turn on and off. The result was a repeating pattern of green, yellow, and red lights.

Morgan's traffic light was an immediate success, and he sold the patent to General Electric

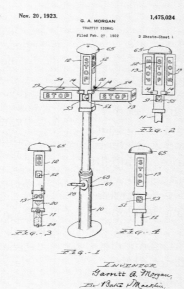

▲ Morgan's patent for the first traffic light

Corporation for $40,000, a large sum in those days. Since then, the traffic light has been the mainstay of urban traffic control.

The technology of traffic lights, however, continues to improve. Some newer models, for example, can change their timing based on the traffic needs for a particular time of day. Some models can even use sensors installed in the street to monitor traffic flow. Other models have

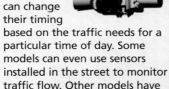

sensors that can be triggered by an ambulance so that they automatically turn green, allowing the ambulance to pass.

The Man

Born in Paris, Kentucky, Garrett Morgan was one of 11 children. Having received only an elementary school education, Morgan educated himself. When he was 14, he left home to seek work in Cleveland, Ohio. He was broke and had few skills, but he was mechanically inclined and quickly taught himself enough about sewing machines to get a job repairing them. He worked very hard, and in 1907 he opened his own sewing-machine repair shop. Just two years later, he bought a house and started a tailoring business.

One night when Morgan was experimenting with a lubricant for the sewing needle, he noticed that the lubricant straightened the fuzz on a fur cloth he had been using. He had accidentally invented hair straightener, the first of many products that Morgan invented.

Think About It

Traffic control is not the only system that uses light as a signal. What are some of these other systems, and what makes light such a good medium for communication?

Answers to *Think About It*

Other systems that utilize light for communication include Morse code (such as when ships communicate with flashing lights), spotlights, advertisements, and crosswalk signals. In addition, the aisles of theaters and airplanes are often equipped with lights to guide foot traffic. Light is a good method of communication because it is not affected by noise, it works well in all lighting conditions, and it is fairly inexpensive. (See also the Extension at right.)

Background

Traffic lights are crucial to our modern transportation needs because in most situations they are far more effective at regulating traffic than other methods. Modern traffic signals are even designed to account for colorblindness. Although colorblind drivers cannot distinguish the color of each light, they can interpret the signal by memorizing the position of each color.

In addition to the traffic light, Morgan also invented a woman's hat fastener, a friction-drive clutch, and a safety hood that protects the wearer from dangerous airborne materials.

The first major test of Morgan's safety hood occurred in 1916 when an explosion in a Cleveland tunnel left 32 men trapped amid smoke and toxic gases. Morgan and his brother heard of the disaster, raced to the scene, donned their safety hoods, and went into the tunnel. Although not all of the men survived, Morgan and his brother retrieved every single man from the tunnel. For his actions, Morgan received a medal from a group of Cleveland citizens as well as an award from the International Association of Fire Engineers.

Extension

Light is often used instead of sound to send signals. Ask: What are some of the advantages and disadvantages of each? Are there any traits shared by both methods? *(Accept all reasonable responses. Both systems can be used to bridge language barriers, as long as the sound signals are not words but rather noises such as traffic whistles or car horns. One advantage of light over sound is that light can usually be detected at greater distances. In a loud urban environment, a sound signal can easily be lost among other noises. A disadvantage of a light signal is that the recipient has to be looking directly at the signal to notice it. If it is loud enough, a sound signal can be heard no matter what direction the listener is facing. However, neither system can be used by all people because many individuals have impaired hearing or sight or both.)*

Unit 8 CONTINUITY OF LIFE

Unit Overview

This unit introduces students to genetics, the study of how characteristics are passed from one generation to the next. In Chapter 22, Lesson 1, students study inherited characteristics and investigate the characteristics of living things. Then, in Lesson 2, students examine the experimental evidence that shows how life comes only from preexisting life. In Lesson 3, students observe living cells and take an imaginary trip inside a cell to witness mitosis. In Lesson 4, students learn that many inherited traits appear in predictable patterns in the offspring of a genetic cross. In Lesson 5, students explore the development of human life from conception to birth.

In Chapter 23, students examine the structure and function of DNA and identify its relationship to chromosomes and genes. Then they investigate the importance of the environment to the development of living things and discuss the future of genetic research.

Using the Themes

The unifying themes emphasized in this unit are **Structures, Cycles,** and **Changes Over Time.** The following information will help you incorporate these themes into your teaching plan. Focus questions that correspond to these themes appear in the margins of this Annotated Teacher's Edition on pages 494, 497, 504, 509, 518, 537, and 546.

The theme of **Structures** is woven throughout the discussion of the DNA double helix. Students learn how the processes of meiosis and mitosis depend on the structure of DNA and the fact that its unique structure allows for self-replication.

Much of the material on reproduction in this unit incorporates the theme of **Cycles.** Students see how genetics and heredity relate to the cycle of life, consisting of birth, reproduction, and death.

Changes Over Time is a dominant theme throughout the unit because of the emphasis placed on the predictable changes that occur through genetic inheritance, mitosis, meiosis, and fetal development. Changes Over Time is also a dominant theme in the final lesson of the unit. Students extrapolate current research trends to predict what exciting and challenging changes might result from future technology.

Using the SourceBook

In Unit 8 of the SourceBook, students explore the science of genetics in depth from the work of Gregor Mendel to the discoveries of chromosomes and cell division. The structure of DNA and its role in making proteins is also examined. Human sexuality, reproduction, and development are also presented.

Bibliography for Teachers

Bishop, Jerry E., and Michael Waldholz. *Genome.* New York City, NY: Simon and Schuster, 1990.

Marks, Jonathan, and R. Brent Lyles. "Rethinking Genes." *Evolutionary Anthropology* 3, no. 4 (1994): 139–146.

Pierce, Benjamin A. *A Practical Guide to Human Heredity.* New York City, NY: John Wiley & Sons, Inc., 1990.

Pollack, Robert. *Signs of Life: The Language and Meanings of DNA.* New York City, NY: Houghton Mifflin Co., 1994.

Bibliography for Students

Aronson, Billy. *They Came From DNA.* New York City, NY: Scientific American Books for Young Readers, 1993.

Asimov, Isaac. *How Did We Find Out About Our Genes?* New York City, NY: Walker Publishing Company, Inc., 1983.

Films, Videotapes, Software, and Other Media

Exploring Genetics and Heredity
CD-ROM (Macintosh, MS-DOS)
Queue, Inc.
338 Commerce Dr.
Fairfield, CT 06430

Genetic Biology
Film or videotape
Coronet/MTI
108 Wilmot Rd.
Deerfield, IL 60015

Heredity
Videotape
Britannica
310 S. Michigan Ave.
Chicago, IL 60604

Mendelbugs: A Genetics Simulation
Software (Apple)
Focus Media, Inc.
P.O. Box 865
Garden City, NY 11530

STV: The Cell
Videodisc
National Geographic Society
Educational Services
1145 17th St. NW
Washington, D.C. 20036

Unit Organizer

Unit/Chapter	Lesson	Time*	Objectives	Teaching Resources
Unit Opener, p. 488				Science Sleuths: Twins or Not? English/Spanish Audiocassettes Home Connection, p. 1
Chapter 22, p. 490	Lesson 1, A Family Likeness, p. 491	2 to 3	1. Identify several ways in which people resemble family members. 2. Construct a family tree and use it to analyze the relationships among members of the same family. 3. Identify several physical traits that unrelated people may have in common. 4. Describe four essential properties of life.	Image and Activity Bank 22-1 Discrepant Event Worksheet, p. 3 Resource Worksheet, p. 4 ▼ Exploration Worksheet, p. 6
	Lesson 2, Recipes for Life, p. 496	1	1. Explain why people once believed in spontaneous generation. 2. Describe an experiment that disproved the theory of spontaneous generation.	Resource Worksheet, p. 9
	Lesson 3, Cells—The Basic Units of Life, p. 498	2	1. Describe the function of the nucleus of a cell. 2. Summarize the process of mitosis. 3. Compare sexual and asexual reproduction.	Image and Activity Bank 22-3 Exploration Worksheet, p. 11 Transparency Worksheet, p. 12 ▼ Transparency 80 Resource Worksheet, p. 14 ▼
	Lesson 4, Mendel's Factors, p. 506	3	1. Describe some of the genetic experiments Mendel performed with pea plants. 2. Compare dominant and recessive traits. 3. Predict the ratio of dominant to recessive traits in the offspring of hybrid plants. 4. Construct a diagram showing the combinations of inherited factors that are possible among the offspring of hybrid parents. 5. Describe the process of meiosis.	Image and Activity Bank 22-4 Resource Worksheet, p. 15 Theme Worksheet, p. 19 Exploration Worksheet, p. 20 Transparency 82 Transparency Worksheet, p. 23 ▼
	Lesson 5, Life Story of the Unborn, p. 516	3	1. Explain how the process of fertilization occurs in humans. 2. Describe the life-support system of an embryo within the uterus. 3. Describe the sequence of events that takes place during the development of a human embryo and fetus. 4. Compare the development of identical and fraternal twins. 5. Describe the birth process.	Image and Activity Bank 22-5 Transparency 84 Transparency Worksheet, p. 25 ▼ Transparency Worksheet, p. 27 ▼ Transparency 87 Resource Worksheet, p. 29
End of Chapter, p. 530				Chapter 22 Review Worksheet, p. 31 Chapter 22 Assessment Worksheet, p. 34
Chapter 23, p. 532	Lesson 1, Living With Instructions, p. 533	2 to 3	1. Identify several different ways that instructions are conveyed. 2. Explain how scientists came to the conclusion that DNA contains the instructions for a living thing. 3. Describe how DNA, chromosomes, and genes are related. 4. Recognize that DNA uses a chemical code to store, convey, and duplicate genetic information.	Image and Activity Bank 23-1 Exploration Worksheet, p. 37
	Lesson 2, Environment or Heredity? p. 540	2	1. Compare genotype with phenotype. 2. Identify examples of environmental and hereditary factors that contribute to an organism's well-being and development. 3. Explain how mistakes in the genetic code affect the development of an organism.	Image and Activity Bank 23-2 Transparency Worksheet, p. 39 ▼ Exploration Worksheet, p. 41 Transparency 89
	Lesson 3, What the Future Holds, p. 544	2	1. Define genetic engineering. 2. Describe some of the current research involving genetics. 3. Explain how advances in genetic research may affect the future.	Math Practice Worksheet, p. 43
End of Chapter, p. 548				Activity Worksheet, p. 45 ▼ Chapter 23 Review Worksheet, p. 46 Chapter 23 Assessment Worksheet, p. 50
End of Unit, p. 550				Unit 8 Activity Worksheet, p. 52 Unit 8 Review Worksheet, p. 53 Unit 8 End-of-Unit Assessment, p. 57 Unit 8 Activity Assessment, p. 63 Unit 8 Self-Evaluation of Achievement, p. 66

* Estimated time is given in number of 50-minute class periods. Actual time may vary depending on period length and individual class characteristics.
▼ Transparencies are available to accompany these worksheets. Please refer to the Teaching Transparencies Cross-Reference chart in the Unit 8 Teaching Resources booklet.

Materials Organizer

Chapter	Page	Activity and Materials per Student Group
22	493	**Exploration 1:** 1 index card per student; paper punch; scissors; 1 knitting needle per class
	498	**Exploration 2:** *Protococcus* (or another alga); microscope; several drops of water; microscope slide and coverslip; eyedropper; small knife; small plastic container with lid
	510	**Exploration 3:** 200 2 cm × 2 cm squares of paper; 2 paper bags or other opaque containers (additional teacher materials for preparing paper squares: 2 sheets of paper or photocopies of page 22 of the Unit 8 Teaching Resources booklet; paper cutter or scissors and metric ruler; see Advance Preparation below.)
23		none

Advance Preparation

Exploration 3, page 510: Cut out 200 squares of paper, about 2 cm × 2 cm each, from 2 sheets of letter-sized paper. Alternatively, the worksheet on page 22 of the Unit 8 Teaching Resources booklet may be photocopied as needed to produce 200 appropriately marked squares. A paper cutter is a convenient way to cut the squares quickly.

Unit Compression

This unit uses a personal approach to explain heredity, reproduction, and human development. Students study their own genetic traits and learn how they developed from conception, through embryonic and fetal stages, to birth. At the same time, a historical perspective of the biology of the cell and of the structure and organization of genetic material within the cell is provided. These topics are essential to the unit.

However, if time is short, parts of Lesson 5 in Chapter 22 could be assigned as homework. Complete omission would sacrifice a discussion of real-world issues related to reproduction as well as information on reproductive biology, however.

The final two lessons in Chapter 23 could be scanned briefly or omitted without undermining the essence of the unit. These lessons introduce subjects, such as genetic engineering, the significance of mutations, genetic disorders, and the influence of the environment on human identity, that fascinate most students. Skipping or skimming these lessons, therefore, is an option to be chosen only if time is very limited.

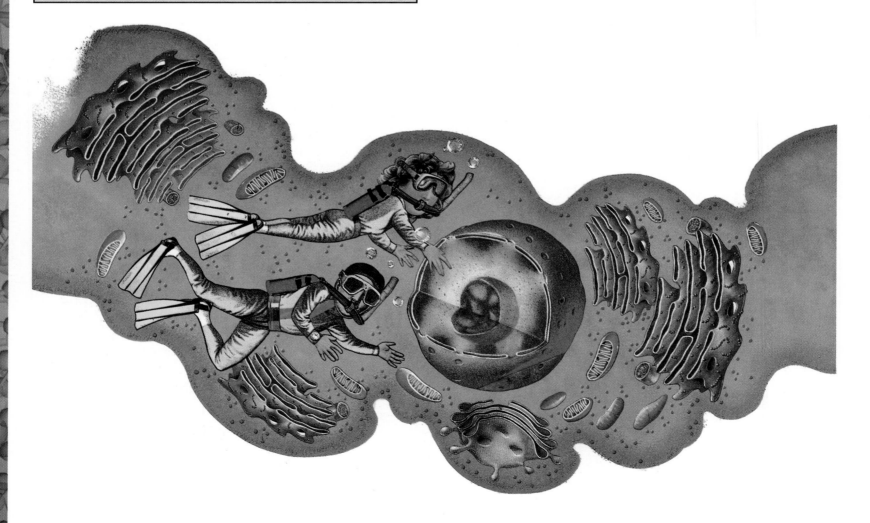

Homework Options

Chapter 22	See Teacher's Edition margin, pp. 492, 494, 497, 499, 501, 505, 507, 511, 513, 517, 521, 525, and 529 Resource Worksheet, p. 4 Resource Worksheet, p. 9 Resource Worksheet, p. 29 SourceBook, pp. S154, S157, and S175
Chapter 23	See Teacher's Edition margin, pp. 534, 536, 541, 545, 546, 547, 549, and 551 Math Practice Worksheet, p. 43 Activity Worksheet, p. 45 SourceBook, pp. S167, S171, and S172
Unit 8	Unit 8 Activity Worksheet, p. 52 Unit 8 SourceBook Activity Worksheet, p. 67

Assessment Planning Guide

Lesson, Chapter, and Unit Assessment	SourceBook Assessment	Ongoing and Activity Assessment	Portfolio and Student-Centered Assessment
Lesson Assessment Follow-Up, see Teacher's Edition margin, pp. 495, 497, 505, 514, 529, 539, 543, and 547 **Chapter Assessment** Chapter 22 Review Worksheet, p. 31 Chapter 22 Assessment Worksheet, p. 34* Chapter 23 Review Worksheet, p. 46 Chapter 23 Assessment Worksheet, p. 50* **Unit Assessment** Unit 8 Review Worksheet, p. 53 End-of-Unit Assessment Worksheet, p. 57*	SourceBook Review Worksheet, p. 69 SourceBook Assessment Worksheets, p. 73* 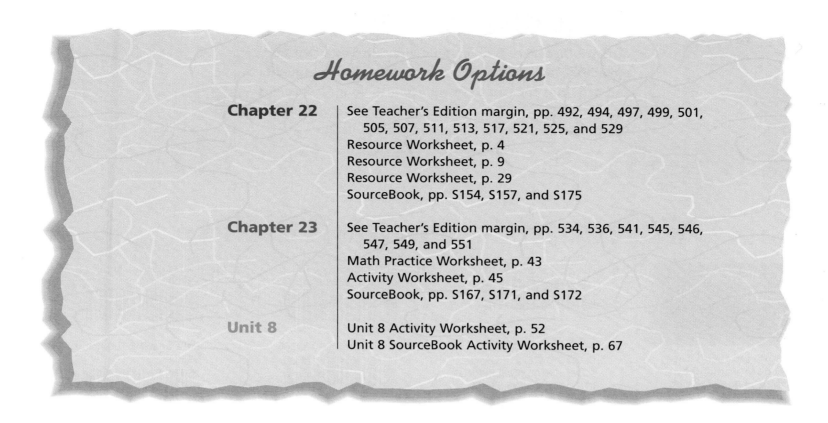	Activity Assessment Worksheet, p. 63* **SnackDisc** Ongoing Assessment Checklists ♦ Teacher Evaluation Checklists ♦ Progress Reports ♦	Portfolio: see Teacher's Edition margin, pp. 512, 523, and 547 **SnackDisc** Self-Evaluation Checklists ♦ Peer Evaluation Checklists ♦ Group Evaluation Checklists ♦ Portfolio Evaluation Checklists ♦

* Also available on the Test Generator software

♦ Also available in the Assessment Checklists and Rubrics booklet

Science Discovery is a versatile videodisc program that provides a vast array of photos, graphics, motion sequences, and activities for you to introduce into your SciencePlus classroom. Science Discovery consists of two videodiscs: Science Sleuths and the Image and Activity Bank.

Using the Science Discovery Videodiscs

Science Sleuths: Twins or Not? Side B

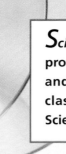

When twins read their father's will, they discover that they have a third sibling, a triplet, with whom they must share their inheritance. Three people claim to be the lost sibling.

Interviews
1. Setting the scene: School teacher; brother **48882** (play ×2)

2. Actress; sister **49840** (play)

3. Geneticist **50506** (play)

4. Applicant #1; dancer **51396** (play)

5. Applicant #2; burger chef **52198** (play)

6. Applicant #3; architect **52714** (play)

Documents
7. Addendum to will **53237** (step)

8. Newspaper ad **53240**

Literature Search
9. Search on the words: BLOOD TYPE, CHROMO-SOME, DNA, FINGERPRINT-ING, HLA, TWINS **53242** (step)

10. Article #1 ("Donor Heart Goes to Wrong Recipient") **53245** (step)

11. Article #2 ("Faulty DNA Fingerprinting Frees Killer") **53248** (step)

12. Article #3 ("Learned or Inherited? Personality Traits of Identical Twins") **53251** (step ×2)

13. Article #4 ("Red Cross Needs Donors") **53255** (step)

14. Article #5 ("Twins and Their Secrets") **53258** (step ×2)

Sleuth Information Service
15. Selected hereditary traits in humans **53262**

16. ABO blood type inheri-tability **53264** (step ×2)

17. Rh inheritability **53268**

18. HLA **53270** (step ×2)

Sleuth Lab Tests
19. Personality profiles of twins and applicants **53274** (step ×4)

20. Medical tests of father, twins, and applicants **53280** (step ×5)

21. Chromosome spread for father, twins, and applicants **53287** (step ×5)

22. DNA analysis, brother **53294**

23. DNA analysis, sister **53296**

24. DNA analysis, Applicant #1 **53298**

25. DNA analysis, Applicant #2 **53300**

26. DNA analysis, Applicant #3 **53302**

Still Photographs
27. Norman and Norma's parents **53304** (step)

28. Norman Baker **53307**

29. Norma Baker **53309**

30. Natalie Backman **53311**

31. Nathan Baxter **53313**

32. Ned Booker **53315**

◀️ Step Reverse ▌▌▌▌   Play ▶ ▌▌▌▌   Pause ❚❚ ▌▌▌▌   Step Forward ▐▶ ▌▌▌▌

A selection of still images, short videos, and activities is available for you to use as you teach this unit. For a larger selection and detailed instructions, see the Videodisc Resources booklet included with the Teaching Resources materials.

22-1 A Family Likeness, page 491

Herders; Masai 4227
The Masai are nomadic herders who live in Kenya and parts of Tanzania.

Nomads; !Kung 4228
The !Kung (Bushmen) people are nomadic hunter-gatherers who live in southwestern Africa.

Merchant; Afghani 4229
Dried fruits, skins, gas, cotton, and wool are Afghanistan's main exports. It is also famous for its carpets.

Lacemaker; Italian 4230
This type of lace-making involved tying knots with threads attached to bobbins (cylindrical pieces of wood). It is a painstaking process that requires great skill.

Metalworker; Iranian 4231
Iran's geographical location (between eastern Europe and the Middle East) historically has made it a crossroads for culture and trade.

Camel driver; Tunisia 4232
Tunisia is a largely desert country in North Africa. Camel caravans have been the only mode of transportation across the sands for centuries.

Man and child; Shanghai, China 4255
The People's Republic of China is the world's third-largest country and is the most populated. It is located in eastern Asia.

Women and children; China 4256
These two older women and young children live in the People's Republic of China, the world's mostpopulated country.

Children; Beijing, China 4257
These young children are at a daycare center in Beijing.

Boatmen and produce; China 4258
Agriculture is the basis of the People's Republic of China's economy. Its economy is dependent on rice. Cotton and tea are grown for export.

22-3 Cells—The Basic Units of Life, page 498

Cell, animal; mitosis (step ×5) 2678–2683
The nucleus of a cell divides into two nuclei containing identical DNA.

22-4 Mendel's Factors, page 506

Cross-pollination 4234
Cross-pollination fertilizes plants with each other's pollen. Each offspring gets half of its DNA from the red parent and half from the white, resulting in two red, three pink, and three white offspring, if one gene does not dominate.

Cross-pollination; Punnett squares (step ×4) 4235–4239
Single trait possible combinations: 4/4 with dominant trait (red) (step) Two traits possible combinations: 9/16 with two dominant traits (red with black spots) (step) Two traits possible combinations: 3/16 with one recessive trait (red with blue spots) (step) Two traits possible combinations: 3/16 with the other recessive trait (white with black spots) (step) Two traits possible combinations: 1/16 with two recessive traits (white with blue spots)

Meiosis (step ×9) 4240–4249

Prophase I: thin strands of DNA tighten up into chromosomes. (step) Metaphase I: chromosome pairs line up. (step) Anaphase I: chromosome pairs move toward the cell's poles. (step) Telophase I: the cell divides in half. (step) Daughter cells–late interphase: the two new cells form nuclei. (step) Prophase II: the nuclei open. (step) Metaphase II: the chromosome pairs line up at the equators. (step) Anaphase II: the chromosome pairs split. (step) Telophase II: the cells divide in half. (step) Telophase II: 4 gametes are formed.

22-5 Life Story of the Unborn, page 516

Reproductive system, female 4191
Eggs are formed in the ovary. They burst out, are caught by the Fallopian tubes, and travel to the uterus.

Reproductive system, male 4192
Sperm is formed in the testes and travels through the vas deferens into the seminal vesicles for storage.

Embryo, human; 40 days 4220
The organ in the middle of this embryo is its heart. Note the tail, which will disappear later, the limb buds, and the eye.

Fetus, human; with dyed skeleton (step) 4221–4222
This fetus is about 11 weeks old. Bone formation begins as cartilage structures start to become bone. Arms, legs, and digits are formed. Note the jawbone and the lack of skull bones. (step) This fetus is between its third and fourth months. Note the more developed skull. The bridge of the nose has formed, appendages are fully formed, and joints are beginning to develop.

Growth; human 4171
Body proportions and age

23-1 Living with Instructions, page 533

Laser applications (step) 2245–2246
Holograms are made by using a split laser beam. One beam produces a wave form of an image that interferes with the other beam. This interference is recorded on a photographic plate. The plate is illuminated and the object is projected. (step) Lasers are used to read barcodes.

X-ray machine 2172
X rays are waves with very short wavelengths that are able to penetrate solid objects.

X ray; broken bone 1937
A broken bone is a failed structure. The high-energy waves of X rays can be used to view bone fractures beneath the skin's surface.

DNA; model 4233
DNA forms the main constituent of the chromosomes of living cells. It has a complicated structure consisting of two long polymeric chains twisted about each other to form a double helix. The patterns determine genes.

23-2 Environment or Heredity, page 540

Moth, peppered 4252
Peppered moths may be either pale or dark. The percentage of dark moths increased through natural selection in areas affected by industrial pollution.

Horse evolution 4251
The fossil record shows the evolution of modern horses from ancient ones.

Unit 8 — CONTINUITY OF LIFE

UNIT FOCUS

Ask students to name several physical characteristics that help to distinguish each of them as individuals. *(Students could identify eye, hair, and skin color; hair texture; height; weight; build; and facial features.)* Then ask how many of these physical characteristics they inherited from their parents. Which were determined by their environment? *(Answers will vary.)* Then ask students to speculate about how we inherited certain physical characteristics from our parents. *(Accept all reasonable responses at this point.)* Point out to students that the combination of traits they acquired from their parents makes each of them unique even though they share some common characteristics with other people.

A good motivating activity is to let students listen to the English/Spanish Audiocassettes as an introduction to the unit. Also, begin the unit by giving Spanish-speaking students a copy of the Spanish Glossary from the Unit 8 Teaching Resources booklet.

Did You Know . . .

The fossilized remains of the first recognizable turtle are estimated to be about 200 million years old. Although the turtle has never been the most common of reptiles, its adaptability has enabled it to develop terrestrial and marine species. It is now the oldest living group of reptiles.

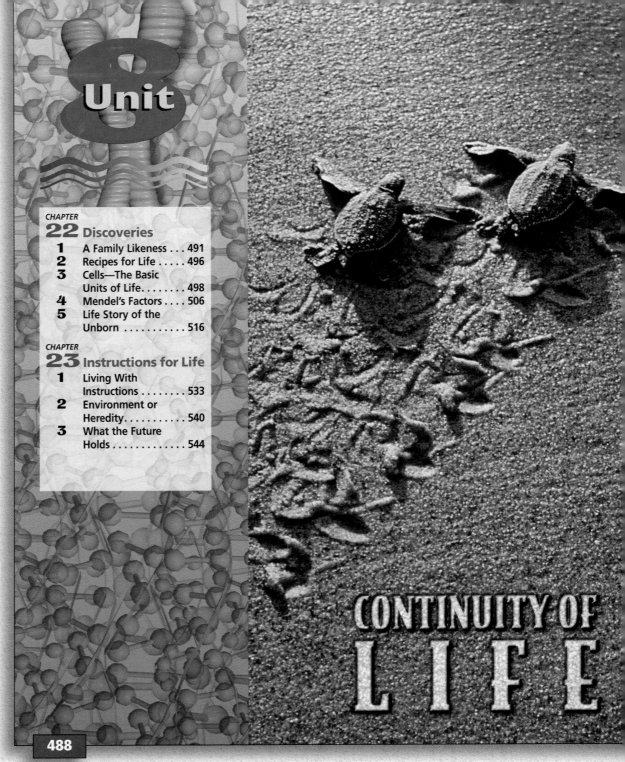

Unit 8

CHAPTER 22 Discoveries
1 A Family Likeness . . . 491
2 Recipes for Life 496
3 Cells—The Basic Units of Life 498
4 Mendel's Factors 506
5 Life Story of the Unborn 516

CHAPTER 23 Instructions for Life
1 Living With Instructions 533
2 Environment or Heredity 540
3 What the Future Holds 544

488

CONTINUITY OF LIFE

Connecting to Other Units

This table will help you integrate topics covered in this unit with topics covered in other units.

Unit 1 Life Processes	Genetic material is central to the life processes and reproduction of all cells and living organisms.
Unit 2 Particles	Mendel's assumption that genetic material was made up of very small particles was essential in developing his theory about inheritance.
Unit 4 Oceans and Climates	Climate, including temperature and moisture levels, usually has a significant effect on the phenotypes of organisms.

As though responding to some invisible signal, newly hatched leatherback turtles head straight for the surf. Unfortunately, many will be snatched up by waiting predators before ever reaching the water. And for those that survive the gauntlet of jaws and beaks, life will continue to be perilous for many months to come. Perhaps one turtle in fifty survives to adulthood.

The tiny turtles would face much greater odds against survival if they did not head immediately to the water. But how do they know to do this? Encoded in the cells of each turtle is a set of behavioral instructions, including the order to make this instinctual trek. These behavioral instructions are inherited by the turtles from their parents, in the same way that their physical characteristics are inherited. Although these turtles appear identical, each is, in fact, unique. The mechanism of heredity is able to recombine the traits of the parents in so many ways that each offspring gets a slightly different set of life instructions. These instructions, responsible for the turtles' similar-yet-different form, will also one day direct surviving females back to the shore where their lives began. There they will lay eggs, passing life instructions to a new generation.

Newly hatched leatherback turtles, Les Hattes Beach, French Guiana

489

Call on volunteers to restate the text on this page in their own words. Point out that, for many animals, some behaviors are instinctive, whereas others must be taught. Encourage students to discuss what types of behavior are instinctive to human babies. (*Sample answer: Human babies breathe, cry, eat, and sleep without being taught to do so.*)

In addition, you may wish to share the following information with students: Genetic studies have helped conservationists learn more about endangered species of sea turtles. Studying the DNA of sea turtles has allowed scientists to confirm that the turtles return to their native beaches to reproduce. Turtles from one nesting site share a set of genes, or a gene pool, that differs from the gene pools of turtles from other nesting sites. This implies that different groups within the same species do not regularly interbreed. It also demonstrates the importance of securing safe nesting sites for each separate population of turtles. In order to ensure the protection of sea turtles, every nesting site and its associated population needs to be treated as a separate entity.

Integrating the Sciences

Earth and Life Sciences

Share the following information with students: Laboratory studies of loggerhead turtles suggest that the hatchlings can sense the Earth's magnetic field. It is thought that this ability not only allows the young sea turtles to locate the surf after hatching, but also allows them to return several years later to nest on the same beach. Other studies indicate that sea turtles may use wave direction and the geological features on the ocean floor to navigate along their migratory routes, which may be hundreds of kilometers long.

Connecting to Other Units, continued

| Unit 6 Sound | Much of the research that has been done in human fetal development has been made possible by ultrasound technology. |
| Unit 7 Light | By using X-ray photography, Franklin, Watson, and Crick were able to collect and analyze data on the structure of DNA. |

Connecting to Other Chapters

Chapter 22
examines Mendel's heredity experiments, cellular reproduction, and human fetal development.

Chapter 23
explores molecular biology, including the structure of DNA, genotype expression, and genetic engineering.

Prior Knowledge and Misconceptions

Your students' responses to the ScienceLog questions on this page will reveal the kind of information—and misinformation—they bring to this chapter. Use what you find out about your students' knowledge to choose which chapter concepts and activities to emphasize in your teaching. After students complete the material in this chapter, they will be asked to revise their answers based on what they have learned. Sample revised answers can be found on page 531.

In addition to having your students answer the questions on this page, you may wish to ask students to write a paragraph describing human reproduction or an explanation of how parents pass on their physical characteristics to their children. Emphasize that there are no right or wrong answers in this exercise. Encourage students to make their responses as complete as possible. Collect the papers, but do not grade them. Instead, read them to find out what students know about reproduction and heredity, what misconceptions they may have, and what about these topics interests them.

CHAPTER
22
Discoveries

1 **Why don't we all look alike?**

How is your body able to mend a cut? **2**

How old do you think you were when your heart began to beat? **3**

ScienceLog

Think about these questions for a moment, and answer them in your ScienceLog. When you've finished this chapter, you'll have the opportunity to revise your answers based on what you've learned.

490

LESSON 1 ~ A Family Likeness

"You got your mom's chin." "You look so much like your aunt." You might have heard your relatives say things like this. Such similarities are called family "likenesses."

Look at yourself carefully in a mirror. Notice details such as the color and shape of your eyes and the shape and size of your nose, eyebrows, mouth, ears, and hands. Then look at other family members (your parents, sisters, brothers, cousins, aunts, uncles, and grandparents) and at family photos for resemblances.

You're the spitting image of your dad!

Your nose is just like Grandpa's.

How do these siblings resemble one another?

491

LESSON 1 ORGANIZER

Time Required
2 to 3 class periods

Process Skills
observing, comparing, inferring

Theme Connection
Structures

New Terms
Family tree—a diagram used to show how individual members of a family are related
Generation—a stage in a family's line of descent that includes all of the individuals born within one life cycle

Traits—characteristics, especially those that are genetically determined

Materials (per student group)
Exploration 1: 1 index card per student; paper punch; scissors; 1 knitting needle per class

Teaching Resources
Discrepant Event Worksheet, p. 3
Resource Worksheet, p. 4
Transparency 78
Exploration Worksheet, p. 6

LESSON 1 ~ A Family Likeness

FOCUS

Getting Started
Ask students to bring in family photos of famous families. Display the photos and ask students to name what traits the family members have in common. Ask students to explain whether the shared characteristics result from genetic or environmental factors. *(Accept all answers without comment at this point.)*

Main Ideas
1. Members of the same family often resemble one another.
2. A family tree may be used to analyze how family members are related to one another.
3. People who are not related may share certain physical traits.
4. All living things demonstrate four essential properties of life.

TEACHING STRATEGIES

Point out to students that they are each related to their family members in different ways. For example, each is a son or a daughter, and each may also be a brother or a sister. (Be sensitive to the fact that some students may not know their biological families.) Direct students' attention to the photographs on page 491. Call on students to suggest how the family members pictured may be related to one another. Involve the class in a discussion of the characteristics that the family members have in common.

Answer to
Caption

 Responses could include color of eyes, skin, and hair; bone structure of the face including shape of face, eyes, nose, and mouth; and texture of skin and hair.

★ **The Discrepant Event Worksheet (Teaching Resources, page 3) is available as an introduction to Lesson 1.**

CHAPTER 22 • DISCOVERIES **491**

1. The following answers describe blood relatives only.
   - Juan is the grandson of Manuel and Mariana and of Caroline and Bill.
   - Juan is the nephew of David, Laura, Teresa, and Diego.
   - Juan is the son of Miguel and Kara
   - Juan is the brother of Elena and Alicia.
   - Juan is the first cousin of Tomás and Rosa.
   - Juan is the first cousin, once removed, of Linda and Ramón.

2. Alicia is Laura's niece. Laura is Alicia's aunt.

3. A generation is a single stage in a family's line of descent. Juan's parents and their brothers and sisters represent one generation. Their children are the next generation.

4. Four generations are included in the diagram.

5. Laura and Nathan's children will be Juan's and Alicia's first cousins.

6. Alicia got her looks from her parents, Miguel and Kara. Her father got his ears from his parents, Manuel and Mariana.

7. Juan has more *ancestors* because he belongs to the generation that came after his father's generation.

A Family-Tree Project

Encourage students to begin work on a family tree. Be aware that family ancestry could be a sensitive issue for some students (e.g., students who are adopted, do not live with their parents, or are part of a single-parent family). Give your students the option of doing a family tree for their own family or for one of the following: a friend's family; a family from a historical novel that is genealogical, such as *The Diary of Miss Jane Pittman* or *Roots;* a royal family; a famous family; or a pedigreed pet.

Homework

The Resource Worksheet that accompanies Family Relations makes an excellent homework assignment (Teaching Resources, page 4). If you choose to use this worksheet in class, Transparency 78 is available to accompany it.

Family Relations

The diagram at the right shows how everyone in Juan's immediate family is related. Use it to answer the following questions:

1. How is Juan related to each of the other people in the diagram?

2. How are Alicia and Laura related?

3. What is meant by a **generation** of a family?

4. How many generations are included in the diagram?

5. If Laura and Nathan have children, how will their children be related to Juan and Alicia?

6. From whom might Alicia get her looks? If she "has her father's ears," from whom did he get them?

7. Who has more *ancestors,* Juan or his father? Explain.

A Family-Tree Project

Now is the time to begin a family history. You could choose your own family or one from which you can get the information you need. The family history will feature a diagram like the one drawn for Juan's family. Include as many generations as possible. Some families are fortunate enough to have four or five generations living at the same time. You may also include relatives who are no longer alive. Go back as many generations as you can by asking questions of family members and by looking through family records. The resulting **family tree,** as this type of diagram is often called, could become quite large.

Preparing a family tree can take quite a bit of time. You can add to it as you proceed through this unit. If you'd like, you could collect photographs to include on the family tree. The photos may help you observe family members' characteristics, or **traits,** more carefully than you could by relying on memory alone. The following Exploration will introduce you to some traits that you know about and others that are probably unfamiliar to you.

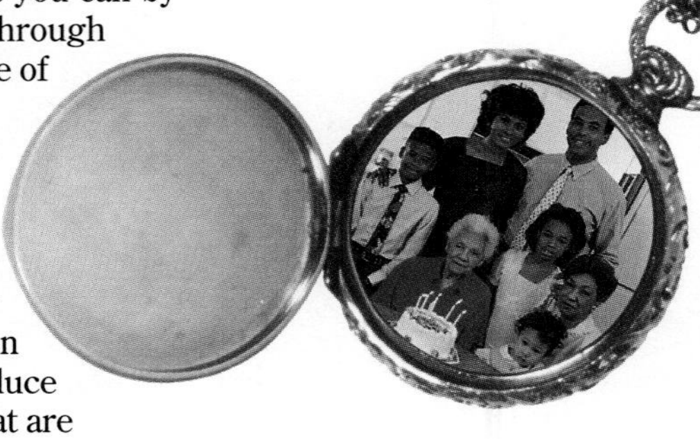

492

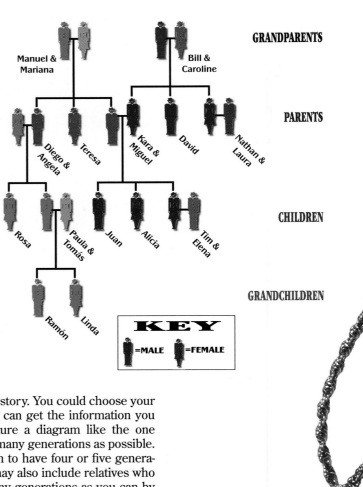

FOUR GENERATIONS OF A FAMILY

GRANDPARENTS

Manuel & Mariana

Bill & Caroline

PARENTS

Diego & Angela

Teresa

Kara & Miguel

David

Nathan & Laura

CHILDREN

Rosa

Paula & Tomás

Juan

Alicia

Tim & Elena

GRANDCHILDREN

Ramón

Linda

KEY

=MALE =FEMALE

A Traits Test

You Will Need

- an index card
- a paper punch
- scissors
- a knitting needle

What to Do

1. Prepare the index card as shown in the diagram.

2. Use the illustrations to help you answer the questions below. For each question, if your answer is *yes,* darken the area by the number, between the hole and the edge of the card. If your answer is *no,* leave the card unchanged. See the example shown.

3. When you have answered all the questions, use scissors to cut out the areas you darkened for step 2. This will open the holes for all the *yes* answers, as shown. Pass in your card.

Determining Your Traits

1. Can you tell the difference between red and green? If you do *not* see a number in the circle, then you have red-green colorblindness.

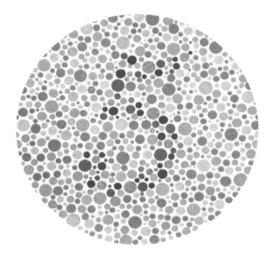
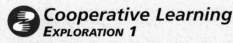

2. Do you have a missing *lateral incisor*? (Lateral incisors are the teeth on either side of your two front teeth. You should have a total of four lateral incisors—two in your upper jaw and two in your lower jaw.)

Exploration 1 continued ▶

493

 Cooperative Learning
EXPLORATION 1

Group size: 2 to 3 students

Group goal: to determine how many traits are common to a population

Positive interdependence: No roles are necessary for this activity. Each group member should prepare his or her own card and must depend on the others for information regarding their genetic traits.

Individual accountability: Each group member should be able to answer any of the questions from Analyzing the Data.

★ **An Exploration Worksheet is available to accompany Exploration 1 (Teaching Resources, page 6).**

EXPLORATION 1

In this Exploration, students should discover that some of the traits are shared by a large majority of the class, some are shared by only two or three people, and some do not appear at all. *Yes* answers mean that students have the dominant trait. *No* answers mean that they have the recessive trait. Note that students will not encounter the terms *dominant* and *recessive* until Lesson 4.

The following steps should help you and your students complete the activity successfully:

1. You may wish to prepare the cards ahead of time to ensure that they align properly. It is important that all of the holes in the cards are aligned. If students prepare their own cards, be sure that they understand that everyone's card should be exactly the same.

2. Review each of the pictures with students. Explain that each one shows a physical trait that some people have and some people do not have. You may wish to point out that these traits are neither good nor bad, but are simply easy to identify.

You may need to clarify the terms *thumb, fingers,* and *pinkie* because some languages do not distinguish between these terms.

3. Suggest that students work in pairs to determine whether they have the physical traits pictured. Monitor the way students use their cards by walking around the classroom to see whether they are correctly filling in the spaces between the holes and the edge of the card.

You may want to demonstrate how students are to cut out the notches in their cards next to their *yes* responses. Be sure they understand that each notch should extend from the edge of the card to the hole next to a number and that only *yes* answers should be notched.

4. When students are finished, collect all of the cards. Ask a different pair of students to perform the knitting-needle step for each trait and to record the results on a class chart. Note that the name of each student that has the dominant trait must be recorded in order to answer the questions in Analyzing the Data.

The Results

Results will vary from class to class. If you have any colorblind students, reassure them that colorblindness is not a disease, but simply a characteristic that they possess. Although the *yes* answers reflect the presence of the dominant trait, these traits may not be the most common among your students because the dominant genes that cause some of these traits are relatively rare in the population.

Answer to
In-Text Question

Ⓐ The *yes* responses will fall out and the *no* responses will remain on the knitting needle.

Answers to
Analyzing the Data

1. Answers will vary. (Because there are over a thousand possible combinations of these 10 characteristics, trying to analyze all of the combinations would be very tedious without the aid of a computer. By looking at just the first four traits, students should get the idea that the more characteristics you look at, the less likely it will be for any two individuals to be identical for all of those characteristics.)

2. It is very unlikely that two students, other than identical twins, will be identical for all of these traits.

3. People share some of the same physical characteristics, but each person is an individual with his or her own set of characteristics.

4. Students should recognize that most people share more generalized characteristics, such as having two arms, two legs, two eyes, two ears, a nose, and so on. Some students may extend the meaning of the term *characteristic* to include breathing, mobility, responding, and other characteristics of living things.

Homework

You may wish to assign the questions in What Is Life? on page 495 as homework.

3. Do you have free ear lobes?

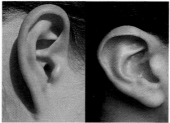

Attached   Free

4. Is the last segment of your thumb straight?

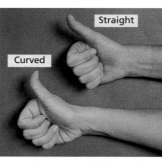

Straight

Curved

5. Do you have dimples?
6. Do you have mid-digit hair?

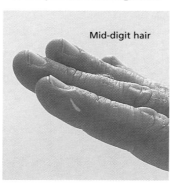

Mid-digit hair

7. Do you have a white forelock—strands of white hair growing just above your forehead?

8. Is your fifth finger (pinky) straight?

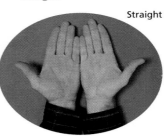

Straight

Bent

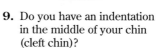

9. Do you have an indentation in the middle of your chin (cleft chin)?

10. Do you have a widow's peak?

Widow's peak

The Results

When all the cards have been collected, stack them together. Insert the knitting needle in each hole, in turn, and then lift the pile of cards. Gently shake the cards until some fall. Which ones will Ⓐ fall? For example, if you answered *yes* to question 1, will your card stay in or fall out of the group? Count how many cards fall and how many remain. Record the numbers in a chart. Repeat this process for each question.

Analyzing the Data

1. Looking at the first four characteristics, how many people in your class answered *yes* for the same two characteristics? for three? for more?

2. Do any two people in your class have the same responses for all of these traits?

3. What do these results tell you about the characteristics of people?

4. What are some characteristics that all people have in common?

Theme Connection

Structures

Make a bulletin board of family pictures of student volunteers. **Focus questions:** What observable physical structures are shared by family members? *(Observable structures may include color of hair, eyes, and skin, build and height, and shape of facial features.)* What shared structures are not observable? *(Unobservable shared structures may include blood type and abnormal heart valves, pancreatic glands, or other organs.)*

Multicultural Extension

Life Expectancies

Remind students that longevity tends to run in families, indicating a genetic link. Also, the average human life span varies greatly in different parts of the world, indicating an environmental link. Suggest that students research the life expectancy of people living in different parts of the world. Students should note cultural factors such as diet and work activities that might contribute to the observed differences. They should arrange their data into a chart and draw conclusions about what the data show.

What Is Life?

Families share likenesses. So do all humans, all mammals, all vertebrates, and all animals. Some characteristics are shared by *all* living things. If you were asked to describe the characteristics of living things, **B** what things would you include? The list of questions below suggests four essential properties that all living things exhibit. See whether you can determine the four properties.

1. When you cut your finger, what does your body do?

2. Why does an earthworm avoid daylight?

3. Cats don't live forever, so why are there so many of them around?

4. Why do we perspire?

Biologists have identified four essential properties of life.

- Self-preservation—staying alive; obtaining energy to live and grow

- Self-regulation—the control of life processes, such as respiration

- Self-organization—forming, grouping, and repairing body cells

- Self-reproduction—making new life of the same kind (such that the offspring of monkeys are monkeys and the offspring of whales are whales)

Now try matching the four questions with the four properties.

This unit is involved with the fourth essential property of life—self-reproduction—and deals with many interesting questions.

- Why can two relatives resemble each other in some ways but still look very different?

- Can two people look exactly alike?

- Who did you get your looks from? Where did they get their looks?

- If someone loses a toe in an accident, is it likely that his or her children will be born with a missing toe?

- How is life passed on?

- How does a human baby develop before birth? Is it possible to predict what a baby will look like before it's born?

- What is a "test-tube" baby?

- Could there ever be another "you"?

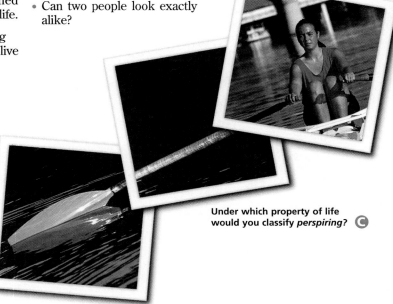

Under which property of life would you classify *perspiring*? C

495

What Is Life?

You may wish to have the class divide into small groups and brainstorm for ideas about the bulleted questions on page 495. Encourage students to write a response for each question in their ScienceLog to review and revise as they progress through the unit.

Answers to
What Is Life?

1. When you cut your finger, your body first bleeds, which cleans the wound; then forms a scab, which protects it from infection; and finally replaces the damaged cells during the healing process. (self-organization)

2. An earthworm avoids daylight to avoid predators and the drying effects of the sun. (self-preservation)

3. There are lots of cats around because they reproduce frequently. (self-reproduction)

4. By perspiring, people eliminate excess heat and maintain a constant body temperature. (self-regulation)

Answer to
In-Text Question and Caption

B Accept all reasonable responses. Students' lists of the characteristics of living things may include reproducing, moving, breathing, etc.

C Perspiring is an example of self-regulation.

FOLLOW-UP

Reteaching

Have students work in groups to make posters or bulletin-board displays to illustrate how plants conduct each of the four essential properties of life. Suggest that they divide the work by having each group choose one essential property to illustrate. Encourage students to be creative by suggesting that they use a variety of materials, including pictures from magazines and newspapers, original art, poems, and their own photographs.

Assessment

Have students write descriptive paragraphs depicting their own, or someone else's, characteristics in as much detail as possible. Paragraphs should describe hair, skin, and eye color, face shape, body shape, and so on. Encourage students to be creative but not immature when describing others. Read some of the paragraphs to the class and ask students to identify the person being described.

Extension

Explain to students that the card system used in Exploration 1 is an early form of a database. Similar systems are still used to sort data from large surveys. Challenge students to devise their own survey based on this card system.

Closure

Discuss how a variety of organisms, such as a grass plant, fish, hawk, and rabbit, carry out the four functions of living things. Discuss how inherited traits allow organisms to carry out these four functions.

LESSON

2 Recipes for Life

FOCUS

Getting Started

Help students recognize the connection between the lesson title and the statements on this page. *(The statements show that people once believed that some organisms were formed from non-living materials.)* Encourage students to suggest their own fanciful recipes for life. *(For instance, fleas come from dog hair, and frogs come from damp grass.)*

Main Ideas

1. All life comes from preexisting life.
2. Francesco Redi performed the crucial controlled experiment that proved life cannot arise from nonliving matter.

TEACHING STRATEGIES

Students should realize that the statements on this page are based on incomplete observations that lead to faulty inferences.

Answers to
In-Text Questions

Ⓐ Sample answer: In some cases, eggs or baby organisms may have been placed in a certain location without people noticing them. When the organisms grew into adults, it might have seemed that the organisms came from nowhere or from the location itself. This was probably the case with the flies on rotting meat, the mice in a pot with dirty shirts and corn, the organisms in a marsh, and the insects that appear on dewy leaves. In other cases, the resemblance between two organisms might have suggested that the two organisms were related. This may explain people's thoughts about the horsehair worms and hairs in a horse's tail and about the barnacle geese and goose barnacles.

Ⓑ Students are likely to be skeptical of these statements. Students should recognize that they would have to design controlled experiments to test the statements.

Throughout history, people have given many different explanations for how living things reproduce. As you read the following statements made centuries ago, suggest why each of the explaⒶnations might have seemed reasonable at the time it was offered.

Hairs from a horse's tail become horsehair worms, which are found in pools.

Flies come from rotting meat.

The barnacle goose grows from goose barnacles found on rocks beside the ocean.

Put one dirty shirt and some grains of corn into an old pot. In 21 days there will be a lively crop of mice.

The emanations rising from the bottom of the marshes bring forth frogs, snails, leeches, herbs, and a good many other things.

Some insects are born on dewy leaves; some are born from the hair and flesh of animals.

What do you think about these explanations? Choose any one of the statements listed above, attributed to seventeenth-century scientist Jan Baptista van Helmont. What could you do to investigate that claim scientifically? Ⓑ

Well into the 1500s, ideas about the reproduction of living things were still being influenced by Aristotle, who lived in Greece 2000 years before van Helmont. Aristotle said that living things could spring from nonliving things by a process called *spontaneous generation*. For example, Aristotle said that eels grow from mud and slime at the bottoms of rivers and oceans.

What happened to change people's ideas about how new generations of living things come about? Read the following story about Francesco Redi, whose experiment eventually changed people's ideas about reproduction.

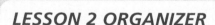
496

LESSON 2 ORGANIZER

Time Required
1 class period

Process Skills
observing, hypothesizing, analyzing, inferring

Theme Connection
Cycles

New Term
Spontaneous generation—the now discredited theory that living organisms can come to life spontaneously from a nonliving source within a short time

Materials (per student group)
none

Teaching Resources
Resource Worksheet, p. 9

Redi's Experiment

In 1628 an English doctor named William Harvey published a book in which he suggested that some living things might come from tiny eggs and seeds instead of from nonliving things (as had been thought previously).

When Francesco Redi, an Italian doctor, read Harvey's book, he decided to investigate whether flies grow spontaneously from rotting meat. His experiment is pictured below.

Redi's Experiment   **Results**

A Meat in open jars   Maggots on the meat

B Meat in tightly closed jars   No maggots on the meat

Think About It

1. What must have happened in part *A*? in part *B*?
2. Where did the maggots in part *A* come from?
3. Redi repeated his experiment, making one important change: he substituted a gauze cloth for the solid lid he had used in part *B*. Why do you think he did this?
4. Why was part *A* so important? What name is given to that part of the experiment?
5. What conclusion could be drawn about spontaneous generation from this experiment?

Although people no longer believe in spontaneous generation, they remain deeply interested in the development of new living things. For example, when you were very young, you probably asked, "Where do babies come from?" In the next section, you'll begin your exploration of this and other questions about reproduction by studying single cells, the basic units of life.

Redi's Experiment

Redi's experiments in 1668 showed that meat covered so that flies could not lay eggs on it never developed maggots. Ask students to write in their ScienceLog what they think Redi's hypothesis might have been. *(If meat is covered, maggots will not grow on it.)*

Answers to
Think About It

1. Maggots developed on the meat in part *A* because flies entered the opened jars and laid eggs on the meat. No maggots appeared in part *B* because flies could not reach the meat to lay eggs on it.

2. The maggots hatched from eggs laid by flies.

3. In Redi's first experiment, two variables were changed by adding lids to the jars: access to the meat and availability of oxygen. By using gauze, it was possible to tell that the flies' access to the meat (and not the availability of oxygen) was the variable that caused the results.

4. Part *A* was important because one needs to compare the effects of changing a variable with a setup in which the variable is left unchanged. Part *A* is called the *control* for the experiment.

5. Spontaneous generation does not occur. Living things come from other living things.

Theme Connection

Cycles

Focus question: If spontaneous generation actually could take place, why wouldn't it make sense to speak of the "life cycle" of a species? *(The life cycle of a species includes reproduction. If spontaneous generation occurred, there would be no need for reproduction so the concept of a life cycle would not be necessary.)*

Homework

As a homework assignment, you may wish to assign the Resource Worksheet that accompanies Redi's Experiment (Teaching Resources, page 9).

FOLLOW-UP

Reteaching

Ask students to write a script for a play titled *Francesco Redi's Experiment*. The play should dramatize Redi's famous experiment that disproved spontaneous generation.

Assessment

Ask students to develop three good quiz questions (with answers) that relate to the most central concepts of the lesson. You may wish to redistribute the questions among students for a real quiz.

Extension

Despite Redi's experiment, some people were still not convinced that life could only come from preexisting life. Have students research Louis Pasteur's experiment with microorganisms in milk in 1864, which finally settled the debate over spontaneous generation.

Closure

Ask students to summarize Redi's experiment and identify Redi's hypothesis, the experimental variable, and the control.

Cells— The Basic Units of Life

FOCUS

Getting Started

Involve students in a brief discussion of how the organisms pictured on this page are alike and how they are different. *(All of them are single-celled, have cell membranes and a nucleus, and live in fresh water. The euglena is green and photosynthesizes, while the others do not. They each have different shapes and move in different ways.)*

Main Ideas

1. The processes in a living cell are controlled by the cell's nucleus.
2. The process by which a cell's nucleus divides to produce two identical nuclei is called *mitosis.*
3. Reproduction can be either sexual or asexual.

TEACHING STRATEGIES

EXPLORATION 2

Before students make their drawings, encourage them to look at each other's slides to compare the cells and structures that they see. You may wish to collect the alga ahead of time so that it is available for students when they begin the Exploration. If so, be sure to keep it in a warm, damp place out of direct sunlight—a closed plastic bag with water sprayed into it is ideal. To save time in class, you may wish to make wet mounts ahead of time or use prepared slides of *Protococcus* or of another alga.

 An Exploration Worksheet accompanies Exploration 2 (Teaching Resources, page 11).

Cells—The Basic Units of Life

You've probably used a microscope to look at single-celled living things like those shown below. They can be found in pond water.

In the following Exploration, you'll look at *Protococcus* —an alga that forms the greenish stain on tree trunks, wooden fences, flowerpots, and buildings.

Euglena

Amoeba

Paramecium

EXPLORATION 2

A Cell Has a Nucleus

You Will Need

- *Protococcus* (or other alga)
- a microscope
- water
- a microscope slide and coverslip
- an eyedropper
- a knife
- a plastic container with lid

What to Do

1. Locate some *Protococcus*, the green "moss" that grows in the places mentioned above. Scrape a small sample into a container. Bring it to the classroom and make a wet mount of it. If you can't find *Protococcus* outdoors, look for algal growth on the glass in an aquarium. Such algae may not be *Protococcus* but will serve your purposes here.
2. Use low power and then high power to examine the algal cells.

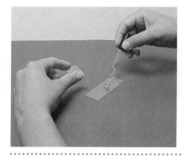
Making a wet mount

3. Draw both a single cell and a group of cells.
4. You'll probably notice that each cell contains several chloroplasts—the parts of the cell that are responsible for photosynthesis. One other structure that should be clearly visible in all of the algal cells is the nucleus. The **nucleus** of a cell controls most of the activities that take place in that cell. Find the nucleus in one of your cells and label it on your drawing. As you will soon discover, the nucleus contains the "recipe" for life. In what process, then, must the nucleus play an important role?

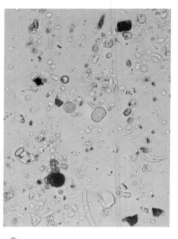
Microscopic view of *Protococcus*

LESSON 3 ORGANIZER

Time Required
2 class periods

Process Skills
observing, hypothesizing, analyzing, inferring

Theme Connection
Changes Over Time

New Terms
Asexual reproduction—reproduction that does not involve the union of sex cells
Cell division—the process by which one cell divides into two new cells

Cytologist—a person who studies cells and their structures
Mitosis—the process by which a cell's nucleus divides, leading to the formation of two identical cells
Nucleus—the structure in a cell that controls most of the cell's activities and contains its chromosomes
Sexual reproduction—reproduction that involves the union on an egg and a sperm

continued

Trouble in the Toy Factory!

Things are not going well at Joy's Toy Factory. It seems that things have gotten totally chaotic. Nobody knows what to do because nobody seems to be in charge. In the old days when the company was small this wasn't a problem. But the company has grown so much recently that it needs some kind of structure.

The toy makers drew up a statement to give to the owner of the factory, asking that a head office be set up. At right is the beginning of the diagram they prepared to describe what a head office could do for them. Copy the diagram into your ScienceLog and complete it to show what the responsibilities

of a head office could include. Take into consideration the problems illustrated in the cartoon, and add any other ideas you have.

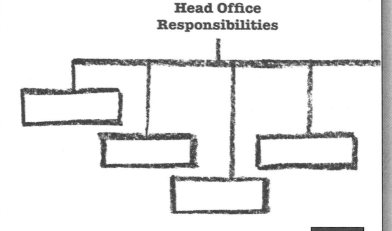

Head Office Responsibilities

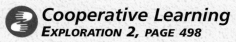
Group size: 3 to 4 students

Group goal: to observe the cell nucleus and infer why it is important

Positive interdependence: Each student should choose a role such as algae handler (to collect and prepare algae for viewing), microscope expert (to adjust the microscope and to see that each group member observes the algae), director (to read the directions and clean the slide), or artist (to draw cells and to check for the group's approval).

Individual accountability: The student who will present the group work to the class can be chosen by drawing lots. Each member of the group must be accountable for knowing the material because students do not know who will be chosen to be the presenter until the end of the exercise.

Answer to
Exploration 2, page 498

3. Accept all reasonable drawings. Students should try to illustrate the interior and the structure of each cell as accurately as possible.

4. The nucleus plays an important role in the process of reproduction.

Answer to
Trouble in the Toy Factory!

Accept all reasonable responses. Responsibilities of the head office might include overseeing the production of toys, routine maintenance of the factory, storing personnel records and toy designs, and planning the construction of new factories to relieve overcrowding.

Homework

You may wish to assign Trouble in the Toy Factory! as homework.

ORGANIZER, continued

Materials (per student group)
Exploration 2: *Protococcus* (or another alga); microscope; several drops of water; microscope slide and coverslip; eyedropper; small knife; small plastic container with lid

Teaching Resources
Exploration Worksheet, p. 11
Transparencies 79–81
Transparency Worksheet, p. 12
Resource Worksheet, p. 14
SourceBook, p. S157

The Cell Is Like a Factory

Encourage students to share any information they may have about *Euglena*. Students may find it interesting to know that there are about 150 species of *Euglena* that live in fresh water that is rich in organic matter. In warm weather, they may become so numerous that they form a green scum on the surface of small ponds and other still bodies of water.

Point out to students that most species of *Euglena* are green because they contain chlorophyll and carry on photosynthesis, but they must also absorb some essential nutrients from the environment. The long, hairlike structure at the end of the *Euglena*'s body is called a *flagellum*. The flagellum twirls in a way that allows the organism to move through the water.

The Cell Is Like a Factory

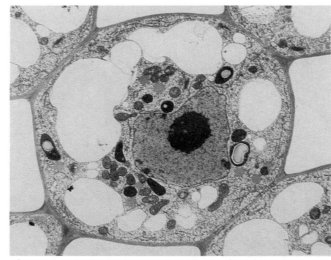

🔺 A root cell of a plant

The photograph above shows that a cell has many parts. Identify Ⓐ as many parts as you can. As in a factory, a lot of complicated action takes place within a cell. It needs direction from a kind of "head office." The head office of a cell is its nucleus. The nucleus in the cell shown above is certainly conspicuous. Just as a factory's head office holds the plans to build a new factory if necessary, the nucleus contains the blueprints for creating new cells.

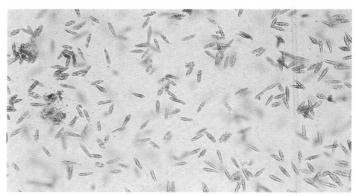

🔺 Euglenas

Consider an organism made of a single cell, such as a *Euglena*. A single-celled organism actually multiplies by dividing! When the organism is ready, its nucleus directs it to split into two identical cells, each of which contains all it requires for all of life's activities. This process is called **cell division**.

Look at the diagram of cell division below.

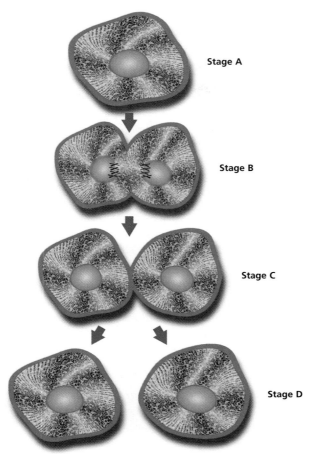

Stage A

Stage B

Stage C

Stage D

1. What happened to the "parent" of the daughter cells?
2. Draw a diagram to show the next cell division. (Draw stage D from the diagram, and add further stages as needed.)
3. Did you see any sign of cell division when you did Exploration 2? Look at your drawings for evidence.

Cell division also takes place in multicellular bodies such as yours. It happens when you need new cells for growth, repair, or other purposes. For example, skin cells are dividing continually to make new skin and to heal cuts.

Next, you will have a closer look at what happens inside a cell during cell division.

501

CROSS-DISCIPLINARY FOCUS

Mathematics
The rule of 69 is a convenient way to estimate how long it takes a population of microorganisms to double, assuming exponential growth. First determine what the percentage growth is in some amount of time. Then divide 69 by that number. Ask: If a bacteria population experiences 5-percent growth each minute, how many minutes will it take for the population to double? *(69 divided by 5 is 13.8, so it would double in a little less than 14 minutes!)*

Homework
You may wish to assign the material on this page as homework.

Not So Simple

Students may find it both interesting and entertaining to have the story of the "cytonauts" presented as a play, with class members taking the roles of the guide, Gloria, Yoon, and a narrator. If time permits, have the "actors" rehearse their parts.

Explain to students that cells must divide once they reach a certain size if growth is to continue. Otherwise, cells will either die or cease to be active when their surface-area-to-volume ratio becomes too small. Point out that having a large surface area allows a cell to exchange materials with its surroundings, taking in food and discarding waste. When a cell's mass increases, its need to take in and discard materials increases. Thus, a cell's surface-area-to-volume-ratio is important because it determines how efficient the cell is at taking in and discarding materials. You may want to use the following example to illustrate what happens to this ratio as a cell increases in size:

A cube that is 1 cm × 1 cm × 1 cm has a surface area of 6 cm² and a volume of 1 cm³. The surface-area-to-volume ratio is thus 6 to 1. If the cube grows to 2 cm in height, width, and depth, it will have a surface area of 24 cm² and a volume of 8 cm³, a ratio of only 3 to 1. Thus, volume increases at a greater rate than surface area. By dividing the cube into two equal halves, the total volume will remain the same, but the total surface area increases by one-third, making the surface-area-to-volume ratio increase to 4 to 1.

When the reading is completed, involve students in a discussion. Check their understanding by asking questions similar to the following:

- Why is it important for a cell to divide after a period of growth? *(As a cell increases in size, its surface-area-to-volume ratio decreases, making the cell less efficient at obtaining necessary materials and eliminating wastes through the membrane. After dividing, the daughter cells have a larger surface-area-to-volume ratio and thus can obtain and eliminate substances more efficiently.)*
- Why are daughter cells identical? *(They consist of two equal portions of the original material from the parent cell.)*
- What cell structure controls the way the material within the parent cell divides? *(The nucleus)*

Not So Simple

Gloria and Yoon visited their town's new high-tech Science Discovery Center. Listen in as they talk to their guide, a microbiologist.

Guide: *During cell division, the materials within the cell, including the nucleus, organize themselves into two equal portions. As a result, the two new daughter cells are identical—and just like the "parent" cell.*

Gloria: *But wait! If you divide one apple into two equal parts, you end up with two half-apples. How can one cell produce two whole cells simply by dividing?*

Guide: *Good point! The split isn't at all simple. The two new cells are actually each a bit smaller than the original one. They'll gradually grow larger until it's time for each of them, in turn, to divide. Each of the resulting cells is a complete living thing that can carry out all of the activities necessary for life. To carry out these activities, they must take in some substances, such as food, and release others, such as wastes.*

Yoon: *So what does that have to do with cells dividing?*

Guide: *The new, smaller cells have a surface area that is large for their volume. This enables them to take in and release substances very well. When the cells grow in size, the increase in volume is greater than the increase in surface area. So their capability to take in and release substances lessens. Eventually, each cell must divide. But before the cell divides, the material in the nucleus doubles!*

Gloria: *No way! You're saying matter can appear from nowhere. That's impossible!*

Guide: *You're right. To see how the material doubles, let's use some special miniaturizing equipment that will make you small enough to enter a cell. So let's strap on the equipment and get going, okay?*

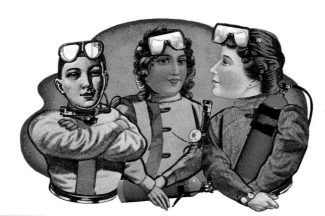

502

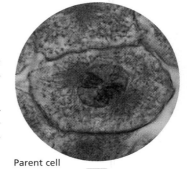

Parent cell

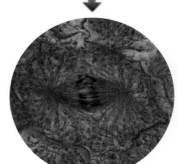

Materials within the nucleus organizing into two equal portions

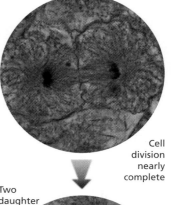

Cell division nearly complete

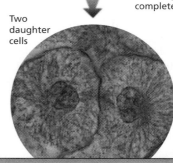

Two daughter cells

- What important process takes place in a cell just before it divides? *(The ladderlike material in the nucleus doubles.)*
- How would you explain what happens when a nucleus duplicates itself? *(Several long, spiral-shaped ladders exist inside the nucleus. Each ladder rung breaks apart near its middle. Immediately, separate components that are floating in the cell fluid move toward the broken rungs and attach themselves, forming two joined pairs of ladders from each single ladder. The membrane around the nucleus*

dissolves. The pairs of ladders separate and move toward opposite sides of the cell. A new membrane forms around each set of ladders, and two new nuclei are created.)
- What is the name of the process that Gloria and Yoon saw? *(Mitosis)*

 A Transparency Worksheet (Teaching Resources, page 12) and Transparencies 79 and 80 are available to accompany Not So Simple.

Yoon: *You mean we are going to be cell explorers?*

Guide: *Right! You'll see what it's like to be a cytonaut. (Cyto means "cell.")*

The miniaturizing equipment worked fast. Gloria and Yoon soon felt like they were underwater, moving through a liquid containing many objects they'd never seen before.

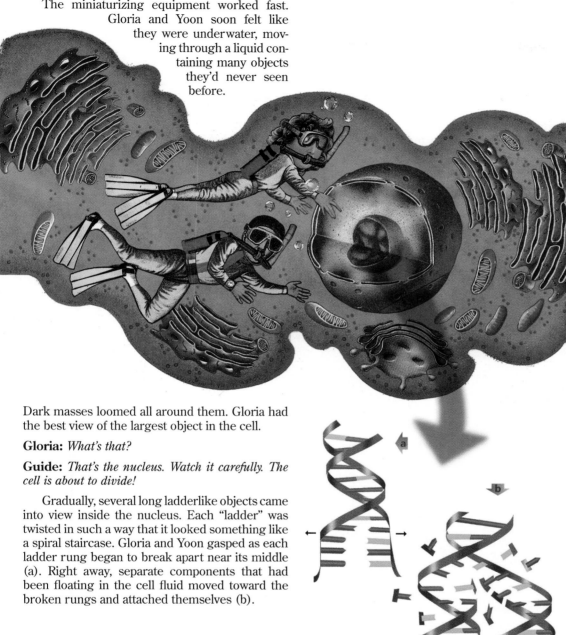

Dark masses loomed all around them. Gloria had the best view of the largest object in the cell.

Gloria: *What's that?*

Guide: *That's the nucleus. Watch it carefully. The cell is about to divide!*

Gradually, several long ladderlike objects came into view inside the nucleus. Each "ladder" was twisted in such a way that it looked something like a spiral staircase. Gloria and Yoon gasped as each ladder rung began to break apart near its middle (a). Right away, separate components that had been floating in the cell fluid moved toward the broken rungs and attached themselves (b).

503

Not So Simple, *continued*

After students have had time to discuss the story, point out the photomicrographs of cell division on page 504. Encourage students to describe what is happening in each picture and to identify its corresponding description in the story.

Remind students that mitosis refers to the duplication of the nucleus. The creation of new cells also requires *cytokinesis,* in which the cytoplasm separates into two new cells after mitosis. Because mitosis immediately precedes cell division in most cases, the term *mitosis* is sometimes used to refer to both nuclear division and cytokinesis. This sequence could more accurately be referred to as *mitotic cell division.* The period between cell divisions is called *interphase.* Cells are in this stage for most of their lives. During this stage, the DNA is duplicated, as are other cell structures such as mitochondria and chloroplasts. Individual chromosomes are not visible at this stage, and there is a membrane around the nucleus of the cell.

Answers to
A Cytologist's Expertise, *page 505*

a. Cells are the basic units of life.

b. Euglenas, paramecia, and amoebas are single-celled living things.

c. Cells can reproduce.

d. Nuclei are control centers of life processes.

e. Cell division produces new cells like the old.

f. Before a cell divides, material in the nucleus makes a copy of itself.

g. Our skin cells undergo cell division for growth and repair.

h. During cell division, the material in the nucleus is equally divided.

 A Resource Worksheet (Teaching Resources, page 14) and Transparency 81 are available to accompany A Cytologist's Expertise on page 505.

Twosomes, *page 505*

Point out that many different organisms, including most flowering plants, have both male and female reproductive organs. These organisms are able to reproduce sexually even though only one parent is involved. Also, because these organisms usually have two copies of every chromosome, the offspring (each of which receives only one of the two copies) may exhibit traits different from those of their parent.

Check students' understanding of the concepts by asking the following question: What is the difference between sexual and asexual reproduction? *(In sexual reproduction, a "male" cell and a "female" cell unite. In asexual reproduction, offspring are produced by cells that are identical to those of the parent.)*

New, complete ladders took shape before their eyes. Now they saw twice as many ladders, but each spiral ladder was joined to its duplicate (c). "Like Siamese twins," Gloria thought.

Guide: *Look! The membrane around the nucleus is dissolving.*

Gloria and Yoon then watched as the spiral ladders lined up across the center of the cell. Suddenly, each "Siamese twin" separated, forming two sets of spiral ladders (d). Everyone watched as one set of spirals moved toward them, while the other set faded from view behind a film.

Guide: *Guess what? You're in one of the new nuclei now. A membrane has enclosed you in this new nucleus, keeping you separate from the other new nucleus, which is now in its own cell. You just witnessed cell division from inside the nucleus! Pretty cool, huh? Let's return to normal size now.*

The photos below represent stages (the formation of new nuclei that Yoon and Gloria saw) of a process called **mitosis**. You'll notice that there aren't any spiral ladders here. That's because the photos show the way you'd see cell division through a microscope. Microscopes simply aren't powerful enough to show these tiny ladders.

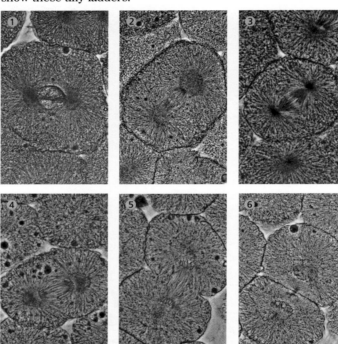

Cell undergoing mitosis

504

Learners Having Difficulty

Instruct students to use several examples of vegetative propagation to demonstrate asexual reproduction. Examples might include the following:

- Cut about 3 cm off the top of a carrot or about 5 cm off the top of a sweet potato, and bury the cutting in damp sand with just the top showing.
- Cut a piece of stem about 10 cm long from a coleus or geranium plant. Cut off any leaves within 4 cm of the cut end, and place the stem in moist sand or a container of water.

Changes Over Time

Focus question: Why are organisms that reproduce sexually better able to adapt to changing environments than are organisms that reproduce asexually? *(In sexual reproduction, DNA from one cell combines with DNA from another cell with different characteristics. Over the course of many generations, the mixture of DNA produces new combinations of traits and gives rise to diversity. There is no possibility for new combinations arising from asexual reproduction, except for those that arise through errors.)*

A Cytologist's Expertise

A **cytologist** studies cells and their structures. How well informed a cytologist are you right now? Check your expertise by doing the following matching exercise. Write eight sensible statements about cells in your ScienceLog by matching the words in Column I with the most appropriate words in Column II.

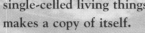

Column I	Column II
a. Cells are	growth and repair.
b. Euglenas, paramecia, and amoebas are	produces new cells like the old.
c. Cells can	control centers of life processes.
d. Nuclei are	the material in the nucleus is equally divided.
e. Cell division	the basic units of life.
f. Before a cell divides, material in the nucleus	reproduce.
g. Our skin cells undergo cell division for	single-celled living things.
h. During cell division,	makes a copy of itself.

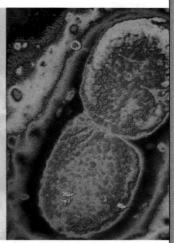

These *E. coli* cells are dividing. *E. coli* is a type of bacteria found in your digestive tract.

Twosomes

Cell division is performed by individual cells. It is a successful form of reproduction for many single-celled organisms, such as bacteria. Multicelled organisms (such as plants and animals) grow and repair damaged tissues by cell division, but they usually reproduce by a more complicated method. This method of reproduction requires two individuals, one male and one female. It is called **sexual reproduction**. In sexual reproduction, two cells—one from the male and one from the female—unite. The offspring, which forms from the union of these cells, inherits a mixture of traits from both parents. It's through this mixing of traits that diversity arises.

By contrast, the kind of reproduction that involves only a single parent is called **asexual reproduction**. Many single-celled organisms reproduce asexually by dividing into two cells. Each cell is then a new individual. Many plants also can reproduce asexually. For example, a new plant can be produced by planting a small stem that has been cut from another plant.

Going Further

Many single-celled organisms can reproduce sexually, but the process differs from sexual reproduction among multicelled organisms. How do single-celled organisms reproduce sexually?

505

LESSON
4 Mendel's Factors

Getting Started

Display any showy flower. Ask students to identify several of the flower's characteristics. Keep track of students' suggestions on the chalkboard. *(Color, size of flower, shape of petals, and so on)* Then ask: Which traits might change in other flowers of this species? *(Color, size, and shape)* Point out that in this lesson students will learn why traits vary among individuals of the same species. Students will also learn how to predict what traits the offspring of a plant will have.

Main Ideas

1. Traits are determined by pairs of inherited factors.
2. When hybrids are crossed, a 3-to-1 ratio of dominant to recessive traits appears in the next generation.
3. Offspring inherit one factor for a specific trait from each parent.

Call on a volunteer to read the riddle at the top of page 506, and have students think of a response. Invite them to share other similar riddles that they know. Then have students read this introductory material silently. When students have finished reading, check their understanding by involving them in a brief discussion of sexual reproduction in flowering plants.

Multicultural Extension

Flowers in the Netherlands

Explain to students that the Netherlands has long been a world leader in the development of flowers, especially tulips and roses. Every day, fresh flowers from the Netherlands are flown to many parts of the world. Challenge students to learn more about plant breeding in the Netherlands and about how it became famous for the thousands of varieties of flowers that have been developed there over the centuries.

"What would you get if you crossed a snake plant with a trumpet vine?"

If you've ever heard jokes like this, you probably already know what "crossed" means. In the case of plants, it means that the pollen from one plant is used to pollinate another plant. When pollen from a flower's *stamens* (a plant's male organs) contacts the ovules in a flower's *pistil* (the female organ), parts of the pollen and ovules unite. Sexual reproduction has occurred, and eventually seeds are formed. We cross plants to produce seeds that will grow into plants with a desirable mixture of characteristics inherited from the parent plants.

By the way, the answer to the joke above is, *a snake in the brass*!

Far-Reaching Labors

What would you get if you crossed a tall pea plant with a short pea plant? (This time, the question is real—not a joke.) The answer is that you would get a second generation of tall pea plants. Why not short plants or plants that fall somewhere in between? In the 1850s, Gregor Mendel, an Austrian monk with a passion for gardening, became intrigued by such questions. He had become curious about pea plants because he'd noticed that in some patches of his pea plants, all the plants grew tall. In other patches, there were only short plants, and in a third type of patch, there were mixtures of tall and short plants. He spent the next 7 years growing pea plants and looking for answers to satisfy his curiosity. In doing so, he made some very important discoveries.

Gregor Mendel ▶

LESSON 4 ORGANIZER

Time Required 3 class periods

Process Skills
analyzing, inferring, predicting

Theme Connection
Changes Over Time

New Terms
Chromosomes—the structures in the nucleus of a cell that carry the hereditary factors, or genes
Cross-pollination—the transfer of pollen from one plant to another plant
Dominant—describes a genetic factor, or trait, that is always expressed when it occurs

Fertilization—the fusion of two cells, one from each parent, that occurs during sexual reproduction
Genes—individual sections of chromosomes that control hereditary traits
Genetics—the study of heredity
Heredity—the passing of traits from one generation to the next
Hybrid—offspring containing one dominant gene and one recessive gene for a trait
Meiosis—the process that results in the formation of sex cells

continued ▶

Tall

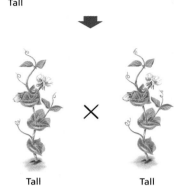

Tall

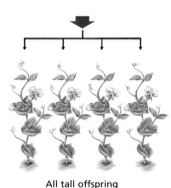

Tall   Tall

Short

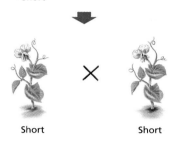

Short   Short

As you read about the stages of Mendel's investigations, search for clues to explain the results.

1. Mendel planted the seeds of tall pea plants in one garden area. He planted the seeds of short pea plants in another area. Why didn't he plant all the seeds in the same area of the garden?

2. He wrapped the flowers of each plant with pieces of cloth. Why?

3. He collected the seeds of each group of plants and planted them the next spring in separate beds. What was he trying to find out?

4. Mendel repeated this procedure many times until he was satisfied that the seeds from tall plants produced only tall plants, and the seeds from short plants produced only short plants. Why did this matter?

5. Mendel then planted the seeds from the pure tall plants and the seeds from the pure short plants. When they bloomed, he transferred pollen from tall plants to the flowers of short plants and later collected the seeds. From which plants (tall or short) did he collect the new seeds?

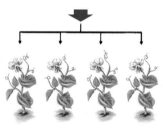

All short offspring

All tall offspring

Pure tall   Pure short

507

Cooperative Learning
FAR-REACHING LABORS PP. 506–508

Group size: 3 or 4 students
Group goal: to describe Mendel's genetic experiments
Positive interdependence: Employ a round-table format in which each student has the same role. One student begins reading page 507 and continues until he or she comes to an in-text question. The reader then gives an answer sheet to the person on his or her left. This student writes down the question and an answer. The group may discuss and agree on an answer before writing it down. This student continues the reading until he or she comes to an in-text question. He or she passes the answer sheet to the left. Students continue the process until all information has been read and discussed and all questions have been answered.
Individual accountability: Each student should be able to construct a concept map using the following terms: *cross-pollination, factors, dominant, recessive, hybrid, genetics,* and *heredity.*

★ A Resource Worksheet is available to accompany Far-Reaching Labors (Teaching Resources, page 15).

Homework

As a homework assignment, have students find biographical information about Gregor Mendel and write a short report on some aspect of his life that they find interesting.

Answers to Far-Reaching Labors are on the next page. ►

ORGANIZER, continued

Pedigree—a diagram showing the appearance of a trait in members of a family
Punnett squares—diagrams that show all the possible combinations of factors inherited by the offspring of two parents
Recessive—describes a genetic factor, or trait, that is expressed when no dominant factor is present
Sex cells—eggs and sperm

Materials (per student group)
Exploration 3: 200 2 cm × 2 cm squares of paper; 2 paper bags or other opaque containers (additional teacher materials for preparing paper

squares: 2 sheets of paper or photocopies of page 22 of the Unit 8 Teaching Resources booklet; paper cutter or scissors and metric ruler; see Advance Preparation on page 487C.)

Teaching Resources
Resource Worksheet, p. 15
Theme Worksheet, p. 19
Exploration Worksheet, p. 20
Transparencies 82 and 83
Transparency Worksheet, p. 23
SourceBook, p. S154

1. Mendel separated the tall and short pea plants so that they could not pollinate one another.

2. By wrapping the flowers with cloth, Mendel ensured that they would not be pollinated by insects bringing pollen from unknown plants. (The procedures outlined in steps 1 and 2 were done to ensure the purity of each strain of plant.)

3. Mendel wanted to find out if the two groups of plants produced offspring that were just like their parents.

4. Mendel needed pure, or true-breeding, plants for his experiments so that he could test one variable at a time—in this case, height.

5. He collected the seeds from the short plants because they were the plants to which he had transferred pollen.

6. Accept all reasonable responses. Sample answer: The pollen determines whether the offspring will be tall or short because pollen from tall plants produced only seeds that grew into tall plants. This could be tested by transferring pollen from short plants to tall ones and then planting the resulting seeds to see whether they grow into plants that are also short. (See number 7 below.)

7. Accept all reasonable responses. One explanation might be that whenever tall plants are crossed with short plants, the resulting seeds will produce only tall plants. Give students a chance to discuss and share their ideas about Mendel's experiments. Students should conclude that Mendel inferred from his results that both plants contributed instructions for height, regardless of which plant supplied the pollen. Mendel also inferred that only tall plants resulted from crossing tall and short plants because the factor for tallness is *dominant* and the factor for shortness is *recessive*.

8. About three-fourths of the plants were tall and about one-fourth were short. This was an important result because it demonstrated that the factor for shortness was still present, but hidden, in the previous generation.

---

6. The next year, the seeds produced by **cross-pollination** were planted. All the resulting plants were tall. What might be inferred from this? How could it be tested?

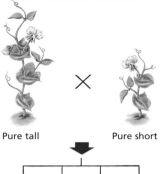

Pure tall   Pure short

All tall

7. Next, Mendel tried transferring pollen from short plants to tall plants, but the results were always the same—all the offspring were tall. How would you explain this?

Pure short   Pure tall

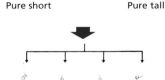

Pure short   Pure tall

All tall

To Mendel, it seemed that each male and female gave its own "instructions" for height to the seeds (and therefore to the next generation of peas). Why do you think he inferred this? Ⓐ He called the "instructions" *factors* and concluded that factors are inherited in pairs. For example, his pure tall parent plants contributed the factor for tallness, and his pure short parent plants contributed the factor for shortness. When two such plants were crossed, somehow tallness always "won." It seemed stronger, so Mendel called it **dominant**. He called the apparently weaker factor for shortness **recessive** because it seemed to recede into the background. Even though the

recessive factor didn't show up in the offspring, Mendel suspected that it was still present, just hidden.

8. Finally, Mendel proceeded one generation further. He crossed the offspring that resulted from crossing a pure tall plant with a pure short plant. The results were surprising. From a total of 1064 plants, Mendel counted 787 tall plants and 277 short plants. He noticed a definite pattern among the test plants. The number of tall plants versus short plants can be expressed as a fraction. What is this fraction (rounded off)? Why was this such an important result?

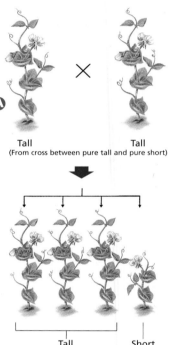

Tall   Tall
(From cross between pure tall and pure short)

Tall   Short

Meeting Individual Needs

Gifted Learners
Point out to students that the life cycles of most plants show a reproductive pattern called *alternation of generations.* This pattern is particularly easy to see in mosses and ferns. Have students find out about the reproductive processes involved in alteration of generations and make a poster diagram to illustrate the processes and identify the chromosome number at each step.

Answer to
In-Text Question

Ⓐ Sample answer: Mendel inferred this because the height of the offspring sometimes resembled the male parent and sometimes resembled the female parent. Thus, the parental instructions seemed to be passed on separately.

Putting It All Together

Mendel's hunch was right. The recessive factor was still there among the offspring that had come from crossing pure tall plants with pure short plants. Mendel gave the name **hybrid** to offspring that contained one dominant factor and one recessive factor. The diagram on the right can be used to show what Mendel did. Write a one-sentence description for each section of the diagram. **B**

Mendel went on to investigate other characteristics of pea plants, such as seed color. He found that the same pattern prevailed. Again and again, three-fourths of the second generation of plants (produced from the hybrids) showed the dominant trait, and one-fourth showed the recessive trait. Mendel was right. The recessive factor was still there in the hybrids, but it was hidden. It reappeared in the next generation.

Mendel experimented for many years. He wrote in one of his reports, "It indeed required some courage to undertake such far-reaching labors."

Mendel's work did not get immediate notice. However, his results eventually became the basis for the science of **genetics,** which is the study of how traits are passed from one generation to the next.

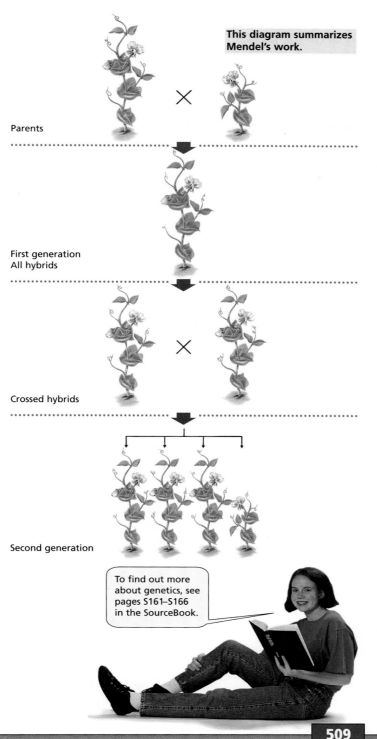

This diagram summarizes Mendel's work.

Parents

First generation
All hybrids

Crossed hybrids

Second generation

To find out more about genetics, see pages S161–S166 in the SourceBook.

509

To find out more about genetics, see pages S161–S166 in the SourceBook.

Answer to
In-Text Question

B • Parents: A tall pea plant is crossed with a short pea plant.
• First generation: The first-generation offspring all exhibit the dominant trait, tallness.
• Crossed hybrids: Two first-generation hybrid plants are crossed.
• Second generation: About three-fourths of the second-generation offspring are tall and about one-fourth are short.

Theme Connection

Changes Over Time

Focus question: How has the study of genetics changed the practice of agriculture? *(Although the selective breeding of plants has been practiced for thousands of years, the study of genetics has revealed the specific mechanisms by which traits are inherited. It has made it possible to transfer a single trait from one species to another and to genetically engineer food crops for higher yields or for desired traits such as resistance to drought, insects, or disease.)* A Theme Worksheet (Teaching Resources, page 19) is available to accompany this Theme Connection.

ENVIRONMENTAL FOCUS

One of the dangers associated with the selective breeding of food crops is a loss of genetic diversity. As farmers attempt to maximize the economic yield of plants, they increasingly rely on a few varieties. This overreliance on a few varieties can have devastating effects, as illustrated by the Irish potato famine of the 1840s. Nearly 2 million Irish starved to death when the few potato varieties on which they depended succumbed to a potato blight. The United States faced a similar problem in 1970 when a fungus, the southern corn leaf blight, destroyed 15 percent of the nation's corn crop because nearly all of the varieties had a common parent. Encourage interested students to research monoculture, the green revolution, and the National Seed Storage Laboratory.

CROSS-DISCIPLINARY FOCUS

Social Studies

Point out to students that a species may migrate from one part of the world to another. Through natural selection and competition for resources, a species adapts to its new environment because those individuals most suited to their environment are the ones that survive and reproduce most successfully. Eventually, a new species may evolve. On the Galápagos Islands, differences in the environments of the various islands influenced groups of a single species to adapt in different ways. As a result, a single species gave rise to many new species. Challenge students to identify such a species, the part of the world from which it migrated, and how the geography of its new environment contributed to the development of the characteristics of the new species. Have them share what they learn by making a diagram for the classroom.

Dominant to Recessive: Why 3 to 1?

Students may be interested to know that some of the other traits of the pea plants that Mendel investigated included smooth or wrinkled seeds, yellow or green seeds, purple or white flowers, green or yellow pods, constricted or inflated pods, and axial or terminal flowers. In each case, the second generation ratio was 3 to 1.

Meeting Individual Needs

Learners Having Difficulty

If students have difficulty understanding how the fractions three-fourths and one-fourth translate into a ratio of 3 to 1, display three pennies and one dime and ask the following questions:

- How many coins are there in all? *(4)*
- What fraction of the coins are pennies? *(3/4)*
- What fraction of the coins are dimes? *(1/4)*
- How many pennies and how many dimes are there? *(There are 3 pennies and 1 dime.)*
- What is the ratio of pennies to dimes? *(The ratio is 3 to 1.)*

When you are satisfied that students understand ratios, have them complete Exploration 3 on this page.

Answers To In-Text Questions

Ⓐ Every sample is subject to random fluctuations in the results. The larger the sample is, the smaller this fluctuation will be, and the closer the resulting ratio will be to 3 to 1.

Ⓑ The bags represent the nuclei of cells from hybrid plants because they contain the genetic factors for plant height.

EXPLORATION 3

Point out to students that marking the squares will be much easier and faster if the task is divided among all members of the group. If using copies of the premarked squares from the Unit 8 Teaching Resources booklet, simply make sure that each bag or container has 50 squares with

Dominant to Recessive: Why 3 to 1?

Mendel observed that when hybrids are crossed, the next generation shows the traits in about the same mathematical relationship each time: three-fourths show the dominant trait, while one-fourth show the recessive trait. This 3-to-1 ratio occurred again and again in Mendel's studies of traits other than plant height. It didn't seem to appear by chance. The larger the number of plants involved in each test, the more clearly defined the 3-to-1 ratio became. (Why?) Ⓐ

Using pieces of paper to represent Mendel's factors, the following Exploration shows how likely the 3-to-1 ratio is. As you do the Exploration, remember that Mendel was working with living things made of cells. In Exploration 3, what part of a living cell might the bag represent? Ⓑ

510

a *T* and 50 squares with a *t*.

Ask students why the bags are labeled as "hybrids." *(Each bag contains both dominant and recessive traits for tallness.)*

Cooperative Learning
EXPLORATION 3

Group size: 2 to 3 students
Group goal: to simulate Mendel's experiments with pea plants and to draw conclusions from the results
Positive interdependence: Each group member should choose a role such as Male Hybrid bag holder (to draw letters

EXPLORATION 3

Crossing Two Factors

You Will Need

- 200 squares of paper (about 2 cm × 2 cm)
- 2 paper bags or other opaque containers

What to Do

1. Mark 100 squares with a capital *T*. Let this stand for the dominant factor for height—tallness.

2. Mark 100 squares with a small *t*. Let this stand for the recessive factor for height—shortness.

3. Drop 50 squares with a *T* and 50 squares with a *t* into a paper bag labeled "Male Hybrid." Then drop 50 squares with a *T* and 50 squares with a *t* into a paper bag labeled "Female Hybrid."

Combinations	Number drawn
TT	
Tt	
tt	

4. Without looking into the bags, select one square from the Male Hybrid bag and one from the Female Hybrid bag. Place them together on a table.

5. Repeat step 4 until all the squares have been used up and all possible combinations of the male and female factors have been made. Set each possible combination *(TT, Tt,* or *tt)* in its own area of the table.

6. Count the number of pairs of *TT* combinations, *Tt* combinations, and *tt* combinations. Record the number of each in your ScienceLog, in a table containing the information shown at left.

from the male bag), Female Hybrid bag holder (to draw letters from the female bag), or tally keeper (to keep an accurate record of all letters drawn). Teams will answer Looking for Meaning on the next page together.

Individual accountability: Randomly choose a presenter from each group to present and explain the team's data to the class.

★ **An Exploration Worksheet is available to accompany Exploration 3 (Teaching Resources, page 20).**

Looking for Meaning

1. How many combinations of squares did you have?

2. In how many pairs did you have at least one *T*? What fraction is this of the total number of squares?

3. In how many pairs did you have only *t*'s? What fraction is this of the total number of pairs?

4. What is the ratio of pairs with at least one *T* to pairs with two *t*'s?

More Peas, Please

Mendel thought that each parent contributed, by chance, one of its factors for height to the peas. Suppose both parents supplied a *T*. The combination of *TT* would result in a tall plant in the next generation. The combination of a *T* from one parent and a *t* from the other parent would also result in a tall plant in the next generation, since *T* is dominant and *t* is recessive. If a plant inherited a *t* from each parent, however, the plant would be short. The diagram below illustrates this.

1. How many different combinations of factors are possible?

2. How does the number of combinations with at least one *T* compare with the number that have only *t*'s?

3. For the cross shown in the diagram below, how many tall plants would you expect for every short plant?

4. What did Mendel actually find in his investigations for this particular cross?

5. Refer again to the diagram outlining Mendel's investigations (page 509). How would you label each plant shown, using combinations of *T* and *t*?

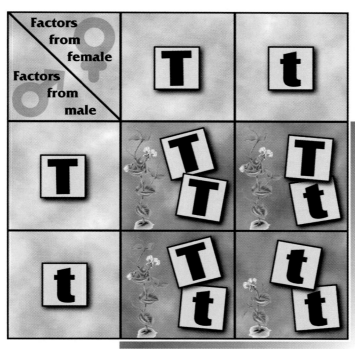

Possible Combinations in the Offspring

511

Answers to
Looking for Meaning

1. There were three possible combinations of squares: *TT*, *Tt*, and *tt*.

2. Although the exact number will vary from group to group, about 75 pairs (or three-fourths of the total number) should contain at least one *T*.

3. Although the exact number will vary from group to group, about 25 pairs (or one-fourth) should have only *t*'s.

4. The ratio will be about 3 to 1, although it will vary slightly from group to group.

More Peas, Please

Be sure students understand that when a factor for tallness *(T)* is present, the offspring will be tall because *T* is dominant. If necessary, point out that this means that even though one of the factors may be for shortness, the offspring will still be tall. The only time the offspring will be short is when both factors are recessive *(t)*.

Answers to
More Peas, Please

1. There are three possible combinations: *TT*, *Tt*, and *tt*.

2. There are three combinations with at least one *T* and one combination with only *t*'s. The ratio is 3 to 1.

3. You would expect about three tall plants for every short plant.

4. Mendel found 787 tall plants and 277 short plants, or a ratio of about 3 to 1. (The exact ratio resulting from Mendel's investigation was 2.84 to 1.)

5. Parents: *TT*, *tt*
First generation: *Tt*
Crossed hybrids: *Tt*, *Tt*
Second generation: *TT*, *Tt*, *Tt*, *tt*

Meeting Individual Needs

Second-Language Learners
There are many opportunities for students to diagram the concepts in this unit rather than to describe them in words. For example, students could diagram mitosis. Later in the unit, students could diagram meiosis and DNA replication. Have students label their diagrams and share them with the class.

Homework
You may wish to assign questions 1–5 in More Peas, Please as homework.

Making Punnett Squares

Point out to students that making Punnett squares is a simple way to determine the various combinations of genes, or factors, that can result from a particular cross. Direct their attention to the Punnett squares on this page and have them follow along as you explain how to fill out the squares.

Point out that the letters across the top will represent the female's genes, or factors, for a particular trait, and the letters on the left will represent the male's genes for the trait. The pair of genes in each box will represent one from the male and one from the female. Show students how to fill in each box by combining the factor to the left of the box with the factor above the box. When there is both a dominant and a recessive factor for a trait, the dominant one is always written first.

Making Punnett Squares

The diagram on page 511 shows all the possible combinations that can be passed on to the offspring of hybrid parent pea plants. Such diagrams are called **Punnett squares**. The passing on of traits from one generation to the next is called **heredity**. Punnett squares are commonly used in studying heredity.

Now it's your turn to practice working with Punnett squares. In your ScienceLog, draw and complete a Punnett square that shows the possible results of mating a black guinea pig with a brown one. A Black is the dominant color, while brown is the recessive color. The black guinea pig in this cross is purebred for a black coat. To indicate the dominant factor for coat color, use the capital letter *B*. For the recessive factor, use the lowercase letter *b*.

What would be the results of crossing two hybrid guinea pigs? In your ScienceLog, draw and complete a Punnett square for such a cross. What ratio of black to brown coat color would you expect the offspring of these hybrids to show? B

Black

Brown

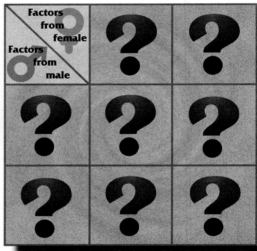

Meeting Individual Needs

Learners Having Difficulty

Allow students to tour the school campus, visit a nursery, or go to a greenhouse to study similarities and differences in a particular species of plant. (Be sure to select a species for which there are many examples.) Differences in height, color, and the number of petals on flowers should be noted. A class discussion can be used to decide whether the differences are based on hereditary or environmental factors. *(Accept all reasonable responses.)*

Predictable Ratios

In Exploration 3, you observed that there were three pairs with at least one *T* for every one pair that consisted only of *t*'s. This is a ratio of 3 to 1. Had you chosen only a few pairs instead of 100, would you have gotten this 3-to-1 relationship? Why or why not? **C**

To arrive at his results, Mendel experimented with many different inherited characteristics of pea plants, such as seed shape and color, pod shape and color, stem length, and flower position. He kept careful records of his experiments and used these records to calculate the ratios in which the traits appeared. In every case, the offspring of hybrid crosses produced the 3-to-1 ratio. One of Mendel's chief contributions was showing that many inherited traits would appear in predictable mathematical patterns.

Thinking Back

Now look over the data you recorded for Exploration 1. All the *yes* answers indicate dominant traits. That is, every *yes* answer you gave shows that you have the dominant expression of that characteristic. Every *no* answer shows that you have the recessive expression of that characteristic. If you show the recessive expression of a characteristic, what kind of factor did each of your parents contribute? Would the same be true if you show the dominant expression of a characteristic? Explain. **D**

Going Further

1. Should you expect there to be a 3-to-1 ratio of dominant to recessive traits for the characteristics you observed in Exploration 1? Why or why not? Suppose everyone in your school completed an index card. For each trait assessed, how do you think the number of dominant to recessive traits would compare?

2. Having a cleft chin is a dominant trait. So is having six fingers. Do you think that three-fourths of the students in your school have cleft chins? How about six fingers? How would you explain your observations?

3. Mendel's pea plants were either tall or short. How does this compare to the heights of your classmates? How might you explain this?

As you can see, studying human inheritance factors is much more complex than studying Mendel's pea-plant factors.

Michael Douglas (top) and his father Kirk show off their famous cleft chins. A cleft chin is a dominant trait. Why, then, don't most people have it?

And Where Did You Get That Nose?

Be sure students understand that a *pedigree* is a family-tree diagram that traces the inheritance of a single trait through several generations of a family. Direct students' attention to the example of the pedigree in the diagram. Check their understanding by asking questions similar to the following:

- How do you know which members of Juan's family have dimples? *(The members with dimples are shown with darker colors.)*
- Are there more family members with dimples or without dimples? *(There are more members with dimples than without dimples.)*
- Do both of Juan's parents have dimples? *(No)*
- Which parent displays the dominant gene for this trait? *(Juan's father)*
- Does at least one person in every generation shown in the pedigree diagram display the dominant expression for this trait? *(Yes)*

- - - - - - - - - - - - - - - -

Answers to
In-Text Questions

Ⓐ A person can possess a recessive trait that neither of the parents possesses. The inheritance of many human characteristics, such as height, intelligence, and eye, skin, and hair color, is not a matter of simple dominance. Most human characteristics are controlled by many genes. Thus, there is much variation in the way that they are expressed. Environment can also alter the expression of genes, and genes can be changed by mutations.

Ⓑ A pedigree can show the individuals in each generation who exhibit a specific trait. Pedigrees are used for studying human traits because making controlled crosses of humans is not practical or ethical, human families are too small to obtain accurate ratios, and the time between generations is too long.

And Where Did You Get That Nose?

Now use what you've learned to add to your family-tree project. Using the questions you answered in Exploration 1, interview or research as many members of the family you chose as possible. Keep a careful record of the information you find.

The diagram on the right is similar to the family tree you drew earlier, but this one traces the inheritance of a particular trait in a family—dimples. Using a separate diagram for each trait, note as many different traits as you can for the family you are studying. Each of these diagrams is called a *pedigree*.

Are there any family members who have characteristics that are not possessed by either of their parents? What can pedigrees Ⓐ tell you about traits that are passed on from generation to generation? Pedigrees are not usually used for studying the inheritance of traits in plants. Why are pedigrees used for studying the inheritance of human traits? Ⓑ

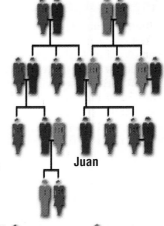

Juan

**K
E
Y** | Female with dimples | Female no dimples |
Male with dimples | Male no dimples |

A pedigree of the dimples trait in Juan's family

A Modern Look at Mendel's Factors

You know a lot now about dominant and recessive factors. By now, too, you've seen how certain physical traits are *inherited*, or passed from one generation to the next. What are the mysterious *factors* that determine these traits? It took scientists quite a while before they discovered an answer to this question.

Toward the end of the nineteenth century, certain structures in the nucleus of a dividing cell were seen through a microscope. Because these structures easily absorbed dye or stain, they were called **chromosomes** (meaning "colored bodies"). Each type of living thing has a specific number of chromosomes in its cells. The chromosomes are in pairs, like Mendel's factors. Human body cells, for example, contain 23 pairs, or a total of 46 chromosomes.

By 1910, Mendel's factors were identified as individual sections of chromosomes. These components were named **genes**, after a Greek word relating to "birth." Each chromosome in a pair carries one of the two genes that determine a trait. Altogether, human chromosomes carry at least 100,000 genes. Genes not only determine an organism's traits, but also provide instructions to the organism's cells as it grows.

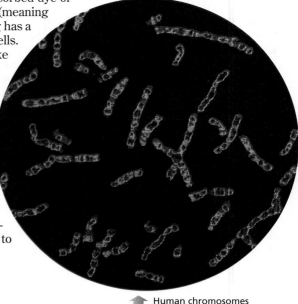

⬆ Human chromosomes

514

FOLLOW-UP

Reteaching

Have students illustrate the processes of mitosis and meiosis and include a short explanation of how the two processes differ. Students should include descriptions of the various stages and should emphasize the number of chromosomes or chromosome pairs that are present during these stages.

Assessment

Have students draw diagrams of three generations, like the one on page 509,

and Punnett squares to show what happens when a pure purple-flowered pea plant and a pure white-flowered pea plant are crossed. Point out that purple flowers are dominant and white flowers are recessive. *(All of the first generation will be purple. The second generation will be 3 purple to every 1 white.)*

Extension

Have students make a Punnett square to show the possible results of crossing two pea plants that are hybrids for seed

Sex Cells

When sexual reproduction takes place, a cell from one parent unites with a cell from the other parent to form a new cell. This process is called *fertilization*. In flowering plants, pollen contains a sperm nucleus (the male contribution), which unites with an ovum, or egg cell (the female contribution), inside an ovule. In the animal kingdom, a sperm from a male unites with an ovum, or egg, from a female. Eggs and sperm are also called **sex cells**.

Now here is a problem! If each human sex cell supplied 46 chromosomes, the offspring would develop from a total of 92 chromosomes. But human cells have only 46 chromosomes. What happens to make this number remain constant, generation after generation? Through a complicated process, sex cells with only 23 *single* chromosomes (one chromosome of each pair) are produced. This way, when the two sex cells unite, the resulting cell has 23 pairs of chromosomes, or a total of 46 chromosomes—the correct number.

The production of sex cells begins like ordinary cell division: the material in the nucleus doubles. The later stages cause the chromosomes to be distributed among four cells, instead of between two cells.

At right is a simplified view of what happens when sex cells form.

If this were a human cell, it would contain 23 pairs of chromosomes. Only three pairs are shown here so that you can more easily see what is happening. Look at the illustration for each step. What is happening to the number of chromosome strands at each step?

In a male, these four cells all become sperm. In a female, only one of the four cells becomes an egg cell. The other three disintegrate.

You have just observed the process that results in the formation of sex cells. This process is called **meiosis**.

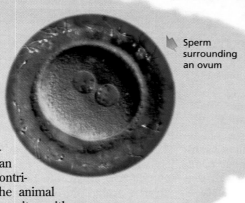

Sperm surrounding an ovum

Electron micrograph of sperm penetrating ovum (background)

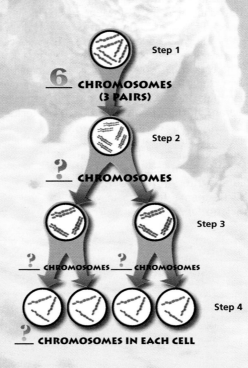

Step 1

6 CHROMOSOMES (3 PAIRS)

Step 2

? CHROMOSOMES

Step 3

? CHROMOSOMES — ? CHROMOSOMES

Step 4

? CHROMOSOMES IN EACH CELL

515

Sex Cells

After students have read page 515 silently, make sure that they understand that sex cells differ from the rest of the cells in the body in that they have only half of the number of chromosomes, one from each pair.

Have students study the diagram. Involve them in a discussion of what is happening to the number of chromosomes at each stage. Explain that until the duplicated chromosomes separate, they are considered to be single chromosomes composed of two chromatids that are joined by a centromere. For this exercise, students should count each chromatid as a separate chromosome.

Students may find it interesting to compare this diagram to the photographs showing mitosis on page 504. Have them compare how the two processes occur. (*In both processes, the number of chromosome strands is doubled and then halved. However, in meiosis a second division takes place that halves the number of chromosome strands a second time.*) Point out to students that in meiosis, the chromosome pairs separate during the first division, and the duplicated chromosomes separate during the second division. This pattern ensures that each sex cell will receive one factor from each pair of factors for a characteristic.

Answers to *Sex Cells*

Step 1: There are six chromosomes, or three pairs, per cell.
Step 2: The number of chromosomes has doubled so that each cell has twelve chromosome strands.
Step 3: The chromosome pairs have separated, and the cells have divided so that each cell has six chromosome strands.
Step 4: The duplicate chromosomes have separated, and the cells have divided again so that each cell has only three chromosomes.

 A Transparency Worksheet (Teaching Resources, page 23) and Transparency 83 are available to accompany Sex Cells.

---

texture—round seeds (R) vs. wrinkled seeds (r)—and seed color—yellow (Y) vs. green (y). Round and yellow are dominant traits. *Answer:*

Factors	RY	Ry	rY	ry
RY	RRYY	RRYy	RrYY	RrYy
Ry	RRYy	RRyy	RrYy	Rryy
rY	RrYY	RrYy	rrYY	rrYy
ry	RrYy	Rryy	rrYy	rryy

Have students determine what the ratio for the combinations of these traits will be. (*9 to 3 to 3 to 1*)

Closure

Ask students to design a pedigree like the one on page 514 in order to illustrate the inheritance of one of the human traits from Exploration 1. Have students integrate the results of the following crosses into their pedigree: pure dominant and pure recessive; two hybrids; pure dominant and hybrid; pure recessive and hybrid. Remind students that the expression of the trait in each generation should approximate the ratio predicted by completing Punnett squares.

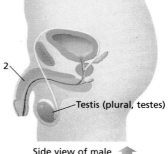

LESSON 5
Life Story of the Unborn

FOCUS

Getting Started

Help students to review what they learned in Lesson 4 about sexual reproduction by asking the following:

- What is the main difference between sex cells and other cells in the body? *(Sex cells have half the number of chromosomes that other cells have.)*
- What are male sex cells called? *(Sperm)*
- What are female sex cells called? *(Egg cells or ova)*
- What process results in the formation of sex cells? *(Meiosis)*
- What is the name of the process in which a sperm and an egg cell unite to form a new cell? *(Fertilization)*

Tell students that they will learn what happens in humans after fertilization takes place.

Main Ideas

1. The development of human life begins with the fertilization of an egg cell by a sperm cell.
2. During pregnancy, a fetus develops at a predictable rate and according to a predictable sequence.

TEACHING STRATEGIES

Have students read the introductory paragraphs silently. Students should be able to recall that human sex cells have only 23 chromosomes; therefore, when the cells unite during fertilization, the resulting fertilized cell has a full complement of 46 chromosomes. Students should also remember that during the process of meiosis, the number of chromosomes in sex cells is reduced by one-half.

 Transparencies 84–87 and two Transparency Worksheets are available to accompany Before You Were You (Teaching Resources, pages 25 and 27).

The development of a new human life is a remarkable process. At five different points in this lesson, you will pause to observe the development of an *embryo*. It all began with the fertilization of an egg cell by a sperm cell. Cell division then began, and from there, all kinds of things started happening.

But first things first. It really begins with two sex cells, each formed through the process of meiosis. Female sex cells (egg cells) are produced in the female's ovaries, while male sex cells (sperm cells) are produced in the male's testes. Each sex cell has only 23 single chromosomes (as opposed to the 23 pairs, or 46 chromosomes, in a body cell). Do you remember **A** why there's a difference? To examine the events that come before fertilization and the development of an embryo, read Before You Were You, and examine the diagrams on this page. For each numbered item, write a brief description of the item's function in your Sciencelog. **B**

Before You Were You

The sperm cells are placed in the vagina (1) by the male reproductive organ, the penis (2). The sperm move upward through the uterus into the Fallopian tube (3), where one sperm cell (4) may fertilize an egg cell (5) that has been released from the ovary (6). The fertilized egg cell begins to divide (7) as it moves into the uterus (8). There, perhaps 1 week after the egg has been fertilized, the embryo embeds itself (9) in the wall of the uterus.

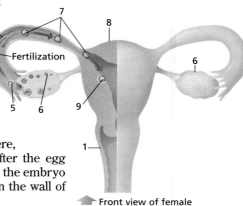

Front view of female reproductive system

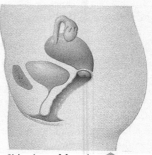

Side view of male reproductive system

Testis (plural, testes)

Side view of female reproductive system

LESSON 5 ORGANIZER

Time Required 3 class periods

Process Skills predicting, organizing, inferring

Theme Connection Cycles

New Terms

Embryo—an unborn child from the second to the eighth week of its development

Fetus—an unborn child from the ninth week following fertilization until birth

In vitro fertilization—a procedure in which an egg is fertilized outside of a woman's body and then is introduced into her uterus

Placenta—the organ that develops within the uterus during pregnancy that allows nutrients and wastes to be exchanged between the mother and the embryo and vice versa

Trimester—a 3-month period in a human pregnancy

Materials (per student group) none

Teaching Resources
Transparencies 84–87
Transparency Worksheet, p. 25
Transparency Worksheet, p. 27
Resource Worksheet, p. 29
SourceBook, p. S175

This diagram shows some details of the embryo in the uterus wall.

1. What do you think happened at (10)?
2. What role might the amniotic cavity (11) play?
3. What observation can you make about the cells of the embryo (12) at this stage?
4. What part do you think the yolk sac (13) plays?
5. Why is it important that the mother's blood vessels (14) be near the embryo?

The life-support system for the embryo (15) is called the **placenta** (16). It is an organ that develops around the embryo and that attaches the embryo to the wall of the uterus. The placenta is rich with blood vessels that come from both the mother and the embryo. However, the blood of the mother does not mix directly with the blood of the embryo. Instead, substances are exchanged between the mother's blood and the embryo's blood across the walls of the blood vessels in the placenta.

As the embryo develops, it grows an umbilical cord (17) that connects the embryo to the placenta. Substances enter and leave the embryo through this cord. What substances are needed by the embryo? What substances must leave the embryo? **C**

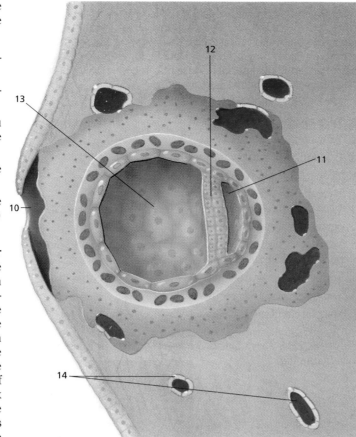

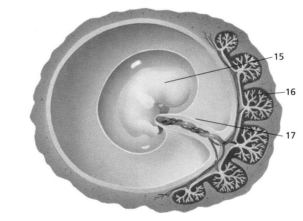

Compare these two diagrams. What necessary part of the life-support system has not yet developed in the upper diagram? **D**

A Sex cells have only 23 chromosomes so that an egg will have 46 chromosomes when fertilized.

B Sample answer:
1. The vagina is where sperm enter the female's reproductive system.
2. The penis delivers sperm from the male.
3. The Fallopian tube is where the sperm and egg unite.
4. The sperm cell contains genetic information from the male.
5. The egg cell contains genetic information from the female.
6. The ovary produces egg cells.
7. The fertilized egg contains all of the genetic information necessary to create a new individual.
8. The uterus is where the fertilized egg embeds itself and develops throughout the pregnancy.
9. The embryo develops from the fertilized egg and grows into a fetus.

C The embryo needs oxygen, nutrients, and water. Waste products, including carbon dioxide, must be carried away from the embryo.

D The umbilical cord has not yet developed.

Homework

You may wish to assign the in-text questions on this page as homework.

Answers to
Before You Were You

1. The fertilized egg made its way into the lining of the uterus.

2. As the cavity grows larger and becomes filled with fluid, it will surround and protect the embryo.

3. The embryo appears to be flat, and all of the cells look alike. (In fact, the cells of the embryo are just beginning to differentiate and take on specific roles at this stage.)

4. Students might assume that the yolk sac supplies nourishment for the embryo until the life-support system, the placenta, develops. This is the function of the yolk in most vertebrates. The yolk sac of a human embryo, however, is the source of the embryo's first blood cells and germ cells.

5. The mother's blood vessels carry nutrients to the embryo and waste products away from the embryo.

Cooperative Learning
YOUR EARLY DEVELOPMENT

Group size: about 5 students

Group goal: to describe the events that occur during the development of a human embryo and fetus

Positive interdependence: Have each group discuss the developments listed in Your Early Development and record their predictions. Each member of the group should choose one of the five episodes on pages 519–527 to teach. Students should regroup so that all common episode members meet together in "expert groups" to read and discuss their episode. Then have students return to their original group to teach their episode to the other group members. As each episode is taught, the groups should reorder their list of developments as necessary. At the end of each episode presentation, discuss as a class the questions that follow each episode. (You may wish to place a time limit on presentations.)

Individual accountability: Each student is responsible for teaching an episode.

Theme Connection

Cycles

Focus question: How does the human female body prepare for fertilization each month? *(Each month an egg is released by the ovary, and the interior of the uterus becomes coated with a protective lining. If the egg is not fertilized, the lining of the uterus is broken down, and the egg is flushed away. The next month, another egg is released, and the cycle begins once again.)*

Your Early Development

Now consider what you already know about your own development. In your ScienceLog, place the following happenings in the order you think they occurred. Decide, for example, what might have happened during the first month, second month, and so on. The list includes abilities and activities.

a. Your fingers and toes took shape.

b. Your heart began to beat.

c. Your memory began.

d. Leg and arm stumps appeared.

e. Your eyes could respond to light.

f. Your mother was aware of your movement.

g. Your baby teeth started to take shape, and tiny buds for your permanent teeth began to develop.

h. You began the cycle of sleeping and waking, and you developed the ability to dream.

i. Your sex was clearly indicated by internal sex organs.

j. You developed fingerprints.

k. Your face and neck began to develop.

As you follow each episode in Adventures of a Life in Progress, refer to your list to see how well you imagined your own beginning!

⬆ A human being in the making

Answer to
Your Early Development

The steps occur in approximately the following sequence:

b. Your heart began to beat—by 22 days.

d. Leg and arm stumps appeared—end of first month.

k. Your face and neck began to develop—second month.

i. Your sex was clearly indicated by internal sex organs—end of second month.

a. Your fingers and toes took shape—end of second month.

g. Your baby teeth started to take shape, and tiny buds for your permanent teeth began to develop—third month.

j. You developed fingerprints—fourth month.

f. Your mother was aware of your movement—fourth month.

e. Your eyes could respond to light—seventh month.

c. Your memory began—seventh month.

h. You began the cycle of sleeping and waking, and you developed the ability to dream—eighth month.

Adventures of a Life in Progress

Episode 1—The First Month

Your life started with the fertilization of an egg cell by a sperm cell. The egg was about 0.1 mm in diameter—even smaller than the period at the end of this sentence. The sperm that fertilized the egg was only one-thirtieth the size of the egg. After fertilization, you began to divide, and things began to happen like clockwork. Later, cells with a specific shape and function began to appear.

By the age of 17 days, blood cells had formed, and shortly thereafter, you had a heart tube. Your heart began to twitch and began the rhythmic beating that must continue until your life's end. By the end of the third week, the cells that would eventually form your sperm or eggs were set aside. In the fourth week, a tube was formed in what became your mid-back region. At the front end, your brain later developed. The back part of the tube later formed your spine. Next came your food canal. By the 25th day, you had a head end and a tail end (and a bit of a tail also). You had no face or neck, and your heart lay close beside where your brain would eventually be.

By the end of the first month, you had tiny bumps where arms and legs would later develop. Your lungs, liver, kidneys, and most other organs had begun to form. Your head had two pouches that would become eyes, two sunken patches of tissue where the nose would form, and a sensitive area behind each eye where the inner ear would form. A lot happened in 1 month. Can you imagine what size and shape your embryo was by then? Ⓐ

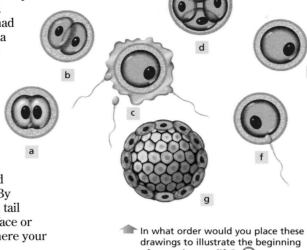

In what order would you place these drawings to illustrate the beginning of a new human life? Ⓑ

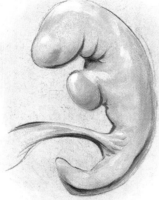

This is what you looked like at 1 month of growth in the womb, shown about 10 times actual size.

Answer to
In-Text Question and Caption

Ⓐ By the end of 1 month, the embryo is about 5 mm long and is shaped somewhat like a peanut.

Ⓑ The order of the drawings is c, f, e, a, b, d, and g.

Integrating the Sciences

Life and Physical Sciences

Have students try to explain how it is possible for substances to pass through the walls of a mother's blood vessels into the embryo's blood vessels in the placenta. Explain that the process of diffusion makes it possible. Review the law of diffusion, which students learned about in Unit 1.

Episode 1—The First Month

The following information may be useful in answering some of the questions that students may have after reading this page:

After one sperm has entered an egg, changes in the egg's surface take place to prevent additional sperm from entering. Once the nuclei of the sperm and egg have united, a nucleus with a complete complement of 46 chromosomes is formed. This new nucleus contains all of the genetic material (DNA) that will then be duplicated and distributed to all the cells of the embryo during cell division.

Shortly after a sperm penetrates an egg, a series of cell divisions called *cleavage* begins. Cleavage in humans is relatively slow in comparison to other animals. For example, the first cleavage occurs about 1 day after fertilization, and subsequent cleavages take place about twice a day for the next 5 or 6 days. (In frogs, cleavage takes place about once every hour.)

After a week's growth, the embryo is a formless group of cells that is about half of a millimeter in diameter. At the end of 2 weeks, tissue differentiation begins to take place. By 3 weeks, the 2.5 mm embryo has begun organ formation. The limb buds appear late in the fourth week. Due to the dominance of the head and neck region, the arm buds seem to be located quite low on the body. Of the two sets of limb buds, the upper pair appears first, begins its differentiation sooner, and is earlier in attaining its final relative size.

As students read about the first month of embryonic development, they may wonder how cells seem to know to differentiate into heart cells, liver cells, lung cells, and so on. In fact, scientists do not know exactly how such specialization and differentiation occurs. However, they do know that all of the many specialized cells that make up an individual are derived from a single fertilized egg cell. During the first few cell divisions, or cleavages, each new cell is still able to form an entire organism. This ability, called *totipotency,* is lost after the first few cleavages.

Jean Hegland recorded in a journal the thoughts she and her husband had as they awaited the birth of their child. She published her thoughts in a book called *The Life Within*. Here is a part of her journal for the first month.

> Then comes heart, and spinal cord, and gut, all primitive, all bulging and shifting like a rose opening in a time-lapsed film, all arising out of that speck of protein that was two cells, that became one, that doubled and divided, transforming from one thing to another as wildly and exactly as objects in a magician's show, where rabbits become scarves that become tulips and then coins, in a perfect, dizzying succession of change. And so these organs, these basic bits of human being, appear like rabbits out of nowhere—out of the intention of the universe—and a new creature takes shape.

What sort of instructions are being followed to make all of the embryo's body parts form at a certain time, in a certain way? Compare the events of the first month with the list of developments you wrote in your ScienceLog. Were your predictions close? **Ⓐ**

Helping the Natural Process
In 1978 a baby's birth made history. That baby's name was Louise Brown, and she was the first "test-tube" baby. Did she really grow in a test tube? What does the term mean? Why was this test-tube procedure done? **Ⓑ**

Sometimes, certain problems interfere with the natural process of fertilization. **In vitro fertilization** is a procedure that allows certain people to have children. Find out what this term means and why the procedure is performed. As you investigate this form of reproductive technology, as it is called, ask yourself some basic questions, such as the following:

1. Why would someone choose to try (or not to try) this procedure?
2. What are the costs, in terms of
   a. risks to parent or baby?
   b. money?
   c. time?
3. How successful is in vitro fertilization likely to be?

Louise Brown—the first "test-tube" baby—with her younger brother, also conceived by in vitro fertilization.

Outline a brief presentation that you could give to someone interested in the procedure. Include the pros and cons of in vitro fertilization and back up your statements with information gathered in your research. End by giving your recommendation for guidelines as to when in vitro fertilization is appropriate and when it is not.

Twin Heartbeats

Sometimes two embryos develop at the same time. The resulting children may be so much alike that even their parents have trouble telling them apart. These are called *identical twins*. Sometimes, though, twins are no more alike than any other siblings (children of the same parents). These are *fraternal twins*. Why are twins sometimes identical and sometimes fraternal? Consider two possibilities.

A The mass of cells from a single fertilized egg separates into two halves early in development. Two babies result.

B Two eggs are released by an ovary. They are fertilized by two different sperm cells. Each is implanted in the uterus and continues to grow. Two babies result.

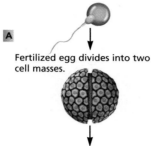

A

Fertilized egg divides into two cell masses.

Two cell masses develop in one sac.

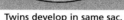

Twins develop in same sac.

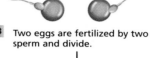

B Two eggs are fertilized by two sperm and divide.

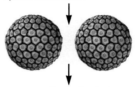

Two masses of cells develop separately, each in its own sac.

Twins develop in separate sacs.

What Do You Think?

Think about the two possibilities above. Do they tell you why twins are sometimes identical and sometimes fraternal?

1. Which instance, *A* or *B*, would produce two identical babies?
2. It is possible at birth to tell whether twins are identical. How?
3. Could identical twins be (a) both boys, (b) both girls, or (c) one boy and one girl? Provide an explanation for your answers.
4. Could fraternal twins be (a) both boys, (b) both girls, or (c) one girl and one boy?

Provide an explanation for your answers.

521

Episode 2—
The Second Month

The following additional developmental landmarks that occur during the second month of embryonic growth are provided for your use in class discussion:

- The development of the human face occurs chiefly between the fifth and eighth weeks. During this period, the face becomes recognizably human.
- The trunk becomes less curved.
- The head becomes more erect.
- A tail develops and then disappears.
- The five major divisions of the brain are formed.
- The cerebral cortex of the brain begins to acquire specialized cells, and the olfactory lobes responsible for interpreting smells are visible.
- The eyelids are present by the end of the second month.
- Parts of the external, middle, and inner ear approach their final form.

Answers to
In-Text Questions

Ⓐ Students will probably predict that the brain develops at a faster rate than other body parts because it will be linked to the control of the organ systems and sensory organs that are also developing quickly.

Ⓑ After eight weeks, the embryo is about 2.5 cm long and has a mass of about 1 g. Events that occurred in the second month (from the list on page 518) include (k) development of face and neck, (i) appearance of internal sex organs, and (a) formation of fingers and toes.

Episode 2—The Second Month

This month was a time of rapid growth. You grew about 5 times longer and about 500 times heavier. Your face and neck developed. Your brain developed at a faster pace than any other part of your body. Why do you think this happened? Ⓐ

By the end of the second month, your sex was clearly indicated by internal sex organs, which could probably have been seen externally too. Bones and muscles began to form. The arm and leg buds that formed last month began to grow into limbs, and your fingers and toes took shape. Gradually, constrictions formed to mark off elbows and wrists, knees and ankles.

Your bones first consisted of a soft substance called cartilage. Your bone structure changed throughout your prenatal (before birth) life, and it will continue to change in your postnatal (after birth) life. In fact, not until you become a mature adult will your skeleton be fully formed.

Ⓑ How big were you by the end of the second month? How many events did you predict correctly?

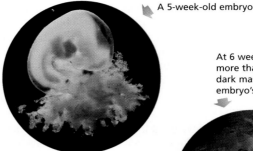

A 5-week-old embryo

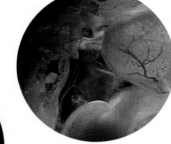

At 6 weeks, you were little more than 1 cm long. The dark masses are this embryo's heart and liver.

As an 8-week-old embryo, you were only the size of a walnut, but you had all of your basic organs and systems. Your head was almost half your total size.

Sometime during this month, this bit of a child, this baby grown to the size of a peanut, will begin to move, to flail its stubby hands and stretch its delicate legs . . . I wonder what it must be like to bend a brand new elbow, to wiggle freshly made toes . . .

And then there are the thousands of things that could go wrong . . .

Each dividing cell, each growing organ must follow the 4-million-year-old plan exactly—the first time. There are no second chances. There is no going back, no catching up, no fixing mistakes. . . .

. . . we have been warned about Thalidomide, radiation, rubella. We have been told about nutrition and exercise, and what the Chinese knew a thousand years ago, that the child of a tranquil mother is calmer, brighter, and weighs more at birth. And so I drink milk for the sake of this creature's bones, eat protein for its mushrooming brain. I avoid alcohol, aspirin, and stress . . .

522

Meeting Individual Needs

Gifted Learners

Have students read Sex Determination on page S165. Have students create a pedigree diagram in which the mother is homozygous for hemophilia and the father does not carry the gene. Have students predict the likelihood that her grandchildren will carry the gene for hemophilia. (*Among her son's children, all of the females and none of the males should carry the gene. All of her daughter's children should have a fifty-percent likelihood of carrying the gene.*)

Opinions and Viewpoints

As you have seen, the developing embryo is very delicate. What are some things that can affect its development? What are the parents' responsibilities to the developing embryo? This subject has prompted quite a bit of research and thought. Use the following questions to help you form your own informed opinions and viewpoints.

1. How dangerous is it for pregnant women to drink alcohol, to smoke, or to abuse drugs? What can happen to the developing embryo as a result of these activities? Find out all you can about this topic by consulting your library for books, magazines, and newspaper articles on the subject. For additional assistance, ask a nurse or doctor.

2. What is rubella? Investigate how a pregnant woman's exposure to this can affect the developing embryo.

3. Thalidomide was a drug used in some countries for a short time to help pregnant women who had nausea during the first months of pregnancy. It had disastrous results.

    a. Do some research about this drug and the tragedy of its use.

    b. What evidence indicates that it was used during the second month of pregnancy?

4. How does stress experienced by the mother affect the developing embryo?

SHE'S ALREADY UP TO A PACK A DAY.

Use tobacco while you're pregnant and your dreams of a healthy child could go up in smoke.

IF YOU'RE PREGNANT, DON'T SMOKE.

523

Opinions and Viewpoints

INDEPENDENT PRACTICE Ask students to select one of the topics, research it, and prepare an oral report for the class. Encourage students who work on the same topic to share what they learn, or you may wish to suggest that students work in small groups.

PORTFOLIO
You may wish for students to put their responses to Opinions and Viewpoints in their Portfolio.

1. More than two alcoholic drinks per day increases the chance of fetal alcohol syndrome. Symptoms of this disorder include facial abnormalities, heart defects, abnormal limb development, and low intelligence. Also, this level of alcohol consumption increases the risk of miscarriage. Even small amounts of alcohol may disrupt normal development.

    Smoking can interfere with the amount of oxygen that an embryo or fetus receives. Babies born to women who smoke tend to be smaller and less healthy than babies born to nonsmoking mothers.

    Any drug (even aspirin) taken during pregnancy may pass from the mother to the developing embryo or fetus. The babies of drug abusers can have serious problems, and these babies tend to weigh less and have a higher chance of death during the first few weeks after birth. They may suffer withdrawal symptoms, such as sleeping and feeding difficulties, trembling, and even seizures. Babies born to women who are intravenous drug users have the added risk of being infected with the AIDS virus.

2. Rubella is a viral infection also known as German measles. It can affect a developing embryo if the mother is infected during her first 4 months of pregnancy. Affected newborns may suffer from deafness, congenital heart disease, mental retardation, cataracts, cerebral palsy, bone abnormalities, and other disorders. About 20 percent of affected babies die in early infancy.

3. a. Thalidomide was never approved for sale in the United States. It caused limb deformities in many of the babies born to women who were given the drug.

    b. The deformities of the arms and legs showed that the drug was taken when the embryonic limbs were beginning to develop.

4. Sample answer: Stress could cause a pregnant woman to lose sleep or rely on alcohol, drugs, or cigarettes to reduce stress. Biochemical effects of stress are also possible.

Episode 3—The Third Month

The following additional developmental landmarks that occur during the third month of growth are provided for your use in class discussion:

- The lips differentiate.
- The intestines withdraw from the umbilical cord and assume their characteristic position.
- The anal canal is formed.
- The kidneys begin to function, and the bladder expands as a sac.
- In males, the rudimentary sex ducts form. The external genitalia attain distinctive features, and the prostate and seminal vesicles appear.
- The lymph glands begin to develop.
- Blood formation begins in bone marrow, and red cells predominate in the blood.
- The fetal pulse can now be detected.
- Nasal passages are partitioned. The bridge of the nose and nasal glands form.
- Hair follicles begin developing on the face.
- The spinal cord attains its definitive internal structure.
- Tear glands are budding.
- Cheeks develop.
- The lungs acquire their distinctive shape.
- The brain attains its general structural features.
- Characteristic organization of the eye is attained.

After 3 months of growing, a fetus is about 7 cm long and has a mass of about 25 g. The mother begins to feel her uterus enlarging.

Multicultural Extension

Ancestor Worship

Long before the study of genetics revealed the transmission of traits from one generation to the next, the belief that ancestors could affect the lives of living descendants was common in many cultures. Ancestor worship remains an integral part of many cultures worldwide. For example, many cultures believe that the souls of their ancestors are reborn in living things or in inanimate objects. Encourage students to report on spiritual beliefs about ancestors as a counterpoint to this unit's discussion of genetics. Students may find it interesting to compare scientific notions about heredity with religious ideas about the influence of ancestors.

Episode 3—The Third Month

From the third month on, the developing child is called a **fetus**. A male fetus undergoes a lot of sexual development during the third month. This could also be called the "tooth month" because your first teeth started to take shape, and tiny buds for your permanent teeth began to develop. Your vocal cords formed during this month. Your digestive system, liver, and kidneys all began to function. Bones and muscles developed, and your face and body began to take on their distinctly human form.

After 3 months of growing, you were about 7 cm long.

... its umbilical cord connects our Riddle to its placenta like an astronaut to his spacecraft or a deep-sea diver to her boat. A placenta is a weird, dense pudding of an organ, just thicker than a beefsteak and roughly round as a dinner plate. On one side, the pale cable of umbilical cord rises from it like a tree trunk from a rooty tangle of blood vessels, and on the underside, those veins have thinned to tiny capillaries that entwine with the capillaries in the wall of my uterus. There, this fetus and I communicate in ... the language of diffusion, so that its needs are met, molecule by molecule, before it knows them. There, its blood rushes constantly to meet mine, to strip it of its lode of oxygen and nutrients, and give me in exchange carbon dioxide, uric acid, wastes.

... After it is born, and the cut stump of its umbilical cord sloughs off, this baby will be scarred forever with the pretty pucker of its navel, a remembrance of the communion it once took from my blood.

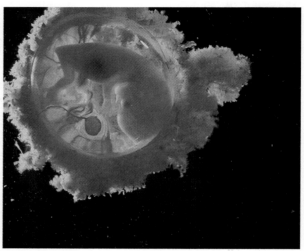

At 11 weeks, each of your hands was about the size of a teardrop, and your body weighed about as much as an envelope and a sheet of paper.

Cross-Disciplinary Focus

Language Arts

Point out to students that "origin tales" are common in many cultures. Some of these tales attempt to explain the origin of life. Challenge students to research some origin tales to share with the class. Suggest that they organize the stories they find into a booklet, think of a title, and design a cover. Display the booklet in the classroom or school library for other students to read. To extend the activity, you may wish to have students write their own origin tales.

Answer to
In-Text Question, page 525

Ⓐ Answers will vary. Point out to students that the first trimester refers to the first three months of pregnancy and includes events (b), (d), (k), (i), (a), and (g) from the list on page 518.

You have now reached the end of the first **trimester** of the developing life. Review what has happened. How does the order of events compare with what you predicted on page 518? Ⓐ

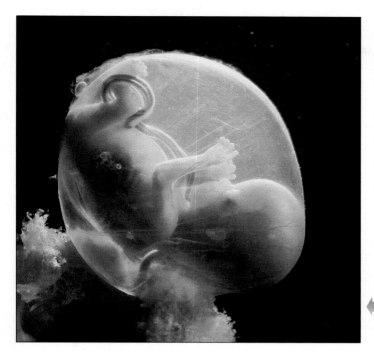

By the 12th week, your eyelids closed and remained shut for about 3 more months.

Analysis and Reflection

1. a. The writer of *The Life Within* calls her unborn child a "Riddle." Why do you suppose she does that?

   b. Explain "the language of diffusion." Where are the membranes that permit diffusion? What substances diffuse? How is this like the process in a hen's egg?

   c. You may have seen the "cut stump" of an umbilical cord on a newborn. What does "sloughs off" refer to?

2. If a pregnant woman smokes cigarettes, do any of the substances from the smoke pass through the placenta into the fetus? Find out about this.

3. Suppose someone said, "A pregnant woman can do whatever she wants to do. She can smoke, drink, take drugs, and not take care of herself, if that's what she decides to do." Would you agree or disagree with the statement? Gather information to support your opinion.

525

Episode 4—
The Second Trimester

Before students begin reading, direct their attention to the title on page 526. Call on a volunteer to explain what is meant by *second trimester.* (*The second 3-month period of pregnancy—the fourth, fifth, and sixth months*) Then have students read the material, referring to the photographs as needed.

The following additional developmental landmarks that occur during the second trimester are provided for your use in class discussion:

Fourth month
- The face looks "human."
- Hair appears on the head.
- Muscles become active spontaneously.
- The body begins to grow faster than the head.
- Gastric and intestinal glands begin to develop.
- The kidneys attain their typical shape and structure.
- In females, the uterus and vagina are recognizable.
- In males, the testes are in position for later descent into the scrotum.
- Most bones are recognizable throughout the body.
- Stretching movements occur and may be felt by the mother.
- The epidermis begins adding layers to form the skin.
- Body hair starts developing.
- Sweat glands appear.

Fifth month
- The body is covered by downy hair called *lanugo.*
- The nose bones begin to harden.
- In females, the vaginal passageway begins to develop.
- The heart beats at twice the adult rate.
- The lungs are formed but do not function.
- The gripping reflex of the hands begins to develop.
- Kicking movements and hiccuping may be felt by the mother.

Sixth month
- The body is lean and better proportioned.
- The colon becomes recognizable.
- The nostrils open.
- The cerebral cortex of the brain is layered in a more typically adult fashion.

Episode 4—The Second Trimester

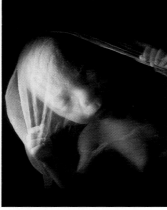

A 4-month-old fetus

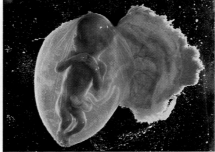

Fetus at $4\frac{1}{2}$ months

In the second trimester of development, your fingerprints and toeprints formed (month 4), followed by oil and sweat glands, scalp hair, nails, and tooth enamel (month 5). Your eyebrows and lashes continued to grow (month 6).

Your mother first became aware of your movement during the fourth month. You were half your birth height (a third of that was the length of your head), and you were wrinkled and dark red in color.

During the fifth month, some of your skin cells died and fell off. They were replaced by new ones. A mixture of the dead cells and a substance from the skin glands formed a protective layer over your skin. Your mass was about half a kilogram. Your head was well balanced, and your back grew straighter.

Your eyes (sealed at the age of 3 months) reopened during the sixth month. You had more taste buds on your tongue and in your mouth than you have now.

They say that already your face, with its newly finished lips, is like no one else's face . . . a twenty-week-old fetus is a porcelain doll of a child, complete with fingernails and eyebrows and a tender fringe of lashes sprouting from the lids of its still-sealed eyes. Fragile, luminous, it seems an unearthly being—more sprite than hearty human—through whose skin shines an intricate lace of veins.

. . . Always when it moves I know it now. It pummels my kidneys, kicks my stomach, punches my lungs. I know when it is resting, and when it wakes and stretches. I know it well by now . . . It knows me too. It dangles . . . inside me . . . listening to all my doings. It hears all that I hear, the sounds intensified and distorted like the . . . racket one hears under the water in a swimming pool.

526

- The internal organs occupy their normal positions.
- Thumb sucking may begin.

By the end of the second trimester, the fetus is about 36 cm long and has a mass of about 850 g.

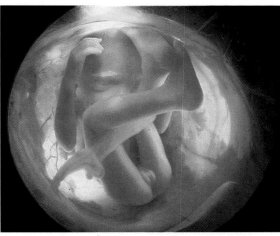

By 22 weeks, your mother could feel your movements, and she may have even been aware of your sleeping and waking cycles.

Thought-Provoking Questions

1. Why do you suppose the features of the fetus are already unique at such an early age?

2. Why do some mothers-to-be read and talk to their growing fetuses?

Episode 5—The Third Trimester

In the seventh month, your eyes could respond to light, and the palms of your hands could respond to a light touch. Fat began to fill in the wrinkles during the eighth and ninth months. By the end of the seventh month, your mass was about 1.4 kg. If you had been born at this point, you could have survived as a premature baby. Your activity increased, and you might have even changed position in the uterus. During this trimester, you were "practice-breathing," even though your lungs were filled with fluid. If your mother smoked just *one* cigarette, the practice-breathing stopped for up to half an hour.

> Medical researchers say that memory begins during the seventh month . . . I wonder what this child will remember . . .
>
> . . . neurologists claim that by the eighth month this fetus will have begun a cycle of waking and sleeping whose rhythms are already influenced by the sun. And . . . when it sleeps, it dreams—or at least its brain waves resemble those of a dreamer.

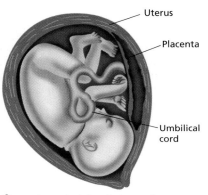

Uterus

Placenta

Umbilical cord

By the end of the ninth month, a fetus turns itself upside down, or into a head-down position. The umbilical cord supplies the fetus with food and oxygen and removes waste until after birth, when the cord is tied and cut.

527

Episode 5— The Third Trimester

The following are some additional developmental landmarks that occur during the third trimester:

Seventh month
- The fetus is lean, wrinkled, and red.
- The eyelids open, and the eyebrows and eyelashes form.
- The nervous system is developed sufficiently so that the fetus practices controlled breathing and swallowing movements.
- The lanugo hair is prominent.
- The brain enlarges and cerebral fissures and convolutions rapidly appear.
- The retinal layers of the eye are complete and light can be perceived.

If delivered at this stage, a baby has at least a 10 percent chance of survival.

Eighth month
- In males, the testes settle into the scrotum.
- The sense of taste is present.
- Weight increase slows.
- Fat is collecting, wrinkles are smoothing, and the body is rounding out from the eighth to the ninth months.

If delivered at this stage, a baby has at least a 70 percent chance of survival.

Ninth month
- Nails reach the fingertips.
- Pulmonary branching is about two-thirds complete. (The lungs do not function normally until birth.)
- The lanugo hair is shed.

The average length of a human pregnancy is 266 days, or 38 weeks. Eventually, there is a decrease in fetal activity, possibly due to a lack of space in the uterus. The fetus turns to a head-down position in preparation for birth. At birth, a human fetus measures about 52 cm and has a mass of about 3.2 kg.

Answers to *Thought-Provoking Questions*

1. The genetic information for specific physical traits is present from the beginning. As a result, these traits begin to appear as soon as the appropriate physical structures begin to take form.

2. Mothers-to-be read and talk to their fetus because, once it has become sufficiently developed in the latter part of a pregnancy, a fetus can hear its mother's voice. Then, when the baby is born, it will recognize and be comforted by the sound of the familiar voice. Talking to the fetus also helps many mothers-to-be feel connected to their baby so that they can comfort and mother their baby even before birth.

Answers to
Going Further

Students will need to do some additional research in order to answer these questions.

1. A new baby usually does not have fully developed tear glands. Therefore, in most cases, there are no tears.

2. A baby's salivary glands are not fully developed at birth. The saliva does not have the ability to break down starch until 3 to 6 months after birth.

3. The eyes of a newborn are sensitive to light, but the muscles cannot focus the eyes on one point. Because of this, a baby may seem cross-eyed until its eye muscles develop adequately.

4. The major changes at birth are that a baby begins to consume food through its mouth and breathe air. Before birth, all substances reached the baby through its umbilical cord. However, a baby continues to grow and change rapidly after birth.

5. Caesarean sections are surgical procedures performed in order to deliver the baby through an incision made through the mother's abdomen and uterus. Caesarean sections are performed when there is a problem with the baby's position or size, when the labor is too lengthy, or when there is some medical emergency. Some risks are involved, however, including maternal infections, excessive blood loss, and injuries to other organs.

Be a Witness, *page 529*
You may wish to give students some guidelines about how to obtain the necessary permissions for these activities. Allow students to work together on this project. Ask them to share their experiences and what they have learned with the class. Encourage class members to ask questions and to discuss and compare experiences.

Birth

At the end of pregnancy, the combined effects of several hormones and the stretching of the uterus cause the uterus to begin to contract with strong muscular movements. With these movements, the baby is pushed out of the uterus. Jean Hegland suggests, "For a baby, contractions may be caresses that express the fluid from its lungs and stimulate its skin in the same way that other mammals . . . do by licking their newborns."

At birth, you gasped, filled your lungs with air, and cried. The placenta and umbilical cord you no longer needed were cut away. When your wound healed, you were left with the scar of the navel. The navel has been called a "permanent reminder of our once-parasitic mode of living."

People have very different life experiences, but everyone shares the prenatal phase, which forms the first chapter in the personal history of each human being.

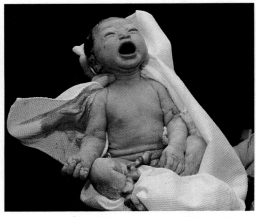
▲ Within seconds after birth, a baby takes its first breaths.

Going Further

1. When a newborn baby cries, are there tears? How do you explain this?

2. How old must a baby be before starch can be added to its diet? Why?

3. What eye development has to take place after birth?

4. Explain what you think Jean Hegland meant when she wrote, "In the years to come, when we celebrate this Riddle's birthday . . . we will be commemorating little more than a change in its means of respiration and nutrition."

5. A sizable percentage of babies in the United States are delivered by *Caesarean section*. Find out what you can about this procedure. How is it performed? What are the advantages of delivery by Caesarean section? the disadvantages?

▲ During pregnancy, the mother's breasts prepare for the production of milk. After the baby is born, breast milk can provide all the nutrients needed for almost the entire first year of a baby's life.

I Want to Know! *page 529*
Students will probably need to do some research to answer these sample questions. Encourage students to write down other questions they may have about human reproduction. These questions could then be answered by a health professional, you, or independent student research.

Answers to
I Want to Know! page 529

1. Yes; the bacteria normally present in human intestines are taken in by the baby with the food that it eats after birth.

2. A fetus could be considered a parasite in that it is completely dependent on its mother for survival. However, parasites usually cause harm to the host (the mother). The health of the mother is usually not jeopardized if proper health care is provided during pregnancy.

Be a Witness

Find out more about mammal birth by watching it or by talking to someone who has seen a mammal birth. Here are some suggestions for what you might do.

1. Plan a visit to a farm, an agricultural fair, or a kennel to watch the birth of a mammal, such as a colt, a calf, or a puppy.

2. Interview a pregnant woman during her pregnancy and later, after her child's birth.

3. Interview a father, both before and after his child's birth. Many fathers are present during the birth of their children.

4. Watch a film of a human birth.

I Want to Know!

Patrick's class arranged for a visit from an obstetrician (a doctor specializing in human birth). Everyone was supposed to do some reading in advance and come up with some good questions. How might the obstetrician have answered them?

1. I read that a newborn has no bacteria living in its digestive system. Is that true? How do we get the bacteria?

2. One book said the fetus lives the life of a parasite. Do you agree? Explain.

3. Are the mother and fetus really separated, such that they don't touch? Would the mother's body really reject the fetus if they were touching?

4. If the mother's blood does not pass through the developing fetus, how does the fetus get what it needs?

5. I read that the fetus is floating and weightless, like an astronaut in space. Can that be so? Is there any advantage to that?

6. Someone told me that viruses and poisons can get into the fetus from the mother. Is that true? Can the AIDS virus do that?

What other questions would you like to ask? Make a list. Arrange to invite a doctor or another health professional to your class. You will likely make a lot of new discoveries.

529

---

3. The placenta separates the mother and fetus, and they do not share the same blood supply. If they were touching, the mother's body might treat the fetus as foreign material and reject it.

4. A mother's blood carries nutrients to the placenta, where they are absorbed by the fetal bloodstream and passed to the fetus through the blood vessels in the umbilical cord.

5. The fetus is floating, or suspended, in the amniotic fluid, which protects it and cushions it against sudden jolts or movements. The sensation of floating is similar to weightlessness, but the fetus is still affected by the force of gravity.

6. Viruses and poisons in a mother's blood travel to the placenta and can be passed to her fetus through the umbilical cord. A fetus may acquire the AIDS virus in this way.

Reteaching

Locate clusters of snail eggs on aquarium plants or on the glass of an aquarium. If snail eggs cannot be obtained from a school aquarium, they can be purchased from a science supply company and easily maintained for future use. Place a cluster of the eggs in a clear glass bowl or petri dish.

Instruct students to observe the development of the snail embryos with a magnifying glass or a microscope and to construct a diagram of their observations. They should also write a paragraph comparing the development of the snail embryos with that of human embryos.

Assessment

Have students work individually or in small groups to summarize the lesson by making a chart that shows human development in the first, second, and third trimesters. The chart should include major developmental landmarks for each stage.

Extension

Explain to students that a developing fetus can be viewed by its parents while it is still in the uterus (with a *sonogram*). Doctors can also collect cells from an unborn infant (by *amniocentesis*) to find out if it has any chromosomal abnormalities that indicate possible deformities. This test also indicates the sex of a fetus. Challenge students to research how these two procedures are performed, what they show, and why they may be requested.

Closure

Have students create a flowchart showing major developmental milestones from fertilization to birth. Details that show what happens during each phase can be included.

Homework

The Resource Worksheet that accompanies I Want to Know! makes an excellent homework assignment (Teaching Resources, page 29).

Answers to *Challenge Your Thinking*, pages 530–531

1. You could draw a family tree for a single-celled organism. However, it should be understood that such an organism could not exist at the same time as its parent. For this reason, there would be no parents, grandparents, or other ancestors at a family reunion for any species of single-celled organisms. Instead, there would be only siblings.

2. Sample answer: Among organisms that reproduce by simple cell division (such as bacteria), every time one cell divides, the number of cells doubles. In other words, cell division results in the multiplication of organisms.

3. Fruit flies develop from eggs laid by other fruit flies, mostly on overripe or rotting fruits and vegetables. Micki can reduce the number of fruit flies in the cupboard by refrigerating, using, or composting fruits and vegetables before they become over-ripe or rotten and by keeping cupboard shelves clean.

4. a. Self-regulation
b. Self-preservation
c. Self-regulation
d. Self-organization
e. Self-reproduction
f. Self-organization

5. a. Sex cells contain one-half of the number of chromosomes that the rest of the body's cells contain.
b. Mitosis is necessary for the growth and repair of tissues. Meiosis produces sex cells, which are needed for reproduction. Fertilization is necessary for the development of a new plant or animal by sexual reproduction.

6. Accept all reasonable answers. Two examples are further lung development and weight gain.

CHALLENGE YOUR THINKING

1. Family Reunion
Could you draw a family tree for single-celled organisms? Could such organisms hold a family reunion? Why or why not?

2. A Contradiction in Terms
Explain to a 9-year-old how this statement can be true: Some living things multiply by division.

3. Spontaneous Generation Revisited
Micki told her mother this morning, "There are a bunch of fruit flies around the bananas in the cupboard, Mom. They just appeared—out of nowhere!" Is Micki right? Did the fruit flies grow from the bananas? Explain what must be happening, and suggest how she can keep fruit flies from becoming more numerous.

530

4. It's Alive!

What life-sustaining characteristics are shown when

a. your body digests food?

b. a fox chases a rabbit?

c. a dog pants?

d. a salamander grows a new tail?

e. a fly lays eggs on food?

f. a wound heals?

5. Sex Cells and Body Cells

a. How do sex cells differ from the rest of the body's cells?

b. Why are the following processes necessary?

- mitosis
- meiosis
- fertilization

6. Playing Catch-Up

A baby born after only 8 months of development has about a 70 percent chance of survival. It has to "catch up" to be as strong as a 9-month-old fetus. What are two improvements in its condition that must be accomplished by a premature baby?

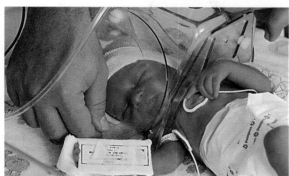

ScienceLog

Review your responses to the ScienceLog questions on page 490. Then revise your original ideas so that they reflect what you've learned.

531

ScienceLog

The following are sample revised answers:

1. A person's appearance is determined by the chromosomes inherited from his or her parents. Because half of these chromosomes come from the mother and the other half come from the father, no person can be identical to one parent. Also, because the combination of chromosomes that a parent passes along may vary greatly, brothers and sisters don't receive the exact same chromosomes from their parents (except in the special case of identical twins).

2. The body mends a cut by replacing damaged cells through the process of mitosis, in which an existing cell divides to form two new cells.

3. A person's heart begins to beat about 3 weeks after fertilization occurs.

★ **You may wish to provide students with the Chapter 22 Review Worksheet that accompanies this Challenge Your Thinking (Teaching Resources, page 31).**

Connecting to Other Chapters

Chapter 22
examines Mendel's heredity experiments, cellular reproduction, and human fetal development.

Chapter 23
explores molecular biology, including the structure of DNA, genotype expression, and genetic engineering.

Prior Knowledge and Misconceptions

Your students' responses to the ScienceLog questions on this page will reveal the kind of information—and misinformation—they bring to this chapter. Use what you find out about your students' knowledge to choose which chapter concepts and activities to emphasize in your teaching. After students complete the material in this chapter, they will be asked to revise their answers based on what they have learned. Sample revised answers can be found on page 549.

In addition to having your students answer the questions on this page, you may wish to assign a "free-write" to assess their prior knowledge. To do this, instruct students to write for 3 to 5 minutes on the relationships among DNA, genes, chromosomes, and heredity. Tell them to keep their pens moving at all times, writing in a stream-of-consciousness fashion. Emphasize that there are no right or wrong answers in this exercise. It might be best to ask students not to put their names on their papers. Collect the papers, but do not grade them. Instead, read them to find out what students know about molecular genetics, what misconceptions they may have, and what is interesting to them about this topic.

CHAPTER
23
Instructions for Life

1 Write down as many ways of giving instructions as you can think of.

2 What kind of cell instructions are necessary when a new life begins?

ScienceLog

Think about these questions for a moment, and answer them in your ScienceLog. When you've finished this chapter, you'll have the opportunity to revise your answers based on what you've learned.

Which has a greater effect on a person, inherited characteristics or the circumstances of the person's life?

3

Nature or Nurture?

532

Living With Instructions

Miguel arrived home looking quite pleased.

"I got a model kit on sale—75 percent off the regular price," he announced, as he hurried to open the cardboard carton. "Look at this," he said, as dozens of parts fell out onto the table.

"Oh no! There are no instructions—not even a diagram," he moaned. "I know it's a model of an airplane, but I'm not sure what it's supposed to look like. The illustration and model number on the outside of the box don't help me much."

Miguel called the store for help. The store manager said, "You'll have to send away for the instructions—to the head office."

What sort of instruction sheets should Miguel expect to get?

Instruction sheets for model kits have illustrations that show what the pieces look like, diagrams and text that explain how to assemble them, and pictures that show what the completed model should look like.

All Kinds of Instructions

Not all instructions include illustrations and diagrams. Some instructions use only words, or a combination of words and numbers. Computers, for example, receive instructions in the form of an alphanumeric code. Earlier, you used holes in cards as a source of information about traits. The holes and cut-away sections did not resemble any of your traits, but they did pass on information. What mechanical means did you use to retrieve, or *translate*, the information? What information did you get? What other examples of instructions can you think of that do not use pictures? Ⓐ

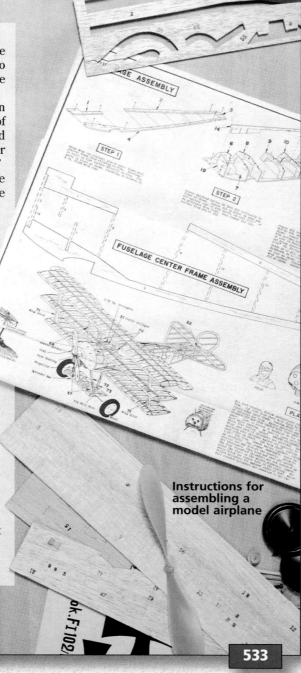

Instructions for assembling a model airplane

533

LESSON 1 ORGANIZER

Time Required
2 to 3 class periods

Process Skills
analyzing, comparing, inferring

Theme Connection
Structures

New Term
DNA—an abbreviation for deoxyribo-nucleic acid, the chemical found in chromosomes that contains the genetic instructions for cells

Materials (per student group)
none

Teaching Resources
Exploration Worksheet, p. 37
SourceBook, p. S167

FOCUS

Getting Started

Display several sets of instructions where students can see them. These may include instructions for putting together a model, assembling an appliance or toy, cooking something, and so on. Ask students to identify features of the instructions, such as diagrams and illustrations, numbered steps, and labeled parts. Then invite them to speculate as to why instructions are important. *(They help us do things correctly.)* Point out that in this lesson students will learn about the instructions that determine the development and functioning of living cells.

Main Ideas

1. Instructions may be conveyed in a variety of ways.
2. Linear programming is one way of providing instructional information.
3. DNA is found in the chromosomes of cells and is the genetic material that determines the development of life.
4. DNA is a form of linear programming that stores the instructions for life in a simple chemical code.

TEACHING STRATEGIES

Answers to
In-Text Questions

Ⓐ In Exploration 1, information was retrieved by passing a knitting needle through the holes in a stack of index cards. Notches allowed the *yes* responses to fall to the floor. These cards provided information about the prevalence of certain human traits among the members of the class. Students may suggest some of the following as examples of instructions that do not use pictures: recipes, electronic signals from satellites or remote-control devices, computer programs, traffic lights, and so on.

Answers to *In-Text Questions*

A Bar codes and computer programs were developed for recent technology. Sheet music and crochet instructions have been used for centuries. Braille was developed in 1824.

Answers to *All Kinds of Instructions*

Bar code

1. A bar code is often used to provide product information, such as the name of an item and its cost. The instructions do not look like the end result.
2. The notation is arranged as a linear series of black bars and white spaces of varying widths. The bars represent binary numbers that stand for numbers, letters, or other characters.
3. The code is scanned by a laser, which sends the information to a computer, where it is translated.

Sheet music

1. Sheet music provides the instructions for playing a tune on a musical instrument. The instructions do not look like the end result.
2. The notation is a series of notes that are arranged on or between an array of five lines.
3. Sheet music is translated into sounds of varying pitch and duration by the musician who reads the notes and plays the musical instrument.

Crochet instructions

1. Crochet instructions are used to direct the types of stitches used, the order in which they are used, and the size and shape of the finished product. The instructions do not look like the end result.
2. The notation is a series of words and abbreviations that represent types of stitches and are arranged in rows.
3. The crocheter uses the words and abbreviations to create the desired fabric.

Braille

1. Braille is a code for numbers and the letters of the alphabet that can be read by touch. The instructions do not look like the final result.
2. The notation is a series of small raised dots on paper.
3. These dots can be read, or translated into the letters of the alphabet and numbers, by touch.

Computer program

1. A computer program provides the instructions that make a computer perform certain functions. The instructions do not look like the end result.
2. The notation is a series of numbers and letters arranged in rows. The information is entered into a computer through a keyboard or on a disc.
3. Computer programs are translated into an internal code that can be interpreted by a computer's central processing unit (CPU). The resulting information may be displayed on a screen or a printer or may be used to operate another machine.

Look at the following examples of sets of instructions. For each set of instructions, consider these questions:

1. What do you think the instructions are about? Do the instructions look like the end result?
2. What notation (set of characters) does each use? How is the notation arranged?
3. How do you think each set is translated so that its instructions can be followed or its information acted upon?

In the examples below, information is provided in a series of symbols or characters arranged in rows or lines. These sets of instructions are forms of *linear programming.* Which sets of instructions were developed for recent technology? Which have been used for centuries? **A**

Sonatina in F Major
Allegro assai
Ludwig van Beethoven (1770–1827)
lives in Tucson, Arizona.
ISBN 0-06-092253-2
90000
22535
0494P
dc in next sc, repeat from * to end of rnd. Then ch 1, and join with sl st to 3rd st of ch-4 first made. 9th rnd: Ch 10, and complete cross st as before, skipping 2 dc between each leg of cross st and inserting hook under ch-1 sp. Ch 3 between each cross st. Skip 2 dc between each cross st. Repeat around (20 cross sts) and join last ch-3 with sl st to 7th st of ch-10 first made. 10th rnd: Sl st in 1st 2 sts of 1st sp, ch 5, * dc in same sp, ch 3, sc in next sp, ch 3, dc in next sp, ch 2, repeat from * to end of rnd. Join last ch-3 with sl st to 3rd st of ch-5 first made. 11th rnd: Ch 6, dc in same sp, * ch 4,
IF UC = 4 THEN GOTO 220
IF UC = 7 THEN GOTO 250
IF UC < 4 OR UC > 4 OR OR UC < 7 OR UC > 7 THEN
GOTO 165
UL$ = "CORMORANTS & SEAGULLS": CL$ =
"UPPER":TC$ = "WATERBIRDS":U = 5:AR = 36521
X$ = "CORMORANTS SEAGULLS, BLUEFISH HERRING
SARDINES SQUID, MOLLUSKS URCHINS STARFISH
WORMS, CRUSTACEANS INSECTS, BACTERIA ZOO-
PLANKTON, ALGAE PHYTOPLANKTON PROTO-
ZOANS,"
"TUNA":CL$ = "FOURTH":TC$ = "FISH

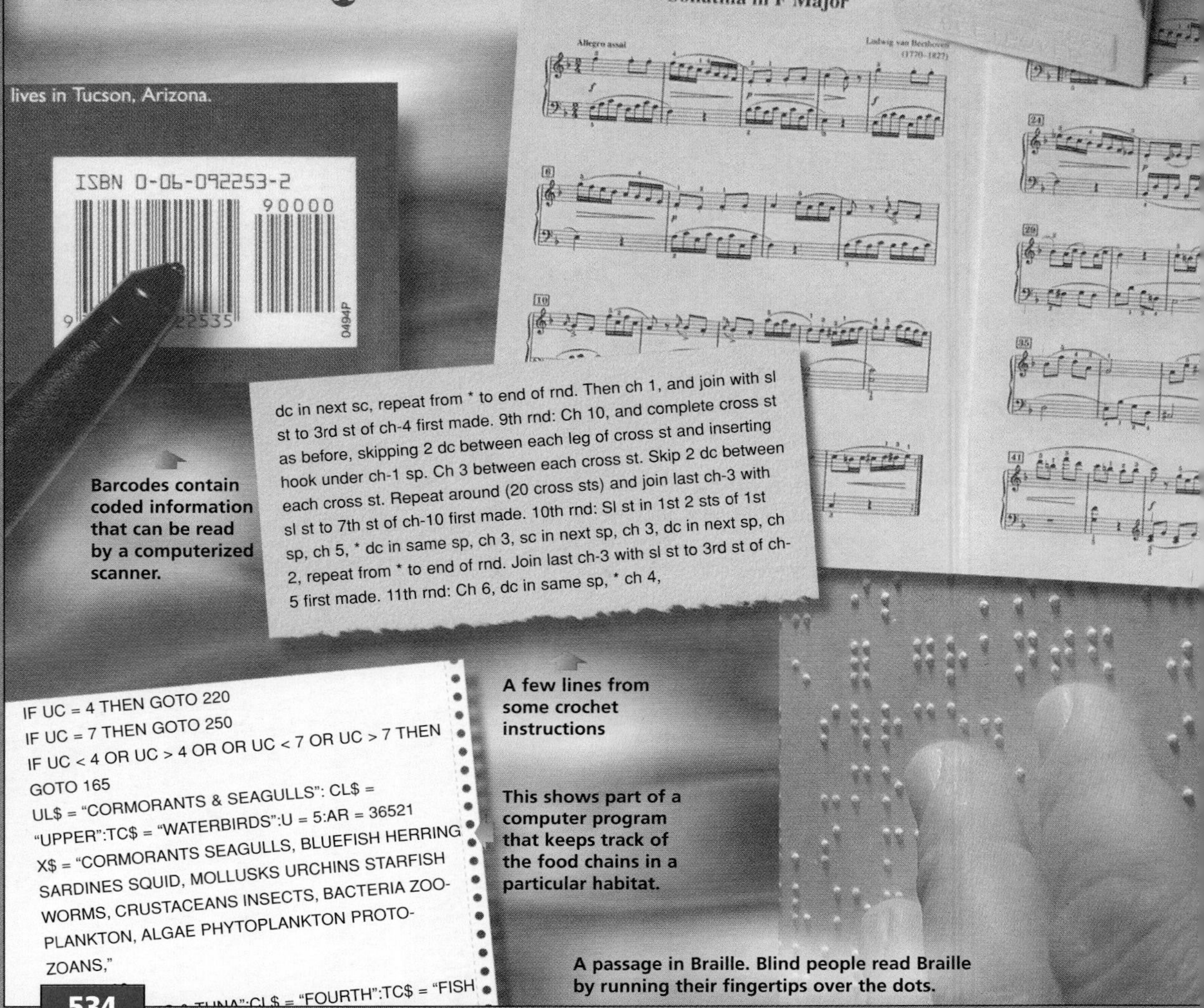

The opening bars of Beethoven's Sonatina in F major

Barcodes contain coded information that can be read by a computerized scanner.

A few lines from some crochet instructions

This shows part of a computer program that keeps track of the food chains in a particular habitat.

A passage in Braille. Blind people read Braille by running their fingertips over the dots.

534

Homework

You may wish to assign question 1 on this page as homework. The next day, questions 2 and 3 can be completed in class.

The Instructions That Shape Life

Like Miguel and his model airplane, scientists often have to work without instructions. For example, biologists spent many years looking for the instructions that guide the development of new living things. They needed to find out whether there was a "head office" where the instructions were kept, what notation was used, how the notation was arranged, and how it was translated. In the next Exploration, you will follow the story of how they found answers to some of these questions.

EXPLORATION 1

Cell Search

What to Do

Read the descriptions below of some important findings made by scientists. Notice how technology helped them as they looked for answers to their questions about reproduction.

A. Body Fabrics

The tissues of the body (such as skin, bone, and muscle) were first described as woven materials, such as linen or rope. During the nineteenth century, improved microscopes allowed biologists to see the cells of living things for the first time.

B. Colored Ribbons

A special dye enabled scientists to observe the nucleus with much greater clarity. The scientists saw certain colored structures in the nucleus that looked like fine, ribbonlike strands. These structures were named *chromosomes,* meaning "colored bodies." The function of the chromosomes was not known at first.

C. Fertile Ground

Scientific techniques allowed biologists to observe living cells and some of their life processes. Using petri dishes and microscopes, they watched fertilization of egg cells (ova) by sperm cells.

In the nineteenth century, improved microscopes advanced the study of living things.

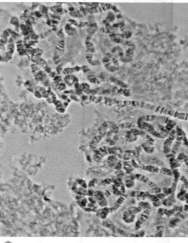

A microscopic view of chromosomes

Exploration 1 continued ▶

535

EXPLORATION 1

After students have had an opportunity to read each of the descriptions, involve them in a discussion of what they have read. Evaluate their understanding by asking questions similar to the following:

- Before people knew about cells, from what did they think tissues of the body were made? *(Woven materials, such as linen or rope)*
- What development allowed biologists to see the cells of living things for the first time? *(The improvement of the microscope)*
- Why is the continual improvement of the microscope so important? *(Improvements in the microscope allow scientists to see smaller and smaller structures within living cells.)*
- What other scientific techniques led to further discoveries? *(Special dyes to stain nuclear material and X-ray techniques to study the shape of molecules)*
- Why is it important that scientists carefully record the results of their investigations? *(It is important so that other scientists can read about and build on previous investigations.)*
- Why do scientists need to share their work? *(Scientists share ideas so that others can benefit from what has been done in the past and from any current research that is occurring in their field. For example, the work done by Rosalind Franklin and other scientists was the springboard for the conclusions drawn by James Watson and Francis Crick.)*

 An Exploration Worksheet (Teaching Resources, page 37) is available to accompany Exploration 1.

Cooperative Learning
EXPLORATION 1, PAGES 535–536

Group size: 3 to 4 students

Group goal: to infer how technology has led to new discoveries concerning reproduction

Positive interdependence: Group members should have roles such as recorder, presenter, and readers. Use a round-table format so that each student has the same role once. One student begins by reading section A. Then the entire group formulates one quiz question about that section. The next student reads section B, and the pattern continues until all sections have been read and all questions have been developed. Students should record all of their group's quiz questions in their ScienceLog. You may wish to have students share their questions with the other groups. Students should then answer the Inferences section together.

Individual accountability: Each student should be able to answer at least one question devised by another group.

Answers to
Inferences

a. C. Fertile Ground
b. E. Shadowy Shapes
c. B. Colored Ribbons
d. D. Inner Division
e. A. Body Fabrics

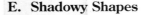

D. Inner Division

Biologists were keenly interested in the behavior of the chromosomes. Chromosomes seemed to appear right before cell division, and then a complex series of movements caused the chromosomes to be distributed equally between the two new cells. Because the same number of chromosomes appeared each time, they concluded that the chromosomes must be duplicated between each division. The careful way in which they were distributed to each new cell led biologists to believe that the chromosomes carried the hereditary instructions for the cells.

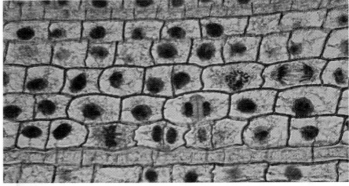
A section of plant tissue containing dividing cells

E. Shadowy Shapes

By the 1950s, scientists were studying the structure of crystals with a photographic technique that uses X rays instead of light. With this technique, Rosalind Franklin exposed crystallized molecules from the nucleus to X rays and then made photographs of the "shadows" cast by those molecules. James Watson and Francis Crick used her photographs to determine the structure of those molecules.

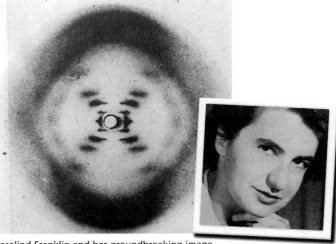
Rosalind Franklin and her groundbreaking image

Inferences

Match each of the following inferences to observations A–E on this and the preceding page.

a. Two parents each contribute part of the instructions that lead to the development of offspring.

b. A model could be constructed to resemble a molecule of material in the nucleus.

c. The chromosomes contain instructions for offspring. These instructions enable living offspring to be constructed from nonliving materials.

d. The nucleus is the "head office" of a cell—the location of the instructions for the development of new life.

e. Body tissues are communities of similar cells joined together to form a structure.

536

Thinking It Over

What answers did biologists have, so far, about the mysteries of reproduction? As you read the following paragraph, think about the words that complete the sentences.

Biologists knew that two sex cells, ____1____ and ____2____, unite to start a new life. They called the combining of these cells ____3____. The center of activity in the formation of the sex cells was always the ____4____, where long, ribbonlike objects called ____5____ took part in the process of cell division. The biologists believed these objects to be the sets of ____6____ for the development of a new individual. They believed these sets were a form of ____7____ programming. They also were able to identify the ____8____ of a molecule of chromosome material.

Substance of Life

What was the material in the nucleus that biologists were viewing? Its shape and structure were identified in 1953 by James Watson and Francis Crick, who received a Nobel Prize for their discovery. The material is called **DNA**—an abbreviation for the chemical that contains the instructions for cells, *deoxyribonucleic acid*. It is found in the chromosomes and is essential for all forms of life.

The models of DNA at right may remind you of the spiral-ladder-shaped structures that Gloria and Yoon saw in the nucleus at the Science Discovery Center. The spiral ladders they watched were molecules of DNA.

Further investigation showed how the DNA molecule stores the instructions for life. There are four chemicals that bond together, two at a time, to form the rungs of the DNA ladder. The instructions for life are based on specific sequences of these bonded pairs. Much is known about how the chemical code is translated into action as new living things develop. However, much also remains to be discovered.

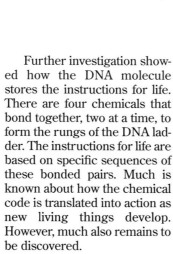

Watson and Crick, with their model of DNA. Their model provided scientists with a valuable tool in the study of heredity.

Model of a DNA molecule

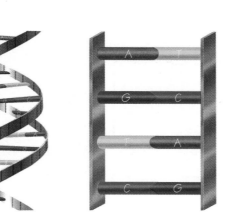

The structure of DNA is much like a coiled ladder or a spiral staircase.

The structure of DNA, uncoiled so that you can see the "ladder rungs" more clearly. The letters C, G, T, and A stand for the chemicals that combine to form the rungs of the "ladder."

537

Homework

You may wish to assign Thinking It Over as homework.

Substance of Life

James Watson's book, *The Double Helix* (New York City, NY: NAL-Dutton, 1969), is an interesting and engaging account of Watson and Crick's discovery of the structure of DNA. Although the reading level may be too high for some students, certain sections of this inexpensive paperback may be useful to you in class discussions.

Theme Connection

Structures
Focus question: How is the structure of the DNA molecule related to its function? *(DNA is like a spiraling ladder. The rungs of the ladder are composed of base pairs of complementary proteins. During the replication phase of mitosis, DNA straightens out and separates where the base pairs meet, resulting in two "half-strands" of DNA. Free nucleotides from the cell then attach to each base in each half-strand. The result is two identical strands of DNA.)*

Integrating the Sciences

Life and Physical Sciences
Ask students to consider the following analogy: Only four chemicals appear in DNA. The various combinations and arrangements of these chemicals are responsible for the endless variety in human traits. Similarly, the 109 elements combine in various ways to make up all the compounds found in our environment.

Mind-Benders

The facts presented in the text are intended to help students understand the incredible complexity of the instructions for life provided by DNA. Have students read the facts silently, and then involve them in a discussion of what the facts mean.

Direct students' attention to the illustration. Point out that each of the letters in the DNA code stands for one of the four chemicals that constitute the rungs of the DNA ladder—A (adenine), T (thymine), G (guanine), and C (cytosine). You may want to point out that although the base pairs are important because they allow DNA to be replicated exactly, the code is actually contained in the linear sequence of nitrogen bases on either side of a DNA double helix. The code is read in groups of three consecutive bases, each of which translates to a particular amino acid. A series of consecutive three-base groups (triplets) translates to a sequence of amino acids for a particular protein.

Be sure students understand that DNA is made of the same chemical components in all living things, and its instructions are carried out in the same manner. Each cell translates the sequence of nitrogen bases in its DNA into sequences of amino acids to form proteins. The proteins made from the DNA code are used to build cell parts and to regulate the chemical activities of the cell. The differences among organisms are determined by the amount of DNA in their cells and by the sequence in which the chemical code is arranged.

Students may not realize that when a DNA molecule replicates itself, it does so at an extremely fast speed. Point out that a DNA molecule may consist of millions of rungs and be twisted hundreds of thousands of times. When a cell divides, the DNA molecule has to untwist and duplicate itself by assembling two new halves in the correct order. All of this may be done in less than 20 minutes—the time it takes the most rapidly dividing cells to divide. Replication occurs at a rate of about 200 base pairs per second. In order to complete replication of a very long molecule in so short a time, a DNA molecule replicates at many points along its length simultaneously.

Mind-Benders

Consider these facts about DNA—the code of life.

- DNA has an alphabet of four chemical "letters" that form the rungs of the spiral ladders. The DNA in a human contains more than a billion of these letters, which would fill about 3000 novels of average length.

- If you could completely uncoil a DNA molecule in one of your cells, it would measure almost 2 m. Each cell contains many strands of DNA, and you have more than 10 trillion cells in your body. All the DNA they contain could stretch to the sun and back.

- Chromosomes contain long strands of DNA, which are "read" in segments called genes. A section of the DNA spiral ladder with 2000 rungs might make up only one gene. Each gene is like a "sentence" in the language of life. Each of your cells contains about 100,000 genes, or "sentences" of instructions.

- A full set of DNA for all 6 billion people alive now could fit on 1 teaspoon and would have a mass of about 1 g.

- The long series of letters shown below is part of the instructions for a protein in human blood. You can see that it is in the form of linear programming. The whole code for the protein would fill up this page. To print the genetic code for one single cell from your body, however, you would need about 1 million pages the size of this page.

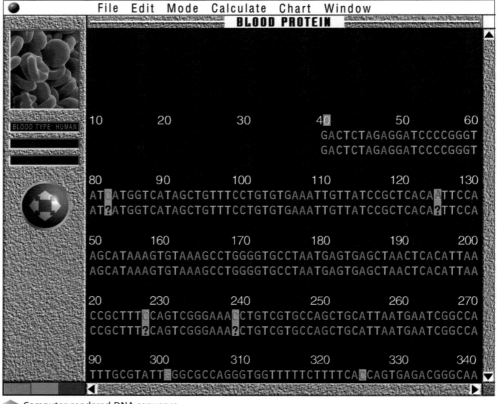

Computer-rendered DNA sequence

CROSS-DISCIPLINARY FOCUS

Art

Have students visually portray the tremendous storage capacity of DNA. Provide old magazines and other art supplies. Divide the class into five groups. Within each group, assign as many of the following roles as required: materials supervisor (to collect and return supplies), researcher (to obtain needed information about DNA), task facilitator (to keep the group on task), presenter (to explain the poster to the class), communicator (to contact other groups for ideas when needed).

Assign each group one of the five Mind-Benders about DNA to illustrate. Students should develop a unique poster that will, as accurately as possible, depict the facts presented in their assigned statement. Suggest such formats as cartoons, collages, or computer-generated graphics. All students should serve as artists in the completion of the poster. Display the completed posters.

Stop and Think

Read the dialogue below and analyze what the characters are saying. Sort out and summarize the thoughts expressed about genes, Ⓐ chromosomes, DNA, and genetic instructions.

Genes are found in chromosomes, right? In fact, aren't they—chromosomes, I mean—made of genes?

Yeah, that's right. I know we have a lot of genes too, something like 100,000 of them, all made of DNA—whatever that is.

Well, DNA is like a genetic blueprint. A book I'm reading says DNA transmits instructions through a kind of alphabet that has four letters.

Right—each letter's a different chemical, and the chemicals join together like ladder rungs. Groups of these are like words…

… and groups of words make up the genetic instructions that tell the cells how to develop!

When you think you understand the structure and organization of genetic material, reconsider the function of DNA. The instructions of the genetic code control the formation of a new life—from the first cell division to the formation of cells of various sorts (such as nerve and muscle cells) to the organization of the cells into an organism. How do you think the instructions Ⓑ account for all the characteristic traits of an organism? Does this mean that every organism is completely programmed by Ⓒ its genes?

What other factors, besides genes, affect who you are? In the Ⓓ following lesson, you'll consider the importance of environment on the formation of life.

Answers to *Stop and Think*

Ⓐ The statements made by the characters provide a framework around which students can summarize the most important facts in the lesson. Suggest that they respond to the questions in the dialogue by writing brief summaries in their ScienceLog. Their summaries should reflect the following ideas: The nuclei of cells contain chromosomes, each of which is made up of many genes. A single human cell, in which there are 46 chromosomes, may contain as many as 100,000 genes, each carrying an instruction that tells the cell how to make some part of a living thing. Each chromosome is made up of a DNA molecule that is shaped like a spiral ladder. Four chemicals combine in a specific way to form the "rungs" of the ladder. The sequence of these chemicals in DNA is the code for genetic instructions. The chemicals of this code are represented by the letters A, G, C, and T. Genes are sections of the DNA molecule.

Ⓑ Students should realize that with so many genes, there are enough instructions to describe all the inherited traits of an organism.

Ⓒ Students may recognize that genes are not the only factors involved in determining all of the characteristics of an individual.

Ⓓ Students may list their parents, where they live, what they eat, and where they go to school as other factors that affect who they are.

FOLLOW-UP

Reteaching

Have students use the illustrations in the text to make models of DNA molecules. Materials such as plastic straws, pipe cleaners, modeling clay, and papier-mâché may be used. Provide space in the classroom for students to display their models.

Assessment

Have students bring examples of instructions to class. Ask them to explain what the instructions are for and why they are useful. Then challenge students to write a short essay on how their instructions are similar to the instructional material found in DNA molecules.

Extension

Every possible combination of three nucleotides (called a *codon*) codes for a specific amino acid or an instruction such as "start code" or "stop code." Have a group of interested students create a poster for the classroom that shows what each nucleotide triplet codes for. *(For instance, guanine-guanine-adenine codes for glycine.)*

Closure

Point out to students that different organisms have different numbers of chromosomes. Suggest that they do some research to discover the number of chromosomes in at least 3 different organisms. Have students compare their information and use it to make a composite chart for the classroom.

LESSON 2 Environment or Heredity?

The instructions have been given and all the pieces have been assembled properly. As a result, a baby has been born. The new human being has 100,000 gene pairs inherited in two sets—one from each parent. The huge number of traits that the gene pairs code for represent a unique combination. With 100,000 gene pairs to juggle, you can see how no one else is—or ever will be— exactly like you. But genes are not the only reason you are who you are. Factors from your environment also greatly affect your unique combination of characteristics.

Family-Tree Time

At this point you can add to your family-tree project. You will need to trace the family's traits back several generations. Talk to the family members you are studying to learn about those traits. In the project, include lists of the following traits:

1. observable traits
2. traits that may be inherited but that will not show up until later in life
3. recessive traits that may have "skipped a generation"

Geneticists call the description of an organism's traits its **phenotype**. Your phenotype is the physical expression of your genes. For example, your phenotype may be for brown eyes.

The actual pair of genes you have inherited for each of your traits is called your **genotype**. If your phenotype is for brown eyes, your genotype may actually include a "brown eye" gene and a "blue eye" gene.

For each individual, the traits you listed in 1, 2, and 3 above will fall under one or both of these categories—*phenotype* and *genotype*. See whether you can identify which traits are part of your phenotype and which are part of your genotype. Can a trait be part of your genotype but not part of your phenotype? Can the reverse be true? **A**

 What are the phenotypes for several traits of the horses shown here? Can you tell what their genotypes are for these traits? **B**

FOCUS

Getting Started

Ask students to suggest some environmental factors that contribute to their physical well-being. *(Examples include good nutrition, medical care, regular exercise, clean air and water, good hygiene, and loving relationships.)* Then have them suggest some hereditary factors that contribute to their physical well-being. *(Examples include longevity, strong physical build, tendency toward good health, and strong and healthy organs.)* Point out that in this lesson they will explore the influence of heredity and environment on development and the expression of genetic traits.

Main Ideas

1. Both heredity and the environment contribute to an organism's development and well-being.
2. If something goes wrong with the genetic code, it is likely to result in the abnormal development of an organism.

TEACHING STRATEGIES

Family-Tree Time

To evaluate students' work on the family-tree project, take a cursory look at their family trees and pedigrees. Reading their material in detail—or requiring too much detail—could constitute an invasion of their privacy.

★ A Transparency Worksheet (Teaching Resources, page 39) and Transparency 88 are available to accompany this lesson.

LESSON 2 ORGANIZER

Time Required
2 class periods

Process Skills
comparing, communicating, classifying, analyzing

New Terms
Genotype—the genetic makeup of an organism
Mutation—a change in the genetic makeup of a cell's chromosomes
Phenotype—the external appearance of an organism

Materials (per student group)
none

Teaching Resources
Transparency Worksheet, p. 39
Exploration Worksheet, p. 41
Transparencies 88 and 89
SourceBook, p. S171

There's just one more piece of information to add to your family-tree project. Read the passage below to discover what that information is, and then complete your family tree. How could the information you gathered for this family tree be valuable to a family? In what ways do you feel more informed about heredity after doing this project? Record your thoughts about these questions in your ScienceLog.

How Important Is Your Environment?

Which has a greater effect on what you are—the genes you inherit, or your environment? What does environment include? Food and nutrition are major factors. Consider a neglected, stray dog that is fortunate enough to be taken in by someone who provides food and good living conditions. The results of good nutrition and care are often dramatic. Would this dog's genotype be the same in each environment? What about its phenotype?

Investigate, as far back as possible, the length of life of people in the family you're studying. Look for patterns. For example, many members of the father's family may have lived to be 80 or 90 years old. Many people believe that longevity, the length of a person's life, is strongly influenced by heredity. In what ways could your environment affect your life span? Identify as many environmental factors as possible.

Add the information about the longevity of your ancestors (or the ancestors of the family you chose) to your family-tree project. How can people alter their environment to try to extend life?

"Environment can affect a person as much as heredity does."

Do you agree or disagree? Begin collecting information that illustrates why this statement is actively debated. Think about the dog described above. Ask people you know about their opinions.

A Field Trip

Visit an experimental nursery and explore the environment. Find out from someone there how they make use of heredity to produce healthy, productive animals or plants. Ask about environmental effects on living things. In your ScienceLog, restate the previous quote for the plants or animals you investigated: Environment can (cannot) affect animals (plants) as much as heredity does.

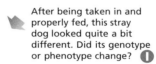 After being taken in and properly fed, this stray dog looked quite a bit different. Did its genotype or phenotype change? **I**

541

CHAPTER 23 • INSTRUCTIONS FOR LIFE **541**

Homework

You may wish to have students complete How Important Is Your Environment? as a homework assignment.

ENVIRONMENTAL FOCUS

Propose the following: Some of the water in a river is diverted, reducing the amount of water and sediment that was carried by the river through a forest. What long-term effects might this have on the phenotypes and genotypes of the forest plants? *(Sample answer: The phenotypes might change because the plants might not grow as well. The genotypes might change if some plants become more likely to survive and reproduce successfully in the changed environment.)*

Answers to
In-Text Questions and Captions,
pages 540–541

A The traits in 1, 2, and 3 can all be stated as genotypes. The traits in 1 and 2 can be stated as phenotypes also. The traits in 3 can be part of a person's genotype but not part of their phenotype (except in the generation in which they appear). A trait can be part of your genotype but not part of your phenotype. If an inherited trait is part of your phenotype, then it must be part of your genotype.

B Students should include all observable physical characteristics in their descriptions of the phenotypes. The genotypes can be determined for recessive traits, but not for the dominant traits, from the photograph alone.

C The information in a family tree could be important to a family in order to understand from whom they inherited certain traits and to predict the likelihood that family members or offspring will inherit certain genetic diseases or disabilities. Students should feel more informed about how traits are inherited and how they inherited their own traits.

D Heredity and environment both affect what a person becomes—the relative importance of each of these factors is not known. The term *environment* includes other factors, such as education and social interactions.

E The dog's phenotype would change, but its genotype would be unaffected.

F A person's lifespan, or longevity, could be affected by environmental factors such as nutrition; amount of sleep; exercise; exposure to disease, radiation, or poisons; drugs; stress; noise; parasites; societal dangers; education; and happiness.

G To extend life, people can develop good health habits and try to avoid or eliminate environmental hazards and poisons.

H Answers will vary. Students may feel that the environment's effect on a person is as great as a person's heredity.

I The dog's phenotype, not its genotype, changed.

Surprising Evidence or Coincidence?

Studies have shown that twins who grow up together are often less alike than twins who are raised apart. This may have several causes. For instance, when twins grow up together, they may try to stand out and be seen as individuals.

Answer to
Surprising Evidence or Coincidence?

The twin study powerfully demonstrates the importance of heredity in human development.

EXPLORATION 2

If some of your students interview twins, have them share what they learn with the class. It would be of value to interview fraternal twins also because they have shared the same environment but are different genetically.

Answers to
Exploration 2

2. Answers will vary. Unlike siblings who are not the same age, fraternal twins experience environmental effects at the same time in their development.

3. Answers will vary. Some studies, for instance, have shown that twins score similarly on certain types of personality tests, even in cases in which the twins have been raised separately.

   The likelihood of having twins is about 1 in every 89 births. For triplets it is 1 in every 7900 births, and for quadruplets it is 1 in every 705,000 births.

4. Students may now have a greater appreciation for how important heredity is in determining the way an organism develops. Both heredity and environment contribute to an organism's development and well-being.

⭐ **An Exploration Worksheet is available to accompany Exploration 2 (Teaching Resources, page 41).**

Surprising Evidence or Coincidence?

The twin brothers shown at right were separated at the age of 4 weeks and were not in touch with each other for 40 years. They grew up in different areas with different families. They were brought together again by a group doing an investigation about heredity and environment. The results surprised everyone. The two men had much in common beyond appearance. They both liked math and hated spelling in school, had similar hobbies, went to the same beach areas in Florida each year, drove the same kind of cars, married women named Betty, named their firstborn sons James Alan and James Allen, had dogs named Toy, had high blood pressure, gained and lost weight at the same stages in their lives, and had the same type of recurring headaches. When reunited they were even wearing similar clothes. How does this study add to your understanding of the role of genes and environment in development?

The twins, moments after their reunion

EXPLORATION 2

How Alike?

1. If you know any identical twins, arrange to interview each one separately. Make up a set of questions in advance about their likes, dislikes, habits, and abilities. Ask each person the same questions, and then compare their answers.

2. If you know any fraternal twins, arrange to ask them the same questions. Their inheritance is not identical, but they have been living in the same environment. What environmental factors affect fraternal twins that do not affect siblings who are not twins?

3. If you don't know any twins, you can do research to discover more about them. Many studies have been done on twins and other multiple births. What are some unusual documented similarities among twins, triplets, or quadruplets? What is the likelihood that a woman will have a multiple birth?

4. What do you think now about the effects of heredity and environment? Which one matters more?

When Heredity Goes Wrong, page 543

Explain that birth defects and miscarriages result from mutations, abnormal chromosomes, and developmental errors. Mutations are permanent changes in genes that can be transmitted to offspring if the mutations are present in the sex cells. Extra or missing genetic instructions may result from the improper separation of chromosomes during cell division. Even perfect instructions may be interpreted improperly, resulting in physiological changes.

⭐ **Transparency 89 is available to accompany When Heredity Goes Wrong on page 543.**

When Heredity Goes Wrong

Most of the time, the instructions for life are precisely followed. But suppose something goes wrong. Although 99 percent of the babies born are the result of normal development, genetic errors can cause problems during development. Some embryos are so damaged that they cannot develop beyond the first stages and are rejected by the mother's body—often before she even knows she is pregnant. It has been estimated that this happens 500,000 times every year in the United States alone.

(A) Why do these things happen? What goes wrong? Could it be that sometimes, in the process of cell division, instructions get improperly duplicated? Perhaps sections of instructions in the sex cells are lost, altered, turned upside down, or placed out of order. Perhaps the instructions are perfect, but there is an error in their translation in the developing embryo.

(B) Study the flowchart, beginning at the top and working down. What is it saying? Write a script to accompany this chart so that someone else will understand it. What would be a good title for the flowchart?

Disaster or Lucky Mistake?

A **mutation** is a permanent change in the genetic code. Mutations may produce changes that are disastrous to living things. On the other hand, without mutations, living things would not have changed, or evolved, to their present form. The remarkable diversity of the living things around you would not exist if mutations had not occurred over time.

What are some examples of nature's "lucky mistakes"? Perhaps this is a subject you would like to investigate.

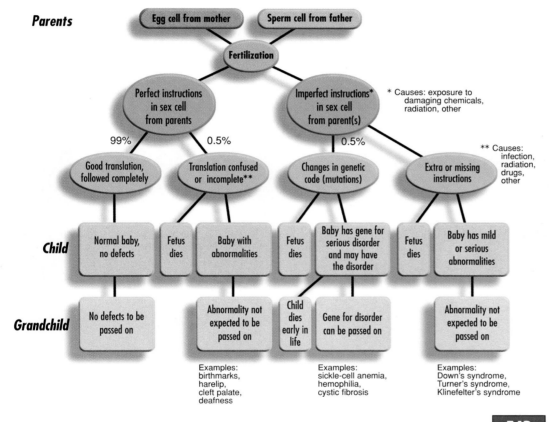

Parents — Egg cell from mother / Sperm cell from father → Fertilization → Perfect instructions in sex cell from parents | Imperfect instructions* in sex cell from parent(s)

* Causes: exposure to damaging chemicals, radiation, other

99% | 0.5% — Good translation, followed completely | Translation confused or incomplete** — 0.5% — Changes in genetic code (mutations) | Extra or missing instructions

** Causes: infection, radiation, drugs, other

Child — Normal baby, no defects | Fetus dies | Baby with abnormalities | Fetus dies | Baby has gene for serious disorder and may have the disorder | Fetus dies | Baby has mild or serious abnormalities

Grandchild — No defects to be passed on | Abnormality not expected to be passed on | Child dies early in life | Gene for disorder can be passed on | Abnormality not expected to be passed on

Examples: birthmarks, harelip, cleft palate, deafness

Examples: sickle-cell anemia, hemophilia, cystic fibrosis

Examples: Down's syndrome, Turner's syndrome, Klinefelter's syndrome

543

Disaster or Lucky Mistake?
Changing the genetic instructions is usually harmful to the organism, but some mutations result in beneficial changes. Such changes are responsible in part for the emergence of new species. Students may realize that most characteristics of any organism they can name originated as mutations.

Reteaching
Compare the genetic instructions of an organism with a recipe for baking a cake. Ask: In a general way, what would be the genotype and the phenotype of the cake? (*The genotype would be the recipe, and the phenotype would be the final appearance of the cake.*) How could environmental factors change the phenotype of the cake? (*Sample answer: If the cake was cooked at the wrong temperature, its appearance would be very different.*)

Assessment
Select teams to debate the following topic: Environment versus heredity—which is more important? One team should support the environment as being the most important influence on phenotype, and the other team should support heredity. Help students come to the conclusion that an organism's potential is programmed in its genes, but that some of its traits can be greatly modified by its environment.

Extension
Point out to students that sometimes

the genotype for a genetic abnormality is present, but its severity may be lessened by controlling environmental factors. One such genetic disease is phenylketonuria (PKU). Challenge students to discover what PKU is and how it can be prevented and controlled.

Closure
Contact the local chapter of the March of Dimes and invite a representative to the classroom to speak about birth defects. The discussion might include the causes of birth defects as well as methods for preventing and treating them.

FOCUS

Getting Started

Ask students to think about some of the practical applications of the study of heredity. Keep track of their responses on the chalkboard. *(Selective breeding produces specific kinds of plants and animals, especially for food. Understanding genetic principles is useful in predicting the likelihood of a disease occurring in a family, in explaining how living things evolve, and in understanding how organisms are alike and how they are different.)* Point out that in this lesson students will explore how the science of genetics may affect the future.

Main Ideas

1. Biologists can manipulate the genetic material of living cells.
2. Advances in the study of genetics may have a profound effect on the future.
3. Genetic engineering involves ethical questions that society will have to answer.

TEACHING STRATEGIES

Have students pause after reading this page. Remind them of the structure of the DNA molecule that they learned about in Lesson 1. Suggest that they return to Lesson 1 and reexamine the diagrams of DNA. Help them to recall that genes are sections of DNA molecules that contain the instructions for specific traits. Then challenge students to predict how genetic engineering might be used in the future. Most students will probably recognize that the manipulation of genes could result in the ability to alter an organism's traits. Some students may wonder if the genes of two different organisms could be combined. Others might think about the ethical issues involved with genetic engineering. Allow as many ideas to be expressed as time permits. Then have students continue reading.

What the Future Holds

While you were still very young, you began to examine the life all around you. You have learned a lot about yourself, your environment, and other living things. You have learned something about the structure of living things and the uses or functions of their parts. You have become especially well acquainted with the plants and animals you use for food. Your eyes alone have provided you with a lot of information about the world around you.

Your experience has mirrored that of scientists through history. Their examination of living things was limited, for a long time, to what their unaided eyes could tell them. After the invention and perfection of the microscope, their main approach changed from studying the whole animal to examining the animal's tiny details. Now biologists are able to study life at the level of the molecule of life—DNA. When you hear the term *molecular biology,* you will know what it means.

Technology has advanced to the point that biologists can now work within the nucleus of a cell. They can manipulate, or move, DNA from one cell to another. This manipulation of DNA is called **genetic engineering**. It has become possible to perform some remarkable procedures. However, the greatest realization seems to be that biologists know enough now to recognize how much more they have to learn about living cells. Molecular biologists and other scientists whose work deals with heredity are called *geneticists.*

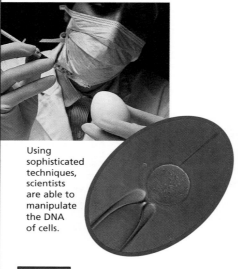

Using sophisticated techniques, scientists are able to manipulate the DNA of cells.

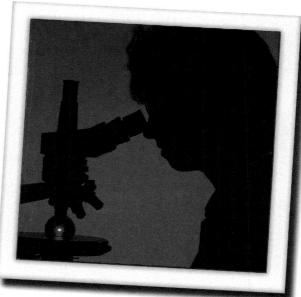

LESSON 3 ORGANIZER

Time Required
2 class periods

Process Skills
analyzing, predicting, inferring

Theme Connection
Changes Over Time

New Terms
Clones—genetically exact organisms that are derived asexually from a single ancestor
Genetic engineering—the manipulation of DNA in living cells

Materials (per student group)
none

Teaching Resources
Math Practice Worksheet, p. 43
SourceBook, p. S172

It is important to note here that there are regulations to control what researchers do with the genetic structure of cells. For example, there are limits to what can be attempted with human fertilized eggs. Biologists themselves want the regulations. Form small groups to discuss why people might fear the results of genetic engineering that involves humans. Do you think the possible dangers posed by genetic engineering outweigh its benefits to human life? Ⓐ

On the Horizon

What should you expect to hear about in the future? Several possibilities are listed below.

- *Improved ways of detecting genetic disorders early in pregnancy.* Amniocentesis is a procedure used to obtain fetal cells for examination. A long needle is used to withdraw some of the fluid surrounding the fetus. This fluid contains skin cells that have been shed by the fetus. Testing of the fetal cells can detect certain genetic disorders. Unfortunately, there is a slight risk of death to the fetus in performing this procedure. A new technique is being developed that requires only a sample of the mother's blood. The technique involves finding the few fetal blood cells that enter the mother's bloodstream through placental leaks. What do you think would be some advantages of using the new technology? Ⓑ

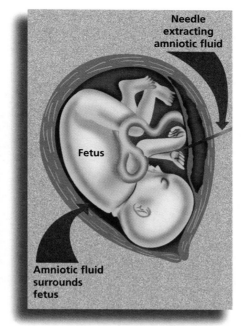

Needle extracting amniotic fluid

Fetus

Amniotic fluid surrounds fetus

Amniocentesis involves withdrawing some of the amniotic fluid that surrounds a fetus and then analyzing cells from that fluid.

A genetic counselor advises couples about the probability of passing on genes for a genetic disorder to their offspring.

On the Horizon

Encourage students to respond to the questions starting on this page. Then have them share and discuss their ideas about the future of genetic research. As discussion motivators, call on volunteers to read what they have written about the people who will be needed in the future to explain the processes of life.

Answers to
In-Text Questions

Ⓐ Accept all reasonable responses. (Encourage students to support their positions with concrete examples. You may also wish to refer to page 555.)

Ⓑ Accept all reasonable responses. Since there is less risk to a fetus, more pregnant women may consent to the procedure and more information could be obtained.

Homework

Ask students to consider how they would want to manipulate genes if they were geneticists. Then have students write a proposal outlining what the benefits of their genetic-engineering work would be.

Answers to
In-Text Questions, pages 546–547

(A) Gene therapy would not have any effect on the descendants of those who have had it because the new genes only affect certain body cells and not the sex cells.

(B) Changing genes in a fertilized egg would affect the descendants of the individual who develops from the egg. Since the gene changes occur before the cells differentiate, the new genes would be part of all the individual's cells, even the sex cells.

(C) Theoretically, human clones are possible. (You may wish to involve students in a discussion of the ethical concerns involved with creating human clones. For example, which individuals would be cloned? Whose genetic material would be used? Could a "super race" of people be created by cloning?)

(D) Students may suggest that poets and musicians will be needed to express the mystery of life. Writers will be needed to explain the science of life. Philosophers will be needed to remind us of the ethical and moral questions involved with genetic research. Artists will be needed to illustrate living things and their processes so that others may know and understand them. Historians will be needed to remind us of the centuries-long search to explain the processes of life.

- *Discoveries about the location of genes that cause birth defects and diseases.* The genes associated with cystic fibrosis and a form of muscular dystrophy have already been located.

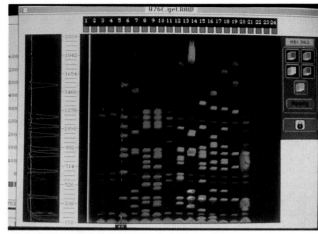

Researchers are gradually putting together a map showing every human gene.

- *Gene therapy.* There is hope that many genetic disorders can be corrected through gene therapy. In gene therapy, some cells are removed from the patient, healthy genes are inserted into them, and the cells are returned to the patient's body. The substance lacking in the person's body is then produced by the cells with the new genes. Would this procedure have any effect on the descendants of a person who had undergone gene therapy? **(A)**

- *Procedures for changing genes in a fertilized egg.* Why might this procedure have more far-reaching effects than the gene therapy described above? Explain. **(B)**

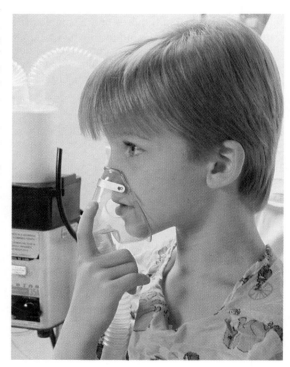

With gene therapy, genetic disorders such as cystic fibrosis could be cured.

Theme Connection

Changes Over Time
Focus question: What kinds of changes to society might you witness in your lifetime as a result of genetic research? *(Answers will vary but should be based on the information presented in this lesson. Answers may include the marketing of genetically engineered foods or the reduction of genetic diseases in humans.)*

Homework

For homework, suggest that interested students research one of the topics on pages 545–547. Have them share what they learn by presenting oral reports to the class.

- *Clones.* When you cut off parts of a mature plant (such as leaves or stems) and grow new plants from them, you are producing **clones**. Clones have the same genetic makeup as the mature plant. They are new plants produced without sexual reproduction.

These plants, which were each grown from a single cell, are all clones of other plants.

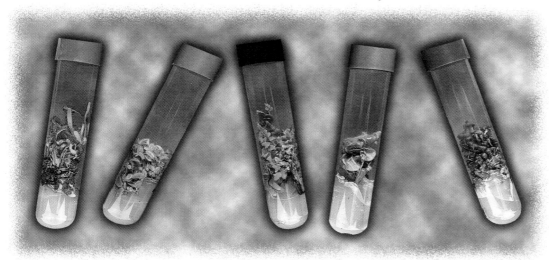

It is now possible to clone certain simple animals. Are human clones possible? This issue is being debated right now. **C**

Keep in mind that human clones would not first appear as full-grown adults. Instead, they would have to mature as every other embryo does—and you have already seen the kind of effect that environment can have on development.

What lies ahead is unknown. The noted scientist Lewis Thomas has said that in the future, people will travel through a "wilderness of mystery." He says, "Science will not be able to explain the full meaning of all that it makes possible." As a result, Thomas believes that every variety of talent will be needed for the future, especially the contributions of poets, artists, musicians, philosophers, historians, and writers. What talents do *you* think will be needed? How could such talents help explain the process of life and how to make use of technology? **D**

Genetic engineering may be very important in the future. To find out more about genetic engineering, read page S172 in your SourceBook.

547

Homework

The Math Practice Worksheet on page 43 of the Unit 8 Teaching Resources booklet makes an excellent homework activity once students have completed Lesson 3.

PORTFOLIO
Students may wish to include their poem or story from the Assessment on this page in their Portfolio.

Reteaching

Have students form debate teams to debate the advantages and disadvantages of one of the technological innovations (such as the genetic engineering of crops) made possible by modern genetics.

Assessment

Have students write poems or stories to express their feelings about what the future will be like as a result of genetic research. Suggest that they use the items listed on pages 545–547 to help them think of some ideas. Encourage students to be creative. Suggest that they include illustrations. Display the finished works for others to read and enjoy.

Extension

Explain to students that when a new molecule of DNA is made by genetic engineering, the resulting molecule may be called *recombinant DNA*. Suggest that they do some research to find out how recombinant DNA is created. Have students share what they learn by making poster diagrams for the classroom. In their diagrams, students may want to use information on the structure of DNA from Lesson 1.

Closure

Ask students to write a response to each of the following hypothetical scenarios:
- Scientists alter the genetic makeup of cattle to increase milk and meat production.
- Scientists manipulate human genes to change the physical features of offspring.
- Scientists alter the chromosomes to change the sex of an unborn baby.

Answers to Challenge Your Thinking, pages 548–549

1. Accept all reasonable answers. Answers could include a director of a plant nursery and a director of forestry management. What each expert would say should include some of the following points:
- Cloning preserves good combinations of genetic traits, which would otherwise be changed and possibly lost by sexual reproduction.
- Cloning produces many individuals with uniform characteristics.
- Cloning is a way to preserve and propagate helpful mutations that occur in body cells.
- One danger of cloning is that all individuals may be lost due to a disease or environmental factor that is detrimental to all.

2. There are several inherited characteristics that are not affected by the environment. For example, eye color and blood type are due to proteins that are not altered by the environment. All acquired characteristics, such as scars or loss of limbs, are not affected by genes.

3. Sample answers:
 a. A *trait* is determined by a pair of genes, each of which is a section of a *DNA* molecule that contains the instructions for producing the trait in the form of *linear programming.*
 b. *X-ray* photographs of DNA allowed *Watson and Crick* to determine that the structure of DNA is a *double helix.*
 c. *Heredity* is the passing of traits from one generation to the next. The genes inherited are an individual's *genotype,* while the expression of the traits makes up an individual's *phenotype.*

4. The following is a sample outline:
 a. Divide 100 radish seeds from one package of seeds into five groups of 20 seeds each.
 b. Give group 1 ideal conditions for the sprouting and growth of seedlings.
 c. Give groups 2, 3, 4, and 5 various combinations of light, warmth, moisture, and soil for the sprouting and growth of seedlings.
 d. Keep records of the sprouting and growth in each group.
 e. Compare results among different groups to determine the effect

---

CHALLENGE YOUR THINKING

1. Coming Soon, to a Theater Near You
The 16-year-old clone was nervous. She had just been reassigned as COPI-864 (Clone, On Parts Inventory), and she was one of the next to be used if transplants were needed—a leg, a kidney, or perhaps her brain. Although most clones had been brainwashed into accepting their roles, COPI-864 plotted her escape.

At home that evening, Diego and Tracy talked about the movie and the whole idea of human clones. They wanted to find out all they could about the possible results of cloning. They were interested not just in humans, but in all sorts of living things.

They started writing down a list of experts who could contribute knowledge and valuable viewpoints to a discussion about cloning. Complete the list they started (shown below) by adding at least five more people.

- Owner of prize cattle or other farm animals
- Director of a zoo for endangered wildlife

What might each expert have to say?

2. Let's See the Evidence!
Are there inherited characteristics that are not affected by the environment? Are there human characteristics that are not affected by genes? Support your answers with evidence.

---

of variation of the environment. Examine group 1 carefully to see if all of the plants in this group are superior, since they had the most nourishing environment.

5. Sample answer:
 a. Row 1: 10W
 Row 2: 2W, 2R, 1W, 2R, 3W
 Row 3: 2W, 1R, 1B, 1W, 1B, 1R, 3W
 Row 4: 4W, 1Y, 5W
 Row 5: 2W, 1R, 1B, 1W, 1B, 1R, 3W
 Row 6: 1G, 1W, 2R, 1W, 2R, 1W, 2G
 Row 7: 2G, 2W, 1G, 2W, 3G

Row 8: 3G, 1W, 1G, 1W, 3G, 1W
Row 9: 1W, 3G, 1W, 3G, 2W
Row 10: 1W, 3G, 1W, 3G, 2W

 b. The instructions are given in lines by symbols and do not resemble the object they describe. Similarly, chromosomes are made up of genes arranged in lines and do not look like the characteristics they produce.

The instructions in the question can be followed to produce the bead pattern. Genetic instructions can be followed to produce a certain set of characteristics in new cells and living things.

3. If X Is to Y . . .

Explain in your own words how the items in each group are related to each other.

a. linear programming, DNA, trait

b. double helix, Watson and Crick, X rays

c. heredity, genotype, phenotype

4. 100 Radish Seeds

Outline how you could use 100 radish seeds to investigate the effects of environment on plants.

5. From Beads to Genes

Below is a design for beading on a costume.

a. Translate it into a form of linear programming.

b. How is your example of linear programming similar to the instructions for life in the DNA molecule?

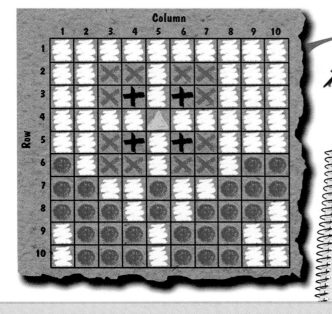

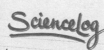

Sciencelog

Review your responses to the ScienceLog questions on page 532. Then revise your original ideas so that they reflect what you've learned.

 You may wish to provide students with the Chapter 23 Review Worksheet that accompanies this Challenge Your Thinking (Teaching Resources, page 46).

Sciencelog

1. Students' answers will vary but may include the following:
   • by spoken word
   • by demonstration
   • by pictures and diagrams
   • by lines of various thicknesses (bar codes)
   • by musical notes
   • by symbols (numbers, letters, etc.)
   • by raised dots (Braille)

2. Development of a new life depends on a set of instructions called genes within the nuclei of cells. Genes are made of a chemical called DNA and are organized in blocks of at least 100,000 genes. These blocks are known as chromosomes.

3. The answer to this question is continually debated by both experts and the general public with no definitive result. It remains open to opinion whether inherited characteristics or the circumstances of a person's life are more important in determining behavior and attitudes.

Homework

As a homework assignment, you may wish to have students complete the Activity Worksheet on page 45 of the Unit 8 Teaching Resources booklet. If you choose to use this worksheet in class, Transparency 90 is available to accompany it.

The Big Ideas

The following is a sample unit summary:

Inheriting family characteristics means that a person has received combinations of genes similar to those of his or her ancestors through the process of sexual reproduction. (1)

All living things are capable of self-preservation, or staying alive; self-regulation, or controlling all life processes; self-reproduction, or making offspring; and self-organization, or forming and repairing body cells. (2)

The nucleus of a cell controls the formation of new cells. Through the process of mitosis, the chromosomes in the nucleus are duplicated and then divided between two cells so that each cell receives an identical set of chromosomes. (3)

Asexual reproduction involves only one parent and occurs by the processes of mitosis and cell division. Sexual reproduction involves two parents and occurs by the union of sex cells produced by meiosis. (4)

Mendel discovered that inherited traits appear in definite ratios among the offspring of a genetic cross. When an individual that is pure for a dominant trait is crossed with an individual that has the recessive trait, all offspring will have the dominant trait. When two hybrids are crossed, a ratio of three dominant to one recessive results. (5)

Characteristics such as the basic body plan, organs, and organ systems develop during the first trimester of pregnancy. Body hair, movements, and further organ differentiation occur during the second trimester. The final development of the nervous system, sense organs, and circulatory system and the smoothing of skin wrinkles occur during the third trimester. (6)

Genetic information is contained in one or more strands of DNA. DNA is a part of the chromosomes, which are found in the nucleus of a cell. Individual units of information are called genes. (7) Chromosomes are a form of linear programming because the DNA they contain is composed of a sequence of chemicals (nitrogen bases), which contain a code. The code is read, from beginning to end, in sequence, and is used to assemble the chemicals that make up a cell. (8)

SOURCEBOOK

If this unit got you interested in genetics, you can find out more about the subject in the SourceBook. Read about how a person's sex is determined by genes, how DNA copies itself, and more.

Here's what you'll find in the SourceBook:

UNIT 8
Foundations of Heredity S154
Chromosomes, Genes, and Heredity S161
DNA—The Material of Heredity S167
Reproduction and Development S173

550

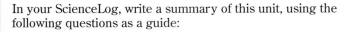

Unit 8

CONTINUITY OF LIFE

The Big Ideas

In your ScienceLog, write a summary of this unit, using the following questions as a guide:

1. What does it mean to "inherit" traits?
2. What are four abilities that all living things share?
3. What part of a cell controls the formation of new cells? How is this accomplished?
4. How do you distinguish asexual reproduction from sexual reproduction?
5. What patterns did Mendel discover in his research?
6. What parts of an embryo develop early in pregnancy? midway through? late in pregnancy?
7. How is genetic information organized?
8. What is meant by the statement, "Chromosomes are a form of linear programming"?
9. How can the environment affect development (both before and after birth)?

Checking Your Understanding

1. Think about the situations presented below. What would be the possible effects on a couple's descendants if
   a. the father had lost a finger due to an accident at the plant where he worked?
   b. the mother (unvaccinated) was exposed to rubella during the early weeks of pregnancy?
   c. the mother abused drugs during pregnancy?
   d. the mother went to a scary movie about vampires during the early weeks of pregnancy?

The environment can alter development both before and after birth. Environmental factors that may harm a fetus include drugs, alcohol, chemicals in tobacco and food, radiation, and viral diseases. Environmental factors can cause mistakes during the development process and can result in abnormalities or birth defects. Newborn babies can also suffer symptoms of alcohol or drug withdrawal and can acquire diseases such as AIDS due to exposure during development. (9)

★ **You may wish to provide students with the Unit 8 Review Worksheet that accompanies this Making Connections (Teaching Resources, page 53).**

e. the mother was exposed to radiation during the early weeks of pregnancy?

f. the father was exposed to rubella during the early weeks of the mother's pregnancy?

g. a woman was vaccinated against rubella before she became pregnant?

h. a woman who abused drugs (of any sort, including alcohol and tobacco) gave them up before she decided to have a child?

i. the father had abused drugs known to cause chromosome damage?

j. the father worked around insecticides suspected of causing genetic mutations?

2. Straight or Curved Thumbs?

a. Draw a Punnett square to show the possible combinations among the children shown in the pedigree at right.

b. Fill in the children's genotypes on the pedigree.

c. Who has straight thumbs and who does not in this family?

d. Use Punnett squares to show the possibility of straight thumbs among the grandchildren.

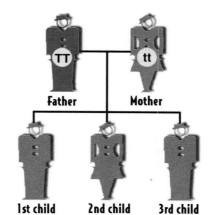

Father   Mother

1st child   2nd child   3rd child

  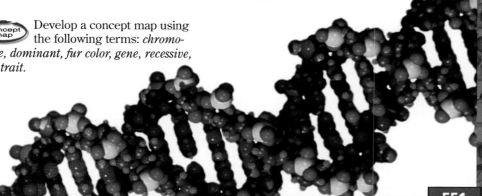

Wife   1st child   Husband   2nd child   Wife   3rd child

3. **concept map** Develop a concept map using the following terms: *chromosome, dominant, fur color, gene, recessive,* and *trait.*

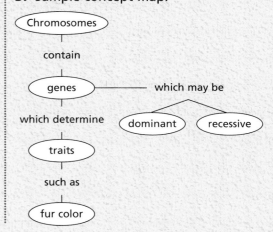

As a homework assignment, you may wish to have students complete the Unit 8 Activity Worksheet on page 52 of the Unit 8 Teaching Resources booklet.

Homework

551

Answers to
Checking Your Understanding,
pages 550–551

a. Loss of the father's finger in an accident would have no effect on his descendants.

b. Exposure to rubella during the early weeks of pregnancy could result in birth defects in the baby.

c. Drug abuse during pregnancy could result in birth defects or addiction of the baby to the drugs used.

d. The mother's watching a scary movie during pregnancy would probably have no effect on the baby.

e. A mother's exposure to radiation during pregnancy can result in birth defects or mutations that could be passed on to successive generations.

f. If he contracted the disease and gave it to the mother, the baby could be born with birth defects.

g. Vaccination could prevent a mother from contracting rubella, thus preventing birth defects.

h. Any time a woman abuses drugs, the drugs may damage a sex cell, which was already partially formed in her ovaries at birth. This damaged sex cell may later be fertilized and develop into an embryo that either is miscarried or becomes a child with birth defects. When the chromosomes of a fetus are damaged, the damage may affect the future generations.

i. His children may be born with birth defects, some of which could be passed on to successive generations.

j. His descendants may inherit any mutations that occur in his sex cells.

2. Answers are on page S214.

3. Sample concept map:

Chromosomes
|
contain
|
genes —— which may be
| /   \
which determine   dominant   recessive
|
traits
|
such as
|
fur color

Background

Until scientists began working with amber, prehistoric DNA could only be extracted from very recent fossils. Then in 1992 DNA from extinct termites found in amber in the Dominican Republic was dated to be about 30 million years old. The following year, in excavating some Lebanese amber, some researchers found DNA from an extinct weevil estimated to have lived 130 million years ago. One group of researchers even used a prehistoric strain of yeast to brew ale!

Research with amber is not without its complexities. Samples become contaminated very easily, making analysis much more troublesome. Another difficulty is identification. For example, in 1994 a research group in Great Britain sheepishly admitted that what they had speculated to be DNA from an extinct mammoth actually belonged to a lab technician.

CROSS-DISCIPLINARY FOCUS

Mathematics

DNA consists of a long, double strand of paired nucleotides, of which there are four types. The information carried by DNA is determined by the order in which these pairs appear in the strand. For instance, if a DNA strand consists of three pairs linked together, there would be $4 \times 4 \times 4$, or 64, possible combinations. Ask: About how many combinations are possible in a DNA strand made up of just 11 pairs? (4^{11}, or about 4.2 million combinations) You may want to tell students that bacterial *E. coli* DNA, one of the simplest known, contains about 2.2 million pairs!

WEIRD Science

Searching for Prehistoric Genes

In the movie *Jurassic Park*, scientists not only duplicated the genetic material of dinosaurs but also re-created the dinosaur itself. A little far-fetched? Maybe, but in fact scientists have retrieved and duplicated prehistoric DNA, and they've done it with the help of an "old sap" called amber.

Trapped in Sap

Golden in color and as clear as glass, amber looks like a gem, but it's really fossilized tree sap.

▲ Many different species of prehistoric plants and animals have been found trapped inside lumps of amber.

Amber has been used in jewelry making for hundreds of years, but today it is popular for another reason. Millions of years ago, microscopic organisms, pollens, insects, and other very small animals were trapped in the sap that oozed from prehistoric trees. As the sap hardened to amber, it mummified its prehistoric victims so well that millions of years later fragments of their DNA, or genetic material, are still preserved.

Extracting the DNA

Genetic researchers are now extracting fragments of prehistoric DNA from the mummified bodies of organisms found in amber. After carefully cleaning the amber to prevent contamination, scientists extract the organic material. In one method, thin slices are cut from the amber until the material can be scooped out. In another method, holes are drilled in the amber, and the material is sucked out with a needle.

The DNA retrieved from prehistoric tissue is almost always badly damaged. Most DNA from living organisms is made up of millions of molecules joined in unbroken strands. The DNA recovered from fossil samples is much shorter—only a few hundred molecules long—and has many missing pieces. However, with new techniques, the extinct DNA can actually be duplicated. The DNA can then be compared with the DNA of a living organism thought to be related to the extinct one.

▼ This scientist is using a special needle to extract genetic material from a chunk of amber.

Hopes for the Future

The search for prehistoric genes is already leading scientists to a better understanding of how living organisms evolved. Some researchers also hope to clone DNA and revive prehistoric species of bacteria, viruses, fungi, and pollen. They believe this will lead to discoveries of new medical drugs, industrial compounds, and natural pesticides. And what about re-creating a dinosaur? Not very likely. The DNA fragments are simply too damaged and incomplete.

Find Out for Yourself

Why is amber such a good preserver of animal and plant tissue? Do some research to find out.

Meeting Individual Needs

Learners Having Difficulty

Have students make a time line showing the major geologic eras and epochs and what types of organisms were common in each one. Then tell students to locate on the time line when each of the organisms mentioned above was encased by amber.

Answers to *Find Out for Yourself*

Amber is a good preserver of animal and plant tissue for two reasons. First, chemicals in amber dehydrate the tissue, which is essential to the preservation of organic material. Second, amber is able to completely flow over and encapsulate the tissue. The amber is airtight and waterproof—a nearly perfect embalmer.

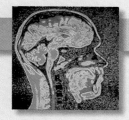

Goats to the Rescue

They're called transgenic (tranz JEHN ik) goats because their cells contain a human gene. They look just like any other goats, but because of their human gene they produce a drug that can save lives.

Life-Saving Genes

Heart attacks are the number one cause of death in the United States. Many heart attacks are triggered by large blood clots that interfere with the flow of blood to the heart. Human blood cells produce a chemical called TPA (tissue plasminogen activator) that dissolves small blood clots. If TPA is given to a person having a heart attack, it can often dissolve the blood clot, stopping the attack and saving the person's life. But TPA is in short supply because it is difficult to produce large quantities in the laboratory. This is where the goats come in. Researchers at Tufts University in Grafton, Massachusetts, have genetically engineered goats to produce this life-saving drug.

Drug-Producing Goats

Producing transgenic goats is a complicated process. First, fertilized eggs are surgically removed from normal female goats. The eggs are then injected with hybrid genes that consist of human TPA genes "spliced" into genes from the mammary glands of a goat. Finally, the altered eggs

▲ A scientist at Tufts University injects human TPA genes into fertilized goat eggs. Above, the first transgenic goat

are surgically implanted into female goats, where they develop into young goats, or kids. Some of the kids that are born actually carry the hybrid gene. When the hybrid kids mature, the females produce milk that contains TPA. Technicians then separate the TPA from the goat's milk for use in heart-attack victims.

◄ The TPA in this milk might help save the life of a heart-attack victim.

The Research Continues

Transgenic research in farm animals such as goats, sheep, cows, and pigs may someday produce drugs more rapidly, cheaply, and in greater quantities than current methods. The way we view the barnyard may never be the same.

Find Out for Yourself

Using chemicals produced by transgenic animals is just one of many gene therapies. Do some research to find out more about gene therapy, how it is used, and how it may be used in the future.

Background

Many proteins that are produced naturally by the human body could be produced by other animals through genetic engineering. This raises the possibility of turning animals into "drug factories." Genetically altered pigs have been used to produce blood and to grow organs that can be used in human transfusions and transplants. Cows have been altered to produce milk that is more like human milk. A dozen different human proteins have been produced in the milk of various transgenic animals.

However, there are serious safety issues concerning these new animal products. Host animals may carry pathogens that are dangerous to humans. For example, sheep and goats sometimes carry a brain disease known as *scrapie.* Research companies that offer transgenic drugs must be sure that pathogens are not present in their products. They must also ensure that any animal proteins that might trigger allergic reactions or other side effects are removed during the purification process. In addition, these companies must be able to prove that their products will function in the same way as their natural counterparts.

Discussion

One concern over transgenic animals is whether it is ethical to tamper with an animal's genetic material. You might want to lead a class discussion on this topic. Ask: How far should we allow genetic engineering to go? Do you consider it to be a valid scientific pursuit? *(Accept all reasonable responses.)*

Multicultural Extension

Fantastic Animals

Many cultural myths include fantastic animals that are mutations or combinations of other animals. Have students research some of these mythical creatures. *(For example, students might choose a satyr, Minotaur, werejaguar, or basilisk.)* Students should discuss what importance these creatures held in the respective cultures.

Answers to
Find Out for Yourself

Another use of gene therapy is to extract the altered cells of a host animal and insert them into the tissue of a human. Such methods are used to treat genetic disorders found in humans, including many types of cancer.

Science Snapshot

Background

Barbara McClintock was born in Hartford, Connecticut, in 1902 and grew up in Brooklyn, New York. She received both a B.S. and a Ph.D. from Cornell University, where she identified all 10 of the chromosomes found in maize. One of the reasons she chose to work with maize was that maize kernels come in a variety of colors and the colors are determined by different genes. She taught at the University of Minnesota before moving on to the Carnegie Institute's Department of Genetics at Cold Spring Harbor, New York, in 1941—a position she held for the next 50 years. In 1944 she became the third woman ever to be elected to the National Academy of Sciences.

McClintock was a very solitary person and preferred to work alone. In 1983, when the news reached her that she had won the Nobel Prize, she reportedly uttered "Oh dear" and walked outside to collect walnuts. She was the first woman to be the sole recipient of a Nobel Prize for physiology or medicine.

Extension

Have students find out who was awarded the Nobel Prize in medicine or physiology for a year of their choice. They should present their findings to the class, including the discoveries that led to each award.

Answers to
Fieldwork

Because scientists had assumed for so long that genes could not move, it was hard for them to accept evidence that genes could move. This is similar to the popular debate over the shape of the Earth. For a long time most people assumed that the Earth was flat. When evidence was presented that the Earth was round, it was hard for many people to accept.

Science Snapshot
Barbara McClintock
(1902–1992)

In 1918, Barbara McClintock's mother was deeply concerned that her daughter would have trouble fitting into society. McClintock had unusual interests and aspirations for a woman of her era. She was about to enter Cornell University, where she was going to become a scientist!

A Relish for Research

In college, McClintock studied genetics, botany, cytology, and zoology. A perfectionist in her research, she soon was recognized as an outstanding researcher in the field of plant genetics. McClintock began working with a type of corn called maize. Her work resulted in some ground-breaking discoveries about genes.

In 1983, Barbara McClintock won the Nobel Prize for her work in genetics.

Jumping Genes

During her research with maize, McClintock noticed that the color of kernels changed unexpectedly from one generation to the next. The responsible genes seemed to move, or "jump," from place to place on the chromosomes. This observation led McClintock to develop a radically new model for gene behavior. The idea that genes could change locations on the chromosome contradicted the prevailing view that genes and chromosomes were stable parts of a cell nucleus.

Scientists were shocked by McClintock's findings, and most of them discounted her ideas for more than 20 years. Over time, however, as more and more research supported her hypothesis, McClintock's model gradually gained acceptance. Finally, in 1983, 65 years after her mother had agonized over McClintock's college plans, McClintock was awarded a Nobel Prize for her work in genetics.

The genes that McClintock discovered are called transposons because they transpose, or shift, on the chromosomes. When such a gene changes location, it often causes a mutation or causes a gene to become inactive. Such genetic changes can lead to new or different characteristics in the offspring of the affected organism.

▶ Transposons cause surprising changes in the patterns of colored corn kernels.

Fieldwork

Many scientists doubted McClintock's findings because a great deal of evidence indicated that genes maintained their positions on the chromosome. Scientists had even constructed maps that showed the relative positions of genes on certain chromosomes. How is this similar to the debate over the shape of the Earth?

554

Supersquash or Frankenfruit?

*T*he fruits and vegetables you buy at the supermarket may not be exactly what they seem. Scientists may have genetically altered these foods to make them look and taste better, contain more nutrients, and have a longer shelf life. So make way for genetically engineered foods because they may already be on the shelves of your local grocery store.

From Bullets to Bacteria

Through genetic engineering, scientists are now able to duplicate one organism's genes and place them into the cells of another species of plant or animal. This technology enables scientists to give plants and animals a variety of new traits. The new traits are then passed along to the organism's offspring and future generations.

▼ A scientist using a "gene gun" to insert DNA into plant cells

Scientists alter plants by inserting DNA with new properties into the plant's cells. The DNA is usually inserted by one of two methods. In the agrobacterial method, the new DNA is placed inside a special bacterium. The bacterium carries the DNA to the plant cell and transfers it into the cell. The DNA particle-gun method works a little differently. In this method, microscopic particles of metal coated with the new DNA are actually fired into the plant cells by way of a special "gene gun."

High-Tech Food

During the past decade, scientists have inserted genes into more than 50 different kinds of plants. Most of the new traits from these genes make the plants more disease resistant or more marketable in some way. For example, scientists have added genes from a caterpillar-attacking bacteria to cotton, tomato, and potato plants. The altered plants produce proteins that kill the caterpillar's crop-eating larvae. Scientists are also trying to

◀ The Flavr-Savr™ tomato was the first genetically altered fruit to reach supermarket shelves.

develop genetically altered peas and red peppers that stay sweeter longer. A genetically altered tomato that lasts longer and tastes better is already in many supermarkets. One day it may even be possible to create a caffeine-free coffee bean.

Are We Ready?

As promising as these genetically engineered foods seem to be, they are not without controversy. Some people are afraid that new, harmful genes could be released into the environment or that foods may be changed in ways that endanger human health. For example, could the nutritional value of some foods be inadvertently changed or reduced in an attempt to make the food look better and last longer? All of these concerns will have to be addressed before the genetically altered food products are widely accepted.

Find Out for Yourself

Are genetically altered foods controversial in your area? Survey a few people to get their opinions about genetically altered foods. Do they think grocery stores should carry these foods? Why or why not?

555

Find Out for Yourself

You may wish to have students develop a questionnaire for people to fill out. (You should approve the questionnaire before students conduct their survey. Be sure to comply with any local guidelines that apply to public survey procedures.) Students can then tally the replies and discuss the results in class. Make sure that students administer the poll only with adult supervision. You might want to have students contact only people whom they know personally.

Background

The Flavr-Savr™ tomato was approved for consumer use by the Food and Drug Administration in 1994. By inserting an extra gene into the tomato, scientists were able to slow down a series of chemical reactions that cause tomatoes to rot. As a result, this genetically altered tomato can be left to ripen on the vine and can stay on the shelf much longer before becoming soft and spoiling. Many consumers think it also has a better flavor.

Consumer reaction to the tomato was not entirely positive, however. After the FDA's approval was announced, some consumer groups quickly began preparing a boycott of the product as well as planning public "tomato squashes," in which the tomatoes were crushed as a protest against genetically engineered foods.

Yet the tomato marked a new era in commercial agriculture. Genetic engineering may eventually produce crops that can tolerate much harsher growing conditions, including the application of herbicides that are currently too strong to be used.

Debate

Discuss with the class the pros and cons of genetically altered foods. Then ask students to form an opinion about the genetic engineering of animals. Is this any different from that of plants? Why or why not? *(Accept all reasonable responses.)* You may wish to divide the class into two teams and organize a debate on the topic of genetically engineered organisms.

Discussion

One controversy over altered food products involves product labeling. Some people feel that genetically altered foods, including individual fruits and vegetables, should be clearly labeled as such. Many of the producers of these foods feel that labeling should be mandated only when there is an obvious need, such as when an allergen is introduced or when the nutritional value of the food is significantly altered. Ask students to share some of their thoughts on this issue.

Life Processes

Unit 1

In This Unit

The Chemistry of Life,
page S2

**Building Biochemicals
With Light,**
page S9

Basic Life Processes,
page S13

Now that you have been introduced to two of the major life processes, photosynthesis and respiration, consider these questions.

1. What are the basic chemicals of life, and what are their primary functions?

2. In what ways are photosynthesis and cellular respiration complementary processes?

3. What is metabolism and how does it relate to digestion?

In this unit you'll explore the chemical basis of life and then see how plants make the sugar that almost all other organisms depend on. You'll also learn more about basic life processes that are characteristic of all living things.

S1

THE CHEMISTRY OF LIFE

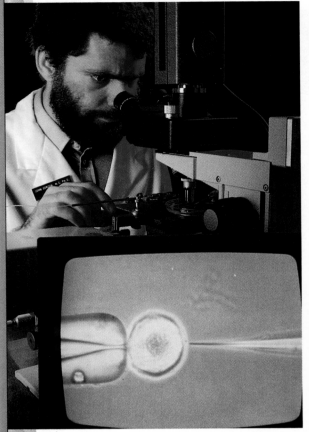

Everything that happens in a living thing is the result of chemical reactions. Just flexing your leg, for example, requires many different chemical changes. Think about this as you bend your leg. Chemical changes in your brain cause nerve cells to send the "bend your leg" message from your brain to your leg. Then another series of chemical changes in your leg causes some of the muscle cells in your leg to contract and others to relax. As a result, your leg bends.

A better understanding of the chemical processes of life may be the key to solving many human problems, from producing enough food for the people of the world to curing diseases such as cancer, heart disease, and AIDS. Currently, scientists are doing extensive research to help them further understand the chemical basis of life as well as the processes that occur in all living things.

The Elements of Life

As you know, all matter—living and nonliving—is made up of particles called *atoms*. Atoms are the smallest particles of elements. But of the 91 naturally occurring elements, just four of them—carbon, hydrogen, oxygen, and nitrogen—make up more than 99 percent of the living matter on Earth. Atoms of these four elements combine with each other by forming very strong and stable chemical bonds.

Carbon atoms form the "backbone" of most biological molecules. Two special abilities of carbon make life as we know it possible. First, carbon atoms can chemically bond with up to four different atoms. Second, carbon atoms can bond to other carbon atoms, making long chains, rings, or branched structures.

Chemists use the term **organic compounds** when referring to most carbon-containing compounds, since they are normally formed by living organisms. Indeed, *organic* means "coming from life." Compounds that do not contain carbon, like water (H_2O)—and a few that do, like carbon dioxide (CO_2)—are called *inorganic compounds*.

 Using an extremely fine syringe, a biologist injects nucleic acid into this mouse cell. This procedure allows scientists to study the biochemistry of the nucleus.

A carbon atom readily bonds with other carbon atoms in various formations. Because carbon forms chemical bonds so readily, numerous carbon-containing compounds exist. Over 4 million compounds of carbon are now known, and the list continues to grow.

Organic compound

A compound that contains carbon and usually is produced by living things

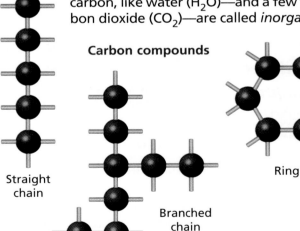

Carbon compounds

Straight chain

Branched chain

Ring

S2

Most organic molecules are very large. Some organic molecules contain thousands or even millions of individual atoms. These atoms are arranged in chemical units called *monomers* that repeat over and over throughout the molecule. Molecules that are made of many monomers are called *polymers*. The prefix *mono* means "one," and the prefix *poly* means "many."

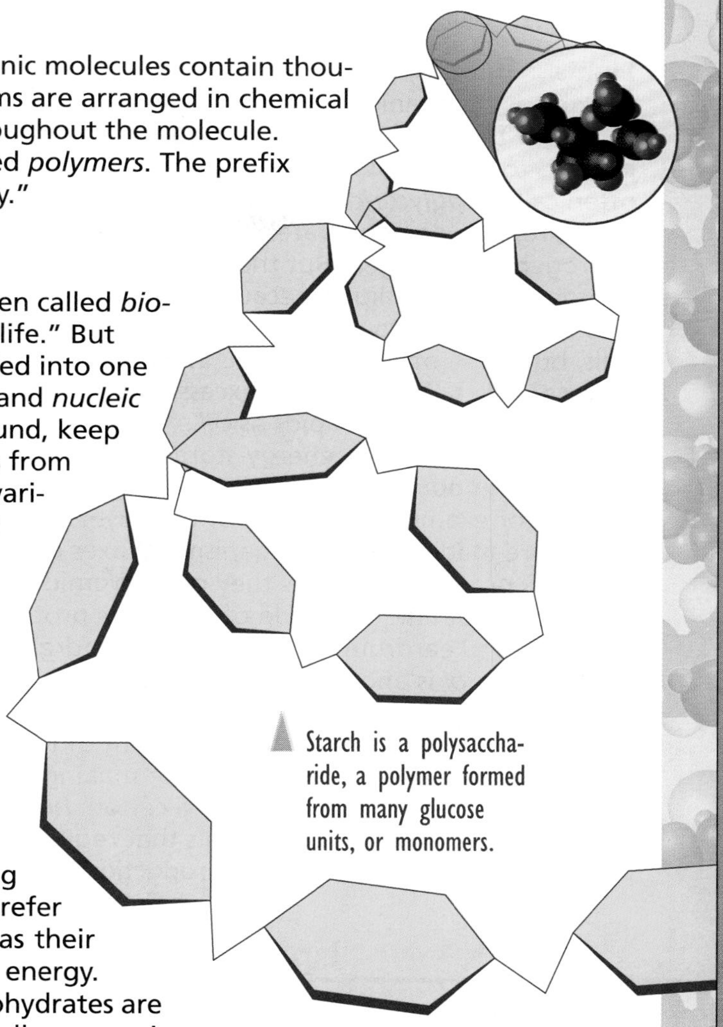

Basic Biochemicals

The organic compounds made by living things are often called *biochemicals* because they are literally the "chemicals of life." But each of the many kinds of biochemicals can be classified into one of four major groups: *carbohydrates*, *lipids*, *proteins*, and *nucleic acids*. As you read about each kind of organic compound, keep in mind that plants synthesize all of these compounds from the inorganic materials carbon dioxide, water, and a variety of minerals. Animals, on the other hand, must eat foods that contain the monomers they use to build the organic compounds that their bodies need.

Carbohydrates Sugars, starch, and cellulose are examples of the organic compounds known as **carbohydrates**. The photo below shows some of the sources of these substances in your diet. Carbohydrates are made entirely of carbon, hydrogen, and oxygen. Almost all carbohydrates contain stored energy that, as you will see later, can be made available to living cells. In fact, most cells prefer to use carbohydrates as their primary source of energy.

Many carbohydrates are polymers of smaller, repeating monomers called *monosaccharides,* or simple sugars. Glucose, a product of photosynthesis, is one type of monosaccharide. When two monosaccharides join together chemically, they form a *disaccharide*, or double sugar. Common table sugar is a disaccharide made of glucose and fructose. Many glucose molecules join together to form the *polysaccharides* starch and cellulose.

Starch is a polysaccharide, a polymer formed from many glucose units, or monomers.

Fruits and honey contain the simple sugars glucose and fructose. Complex carbohydrates, such as starches, are found in bread, cereal, and pasta.

Carbohydrates

Organic molecules composed of one or more monosaccharides that are often used for energy by cells

S3

Lipids   Fats, oils, waxes, and steroids are examples of the group of organic compounds called **lipids**. Most lipids consist of three long *fatty-acid molecules* chemically bonded to a *glycerol molecule*. These kinds of lipids are often called *triglycerides* because of their three-chain structure. Like carbohydrates, lipids consist of carbon, hydrogen, and oxygen. But there are usually many more hydrogen atoms in lipids because of their very long carbon chains. Lipids contain energy that can be used by cells, but more often lipids serve as excess-energy storage molecules. Animals store excess lipids primarily as *fat*; plants store excess lipids as *oil*.

In addition to their energy-storage function, lipids play other important roles in living things. Cell membranes, for example, are made of two layers of lipids. *Waxes* are made by many organisms. Waxes on the leaves of plants protect the tissues they cover from dehydration, and those that line the inside of your ears protect your ear canal and eardrum from dirt and microorganisms.

Cholesterol is another type of lipid that you have probably heard much about. Although excess cholesterol can lead to heart disease, a small amount of it is necessary in animals. Cholesterol is used for making cell membranes, nerve and brain tissues, and certain *hormones*. Hormones are important chemicals that regulate life processes such as growth and reproduction. The photo below shows some of the sources of lipids in your diet.

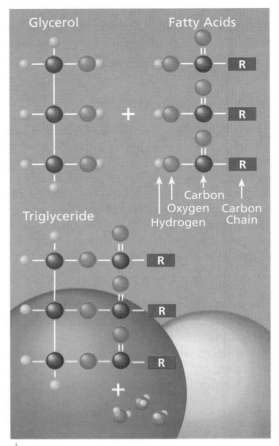

▲ Most lipids are made by joining three fatty-acid molecules to a glycerol molecule.

Lipids

Organic molecules composed of fatty acids and glycerol that store energy and make up cell membranes in living things

▼ Lipids are found in each of the items shown here.

Proteins

Large organic molecules composed of amino acids that either act as structural materials in an organism or regulate the chemical activities of an organism

◀ Proteins are found in dairy products, nuts, meat, and fish.

Proteins   Most of the different kinds of organic molecules found within living things are **proteins**. The photo above shows some of the sources of protein in your diet. All proteins contain carbon, hydrogen, oxygen, and nitrogen, and many also contain sulfur. Proteins are complex polymers of smaller molecules called *amino acids*.

There are 20 different kinds of amino acids. These amino acids make up all the different proteins in living things in much the same way that the 26 letters of the alphabet make up all the different words in the English language. In the same way that words with different meanings are spelled with different letters, proteins (with different chemical properties and functions) are made by varying the kinds, the number, and the sequence of amino acids. For example, two proteins that differ by only one amino acid can have completely different properties, just as the words *live* and *love* have completely different meanings. The longest words in the English language contain only a few dozen letters, but the longest proteins have over 1000 amino acids. Just think how many different words you could make if words had more than 1000 letters! Because of their great size, many different kinds of proteins are possible.

Proteins are very important to the structure and function of living things. Some proteins, such as *collagen* in skin and *myosin* in muscle cells, form strong, elastic fibers. Other proteins, called *enzymes*, are important to the chemical activities of cells. Enzymes are involved in nearly every chemical reaction in a living cell. Still other proteins, such as hemoglobin in the blood, carry substances to places in the body where they are needed. Cell membranes also contain many *carrier proteins* that transport particles into and out of cells.

▲ Insulin, with only 51 amino acids, is one of the smallest proteins in humans. Each color symbolizes a different amino acid contained in the insulin molecule. Two "sulfur bridges" hold the two strings of amino acids together.

DID YOU KNOW...

that by rearranging the sequence of just 20 different amino acids, it is possible to create 2.43×10^{18} different proteins? Although not *all* combinations of amino acids are used to create proteins in organisms, you can now see why proteins are the most diverse of all biological molecules.

S5

Nucleic acids

Large organic molecules composed of nucleotides that store the information for building proteins

Nucleic Acids The largest and most complex organic molecules in living things are the nucleic acids—DNA (deoxyribonucleic acid) and RNA (ribonucleic acid). Nucleic acids are very long chains of monomers called *nucleotides*. Nucleotides are composed of the elements carbon, hydrogen, oxygen, nitrogen, and phosphorus. To give you an idea of how long some nucleic acids are, your body contains about 25 billion kilometers (more than four times the distance to Pluto) of DNA molecules alone!

DNA is located in the nucleus of each cell. It is a very important molecule because it contains the instructions for putting the amino acids of proteins together. These instructions are in the form of a code called the *genetic code*. While DNA functions as the "recipe book" for proteins, RNA does the actual physical work of synthesizing the proteins.

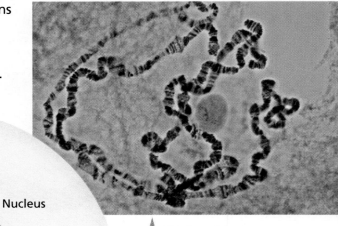

Each group of three nucleotides in a DNA molecule is a "code word" for a specific amino acid.

RNA copies the protein-building information from DNA and moves out of the nucleus and into the cytoplasm of the cell.

RNA nucleotides act as a pattern for lining up the amino acids required for building proteins.

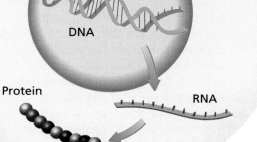

Nucleus

DNA

Protein

RNA

▲ This is a close-up of DNA from a fruit fly. DNA contains information needed to make proteins.

Water

You could live about 5 weeks without eating a bite of food. However, you could live only about 5 days without drinking water. Water is the most important inorganic substance in living things, making up from 50 to 98 percent of the mass of a living cell. About 65 percent of the total volume of your body is water. If you were to lose only 15 percent of your body's normal amount of water, it could be fatal.

We share our need for water with all living things. For one thing, water dissolves the chemicals that participate in the many chemical reactions that occur inside organisms. Among its many important properties, water has a high capacity to absorb and give off heat without great changes in its temperature. Water's *heat capacity*, therefore, keeps living things from both freezing and overheating. As the only common substance that is a liquid at most of the temperatures found on Earth, water is also the habitat of thousands of living species.

S6

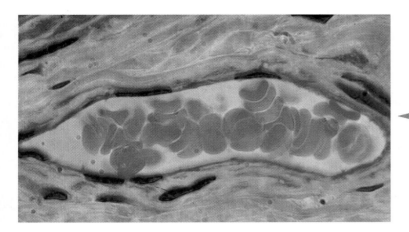

◄ Without water, there can be no life. These blood and muscle cells exist in solutions that are mostly water.

Polar molecule

A molecule that carries unevenly distributed electrical charges

A Polar Substance   Many of the properties of water are due to its chemical structure. A water molecule forms when two hydrogen atoms chemically bond with an oxygen atom. The negatively charged electrons of the molecule spend more time near the oxygen nucleus than they spend near the hydrogen nuclei. As a result of this arrangement of electrons, the oxygen side of a water molecule has a slight negative charge, and the hydrogen side has a slight positive charge. Molecules, such as water, that have positive and negative poles are said to be **polar molecules**.

Polar molecules are attracted to other molecules that have electric charges. Because water is polar, it makes an excellent *solvent*. A solvent is a substance that can dissolve another substance.

Adhesion and Cohesion   Water molecules tend to stick to things they cannot dissolve, such as glass, soil particles, and cells. This property of water is called *adhesion*. As you know, adhesion is partly responsible for water's ability to rise inside small tubes such as the vessels in plants that conduct water from the roots to the leaves. Water molecules are attracted to each other as well, due to their polarity. This property, called *cohesion*, keeps water columns from breaking as they rise inside plants.

◄ The unequal distribution of positive and negative charges in a water molecule results in its bent shape.

When crystals of common table salt (NaCl) are dropped into water, the water molecules are attracted to the individual sodium and chloride ions that make up the salt. The attraction of the water molecules pulls the ions away from the salt crystals, and each ion becomes surrounded by water molecules. As a result, the salt disappears into the water, or dissolves. ►

Chloride ion (Cl^-)

Sodium ion (Na^+)

Salt (NaCl)

Water molecule (H_2O)

Water's Chemical Role Water plays an extremely important part in the chemical reactions that occur in living things. First of all, dissolved substances react much more rapidly than undissolved substances. In fact, many substances will not react at all unless they are dissolved in water. Water dissolves most of the substances that participate in the chemical reactions of life. Second, water is often a participant in chemical reactions. For example, splitting apart large organic polymers requires the addition of water. This type of reaction is called *hydrolysis*, which means "splitting with water." On the other hand, large organic molecules are made by a reaction known as a *condensation* reaction. In a condensation reaction, water is produced when monomers are joined to form polymers.

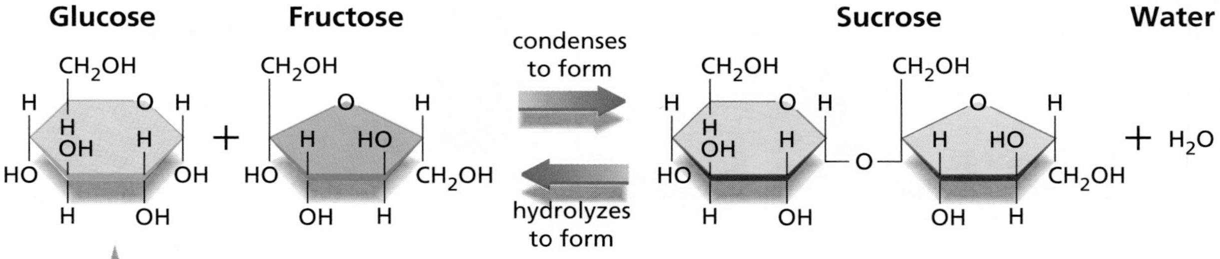

Glucose	Fructose		Sucrose	Water

condenses to form →

hydrolyzes to form ←

$+ H_2O$

A condensation reaction between glucose and fructose produces sucrose (table sugar) and water. This reaction is reversible by hydrolysis, which uses water to break sucrose down into glucose and fructose.

SUMMARY

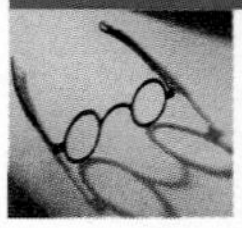

Carbon is the element that forms the backbone of most of the important compounds of living things. These organic compounds include carbohydrates, lipids, proteins, and nucleic acids. Carbohydrates are made from monosaccharides, and lipids are made from fatty acids and glycerol molecules. Both are used as energy-storage compounds. Proteins are made by joining amino acids together. Some proteins are structural materials, while others, called enzymes, make possible the chemical reactions necessary for life. Nucleic acids are made of long chains of nucleotides. DNA is a nucleic acid that contains the instructions for making proteins. RNA copies information from DNA and uses it to build proteins. Water is the most abundant inorganic substance in living things. It dissolves many materials and participates in chemical reactions.

BUILDING BIOCHEMICALS WITH LIGHT

All of the organisms on Earth are ultimately dependent on the sun. Without the sun's warming rays, the Earth would soon become a frozen wasteland, a sphere of ice on which no living thing could survive. But the sun brings more than just warmth to the Earth. The sun also supplies the energy that plants need to build the organic molecules other living things depend on.

Chlorophyll

A green chemical that absorbs energy from the sun during photosynthesis

Photosynthesis

During *photosynthesis*, plants use sunlight, water (H_2O), and carbon dioxide (CO_2) to make organic molecules and oxygen (O_2). Plants use some of these organic molecules as building blocks for making lipids, nucleic acids, proteins, and carbohydrates. Remember that carbohydrates are a major source of energy for most living things. A basic formula used to represent any kind of carbohydrate produced by a plant is CH_2O.

The Place   Photosynthesis begins when the sun's energy falls on plant organelles called *chloroplasts*. Chloroplasts are found in cells inside the leaves of plants. A chemical called **chlorophyll** inside the chloroplasts absorbs the sunlight.

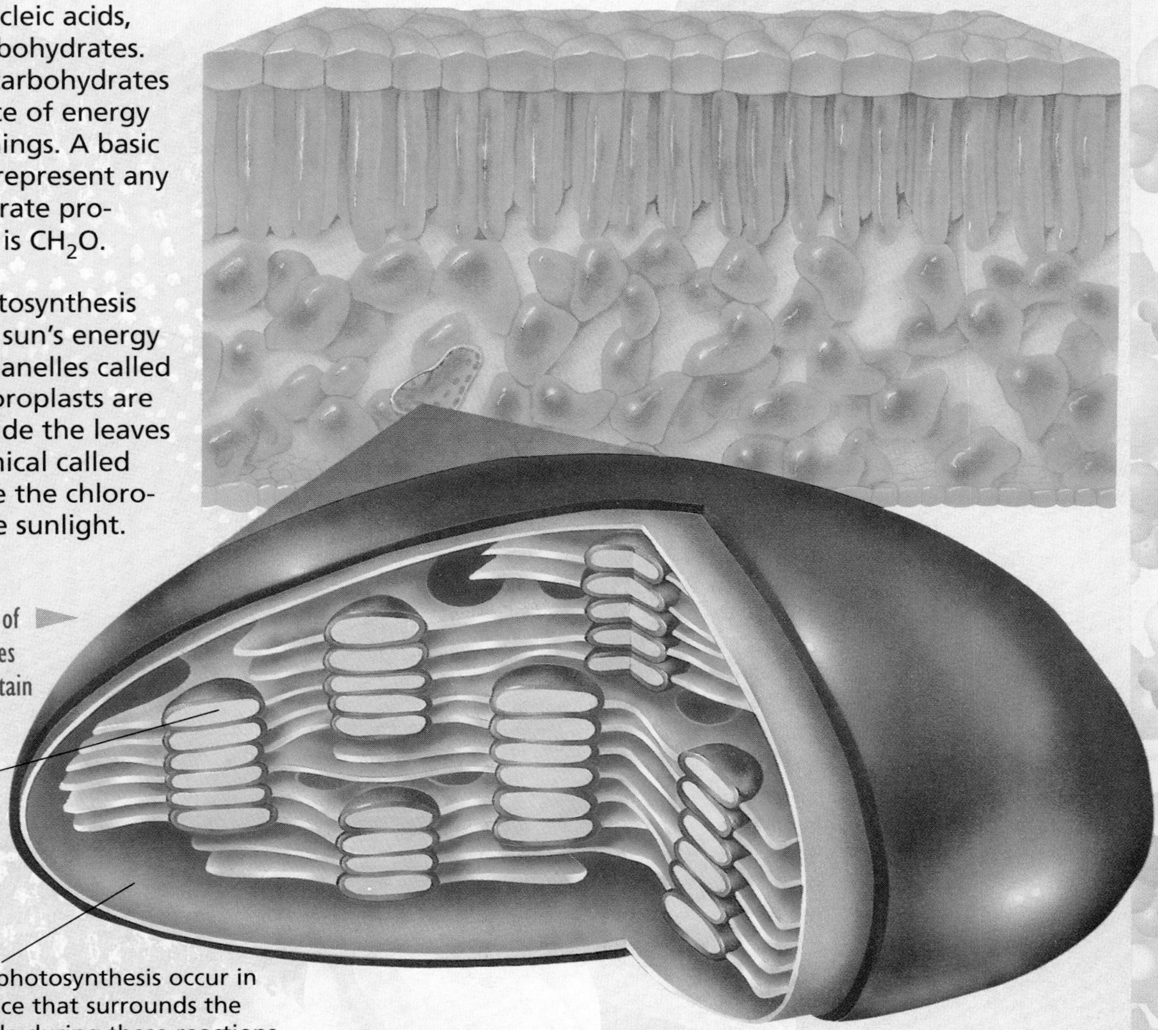

Inside the chloroplasts of plant cells are structures called *grana*, which contain chlorophyll molecules.

Many of the reactions of photosynthesis occur in the grana.

Other reactions of photosynthesis occur in the cellular substance that surrounds the grana. Sugar is made during these reactions.

S9

The Process   The overall chemical equation that describes photosynthesis is shown below. The equation shows that a basic carbohydrate is made from a molecule of carbon dioxide and a molecule of water, with a molecule of oxygen being released as a byproduct. During some of the reactions of photosynthesis, the water molecules are split. As a result, oxygen is produced. This oxygen eventually leaves the plant and enters the atmosphere by passing through the *stomata* in the leaves. The carbon dioxide and the hydrogen from water are used in a separate series of reactions that occur in the cellular material that surrounds the grana inside chloroplasts. These reactions lead to the formation of carbohydrates and other organic molecules.

The Products   Some of the organic compounds that plants produce, such as carbohydrates, contain energy that cells use. The carbohydrates made by plants include glucose, sucrose, and starch. Some plants have specialized parts in which starch is stored. Potatoes are one example of such a part. One polysaccharide that plants make, called *cellulose*, is not used by plants for energy. Plants use cellulose to make their cell walls.

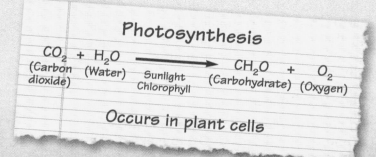

Photosynthesis

$$CO_2 + H_2O \xrightarrow[\text{Chlorophyll}]{\text{Sunlight}} CH_2O + O_2$$

(Carbon dioxide)   (Water)   (Carbohydrate)   (Oxygen)

Occurs in plant cells

DID YOU KNOW...

that of all the energy from the sun that reaches the Earth each day, only about 1 percent is used for photosynthesis? Of the remaining 99 percent, about one-third is immediately reflected back into space, and the rest is absorbed by the Earth and changed into heat.

O_2

CH_2O

CO_2

H_2O

Plants take in carbon dioxide through their leaves and water through their roots. These two chemicals, along with chlorophyll and the energy in sunlight, are used by plants to form carbohydrates and oxygen.

S10

Reaping the Benefits of Photosynthesis

The carbohydrates that a plant makes—sugar and starch—are used as energy sources by the plant. An animal that eats the plant takes in these molecules and also uses them for energy to carry out its own life processes. Any extra energy is stored in the animal's body as fat or as animal starch, which is also called *glycogen*. As one animal eats another animal, the energy that came to the plants from the sun is passed on again and again. In this way, the energy that is used by almost all living things on Earth can be traced back to sunlight. But how do plants and the animals that eat plants actually get energy from carbohydrates? The answer lies in a chemical process called *cellular respiration*.

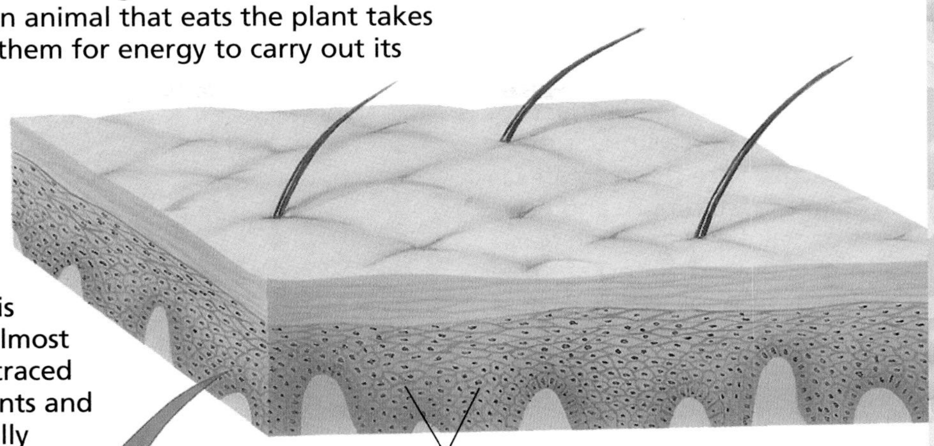

Skin cells

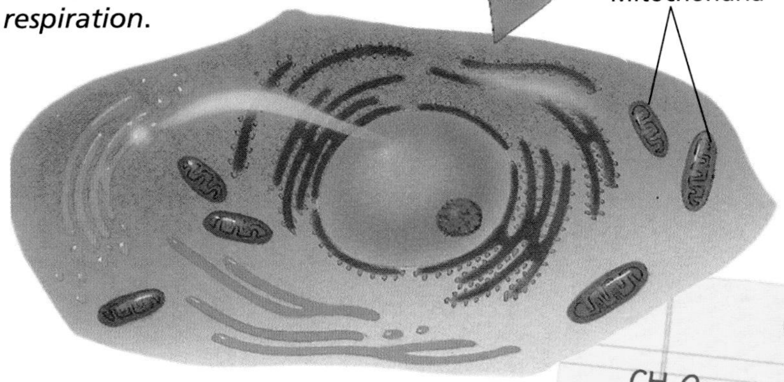

Mitochondria

Some of the reactions of cellular respiration occur in the cytoplasm of cells. Other reactions of cellular respiration occur inside organelles called *mitochondria*.

Cellular Respiration

$$CH_2O + O_2 \longrightarrow CO_2 + H_2O + Energy$$

(Carbohydrate) (Oxygen) (Carbon dioxide) (Water) (ATP)

Occurs in both plant and animal cells

Getting Energy From Glucose As you know, plant and animal cells rely on organic molecules, primarily glucose, for energy. When an animal ingests plant material, it digests the starch in the plant, breaking it down into glucose molecules. Plants must also convert their own starch into glucose to get energy out of it. The glucose then fuels cellular respiration, the chemical process that transforms the energy in glucose into energy that can be used by the organism. As in photosynthesis, many different chemical reactions occur during cellular respiration. The overall process of cellular respiration can be represented by the chemical equation shown above.

During cellular respiration, the energy contained in an organic compound, such as glucose, is released. But this released energy cannot be used directly by cells. Instead, the energy is used to form another molecule called **ATP**, which, chemically speaking, is a type of nucleotide. As you know, most nucleotides are used to build nucleic acids. But the function of ATP is different. ATP molecules are "packets of energy" that cells use to live. The energy contained in ATP is used by organisms in a variety of life processes, including movement, building organic polymers, and cell division.

ATP

*An organic molecule (**a**denosine **tri**phosphate) used to deliver energy for life processes*

S11

A Precarious Balance

If you compare the equation for photosynthesis with the equation for cellular respiration, you will see that one is basically the reverse of the other. Products of photosynthesis (carbohydrates and oxygen) are the reactants of cellular respiration, and products of cellular respiration (carbon dioxide and water) are the reactants of photosynthesis. Because of this relationship between photosynthesis and cellular respiration, it has been hypothesized that plants and animals could coexist in a sealed environment such as a glass globe. In fact, artificial ecosystems have been created and do last quite a long time. Perhaps the most famous of these was the Biosphere II project, shown to the right. This project was an attempt to demonstrate that plants and animals (including humans) could live together in a closed environment for an extended period of time.

From 1991 to 1993, eight "biospherians" lived under an air-tight glass dome that contained a variety of plant and animal species. During their stay in Biosphere II, they raised their own livestock, grew their own crops, and collected scientific data. Today, the facility is being used as a research center. In it, scientists conduct experiments to further refine our understanding of the intricate balance between plant and animal life and the transfer of energy within an ecosystem.

Atmospheric oxygen   Atmospheric carbon dioxide   Energy   Energy

Life on Earth is possible because of the interaction between photosynthesis and cellular respiration. If plant life disappeared from the Earth, most other organisms would soon cease to exist due to a lack of food and oxygen.

SUMMARY

Plants make organic molecules such as carbohydrates by carrying out photosynthesis. Photosynthesis occurs inside plant organelles called chloroplasts. Photosynthesis uses water, carbon dioxide, and energy from the sun to produce organic compounds and oxygen. Plants and animals use glucose in cellular respiration. During cellular respiration, glucose and oxygen are used to produce energy in the form of ATP. Carbon dioxide and water are produced as byproducts. Photosynthesis and cellular respiration are interdependent processes because the products of photosynthesis are the reactants of cellular respiration, and the products of cellular respiration are the reactants of photosynthesis.

S12

BASIC LIFE PROCESSES

The many chemical activities of living things can be grouped according to several major processes. Among these processes are *obtaining raw materials*, *metabolism*, *excretion*, and *regulation*. You may have observed processes that resemble these in nonliving things. For example, streams gather and transport sediments, while an air conditioner adjusts the air temperature in a room. However, all organisms—and the cells of which they are made—perform not just one, but all, of these processes. As you read about them, notice how each process is related to and depends on the others.

Even though a computer has complex organization, responds to stimuli, and has moving parts, it is not alive. What can the mouse do that the computer cannot do?

Obtaining Raw Materials

To remain alive, living things must take in raw materials such as food, water, and mineral nutrients. *Food* is any substance that provides organisms with organic compounds. When you eat a snack or a meal, you supply your body with new organic molecules. Some organisms, such as most plants, algae, and certain bacteria, take in only inorganic substances. From these, they make the organic substances they need by carrying out photosynthesis. Organisms that make their own food, called *producers*, form the basis of all food chains. Organisms called *consumers* must eat other organisms to obtain organic substances.

The movement of materials into and out of cells is regulated by the cell membrane, which allows certain substances to enter and keeps others out.

The materials obtained by both producers and consumers must be transported to the individual cells of the organism, where they are used. However, in order for cells to take in materials, the materials must first pass through cell membranes. Movement through a cell membrane is accomplished by two processes: diffusion and active transport.

S13

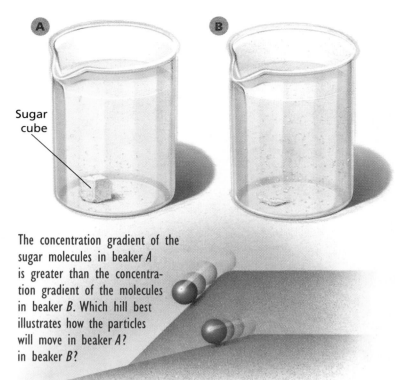

Sugar cube

The concentration gradient of the sugar molecules in beaker A is greater than the concentration gradient of the molecules in beaker B. Which hill best illustrates how the particles will move in beaker A? in beaker B?

Concentration gradient

The difference between the concentrations of a substance in two areas

Diffusion and Osmosis The movement of particles that occurs when the concentration of a dissolved substance differs in two neighboring areas is called *diffusion*. A difference in concentration can be pictured as a hill. Just as a ball would roll down the slope of a hill, from high to low, the movement of diffusing molecules and ions also occurs from high to low—high concentration to low concentration. In other words, the particles move down (or with) the concentration gradient. The **concentration gradient** is simply the difference between the concentrations of a substance in two areas. The steeper the slope of the hill is, the faster the ball rolls. The same is true for diffusion—the greater the concentration gradient is, the faster the process of diffusion occurs.

Of course, molecules and ions do not really roll downhill. However, they are constantly moving back and forth between areas of different concentration. Since there are more particles to move from high to low, and fewer to move in the opposite direction, an area that is less concentrated gains more particles than it loses. As a result, the less concentrated area experiences an overall increase in molecules or ions of that kind.

Osmosis follows the same principle as diffusion but refers only to the diffusion of water through semipermeable membranes, such as those surrounding cells. The concentration of water molecules depends on the number of dissolved particles contained in the water. There are three types of environments that can exist around cells—hypotonic, hypertonic, and isotonic. The number of dissolved particles and the concentration of water molecules is different in each environment.

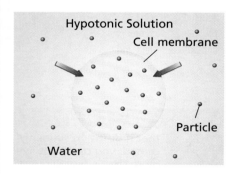

Hypotonic Solution
Cell membrane
Particle
Water

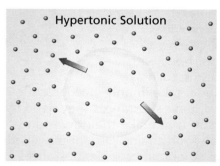

Hypertonic Solution

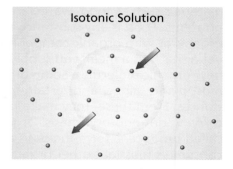

Isotonic Solution

In a hypotonic environment there are more water molecules (fewer dissolved particles) outside the cell than inside. Therefore, the water molecules tend to move into the cell in order to equalize the concentration.

In a hypertonic environment there are fewer water molecules (more dissolved particles) outside the cell than inside. Therefore, water tends to leave the cell.

In an isotonic environment the concentration of water molecules is equal on both sides of the membrane. Therefore, there is no net movement of water in either direction.

S14

Diffusion and osmosis are things that you might say "just happen" to a cell. The cells themselves do not take an active part in obtaining materials by either process. However, only certain materials can enter a cell by these means. First of all, the material entering a cell must be in higher concentration outside the cell. Second, the materials must be able to dissolve in water and must have molecules or ions small enough to pass through the tiny openings in a cell membrane.

Active Transport   Sometimes a cell may need a substance that is already more concentrated inside the cell than it is outside the cell. Such substances cannot get in by diffusion. However, there *is* a method by which these substances can enter. This method is called **active transport**. Unlike diffusion, active transport requires energy. Where do you think this energy comes from? If you said from the ATP molecules made during cellular respiration, you are absolutely correct.

To understand why active transport uses energy but diffusion and osmosis do not, imagine that you are riding your bicycle and you come to a hill. If the hill goes down, you can relax because all you have to do is steer. You don't get tired because you don't use energy to roll downhill. But if the hill goes up, you have to use energy to pedal up the hill. As long as the hill goes up, you must continue to use energy.

Active transport

The movement of materials into and out of cells against a concentration gradient, requiring the use of energy

▲ Pedaling uphill is like active transport.

◄ Coasting downhill is similar to diffusion and osmosis.

Substances entering a cell by diffusion or osmosis are like your bicycle rolling downhill. Because both diffusion and osmosis occur with (or down) the concentration gradient, no energy is needed. On the other hand, active transport moves materials *against* (or up) the concentration gradient. Just as you must expend energy to ride your bicycle uphill, a cell must use energy to move materials against a concentration gradient.

S15

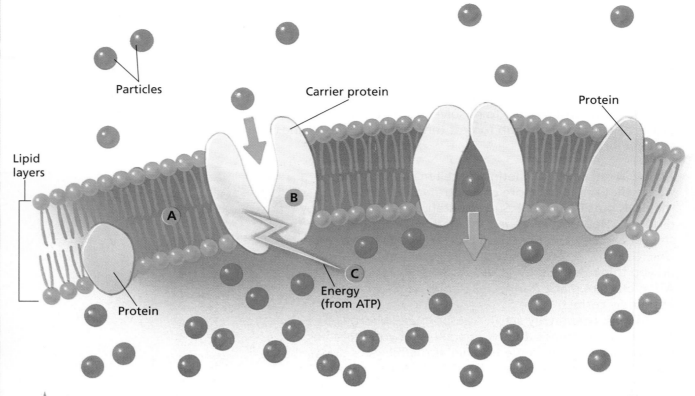

Particles

Carrier protein

Protein

Lipid layers

B

A

C

Energy (from ATP)

Protein

▲ The cell membrane (A) is composed of a double layer of lipids in which many kinds of proteins are embedded. Carrier proteins (B) act like gates. When a particle has to go against its concentration gradient through one of these gates, a chemical reaction inside the cell breaks down ATP to supply the needed energy. The energy released from the ATP (C) is used to change the shape of the carrier protein and move the particle through the membrane.

The secret to active transport is in the structure of the cell membrane. Large protein molecules called *carrier molecules* are positioned in the lipid layers of the membrane. These molecules change their shape and move ions or molecules into and out of the cell. Changing the shape of the carrier protein is what requires energy.

Metabolism

Living things are able to organize and reorganize the substances they take in for specific

functions. This involves breaking down large organic molecules into smaller molecules, as well as joining small molecules together to make larger ones. Together, all the chemical breaking-down and building-up processes of organisms are referred to as **metabolism**.

There are three basic chemical processes that make up metabolism: digestion, cellular respiration, and synthesis. In *digestion*, large food molecules are broken down into simpler compounds. Some of these compounds are then used in the process of *cellular respiration*, which supplies the energy for cell activities. Although the simple sugar glucose is the primary fuel for cellular respiration, fatty acids and amino acids can be used as well. In fact, if an

organism is suffering from starvation, it will break down any available organic compound to obtain energy.

Some of the products of digestion, such as sugars, amino acids, and nucleotides, are combined in new ways to make the specific carbohydrates, lipids, proteins, and nucleic acids that an organism needs for its activities. This process of putting smaller molecules together to make larger molecules is called *synthesis*. Photosynthesis is a great example of synthesis. As the name implies, it is a process that uses light to synthesize a product—organic molecules. From the raw materials obtained from its environment, an organism can synthesize many of the substances that are necessary for life.

Metabolism

All of the chemical reactions of an organism

S16

Excretion

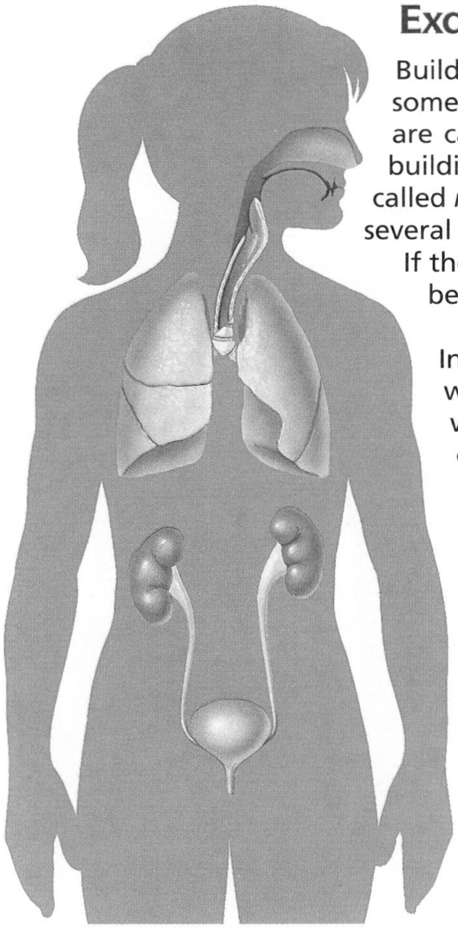

Building things up and breaking things down usually produces some materials that are unusable. These unusable materials are called *wastes*. Wastes that are produced by the chemical building-up and breaking-down processes of metabolism are called *metabolic wastes*. They include carbon dioxide, water, several nitrogen-containing compounds, and inorganic salts. If these wastes build up in a cell or an organism, they may become harmful, or toxic.

The removal of metabolic wastes is called **excretion**. Individual cells and unicellular organisms eliminate these wastes primarily by diffusion. In large multicellular animals, wastes are usually gathered from the area around individual cells by the fluids of the *circulatory system*. These wastes are then filtered out of the fluid for removal from the body. In humans, organs such as the lungs, skin, and kidneys play an important role in excretion. In plants, the primary metabolic waste product is oxygen, which diffuses out of the leaves through the stomata.

Excretion

The removal of metabolic wastes

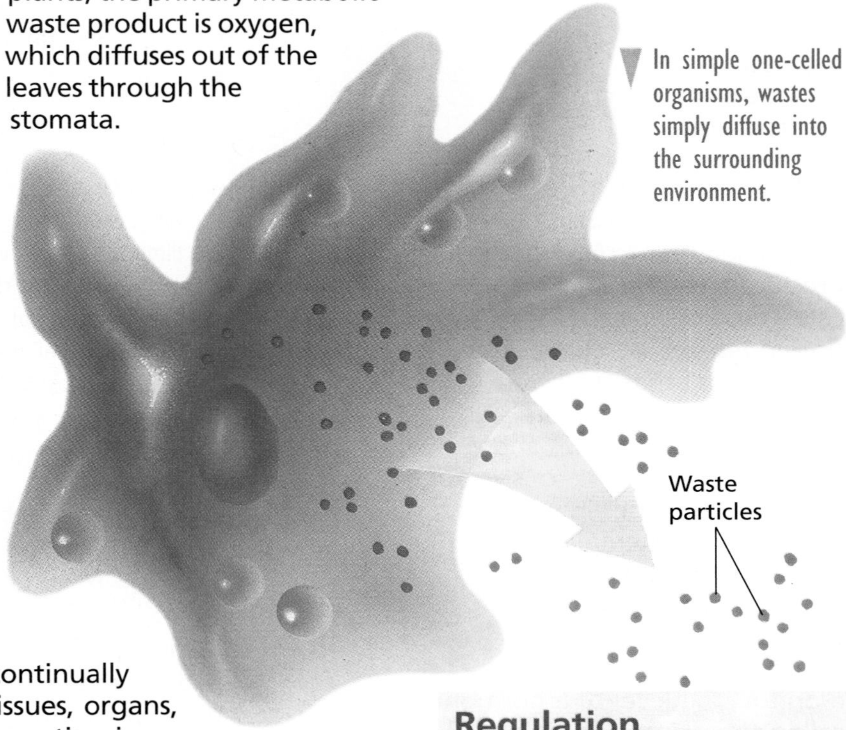

In simple one-celled organisms, wastes simply diffuse into the surrounding environment.

In complex animals, such as humans, special organs rid the body of metabolic wastes. Lungs remove carbon dioxide and water from the body. Skin excretes water, nitrogen-containing compounds, and salts in the form of perspiration. Kidneys remove nitrogen-containing compounds, salts, and water, which are excreted in the form of urine.

Waste particles

Regulation

To remain alive, all organisms must continually monitor the activities of their cells, tissues, organs, and organ systems and keep them operating in a coordinated manner. This life process is called **regulation**. In order to regulate its activities, the individual parts of an organism must be able to send and receive messages. Plants, for example, regulate their activities by producing chemical substances called *growth regulators*. These chemicals initiate such changes as cell specialization, growth, tropisms (responses to light, gravity, or touch), and flower formation. Animals, on the other hand, have two highly developed systems that work individually and together to regulate the activities of their bodies.

Regulation

The coordination of the internal activities of an organism

S17

One of these systems, the *nervous system,* acts very quickly, controlling adjustments that must be made immediately. For example, fleeing a predator is a response that requires quick action. The messages of the nervous system are passed along by transporting ions into and out of *nerve cells.* Such *electrochemical messages* are transmitted from nerve cell to nerve cell at a speed of 100 m/s.

The other regulatory system in animals is called the *endocrine system.* The endocrine system acts more slowly than the nervous system. It co-ordinates body activities and controls gradual changes such as growth and development. The organs of the endocrine system, called *endocrine glands,* produce a variety of chemicals. These chemicals, called *hormones,* act as "chemical messengers" that travel through the bloodstream to affect certain tissues and organs.

There are cases in which the nervous system and the endocrine system work together to regulate activities in the body. For example, the nervous system monitors the level of a hormone called *thyroxine* in the bloodstream. Thyroxine, which is produced by the thyroid gland, increases the rate of cellular respiration in cells. The thyroid gland is signaled to produce more thyroxine when the body's cells are producing too little ATP and to produce less thyroxine when ATP levels rise too high.

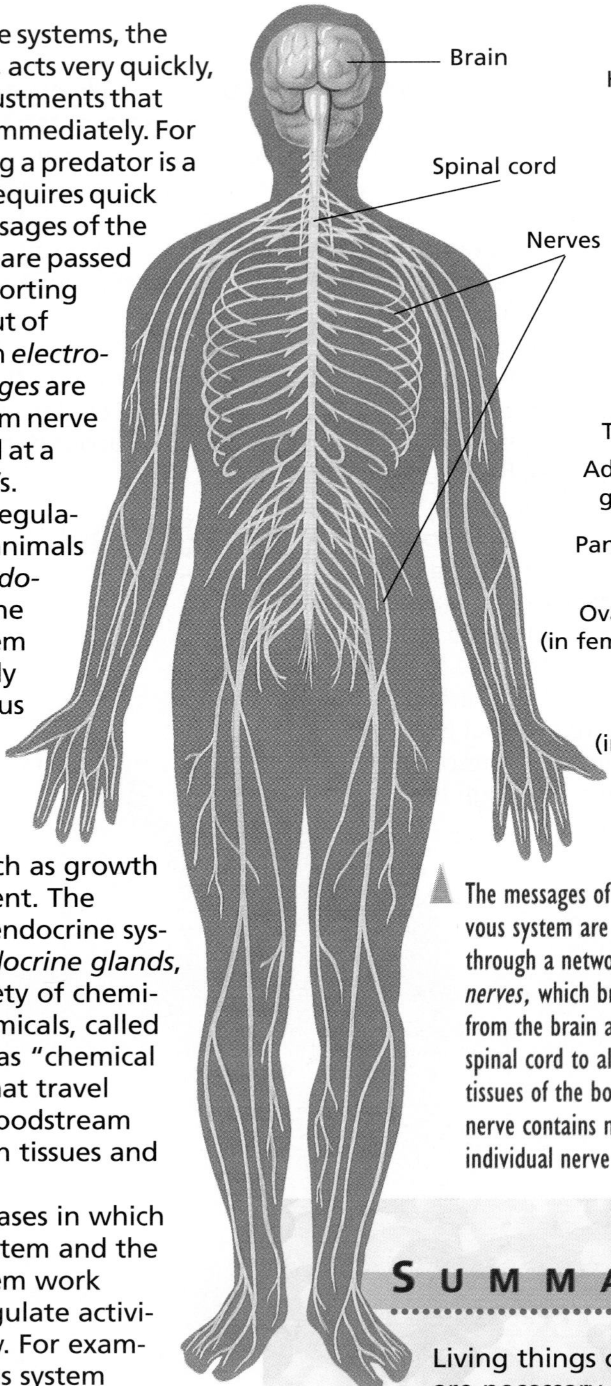

S18

The messages of the nervous system are sent through a network of *nerves,* which branch from the brain and spinal cord to all the tissues of the body. Each nerve contains many individual nerve cells.

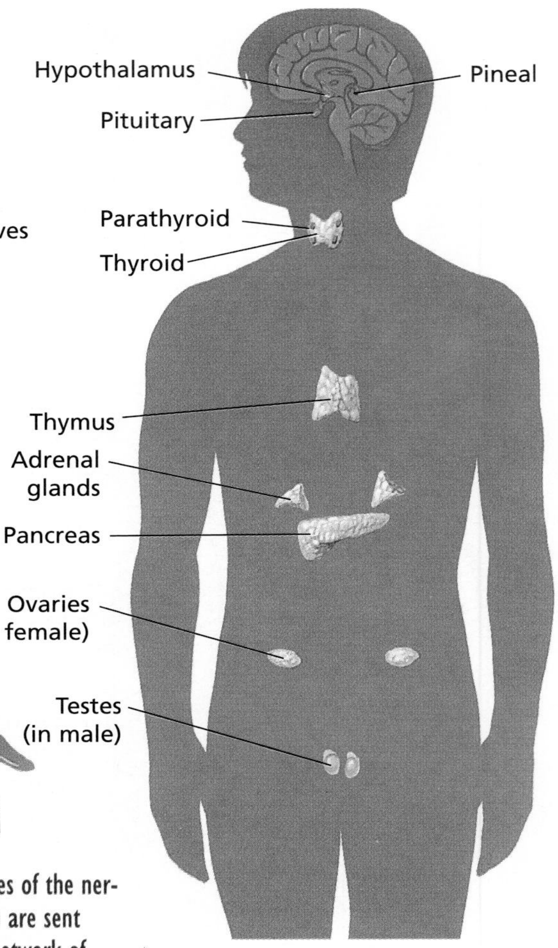

Endocrine glands release chemicals called hormones. Some hormones are proteins, and others are lipids.

SUMMARY

Living things carry out certain processes that are necessary for maintaining life. First, organisms must take in raw materials. The materials are then transported to the cells, where they pass through membranes by diffusion or active transport. These materials are processed by the various chemical reactions of metabolism to produce the energy and the many different molecules that organisms need to live. Then, by the process of excretion, metabolic wastes and excess water are removed from cells and organisms. The various functions of an organism are regulated by slow-acting chemical messengers and (in animals) a quick-acting nervous system.

Labels on the nervous system diagram: Brain, Spinal cord, Nerves

Labels on the endocrine system diagram: Hypothalamus, Pineal, Pituitary, Parathyroid, Thyroid, Thymus, Adrenal glands, Pancreas, Ovaries (in female), Testes (in male)

Concept Mapping

The concept map shown here illustrates major ideas in this unit. Complete the map by supplying the missing terms. Then extend your map by answering the additional question below. Write your answers in your ScienceLog. **Do not write in this textbook.**

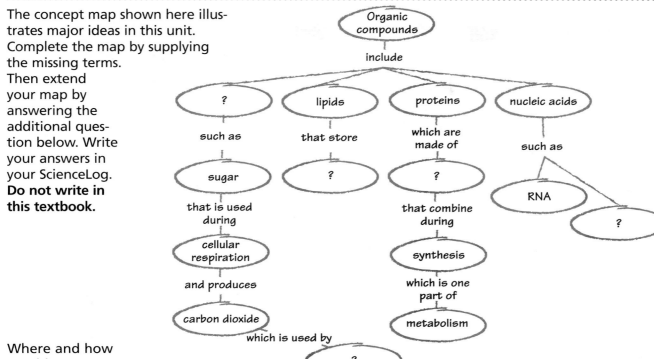

Where and how would you connect the terms *carbon*, *chloroplasts,* and *ATP*?

Checking Your Understanding

Select the choice that most completely and correctly answers each of the following questions.

1. Which biochemical is most readily used as an energy source by cells?
   a. protein
   b. amino acid
   c. carbohydrate
   d. glucose

2. Which is NOT an organic compound?
   a. ATP
   b. DNA
   c. H_2O
   d. $C_6H_{12}O_6$

3. Water is important to living things because
   a. it participates in hydrolysis reactions.
   b. it contains carbon atoms.
   c. it is an uncharged molecule.
   d. it is made of amino acids.

4. Energy from the sun reaches humans by way of
   a. starch.
   b. glucose.
   c. ATP.
   d. All of the above

5. Which is NOT true of metabolic wastes?
   a. They are excreted.
   b. They are produced only by animals.
   c. They may result from cellular respiration.
   d. They may be used by other organisms.

S19

Interpreting Graphs

A cell is placed in a salt solution. Which graph (*A* or *B*) indicates that the solution is hypertonic? Explain your answer.

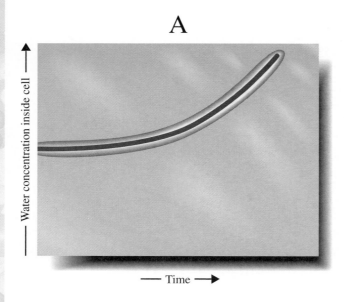

A

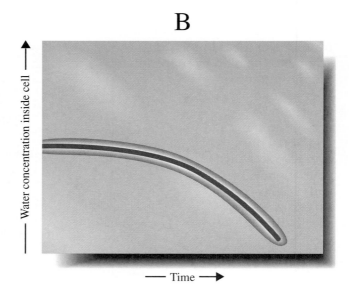

B

Critical Thinking

Carefully consider the following questions, and write a response in your ScienceLog that indicates your understanding of science.

1. While doing research, an organic chemist determines that a certain polymer contains carbon, hydrogen, oxygen, nitrogen, and phosphorus. The chemist then hydrolyzes a fresh sample of the compound. Name the monomers that result from the hydrolysis.

2. Let's say that a scientist is able to make water that has cohesive, but not adhesive, properties. Why would using this special water to water the plants in Biosphere II eventually have an effect on *all* of the life in this facility?

3. One theory suggests that the extinction of the dinosaurs resulted from a giant meteorite that crashed into the Earth about 65 million years ago. Scientists think that the tremendous impact could have thrown a giant cloud of dust into the atmosphere, preventing sunlight from reaching Earth's surface for many months. Explain how this could have caused the extinction of the dinosaurs.

4. Explain the following statement: A cell low in ATP is probably starving for raw materials.

5. Does your body's cellular respiration produce metabolic wastes at the same rate during the day as it does at night when you are sleeping? Explain your answer.

Portfolio Idea

Imagine that you are a reporter who is interviewing a water molecule and a glucose molecule. During the interview, each molecule boasts that *it* is the most important molecule on Earth. Use all the information that you gain from your interview to write a story that supports one of the molecules as being the most important. Be creative, but use facts and logic in reaching your conclusion.

S20

Particles

Unit 2

IN THIS UNIT

Particles of Matter,
page S22

Particles in Motion,
page S27

Particles of Particles,
page S35

Now that you have been introduced to the concept of particles, consider the following questions.

● **1.** What is a *mole*? How is it used in measuring matter?

2. How can the kinetic molecular theory of matter be used to explain certain physical properties of matter?

3. Are there particles of matter that are smaller than atoms? What are these particles called?

4. In what way does the particle model of matter fail? How could you account for this failure?

In this unit, you will take a closer look at particles of matter and theories about the nature of particles.

S21

PARTICLES OF MATTER

Scientists often describe matter as consisting of particles. But just what do they mean by the term *particle*? A **particle** is any small unit that is part of a larger whole. As you know, an atom is the smallest particle of an element, and a proton is a particle within an atomic nucleus. There are even particles inside protons called *quarks*. When atoms interact, other types of particles, such as molecules and ions, are formed. Light consists of particles, or packets of energy, called *photons*. And according to theory, even the force of gravity may consist of particles called *gravitons*.

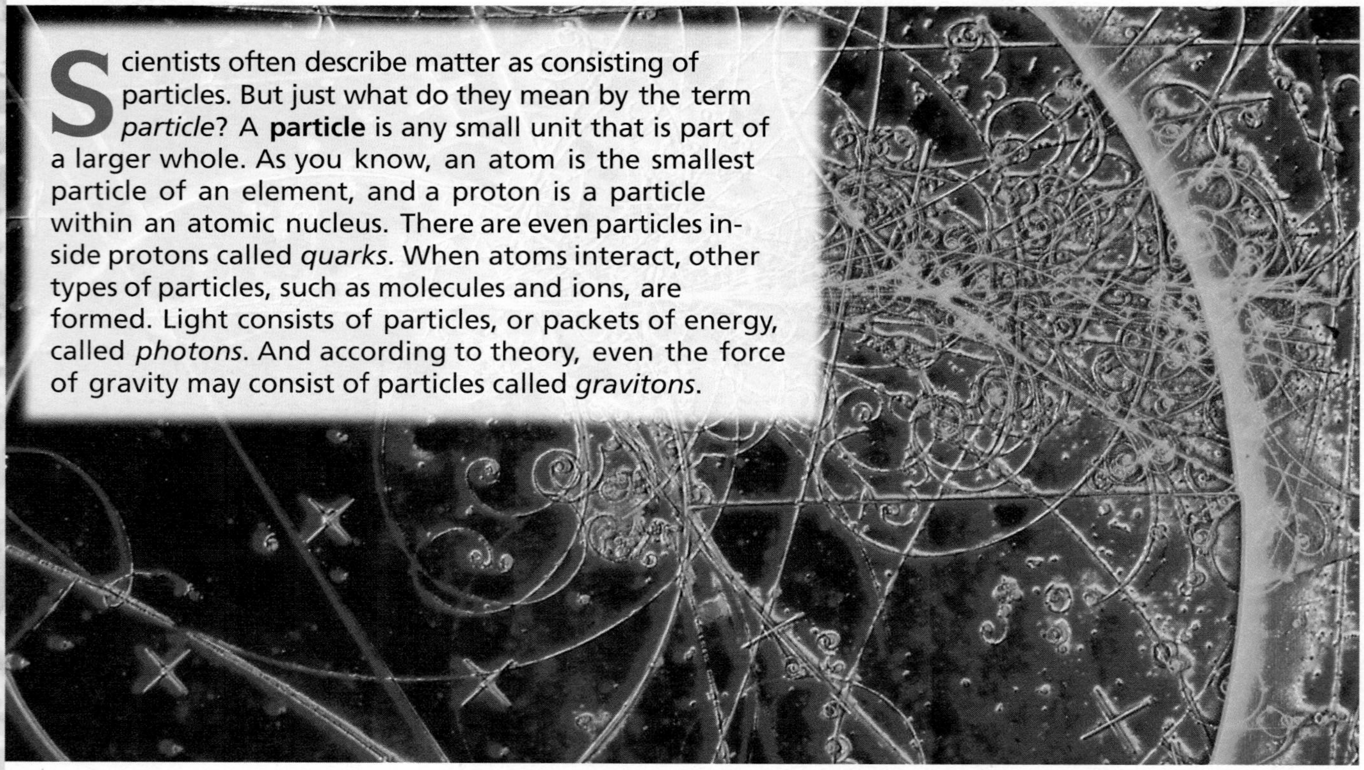

▲ This artificially colored photograph shows the tracks of many subatomic particles in a bubble chamber. These particles appear when atoms collide and break apart.

Particle

Any small unit (of matter or energy) that, with others, forms a larger whole

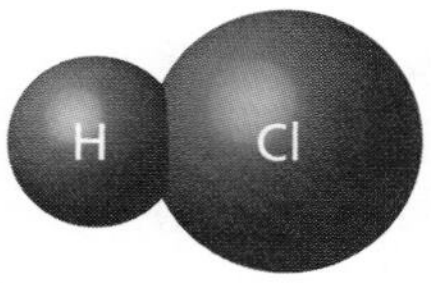

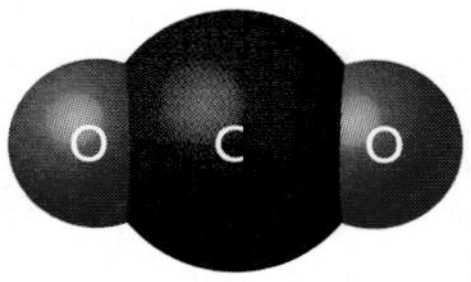

▲ These models represent molecules of different chemical compounds.

Molecules and Ions

You frequently hear the word *molecule* used to describe any particle composed of two or more atoms. We often say that all compounds—substances made from two or more elements—are made of molecules. This, however, is not quite accurate. In the scientific sense, the term *molecule* refers only to neutral groups of atoms that are joined by a particular type of chemical bond.

You may recall that there are two distinct types of chemical bonds—ionic and covalent. An ionic bond forms between atoms that have opposite electrical charges. A covalent bond forms when two or more atoms share electrons. The particles we call *molecules* consist of atoms that are joined by covalent bonds and have no net electric charge. For example, water consists of water molecules in which two hydrogen atoms share electrons with one oxygen atom. Thus, the chemical formula for a molecule of water is H_2O.

On the other hand, a substance such as sodium chloride (table salt) does not consist of molecules. Rather, it contains particles called *ions*, which form when electrons move from one atom to another. Atoms that lose electrons take on a positive charge, and those that gain electrons take on a negative charge. Two ions with opposite charges attract each other, forming an ionic bond.

S22

By magnifying grains of table salt, you can see that salt crystals are shaped like cubes.

In a crystal of table salt, opposing electric charges hold positive sodium ions (Na^+) and negative chloride ions (Cl^-) together. Chloride ions surround each sodium ion, and sodium ions surround each chloride ion, as shown in the illustration to the right. For every Na^+ ion in a salt crystal there is one Cl^- ion. However, since the ions in a salt crystal do not *share* electrons, their groupings are not considered molecules. The designation NaCl is simply a way of writing formulas and equations in which sodium chloride takes part. Therefore, the symbol NaCl is called a *formula unit.*

Can you see how useful the term *particle* might be? You can talk about particles of sugar (sugar molecules) and particles of salt (Na^+ and Cl^- ions), as well as the properties they share, without having to consider the details. This comes in handy in a number of areas of study.

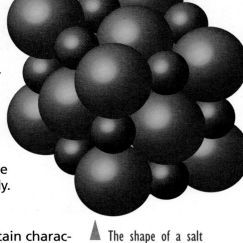

The shape of a salt crystal is determined by the arrangement of its ions.

Counting Particles

All types of particles, no matter what they are, share certain characteristics. One of these is the ability to be counted. Particles can be counted either individually or in groups. But since particles of matter are so small, we need a special way to count them.

The Mole   When you go to the store to buy eggs, you usually buy them by the dozen. When you buy a dozen eggs, you know that you are getting 12 eggs. The word *dozen* is a counting term for "groups of 12." Another counting term you may be familiar with is the *gross,* which is 12 dozen. When you buy a gross of oranges you get 144 oranges. It does not matter whether you buy eggs or oranges, because the number of items in a dozen or a gross is always the same. If you buy a dozen, you get 12. If you buy a gross, you get 144.

In science, a very useful counting term for particles of matter is called the *mole* (mol). No, we are not talking about little furry animals! A **mole** is a unit for measuring the *amount of a substance.* Just as there are always 12 items in a dozen—no matter what you are counting—a mole of any substance always contains the same number of particles.

Mole

The SI unit for the amount of a substance

S23

Avogadro's number

The number of particles in 1 mol of a substance (6.022137 x 10²³)

Avogadro's Number Because the basic particles of matter are so small, the number of particles in a mole is very large. But just how many particles are there in a mole? It has been determined that one mole of any substance contains 6.022137×10^{23} particles. This number is called **Avogadro's number** in honor of the Italian scientist Amedeo Avogadro, who first suggested the idea of molecules. Written out, Avogadro's number is 602,213,700,000,000,000,000,000.

Imagine getting the help of all 6 billion people on Earth to count the number of particles in only 1 mol of a substance. If each person counted 1 particle per second, it would take over 3 million years to count all the particles in a mole! To put it another way, if you made $40,000 every *second* at your job and you had been working since Earth formed 4.5 billion years ago, you would not yet have earned Avogadro's number of pennies.

Using the Mole Using the mole is like using any other standard quantity, such as a dozen or a gross. Suppose you were in charge of preparing breakfast for a group of 36 students and you wanted to serve breakfasts with 2 eggs and 1 sausage patty on each plate. You could order, say, 6 dozen eggs and 3 dozen sausage patties. This way, you would have just enough eggs and sausage patties so that none would be left over after 36 breakfast plates were filled.

In a similar way, chemists usually determine the amounts of the chemicals they will need for a particular chemical reaction in terms of moles. For example, to make water molecules, you need twice as many hydrogen atoms as oxygen atoms. Therefore, you could use 2 mol of hydrogen atoms and 1 mol of oxygen atoms. Just as you were sure you could serve 36 identical breakfast plates by buying eggs and sausage patties in dozens, chemists ensure that they get the correct ratio of atoms by using moles. But how can moles be translated into units you are already familiar with? To understand the answer to this question, you must look at the periodic table.

◀ If you could stack 6.02×10^{23} pennies on top of one another, the stack would reach to the other side of the Milky Way galaxy.

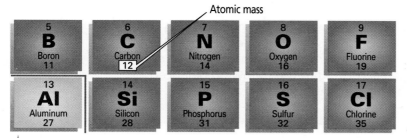

Atomic mass

| 5
B
Boron
11 | 6
C
Carbon
[12] | 7
N
Nitrogen
14 | 8
O
Oxygen
16 | 9
F
Fluorine
19 |
| 13
Al
Aluminum
27 | 14
Si
Silicon
28 | 15
P
Phosphorus
31 | 16
S
Sulfur
32 | 17
Cl
Chlorine
35 |

▲ By finding the atomic mass of an element on the periodic table, you can determine the number of grams in 1 mol of that element.

Moles to Grams
If you look at the periodic table, you will find that hydrogen (H) has an atomic mass of 1. This value tells you the number of grams in 1 mol of that element. In other words, there is 1 g of hydrogen in 1 mol of hydrogen atoms. Likewise, the atomic mass of oxygen (O) is 16. Thus, 1 mol of oxygen has a mass of 16 g. As you might expect, 2 mol of hydrogen atoms would be twice as much as 1 mol, or 2 g of hydrogen. *To find the number of grams in 1 mol of any particular element, just look up its atomic mass and use it as the number of grams.*

Let's again use eggs to illustrate how moles relate to grams. Suppose that you have three sizes of eggs—small (30 g), medium (40 g), and large (50 g). If you have a dozen of the small eggs, you would have a total of 360 g of eggs. A dozen medium eggs would be 480 g, and a dozen large eggs would be 600 g. You could order 360 g of small eggs or 600 g of large eggs and expect to get 1 dozen in either case.

In the same way, scientists use grams to determine the number of moles that they want to use. If you know the atomic mass of the atoms of an element, you can determine how many grams of that element you need to get 1 mol (or 6.02×10^{23} atoms), as the photos on this page illustrate.

Moles of Compounds
Atoms of elements are *not* the only particles that you can count with moles, or Avogadro's number. You can also have a mole of ions or a mole of molecules. Therefore, you can measure amounts of *compounds* in moles. Recall that a water molecule (H_2O) has two atoms of hydrogen and one atom of oxygen. Therefore, 1 mol of water contains 2 mol of hydrogen and 1 mol of oxygen. Still, there are only 6.02×10^{23} *molecules* in 1 mol of water.

To determine the number of grams in 1 mol of water, you would first find the atomic masses of the hydrogen and oxygen atoms. Then you would add the atomic masses of each of the atoms in one molecule. For a water molecule (H_2O), your calculation would be as follows: $1 + 1 + 16 = 18$. Therefore, by applying the same rule that you used for moles of elements, you would need 18 g of water to get 1 mol of water molecules.

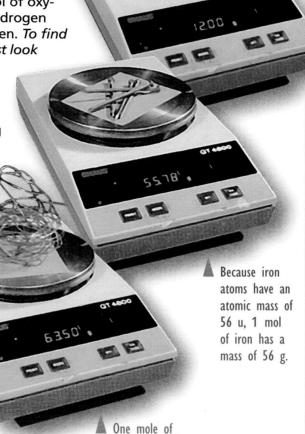

▼ Carbon atoms have an atomic mass of 12 u. Therefore, 1 mol of carbon has a mass of 12 g.

▲ Because iron atoms have an atomic mass of 56 u, 1 mol of iron has a mass of 56 g.

▲ One mole of copper, with an atomic mass of 64 u, has a mass of 64 g.

S25

The Mole in Chemistry Knowing the mass of a mole of any substance is very important to chemists. Again, suppose that you wanted to combine hydrogen with oxygen to get water. In nature, hydrogen gas occurs as H_2 molecules, and oxygen gas occurs as O_2 molecules. Therefore, to make water, you must combine hydrogen and oxygen according to the following balanced chemical equation:

$$2H_2 + O_2 \rightarrow 2H_2O$$

The equation states that two *molecules* of H_2 react with each *molecule* of O_2. That is, you will need twice as many H_2 molecules as O_2 molecules for your reaction. This is where moles come in handy. If you use twice as many moles of H_2 as you do of O_2, you will have the right proportions of the two chemicals to make water. Suppose that you use 10 mol of H_2 and 5 mol of O_2; how many grams of each would you need?

▲ Two molecules (or moles) of H_2 plus one molecule (or mole) of O_2 yields two molecules (or moles) of water.

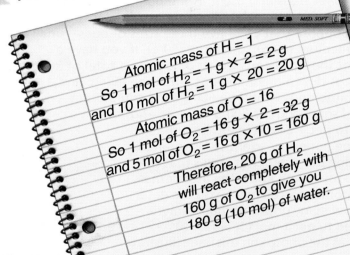

Atomic mass of H = 1
So 1 mol of H_2 = 1 g × 2 = 2 g
and 10 mol of H_2 = 1 g × 20 = 20 g

Atomic mass of O = 16
So 1 mol of O_2 = 16 g × 2 = 32 g
and 5 mol of O_2 = 16 g × 10 = 160 g

Therefore, 20 g of H_2 will react completely with 160 g of O_2 to give you 180 g (10 mol) of water.

SUMMARY

Scientists often use the term *particle* when discussing matter and energy. Molecules are particles made up of atoms held together by covalent bonds. Ions are particles formed when atoms gain or lose electrons. The simplest grouping of ions in a compound formed by ionic bonds is called a formula unit. *Mole* is a term used for counting the basic particles of matter, whether they be atoms, molecules, or ions. Avogadro's number is the number of particles in a mole of a substance. A mole of any one substance has exactly the same number of particles as a mole of any other substance. The mass of a mole of any substance is the atomic mass of its particles given in grams.

PARTICLES IN MOTION

Joseph Black, the Scottish chemist who in the 1750s first identified carbon dioxide gas, devoted much of his life to investigating the effects of heat on matter. Among the things he studied were the freezing, melting, and boiling points of different substances. In one experiment, Black observed how the temperature of water changes as it is heated to boiling. Of this experiment he wrote:

The liquid gradually warms, and at last attains the temperature, which it cannot pass without assuming the form of vapor . . . However long and violently we boil a liquid, we cannot make it hotter than when it began to boil. The thermometer always points to the same degree, namely, the vapor point of that liquid. Hence the vapor point of a liquid is often called its boiling point.

Why is it impossible, under ordinary conditions, to raise the temperature of boiling water above 100°C? How do particles of liquid water differ from particles of water vapor? What holds particles of boiling water together? Forces of attraction between particles of matter and the *kinetic molecular theory of matter* can be used to answer these and other questions about the behavior of matter.

Forces of Attraction

Have you ever wondered what holds a piece of paper, a steel beam, or a raindrop together? As you know, many different forces draw particles of matter together. Electrons stay in orbit about an atom's nucleus because the negatively charged electrons are attracted by the positively charged nucleus. Water from a faucet falls toward the Earth because of the pull of gravity. Yet when you pour water into a clean glass, some of it creeps up the inside of the glass, *against* the pull of gravity. Two forces of attraction—cohesion and adhesion—are responsible for the behavior of the water in a glass.

If you look closely, you will ► see that the surface of the water in a clean glass graduated cylinder is not level.

1 Because water tends to adhere (stick) to glass, it moves up the side of the cylinder.

2 The surface of the water forms a curve called a *meniscus*.

3 Because water molecules cohere (stick to each other), the water's surface remains unbroken.

Cohesion

The force of attraction between like particles

Adhesion

The force of attraction between unlike particles

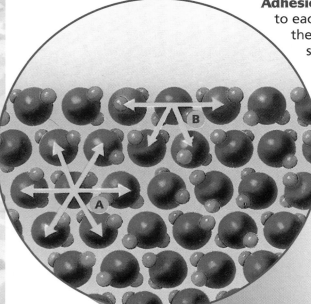

1 A molecule beneath the surface of the water (position **A**) is equally attracted to other water molecules on all sides.

2 A molecule on the surface of the water (position **B**) is attracted to other water molecules beneath it as well as next to it, but there are no water molecules to attract it upward.

3 Thus, the molecules at the surface are pulled closer together, causing them to act like a thin elastic film.

DID YOU KNOW...

that surface tension is what makes raindrops round?
In a drop of water, all surface particles are pulled toward the center of the drop, giving it a spherical shape.

S28

Cohesion Particles of matter that are alike, such as water molecules, tend to stick together. This force of attraction between like particles is called **cohesion.** In solids, cohesion may sometimes be very strong. Thus, cohesion holds solids, such as sheets of paper and steel beams, in a definite shape. Cohesion also holds together the particles of liquids. However, the cohesion between the particles of liquids is not usually as strong as the cohesion between the particles of solids. Imagine trying to hold up a sheet of water in the same way you hold up a sheet of paper. Because there is less cohesion between water molecules than there is between paper particles, a sheet of water would not hold together.

Adhesion Particles of different kinds of matter may also stick to each other. For example, when water touches clean glass, the water and glass particles attract one another much more stongly than water particles attract each other. The force of attraction between the particles of two different substances is called **adhesion.** There is a great deal of adhesion between glues and the substances to which they stick.

Surface Tension Have you ever watched water dripping from a leaky faucet? It does not drip molecule by molecule. Rather, the water molecules stay together until a large drop forms. As the drop falls, it becomes a sphere. But without a container, what keeps drops of liquid water in this shape? The spherical shape of a falling water drop results from a property of liquids called *surface tension.* Surface tension is the tendency of the particles at the surface of a liquid to pull together. As the diagram to the left shows, *surface tension* results from cohesion between a liquid's molecules. Because of surface tension, some insects are able to walk across the surface of a pond without sinking into the water.

▼ Surface tension allows this water strider to walk on water.

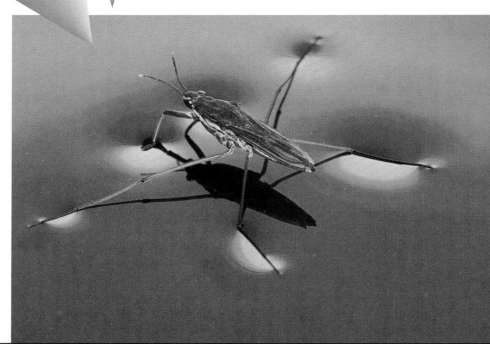

The Kinetic Molecular Theory of Matter

One of the most important theories of modern science is the *kinetic molecular theory of matter*, which states that *matter is made of very tiny particles that are in constant motion*. The particles that make up matter are, of course, atoms, molecules and ions. Because they are constantly moving, these particles have kinetic energy—the energy of motion. The faster the particles move, the more kinetic energy they have.

States of Matter   The kinetic molecular theory can be used to explain the difference between the three states of matter—solids, liquids, and gases. In a solid, the atoms or molecules do not have enough kinetic energy to overcome the cohesion that holds them together. Therefore, the particles of a solid cannot move about freely; they simply vibrate back and forth about fixed positions. For this reason, solids have a definite shape and a definite volume.

The atoms or molecules in a liquid have more kinetic energy than they do when in a solid state. Although the particles of a liquid still cling together because of cohesion, they are free to move around or slide past each other. The motion of these particles resembles the way that marbles in a plastic bag would roll around if you squeezed the bag. Because the particles of a liquid are held together by cohesion, liquids have a definite volume. But because the particles of a liquid can change position, liquids have no definite shape. Liquid water, for example, can be poured from a tall, narrow glass into a short, wide glass, but it will occupy the same volume in both glasses.

Gas particles have enough kinetic energy to overcome the forces of cohesion altogether. The atoms or molecules of gases fly off in straight lines until they collide with other particles or the walls of a container. For this reason, gases expand until they fill the container they occupy. Thus, they do not have definite shapes or volumes.

The atoms or molecules in a solid vibrate in fixed positions. In many solids, the particles have a very orderly arrangement that gives the solid a crystal shape.

Particles in a gas have a lot of kinetic energy and little cohesion, and they can move freely from place to place. On average, the distance between the particles in a gas is very great compared with the size of the particles.

Liquid molecules can move around but are always in contact with other molecules. The shape of the container in which a liquid is held determines the shape of the liquid.

S29

Diffusion

The mixing of the particles of one substance with the particles of another substance because of the motions of their particles

Diffusion   We can see that atoms and molecules are in motion when we observe **diffusion**, which is the mixing of the particles of two or more substances as a result of their motion. You have already learned that diffusion is an important way that living systems distribute materials. Although you cannot see diffusion happening inside your body, you experience it every time you smell something. Because gas particles travel much farther between collisions than do particles of solids or liquids, diffusion is more rapid in gases. For example, if you open a bottle of strong perfume inside a room, the perfume molecules will quickly diffuse through the air. Air molecules help to distribute the perfume molecules, and you can soon smell the perfume all over the room.

Another way to observe diffusion is to place a drop of ink into a glass of water. The ink particles move among the water molecules until they are equally distributed throughout the glass. As you might predict by using the kinetic molecular theory, the temperature of the water influences the rate of diffusion. Because particles move faster at higher temperatures, diffusion is more rapid at higher temperatures.

Diffusion even occurs between the particles of solids. However, because the particles of solids are not nearly as free to move around, diffusion occurs very slowly in solids. For example, if you were to stack a lead brick on top of a gold brick, a very slow diffusion would occur between these two solids. After several months, particles of one solid could be detected in the other.

▲ Over time, ink gradually mixes with water to form a uniform mixture. This is called *diffusion*.

Changes of State We can also use the kinetic molecular theory of matter to explain how matter changes states. Recall Joseph Black's discovery that the temperature of a liquid at its boiling point remains constant even when it is heated further. As you know, temperature is a measure of the average amount of kinetic energy of the particles in a substance. Thus, when the temperature of a substance increases, it gains kinetic energy and its particles move faster. However, the molecules in a boiling liquid *cannot* move faster unless they break free of the cohesion that holds them together. Therefore, the temperature of boiling water stays the same even when you add more heat. Additional heat provides the energy needed to overcome the cohesion of water molecules. These molecules escape the liquid as a gas. Until all molecules have escaped, the temperature of the water stays at its boiling point.

Likewise, adding heat to a solid increases the kinetic energy of its particles. When the particles acquire enough energy to move around each other freely, the solid melts. Once at its melting point, however, the temperature will remain constant until all the solid has melted. After melting, the particles of the solid can move even faster when more heat is added. As a result, the kinetic energy of the particles and the temperature of the substance can increase.

What happens when gases and liquids cool? As heat leaves a substance, its particles lose kinetic energy and slow down. Once the particles lose enough kinetic energy, cohesion begins to draw the particles together. When the particles of a gas cohere (stick together), they get much closer together, and we say that the gas *condenses*. Heat is given off in the process. When the particles of a liquid lose their ability to move past each other freely, we say that the liquid freezes.

Changing States of Water

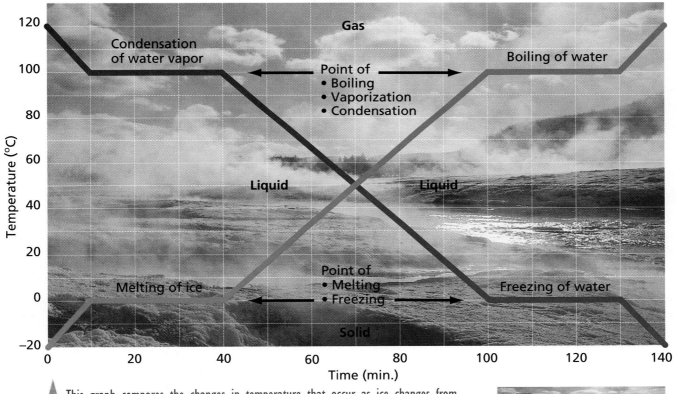

This graph compares the changes in temperature that occur as ice changes from a solid to a liquid and then to a gas, and as water vapor changes from a gas to a liquid to a solid. Notice that the boiling, vaporization, and condensation points of a substance are the same, as are the melting and freezing points.

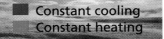

S31

Applications of the Kinetic Molecular Theory of Matter

Conditions that affect the way particles move, such as temperature and pressure, may change the volume occupied by matter. Because the particles in solids and liquids are already quite close together, the volumes of solids and liquids change very little as temperature changes. In most cases, pressure has even less effect on the volume of solids and liquids. However, any change in temperature or pressure greatly affects the volume of a gas. The particles of a gas will readily spread out or move together to fit the conditions. For example, if you leave a balloon full of air in bright sunlight, the balloon expands as the air inside warms up. If you push down on the handle of a bicycle pump, the volume of the air inside the pump decreases and the pressure increases. Several scientific laws describe the effect of temperature and pressure on matter. Let's see how the kinetic molecular theory of matter can be used to explain these laws.

Boyle's Law   Robert Boyle was an Irish scientist who lived during the seventeenth century, around the same time as Isaac Newton. He studied how gases behave when the pressure on them changes. Robert Boyle experimented with several gases, such as oxygen and carbon dioxide. In each case, he found that *doubling* the pressure always reduced the volume of a gas by *one-half*, if its temperature remained constant. His observations led to what is now called **Boyle's law**, which can be stated as follows: *The original volume (V_1) occupied by a certain amount of gas multiplied by its pressure (p_1) is equal to its new volume (V_2) multiplied by its new pressure (p_2)*, or

$$V_1 \times p_1 = V_2 \times p_2$$

For example, suppose that the air trapped by the piston seen on this page has a volume of 200 mL (V_1) when the pressure is equal to 1 atm (p_1). If the total pressure increases to 2 atm (p_2), its new volume (V_2) will be 100 mL.

$$200 \text{ mL} \times 1 \text{ atm} = V_2 \times 2 \text{ atm}$$

$$V_2 = \frac{200 \text{ mL} \cdot 1 \text{ atm}}{2 \text{ atm}}$$

$$V_2 = 100 \text{ mL}$$

1 atm = *atmospheric pressure at sea level* (101.3 kPa)

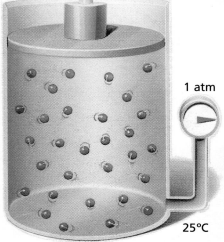

1 atm

25°C

A gauge measures the pressure exerted on a gas by a piston. As pressure increases, volume decreases in proportion to the increase in pressure.

2 atm

25°C

3 atm

25°C

S32

The kinetic molecular theory of matter can be used to explain Boyle's law in the following way: As the force that causes the pressure pushes down on the air inside the cylinder, the air inside the cylinder pushes upward against the piston with an equal force. Otherwise, the piston would go all the way down to the bottom of the cylinder. The pressure exerted on the bottom of the piston to hold it up is a result of the constant bombardment by the trapped air particles inside. Each time the volume occupied by the air inside the cylinder is reduced by *half*, the number of particles per unit of volume *doubles*. Thus, twice as many particles strike the surface of the piston per second and therefore exert twice the force.

Pascal's Law   Blaise Pascal was a French mathematician, scientist, and philosopher who also lived during the seventeenth century. Among other things, he is known for his work on the effect of pressure on liquids. Pascal observed that pressure applied to a liquid in a closed container is felt throughout the container. This observation is described in *Pascal's law*: *Whenever pressure is increased at any point, an equal change in pressure occurs throughout the liquid.*

Hydraulic systems, such as the hydraulic lift shown below, work on the principle of Pascal's law. A hydraulic lift has an *input piston*, which is used to apply pressure to a liquid in a closed container, and an *output piston*, on which the pressure can act. Because the output piston has a larger surface area than the input piston, a small force exerted on the input piston creates a large force on the output piston. The brake system of an automobile uses the same principle. A small amount of pressure applied to the brake pedal is transmitted by the brake fluid to the brake pads, where the force is multiplied many times—enough to bring the car to a stop.

Pascal's law can also be explained by the kinetic molecular theory of matter. Liquids, like gases, are fluids. A *fluid* is any substance that can flow and change shape. Fluids have these characteristics because their particles can move about when pressure is applied to them. Unlike the particles of gases, however, the particles of liquids cannot be compressed (pushed closer together) because they are already close together. Thus, when pressure is applied to a liquid in a closed container, it cannot be compressed and it cannot change shape. As a result, the pressure applied at one point causes the liquid to push on all sides of its container.

$\frac{100 \text{ N}}{\text{cm}^2}$

$\frac{100 \text{ N}}{\text{cm}^2}$

According to Pascal's law, pressure applied to the cork in a bottle full of liquid is dispersed equally throughout the liquid in all directions.

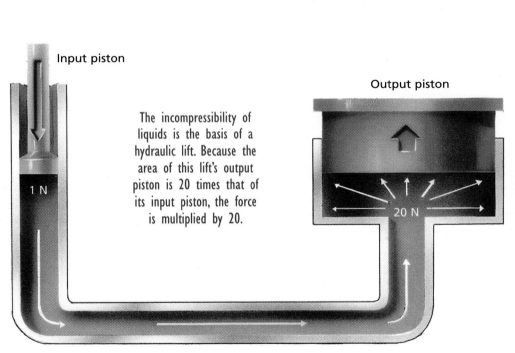

Input piston

Output piston

The incompressibility of liquids is the basis of a hydraulic lift. Because the area of this lift's output piston is 20 times that of its input piston, the force is multiplied by 20.

1 N

20 N

S33

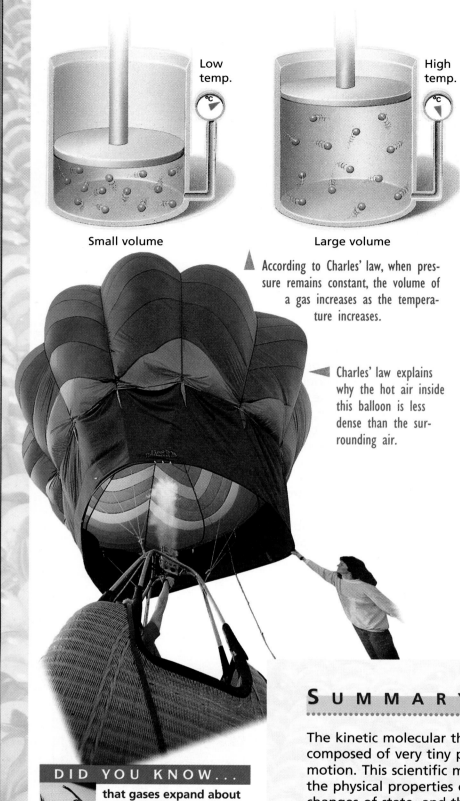

Low temp. °C

High temp. °C

Small volume

Large volume

▲ According to Charles' law, when pressure remains constant, the volume of a gas increases as the temperature increases.

◀ Charles' law explains why the hot air inside this balloon is less dense than the surrounding air.

Charles' Law In the late eighteenth century, the French scientist Jacques Charles observed that the volume of air increases steadily as its temperature increases. This observation is described by **Charles' law**: *The volume of a gas increases as the temperature increases, if the pressure remains the same.*

When using the Kelvin temperature scale, this relationship can be expressed as a direct proportion, which is written as

$$\frac{V_1}{T_1} = \frac{V_2}{T_2}$$

In other words, if you double the Kelvin temperature of a gas, its volume doubles. As is the case when you use Boyle's law, if you know three values, you can calculate the fourth.

How does the kinetic molecular theory apply to Charles' law? If the temperature of a gas increases, the gas particles move faster and collide with the walls of their container more often. Because of this, they exert more force. If pressure is to remain constant, the particles must travel greater distances, requiring the container to expand and the volume to increase.

S34

SUMMARY

The kinetic molecular theory states that matter is composed of very tiny particles that are in constant motion. This scientific model helps explain many of the physical properties of matter, as well as diffusion, changes of state, and the effects of changes in temperature and pressure on volume. Boyle's law describes how the volume of a gas changes as the pressure changes. Pascal's law describes the effects of applying pressure on a liquid in a closed container. Charles' law describes how the volume of a gas changes as the temperature changes.

PARTICLES OF PARTICLES

Each element consists of its own unique atoms—the smallest particles of matter that we usually deal with in the field of chemistry. But are atoms the smallest particles found in nature? The answer is no. Among the other particles that you have come across in your investigations of matter are protons, neutrons, and electrons. These particles are studied by physicists.

The current view of the universe considers *matter* to be the most basic entity (thing). This means that every substance that exists must be composed of fundamental particles of matter. Until the late nineteenth century, matter was thought to be formed of indivisible atoms. Then, due to the work of J. J. Thomson, Ernest Rutherford, and James Chadwick, new models of the atom—composed of protons, neutrons, and electrons— were proposed.

More recently, even smaller particles have been identified. Protons and neutrons, for example, are now thought to be made of three smaller particles called **quarks.** Although quarks have never been isolated, physicists have inferred that they exist from the interactions of atoms in huge machines called particle accelerators.

▲ Atoms consist of smaller pieces called subatomic particles.

Quark

A theoretical particle that is thought to make up protons and neutrons

▲ This model of a helium nucleus shows that protons and neutrons are made of three smaller particles called quarks.

Particle Interactions

Scientists have used their knowledge of subatomic particles and the forces involved in their interactions to explain the behavior of matter and the particles of which it is composed. According to present theories, four fundamental forces help to hold everything together.

◄ Physicists use this particle accelerator, at Fermilab in Batavia, Illinois, to explore the interior of atoms.

S35

Gravitational Force   The primary interaction that affects stars, planets, and most of the visible objects around you is *gravitational force.* Gravity pulls everything near the Earth toward the center of the Earth and keeps you from flying off into space. The strength of this pull is directly proportional to the masses of the objects that are attracted and is inversely proportional to the square of the distance between them—the greater the distance, the less the attractive force.

Electromagnetic Force   At the level of atoms and molecules, interactions that involve *electromagnetic force* have a greater influence than does gravity. Electromagnetic interactions occur between particles that are either electrically charged or magnetic. Remember that like electric charges repel each other, and opposite electric charges attract each other. Magnetic fields work in a similar way. Electrons are held in orbit about atomic nuclei because of the electrical attraction between negatively charged electrons and positively charged nuclei. Atoms stick together to form molecules for the same reason. The theory that explains electromagnetic force is called *quantum electrodynamics,* or QED.

Strong Force   The *strong force* is what holds an atomic nucleus together. As you know, atomic nuclei contain the small particles called protons and neutrons. Each of these particles is thought to be made of still smaller particles called quarks. The strong force holds the three quarks inside each proton and neutron together. The strong force also holds the nucleus itself together. This means that the strong force is stronger than the electromagnetic force that tends to repel protons, which have like electrical charges.

Weak Force   The fourth fundamental force is called the *weak force.* The weak force is responsible for several types of radioactive decay. This force also helps to hold particles such as neutrons together. In weak-force interactions, particles called *neutrinos* are either produced or absorbed. Neutrinos are very small. In fact, physicists are still trying to determine whether they have any mass at all. As indicated by their name, neutrinos do not have any charge.

▼ Some scientists think that there are four forces that hold everything together. Each force operates at a different level of organization, as shown here.

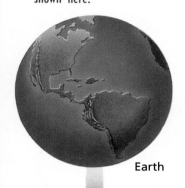

Earth

Gravitational force

Moon

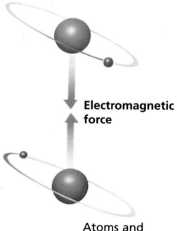

Electromagnetic force

Atoms and molecules

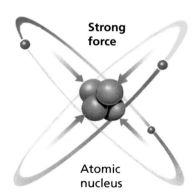

Strong force

Atomic nucleus

Weak force

Neutron

S36

Quantum Theories

The theories explaining the electromagnetic, strong, and weak forces are called *quantum* theories. These theories assume that only discrete packages, or *quanta*, of energy can be transferred when particles interact. For example, photons are quanta of light energy. The transfer of quanta of energy can be compared to buying eggs at a grocery store. You can buy 1 egg or a dozen eggs, but you cannot buy 2.7 eggs. The same is true of photons of light. An object might absorb 2 or 3 photons, but it cannot absorb 2.7 photons.

Because electromagnetic, strong, and weak forces can all be described by quantum theories, many scientists have tried to combine these forces into a theory that explains them all. One theory, called the *electroweak theory*, combines the electromagnetic and weak forces. The electroweak theory explains everything that can be explained by considering electromagnetic force and weak force as separate forces. This theory has even successfully predicted the existence of some previously unknown particles.

An egg can be divided only by breaking the shell. Quanta, however, are indivisible.

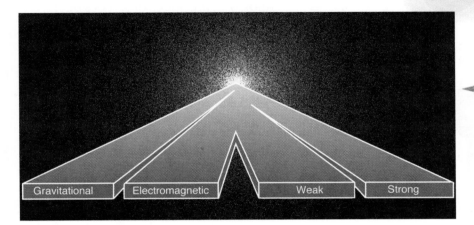

| Gravitational | Electromagnetic | Weak | Strong |

Scientists theorize that in the first few seconds after the big bang, the four fundamental forces separated from one original unifying force. Gravitational force was the first to separate from the other three. Next, the strong force separated from the electroweak force. Finally, the electroweak force separated into the electromagnetic and weak forces.

Since the electroweak theory was proposed, a number of theories that describe the strong force and the electroweak force in a single theory have also appeared. As a group, these are called *grand unified theories*, or *GUTs.* All GUTs share some features, including the prediction that electroweak and strong forces become a single force at extremely high energies when particles are very close together. Eventually, physicists may be able to explain all four fundamental forces with a single "theory of everything."

Where Do We Go From Here?

Before the mid-twentieth century, the *law of conservation of mass* (matter cannot be created or destroyed) and the *law of conservation of energy* (the amount of energy in the universe is constant) were thought to always hold true. But we now know that matter can be transformed into energy. Every second, for example, 657 million tons of the sun's hydrogen is converted into 653 million tons of helium. The "missing" 4 million tons of matter is converted into radiant energy, some of which we see as sunlight.

Hydrogen atoms combine by nuclear fusion to form helium. Smaller particles and a great deal of energy are released by this reaction.

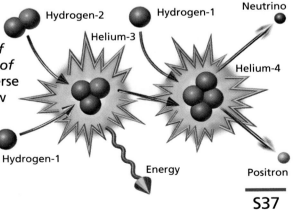

Hydrogen-2
Hydrogen-1
Neutrino
Helium-3
Helium-4
Hydrogen-1
Energy
Hydrogen-1
Positron

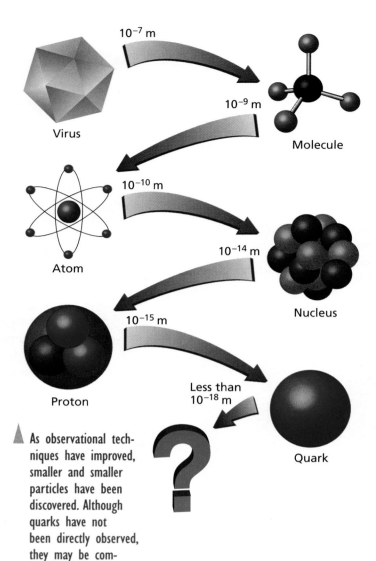

10^{-7} m

Virus

10^{-9} m

Molecule

10^{-10} m

Atom

10^{-14} m

Nucleus

10^{-15} m

Proton

Less than 10^{-18} m

Quark

▲ As observational techniques have improved, smaller and smaller particles have been discovered. Although quarks have not been directly observed, they may be composed of something even smaller.

In 1905, Albert Einstein proposed a mathematical relationship between mass and energy that he expressed with his equation $E = mc^2$. This famous equation states that *energy* is equal to *mass* multiplied by the *speed of light* squared. To account for the creation of energy from mass, a new law combining the two previous laws was needed. This law has become known as the **law of conservation of mass and energy**. It states: *The total amount of matter* and *energy in the universe does not change.* Whether mass is converted into energy or energy is converted into mass, the total amount of mass and energy remains the same.

Our idea of the atom has also changed. The "indivisible" atom has been subdivided far beyond protons, neutrons, and electrons. It has been broken into quarks, leptons, bosons, and many other pieces. In fact, over 200 subatomic particles have been cataloged. Just how far can this go? If matter can become energy, and energy can become matter, can we consider any particle of matter as fundamental? Perhaps there is a "particle" common to both matter and energy that can answer such questions.

DID YOU KNOW...

that the top quark is nearly as massive as a gold atom?
After 20 years of searching, scientists were surprised to find that this quark, whose discovery was confirmed in 1995, has so large a mass.

SUMMARY

According to current theories, matter consists of particles that are affected by four forces—gravitational, electromagnetic, strong, and weak—that hold everything in the universe together. Each force operates at a different level of matter, from galaxies to subatomic particles. Quantum theories, which include all but the force of gravity, assume that quanta of energy are transferred when particles interact. Grand unified theories attempt to combine three fundamental forces, while a "theory of everything" would combine all four. The relationship between mass and energy is $E = mc^2$. According to the law of conservation of mass and energy, the total amount of matter and energy in the universe does not change.

Unit CheckUp

Concept Mapping

The concept map shown here can be used to illustrate major ideas in this unit. Complete the map by placing terms from the list in the appropriate positions. Then extend your map by answering the additional question below. Use your ScienceLog. **Do not write in this textbook.**

Where and how would you connect the terms *formula unit, electromagnetic force,* and *the kinetic molecular theory of matter*?

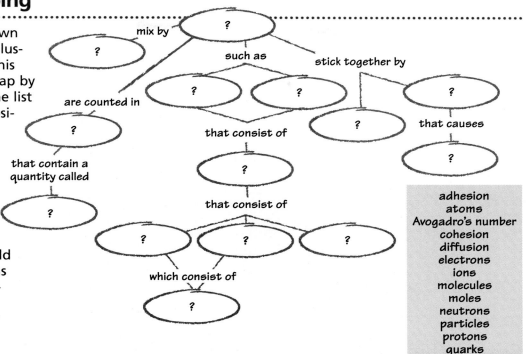

adhesion
atoms
Avogadro's number
cohesion
diffusion
electrons
ions
molecules
moles
neutrons
particles
protons
quarks
surface tension

Checking Your Understanding

Select the choice that most completely and correctly answers each of the following questions.

1. All of the following are particles of matter except
   a. molecules.
   b. ions.
   c. atoms.
   d. photons.

2. If you wanted to indicate the number of atoms in a certain amount of a substance, you could use the SI unit called a
   a. mole.
   b. formula unit.
   c. gram.
   d. liter.

3. Adhesion
   a. is the force of attraction between unlike particles.
   b. increases as the kinetic energy of molecules increases.
   c. is greatest in solids.
   d. decreases as the temperature of a substance decreases.

4. According to Boyle's law,
   a. the volume of a gas increases as its temperature increases.
   b. the volume of a gas decreases as the pressure increases.
   c. the pressure applied to a liquid in a closed container is evenly distributed.
   d. the temperature of a substance increases as the pressure on it increases.

5. The force between particles that acts over the greatest distance is
   a. weak force.
   b. strong force.
   c. gravitational force.
   d. electromagnetic force.

S39

Interpreting Photos

Explain how mass and energy are conserved in the process seen in this photo.

Critical Thinking

Carefully consider the following questions, and write a response in your ScienceLog that indicates your understanding of science.

1. Suppose that your teacher asks you to bring 5 mol of sodium bicarbonate (NaHCO$_3$) to your lab table. What information would you need to know in order to determine how much sodium bicarbonate to obtain? Look up this information, and calculate the amount you will need. Show your work.

2. The force of cohesion between atoms of liquid mercury is greater than the force of cohesion between water molecules. Which substance, mercury or water, has the greatest surface tension? Explain your answer.

3. The condensation point of a gas is the same as its boiling point. Likewise, the freezing point of a liquid is the same as its melting point. How would you explain these observations using the kinetic molecular theory of matter?

4. Suppose that 100 N of force are required to raise an object with a hydraulic lift, but that you can apply only 0.1 N of force. How could you alter the hydraulic lift shown on page S33 to accomplish this task? Explain.

5. What will the volume of a 2 L balloon be after the temperature falls from 300 K to 270 K? Show your work.

Portfolio Idea

Imagine that you could be shrunk to the size of the smallest subatomic particle and go on a voyage inside an atom. Write a short story describing your adventure into the interior of an atom. Be sure to describe all of the particles and forces that you would encounter on your journey and to explain how these particles and forces interact to produce what we know as matter.

S40

Unit 3

IN THIS UNIT

Force, Work, and Power,
   page S42

Simple Machines,
   page S47

Machines at Work,
   page S52

MACHINES, WORK & ENERGY

Now that you have been introduced to machines and how they relate to work and energy, consider the following questions.

● **1.** What simple machines make up the mechanical systems of your home?

2. How is the input energy supplied to each system? What does the output energy do?

3. What is the mechanical advantage and efficiency of each mechanical system?

In this unit, you will take a closer look at the physics behind machines and at how machines affect our daily lives.

S41

FORCE, WORK, AND POWER

When the forces applied by each of these arm wrestlers are equal, neither arm wrestler can move the arm of the other.

Work: Force Over a Distance

In everyday language, you might say that sitting at your desk reading this book is "work." But in a scientific sense, reading is not work. To a scientist, **work** involves using a *force* to move an *object* over a *distance*. Thus, work is related to motion. In order for work to be done, an object must be moved over some distance.

Think about the last time you played tug of war. When you win a tug-of-war contest, you do work in the scientific sense. The purpose of such a contest is to move an object (the other team) over a distance (across the center line). Your team must apply a force (pull on the rope) to move the other team. From your team's perspective, this force is the *effort*. As your team pulls on the rope, the opposing team resists your team's force by pulling in the opposite direction.

If neither team moves in a tug-of-war contest, nobody is doing any work. But if your team's effort is greater than the

Do you realize that you are constantly under the influence of a far-reaching force? Even though you might not feel it, this force is pulling on you right now. The force is, of course, gravity. As you have learned, the size of the gravitational force pulling on you is equal to your weight. Fortunately, another force usually balances your weight. When you are standing or sitting, for example, an equal force pushes up on you from below. This force, supplied by the ground or the seat of the chair, acts in the opposite direction of the gravitational force. In situations involving balanced forces, there is no change in motion.

When a force is not balanced, however, something will change. For example, if you push hard enough on the back of a chair, the chair will probably move. The chair starts to move because the force you supply is *unbalanced* (not opposed by an equal force).

Work

The result of using a force to move an object over a distance

DID YOU KNOW...

that by studying this textbook you are doing very little work? Because the only motion involved in reading is that of your eyes, total work is minimal. Why, then, does your head hurt?

In the scientific sense, climbing stairs is a lot more work than holding a barbell above your head.

S42

1 As long as the efforts of both teams are equal, all forces are balanced and nobody moves. These people might be pulling and straining, yet neither team is doing *work* in the scientific sense unless one team moves some distance.

2 By using more force than the opposing team, your team can move the opposing team across the line. Now your team has done *work* in the scientific sense.

other team's effort and causes them to *move*, your team has done work. To find out how much work your team did in winning the contest, you must know how much force your team used and how far the other team moved.

To find how much work your team did, multiply the *force* that your team applied by the *distance* over which the force was applied. In other words, find the work that your team did by using the following equation:

$$\text{Work} = \text{Force} \times \text{Distance},$$
$$\text{or}$$
$$W = F \times d$$

Because we measure force in newtons and distance in meters,

the unit of work is the *newton-meter* (N·m). This unit is also called a *joule* (J).

Let's assume that your team applied a force of 10,000 N more than the other team and that your team pulled the other team a distance of 3 m. The work your team did can be calculated as follows:

$$W = F \times d$$
$$W = 10{,}000 \text{ N} \times 3 \text{ m}$$
$$W = 30{,}000 \text{ N·m,}$$
$$\text{or}$$
$$30{,}000 \text{ J}$$

Therefore, your team did a total of 30,000 J of work in winning the tug of war.

◄ By definition, a joule is a force of 1 N acting over a distance of 1 m, or 1 N·m. This person is doing 1 J of work. As you can see, 1 J is not a very large amount of work.

S43

In the previous example, we assumed that the force stayed the same throughout the tug-of-war contest. In most situations, however, the applied force is not constant. For example, suppose you are pulling on a spring. The farther the spring stretches, the greater the force you must apply. In cases where the amount of applied force changes, *average force* is used to calculate the work that is done.

Power

The rate at which work is done

As you pull back on the string of an archer's bow, you must apply different amounts of force. You apply a small force when you start pulling back on the bowstring, but you must apply more force as you continue to pull. In this situation, you would use average force to calculate work.

Power: Work Over Time

Suppose your team is challenged to a rematch of the tug-of-war contest. This time you beat the other team even faster than you did the first time. In the first contest, it took your team 5 minutes to pull the other team 3 m. Now it takes your team only 2 minutes. Assuming that the force your team used was the same in both games, how do you think this time difference affected the amount of work your team did? Actually, the work was exactly the same in both cases.

The amount of work that you do does not depend on how fast you do it. Notice that in the equation for work, there is no variable for time. To describe how fast a certain amount of work is done, we use the term **power**. The faster an amount of work is done, the greater the power that was produced. Power equals *work* divided by *time* (in seconds). This relationship is given in the following equation:

$$P = \frac{W}{t}$$

These runners will do about the same amount of work, but they do not have the same power. Which runner has more power? Why?

S44

James Watt defined power in terms of a workhorse. One *horsepower* (hp) was the power of a horse lifting 550 lb. 1 ft. in 1 second. One horsepower = 746 W.

Because work equals force times distance, the equation for calculating power can also be written as follows:

$$P = \frac{(F \times d)}{t}$$

If you exert a force of 1 N over a distance of 1 m for 1 second, the power would be calculated as follows:

$$P = \frac{(1 \text{ N} \times 1 \text{ m})}{1 \text{ s}}$$

$$P = 1 \frac{\text{N·m}}{\text{s}} = 1 \frac{\text{J}}{\text{s}} = 1 \text{ W}$$

You can see from the equation that power is measured in *joules per second* (J/s), or *watts* (W). The watt is the SI unit of power. It was named after James Watt, the inventor of the sliding-valve steam engine. To generate 1 W of power, you must do 1 J of work in 1 second. Because a watt is rather small, however, the *kilowatt* is often used to indicate power. One kilowatt of power equals 1000 W. To generate 1000 W, or 1 kW, of power, you must do 1000 J of work in 1 second.

How much power did your tug-of-war team have in each of the two matches? Recall that your team did 30,000 J of work in both of the matches. Therefore, you would calculate the power of your team in each match (P_1 and P_2) as follows:

$$P_1 = \frac{W}{t} = \frac{30,000 \text{ J}}{5 \text{ min}} \times \frac{\text{min}}{60 \text{ s}}$$

$$P_1 = 100 \text{ J/s, or } 100 \text{ W}$$

$$P_2 = \frac{30,000 \text{ J}}{2 \text{ min}} \times \frac{\text{min}}{60 \text{ s}}$$

$$P_2 = 250 \text{ J/s, or } 250 \text{ W}$$

Note: The term min/60 s is a conversion factor for changing minutes into seconds.

What common household appliances have power ratings similar to these?

S45

Electricity and Energy Use

Work is done by electrical appliances as they transform the electrical energy in electricity into other forms of energy. Because the watt is the unit used to measure the rate at which objects do work, the rate at which electrical appliances, such as light bulbs and hair dryers, transform energy is expressed in watts.

The total amount of energy an appliance uses depends on how long it operates. The units for total energy consumption are *watt-hours* (Wh) or *kilowatt-hours* (kWh). For example, a 60 W light bulb that burns for 1 hour uses 60 W × 1 h, or 60 Wh, of electricity. When several electrical appliances operate at the same time, a great deal of electricity may be used. Therefore, the amount of electrical energy used in a home is usually measured in kilowatt-hours (where 1 kWh = 1000 Wh).

As the following calculation shows, a kilowatt-hour is actually a unit of energy, not a unit of power:

$$1 \text{ kWh} = 1000 \text{ W} \times 1 \text{ h}$$

$$1 \text{ kWh} = \frac{1000 \text{ J}}{\text{s}} \times 3600 \text{ s}$$

$$1 \text{ kWh} = 3{,}600{,}000 \text{ J}$$

When you pay your electric bill, you are really paying for energy. For example, a 60 W light bulb consumes 216,000 J of energy in 1 hour:

$$60 \text{ J/s} \times 3600 \text{ s} = 216{,}000 \text{ J}$$

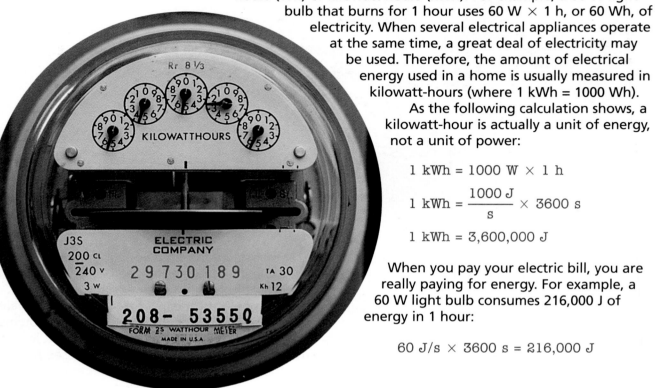

▲ An electric meter, such as this one, measures the number of kilowatt-hours of electrical energy that you use. On an average day, a house may use from 60 to 100 kWh of electrical energy.

SUMMARY

When a force causes an object to move over a distance, work is done. To calculate work, which is measured in joules, multiply force by distance. Power is the rate at which work is done. To calculate power, which is measured in joules per second, or watts, divide work by time. Electrical energy is usually measured in kilowatt-hours, which are units of energy, not power.

SIMPLE MACHINES

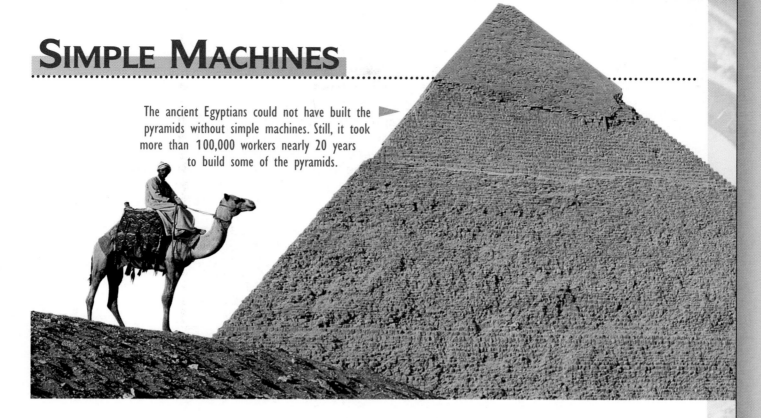

The ancient Egyptians could not have built the pyramids without simple machines. Still, it took more than 100,000 workers nearly 20 years to build some of the pyramids.

Can you imagine how much work was done in building the ancient Egyptian pyramids? Each of the stone blocks used to build the pyramids has a mass of thousands of kilograms. Yet because it took many years to build a pyramid, the power involved was not very great. You have already seen that people generate very little power in a tug-of-war contest. In the two previous examples, your team generated only 100 W and 250 W—about the same amount of power as a light bulb.

As you can see, humans cannot apply much force, and human power is not very effective in doing large amounts of work quickly. This is why humans invented machines. For example, moving a heavy boulder probably requires more force than you alone could normally supply. You might not be able to move it at all by simply pushing on it. But by applying the same force to a metal bar placed under the boulder, you can produce a force large enough to move the heavy boulder. The bar is a type of *simple machine*. You do work on the bar by applying a force to one end and moving that end. The other end of the bar, in turn, does work on the boulder by lifting it some distance.

A **simple machine** is a device that changes the size or direction of a force. While a simple machine can make your task easier, it cannot do more work than is put into it. Remember, the amount of work depends on the force applied to an object and the distance the object moves. If you want to decrease the force needed to do work, you must increase the distance over which the force is applied. On the other hand, if you want to decrease the distance over which work is done, you must increase the force applied. These relationships are true for all machines. Among the devices that we think of as simple machines are the screw, wedge, wheel and axle, and pulley. However, these simple machines are forms of just two basic machines—the *inclined plane* and the *lever*.

This metal bar is a simple machine that enables you to apply more force to an object.

Simple machine

A device that changes the size or direction of a force

S47

The Inclined Plane

An *inclined plane* is a simple machine with no moving parts. All simple machines, including the inclined plane, work on the same basic principle. An inclined plane, such as the ramp shown in the photograph to the left, simply increases the distance over which an applied force acts. As the length of an inclined plane increases, the force needed to move a load to the same height decreases. If you shorten the length of an inclined plane (make it steeper), more force is required. Again, the amount of work done is the same in both cases. It is only a matter of which is more important in getting the job done— the amount of force you must use or the distance you must travel.

Ramps are examples of inclined planes connecting one level to another.

Because an inclined plane forms a sloping surface, a load must move through a greater distance to rise to a certain height. Therefore, you can move a load up an inclined plane with less force than would be required to lift it straight up to the same height.

The Screw   Though they are simple machines in their own right, screws are really forms of the inclined plane. By looking at the photograph below, you can see that a screw is an inclined plane wrapped around a cylinder. Screws are very useful because the length of the inclined plane can be spread over a large distance. The closer together the threads on a screw are, the longer the inclined plane is. It will take more turns to fully tighten such a screw, but it will take less force to do it.

▼ A screw is an inclined plane wrapped around a cylinder.

S48

The Wedge Two inclined planes placed back to back make a wedge. When a wedge is used to split a log, as shown in the photo to the right, the two inclined planes move the resistance (the wood) on either side apart. As you might imagine, a short, wide wedge would split a log without being driven very far. However, the wedge would have to be pounded with a great deal of force. A long, narrow wedge would be a lot easier to pound but would have to be driven farther. There are probably many wedges on your kitchen counter. A knife is really a very narrow wedge.

◄ A wedge is two inclined planes placed back to back. Wedges are used to split an object in two.

The Lever

A pole or bar used to move a boulder is an example of a *lever*. The diagram below shows the basic parts of a lever. Although a lever might seem to do more work than is put into it, this is not the case. A lever only *transfers* effort. When the effort arm is longer than the resistance arm, applying a small force to the end of the effort arm results in a larger force at the end of the resistance arm. This is because the effort force acts over a greater distance than the resistance force. As in all machines, the work done by a lever can never exceed the work put into the lever. A lever merely changes the way that work is done.

DID YOU KNOW...

that principles of simple machines were developed by Greek mathematician Archimedes in about 250 B.C.? Archimedes once suggested that given a place to stand, he could move the Earth with a lever.

A typical lever

A A lever always turns on a fixed point called a *fulcrum*. The fulcrum is the point that supports the lever.

B The *effort arm* is the part of a lever between the fulcrum and the point where force is applied.

C The *resistance arm* is the part of a lever between the fulcrum and the load.

S49

As seen below, there are three classes of levers, based on the location of the fulcrum. Notice that each class of lever changes the way work is done in a different way. A first-class lever reverses the direction of the applied force. If the effort and resistance arms are not equal, a first-class lever will also change the size and distance of an applied force. Second- and third-class levers change the size and distance but not the direction of an applied force. You probably have seen and used many examples of all three kinds of levers.

Three Classes of Levers

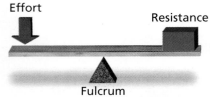

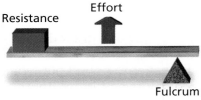

1 A first-class lever has the fulcrum between the resistance and the effort and *reverses* the direction of the applied force. Because the effort and resistance arms of this particular lever are equal, the effort and resistance are equal and so are the distances that each moves.

2 A second-class lever has the resistance between the fulcrum and the effort and *does not change* the direction of the applied force. A small amount of effort applied over a long distance moves a large resistance over a short distance.

3 A third-class lever has the effort between the fulcrum and the resistance and *does not change* the direction of the applied force. A large amount of effort applied over a short distance moves a small resistance over a long distance.

▲ A screwdriver is a wheel and axle. This person is using a screwdriver to increase the amount of force applied to a screw.

The Wheel and Axle   Have you ever tried to turn the shaft of a doorknob when the knob was missing? If so, you know how much harder it was to open the door. Imagine trying to make a car turn a corner without a wheel on the steering column. A doorknob and a steering wheel are common examples of the *wheel and axle*, a machine that is similar to a lever. A wheel and axle has a fulcrum at the center of the axle. The resistance arm of a wheel and axle is the radius of the axle, and the effort arm is the radius of the wheel. Effort applied to the wheel is transferred to the axle.

Have you noticed that large trucks and tractors have very large steering wheels? In a wheel and axle, the size of the wheel determines how much force acts on the axle. The larger the wheel, the more distance it covers in turning the axle. Therefore, a small force used to turn a large steering wheel will cause a large force to act on the axle of the steering wheel. This large force is transferred to the turning mechanism, making it much easier to turn a heavy truck.

S50

The Pulley A pulley is similar to a wheel and axle. However, the axle on a pulley does not turn. Instead, the wheel of a pulley turns as a rope runs over it. A pulley can be used in two different ways—it can be *fixed* to something or it can *move* along the rope. With a *fixed pulley*, the load (resistance) hangs down on one side, and the effort is applied to the other side. A fixed pulley does not change the effort force or the distance it moves; it simply changes the direction of the force. With a *movable pulley*, the resistance hangs from the axle of the pulley, and the pulley hangs on a rope. A movable pulley does not change the direction of the effort force, but it changes both the size of the force and the distance over which it acts.

A A fixed pulley changes the direction of the effort force but does not multiply effort force. Such pulleys can be useful because it is often easier to pull downward than to lift upward. When you pull down on the rope of a fixed pulley, the load rises. Notice that one length of rope supports the entire load (resistance). Because the effort and resistance distances are the same, there is no gain in force.

B A movable pulley reduces the effort force but does not change the direction of effort force. Notice that two lengths of rope support the load (resistance). Because the fixed rope supports half the weight of the load, it takes half as much effort to lift the load. Since force decreases, distance increases. The effort distance is twice the resistance distance.

SUMMARY

Machines help people overcome the limitations of human strength. The inclined plane and the lever are the most basic machines. All other simple and complex machines are forms of these two machines. Simple machines change the size or direction of a force. They do not increase the amount of work done. To decrease the force needed to accomplish work, the distance over which the force is applied must increase. Conversely, if distance decreases, force must increase.

DID YOU KNOW...

that several pulleys can be combined to multiply force even more?
Movable pulleys and fixed pulleys work together in a *block and tackle.* By moving the end of the rope a long distance with a small force, a very large force acts to lift a heavy object a small distance.

S51

MACHINES AT WORK

Any tool that helps us do work is a machine. Pencil sharpeners, drills, screwdrivers, and crowbars are machines. Even a baseball bat is a machine. Machines help do work in three ways. First, a machine can change the force applied to an object. Second, a machine can change the direction of a force. Third, a machine can change the distance through which a force moves. You already know that work equals force multiplied by distance. Because of this relationship, machines can reduce the force needed to do a given amount of work, thus making work easier.

Work Is Work

Although machines can make it easier to perform a task, they do not reduce the amount of work needed to do the job. For example, it takes the same amount of work to lift a heavy crate regardless of how the crate is lifted. You might lift the crate without the aid of a machine, or you might use a simple machine such as an inclined plane or a pulley to lift it. If you use a machine, you may exert less force, but you must exert that force over a longer distance. No matter how you do it, the amount of work required to lift the crate to a certain height is always the same. In other words, the amount of work you put into a machine (work input) can never be less than the work that comes out (work output).

In an **ideal machine**, *work input* always equals *work output*. Since work equals force multiplied by distance, we can set up the following equations:

$$\text{work input} = \text{work output}$$

$$(\text{effort force}) \times (\text{effort distance}) = (\text{resistance force}) \times (\text{resistance distance})$$

$$F_e \times d_e = F_r \times d_r$$

Now suppose you are faced with a task in which it is important to minimize the force you must apply. For example, suppose you want to lift a 360 N load 0.5 m off the ground with a lever. If you can only apply a force of 90 N, how far would the opposite end of the lever have to move?

$$F_e \times d_e = F_r \times d_r$$

$$90 \text{ N} \times d_e = 360 \text{ N} \times 0.5 \text{ m}$$

$$d_e = \frac{360 \text{ N} \times 0.5 \text{ m}}{90 \text{ N}}$$

$$d_e = 2 \text{ m}$$

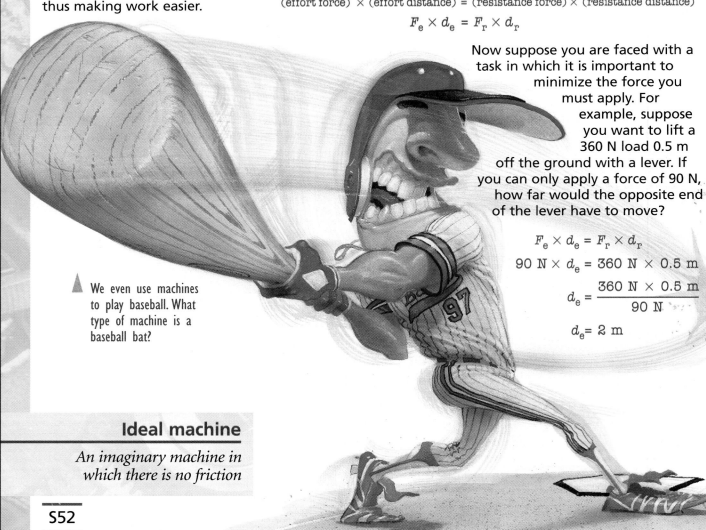

▲ We even use machines to play baseball. What type of machine is a baseball bat?

Ideal machine

An imaginary machine in which there is no friction

S52

You can see from this example that in order for the *effort force* to be less than the *resistance force,* the *effort distance* must be greater than the *resistance distance.* In this case, the resistance force (F_r) is four times greater than the effort force (F_e) you applied. However, your effort force moved four times farther than the load was lifted. As the figure to the right illustrates, a machine can change the size of a force, but it cannot supply more work than is put into it.

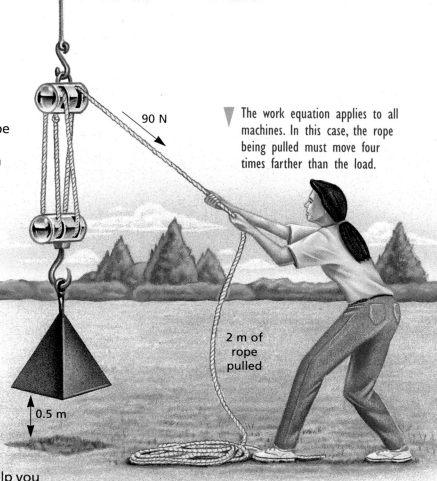

90 N

2 m of rope pulled

0.5 m

▼ The work equation applies to all machines. In this case, the rope being pulled must move four times farther than the load.

Machines and Mechanical Advantage

If machines do not increase the amount of work you can do, what would make one machine better than another? Besides the fact that you can choose whether to increase force or distance, what else might help you decide to use a particular machine? The usefulness of a machine can be rated by comparing the amount of force you get out with the amount of force you put in.

You can find the amount by which an ideal machine increases force by simply dividing the output (resistance) force by the input (effort) force. The resulting number is called the **mechanical advantage** (MA) of the machine. Mechanical advantage is a ratio of the output force of a machine compared with the input force. For example, suppose an automobile mechanic uses an engine hoist to pull an engine from a car. If the engine weighs 4000 N and the mechanic uses a force of only 50 N to lift it, the mechanical advantage of the hoist can be figured as follows:

$$MA = \frac{\text{Output force}}{\text{Input force}} = \frac{4000\text{ N}}{50\text{ N}} = 80$$

The mechanical advantage of the hoist is 80. This means that while raising the engine, the machine applied 80 times more force to the engine than the mechanic applied to the machine.

Mechanical advantage can also be *less* than one. By using a machine with a mechanical advantage of less than 1, you exchange distance or speed for force. For example, when you turn the handle of an eggbeater, the blades turn much faster than the handle.

Mechanical advantage

The factor by which a machine multiplies effort in order to equal resistance

The mechanical advantage of an eggbeater is less than 1. The blades turn faster than the handle, increasing speed (or distance over time) rather than force. ▶

S53

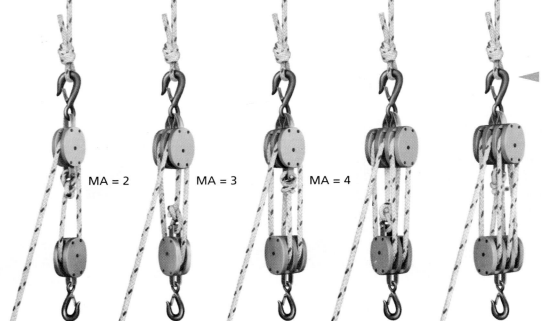

MA = 2   MA = 3   MA = 4

The easiest way to determine the mechanical advantage of a pulley system is to count the ropes that support the resistance force, or the load. What is the mechanical advantage of the two pulley systems on the right?

Computing Mechanical Advantage Mechanical advantage can sometimes be calculated without knowing the input and output forces. Levers and their related machines, the wheel and axle and the pulley, increase effort force because their effort arms are longer than their resistance arms. Thus, you can determine a lever's mechanical advantage by simply relating effort distance to resistance distance. This is done by dividing the length of the effort arm by the length of the resistance arm. For example, if the length of the effort arm of the lever you plan to use is 2.0 m and the length of the resistance arm is 0.2 m, you would find the mechanical advantage as follows:

$$MA = \frac{\text{Effort arm}}{\text{Resistance arm}} = \frac{2.0 \text{ m}}{0.2 \text{ m}} = 10$$

Your lever has a mechanical advantage of 10. In other words, this lever multiplies your effort force by 10.

The mechanical advantage of an inclined plane can be found by dividing its length by its height. For example, if the length of an inclined plane is 3.0 m and its height is 0.6 m, you would find its mechanical advantage as follows:

$$MA = \frac{\text{Length}}{\text{Height}} = \frac{3.0 \text{ m}}{0.6 \text{ m}} = 5$$

So the force needed to push a box up this inclined plane would ideally be one-fifth of the force you would need to lift the box directly. The same equation can be used to calculate the mechanical advantage of a wedge or screw.

The Effect of Friction Friction, as you have learned, is a force that results when two surfaces rub together. It opposes the motion of two objects in contact with each other. For example, suppose you use an inclined plane to lift a heavy box. Dividing the length of the inclined plane by its height will give its *ideal mechanical advantage*. However, because there is friction between the box and the surface of the inclined plane, some of the force you use is needed to overcome the friction. Therefore, the *actual mechanical advantage* of the inclined plane is less than what you calculated.

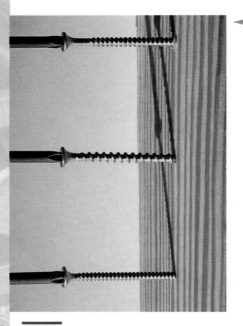

The closer together the threads on a screw are, the greater its mechanical advantage is. Which of these screws has the greatest mechanical advantage?

S54

The actual mechanical advantage of this inclined plane is less than its ideal mechanical advantage because some force is used to overcome friction.

Imagine that you are moving a 500 N box up an inclined plane, or ramp, into a truck. Suppose the ideal mechanical advantage of the ramp is 5. This means that you should only have to apply a 100 N force to move the box up the ramp. Unfortunately, friction between the box and the ramp increases the input force you must exert to move the box. Therefore, the actual mechanical advantage of the ramp is less than 5. This is true in every real situation when a machine does work. The greater the friction is, the smaller the actual mechanical advantage will be.

The Efficiency of Machines

Because all machines have at least two surfaces in contact, friction is a major consideration in the design and application of machines. And since all machines have some friction, the actual mechanical advantage of a machine is always less than its ideal mechanical advantage. In other words, a real machine always puts out less work than is put into it. The difference between the work that comes out of a machine and the work put into it is a measure of the machine's **efficiency**, which is expressed as a percentage. The smaller the difference, the greater the efficiency.

One way to determine the efficiency of a mechanical system is to divide the work output of the system by the work input. For example, suppose you use 16 kJ (kilojoules) of energy to pedal a bicycle. Of this amount, 5 kJ is used to overcome friction. The efficiency of the bicycle would be calculated as follows:

$$\text{Efficiency} = \frac{\text{Work output}}{\text{Work input}} = \frac{16 \text{ kJ} - 5 \text{ kJ}}{16 \text{ kJ}} = \frac{11 \text{ kJ}}{16 \text{ kJ}} = 0.69 = 69\%$$

Therefore, your bicycle has an efficiency of 69 percent. This means that only 69 percent of the work you do in pedaling your bicycle is used by the bicycle to move forward. Efficiency, then, is a measure of how well machines overcome friction.

Every new appliance sold in the United States now carries a notice indicating the efficiency of the machine. This translates into savings—the higher the efficiency of the appliance, the lower your utility bill.

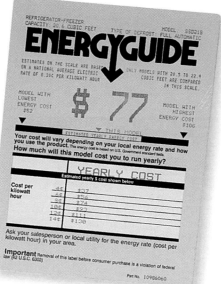

Efficiency

The ratio of the work done by a machine to the work put into it

S55

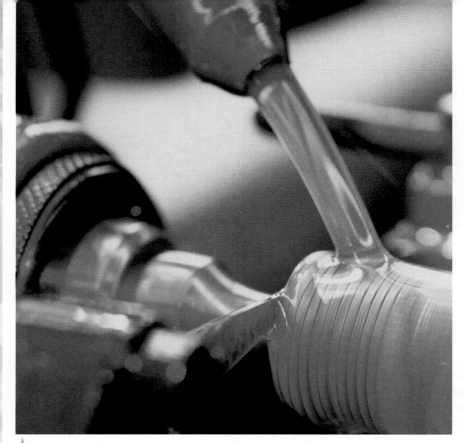

The oil on this rotating spindle helps reduce friction, thereby increasing efficiency.

Increasing Efficiency One way to improve the efficiency of a machine is to reduce friction. Friction can be reduced in many ways. For example, oil or grease makes the surfaces of rotating parts turn more easily. Bearings, such as the ball bearings inside bicycle wheels, are also used to reduce friction between rotating parts.

Many people throughout history have tried to invent machines that would be 100 percent efficient. These machines were called perpetual-motion machines. They were designed to produce at least as much energy as they consumed, or to accomplish as much work as was put into them. In other words, a perpetual-motion machine is one that would continue operating forever with no additional input of energy. However, no one has ever succeeded in inventing a perpetual-motion machine because no one has ever been able to totally eliminate the loss of energy or work due to friction.

SUMMARY

Machines make work easier, but they do not decrease the amount of work needed for any task. The mechanical advantage of a machine is the relationship between its output force and its input force, or between its effort distance and resistance distance. Because of friction, the work that comes out of a machine can never equal the work that is put into it. Efficiency is the ratio of work output to work input. Despite many efforts to build a perpetual-motion machine, no one has been able to make output work equal input work because of the presence of friction.

S56

Unit CheckUp

Concept Mapping

Starting with the terms supplied below, construct a concept map that illustrates major ideas from this unit. Arrange the terms in an appropriate manner and connect them with linking words. Then extend your concept map by adding as many additional terms from the unit as you can. Use your ScienceLog. **Do not write in this textbook.**

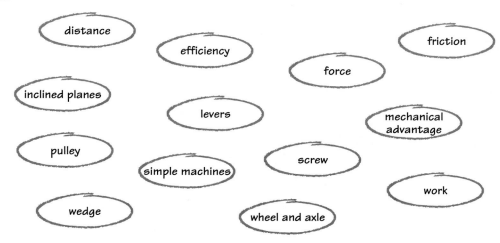

Checking Your Understanding

Select the choice that most completely and correctly answers each of the following questions.

1. The quantity that is determined by the distance moved and the force used is called
   a. power.
   b. work.
   c. friction.
   d. newtons.

2. Which of the following would constitute a power rating?
   a. meters per second
   b. joules per second
   c. watts per hour
   d. watts per meter

3. Which of the following simple machines would be considered a type of lever?
   a. ramp
   b. screw
   c. wedge
   d. wheel and axle

4. The screw is a simple machine that works like the
   a. lever.
   b. inclined plane.
   c. wheel and axle.
   d. pulley.

5. With an ideal machine, the work you get out of it equals
   a. the distance it must move.
   b. the force you use on it.
   c. the work you put into it.
   d. the resistance it has.

S57

Interpreting Photos

What type of simple machine is shown in the photo below? How would the mechanical advantage of this machine be calculated? Identify any sources of friction that could decrease the efficiency of this machine.

Critical Thinking

Carefully consider the following questions, and write a response in your ScienceLog that indicates your understanding of science.

1. Describe how you would determine the amount of power involved in walking up a flight of stairs. Use actual values in your description.

2. How much energy does a 100 W light bulb use in 24 hours? Show all your work.

3. How does a knife function as a double inclined plane? Why does sharpening a knife make it cut better?

4. Why can you use either force or distance to calculate ideal mechanical advantage? Use formulas to support your answer.

5. Some people think that levers are more efficient than inclined planes. Do you think they are right? Explain your answer.

P o r t f o l i o   I d e a

Locate several examples of non-electric mechanical devices in and around your home. Then produce a technical brochure for one or more of these devices. Each brochure should have a photograph or drawing of the device; a schematic diagram showing the types of simple machines that make up the device (levers, pulleys, wheels and axles, inclined planes, screws, and wedges); and a written description of the device, how it works, and how it makes your task easier. Also try to determine the mechanical advantage for each simple machine that you identify in your mechanical devices.

S58

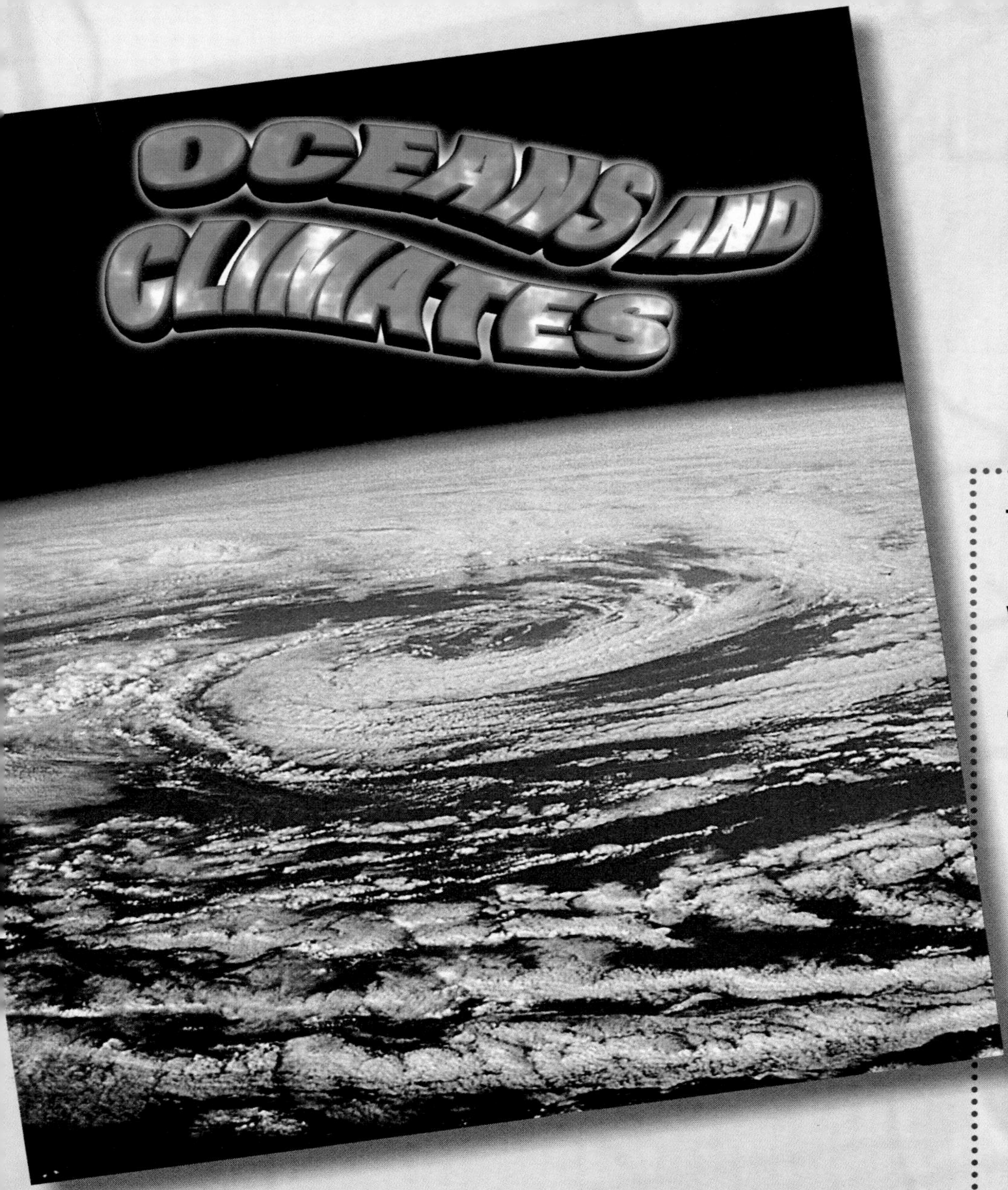

OCEANS AND CLIMATES

IN THIS UNIT

The Atmosphere,
page S60

Weather and Climate,
page S66

Ocean Resources,
page S75

Now that you have been introduced to oceans and climates, consider the following questions.

● **1.** How does the composition of the atmosphere affect life on Earth as we know it? How has the atmosphere changed over time?

2. In what ways do heat and moisture control the weather patterns in the atmosphere?

3. How are the resources of the ocean utilized, and in what ways must these resources be protected?

In this unit, we will take a closer look at the atmosphere and its structure, and at the ocean's resources.

S59

THE ATMOSPHERE

Unlike Earth, Mercury has no atmosphere.

You could, perhaps, live about 5 weeks without food and about 5 days without water. But you would last less than 5 minutes without air. In a sense, then, Earth's blanket of air, called the **atmosphere**, could be considered its most important resource.

Earth's atmosphere is one of its most distinguishing features. Life could not exist without it. In addition to supplying the gases required for life processes, the atmosphere protects organisms from exposure to harmful radiation and from bombardment by meteors. The atmosphere also gives our planet a moderate climate. Without an atmosphere, the Earth would have a climate like Mercury, with extremely hot days and terribly cold nights. There would be no wind, no clouds, and no liquid water. Without an atmosphere, our planet would be nothing like it is today. However, the atmosphere was once very different from the way it is now. In fact, you would not be able to breathe the air that was present in the atmosphere during most of Earth's history. How, then, did the present atmosphere change to become capable of supporting life as we know it?

Atmosphere

The layer of gases that surrounds the Earth

Evolution of the Atmosphere

Like living things and the physical features of the Earth, the atmosphere is thought to have evolved over several billion years to reach its present form. The Earth's earliest atmosphere probably formed shortly after the planet condensed from a swirling cloud of gas and

According to one scientific model, the Earth's present atmosphere took billions of years to develop. Its first atmosphere, consisting of gases left over from the planet's formation, was scattered by the solar wind.

A second atmosphere then formed from gases released by volcanoes. The action of lightning or ultraviolet radiation on atmospheric gases formed complex organic compounds that eventually developed into primitive living things.

Organisms capable of photosynthesis eventually evolved. Oxygen, a byproduct of photosynthesis, began to accumulate in the atmosphere.

S60

dust about 4.5 billion years ago. This early atmosphere most likely consisted of gases left over from the Earth's formation: hydrogen, ammonia, methane, and water. But once the sun (which probably formed about the same time) began to shine, the high-energy radiation streaming from it most likely scattered these gases into space.

The Primitive Atmosphere Evidence suggests that after the original atmosphere was scattered, a new atmosphere formed from gases that were trapped inside the rocks of Earth's crust when it first solidified. Most of these gases—which included hydrogen, water vapor, carbon dioxide, and nitrogen—were released during the volcanic eruptions that were common during Earth's early history. But as this new atmosphere was forming, it was also evolving. Hydrogen, which is too light to be held by gravity, escaped into space. And as Earth and its atmosphere cooled, much of the water vapor condensed and fell as rain. As a result, the first oceans formed when rainwater collected in low places on the Earth's surface.

Conditions in this primitive atmosphere may have allowed certain important chemical reactions to occur. For example, energy from ultraviolet solar radiation and the electrical discharges of lightning could have caused some of the gases in the atmosphere to combine into a variety of complex organic compounds. These compounds collected in the oceans. Scientists think that the first life-forms on Earth developed from these organic compounds more than 3.5 billion years ago.

Oxygen Enters the Picture The early atmosphere contained almost no free (uncombined) oxygen. Oxygen (O_2) was introduced to the atmosphere by two processes. A small amount of oxygen entered the atmosphere when water molecules were split by lightning and solar radiation. However, most of the oxygen on Earth today is the result of photosynthesis. During photosynthesis, carbon dioxide and water are used to make sugar. Oxygen is a byproduct of that process. Evidence suggests that primitive organisms evolved the ability to carry out photosynthesis fairly early in the Earth's history.

Some of the free oxygen in the atmosphere formed the gas ozone (O_3), which collected in a layer high above the Earth. This was an important development because this layer absorbed most of the sun's high-energy ultraviolet radiation, which is harmful to living things. Once the layer stabilized, the rapid evolution of early life-forms began. Increasing numbers of photosynthetic bacteria and algae began to release a steady supply of oxygen gas into the atmosphere and the ocean.

At some point, oxygen became common enough that an ozone shield formed. Life evolved rapidly.

When the oxygen level in the atmosphere reached a certain point, larger life-forms became possible. On a huge scale, organisms used up carbon dioxide to form calcium carbonate shells.

Oxygen continued to accumulate. Today, it makes up more than 20 percent of the atmosphere.

S61

Over hundreds of millions of years, oxygen accumulated in the atmosphere and oceans. By the beginning of the Cambrian geologic period, about 570 million years ago, enough oxygen had become dissolved in the oceans to support larger life-forms. An evolutionary explosion resulted. Most of these primitive animals had shells made of *calcium carbonate*—a compound formed by the reaction of carbon dioxide and the element calcium, which dissolves in water. Much of the carbon dioxide that had been part of Earth's early atmosphere was used by these animals to make their shells. Over millions of years, the shells of dead organisms collected on the ocean floor and were compressed and cemented together to form limestone. Today, most of the carbon dioxide from the primitive atmosphere remains chemically bound in massive deposits of limestone.

The Atmosphere Today

For some time after the Cambrian period, oxygen continued to accumulate in the atmosphere. Eventually, the concentration of oxygen in the atmosphere stabilized near today's levels. Today, the proportions of oxygen, carbon dioxide, and nitrogen in the atmosphere are fairly stable, kept in equilibrium by cycles that exchange gases among the living and nonliving parts of the environment. Currently, the atmosphere consists of about 78 percent nitrogen, 21 percent oxygen, and 0.9 percent argon, with traces of water vapor, carbon dioxide, and other gases. This mixture of atmospheric gases is called **air.**

▲ Early corals and other organisms used much of the carbon dioxide dissolved in ocean water to make their skeletons or shells.

Air

The mixture of gases in the atmosphere

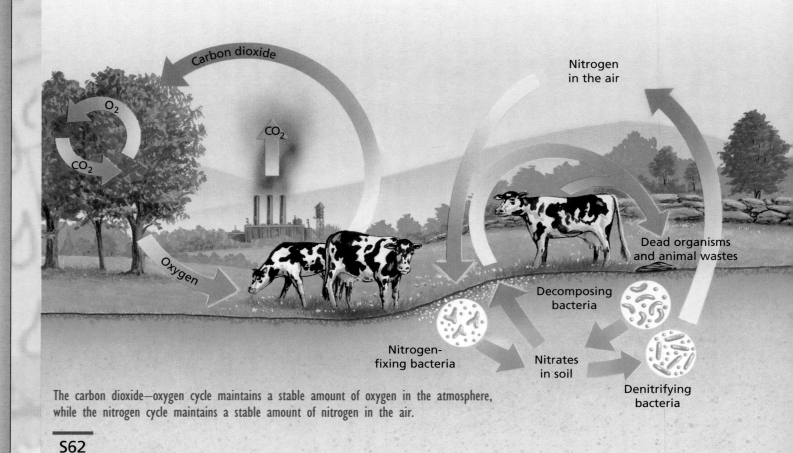

The carbon dioxide—oxygen cycle maintains a stable amount of oxygen in the atmosphere, while the nitrogen cycle maintains a stable amount of nitrogen in the air.

S62

Pressure Changes If you have ever traveled up a mountain, you might have noticed that your ears "popped." This happens because the air pressure outside your ears gradually decreases as you climb, while the air pressure inside your ears remains the same. Your ears pop when the pressure difference between the outside and inside of your ears suddenly equalizes. But why does air pressure change as you climb? Examine the diagram on this page for some clues.

Temperature Changes The temperature also changes as you ascend. If you measured the change, you would find that the temperature falls about 6.5°C for each 1000 m of altitude gained. For example, if you ascended in a balloon to an altitude of 3 km, you would find the temperature about 20°C lower than at sea level.

Layers of the Atmosphere The Earth's atmosphere consists of several layers. Each layer is characterized by either a rise or a fall in air temperature. A steady decrease in pressure also takes place the farther up you go. The lowest and densest layer, the *troposphere*, lies next to the surface. By mass, the troposphere contains about 90 percent of the atmosphere. It is also the layer in which almost all of Earth's weather occurs. The sun's energy heats the Earth's surface, which in turn heats the air next to it. Warm air near the surface becomes less dense and rises. This is balanced by cool air sinking to the surface somewhere else. The constant movement of air in the troposphere drives the Earth's wind and weather systems.

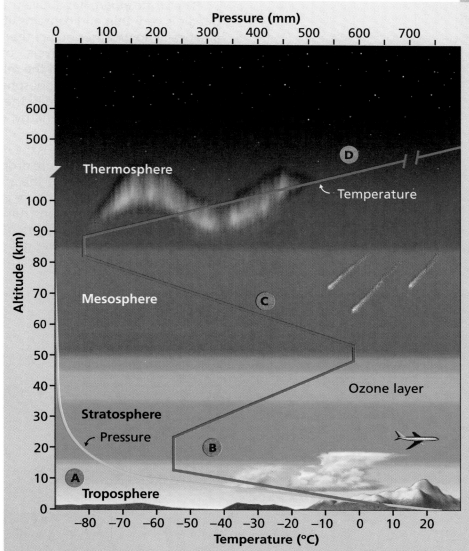

The layers of the atmosphere are marked by changes in temperature and pressure. The temperature of the air changes with altitude. At first it drops, but at higher altitudes, the absorption of high-energy solar radiation by ozone molecules causes the temperature to rise. Air pressure is simply the weight of all the air overlying a given area, and it is caused by gravity pulling the atmosphere toward the surface. At sea level, the weight of the air around you exerts a pressure of about 100,000 N/m^2! At higher altitudes, less of the atmosphere is overhead, so air pressure is less.

A The troposphere is the lowest layer of the atmosphere. Virtually all weather takes place there. Atmospheric pressure is greatest in the troposphere and declines with altitude.

B Temperature in the stratosphere remains constant for the first 10 km or so. It then begins to rise due to the presence of ozone, most of which is located in the middle to uppper stratosphere.

C In the mesosphere, the temperature drops steadily with increasing altitude as the ozone concentration and the atmospheric pressure decrease.

D The thermosphere blends gradually into the vacuum of space. The absorption of high-energy solar radiation, however, causes temperatures in the ultra-thin air of the thermosphere to soar.

S63

Jet stream

A belt of high-speed wind circling the Earth between the troposphere and the stratosphere

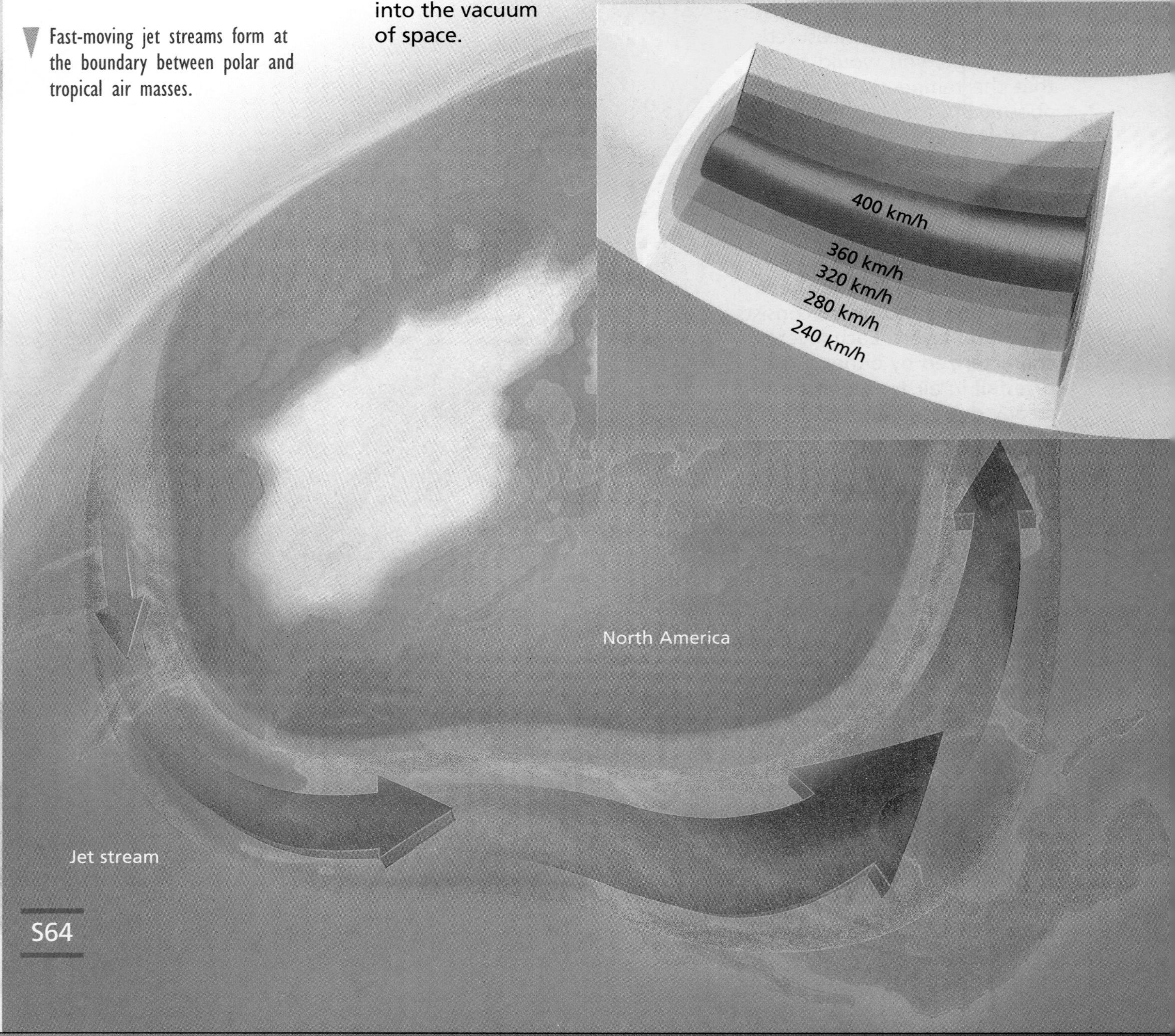

The *stratosphere* lies above the troposphere. In the stratosphere, the air is very thin and contains little moisture or dust. As a result, practically no weather occurs there. At the base of the stratosphere are broad, fast-moving "rivers" of air, called **jet streams**. The jet streams circle the planet in the mid- and upper latitudes, affecting weather patterns in the troposphere below. The stratosphere also contains most of the ozone in the atmosphere. Ozone molecules absorb some of the sun's energy, causing the temperature in the stratosphere to increase.

Above the stratosphere is the *mesosphere*. Once again, throughout this layer the temperature drops steadily with increasing altitude.

The uppermost layer of the atmosphere is the *thermosphere*. The "thermos" in thermosphere is a reference to the high temperatures found there. Special instruments have recorded temperatures in the thermosphere as high as 2000°C. In the upper part of the thermosphere, or *exosphere*, gas molecules are so far apart that little interaction occurs among them, and the Earth's atmosphere gradually fades into the vacuum of space.

▼ Fast-moving jet streams form at the boundary between polar and tropical air masses.

400 km/h

360 km/h
320 km/h
280 km/h
240 km/h

North America

Jet stream

S64

The Ionosphere Within the uppermost part of the mesosphere and the lower part of the thermosphere is a region called the *ionosphere.* It begins at a height of about 60 km and continues up to about 400 km. The ionosphere is so named because it contains electrically charged particles called *ions.* These ions are formed when atoms and molecules in the atmosphere absorb solar radiation and lose some of their electrons. Colorful light displays, known as *auroras,* result from the recapture of electrons by ions in the ionosphere. Auroras, however, are usually visible only near the poles. Ions, which are guided by Earth's magnetic field, are concentrated there because the magnetic field is much stronger near the poles.

Have you noticed that AM radio signals often travel greater distances at night than during the day? The ionosphere is responsible for this effect because it reflects these radio waves. At night, the lower layers of the ionosphere dissipate because ions cannot form in the absence of the sun's rays. Thus, radio waves are able to travel farther because they are reflected off a higher level at night.

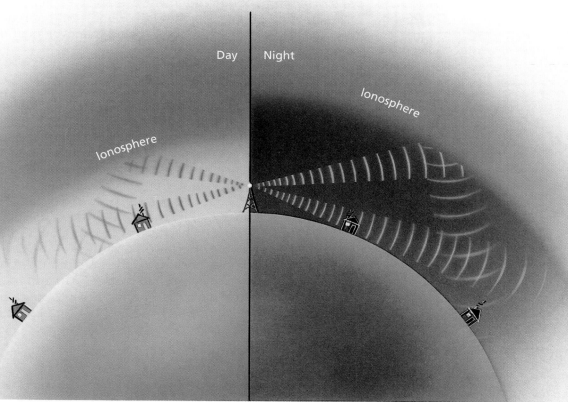

Day   Night

Ionosphere

Ionosphere

▲ AM radio waves are able to travel around the curve of the Earth because they are reflected by the ionosphere. These radio waves travel long distances by bouncing back and forth between the Earth and the ionosphere. But every time a radio wave is reflected, some of its energy is lost, so it eventually fades out. At night, the lowest layer of the ionosphere dissipates, allowing the waves to travel farther before being reflected back to the Earth. This is why AM radio waves can travel farther at night.

SUMMARY

The atmosphere is one of Earth's most vital resources. The Earth's atmosphere most likely originated as gases released by volcanic action. The atmosphere evolved over millions of years through chemical reactions and the actions of living things. Living things were the primary producers of the oxygen in the atmosphere of today. The composition of the atmosphere is now fairly stable. The atmosphere is divided into several invisible, yet distinct, layers.

S65

WEATHER AND CLIMATE

Weather

The condition of the atmosphere at a given time and place

Climate

The average weather for a particular place over many years

If you live along the Gulf Coast, in Hawaii, or in Florida, you know that it will be warm and humid for most of the year. If you live in the Northeast or the Midwest, your summers will be warm and humid, but winters will be cold and snowy. If you live in the Southwest, it will be warm and dry most of the year. If you live in the Pacific Northwest, it will be cool and rainy most of the year.

Weather is the condition of the atmosphere at a particular time and place. Specific weather conditions, such as temperature, humidity, clouds, winds, and precipitation, usually change from day to day and often from hour to hour. The average weather conditions over a long period of time constitute the **climate** of that region. Weather and climate both are created by the sun's uneven heating of Earth's atmosphere.

Heating the Atmosphere

Although the sun releases a huge amount of radiant energy into space, only about three-billionths of this energy is received by the Earth. Yet even this tiny fraction of the sun's radiation contains a very large amount of energy. In fact, the total amount of energy that reaches the Earth is thousands of times greater than all the energy used by every country on Earth combined. So what happens to all this energy? About 20 percent of the solar radiation that reaches Earth is absorbed by the gases of the atmosphere. Another 33 percent is reflected back into space by clouds, snow, and ice. The remaining 47 percent is absorbed at the Earth's surface. As you know, because of the greenhouse effect, some of the energy absorbed by the Earth and re-emitted as heat does not escape back into space but is absorbed by atmospheric gases. As a result, the atmosphere warms up.

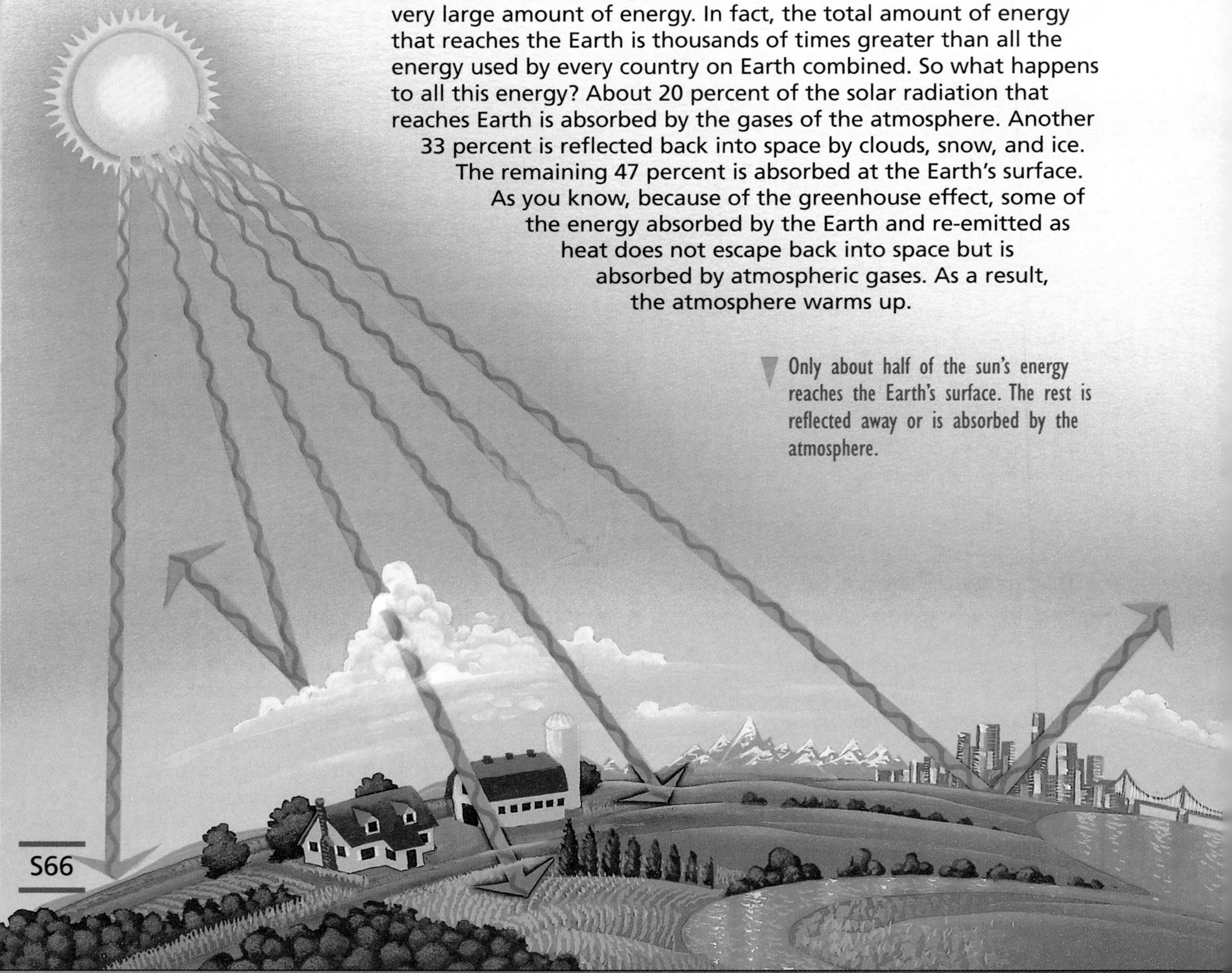

▼ Only about half of the sun's energy reaches the Earth's surface. The rest is reflected away or is absorbed by the atmosphere.

S66

However, the solar radiation striking the Earth is not evenly distributed. Because Earth is a sphere, the sun's rays strike different places at different angles. This uneven distribution of solar radiation causes unequal heating of the surface. Since the troposphere is heated by the Earth's surface, it too is heated unevenly. For example, air over the equator is heated much more than air over the poles. This causes the air over the equator to be less dense and to have a lower pressure than the air over the poles.

▼ The sun's rays strike the Earth more directly at the equator and more obliquely at the poles.

DID YOU KNOW...

that the troposphere is about twice as thick at the equator as at the poles? In the equatorial regions, strong updrafts caused by intense solar heating push the boundary between the troposphere and stratosphere upward. But in the calm polar regions the boundary drops down to within 5 km of the surface.

S67

Air Circulation

In general, air flows from a region of high pressure toward a region of low pressure. Thus, cold dense air from the poles flows toward the equator along the Earth's surface. At the equator, the air becomes warmer, expands, and rises. The heated air flows away from the equator along the top of the troposphere. At the poles, the air cools, becomes denser, sinks, and starts a new cycle as it flows back toward the equator. The first illustration on this page shows this idealized pattern of air circulation.

However, the Earth's pattern of air flow actually consists of several smaller *wind belts.* As equatorial air flows toward the poles along the top of the troposphere, it gradually cools. About one-fourth of the way to the poles, most of this air sinks back to the surface. When it strikes the surface, the mass of sinking air divides. Some of it flows back toward the equator, while the rest continues on toward the poles.

Because the Earth rotates, the path taken by air moving between the equator and the poles is not a straight line. Rather, it appears to be deflected. This phenomenon, called the **Coriolis effect**, makes objects moving north or south in the Northern Hemisphere drift to the right when one faces the direction of their movement. In the Southern Hemisphere, the drift is to the left. In the Northern Hemisphere, the Coriolis effect causes air that is moving toward the equator to blow from northeast to southwest. Such is the case with the wind belts known as the *trade winds* and the *polar easterlies.* Air moving toward the poles creates winds that blow from southwest to northeast. These winds are called *prevailing westerlies.* Notice how the direction of the winds in the wind belts differs between the Northern and Southern Hemispheres.

▲ At the equator, air is warmed and rises. At the poles, air is cooled and sinks.

Much of the warmed air loses ▶ energy and sinks at about 30 degrees latitude. This breaks up the equator-to-pole circulation into smaller convection cells.

Coriolis effect

The apparent bending of the path of a moving object due to Earth's rotation

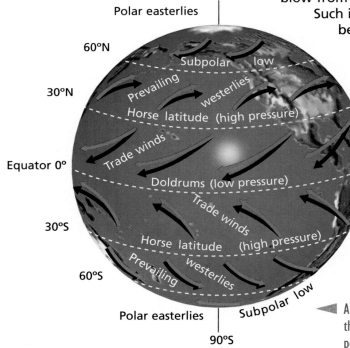

◀ A combination of the Coriolis effect and the circulation of air to and from the poles causes the global wind belts.

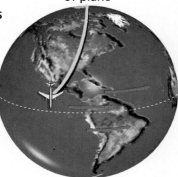

Curved path of plane

Earth spinning

▲ Because of the Coriolis effect, objects traveling in a southerly direction in the Northern Hemisphere appear to curve to the right.

Water Enters the Atmosphere

When water is heated by energy from the sun, some of it evaporates and enters the atmosphere as water vapor. All bodies of water contribute water vapor to the atmosphere. Plants also contribute water vapor to the air. The oceans, however, are by far the most important source of water vapor.

Water vapor in the air is called *humidity*. Air, however, can hold only a certain amount of water vapor. This amount is determined by the temperature of the air. Warm air can hold more water vapor than cold air. Air that is holding as much water vapor as possible is said to be *saturated*.

But air does not usually contain all the water vapor that it could possibly hold. The term **relative humidity** refers to the amount of water vapor that is currently in the air, as compared with the maximum amount the air could possibly hold. In other words, relative humidity is the percent to which air is saturated with water vapor. For example,

if a certain volume of air could hold 40 g of water but has only 20 g, the relative humidity would be 50 percent. If the same mass of air had 30 g of moisture, the relative humidity would be 75 percent. When this air mass contains 40 g of water, its relative humidity is 100 percent, and it is saturated.

Relative humidity changes as air temperature changes. The amount of water vapor itself does not change, but the amount of water vapor that air can hold does change. For example, as air temperature drops, the air can hold less water vapor. Therefore, the relative humidity rises. When air is warmed, it can hold more water vapor. Therefore, the relative humidity drops.

This graph shows the maximum amount of water vapor that air can hold over a range of air temperatures. As you can see, the ability of air to hold moisture increases sharply with increases in temperature.

Relative humidity

The amount of water vapor in air compared with the amount of water vapor that air could hold at that temperature

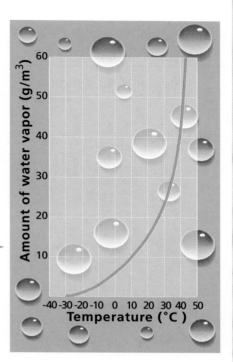

DID YOU KNOW...

that a huge amount of water evaporates from the Earth's surface every day? The amount that evaporates, 1100 km³ (1.1 quadrillion L), would cover the continental United States to a depth of about 12 cm.

This graph shows how the relative humidity might vary throughout a typical sunny day. As the temperature changes, so does the relative humidity, even though the actual amount of moisture in the air stays constant.

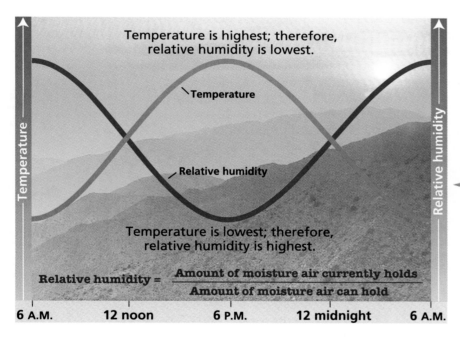

Temperature is highest; therefore, relative humidity is lowest.

Temperature

Relative humidity

Temperature is lowest; therefore, relative humidity is highest.

$$\text{Relative humidity} = \frac{\text{Amount of moisture air currently holds}}{\text{Amount of moisture air can hold}}$$

6 A.M.   12 noon   6 P.M.   12 midnight   6 A.M.

S69

Water Leaves the Atmosphere

Each day about 1100 km³ of water move from the Earth's surface into the atmosphere. And each day, about the same amount of water returns to the surface as precipitation. Yet the moisture does not return evenly to the Earth. Some regions are drenched while others receive little moisture for years on end. Why does this happen? Where will it rain today? Weather forecasters attempt to answer these challenging questions all the time, often unsuccessfully. A discussion of some of the factors involved in the return of water to Earth's surface follows.

Condensation   The temperature at which the relative humidity reaches 100 percent is called the *dew point.* If the temperature continues to drop, water vapor in the air will change back into a liquid. This process is called **condensation.** Water vapor may condense directly on cool objects, such as rocks, metal, windows, or plants. This moisture is called *dew.* If the dew point is below 0°C, *frost* will form. Frost is composed of solid ice crystals that form directly on an object when water vapor condenses below its freezing temperature. Frost is not frozen dew.

There are several ways that water can be cooled to the dew point. Air may be cooled as it is pushed upward to rise over mountains. It may also be pushed upward by denser air moving under it. Visible water droplets will form in the air when it is cooled to the dew point, provided there are tiny particles such as dust or salt crystals in the air. The particles give water vapor a surface on which to condense. The visible droplets that result form clouds.

Precipitation   Cloud droplets are so small that even slight air movements keep them floating. When cloud droplets collide and merge, however, larger, heavier drops form and may fall as *rain* or *drizzle.* Whenever the temperature in a cloud is below 0°C, ice crystals may form. Ice crystals grow as they fall, forming snowflakes. These may fall as *snow* if the air is cold enough all the way to the ground. However, if snowflakes pass through a layer of warm air, they will melt and become rain. Sometimes, rain passing through a cold layer of air freezes as it falls. It is then called *sleet.* In violent storms, strong updrafts may carry raindrops high enough to freeze. As they fall, a new layer of water collects on them and freezes as they rise again. Often, they are carried up and down several times, causing multiple layers of ice to accumulate. This forms *hail.* Sometimes hail can grow large enough to damage objects that it strikes on the ground.

Condensation

The changing of a gas into a liquid

▲ At the dew point, the moisture in the air condenses. When conditions are right, fog—a cloud very near the Earth's surface—is formed.

Relative sizes of ▶ condensed water vapor

Cloud droplet

Drizzle droplet

Rain drop

Clouds In 1803 an Englishman named Luke Howard developed a classification system to describe various types of clouds. Cloud formation depends on certain conditions in the atmosphere, such as air temperature, air pressure, humidity, and air currents (wind). Howard identified and named three basic cloud types based on their appearance: *cumulus*, from the Latin for "heap"; *cirrus*, from the Latin for "lock of hair"; and *stratus*, from the Latin for "layer."

If you have done much cloud gazing, you know that clouds do not always fit into one of the three categories above but appear to be combinations of the three basic types. Howard suggested names for these combinations as well. For example, sometimes cumulus clouds become so crowded together that they almost form a layer across the sky. Howard called these clouds *stratocumulus*.

The modern classification system for clouds also takes into account the altitude at which clouds form and whether they produce rain. Clouds that form at altitudes above 7 km are given the prefix *cirro,* and those that form between 2 and 7 km above the ground are given the prefix *alto.* Clouds that produce rain or snow are called *nimbus clouds*. *Nimbus* comes from the Latin word meaning "rainy cloud." The illustration below shows some of the characteristics of these basic cloud types.

DID YOU KNOW...

that a thunderstorm once produced hail larger than softballs?
Hailstones produced by this storm over Potter, Nebraska, in 1928 fell with such force that they became embedded in the ground!

Classification of clouds according to height and form

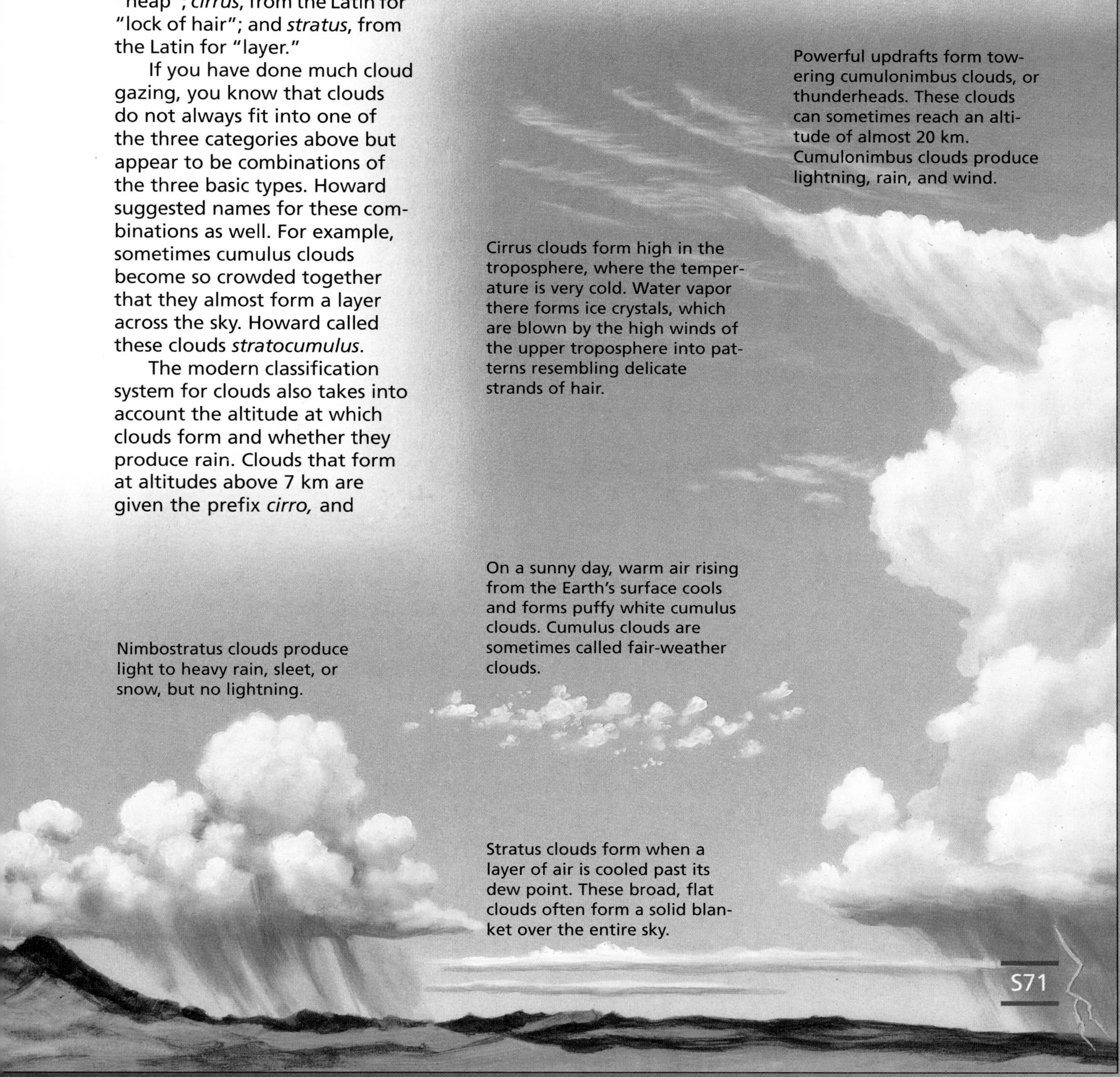

Powerful updrafts form towering cumulonimbus clouds, or thunderheads. These clouds can sometimes reach an altitude of almost 20 km. Cumulonimbus clouds produce lightning, rain, and wind.
Cirrus clouds form high in the troposphere, where the temperature is very cold. Water vapor there forms ice crystals, which are blown by the high winds of the upper troposphere into patterns resembling delicate strands of hair.
On a sunny day, warm air rising from the Earth's surface cools and forms puffy white cumulus clouds. Cumulus clouds are sometimes called fair-weather clouds.
Nimbostratus clouds produce light to heavy rain, sleet, or snow, but no lightning.
Stratus clouds form when a layer of air is cooled past its dew point. These broad, flat clouds often form a solid blanket over the entire sky.
S71

Air Masses

When air remains over one part of the Earth's surface for a long period of time, it takes on the characteristic temperature and humidity of that region. Such a body of air is called an **air mass.** In general, air masses that form over land are dry, and those that form over oceans are humid. Air masses that form in polar regions are cold, and those that form in regions close to the equator are warm.

Only regions that have light winds and consistent surface features over a large area can form air masses. For example, the Earth's surface at the equator is mostly covered by oceans. At the poles, vast areas are covered by the polar icecaps. Regions such as these are called *source areas.*

Fronts

As air masses move, they encounter each other. When air masses meet, little mixing takes place. Instead, one air mass pushes the other air mass out of the way, replacing it. The boundary between air masses is usually fairly sharp. Such a boundary is called a **front.** Temperature, pressure, and wind direction usually change significantly with the passage of a front. Most fronts form when one air mass moves into a region occupied by another air mass.

Air mass

A large body of air with uniform temperature and humidity

Front

The boundary between two air masses

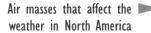 Air masses that affect the weather in North America

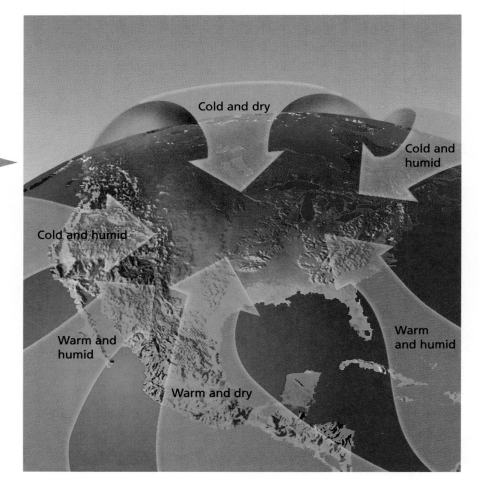

Cold and dry

Cold and humid

Cold and humid

Warm and humid

Warm and humid

Warm and dry

Cold Fronts If a cold air mass invades warmer air, a *cold front* is created. Since colder air is denser, it pushes under warmer air like a wedge. The lifting of a warm air mass usually causes cumulus clouds to form along the cold front. These clouds often bring stormy weather and heavy rain. Cold fronts tend to move quickly because dense, cold air easily pushes into warmer air. The turbulent weather associated with a cold front usually passes quickly. The air mass behind a cold front normally brings cool or cold, dry weather.

Warm Fronts When a warm air mass pushes into a cold air mass, a warm front is formed. Because it is less dense, the invading warm air rides up and over the colder air. As the warm air is lifted, cirrus clouds are formed high in the troposphere. As time passes, the cloud cover becomes lower and thicker until a solid sheet of stratus clouds covers the sky. A long period of gentle rain may occur, followed by slow clearing and rising temperatures. The weather changes that accompany a warm front are not as dramatic as those that accompany a cold front.

Severe Weather Sometimes the interaction of air masses causes very turbulent weather to occur. For example, thunderstorms are common along cold fronts. The thunderstorms occur when warm, humid air is lifted very rapidly by the cold air mass. Towering *cumulonimbus* clouds, or thunderheads, develop as a result. Rain, lightning, and wind accompany the thunderstorms. Sometimes, cold fronts are preceded by a solid line of thunderstorms, called a *squall line.* While severe weather is common along cold fronts, it is rare along warm fronts. This is because warm fronts, unlike cold fronts, are a zone of gradual transition between air masses.

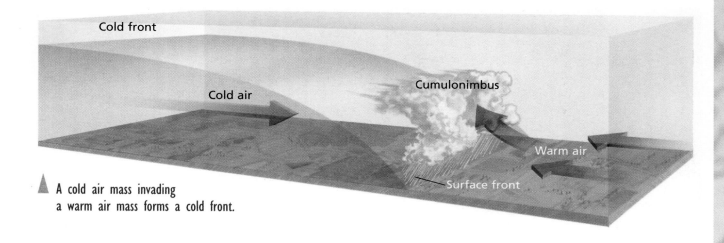

▲ A cold air mass invading a warm air mass forms a cold front.

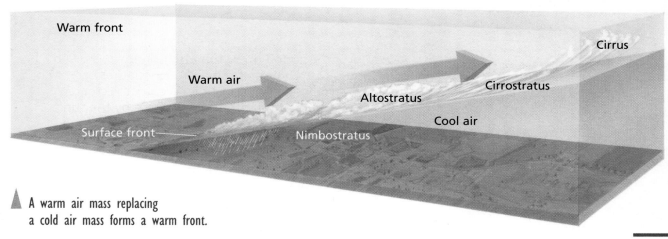

▲ A warm air mass replacing a cold air mass forms a warm front.

S73

About 5 percent of all thunderstorms develop to the point that they become classified as severe. Severe thunderstorms produce heavy rain, dangerous lightning, and damaging winds. They may also produce hail. Occasionally, small but extremely violent phenomena called *tornadoes* accompany severe thunderstorms. A tornado is a violently whirling column of air, usually seen as a funnel-shaped cloud extending from the base of a thunderhead. Tornadoes can be anywhere from 3 m to 3 km across. Scientists are not completely sure how tornadoes form, but they do know the kinds of weather conditions that are likely to produce them. These conditions are typically found along cold fronts, where strongly contrasting air masses—one cool and dry, the other warm and moist—collide. Such conditions are most common in the spring.

Tornadoes are common in only a few places on Earth. The vast majority occur in the United States, which averages more than 800 every year. Most of these occur in "Tornado Alley," a corridor running from Texas to Michigan. Although tornadoes usually last only a few minutes, they can be extremely destructive.

▲ A towering thunderhead

▼ Tornadoes are the most concentrated form of wind energy. Their fierce winds, sometimes more than 400 km/h, can destroy almost everything in their path.

SUMMARY

Weather is the condition of the atmosphere at a given time and place. Climate is the average weather of a region over a long period of time. Weather is created by the uneven heating of the atmosphere by the sun. There is a general pattern of circulation between the equator and the poles. This pattern is broken up into several zones. The rotation of the Earth deflects the winds from a straight-line path. Water vapor in the atmosphere condenses when it cools, forming clouds and precipitation. The motion and interaction of air masses cause many weather phenomena.

S74

OCEAN RESOURCES

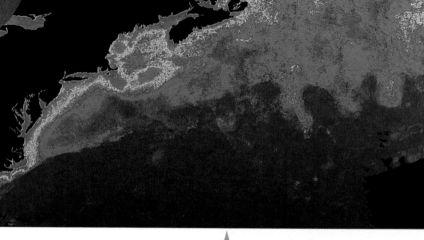

Seen from space, perhaps the Earth's most compelling feature is the enormous, deep blue oceans that seem to dominate the planet. The oceans are ancient; life most likely began there some 3.5 billion years ago. For more than 3 billion years, living things were found only in the oceans; the land was barren. The ocean is and has been home to a vast number of living things. Today, the oceans provide a wealth of important living and nonliving natural resources to the Earth's dominant life-form—humans.

Food From the Sea

Oceans cover almost two-thirds of the Earth's surface. Living things can be found in every part of the oceans. In theory, oceans could produce as much food as the land. Then why does less than 10 percent of the world's food supply come from the ocean? Part of the answer is the type of food we collect there.

This image shows the concentration of living things in the ocean. Most of the ocean's living things are microscopic plankton (inset).

Plankton Most of the mass of the organisms (biomass) in the ocean is in the form of **plankton**. Most plankton are so small that they can be seen clearly only with a microscope. Microscopic plants, called *phytoplankton,* are found mostly in the upper few meters of the sea, where there is enough light for photosynthesis. Tiny animals, called *zooplankton,* feed on the phytoplankton. Both kinds of plankton are eaten by larger animals that, in turn, become food for even larger fish and other marine animals. Thus, phytoplankton are the beginning of a food chain that supports the ocean's animal life. Plankton are not evenly dispersed throughout the ocean. Because of its remoteness from land, much of the ocean is poor in nutrients and is therefore a biological desert. Such regions are poor even in plankton.

Plankton

The small plants and animals floating near the surface of the ocean

Fish Fish are the main food we take from the ocean. They are caught primarily in areas of the ocean that lie over continental shelves. There the water is rich in the phytoplankton that are the basis of most ocean food chains. Some areas near the coasts act like natural fish farms. This is partly because of *upwelling*, which is the rising of cold water from below to replace surface waters that are removed by steady winds and evaporation. Upwelling brings up nutrients from the ocean bottom that are needed for the growth of phytoplankton. Upwelling also occurs in parts of the open ocean.

S75

Today, humans catch nearly one-third of all the fish large enough to be taken in nets. Unfortunately, careless overharvesting has taken a toll on the population of certain fishes used as food. Some kinds of fish that were once plentiful are now scarce. Laws have been enacted to protect some endangered fish populations. However, in the absence of strong enforcement, these laws are often broken. Scientists are trying to determine the conditions in the ocean that affect fish populations. This information may help researchers to learn how many fish can be harvested without doing irreparable damage. Using that information could help ensure a constant supply of fish.

Other Ocean Food Sources

Like fish, shrimp and shellfish are also important sources of food. Seaweeds, especially certain red and brown algae that grow anchored to the ocean floor, are another food resource. In many countries, the seaweed *kelp* is a popular food. Substances extracted from algae, such as algin, agar, and carrageenan, are used as stabilizers in foods such as ice cream, cheese spreads, and salad dressings. One potential source of food is *krill*. This small, shrimplike animal is found in large numbers in the cold waters of the Southern Hemisphere. Krill can be used directly for food or as part of other foods, just as shrimp are.

Kelp, which forms "forests" up to 30 m tall, grows in cold and rocky coastal waters. It is both a food source and a source of raw materials for humans. It is also a key player in aquatic ecosystems, providing food and shelter for dozens of species of marine animals.

Overuse of a resource can sometimes destroy it. Redfish were almost eliminated from the Gulf of Mexico because of their popularity as food fish. Because of strict limits on the allowable catch, redfish have made a dramatic comeback.

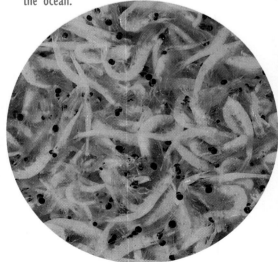

Krill are small shrimplike animals that are found in huge numbers in some parts of the ocean.

S76

Other Ocean Resources

For years, proposals to mine the oceans for their mineral wealth have been discussed. Because rivers, streams, and undersea hot springs dump huge amounts of dissolved material into the ocean, the ocean floor and even the water itself are rich in minerals. However, any ocean mining operation would face considerable difficulties. For example, to recover useful amounts of minerals from the water, large quantities of sea water would have to be processed, at great expense. It is far too costly to extract even such a valuable mineral as gold from sea water. Only a few highly concentrated minerals—sodium, bromine, and magnesium salts—can now be economically extracted from sea water.

Another option is to mine the sea floor for its mineral wealth. The mid-ocean ridges are honeycombed with mineral lodes deposited by hot springs. In addition, vast areas of the ocean floor are littered with *manganese nodules.* Trillions of tons of these egg- to potato-sized lumps of high-grade ore are thought to exist. Nodules contain high concentrations of manganese, iron, nickel, copper, and cobalt. They also contain smaller amounts of zinc, silver, gold, and lead. As rich a mineral source as manganese nodules are, though, mining them is not currently economical. Because these nodules are found only at great depths and far from shore, mining them would be difficult and very expensive.

Fuel From the Sea One of the most important sources of energy from the sea is *petroleum.* When marine organisms die, their remains fall to the ocean floor along with other debris. Since the water at the ocean bottom does not contain much oxygen, organic matter does not readily decay. Instead, it accumulates along with other, inorganic sediments. As more sediments accumulate, the buried organic matter is eventually chemically converted into petroleum. The largest known petroleum deposits on Earth formed when the African continent drifted northward and pushed organic-rich sediments against Asia.

Vast areas of the ocean floor are carpeted with deposits of manganese nodules. Manganese, nickel, cobalt, and other valuable metals are concentrated in nodules such as these.

Artist's conception of an undersea mining operation

These sediments ultimately became the large petroleum reserves of the Middle East. Other large reserves are found on the north shore of Alaska, beneath the North Sea, and along the coast of the Gulf of Mexico.

Protecting Ocean Resources

The technical difficulty of obtaining ocean resources is not the only problem to be overcome in utilizing the ocean's resources. Deciding who owns these resources may ultimately be an even bigger problem, one that could lead to serious conflict if not solved. Many questions remain to be settled: How far into the ocean do the borders of a country extend? Do nations have the right to collect these resources from the ocean beyond their borders? Do landlocked countries have any rights to ocean resources? Until there is agreement among nations, there cannot be full or fair use of the ocean's resources.

Some people would say that we should be less concerned about how to use the oceans and more concerned about how to preserve them. For example, pollution is a serious problem that threatens the vitality of the ocean. The chief sources of ocean pollution are sewage and garbage, industrial waste, and agricultural chemicals and wastes. These pollutants not only are detrimental to ocean life in the areas where they are dumped, but also can be spread by currents to affect the entire ocean. Oil spills from tankers and drilling platforms can also cause major harm. One accident can damage a coastal ecosystem for many years.

Because the oceans belong to no single nation, preserving them requires the cooperation of all nations. This will not be an easy task, but humanity must act soon to safeguard this valuable resource or face the consequences.

▲ Huge, costly drilling platforms are used to extract the oil trapped beneath the ocean floor.

▼ Pollution threatens the vitality of the oceans.

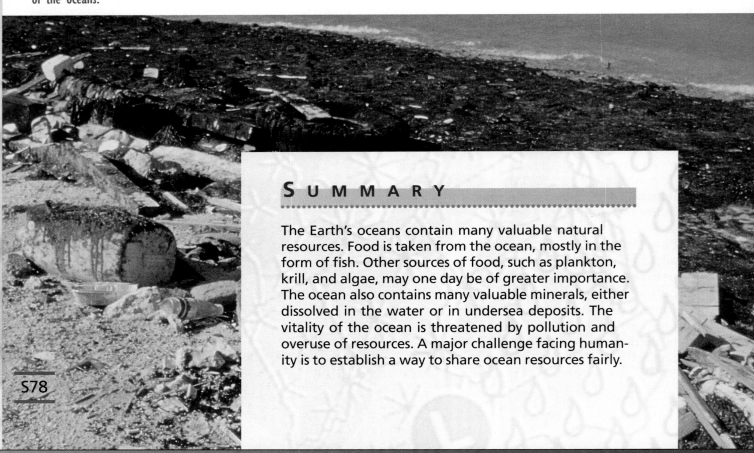

S U M M A R Y

The Earth's oceans contain many valuable natural resources. Food is taken from the ocean, mostly in the form of fish. Other sources of food, such as plankton, krill, and algae, may one day be of greater importance. The ocean also contains many valuable minerals, either dissolved in the water or in undersea deposits. The vitality of the ocean is threatened by pollution and overuse of resources. A major challenge facing humanity is to establish a way to share ocean resources fairly.

S78

Concept Mapping

The concept map below illustrates major ideas in this unit. Complete the map by supplying the missing terms. Then extend your map by answering the additional question below. Write your answers in your ScienceLog. **Do not write in this textbook.**

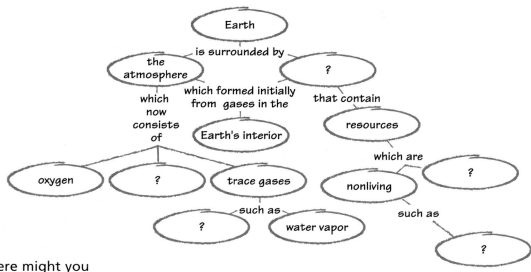

How and where might you connect the terms *argon, petroleum,* and *food*?

Checking Your Understanding

Select the choice that most completely and correctly answers each of the following questions.

1. The Earth's early atmosphere
   a. was probably much like the atmosphere today.
   b. contained little oxygen.
   c. enabled living things to evolve rapidly.
   d. consisted largely of ozone.

2. Which layer of the atmosphere is capable of supporting life?
   a. troposphere
   b. stratosphere
   c. mesosphere
   d. thermosphere

3. Which is an example of weather?
   a. a squall line
   b. a short growing season
   c. scant year-round rainfall
   d. hot, humid summers

4. Outside, the temperature cools to the dew point, causing
   a. rain to fall.
   b. condensation to occur.
   c. dew to evaporate.
   d. the relative humidity to exceed 100 percent.

5. Mineral resources from the ocean
   a. are off-limits to all nations by international agreement.
   b. belong to all nations by international agreement.
   c. are retrieved only with great difficulty.
   d. have largely replaced land-based mineral stores.

S79

Interpreting Illustrations

The photo here shows a home weather station. According to the readings on the station, can you tell what season of the year it is? What is the current air temperature? What is the relative humidity? How much rain has fallen since the rain gauge was last emptied? In which direction is the wind blowing, and how fast is it blowing?

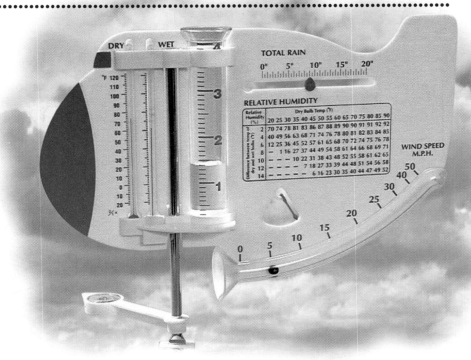

Critical Thinking

Carefully consider the following questions, and write a response in your ScienceLog that indicates your understanding of science.

1. How are volcanoes connected to the formation of the atmosphere and oceans?

2. How did the evolution of living things affect the evolution of the atmosphere?

3. Large thunderstorms sometimes form anvil-shaped clouds like the one shown below. How might the jet stream contribute to this type of cloud's characteristic shape?

4. How do you think the Coriolis effect would change (if at all) if the Earth were to rotate twice as fast? Explain.

5. Without the atmosphere, the Earth's surface would be very different. Describe a few major ways in which the atmosphere and the surface interact.

Portfolio Idea

Use your knowledge of air masses to put together a weather forecast for your area. Start by examining weather maps for the last 3 days. (Your local newspaper may have these.) Then use your knowledge of weather systems to predict the changes that will take place in your area over the next 48 hours. Do NOT use existing forecasts; make your own. At the end of the 48-hour period, compare your predictions, and those of your classmates, with the actual results.

Electromagnetic Systems

Unit 5

SourceBook

IN THIS UNIT

Two Kinds of Electricity,
page S82

Magnets and Magnetic
Effects,
page S91

Using Electricity,
page S97

Now that you have been introduced to electromagnetic systems, take a closer look at the basics of these systems by answering the following questions. ····

1. What is static electricity? How is it different from current electricity?

2. How do the properties of certain materials cause magnetism?

3. What is the relationship between magnetic effects and electric effects?

4. What kinds of devices make and use electricity? How do they work?

In this unit, you will examine electricity and magnetism, the relationship between them, and the methods and machines used to make, distribute, and harness electricity.

S81

TWO KINDS OF ELECTRICITY

When you think of electricity, you probably think of something that comes from a wall outlet and operates machines and appliances. That certainly is electricity—electricity in motion. But have you ever walked across a carpet and then felt a shock when you touched a metal doorknob?

Do your clothes sometimes stick together after being in the dryer? Have you ever rubbed a balloon on your shirt or hair and then stuck the balloon to a wall? Does your hair ever crackle and cling when you comb it? Such experiences are the result of *static electricity*.

The effects of static electricity can be demonstrated by this simple activity.

Static Electricity

Since ancient times people have noticed something curious about a natural substance called *amber*. Amber is ancient tree sap that has hardened over the centuries into a solid. Amber ranges in color from yellow to brown and sometimes contains the remains of insects. The ancient Greeks noticed that when amber is rubbed with cloth, it attracts objects, such as bits of straw and paper. The Greek word for amber is ηλεκτρον (*elektron*), and amber's attraction for small objects is due to what we now call *electric charges*.

◄ When amber is rubbed with a cloth, bits of cloth and paper are attracted to it.

Electric Charges Benjamin Franklin, who was a scientist as well as a statesman, studied electric charges. He is famous for having demonstrated that lightning is a form of electricity. In his day, it was known that there were different types of electricity. It was also known that electric charges could be created by rubbing two materials together. Franklin hypothesized that there are only two kinds of electric charges—positive (+) and negative (–). Franklin defined a *negative charge* as the type of charge given to a rubber rod when it is rubbed with fur or wool. He defined a *positive charge* as the charge given to a glass rod that has been rubbed with a silk cloth. An object that has neither a positive nor a negative charge is considered *neutral*. But how do objects become electrically charged?

Franklin's experiment, while very dangerous, demonstrated that lightning is a form of electricity.

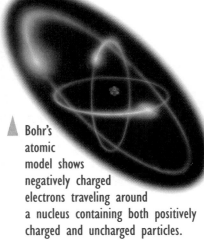

Bohr's atomic model shows negatively charged electrons traveling around a nucleus containing both positively charged and uncharged particles.

You have learned that all substances are made of atoms. According to current scientific models, atoms are composed of three basic particles—negatively charged electrons, positively charged protons, and neutrons that have no charge. When atoms have the same number of electrons and protons, they are electrically neutral. Normally, you are neutral because the negative electrons in your body are balanced by an equal number of positive protons.

Producing Charges Rubbing two materials together causes electrons to be torn away from one material and added to the other. For example, clothes taken from a dryer often cling together because they are electrically charged. This charge results from the different kinds of cloth rubbing together as the clothes are tumbled in the dryer. Clothes that have picked up electrons take on a negative charge. Those that have lost electrons take on a positive charge. Likewise, running a comb through your hair may cause electrons to move from your hair to the comb. In this case, the comb takes on a negative charge because it has acquired extra electrons. Your hair is left with a positive charge because it has lost electrons. When you walk across a carpet, the friction of your shoes on the carpet may cause you to pick up extra electrons, giving your body an overall negative electric charge.

Has this ever happened to you? Differing electric charges make clothes cling together.

S83

Static electricity

Electric charges that accumulate on an object

Electric charges can make your hair stand on end. The electric charge here was produced by a Van de Graaff generator.

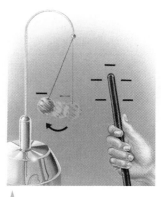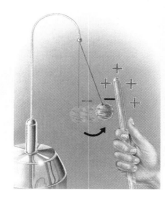

A rubber rod that has been rubbed repels a negatively charged ball (left). A glass rod that has been rubbed attracts the ball (right).

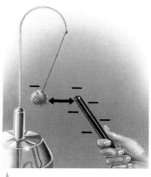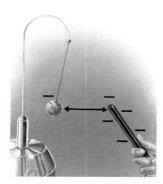

The electric force between objects depends on their charges and the distance between the objects. The precise relationship is given by Coulomb's law.

Electric force

The force that causes two like-charged objects to repel each other or two unlike-charged objects to attract each other

An electric charge results when an object either gains or loses electrons. Because this type of electric charge builds up on objects and does not move, as does electricity through a wire, it is called **static electricity**. The word *static* means "staying in one place." Any substance can be given a static electric charge.

Electric Forces and Fields   When two charged objects are brought close together, they either attract or repel each other. The behavior of objects with electric charges can be described by a simple rule: *Like charges repel, and unlike charges attract.* This is called the ***law of electric charges***.

Recall that force is always needed to cause an object to move. Thus, when two electrically charged objects cause each other to move, a force must be involved. This force is called **electric force**. Electric forces cause charged objects to move apart or come together depending on whether their charges are alike or different.

Two things seem to affect the strength of an electric force. First, the more an object is rubbed to give it an electric charge, the stronger the electric force that is produced. In other words, the strength of the electric force between two objects increases as the amount of electric charge on those objects increases.

The distance between two charged objects also affects the strength of an electric force. Charged objects have a greater effect on each other as they come closer together. For example, if the distance between two charged objects is halved, the strength of the force becomes four times greater. In the same way, doubling the distance between two charged objects reduces the force to one-fourth of what it was. These two relationships are expressed in *Coulomb's law*: *The force between two charged objects is directly proportional to the product of their charges and inversely proportional to the square of the distance between them*. How is this law similar to Newton's law of gravitation?

Around every charged object there is a space in which the effects of the electric force may be observed. Such a region of space is called an **electric field**, and it surrounds the charged object in all directions. As the distance from a charged object increases, the strength of its field becomes weaker. At a great enough distance, the electric field is too weak to be noticeable.

Electric field

The region of space around an electrically charged object in which the effects of the electric force may be observed

▲ The lines indicate the field existing between two opposite charges (left) and two similar charges (right).

Electricity on the Move

A bolt of lightning has tremendous power. In fact, a single flash of lightning can heat the air to a temperature higher than the surface of the sun. Lightning has melted holes in church bells and welded chains into iron bars. Potatoes in a field struck by lightning have been cooked in the ground. But for all its great power, lightning is basically a spark—very much like the spark that jumps from your hand to a doorknob, only much larger.

Lightning begins as static electricity that is caused by charges building up in storm clouds. These excess charges are rapidly transferred to the ground and result in the spark we see as a bolt of lightning.

S85

Scuffing your feet across a carpet may give your body an overall electric charge. But you don't notice this static charge unless you touch a metal object such as a doorknob. Then you suddenly feel a small jolt as extra electrons jump from your hand to the metal object. When you touch a doorknob, you complete a path over which the excess electrons can travel. As electrons travel through a conducting path, they become electricity in motion. Although you see a spark when this happens, you cannot actually see the electric flow itself. As with lightning, a spark results from the rapid movement of electric charges, which heats the air and causes it to glow.

Electric current

Electrons moving from one place to another

Electric Current

The electron model can be used not only to explain how objects become electrically charged, but also to help explain how electrons move from one place to another. In some ways, electrons behave like water, which will flow through a pipe when pressure of a pump is applied. In a similar way, a form of "electric pressure" can cause electrons to "flow" from an area where they are in excess to an area where they are lacking. For example, electrons will travel from the negative pole to the positive pole of a battery. However, in order for electrons to flow, there must be a path. A *conductor*, such as a wire, forms that path. If the conductor touches only the negative pole of the battery, the electrons will pile up on the conductor—static electricity. But if the negative and positive poles of a battery are connected by a conductor, electrons will flow through the conductor from one terminal to the next. This flow of electrons from one place to another is called an **electric current**.

Electric charges can build up on your body when you walk across a rug. When you touch a doorknob, the charges can jump from your hand to the metal knob.

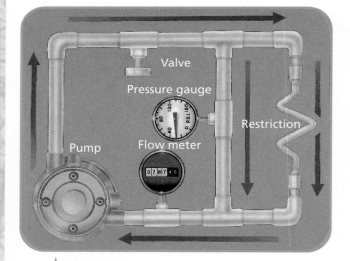

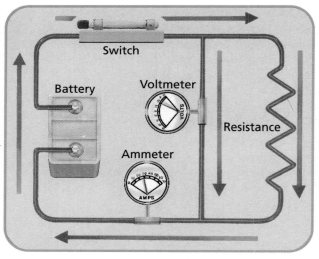

The flow of water in a system of pipes (left) can be compared to the flow of electricity through a circuit (right).

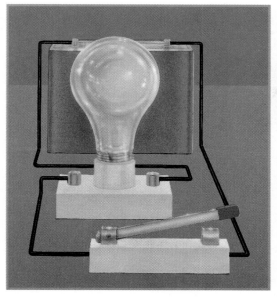

An electric circuit must form a continuous path in order for electrons to travel.

Describing Electric Current

Water will not flow in a pipe unless a force pushes or pulls it along. That force can be gravity or it can be pressure supplied by a pump. For example, a pump is often used to lift water to the top of a water tower. The water at the top of the tower has *potential* energy. When the water is allowed to flow from the tower, its potential energy is converted into the kinetic energy of motion, which can do work.

In the same way, potential energy can be stored by electric charges. Due to the electrical force of attraction between opposite charges, it takes work to pull them apart. Once these charges are separated, they have potential energy. When the separated charges are allowed to recombine, this potential energy is converted into other forms of energy—heat, light, and sound, for example.

Water stored in a water tower is similar to electric charges stored in a battery.

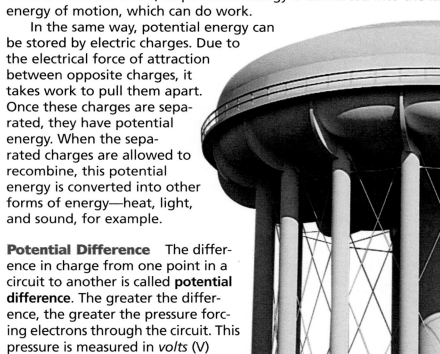

Potential Difference   The difference in charge from one point in a circuit to another is called **potential difference**. The greater the difference, the greater the pressure forcing electrons through the circuit. This pressure is measured in *volts* (V) and is referred to as *voltage*. The greater the voltage, the more work that can be done.

Potential difference

The difference in charge between two points in a circuit

S87

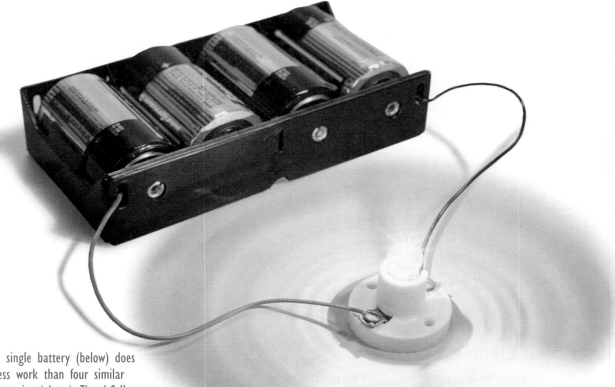

A single battery (below) does less work than four similar batteries (above). The 6.0 V combination has a potential difference four times greater than the 1.5 V battery.

Potential difference can be compared to water pressure. For example, water will not flow through a hose unless there is pressure. This pressure is supplied by a pump. It can also be supplied by gravity if the water source is higher than the outlet. In a like manner, an electric charge will not flow unless there is a difference in electric potential. To keep a charge flowing continuously, there must be a way of maintaining a potential difference. This is often done with a battery. A chemical reaction in the battery maintains the potential difference needed to keep the charges flowing.

If the flow of electrons is compared to water running down a hill, then the voltage can be thought of as the height of the hill. For example, an ordinary battery, such as the one shown in the photo above, maintains a potential difference of 1.5 volts. Think of it as a hill 1.5 V high. Four such batteries wired together in series create a hill 6.0 V high.

Current The amount of electric charge passing a point in 1 second is a measure of the **current**. The unit of electric current is the *ampere* (A). One ampere equals a charge of 6.25×10^{18} electrons moving past a point in 1 second. Current is like the rate of flow in water, which is often measured in liters per minute. The electrical devices you are familiar with require different amounts of current. Some devices, such as space heaters or hair dryers, use several amperes or even more. Other devices, such as calculators or portable radios, use only a few milliamperes.

Current

The flow of electric charge through a conductor

S88

Resistance Water moving through a hose encounters friction with the walls of the hose. The longer and narrower the hose, and the rougher its inner wall, the greater the friction will be. The greater the friction, the more the flow is reduced. Similarly, when electrons move through a material, they meet resistance. **Resistance** is an opposition to electric current that limits the flow of electrons in a circuit. Resistance is measured in *ohms* (Ω). A circuit has a resistance of 1 Ω when a potential difference of 1 V produces a current of 1 A.

You may recall that conductors have low resistances, but insulators have high resistances. For example, copper wires, such as those in your home, offer very little resistance to electric current, while rubber and glass offer a very large resistance. The amount of current in any electric circuit is determined by *both* the voltage and the resistance in the circuit. Think again of water flowing down a hill through a pipe. The amount of water that will pass through the pipe is determined by both the height of the hill *and* the size of the pipe. A narrow pipe offers more resistance to the flow of water than a large pipe, and it will therefore carry less water at a given pressure. A narrow pipe is similar to an electric circuit with high resistance. When an electric circuit has high resistance, the current in the circuit is reduced for any given voltage.

Resistance can be thought of as electrical friction. When electric current meets resistance, heat is produced. Consider the wires in a toaster oven. These wires are made of metal that has a relatively high resistance. As electricity flows through the wires, it encounters resistance, or electrical friction. The wires heat up as a result. This is similar to what happens when you rub your hands together. The harder you press them together or the faster you rub them, the warmer they get.

Resistance

The opposition to current in an electric circuit

▼ The resistance to flow in these pipes is mainly a function of their diameter.

◀ The wires in a toaster oven provide a lot of resistance to electric current.

S89

Ohm's Law The voltage, current, and resistance in an electric circuit are related to each other by a rule known as ***Ohm's law***. This relationship was discovered by a German schoolteacher, Georg Ohm, in the early 1800s. Ohm experimented with electric circuits made of wires having different amounts of resistance. By doing so, he discovered a general rule that describes the relationship among voltage (E), current (I), and resistance (R) in a circuit. Ohm's law can be written in equation form as follows:

$$\text{current} = \frac{\text{voltage}}{\text{resistance}}$$

or

$$I = \frac{E}{R}$$

For example, suppose an automobile with a 12 V battery has headlights with a resistance of 4 Ω. When the lights are on, the current needed to run the lights is

$$I = \frac{E}{R} = \frac{12 \text{ V}}{4 \text{ }\Omega} = 3 \text{ A}$$

By mathematically rearranging the terms, Ohm's law can also be written in other ways, for example:

$$E = IR \qquad \text{or} \qquad R = \frac{E}{I}$$

In short, if you know two of the values for a circuit, you can always find the third value by using Ohm's law.

▲ Using this illustration may help you recall the alternative forms of Ohm's law.

S U M M A R Y

Electric phenomena can be explained in terms of positive and negative electric charges. An electric charge results when something either gains or loses electrons. Like charges repel each other, and unlike charges attract each other. Static electricity refers to stationary electric charges. The force between two charged objects depends on the amount of charge and the distance between the objects. An electric current is the flow of electrons as they move from a place where there are more of them to a place where there are fewer. Resistance is an opposition to electric current. Ohm's law states that the electric current increases with increasing voltage but decreases with increasing resistance.

MAGNETS AND MAGNETIC EFFECTS

You can see the effects of electricity all around you, and you have learned that electricity is related to magnetism. But what is magnetism, and what produces magnetic forces? Over 2000 years ago, the Greeks discovered a naturally occurring material with unusual properties. This material, called lodestone, attracted pieces of iron. In honor of the region of Magnesia, where the lodestone was found, the term *magnet* was given to any material that attracted iron and other metals. Common magnets are specially treated pieces of steel that can attract iron. They also attract steel (which is largely iron), nickel, and cobalt.

▲ Lodestone is a naturally occurring magnetic material.

▲ Magnets come in many different shapes and sizes.

Describing Magnetism

Magnetism is not evenly distributed throughout a magnet. If you dip a bar magnet into a box of nails, you will find that the nails cling mostly to the ends of the bar. The ends of a bar magnet are called *poles*. Poles are the points on a magnet at which its magnetic attraction is the strongest.

▲ Attractive force is strongest at the poles of a magnet.

Magnetic Poles   A simple demonstration illustrates an important fact about magnetic poles. If you allow a bar-shaped magnet to swing freely from a string, it will always point in a north-south direction. One pole points north; the other pole points south. If you disturb the magnet, after a moment or two it will return to this same orientation. Because magnets always act this way, the pole of the magnet that points north is called its north-seeking pole or *north pole*. The opposite end is called the *south pole* of the magnet. A compass is simply a magnet that is free to turn so that its north pole always points north.

▲ Because a magnet always aligns itself in a north-south direction, its two ends are called its north and south poles.

S91

Magnetic force

The attraction or repulsion between the poles of magnets

Magnetic field

A region of space around a magnet in which magnetic forces are noticeable

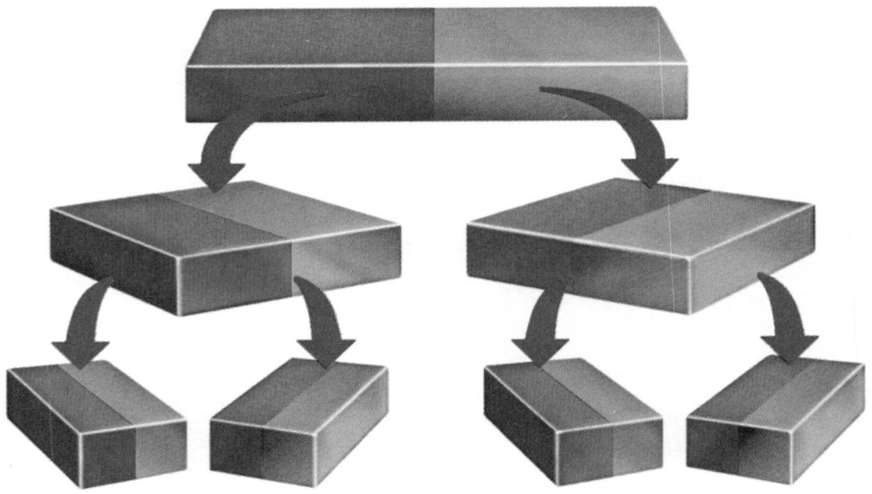

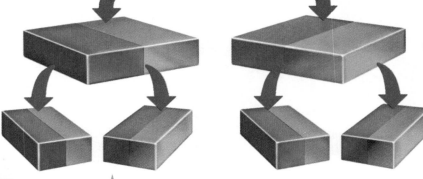

▲ If you continue cutting a magnet, each piece will still have two poles.

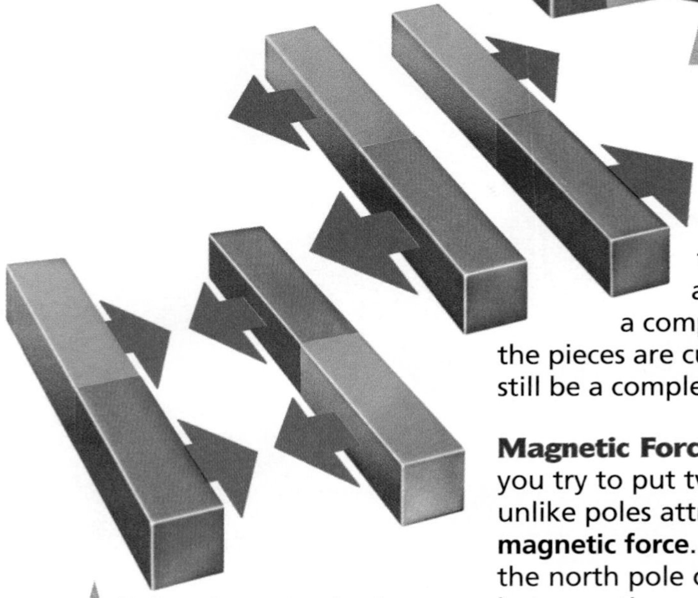

▲ Similar poles repel each other, whereas opposite poles attract, even at a distance.

▼ This horseshoe magnet has a circular magnetic field. Like all magnets, the field is strongest at the magnetic poles.

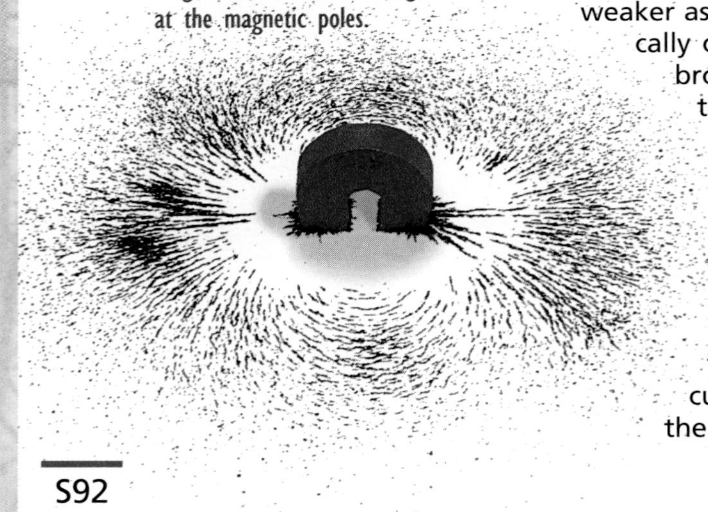

What do you suppose would happen if you cut a bar magnet in half? You might think that the north pole would be in one piece and the south pole would be in the other piece. The actual result is shown above. Each piece becomes a complete magnet with both north and south poles. If the pieces are cut into even smaller ones, the smallest piece will still be a complete magnet with a north and a south pole.

Magnetic Forces   You probably know what happens when you try to put two magnets together—like poles repel and unlike poles attract. This attraction and repulsion is called **magnetic force**. If the north pole of one magnet is brought near the north pole of another magnet, a repulsive force develops between them. If the north pole of one magnet is brought near the south pole of a second magnet, an attractive force develops. In this way, magnetic force is similar to electric force. Magnetic force is similar to electric force in other ways as well. First, both kinds of force work at a distance. In other words, two magnets do not need to touch in order to attract or repel each other. Furthermore, both magnetic force and electric force become weaker as the distance between two magnets or two electrically charged objects increases. As two magnets are brought closer together, the push or pull between them becomes much greater.

Magnetic Fields   You can feel the force between two magnets as they either attract or repel one another. The region of space in which magnetic forces are noticeable around a magnet is called the **magnetic field**. You can see the shape of a magnetic field by sprinkling iron filings around a magnet. The filings line up along curving lines running from one end of the magnet to the other. These lines are known as *lines of force*.

S92

The needle of each compass shows the orientation of the magnetic field around the magnet.

As you move a compass around a magnet, the compass needle turns. This happens because the magnetic compass needle is affected by the magnetic field produced by the magnet. At any location in the field, the compass needle comes to rest in line with the magnetic lines of force. These lines, however, are only imaginary. Like the lines of latitude and longitude on a map, magnetic lines of force are "inventions" that help us understand how things work. They can be used to predict which way a compass needle will point.

Magnetic lines of force never cross each other. When two or more magnets produce fields that overlap, the result is a combined field. If opposite poles are aligned, most of the lines run from the north pole of one magnet to the south pole of the other.

The Source of Magnetism

Although people have known about magnets for centuries, it was not until the early nineteenth century that theories were developed to explain magnetism. The currently accepted theory suggests that magnetism is actually a property of electric charges in motion. You already know that the electrons of an atom are in constant motion. They not only orbit the nucleus but also "spin" about an axis. If the electron spins one way, it causes one type of magnetic force. If the electron spins the other way, it causes the opposite magnetic force. A pair of electrons spinning in opposite directions has no net magnetism because the forces cancel each other. Therefore, the magnetic properties of an atom are related to how many unpaired electrons it has. The motion of unpaired electrons can give an atom north and south poles, making it act like a tiny magnet.

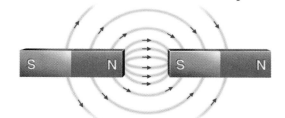

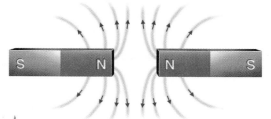

Magnetic lines of force never overlap, even when the poles of two magnets are brought close to one another, as shown here.

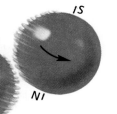
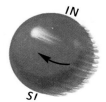

Spinning electrons act like tiny magnets. While paired electrons cancel each other out, unpaired electrons cause the atom to be magnetic.

S93

Within materials that have magnetic properties, such as an iron bar, atoms are grouped together in clusters called **domains**. The atoms within each domain have about the same magnetic alignment. In other words, all of their individual north and south poles point in the same direction. In an unmagnetized piece of iron, these small domains are arranged in a random manner. Because of this, their magnetic properties cancel each other. But when you bring the south pole of a magnet near the iron bar, the magnet attracts the north poles of the atoms in the iron bar and causes most of the domains to line up in the same general direction. When most of the domains are aligned, it makes the whole iron bar into a single large magnet.

Domains

In magnetic materials, clusters of atoms in which the magnetic fields of most atoms are aligned in the same direction

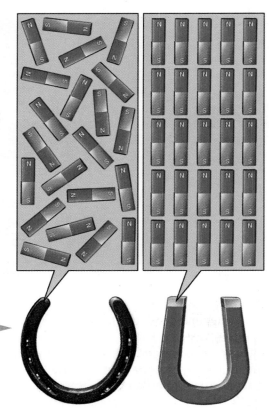

In most materials, such as this horseshoe, magnetic domains are randomly arranged. In magnets, however, the domains are aligned and produce a magnetic effect. ▶

▲ Some alnico magnets are strong enough to support your weight.

Types of Magnets

The ancient Chinese discovered that a piece of lodestone will always point in the same direction when allowed to move freely. As a result, lodestone was used as the first compass. The Chinese also found that iron can be magnetized by rubbing it with lodestone. Lodestone is actually a naturally magnetic form of iron oxide called *magnetite*. Artificial magnets, on the other hand, are made of metals such as iron, cobalt, and nickel. Materials that are easily magnetized but quickly lose their magnetism are called *temporary magnets*. Some harder metals, such as steel, are harder to magnetize but tend to retain their magnetism better. A magnet made of such material is called a *permanent magnet*.

Artificial Magnets   There are only five elements that can be made into magnets—iron, cobalt, nickel, gadolinium, and dysprosium. In their pure forms, however, these metals can be magnetized only temporarily. Excess heat or a sudden blow may cause their domains to become disorganized. To make a permanent magnet, you need an *alloy*. An alloy is a mixture of two or more metals, or a mixture of metals and nonmetals. The most common material used for making permanent magnets is steel, an alloy of iron and carbon. Strong permanent magnets can also be made of an alloy called *alnico*, which contains iron, aluminum, nickel, cobalt, and copper.

The best material known for permanent magnets is an alloy called *magnequench*. First fabricated in 1985, magnequench consists of the elements iron, neodymium, and boron. The magnetic field produced by this material is 10 times stronger than that produced by the best alnico magnets.

S94

Electromagnets   As you know, electromagnets are temporary magnets made by using an electric current. Due to the close relationship between electricity and magnetism, magnets can be used to produce electricity (as in a generator), and electricity can be used to make temporary magnets. These magnets can be turned on and off simply by flipping a switch. Powerful electromagnets, such as the one shown to the right, have been constructed for a variety of uses.

The Earth as a Magnet

In 1600, William Gilbert, who was the physician of Queen Elizabeth of England, proposed that the reason a suspended lodestone lines up in a north-south direction is that the Earth itself is a giant lodestone! To test his theory, Gilbert performed an experiment in which he carved a natural lodestone into the shape of a sphere. He found that a compass placed on the surface of the round lodestone acts just as a compass does at different places on the Earth's surface.

In fact, the Earth behaves as if a giant bar magnet were running through its center. Because the north pole of a magnet (by definition) is the pole that points northward, the magnetic pole at that location on Earth must in fact be a magnetic south pole. This follows from the behavior of magnetic poles—unlike poles attract. So when you use an ordinary compass, you are actually making use of two magnets. One of these magnets is small—the needle of the compass. The other magnet is very large. That magnet is the Earth itself.

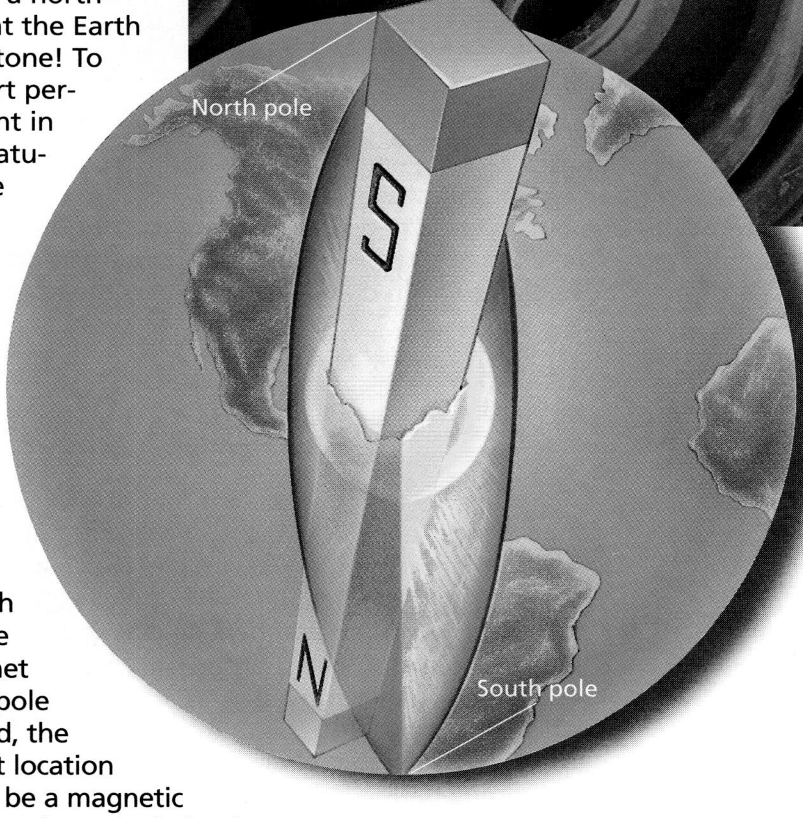

North pole

South pole

An electromagnet attached to this crane can lift hundreds of kilograms at a time.

The Earth itself is a magnet, with north and south poles. The north magnetic pole of the Earth is actually a magnetic south pole.

S95

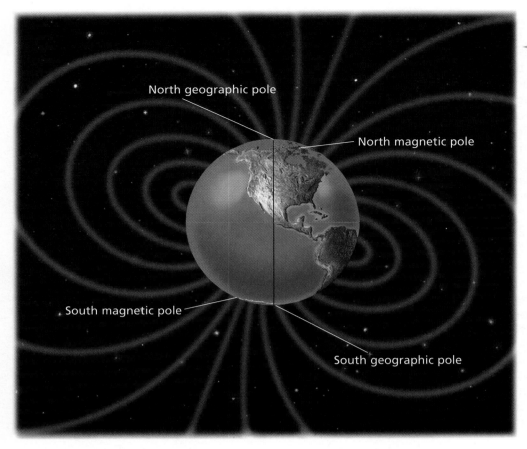

North geographic pole

North magnetic pole

South magnetic pole

South geographic pole

The Earth has a magnetic field that is strongest at the poles. Notice that the Earth's geographic and magnetic poles are not in the same location.

Like all magnets, the Earth is surrounded by a magnetic field. Recall that you can see evidence of a magnetic field by sprinkling small pieces of iron on a piece of paper over a magnet. If there were some way to do this for the Earth, you would have a picture of this magnetic field. It is a little-known fact that the magnetic poles drift as much as several kilometers every year. Since it was first discovered in 1831, the north magnetic pole has drifted about 1000 km to the northeast.

DID YOU KNOW...

that the Earth's magnetic field sometimes reverses polarity?
Every 500,000 years or so, the field collapses and then flips. The Earth's north pole becomes its south pole and vice versa!

SUMMARY

A magnet is any material that attracts iron and certain other metals. A magnet always has both a south pole and a north pole. As with electric charges, like magnetic poles repel and unlike poles attract. Magnetic fields surround magnets much like electric fields surround electrically charged objects. The magnetic properties of an atom are related to the movement of its electrons and to the number of unpaired electrons it has. In a magnetized bar of iron, the magnetic domains line up with like poles pointing in the same direction. Artificial magnets are made from alloys, and electromagnets are produced by the action of an electric current. The Earth also has a magnetic field, and it behaves as if a giant bar magnet were running through its center.

S96

USING ELECTRICITY

To use electricity, you need four basic things: a source of electricity, a way to manipulate voltage, a device for converting the electricity into useful work, and a pathway for the current. We will now take a look at each of these requirements.

Starting the Flow of Current

Electricity must be generated before it can be used. To start an electric current flowing, an electric generator can be used to change mechanical energy into electrical energy. Generators come in a variety of sizes. Some, such as the generator that operates a bicycle lamp, could fit in the palm of your hand. Others, such as those that supply cities with electricity, are enormous devices weighing tons. We know what a generator is, but how does it work?

Basically, an electric generator consists of a coil of wire, called an *armature*, and a permanent magnet. A simple generator is shown above.

In most working generators, the field structure rotates and the armature stays fixed. Furthermore, the armature consists of hundreds of turns of wire, rather than the single coil shown here. In all generators, the part that rotates (whether it is the armature or the magnet) is called the *rotor*. The part that remains fixed is called the *stator*.

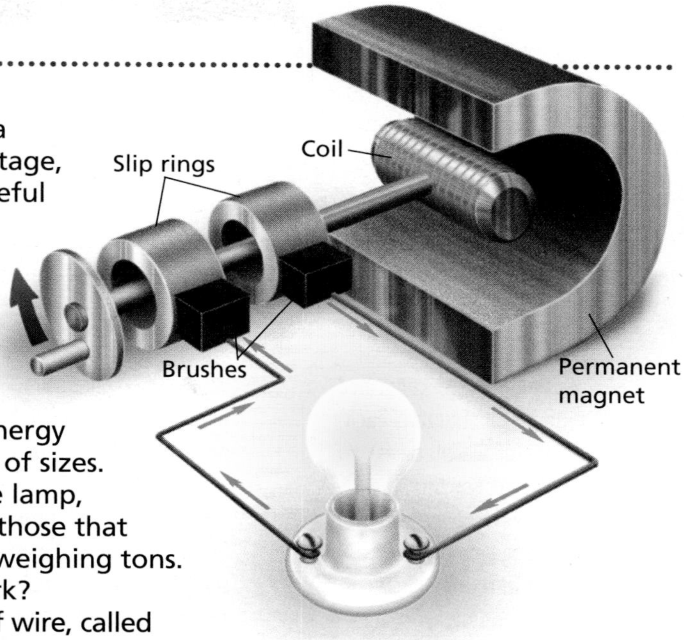

Alternators As you know, household current is alternating current, called simply AC. A device that generates alternating current is called an **alternator**, or AC generator. You may have heard this term before, probably in reference to the small electric generator found in automobiles. Automotive alternators convert a small fraction of the engine's mechanical energy into electrical energy to operate the car's lights, radio, and gauges. The operation of an alternator is summarized in the illustration below.

Slip rings
Coil
Brushes
Permanent magnet

In a simple generator, the rotating armature sweeps through lines of magnetic force produced by the magnet. The changing magnetic field sets electrons within the armature in motion, causing an electric current to flow.

Alternator

A mechanical device for generating alternating current

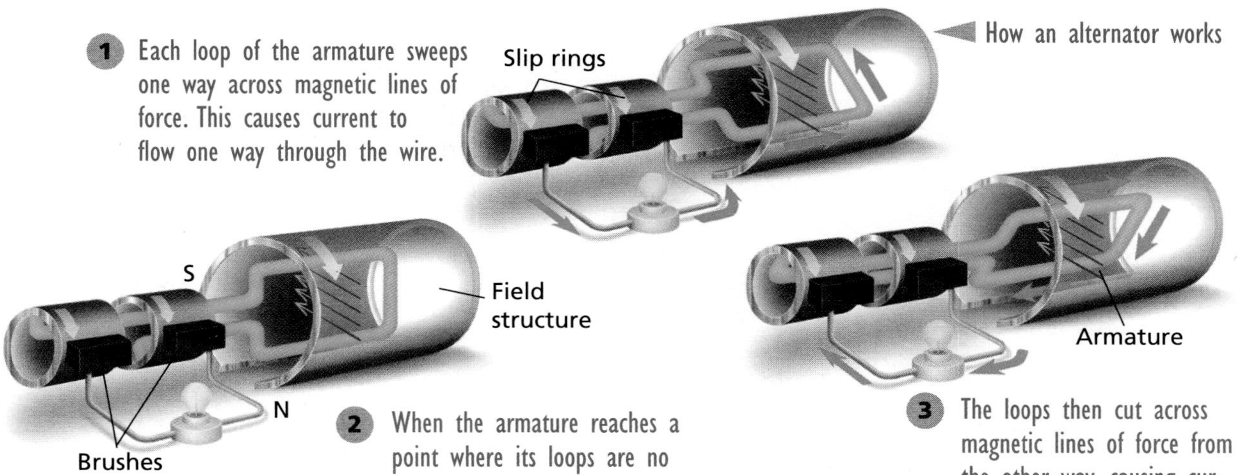

1 Each loop of the armature sweeps one way across magnetic lines of force. This causes current to flow one way through the wire.

Slip rings

How an alternator works

S

Field structure

Brushes

N

2 When the armature reaches a point where its loops are no longer crossing lines of force, the current drops to zero.

Armature

3 The loops then cut across magnetic lines of force from the other way, causing current to flow in the opposite direction through the circuit.

S97

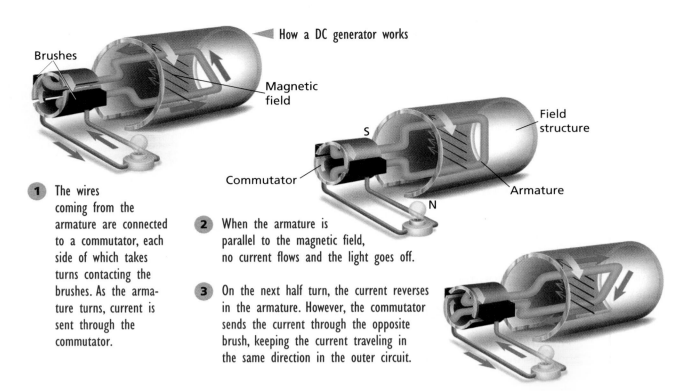

◀ How a DC generator works

Brushes

Magnetic field

S

Commutator

Field structure

Armature

N

1 The wires coming from the armature are connected to a commutator, each side of which takes turns contacting the brushes. As the armature turns, current is sent through the commutator.

2 When the armature is parallel to the magnetic field, no current flows and the light goes off.

3 On the next half turn, the current reverses in the armature. However, the commutator sends the current through the opposite brush, keeping the current traveling in the same direction in the outer circuit.

Being Direct Not all generators produce alternating current. A DC generator produces direct current. The diagram above summarizes its operation. You might notice a similarity between the AC and DC generators: in both, the current in the coil drops to zero and then reverses direction with every half turn. To compensate, the DC generator has a mechanical device called a *commutator*, which switches the connections to the circuit to keep the current flowing in the same direction.

Transforming Electricity

In getting electricity to do what you want it to do, it is often necessary to *transform* it—raise or lower its voltage. Since voltage is a measure of

the amount of "push" that an electric current has, the higher the voltage, the stronger the push.

As you already know, most of our electricity comes in the form of alternating current (AC). You may wonder why this type of current is used instead of direct current (DC), which flows only in one direction. After all, wouldn't the constantly changing current cause problems?

The primary reason AC is used is that it is easy to change its voltage. The device used to do this is called a *transformer*, shown below. In its simplest form, a transformer consists of coils of wire wound around an O-shaped iron core. Alternating current from the source flows through the *primary coil*. With each surge of current, the iron core becomes an electromagnet. Every time the current alternates, or switches direction, the electromagnet changes polarity. This causes the other coil of the transformer to respond as though it were rotating between the poles of a magnet. Current is therefore produced in the *secondary coil*.

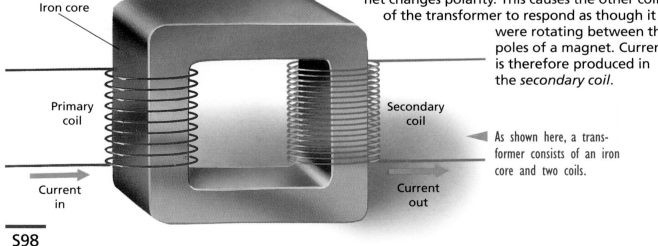

Iron core

Primary coil

Secondary coil

Current in

Current out

◀ As shown here, a transformer consists of an iron core and two coils.

S98

Wind-Up Voltage The relative amount of voltage produced in the secondary coil of a transformer depends on the number of turns of wire in each of the two coils. If the primary coil has more turns than the secondary coil, the voltage output will be lowered. This type of transformer is called a *step-down* transformer. If, on the other hand, the secondary coil has more turns than the primary coil, the voltage will increase. This is called a *step-up* transformer.

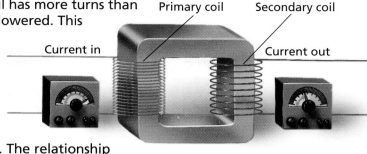

You can figure out the voltage in either of the two coils if you know the voltage in one coil and the number of turns in both coils. The relationship between voltage and turns in a transformer is given by the following equation:

$$\frac{\text{Voltage in the primary}}{\text{Voltage in the secondary}} = \frac{\text{Turns in the primary}}{\text{Turns in the secondary}}$$

$$\text{or} \quad \frac{E_P}{E_S} = \frac{T_P}{T_S}$$

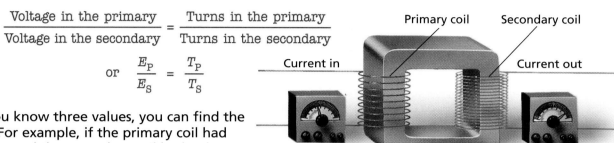

If you know three values, you can find the fourth. For example, if the primary coil had 100 turns and the secondary coil had only 50 turns, the voltage would be transformed from high to low voltage. If this transformer were plugged into a standard wall socket at 120 V, the output from the secondary coil would be 60 V.

A step-down transformer (top) and a step-up transformer (bottom)

Effects on Current If you transform the electricity by increasing or decreasing the voltage, there is a corresponding effect on current. A transformer can change the voltage (force) in a circuit, but the total electrical energy (work) must remain the same. In other words, when voltage is stepped up, current is stepped down. Remember that electrical energy is a function of both voltage and current. The change in current can be calculated as follows:

$$\frac{\text{Voltage in the primary}}{\text{Voltage in the secondary}} = \frac{\text{Current in the secondary}}{\text{Current in the primary}}$$

$$\text{or} \quad \frac{E_P}{E_S} = \frac{I_S}{I_P}$$

If, for example, the voltage is boosted from 100 V to 1000 V and the current in the primary coil is 2 A, the current in the secondary coil will be 0.2 A.

From the generating plant, electricity is usually transmitted over special lines at very high voltages, up to 765,000 V. This high voltage produces very low currents and therefore reduces losses caused by resistance in the power lines. Electrical substations reduce this high voltage to around 100,000 V for distribution to local power stations. A substation reduces the voltage to around 15,000 V for distribution within neighborhoods. And near your house, a small transformer further reduces the voltage to about 110 V, which is what most of your appliances use.

S99

Electricity Into Motion

A **motor** is a device that changes electrical energy into mechanical energy that can be used to do work. There are three basic types of motors: AC motors, DC motors, and universal motors. As their names suggest, AC motors work with alternating current, and DC motors work with direct current. Universal motors will work with either AC or DC.

How Motors Work   Motors operate on the same general principle as generators, only in reverse. In fact, some motors can be used as generators if they are turned by an outside force. Like a generator, an electric motor consists of an armature and a permanent magnet. The magnet creates a magnetic field in which the armature turns. The rotating magnetic field of the armature interacts with the magnetic field of the permanent magnet to cause the rotational motion of the motor.

1 In a motor, the armature becomes an electromagnet when it is supplied with electricity from an outside source. Because like poles of magnets repel and unlike poles of magnets attract, the armature assembly turns in response to the magnetic field of the permanent magnet.

▼ Like a generator, a motor consists of a permanent magnet and an armature. In a real motor, the armature would be attached to a drive shaft that turns and does work.

2 As the armature turns, however, the commutator causes the current passing through the armature to change direction. The polarity of the electromagnet is thus reversed, and the armature is forced to turn again. These continuously changing magnetic fields keep the armature turning for as long as the motor is in operation.

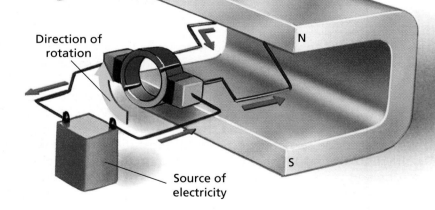

Brushes

Commutator

Armature

N

S

Permanent magnet

Direction of rotation

N

S

Source of electricity

DID YOU KNOW...

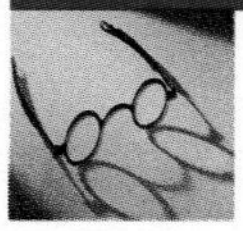

that electric motors are among the most efficient devices available for converting energy?
Many electric motors are 95 percent efficient. That is, they convert 95 percent of the electrical energy they use into mechanical energy. Friction and resistance claim the other 5 percent.

Circuit

A continuous path through which electricity flows

Pathways for Current

As you know, in order for electricity to flow, there must be an unbroken pathway that electrons can move through. Such a pathway is called a **circuit**. Circuits have three parts: a source of electric current, an output device, and a connection.

The source is typically a generator or a battery. When you plug in a blow-dryer, the source is not the wall socket but the distant generator that supplies your house with current. Output devices, such as lamps and motors, use the energy carried by the circuit to perform useful functions. A lamp, for example, converts electricity into light (and heat); a motor converts electrical energy into rotary motion. The connection, such as a wire, joins the components of a circuit together, providing a pathway for the electrons. As long as the pathway is intact, electricity will flow. If the connection is broken, however, electricity will cease to flow and the output device will quit working.

S101

Controlling Circuits For electricity to be useful, the current must be controlled in some way. For example, there should be a way to turn an electrical device off. The simplest device for controlling electric current is a *switch*. A switch stops the flow of electric current by introducing a gap into the circuit. Another common current-control device is the *rheostat*. A rheostat allows the amount of current flowing through a circuit to be varied at will by adding resistance to the current. The dimmer switch on a lamp and the volume knob on a stereo are examples of rheostats.

Circuits may be simple or quite complex. Some circuits, such as those in a computer, may contain hundreds or thousands of individual components.

As you know, there are two basic types of circuits—*series* and *parallel*. A series circuit has a single current pathway. The components of a series circuit are wired "in series," meaning that each component is located on the same current pathway. The entire electric current passes through every component in the circuit, one after another.

A parallel circuit, on the other hand, has at least two current pathways. Typically, a parallel circuit consists of branches coming off of a main current pathway. Each of the branches goes through an electrical component or output device. An electric current passing through a parallel current divides. Some current continues along the main circuit and some diverts into the branching circuits. After passing through the branches, the separate currents rejoin the common current pathway.

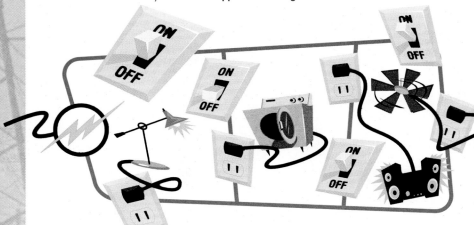

In a series circuit, all current follows one path. If the circuit is broken (or the switch is thrown), all of the appliances will go off.

In a parallel circuit, the current follows several branching paths. If the circuit is broken at any one branch, only the appliances on that branch will go off. The current will follow the alternate paths.

S U M M A R Y

Generators convert mechanical energy into electrical energy. Alternators generate alternating current, and DC generators generate direct current. Transformers allow for regulating the voltage in AC circuits. Motors convert electrical energy into mechanical energy. Circuits are the pathways through which electric current flows. Series circuits consist of a single current pathway, whereas parallel circuits consist of two or more current pathways.

Concept Mapping

The concept map shown here illustrates major ideas in this unit. Complete the map by supplying the missing terms. Then extend your map by answering the additional question below. Use your ScienceLog. **Do not write in this textbook.**

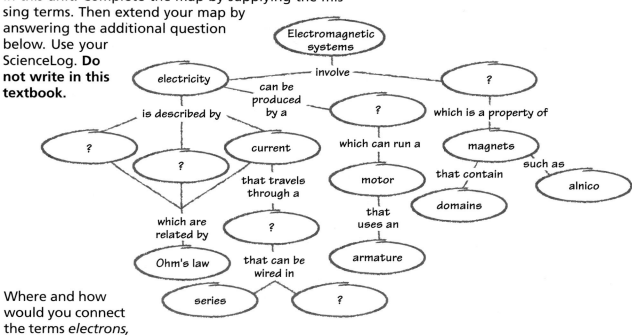

Where and how would you connect the terms *electrons, poles,* and *transformer*?

Checking Your Understanding

Select the choice that most completely and correctly answers each of the following questions.

1. An electric current results from
   a. the flow of positively charged particles through a conductor.
   b. the flow of electrons between areas of greater and lesser charge.
   c. the movement of ions through wires.
   d. the flipping of a switch.

2. Which of the following correctly describes the behavior of charged particles?
   a. Negatively charged particles attract positively charged particles.
   b. Negatively charged particles attract other negatively charged particles.
   c. Positively charged particles attract other positively charged particles.
   d. Negatively charged particles attract both positively charged and negatively charged particles.

3. Voltage can be compared to
   a. flow rate.
   b. current velocity.
   c. pressure.
   d. friction.

4. "The greater the distance between two objects, the weaker their interaction." This statement applies to
   a. only an electric force.
   b. only a magnetic force.
   c. both an electric and a magnetic force.
   d. neither an electric nor a magnetic force.

5. A generator converts
   a. electrical potential into electrons.
   b. electrical energy into mechanical energy.
   c. direct current into alternating current.
   d. mechanical energy into electrical energy.

S103

Critical Thinking

Carefully consider the following questions, and write a response in your ScienceLog that indicates your understanding of science.

1. Suppose that you have an iron nail and a compass. How could you demonstrate whether the nail was magnetized or not?

2. A certain transformer has a primary coil with 30 turns of wire and a secondary coil with 10 turns; the voltage in the secondary coil is 30 V. What is the voltage in the primary coil? Show your work.

3. Suppose you coupled a generator to an electric motor so that the motor drove the generator and the generator supplied the motor with electricity. Would this system run itself indefinitely? Why or why not?

4. Why is AC, but not DC, easily transformed? What would happen if you passed DC through a transformer? Explain.

5. All magnets have a *Curie point*, the temperature at which the materials cease to be magnetic. Above that temperature, no magnetism is present. As you know, there is a relationship between temperature and molecular behavior. How does this relationship help to explain the Curie point?

Interpreting Photos

Some of the bulbs in this string of lights are not working. What type of circuit—series or parallel—would account for the on-and-off pattern of bulbs shown here? Explain.

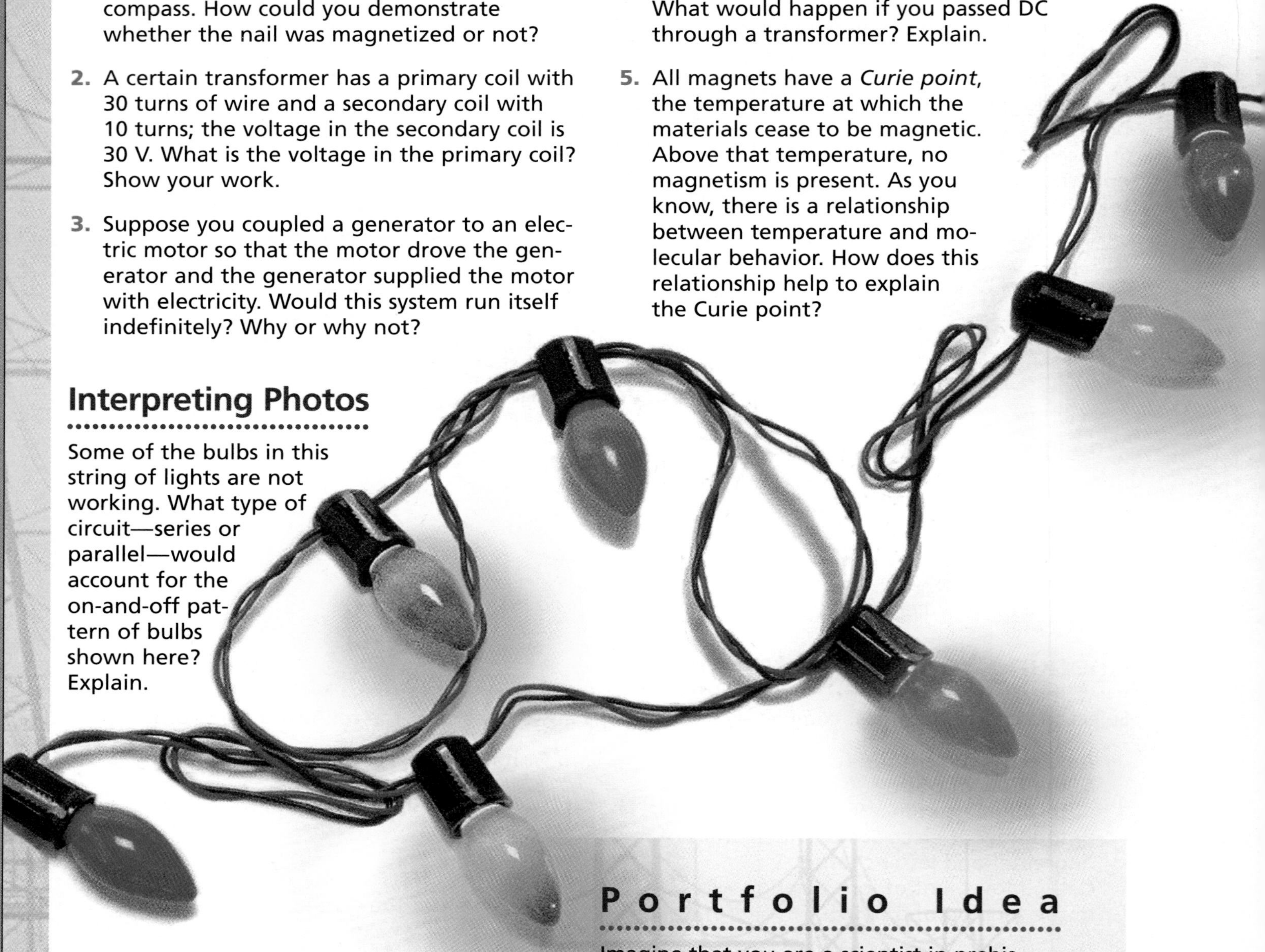

Portfolio Idea

Imagine that you are a scientist in prehistoric times, and you have just discovered magnetism. You are describing its properties to a colleague in a letter. Since you have discovered it well before the modern age of science, most of the words associated with magnetism do not yet exist. Describe your discoveries about magnets and magnetism to your colleague without using the terms *magnet*, *magnetism*, *pole*, *magnetic field*, or *lines of force*.

S104

SOUND

Unit 6

IN THIS UNIT

Waves,
    page S106

Sound Waves,
    page S115

Sound Effects,
    page S121

Now that you have been introduced to sound and the nature of its qualities, consider the following questions.

1. How do waves transmit energy from one place to another? How can you observe this with sound waves?

2. Does the fact that sound travels as waves affect how we hear it?

3. In what ways do we use sound, and how can we protect ourselves from sound levels that are too high?

In this unit, you will take a closer look at the physical basis for sound—waves—and at how modern technology uses sound.

S105

WAVES

You have been learning about sound waves. But what, exactly, are waves? There are many kinds of waves in addition to sound waves. You have probably heard of water waves, radio waves, microwaves, earthquake waves, and electromagnetic waves, as well as other types of waves. Because sound is made of waves, it has certain properties that are characteristic of all waves. So let's take a step back and investigate the phenomenon of waves in general.

If you have ever dropped a stone into a quiet pond, you produced many waves, or disturbances, in the surface of the water. A **wave** is a disturbance that transmits energy as it travels through matter or empty space. The material through which a wave travels is called a *medium*. The waves on the pond are produced when kinetic energy from the stone is transferred to the medium, which is the water. The waves then carry the energy outward through the medium, from the point where the stone hit the water.

It is important to understand that the energy of the wave moves *through* the medium; the medium itself does not travel with the wave's energy. It is like a long line of people standing close together. Imagine that the person at the end of the line pushes the next person in line, and that person pushes the next, and so on. Can you see how the disturbance can pass to the head of the line without any person taking a step forward?

Wave

A disturbance that transmits energy as it travels through matter or space

▲ Energy from a stone radiates outward in the form of waves.

Notice that although ▶ the wave continues traveling to the right, the floating object (as well as the water itself) remains in one place, making small bobbing motions.

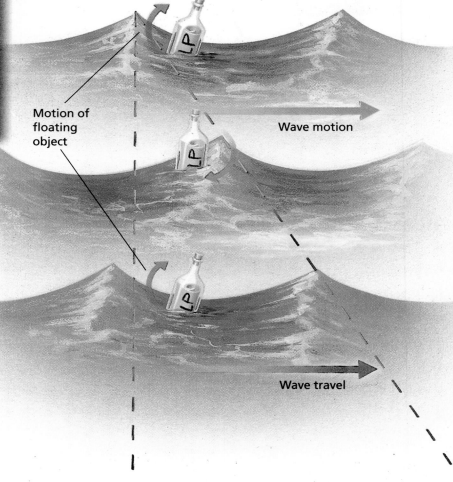

Motion of floating object

Wave motion

Wave travel

Types of Waves

A liquid, such as water, is one medium for waves. But gases, like those in the atmosphere, and solids, such as iron, can also be mediums for waves. We can study the action of waves without getting wet by considering waves in a rope. The rope acts as the medium for the waves. To study waves in more detail, look at the illustration below.

Transverse Waves   The waves produced by the up-and-down motion of a medium are called **transverse waves**. *Transverse* means "moving across." In a transverse wave, the material of the medium moves across (at right angles to) the direction in which the wave travels. Look at the illustration again. Notice that although the wave travels

to the right, no part of the rope moves to the right. Look at the colored markers on the rope. If you follow the motion of any one of them from one drawing to the next, you will see that each marker moves up and down. Every point on the rope follows the motion of the student's hand.

The top of a transverse wave is called a *crest*. Each crest is followed by a depression called a *trough*. Transverse waves consist of a series of crests and troughs that follow each other through a medium. The motion of the particles of the rope is perpendicular to the direction in which the wave is traveling. If the rope were shaken from side to side, would the wave still be transverse?

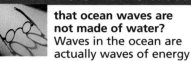
Transverse wave

A wave in which the motion of the particles of the medium is perpendicular to the path of the wave

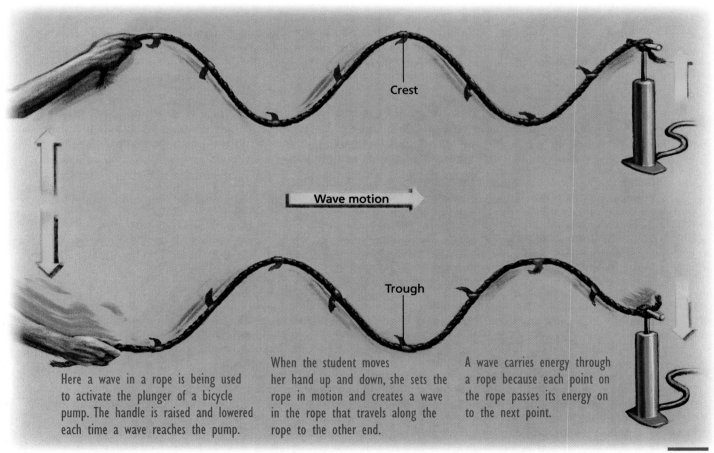

Crest

Wave motion

Trough

Here a wave in a rope is being used to activate the plunger of a bicycle pump. The handle is raised and lowered each time a wave reaches the pump.

When the student moves her hand up and down, she sets the rope in motion and creates a wave in the rope that travels along the rope to the other end.

A wave carries energy through a rope because each point on the rope passes its energy on to the next point.

S107

Longitudinal wave

A wave in which the particles of the medium move back and forth, parallel to the direction of motion of the wave

Longitudinal Wave A different kind of wave motion can be demonstrated by using a coiled spring like the one shown in the illustration below. In this case, a student produces a wave by moving her hand toward and away from a bicycle pump with a motion that is parallel to the length of the spring. The particles of the spring also move toward and away from the pump. Waves in which the particles of the medium move back and forth parallel to the path of the wave are called **longitudinal waves**.

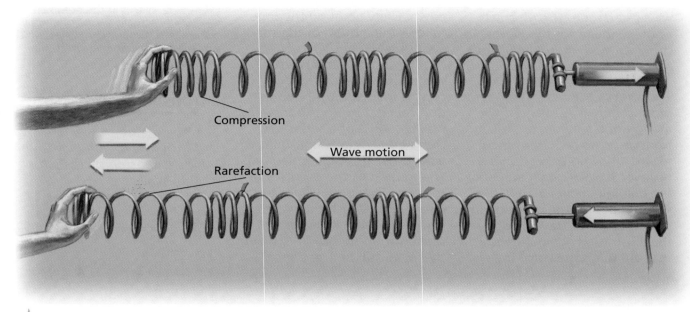

Compression

Wave motion

Rarefaction

▲ Here a spring is being used to activate the plunger of a bicycle pump. Each time a complete wave reaches the pump, it moves the pump handle in and out.

Notice that longitudinal waves do not have crests and troughs. Every time the student pushes the spring forward, she pushes some of the coils of the spring closer together. In other words, the student causes a *compression* in the spring. In each compression of a longitudinal wave, the particles of the medium are more dense than they are in the surrounding material. When the student pulls her hand back, she separates the coils. This causes an expansion in the spring, where the particles of the medium become less dense. This part of a longitudinal wave is called a *rarefaction*. Longitudinal waves consist of alternating compressions and rarefactions moving through a medium.

The crest of a transverse wave can be compared to the compression of a longitudinal wave, while a trough can be compared to a rarefaction.

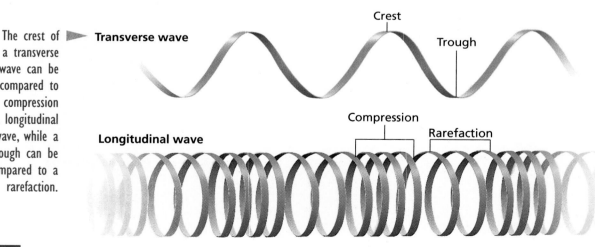

Transverse wave

Crest

Trough

Longitudinal wave

Compression

Rarefaction

S108

Properties of Waves

Just as you can describe matter by various physical properties, such as color, size, mass, and density, waves also have certain properties by which they can be described. Although they are produced in many different ways and cause many different effects, all waves can be described by four basic properties.

Amplitude   You can make a small wave in a rope by flicking your wrist while holding one end of the rope. You could make a much larger wave by moving your whole arm to shake the rope. Scientists use the term *amplitude* when referring to the size of a wave. **Amplitude** is the maximum distance that a certain point on a wave moves from a rest position. The amplitude of any given wave depends on the amount of energy in the wave. The greater the energy that the wave carries, the greater the wave's amplitude is.

Amplitude

The maximum distance that a wave rises or falls from a normal rest position

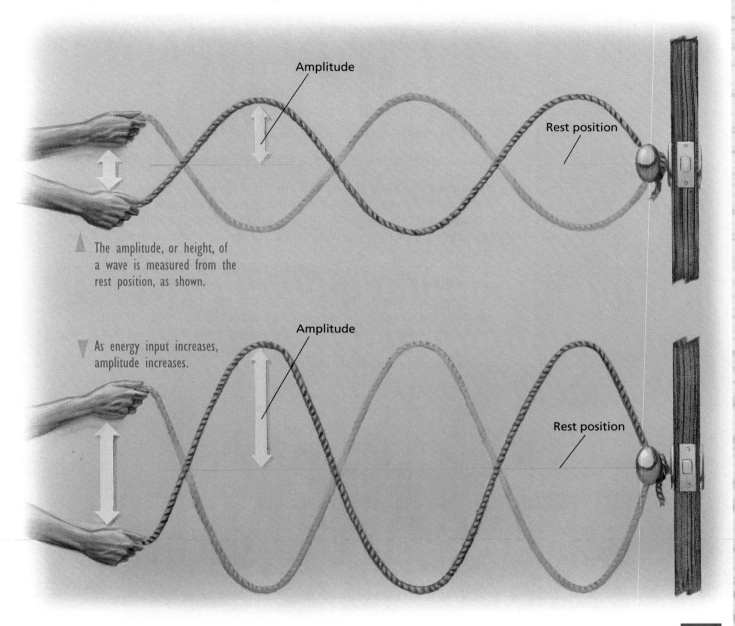

The amplitude, or height, of a wave is measured from the rest position, as shown.

As energy input increases, amplitude increases.

S109

Wavelength Waves are characterized not only by their amplitude, but also by their *wavelength*. **Wavelength** is the distance between a particular point on one wave and the identical point on the next wave. The easiest way to find the wavelength of a transverse wave is to measure the distance from one crest to the next. The easiest way to find the wavelength of a longitudinal wave is to measure the distance between two compressions.

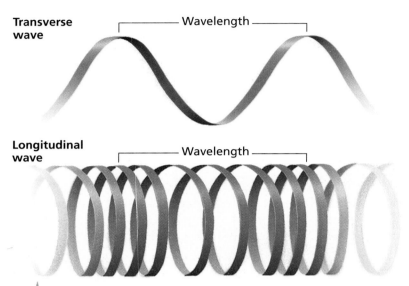

Transverse wave

Wavelength

Longitudinal wave

Wavelength

▲ The wavelengths of transverse and longitudinal waves are illustrated here.

Frequency If you shake one end of a rope very slowly, you will not produce many vibrations in the rope. However, if you shake the rope rapidly, you will create a large number of vibrations. The number of vibrations produced in a given amount of time is called **frequency**. Frequency indicates the rate at which vibrations are produced. It is usually measured in hertz (Hz), the number of vibrations per second.

There is an important relationship between the frequency of a wave and its wavelength. As the frequency increases, the wavelength decreases. You can see this by watching the vibrations in a rope as you shake it at different rates. The faster you shake the rope, the closer together the crests will be.

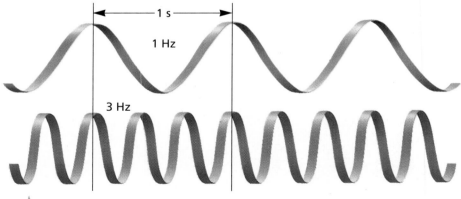

1 s

1 Hz

3 Hz

▲ The wave on the bottom has a wavelength one-third that of the wave on the top, while it has a frequency three times as great.

▲ Objects that vibrate with constant frequencies can be used to measure time. The pendulum of a grandfather clock, for example, might swing back and forth once each second—a frequency of 1 Hz—while the quartz crystals used in most digital watches vibrate at a frequency of 32,768 Hz. Frequencies also determine phenomena such as the pitch of musical tones and the color of light.

S110

Wave Speed Another basic property of waves is the speed at which they travel. The speed of a wave can be determined by observing a certain point on the wave as it moves. In longitudinal waves, you could observe a single compression or rarefaction as it travels through the medium. In transverse waves, you could observe a single crest or trough. For example, the speed of a wave through water can be found by observing one crest and measuring the distance it moves in a certain amount of time. Wave speed is usually given in meters per second (m/s).

Wave speed can be calculated if you know the wavelength and frequency of a wave. The relationship between speed, frequency, and wavelength is expressed in the following formula:

$$\text{speed} = \text{frequency} \times \text{wavelength}$$

or

$$v = f\lambda$$

For example, suppose that you are fishing from an anchored boat. Several waves pass the boat. You could find the speed of those waves by calculating the number of waves that pass in 1 second (frequency) and measuring the length of 1 wave. If the frequency is 2 waves/s (2 Hz) and the wavelength is 0.5 m, the speed is

$$v = f\lambda$$
$$v = 2 \text{ waves/s} \times 0.5 \text{ m/wave}$$
$$v = 1 \text{ m/s}$$

The same formula can be used for sound waves. The speed of sound (*v*) is known to be about 346 m/s. So if you can determine either the frequency (*f*) or the wavelength (λ) of a sound, you can calculate the other value.

Wavelength

The distance between two identical points on neighboring waves

Frequency

The number of waves produced in a given amount of time

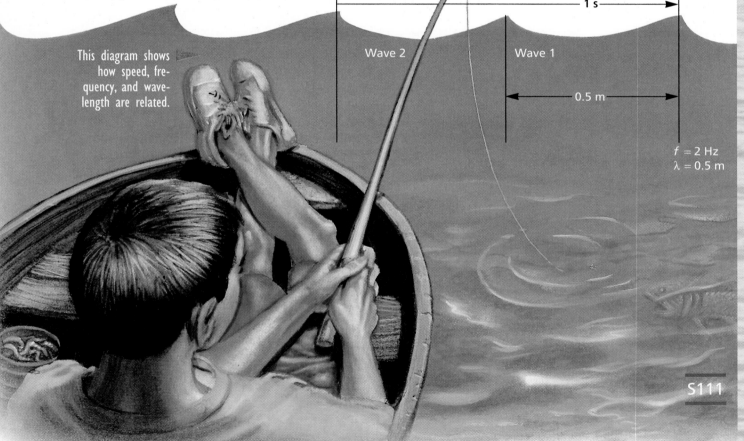

This diagram shows how speed, frequency, and wavelength are related.

1 s

Wave 2

Wave 1

0.5 m

f = 2 Hz
λ = 0.5 m

Characteristics of Waves

There are certain characteristics that all waves have in common. *Reflection*, for example, is displayed by waves when they encounter obstacles. Other characteristics include *refraction*, *diffraction*, and *interference*.

Reflection

The process in which a wave bounces back after striking a barrier that does not absorb all of the wave's energy

Reflection   Have you ever watched or played a game of pool? A skillful player knows how to use the side rail of the table to his or her advantage. Success often depends on getting the balls to bounce at the desired angles. Normally, a ball will bounce off the side rail of the table at the same angle at which it strikes it.

When playing pool, you must consider the laws of reflection.

Waves bounce off barriers in the same manner that a pool ball does. This is called **reflection**. To demonstrate that waves are reflected, try these experiments. Tie one end of a rope to a doorknob. Hold the other end of the rope and send a transverse wave pulse through the rope. You will see that when the wave reaches the doorknob, it bounces back. Longitudinal waves are also reflected by barriers. Hold a coiled spring at one end. If you send a longitudinal wave down the spring, the wave will come back when it reaches the end of the spring. To demonstrate the reflection of water waves, make a wave in a sink or tub full of water by tapping your finger on the surface of the water. When the wave reaches the side of the sink or tub, you will see it bounce back. For an example of the reflection of sound waves, recall how sound is reflected from walls or the sides of buildings to produce echoes.

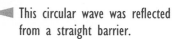

This circular wave was reflected from a straight barrier.

S112

If waves approach the shore at an angle, the ends closest to the shore slow down first, and the waves are refracted.

Refraction The bending of a wave when it passes from one medium to another is called **refraction**. Refraction is due to a change in the speed of a wave. You can observe the refraction of waves at the ocean shore. As waves in the ocean approach the shore, they enter shallow water and slow down. This is because waves travel faster in deep water than in shallow water. If a wave approaches the shore at an angle, the end closest to the shore slows down first. As a result, the wave bends toward the beach.

Refraction can be compared to the column of a marching band that marches at an angle from hard pavement into a patch of soft mud. The band members at one side of the front row will encounter the mud first and, finding it harder to march, will slow down and march with shorter strides. The others in the front row will continue at their initial speed until each, in turn, enters the mud. In effect, each row in the column of marchers will turn toward the mud. The opposite effect will occur when the column returns from the mud to the pavement.

Refraction

The process in which a wave changes direction because its speed changes

Diffraction

The bending of waves as they pass the edges of objects

Diffraction Waves can also go around corners. Water waves clearly illustrate this behavior. The image to the right shows ripples from a vibrating source touching the water surface in a "ripple tank." Straight water waves (coming from the top of the picture) bend around the edge of the barrier. This is called **diffraction**. All waves, including sound waves, exhibit diffraction. For example, you can hear a sound from around a corner, even if there is nothing to reflect the sound to you.

Straight waves can be bent around the edges of a barrier.

S113

Interference

The result of two or more waves overlapping

Interference   Have you ever thrown two stones into a still pond at the same time and seen the ripples from the two splashes pass through each other?

Waves that meet in this manner affect each other by a process called **interference**. When two troughs meet, they combine to make a single, larger trough. Likewise, when two crests meet, they combine to make a single, larger crest. The result is deeper troughs and higher crests. This phenomenon is called *constructive interference*. On the other hand, when a crest and a trough of equal amplitude meet, they cancel each other out. This is called *destructive interference*. Interference is a characteristic of all waves, including sound waves.

When two waves cross each other's path, they affect each other's amplitude.

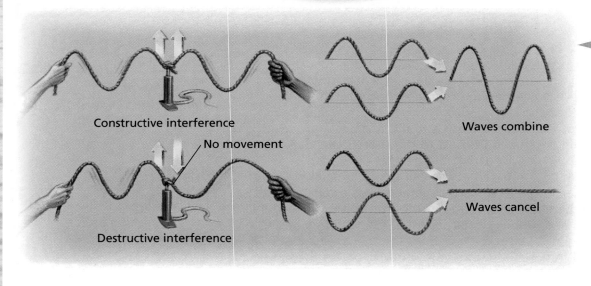

Constructive interference

No movement

Destructive interference

Waves combine

Waves cancel

Waves that undergo constructive interference (top) produce a single, stronger wave. Waves that undergo destructive interference (bottom) tend to cancel each other out.

DID YOU KNOW...

that the property of destructive interference is used to reduce noise in the workplace? Special instruments pick up the background noise in office buildings and factories, invert the signals, and output sound waves that cancel the background noise. This greatly reduces noise levels of the working environment.

SUMMARY

A wave transmits energy through a medium or through empty space. Both transverse waves and longitudinal waves exhibit the properties of amplitude, wavelength, frequency, and speed. Reflection, refraction, diffraction, and interference are characteristics common to all waves. Reflection is the bouncing back of a wave when it meets a barrier. Refraction is the bending of a wave when it passes at an angle from one medium to another, either slowing down or speeding up. Diffraction is the bending of straight waves around the edge of a barrier. When two or more waves meet, they interfere with each other.

S114

SOUND WAVES

Without air, there would be no familiar sounds. In fact, we would hear no sounds at all. This is because air is a medium for sound. Sound moves through air in the same way that ripples move through water. A sound begins when a vibrating object, such as a guitar string, pushes on the air particles around it. These particles then push on other particles, causing a disturbance that is transmitted outward from the source. As sound waves move through the air, the air particles are alternately crowded together and spread apart. This back-and-forth motion of the particles of air is similar to the way longitudinal waves move through a spring. In other words, sound consists of longitudinal waves.

When air is removed from the bell jar, the alarm clock will not be heard. No air molecules are there to transfer the energy as a wave.

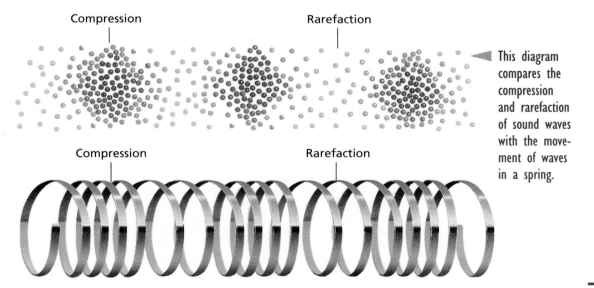

Compression   Rarefaction

Compression   Rarefaction

This diagram compares the compression and rarefaction of sound waves with the movement of waves in a spring.

S115

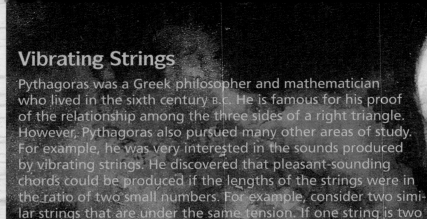

Vibrating Strings

Pythagoras was a Greek philosopher and mathematician who lived in the sixth century B.C. He is famous for his proof of the relationship among the three sides of a right triangle. However, Pythagoras also pursued many other areas of study. For example, he was very interested in the sounds produced by vibrating strings. He discovered that pleasant-sounding chords could be produced if the lengths of the strings were in the ratio of two small numbers. For example, consider two similar strings that are under the same tension. If one string is two times the length of the other, they will produce two notes that are one octave apart (for example, from C to C). If the lengths of the strings are in the ratio of two to three, then a fifth chord is produced (for example, from C to G). These chords are considered to be very pleasing to the ear.

Pythagoras was so impressed by his discovery that he made it the basis for a whole school of thought. He believed that the beauty of the universe was related to the beauty of these chords. For example, he thought that the movements of the planets and stars were based on the same ratios that he discovered in vibrating strings. The phrase "music of the spheres" comes from this idea. Later scientists disproved Pythagoras's theory concerning the movements of stars and planets. However, his law of vibrating strings, which came from direct observation, has withstood the test of time.

DID YOU KNOW...

...that Pythagoras was the leader of a secret society? The investigations he and his colleagues did with numbers and ratios were jealously guarded as wonders that the common people would not understand.

▲ Pythagoras thought that the sun and the planets moved in perfect harmony, one that could be expressed with simple numbers.

Standing Waves

An important characteristic of vibrating strings is the patterns of waves that they form. Suppose you send a wave through a rope and the wave is reflected by a barrier. If you send only one crest, it will travel to the barrier and then be reflected back toward you. The crest will encounter no interference. However, if you continue to produce waves, the waves traveling toward the barrier will soon encounter waves reflected from the barrier. In this situation, the waves will interfere with one another.

Because the strings on a musical instrument like the guitar are fixed at both ends, they reflect from both ends. Waves traveling in opposite directions that have matching characteristics—amplitude, frequency, and speed—combine to form **standing waves**.

The points on standing waves that have no vibration are called *nodes*. Nodes are caused by destructive interference between waves moving in opposite directions. The points on standing waves that vibrate with the greatest amplitude are called *antinodes*. Antinodes are the result of constructive interference between waves moving in opposite directions.

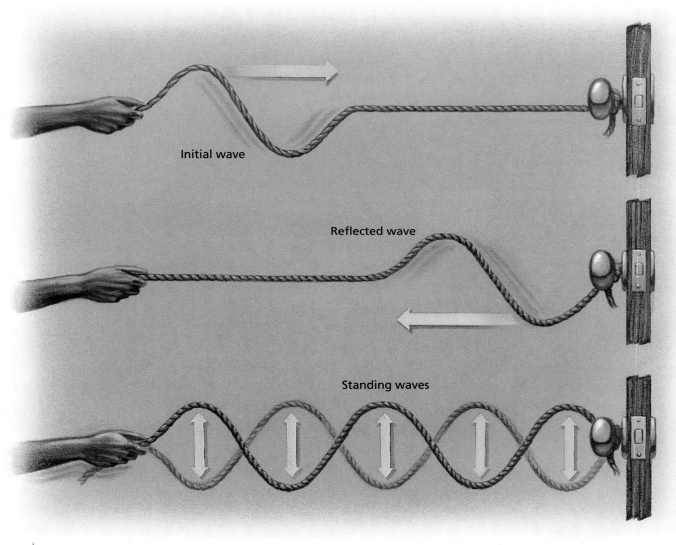

Initial wave

Reflected wave

Standing waves

▲ When reflected waves meet waves moving in the opposite direction, standing waves are formed.

S117

Standing waves look very different from traveling waves. Standing waves do not appear to move through the medium. Instead, the waves cause the medium to vibrate in a series of stationary *loops*. Each loop is separated from the next by a node, or point of no movement. The distance from one node to the next is always one-half of a wavelength.

The following diagram shows the vibrations of a string. In each example, a standing wave has been produced. Notice that both ends of the string are fastened tightly and cannot move. Therefore, there must be a node at each end of the string.

Standing waves can also be formed by longitudinal waves. Standing waves in a column of air cause the sound you hear when you blow across the top of a soda bottle. Columns of standing waves are also responsible for the sounds of flutes, clarinets, and other wind instruments.

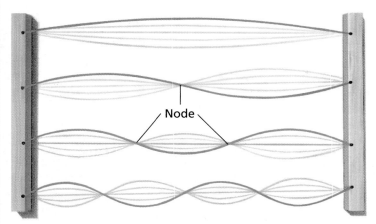

Node

1 The frequency of the first vibration has one loop in the wave. Since the loop is one-half wavelength long, the wavelength is twice the length of the string.

2 In this wave, the string has a total of three nodes. The entire string is equal in length to one wavelength.

3 There are four nodes in this wave. Here the string is equal to the length of one and one-half wavelengths.

4 Five nodes are present in this wave. The entire length of the string includes two wavelengths.

Each of these wavelengths occurs at a different frequency. When you pluck a string, it produces all of these frequencies at the same time. The lowest frequency is called the *fundamental*; higher frequencies are called *overtones*. The number and relative intensity of the overtones give the characteristic sound to the instrument.

DID YOU KNOW...

that standing waves can cause glass to break and bridges to collapse? When standing waves match the natural vibrations of a material object, they may become amplified, often enough to cause the object to shatter. Opera singers have been known to break goblets, and in 1940, the Tacoma Narrows Bridge was destroyed by winds that induced destructive standing vibrations in the roadway.

▼ When guitar strings are plucked, a variety of standing waves are produced, blending to form a guitar's distinctive sound quality.

The Doppler Effect

Have you ever heard the blast of the warning horns as a train passed through a railroad crossing? Did you notice how the sound of the horns changed as the engine passed by? You probably heard a sudden drop in the pitch of the sound. But why?

The pair of photographs below shows the pattern of sound waves produced by a moving sound source. Notice that as the horns on the train move forward, the sound waves in front of the train are closer together, giving these waves a shorter wave-length and thus a higher frequency. The sound waves in back of the train are spread farther apart, so they have a longer wavelength and a lower frequency. This change in the apparent frequency of the waves from in front of to behind a moving wave source is called the **Doppler effect**.

As sound waves are crowded toward you, the frequency at which they reach your ears increases. Remember that higher frequency sound waves have a higher pitch. Therefore, as a train approaches you, you hear a higher pitch than you would if the train were standing still. Once the train has passed by, the apparent pitch of each horn decreases. However, the actual pitch does not change. The engineer on board the train always hears the same sound.

Doppler effect

A change in the apparent frequency of waves caused by the motion of either the observer or the source of the waves

▲ As the train approaches, the sound waves are "bunched up."

▲ The pitch of the horn seems to get lower after the train passes by because the sound waves are spread out.

The Doppler effect also applies to situations in which the observer, rather than the wave source, is moving. Think of watching water waves from a moving boat. When the boat heads into the waves, the waves will hit the boat more often. The frequency of the waves will appear to increase. When the boat moves in the same direction as the waves, the waves will hit the boat less often. The frequency of the waves will appear to decrease. To someone watching the waves from the shore, the actual frequency of the waves will not change at all.

The Doppler effect is a property of all waves, not just sound and water waves. Light waves from distant galaxies also display the Doppler effect, in the form of a shift toward the red end (longer wavelengths) of the spectrum of light. This *red shift* indicates that these galaxies are moving away from us at high speed.

As indicated by this artist's conception, light from faster-moving galaxies appears shifted more toward the red end of the spectrum.

S120

SUMMARY

There would be no sound without air or some other transmitting medium. Sounds in air begin when vibrating objects push on air particles, causing back-and-forth motions of these particles, or longitudinal waves. Standing waves are formed when similar waves traveling in opposite directions interfere with each other. They form a pattern in which the medium vibrates with nodes and antinodes. The Doppler effect—a property of all waves—is a change in the apparent frequency of waves caused by the movement of the wave source, its observer, or both.

SOUND EFFECTS

In addition to the sounds we can hear, there are sounds we cannot hear. These sounds can be used in a variety of ways, from locating earthquakes and measuring the depth of the ocean to ensuring good health. Whether produced naturally or artificially, audible and inaudible sounds are an important aspect of our daily lives.

Infrasonic waves

Sound waves at a frequency below that which humans can hear

Infrasonic Waves

Sounds with frequencies below 20 Hz consist of **infrasonic waves**. These frequencies are too low for human ears to detect. Earthquakes produce infrasonic waves that move through the solid material of the Earth. Two types of waves originate from the focus of an earthquake—primary and secondary waves. *Primary waves*, or P waves, are longitudinal waves that travel by compressing rock and soil in front of them and stretching rock and soil behind them. These waves move quickly and are the first ones to reach earthquake-recording stations. It is for this reason that they are called primary waves. *Secondary waves*, or S waves, are transverse waves. These waves travel more slowly than P waves and cause the ground to move up and down or from side to side.

Primary and secondary earthquake waves come from the focus of an earthquake, which is often far below the epicenter where surface waves originate.

The interaction of P and S waves with the Earth's surface produces a third type of wave, called a *surface wave*. These waves cause the surface of the Earth to shake and roll. Surface waves are unique in that they originate from the epicenter of an earthquake, not from its focus. Their rolling action, which is like the rising and falling of ocean waves, causes many buildings to collapse.

The instrument used to record earthquake waves is called a *seismograph*. A seismograph consists of a rotating drum wrapped with paper and a pen attached to a suspended weight. The pen presses gently against the paper-wrapped drum. The recording from a seismograph is called a seismogram.

Since P waves travel fastest, they are the first waves recorded on the seismogram. The difference in travel time between the arrival of P and S waves is used to determine the distance between the recording station and the epicenter of the earthquake.

Seismologists keep detailed records of earthquake activity.

S121

Volcanic eruptions and tornadoes also produce infrasonic waves that travel for thousands of kilometers through the atmosphere. When Krakatau erupted in one of the largest volcanic events in recorded history, the infrasonic waves it produced could still be detected after they had circled the Earth several times.

Infrasonic waves are not always connected with disasters. Scientists use infrasonic waves to investigate the inner structure of the sun. *Helioseismology,* the study of solar "earthquakes," is based on the assumption that sound waves generated within the sun cause a pattern of peaks and valleys on the sun's surface. Because the movement of waves through a medium depends on factors such as elasticity and density, helioseismologists can learn about processes inside the sun by analyzing the surface wave patterns.

Ultrasonic Waves

Insects and other animals can detect sounds that are much higher than those detected by the human ear. Sounds in this upper range are called ultra-high-frequency sounds, or **ultrasonic waves.** Ultrasonic waves have been put to use in a variety of ways, from industrial cleaners to medical diagnostic and treatment instruments.

Many ships have a navigation system that uses echoes of ultrasonic waves to find the depth of the water. This system is called *sonar,* which stands for *so*und *na*vigation *r*anging. This system works in very much the same way that echolocation works in bats. A sonar device sends short pulses of ultrasonic waves through the water. When the sound waves hit the sea floor or other underwater objects, some of the waves are reflected back to the ship as an echo. A receiver detects the echo, and the time delay indicates the distance. Some autofocus cameras operate in the same way.

Ultrasonic waves can also be used to clean jewelry, machine parts, and electronic components. A device called an *ultrasonic cleaner* consists of a container that holds a bath of water and a mild detergent. Sound waves are sent through the bath, causing intense vibrations in the water that remove dirt from the items placed in the bath.

Ultrasonic waves

Sound waves at frequencies too high for humans to hear

▲ Bats use ultrasonic waves to locate prey in the dark.

Submarine pilots rely completely on sonar in order to navigate underwater. ▶

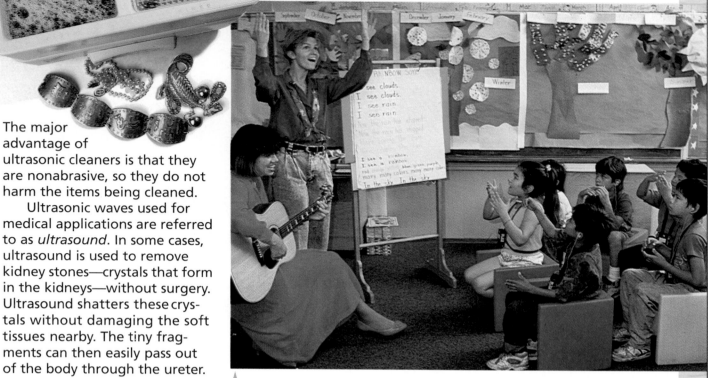

An ultrasonic cleaner, such as the one shown here, uses high-frequency sound waves to clean jewelry.

The major advantage of ultrasonic cleaners is that they are nonabrasive, so they do not harm the items being cleaned.

Ultrasonic waves used for medical applications are referred to as *ultrasound*. In some cases, ultrasound is used to remove kidney stones—crystals that form in the kidneys—without surgery. Ultrasound shatters these crystals without damaging the soft tissues nearby. The tiny fragments can then easily pass out of the body through the ureter.

Ultrasound also provides a way to "see" inside the body. Ultrasonic waves bounce off high-density tissues and are converted into electrical signals that are fed into a computer. The computer uses these signals to form a type of picture called a *sonogram*. Using this technology, doctors can locate tumors and gallstones. They can also examine a developing baby inside its mother to determine whether it is forming normally and if it is in the proper position.

Your first baby picture may have been a sonogram taken by ultrasonic waves.

By using electronic technology, people with hearing loss can interact with sound.

Sounds of Silence

Most people can hear a wide range of sounds. This is not the case for individuals who are hard of hearing. However, even a person who is completely deaf can detect certain sound waves in the environment. Since sound waves cause vibrations in solid materials, these vibrations can be felt through the sense of touch. Low-frequency sounds, especially, can be felt this way. Deaf people can dance to the beat of a loud band by sensing the strong beat of the bass guitar and drums. For people with impaired hearing, hearing aids amplify frequencies that are hard to hear. Modern electronic technology has produced hearing aids so small that they can be inserted completely into the ear canal.

S123

Hearing loss can result from exposure to music played at high volumes or from the extended use of personal stereo headphones.

Loud sounds and even loud music can permanently damage your hearing. Sound intensity is measured in *decibels* (dB). The higher the decibel level, the more intense the sound. Generally, the more intense a sound, the louder it is. Above a certain level of intensity, noise can harm your health. A quiet conversation, for example, is only about 40 dB, a harmless level of sound. Hearing damage may begin, however, when a person is exposed to noise levels of about 75 dB for 8 hours a day. A noise of more than 115 dB causes immediate pain. Loud music and crowd noise at rock concerts can reach 100 to 130 dB. Repeated exposure to high noise levels at rock concerts and the long-term use of personal stereos may cause permanent hearing impairment. Do not let the world of sound fade away due to carelessness!

S124

Sound Intensity Levels

Type of sound	Intensity (dB)	Hearing damage
Whisper	10–20	None
Soft music	30	None
Conversation	60–70	None
Vacuum cleaner	70	After long exposure
Heavy street traffic	70–80	After long exposure
Construction equipment	100	Progressive
Thunder	110–120	Progressive
Loud rock concert	110–130	Immediate and irreversible
Jet engine at takeoff	120–150	Immediate and irreversible

SUMMARY

Sound waves, both audible and inaudible, can be used in many ways. Infrasonic waves are very-low-frequency sound waves. They are produced by several natural phenomena, such as earthquakes, and are even used to study the sun's interior. Ultrasonic waves are used by sonar devices, ultrasonic cleaners, and some medical instruments. People who are hearing-impaired cannot hear the sounds that are constantly around us, but they can sense strong vibrations. Loud sounds or music can cause permanent hearing loss.

Concept Mapping

Starting with the terms supplied here, construct a concept map that illustrates major ideas from this unit. Arrange the terms in an appropriate manner and connect them with linking words. Then extend your map by adding as many additional terms from the unit as you can. Use your ScienceLog. **Do not write in this textbook.**

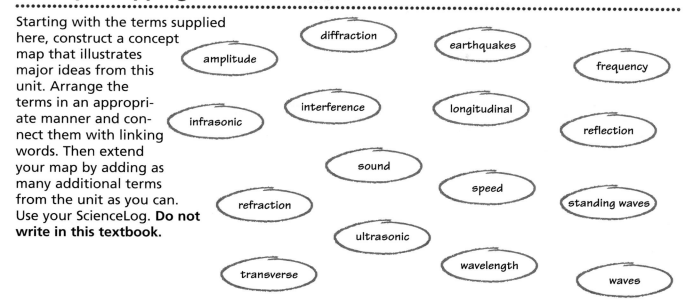

Checking Your Understanding

Select the choice that most completely and correctly answers each of the following questions.

1. When a wave moves the particles of its medium at right angles to the direction of the wave itself, the wave is a
   a. transverse wave.
   b. longitudinal wave.
   c. sound wave.
   d. wavelength.

2. The relationship between frequency and wavelength is expressed by which of the following?
   a. As frequency increases, wavelength increases.
   b. As frequency decreases, wavelength decreases.
   c. As frequency increases, wavelength decreases.
   d. As frequency decreases, wavelength stays the same.

3. The action of waves bending around a corner is called
   a. reflection.
   b. refraction.
   c. diffraction.
   d. interference.

4. A train sounding its horn moves toward you at a constant speed. As it passes the location where you are standing, the pitch of the sound gets lower because
   a. the frequency of the sound produced decreases.
   b. the wavelength of the sound produced increases.
   c. the frequency of the sound heard increases.
   d. the wavelength of the sound heard increases.

5. Which of the following would pose the most immediate danger to your hearing?
   a. a singing trio that couldn't stay in tune
   b. a supersonic jet flying over your house
   c. sitting in the third row of a loud rock concert
   d. operating a vacuum cleaner for 30 minutes a day

S125

Interpreting Graphs

According to the graph shown here, which animal's range of hearing corresponds to the longest bar on the graph? Which animal actually has the greatest range of hearing? How can you explain this apparent discrepancy?

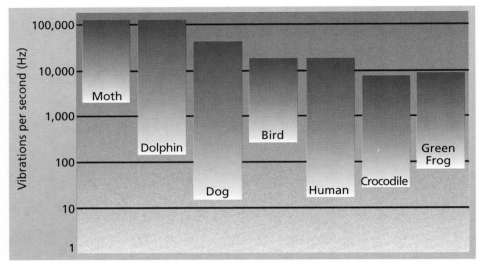

Critical Thinking

Carefully consider the following questions, and write a response in your ScienceLog that indicates your understanding of science.

1. In what ways are sound waves like other waves? In what ways are they different?

2. A popular science-fiction film used the following statement in its advertisements: "In space, no one can hear you scream . . ." Explain this statement in terms of what you have learned about sound waves.

3. Assume that sound travels through air at 346 m/s. If you were able to produce a sound wave with a frequency of 440 Hz, what would the wavelength of the sound wave be?

4. Suppose that you were at an auto-racing track. If you could listen only to the sound of the cars as they passed, would there be a way to estimate which car was moving the fastest? Explain.

5. If a tree falls in a forest and no one is there to hear it, does the tree make a sound when it hits the ground? Explain your answer.

Portfolio Idea

When new highways are constructed through residential areas, noise-abatement walls are often built along the roadway. These walls help reduce the noise of traffic entering neighborhoods, noise that can often reach levels exceeding 70 dB. Do some research to find out how these walls work and in what cities they have been used. Also investigate some of the problems associated with using or not using noise-abatement walls along busy highways.

LIGHT

Unit 7

IN THIS UNIT

Light: A Split Personality?
page S128

Visible Light,
page S133

Invisible Light,
page S137

High-Tech Light,
page S144

Now that you have been introduced to light, its relationship to heat and color, and how images form, consider the following questions.

● **1.** How does light behave—as a stream of particles, a series of waves, or both?

2. What causes us to see the different colors of the objects around us?

3. What is "invisible" light?

4. How does modern technology make use of the properties of light?

In this unit, you will take a closer look at the composition of light and at some of the high-tech ways in which light is used.

S127

LIGHT: A SPLIT PERSONALITY?

E ver since scientists first began investigating light, there have been conflicting ideas about it, not only about what gives light color, but also about the very makeup of light. Light seems to act in different ways under different circumstances. Various theories of light have been advanced to explain some of its properties. But, as you will see, none of these theories explains all the known properties of light.

The Particle Theory

Isaac Newton, whose laws of motion and gravity are the basis of traditional physics, wrote a scientific paper on light in 1672. Newton performed many experiments with light at his home in England. Besides using prisms to study color, he also studied how light is affected by objects in its path. He wanted to learn how light creates shadows, how mirrors reflect light, and how various materials refract light. Newton explained the results of his experiments by assuming that rays of light consist of streams of tiny particles. He called these particles "corpuscles."

Newton developed a theory that explained all the phenomena of light known in his day. For example, Newton argued that light particles are reflected in the same way that a ball bounces off a wall. In both cases, the angle of incidence equals the angle of reflection. Also, light changes its direction when it enters a new medium. This behavior of light is known as *refraction*. Newton compared the refraction of light to the behavior of a ball rolling down an incline.

Newton's most compelling evidence in favor of a particle theory of light, however, was that light seems to travel in a straight line. For example, an object placed in the path of light casts a shadow; in other words, the object stops the light. In the same way, a rolling ball can be stopped by a barrier placed in its path.

Newton thought that light consists of streams of tiny particles.

The ball rolls faster down the incline, and its direction changes.

The ball slows down when it leaves the incline, and its direction changes again.

The path of a rolling ball bends as the ball rolls down an incline, as seen by the changes in angles. Because light bends in a similar way, Newton saw this as evidence that light consists of particles. However, while the ball in this model speeds up, refracted light slows down.

S128

Meanwhile, the Dutch physicist Christian Huygens theorized that light consists of waves. To support this idea, he pointed out that one beam of light passes through another without either beam being disturbed. Waves can do this; streams of particles presumably cannot. A wave theory could also explain reflection and refraction of light because both phenomena are common to waves such as water waves. But Huygens could not explain why light seems to travel only in straight lines, whereas sound waves can travel around barriers. The bending of waves around barriers is called *diffraction*. If both sound and light consist of

waves, why can you hear, but not see, around corners? If light consists of waves, it should also be diffracted around barriers. Because there was not enough evidence supporting Huygens's wave theory, Newton's particle theory of light became the accepted theory for the next 100 years.

◄ Huygens thought that light, like sound, consists of waves.

The Wave Theory

In 1801, another English scientist, Thomas Young, discovered that light, like sound, *can* be diffracted. This was a powerful argument in favor of the wave theory

of light because diffraction is a property of waves but not of particles. An example of the diffraction of water waves is seen in the photos to the left, below.

To see whether light behaves as particles or as waves, Young did a simple experiment. He made two narrow, parallel slits in a card. He then placed a light source on one side of the card and a screen on the other. Young reasoned that if light consists of particles, two bright bands of light should appear on the screen. But if light consists of waves, the light should be diffracted as it passes through the slits. Because each slit acts as a separate circular wave source, the waves from these two wave sources would *interfere* with each other, resulting in a pattern of bright and dark bands on the screen. This is exactly what happened in Young's experiment, as the diagram below shows.

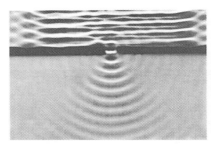

▲ When water waves pass through a sufficiently small opening, they are diffracted, or bent around the edges of the opening. When water waves pass through a large opening, little or no diffraction takes place.

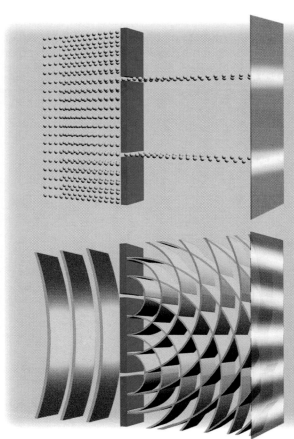

If light consists of particles (left), no interference will occur. If light consists of waves (below), diffracted waves will interfere with one another.

Bright bands appear where the crests and troughs of waves from two wave sources coincide.

Dark bands appear where the crests of waves from one wave source coincide with the troughs of waves from the other wave source.

S129

Huygens may not have detected any diffraction of light because, in order for a wave to be diffracted, the opening through which it passes must not be much larger than its wavelength. Because the wavelengths of visible light are very small (about 0.0005 mm), an opening must be very narrow in order to diffract light. The slits Young used were narrow enough to cause the diffraction of light and the interference of light waves. As a result of Young's experiments, the wave theory eventually replaced Newton's particle theory as the accepted model of light.

Broadening the Spectrum

In 1864, the wave theory of light received a major boost from the Scottish physicist James Clerk Maxwell. While studying the relationship between electricity and magnetism, Maxwell showed that light consists of **electromagnetic waves**. Using mathematics, he showed that electromagnetic waves, which have an electric field and a magnetic field at right angles to each other, would travel at a constant velocity. Maxwell was surprised to find that this calculated velocity is the same as the velocity of light! Although he was not looking for a theory of light, the coincidence was striking. Thus, Maxwell suggested that light is an electromagnetic wave.

The German scientist Heinrich Hertz performed an experiment that verified Maxwell's theory. Hertz generated electromagnetic waves with a wavelength of approximately 1 m. Later experiments showed that a wide range of electromagnetic waves could be generated. Electromagnetic waves can have wavelengths that range from thousands of kilometers to billionths of a centimeter. These waves are a type of energy called *electromagnetic radiation*. Visible light, whose wavelength (and frequency) is approximately in the middle of the range of all electromagnetic waves, is only part of a much broader spectrum of electromagnetic radiation.

Electromagnetic waves

Waves that carry both electric and magnetic energy and move through a vacuum at the speed of light

DID YOU KNOW...

that the speed of electromagnetic waves is not the same in all materials? Electromagnetic waves move more slowly in denser materials. The degree to which a wave is slowed depends on its wavelength.

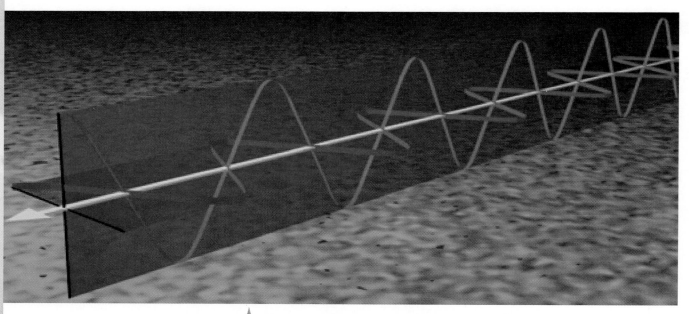

▲ Notice that the electric and magnetic fields of an electromagnetic wave are at right angles to each other and to the direction of travel.

S130

The Particle Theory Revisited

After the discovery of electromagnetic waves, many scientists were satisfied that they knew all there was to know about light. However, in 1887, Heinrich Hertz discovered the **photoelectric effect**. Hertz found that some metals emit electrons when light hits a thin plate of metal. However, the light must be above a certain *threshold frequency* for each kind of metal. If the frequency of the light hitting a particular metal plate is slightly less than the threshold frequency, no electrons are emitted, even if the light intensity is great. On the other hand, if the frequency of the light is slightly greater than the threshold frequency, electrons are emitted no matter how weak the light is.

This discovery presented a problem—the wave theory of light could not explain the photoelectric effect. If light consists of waves, then any frequency of light should give electrons enough energy to escape from a metal plate, as long as the intensity of the light is great enough. According to the wave theory, the energy of light would be spread out over the front edge of the light wave. Thus, as the intensity of the light increased, so would the amount of energy spread over the wave front. But the characteristic threshold frequency of each material showed that this is not the case.

Then, in 1905, Albert Einstein offered an explanation of the photoelectric effect. Using the ideas of the German physicist Max Planck, Einstein proposed that light is composed of tiny packets (particles) of light energy. These packets of energy were later called **photons**. According to Einstein, each photon has a definite amount of energy that depends on the frequency of the light. To see how the photoelectric effect works, look at the diagram above.

The photoelectric effect has many practical applications. When many photons strike a light-sensitive metal plate, the many electrons emitted can initiate an electric current. Electric currents produced by the photoelectric effect operate photographic light meters, produce the sound in a movie soundtrack, and turn street lights on and off.

Photoelectric effect

The emission of electrons by a substance when illuminated by light of a sufficient frequency

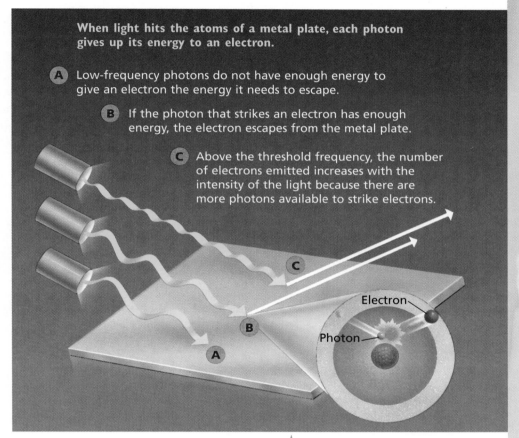

When light hits the atoms of a metal plate, each photon gives up its energy to an electron.

A Low-frequency photons do not have enough energy to give an electron the energy it needs to escape.

B If the photon that strikes an electron has enough energy, the electron escapes from the metal plate.

C Above the threshold frequency, the number of electrons emitted increases with the intensity of the light because there are more photons available to strike electrons.

Electron

Photon

The photoelectric effect occurs when some metals emit electrons as light shines on them.

Photon

A tiny package of electro-magnetic energy

S131

The Particle-Wave Theory

Newton was the first to suggest that light consists of particles because it travels in a straight line. But in 1801, Young showed that light behaves as waves because it is diffracted and shows interference. In 1864, Maxwell discovered that light waves and other electromagnetic waves behave in the same way. However, to explain the photoelectric effect, Albert Einstein proposed in 1905 that light *does* sometimes behave as particles. Thus, it seems that neither waves nor particles alone can explain the behavior of light. What is light then—a wave or a stream of photons?

Today scientists realize that the wave model is the best explanation for some behaviors of light, while the particle model is the best explanation for other behaviors of light. This has led to what is called the *particle-wave theory* of light. How can light exhibit the qualities of both particles and waves, yet not be purely one or the other? Perhaps scientists have yet to determine the true nature of light. In other words, a third explanation may await discovery.

When light passes through a large opening (left), it behaves as the particle model predicts. When light passes through a small opening (right), it behaves as the wave model predicts.

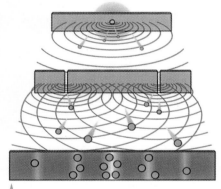

Light has both particle and wave characteristics. Individual photons can be detected as they hit the screen, but a wavelike interference pattern is also evident.

S U M M A R Y

The photoelectric effect and the fact that light travels in a straight line suggest that light consists of a stream of particles. However, behaviors such as diffraction and interference suggest that light is a wave. Other behaviors of light, such as reflection and refraction, can be explained by either a wave theory or a particle theory of light. Light waves are electromagnetic waves, which can travel through a vacuum. Individual packets of light energy are called photons. The amount of energy in a photon of light increases as the frequency of the light increases. Today, physicists explain the properties of light with a combined particle-wave theory.

S132

VISIBLE LIGHT

isible light consists of all the electromagnetic waves that humans can see. Within the entire range of electromagnetic waves, visible light consists of only a very narrow range of wavelengths. These wavelengths have a range of about 400–760 nanometers (nm), or billionths of a meter. Because our eyes are able to detect visible light waves and our brains are able to interpret them, we can see the objects that surround us.

Sources of Light

Light comes from many different sources. The sun and stars are natural sources of light. Candles and electric lamps are artificial sources of light. However, most light originates from its source in the same way—by the excitement of electrons. Electrons orbit atoms in regions called energy levels. The electrons in the energy level closest to an atom's nucleus have less energy than the electrons in energy levels that are farther away from the nucleus. If an electron absorbs energy, it jumps to a higher energy level, as the diagram above shows. When the electron drops back to its normal energy level, it emits a photon of light. The amount of energy in the photon emitted is determined by how far the electron jumps. Many types of energy can excite electrons. For example, energy that comes from nuclear reactions inside the sun produces light by exciting electrons in the sun's outer layers. Electricity

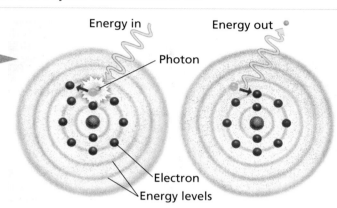

When an electron absorbs the energy of a photon (left), it rises to a higher energy level. As it falls (right), it emits a photon of light.

excites the electrons in a light bulb. Light excites electrons in luminescent materials, causing them to glow in the dark.

Electrons in the luminescent paint on the dial of this timer are excited by light and then slowly give off their own light, enabling you to see how much time remains.

Color

As you can see by looking at a rainbow or at the objects around you, you interpret visible light in many different colors. Nearly 300 years ago, the English physicist Isaac Newton observed all the colors of a rainbow by passing sunlight through a prism. The colors that we see are now known as the **visible spectrum**. Newton also noted that when all the wavelengths of light in the visible spectrum are recombined, white light results. Thus, he concluded that white light contains all the wavelengths of visible light.

A rainbow reveals that visible light consists of many colors.

Visible spectrum

All the colors of visible light

S133

Each color of light has a different wavelength. The colors of light separate when passing through a prism because the waves with shorter wavelengths are bent more than those with longer wavelengths.

But why does a prism separate the colors of light? As you know, light slows down when it passes from air into a denser medium, such as the glass of a prism. As a beam of light slows down, its path of travel is bent (refracted). However, waves with long wavelengths are not bent as much as waves with shorter wavelengths. Therefore, each color of light is bent at a different angle when it passes through a prism, as the illustration above shows.

Now imagine a bowl of fruit sitting on a table. The bowl contains an apple, a banana, and a plum. When white light shines on the fruit, you see that the apple looks red, the banana looks yellow, and the plum looks purple. Since you know that white light is a mixture of all colors, what do you think happened to the other colors of light that fell on each fruit? The fruits contain substances called *pigments*, which absorb some wavelengths of light and reflect others. The color of each fruit results from the wavelengths of light that it reflects. In fact, the wavelengths of reflected light determine the perceived color of each object that we see.

White light has three *primary colors*: red, green, and blue. These colors, seen in the diagram to the left, are called the primary colors because all the colors we see can be made from them. You are able to see the colors of light reflected by an object because your eyes have color receptors called *cones*. There are three different types of cones, each type sensitive to one of the primary colors of light. When two primary colors of light mix, a new color that is a combination of these two colors is produced.

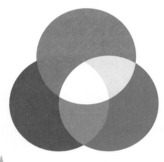

Red, green, and blue are the primary colors of light. When two of these colors overlap, one of the secondary colors (magenta, yellow, or cyan) is produced.

The colorful picture of the balloon above is produced by overlaying three colors of film—magenta, yellow, and cyan—and a piece of black film, which provides contrast.

Rays and the Behavior of Light Waves

The wave theory can explain much about the behavior of light. As you know, light waves appear to travel out from a source as a series of expanding circles. But every point on each circle is traveling directly away from the source along a straight line called a *ray*. A ray is, in effect, the radius of the circle. Light rays are not real, however. They are simply imaginary lines that represent the path that light waves take in a particular direction.

The *ray model* of light is a very useful way to describe the behavior of light waves—how they are refracted, how they are reflected, and how they form images (likenesses) of objects.

Lenses and Refraction   As you know, when light travels from one medium to another, it is refracted, or bent. *Lenses* are used to refract light rays so that images of an object can be formed. A lens is a curved piece of glass, plastic, or some other transparent material. Lenses refract light because the light moves slower through the material of the lens than it does through the air.

There are two basic types of lenses—convex and concave, as shown in the photos to the right. A *convex lens*, which is thicker in the center than at the edges, causes light rays to converge. A *concave lens*, which is thinner in the center than at the edges, causes light rays to diverge.

By using the ray model, the way that refracted light forms images of objects can be explained. A **real image** forms where light rays converge, or are *focused*. A **virtual image** forms not where light rays converge, but where they appear to originate. Because concave lenses cannot make light rays converge, they cannot form real images. However, a virtual image forms in front of a concave lens, where light rays appear to originate.

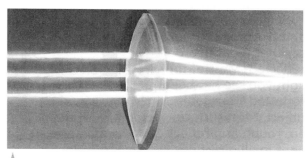

▲ Convex lenses cause light rays to converge, or come together.

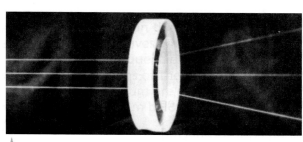

▲ Concave lenses cause light rays to diverge, or spread apart.

◀ A convex lens forms a real image where light passing through the lens converges. A convex lens can also form a virtual image at the location where light passing through the lens appears to originate.

◀ A concave lens forms a virtual image where light rays appear to originate. A real image cannot be made because light rays from the object never converge.

Real image

An image that forms where light rays coming from an object converge; it can be projected on a screen

Virtual image

An image that is not formed by the convergence of light rays; it cannot be projected on a screen

S135

Power Waves *Power waves* are very low frequency electromagnetic waves that are produced by electric generators. These waves are produced as electric current moves through transmission lines. They cause static on your car radio when you drive near them. Their frequencies range between 10 Hz and 100 Hz, with wavelengths of thousands of kilometers.

Radio Waves When you turn on a radio, you hear the effects of another type of electromagnetic wave. *Radio waves* consist of a wide range of electromagnetic waves with frequencies between 10^4 and 10^{12} Hz. Ordinary AM radio broadcasts use the lower-frequency waves between 535 kilohertz (kHz) and 1605 kHz. (One kilohertz is 1000 waves per second.) AM radio waves are reflected back to the Earth's surface by the layer of the atmosphere called the *ionosphere*. Thus, AM radio broadcasts can be received far away from the radio station. At night, you might be able to pick up AM radio stations from halfway across the country. FM radio waves normally use higher frequencies, between 88.1 megahertz (MHz) and 107.9 MHz. (One megahertz is 1,000,000

waves per second.) FM radio waves are not reflected by the ionosphere, so they cannot be received past the curve of the Earth like AM waves. Television (TV) broadcasts use high-frequency radio waves that also travel only in straight lines. Satellites are used to relay radio and television waves around the world.

Radar waves are radio waves of even higher frequencies than TV waves. Waves at these frequencies are reflected by many materials, especially metals. Reflected radar waves function like reflected sound waves, which are heard as an echo. A radar echo can be used to "see" objects in the dark, through

These air traffic controllers use radar to scan the sky for aircraft in the vicinity.

fog, or at a great distance. For this reason, radar is used for navigation on ships and planes.

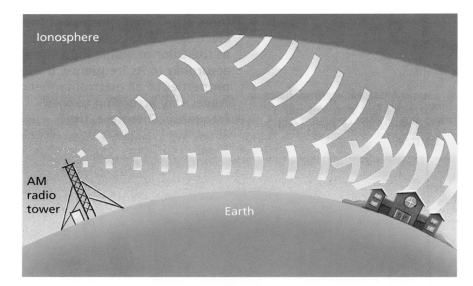

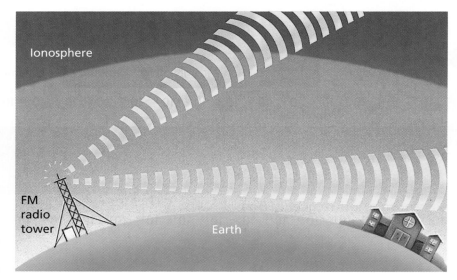

AM radio waves (above) have longer wavelengths and are reflected by the ionosphere. FM waves (left) have shorter wavelengths, which allow them to pass through the ionosphere.

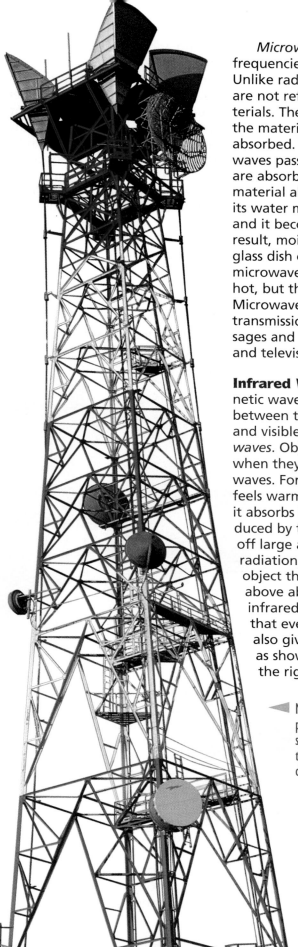

Microwaves have the highest frequencies of any radio waves. Unlike radar waves, microwaves are not reflected by most materials. They either pass through the materials or are easily absorbed. For example, microwaves pass through glass but are absorbed by water. When a material absorbs microwaves, its water molecules move faster and it becomes hotter. As a result, moist food or water in a glass dish can be heated in a microwave oven. The food gets hot, but the glass dish does not. Microwaves are also used in the transmission of telephone messages and long-distance radio and television broadcasts.

Infrared Waves Electromagnetic waves with frequencies between those of microwaves and visible light are *infrared waves*. Objects become warmer when they absorb infrared waves. For example, your skin feels warmer in sunlight because it absorbs infrared waves produced by the sun. The sun gives off large amounts of infrared radiation into space. But any object that has a temperature above absolute zero radiates infrared waves. This means that everything around you also gives off infrared waves, as shown in the image to the right.

◀ Microwave towers relay telephone conversations and television programs to areas where the installation of cables is difficult or impossible.

DID YOU KNOW...

that FM radio waves, television waves, and microwaves are all line-of-sight methods of communication? Because they are not reflected by the ionosphere, these waves can be picked up only by receivers that are above the horizon. Therefore, FM radio, television, and microwave towers are usually very tall so that their signals can be received as far away as possible.

▲ The lighter colors in this infrared image indicate higher temperatures.

Visible Light Waves *Visible light* is the very narrow band of frequencies that can be seen with the human eye. We see different frequencies within this narrow band as different colors. As you know, the lowest frequencies of visible light are seen as red; the highest frequencies are seen as blue or violet. White light contains all the frequencies of visible light.

S139

Ultraviolet Light   If you spend a lot of time outdoors in the sun, you should be aware of *ultraviolet light*. The waves in this part of the electromagnetic spectrum have frequencies just above those of visible light. The ultraviolet light in sunlight causes sunburn. Too much ultraviolet light can be dangerous because it may cause harmful mutations and skin cancer. In limited amounts, ultraviolet light can be beneficial. For example, when skin cells are exposed to ultraviolet light, they begin to produce vitamin D, a substance necessary for healthy teeth and bones. Ultraviolet light also kills germs and is often used for this purpose in hospitals.

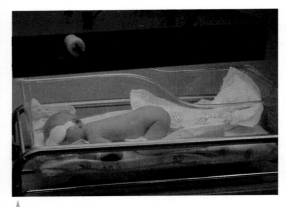

▲ Ultraviolet light is used to treat jaundice in newborn babies.

X Rays   *X rays* are high-frequency electromagnetic waves that are very useful in the field of medicine. They easily pass through skin and other tissues, but not through bone. X rays can produce an image on film or a picture on a screen. This means that X rays can be used to see inside your body. A doctor or dentist has probably made X-ray photographs of parts of your body. However, too much exposure to X rays can kill living cells. Therefore, machines that produce X rays must be used only by trained personnel.

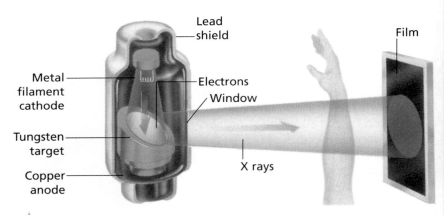

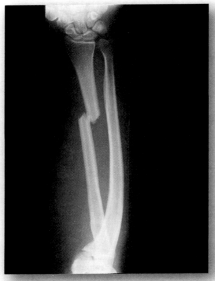

Lead shield

Film

Metal filament cathode

Electrons

Window

Tungsten target

X rays

Copper anode

▲ X rays are produced in a special tube when electrons strike a tungsten target. The reflected X rays pass through the object to form an image on film.

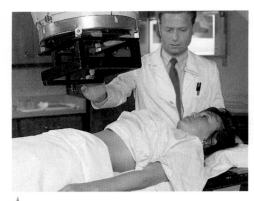

▲ Gamma rays are used to treat patients with cancer.

Gamma Rays   *Gamma rays* are very high frequency waves similar to X rays. They are much more dangerous, however, because they pass through matter more easily. Nevertheless, gamma rays are useful in fighting cancer. Beams of gamma rays can be aimed at the location of the cancer. The rays pass deep into the body to reach and kill the cancer cells.

S140

Quicker Than the Eye

Because visible light is the easiest type of electromagnetic radiation to detect, most early investigations were done with light. However, many of the characteristics of light apply to other forms of electromagnetic radiation. One of the most impressive characteristics of light, and of all other electromagnetic radiation, is its incredible speed.

Speed of Light   How long does it take light from a lamp to get to your eyes? The first scientists who tried to measure the speed of light found that light travels at a speed that is far too fast to measure by ordinary means. In the sixteenth century, the Italian scientist Galileo attempted to measure the time it took light to travel about 2 km. However, light travels this distance almost instantly. Galileo had no way to measure such a short period of time. In about 1676, however, the Danish astronomer Ole Rømer succeeded in calculating the speed of light. By studying the eclipses of the moons of Jupiter, he was able to figure out how long it takes for light to travel from Jupiter to Earth. His calculations turned out to be quite accurate.

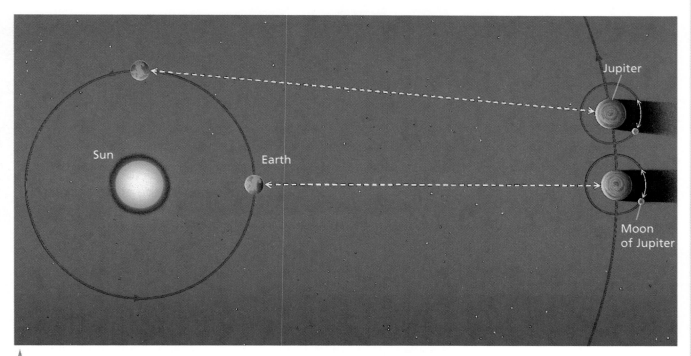

▲ Light from an eclipse of one of Jupiter's moons takes longer to reach the Earth when the Earth is farther away from Jupiter. By comparing this difference in time, Rømer estimated the speed of light.

Modern measurements, of course, are much more accurate than Rømer's calculations. Today, the value used for the speed of light has been determined to eight decimal places. In a vacuum, light travels at a speed of 2.99792458×10^8 m/s, or about 3×10^8 m/s. At this speed, a beam of light could travel from Los Angeles to Atlanta in less than one-hundredth of a second. All forms of electromagnetic radiation travel through space at the speed of light.

S141

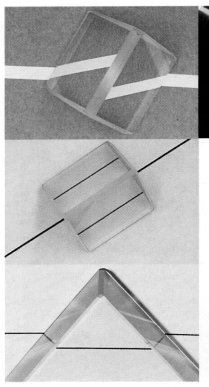

Light is refracted when it travels from one medium to another because it has a different speed in each medium.

Index of refraction

The ratio of the speed of light in a vacuum to its speed in another medium

Speed Limits   The speed of light noted on the previous page is only valid for light in a *vacuum*. In air, the speed of light is slightly slower. Also, when light moves from one material into another, its speed changes. For example, when light moves from air into water, it slows down. Light waves travel about 25 percent slower in water than they do in air. On the other hand, light moving from water into air speeds up. This change of speed causes light entering or leaving water to be refracted along a new path.

Some materials slow light down more than water does. The more that light slows down when it passes from one medium to another, the more it is bent. The degree to which a material can bend light is given by its **index of refraction**. To find the index of refraction for any material, simply divide the speed of light in a vacuum by the speed of light in that material. For example, you can find the index of refraction for a diamond by doing the following calculation:

$$\text{index of refraction} = \frac{\text{speed of light in a vacuum}}{\text{speed of light in diamond}}$$

$$= \frac{3.00 \times 10^8 \text{ m/s}}{1.24 \times 10^8 \text{ m/s}}$$

$$= 2.42$$

S142

The index of refraction of some common materials is given in the table to the right. The larger the index of refraction, the greater the change in the path of the light rays.

Infrared binoculars demonstrate that electromagnetic waves other than visible light are also affected by the materials through which they travel. Images form in infrared binoculars because their lenses refract infrared waves in much the same way that the lenses of a normal set of binoculars refract visible light. Radio waves can also undergo refraction. Because the density of the atmosphere changes at different altitudes, radio waves are gradually refracted. This refraction sometimes carries FM radio waves past the horizon.

Indices of Refraction	
Substance	Index of refraction
air	1.00
ice	1.31
water	1.33
quartz	1.46
glass	1.52
amber	1.54
ruby	1.76
diamond	2.42

By using infrared binoculars, objects in complete darkness become visible.

SUMMARY

Visible light is only one part of a larger range of electromagnetic radiation called the electromagnetic spectrum. Electromagnetic waves are classified according to frequency into power waves, radio waves, infrared waves, visible light, ultraviolet light, X rays, and gamma rays. The speed of light (and other electromagnetic radiation) in a vacuum is about 3.00×10^8 m/s. As light enters another medium, its speed decreases. This change of speed causes light to be refracted by an amount determined by each material's index of refraction.

DID YOU KNOW...

that the sparkle of a cut diamond results from its index of refraction?
A diamond slows light to less than half of its speed in air. Therefore, diamonds have a very high index of refraction. As a result, light rays that enter a diamond are refracted and separated into different colors, just as they are by a prism.

S143

HIGH-TECH LIGHT

The unique properties of light have led to many innovations that affect our daily lives. From the sunglasses you wear at the beach to the lasers that scan prices at the grocery store to the fiber-optic cables that carry your phone conversations with your friends, the technology of light is all around you. Both the wave and particle models of light have inspired important advances in technology and improved our understanding of the universe. In this section, you will look at a few of these technological advances and discover how we use some of the properties of light.

Polarized light

Light consisting of light waves that vibrate in only one plane

Polarization of Light

Picture yourself on the shore of a lake on a beautifully clear day. Sunlight is reflected off the sparkling lake, making you squint. You put on a pair of polarizing sunglasses, and the glare disappears. Now you can even see into the water. How can a pair of sunglasses make such a difference? Sunglasses that eliminate glare contain polarizing filters.

Like all electromagnetic waves, light waves are transverse waves, which vibrate at right angles to their direction of travel. In a beam of light, however, some of the waves vibrate up and down, some vibrate from side to side, and others vibrate at all the angles in between. As light waves pass through a polarizing filter, all waves are blocked except those that vibrate in the direction allowed

by the filter. The diagram below shows how polarizing filters work.

Light that vibrates in only one plane or orientation is called **polarized light.** The light causing the glare at the surface of the lake is polarized horizontally when it is reflected from the surface of the water. Your sunglasses have a vertical polarizer that blocks the glare from the lake so that it does not reach your eyes. Because the rest of the light in your environment is not polarized, there are plenty of light waves that vibrate vertically and can pass unobstructed through your sunglasses. You can also see the effect of a polarizing filter on glare by looking into a store window, like the one in the photos on the next page.

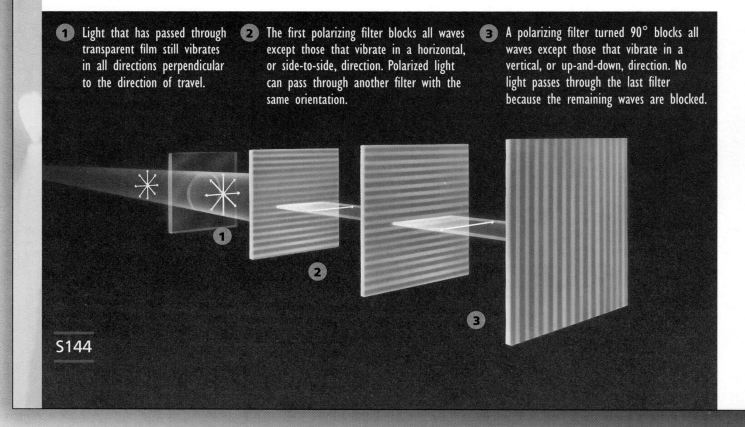

1. Light that has passed through transparent film still vibrates in all directions perpendicular to the direction of travel.

2. The first polarizing filter blocks all waves except those that vibrate in a horizontal, or side-to-side, direction. Polarized light can pass through another filter with the same orientation.

3. A polarizing filter turned 90° blocks all waves except those that vibrate in a vertical, or up-and-down, direction. No light passes through the last filter because the remaining waves are blocked.

S144

While you cannot detect polarized light with your eyes alone, you can find out whether light is polarized by using a polarizing filter. To do so, look at the light through a polarizing filter and then rotate the filter 90 degrees. If the light is polarized, it will get dimmer or brighter as you rotate the filter.

Lasers

American industries spend billions of dollars every year making and using laser light for hundreds of purposes. From medical applications and space research to home entertainment, lasers have made a large impact on modern technology. But how is laser light different from ordinary light?

▲ The glare from this window makes it difficult to see the inside of the store.

▲ Because this photograph was taken through a polarizing filter, much of the glare has been eliminated.

DID YOU KNOW...

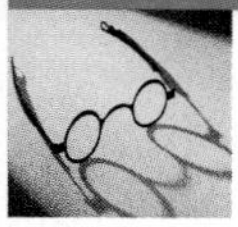

that light technology is being used in research on nuclear fusion? Powerful laser beams are being used in attempts to heat the fuel for nuclear fusion to the extremely high temperatures needed to cause atomic nuclei to fuse.

◄ Lasers can be used to measure the distance from locations on Earth to satellites in orbit around the Earth.

S145

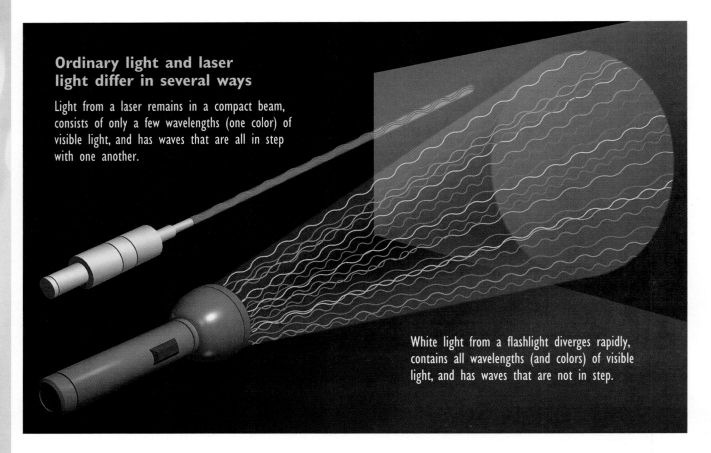

Ordinary light and laser light differ in several ways

Light from a laser remains in a compact beam, consists of only a few wavelengths (one color) of visible light, and has waves that are all in step with one another.

White light from a flashlight diverges rapidly, contains all wavelengths (and colors) of visible light, and has waves that are not in step.

Properties of Laser Light

The beams of light from an ordinary flashlight and a low-power laser can have about the same amount of total energy. But there is a big difference between them. You can see some of these differences in the diagram above.

First, a flashlight beam that is about 3 cm across when it leaves the flashlight spreads to a width of about 40 cm by the time it reaches the other end of a room. By contrast, a beam of laser light that is 5 mm wide when it leaves the laser makes the same size spot when it

reaches the other end of the room—5 mm. Laser light spreads out so little that a laser beam from Earth can be reflected from mirrors left on the moon by American astronauts.

Second, a flashlight's white light contains most wavelengths of visible light. The light in a laser beam, however, contains only a few, similar wavelengths. For instance, a helium-neon laser produces visible light with long wavelengths, and therefore the laser beam looks red. Other types of lasers produce light with different wavelengths (and colors).

Third, the crests and troughs of light waves in ordinary light overlap. However, the light waves in a laser beam are *in step*. That is, crests travel with crests and troughs travel with troughs. Light consisting of waves that are in step is called **coherent light**.

How Lasers Work
A laser is a device that produces coherent light. The most commonly used laser is the low-power, helium-neon laser. It produces a thin, straight beam of red light. This light has the same wavelength as the light from red neon signs (like those you see in store windows) and originates in much the same way.

The tubing used in a neon sign is filled with neon gas at low pressure. As an electric current passes through the tube, electrons in the neon atoms absorb energy. When an electron has extra energy, it is said to be in an *excited state*. However, electrons cannot stay in an excited state, and they quickly release energy by emitting a photon of light. This causes the neon gas in the tube to glow. When atoms give off photons of light energy, their electrons are said to return to the *ground state*.

Coherent light

Light consisting of waves that are in step

S146

Every photon released by excited neon atoms has the same wavelength. However, in a neon tube, the photons are released in all directions and at different times.

Because all of the photons released by neon atoms have the same wavelength, a neon tube can be converted into a laser. When a photon of red light strikes an already excited neon atom, the atom immediately releases another, identical photon of red light. This process, seen in the diagram to the right, is called **stimulated emission**. If each of the two photons produced by stimulated emission from an excited neon atom strikes another excited neon atom, four photons result, then eight, then sixteen, and so on, as the diagram below shows. In a microsecond, so many photons are produced that the neon tube glows brightly. In a laser, the two ends of the tube are coated with a reflecting surface

that causes any photons traveling down the tube to bounce back and forth between the ends. The bouncing photons stimulate emission from other atoms that they strike. Soon an enormous number of photons flow back and forth in the tube, all in step with one another. One of the mirrored ends has a thinner coating, which allows some of the light to pass through. From this end, 5 to 10 percent of the coherent light escapes. This light is the laser beam.

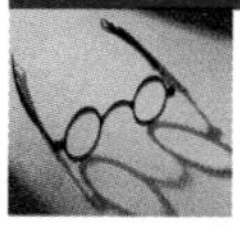

Stimulated emission

A process of forcing identical photons into step in order to produce coherent light

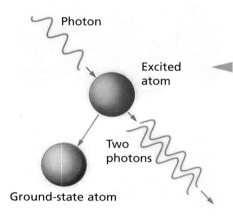

Photon

Excited atom

Two photons

Ground-state atom

◄ When a photon strikes an already excited atom, the excited atom will emit an additional photon. The two photons will then leave the atom together, in step with one another.

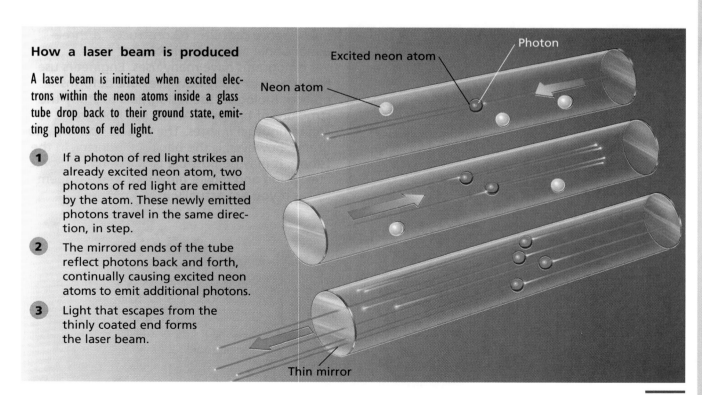

How a laser beam is produced

A laser beam is initiated when excited electrons within the neon atoms inside a glass tube drop back to their ground state, emitting photons of red light.

1. If a photon of red light strikes an already excited neon atom, two photons of red light are emitted by the atom. These newly emitted photons travel in the same direction, in step.

2. The mirrored ends of the tube reflect photons back and forth, continually causing excited neon atoms to emit additional photons.

3. Light that escapes from the thinly coated end forms the laser beam.

Photon

Excited neon atom

Neon atom

Thin mirror

S147

An **optical fiber** is a long, thin light pipe. Optical fibers are made of glass and encased in plastic. Light travels more slowly in glass than it does in plastic, making total internal reflection possible. The fiber can be bent and twisted into many shapes without losing the light.

Fiber-optic telephone cables also consist of optical fibers encased in plastic. When you talk on the telephone, your voice is converted into an electrical signal. At a local telephone station, this signal is changed into a digital code. The electric bits of the code are used to trigger a tiny laser. The laser translates this code into a series of flashes of infrared light, which travel

▲ Total internal reflection keeps a beam of light within a thin glass fiber.

through the optical fiber. When you speak, you leave spaces between your words and sentences. Therefore, the code for your conversation takes up only a small portion of the light signal. The spaces between the flashes of your conversation are filled with bits from other conversations. Because each fiber can carry 46 million bits per second, a cable of 12 fibers can carry nearly 50,000 telephone conversations at once.

▲ The laser used to send light in an optical fiber is very small. One such laser is shown here—placed next to a quarter for scale.

Optical fiber

A long, thin light pipe that uses total internal reflection to carry light

S U M M A R Y

The unique properties of light have led to innovations that affect our daily lives. Polarizing sunglasses eliminate glare because they contain filters that allow only light waves vibrating in one plane to pass. A laser is a device that produces coherent light by forcing identical photons into step through a process called stimulated emission. Laser light can be used to make holograms. Optical fibers carry information in beams of light that are guided by total internal reflection. Telephone systems use both optical fibers and lasers to greatly enhance our ability to communicate.

Unit 8

CONTINUITY OF LIFE

IN THIS UNIT

Foundations of Heredity,
page S154

Chromosomes, Genes, and Heredity,
page S161

DNA—The Material of Heredity,
page S167

Reproduction and Development,
page S173

Now that you have been introduced to heredity and the processes of reproduction and development, consider these questions.

● **1.** What evidence did the discoveries of mitosis and meiosis provide in support of Mendel's laws of heredity?

2. How does our modern understanding of genetics differ from Mendel's concept?

3. How does the structure of DNA make it an ideal molecule for the transmission of inherited traits?

4. What is asexual reproduction and how does it occur?

In this unit, you'll take a closer look at some patterns of inheritance and DNA, the molecule of heredity. You will also learn about the different ways that organisms reproduce.

S153

FOUNDATIONS OF HEREDITY

▲ What principle of reproduction does this litter seem to violate?

The process of *reproduction* ensures the continuation of a species by forming new individuals of that species. The organisms that result from reproduction resemble their parents—at least in the most basic ways. Live-oak trees produce offspring that are also live-oak trees, and human beings produce other human beings. In other words, each species produces offspring that belong to that species.

Although offspring generally resemble their parents, they also may have differences. External differences are easiest to see. For example, there can be differences in height, coloration, and shape of body parts. But there may be internal differences as well. For instance, most people have type O or type A blood, while others have type B or type AB blood. Such differences, external and internal, are called *variations*.

For thousands of years, people knew that many characteristics are passed from parents to offspring. With this knowledge, people selected and bred certain plants and animals to bring out desirable characteristics. Yet it was not until the twentieth century that we discovered *how* those characteristics, or *traits*, pass from parents to offspring.

Heredity

The passing of traits from parents to offspring

Mendel's Discoveries

Gregor Mendel was an Austrian botanist who lived from 1822 to 1884. He is famous because he discovered the basic principles of **heredity**. Mendel became interested in plants while growing up on his family's farm. At the age of 21, he entered a monastery, where he studied to become a priest and a teacher. Later, the monastery sent Mendel to the University of Vienna, where he studied science and mathematics. After his return to the monastery, Mendel taught natural science at a nearby high school.

◄ Sometimes organisms are selectively bred "just for show." These fancy goldfishes are called lionhead goldfishes. You may have seen them in a pet store, or perhaps you have some in your aquarium at home.

Mendel also applied what he had learned at the university to his interest in plants. In the monastery garden, he did research with pea plants that led to many important discoveries about heredity. Much of Mendel's success was due to his understanding of mathematics. Mendel analyzed his results by using the mathematical principles he had learned at the University of Vienna. He was among the first to analyze biological experiments using mathematics.

Crossing Pea Plants Mendel studied seven pea-plant traits, each with two contrasting forms. No matter which trait Mendel was studying (stem length, flower position, seed shape, etc.), he always began his experiments by *cross-pollinating* purebred plants that exhibited contrasting forms of the trait. A **purebred** plant produces offspring that all have one form of a particular trait. For example, a purebred tall pea plant produces only tall pea plants, generation after generation. And a purebred short pea plant produces only short pea plants.

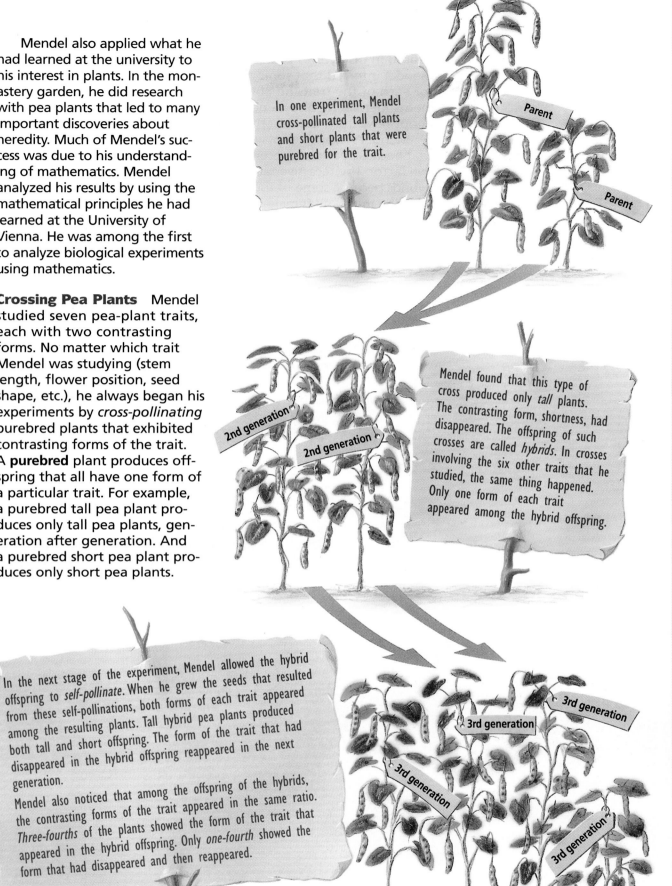

In one experiment, Mendel cross-pollinated tall plants and short plants that were purebred for the trait.

Parent

Parent

Mendel found that this type of cross produced only *tall* plants. The contrasting form, shortness, had disappeared. The offspring of such crosses are called *hybrids*. In crosses involving the six other traits that he studied, the same thing happened. Only one form of each trait appeared among the hybrid offspring.

2nd generation

2nd generation

In the next stage of the experiment, Mendel allowed the hybrid offspring to *self-pollinate*. When he grew the seeds that resulted from these self-pollinations, both forms of each trait appeared among the resulting plants. Tall hybrid pea plants produced both tall and short offspring. The form of the trait that had disappeared in the hybrid offspring reappeared in the next generation.

Mendel also noticed that among the offspring of the hybrids, the contrasting forms of the trait appeared in the same ratio. *Three-fourths* of the plants showed the form of the trait that appeared in the hybrid offspring. Only *one-fourth* showed the form that had disappeared and then reappeared.

3rd generation

3rd generation

3rd generation

3rd generation

S155

Dominant

Describes the form of a genetic factor that is always expressed

Recessive

Describes the form of a genetic factor that is not expressed when the dominant factor for the trait is present

Mendel's Factors Seeing that one form of each trait had disappeared in his hybrids but reappeared in the next generation, Mendel inferred that some "distinct element" must be responsible for each of the traits he observed. He called these distinct elements *factors*. He also came to several important conclusions about how traits are inherited, as illustrated below.

Mendel's hybrids displayed only the dominant form of each trait because each had received one dominant factor and one recessive factor. The recessive factors reappeared in the offspring of the hybrids because some of them received the factor for shortness from both the male and female reproductive cells. Only when two recessive factors were paired together was the recessive factor expressed.

Mendel published the results of his work in 1866. However, no one paid much attention to them at the time. But over the next 35 years, important discoveries about cells were made. Those discoveries prepared the scientific community for the rediscovery of Mendel's work in 1900. Unfortunately, Mendel had already died.

2
The factors for a trait separate when reproductive cells are formed. As a result, reproductive cells receive and carry only one factor for each trait. When two reproductive cells combine to produce a new individual, there is again a pair of factors for each trait. This principle can be called the *law of segregation*.

3
There are different forms of the factor for each expression of a trait. In peas, for example, there are two kinds of factors for the trait of plant height—one for tallness and the other for shortness.

1
Each parent contributes one factor for each trait to its offspring. Thus, individuals have a *pair* of factors for each trait.

4
One of the two factors for a trait prevents the expression of the other. For example, the factor for tallness prevents the expression of the factor for shortness. The factor that prevents the expression of the other is the **dominant** factor. The factor whose expression is prevented is the **recessive** factor.

S156

Cell Division

During the 1860s, several scientists began to examine living cells that were in the process of dividing. The scientists noticed certain structures in the nucleus of each cell undergoing the division process. These structures were named *chromosomes* because they absorbed a colored dye that made them more visible under the microscope. *Chroma* means "color" in Greek. The scientists also noticed that when a cell divided, equal numbers of chromosomes were distributed to each new cell. They also observed that each new cell ended up with the same number of chromosomes as the original cell.

Later, scientists who were studying the formation of reproductive cells observed a second kind of cell division. During the process of reproductive-cell formation, two successive divisions occur. The result is four new cells, each having only one-half of the original cell's number of chromosomes. Let's take a closer look at these two types of cell division.

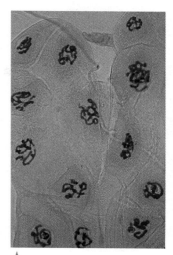

▲ The darkly stained objects in this photograph are chromosomes in the cells of a fruit fly.

Mitosis   Look at your hand. Each square centimeter of skin is made of more than 150,000 cells. Yet many of these cells will be gone by tomorrow. These surface skin cells are actually dead, and normally they are shed or get washed away. But don't worry, you'll never run out of skin, because brand-new skin cells are continually being produced. These new cells will replace the skin cells you lose. In other parts of your body, cells are also reproducing rapidly. For example, as you grow taller, your bones grow in length by adding new bone cells at the ends of the bones.

The chromosomes in the nucleus of a cell are the "blueprints" for that cell. They contain all the information needed to build new cell materials and to control the cell's activities. Suppose you want to have two identical houses built at the same time. You would not cut the blueprints in two and give one-half to each builder. Instead, you would make an exact copy, or duplicate, of the blueprints so that each builder could have a full set of instructions. The same is true for the blueprints of a cell. Each new cell must carry a full set of chromosomes. Therefore, chromosomes must be duplicated before the nucleus of a cell divides. This duplication process occurs during a certain stage in the life of a cell.

A *parent* cell reproduces by dividing into two new cells.

The new cells are called *daughter* cells.

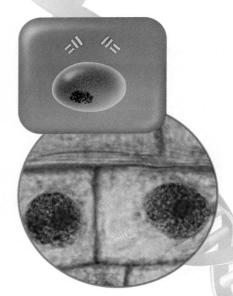

A cell spends most of its life in a stage called *interphase*. This is the time when the cell *seems* to be at rest. Actually, the cell is not resting at all. It's really very busy—growing and carrying out all of its normal life activities. Toward the middle of interphase, each chromosome in the cell duplicates. This doubling of the number of chromosomes in a cell is a sign that cell division is about to begin.

The first part of a cell that divides is the nucleus. This division of the nucleus is called **mitosis**. During mitosis, the division of the nucleus occurs in four consecutive stages that enable each daughter cell to receive one copy of each chromosome. Scientists call these stages—from first to last—*prophase*, *metaphase*, *anaphase*, and *telophase*.

Late Interphase Each chromosome has been duplicated. Once chromosomes have been duplicated, they exist as two strands joined together by a structure called a *centromere*.

Prophase Each of the chromosomes inside the nucleus begins to thicken. As the chromosomes become visible, the membrane around the nucleus disappears. Slender fibers, called *spindle fibers*, extend from opposite sides of the cell. A spindle fiber from each side attaches to the centromere of each duplicated chromosome.

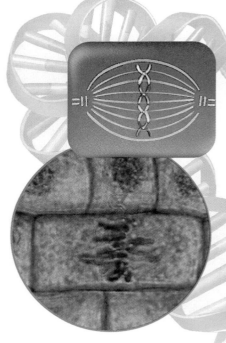

Metaphase The duplicated chromosomes line up across the middle of the cell. Then the centromeres divide, separating the duplicate chromosomes.

Anaphase The spindle fibers shorten and pull the duplicate chromosomes through the cytoplasm toward opposite sides of the cell.

Mitosis

The process by which a cell's nucleus divides, leading to the formation of two identical cells

Meiosis Many organisms reproduce by uniting two sex cells—one male and one female—during a process called *fertilization.* As a result, a fertilized cell, or *zygote,* forms. The zygote is the first cell of a new organism. It will divide by mitosis many times and develop into a new organism.

If sex cells contained the same number of chromosomes as other body cells, the number of chromosomes in a zygote would end up being double the normal amount. For example, every human body cell, such as a skin cell or a liver cell, has 46 chromosomes in its nucleus. If each human sex cell (egg or sperm) also contained 46 chromosomes, then a human zygote would contain 92 chromosomes. This means that every body cell of the new human being that developed from that zygote would also contain 92 chromosomes. This simply does not happen. Each species has a

characteristic number of chromosomes in each of its cells; that number remains stable from one generation to the next. Scientists realized that for this to be the case, the number of chromosomes in sex cells would have to be exactly half the number of chromosomes in body cells.

While studying the formation of sex cells, scientists discovered a cell-division process that

This tiny sperm cell is attempting to fertilize a much larger egg cell. This particular sperm cell and egg cell are from a sea urchin.

results in daughter cells with half as many chromosomes as the parent cell. This process is called **meiosis**. In meiosis, the nucleus divides in stages that are very similar to those of mitosis, with one major exception. In meiosis, the duplicated chromosomes do not line up randomly during metaphase of the first division, as they do in mitosis. Instead, each duplicated chromosome lines up across from another duplicated chromosome of the same size and shape. Seeing this, researchers realized that chromosomes occur *in pairs.*

Meiosis

The process by which a parent cell divides twice, leading to the formation of four sex cells, with half the original number of chromosomes

Telophase The separated chromosomes cluster at the opposite sides of the cell. A nuclear membrane forms around each group of chromosomes. At the same time, the chromosomes untwist, becoming longer and thinner.

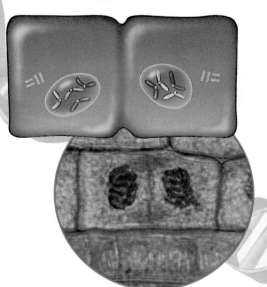

Cytokinesis At the end of telophase, the cytoplasm is divided between the two new cells, and they separate.

DID YOU KNOW...

that some human genetic disorders are caused by mistakes that happen during meiosis?
Sometimes sex cells are formed that have an extra number 21 chromosome. Down syndrome, a condition characterized by varying degrees of mental retardation, occurs in an offspring that develops from the union of one of these abnormal sex cells with a normal sex cell.

Cells that contain pairs of chromosomes are described as being *diploid*. Diploid cells are represented by the symbol 2*n*. The pairs of chromosomes separate during the first division of meiosis, and then the duplicates separate during a second division. The result is that each of the four daughter cells carries only one chromosome from each pair. Cells with a half-set of chromosomes are called *haploid* cells and are represented by the symbol *n*. Although he had never observed meiosis itself, Mendel had described this same pattern in explaining the inheritance of factors. Traits are determined by a pair of factors, the factors of each pair separate during formation of sex cells, and only one factor for each trait is transmitted by each parent.

Meiosis

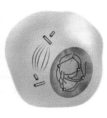

In meiosis, each duplicated chromosome lines up across from another duplicated chromosome of the same size and shape.

Pairs of duplicated chromosomes separate during the first division.

The duplicates separate during the second division.

Each resulting sex cell has only one of each kind of chromosome.

SUMMARY

Reproduction is the means by which a species continues. During reproduction, characteristics are passed from parents to their offspring. Gregor Mendel was the first to describe the basic principles by which traits are inherited. He theorized that a trait is determined by a pair of factors. These factors separate during the formation of sex cells. Each parent transmits one factor for a trait to its offspring. Factors occur in different forms. A dominant factor prevents the expression of a recessive factor. Cells reproduce themselves through a process called cell division. During mitosis, the chromosomes in the nucleus of a parent cell are doubled and then divided so that two daughter cells receive the same kind and number of chromosomes as the parent cell. During meiosis, pairs of like chromosomes separate so that four daughter cells are produced, each having half as many chromosomes as the parent cell.

S160

CHROMOSOMES, GENES, AND HEREDITY

Without knowing what they were or where they were located, Mendel reasoned that "factors" inherited from parents are responsible for the characteristics of an organism. Many years later, the behavior of chromosomes was observed during the formation of reproductive cells. However, the significance of this behavior was not immediately recognized. When Mendel's work was rediscovered in 1900, a new science, called **genetics**, was born. This name, which comes from a Greek word meaning "beginning," was first suggested at a scientific meeting on heredity in 1906.

The Chromosome Theory

In 1903, Walter S. Sutton, a researcher at Columbia University, was conducting a microscopic examination of meiosis in grasshoppers. Having read Mendel's paper about the inheritance of traits in pea plants, Sutton realized that the behavior of Mendel's factors could be explained by the behavior of chromosomes during meiosis. He inferred that, because of this relationship, these factors must somehow be related to chromosomes. He also realized that for every organism, there are many more traits than there are chromosomes. Sutton reasoned that more than one hereditary factor must be located on each chromosome. Sutton's ideas became known as the *Chromosome theory*, which simply states that *hereditary factors are found on chromosomes*. Today, scientists call these hereditary factors *genes*.

The term *gene* was first used to refer to a hereditary factor in 1909. Over the next 50 years, *geneticists* studied the way in which many genes are inherited. They even identified the exact locations of specific genes on the chromosomes of many organisms. However, the chemical nature of a gene and how it works remained a mystery until the 1950s. Even today, there is still much more to be learned about genes and chromosomes.

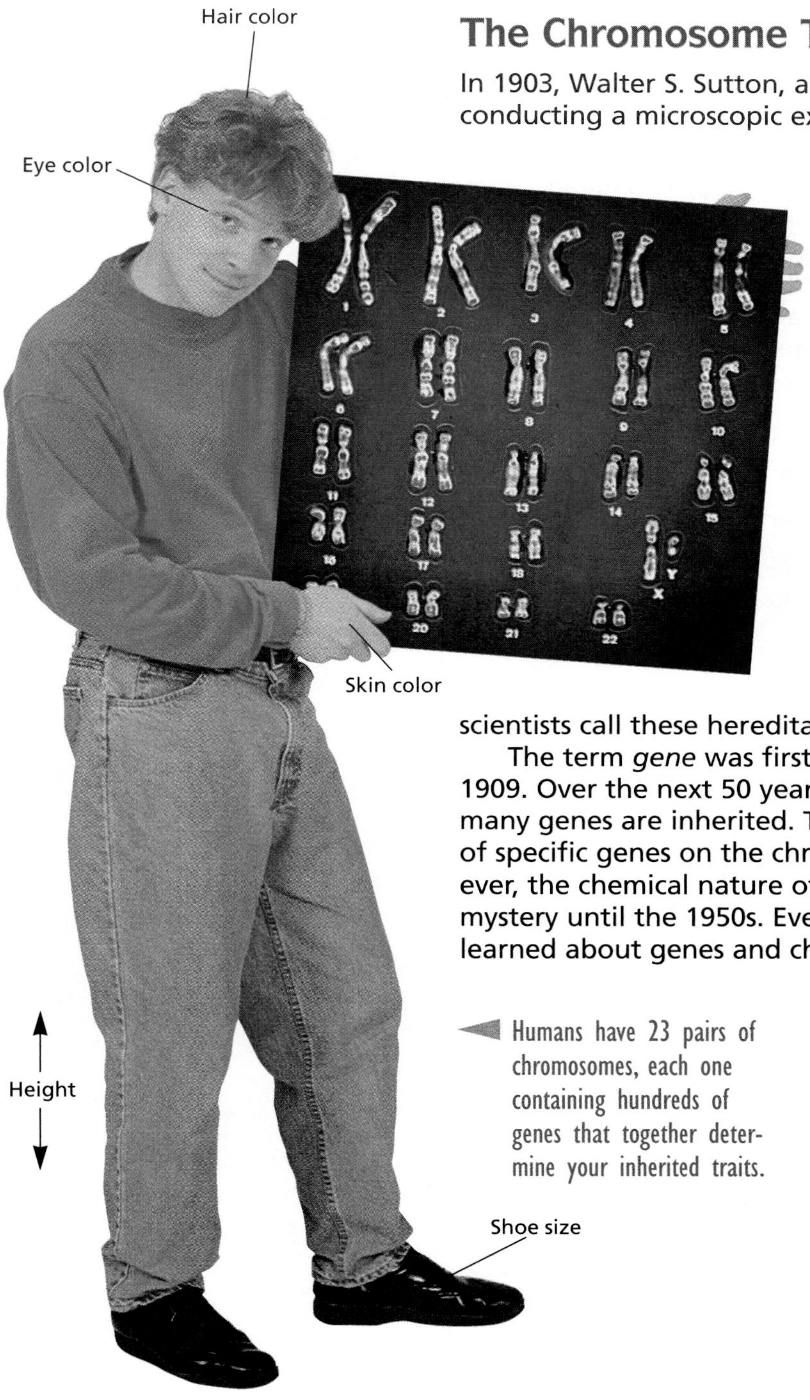

◄ Humans have 23 pairs of chromosomes, each one containing hundreds of genes that together determine your inherited traits.

Hair color

Eye color

Skin color

Height

Shoe size

Genetics

The science in which heredity is studied

S161

Working With Genes

You can represent individual genes by using letters. Capital letters are used to indicate dominant genes. For example, *T,* is used to represent the gene for tallness in pea plants, and *R,* is used to represent the gene for round seed shape in pea plants. Remember that both of these traits are dominant in pea plants. The lowercase version of the *same* letter indicates the recessive gene for the same trait. The gene for shortness, then, is *t,* and the gene for wrinkled seed shape is *r.*

Since traits are produced by *pairs* of genes, a pair of letters can be used to represent the genes that a particular individual has for a trait. For example, pea plant height is determined by one of three possible gene pairs—*TT, Tt,* or *tt.* Pea-seed shape can be determined by *RR, Rr,* or *rr.* The specific pair of genes that an organism has for a trait is called its **genotype**. The physical description of a genotype, or what the organism actually looks like, is called a **phenotype**. Round seeds, wrinkled seeds, tall plants, or short plants are terms that describe phenotype. If both genes for a trait are identical (*TT* or *tt*), the genotype is said to be *homozygous,* or pure, for that trait (*homo* means "same"). If the two genes are different (*Tt*), the genotype is said to be *heterozygous,* or hybrid (*hetero* means "different").

Today, you can make the same genetic crosses that Mendel made about 150 years ago. But you don't need to find garden space to do it. You just have to know how to use a special chart called a *Punnett square.*

Genotype

The genetic makeup of an organism according to its genes

Phenotype

The external appearance of an organism as determined by its genotype

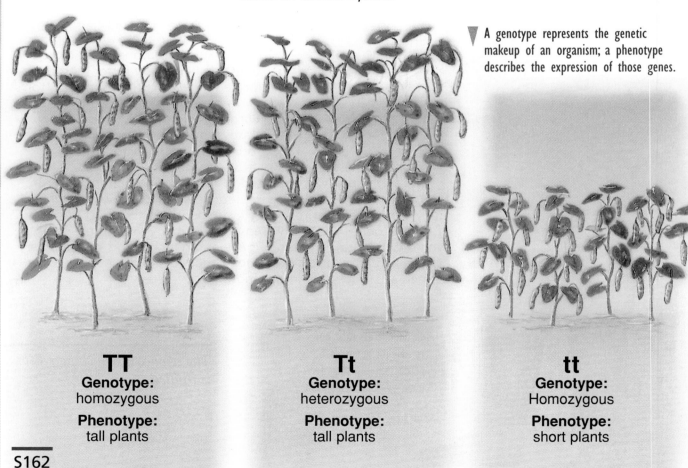

A genotype represents the genetic makeup of an organism; a phenotype describes the expression of those genes.

TT
Genotype:
homozygous

Phenotype:
tall plants

Tt
Genotype:
heterozygous

Phenotype:
tall plants

tt
Genotype:
Homozygous

Phenotype:
short plants

S162

Constructing a Punnett Square The Punnett-square method for displaying the possible results of a genetic cross was developed by the British scientist Reginald Punnett in the early twentieth century. Follow the steps below to see how a Punnett square is used to predict the outcome of a cross between homozygous tall pea plants and homozygous short pea plants.

1 First draw a square and divide it into two rows and two columns.

2 Now write symbols for the two genes of one parent's genotype at the top of the square, one above each column. Then write the symbols for the two genes of the other parent's genotype down the left side of the square, one beside each row. These letters represent the gene content of the parents' reproductive cells.

3 Finally, copy the gene symbol from the top of each column into the boxes below, and copy the gene symbol from the left side of each row into the boxes of that row. Typically, the dominant gene is written first, followed by the recessive gene. All of the possible gene combinations from the cross now appear in the boxes. These are the possible genotypes of the offspring.

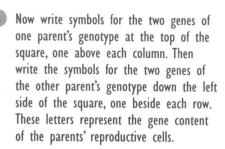

4 Notice that all of the genotypes in the square are heterozygous. Since tallness *(T)* is dominant to shortness *(t)*, what will the phenotype of all the offspring be?

5 Now take a couple of the offspring and cross them in another Punnett square. What is the phenotype of each member of the second generation? What is the phenotype ratio?

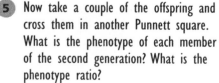

S163

Pedigree

A chart that traces the inheritance of a genetic trait through several generations of a family

Using a Pedigree Suppose you decide to raise rabbits with black fur (B), which is dominant to brown fur (b). You go to a pet store to buy several pairs of black rabbits. But how will you know if the rabbits you buy will produce only offspring with black fur? You cannot tell just by looking at them, because a black rabbit can have either one of two different genotypes—homozygous BB or heterozygous Bb. If some of the rabbits you buy have the Bb genotype, your rabbit warren will eventually contain some brown rabbits. To be sure that you buy only homozygous black rabbits, you must obtain a **pedigree** for each rabbit.

▼ A pedigree is a chart that traces the genetic history of a particular individual. To be useful, a pedigree usually must go back several generations and show as many family members as possible.

If there are no brown rabbits among the ancestors of a particular rabbit for several generations, you can be fairly certain that the rabbit is homozygous for black fur.

But if there are brown ancestors in the rabbit's recent past, there is a chance that the rabbit is heterozygous for fur color.

■ Male

● Female

1st Generation

2nd Generation

3rd Generation

Based on the information given in this pedigree, would it be wise for you to buy this black rabbit as a breeder for your rabbit warren?

S164

Sex Determination

What determines whether an individual is male or female? In the early 1900s, biologist Thomas Hunt Morgan found an answer to this question. He made his discovery while conducting genetic experiments with fruit flies, the tiny insects that fly around overripe bananas.

Morgan noticed that the cells of female flies have four pairs of chromosomes that are alike in shape and size. In male flies, however, only three of the pairs match; the fourth pair does not. One chromosome of this pair is of normal size, while the other is much shorter. Morgan believed that this mismatched fourth pair was responsible for the sex of the male fruit fly. Morgan called the nonmatching chromosome (the short one) in the male the *Y chromosome*. The other chromosome in the pair he called the *X chromosome*.

Further research has revealed that sex in most animals, including humans, is determined by the inheritance of X and Y chromosomes. Females have two X chromosomes, but males have one X and one Y chromosome. Some conditions in humans, such as hemophilia and colorblindness, are caused by abnormal genes that are located on the X chromosome.

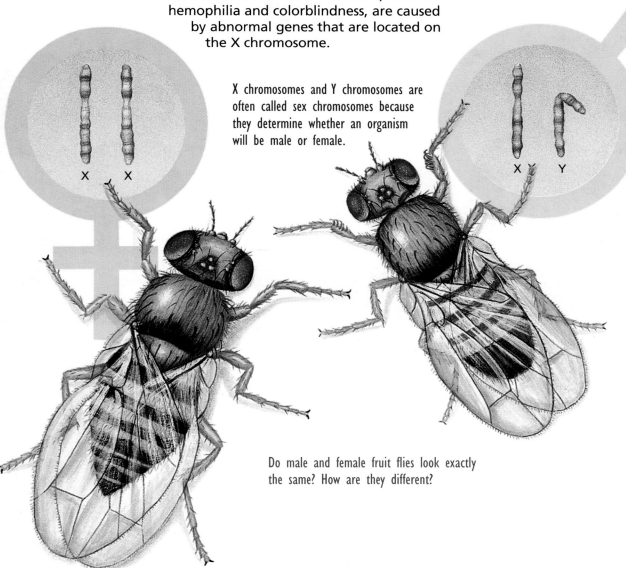

X chromosomes and Y chromosomes are often called sex chromosomes because they determine whether an organism will be male or female.

X X

X Y

Do male and female fruit flies look exactly the same? How are they different?

Other Patterns of Inheritance

Mendel described only one pattern of inheritance. The traits he studied in peas are each controlled by one pair of genes. For each of these traits there are two forms of the gene—one dominant and one recessive. Modern geneticists refer to this pattern of inheritance as *simple dominance*. However, simple dominance is only one way that genetic traits are inherited.

For some traits, there are no dominant or recessive genes. Instead, *both* genes in a heterozygous pair are expressed. This pattern of inheritance is called *incomplete dominance*. Flower color in snapdragons, for example, shows incomplete dominance. Crosses made between a red-flowered plant and a white-flowered plant produce all pink-flowered plants. They would not produce red-flowered plants or white-flowered plants, as you would expect if one color were completely dominant over the other.

Many traits in humans are determined by several pairs of genes. This pattern of inheritance is called *polygenic inheritance*. The effect of each pair of genes is added together to produce the phenotype of the individual. Skin color, for example, is a trait produced by the action of several pairs of genes.

▲ When plants with red flowers are crossed with white-flowered plants, the resulting offspring have pink flowers.

These students exhibit a wide range of skin colors.

S U M M A R Y

Hereditary factors called genes are located on chromosomes. Dominant genes are indicated by capital letters, while recessive genes are indicated by lowercase letters. A Punnett square is used to determine the genotypes and phenotypes of the offspring from genetic crosses. A pedigree is used to trace an individual's genetic history. In many animals, sex is determined by the inheritance of X and Y chromosomes. In addition to simple dominance, there are other patterns of inheritance, including incomplete dominance and polygenic inheritance.

S166

DNA—THE MATERIAL OF HEREDITY

DNA (deoxyribonucleic acid) was discovered in 1869. Because scientists found it in the nucleus of cells, they dubbed it a "nucleic acid." By the 1940s, scientists knew that chromosomes were made of both DNA and protein. But they didn't know whether it was the DNA or the protein that was the genetic material of cells. It was not until the 1950s that scientists were able to demonstrate that DNA is the material responsible for heredity. Yet, how does DNA work? And what is it made of?

Structure of DNA

Have you ever put together a model of an airplane or an automobile? Would you be able to put all the tiny pieces of the model together correctly if you lost the instructions? You might if you knew what the model was supposed to do. Back in 1953, the American biologist James Watson and the English physicist Francis Crick knew what DNA was supposed to be able to do—duplicate itself and act as the genetic material of a cell. Armed with this knowledge, and with the information collected by many other scientists, they were able to build a three-dimensional model of DNA.

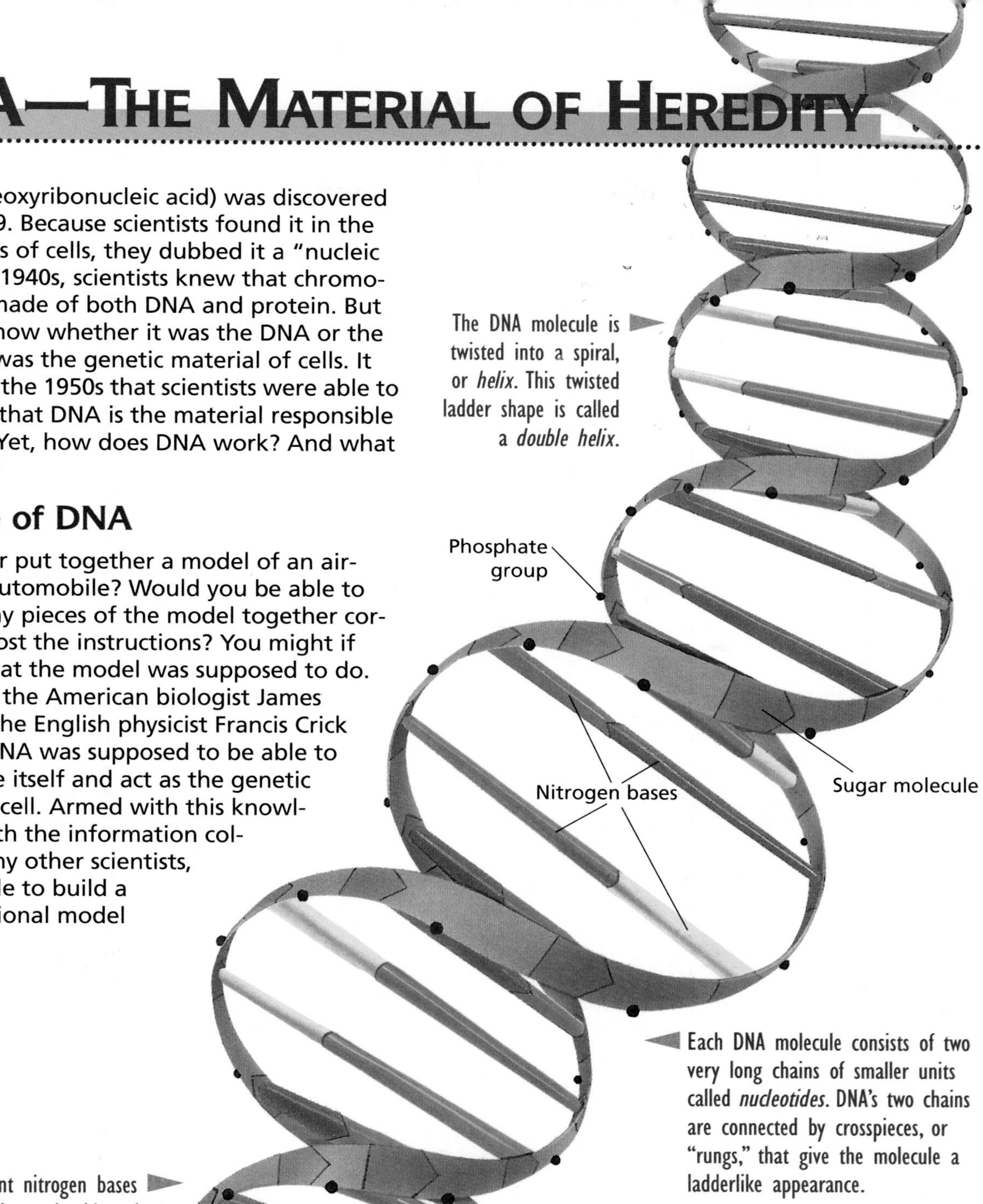

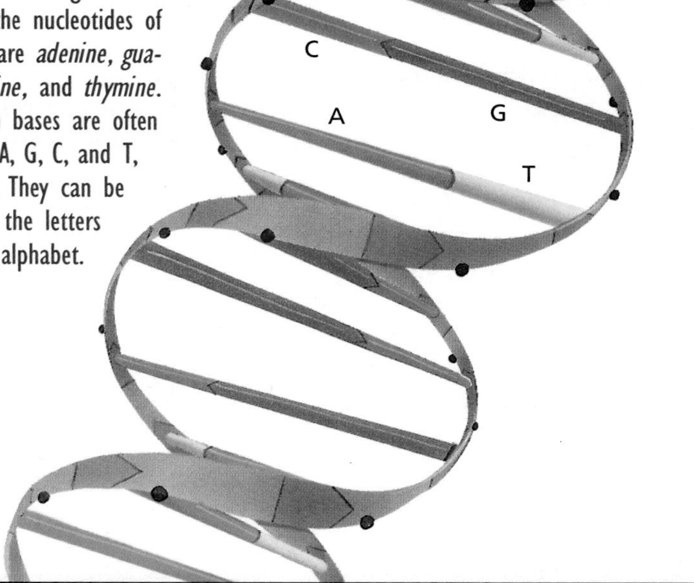

The DNA molecule is twisted into a spiral, or *helix*. This twisted ladder shape is called a *double helix*.

Phosphate group

Nitrogen bases

Sugar molecule

Each DNA molecule consists of two very long chains of smaller units called *nucleotides*. DNA's two chains are connected by crosspieces, or "rungs," that give the molecule a ladderlike appearance.

Phosphate group

Nitrogen base

Sugar molecule

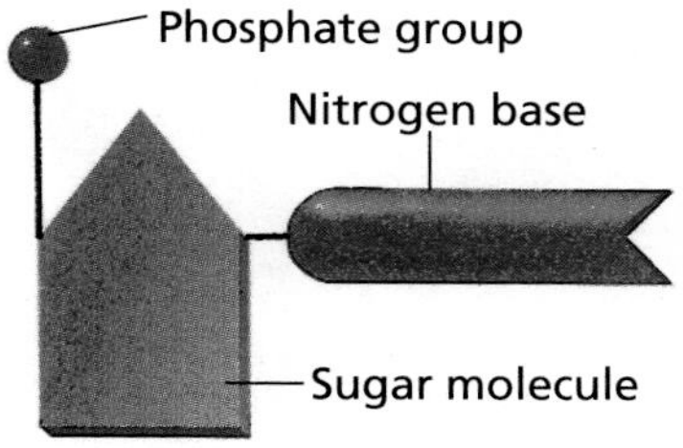

Each nucleotide in a DNA molecule is made of three parts—a sugar molecule, a phosphate group, and a molecule called a *nitrogen base*.

Four different nitrogen bases are found in the nucleotides of DNA. They are *adenine*, *guanine*, *cytosine*, and *thymine*. The nitrogen bases are often abbreviated as A, G, C, and T, respectively. They can be thought of as the letters of the genetic alphabet.

C

A

G

T

S167

DID YOU KNOW...

that the DNA molecules from *one* of your cells placed end to end are probably longer than you are tall? Stretched out to their full length, they would be about 2 m long!

Functions of DNA

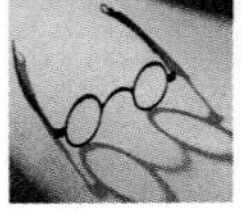

DNA has two major functions. It undergoes duplication, also called *replication*, and it directs *protein synthesis*. In order to understand how DNA accomplishes either of these processes, you must first know the way that the nucleotides are arranged to form the "sides" and the "rungs" of the DNA "ladder."

Nucleotides are bonded together in a specific way to form the double helix. The sides are formed by bonding the sugar of one nucleotide to the phosphate of the next nucleotide in a continuous chain. The nitrogen bases stick out from the sugar and phosphate sides. Each of the nitrogen bases of one side is bonded to a base from the other side. Look again at the DNA model on the previous page. The resulting nitrogen-base pairs form the rungs of the ladder. But each base can be paired with only one other base. Adenine (A) always pairs with thymine (T), and guanine (G) always pairs with cytosine (C).

DNA Replication Replication is the making of an exact copy of a DNA molecule. Replication occurs during the interphase portion of a cell's life cycle. The process of replication begins when the two chains of a DNA molecule begin to separate.

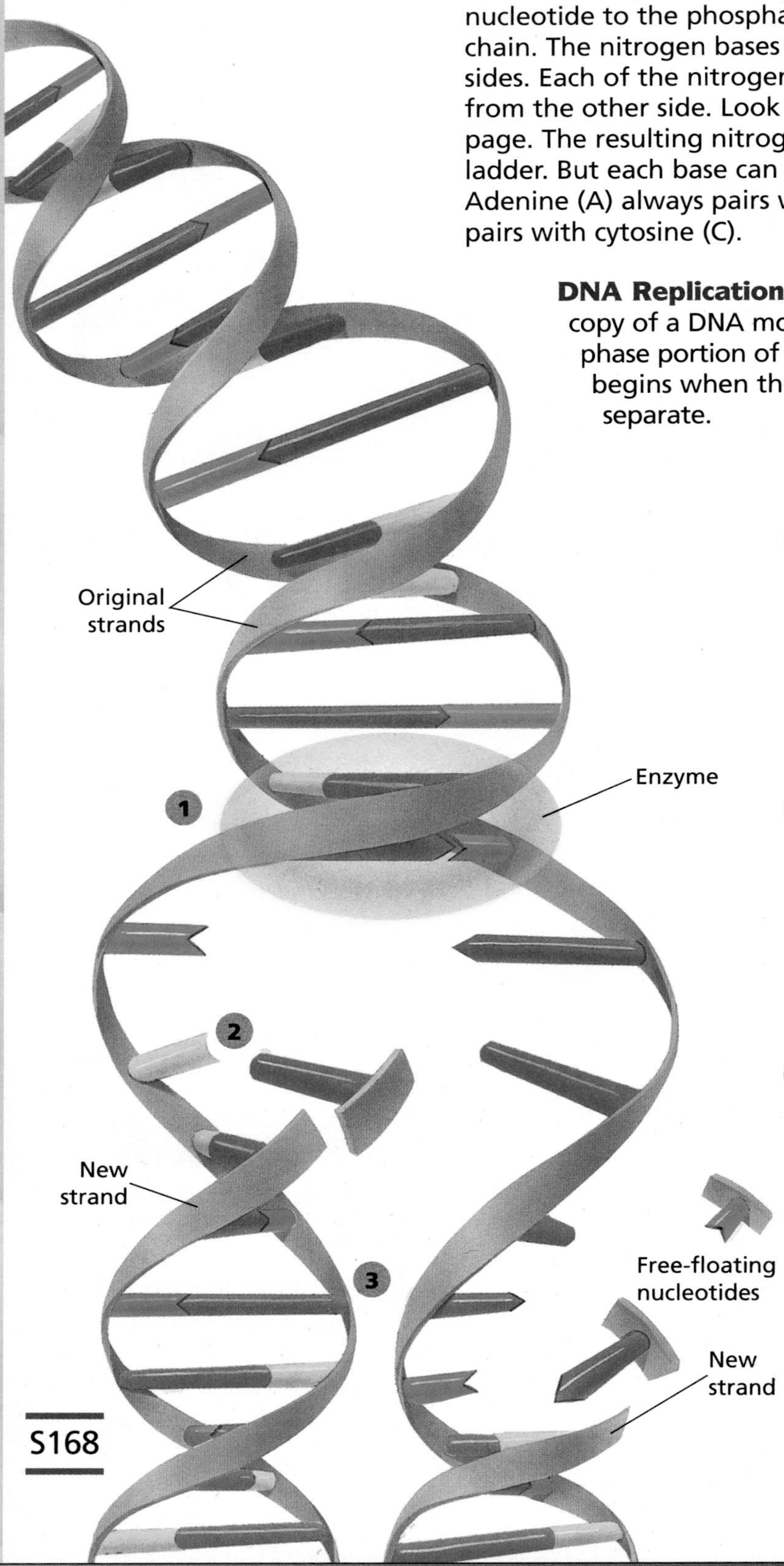

Original strands
Enzyme
1
2
New strand
3
Free-floating nucleotides
New strand

1. The nitrogen-base pairs are pulled apart by a type of protein called an *enzyme*. The separation of the two sides of the DNA molecule is often compared to the unzipping of a zipper. When the chains have separated, individual nucleotides floating freely in the nucleus line up across from the nitrogen bases of the separated chains.

2. New base pairs form as the nitrogen bases of the new nucleotides bond to the nitrogen bases on each half of the old molecule. Remember, there are only two kinds of nitrogen-base pairs that can be made—C always pairs with G, and A always pairs with T. As a result, the base-pair sequence of the original DNA reappears.

3. Once replication is complete, each side of the original DNA has produced a new DNA molecule. The two new DNA molecules are identical to each other and to the original molecule.

Protein Synthesis   The DNA in each of the chromosomes of an organism has a different sequence of nitrogen-base pairs. Genes are no more than sections of a DNA molecule. One gene may consist of thousands of nitrogen-base pairs. The individual genes of a DNA molecule contain instructions for making specific proteins. This is an extremely vital role of DNA, since the primary structural and regulatory chemicals of living things are proteins. Some of the many different proteins made by an organism are used to build its various parts. Other proteins, called enzymes, assist in making the variety of other substances that an organism produces. You just read about how one enzyme functions to unzip DNA during replication.

Each "word" in the genetic code, as found in RNA, is composed of three "letters," or nitrogen bases. Sixty-four such three-letter "words," or triplets, can be made.

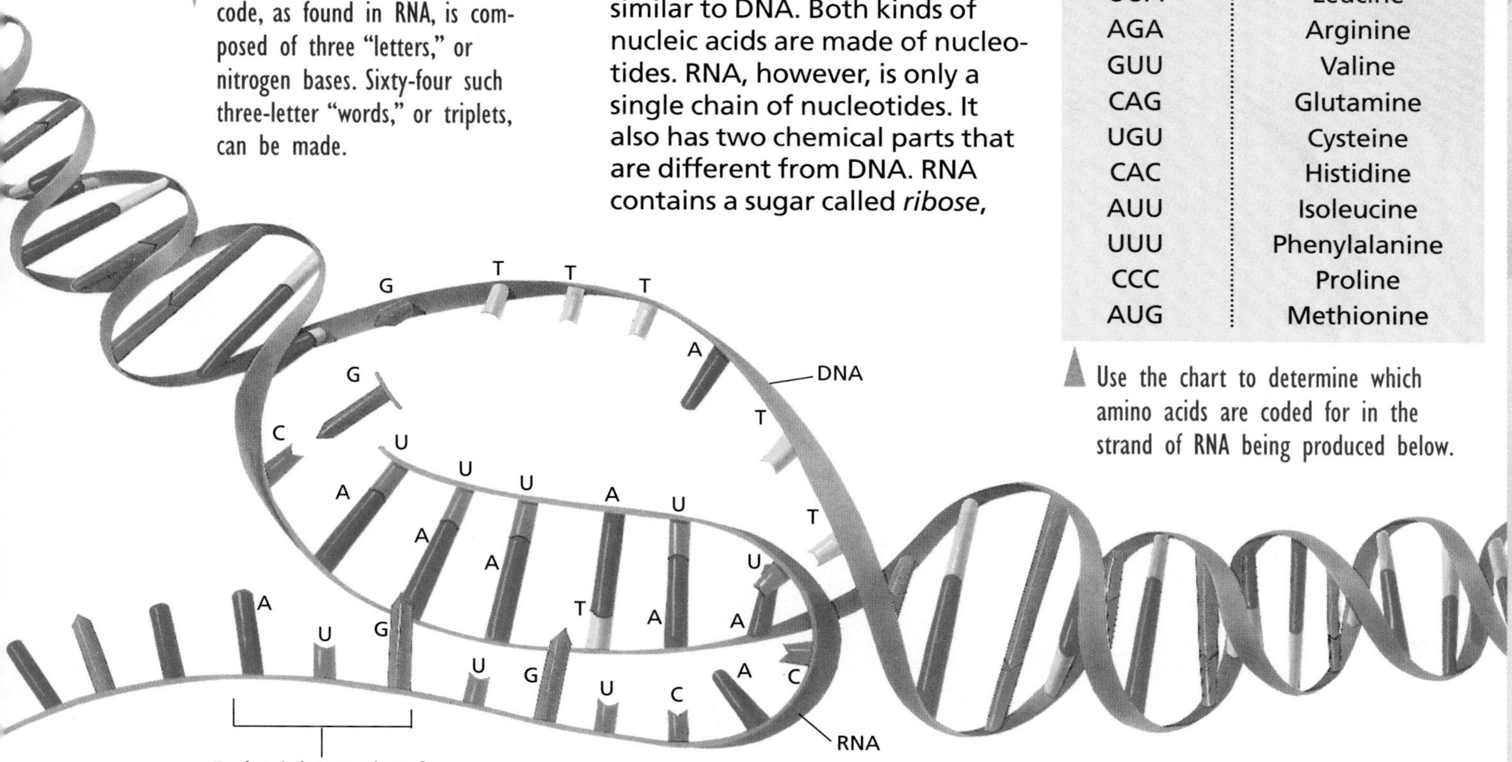

The Genetic Code

The instructions that DNA contains for making proteins are not written in words that are familiar to you. Instead, they are written in a "language" that uses the nitrogen bases A, T, G, and C as "letters." This language is called the **genetic code**.

Recall that amino acids are the molecules that make up proteins. The string of code words in a gene causes many amino acids to come together in a certain order to make a particular protein. But this cannot be a direct process because proteins are made in the cytoplasm of a cell, while DNA is found only in the nucleus. Instead, proteins are made in a more indirect way, by way of another nucleic acid, called RNA. RNA copies the information for making a protein from DNA. The RNA then brings the protein "recipe" to the cytoplasm.

RNA, or *ribonucleic acid*, is similar to DNA. Both kinds of nucleic acids are made of nucleotides. RNA, however, is only a single chain of nucleotides. It also has two chemical parts that are different from DNA. RNA contains a sugar called *ribose*,

while DNA contains a similar sugar called *deoxyribose*. As you can see, the molecules' names are based on their sugar parts. Instead of thymine, RNA has a nitrogen base called *uracil* (U). As with DNA, the nitrogen bases of RNA also form pairs, but there is one difference. The U of RNA pairs with A, since U takes the place of T in RNA.

Genetic code

The language in which the instructions for proteins are written in DNA

Some RNA Codes and Their Associated Amino Acids

Triplet	Amino acid
AAA	Lysine
GUG	Valine
UUA	Leucine
AGA	Arginine
GUU	Valine
CAG	Glutamine
UGU	Cysteine
CAC	Histidine
AUU	Isoleucine
UUU	Phenylalanine
CCC	Proline
AUG	Methionine

Use the chart to determine which amino acids are coded for in the strand of RNA being produced below.

Each triplet "codes" for a particular amino acid.

From DNA to RNA to Protein   Prior to protein synthesis, a special RNA molecule, called *messenger RNA* (mRNA), is synthesized from the DNA of a gene and carries the protein-making instructions from the DNA in the nucleus to a ribosome in the cytoplasm. Ribosomes are the structures on which proteins are assembled. In the cytoplasm, there is another kind of RNA, known as *transfer RNA* (tRNA). Each tRNA molecule (which has a three-looped, or cloverleaf, shape) carries a particular amino acid. For example, one kind of tRNA carries the amino acid *leucine*, while another carries *valine*. The amino acids carried by the tRNA molecules line up end to end at the ribosome. Then the amino acids are joined together by chemical bonds to form a protein.

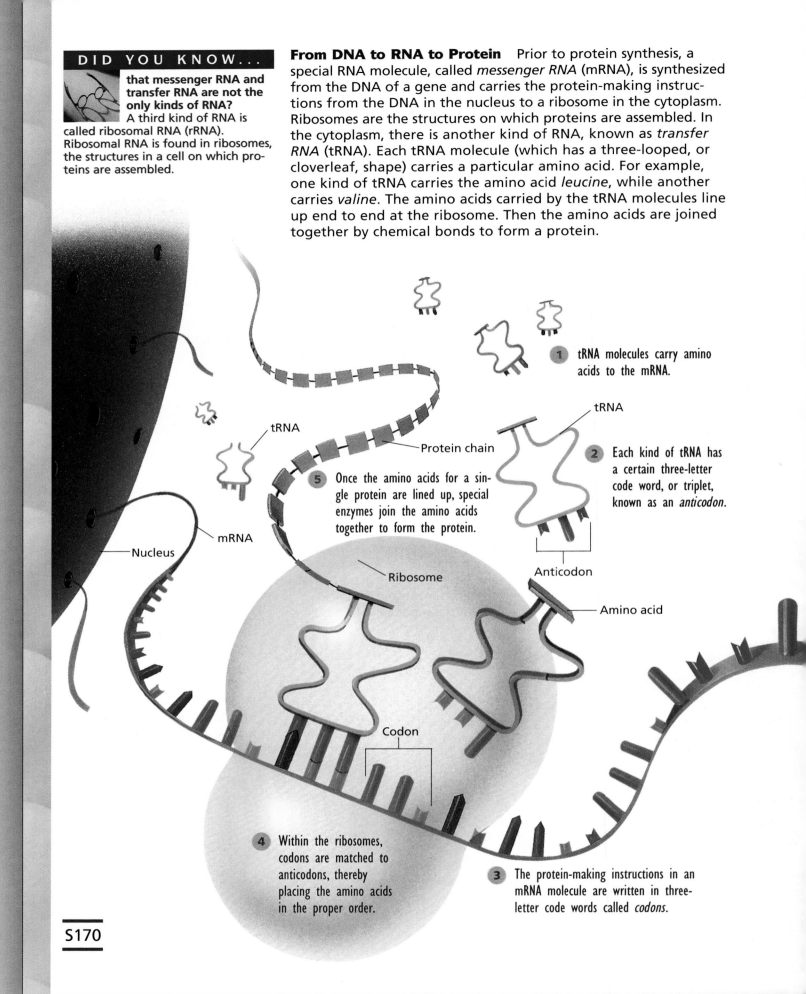

1 tRNA molecules carry amino acids to the mRNA.

tRNA

2 Each kind of tRNA has a certain three-letter code word, or triplet, known as an *anticodon*.

tRNA

Protein chain

5 Once the amino acids for a single protein are lined up, special enzymes join the amino acids together to form the protein.

mRNA

Nucleus

Ribosome

Anticodon

Amino acid

Codon

4 Within the ribosomes, codons are matched to anticodons, thereby placing the amino acids in the proper order.

3 The protein-making instructions in an mRNA molecule are written in three-letter code words called *codons*.

Changes in DNA

Right before a cell divides, its DNA is replicated so that each new cell receives a copy of the complete set of genetic instructions for an organism. Since the nitrogen bases of DNA pair only in certain ways, the copies are normally exact. However, mistakes during copying are sometimes made. Also, factors from the environment, such as chemicals and radiation, can cause accidents that destroy parts of a DNA molecule. In both cases, the result is a change in the nitrogen-base sequence of a DNA molecule. Thus, its genetic instructions may change as well. Such a change is called a *mutation*.

Mutations are important because they are the source of new genetic variations. As such, they have played a major role in the evolution of life on Earth. Although mutations can be beneficial, they are more frequently harmful. Fortunately, a mutation usually affects only one cell, and so there is essentially no effect on the entire organism. Only mutations that occur in reproductive cells can affect an entire organism. Remember that during fertilization, reproductive cells come together to form a new organism. In this way, these mutations can be introduced

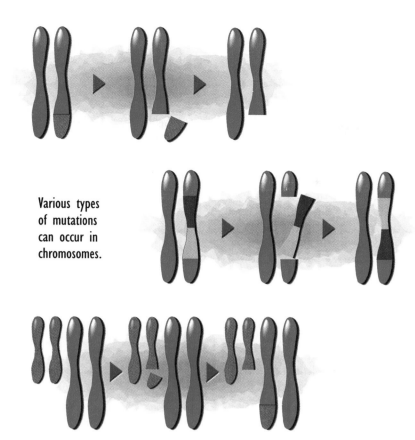

Various types of mutations can occur in chromosomes.

into offspring. There are two basic types of mutations—*gene mutations* and *chromosome mutations*.

Gene Mutations   A gene mutation is a change in the structure of a single gene and is therefore also called a point mutation. Most often, this kind of mutation affects only one nucleotide, or maybe several nucleotides, in the gene. But still, a change as small as even one base in the sequence of bases that make up a gene can have a significant effect. It can cause the gene to produce an altered protein or, sometimes, no protein at all.

Chromosome Mutations
Chromosome mutations are changes caused by the breaking of chromosomes. Normally, chromosome breaks are repaired. But if the parts of a broken chromosome are improperly rejoined or if some parts are lost, major changes in the genetic instructions of a cell can occur. Chromosome mutations affect many genes. As a result, they usually have a far more serious effect on a developing embryo than gene mutations. In fact, most are fatal and the offspring is never born.

◄ This black rat snake is white because a gene mutation has resulted in the inability to produce normal pigments.

S171

Genetic Engineering

Humans have learned how to engineer changes in DNA. Early in the 1970s, scientists discovered several bacterial enzymes that could be used to cut DNA molecules into pieces. By using these enzymes, scientists are able to remove a gene out of a cell from one organism and place it into a cell from another organism. The gene that was removed combines with the DNA in the new cell. That cell then has some new hereditary instructions. The new DNA formed by adding DNA from one organism to the DNA of another organism is called **recombinant DNA**.

Bacteria containing recombinant DNA are now used to make substances that are vital to human beings. One such substance is insulin, which is needed by many people who suffer from a disease called diabetes.

Most recently, recombinant DNA has been used to replace incomplete or undesirable DNA instructions in plants and animals. Scientists have been able to improve certain food crops and farm animals by adding genes for higher nutritional value and disease resistance. In a new kind of medical procedure called *gene therapy*, doctors use recombinant DNA techniques in attempts to cure certain human genetic diseases.

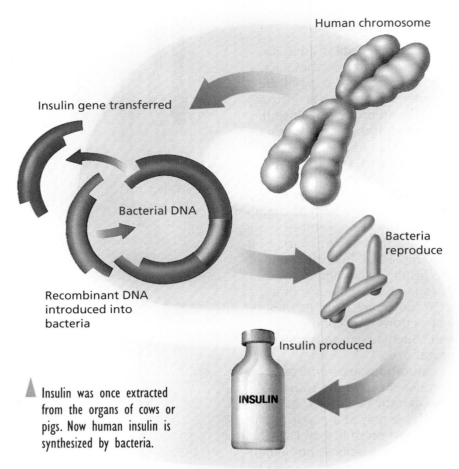

Human chromosome

Insulin gene transferred

Bacterial DNA

Recombinant DNA introduced into bacteria

Bacteria reproduce

Insulin produced

INSULIN

▲ Insulin was once extracted from the organs of cows or pigs. Now human insulin is synthesized by bacteria.

Recombinant DNA

A molecule made from the DNA of different organisms

S U M M A R Y

DNA is the material responsible for heredity. A molecule of DNA is a double strand of nucleotides that is twisted into a spiral shape called a double helix. The hereditary information carried by DNA is in a code that is contained in the sequence of the four nitrogen bases adenine, thymine, cytosine, and guanine. Because only certain bases can bond together in pairs, a DNA molecule can be copied by the process of replication. The instructions of the genetic code are used to put amino acids together to make proteins. Messenger RNA molecules take the instructions for making proteins from the nucleus into the cytoplasm, where proteins are made. Transfer RNA molecules carry amino acids to the messenger RNA. A change in the sequence of bases in a DNA molecule is called a mutation. A single gene or an entire chromosome may be changed by a mutation. Humans can alter DNA by taking DNA pieces from one cell and placing them into another cell, thus making recombinant DNA.

S172

REPRODUCTION AND DEVELOPMENT

You now know about chromosomes and genes and the role that meiosis plays in passing traits from one generation to the next. The sperm cells and egg cells that result from meiosis combine during a form of reproduction called **sexual reproduction.** But not all organisms reproduce by using sex cells, also called *gametes.* Some organisms, such as bacteria and sponges, produce offspring by **asexual reproduction,** which doesn't rely on the formation of sperm and eggs.

Asexual Reproduction

If you watch live bacteria under a microscope long enough, you should eventually see some of them divide in two. These dividing bacteria are reproducing asexually—reproducing without using gametes. Many single-celled organisms, such as bacteria, reproduce by simply splitting. This form of asexual reproduction is called fission. *Fission* means "splitting in half." In fission, one parent cell produces two identical daughter cells. Similarly, asexual reproduction in some protists occurs by mitosis. But fission and mitosis are not the only ways that organisms can reproduce asexually. Nor does asexual reproduction occur only in bacteria and protists.

▲ This bacterium is undergoing fission, a type of asexual reproduction.

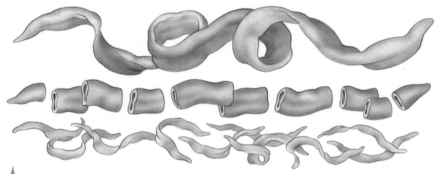

▲ Some simple worms reproduce by *fragmentation.* In fragmentation, the worm's body breaks up into many pieces. Each piece grows into a complete new worm.

Asexual reproduction

Producing new individuals without using gametes

Sexual reproduction

Producing new individuals by uniting gametes

The hydra, a relative of the jellyfish, reproduces by *budding.* In budding, a small knob of tissue grows from the body wall of the hydra. Before long, this outgrowth develops into a new hydra. Usually the offspring detaches from the parent and begins living on its own. ▶

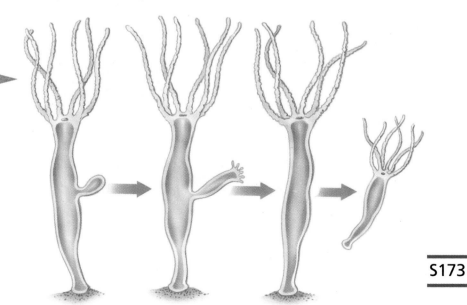

S173

Sexual Reproduction

Many protists, fungi, plants, and animals reproduce sexually, using gametes. During the life cycle of a sexually reproducing organism, there are two different cellular phases. In one phase, cells are diploid (contain pairs of chromosomes). In the other phase, they are haploid (contain a half-set of chromosomes). Remember that haploid cells result from meiosis.

The Plant Life Cycle   Plants have a life cycle that is described as an *alternation of generations*. This description comes from the fact that plants alternate between a multicellular diploid stage and a multicellular haploid stage. The diploid plant is called a *sporophyte*. Sporophytes are so named because they produce haploid cells, called *spores,* by meiosis. Some of the spores that are produced by a sporophyte undergo mitosis and develop into multicellular *gametophytes*, which are haploid. The gameto- phytes produce haploid sperm and egg cells, which unite to form another sporophyte. Let's take a look at a fern and see how it alternates between a sporophyte stage and a gametophyte stage during its life cycle.

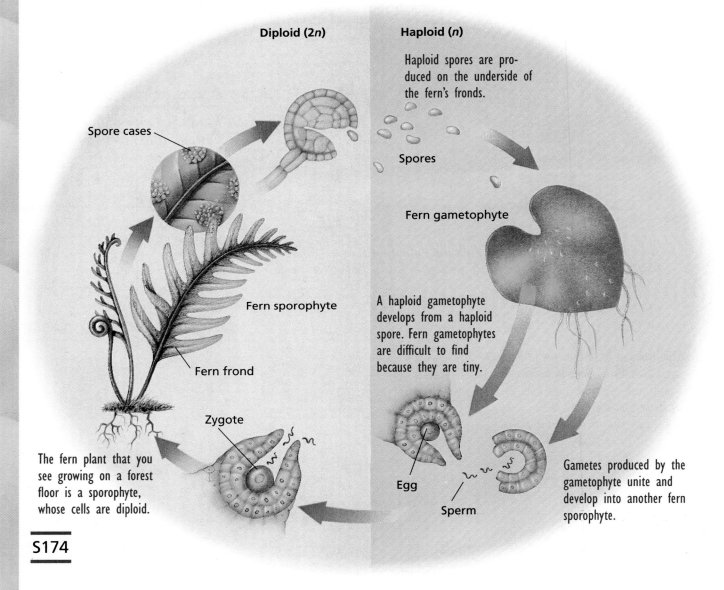

Diploid (2*n*)

Haploid (*n*)

Haploid spores are pro- duced on the underside of the fern's fronds.

Spores

Spore cases

Fern gametophyte

Fern sporophyte

A haploid gametophyte develops from a haploid spore. Fern gametophytes are difficult to find because they are tiny.

Fern frond

Zygote

The fern plant that you see growing on a forest floor is a sporophyte, whose cells are diploid.

Egg

Sperm

Gametes produced by the gametophyte unite and develop into another fern sporophyte.

S174

The Animal Life Cycle The life cycle of an animal is different from a plant's life cycle. In animals, the only haploid cells that exist are gametes. All the other cells in an animal's life cycle are diploid. The diploid zygote and the diploid individual that develops from the zygote are the most obvious parts of the animal life cycle.

These male and female salmon are breeding in a stream in northern California. The salmon are diploid and the gametes that they shed are haploid.

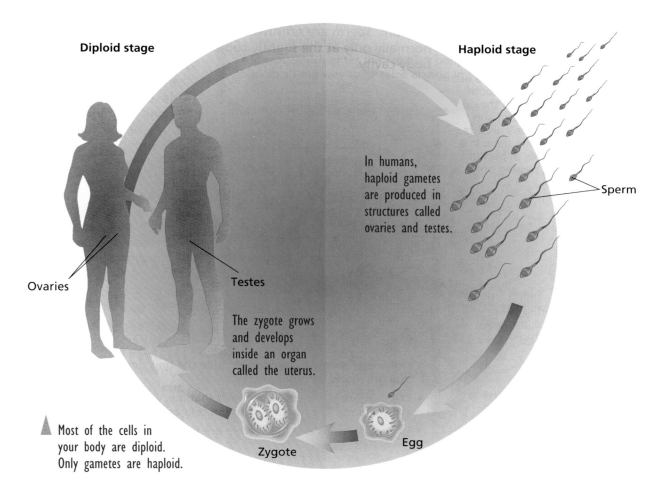

Diploid stage

Ovaries

Testes

The zygote grows and develops inside an organ called the uterus.

Most of the cells in your body are diploid. Only gametes are haploid.

Zygote

Haploid stage

In humans, haploid gametes are produced in structures called ovaries and testes.

Sperm

Egg

Human Reproduction

You are born with all the parts of your reproductive system, but it doesn't fully develop until you reach puberty. **Puberty** is the stage of life during which a person matures sexually. It usually begins sometime between the ages of 10 and 14. During puberty, males begin producing sperm cells, and females begin producing egg cells. The production of these gametes is mostly under the control of chemicals called *hormones*. Hormones are made by structures called *endocrine glands*. Hormones enter the bloodstream and cause changes in the body by affecting specific tissues and organs.

Puberty

The stage of physical development when sexual reproduction first becomes possible

S175

Conductor a material through which electric current flows easily (301)

Conservation of energy the law stating that energy can be transferred or changed from one form to another but cannot be created or destroyed (181)

Continental climate a climate typical of large land-masses, characterized by relatively large daily and seasonal temperature changes (256)

Control an experimental setup or subject used as a standard of comparison with another setup or subject that is identical except for one variable (216)

Controlled experiment an experiment performed to evaluate another experiment that is identical except for one variable (216)

Converging lens a lens that is thicker at its center than at its edge. Light rays bend toward each other as they pass through a converging lens. (469)

Convex mirror a mirror in which the reflective surface curves outward (467)

Coriolis effect the effect, caused by the rotation of the Earth, whereby moving objects (including winds and ocean currents) veer to the right in the Northern Hemisphere and to the left in the Southern Hemisphere (273, S68)

Coulomb a unit of measure of electrical charge, equal to the charge of 6.24×10^{18} (6.24 billion billion) electrons (343)

Cross-pollination the transfer of pollen from a flower of one plant to the flower of another plant (508)

Current a flow of electric charge through a conductor. Current is measured in amperes. (343, S88)

Cytologist a person who studies the structure, function, behavior, growth, and reproduction of cells (505)

D

DNA (deoxyribonucleic acid) the chemical in chromosomes which carries the genetic information that instructs the cells of living organisms (537)

Daughter cells the two cells that result from the division of a single parent cell (501)

Decibel (dB) a unit of measure of the loudness of sound (405)

Density the mass per unit volume of a given substance (125, 245)

Density current currents caused by density differences in liquids or gases. The denser substance flows downward, beneath the less dense substance. (248)

Dew point the temperature at which water vapor in the air begins to condense (231)

Diffraction the bending of waves as they pass the edges of objects (S113)

Diffuse reflection scattered reflection; that is, light reflected in more or less random directions (455)

Diffusion the uniform intermingling of particles of one substance with particles of another substance because of the motion of both types of particles. As a result of diffusion, particles move from regions of greater concentration to regions of lesser concentration. (36, S30)

Digestion the process by which an organism breaks down food into substances that can be used by individual cells (55)

Digestive system the organs associated with the intake, digestion, and absorption of food (65)

Direct current an electric current that flows continuously in one direction only (321)

Doldrums a region of the ocean near the equator characterized by calm weather and very light winds (275)

Domains in magnetic materials, clusters of atoms in which the magnetic fields of most atoms are aligned in the same direction (S94)

Dominant describing a genetic factor, or trait, that is always expressed. The presence of a dominant trait prevents a recessive trait from being expressed in offspring. (508, S156)

Doppler effect a change in the apparent frequency of waves caused by the motion of either the observer or the source of the waves (S119)

Dry cell a type of chemical cell in which the electrolyte is a paste rather than a liquid (310)

E

Echolocation the determination of the positions of objects by emitting sound that is reflected back to the sender as echoes. Bats use echolocation to navigate. (389)

Efficiency the ratio of the work done by a machine to the work put into it (176, S55)

El Niño effect a variation in worldwide weather patterns that recurs every 3–5 years and that is caused by changes in the wind conditions over the eastern Pacific Ocean (276)

Electric current the flow of electrons from one place to another (300, 303, S86)

Electric field the region of space around an electrically charged object in which the effects of the electric force may be observed (S85)

Electric force the force that causes two like-charged objects to repel each other or two unlike-charged objects to attract each other (S84)

Electricity the energy of negative particles (electrons) flowing in a conductor (293)

Electrode a terminal that conducts electrons into or away from the electrolyte in a battery (305)

Electrolyte a substance that conducts electricity; usually dissolved in water or some other solvent (305)

Electromagnet a magnet produced by an electric current flowing through a coil of insulated wire wrapped around a core of steel or iron (338)

Electromagnetic spectrum the entire range of electromagnetic waves, of which visible light is a small part (S137)

Electromagnetic waves waves that carry both electric and magnetic energy and that move through a vacuum at the speed of light (S130)

Electron a negatively charged particle found in atoms (130, 302)

Element a substance that consists of only one kind of atom and that cannot be separated into other substances by ordinary chemical changes (95)

S182

Embryo a developing organism in the early stages of growth; a human from the second to the eighth week of development (516, S178)

Endothermic describes processes in which heat is absorbed (115)

Energy the capacity to do work (153)

Energy converter a physical system in which energy is changed from one form into another (166, 427)

Evaporation the process by which a liquid becomes a gas (34, 104)

Evaporative cooling the removal of heat from the surroundings of a liquid that is undergoing evaporation (116)

Exothermic describes processes in which heat is given off (115)

Excretion the removal of metabolic wastes (S17)

Excretory system the organs associated with collecting and eliminating metabolic wastes (65)

Exponent a small number placed at the upper right of another number that tells how many times the lower number is to be multiplied by itself. For example, $10^3 = 10 \times 10 \times 10$; 3 is the exponent. (80)

F

Family tree a diagram for a family that shows parents and offspring for a number of generations (492)

Fertilization the union of a sperm cell with an egg cell, or ovum, resulting in the development of a new organism (515)

Fetus a developing organism that is in the later stages of development but is still in the womb or egg; a human from the ninth week of development until birth (524, S178)

Filter a process or device used for screening out something. An example of a filter is a colored piece of glass that absorbs certain colors of light while permitting other colors of light to pass through. (431)

Flywheel a heavy rotating wheel that stabilizes the speed of the device to which it is attached (187)

Focal length the distance from the focal point of a lens or mirror to the lens or mirror itself (471)

Focal point the point at which light beams passing through a lens or reflecting off a mirror converge (471)

Forced vibration the vibration produced in an object when it comes in contact with another object that is already vibrating (403)

Fraternal twins two individuals who are born at the same time from the same mother but who developed from different fertilized eggs. Such individuals have different genetic makeups. (521)

Frequency the number of repetitions in a given interval of time, such as the number of vibrations of an object per second or the number of complete waves passing a point per second (371, S111)

Front the boundary between two air masses (253, S72)

Fuse a safety device containing a strip of metal that melts when too much current passes through a circuit containing the device (349)

G

Galvanometer an instrument that measures small amounts of electric current (306)

Gear a toothed wheel that meshes with another toothed wheel in order to transmit force (187)

Generation a stage in a family's line of descent that includes all the individuals born within one life cycle (492)

Genes individual components of a chromosome which carry the factors that pass hereditary traits from parents to offspring (514)

Gene therapy the process of replacing defective genes in cells by injecting healthy genes into the cells (546)

Genetic code the language in which the instructions for proteins are written in DNA (S169)

Genetic engineering the manipulation of DNA by splicing and recombining it in order to produce new characteristics in species (544)

Geneticist a person who specializes in the study of genetics (544)

Genetics the science in which heredity is studied (509, S161)

Genotype the genetic makeup of an organism according to its genes (540, S162)

Global warming the apparent overall warming trend of the Earth's atmosphere (218)

Greenhouse effect the warming of the Earth's surface and lower atmosphere caused by heat rising from the surface of the Earth and being re-radiated back toward Earth by gases in the atmosphere (220)

Guard cells the kidney-shaped cells found on either side of each stoma on a plant leaf. Guard cells control the passage of gases into and out of the plant by causing the stomata to open and close. (20)

Gulf Stream a warm ocean current that flows from the tropical Atlantic Ocean northward along the eastern coast of North America (270)

H

Harmonic a component of a musical tone, the frequency of which is a multiple of the fundamental frequency (413)

Heredity the passing of traits from parents to offspring (512, S154)

Hertz (Hz) the basic unit of measure of frequency, equal to 1 cycle per second (321, 396)

Horse latitudes regions of the ocean at about 35°N and S that are known for their lack of winds (275)

Hurricane a severe tropical cyclonic storm with sustained winds in excess of 118 kilometers per hour (277)

Hybrid an offspring that contains one dominant gene and one recessive gene for a given trait (509)

Hydrometer an instrument that measures the density of liquids (249)

S183

Hypertonic solution a solution with a higher concentration of a dissolved substance than another solution **(52)**

Hypotonic solution a solution with a lower concentration of a dissolved substance than another solution **(52)**

I

Ideal machine an imaginary machine in which there is no friction **(S52)**

Identical twins two individuals produced from the same fertilized egg. Such individuals have an identical genetic makeup. **(521)**

Image the visual impression of an object produced by reflection in a mirror or refraction by a lens **(463)**

Impermeable not allowing substances to pass through **(41)**

Incident beam a beam of light that falls on or strikes something, such as a beam of light falling on a mirror **(452)**

Inclined plane a simple machine consisting of a plane set at an angle to a horizontal surface, forming a ramp **(174)**

Index of refraction the ratio of the speed of light in a vacuum to its speed in another medium **(S142)**

Inference a hypothesis, drawn from observations, that attempts to explain or to make sense of the observations **(81)**

Infrasonic waves sound waves at frequencies below those which humans can hear **(S121)**

Insulator a material through which electric current passes with difficulty or not at all **(301)**

Insulin the chemical substance released into the blood by the pancreas that enables the body to use sugar as a fuel in the process of respiration **(64)**

Interference the result of two or more waves overlapping **(S114)**

In vitro fertilization fertilization that takes place outside a living organism. The fertilized egg is maintained in an artificial environment until it can be placed into the uterus of a female for normal embryonic development. **(520)**

Isobar a line on a weather map that connects points of equal pressure **(279)**

Isotonic solution a solution with the same concentration of a dissolved substance as another solution. The term often refers to a solution of salt and water with the same salt concentration as blood. **(45)**

J

Jet stream a belt of high-speed wind circling the Earth between the troposphere and the stratosphere **(S64)**

Joule (J) a unit of measure of work, equal to the amount of work done when a force of 1 newton is exerted over a distance of 1 meter **(155)**

K

Kinetic energy the energy of motion **(162)**

L

Larynx the upper part of the trachea that contains the vocal cords and produces vocal sounds; the "voice box" **(368)**

Lens a piece of glass or other transparent material that is curved on one or both sides and that is used to refract light **(469)**

Lever a simple machine consisting of a bar that pivots about a fixed point; used to transmit or increase force or motion **(178)**

Light electromagnetic radiation in the wavelength range including infrared, visible, ultraviolet, and X rays; often used to refer specifically to the range that is visible to humans **(425)**

Lipids organic molecules composed of fatty acids and glycerol that store energy and make up cell membranes in living things **(S4)**

Longitudinal wave a wave, such as a sound wave, in which the particles of the medium move back and forth, parallel to the direction of motion of the wave **(S108)**

M

Machine a device that helps to do work by multiplying force or distance **(149)**

Magnetic field a region of space around a magnet in which magnetic forces are noticeable **(S92)**

Magnetic force the attraction or repulsion between the poles of magnets; the force that electric currents exert on each other **(317, S92)**

Magnetic lines of force invisible curved paths, between the poles of a magnet, along which a magnet exerts its force **(317)**

Maritime climate a climate typical of the ocean, characterized by relatively small daily and seasonal temperature changes **(256)**

Mechanical advantage the factor by which a machine multiplies force **(177, S53)**

Mechanical energy the kinetic energy of moving objects **(164)**

Mechanical system a machine composed of more than one simple machine **(150)**

Medium a substance, such as air or water, through which a wave travels **(380)**

Megaphone a cone-shaped device used to direct or amplify the voice **(403)**

Meiosis a process, occurring within the nucleus during cell division, that leads to the formation of sex cells with half of the organism's usual number of chromosomes **(515, S159)**

Metabolism all of the chemical reactions of an organism **(S16)**

Mitosis the process by which a cell's nucleus divides, leading to the formation of two identical cells **(504, S158)**

Model a representation of a phenomenon that simulates the structure, function, or effect of the phenomenon and that allows predictions to be made about the phenomenon **(83)**

S184

Mole the SI unit for the amount of a substance (S23)

Molecular biology the branch of biology that deals with the organization of living matter and genetic inheritance at the molecular level (544)

Molecule the smallest unit of a substance that has all of the physical and chemical properties of the substance and that is composed of two or more atoms (95)

Motor a device for converting electrical energy into mechanical energy (S100)

Mutation a change in the DNA of a gene (543)

N

Negative electrode the terminal that conducts electrons into the electrolyte in a chemical cell (313)

Nervous system the network of structures, including the brain and spinal cord, that control the actions and reactions of the body (65)

Neutron a particle with no charge that is found in the nucleus of an atom (135)

Newton (N) the SI unit of measure for force. One newton (1 N) is approximately equal to the weight (gravitational force) of a 100 g mass. (260)

Noise undesired sound, especially nonmusical sound that includes a random mix of frequencies (360)

Nucleic acids large organic molecules composed of nucleotides that store the information for building proteins (S6)

Nucleus the central core of an atom that contains protons and neutrons and that accounts for most of the mass of the atom (133); the central mass of protoplasm that is found in most plant and animal cells, controls the activities of the cell, and contains genes (498)

O

Observation direct evidence obtained through use of the senses; also, the act of obtaining direct evidence through use of the senses (81)

Octave the musical distance, or interval, between a musical tone and one of twice the frequency, such as from middle C to the C just above it (400)

Offspring the child or children of a parent or parents; in the case of one-celled organisms, the new cells are the offspring of the original cell (505)

Optical fiber a long, thin light pipe that uses total internal reflection to carry light (S150)

Organ part of an animal or plant that performs a specialized function (17)

Organic compound a compound that contains carbon and usually is produced by living things (S2)

Oscilloscope an electronic instrument that shows waves as curves on a screen (409)

Osmosis the movement of water across a semipermeable membrane from areas where water particles are more concentrated to areas where they are less concentrated (43)

Ovum a female sex cell produced as a result of meiosis; also known as an egg (515)

P

Pancreas the gland between the stomach and small intestine that enables the body to use sugar as a fuel in the process of respiration by releasing insulin into the blood (64)

Parallel circuit an electrical circuit arranged in such a way that the current passes through more than one pathway simultaneously (331)

Particle any small unit of matter that, with others, forms a larger whole (88, S22)

Pascal a unit of measure of pressure, equal to 1 newton per square meter (266)

Pedigree a chart that traces the inheritance of a genetic trait through several generations of a family (514, S164)

Permeable allowing substances to pass through (41)

Petiole the stalk that fastens a leaf to the stem of a plant (18)

Phenotype the external appearance of an organism as determined by its genotype (540, S162)

Photoelectric effect the emission of electrons by a substance when illuminated by light of a sufficient frequency (S131)

Photon a tiny package of electromagnetic energy (S131)

Photosynthesis the process by which green plants and plantlike organisms use energy from sunlight to convert water and carbon dioxide into sugars that can be used for food (14)

Piezoelectric effect the production of electric currents by certain crystals when they are squeezed or stretched (323)

Pigment a material that gives a substance its color by absorbing some colors of light and reflecting others (24)

Pitch the highness or lowness of a sound, determined by the frequency of the sound wave (361)

Placenta an organ that develops around an embryo and attaches the embryo to the wall of the mother's uterus. The placenta exchanges nutrients, wastes, and gases between the blood of the mother and the blood of the embryo. (517)

Plane mirror a flat mirror; reflects light to produce virtual images (463)

Plankton the small plants and animals floating near the surface of the ocean (S75)

Polar molecule a molecule that carries unevenly distributed electrical charges (S7)

Polarized light light consisting of light waves that vibrate in only one plane (S144)

Positive electrode the terminal that conducts electrons away from the electrolyte in a chemical cell (313)

Potential difference the difference in charge between two points in a circuit (S87)

Potential energy stored energy, such as the energy in a stretched or compressed spring or in an object raised above the surface of the Earth (159)

Power the rate at which work is done (157, S44)

S185

 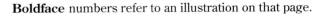

Boldface numbers refer to an illustration on that page.

A

Absorption spectrum, 26, **26**
Abyssal plains, 240
Acids
  amino, S5
  nucleic, S6
Adhesion, 50, S7, S28
Aeronautical engineering
  noise control and, 419
  Richard Linn and, 419, **419**
AIDS (acquired immune
  deficiency syndrome), 142
Air
  circulation, S68
  defined, S62
Air mass
  defined, 252, S72
  motion of, 253, S72
  types of, 252, **252**, S72,
    S72
Alternating current, 320
Alternator, S97, **S97**
Amino acids, S5
Amniocentesis, 545, **545**
Ampere
  defined, 343
  human body and, 347
  understanding of, 345
Amphibians
  effect of pollution on, 72
  ozone and, 72
  shrinking population of, 72
Amplitude (of waves), 371,
  S109, **S109**
Anaphase, S158, **S158**
Animals
  carnivorous, 9
  life cycle of, S175
Anvil (of ear), 393
Archimedes, S49
Art
  machines as, 209, **209**
Artificial magnets, S94
Asexual reproduction, 505,
  S173, **S173**
Atmosphere, S60–S65
  air in, S62
  defined, 251, S60, **S60**
  density currents in, 251–252
  evolution of, S60–S61,
    S60–S61
  heating of, S66, **S66**
  layers of, **260**, S63, **S63**
  oxygen levels in, S61–S62,
    S61
  pressure unit, 268, S32
  temperature changes in, S63
Atmospheric pressure, 261, S63
Atoms, S2, **S2**
  Albert Einstein and, 140

  defined, 92
  John Dalton and, 127–128
  models of, 127–135
ATP (adenosine triphosphate),
  S11
Auditory nerve, 394
Avogadro's number, S24
Axle
  wheel and, 182, S50, **S50**

B

Barometer, mercury, 266
Battery
  defined, 310
  examples of, **313**
Beam
  incident, 452–454
  object, S148
  reference, S148
  reflected, 452–454
Belts
  gears and, 190
  wheels and, 190
Bicycles, 208
Biochemicals, S3
  types of, S3–S6
Biomass, 67
Blade (of leaf), 18, **18**
Body systems
  circulatory, 59, **59,** 65
  digestive, 65
  excretory, 65
  nervous, 65
  respiratory, 65
Bohr, Niels, 133, **133**
Bohr model, 133–135,
  133–135
Boyle, Robert, 267, S32
Boyle's law, S32–S33
Breathing
  rate, 59–60
  respiration and, 58
  structures, 58–59, **59**

C

Caesarean section, 528
Capillary action, 48, **48**
Carbohydrates, S3, **S3**
Carbon dioxide
  as component of air, 13
  limewater test for, 13, **13**
  sources of, 224
  starch and, 14
  use of by plants, 13–14
Carnivorous animals, 9
Cars
  electric, 353, **353**

Cell division, 500, **501–502,**
  504
Cells
  blood, **59**
  chemical, 305, 310–314
  division, 500–504,
    501–502, 504,
    S157–S160, **S157–S160**
  dry, 310–313
  examples of, 7, **22, 498,**
    515, **515, S11, S157,**
    S159
  membranes of, 41
  of leaves, 20–21, **20–21**
  parts of, 498, **498,** 500,
    500
  solar, 322, **322**
  wet, 311
Cellular respiration, 57, S11
Charged particles, 301–303
  discovery of, 302
  J. J. Thomson and, 302
  negatively charged, 301–303
  positively charged, 301–303
  static electricity and, 303
  uncharged, 303
Chemical cell
  defined, 305
  technology, 311–314
Chemical energy, 160
  is a form of potential energy,
    160
Chemical role of water, S8
Chlorophyll, 23, **24**
  defined, S9
  function of, 23
  photosynthesis and, 24,
    S9–S11
Chloroplasts, S9, **S9**
Chromosomes, S157, **S157**
  defined, 514
  human, 514, **514**
  mutations in, S171
  sexual reproduction and,
    515, S159–S160,
    S159–S160
Chromosome theory
  Walter S. Sutton and, S161
Chu, Paul, **352**
  superconductors and, 352
Circuits (electric), 296
  components, **327**
  controlling, 335–337, S102
  defined, S101
  examples of, **336–337,**
    S86–S87
  symbols, **328**
  types of, 331, S102, **S102**
Circulatory system, 59, **59**
Circumstantial evidence, 81–82
Cirrus clouds, S71, **S71**
Climate, 256
  defined, S66

  types of, 256
Clones, 547, **547**
Clouds
  formation of, 34, **34,** S71
  types of, S71, **S71**
Cochlea (of ear), **392,** 394
Coherent light, S146
Cohesion, 49, 50, S7, S28
Cold front, 253, **253,** S73, **S73**
Colors, S133–S134
  complementary (light), 438
  light and, 436–438
  primary (light), 438, **438,**
    S134, **S134**
  primary (paint), 458–459
  secondary (light), 438, **438,**
    S134, **S134**
  visible spectrum and, S133
Complementary colors of light,
  438
Compounds, 92
  defined, 95
  organic, S2
  inorganic, S2
Compression (of a wave), S108,
  S108, S115
Concave lenses, S135, **S135**
Concave mirrors, 473–476,
  S136, **S136**
Condensation, 34, S70
Condensation reaction, S8
Conductors, electrical, 301
Coniferous trees, 25
Constructive interference,
  S114, **S114**
Continental climate, 256
Convex lenses, S135, **S135**
Convex mirrors, 467–468, S136,
  S136
Coriolis effect, 272
  defined, 273, S68
Coriolis, Gustave-Gaspard de,
  271
  wind and, 271
Coulomb
  defined, 343
  understanding of, 345
Cross-pollination, 508, S155
Cumulonimbus clouds, S71,
  S71
Cumulus clouds, S71, **S71**
Current
  alternating, 320
  deep-ocean, 248, **248**
  defined, 343, S88
  direct, 321
  effects of voltage on, S99
  electric, 300, 303, S86
  electricity and, S97
Cytokinesis, S159, **S159**

D

Dalton, John, **92**
  particle theory and, 92–96
Decibels
  defined, 405
  scale of, 406
Democritus, **92**
  particle theory and, 92–95
Density, 102, 125, 245
Density currents, 248, 251–252, **251**
Design
  industrial, 209
Destructive interference, S114, **S114**
Dew point
  defined, 231
  measurement of, 232
Diffraction (of waves), S113, **S113**
Diffuse reflection, 455, **455**
Diffusion, 36–40, S30, **S30**
  law of, 38
  model of, **37**
  through a membrane, 39–40
Digestion, 55–56, **56**
Digestive system, 65
Direct current, 321
Distance
  work and, S42–S44
DNA, 537–539, **537,** S167–S172, **S167–S170**
  changes in, S171
  functions of, S168
  genetic engineering and, 544, **544**
  nucleic acids and, S6, **S6**
  prehistoric genes and, 552
  recombinant, S172
  replication, S168
  structure of, **537,** S167
Dogs
  hearing aids for, 421, **421**
Doldrums, 275
Domains, S94, **S94**
Dominance
  incomplete, S166
  simple, S166
Dominant factors, 508–511, S156
Doppler effect, S119–S120
Dry cells, 310–313

E

Eardrum, 393, **393**
Ear, human, 392–394, **392–394**
Eastern meadowlark, **368**
Ebola virus, 142, **142**
Echoes
  sound and, 385–390
Echolocation, 389–390

Ecology
  desert lizards and, 74
Efficiency
  defined, 176, S55
  of machines, 176, 180–181, S55–S56
Einstein, Albert, 140, **140**
  particle-wave theory and, S132
  photoelectric effect and, S131
Electric cars, 353, **353**
Electric current, 300, 303, S86
  defined, 343, S88
  description of, S87
  detection of, 306
  effects of voltage on, S99
Electric field, S85, **S85**
Electric force
  defined, S84
  law of electric charges and, S84
Electricity
  connection to magnetism, 298
  energy use and, S46
  from chemicals, 310–314
  from magnetism, 315–321
  kinds of, S82–S90
  measurement of, 343–346
  motors and, S100
  static, 303, S82–S84
  transformation of, S98–S99
Electrodes, 305, **305**
  positive and negative, 305, 313
Electrolyte, 305, **305**
Electromagnet, 338–342, S95, **S95**
  construction of, 338
Electromagnetic spectrum, S137–S140, **S137**
Electromagnetic waves
  characteristics of, S130
  defined, S130
  example of, **S130**
  James Clerk Maxwell and, S130
  types of, **S137,** S138–S140, **S138–S140**
Electron microscope, 7
Electrons, 130–135
  defined, 302
Elements, 92–96
El Niño effect, 276, **276**
Embryo, 516–517, **517**
  defined, S178
Endothermic change, 115
Energy
  changes in, 164–165
  electricity and, S46
  forms of, 159–166
  in food, 5–9, 57–65
  machines and, 180–185
  transferring, 186–192
  work and, 153, 158–163
Engineering, genetic, 544
Environment
  human development and, 540–542

Eustachian tube, 393, **393**
Evaporative cooling, 116
Excretory system, 65
Exosphere, **260,** S64
Exothermic change, 115
Exponents, 80

F

Factors (genes)
  dominant, 508–511, S156
  recessive, 508–511, S156
Family tree, 492, **492, 514**
Female reproductive system, S177, **S177**
Fertilization, 515, **515**
  in vitro, 520
Fetus
  defined, 524, S178, **S178**
  description of, 524–527, **524–527**
Fiber optics, S149–S150
Filter
  light and, 431–435, 439
Fire, whirlwinds of, 284, **284**
Fire detection
  William Grosshandler and, 420
Fish (as a food source), S75–S76, **S76**
Food in space, 73
Food sources
  ocean, S75–S76, **S75–S76**
Force
  average, S44
  electric, S84, **S84**
  work and, S42–S44
Forced vibration, 403
Fourier, Jean, **220**
  greenhouse effect and, 220
Franklin, Benjamin
  lightning and, S83, **S83**
Franklin, Rosalind, **536**
  nuclear X rays and, 536, **536**
Fraternal twins, 521, **521**
Frequency (of waves), 371, S111
  wave speed and, S111
Friction
  effect on machines, S54–S55
  similarity to resistance, 330
Front, 253–255, **253–255,** S72–S73, **S73**

G

Galvanometers, 306
Gamma rays, S140
Gearbox, 192, **192**
Gears, 187–192
  belts and, 190
  kinds of, 191, **191**
Generation, family, 492

Generators, 318–319, **318–319,** S97–S98, **S97–S98**
Genes, 514, S161–S163
  Barbara McClintock and, 554
  mutations in, S171
Gene therapy, 546, S172
Genetic code, S169
Genetic disorders
  detection of, 545
Genetic engineering
  defined, 544, S172
  of food, 555
Genetics, 509, S161
Genotype, 540
  defined, S162
Gilbert, William
  magnetism and, S95
Glands
  pancreas, 64, **64**
  pituary, 64, **64**
  thyroid, 64, **64**
Glenn, John
  food in space and, 73
Global warming, 218, 225
Goats
  drug-producing, 553
  transgenic, 553
Grasshoppers
  sound and, 369
Greenhouse effect, 217–225
  defined, 220
  Jean Fourier and, 220
  planets and, 220
Gropius, Walter
  industrial design and, 209
Grosshandler, William
  fire detection and, 420
Guard cells, 20–21, **21**
Guericke, Otto von
  Magdeburg experiment and, 265
Gulf Stream, 270

H

Hair cells (of ear), 394, **394**
Hammer (of ear), 393
Harmonic (sounds), 413
Harvey, William
  Francesco Redi and, 497
  spontaneous generation and, 497
Hawaiian Islands, 242, **242**
Hearing
  process of, 393–394, **393–394**
Hearing aids
  for dogs, 421, **421**
Heat capacity
  of water, 29
Hensler, Mike, **207**
  Surf Chair and, 207, **207**

Heredity
defined, 512, S154
Gregor Mendel and, S154–S156
human development and, 543
Punnett squares and, 512, **512**
Hertz (unit of frequency), 321, 396
Hertz, Heinrich
electromagnetic waves and, S130
James Clerk Maxwell and, S130
Holograms, S148, **S148**
Holography, S148, **S148**
Horse latitudes, 275
Human development, 516–528, **516–528**
stages of, S178
Humidity, S69
Humpback whale, **368**
Hurricanes, 277–279, **277–279**
Huygens, Christian
particle theory and, S129
Hydrolysis, S8
Hydrometer, 249–250, **249–250**
Hydrothermal vents, 241, **241**
Hypertonic solution, 52, **52**
Hypotonic solution, 52, **52**

I

Ideal machine, S52
Identical twins, 521, **521**
Images, 463
real, 466, S135
virtual, 466, S135
Impermeable membranes, 41
Inclined plane
defined, 174, S48
examples of, **S48–S49**
Index of refraction, S142–S143
Industrial design, 209, **209**
Inference
differentiating from observation, 81–82
Infrared waves, S139
Infrasonic waves, S121–S122
Inheritance
patterns of, S166
Inorganic compound, S2
Insulators (nonconductors), 301
Interference (of waves), S114, **S114**
Interphase, S158, **S158**
In vitro fertilization, 520
Iodine
in search for starch, 7–9, **7**
Ionosphere, S65, **S65**
Ions, S22–S23, **S23**
Isobars, 279, **279**
Isotonic solution, 45, **52**

J

Jet stream, S64, **S64**
Joule, James Prescott, 155, **181**
and efficiency, 181
Joule (unit of work), 155, S43

K

Kelp (as a food source), S76, **S76**
Kemph, Martine, **207**
voice-controlled wheelchair and, 207, **207**
Kilojoule, 155
Kinetic energy, 162
Krill (as a food source), S76, **S76**

L

Larynx, 368
Laser beam
how produced, **S147**
Lasers, S145–S148, **S145–S146**
Lavoisier, Antoine
role of oxygen and, 92
Law
of diffusion, 38
of electric charges, S84
Leaves, 18–21
cross section of, **21**
parts of, 18, **18**
types of, **19, 25**
Lenses
compared with mirrors, 475
concave, S135, **S135**
converging, 469–472, **471, 472**
convex, S135, **S135**
defined, 469, S135
refraction and, S135
Levers, 178, S49–S50
classes of, S50, **S50**
parts of, S49
Life cycles
of animals, S175
of plants, S174, **S174**
Light
absorption of, 443
behavior of, 443–450
coherent, S146
color and, 436–439
complementary colors of, 438
filters and, 431–435, 439
lasers, S145–S148, **S146**
Newton's theory on, 434
polarized, S144–S145, **S144**
primary colors of, 438, **438**, S134, **S134**
prisms and, 431–434
scattering of, 443
secondary colors of, 438, **438**, S134, **S134**

sources of, S133
speed of, S141–S142
temperature and, 430
transmission of, 443
ultraviolet, S140, **S140**
visible, S133–S136, S139
Lightning
as a form of electricity, S85
Light pollution, 486, **486**
Limewater
carbon dioxide test, 13, **13**, 61
defined, 13
Linn, Richard, **419**
aeronautical engineering and, 419
noise control and, 419
Lipids, S4, **S4**
Lizards, desert, 74, **74**
Longitudinal wave, S108, **S108**
Loudness, 402–408
measurement of, 405

M

Machines
as art, 209, **209**
efficiency of, 176, 180–181, S55–S56
energy and, 180–185
examples of, 147–151, **147–151, 169, 171,** 182, 185, **185**, S47–S51, **S47–S51**
ideal, S52
mechanical advantage and, 177, S53–S55
micromachines, 206
simple, 150, **150**, 178, S47–S51, **S47–S51**
Magdeburg experiment, 265
Magnet
defined, S91, **S91**
Earth as a, S95–S96, **S95–S96**
examples of, **S91–S92**
in medicine, 354
types of, S94–S95, **S94–S95**
Magnetic field, S92–S93, **S92**
Magnetic force, 317, S92, **S92**
Magnetic lines of force, 317, S92–S93, **S93**
Magnetic poles, S91–S92, **S91–S92**
Magnetic resonance imaging (MRI), 354, **354**
Magnetism
connection to electricity, 315–321, S97, **S97,** S100, **S100**
description of, S91–S93
source of, S93–S94, **S93–S94**
Male reproductive system, S176, **S176**
Mariana Trench, 242, **242**
Trieste (bathyscaph) and, 268, **268**

Maritime climate, 256
Marsili, Count Luigi
Mediterranean Sea and, 243–247
Mass
volume and, 121
Maxwell, James Clerk
electromagnetic waves and, S130
Heinrich Hertz and, S130
particle-wave theory and, S132
McClintock, Barbara, **554**
study of genes and, 554
Mechanical advantage, 177, S53–S55
computation of, S54
effect of friction on, S54–S55
Mechanical energy, 164
Mechanical systems, 150–151, **150–151,** 193–197, **193–197**
Mediterranean Sea, 243–247, **247**
Count Luigi Marsili and, 243–247
Medium, 380
Megaphone, 403, **403, 404**
Meiosis, 515, **515**, S159–S160, **S160**
Membrane
impermeable, 41
permeable, 41
semipermeable, 41–42
Mendel, Gregor, **506**, S156
factors and, 506–515, S156
heredity and, S154–S156
patterns of inheritance and, S166
reproduction of plants and, 507–509, **507–509,** S155, **S155**
Mendel's factors
dominant, 508, S156
recessive, 508, S156
Mercury barometer, 266
Mesosphere, **260**, S63, **S63,** S64
Metaphase, S158, **S158**
Meteorology, 287
Micromachines, 206
Microscope
electron, 7
scanning tunneling, 77, 206
Mid-ocean ridges, 241, **241**
Mirrors
compared with lenses, 475
concave, 473–476, **473, 475, 476**, S136, **S136**
convex, 467–468, **467–468**, S136, **S136**
in space, 484, **484**
plane, 463–466, **463, 465,** S136, **S136**
reflection and, S136, **S136**
Mitosis, 504, **504**, S157–S158
stages in, **S158–S159**
Model, 83–84

Molecules
    defined, 95, S22
    ions and, S22–S23
    models of, **97–98, S22**
    polar, S7
Moles, S23–S26
    Avogadro's number and, S24
    in chemistry, S26
Morgan, Garrett, **487**
    traffic lights and, 487
Morgan, Thomas Hunt
    sex determination and, S165
Motors, 341–342, **342,** S100,
    S100
Mount St. Helens, 141
Music
    synthesizers and, 418, **418**
Mutations, 543
    in chromosomes, S171
    in genes, S171

N

Neon art
    485, **485**
Nervous system, 65
Neutron, 135
Newton (unit of weight), 260
Newton, Isaac, **434**
    particle theory of light and,
        S128
    theory of light and color
        and, 434
Nimbostratus clouds, S71, **S71**
Noise, 360
    effect on health, 360, 407
Noise control
    aeronautical engineering
        and, 419
    Richard Linn and, 419, **419**
Noise pollution, 407–408
Nucleic acids, S6

O

Observation
    differentiating from infer-
        ence, 81–82
Ocean
    abyssal plains, 240
    currents in, 248, **248**
    food sources in, S75–S76,
        S75–S76
    hydrothermal vents, 241,
        241
    Mariana Trench, 242, **242**
    mid-ocean ridges, 241, **241**
    movement of, 243–250
    pressure in, 263–264
    resources in, S75–S78
    undersea canyons, 240, **240**
Octave, 400
Ohm's Law, S90
Optical fiber, S150
Organic compound, S2

Organs
    defined, 17
    of plants, 17, **17**
Oscilloscope, 409–411, **409**
Osmosis, 43–46
    defined, 43
Oval window (of ear), 394
Ozone
    defined, 72
    effect on amphibian popula-
        tion, 72

P

Paint, primary colors of,
    458–459
Pancreas, 64, **64**
Parallel circuits, 331, S102,
    S102
Particles
    changes of state and,
        111–114
    charged, 301–303
    counting of, S23–S26
    mass and volume and,
        119–123
    motion of, S27–S34
    of gases, 103–104
    of liquids, 103–104
    of solids, 103–104
    size of, 99–102
    subatomic, S35–S38, **S35**
    temperature and, 108–110
Particle theory, 92–98
    Democritus and, 92–95
    John Dalton and, 92–96
Particle theory of light,
    S128–S129, S131
Particle-wave theory, S132
Particulates, 141
Pascal (unit of pressure), 266
Pascal, Blaise, 266, **266**
Patterns of inheritance, S166
Pedigree, 514, **514,** S164,
    S164
Permeable membranes, 41
Petiole (of leaf), 18, **18**
Petroleum, S77–S78
Phenotype, 540, S162
Photoelectric effect, S131,
    S131
Photon, S131, **S131**
Photosynthesis
    as a food-making process, 16
    benefits of, S11
    defined, 14, S9
    process of, S9–S10
Pianka, Eric, **74**
    zoology and, 74
Piezoelectric
    crystals, 323, **323,** 420,
        420
    effect, 323
Pinna (of ear), 393
Pitch, 361, 400–401
Pituitary gland, 64, **64**
Placenta, 517, **517**

Plane mirrors, 463–466, **463,
    465,** S136, **S136**
Planets
    comparison of four, 214–215,
        214–215
Plankton
    as a food source, S75, **S75**
    types of, S75
Plants
    life cycle of, S174, **S174**
    Mendel and reproduction of,
        507–509, **507–509,**
        S155, **S155**
    organs of, 17, **17**
    purebred, S155
    structure of, 17, **17**
Plimsoll marks, 257, **257**
Pointillism, 143, **143**
Polar easterlies, S68
Polar molecule, S7
Polarization of light, S144–S145,
    S144
Poles, magnetic, S91–S92,
    S91–S92
Pollution
    effect on amphibian popula-
        tion, 72
    light, 486, **486**
    noise, 407–408
Potential difference, S87–S88
Potential energy, 159–161
Power, 157, S44–S45
Precipitation, S70
    types of, S70–S71,
        S70–S71
Prehistoric genes
    DNA and, 552
Pressure
    atmospheric, 261, S63
    measuring, 266, **266,** 268
    ocean, 263–264
Prevailing westerlies, S68
Primary colors
    of light, 438, **438,** S134,
        S134
    of paint, 458–459
Prisms
    light and, 431–434
Prophase, S158, **S158**
Proteins, S5, **S5**
Protons, 130, **130**
    defined, 302
Puberty, S175
Pulley, S51
    examples of, **171–172**
    types of, S51
Pulse rate
    testing of, 60
Punnett squares, 512, **511,
    512,** S163, **S163**

R

Radiometer, 426, **426**
Rails, electric, 355
Rain forest, 3, **2–3,** 32
    transpiration in, 32–33

Rarefaction (of a wave), S108,
    S108, S115
Rate
    breathing, 60
    pulse, 60
Real images, 466, S135
Recessive factors, 508–511,
    S156
Recombinant DNA, S172
Redi, Francesco
    spontaneous generation and,
        497
    William Harvey and, 497
Reflection
    diffuse, 455, **455**
    mirrors and, 452–455, S136,
        S136
    (of waves) defined, S112,
        S112
    specular, 454, **455**
    total internal, 478, **478**
Refraction, 477–479, **479**
    index of, S142–S143
    lenses and, S135
    (of waves) defined, S113,
        S113
Regenerative breaking system
    in electric cars, 353
Relative humidity, S69
Reproduction, 505, 515–516
    asexual, 505, S173, **S173**
    in humans, S175–S177
    sexual, 505, 515–516,
        S173–S175
Reproductive system
    female, S177, **S177**
    male, S176, **S176**
Resistance, 329–330, S89
    applications of, 330
    examples of, **S89**
Resistor, 330
Resonance, 404
Respiration, 57–65
    control of, 64
    defined, 57
    products of, 61–63
Respiratory system, 65
Rheostat, 330, **330,** S102
Ride, Sally, **73**
    food in space and, 73
Robots
    cell-sized, 206, **206**
Root hairs, 35, **35,** 38, **38**
Root pressure, 47–48, **48**
Round window (of ear), 394
Rumford, Count, **181**
    and efficiency, 181
Rutherford, Ernest, 130, **130,**
    132–133

S

Sargasso Sea, 273, **273**
Scanning tunneling microscope,
    77, 206
Screw, S48, **S48**
Sea water
    composition and density of,
        249

S191

Secondary colors of light, 438, **438**, S134, **S134**
Self-pollination, S155
Semicircular canals (of ear), 393
Semipermeable membranes, 41–42
Series circuits, 331, S102, **S102**
Seurat, Georges
  pointillism and, 143
Sex
  cells, 515, **515**
  determination of, S165
Sexual reproduction
  defined, 505, S173
  in animals, S175, **S175**
  in plants, S174, **S174**
Simple machines, 178, S47–S51, **S47–S51**
  defined, S47
  examples of, 150, **150**
Solar cells, 322, **322**
Solutions
  hypertonic, 52, **52**
  hypotonic, 52, **52**
  iodine, 7, **7**
  isotonic, 45, 52, **52**
  sodium bicarbonate, **14**
  sodium hydroxide, **14**
Solvents, S7
Sonar, 390
Sonic boom, 384
Sonogram, S123, **S123**
Sound
  classification of, 361
  descriptions of, 360
  echoes, 385–390
  effect on health, 407–408
  in nature, 368–369
  intensity levels of, S124
  loudness and, 402–408
  of vibrating strings, S116
  pitch and, 361, 400–401
  quality of, 409–413
  speed of, 383–386
  transmission of, 376–382
  vibrations and, 370–372, 403–404
  waves, 378–381, **378**, S115–S120, **S115**, **S118**, **S119**
Sound meter, 405, **405**
Spectrum
  electromagnetic, S137–S140, **S137**
  of light, 431
  visible, S133
Specular reflection, 454, **455**
Speed of light, S141–S142
Spontaneous generation, 496–497
Standing waves, S117–S118, **S117–S118**
Starch
  carbon dioxide and, 14
  composition of, 13
  defined, 7
  in green leaves, 8–9
  and light, 10
  storage of, 7
Static electricity, 303, S82–S84
Stationary front, 254, **254**

Stimulated emission, S147, **S147**
Stirrup (of ear), 393, **394**
Stomata, 20–21, **20–21**
Stratosphere, **260**, S63–S64, **S63**
Stratus clouds, S71, **S71**
Submarines, 285, **285**
Superconductors
  Paul Chu and, 352
Surf Chair, 207, **207**
  Mike Hensler and, 207, **207**
Sutton, Walter S.
  chromosome theory and, S161
Switch, 336–337, **336–337**
Synthesizers, 418, **418**
Syrinx, 368

T

Telophase, S158, **S159**
Temperature
  light and, 430
  on Earth's surface, **226**
  particles and, 108–110
  state and, 111
Thermocouples, 323, **323**
Thermosphere, **260**, S63, **S63**, S64
Thomson, J.J., **129**
  discovery of charged particles and, 302
  view of atoms and, 129
Thyroid gland, 64, **64**
Thyroxin, 64
Tornadoes, S74, **S74**
Torricelli, Evangelista, 266, **266**
Total internal reflection, 478, **478**
Trade winds, 252, 269, **269**, S68, **S68**
Traffic lights, 487, **487**
  Garrett Morgan and, 487, **487**
Traits, 492–494
Transgenic goats, 553
Translucence, 448
Transparency, 447
Transpiration, 31–34, 50
  defined, 32
Transverse wave, S107, **S108**
Trees, coniferous, 25
Trieste (bathyscaph)
  Mariana Trench and, 268, **268**
Trimester
  defined, 525
  first, 519–524
  second, 526
  third, 527
Troposphere, **260**, S63, **S63**
Tsunamis, 286
Turtles
  leatherback, 489, **488–489**

Twins
  fraternal, 521, **521**
  identical, 521, **521**

U

Ultrasonic waves, S122–S123
  medical application of (ultrasound), S123
  used for sonograms, S123
Ultraviolet light, S140, **S140**
Undersea canyons, 240, **240**

V

Van Helmont, Jan Baptista
  reproduction of living things and, 496
  water experiment and, 11–12
Vegetarian diets, 75
Vein (of leaf), 18, **18**
Vibrating strings, S116
Vibrations, 370–372, **370**
  defined, 370
  forced, 403
Virtual images, 466, S135
Viruses, 142, **142**
Visible spectrum, S133
Visible light waves, S139
Voice-controlled wheelchair, 207, **207**
  Martine Kemph and, 207, **207**
Volume
  mass and, 121
Volts, 344
  human body and, 347
  understanding of, 345
Voltage, 312
  effects on current, S99
  electrical, 344

W

Waring Blender®, **209**
Warm front, 253, **253**, S73, **S73**
Water
  chemical role of, S8
  cycle, 31–34, **34**
  facts about, 29
  movement, 47–51
  root pressure and, 47
Water cycle, 31–34, **34**
Watery liquid (of ear), 394, **394**
Watt (unit of power), S45
  understanding of, 345
Watt, James
  power and, S45
Waves, S106–S114
  characteristics of, S106, S112–S114, **S112–S114**
  compression, S108, **S108**, **S115**
  defined, S106

Doppler effect, S119–S120
electromagnetic, S130, **S130**
infrared, S139
infrasonic, S121–S122
power, S138
properties of, S109–S111, **S109–S111**
radio, S138–S139, **S138**
rarefaction, S108, **S108**, **S115**
sound, S115–S120
speed of, S111
types of, S107–S108, **S108**, S117–S118, **S117–S118**
ultrasonic, S122–S123
visible light, S139
Wavelength, 379, S110, **S110**, S111
Wave theory of light, S129–S130, S131
  Thomas Young and, S129
Weather, 256, S66
  maps, 254, **254–255**
Wedge, S49, **S49**
Westerlies, 269, **269**
Wet-bulb depression, 231
Wet-bulb thermometer, 117, 229
Wet cell, 311, **311**
Wheel
  belts and, 190
  gears and, 190
Wheel and axle (machines)
  defined, 182, S50
  examples of, 185, **185**, **S50**
Wheelchair
  Surf Chair, 207, **207**
  voice-controlled, 207, **207**
Whirlwinds of fire, 284, **284**
Wind
  belts, S68, **S68**
  Gustave-Gaspard de Coriolis and, 271
  surface currents and, 271
Work, 152–157, S42–S44
  calculation of, 155
  defined, 153, S42
  distance and, S42–S43, **S43**
  energy and, 158–163
  force and, S42–S44
  input, 173, 176, S52
  output, 173, 176, S52

X

X rays, S140, **S140**

Y

Young, Thomas
  wave theory of light and, S129

Z

Zoology
  Eric Pianka and, 74

Photo Credits

Abbreviated as follows: (t) top; (c) center; (b) bottom; (l) left; (r) right; (bckgd) background, (bdr) border.

All HRW photos by Sam Dudgeon except where noted.

FRONT COVER: (t), Richard Kaylin/Tony Stone Images; (br), Carl Roessler/Animals Animals

BACK COVER: (bc), Jeff Smith/FotoSmith/Reptile Solutions of Tucson

TITLE PAGE: (bl), Jeff Smith/FotoSmith/Reptile Solutions of Tucson

TABLE OF CONTENTS: Page iv(tr), John Langford/HRW; (cr), Renee Lynn/Photo Researchers; (bl), John Langford/HRW; v(tc), John Langford/HRW; (bl), (br), Michelle Bridwell/HRW; vi(tl), John Langford/HRW; (tr), HRW; vii(tc), Stephen Dalton/Photo Researchers; (tr), John Langford/HRW; (cr), Scott Van Osdol/HRW; (bl), John Langford/HRW; viii(t), John Langford/HRW; (tr), Jennifer Dix/HRW; (br), John Langford/HRW; ix(tl), Peter Van Steen/HRW; (tc), Comstock; (cr), Peter Van Steen/HRW; (bl), John Langford/HRW; (br), Howard Sochurek/Stock Market; x(t), Lennart Nilsson/From *The Incredible Machine*, National Geographic Society; (bc), Runk/Schoenberger/Grant Heilman Photography; (br), C & M Denis-Huot/Peter Arnold; xi(t), Bill Ross/Westlight; (cr), Bill Beatty/Visuals Unlimited; (br), Philippe Plailly/Science Photo Library/Photo Researchers; xii(cl), Patrick R. Dunn/HRW; (br), HRW; xiii(tl), Carlos Austin; (c), Michelle Bridwell/HRW; xiv(l), (br), HRW; xv(tl), Michelle Bridwell/HRW; (cr), (bl), HRW; xvi(cr), Jana Birchum/HRW; (bl), Michelle Bridwell/HRW; xvii(bc), Daniel Schaefer/HRW.

UNIT 1: Page 2-3(bkgd), Gary Braasch/Tony Stone Images; 2(tl), Grant Heilman/Grant Heilman; 4(c), Renee Lynn/Photo Researchers; (bl), (br), HRW; (bkgd), Index Stock Photography; 5(tl), John Langford/HRW; (bl), (bc), (b), Michelle Bridwell/HRW; 6(tl), Lance Nelson/Stock Market; 6(tc), (tr), (br), John Langford/HRW; 7(l), Manfred Kage/Peter Arnold; 9(tl), (r), (bl), John Langford/HRW; 11(bl), Runk/Schoenberger/ Grant Heilman; (bc), Grant Heilman/Grant Heilman; 12(tl), (tr), (cr), HRW; (bc), Cindy Bland Verheyden/HRW; 18(bl), (bc), Runk/Schoenberger/Grant Heilman; 19, HRW (sycamore,lily); Runk/Schoenberger/Grant Heilman(elm, grass,maple,oak); Jerome Wexler/Photo Researchers(carrot); Fritz Pölking/Peter Arnold(beech); Ed Reschke/Peter Arnold(horse chestnut); 20(bl), Spike Walker/Tony Stone Images; 21(c), (bl), John Langford/HRW; (bkgd), Diamar Portfolios Landscapes & Scenery CD ROM by Diamar Interactive Corp.; 23(b), HRW; 25, HRW(sycamore,pine,orange), (tr); Runk/Schoenberger/ Grant Heilman(maple); (bl), Cindy Bland Verheyden/HRW; 26(t), Michelle Bridwell/HRW; 27(bl), Jim Frank/Photonica; 28(l), John Langford/HRW; (tr), David Nunuk/Science Photo Library/Photo Researchers; (br), Andrew Syred/Science Photo Library/Photo Researchers; 29(bkgd), Kevin Kelley/Tony Stone Images; 30(tl), Manfred Kage/Peter Arnold; (tc), Charles Philip/Westlight; (cl), Alan Carey/Photo Researchers; (c), Stephen Dalton/Photo Researchers; (tcr), Renee Lynn/Photo Researchers; (bcr), Index Stock; (bl), Rod Planck/Photo Researchers; (br), R. W. Jones/Westlight; (bkgd), Ralph Clevenger/Westlight; 31(bkgd), John Langford/HRW; 32(tr), Michael Fogden/DRK Photo; 36(tr), (cl), (bl), (bc), (br), John Langford/HRW; (b), Superstock; 37(cr), Michelle Bridwell/HRW; (bl), (bc), (br), John Langford/HRW; 45(tr), John Langford/HRW; (c), Robert Wolf; (cr), Phil A. Harrington/Peter Arnold; (br), Michael Mitchell; 46(t), Queen's Printer for Ontario, 1993, Reproduced with permission/Ontario Ministry of Agriculture and Food; (b), David Phillips/HRW; 47(r), Tom & Pat Leeson; 50(b), Michelle Bridwell/HRW; 51(tr), John Langford/HRW; 53(t), HRW; (bl), John Langford/HRW; 54(tr), Manoj Shah/Tony Stone Images; (br), NASA/Science Source/Photo Researchers; 57(tc), Nawrocki Stock Photo; (cl), Craig Lovell/Viesti Associates; (cr), Lori Adamski Peek/Tony Stone Images; (bc), HRW; (br), Paula Gaetani/Stock Shop; 58(bl), Stuart Westmorland/ Tony Stone Images; (bc), Rod Planck/Tony Stone Images; (br), Leonard Lee Rue III/Tony Stone Images; 59(br), CNRI/Science Photo Library/Photo Researchers; 60(c), (bl), (bc), (br), John Langford/HRW; 61(r), (cl), (br), John Langford/HRW; 62(tr), John Langford/HRW; 64(t), John Langford/HRW; 65(br), John Langford/HRW; 66(tr), NASA/Science Source/Photo Researchers; 67(bl), Ron Sherman/Tony Stone Images; (br), Peter May/Peter Arnold; 69(tl), (tc), (tr), (cr), HRW; 70(cl), Ken Eward/Science Source/Photo Researchers; (tr), Gary Braasch/Tony Stone Images; 72(cr), John Swedberg/Bruce Coleman; (bl), Zefa Germany/Stock Market; (tr), Michael Fogden/Bruce Coleman; 73(tl-ICON), Rogge/Stock Market; (tr), NASA; 74 (tl-ICON), Tony Stone Images; (tr), Charles C. Place/Image Bank; (bl), Dr. Eric Pianka; 75(tl-ICON), Howard Sochurek/Stock Market; (tc), Ken Karp/HRW; (tr), (cl), HRW; (bl), Superstock; (br), Stock Editions/HRW.

UNIT 2: Page 76-77(bkgd), Lawrence Berkeley Laboratory/Science Source/Photo Researchers; 76(tl), John Raffo/Science Source/Photo Researchers; 78(tl), Kenneth Love/National Geographic Society; (bl), Ken Lax/Stock Shop; (bkgd), Patrice Loiez, Cern/Science Photo Library/Science Source/Photo Researchers; 79(cl), Michelle Bridwell/HRW; 81(tr), Christopher Springmann/Stock Market; 82(tl), David Phillips/HRW; (br), John Langford/HRW; 83(tr), David Phillips/HRW; (bl), John Langford/HRW; 84(cr), John Langford/HRW; 85(tc), John Langford/HRW; (b), S. J. Krasemann/Peter Arnold; 86(tl), Neal Graham/Stock Shop; (tr), Steve Vidler/Nawrocki Stock Photo; (cl), B. Rubel/FPG Int'l.; (bc), Steven Ferry; (bkgd), HRW; 87(tr), Michelle Bridwell/HRW; 88(br), HRW; 90(c), HRW; 92(tr), (cr), Culver Pictures; (bc), Vaughn Fleming/Science Photo Library/Photo Researchers; (br), Frank Lawrence Stevens/Nawrocki Stock Photo; (bl), Tom McHugh/Photo Researchers; 93(cl), Culver Pictures; 94(tl), F. & A. Michler/Peter Arnold; (tr), Bob Pizaro/Comstock; 95(bl), (bc), David Phillips/HRW; 96(tr), (tc), David Phillips/HRW; 100(cr), Rossane Olson/Tony Stone Images; 101(tr), Garry Hunter/Tony Stone Images; (other), Specular Replicas CD ROM by Specular Int'l. (helicopters); Jeff Greenberg/Peter Arnold (city); 102(bc), David Phillips/HRW; (bkgd), Tom Bean/Stock Market; 103(bl), Michelle Bridwell/HRW; 107(cr), (bl), John Langford/HRW; 114(br), Simon Fraser/Science Photo Library/Photo Researchers; 116(bc), Tom Lyle/Stock Shop; (bkgd), David Phillips/HRW; 118(tr), John Langford/HRW; (other), Superstock (cloud); Clyde H. Smith/Peter Arnold(dark sky); Peter Arnold (lake); Robert MacKinlay/Peter Arnold(hills); 122(tc), (tr), (br), John Langford/HRW; 123(cl), (c), (tr), John Langford/HRW; (cr), Michelle Bridwell/HRW; 124(tr), Specular Replicas CD ROM by Specular Int'l.; 125(bc), Specular Replicas CD ROM by Specular Int'l.; 126(bc), John Langford/HRW; 129(tr), Bettmann Archive; (b), Michelle Bridwell/HRW; 130(bl), Science Photo Library/Photo Researchers; 133(bl), Bettmann Archive; (br), Michelle Bridwell/HRW; 135(tr), A. Barrington Brown/Science Photo Library/Photo Researchers; 137(cr), Michelle Bridwell/HRW; 138(cl), Michelle Bridwell/HRW; (tr), Lawrence Berkeley Laboratory/Science Source/Photo Researchers; 139(cl), Michelle Bridwell/HRW; 140(tl), UPI/Bettmann Archive; (bc), Bettmann Archive; 141(tl-ICON), Tony Stone Images; (c), D. Dzurisin/U.S. Geological Survey/Westlight; (bl), Paul Chauncey/Stock Market; 142(tl-ICON), Howard Sochurek/Stock Market; (tc), Richard J. Green/Photo Researchers; 143(tr), (cr), Georges Seurat, French, 1859-1891, *A Sunday on La Grande Jatte* 1884, oil on canvas, 1884-86, 207.5X308 cm, Helen Birch Bartlett Memorial Collection, 1926.224/©1994, The Art Institute of Chicago, All Rights Reserved.

UNIT 3: Page 144-145(bkgd), Vicki Miller/Featherstone Kite courtesy of Mid-America Museum, Hot Springs, Ark.; 144(tl), Steven Gottlieb/FPG Int'l.; (l), Stuart Cohen/Comstock; 146(tr), (bl), John Langford/HRW; 147(r), Superstock; (cr), Don Morley/Tony Stone Images; (bl), Peter Le Grand/Tony Stone Images; (cl), David Phillips/HRW; 148(tc), HRW; (c), Randy Duchaine/Stock Market; (cr), (b), David Phillips/HRW; (bkgd), Superstock; 149(tc), Tony Stone Images; (br), John Langford/HRW; (cr), Peter Grumann/Image Bank; 150(tc), (tl), David Phillips/HRW; (cr), Steven Ferry; (bl), NASA; 151(tr), HRW; (bl), Steven Ferry; (br), John Langford/HRW; 152(tr), Patti McConville/Image Bank; (cr), Guy Gillette/Photo Researchers; (bl), Jose L. Pelaez/Stock Market; (br), Superstock; 153(t), Comstock; (br), Michelle Bridwell/HRW; 154(tr),(cr), (br), (bl), (bc), John Langford/HRW; 155(tl), (cl), (bl), (br), (bc), John Langford/HRW; 156(b), David Phillips/HRW; 157(r), David Phillips/HRW; (bl), John Langford/HRW; 159(tr), (bl), David Phillips/HRW; 160(t), (cr), David Phillips/HRW; (bl), Eugen Gebjardt/FPG Int'l.; (br), Superstock; (bkgd), The Harold E. Edgerton 1992 Trust/courtesy of Palm Press, Inc.; 161(tl), David Madison; (cl), Michelle Bridwell/HRW; (tr), (cr), John Langford/HRW; (bl), David Phillips/HRW; (br), Superstock; 163(t), (c), (b), John Langford/HRW; 164(cr), John Langford/HRW; (b), David Phillips/HRW; 165(b), The Harold E. Edgerton 1992 Trust/courtesy of Palm Press, Inc.; 166(r), Barbara Van Cleve/Tony Stone Images; (br), Scott Van Osdol/HRW; (bl), David Phillips/HRW; (cr), Steven Ferry; (bkgd), Tony Stone Images; 167 (t), David Phillips/HRW; (br), HRW; 169(tl), David Phillips/HRW; (tr), (cr), Ron Kimball; (bl), Cindy Bland Verheyden/HRW; 172(l), Michelle Bridwell/HRW; 173(br), Michelle Bridwell/HRW; 175(br), John Langford/HRW; 178(tc), Michelle Bridwell/HRW; (tr), Mike & Carol Werner/Comstock; (cr), David Phillips/HRW; (b), J. Barry O'Rourke/Stock Market; 179(tl), David Phillips/HRW; (tr), Steven Ferry; (cr), (br), Michelle Bridwell/HRW; (b), Superstock; 181(br), HRW; (tr), (cr), Culver Pictures; 185(br), David Phillips/HRW; 186(tr), David Phillips/HRW; (br), Norco Products, Ltd.; 187(bl), David Phillips/HRW; (bkgd), Russ Kinne/Comstock; 190(br), Michael Melford/Image Bank; 193(tr), (br), HRW; 194(tr), (bc), HRW; (cr), Michelle Bridwell/HRW; 195(t), (b), HRW; (c), Specular Replicas CD ROM by Specular Int'l.; 197(cl), Michelle Bridwell/HRW; 198(bl), Brian Smith/Black Star; (cl), Steve Chenn/Westlight; (cr), Comstock; (br), Superstock; 199(br), David Phillips/HRW; 201(tr), (cr), Michelle Bridwell/HRW; 204(cl), John Langford/HRW; (tr), Vicki Miller/Featherstone Kite courtesy of Mid-America Museum, Hot Springs, Ark.; 206(tl-ICON), Rogge/Stock Market; (bl), Peter Menzel; (br), IBM Corporation, Research Division Almaden Research Center; 207(tl-ICON), Tony Stone Images; (tr), A.W. Stegmeyer/Upstream; (bl), Kathy Raddatz/*People Weekly* 1986; 208(tl-ICON), Ron Kimball; (tr), Loren Callahan; (cl), Michelle Bridwell/HRW; (bl), Barry Gregg/Advanced Transportation Products; 209(tr), Museum of Modern Art, New York "Tizio Lamp" by artist Richard Sapper, 1972, black metal; manufactured by Artemide; (bl), R. Beaudry/Waring Products Division.

UNIT 4: Page 210-211(bkgd), NASA; 210(l), Superstock; (tl), Warren Faidley/International Stock; 211(t), Superstock; 212(tl), Photri/Stock Market; (tc), Superstock; (t), US Geological Survey/Science Photo Library/Photo Researchers; (bl), Rod Planck/Photo Researchers; (bc), Superstock; 213(c), Keith Gunnar/FPG Int'l.; (bl), Zefa-Hummel/Stock Market; (br), Dallas & John Heaton/Westlight; (b), Superstock; 214(tc), NASA/Science Source/Photo Researchers; (bc), Photri/Stock Market; 215(tc), Jack Zehrt/FPG Int'l.; (bc), US Geological Survey/Science Library/Photo Researchers; 217(br), David Phillips/HRW; 218(b), Mountain High Maps CD ROM by Digital Wisdom and Quarto Publishing; 219(bl), HRW; (bkgd), Superstock; (bkgd), City Builder CD ROM by Dedicated Digital Imagery; 220(tc), Specular Replicas CD ROM by Specular Int'l.; (br), Science Photo Library/Photo Researchers; 221(cl), Ron Watts/Westlight; (b), E. Nagele/FPG Int'l.; 224(t), Mide Dobel/MasterFile; (cr), E.R. Degginger/Photo Researchers; (bc), Stephen J. Krasemann/ Photo

Goldberg/Sygma; (r), HRW; 555(tl-ICON), (tc), HRW; (bl), Cornell University New York State Center for Advanced Technology.

ART CREDITS

Abbreviated as follows: (t) top; (b) bottom; (l) left; (r) right; (c) center.

S195

Illustration; 344(t), Mark Persyn Illustration; 347, Blake Thornton/Rita Marie & Friends; 348, Blake Thornton/Rita Marie & Friends.

UNIT 6: Page 361, Blake Thornton/Rita Marie & Friends; 364-365, Boston Graphics, Inc.; 366, Howard Fullmer/Cary & Company; 369, Jack Graham/Carol Chivlovsky Design, Inc.; 370-371, Boston Graphics, Inc.; 377(r), Patti Bonham/Washington-Artists' Represents, Inc.; 381, Darrel Tank/Sweet Represents; 385, Boston Graphics, Inc.; 386, Frank Demes; 388-389, Tim Ladwig/Suzanne Craig Represents; 392, Boston Graphics, Inc.; 393, Boston Graphics, Inc.; 393-394, Martens & Keiffer/Carol Chivlovsky Design, Inc.; 397(t), Michael Koester/Woody Coleman Presents; 397(c), Jack Graham/Carol Chivlovsky Design, Inc.; 397(b), David Chen/James Conrad Artist Representative; 399, Tim Ladwig/Suzanne Craig Represents; 402(t), Michael Koester/Woody Coleman Presents; 406, Valerie Marsella; 407, Martens & Keiffer/Carol Chivlovsky Design, Inc.; 409, Stephen Durke/Washington-Artists' Represents, Inc.; 412(c), Stephen Durke/Washington-Artists' Represents, Inc.; 414, Aletha Reppel/Suzanne Craig Represents.

UNIT 7: Page 428, Darrel Tank/Sweet Represents; 431-435, Patti Bonham/Washington-Artists' Represents, Inc.; 437, Jim Pfeffer; 443-445, Darrel Tank/Sweet Represents; 446(t), Darrel Tank/Sweet Represents; 446(b), Tim Ladwig/Suzanne Craig Represents; 450(t), Patti Bonham/Washington-Artists' Represents, Inc.; 450(b), Stephen Durke/Washington-Artists' Represents, Inc.; 451, Stephen Durke/Washington-Artists' Represents, Inc.; 453, Stephen Durke/Washington-Artists' Represents, Inc.; 454, Tim Ladwig/Suzanne Craig Represents; 456, Blake Thornton/Rita Marie & Friends; 468(t), Darrel Tank/Sweet Represents; 468(b), Stephen Durke/Washington-Artists' Represents, Inc.; 471(t), Darrel Tank/Sweet Represents; 471(b), Lori Anzalone/Jeff Lavaty Artist Agent; 472(l), David Fischer; 472(r), Martens & Keiffer/Carol Chivlovsky Design, Inc.; 474, Darrel Tank/Sweet Represents; 475, Patti Bonham/Washington-Artists' Represents, Inc.; 477, Stephen Durke/Washington-Artists' Represents, Inc.; 478, Uhl Studio; 479, Stephen Durke/Washington-Artists' Represents, Inc.

UNIT 8: Page 490, Martens & Keiffer/Carol Chivlovsky Design, Inc.; 496(t), Barbara Kiwak/Barbara Gordon Associates Ltd.; 497, Joani Pakula; 499, Jack Graham/Carol Chivlovsky Design, Inc.; 502, Aletha Reppel/Suzanne Craig Represents; 503(t), Aletha Reppel/Suzanne Craig Represents; 503(b), Bill Geisler; 504(t), Bill Geisler; 504(b), Aletha Reppel/Suzanne Craig Represents; 507-509, Joani Pakula; 516-517, David Fischer; 519, Bill Geisler; 521, Bill Geisler; 527, Bill Geisler; 530(t), Keith Locke/Suzanne Craig Represents; 530(b), Jack Graham/Carol Chivlovsky Design; 531, Don Sullivan; 537, Bill Geisler; 545, Bill Geisler.

SOURCEBOOK ILLUSTRATION CREDITS

UNIT 1: Page S2(br), Academy Artworks, Inc.; S3(tr), Academy Artworks, Inc.; S4(tr), Academy Artworks, Inc.; S5(tc), Academy Artworks, Inc.; S6(cl), Academy Artworks Inc.; S7(c), Academy Artworks Inc.; (br), Mark Persyn; S8(c), Academy Artworks, Inc.; S9(cr), Martens & Keiffer / Carol Chislovsky Design Inc. Design Inc.; S10, Wendy-Smith Griswold/ Melissa Turk & The Artist Network & The Artist Network; S11(tr), Martens & Keiffer / Carol Chislovsky Design Inc. Design Inc.; S12(ct) Greg Harris / Cornell & McCarthy; S14(tl), Uhl Studio, S14(cb), Academy Artworks, Inc.; S16(tc), Martens & Keiffer / Carol Chislovsky Design Inc. Design Inc.; S17(tl), Martens & Keiffer / Carol Chislovsky Design Inc. Design Inc.; S17(cr), Martens & Keiffer / Carol Chislovsky Design Inc. Design Inc.; S18(c), Martens & Keiffer / Carol Chislovsky Design Inc. Design Inc.; S18(tr), Martens & Keiffer / Carol Chislovsky Design Inc. Design Inc.; S20(tc), Glasgow & Associates.

UNIT 2: Page S22(bl), Academy Artworks, Inc.; S23(cr), Academy Artworks, Inc.; S24(l), Richard Wehrman; S26(tl), John Francis; S26(cr), Mark J. Persyn; S28(cl), John Francis; S29(tr), Academy Arts Inc., S29(cr), Academy Arts Inc.; S29(bl), Academy Arts Inc.; S31(b), Liaison Production Services; S32(b), Uhl Studio; S33(tr), John Francis; S33(b), John Francis; S34(tl), Uhl Studio; S35(tc), Uhl Studio; S36(b), Uhl Studio; S38(tl), Academy Arts, Inc.

UNIT 3: Page S43(t), Bob Dorsey; Page S43(b), Bob Dorsey; S45(t), David Merrell / Suzanne Craig Represents; S48(c), Joe McDermott / Koralik Associates; S49(b), Joe McDermott / Koralik Associates; S50(t),Patti Bonham / Washington; S51(t), Darrel Tank; S52(b), Gary Locke / Suzanne Craig; S53(tr),Joe McDermott / Koralik Associates; S54(t), Darrel Tank.

UNIT 4: Page S60(b), Steven Adler / Jeff Lavaty Artist Agent; S61, Steven Adler / Jeff Lavaty Artist Agent; S62(b), Greg Harris / Cornell & McCarthy; S63, Stephen Durke / Washington-Artists' Represents, Inc.; S64(all), Craig Attebury / Jeff Lavaty Artist Agent; S65(c), John Francis; S66, Richard Wehrman; S67, Stephen Durke / Washington-Artists' Represents,Inc.; S68(all), Stephen Durke / Washington-Artists' Represents, Inc.; S69(cr), Glasgow & Associates; S69(b), Liaison Production Services; S70(b), Stephen Durke / Washington-Ariitsts' Represents, Inc.; S71, Todd Lockwood / Carol Guenzi Agents; S72, Glasgow & Associates; S73, Craig Attebury / Jeff Lavaty Artist Agent.

UNIT 5: Page S83(br), Keith Locke / Suzanne Craig Represents; S84(lc), Joani Pakula; S85(tr), Mark Persyn; S86(b), Don Brautigam / Bill Erlacher, Artists Associates; S87(t), Don Brautigam / Bill Erlacher, Artists Associates; S90(tl), Blake Thornton / Rita Marie & Friends; S92(t), Darrel Tank; S93(rc)(rb), Mark Persyn; S94(tr), Mark Persyn; S95(cb), Brian Harrold / American Artists Rep., Inc.; S96, Mark Persyn; S97(all), Jim Pfeffer; S98(t), Jim Pfeffer; S98(b), Tony Randazzo / American Artists Rep., Inc.; S99, Tony Randazzo / American Artists Rep., Inc.; S100, Tony Randazzo / American Artists Rep., Inc.; S101, Blake Thornton / Rita Marie & Friends; S102, Blake Thornton / Rita Marie & Friends; S104, Digital Art.

UNIT 6: Page S106(rc), Uhl Studio; S107, Todd Lockwood / Carol Guenzi Agents; S108(t), Todd Lockwood / Carol Guenzi Agents; S108(b), Uhl Studio; S109, Todd Lockwood / Carol Guenzi Agents; S110(t)(b), Uhl Studio; S111(b), Todd Lockwood / Carol Guenzi Agents; S114(c), Todd Lockwood / Carol Guenzi Agents; S115(b), Uhl Studio; S116, Paul Hess / Garden Studio Illustrators; S117, Todd Lockwood / Carol Guenzi Agents; S118(tl), Uhl Studio; S120, Uhl Studio; S121(tr),Wendy Smith-Griswold / Melissa Turk & The Artist Network.

UNIT 7: Page S128(tl), Joani Pakula; S128(bc), Keith Locke / Suzanne Craig Represents; S129(tl), Joani Pakula; S130, Liaison Production Services; S131, Brian Harrold / American Artists Rep., Inc.; S132(c), Brian Harrold / American Artists Rep., Inc.; S133(tr), Brian Harrold / American Artists Rep., Inc.; S135(bl), Liaison Production Services; S136, Liaison Production Services; S137, Digital Art; S138(c)(b), Brian Harrold / American Artists Rep., Inc.; S141, Uhl Studio; S144, John Francis; S146, Digital Art; S147(b), George Ladas / Melanie Kirsch Artist Represents; S148(t), Digital Art; S152, Liaison Production Services.

UNIT 8: Page S155, John Francis; S156, Joel Spector; S157(b), David Fischer; S158, Jean E. Calder; S159(b), Jean E. Calder; S160, Morgan Cain & Associates; S162, John Francis; S163, Michael Krone; S164, Joyce Kitchell / James Conrad; S165, Wendy Smith-Griswold / Melissa Turk & The Artist Network; S167(all), Martens & Keiffer / Carol Chislovsky Design Inc.; S168, Martens & Keiffer / Carol Chislovsky Design Inc.; S169, Martens & Keiffer / Carol Chislovsky Design Inc.; S170, Martens & Keiffer / Carol Chislovsky Design Inc.; S172, Jean E. Calder; S173(cl)(br), Morgan Cain & Associates; S174, Jean E. Calder; S175(c), Morgan Cain & Associates; S176(b), David Fischer; S177(br), Martens & Keiffer / Carol Chislovsky Design Inc. Design Inc.

Acknowledgment

For permission to reprint copyrighted material grateful acknowledgment is made to the following source:

The Humana Press, Inc.: From *The Life Within: Celebration of a Pregnancy* by Jean Hegland. Copyright © 1991 by the Humana Press, Inc.

S196

PHOTO CREDITS

TITLE PAGE: Page T5(tr), HRW photo by Daniel Schaefer; (br), Carlos Austin

OWNER'S MANUAL: Page T15, T17, T18, T22, T24, T25(b), T28, T29, T36, T41, T44(t),T45, T46, T55(r), T56, T58, HRW photo by Sam Dudgeon; T19, T42, T44(b), T49, T57, HRW photo by Daniel Schaefer; T16, T50, T51, HRW photo by Tomas Pantin; T25(t), T26, T33, T35, T38, T39, T47, T48, HRW photo by John Langford; T55(l), HRW photo by Jack Newkirk; T20, T23, T30, HRW photo by Scott Van Osdol; T31, David Young Wolff/Photo Edit; T43, Michelle Bridwell/Frontera Fotos

UNIT 1: Page 1D, HRW photo by Michelle Bridwell

UNIT 2: Page 75D, HRW photo by Daniel Schaefer

UNIT 3: Page 143A, HRW photo by Sam Dudgeon; 143C(tr) HRW photo by David Phillips; (br) Mike & Carol Werner/Comstock; 143D, HRW photo by Daniel Schaefer

UNIT 4: Page 209D, HRW photo by John Langford

UNIT 5: Page 287D, HRW photo by John Langford

UNIT 6: Page 355C, HRW photo by Jennifer Dix; 355D(bc), HRW photo by Sam Dudgeon; (tl) HRW photo by John Langford

UNIT 7: Page 421A, Comstock; 421D, HRW photo by Sam Dudgeon

UNIT 8: Page 487A, Dan Richardson/Medichrome/The Stock Shop; 487D, HRW photo by John Langford

ANSWERS TO TABLES AND DIAGRAMS: Page S208(tl), Ron Watts/Westlight; (c), E. Nagele/FPG Int'l.; S213, Biophoto Associates, Science Source/Photo Researchers

ART CREDITS

All work, unless otherwise noted, is contributed by Holt, Rinehart and Winston.

Page 1A, Richard Wehrman; 75F, Todd Lockwood/Carol Guenzi Agents; 287A, Howard Fullmer/Cary & Company; 287F, Mark Persyn Illustration; 355A, Valerie Marsella; 355C, Martens & Keiffer/Carol Chivlovsky Design, Inc.; 487A, Bill Geisler; 487C, Aletha Reppel/Suzanne Craig Represents

S197

Concept Mapping

Sample concept map:

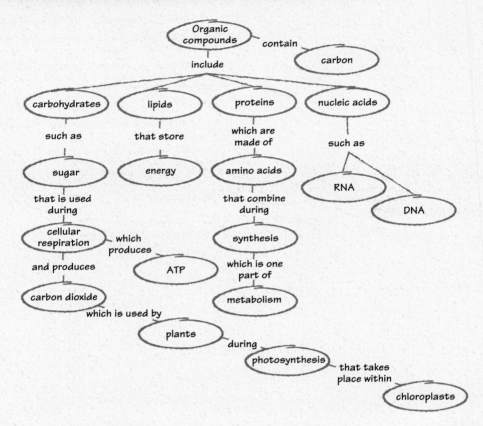

Checking Your Understanding

1. d. glucose
2. c. H_2O
3. a. it participates in hydrolysis reactions.
4. d. All of the above
5. b. They are produced only by animals.

Interpreting Graphs

Graph B. This graph indicates that water is moving out of the cell. Water moves out of cells (from an area of higher water concentration to an area of lower water concentration) when they are placed in hypertonic solutions.

Critical Thinking

1. Nucleotides
2. Without its adhesive properties, it would be impossible for water to travel up the vessels of plants and into their leaves. Without water in the leaves, photosynthesis could not occur. The other organisms in Biosphere II, like the scientists and other animals, that depend on photosynthesis for organic compounds and oxygen would suffer and eventually die or be forced to leave the facility.
3. Without sunlight, plants could not have carried out photosynthesis. Without photosynthesis, animals, including the dinosaurs, could have eventually perished from lack of food or oxygen. Another possibility is that the blocked sunlight caused a sudden drop in temperature on Earth, and the dinosaurs were unable to adapt.
4. Some of the raw materials that cells need must be actively transported into the cells by proteins located in their cell membranes. These proteins depend on the energy contained in ATP molecules to function.
5. No. You would expect metabolic wastes to be produced at a greater rate during the day, when your body is active and its demand for energy (ATP) is high. At night, when you are sleeping, your body requires much less ATP, so the rate of cellular respiration decreases.

Portfolio Idea

Stories will vary, but most students will probably come to the conclusion that water is more important than glucose. Facts in support of water include the following: Its heat capacity keeps living things from both freezing and overheating. It's the habitat of millions of living things. It dissolves most of the substances that participate in the chemical reaction of life. It participates in many of the chemical reactions that occur in living things. It is used during photosynthesis to synthesize organic compounds, including glucose. Facts in support of glucose include the following: It is the organic molecule that most cells use for energy, without which they would suffer.

Concept Mapping

Sample concept map:

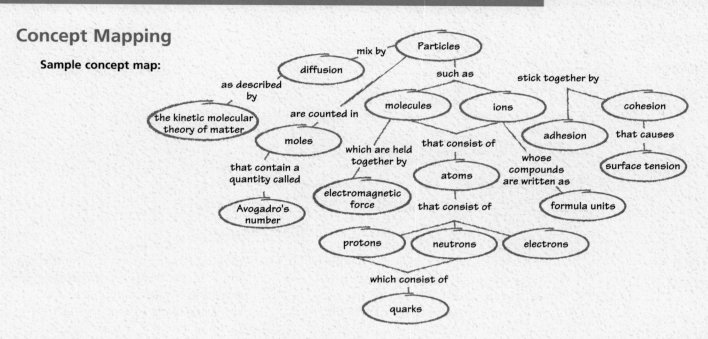

Checking Your Understanding

1. d. photons.
2. a. mole.
3. a. is the force of attraction between unlike particles.
4. b. the volume of a gas decreases as the pressure increases.
5. c. gravitational force.

Interpreting Photos

Some of the mass of a burning log and the oxygen it consumes is converted to the mass of the carbon dioxide and water vapor given off, as well as to the mass of the ash that remains. The remainder of the mass in the burning log is converted to heat and light energy.

Critical Thinking

1. First you would need to know the atomic mass of each atom in the sodium bicarbonate molecule. By consulting a periodic table, you would find the following atomic masses: Na is 23, H is 1, C is 12, and O is 16. Because there is one atom each of Na, H, and C and three atoms of O, you would perform the following addition problem: atomic mass of $NaHCO_3$ = 23 + 1 + 12 + 3(16) = 23 + 1 + 12 + 48 = 84. This means that there are 84 g of $NaHCO_3$ in a mole of sodium bicarbonate. To obtain 5 mol, you would perform the following multiplication problem: mass of 5 mol $NaHCO_3$ = 84 g × 5 = 420 g.

2. Mercury has more surface tension than water. The amount of surface tension is determined by the strength of cohesion between the particles of a substance.

3. The particles of matter must gain or lose energy to change state. As particles gain enough energy (reach a certain temperature), they begin to move apart. This happens at a specific melting point or boiling point. As the particles lose energy, the particles come together at the same temperature—the freezing or condensation point.

4. Because the force of the output piston must lift with a force 1000 times greater than the force exerted on the input piston, you would have to alter the sizes of the input and output pistons so that the output piston has a surface area that is 1000 times greater than that of the input piston. For example, you could decrease the surface area of the input piston to 0.1 cm² and increase the surface area of the output piston to 100 cm².

5. To determine the final volume of the balloon, which would change as the volume of the gas inside it changes, you would use Charles's law.

$$\frac{V_1}{T_1} = \frac{V_2}{T_2}$$

$$\frac{2 \text{ L}}{300 \text{ K}} = \frac{V_2}{270 \text{ K}}$$

$$V_2 = \frac{(270 \text{ K} \times 2 \text{ L})}{300 \text{ K}} = \frac{540 \text{ L}}{300} = 1.8 \text{ L}$$

Portfolio Idea

Students' accounts will vary; however, they should point out the various particles that are found in an atom. See the illustration on page S38 for examples of particles that students might mention. Students should also clearly describe the particles and forces that they would encounter.

Concept Mapping

Sample concept map:

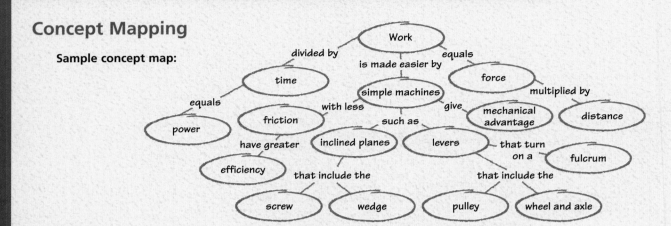

Checking Your Understanding

1. b. work.
2. b. joules per second
3. d. wheel and axle
4. b. inclined plane.
5. c. the work you put into it.

Interpreting Photos

The photo shows a wheel and axle that is used to pull a boat onto a boat trailer. The wheel is formed by the handle, and the axle is formed by the wheel used to wind the cable. The mechanical advantage of this wheel and axle could be calculated by dividing the radius of the wheel, or effort arm (from the center to the handle), by the radius of the axle, or resistance arm (from the center to the outside of the take-up wheel). Sources of friction include the surfaces in contact between the axle and the base, between the cable on the guide wheel and the take-up wheel, and between the boat on the trailer and lake water and, though very small, the air resistance.

Critical Thinking

1. First you must find your mass (for example, 60 kg) and convert it into weight in newtons (60 kg × 10 N/kg = 600 N). Then you must measure the height of the staircase (for example, 5 m). Next you would use the formula $W =$ $F \times d$ (in this case, d = height of the staircase) to calculate how much work you would do in walking to the top of the staircase (W = 600 N × 5 m = 3000 J). Finally, you would perform the experiment to find out how fast you can walk to the top of the staircase (for example, 15 seconds). You would then use the formula $P = W/t$ to calculate your power, which would be given in watts (P = 3000 J ÷ 15 seconds = 200 W).

2. 100 W = 100 J/s. Therefore, 100 J/s × 3600 s/hr. × 24 hr. = 8,640,000 J. The following is another way to solve this problem: In 24 hours, a 100 W light bulb uses 100 W × 24 hr. ÷ 1000 Wh/kWh = 2.4 kWh. So, 2.4 kWh × 3,600,000 J/kWh = 8,640,000 J.

3. Both sides of a knife have a sloping surface. Thus, a knife, like a wedge, is two inclined planes placed back to back. Sharpening decreases the angle (and increases the distance) of one or both sides of a wedge. This increases the mechanical advantage of the wedge (knife) and makes it possible to use less force to push the knife into the material it is cutting.

4. Mechanical advantage is usually thought of as the amount by which a machine increases effort force. It can be calculated by dividing the output force by the input force. However, if force increases, distance must decrease for work to stay the same, as the formula $W = F \times d$ indicates. And if force decreases, distance must increase for work to stay the same. Because the distance you move something will affect the amount of force needed to move it, mechanical advantage can also be calculated by dividing resistance distance by effort distance.

5. Answers will vary. A simple lever (any class) is considered by many to be the most efficient machine because friction has the least amount of effect on a lever. The only place where friction occurs in a lever is the point where the fulcrum touches the arms of the lever.

Portfolio Idea

Results will vary depending on the devices chosen by the students. Be sure each brochure includes a photo or drawing of the device, a schematic diagram showing the types of simple machines in the device, calculations of the mechanical advantage of the machines or device, and a written description of the device, how it works, and how it makes work easier. *Caution students to not take apart electrical devices and to use care when handling any mechanical device.*

Concept Mapping

Sample concept map:

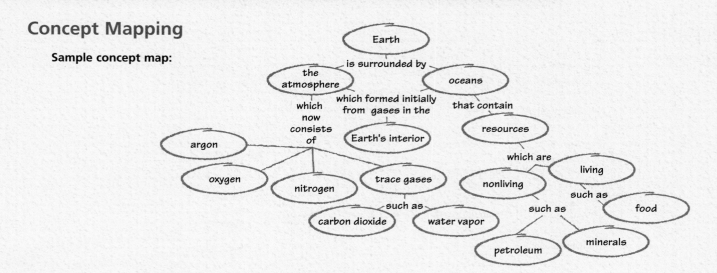

Checking Your Understanding

1. b. contained little oxygen.
2. a. troposphere
3. a. a squall line
4. b. condensation to occur.
5. c. are retrieved only with great difficulty.

Interpreting Illustrations

Because weather conditions depend on location, students would not be able to determine the season from the weather station. Current readings show an air temperature of 73°F with a relative humidity of about 57% (from the chart, using a dry-bulb/wet-bulb temperature difference of 10°F and a dry-bulb temperature of 73°F). There are 1.4 in. of rain in the rain gauge (a total of 9 in. recorded). Assuming that the pointer is pointing north, the wind is blowing from the northwest at a speed of 5 mph.

Critical Thinking

1. After the original atmosphere was scattered, a new atmosphere is thought to have formed from gases, including water vapor, released by many active volcanoes. This water vapor later condensed to form the oceans.

2. As living organisms evolved the ability to carry out photosynthesis, oxygen was added to the atmosphere. Ozone (a form of oxygen) then formed a shield that protected living organisms from solar radiation. Other organisms used up a large portion of the carbon dioxide to form calcium carbonate shells.

3. Towering thunderheads occasionally reach to the top of the troposphere where the jet stream flows. These high-speed winds blow the tops of tall thunderheads into the anvil shape shown.

4. If the Earth were to rotate twice as fast, the Coriolis effect would be enhanced. As an object (such as wind) traveled north or south, twice as much ground would move beneath it, and the path would be more curved.

5. Answers will vary. Possible answers include the following: Because of the atmosphere that surrounds it, the surface of the Earth undergoes weathering and erosion. In the absence of a substantial atmosphere, such as on the moon, weathering and erosion would not occur. The oxygen in the atmosphere causes chemical weathering of the Earth's surface through oxidation, and precipitation causes the erosion of soil. The oceans, which condensed from the atmosphere, change the Earth's surface through waves, tides, and their effect on climate and weather.

Portfolio Idea

Students' predictions may vary according to their interpretation of the weather maps. They should be able to explain why they made their predictions and to describe the assumptions or methods they used in making those predictions. Have students check the accuracy of their predictions against subsequent observations or future weather maps or reports.

Concept Mapping

Sample concept map:

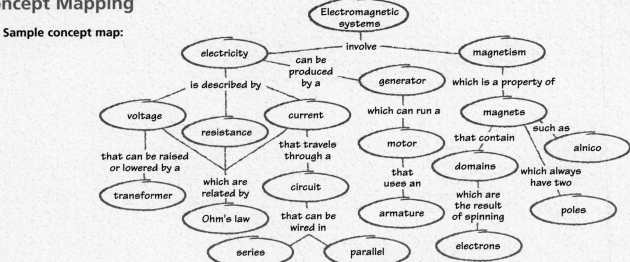

Checking Your Understanding

1. b. the flow of electrons between areas of greater and lesser charge.
2. a. Negatively charged particles attract positively charged particles.
3. c. pressure.
4. c. both an electric and a magnetic force.
5. d. mechanical energy into electrical energy.

Critical Thinking

1. Sample answer: Electric currents are caused by moving electrons. Because magnetism is a property of moving electrons, an electric current will produce a magnetic field in any material through which it flows that is capable of being magnetized.
2. Using the formula $\frac{E_p}{E_s} = \frac{T_p}{T_s}$, voltage in the primary circuit is calculated by cross-multiplying as follows:

   $E_p = \frac{30 \text{ V} \times 30 \text{ turns}}{10 \text{ turns}} = 90$ V
3. No; friction and electrical resistance, no matter how small, will eventually remove enough energy from the system to cause it to stop.

4. AC is easily transformed because it is easy to change its voltage. Nothing would happen if you passed DC through a transformer; DC would not produce a changing magnetic field, so no current would be produced in the secondary coil.
5. As the temperature of a material rises, its atoms move more vigorously. At the Curie point, the arrangement of the atoms and molecules in a material are affected by this motion in such a way that domains cannot be formed. Without domains, in which atoms are aligned magnetically, there can be no magnetism in the material.

Interpreting Photos

A parallel circuit would account for the on-off pattern of bulbs. Because two bulbs are not working, one or both must be faulty. If the wiring were composed of a series circuit, one faulty bulb would cause *all* of the other bulbs to go out. Two series circuits would show every second bulb out due to a single faulty bulb, but that pattern is not indicated. A burned-out bulb in a parallel circuit does not affect any other light in the string. If the string of lights is wired in parallel, both unlit bulbs must be faulty.

Portfolio Idea

Students' letters should be imaginative as well as scientifically accurate. Check to make sure that they understand magnetism enough to describe it with alternative terms.

Concept Mapping

Sample concept map:

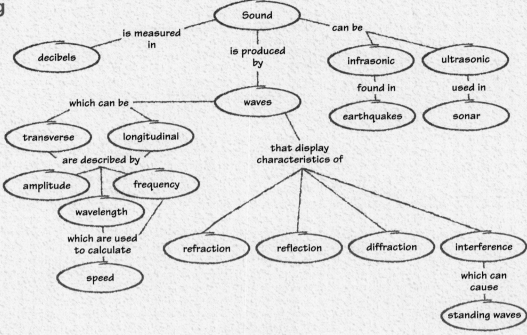

Checking Your Understanding

1. a. transverse wave.
2. c. As frequency increases, wavelength decreases.
3. c. diffraction.
4. d. the wavelength of the sound heard increases.
5. c. sitting in the third row of a loud rock concert

Interpreting Graphs

The dog's hearing range is the longest bar on the graph, even though the dophin has the greatest hearing range, because the graph does not have a constant scale. The increments between the grid lines get larger going up the vertical axis. For example, the area between 1 and 10 Hz can be divided into increments of 1, and the area between 1000 and 10,000 can be divided into increments of 1000. Therefore, the bar representing the dolphin is shorter than that of the dog but covers a greater range of frequencies.

Critical Thinking

1. Sample answer: Like other waves, sound waves transmit energy from one place to another. Sound waves, however, require a medium (such as water) in which to travel, while light waves (electromagnetic radiation) do not. Sound waves are longitudinal; water waves and light waves are transverse.
2. Because sound waves require a medium (such as air) in which to travel, a scream could not be transmitted through the vacuum of space.
3. $v = \lambda f$
   $\lambda = v \div f$
   $\lambda = 346 \text{ m/s} \div 440 \text{ waves/s}$
   $\lambda = 0.786 \text{ m/wave}$
   The wavelength of a wave with a frequency of 440 Hz would be 0.786 m.
4. The fastest car could be identified by listening for the Doppler effect. The faster a car travels, the more it compresses the sound waves in front of it and stretches those behind it. This would cause a greater difference in the pitch of its engine noise. The car with the greatest difference in pitch would be the fastest.
5. In scientific terms, the answer to this famous question depends on one's definition of sound. A falling tree would certainly cause a disturbance, or sound waves, in the media around it. But sound could also be defined as the waves that reach the ear and are interpreted by the brain. According to the second definition, if no ear (or comparable structure) receives these waves, the energy produced by the falling tree would not be considered "sound."

Portfolio Idea

Accept all reasonable responses. (Students may do library research, interview city employees, and conduct surveys in affected neighborhoods. Caution students to stay clear of the highways themselves. Students should report to the class their findings regarding the use of noise-abatement walls.)

Concept Mapping

Sample concept map:

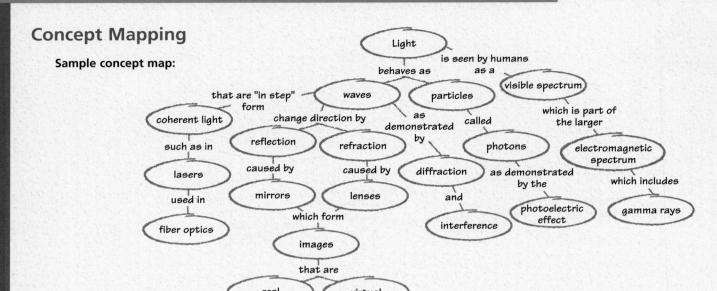

Checking Your Understanding

1. b. photoelectric effect
2. d. photons are emitted as excited electrons fall to lower energy levels.
3. a. red, blue, and green
4. a. X rays
5. c. polarizing lenses

Interpreting Illustrations

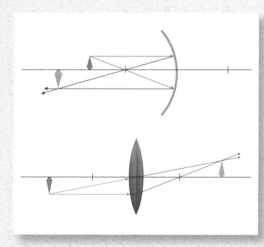

Critical Thinking

1. Answers will vary. The following is a sample answer: Light can be explained as streams of particles because it travels in a straight line. However, light can also be explained as waves because it is diffracted and can show interference.
2. Yellow; a white wall would reflect both red and green light, which produces yellow when combined.
3. No. The sounds transmitted by the station have been translated by microphones into signals that can be carried by electromagnetic radio waves over large distances. Once these signals reach your radio, they are retranslated by speakers into sounds that you recognize.
4. Take a lens that you know is polarizing and look through it while wearing the sunglasses. Rotate the polarizing lens 90°. If the light you see becomes brighter or dimmer as you turn the lens, the sunglasses have polarizing lenses. This test works because light coming through two polarizing lenses is blocked when the lenses are oriented 90° to each other. (Diagrams should be similar to the illustration on page S144.)
5. Answers will vary. The analogies should show that laser light is one color and is coherent (the light waves vibrate in step), while the white light of a flashlight consists of all the colors of light and the light is not coherent (the light waves do not vibrate in step).

Portfolio Idea

Answers will vary. In medicine, lasers are used in many types of surgery because they can be focused to a tiny point and can seal broken blood vessels while they cut or destroy tissue. For example, eye surgery is often performed with the aid of an argon laser, and cancer surgery can be performed with the aid of infrared-producing carbon dioxide and YAG (yttrium aluminum garnet) lasers. In the entertainment industry, infrared lasers are used in compact disc (CD) players and videodisc players, while a variety of visible-light lasers are used in light shows.

Concept Mapping

Sample concept map:

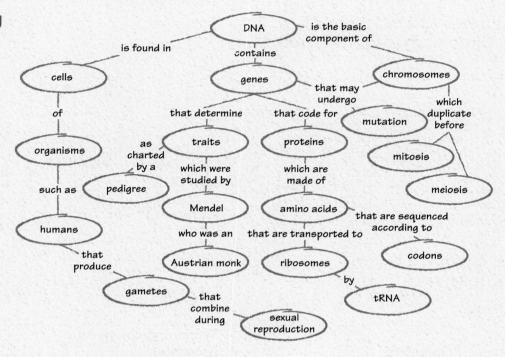

Checking Your Understanding

1. b. all tall.
2. c. genes.
3. d. Both *a* and *c*
4. a. Four daughter cells form.
5. b. TACGTAC

Interpreting Photos

The X chromosome probably contains the mutation. The sex of the fly is probably male. With only one X chromosome, a male would have to have only one mutated gene to have white eyes. White-eyed females occur less frequently because they must have two mutated genes (one on each of their X chromosomes) for the mutation to be expressed. This means that white-eyed males are more common than white-eyed females.

Critical Thinking

1. The Punnett square would look like the following:

	Y	y
Y	YY	Yy
y	Yy	yy

Phenotypes: 75 percent yellow, 25 percent green
Genotypes: 25 percent *YY*, 50 percent *Yy*, 25 percent yy

2. A skin cell from the organism would have 12 chromosomes. Meiosis produces sex cells that have half the number of chromosomes of regular body cells. Since each sex cell had 6 chromosomes, a body cell would have twice as many—12.

3. TCTTAATTT; UCU, UAA, UUU
4. Offspring that result from asexual reproduction have the exact same genes as their parents. No genetic variation occurs that might allow the organism to survive better in a changed environment. Offspring that result from sexual reproduction could have genetic variations that their ancestors did not have. These variations would result in some offspring that are better adapted to a changed environment, thus providing a better chance of survival for that type of organism.

Portfolio Idea

Answers will vary, but the summaries should reflect Mendel's ideas about genetics and heredity and Sutton's ideas about chromosomes and genes.

Answers to Tables and Diagrams

Unit 1, Making Connections, page 71
Checking Your Understanding, question 5

Sample concept map:

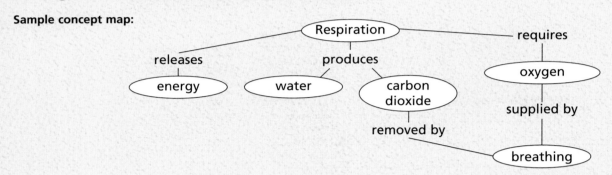

Unit 2, Chapter 4, page 81
Observation vs. Inference

Sample completed table:

Inference	Supporting observation (evidence)
1. The defendant committed the robbery.	The defendant was seen in the area of the robbery.
2. The defendant cut his hand while breaking into the store.	The defendant had cut his right hand. There was a broken window at the store.
3. The blood on the doorknob was that of the defendant.	The blood type found on the doorknob matched that of the defendant.
4. The defendant left a fingerprint during the robbery.	The defendant's fingerprint was found on the countertop.
5. The money the defendant spent the next day was from the robbery.	The defendant was observed spending more than the usual amount of money.
6. The defendant is likely to have committed an additional crime.	The defendant has a past criminal record.
7. The defendant is as innocent as the others at the scene.	Many people were seen in the area of the robbery.
8. The defendant left his fingerprint on the countertop while making a purchase.	The defendant bought a newspaper at the store hours before the robbery.

Unit 2, Chapter 6, page 110
Lesson 1 Follow-Up, Assessment

Graph of Leilani's data:

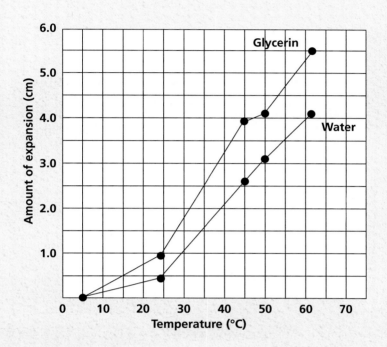

Unit 2, Chapter 6, page 124
Challenge Your Thinking, question 2

Sample completed diagram:

The Matter-Go-Round

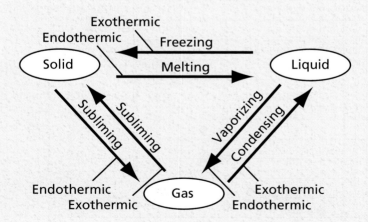

Unit 4, Chapter 10, page 221
Exploration 2, Part 1

Sample labeled illustration:

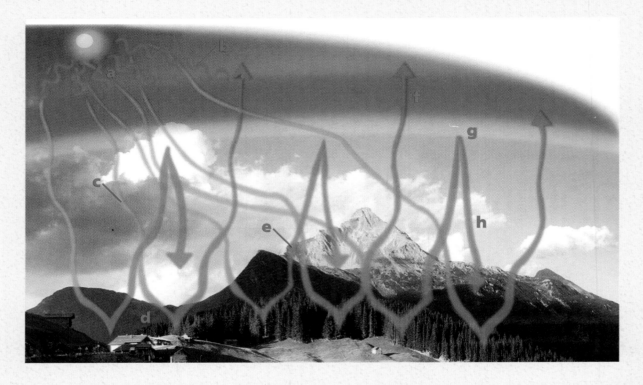

Unit 4, Chapter 10, page 223
Exploration 2, Part 4

Sample graph:

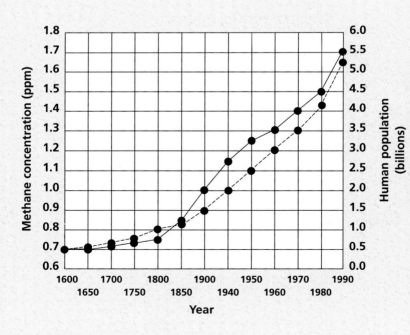

Unit 4, Chapter 10, page 225
Global Warming—How Much?

Sample completed web of changes:

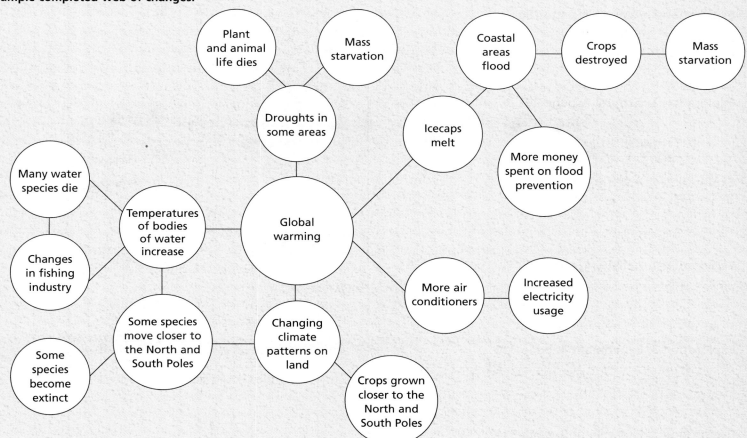

Unit 4, Chapter 12, page 260
Answer to In-Text Question Ⓐ

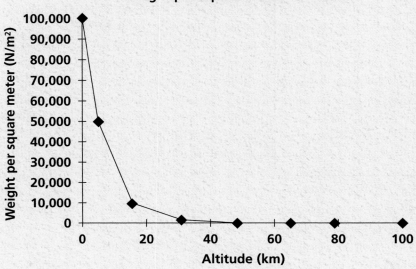

Weight per Square Meter vs. Altitude

Unit 5, Chapter 15, page 328
Circuit Symbols

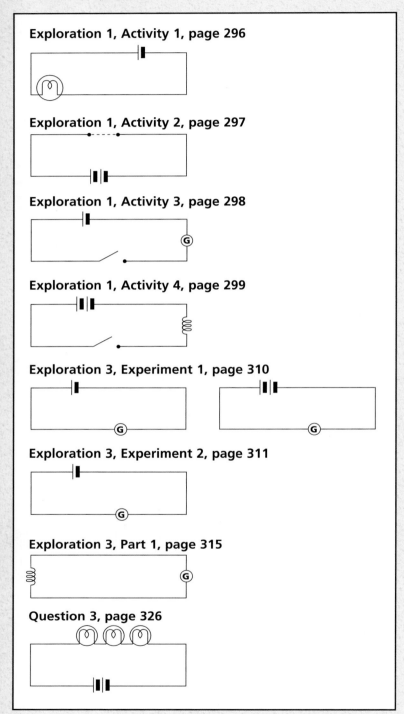

Exploration 1, Activity 1, page 296

Exploration 1, Activity 2, page 297

Exploration 1, Activity 3, page 298

Exploration 1, Activity 4, page 299

Exploration 3, Experiment 1, page 310

Exploration 3, Experiment 2, page 311

Exploration 3, Part 1, page 315

Question 3, page 326

Unit 5, Chapter 15, page 331
Exploration 2, Part 2, questions 1–5

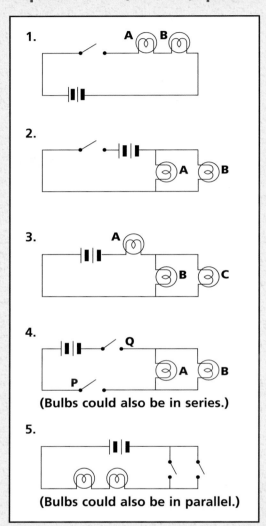

1.

2.

3.

4.
(Bulbs could also be in series.)

5.
(Bulbs could also be in parallel.)

Answers to Tables and Diagrams

Unit 5, Chapter 15, page 337
Meeting Individual Needs

Sample circuit diagrams:

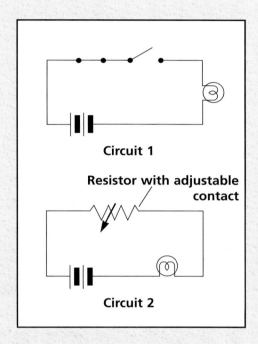

Unit 5, Chapter 15, page 337
Exploration 3, Part 2, questions (a)–(c)

Sample circuit diagrams:

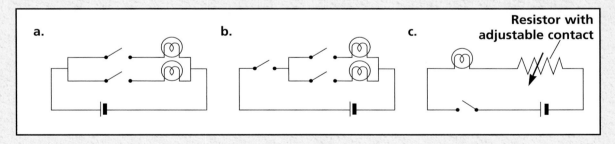

Unit 7, Chapter 19, page 433
Exploration 3, Experiment 5

Sample setup:

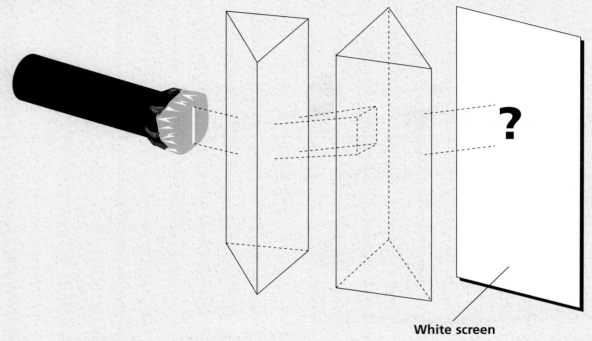

White screen

Unit 7, Chapter 21, page 471
Exploration 3, Part 2, question 2

Sample diagram:

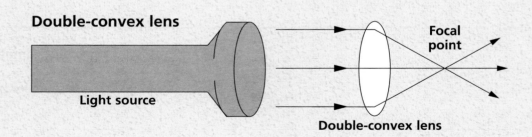

Double-convex lens

Light source

Focal point

Double-convex lens